SAFETY ANALYSIS OF FOODS OF ANIMAL ORIGIN

SAFETY ANALYSIS OF FOODS OF ANIMAL ORIGIN

Edited by
Leo M.L. Nollet and Fidel Toldrá

CRC Press
Taylor & Francis Group
Boca Raton London New York

CRC Press is an imprint of the
Taylor & Francis Group, an **informa** business

CRC Press
Taylor & Francis Group
6000 Broken Sound Parkway NW, Suite 300
Boca Raton, FL 33487-2742

First issued in paperback 2017

© 2011 by Taylor and Francis Group, LLC
CRC Press is an imprint of Taylor & Francis Group, an Informa business

No claim to original U.S. Government works

ISBN-13: 978-1-4398-4817-3 (hbk)
ISBN-13: 978-1-138-11221-6 (pbk)

Library of Congress Cataloging-in-Publication Data

Safety analysis of foods of animal origin / editors, Leo M.L. Nollet, Fidel Toldrá.
    p. cm.
    "A CRC title."
    Includes bibliographical references and index.
    ISBN 978-1-4398-4817-3 (hardcover : alk. paper)
    1. Meat inspection. 2. Meat--Analysis. 3. Dairy inspection. 4. Dairy products--Analysis. I. Nollet, Leo M. L., 1948- II. Toldrá, Fidel.

TS1975.S24 2011
664'.9072--dc22
        2010026750

Visit the Taylor & Francis Web site at
http://www.taylorandfrancis.com

and the CRC Press Web site at
http://www.crcpress.com

Contents

Preface...ix

Editors..xi

Contributors.. xiii

PART I   MEAT, PROCESSED MEATS AND POULTRY

1   Methods to Predict Spoilage of Muscle Foods 3
    GERALDINE DUFFY, ANTHONY DOLAN, AND CATHERINE M. BURGESS

2   Microbial Foodborne Pathogens 21
    MARIOS MATARAGAS AND ELEFTHERIOS H. DROSINOS

3   Parasites...59
    ANU NÄREAHO

4   Mycotoxin Analysis in Poultry and Processed Meats....................................77
    JEAN-DENIS BAILLY AND PHILIPPE GUERRE

5   Detection of Genetically Modified Organisms in Processed Meats and Poultry 125
    ANDREA GERMINI AND ALESSANDRO TONELLI

6   Detection of Adulterations: Addition of Foreign Proteins 155
    MARÍA CONCEPCIÓN GARCÍA LÓPEZ AND MARÍA LUISA MARINA ALEGRE

7   Detection of Adulterations: Identification of Animal Species...................187
    JOHANNES ARJEN LENSTRA

8   Detection of Irradiated Ingredients 207
    EILEEN M. STEWART

9   Growth Promoters...229
    MILAGRO REIG AND FIDEL TOLDRÁ

10   Antibiotic Residues in Muscle Tissues of Edible Animal Products 249
    ERIC VERDON

11 Determination of Persistent Organic Pollutants in Meat......................349
ANNA LAURA IAMICELI, IGOR FOCHI, GIANFRANCO BRAMBILLA, AND
ALESSANDRO DI DOMENICO

12 Biogenic Amines...399
M. CARMEN VIDAL-CAROU, M. LUZ LATORRE-MORATALLA, AND
SARA BOVER-CID

13 Nitrosamines 421
SUSANNE RATH AND FELIX GUILLERMO REYES REYES

14 Polycyclic Aromatic Hydrocarbons...441
PETER ŠIMKO

PART II FISH AND SEAFOODS

15 Assessment of Seafood Spoilage and the Microorganisms Involved........................463
ROBERT E. LEVIN

**16 Detection of the Principal Foodborne Pathogens in Seafoods and
Seafood-Related Environments** 485
DAVID RODRÍGUEZ-LÁZARO AND MARTA HERNÁNDEZ

17 Parasites...507
JUAN ANTONIO BALBUENA AND JUAN ANTONIO RAGA

18 Techniques of Diagnosis of Fish and Shellfish Virus and Viral Diseases 531
CARLOS PEREIRA DOPAZO AND ISABEL BANDÍN

19 Marine Toxins 577
CARA EMPEY CAMPORA AND YOSHITSUGI HOKAMA

20 Detection of Adulterations: Addition of Foreign Proteins 603
VÉRONIQUE VERREZ-BAGNIS

21 Detection of Adulterations: Identification of Seafood Species.............................615
ANTONIO PUYET AND JOSÉ M. BAUTISTA

22 Spectrochemical Methods for the Determination of Metals in Seafood.................641
JOSEPH SNEDDON AND CHAD A. THIBODEAUX

23 Food Irradiation and Its Detection 663
YIU CHUNG WONG, DELLA WAI MEI SIN, AND WAI YIN YAO

24 Veterinary Drugs..687
ANTON KAUFMANN

25 Analysis of Dioxins in Seafood and Seafood Products............................707
LUISA RAMOS BORDAJANDI, BELÉN GÓMARA, AND MARÍA JOSÉ GONZÁLEZ

26 Environmental Contaminants: Persistent Organic Pollutants............................727
MONIA PERUGINI

27 Biogenic Amines in Seafood Products..743
CLAUDIA RUIZ-CAPILLAS AND FRANCISCO JIMÉNEZ-COLMENERO

28 Detection of GM Ingredients in Fish Feed...761
KATHY MESSENS, NICOLAS GRYSON, KRIS AUDENAERT, AND MIA EECKHOUT

PART III   MILK AND DAIRY FOODS

29 Microbial Flora...781
EFFIE TSAKALIDOU

30 Spoilage Detection 799
MARIA CRISTINA DANTAS VANETTI

31 PCR-Based Methods for Detection of Foodborne Bacterial Pathogens in Dairy
Products 811
ILEX WHITING, NIGEL COOK, MARTA HERNÁNDEZ,
DAVID RODRÍGUEZ-LÁZARO, AND MARTIN D'AGOSTINO

32 Mycotoxins and Toxins 823
CARLA SOLER, JOSÉ MIGUEL SORIANO, AND JORDI MAÑES

33 Detection of Adulterations: Addition of Foreign Lipids and Proteins 851
SASKIA M. vAN RUTH, MARIA G. E. G. BREMER, AND ROB FRANKHUIZEN

34 Detection of Adulterations: Identification of Milk Origin.......................865
GOLFO MOATSOU

35 Analysis of Antibiotics in Milk and Its Products 887
JIAN WANG

36 Chemical Contaminants: Phthalates.. 907
JIPING ZHU, SUSAN P. PHILLIPS, AND XU-LIANG CAO

37 Environmental Contaminants...929
SARA BOGIALLI AND ANTONIO DI CORCIA

38 Allergens..951
VIRGINIE TREGOAT AND ARJON J. vAN HENGEL

Index 969

Preface

Safety of food, in general, and safety of foods of animal origin are of great importance for consumers, the food industry, and health authorities. Consumers expect healthy and safe food. Producers need to meet certain standards, guidelines, or directives imposed by local, governmental, continental, and global authorities. Producers and authorities have to rely on adequate methods of analysis for an accurate detection, including the choice of an adequate sample preparation method, which is also of high relevance.

Analysis and detection methods are evolving fast toward miniaturization, automation, and increased limits of detection. Contamination of foods may be of different origins, either biological (bacteria, viruses, or parasites and products of organisms, e.g., marine toxins or mycotoxins) or chemical (residues of growth promoters, antibiotics, food contact materials, persistent organic materials, environmental contaminants, and many others). The more man interferes with the production of raw materials and end products, the more the detection methods required to check the authenticity of foods, the addition of foreign compounds, the use of irradiation, and the presence of genetically modified organisms.

This book, *Safety Analysis of Foods of Animal Origin*, is divided into three parts: Part I deals with meat, processed meats, and poultry; Part II with fish and seafood products; and Part III with milk and dairy products.

In all three parts, selected chapters (Chapters 1 through 3, 15, 17, and 29) deal first with the safety aspects of biological agents and products of different organisms. They also deal with methods to control the presence of bacteria, viruses, or parasites in food (Chapters 5, 16, 18, 30, and 31). Sometimes it is not the biological agent that is hazardous, but rather its products—this is covered in Chapters 4, 12, 19, 28, and 32.

Authenticity is a very important factor in the food industry and for consumers. Aspects like adulteration, addition of foreign compounds, irradiation, and genetically modified organisms are discussed in depth in several chapters (Chapters 5 through 8, 20 through 23, 28, 33, and 34).

Residues in foods may be from internal or external sources. Several chapters discuss residues of growth promoters, antibiotics, persistent organic pollutants, biogenic amines, *n*-nitroso compounds, and polycyclic aromatic compounds (Chapters 9 through 14, 24 through 27, and 35 through 38). This list of compounds is not exhaustive.

In each chapter, the authors start with a discussion of the parameter in question. Sample preparation and cleanup methods are reviewed in depth. This is followed by a detailed overview of different separation and detection methods. Special attention is given to limits of detection and reliability of methods. Finally, a brief summary covers the presence of these parameters in different end products, regions, and countries.

All the chapters have been written by renowned scientists who are experts in their fields of research. Only the most recent techniques and related literature have been dealt with. We would like to thank all the contributing authors for their great efforts in producing this excellent work.

There are two ways of spreading light: to be the candle or the mirror that reflects it. (Edith Wharton)

Leo M.L. Nollet
Fidel Toldrá

Editors

Dr. Leo M.L. Nollet is an editor and associate editor of several books. He has edited the first and second editions of *Food Analysis by HPLC* and the *Handbook of Food Analysis* (which is a three-volume book) for Marcel Dekker, New York (now CRC Press of Taylor & Francis Group). He has also edited the third edition of the book, *Chromatographic Analysis of the Environment* (CRC Press, Boca Raton, Florida) and the second edition of the *Handbook of Water Analysis* (CRC Press, Boca Raton, Florida) in 2007. He coedited two books with F. Toldrá that were published in 2006: *Advanced Technologies for Meat Processing* (CRC Press, Boca Raton, Florida) and *Advances in Food Diagnostics* (Blackwell Publishing, New York). He also coedited *Radionuclide Concentrations in Foods and the Environment* with M. Pöschl in 2006 (CRC Press, Boca Raton, Florida).

Dr. Nollet has coedited several books with Y. H. Hui and other colleagues, including the *Handbook of Food Product Manufacturing* (Wiley, New York, 2007); the *Handbook of Food Science, Technology and Engineering* (CRC Press, Boca Raton, Florida, 2005); and *Food Biochemistry and Food Processing* (Blackwell Publishing, New York, 2005). He has also edited the *Handbook of Meat, Poultry and Seafood Quality* (Blackwell Publishing, New York, 2007).

Dr. Nollet has worked on the following six books on analysis methodologies with F. Toldrá for foods of animal origin, several of these books have already been published by CRC Press, Boca Raton, Florida:

Handbook of Muscle Foods Analysis
Handbook of Processed Meats and Poultry Analysis
Handbook of Seafood and Seafood Products Analysis
Handbook of Dairy Foods Analysis
Handbook of Analysis of Edible Animal By-Products
Handbook of Analysis of Active Compounds in Functional Foods

He has also worked with Professor H. Rathore on the *Handbook of Pesticides: Methods of Pesticides Residues Analysis*, which was published by CRC Press, Boca Raton, Florida, in 2009.

Dr. Fidel Toldrá is a research professor in the Department of Food Science at the Instituto de Agroquímica y Tecnología de Alimentos (CSIC), Valencia, Spain, and serves as the European editor of *Trends in Food Science & Technology*, the editor in chief of *Current Nutrition & Food Science*, and as a member of the Flavorings and Enzymes Panel at the European Food Safety Authority. In recent years, he has served as an editor or associate editor of several books. He was the editor of *Research Advances in the Quality of Meat and Meat Products* (Research Signpost,

Trivandrum, Kerala, India, 2002) and the associate editor of the *Handbook of Food and Beverage Fermentation Technology* and the *Handbook of Food Science, Technology and Engineering* published in 2004 and 2006, respectively, by CRC Press, Boca Raton, Florida. He coedited two books with L. Nollet that were published in 2006: *Advanced Technologies for Meat Processing* (CRC Press, Boca Raton, Florida) and *Advances in Food Diagnostics* (Blackwell Publishing, New York). Both he and Nollet are also associate editors of the *Handbook of Food Product Manufacturing* published by John Wiley & Sons, New York, in 2007. He has also edited *Meat Biotechnology* and *Safety of Meat and Processed Meat* (Springer, Berlin, Germany, 2008 and 2009, respectively) and has authored *Dry-Cured Meat Products* (Food & Nutrition Press—now Wiley-Blackwell, New York, 2002).

Dr. Toldrá has worked on the following six books on analysis methodologies with L. Nollet for foods of animal origin, several of these books have already been published by CRC Press, Boca Raton, Florida:

> *Handbook of Muscle Foods Analysis*
> *Handbook of Processed Meats and Poultry Analysis*
> *Handbook of Seafood and Seafood Products Analysis*
> *Handbook of Dairy Foods Analysis*
> *Handbook of Analysis of Edible Animal By-Products*
> *Handbook of Analysis of Active Compounds in Functional Foods*

Dr. Toldrá was awarded the 2002 International Prize for Meat Science and Technology by the International Meat Secretariat. He was elected as a fellow of the International Academy of Food Science & Technology in 2008 and as a fellow of the Institute of Food Technologists in 2009. He recently received the 2010 Distinguished Research Award from the American Meat Science Association.

Contributors

Kris Audenaert
Faculty of Biosciences and Landscape
    Architecture
Department of Plant Production
University College Ghent
Ghent University Association
Ghent, Belgium

Jean-Denis Bailly
Mycotoxicology Research Unit
National Veterinary School of Taulouse
Toulouse, France

Juan Antonio Balbuena
Cavanilles Institute of Biodiversity and
    Evolutionary Biology
University of Valencia
Valencia, Spain

Isabel Bandín
Departamento de Microbiología y
    Parasitología
Instituto de Acuicultura
Universidad de Santiago de Compostela
Santiago de Compostela, Spain

José M. Bautista
Faculty of Veterinary Sciences
Department of Biochemistry and Molecular
    Biology
Universidad Complutense de Madrid
Ciudad Universitaria
Madrid, Spain

Sara Bogialli
Dipartimento di Chimica
Università La Sapienza
Rome, Italy

Luisa Ramos Bordajandi
Instrumental Analysis and Environmental
    Chemistry Department
Spanish National Research Council
General Organic Chemistry Institute
Madrid, Spain

Sara Bover-Cid
Meat Technology Center
Institute for Food and Agricultural Research
    and Technology
Girona, Spain

Gianfranco Brambilla
Istituto Superiore di Sanitá
Roma, Italy

Maria G. E. G. Bremer
RIKILT
Institute of Food Safety
Wageningen University and Research Centre
Wageningen, the Netherlands

Catherine M. Burgess
Ashtown Food Research Centre
Dublin, Ireland

Cara Empey Campora
Department of Pathology
John A. Burns School of Medicine
University of Hawaii
Honolulu, Hawaii

Xu-Liang Cao
Food Research Division
Health Canada
Ottawa, Ontario, Canada

Nigel Cook
Food and Environment Research Agency
York, United Kingdom

Martin D'Agostino
Food and Environment Research Agency
York, United Kingdom

Antonio Di Corcia
Dipartimento di Chimica
University of Roma La Sapienza
Rome, Italy

Anthony Dolan
Ashtown Food Research Centre
Dublin, Ireland

Alessandro di Domenico
Istituto Superiore di Sanitá
Rome, Italy

Carlos Pereira Dopazo
Departamento de Microbiología y
    Parasitología
Instituto de Acuicultura
Universidad de Santiago de Compostela
Santiago de Compostela, Spain

Eleftherios H. Drosinos
Laboratory of Food Quality Control and
    Hygiene
Department of Food Science & Technology
Agricultural University of Athens
Athens, Greece

Geraldine Duffy
Ashtown Food Research Centre
Dublin, Ireland

Mia Eeckhout
Faculty of Biosciences and Landscape
    Architecture
Department of Food Science and Technology
University College Ghent
Ghent University Association
Ghent, Belgium

Igor Fochi
Istituto Superiore di Sanitá
Rome, Italy

Rob Frankhuizen
RIKILT
Institute of Food Safety
Wageningen University and Research Centre
Wageningen, the Netherlands

Andrea Germini
Faculty of Agriculture
University of Parma
Parma, Italy

Belén Gómara
Instrumental Analysis and Environmental
    Chemistry Department
Spanish National Research Council
General Organic Chemistry Institute
Madrid, Spain

María José González
Instrumental Analysis and Environmental
    Chemistry Department
Spanish National Research Council
General Organic Chemistry Institute
Madrid, Spain

Nicolas Gryson
Faculty of Biosciences and Landscape
    Architecture
Department of Food Science and Technology
University College Ghent
Ghent University Association
Ghent, Belgium

Philippe Guerre
Mycotoxicology Research Unit
National Veterinary School of Toulouse
Toulouse, France

Arjon J. van Hengel
European Commission
Joint Research Centre
Institute for Reference Materials and
    Measurements
Geel, Belgium

Marta Hernández
Laboratory of Molecular Biology and
    Microbiology
Instituto Tecnologico Agrario de Castilla y
    León
Valladolid, Spain

Yoshitsugi Hokama
Department of Pathology
John A. Burns School of Medicine
University of Hawaii
Honolulu, Hawaii

Anna Laura Iamiceli
Istituto Superiore di Sanitá
Rome, Italy

Francisco Jiménez-Colmenero
Department of Meat and Fish Science and
    Technology
Consejo Superior de Investigaciones
    Científicas
Instituto del Frío
Ciudad Universitaria
Madrid, Spain

Anton Kaufmann
Official Food Control Authority of the Canton
    of Zurich
Kantonales Labor Zurich
Zurich, Switzerland

M. Luz Latorre-Moratalla
Department of Nutrition and Bromatology
Faculty of Pharmacy
University of Barcelona
Barcelona, Spain

Johannes Arjen Lenstra
Faculty of Veterinary Medicine
Institute for Risk Assessment Sciences
Utrecht University
Utrecht, the Netherlands

Robert E. Levin
Department of Food Science
University of Massachusetts
Amherst, Massachusetts

María Concepción García López
Faculty of Chemistry
Department of Analytical Chemistry
University of Alcalá
Madrid, Spain

María Luisa Marina Alegre
Faculty of Chemistry
Department of Analytical Chemistry
University of Alcalá
Madrid, Spain

Jordi Mañes
Faculty of Pharmacy
Laboratory of Toxicology and Food Chemistry
University of Valencia
Valencia, Spain

Marios Mataragas
Laboratory of Food Quality Control and
    Hygiene
Department of Food Science & Technology
Agricultural University of Athens
Athens, Greece

Kathy Messens
Faculty of Biosciences and Landscape
    Architecture
Laboratory AgriFing
Department of Food Science and Technology
University College of Ghent
Ghent University Association
Ghent, Belgium

Golfo Moatsou
Laboratory of Dairy Technology
Department of Food Science and Technology
Agricultural University of Athens
Athens, Greece

Anu Näreaho
Witold Stefanski Institute of Parasitology
Polish Academy of Sciences
Warsaw, Poland

Monia Perugini
Department of Food Science
University of Teramo
Teramo, Italy

Susan P. Phillips
Departments of Family Medicine and
    Community Health and Epidemiology
Queen's University
Kingston, Ontario, Canada

Antonio Puyet
Faculty of Veterinary Sciences
Department of Biochemistry and Molecular
    Biology
Universidad Complutense de Madrid
Ciudad Universitaria
Madrid, Spain

Juan Antonio Raga
Cavanilles Institute of Biodiversity and
    Evolutionary Biology
University of Valencia
Valencia, Spain

Susanne Rath
Grupo de Toxicologia de Alimentos e
    Fármacos
Departamento de Química Analítica
Instituto de Química
Universidade Estadual de Campinas
Campinas, Brazil

Milagro Reig
Institute of Food Engineering for
    Development
Universidad Politécnica de Valencia
Valencia, Spain

Felix Guillermo Reyes Reyes
Department of Food Science
University of Campinas
Campinas, Sao Paulo, Brazil

David Rodríguez-Lázaro
Food Safety and Technology Research Group
Instituto Tecnologico Agrario de Castilla y
    León
Valladolid, Spain

Claudia Ruiz-Capillas
Department of Meat and Fish Science and
    Technology
Consejo Superior de Investigaciones
    Científicas
Instituto del Frío
Ciudad Universitaria
Madrid, Spain

Saskia M. van Ruth
RIKILT
Institute of Food Safety
Wageningen, the Netherlands

Peter Šimko
Institute of Food Science and Biotechnology
Faculty of Chemistry
Brno University of Technology
Brno, Czech Republic

Della Wai Mei Sin
Analytical and Advisory Services Division
Government Laboratory
Hong Kong, People's Republic of China

Joseph Sneddon
Department of Chemistry
McNeese State University
Lake Charles, Louisiana

Carla Soler
Faculty of Pharmacy
Laboratory of Food Technology
University of Valencia
Valencia, Spain

José Miguel Soriano
Faculty of Pharmacy
Laboratory of Food Chemistry and Toxicology
University of Valencia
Valencia, Spain

Eileen M. Stewart
Agriculture, Food and Environmental Science Division
Agri-Food and Bioscience Institute
Belfast, Ireland

Chad A. Thibodeaux
Department of Chemistry
McNeese State University
Lake Charles, Louisiana

Fidel Toldrá
Department of Food Science
Consejo Superior de Investigaciones Científicas
Instituto de Agroquímica y Tecnología de Alimentos
Valencia, Spain

Alessandro Tonelli
Faculty of Agriculture
University of Parma
Parma, Italy

Virginie Tregoat
European Commission
Joint Research Centre
Institute for Reference Materials and Measurements
Geel, Belgium

Effie Tsakalidou
Laboratory of Dairy Research
Department of Food Science and Technology
Agricultural University of Athens
Athens, Greece

Maria Cristina Dantas Vanetti
Departamento de Microbiologia
Universidade Federal de Viçosa
Viçosa, Brazil

Eric Verdon
E.V. Community Reference Laboratory for Antimicrobial and Dye Residues in Food from Animal Origin
Fougéres, France

Véronique Verrez-Bagnis
Institut français de recherche pour l'exploitation de la mer
Nantes, France

M. Carmen Vidal-Carou
Department of Nutrition and Bromatology
Faculty of Pharmacy
University of Barcelona
Barcelona, Spain

Jian Wang
Calgary Laboratory
Canadian Food Inspection Agency
Calgary, Alberta, Canada

Ilex Whiting
Food and Environment Research Agency
York, United Kingdom

Yiu Chung Wong
Analytical and Advisory Service Division
Government Laboratory
Homantin Government Office
Hong Kong, People's Republic of China

Wai Yin Yao
Analytical and Advisory Services Division
Government Laboratory
Hong Kong, People's Republic of China

Jiping Zhu
Chemistry Research Division
Safe Environments Programme
Health Canada
Ottawa, Ontario, Canada

MEAT, PROCESSED MEATS AND POULTRY

Chapter 1

Methods to Predict Spoilage of Muscle Foods

Geraldine Duffy, Anthony Dolan, and Catherine M. Burgess

Contents

1.1 Introduction 4
1.2 Culture-Based Methods 5
    1.2.1 Agar Plate Count Methods 5
    1.2.2 Alternative Culture Methods 5
1.3 Direct Epifluorescent Filtration Technique 6
1.4 ATP Bioluminescence Methods 8
1.5 Electrical Methods 9
1.6 *Limulus* Amoebocyte Lysate Assay 9
1.7 Spectroscopic Methods 10
1.8 Developmental Methods 10
    1.8.1 Flow Cell Cytometry 11
    1.8.2 Molecular Methods 11
        1.8.2.1 Polymerase Chain Reaction 11
        1.8.2.2 Fluorescent *In Situ* Hybridization 13
1.9 Electronic Nose 14
1.10 Time–Temperature Integrators 14
1.11 Conclusion 15
References 16

1.1 Introduction

All animals, birds, fish, etc. contain a host of microorganisms in their intestinal tract and on their exposed outer skins, membranes, etc. During the slaughter and processing of the live organism into food, the muscle surface can become contaminated with microorganisms. Microbial contamination on the food and its composition/diversity is dependent on both the microbial load of the host organism and the hygiene practices employed during slaughter, processing, and distribution [1]. For example, during beef slaughter cross-contamination of microbial flora from the bovine hide, feces, and gut contents are recognized as the main cause of microflora on the beef carcass [2]. Among the principal genera of bacteria that are present on postslaughter muscle surfaces are *Pseudomonas* spp., *Acinetobacter* spp., *Aeromonas* spp., *Brochothrix thermosphacta*, members of the lactic acid bacteria (LAB) such as *Lactobacillus* and *Leuconostoc*, as well as many members of the Enterobacteriaceae including *Enterobacter* and *Serratia* spp. [3–7].

From a microbiological standpoint, muscle foods have a particularly unique nutritional profile, with intrinsic factors such as a neutral pH, high water content, a high protein content, and fat providing an excellent platform for microbial growth. Thus, during storage of muscle foods, favorable environmental conditions (temperature, pH, a_w, etc.) will allow the microflora to grow. As the microorganisms grow they metabolize the food components into smaller biochemical constituents, many of which emit unacceptable flavors, odors, colors, or appearance [6,8]. Spoilage may be defined as the time when the microorganisms reach a critical level, usually at around $\log_{10}$ 7–8 colony forming units (CFU) g^{-1}, to induce sufficient organoleptic changes to render the food unacceptable to the consumer. The particular species of bacteria that contaminate the muscle, along with the environmental conditions, will determine the spoilage profile of the stored muscle food [5]. Under aerobic storage conditions, certain species of the genus *Pseudomonas* are generally considered to significantly contribute to spoilage. This is due to the organisms' ability to utilize amino acids and grow well at refrigeration temperatures. Although it is a facultative anaerobe, under anaerobic conditions, the bacterium *B. thermosphacta* is considered a dominant member of the spoilage flora of meat products, producing lactic acid and ethanol as by-products of glucose utilization [9]. Recently, the use of modified atmosphere packaging (MAP) has gained popularity as a method of preservation. Gas mixtures containing variable O_2 and CO_2 concentrations are used to inhibit the growth of different spoilage-related bacteria. Under certain MAP conditions, lactic acid bacteria dominate and are prolific spoilers [10].

The storage period under a particular set of environmental conditions until the spoilage microflora reaches a threshold level is known as shelf life. To extend the shelf life of muscle foods, a range of procedures to prevent or retard microbial growth are deployed. When storing fresh muscle foods, where only chill storage temperatures (<5°C) are employed to retard microbial growth, the shelf life can be measured in days. Modified atmosphere or vacuum packaging can extend shelf life to several weeks or months. Extension of shelf life beyond this period requires the use of more robust and invasive preservation techniques such as freezing, mild or severe heat treatment (canning), reducing water activity (a_w), altering pH (acidic or alkaline), or the use of chemical or biological preservatives. However, all of above preservation processes generally have an unwanted deleterious influence on the organoleptic quality of the food. Therefore, there is an ever increasing move away from heavily preserved food to fresh and minimally preserved foods with a limited shelf life, imposing a greater need for industry to be able to accurately predict when spoilage of the food will occur. As there is a direct correlation between microbial load and spoilage, food hygiene regulators and industry set microbiological guidelines and criteria for specific foodstuffs, which are used to predict spoilage and determine shelf life. A number of direct and indirect techniques are available to assess the microbial load or its metabolites in food at the point of food production, which will give a predicted shelf life under a defined set of storage conditions. This chapter will review a selection of commonly used and emerging technologies that are used to directly or indirectly enumerate the total microbial load and predict spoilage.

1.2 Culture-Based Methods

Microbial cultural assays are generally dependent on the growth of a microbial population to form colonies on an agar plate, which are visible to the analyst. Specific conditions such as temperature, moisture content, atmosphere, and nutrient availability on solid media (agar) are used to induce this growth.

1.2.1 Agar Plate Count Methods

The gold standard method to assess microbial numbers remains the aerobic standard plate count (SPC). This cultural method has been widely and successfully used for many years in the food, pharmaceutical, and medical sectors. Serial dilutions of the sample material are prepared, plated onto agar (plate count agar), and incubated under specific conditions. When visible colonies appear, the number of CFU per gram of food can be readily calculated. The Association of Official Analytical Chemists (AOAC) Official Method 966.23 [11] and the International Organization for Standardization (ISO) (No. 4833:2003) [12] have standardized the test protocol. All alternative methods must generally be correlated or validated against these methods.

Although "gold standard" indicates the method is perfect, there are in fact some drawbacks to the method. The SPC result is often referred to as "total viable count" implying that "all" viable microorganisms will be incorporated in results of the assay. This is not so, as certain microorganisms, referred to as viable but nonculturable (VBNC) [13], may have growth requirements not met by the incubation conditions. The failure of the assay to account for these organisms may lead to an underestimation of the *true* microbial load. From a practical perspective the method is also very slow and labor intensive, requiring 3 days for the colonies to form and thus, a result to be obtained. For products with a short shelf life this delay is very impractical and a product may be in retail distribution before microbial counts are obtained.

1.2.2 Alternative Culture Methods

There are alternative agar-based methods, such as Petrifilm® (3M Microbiology Products, USA), that are AOAC accredited (Method 990.12) [14] and show comparable counts to SPC for a wide variety of meat samples [15]; although some problems have been noted [16]. Another product, SimPlate® (IDEXX Labs Inc., USA), has also been applied to meat muscle with relative success and is an approved AOAC method [17,18].

There is also an automated method based on a liquid media–based most probable number (MPN) technique (TEMPO®, bioMériuex, France). The system is based on wells containing a traditional culture media formula with a fluorescent indicator. Each well corresponds to an MPN dilution tube. Once the sample is distributed in the wells, the microorganisms metabolize the culture media producing a fluorescent signal. The system uses an MPN calculation to assess the number of microorganisms in the original sample. Apart from the obvious advantage that this type of automated instrumentation offers, the TEMPO system has a reduced incubation time (≤48 h) compared with the ISO SPC method that takes 3 days. When applied to meat samples, the technology shows a high correlation with the SPC ($r = .99$) [19].

1.3   Direct Epifluorescent Filtration Technique

An alternative approach to culture is to directly extract the microorganisms from the muscle food by membrane filtration. When concentrated onto the membrane surface, the microorganisms can be stained using a fluorescent dye and the cells then detected and enumerated using epifluorescent microscopy.

The first step in this direct epifluorescent filtration technique (DEFT) is the use of membrane filtration to recover the bacteria from the food and this step poses some challenges in relation to muscle foods. When membrane filtration is used to recover microorganisms from muscle foods, they must be first placed in a liquid media and homogenized, stomached, or pulsified (Microgen Bioproducts, UK) [20] to remove the bacteria from the food surface or matrix into the liquid diluent. A problem encountered is that food particles in the liquid have a tendency to clog the pores of the membrane during filtration. This may mean that the required volume cannot be filtered and that any food debris on the membrane can interfere with the enumeration of bacterial cells. Some approaches to improve filterability of muscle foods have been employed to physically or chemically remove as much of the food suspension as possible before filtration. These have included the use of low-speed centrifugation, appropriate surfactants such as Tween 80 and sodium dodecyl sulfate (SDS), and the proteolytic enzyme, Alcalase [21].

Once the microorganisms are concentrated onto the membrane surface, the membrane is overlaid with a fluorescent dye such as acridine orange and mounted on a glass slide. The microorganisms are viewed using fluorescent microscopy, and the total numbers of organisms in a defined number of fields of view are counted. The microscopic count is used to predict the "gold standard" plate count using a calibration curve relating the DEFT count to the aerobic plate count.

DEFT (Figure 1.1) has been applied to the estimation of microbial numbers in a range of muscle foods (Table 1.1). Although acridine orange is the most commonly used fluorescent dye, it

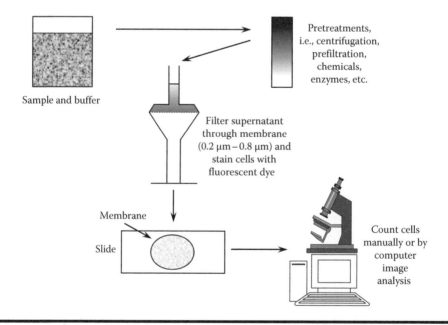

Figure 1.1   Flow diagram of a direct epifluorescent filtration technique (DEFT) for enumeration of microorganisms from muscle foods.

Table 1.1 Correlation of Direct Epifluorescent Filtration Technique (DEFT) with the Standard Aerobic Plate Count (SPC) for Enumeration of Microorganisms in a Range of Muscle Foods

Muscle Food	Treatment of Sample before Filtration through Membrane (0.4–0.8 µm)	Fluorescent Dye	Correlation with SPC	Reference
Fresh meat	Stomached 2 min	Acridine Orange	$r = .91$	[97]
Canned hams	Stomached 2 min, prefiltration through glass microfiber filter	Acridine Orange	Poor	[98]
Raw ground beef	Stomached 2 min, prefiltered through nylon filter, Triton X, and bactotrypsin	Acridine Orange	$r = .79$	[99]
Raw beef pieces	Stomached 30 s, prefiltered through glass microfiber filter, Triton X	Acridine Orange	$r^2 = .91$	[100]
Raw pork mince	Stomached 30 s, Tween 80, SDS, Alcalase 0.6 L	Acridine Orange	$r = .97$	[101]
Raw beef mince	Stomached 30 s, low-speed centrifugation, Tween 80, SDS, Alcalase 0.6 L	Acridine Orange	$r = .97$	[102]
Lamb carcasses	Stomached 30 s, low-speed centrifugation, Tween 80, SDS, Alcalase 2.5 L	Acridine Orange	$r^2 = .87$	[103]
Minced beef	Stomached 30 s, low-speed centrifugation, Tween 80, SDS, Alcalase 2.5 L	Acridine Orange	$r^2 = .97$	[21]
Processed meat (minced beef, cooked ham, bacon rashers, frozen burgers)	Stomached 30 s, low-speed centrifugation, Tween 80, SDS, Alcalase 2.5 L	*Bac*Light	$r^2 = (.90, .87, .82, .80)$	[22]

does not distinguish between live and dead cells and so may overestimate the bacterial load in processed meat samples containing large numbers of dead cells. To overcome this problem, a viability stain *Bac*Light® (Molecular Probes Inc., The Netherlands) was reported to successfully distinguish between live and dead cells and in a DEFT gave a good correlation with the SPC for microorganisms in processed meat ($r^2 = .87–.93$) [22]. The DEFT takes approximately 15–20 min to analyze one sample and so at most 20 samples can be analyzed manually in a working day.

The DEFT has been successfully automated for high throughput enumeration of microorganisms in milk samples [23]. Commercial systems for analysis of milk include the Bactoscan® (Foss,

Denmark) and Cobra® systems (Biocom, France). However, the DEFT has not been automated for muscle foods. This has hugely impacted its uptake commercially by this industry sector as, apart from the small number of samples that can be analyzed manually daily, the approach is labor intensive and requires significant operator skills. Manual enumeration is particularly difficult when there are very high or low numbers of microorganisms on the slide or when there is particulate debris on the slide. Future developments to make this approach commercially suitable for muscle foods may incorporate the initial membrane filtration approach to extract the microorganisms with an automated detection system. A solid-phase cytometry method has been proposed by D'Haese and Nelis [24] and could use a laser beam to detect microorganisms recovered onto a membrane filter. This method would potentially be very rapid and automated but a potential problem could arise from any food debris remaining on the membrane surface.

1.4 ATP Bioluminescence Methods

Enzyme-mediated light production, bioluminescence, is a widespread phenomenon in nature [25]. Bioluminescent organisms are widely distributed throughout the oceans and include bacteria, sea anemones, worms, crustaceans, and fish. Fireflies and glow worms are the best-known terrestrial organisms producing light. The principles of firefly bioluminescence were discovered over 40 years ago [26]. In the firefly bioluminescence reaction, adenosine 5′-triphosphate (ATP) reacts with the enzyme luciferase and the substrate luciferin producing a photon of light. ATP is a high-energy substance found only in living cells. It takes part in all metabolic pathways and therefore its concentration in all cells including bacterial cells is strictly regulated. When luciferin and luciferase are added to a cell suspension the amount of light emitted is proportional to the amount of ATP present. The amount of light can be measured using a photometer to give an indirect indication of the microbial population density.

$$\text{Luciferin} + \text{ATP} + \xrightarrow[\text{Mg}^{2+}]{\text{Luciferase}} \text{AMP} + \text{CO}_2 + \text{pyrophosphate} + \text{oxyluciferin} + \text{photon}$$

The firefly bioluminescence reaction has been exploited as a rapid and sensitive method for measuring cell numbers, including microbial cells.

The ATP bioluminescent assay has been widely applied to assess hygiene, based on detection of all ATP present [27,28]. However, a major problem in the use of bioluminescence to predict the microbial SPC of foods is interference from nonmicrobial ATP. If an accurate estimation of the microbial load is to be obtained, nonmicrobial somatic ATP must be destroyed before the bioluminescence test is carried out. The most common approach is the enzymatic destruction of nonmicrobial ATP, followed by release and estimation of residual ATP from the microbial cells [29,30]. Another approach is to separate the microorganisms from the rest of the material and estimate the ATP in the microbial fraction. Stannard and Wood [31] used this approach to estimate bacterial numbers in minced beef. The results show a linear relationship ($r = .94$) between colony counts and microbial ATP content in raw beef. An ATP bioluminescence test was shown by Siragusa et al. [32] to be an adequate means to assess the microbial load of poultry carcasses. This assay utilized differential extraction and filtration to separate somatic ATP from microbial ATP in a very rapid time frame. The assay required approximately 5 min to complete: approximately 3.5 min to sample and 90 s analytical time. The correlation coefficient (r) between aerobic colony counts and the ATP test was .82. Ellerbroek and Lox [33] used an ATP bioluminescence approach to investigate the total bacterial counts on poultry neck and carcasses. The correlation between the bioluminescence method and the total viable counts of neck skin samples was $r = .85$, whereas a lower correlation was reported between the bioluminescence count and the total viable counts on the carcass ($r = .66$).

Commercially available bioluminescent systems include Celsis® (Celsis International plc., UK) and Bactofoss® (Foss) but their application to date has been aimed at hygiene testing and liquid foods rather than muscle foods.

1.5 Electrical Methods

Electrical methods for assessing bacterial numbers include impedance and conductance. Impedance is the opposition to flow of an alternating electrical current in a conducting material [34]. The conductance of a solution is the charge carrying capacity of its components and capacitance is the ability to hold a charge [34].

When monitoring the growth of microorganisms, the conducting material is a microbiological medium. As microorganisms grow they utilize nutrients in the medium, converting them into smaller more highly charged molecules, for example, fatty acids, amino acids, and various organic acids [35]. If electrodes are immersed in the medium and an alternating current is applied, the metabolic activity of the microorganisms results in detectable changes in the flow of current. Typically, impedance decreases while conductivity and capacitance increase [35]. When the microbial population reaches a threshold of 10^6–10^7 CFU mL^{-1} an exponential change in impedance can be observed [34]. The elapsed time until this exponential change occurs is defined as impedance detection time and is inversely proportional to the initial microbial numbers in the sample.

The most commonly used application of impedance is shelf-life testing. This test determines whether a sample contains above or below a predetermined concentration of microorganisms. Impedance testing has been used in conjunction with a calibration curve with the SPC for a number of products including raw milk ($r = -.96$) [36], frozen vegetables (92.6% agreement between methods) [37] and meat [38]. A conductance method was used to predict microbial counts on fish [39] with a correlation of $r = -.92$ to $-.97$ using brain heart infusion.

Of all developed alternative methods to predict microbial load, the impedance technique has been most widely accepted within the food industry. Commercially available automated systems include the Malthus® (Malthus Instruments Ltd., UK) system, which measures conductance, and the Bactometer® (Bactomatic Inc., USA) system, which can measure impedance, conductance, and capacitance. Both systems can measure several hundred samples simultaneously and have detection limits of ≥1.0 CFU mL^{-1}. Using these systems to predict the SPC count on meat, correlations of $r = -.83$ and $r = -.80$ were reported for the Bactometer and Malthus machines, respectively [40]. In the muscle food sector, uptake has been in the processing sector rather than for raw foods.

1.6 *Limulus* Amoebocyte Lysate Assay

Gram-negative bacteria are important food spoilage organisms in muscle foods [41]. They differ from gram-positive bacteria in that their cell wall contains lipopolysaccharides (LPS). Based on this difference a *Limulus* amoebocyte lysate (LAL) assay method that targets LPS has been developed. LPS contains an endotoxin that activates a proteolytic enzyme found in the blood cells (amoebocytes) of the horseshoe crab (*Limulus polyphemus*). The enzyme activates a clotting reaction, which results in gel formation. The concentration of LPS is determined by making serial dilutions of the sample and noting the greatest dilution at which a gel is formed within a given time [41]. The reaction has been used to develop a colorometric assay.

LAL has been applied to the evaluation of microbial contamination on pork carcasses [42]. Although the test correlated well with coliform numbers, it did not correlate well with total numbers of organisms, indicating its limited usefulness as a spoilage indicator. However, more recently a chromogenic LAL was reported by Siragusa et al. [43] to rapidly predict microbial contamination on beef carcasses. A high correlation ($r^2 = .90$) was reported with the standard aerobic plate count.

1.7 Spectroscopic Methods

Various spectroscopic methods have been proposed as rapid, noninvasive methods for the detection of microbial spoilage in muscle foods. Such methods are based on the measurement of biochemical changes that occur in the meat as a result of the decomposition and formation of metabolites caused by the growth and enzymatic activity of microorganisms, which eventually results in food spoilage.

Fourier transform infrared (FT-IR) spectroscopy involves the observation of vibrations in molecules when excited by an infrared beam. An infrared absorbance spectrum gives a fingerprint-like spectral signature, which is characteristic of any chemical or biochemical substance [44]. Such a method is therefore potentially useful to measure biochemical changes in muscle foods due to microbial growth and could be used as an indicator for spoilage. FT-IR spectroscopy has been successfully employed to discriminate, classify, and identify microorganisms. Some examples include discrimination between *Alicyclobacillus* strains associated with spoilage in apple juice [45], the discrimination of *Staphylococcus aureus* strains from different staphylococci [46], and the setting up of a spectral database for the identification of coryneform strains [47]. Mariey et al. [48] provide a review of many other characterization methods using FT-IR. This also gives an overview of the statistical methods used to interpret spectroscopic data.

In addition to these discriminatory uses, FT-IR has shown promise for use as a spoilage detection method. FT-IR has been used to predict spoilage of chicken breasts in a rapid, reagentless, noninvasive manner [49]. The metabolic snapshot correlated well with the microbial load. Ellis et al. [50] applied FT-IR to predict microbial spoilage of beef; although the correlation with the microbial load was less accurate than for poultry.

Another spectroscopic method that has been used in recent times for detection of microbial spoilage is short-wavelength-near-infrared (SW-NIR) diffuse reflectance spectroscopy (600–1100 nm). It has the advantage over FT-IR in that it is useable through food packaging and can be used to examine bulk properties of a food due to its greater pathlength [51]. This technique was applied to predict spoilage of chicken breast muscle and the results showed that SW-NIR could be used in a partial least squares model to predict microbial load [51]. Lin et al. [52] have used this technique with success in predicting spoilage of rainbow trout.

1.8 Developmental Methods

There are a number of emerging methods and technologies that are being shown to be suitable for the rapid and specific identification or enumeration of microorganisms from clinical or liquid samples. Although most have not yet been applied to predict the total microbial flora or spoilage of muscle foods, they have the potential with further development to be applied in the future. Some of these technologies are summarized in the following sections.

1.8.1 Flow Cell Cytometry

Flow cell cytometry is a technique that can be used to detect and enumerate cells as they are passed on an individual cell basis, suspended in a stream of fluid, past a laser beam. A flow cytometer typically has several key components including a light or excitation source, a laser that emits light at a particular wavelength, and a liquid flow that moves liquid-suspended cells through the instrument past the laser and a detector, which is able to measure the brief flashes of light emitted as cells flow past the laser beam. Thus, individual cells can be detected and counted by the system. The technique has been successfully applied to the enumeration of microorganisms in raw milk [53] and milk powder [54], but it has not yet been applied to muscle foods. As described in Section 1.3, a solid-phase cytometry method could potentially be applied to muscle foods, based on the assumption that the microorganisms could be successfully extracted from the food onto a filter and the filter then scanned by a laser beam [24].

1.8.2 Molecular Methods

Major advances in biotechnology have rapidly progressed the use of genetic tools for microbial detection. In particular, developments in the level of genomic information available for foodborne pathogens have been widely exploited in methods to detect and genetically characterize microorganisms. Genetic tools are now commonly used to detect specific pathogens or groups of spoilage microorganisms. However, to date the use of molecular technology to detect and enumerate, in a single assay, all microorganisms in a food sample is limited by the huge diversity of microorganisms likely to be present and identification of a common gene target present in all the foodborne microorganisms. The use of the 16S ribosomal RNA (rRNA) gene has been reported for this purpose [55]. If technological complexity can be overcome, this approach has enormous potential as a very rapid and specific test to predict spoilage in muscle foods.

1.8.2.1 Polymerase Chain Reaction

Nucleic acid methods that include an amplification step for the target DNA/RNA are now routinely employed in molecular biology. These amplification methods can increase the target nucleic acid material more than a billion fold and are particularly important in the arena of food microbiology where one of the major hurdles is the recovery and detection of very low numbers of a particular species. The most popular method of amplification is the polymerase chain reaction (PCR) technique (Figure 1.2). In PCR, a nucleic acid target (DNA) is extracted from the cell and denatured into single-stranded nucleic acid. An oligonucleotide primer pair specific for the selected gene target, along with an enzyme (usually *Taq* polymerase, a thermostable and thermoactive enzyme originally derived from *Thermus aquaticus*) in the presence of free deoxynucleoside triphosphates (dNTPs), is used to amplify the gene target exponentially, resulting in a double replication of the starting target material. This reaction is carried out in an automated, programmable block heater called a thermocycler, which provides the necessary thermal conditions needed to achieve amplification. Following amplification, the PCR products are separated by gel electrophoresis, stained with ethidium bromide, and visualized using ultraviolet light. This type of PCR, sometimes referred to as conventional PCR, can be used for the identification of specific groups of spoilage bacteria in meat including lactic acid bacteria [56,57].

A quantitative method using electrochemiluminescence to measure the PCR product was applied to predict the spoilage bacterial load on aerobically stored meat [58]. The correlation of

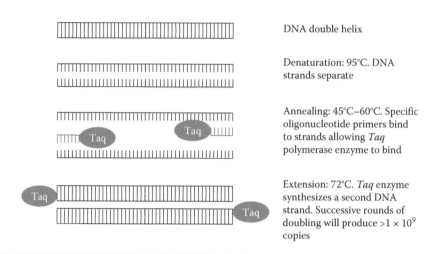

	DNA double helix
	Denaturation: 95°C. DNA strands separate
	Annealing: 45°C–60°C. Specific oligonucleotide primers bind to strands allowing *Taq* polymerase enzyme to bind
	Extension: 72°C. *Taq* enzyme synthesizes a second DNA strand. Successive rounds of doubling will produce >1 × 10⁹ copies

Figure 1.2   Diagram showing the main events in a typical polymerase chain reaction (PCR).

this method with the SPC was $r = .94$. Gutierrez et al. [55] combined conventional PCR with an enzyme-linked immunosorbent assays (ELISA) to allow enumeration of microorganisms. On applying this technique to the detection of the microbial load in meat samples, a good correlation was achieved between the SPC counts and the PCR-ELISA ($r = .95$). The authors did express concerns about the complexity of the assay and about its suitability as a routine assay.

Recently, a more advanced quantitative PCR technology, in the form of real-time PCR (RT-PCR), has entered and revolutionized the area of molecular biology [59]. RT-PCR allows continuous monitoring of the amplification process through the use of fluorescent double-stranded DNA intercalating dyes or sequence-specific probes [60]. The amount of fluorescence after each amplification cycle can be measured and visualized in real time on a computer monitor attached to the RT-PCR machine. A number of dye chemistries have been reported for use in RT-PCR, from DNA-binding dyes such as SYBR® Green (Molecular Probes Inc.) to more complex fluorescent probe technologies such as TaqMan® (Roche Molecular Systems, UK), molecular beacons, and HybProbes® (Roche Molecular Systems) [61]. Whatever signal chemistry is used, RT-PCR not only allows quick determination of the presence/absence of a particular target, but can also be used for the quantification of a target that may then be related to microbial counts.

To quantify microorganisms by RT-PCR, a set of standards of known concentration must first be analyzed. The standards may be of known CFU per milliliter or known gene copy number and then related to the C_T (threshold cycle) of the reaction to generate a standard curve, which can be used to quantify unknown samples. An important factor to be considered when quantifying bacteria is that relying on a DNA-based RT-PCR will lead to a count that comprises of live, dead, and VBNC bacteria, which could potentially lead to an overestimation of the numbers present. A way to overcome this is by coupling RT-PCR with reverse transcription. This technique transcribes RNA (present only in viable cells) into complementary DNA (cDNA), which can then be employed in a RT-PCR reaction. Because the cDNA originates from RNA the quantification will be based on viable cells only, leading to a more accurate determination of the number of metabolically active bacteria.

To date, RT-PCR has been mainly used for the sensitive and rapid detection of a wide range of pathogens, such as *Salmonella* spp., *Escherichia coli*, and *Listeria* on meat [62–65], and a multiplex assay that has the ability to detect more than one pathogen at the same time has been described [66]. The quantitative feature of RT-PCR has been examined for the enumeration of the spoilage organism *Lactobacillus sakei* in meat products, and the application of a live staining method in combination with real-time technology for quantitative analysis has been reported using model organisms [67,68]. There is huge potential for this technology to quantify total microorganisms using an RNA gene target common to all microflora likely to be present.

1.8.2.2 Fluorescent In Situ Hybridization

In situ hybridization (ISH) using radiolabeled DNA was first reported by Pardue and Gall [69] and John et al. [70] for direct examination of cells. It was applied to bacteria for the first time in 1988 [71], and with the advent of fluorescent labels the technique became more widely used [72]. Fluorescent *in situ* hybridization (FISH) is a technique that specifically detects nucleic acid sequences in a cell using a fluorescently labeled probe that hybridizes specifically to its complementary target gene within the intact cell. The target gene is the intercellular rRNA in the microorganism as these genes are relatively stable, occur in high copy numbers, and have variable and conserved sequence domains, which allows for the design of discriminatory probes either specific to an individual species or to particular genera [73]. FISH generally involves four steps, fixation of the sample, permeabilization of cells to release the nucleic acids, hybridization with the fluorescent labeled probe, and detection by fluorescent microscopy. Traditionally, FISH methods have been implemented using DNA oligonucleotide probes. A typical oligonucleotide probe is between 15 and 30 base pairs in length. Short probes have easier access to the target but also may have fewer labels [74]. There are a number of ways in which probes can be labeled. Direct labeling is most commonly used where the fluorescent molecule is directly bound to the oligonucleotide either chemically during synthesis or enzymatically using terminal transferase at the 3′-end. This method is considered to be the fastest, cheapest, and most convenient [75]. Sensitivity of FISH assays can be increased using indirect labeling, where the probe is linked to a reporter molecule that is detected by a fluorescent antibody [76] or where the probe is linked to an enzyme and a fluorescent substrate can be added [77]. A more recent development in probe technology is the development of peptide nucleic acid (PNA) probes. PNAs are uncharged DNA analogs in which the negatively charged sugar–phosphate backbone is replaced by an achiral sugar–phosphate backbone formed by repetitive units of *N*-(2-aminoethyl) glycine [78]. PNA probes can hybridize to target nucleic acids more rapidly and with higher affinity and specificity than DNA probes [79].

FISH technology has been applied to the detection of bacterial pathogens in clinical and food samples [80,81] and can allow direct identification and quantification of microbial species. A FISH assay has been developed for the *Pseudomonas* genus, which is important in milk spoilage, allowing for the specific detection and enumeration of this group of organisms in milk much more rapidly than a cultural method [82]. In the wine industry, lactic acid bacteria can be detrimental or beneficial depending on the species, and when they develop in the process. A FISH technique has been described as the one that utilizes probes to differentiate between different LAB genera so that it is possible to identify potential spoilage strains from the species responsible for successful fermentation [83]. FISH technology has potential as a method to enumerate all microorganisms in a food sample using a common gene target but could be limited in its uptake by its current reliance on microscope-based detection.

1.9 Electronic Nose

It is well known that microorganisms produce a range of volatiles as they grow on food and can be used to identify particular species of microorganisms that have a unique volatile fingerprint or potentially, to determine the total level of microbial contamination on a food and predict spoilage. Gardener and Bartlett [84] defined an electronic nose as "an instrument which comprises an array of electronic, chemical sensors with partial specificity and an appropriate pattern recognition system, capable of recognising simple or complex odours." An electronic nose normally consists of a vapor-phase flow over the sensor, interaction with the sensor, and analyses of the interaction using computer software. The field of sensor development is highly active and includes a range of sensor types based on metal oxide, metal oxide silicon, piezoelectric, surface acoustic waves, optical, and electrochemical premises [85].

Blixt and Borch [86] investigated the use of an electronic nose to predict the spoilage of vacuum-packaged beef. The volatile compounds were analyzed using an electronic nose containing a sensory array composed of 10 metal oxide semiconductor field-effect transistors, four Tagushi type sensors, and one CO_2-sensitive sensor. Two of the Tagushi sensors performed best and correlated well with evaluation of spoilage by a sensory panel. They did not attempt to correlate the results with microbial counts.

Du et al. [87] used an electronic nose (AromaScan) to predict spoilage of yellowfin tuna fish. The change in fish quality as determined by AromaScan (AromaScan plc., UK) followed increases in microbiological counts in tuna fillets, indicating that electronic nose devices can be used in conjunction with microbial counts and sensory panels to evaluate the degree of decomposition in tuna during storage.

1.10 Time–Temperature Integrators

One of the key contributors to the spoilage of fresh muscle foods is a breakdown in the chill chain during distribution. The prediction of spoilage and the application of an optimized quality and safety assurance scheme for chilled storage and distribution of fresh meat and meat products would be greatly aided by the continuous monitoring of temperature during distribution and storage.

A time–temperature integrator (TTI) is defined as a small, inexpensive device that can be incorporated into a food package to show a visible change according to the time and temperature history of the stored food [88]. TTIs are devices that contain a thermally labile substance, which can be biological (microbiological or enzymatic), chemical, or physical. Of these groups, biological TTIs are the best studied. TTIs can be used to determine both whether a heat treatment has worked effectively and whether temperature abuse has occurred during storage. Different types of TTIs have been used for determining the effectiveness of heat treatment. α-Amylase from *Bacillus* species has been evaluated in a number studies [88–91] and a recent report describes the use of amylase from the hyperthermophile *Pyrococcus furiosus* as a sterilization TTI [92].

TTIs can be used to monitor temperature abuse during storage, transportation, and handling and thus to monitor such abuses that may lead to a shortened shelf life and spoilage. This type of TTI must (1) be easily activated and sensitive; (2) provide a high degree of precision; (3) have tamper-evident characteristics; (4) have a response that is irreversible, reproducible, and correlated with food quality changes; (5) have determined physical and chemical characteristics; and (6) have an easily readable response [93].

Giannakourou et al. [94] demonstrated that TTI readings using a commercially available enzyme could be adequately correlated to the remaining shelf life of the product (in this case marine-cultured gilt-head seabream) at any point of its distribution. The same TTI has also shown positive results for fresh chicken storage [95]; although the TTI predictions would be inaccurate following an extreme instance of temperature abuse, with the TTI indicator changing color before the product had actually spoiled [96].

Numerous time–temperature indicators are commercially available for all types of food stuffs. However, it is necessary to validate the TTI of choice with the product and process of choice before correlations can be made between the TTI and the potential for spoilage of the product.

1.11 Conclusion

Muscle foods pose considerably more challenges than other foods for development and successful application of spoilage detection methods. These include a highly complex food tissue matrix in which the microorganisms may be embedded and strongly attached, and from which they must be detached to detect and enumerate the microorganisms. This means the food must generally be placed in a liquid diluent and then physically manipulated to release the microorganisms into the liquid. This dilution effect obviously creates a need for a more sensitive detection method than a sample to which the method could be applied directly, such as a liquid food (i.e., milk). In addition, the microflora is generally quite diverse and the dominant flora is very much dependent on the storage environment. At the early stage of the food process the levels of microorganisms on the raw muscle food may be as low as log 2.0 CFU g^{-1}, thus posing additional challenges for the sensitivity of the detection method.

The gold standard method to predict spoilage remains the aerobic SPC, but it is still required by the industry that alternative methods are validated against this method. However, as previously described in this chapter, the SPC is far from perfect and more rapid methods to predict spoilage are urgently needed by the muscle foods sector (fish, meat, and poultry industries). These alternative spoilage methods must generally give results that are comparable and validated against "gold standard" cultural microbial methods. The methods must be sensitive, rapid, suited to online use, and at least semiautomated. They must be suited to routine use, without the need for highly skilled operators, as high staff turnover is often a major issue in the muscle food industry, and it is neither practical nor possible to keep retraining staff to carry out a test that is highly complex.

The muscle food sector has undoubtedly been the slowest sector in the food industry to take on board alternative technologies for spoilage prediction. However, they are now being compelled by their customers, regulatory authorities, and consumers to more accurately predict shelf life. This is even more pertinent with the continued market move toward chilled prepared foods with minimal preservatives and a short shelf life. Although there has been much research and development in the area of rapid spoilage detection methods, recent research on rapid microbial methods has tended to focus more on methods for identification of specific species of microorganisms rather than on the total microbial load. However, some of the emerging technologies developed, albeit for other applications, have enormous potential to be further developed for enumeration of the total microbial load and to predict spoilage. More research efforts should now be refocused in this direction using the newer technologies to overcome the hurdles that have to date prevented the widespread uptake of rapid methods to predict spoilage by the muscle food sector.

References

1. Bell, R.G., Distribution and sources of microbial contamination on beef carcasses. *J Appl Microbiol* 82, 292, 1997.

2. Elder, R.O., Keen, J.E., Siragusa, G.R., Barkocy-Gallagher, G.A., Koohmaraie, M. and Laegreid, W.W., Correlation of enterohemorrhagic *Escherichia coli* O157 prevalence in feces, hides, and carcasses of beef cattle during processing. *Proc Natl Acad Sci USA* 97, 2999, 2000.

3. Ercolini, D., Russo, F., Torrieri, E., Masi, P. and Villani, F., Changes in the spoilage-related microbiota of beef during refrigerated storage under different packaging conditions. *Appl Environ Microbiol* 72, 4663, 2006.

4. Hinton, A., Jr., Cason, J.A. and Ingram, K.D., Tracking spoilage bacteria in commercial poultry processing and refrigerated storage of poultry carcasses. *Int J Food Microbiol* 91, 155, 2004.

5. Borch, E., Kant-Muermans, M.L. and Blixt, Y., Bacterial spoilage of meat and cured meat products. *Int J Food Microbiol* 33, 103, 1996.

6. Huis in't Veld, J.H.J., Microbial and biochemical spoilage of foods: an overview. *Int J Food Microbiol* 33, 1, 1996.

7. Gustavsson, P. and Borch, E., Contamination of beef carcasses by psychrotrophic *Pseudomonas* and *Enterobacteriaceae* at different stages along the processing line. *Int J Food Microbiol* 20, 2, 1993.

8. Dainty, R.H., Edwards, R.A. and Hibbard, C.M., Time course of volatile compound formation during refrigerated storage of naturally contaminated beef in air. *J Appl Bacteriol* 59, 303, 1985.

9. Pin, C., Garcia de Fernando, G.D. and Ordonez, J.A., Effect of modified atmosphere composition on the metabolism of glucose by *Brochothrix thermosphacta*. *Appl Environ Microbiol* 68, 4441, 2002.

10. Chenoll, E., Macian, M.C., Elizaquivel, P. and Aznar, R., Lactic acid bacteria associated with vacuum-packed cooked meat product spoilage: population analysis by rDNA-based methods. *J Appl Microbiol* 102, 498, 2007.

11. AOAC, *Official Method of Analysis*, 15th ed., AOAC, Washington, DC, 1990.

12. International Organization for Standardization, *EN ISO 4833:2003*, Microbiology of food and animal feeding stuffs—horizontal method for the enumeration of microorganisms—colony count technique at 30°C, International Organization for Standardization, Geneva, Switzerland, 2003.

13. Besnard, V., Federighi, M., Declerq, E., Jugiau, F. and Cappelier, J.M., Environmental and physicochemical factors induce VBNC state in *Listeria monocytogenes*. *Vet Res* 33, 359, 2002.

14. AOAC, *Official Method of Analysis*, 16th ed., AOAC, Washington, DC, 1995.

15. Park, Y.H., Seo, K.S., Ahn, J.S., Yoo, H.S. and Kim, S.P., Evaluation of the Petrifilm plate method for the enumeration of aerobic microorganisms and coliforms in retailed meat samples. *J Food Prot* 64, 1841, 2001.

16. Dawkins, G.S., Hollingsworth, J.B. and Hamilton, M.A., Incidences of problematic organisms on Petrifilm aerobic count plates used to enumerate selected meat and dairy products. *J Food Prot* 68, 1506, 2005.

17. Beuchat, L.R., Copeland, F., Curiale, M.S., Danisavich, T., Gangar, V., King, B.W., Lawlis, T.L., Likin, R.O., Okwusoa, J., Smith, C.F. and Townsend, D.E., Comparison of the SimPlate total plate count method with Petrifilm, Redigel, conventional pour-plate methods for enumerating aerobic microorganisms in foods. *J Food Prot* 61, 14, 1998.

18. Feldsine, P.T., Leung, S.C., Lienau, A.H., Mui, L.A. and Townsend, D.E., Enumeration of total aerobic microorganisms in foods by SimPlate Total Plate Count-Color Indicator methods and conventional culture methods: collaborative study. *J AOAC Int* 86, 257, 2003.

19. Paulsen, P., Schopf, E. and Smulders, F.J., Enumeration of total aerobic bacteria and *Escherichia coli* in minced meat and on carcass surface samples with an automated most-probable-number method compared with colony count protocols. *J Food Prot* 69, 2500, 2006.

20. Kang, D.-H., Dougherty, R.H. and Fung, D.Y.C., Comparison of Pulsifier and Stomacher to detach microorganisms from lean meat tissues. *J Rapid Methods Auto Microbiol* 9, 27, 2001.

21. Duffy, G., Sheridan, J.J., McDowell, D.A., Blair, I. and Harrington, D., The use of Alcalase 2.5L in the acridine orange direct count technique for the rapid enumeration of bacteria in beef mince. *Lett Appl Microbiol* 13, 198, 1991.

22. Duffy, G. and Sheridan, J.J., Viability staining in a direct count rapid method for the determination of total viable counts on processed meats. *J Microbiol Methods* 31, 167, 1998.

23. Hermida, M., Taboada, M., Menendez, S. and Rodriguez-Otero, J.L., Semi-automated direct epifluorescent filter technique for total bacterial count in raw milk. *J AOAC Int* 83, 1345, 2000.

24. D'Haese, E. and Nelis, H.J., Rapid detection of single cell bacteria as a novel approach in food microbiology. *J AOAC Int* 85, 979, 2002.

25. Thore, A., Technical aspects of the bioluminescent firefly luciferase assay of ATP. *Sci. Tools* 26, 30, 1979.

26. McElroy, W.D., The energy source for bioluminescence in an isolated system. *Proc Natl Acad Sci USA* 33, 342, 1947.

27. Davidson, C.A., Griffith, C.J., Peters, A.C. and Fielding, L.M., Evaluation of two methods for monitoring surface cleanliness-ATP bioluminescence and traditional hygiene swabbing. *Luminescence* 14, 33, 1999.

28. Aycicek, H., Oguz, U. and Karci, K., Comparison of results of ATP bioluminescence and traditional hygiene swabbing methods for the determination of surface cleanliness at a hospital kitchen. *Int J Hyg Environ Health* 209, 203, 2006.

29. Bossuyt, R., Determination of bacteriological quality of raw milk by an ATP assay technique. *Milchwissenschaft* 36, 257, 1981.

30. Bossuyt, R., A 5 minute ATP platform test for judging the bacteriological quality of raw milk. *Neth Milk Dairy J* 36, 355, 1982.

31. Stannard, C.J. and Wood, J.M., The rapid estimation of microbial contamination of raw meat by measurement of adenosine triphosphate (ATP). *J Appl Bacteriol* 55, 429, 1983.

32. Siragusa, G.R., Dorsa, W.J., Cutter, C.N., Perino, L.J. and Koohmaraie, M., Use of a newly developed rapid microbial ATP bioluminescence assay to detect microbial contamination on poultry carcasses. *J Biolumin Chemilumin* 11, 297, 1996.

33. Ellerbroek, L. and Lox, C., The use of neck skin for microbial process control of fresh poultry meat using the bioluminescence method. *Dtsch Tierarztl Wochenschr* 111, 181, 2004.

34. Firstenberg-Eden, R., Electrical impedance method for determining microbial quality of foods. In: *Foodborne Microorganisms and Their Toxins: Developing Methodology*, Pierson, M.D. and Stern, N.J., Eds., Marcel Dekker, New York, NY, 679, 1985.

35. Dziezak, J.D., Rapid methods for microbiological analysis of food. *Food Technol* 41, 56, 1987.

36. Firstenberg-Eden, R. and Tricarico, M.K., Impedimetric determination of total, mesophilic and psychrotrophic counts in raw milk. *J Food Sci* 48, 1750, 1983.

37. Hardy, D., Kraeger, S.J., Dufour, S.W. and Cady, P., Rapid detection of microbial contamination in frozen vegetables by automated impedance measurements. *Appl Environ Microbiol* 34, 14, 1977.

38. Firstenberg-Eden, R., Rapid estimation of the number of microorganisms in raw meat by impedance measurements. *Food Technol* 37, 64, 1983.

39. Ogden, I.D., Use of conductance methods to predict bacterial counts in fish. *J Appl Bacteriol* 61, 263, 1986.

40. Bollinger, S., Casella, M. and Teuber, M., Comparative impedance evaluation of the microbial load of different foodstuffs. *Lebensm Wiss Technol* 27, 177, 1994.

41. Heeschen, W., Sudi, J. and Suhren, G., Application of the *Limulus* test for detection of gram negative micro-organisms in milk and dairy products. In: *Rapid Methods and Automation in Microbiology and Immunology*, Habermahl, K.O., Ed., Springer-Verlag, Berlin, 638, 1985.

42. Misawa, N., Kumamoto, K., Nyuta, S. and Tuneyoshi, M., Application of the *Limulus* amoebocyte lysate test as an indicator of microbial contamination in pork carcasses. *J Vet Med Sci* 57, 351, 1995.

43. Siragusa, G.R., Kang, D.H. and Cutter, C.N., Monitoring the microbial contamination of beef carcass tissue with a rapid chromogenic *Limulus* amoebocyte lysate endpoint assay. *Lett Appl Microbiol* 31, 178, 2000.

44. Gillie, J.K., Hochlowski, J. and Arbuckle-Keil, G.A., Infrared spectroscopy. *Anal Chem* 72, 71R, 2000.

45. Lin, M., Al-Holy, M., Chang, S.S., Huang, Y., Cavinato, A.G., Kang, D.H. and Rasco, B.A., Rapid discrimination of *Alicyclobacillus* strains in apple juice by Fourier transform infrared spectroscopy. *Int J Food Microbiol* 105, 369, 2005.

46. Lamprell, H., Mazerolles, G., Kodjo, A., Chamba, J.F., Noel, Y. and Beuvier, E., Discrimination of *Staphylococcus aureus* strains from different species of *Staphylococcus* using Fourier transform infrared (FT-IR) spectroscopy. *Int J Food Microbiol* 108, 125, 2006.

47. Oberreuter, H., Seiler, H. and Scherer, S., Identification of coryneform bacteria and related taxa by Fourier-transform infrared (FT-IR) spectroscopy. *Int J Syst Evol Microbiol* 52, 91, 2002.

48. Mariey, L., Signolle, J.P., Amiel, C. and Travert, J., Discrimination, classification, identification of microorganisms using FT-IR spectroscopy and chemometrics. *Vib Spectrosc* 26, 151, 2001.

49. Ellis, D.I., Broadhurst, D., Kell, D.B., Rowland, J.J. and Goodacre, R., Rapid and quantitative detection of the microbial spoilage of meat by Fourier transform infrared spectroscopy and machine learning. *Appl Environ Microbiol* 68, 2822, 2002.

50. Ellis, D.I., Broadhurst, D. and Goodacre, R., Rapid and quantitative detection of the microbial spoilage of beef by Fourier transform infrared spectroscopy and machine learning. *Anal Chim Acta* 514, 193, 2004.

51. Lin, M., Al-Holy, M., Mousavi-Hesary, M., Al-Qadiri, H., Cavinato, A.G. and Rasco, B.A., Rapid and quantitative detection of the microbial spoilage in chicken meat by diffuse reflectance spectroscopy (600–1100 nm). *Lett Appl Microbiol* 39, 148, 2004.

52. Lin, M., Mousavi, M., Al-Holy, M., Cavinato, A.G. and Rasco, B.A., Rapid near infrared spectroscopic method for the detection of spoilage in rainbow trout (*Oncorhynchus mykiss*) fillet. *J Food Sci* 71, S18, 2006.

53. Holm, C., Mathiasen, T. and Jespersen, L., A flow cytometric technique for quantification and differentiation of bacteria in bulk tank milk. *J Appl Microbiol* 97, 935, 2004.

54. Flint, S., Walker, K., Waters, B. and Crawford, R., Description and validation of a rapid (1 h) flow cytometry test for enumerating thermophilic bacteria in milk powders. *J Appl Microbiol* 102, 909, 2007.

55. Gutierrez, R., Garcia, T., Gonzalez, I., Sanz, B., Hernandez, P.E. and Martin, R., Quantitative detection of meat spoilage bacteria by using the polymerase chain reaction (PCR) and an enzyme linked immunosorbent assay (ELISA). *Lett Appl Microbiol* 26, 372, 1998.

56. Yost, C.K. and Nattress, F.M., The use of multiplex PCR reactions to characterize populations of lactic acid bacteria associated with meat spoilage. *Lett Appl Microbiol* 31, 129, 2000.

57. Goto, S., Takahashi, H., Kawasaki, S., Kimura, B., Fujii, T., Nakatsuji, M. and Watanabe, I., Detection of *Leuconostoc* strains at a meat processing plant using polymerase chain reaction. *Shokuhin Eiseigaku Zasshi* 45, 25, 2004.

58. Venkitanarayanan, K.S., Faustman, C., Crivello, J.F., Khan, M.I., Hoagland, T.A. and Berry, B.W., Rapid estimation of spoilage bacterial load in aerobically stored meat by a quantitative polymerase chain reaction. *J Appl Microbiol* 82, 359, 1997.

59. Bellin, T., Pulz, M., Matussek, A., Hempen, H.G. and Gunzer, F., Rapid detection of enterohemorrhagic *Escherichia coli* by real-time PCR with fluorescent hybridization probes. *J Clin Microbiol* 39, 370, 2001.

60. Wittwer, C.T., Herrmann, M.G., Moss, A.A. and Rasmussen, R.P., Continuous fluorescence monitoring of rapid cycle DNA amplification. *Biotechniques* 22, 130–131, 134, 1997.

61. Mackay, I.M., Real-time PCR in the microbiology laboratory. *Clin Microbiol Infect* 10, 190, 2004.

62. Josefsen, M.H., Krause, M., Hansen, F. and Hoorfar, J., Optimization of a 12-hour TaqMan PCR-based method for detection of *Salmonella* bacteria in meat. *Appl Environ Microbiol* 73, 3040, 2007.

63. Navas, J., Ortiz, S., Lopez, P., Jantzen, M.M., Lopez, V. and Martinez-Suarez, J.V., Evaluation of effects of primary and secondary enrichment for the detection of *Listeria monocytogenes* by real-time PCR in retail ground chicken meat. *Foodborne Pathog Dis* 3, 347, 2006.

64. Perelle, S., Dilasser, F., Grout, J. and Fach, P., Screening food raw materials for the presence of the world's most frequent clinical cases of Shiga toxin-encoding *Escherichia coli* O26, O103, O111, O145 and O157. *Int J Food Microbiol* 113, 284, 2007.

65. Holicka, J., Guy, R.A., Kapoor, A., Shepherd, D. and Horgen, P.A., A rapid (one day), sensitive real-time polymerase chain reaction assay for detecting *Escherichia coli* O157:H7 in ground beef. *Can J Microbiol* 52, 992, 2006.

66. Nguyen, L.T., Gillespie, B.E., Nam, H.M., Murinda, S.E. and Oliver, S.P., Detection of *Escherichia coli* O157:H7 and *Listeria monocytogenes* in beef products by real-time polymerase chain reaction. *Foodborne Pathog Dis* 1, 231, 2004.

67. Rudi, K., Moen, B., Dromtorp, S.M. and Holck, A.L., Use of ethidium monoazide and PCR in combination for quantification of viable and dead cells in complex samples. *Appl Environ Microbiol* 71, 1018, 2005.

68. Guy, R.A., Kapoor, A., Holicka, J., Shepherd, D. and Horgen, P.A., A rapid molecular-based assay for direct quantification of viable bacteria in slaughterhouses. *J Food Prot* 69, 1265, 2006.

69. Pardue, M.L. and Gall, J.G., Molecular hybridization of radioactive DNA to the DNA of cytological preparations. *Proc Natl Acad Sci USA* 64, 600, 1969.

70. John, H.A., Birnstiel, M.L. and Jones, K.W., RNA–DNA hybrids at the cytological level. *Nature* 223, 582, 1969.

71. Giovannoni, S.J., DeLong, E.F., Olsen, G.J. and Pace, N.R., Phylogenetic group-specific oligodeoxy-nucleotide probes for identification of single microbial cells. *J Bacteriol* 170, 720, 1988.

72. Amann, R.I., Krumholz, L. and Stahl, D.A., Fluorescent-oligonucleotide probing of whole cells for determinative, phylogenetic, and environmental studies in microbiology. *J Bacteriol* 172, 762, 1990.

73. Amann, R., Fuchs, B.M. and Behrens, S., The identification of microorganisms by fluorescence *in situ* hybridisation. *Curr Opin Biotechnol* 12, 231, 2001.

74. Bottari, B., Ercolini, D., Gatti, M. and Neviani, E., Application of FISH technology for microbiological analysis: current state and prospects. *Appl Microbiol Biotechnol* 73, 485, 2006.

75. Moter, A. and Gobel, U.B., Fluorescence in situ hybridization (FISH) for direct visualization of microorganisms. *J Microbiol Methods* 41, 85, 2000.

76. Zarda, B., Amann, R., Wallner, G. and Schleifer, K.H., Identification of single bacterial cells using digoxigenin-labelled, rRNA-targeted oligonucleotides. *J Gen Microbiol* 137, 2823, 1991.

77. Schonhuber, W., Fuchs, B., Juretschko, S. and Amann, R., Improved sensitivity of whole-cell hybridization by the combination of horseradish peroxidase-labeled oligonucleotides and tyramide signal amplification. *Appl Environ Microbiol* 63, 3268, 1997.

78. Lehtola, M.J., Loades, C.J. and Keevil, C.W., Advantages of peptide nucleic acid oligonucleotides for sensitive site directed 16S rRNA fluorescence in situ hybridization (FISH) detection of *Campylobacter jejuni*, *Campylobacter coli* and *Campylobacter lari*. *J Microbiol Methods* 62, 211, 2005.

79. Jain, K.K., Current status of fluorescent in-situ hybridisation. *Med Device Technol* 15, 14, 2004.

80. Stender, H., Lund, K., Petersen, K.H., Rasmussen, O.F., Hongmanee, P., Miorner, H. and Godtfredsen, S.E., Fluorescence *in situ* hybridization assay using peptide nucleic acid probes for differentiation between tuberculous and nontuberculous *Mycobacterium* species in smears of *Mycobacterium* cultures. *J Clin Microbiol* 37, 2760, 1999.

81. Fang, Q., Brockmann, S., Botzenhart, K. and Wiedenmann, A., Improved detection of *Salmonella* spp. in foods by fluorescent in situ hybridization with 23S rRNA probes: a comparison with conventional culture methods. *J Food Prot* 66, 723, 2003.

82. Gunasekera, T.S., Dorsch, M.R., Slade, M.B. and Veal, D.A., Specific detection of *Pseudomonas* spp. in milk by fluorescence in situ hybridization using ribosomal RNA directed probes. *J Appl Microbiol* 94, 936, 2003.

83. Blasco, L., Ferrer, S. and Pardo, I., Development of specific fluorescent oligonucleotide probes for in situ identification of wine lactic acid bacteria. *FEMS Microbiol Lett* 225, 115, 2003.

84. Gardener, J.W. and Bartlett, P.N., *Electronic Noses—Principles and Applications*, Oxford University Press, Oxford, 1999.

85. Magan, N. and Sahgal, N., Electronic nose for quality and safety control. In: *Advances in Food Diagnostics*, Nollet, L.M.L., Toldrá, F. and Hui, Y.H., Eds., Blackwell Publishing, Oxford, 119, 2007.

86. Blixt, Y. and Borch, E., Using an electronic nose for determining the spoilage of vacuum-packaged beef. *Int J Food Microbiol* 46, 123, 1999.

87. Du, W.X., Kim, J., Cornell, J.A., Huang, T., Marshall, M.R. and Wei, C.I., Microbiological, sensory, and electronic nose evaluation of yellowfin tuna under various storage conditions. *J Food Prot* 64, 2027, 2001.

88. Van Loey, A., Hendrickx, M., Ludikhuyze, L., Weemaes, C., Haentjens, T., De Cordt, S. and Tobback, P., Potential *Bacillus subtilis* alpha-amylase-based time–temperature integrators to evaluate pasteurization processes. *J Food Prot* 59, 261, 1996.

89. Mehauden, K., Cox, P.W., Bakalis, S., Simmons, M.J.H., Tucker, G.S. and Fryer, P.J., A novel method to evaluate the applicability of time temperature integrators to different temperature profiles. *Innov Food Sci Emerg Technol* 8, 507–514, 2007.

90. Guiavarch, Y., Zuber, F., van Loey, A. and Hendrickx, M., Combined use of two single-component enzymatic time–temperature integrators: application to industrial continuous rotary processing of canned ravioli. *J Food Prot* 68, 375, 2005.

91. Guiavarc'h, Y., Van Loey, A., Zuber, F. and Hendrickx, M., Development characterization and use of a high-performance enzymatic time–temperature integrator for the control of sterilization process' impacts. *Biotechnol Bioeng* 88, 15, 2004.

92. Tucker, G.S., Brown, H.M., Fryer, P.J., Cox, P.W., Poole II, F.L., Lee, H.S. and Adams, M.W.W., A sterilisation time–temperature integrator based on amylase from the hyperthermophilic organism *Pyrococcus furiosus. Innov Food Sci Emerg Technol* 8, 63, 2007.

93. Ozdemir, M. and Floros, J.D., Active food packaging technologies. *Crit Rev Food Sci Nutr* 44, 185, 2004.

94. Giannakourou, M.C., Koutsoumanis, K., Nychas, G.J. and Taoukis, P.S., Field evaluation of the application of time temperature integrators for monitoring fish quality in the chill chain. *Int J Food Microbiol* 102, 323, 2005.

95. Moore, C.M. and Sheldon, B.W., Use of time–temperature integrators and predictive modeling to evaluate microbiological quality loss in poultry products. *J Food Prot* 66, 280, 2003.

96. Moore, C.M. and Sheldon, B.W., Evaluation of time–temperature integrators for tracking poultry product quality throughout the chill chain. *J Food Prot* 66, 287, 2003.

97. Pettipher, G.L. and Rodrigues, U.M., Rapid enumeration of microorganisms in foods by the direct epifluorescent filter technique. *Appl Environ Microbiol* 44, 809, 1982.

98. Liberski, D.J.A., Bacteriological examinations of chilled, cured canned pork hams and shoulders using a conventional microbiological technique and the DEFT method. *Int J Food Microbiol* 10, 19, 1990.

99. Qvist, S.H. and Jakobsen, M., Application of the direct epifluorescent filter technique as a rapid method in microbiological quality assurance in the meat industry. *Int J Food Microbiol* 2, 139, 1985.

100. Walls, I., Sheridan, J.J. and Levett, P.N., A rapid method of enumerating microorganisms from beef, using an acridine orange direct-count technique. *Irish J Food Sci Technol* 13, 23, 1989.

101. Sheridan, J.J., Walls, I. and Levett, P.N., Development of a rapid method for enumeration of bacteria in pork mince. *Irish J Food Sci Technol* 14, 1, 1990.

102. Walls, I., Sheridan, J.J., Welch, R.W. and McDowell, D.A., Separation of microorganisms from meat and their rapid enumeration using a membrane filtration-epifluorescent microscopy technique. *Lett Appl Microbiol* 10, 23, 1990.

103. Sierra, M.L., Sheridan, J.J. and McGuire, L., Microbial quality of lamb carcasses during processing and the acridine orange direct count technique (a modified DEFT) for rapid enumeration of total viable counts. *Int J Food Microbiol* 36, 61, 1997.

Chapter 2

Microbial Foodborne Pathogens

Marios Mataragas and Eleftherios H. Drosinos

Contents

2.1 Introduction 22
2.2 Cultural Methods 22
    2.2.1 Enumeration Methods 23
        2.2.1.1 Plate Count 23
        2.2.1.2 Most Probable Number 24
    2.2.2 Detection Methods 24
2.3 Alternative or Rapid Microbiological Methods 24
    2.3.1 Methods with a Concentration Step 25
    2.3.2 Detection and Enumeration Methods 25
2.4 *Listeria monocytogenes* 27
    2.4.1 Detection of *Listeria monocytogenes* 27
    2.4.2 Enumeration of *Listeria monocytogenes* 29
    2.4.3 Confirmation of *Listeria monocytogenes* 30
2.5 *Escherichia coli* O157:H7 32
    2.5.1 Detection of *Escherichia coli* O157:H7 32
    2.5.2 Enumeration of *Escherichia coli* O157:H7 34
    2.5.3 Confirmation of *Escherichia coli* O157:H7 34
2.6 *Salmonella* spp. 35
    2.6.1 Detection and Confirmation of *Salmonella* spp. 36
2.7 *Staphylococcus aureus* 38
    2.7.1 Enumeration and Confirmation of *Staphylococcus aureus* 38

2.8 *Yersinia enterocolitica* 40
    2.8.1 Detection and Confirmation of *Yersinia enterocolitica*... 40
2.9 *Bacillus cereus* 42
    2.9.1 Enumeration and Confirmation of *Bacillus cereus*... 42
2.10 *Clostridium perfringens* 44
    2.10.1 Enumeration and Confirmation of *Clostridium perfringens* 45
2.11 *Campylobacter jejuni*..47
    2.11.1 Detection and Confirmation of *Campylobacter jejuni*..47
References... 49

2.1 Introduction

Prevention of foodborne infections and intoxications are of paramount importance today. Hazard analysis and critical control point (HACCP)-type food safety management systems are applied by food enterprises to achieve this goal. Validation of all control measures requires, among other activities, microbiological testing of food and environmental samples. The presence of pathogenic bacteria on raw meat (beef, lamb, and pork) and poultry is the result of their contamination from the live animal, equipment, employees, and environment. *Salmonella, Listeria monocytogenes, Staphylococcus aureus, Yersinia enterocolitica, Escherichia coli* (mainly *E. coli* O157:H7), *Campylobacter jejuni*, and *Clostridium perfringens* often occur on raw meat and poultry. These pathogens have been implicated in foodborne outbreaks associated with the consumption of meat and poultry. *C. jejuni* frequently occurs on poultry meat, whereas *E. coli* is rarely found on this type of meat. However, beef has been implicated in many foodborne outbreaks associated with *E. coli. Salmonella* and *L. monocytogenes* may be found on all types of meat, including beef, lamb, pork, and poultry, and *Y. enterocolitica* is usually present on pork meat surfaces [1,2]. Psychrotrophic pathogens such as *L. monocytogenes* and *Y. enterocolitica* are of great concern because they are able to reach high numbers at refrigerated temperatures, especially when products are kept under abused temperatures (>7–8°C) for extended periods of time [3]. *S. aureus* and *C. perfringens* are also of great concern due to toxin production in food as a result of their growth. For more detailed information on the protocols and the culture media (including their preparation), for both cultural and rapid microbiological methods, reference works should be consulted [1,4–6]. The analytical essentials of microbiological examination of foods, as documented by the late Professor Mossel, are important elements of background information for the person performing the analysis [7].

2.2 Cultural Methods

Cultural or traditional methods are simple and relatively inexpensive, but they are time consuming. A food sample (usually 25 g) is homogenized in a stomacher bag with 225 mL of diluent using a stomacher machine to prepare a 1:10 dilution. Diluent must be correctly prepared in terms of buffer capacity and osmotic pressure (saline peptone water [SPW], 0.1% peptone and 0.85% NaCl); otherwise the microbial cells of the target microorganism may be stressed, influencing the final result. The sample withdrawn for microbiological analysis should be representative and randomly selected from different areas of the food to assure, in some degree, detection of the target microorganism if this is not uniformly distributed in the food, which very often is the case for solid foods. Information on the statistical basis of sampling plans and practical aspects of sampling and analysis are provided by Jarvis [8]. Further decimal dilutions may be required depending on the population level of the target microorganism present in the food. An adequate volume of sample from the appropriate dilution is spread (0.1 mL), poured (1.0 mL), or streaked on selective agars to differentiate or enumerate the target microorganism. Nonselective agars may also be used to perform confirmatory biochemical and serological tests. In some cases, an enrichment and, if it is necessary, a preenrichment step may be included to suppress the growth of other microorganisms, allowing at the same time the recovery of injured cells of the target microorganism.

Laboratory media used to subculture the microorganisms present in the food sample are divided into three categories: elective, selective, and differential [9]. Elective media are those that contain agents (e.g., microelements) that support the growth of the target microorganism but do not inhibit the growth of the accompanying microflora. The latter is achieved by the use of the selective media, which contain inhibitory agents, such as inorganic salts, triphenylmethane dyes, surface-active agents, and antibiotics. These agents inhibit the growth of the nontarget microorganisms as well as, in some cases, the growth of the microorganism under examination but in lesser degree. Differential media contain agents that allow the differentiation of the microorganisms (e.g., chromogenic media). These media contain chromogenic ingredients that produce a specific color or reaction due to bacterial metabolism. These agents react with the colonies, changing the color of the media. Usually, the media contain all the preceding agents to ensure proper identification of the target microorganism. For instance, the chromogenic media Agar Listeria Ottavani & Agosli (ALOA) agar [10] and RAPID' L. mono Listeria Agar (RAPID' L. mono) [11] use the following properties to differentiate *Listeria* spp. and *L. monocytogenes* from the other Listeriae species. ALOA contains a chromogenic compound which colors the Listeriae colonies due to its degradation from the enzyme β-glucosidase. This enzyme is produced from all *Listeria* species. The differentiation of pathogenic *Listeria* from the nonpathogenic species is based on the formation of phosphatidylinositol phospholipase C (PI-PLC). This compound hydrolyzes a specific substrate added to the growth medium, resulting in a turbid halo (ALOA) or a specific color of colonies (RAPID' L. mono) [12].

Petrifilm method (3M, Minneapolis, Minnesota) is another method that uses a plastic film together with the appropriate medium in dried form. It is used mainly for coliforms (red colonies with gas bubbles) and *E. coli* (blue colonies with gas bubbles). One milliliter of sample is added directly to the plates to rehydrate the medium. Plates are then incubated and counted. Validation and collaborative studies have found the Petrifilm method to be not significantly different from the traditional methods [6,13,14].

2.2.1 Enumeration Methods

In general, two enumeration methods are used most often—the plate count and most probable number, the latter method being used for certain microorganisms, such as coliforms [15] and *E. coli* [16].

2.2.1.1 Plate Count

Plate count is the most popular cultural enumeration method. The procedure involves homogenization of the food sample, dilution, plating on various media, and incubation at selected temperatures according to which microorganism is under examination. After incubation for a sufficient period of time, counting of the specific colonies of the target microorganism is performed. If confirmation of the target microorganism is required, then a number of randomly selected colonies are obtained. The ratio of the colonies confirmed as the target microorganism to the total colonies tested should be calculated to ascertain the number of viable cells per gram of food sample. For instance, if the mean number of presumable *C. perfringens* colonies from two pour agar plates is 20 at the second dilution (10^{-2}) and the confirmed *C. perfringens* colonies of 10 randomly selected (5 per plate) are 8, then the number of viable *C. perfringens* cells per gram of food sample will be $20 \times 10^2 \times (8/10) = 1.6 \times 10^3$ [1]. A recent critical review of the uncertainty in the enumeration of microorganisms in foods is given by Corry et al. [17].

2.2.1.2 Most Probable Number

The number of viable cells in a food sample is assessed based on probability tables. The food sample is diluted (10-fold dilutions), and then samples from each dilution are transferred to three tubes containing a growth medium (broth). After incubation of tubes, turbidity is measured and the tubes showing turbidity (growth) are compared to probability tables to find the population level of the target microorganism present in the food [1].

2.2.2 Detection Methods

Detection methods are used to determine the presence or absence of a specific pathogen. These methods include additional steps (for example, preenrichment and enrichment) to allow the increase of pathogens to a detectable population and recovery of injured cells, because the target microorganism may be present in very low levels in comparison with the population levels of the dominant microflora.

Sublethal exposure of microbial cells during processing of foods may lead to the inability of the microorganisms to form visible colonies on plate count agars. Although cells may remain undetected on selective agars, they are still viable (but not culturable), and under conditions that favor their growth may recover and become active. This is of great importance for foodborne pathogens that may lead to a food poisoning outbreak. Therefore, additional steps such as the previously mentioned enrichment steps are included in the analytical procedures to allow the resuscitation/repairing of the injured cells. There are many factors that influence the resuscitation of injured cells, such as composition and characteristics of the medium and environmental parameters [18]. Therefore, the analytical methods for the detection of the microorganisms are constructed in such a way as to allow maximum performance (recovery of stressed cells).

Usually, 25 g of food sample is aseptically weighted in a stomacher bag, homogenized in an enrichment broth (225 mL), and incubated for a certain period of time at a known temperature. After incubation, a sample from the broth is streaked on a selective agar plate using a bacteriological loop. If the examined microorganism is present, it is indicated by its characteristic colonies forming on the agar. To confirm the microorganism at strain level, some additional biochemical or serological tests may be needed. These tests are performed on a pure culture; therefore, colonies from the selective agar plates are purified (streaking) on nonselective agar plates, for example, nutrient agar or brain heart infusion (BHI) agar.

2.3 Alternative or Rapid Microbiological Methods

Rapid microbiological methods are much faster, but one disadvantage is that they are expensive. Thus, a careful look at the requirements of a laboratory or a food industry is required before the adoption of a method. These methods also include an enrichment step called a concentration step, aiming to separate and concentrate the target microorganism or toxin. In this way, the detection time is made shorter and specificity is improved.

2.3.1 Methods with a Concentration Step

Methods that concentrate the target microorganism or toxin are

1. The *immunomagnetic separation* (IMS), in which antibodies linked to paramagnetic particles are added and the target microorganism is trapped because of the interaction between antigen and antibody. Commercial kits are available for IMS of various foodborne pathogens, such as *L. monocytogenes, Salmonella* spp., and *E. coli* O157:H7 (Dynabeads™, Dynal Biotech, Oslo, Norway). The IMS for *Salmonella* (10-min duration) has been proved to successfully replace the enrichment step (overnight incubation) of the standard procedure for the detection of *Salmonella*, shortening the time needed to obtain results.
2. The *metal hydroxide–based bacterial concentration* technique, in which metal (hafnium, titanium, or zirconium) hydroxide suspensions react with the opposing charge of the bacterial cells. The cells are then separated by centrifugation, resuspended, and plated.
3. The *hydrophobic grid membrane filter*, which is a filtration method similar to the method used for water. The food sample first is filtered to remove large particles (>5 μm) and then is filtered through a grid membrane on which the microorganisms are retained. The membrane is placed on a selective agar and after an appropriate incubation period, the colony counts are calculated.
4. The *direct epifluorescent technique* (DEFT), used for enumerating viable bacteria in milk and milk products. Microorganisms' cells are concentrated through filtration on a membrane and then retained microorganisms are colored, usually with acridine orange (fluorescent dyes) and counted. Viable cells are red (acridine orange fluoresces red with ribonucleic acid [RNA]) and nonviable green (acridine orange fluoresces green with deoxyribonucleic acid [DNA]) [6,14,19,20].

2.3.2 Detection and Enumeration Methods

Some of the most widely used methods for the identification and detection of foodborne pathogens are the following:

1. *Polymerase chain reaction (PCR)–based methods coupled to other techniques*—most probable number counting method (MPN-PCR) [21], surface plasmon resonance, and PCR acoustic wave sensors [22], LightCycler real-time PCR (LC-PCR), PCR enzyme-linked immunosorbent assay (PCR-ELISA) [23], sandwich hybridization assays (SHAs), and fluorescent *in situ* hybridization (FISH) detection test [24]. From these methods, ELISA has been widely used for pathogen detection and identification, especially for *Salmonella* spp. and *L. monocytogenes*. The detection limit is 10^4 colony forming units (CFU)/g; therefore, a cultural enrichment step is required before testing. Specific antibodies for the target microorganism, contained in microtiter plates, react with the antigen, which is detected using a second antibody conjugated to an enzyme (horseradish peroxidase or alkaline phosphatase) to give a colorimetric reaction after the addition of substrate.
2. *Adenosine triphosphate (ATP) bioluminescence*, which can be used as an indicator of microbial contamination in foods and processing plants. This method detects the presence of

bacterial ATP. In a buffer containing magnesium, luciferase is added to a sample along with luciferin. The latter is oxidized (oxyluciferin) and the photons of light produced are measured by a luminometer. A standard curve is made to calculate the contamination level; the sensitivity of the method is 10^4 cfu/mL.

3. *Reversed passive latex agglutination*, which is used for the detection of toxins such as shiga toxins from *E. coli*. Latex beads containing antibodies (rabbit antiserum) specific for the target microorganism react with the target antigen if present. The particles agglutinate and a V-shaped microtiter well has a diffused appearance. If the antigen is not present, then a dot will appear.

4. *Impedance or conductance technique*, frequently used for enumeration. This method rapidly detects the growth of a specific microorganism based on the production of charged metabolites (direct method) or based on the carbon dioxide liberation (indirect method). In the first method, detection is measured by the change in the conductivity of the culture medium because of the accumulation of various products produced by the microorganism, such as organic acids. These changes are recorded at constant time intervals. "Time to detection" is the time needed in order for the conductance value to be changed. Because the time to detection is dependent on the inoculum size, a calibration curve is made for a known wide range of population levels of the desired microorganism. Using this calibration curve, the calculation of the population level of an unknown sample is simple after the automatic determination of the time to detection by the equipment. In the other method, the sample is distinguished from the potassium hydroxide bridge by a headspace in the test tube. The carbon dioxide produced during the microbial growth in the headspace reacts with potassium hydroxide, forming potassium carbonate, which is less conductive. Conductance decrease is the recorded parameter [6,14,20].

Genotypic, molecular methods are useful in identifying bacteria either as a complement or an alternative to phenotypic methods; besides enhancing the sensitivity and specificity of the detection process, they reduce much of the subjectivity inherent in interpreting the results. DNA is invariant throughout the microbial life cycle and after short-term environmental stress factors. This is the reason that molecular methods targeting genomic DNA are generally applicable [25]. Restriction fragment length polymorphism (RFLP) of total genomic DNA represents a technique belonging to the first-generation molecular methods [26] widely used in microbial differentiation. Southern blot hybridization tests, which enhance the result of agarose gel electrophoresis by marking specific DNA sequences, have also been used. Second-generation molecular techniques (known as PCR-based technologies), such as PCR-RFLP and randomly amplified polymorphic DNA-PCR (RAPD-PCR), have been used for differentiation and identification of microbial isolates [25]. Recent advances in PCR technology, namely real-time PCR [27], enable results to be obtained within a few hours [28]. Quantification of microorganisms is of major importance, especially in the case of toxigenic bacteria, since their concentration determines toxin production [25]. Biosensor technology promises equally reliable results in much shorter times, and is currently gaining extreme interest. Many biosensors rely on either specific antibodies or DNA probes to provide specific results [28].

The current trend is toward culture-independent PCR-based methods, which, unlike the previously mentioned ones, are believed to overcome problems associated with selective cultivation and isolation of microorganisms from natural samples. The most commonly used method among the culture-independent fingerprinting techniques is PCR followed by denaturing gradient gel electrophoresis (DGGE). PCR-DGGE provides information about the variation of the PCR products of the same length but with different sequences on differential mobility in an acrylamide gel matrix of increasing denaturant concentration [25,29].

2.4 *Listeria monocytogenes*

L. monocytogenes is widely distributed in the environment and can be found in many food commodities [3,30]. It is a very persistent microorganism that survives on surfaces and equipment of food processing units in conditions of insufficient cleaning [31–35]. Postprocessing contamination from the plant environment (equipment, personnel, floors, etc.) is the most frequent reason for its presence on meat surface. Cross-contamination may also occur at the retail outlet, as well as in the home, especially when the products have been mishandled and improper hygiene practices have been followed [35,36]. Various foods have been associated with *L. monocytogenes* outbreaks. Milk and dairy products (e.g., cheese), meat (including poultry) and meat products, vegetables, and fish and fish products have been implicated in outbreaks of foodborne *L. monocytogenes* [37]. The pathogen is usually killed during cooking, but it is capable of growing in foods stored at refrigeration temperatures (psychrotrophic microorganism) [38,39]. High salt concentrations and acid conditions do not permit *L. monocytogenes* growth [39]. However, it may survive even under these stressful environmental conditions [40,41]. Therefore, consumption of raw products or manufacturing of products without a killing step (e.g., cooking) with products that support pathogen growth—those with, for example, high initial pH, low salt content, or high water activity—or that are stored at refrigeration temperatures for a long period of time may increase the potential of listeriosis infection involving *L. monocytogenes* [39,42]. *L. monocytogenes* is a significant hazard, particularly for the elderly, immunocompromised people, infants, and pregnant women.

2.4.1 *Detection of Listeria monocytogenes*

The method for cultural detection of *L. monocytogenes* in raw meat and poultry is shown in Figure 2.1 [43]. Two enrichment steps are employed in the method to detect *Listeria* presence. With enrichment, it is feasible to detect low numbers of *Listeria,* as few as as one cell per 25 g of food, because the microorganism is allowed to grow to a level of ca. 10^4–10^5 cfu/g. The first enrichment step includes half Fraser broth (half-concentrated Fraser broth) containing only half concentration of the inhibitory agents (antibiotics), because these agents may have a negative effect on stressed or injured *Listeria* cells [44,45]. Antibiotics (acriflavin and nalidixic acid) are used to suppress the growth of the accompanying microflora, which may outgrow *Listeria* due to its slow growth. *Listeria* presence on the selective agar plates is observed by the formation of characteristic colonies. They are gray-green with a black center surrounded by a black zone on PALCAM agar [46]. Aesculin and ferrous iron are also added to the Fraser broth in conjunction with antibiotics to allow detection of β-D-glycosidase activity by *Listeria*, causing blackening of the medium [45].

Molecular methods that monitor the incidence of *Listeria* spp. in foods are also applied. Suggested techniques include fluorescent antibody assay, enzyme immunoassay, flow cytometry (FCM), and DNA hybridization [47]. DNA hybridization is the simplest molecular method used for the detection of *Listeria* spp. and *L. monocytogenes* in foods. The presence of a target sequence is detected using an oligonucleotide probe of a sequence complementary to the target

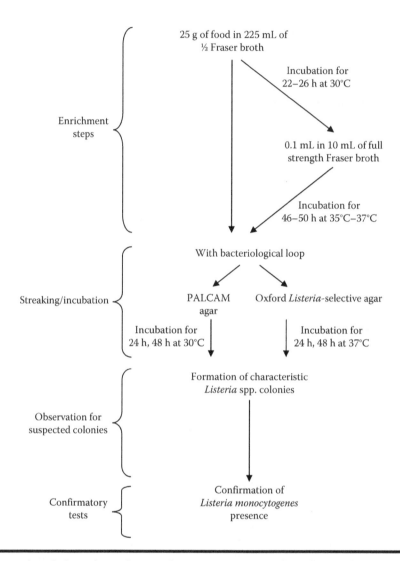

Figure 2.1 Cultural detection scheme of *L. monocytogenes* based on ISO standard. (Based on ISO. 1996. International Standard, ISO 11290-1: *Microbiology of food and animal feeding stuffs—Horizontal method for the detection and enumeration of Listeria monocytogenes—Part 1: Detection method.* Geneva: International Organization for Standardization.)

DNA sequence, containing a label for detection. Radioactive isotopes, biotinylated probes, probes incorporating digoxygenin, or fluorescent markers allow detection of target sequences [45]. PCR combined with DNA hybridization in a microtiter plate is a convenient and highly sensitive and specific approach for detection of *Listeria* spp. in a high-throughput 96-well format [48]. Commercially available DNA hybridization tests are routinely used for food testing and have been proven to be extremely sensitive and accurate. In contrast to DNA hybridization, in which large amounts of DNA or RNA are necessary for detection, PCR provides amplification results starting from very small amounts of target DNA [45]. Detection using PCR is carried out after selectively enriching samples for 24–48 h. Multiplex PCR allows the simultaneous detection of more than one pathogen in the same sample, such as *L. monocytogenes* and *Salmonella* [49,50] or *L. monocytogenes* and other *Listeria* species [51,52]. This approach is most attractive for food analysis, where testing time, reagents, and labor costs are reduced. To detect only living pathogens, RNA can be used instead of DNA. The presence of specific RNA sequences is an indication of live cells. When an organism dies, its RNA is quickly eliminated, whereas DNA can last for years, depending on storage conditions. Klein and Juneja [53] used reverse transcription-PCR (RT-PCR) to detect live *L. monocytogenes* in pure culture and artificially contaminated cooked ground beef. DNA microarrays are a recent technique that has found applicability in the detection of *L. monocytogenes*. Call et al. [54] used probes specific for unique portions of the *16S rRNA* gene in *Listeria* spp. to demonstrate how each *Listeria* species can be differentiated by this method. In this procedure, PCR is first performed using universal primers to amplify all the *16S rRNA* genes present in a sample. The various amplified DNA fragments bind only to the probes for which they have a complementary sequence. Because one of the oligonucleotides used in the PCR contains a fluorescent label, the spots where the amplified DNA has bound fluoresce. Pathogens are identified by the pattern of fluorescing spots in the array [55]. Lampel et al. [56] and Sergeev et al. [57] claim that in pure culture the detection limit of the array is 200 *L. monocytogenes* cells. Sergeev et al. [57] also noted that the array is appropriate for detection of pathogens in food and environmental samples. Microarrays are able to identify a number of pathogens or serotypes at once, but they still require culture enrichment and PCR steps to improve sensitivity and specificity of detection [55].

2.4.2 Enumeration of Listeria monocytogenes

Cultural enumeration method of *L. monocytogenes* based on the International Organization for Standardization (ISO) method [58] is displayed in Figure 2.2. The method has a detection limit ≥100 cfu/g. If numbers of *Listeria* lower than 100 cfu/g are expected, then the following procedure might be applied, which allows detection equal to or above 10 cfu/g. One milliliter of sample from the first 1:10 dilution is spread on three PALCAM agar plates (0.333 mL on each agar plate) and after incubation the colonies on all three plates are measured as a single plate. However, if even lower *Listeria* concentration is expected (1 cfu/g), then the first dilution is made with 1 part of food sample and 4 parts of diluent (1:5) (SPW or half Fraser broth). SPW (0.1% peptone and 0.85% NaCl) or half Fraser broth have large buffer capacity, which favors the growth and repair of stressed or injured cells.

Traditional PCR methods are able to detect the presence of a pathogen but are not able to quantify the level of contamination. One way to approach this problem is the use of competitive PCR. In this method, a competitor fragment of DNA which matches the gene to be amplified is introduced into the sample. In general, the competitor fragment is synthesized as a deletion mutant that can be amplified by the same primers being used to amplify the target DNA. The competitor fragment is distinguished from the pathogen gene fragment by its smaller size [55]. To determine the level of pathogen contamination, DNA purified from the food sample is serially diluted and added to a constant amount of competitor DNA. PCR is performed and the intensity of the pathogen's gene signal is compared to that of the competitor DNA on an agarose gel. The number of cells in the original sample can be estimated by comparing the intensity of the two DNA fragments (target versus competitor) using a standard curve [59]. Choi and Hong [60] used a variation of competitive PCR based on the presence of a restriction endonuclease site in the amplified gene for *L. monocytogenes* detection. The method was

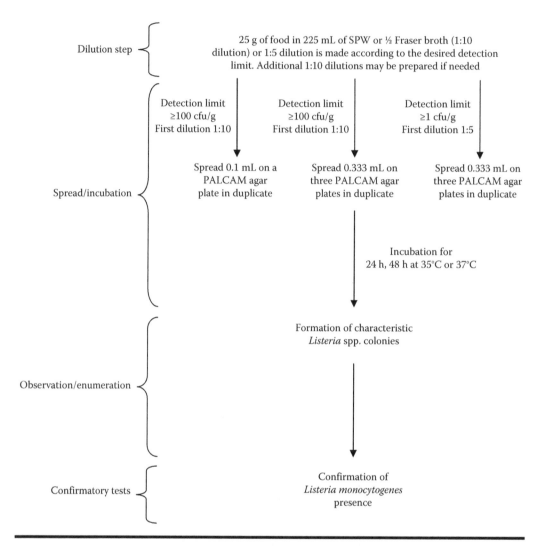

Dilution step

25 g of food in 225 mL of SPW or ½ Fraser broth (1:10 dilution) or 1:5 dilution is made according to the desired detection limit. Additional 1:10 dilutions may be prepared if needed

Detection limit
≥100 cfu/g
First dilution 1:10

Detection limit
≥100 cfu/g
First dilution 1:10

Detection limit
≥1 cfu/g
First dilution 1:5

Spread/incubation

Spread 0.1 mL on a PALCAM agar plate in duplicate

Spread 0.333 mL on three PALCAM agar plates in duplicate

Spread 0.333 mL on three PALCAM agar plates in duplicate

Incubation for
24 h, 48 h at 35°C or 37°C

Formation of characteristic *Listeria* spp. colonies

Observation/enumeration

Confirmatory tests

Confirmation of *Listeria monocytogenes* presence

Figure 2.2 Cultural enumeration of *L. monocytogenes* based on ISO method. (Based on ISO. 1998. International Standard, ISO 11290-2: *Microbiology of food and animal feeding stuffs—Horizontal method for the detection and enumeration of Listeria monocytogenes—Part 2: Enumeration method*. Geneva: International Organization for Standardization.)

completed within 5 h without enrichment and was able to detect 10^3 cfu/0.5 mL milk using the *hlyA* gene as target. The detection limit could be reduced to 1 cfu if culture enrichment for 15 h was conducted first.

2.4.3 Confirmation of Listeria monocytogenes

L. monocytogenes presence is confirmed by the use of various biochemical tests. The tests are performed on purified cultures. From the PALCAM or Oxford agars, five suspected and randomly chosen colonies are isolated and streaked on tryptone soya agar containing 0.6% yeast extract (TSYEA). *Listeria* species are easily identified by Gram staining, motility, catalase, and oxidase reactions. *Listeria* spp. is Gram-positive, small rods, motile, catalase-positive, and oxidase-negative. The motility test should be performed in a semisolid TSYEA tube (TSYE broth or TSYEB supplemented with 0.5% agar) incubated at 25°C because at incubation temperatures above 30°C the motility test is negative (nonmotile). The tube is inoculated by stabbing and is observed for growth around the stab (a characteristic umbrella-like shape of turbidity is formed) [61]. Sugar fermentation, hemolysis, and the Christie–Atkins–Munch–Petersen (CAMP) test may be used to differentiate the *Listeria* species (Figure 2.3). *L. monocytogenes*, *L. ivanovii*, and *L. seeligeri* are β-hemolytic species on horse or sheep blood agar. The CAMP test distinguishes the three species of *Listeria* and should be done on sheep blood agar. An enhanced β-hemolysis zone is observed close to *S. aureus* NCTC 1803 when either *L. monocytogenes* or *L. seeligeri* are streaked on blood agar. *L. seeligeri* shows a less enhanced β-hemolysis zone than *L. monocytogenes*. *L. ivanovii* shows a wide enhanced β-hemolysis zone with *Rhodococcus equi* NCTC 1621. The plates are incubated at 37°C for no longer than 12–18 h. The *Listeria* isolates streaked on blood agar for the CAMP test are derived from the hemolysis plates used to examine the β-hemolysis property. The *Listeria*

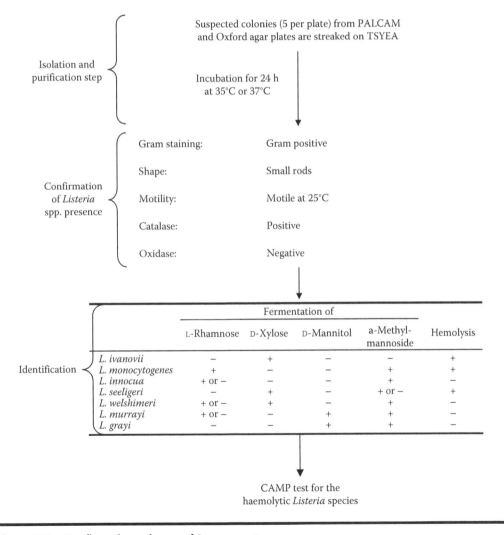

Figure 2.3 Confirmation scheme of *L. monocytogenes*.

streaks should not touch the streaks of the *S. aureus* and *R. equi* control strains. The control strains are streaked parallel to each other and the suspected *Listeria* isolated in between the two streaks [45,61]. Alternatively, various commercial identification kits such as API 10 Listeria (BioMerieux, Marcy Etoile, France) might be used instead of traditional biochemical tests, which are time consuming. Finally, the previous selective agars, PALCAM and Oxford, may be substituted by other selective chromogenic media such as ALOA agar and RAPID' L. mono, as mentioned earlier in Section 2.2, which allow the direct differentiation between *Listeria* species by specific reactions on the agar plates [12,62]. In this way, the direct detection or enumeration of a specific *Listeria* species is feasible from the dilutions of the original sample.

2.5 *Escherichia coli* O157:H7

Pathogenic *E. coli* includes a variety of types having different pathogenicity based on the virulence genes involved. The different types of pathogenic *E. coli* are the enteropathogenic *E. coli* (EPEC), the enteroinvasive *E. coli* (EIEC), the enterotoxigenic *E. coli* (ETEC), the enteroaggregative *E. coli* (EAEC), and the enterohemorrhagic *E. coli* (EHEC) [63]. The latter belongs to verocytotoxigenic *E. coli* (VTEC), which produces verocytotoxins or shiga toxins. VTEC *E. coli* are of great concern because they include the most predominant foodborne pathogen *E. coli* O157:H7. The letters and numbers, for example, O157:H7, refer to the microorganism serogroup. The somatic antigens are designated with the letter "O" and the flagella antigens with the letter "H" [64]. *E. coli* O157:H7 can be found on raw and processed meat [65–68]. Most often it has been isolated from beef, which is believed to be the main vehicle for outbreaks associated with pathogenic *E. coli* O157:H7. The source of contamination of meat is usually the bovine feces or the intestinal tube during slaughtering. Their contact with muscle tissue results in meat contamination [64]. Heat treatment and fermentation processes are sufficient for producing a safe finished product. However, if these processes are not adequate, then *E. coli* O157:H7 may survive during manufacturing if the microorganism is present in the raw material [69–71]. Factors other than process may play significant roles in producing safe products, including the implementation of good manufacturing practices (GMP) or good hygiene practices (GHP) to avoid postprocess contamination [35,36,71]. For the detection of EPEC, EIEC, ETEC, and EAEC there is no standard sensitive procedure and usually the food sample is diluted in BHI broth, incubated at 35°C for 3 h to allow microbial cells to resuscitate. Then an enrichment step (at 44°C for 20 h) in tryptone phosphate broth and plating on Levine eosin–methylene blue agar and MacConkey agar are performed. Lactose-positive (typical) and lactose-negative (nontypical) colonies are collected for characterization using various biochemical, serological, or PCR-based tests [13].

2.5.1 *Detection of Escherichia coli O157:H7*

The cultural method for detecting and identifying *E. coli* O157:H7 [72] is shown in Figure 2.4. Pathogenic *E. coli* O157:H7 does not ferment sorbitol and does not possess β-glucuronidase, produced by almost all other *E. coli* strains [73]. The selective media exploit these attributes to distinguish the pathogenic *E. coli* O157:H7 from other, nonpathogenic *E. coli* strains. The method includes an enrichment step using a selective enrichment broth (tryptone soya broth [TSB] supplemented with novobiocin) to resuscitate the stressed cells and suppress the growth of the background flora.

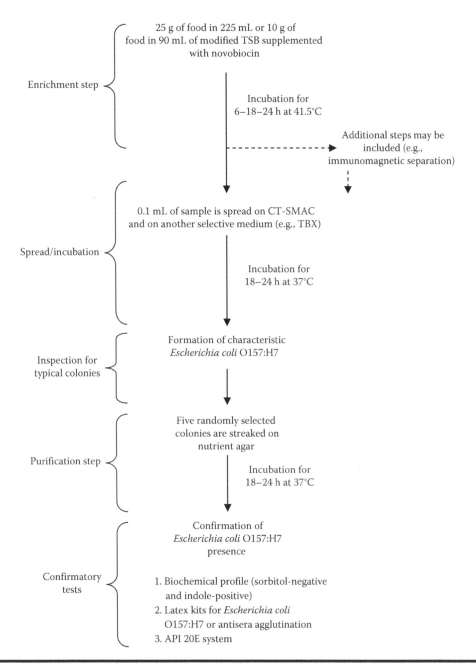

Figure 2.4 Cultural detection of *E. coli* O157:H7 based on ISO method. (Based on ISO. 2001. International Standard, ISO 16654: *Microbiology of food and animal feeding stuffs—Horizontal method for the detection of Escherichia coli O157.* Geneva: International Organization for Standardization.)

Before plating onto agar plates, intermediate steps may be involved. The cell antigen O157:H7 is characteristic of the microorganism pathogenicity and therefore the IMS method (manufacturer instructions are followed to implement this technique) increases the detection of *E. coli* O157:H7 [74]. *E. coli* O157:H7 are captured on immunomagnetic particles, washed with sterile buffer, resuspended using the same buffer, and a sample of the washed and resuspended magnetic particles is inoculated on a selective medium to obtain isolated colonies.

The selective agar used to subculture the sample is the modified MacConkey agar containing sorbitol instead of lactose, as well as selective agents such as potassium tellurite and cefixime (CT-SMAC) [75] and the tryptone bile glucuronic medium (TBX) [63,64]. Because sorbitol-negative microorganisms other than *E. coli* O157:H7 may grow on the agar plates (such as *Proteus* spp. and some other *E. coli* strains), the addition of cefixime (which inhibits *Proteus* spp. but not *E. coli*) and tellurite (which inhibits *E. coli* strains other than *E. coli* O157:H7) substantially improves the selectivity of the medium [13]. CT-SMAC agar medium has been found the most effective for the detection of shiga toxin–producing *E. coli* O157:H7 [76]. Typical *E. coli* O157:H7 colonies are 1 mm in diameter and are colorless (sorbitol-negative) or pellucid with a very slight yellow-brown color. However, because sometimes *E. coli* O157:H7 forms colonies similar to other *E. coli* strains (pink to red surrounded by a zone), further purification (streaking) on nutrient agar and confirmation of the typical and nontypical colonies is required.

Biochemical methods require time; hence, PCR-based protocols, including multiplex PCR (MPCR), have been developed. Detection of STEC strains by MPCR was first described by Osek [77]. A protocol was developed using primers specific for genes that are involved in the biosynthesis of the O157 *E. coli* antigen (*rfb* O157), and primers that identify the sequences of shiga toxins 1 and 2 (*stx*1 and *stx*2) and the intimin protein (*eae*A) involved in the attachment of bacteria to enterocytes [25]. The different strains were identified by the presence of one to four amplicons [77]. More protocols have been developed and applied in the detection and identification of *E. coli* in feces and meat (pork, beef, and chicken) samples [78,79]. Later, Kadhum et al. [80] designed an MPCR to determine the prevalence of cytotoxic necrotizing factors and cytolethal distending toxin–producing *E. coli* on animal carcasses and meat products, from Northern Ireland, in a preliminary investigation into whether they could be a source of human infection.

2.5.2   Enumeration of Escherichia coli O157:H7

The cultural enumeration method of *E. coli* O157:H7 based on the ISO standard method [81] is presented in Figure 2.5. The key step in the case of stressed cells is the additional incubation period required (at 37°C for 4 h) before incubation at 44°C for 18–24 h. Typical *E. coli* O157:H7 colonies have a blue color, and plates with colonies (blue) less than 150 and less than 300 in total (typical and nontypical) are counted. The detection limit of the method is a population of 10 cfu/g.

2.5.3   Confirmation of Escherichia coli O157:H7

To confirm the presence of *E. coli* O157:H7, the following tests should be carried out. *E. coli* O157:H7 is negative to sorbitol, unlike most nonpathogenic *E. coli* strains, and indole positive. After defining the biochemical profile of the suspected colonies, latex kits for *E. coli* O157:H7 or

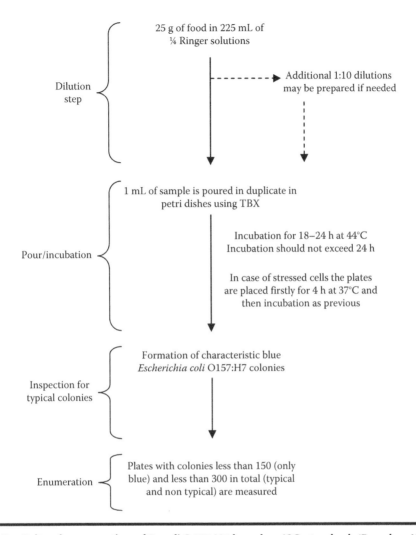

Dilution step
25 g of food in 225 mL of ¼ Ringer solutions

Additional 1:10 dilutions may be prepared if needed

Pour/incubation
1 mL of sample is poured in duplicate in petri dishes using TBX

Incubation for 18–24 h at 44°C
Incubation should not exceed 24 h

In case of stressed cells the plates are placed firstly for 4 h at 37°C and then incubation as previous

Inspection for typical colonies
Formation of characteristic blue *Escherichia coli* O157:H7 colonies

Enumeration
Plates with colonies less than 150 (only blue) and less than 300 in total (typical and non typical) are measured

Figure 2.5   Cultural enumeration of *E. coli* O157:H7 based on ISO standard. (Based on ISO. 2001. International Standard, ISO 16649-2: *Microbiology of food and animal feeding stuffs—Horizontal method for the enumeration of beta-glucuronidase-positive Escherichia coli—Part 2: Colony-count technique at 44°C using 5-bromo-4-chloro-3-indolyl beta-D-glucuronide*. Geneva: International Organization for Standardization.)

antisera agglutination can be used to confirm *E. coli* O157:H7. Commercial kits such as API 20E (BioMerieux, Marcy Etoile, France) constitute an alternative for *E. coli* O157:H7 confirmation. *E. coli* O157:H7 toxins can be detected using reversed passive latex agglutination and cultured vero cells. Polymyxin B may be used in the culture to facilitate shiga toxin release [6].

2.6   *Salmonella* spp.

Salmonella spp. has been isolated from all types of raw meat including poultry, pork, beef, and lamb. All these products have been implicated in outbreaks of *Salmonella* spp. Most often, however, *Salmonella* spp. occurs in poultry and pork meat. The main source of contamination of the raw meat is the transfer of the microorganism from feces to the meat tissue during slaughtering and the following processing [82]. Postprocess contamination may also occur and, therefore, the GHP regarding equipment and personnel are essential.

2.6.1 Detection and Confirmation of Salmonella spp.

The cultural method for detecting and identifying *Salmonella* spp. [83] is depicted in Figure 2.6. The microbiological criterion for *Salmonella* spp. is "absence in 25 g." The method includes two enrichment steps—a preenrichment step to allow injured cells to resuscitate and a selective enrichment step to favor the growth of *Salmonella* cells. In the first step, a nonselective but nutritious medium is used (buffered peptone water); in the second step, the selective medium contains selective agents to suppress the growth of accompanying microflora. Two different selective media are used in the second step because the culture media have different selective characteristics against the numerous *Salmonella* serovars [20]. Time and temperature of incubation during the preenrichment and selective enrichment steps play a significant role in the selectivity of the media. One of the selective media used in the second enrichment step has historically been a selenite cystine broth that contains a very toxic substance (sodium biselenite), and for this reason its use has been replaced by other media such as a Müller-Kauffmann tetrathionate/novobiocin (MKTTn) broth. Rappaport-Vassiliadis soya peptone (RVS) broth is the standard Rappaport-Vassiliadis (RV) broth but with tryptone substituted by soya peptone because it has shown better performance than the standard broth [13]. The next step is plating of the samples on selective differential agars containing selective agents such as bile salts and brilliant green, which have various diagnostic characteristics (e.g., lactose fermentation, H_2S production, and motility) to differentiate *Salmonella* spp. from the other microflora such as *Proteus* spp., *Citrobacter* spp., and *E. coli*. The Oxoid Biochemical Identification System (OBIS) *Salmonella* test (Oxoid, Basingstoke, U.K.) is a rapid test to differentiate *Salmonella* spp. from *Citrobacter* spp. and *Proteus* spp. The principle of the test is based on the determination of pyroglutamyl aminopeptidase (PYRase) and nitrophenylalanine deaminase (NPA) activity, to which *Salmonella* spp. is negative, *Citrobacter* spp. is PYRase-positive and NPA-negative, and *Proteus* spp. NPA-positive and PYRase-negative. Selective agars differ in their selectivity toward *Salmonella,* and for this reason a number of media are used in parallel (xylose lysine desoxycholate [XLD] or xylose lysine tergitol-4 [XLT-4] and phenol red/brilliant green agar). The last steps include biochemical and serological confirmation of suspected *Salmonella* colonies to confirm the identity and to identify the serotype of the isolates [13,84]. *Salmonella* spp. is lactose-negative, H_2S-positive, and motile. However, lactose-positive strains have been isolated from human infections, and an additional selective medium agar may therefore be needed. Bismuth sulfite agar is considered as the most suitable medium for such strains [13,85,86].

The most frequently isolated serovars from foodborne outbreaks are *S. typhimurium* and *S. enteritidis.* Traditional phenotypic methods such as biotyping, serotyping, and phage typing of isolates, as well as antimicrobial susceptibility testing, provide sufficient information for epidemiological purposes. Molecular genetic methods have revolutionized the fingerprinting of microbial strains. However, not all of them have been internationally standardized, and problems in interpreting the results of different laboratories might occur. Nevertheless, the accuracy and speed at which results are obtained have rendered them more and more applicable.

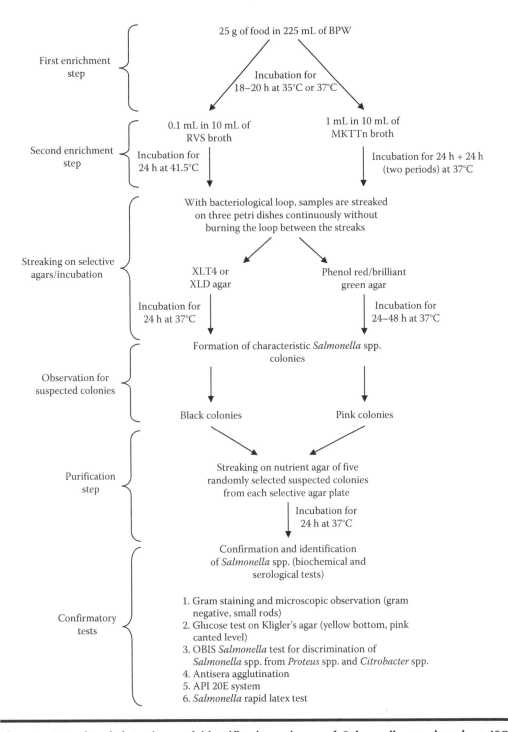

Figure 2.6 Cultural detection and identification scheme of *Salmonella* spp. based on ISO method. (Based on ISO. 1993. International Standard, ISO 6579: *Microbiology of food and animal feeding stuffs—Horizontal method for the detection of Salmonella spp.* Geneva: International Organization for Standardization.)

The assay generally used to identify *Salmonella* serovars is represented by a serological method which requires the preparation of specific antibodies for each serovar and is thus extremely complex and time consuming [25]. Plasmids are characteristic of *Salmonella* and therefore plasmid analysis can often be used to differentiate strains [87]. A faster alternative involves PCR approaches. On the basis of primers designed for detecting O4, H:i, and H:1,2 antigen genes from the antigen-specific genes *rfbJ*, *fliC*, and *fljB* (coding for phase 2 flagellin), respectively, Lim et al. [88] described an MPCR for the identification of *S. typhimurium,* whose presence was associated with the appearance of three amplification products. MPCR targeted to the *tyv* (CDP-tyvelose-2-epimerase), *prt* (paratose synthase), and *invA* (invasion) genes were designed to identify *S. enterica* serovar Typhi and *S. enterica* serovar Paratyphi A by the production of three or two bands, respectively [89]. PCR amplifications of the 16S–23S spacer region of bacterial rRNA as well as specific monoclonal antibodies to the lipopolysaccharide of *S. typhimurium* DT104 have been used [90].

2.7   *Staphylococcus aureus*

Reservoirs of the *S. aureus* microorganism are the animals in which it is part of their normal microflora. Food contamination with *S. aureus* may occur through humans, who also carry staphylococci. Food poisoning by *S. aureus* is the result of ingestion of food containing staphylococcal enterotoxin(s). Enterotoxin is a heat-stable substance, and high cell numbers are required to produce sufficient amounts of toxin. Temperatures above 15°C favor the rapid growth of the microorganism and the production of enterotoxin. The minimum temperatures for microorganism growth and enterotoxin production are 7 and 10°C, respectively. Attention is required in the implementation of GMP and GHP to minimize the contamination of raw materials with *S. aureus* and to avoid postprocess contamination of processed meat products since staphylococci are part of the natural microflora of humans and animals [91].

2.7.1   *Enumeration and Confirmation of Staphylococcus aureus*

The cultural enumeration method of *Staphylococcus* spp. based on ISO [92] is shown in Figure 2.7. The method has a detection limit ≥100 cfu/g. If lower numbers of staphylococci than 100 cfu/g are expected, then the procedure followed for *L. monocytogenes* enumeration may be applied. Low numbers of *S. aureus* are of little significance because extensive growth is needed in order for the microorganism to produce sufficient amounts of enterotoxin, and therefore an enrichment step is not required for its isolation. The most widely used and accepted medium for *S. aureus* is the Baird-Parker (BP) agar [93] (egg yolk–glycine–potassium tellurite–sodium pyruvate). Sodium pyruvate assists the resuscitation of stressed cells, while potassium tellurite, glycine, and lithium chloride enhance the medium's selectivity. *Staphylococcus* spp. forms black colonies (tellurite reduction), and *S. aureus* colonies are also surrounded by a halo (clearance of egg yolk due to lipase activity). Plates having 15–300 colonies in total (*Staphylococcus* spp. and *S. aureus,* if present) are measured. A coagulase test, reversed-passive latex agglutination test, or ELISA methods for enterotoxin detection may be used as confirmatory tests for *S. aureus* presence. The coagulase test is considered positive for enterotoxin presence only in case of a strong positive reaction. API Staph (BioMerieux, Marcy Etoile, France) may be also used to identify the isolated colonies from the agar plates [94].

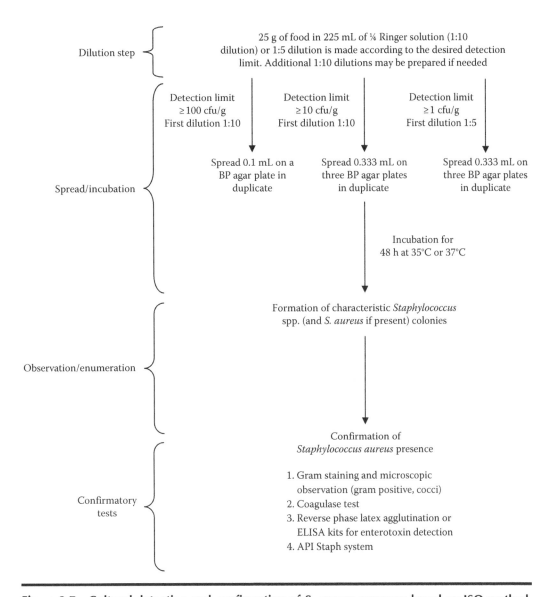

Dilution step

25 g of food in 225 mL of ¼ Ringer solution (1:10 dilution) or 1:5 dilution is made according to the desired detection limit. Additional 1:10 dilutions may be prepared if needed

Detection limit ≥ 100 cfu/g First dilution 1:10

Detection limit ≥ 10 cfu/g First dilution 1:10

Detection limit ≥ 1 cfu/g First dilution 1:5

Spread/incubation

Spread 0.1 mL on a BP agar plate in duplicate

Spread 0.333 mL on three BP agar plates in duplicate

Spread 0.333 mL on three BP agar plates in duplicate

Incubation for 48 h at 35°C or 37°C

Formation of characteristic *Staphylococcus* spp. (and *S. aureus* if present) colonies

Observation/enumeration

Confirmation of *Staphylococcus aureus* presence

Confirmatory tests

1. Gram staining and microscopic observation (gram positive, cocci)
2. Coagulase test
3. Reverse phase latex agglutination or ELISA kits for enterotoxin detection
4. API Staph system

Figure 2.7 Cultural detection and confirmation of *S. aureus* presence based on ISO method. (Based on ISO. 1999. International Standard, ISO 6888-1: *Microbiology of food and animal feeding stuffs—Horizontal method for the enumeration of coagulase-positive staphylococci (Staphylococcus aureus and other species)—Part 1: Technique using Baird-Parker agar medium.* Geneva: International Organization for Standardization.)

Molecular techniques have been applied in the case of *S. aureus* to quickly determine its presence and identification. Occasionally, isolates of *S. aureus* give equivocal results in biochemical and coagulase tests [95]. Most *S. aureus* molecular identification methods have been PCR-based. Primers targeted to the nuclease (*nuc*), coagulase (*coa*), protein A (*spa*), *femA* and *femB*, *Sa442*, 16S rRNA, and surface-associated fibrinogen-binding genes have been developed [96,97].

S. aureus food poisoning is caused by ingestion of preformed toxins (*Staphylococcus aureus* enterotoxins [SEs]) produced in foods. It has been reported that nearly all SEs are superantigens and are encoded by mobile genetic elements including phages, plasmids, and pathogenicity islands [98,99]. Several methods for SE detection from isolated strains and foods have been described in the recent years; these include biological, immunological, chromatographical, and molecular assays [100,101]. The four SEs originally described can be detected with commercial antisera or by PCR reactions [102,103].

Detection and identification of methicillin-resistant *S. aureus* (MRSA) has gained great attention since in immunocompromised patients it can cause serious infections which may ultimately lead to septicaemia. Since MRSA strains mainly appear in nosocomial environments, most of the techniques developed for their detection are focused on clinical or blood isolates [104]. Such techniques include DNA probes [27,105], peptide nucleic acid probes [106], MPCR [97], real-time PCR [107–109], LightCycler PCR [108,109], and a combination of fluorescence *in situ* hybridization and FCM [110]. Recent advances include the development of segment-based DNA microarrays [104]. Although, as mentioned earlier, MRSA strains are mainly encountered in nosocomial environments, food can be considered an excellent environment for introducing pathogenic microorganisms in the general population, especially in immunocompromised people and in the intestinal tract, transfer of resistant genes between nonpathogenic and pathogenic or opportunistic pathogens could occur [111]. A community-acquired case was reported in 2001, in which a family was involved in an outbreak after ingesting MRSA with baked port meat contaminated by the handler [112]. Therefore, the techniques applied in different samples might have applicability in food products.

2.8 *Yersinia enterocolitica*

Infections with *Y. enterocolitica* involve meat and meat products. In particular, pork meat has been implicated in *Y. enterocolitica* outbreaks (yersiniosis). Not all *Y. enterocolitica* strains cause illness. The most common serotypes causing yersiniosis are the serotypes O:3, O:9, O:5,27, and O:8. Because contamination of meat with high numbers of *Y. enterocolitica* may occur during preprocess (e.g., slaughtering), precautionary measures such as GHP are essential [113]. Contamination with *Y. enterocolitica* is a serious concern due to its ability to grow at refrigerated temperatures (4°C) [13,91].

2.8.1 *Detection and Confirmation of Yersinia enterocolitica*

The cultural method for detecting *Y. enterocolitica* [114] is presented in Figure 2.8. The method involves elements of the methods from Schiemann [115,116], the Nordic Committee on Food Analysis [117], and Wauters et al. [118]. If specific serotypes are considered (e.g., O:3), then two isolation procedures are proposed to run in parallel [13]. The procedure involving enrichment with irgasan–ticarcillin–potassium chlorate (ITC) broth is selective for serotype O:3 and possibly O:9. However, poor recovery of the serotype O:9 from ground pork using ITC has been found by De Zutter et al. [119]. After enrichment with ITC, plating of the samples should be done on *Salmonella–Shigella* sodium deoxycholate calcium chloride (SSDC) instead of cefsulodin irgasan novobiocin (CIN) because the latter medium is inhibitory for the serotype O:3. Furthermore, the isolation and identification of *Y. enterocolitica* from ground meat on CIN medium agar has

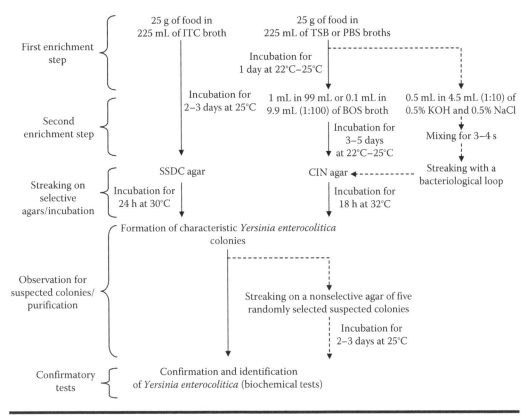

First enrichment step

25 g of food in 225 mL of ITC broth

25 g of food in 225 mL of TSB or PBS broths

Incubation for 1 day at 22°C–25°C

Second enrichment step

Incubation for 2–3 days at 25°C

1 mL in 99 mL or 0.1 mL in 9.9 mL (1:100) of BOS broth

0.5 mL in 4.5 mL (1:10) of 0.5% KOH and 0.5% NaCl

Incubation for 3–5 days at 22°C–25°C

Mixing for 3–4 s

Streaking on selective agars/incubation

SSDC agar

Incubation for 24 h at 30°C

CIN agar

Incubation for 18 h at 32°C

Streaking with a bacteriological loop

Observation for suspected colonies/ purification

Formation of characteristic *Yersinia enterocolitica* colonies

Streaking on a nonselective agar of five randomly selected suspected colonies

Incubation for 2–3 days at 25°C

Confirmatory tests

Confirmation and identification of *Yersinia enterocolitica* (biochemical tests)

Figure 2.8 Cultural detection of *Y. enterocolitica* based on ISO method. (Based on ISO. 1994. *International Standard, ISO 10273: Microbiology of food and animal feeding stuffs—Horizontal method for the detection of presumptive pathogenic Yersinia enterocolitica.* Geneva: International Organization for Standardization.)

been proved to cause problems because many typical *Yersinia*-like colonies may grow [120]. After enrichment (primary) with TSB or peptone sorbitol bile salts (PBS) broth (peptone buffered saline with 1% sorbitol and 0.15% bile salts), an alkali treatment (potassium hydroxide [KOH]) may be used to increase recovery rates of *Yersinia* strains instead of secondary enrichment with bile oxalate sorbose (BOS) [121]. This method should not be used with the procedure involving the ITC broth as a selective enrichment step [122]. On SSDC agar, the *Yersinia* colonies are 1 mm in diameter, round, and colorless or opaque. On CIN agar, the colonies have a transparent border with a red circle in the center (bull's eye).

Yersinia strains and *Y. enterocolitica* serotypes may be distinguished using biochemical tests. *Y. enterocolitica* may be identified using urease and citrate utilization tests, and fermentation of the following sugars: sucrose, raffinose, rhamnose, α-methyl-D-glucoside, and melibiose. *Y. enterocolitica* is urease and sucrose positive, but negative in the other tests. The most frequently used tests to identify pathogenic *Y. enterocolitica* strains are calcium-dependent growth at 37°C, Congo red binding on Congo red magnesium oxalate (CR-MOX) agar, or low-calcium Congo red BHI agarose agar (CR-BHO), which determine the Congo red dye uptake, pyrazinamidase activity, and salicin–esculin fermentation [115,122–126]. Because the last two tests are not plasmid dependent as are the other tests, the pyrazinamidase, salicin, and esculin tests are considered the most reliable biochemical screening tests for pathogenicity because plasmids may be lost during subculture. Before testing, suspected colonies may be subcultured on a non-selective medium incubated at 25°C to reduce the risk of plasmid loss [122]. Pathogenic strains are negative to these three tests. Esculin fermentation and pyrazinamidase activity tests should be conducted at 25°C, whereas salicin fermentation is conducted at 35 or 37°C. Commercial kits for *Y. enterocolitica* identification such as API 20E (BioMerieux, Marcy Etoile, France) also may be used as an alternative that has been proved to be suitable for routine laboratory diagnostics [120].

From a food hygiene point of view, *Y. enterocolitica* is of major importance and is a very heterogeneous species. Nonpathogenic strains may contaminate food products to the same extent as pathogenic *Y. enterocolitica*, and a principal goal for nucleic acid–based methods has been to separate this group of pathogenic bacteria. Both polynucleotide and oligonucleotide probes, as well as PCR-based methods, have been applied for its detection and quantification in meat and meat products [127,128]. Nested-PCR has also been developed for its detection in meat food products and can satisfactorily detect pathogenic *Y. enterocolitica* even in the presence of a high background of microflora [129]. Comparative genomic DNA (gDNA) microarray analysis has recently been developed to differentiate between nonpathogenic and pathogenic biotypes [130].

2.9 *Bacillus cereus*

B. cereus can be found in meat and especially in dishes containing meat. Outbreaks attributed to *B. cereus* infections have also been associated with cooked meats. Its presence in food is not considered significant since high numbers (>10^5–10^6 cfu/g) are needed to cause a diarrheal or emetic syndrome. The two types of illness are caused by an enterotoxin (diarrheagenic or emetic) produced by the microorganism. Because other *Bacillus* species are closely related physiologically to *B. cereus,* including *B. mycoides, B. thuringiensis,* and *B. anthracis,* further confirmatory tests are required to differentiate typical *B. cereus* (egg yolk reaction, inability to ferment mannitol) from the other species [131].

2.9.1 *Enumeration and Confirmation of Bacillus cereus*

The presence of low numbers of *B. cereus* is not considered significant, and thus an enrichment step is not needed unless *B. cereus* growth is likely to occur (Figure 2.9). However, if enrichment must be applied, this can be done using BHI broth supplemented with polymyxin B and sodium chloride [132]. To enhance selection of *B. cereus,* the following attributes of the microorganism are employed: its resistance to the antibiotic polymyxin, the production of phospholipase C causing turbidity around colonies grown on agar containing egg yolk, and its inability to ferment mannitol. The media used for selection are usually the mannitol–egg yolk–polymyxin (MYP) [133] and the Kim-Goepfert (KG) agars [134]. Because of the similarity in composition and functionality of the KG medium with the polymyxin pyruvate egg yolk mannitol bromothymol blue agar (PEMBA) [132,135], the latter medium may be used instead of KG [131].

Colonies on MYP agar have a surrounding precipitate zone (turbidity) and both colonies and zone are pink (no fermentation of mannitol). On PEMBA agar, the colonies are peacock blue

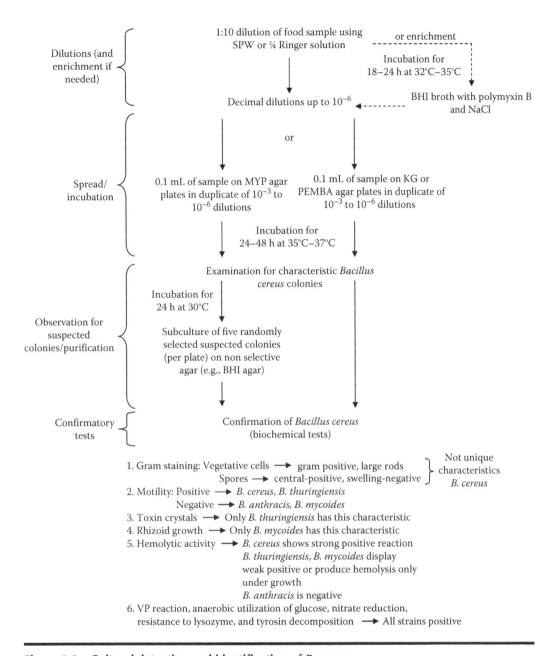

Figure 2.9 Cultural detection and identification of *B. cereus*.

with a blue egg yolk precipitation zone. Finally, on KG agar the colonies are translucent or white cream. Plates having 10–100 colonies per plate are counted instead of 30–300 colonies per plate because turbidity zones may overlap each other and measurement of the colonies with a precipitation zone may be difficult. For low numbers (<100–1000 cfu/g) of *B. cereus* in the food sample, the MPN technique may be used. A suitable medium for this purpose is the trypticase soy polymyxin broth. Each of three tubes of 1:10, 1:100, and 1:1000 is inoculated with 1 mL of sample and the tubes are incubated at 30°C for 48 h and examined for tense turbidity. Confirmation of *B. cereus* presence is required before determining the MPN [131–132]. If only spores are to be counted, the sample is heated (the initial 1:10 dilution is heated for 15 min at 70°C) or treated with alcohol (1:1 initial dilution in 95% ethyl alcohol for 30 min at room temperature) to kill the vegetative cells, and the detection and identification scheme is followed (Figure 2.9). Potential emetic strains can be identified using the identification kit from BioMerieux called API 50CHB (BioMerieux, Marcy Etoile, France) [136]. Before testing isolated colonies for *B. cereus* identity, the culture should be purified on a nonselective agar (e.g., BHI agar) to promote sporulation. Isolated colonies grown on KG agar, used as a selective agar, may be tested directly because KG medium favors sporulation.

ELISA and reverse passive latex agglutination (RPLA) tests are commercially available for *Bacillus* diarrheal enterotoxin. No tests have been developed for emetic enterotoxin due to purification problems, although tissue culture assay using HEp-2 cells may be useful for the detection and purification of the emetic toxin [131,136].

Several molecular techniques have also been developed for the detection and characterization of *B. cereus* derived from food products. Immunological methods for semiquantitative identification of enterotoxins are available (ELISA, RPLA), which demand at least 2 days to obtain a result, since enterotoxin expression during growth is necessary [137]. Although genetic probes are also applied for detection of *B. cereus*, the information provided would involve the presence of the gene and not the level of enterotoxin production. It seems that the production of enterotoxins from enterotoxin-positive strains is too low to cause food poisoning [138]. A good choice for the detection of *B. cereus* would be the use of probes directed to the phospholipase C genes, which are present in the majority of the strains. Different confirmatory tests exist for *B. cereus*. For enterotoxic *B. cereus*, molecular diagnostic (PCR-based) [139,140], biochemical, and immunological assays [139,141,142] are commercially available. Three methods for detection of the emetic toxin have been described during the past years—a cytotoxicity assay, liquid chromatography-mass spectrometry (LC-MS) analysis, and a sperm-based bioassay [143,144]. They have, however, proved difficult to use for routine applications and are not specific enough. Recently, a novel PCR-based detection system has been developed based on the emetic toxin cereulide gene [145].

The latest trend is toward the development of molecular tools that would be able to characterize virulence mechanisms of bacterial isolates within minutes [146]. The next generation assays, such as biosensors and DNA chips, have already been developed [147]. They can be classified in high-density DNA arrays [148] and low-density DNA sensors [149]. An automated electrochemical detection system, which allows simultaneous detection of presently described toxin-encoding genes of pathogenic *B. cereus* [146], and a nanowire labeled direct-charge transfer biosensor capable of detecting *Bacillus* species have also been developed [150].

2.10 *Clostridium perfringens*

Foods usually associated with *C. perfringens* infections are cooked meat and poultry. Its presence in raw meats and poultry is not unusual. The illness (diarrhea) is caused by a heat-sensitive enterotoxin produced only by sporulating cells. Usually, large numbers of the microorganism are required to cause illness. As a consequence, the microorganism is enumerated using direct plating without enrichment. Also, *C. perfringens* does not sporulate in food and therefore there is no need to heat the sample before enumerating the microorganism [151].

2.10.1 *Enumeration and Confirmation of* Clostridium perfringens

The selective media used for enumeration of *C. perfringens* contain antibiotics to inhibit other anaerobic microorganisms, along with iron and sulfite because *Clostridia* reduce the latter to sulfide, which reacts with iron to form a black precipitate (black colonies) characteristic of clostridia. The most commonly used and useful medium to recover *C. perfringens* is the egg yolk free tryptose sulfite cycloserine (EY-free TSC) agar (Figure 2.10) [152]. EY-free TSC agar is used in pour plates. Cycloserine is added to inhibit growth of *Enterococci*. Because other sulfite-reducing clostridia that produce black colonies may grow on EY-free TSC agar, further confirmatory tests are needed to identify the presence of *C. perfringens* (Figure 2.11). If low numbers are expected, the

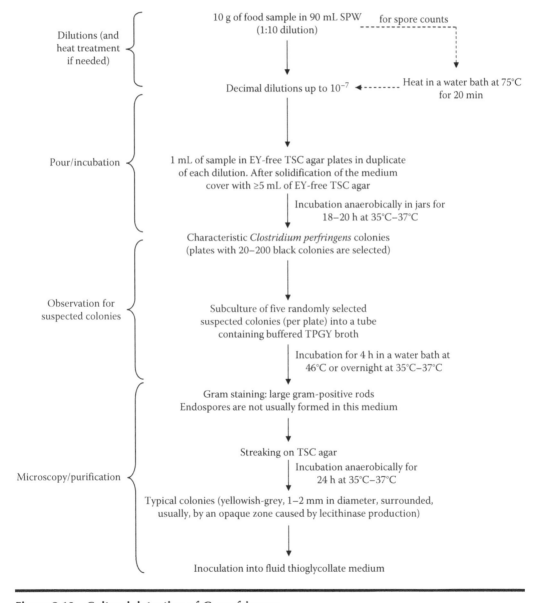

Figure 2.10 **Cultural detection of** *C. perfringens.*

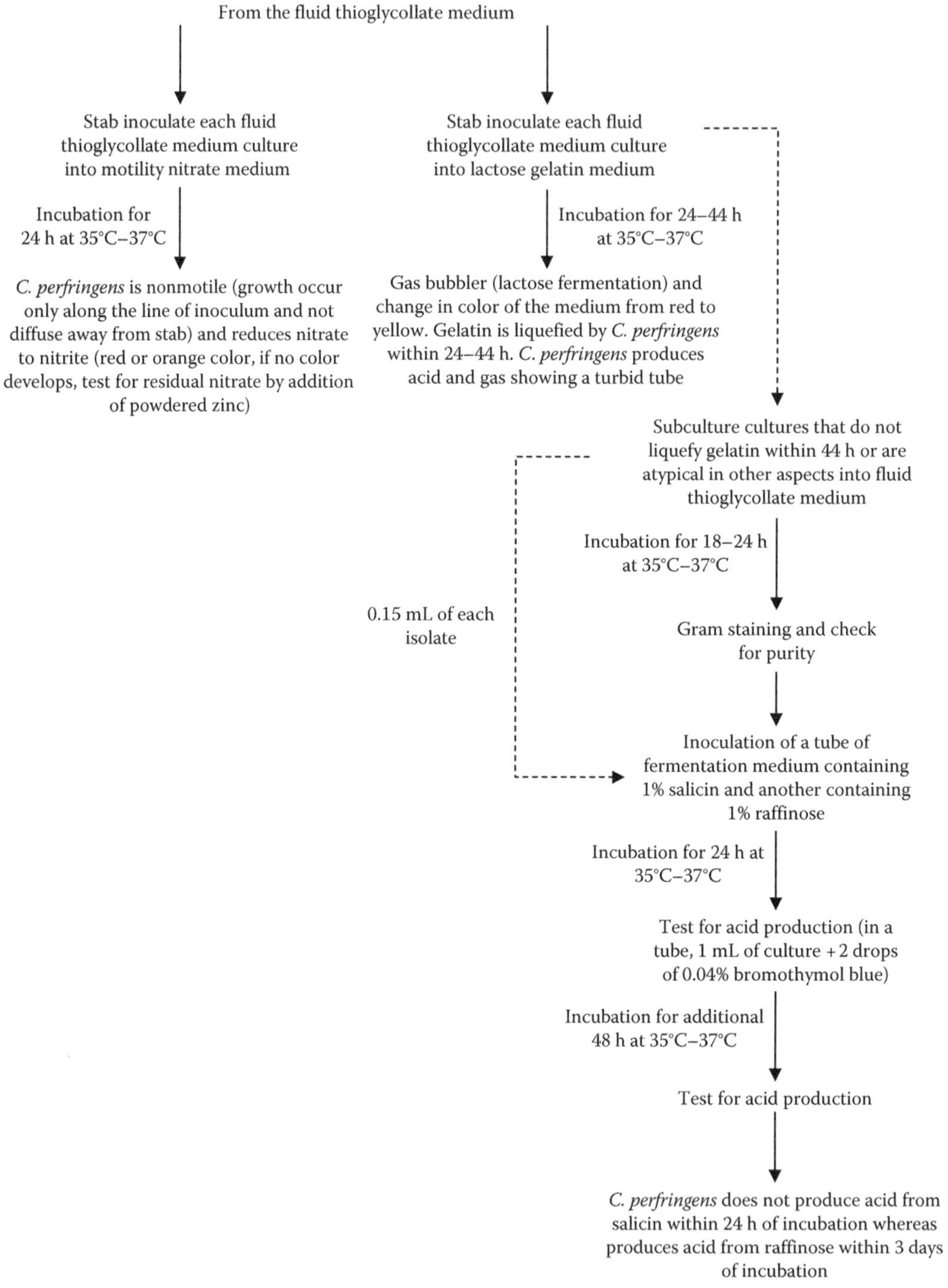

From the fluid thioglycollate medium

Stab inoculate each fluid thioglycollate medium culture into motility nitrate medium

Incubation for 24 h at 35°C–37°C

C. perfringens is nonmotile (growth occur only along the line of inoculum and not diffuse away from stab) and reduces nitrate to nitrite (red or orange color, if no color develops, test for residual nitrate by addition of powdered zinc)

Stab inoculate each fluid thioglycollate medium culture into lactose gelatin medium

Incubation for 24–44 h at 35°C–37°C

Gas bubbler (lactose fermentation) and change in color of the medium from red to yellow. Gelatin is liquefied by *C. perfringens* within 24–44 h. *C. perfringens* produces acid and gas showing a turbid tube

Subculture cultures that do not liquefy gelatin within 44 h or are atypical in other aspects into fluid thioglycollate medium

Incubation for 18–24 h at 35°C–37°C

0.15 mL of each isolate

Gram staining and check for purity

Inoculation of a tube of fermentation medium containing 1% salicin and another containing 1% raffinose

Incubation for 24 h at 35°C–37°C

Test for acid production (in a tube, 1 mL of culture + 2 drops of 0.04% bromothymol blue)

Incubation for additional 48 h at 35°C–37°C

Test for acid production

C. perfringens does not produce acid from salicin within 24 h of incubation whereas produces acid from raffinose within 3 days of incubation

Figure 2.11 Identification scheme of *C. perfringens*.

MPN technique or enrichment using buffered trypticase peptone glucose yeast extract (TPGY) broth may be used. Two grams of food sample is inoculated into 15–20 mL of medium in a tube. The tube is incubated at 35–37°C for 20–24 h. With a bacteriological loop a sample from the positive tubes (turbidity and gas production) is streaked on EY-free TSC agar plates [151]. Enterotoxin of *C. perfringens* can be detected using commercial kits such as ELISA and RPLA.

A nonisotopic colony hybridization technique has been developed for the detection and enumeration of *C. perfringens*; this proved to be more sensitive than the conventional culture methods [153]. It provides quantitative assessment of the presence of potentially enterotoxigenic strains of *C. perfringens* as determined by the presence of the enterotoxin A gene, and the results are acquired within 48 h. A multiplex PCR assay has also been developed for the detection of *C. perfringens* type A [154] and has been evaluated in relation to American retail food by Wen et al. [155]. Methods similar to the ones described earlier for *B. cereus* [137] have also been applied.

2.11 *Campylobacter jejuni*

Campylobacter species are part of intestinal tract microflora of animals and thus may contaminate foods such as meat, poultry, and their products. The most frequent *Campylobacter* species implicated in illnesses is *C. jejuni*. The microorganism is Gram-negative, motile, and oxidase-positive, forming curved rods. Poultry is considered the most important vehicle of *Campylobacter* illness; several outbreaks have been associated with poultry [156,157]. *C. coli* and *C. lari* have also been isolated from poultry and recognized as potential hazards to human health, causing illness, though less frequently than *C. jejuni* [158].

2.11.1 *Detection and Confirmation of Campylobacter jejuni*

In general, *Campylobacter* species are sensitive microorganisms and are stressed during processing, and therefore an enrichment step is needed to resuscitate injured cells. Also, the microorganism fails to grow under normal atmospheric conditions since *Campylobacter* is microaerophilic and capnophilic, and gas jars should be used to provide the right gas atmosphere (5% oxygen, 10% carbon dioxide, and 85% nitrogen). Because of its sensitivity to oxygen, food samples should be kept before analysis in an environment without oxygen (100% nitrogen) with 0.01% sodium bisulfite and under refrigeration. Wang's medium may be used for this purpose [159].

The cultural detection of *Campylobacter* spp. [160] is shown in Figure 2.12. Usually, 10 g of food sample (ground beef) are added to 90 mL of enrichment broth. Sampling of poultry carcasses and large pieces of foods may be performed by the surface rinse technique. The sample is placed in a sterile stomacher bag with 250 mL of Brucella broth and the surface is rinsed by shaking and massaging. The broth (rinse/suspension) is filtered and centrifuged at $16,000 \times g$ for 10 min at 4°C. The supernatant fluid is discarded and the pellet is suspended in 2–5 mL of enrichment broth. After enrichment or during the direct plating without enrichment, two selective agars are used, specifically, Karmali agar and one of the following agars: Butzler agar, Campy-BAP or Blaser agar, *Campylobacter* charcoal differential agar (CCDA)-Preston blood-free agar, and Skirrow agar. It has been found that CCDA-Preston blood-free medium has excellent selectivity and is good for quantitative recovery of *C. jejuni* [159]. The oxygen tolerance of *Campylobacter* may be enhanced by adding to the growth media 0.025% of each of the following: ferrous sulfate, sodium metabisulfite, and sodium pyruvate (FBP supplement) [161].

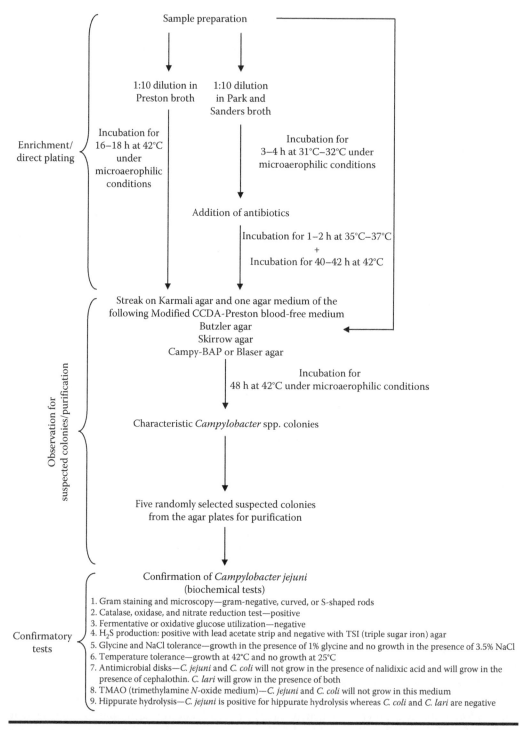

Figure 2.12 Cultural method for detecting and identifying *C. jejuni* based on ISO standard. (Based on ISO. 1995. International Standard, ISO 10272: *Microbiology of food and animal feeding stuffs—Horizontal method for detection and enumeration of Campylobacter spp.* Geneva: International Organization for Standardization.)

Purification of the culture is made as follows for conducting confirmatory tests: Colonies from the selective agar plates are transferred to a Heart Infusion agar with 5% difibrinated rabbit blood (HIA-RB), and plates are incubated at 42°C for 24 h under microaerophilic conditions. The culture is transferred to 5 mL of HIB and the density of the cells is adjusted to meet the McFarland no. 1 turbidity standard (BioMerieux, Marcy Etoile, France). This cell suspension is used further for biochemical testing in tubes or on agar plates [159]. Finally, the commercial kit API Campy (BioMerieux, Marcy Etoile, France) may be used as an alternative for differentiation of *Campylobacter* spp.

Polynucleotide and oligonucleotide probes have been used for the detection of *C. jejuni*; they are reviewed by Olsen et al. [137]. A rapid and sensitive method based on PCR for the detection of *Campylobacter* spp. from chicken products, described by Giesendorf et al. [162], provided results within 48 h with the same sensitivity as the conventional method. Konkel et al. [163] developed a detection and identification method based on the presence of the *cad*F virulence gene, an adhesin to fibronectin, which aids the binding of *C. jejuni* to the intestinal epithelial cells. This method may be useful for the detection of the microorganism in food products, since it does not require bacterial cultivation before its application. Further techniques have been developed since then with the incorporation of an enrichment step before the PCR and real-time PCR amplification, respectively [164,165]. A more recent evaluation of a PCR assay for the detection and identification of *C. jejuni* in poultry products reduced the time of analysis to 24 h or less depending on the necessity of the enrichment step [166]. This method did not seem to be appropriate for ready-to-eat products but was proven to be useful in naturally contaminated poultry samples. Further improvements and trends include multiplex PCRs, reviewed by Settanni and Corsetti [25] as well as real-time nucleic acid sequence-based amplifications with molecular beacons [167].

References

1. Downes, F. P. and Ito, K. 2001. *Compendium of Methods for the Microbiological Examination of Foods*, fourth edition. Washington: American Public Health Association.
2. Van Nierop, W., Duse, A. G., Marais, E. et al. 2005. Contamination of chicken carcasses in Gauteng, South Africa, by *Salmonella*, *Listeria monocytogenes* and *Campylobacter*. *International Journal of Food Microbiology* 99:1–6.
3. National Advisory Committee on Microbiological Criteria for Foods. 2005. Considerations for establishing safety-based consume-by date labels for refrigerated ready-to-eat foods. *Journal of Food Protection* 68:1761–1775.
4. Mossel, D. A. A., Corry, J. E. L., Struijk, C. B., and Baird, R. M. 1995. *Essentials of the Microbiology of Foods: A Textbook for Advanced Studies*. Chichester: Wiley.
5. Atlas, R. M. 2006. *Handbook of Microbiological Media for the Examination of Food*, second edition. Boca Raton, FL: CRC Press.
6. Anonymous. 2007. *The Compendium of Analytical Methods*, Ottawa. http://www.hc-sc.gc.ca/fn-an/res-rech/analy-meth/microbio/index_e.html. Accessed 10 January 2007.
7. Mossel, D. A. A. 1991. Food microbiology: an authentic academic discipline with substantial potential benefits for science and society. *Journal of the Association of Official Analytical Chemists* 74:1–13.
8. Jarvis, B. 2000. Sampling for microbiological analysis. In *The Microbiological Safety and Quality of Food*, eds. B. M. Lund, A. C. Baird-Parker, and G. W. Gould, pp. 1691–1733. Gaithersburg, MD: Aspen Publishers.
9. Betts, R. and Blackburn, C. 2002. Detecting pathogens in food. In *Foodborne Pathogens: Hazards, Risk Analysis and Control*, eds. C. de W. Blackburn and P. J. McClure, pp. 13–52. Boca Raton, FL: CRC Press.

10. Ottoviani, F., Ottoviani, M., and Agosti, M. 1997. A esperienza su un agar selettivo e differenziale per *Listeria monocytogenes*. *Industrie Alimentari* 36:1–3.

11. Foret, J. and Dorey, F. 1997. Evaluation d'un nouveau milieu de culture pour la recherche de *Listeria monocytogenes* dans le lait cru. *Sciences Des Aliments* 17:219–225.

12. Becker, B., Schuler, S., Lohneis, M., Sabrowski, A., Curtis, G. D. W., and Holzapfel, W. H. 2006. Comparison of two chromogenic media for the detection of *Listeria monocytogenes* with the plating media recommended by EN/DIN 11290–1. *International Journal of Food Microbiology* 109:127–131.

13. de Boer, E. 1998. Update on media for isolation of *Enterobacteriaceae* from foods. *International Journal of Food Microbiology* 45:43–53.

14. Gracias, K. S. and McKillip, J. L. 2004. A review of conventional detection and enumeration methods for pathogenic bacteria in food. *Canadian Journal of Microbiology* 50:883–890.

15. ISO. 2006. International Standard, ISO 4831: *Microbiology of food and animal feeding stuffs— Horizontal method for the detection and enumeration of coliforms—Most probable number technique.* Geneva: International Organization for Standardization.

16. ISO. 2005. International Standard, ISO 7251: *Microbiology of food and animal feeding stuffs—Horizontal method for the detection and enumeration of presumptive Escherichia coli—Most probable number technique.* Geneva: International Organization for Standardization.

17. Corry, J. E. L., Jarvis, B., Passmore, S., and Hedges, A. 2007. A critical review of measurement uncertainty in the enumeration of food micro-organisms. *Food Microbiology* 24:230–253.

18. Besse, N. G. 2002. Influence of various environmental parameters and of detection procedures on the recovery of stressed *L. monocytogenes*: a review. *Food Microbiology* 19:221–234.

19. Otero, A., Garcia-Lopez, M. L., and Moreno, B. 1998. Rapid microbiological methods in meat and meat products. *Meat Science* 49:S179–S189.

20. Forsythe, S. J. 2000. *The Microbiology of Safe Food*. Malden, MA: Blackwell Science.

21. Blais, B. W., Leggate, J., Bosley, J., and Martinez-Perez, A. 2004. Comparison of fluorogenic and chromogenic assay systems in the detection of *Escherichia coli* O157 by a novel polymyxin-based ELISA. *Letters in Applied Microbiology* 39:516–522.

22. Deisingh, A. K. and Thomson, M. 2004. Strategies for the detection of *Escherichia coli* O157:H7 in foods. *Journal of Applied Microbiology* 96:419–429.

23. Perelle, S., Dilasser, F., Malorny, B., Grout, J., Hoorfar, J., and Fach, P. 2004. Comparison of PCR-ELISA and Light Cycler real-time PCR assays for detecting *Salmonella* spp. in milk and meat samples. *Molecular and Cellular Probes* 18:409–420.

24. Lehtola, M. J., Loades, C. J., and Keevil, C. W. 2005. Advantages of peptide nucleic acid oligonucleotides for sensitive site directed 16S rRNA fluorescence *in situ* hybridization (FISH) detection of *Campylobacter jejuni*, *Campylobacter coli* and *Campylobacter lari*. *Journal of Microbiological Methods* 62:211–219.

25. Settanni, L. and Corsetti, A. 2007. The use of multiplex PCR to detect and differentiate food- and beverage-associated microorganisms: a review. *Journal of Microbiological Methods* 69:1–22.

26. Rossello-Mora, R. and Amann, R. 2001. The species concept for prokaryotes. *FEMS Microbiology Reviews* 25:39–67.

27. Levi, K., Smedley, J., and Towner, K. J. 2003. Evaluation of a real-time PCR hybridization assay for rapid detection of *Legionella pneumophila* in hospital and environmental water samples. *Clinical Microbiology and Infection* 9:754–758.

28. Lazcka, O., Del Campo, F. J., and Xavier Munoz, F. 2007. Pathogen detection: a perspective of traditional methods and biosensors. *Biosensors and Bioelectronics* 22:1205–1217.

29. Muyzer, G., de Waal, E. C., and Uitterlinden, A. G. 1993. Profiling of complex microbial populations by denaturing gradient gel electrophoresis analysis of polymerase chain reaction-amplified genes coding for 16S rRNA. *Applied and Environmental Microbiology* 59:695–700.

30. National Advisory Committee on Microbiological Criteria for Foods. 1991. *Listeria monocytogenes*. *International Journal of Food Microbiology* 14:185–246.

31. Samelis, J. and Metaxopoulos, J. 1999. Incidence and principal sources of *Listeria* spp. and *Listeria monocytogenes* contamination in processed meats and a meat processing plant. *Food Microbiology* 16:465–477.

32. Tompkin, R. B. 2002. Control of *Listeria monocytogenes* in the food processing environment. *Journal of Food Protection* 65:709–725.

33. Barbalho, T. C. F., Almeida, P. F., Almeida, R. C. C., and Hofer, E. 2005. Prevalence of *Listeria* spp. at a poultry processing plant in Brazil and a phage test for rapid confirmation of suspect colonies. *Food Control* 16:211–216.

34. Wilks, S. A., Michels, H. T., and Keevil, C. W. 2006. Survival of *Listeria monocytogenes* Scott A on metal surfaces: implications for cross-contamination. *International Journal of Food Microbiology* 111:93–98.

35. Gibbons, I., Adesiyun, A., Seepersadsingh, N., and Rahaman, S. 2006. Investigation for possible source(s) of contamination of ready-to-eat meat products with *Listeria* spp. and other pathogens in a meat processing plant in Trinidad. *Food Microbiology* 23:359–366.

36. Reij, M. W., den Aantrekker, E. D., and ILSI Europe Risk Analysis in Microbiology Task Force. 2004. Recontamination as a source of pathogens in processed foods. *International Journal of Food Microbiology* 91:1–11.

37. Jay, J. M. 1996. Prevalence of *Listeria* spp. in meat and poultry products. *Food Control* 7:209–214.

38. Mataragas, M., Drosinos, E. H., and Metaxopoulos, J. 2003. Antagonistic activity of lactic acid bacteria against *Listeria monocytogenes* in sliced cooked cured pork shoulder stored under vacuum or modified atmosphere at 4 ± 2°C. *Food Microbiology* 20:259–265.

39. Mataragas, M., Drosinos, E. H., Siana, P., Skandamis, P., and Metaxopoulos, I. 2006. Determination of the growth limits and kinetic behavior of *Listeria monocytogenes* in a sliced cooked cured meat product: validation of the predictive growth model under constant and dynamic temperature storage conditions. *Journal of Food Protection* 69:1312–1321.

40. Glass, K. A. and Doyle, M. P. 1989. Fate of *Listeria monocytogenes* in processed meat during refrigerated storage. *Applied and Environmental Microbiology* 55:1565–1569.

41. Gandhi, M. and Chikindas, M. L. 2007. Listeria: a foodborne pathogen that knows how to survive. *International Journal of Food Microbiology* 113:1–15.

42. Drosinos, E. H., Mataragas, M., Veskovic-Moracanin, S., Gasparik-Reichardt, J., Hadziosmanovic, M., and Alagic, D. 2006. Quantifying nonthermal inactivation of *Listeria monocytogenes* in European fermented sausages using bacteriocinogenic lactic acid bacteria or their bacteriocins: a case study for risk assessment. *Journal of Food Protection* 69:2648–2663.

43. ISO. 1996. International Standard, ISO 11290-1: *Microbiology of food and animal feeding stuffs— Horizontal method for the detection and enumeration of Listeria monocytogenes—Part 1: Detection method*. Geneva: International Organization for Standardization.

44. Patel, J. R. and Beuchat, L. R. 1995. Enrichment in Fraser broth supplemented with catalase or Oxyrase®, combined with the microcolony immunoblot technique, for detecting heat-injured *Listeria monocytogenes* in foods. *International Journal of Food Microbiology* 26:165–176.

45. Gasanov, U., Hughes, D., and Hansbro, P. M. 2005. Methods for the isolation and identification of *Listeria* spp. and *Listeria monocytogenes*: a review. *FEMS Microbiology Reviews* 29:851–875.

46. Van Netten, P., Perales, I., van de Moosdijk, A. Curtis, G. D. W., and Mossel, D. A. A. 1989. Liquid and solid selective differential media for the detection and enumeration of *L. monocytogenes* and other *Listeria* spp. *International Journal of Food Microbiology* 8:299–316.

47. Klinger, J. D., Johnson, A., and Croan, D. et al. 1988. Comparative studies of nucleic acid hybridization assay for *Listeria* in foods. *Journal of the Association of Official Analytical Chemists* 71:669–673.

48. Cocolin, L., Manzano, M., Cantoni, C., and Comi, G. 1997. A PCR-microplate capture hybridization method to detect *Listeria monocytogenes* in blood. *Molecular and Cellular Probes* 11:453–455.

49. Li, X., Boudjellab, N., and Zhao, X. 2000. Combined PCR and slot blot assay for detection of *Salmonella* and *Listeria monocytogenes*. *International Journal of Food Microbiology* 56:167–177.

50. Hsih, H. Y. and Tsen, H. Y. 2001. Combination of immunomagnetic separation and polymerase chain reaction for the simultaneous detection of *Listeria monocytogenes* and *Salmonella* spp. in food samples. *Journal of Food Protection* 64:1744–1750.

51. Bubert, A., Kohler, S., and Goebel, W. 1992. The homologous and heterologous regions within the *iap* gene allow genus- and species-specific identification of *Listeria* spp. by polymerase chain reaction. *Applied and Environmental Microbiology* 58:2625–2632.

52. Wesley, I. V., Harmon, K. M., Dickson, J. S., and Schwartz, A. R. 2002. Application of a multiplex polymerase chain reaction assay for the simultaneous confirmation of *Listeria monocytogenes* and other *Listeria* species in turkey sample surveillance. *Journal of Food Protection* 65:780–785.

53. Klein, P. G. and Juneja, V. K. 1997. Sensitive detection of viable *Listeria monocytogenes* by reverse transcription-PCR. *Applied and Environmental Microbiology* 63:4441–4448.

54. Call, D. R., Borucki, M. K., and Loge, F. J. 2003. Detection of bacterial pathogens in environmental samples using DNA microarrays. *Journal of Microbiological Methods* 53:235–243.

55. Churchill, R. L. T., Lee, H., and Hall, C. J. 2006. Detection of *Listeria monocytogenes* and the toxin listeriolysin O in food. *Journal of Microbiological Methods* 64:141–170.

56. Lampel, K. A., Orlandi, P. A., and Kornegay, L. 2000. Improved template preparation for PCR-based assays for detection of food-borne bacterial pathogens. *Applied and Environmental Microbiology* 66:4539–4542.

57. Sergeev, N., Distler, M., Courtney, S. et al. 2004. Multipathogen oligonucleotide microarray for environmental and biodefense applications. *Biosensors and Bioelectronics* 20:684–698.

58. ISO. 1998. International Standard, ISO 11290-2: *Microbiology of food and animal feeding stuffs—Horizontal method for the detection and enumeration of Listeria monocytogenes—Part 2: Enumeration method*. Geneva: International Organization for Standardization.

59. Schleiss, M. R., Bourne, N., Bravo, F. J., Jensen, N. J., and Bernstein, D. I. 2003. Quantitative–competitive PCR monitoring of viral load following experimental guinea pig cytomegalovirus infection. *Journal of Virological Methods* 108:103–110.

60. Choi, W. S. and Hong, C. H. 2003. Rapid enumeration of *Listeria monocytogenes* in milk using competitive PCR. *International Journal of Food Microbiology* 84:79–85.

61. Prentice, G. A. and Neaves, P. 1992. The identification of *Listeria* species. In *Identification Methods in Applied and Environmental Microbiology*, eds. R. G. Board, D. Jones, and F. A. Skinner, pp. 283–296. Oxford: Blackwell Scientific.

62. Beumer, R. R. and Hazeleger, W. C. 2003. *Listeria monocytogenes*: diagnostic problems. *FEMS Immunology and Medical Microbiology* 35:191–197.

63. Fratamico, P. M., Smith, J. L., and Buchanan, R. L. 2002. *Escherichia coli*. In *Foodborne Diseases*, eds. D. O. Cliver and H. P. Riemann, pp. 79–102. London: Academic Press.

64. Bell, C. and Kyriakides, A. 2002a. *Pathogenic Escherichia coli*. In *Foodborne Pathogens: Hazards, Risk Analysis and Control*, eds. C. de W. Blackburn and P. J. McClure, pp. 280–306. Boca Raton, FL: CRC Press.

65. Smith, H. R., Cheasty, T., Roberts, D. et al. 1991. Examination of retail chickens and sausages in Britain for vero cytotoxin-producing *Escherichia coli*. *Applied and Environmental Microbiology* 57:2091–2093.

66. Heuvelink, A. E., Wernars, K., and de Boer, E. 1996. Occurrence of *Escherichia coli* O157 and other verocytotoxin-producing *E. coli* in retail raw meats in the Netherlands. *Journal of Food Protection* 59:1267–1272.

67. Bolton, F. J., Crozier, L., and Williamson, J. K. 1996. Isolation of *Escherichia coli* O157 from raw meat products. *Letters in Applied Microbiology* 23:317–321.

68. Chapman, P. A., Siddons, C. A., Cerdan Malo A. T. et al. 2000. A one year study of *Escherichia coli* O157 in raw beef and lamb products. *Epidemiology and Infection* 124:207–213.

69. Getty, K. J. K., Phebus, R. K., Marsden, J. L. et al. 2000. *Escherichia coli* O157:H7 and fermented sausages: a review. *Journal of Rapid Methods and Automation in Microbiology*, 8:141–170.

70. Pond, T. J., Wood, D. S., Mumin, I. M., Barbut, S., and Griffiths, M. W. 2001. Modeling the survival of *Escherichia coli* O157:H7 in uncooked, semidry, fermented sausage. *Journal of Food Protection* 64:759–766.

71. Adams, M. and Mitchell, R. 2002. Fermentation and pathogen control: a risk assessment approach. *International Journal of Food Microbiology* 79:75–83.

72. ISO. 2001. International Standard, ISO 16654: *Microbiology of food and animal feeding stuffs—Horizontal method for the detection of Escherichia coli O157*. Geneva: International Organization for Standardization.

73. Farmer, J. J. and Davis, B. R. 1985. H7 antiserum-sorbitol fermentation medium: a single tube screening medium for detecting *Escherichia coli* O157:H7 associated with hemorrhagic colitis. *Journal of Clinical Microbiology* 22:620–625.

74. Bolton, F. J., Crozier, L., and Williamson, J. K. 1995. Optimisation of methods for the isolation of *Escherichia coli* O157 from beefburgers. *PHLS Microbiology Digest* 12:67–70.

75. Zadik, P. M., Chapman, P. A., and Siddons, C. A. 1993. Use of tellurite for the selection of vero cytotoxigenic *Escherichia coli* O157. *Journal of Medical Microbiology* 39:155–158.

76. Heuvelink, A. E., Zwartkruis-Nahuis, J. T. M., and de Boer, E. 1997. Evaluation of media and test kits for the detection and isolation of *Escherichia coli* O157 from minced beef. *Journal of Food Protection* 60:817–824.

77. Osek, J. 2002. Rapid and specific identification of Shiga toxin producing *Escherichia coli* in faeces by multiplex PCR. *Letters in Applied Microbiology* 34:304–310.

78. Kim, J. Y., Kim, S. Y., Kwon, N. H. et al. 2005. Isolation and identification of *Escherichia coli* O157:H7 using different detection methods and molecular determination by multiplex PCR and RAPD. *Journal of Veterinary Science* 6:7–19.

79. Muller, D., Hagedorn, P., Brast, S. et al. 2006. Rapid identification and differentiation of clinical isolates of enteropathogenic *Escherichia coli* (EPEC), atypical EPEC, and shiga toxin-producing *Escherichia coli* by a one-step multiplex PCR method. *Journal of Clinical Microbiology* 44:2626–2629.

80. Kadhum, H. J., Ball, H. J., Oswald, E., and Rowe, M. T. 2006. Characteristics of cytotoxic necrotizing factor and cytolethal distending toxin producing *Escherichia coli* strains isolated from meat samples in Northern Ireland. *Food Microbiology* 23:491–497.

81. ISO. 2001. International Standard, ISO 16649-2: *Microbiology of food and animal feeding stuffs—Horizontal method for the enumeration of beta-glucuronidase-positive Escherichia coli—Part 2: Colony-count technique at 44 degrees C using 5-bromo-4-chloro-3-indolyl beta-D-glucuronide*. Geneva: International Organization for Standardization.

82. Bell, C. and Kyriakides, A. 2002b. *Salmonella*. In *Foodborne Pathogens: Hazards, Risk Analysis and Control*, eds. C. de W. Blackburn and P. J. McClure, pp. 307–335. Boca Raton, FL: CRC Press.

83. ISO. 1993. International Standard, ISO 6579: *Microbiology of food and animal feeding stuffs—Horizontal method for the detection of Salmonella spp*. Geneva: International Organization for Standardization.

84. Gray, J. T. and Fedorka-Cray, P. J. 2002. *Salmonella*. In *Foodborne Diseases*, eds. D. O. Cliver and H. P. Riemann, pp. 55–68. London: Academic Press.

85. D'Aoust, J. Y., Sewell, A. M., and Warburton, D. W. 1992. A comparison of standard cultural methods for the detection of foodborne *Salmonella*. *International Journal of Food Microbiology* 16:41–50.

86. Ruiz, J., Nunez, J., Diaz, J., Sempere, M. A., Gomez, J., and Usera, M. A. 1996. Comparison of media for the isolation of lactose-positive *Salmonella*. *Journal of Applied Bacteriology* 81:571–574.

87. Lukinmaa, S., Nakari, U. M., Eklund, M., and Siitonen, A. 2004. Application of molecular genetic methods in diagnostics and epidemiology of food-borne bacterial pathogens. *Acta Pathologica, Microbiologica et Immunologica Scandinavica* 112:908–929.

88. Lim, Y. H., Hirose, K., Izumiya, H. et al. 2003. Multiplex polymerase chain reaction assay for selective detection of *Salmonella enterica* serovar *typhimurium*. *Japanese Journal of Infectious Diseases* 56:151–155.

89. Ali, K., Zeynab, A., Zahra, S., and Saeid, M. 2006. Development of an ultra rapid and simple multiplex polymerase chain reaction technique for detection of *Salmonella typhi*. *Saudi Medical Journal* 27:1134–1138.

90. Pritchett, L. C., Konkel, M. E., Gay, J. M., and Besser, T. E. 2000. Identification of DT104 and U302 phage types among *Salmonella enterica* serotype *typhimurium* isolates by PCR. *Journal of Clinical Microbiology* 38:3484–3488.

91. Sutherland, J. and Varnam, A. 2002. Enterotoxin-producing *Staphylococcus, Shigella, Yersinia, Vibrio, Aeromonas* and *Plesiomonas*. In *Foodborne Pathogens: Hazards, Risk Analysis and Control*, eds. C. de W. Blackburn and P. J. McClure, pp. 385–415. Boca Raton, FL: CRC Press.

92. ISO. 1999. International Standard, ISO 6888-1: *Microbiology of food and animal feeding stuffs— Horizontal method for the enumeration of coagulase-positive staphylococci (Staphylococcus aureus and other species)—Part 1: Technique using Baird-Parker agar medium*. Geneva: International Organization for Standardization.

93. Baird-Parker, A. C. 1962. An improved diagnostic and selective medium for the isolating coagulase positive Staphylococci. *Journal of Applied Bacteriology* 25:12–19.

94. Wong, A. C. L. and Bergdoll, M. S. 2002. *Staphylococcal food poisoning*. In *Foodborne Diseases*, eds. D. O. Cliver and H. P. Riemann, pp. 231–248. London: Academic Press.

95. Brown, D. F. J., Edwards, D. I., Hawkey, P. M. et al. 2005. Guidelines for the laboratory diagnosis and susceptibility testing of methicillin-resistant *Staphylococcus aureus* (MRSA). *Journal Antimicrobial Chemotherapy* 56:1000–1018.

96. Geha, D. J., Uhl, J. R., Gustaferro, C. A., and Persing, D. H. 1994. Multiplex PCR for identification of methicillin-resistant Staphylococci in the clinical laboratory. *Journal of Clinical Microbiology* 32:1768–1772.

97. Mason, W. J., Blevins, J. S., Beenken, K., Wibowo, N., Ojho, N., and Smeltzer, M. S. 2001. Multiplex PCR protocol for the diagnosis of staphylococcal infection. *Journal of Clinical Microbiology* 39:3332–3338.

98. Alouf, J. E. and Muller-Alouf, H. 2003. Staphylococcal and streptococcal superantigens: molecular, biological and clinical aspects. *International Journal of Medical Microbiology* 292:429–440.

99. Orwin, P. M., Fitzgerald, J. R., Leung, D. Y., Gutierrez, J. A., Bohach, G. A., and Schlievert, P. M. 2003. Characterization of *Staphylococcus aureus* enterotoxin L. *Infection and Immunity* 71:2916–2919.

100. Martin, M. C., Fueyo, J. M., Gonzalez-Hevia, M. A., and Mendoza, M. C. 2004. Genetic procedures for identification of enterotoxigenic strains of *Staphylococcus aureus* from three food poisoning outbreaks. *International Journal of Food Microbiology* 94:279–286.

101. Nakano, S., Kobayashi, T., Funabiki, K., Matsumura, A., Nagao, Y., and Yamada, T. 2004. PCR detection of *Bacillus* and *Staphylococcus* in various foods. *Journal of Food Protection* 67:1271–1277.

102. McLauchlin, J., Narayanan, G. L., Mithani, V., and O'Neill, G. 2000. The detection of enterotoxins and toxic shock syndrome toxin genes in *Staphylococcus aureus* by polymerase chain reaction. *Journal of Food Protection* 63:479–488.

103. Martin, G. S., Mannino, D. M., Eaton, S., and Moss, M. 2003. The epidemiology of sepsis in the United States from 1979 through 2000. *The New England Journal of Medicine* 348:1546–1554.

104. Palka-Santini, M., Pützfeld, S., Cleven, B. E., Krönke, M., and Krut, O. 2007. Rapid identification, virulence analysis and resistance profiling of *Staphylococcus aureus* by gene segment-based DNA microarrays: application to blood culture post-processing. *Journal of Microbiological Methods* 68:468–477.

105. Poulsen, A. B., Skov, R., and Pallesen, L. V. 2003. Detection of methicillin resistance in coagulase-negative staphylococci and in staphylococci directly from simulated blood cultures using the EVI-GENE MRSA Detection Kit. *Journal of Antimicrobial Chemotherapy* 51:419–421.

106. Oliveira, K., Brecher, S. M., Durbin, A. et al. 2003. Direct identification of *Staphylococcus aureus* from positive blood culture bottles. *Journal of Clinical Microbiology* 41:889–891.

107. Tan, T. Y., Corden, S., Barnes, R., and Cookson, B. 2001. Rapid identification of methicillin-resistant *Staphylococcus aureus* from positive blood cultures by real-time fluorescence PCR. *Journal of Clinical Microbiology* 39:4529–4531.

108. Wellinghausen, N., Wirths, B., Essig, A., and Wassill, L. 2004. Evaluation of the Hyplex BloodScreen Multiplex PCR-Enzyme-linked immunosorbent assay system for direct identification of gram-positive cocci and gram-negative bacilli from positive blood cultures. *Journal of Clinical Microbiology* 42:3147–3152.

109. Wellinghausen, N., Wirths, B., Franz, A. R., Karolyi, L., Marre, R., and Reischl, U. 2004. Algorithm for the identification of bacterial pathogens in positive blood cultures by real-time LightCycler polymerase chain reaction (PCR) with sequence-specific probes. *Diagnostic Microbiology and Infectious Disease* 48:229–241.

110. Kempf, V. A., Mandle, T., Schumacher, U., Schafer, A., and Autenrieth, I. B. 2005. Rapid detection and identification of pathogens in blood cultures by fluorescence in situ hybridization and flow cytometry. *International Journal of Medical Microbiology* 295:47–55.

111. Sorum, H. and L'Abee-Lund, T. M. 2002. Antibiotic resistance in food-related bacteria-a result of interfering with the global web of bacterial genetics. *International Journal of Food Microbiology* 78:43–56.

112. Jones, T. F. Kellum, M. E., Porter, S. S., Bell, M., and Schaffner, W. 2002. An outbreak of community-acquired foodborne illness caused by methicillin-resistant *Staphylococcus aureus*. *Emerging Infectious Diseases* 8:82–84.

113. Nesbakken, T., Eckner, K., Hoidal, H. K., and Rotterud, O. J. 2003. Occurrence of *Yersinia enterocolitica* and *Campylobacter* spp. in slaughter pigs and consequences for meat inspection, slaughtering, and dressing procedures. *International Journal of Food Microbiology* 80:231–240.

114. ISO. 1994. International Standard, ISO 10273: *Microbiology of food and animal feeding stuffs— Horizontal method for the detection of presumptive pathogenic Yersinia enterocolitica*. Geneva: International Organization for Standardization.

115. Schiemann, D. A. 1982. Development of a two-step enrichment procedure for recovery of *Yersinia enterocolitica* from food. *Applied and Environmental Microbiology* 43:14–27.

116. Schiemann, D. A. 1983. Alkalotolerance of *Yersinia enterocolitica* as a basis for selective isolation from food enrichments. *Applied and Environmental Microbiology* 46:22–27.

117. Nordic Committee on Food Analysis. 1987. *Yersinia enterocolitica. Detection in Food*. Method no. 117, second edition. Norway: Esbo Nordic Committee on Food Analysis.

118. Wauters, G., Goossens, V., Janssens, M., and Vandepitte, J. 1988. New enrichment method for isolation of pathogenic *Yersinia enterocolitica* O:3 from pork. *Applied and Environmental Microbiology* 54:851–854.

119. De Zutter, L., Le Mort, L., Janssens, M., and Wauters, G. 1994. Short-comings of irgasan ticarcillin chlorate broth for the enrichment of *Yersinia enterocolitica* biotype 2, serotype 9 from meat. *International Journal of Food Microbiology* 23:231–237.

120. Arnold, T., Neubauer, H., Nikolaou, K., Roesler, U., and Hensel, A. 2004. Identification of *Yersinia enterocolitica* in minced meat: a comparative analysis of API 20E, *Yersinia* identification kit and a 16S rRNA-based PCR method. *Journal of Veterinary Medicine B* 51:23–27.

121. Logue, C. M., Sheridan, J. J., Wauters, G., Mc Dowell, D. A., and Blair, I. S. 1996. *Yersinia* spp. and numbers, with particular reference to *Y. enterocolitica* bio/serotypes, occurring on Irish meat and meat products, and the influence of alkali treatment on their isolation. *International Journal of Food Microbiology* 33:257–274.

122. Weagant, S. D. and Feng, P. 2001. Yersinia. In *Compendium of Methods for the Microbiological Examination of Foods*, eds. F. P. Downes and K. Ito, pp. 421–428. Washington: American Public Health Association.

123. Gemski, P., Lazere, J. R., and Casey, T. 1980. Plasmid associated with pathogenicity and calcium dependency of *Yersinia enterocolitica*. *Infection and Immunity* 27:682–685.

124. Kandolo, K. and Wauters, G. 1985. Pyrazinamidase activity in *Yersinia enterocolitica* and related organisms. *Journal of Clinical Microbiology* 21:980–982.

125. Farmer, J. J., Carter, G. P., Miller, V. L., Falkow, S., and Wachsmuth, I. K. 1992. Pyrazinamidase, CR–MOX agar, salicin fermentation-esculin hydrolysis, and D-xylose fermentation for identifying pathogenic serotypes of *Yersinia enterocolitica*. *Journal of Clinical Microbiology* 30:2589–2594.

126. Bhaduri, S. and Cottrell, B. 1997. Direct detection and isolation of plasmid-bearing virulent serotypes of *Yersinia enterocolitica* from various foods. *Applied and Environmental Microbiology* 63:4952–4955.

127. Johannessen, G. S., Kapperud, G., and Kruse, H. 2000. Occurrence of pathogenic *Yersinia enterocolitica* in Norwegian pork products determined by a PCR method and a traditional culturing method. *International Journal of Food Microbiology* 54:75–80.

128. Lambertz, S. T., Granath, K., Fredriksson-Ahomaa, M., Johansson, K. E., and Danielsson-Tham, M. L. 2007. Evaluation of a combined culture and PCR method (NMKL-163A) for detection of presumptive pathogenic *Yersinia enterocolitica* in pork products. *Journal of Food Protection* 70:335–340.

129. Lucero-Estrada, C. S. M., del Carmen Velazquez, L., Di Genaro, S., and de Guzman, A. M. S. 2007. Comparison of DNA extraction methods for pathogenic *Yersinia enterocolitica* detection from meat food by nested PCR. *Food Research International* 40:637–642.

130. Howard, S. L., Gaunt, M. W., Hinds, J., Witney, A. A., Stabler, R., and Wren, B. W. 2006. Application of comparative phylogenomics to study the evolution of *Yersinia enterocolitica* and to identify genetic differences relating to pathogenicity. Journal of Bacteriology 188:3645–53.

131. Bennett, R. W. and Belay, N. 2001. *Bacillus cereus*. In *Compendium of Methods for the Microbiological Examination of Foods*, eds. F. P. Downes and K. Ito, pp. 311–316. Washington: American Public Health Association.

132. Shinagawa, K. 1990. Analytical methods for *Bacillus cereus* and other *Bacillus* species. *International Journal of Food Microbiology* 10:125–42.

133. Mossel, D. A. A., Koopman, M. J., and Jongerius, E. 1967. Enumeration of *Bacillus cereus* in foods. *Applied Microbiology* 15:650–653.

134. Kim, H. U. and Goepfert, J. M. 1971. Enumeration and identification of *Bacillus cereus* in foods. I. 24-hour presumptive test medium. *Applied Microbiology* 22:581–587.

135. Holbrook, R. and Anderson, J. M. 1980. An improved selective and diagnostic medium for the isolation and enumeration of *Bacillus cereus* in foods. *Canadian Journal of Microbiology* 26:753–759.

136. Griffiths, M. W. and Schraft, H. 2002. *Bacillus cereus* food poisoning. In *Foodborne Diseases*, eds. D. O. Cliver and H. P. Riemann, pp. 261–270. London: Academic Press.

137. Olsen, J. E., Aabo, S., Hill, W. et al. 1995. Probes and polymerase chain reaction for detection of food-borne bacterial pathogens. *International Journal of Food Microbiology* 28:1–78.

138. Granum, P. E., Tomas, J. M., and Alouf, J. E. 1995. A survey of bacterial toxins involved in food poisoning: a suggestion for bacterial food poisoning toxin nomenclature. *International Journal of Food Microbiology* 28:129–144.

139. Hansen, B. M., Leser, T. D., and Hendriksen, N. B. 2001. Polymerase chain reaction assay for the detection of *Bacillus cereus* group cells. *FEMS Microbiology Letters* 202:209–213.

140. Manzano, M., Cocolin, L., Cantoni, C., and Comi, G. 2003. *Bacillus cereus, Bacillus thuringiensis* and *Bacillus mycoides* differentiation using a PCR-RE technique. *International Journal of Food microbiology* 81:249–254.

141. Pruss, B. M., Dietrich, R., Nibler, B., Martlbauner, E., and Scherer, S. 1999. The hemolytic enterotoxin HBL is broadly distributed among species of the *Bacillus cereus* group. *Applied and Environmental Microbiology* 65: 5436–5442.

142. Stenfors, L. P., Mayr, R., Scherer, S., and Granum, P. E. 2002. Pathogenic potential of fifty *Bacillus weihenstephanensis* strains. *FEMS Microbiology Letters* 215:47–51.

143. Haggblom, M. M., Apetroaie, C., Andersson, M. A., and Salkinoja-Salonen, M. S. 2002. Quantitative analysis of cereulide, the emetic toxin of *Bacillus cereus*, produced under various conditions. *Applied and Environmental Microbiology* 68:2479–2483.

144. Anderson, M. A., Jaaskelainen, E. L., Shaheen, R., Pirhonen, T., Wijnands, L. M., and Salkinoja-Salonen, M. S. 2004. Sperm bioassay for rapid detection of cereulide-producing *Bacillus cereus* in food and related environments. *International Journal of Food Microbiology* 94:175–183.

145. Horwood, P. F., Burgess, G. W., and Oakey, H. J. 2004. Evidence for non-ribosomal peptide synthetase production of cereulide (the emetic toxin) in *Bacillus cereus. FEMS Microbiology Letters* 236:319–324.

146. Liu, Y., Elsholz, B., Enfors, S. O., and Gabig-Ciminska, M. 2007. Confirmative electric DNA array based test for food poisoning *Bacillus cereus. Journal of Microbiological Methods* 70:55–64.

147. Homs, M. C. I. 2002. DNA sensors. *Analytical Letters* 35:1875.

148. Chee, M., Yang, R., Hubbell, E. et al. 1996. Accessing genetic information with high-density DNA arrays. *Science* 274:610–614.

149. Albers, J., Grunwald, T., Nebling, E., Piechotta, G., and Hintsche, R. 2003. Electrical biochip technology-a tool for microarrays and continous monitoring. *Analytical and Bioanalytical Chemistry* 377:521–527.

150. Pal, S., Alocilja, E. C., and Downes, F. P. 2007. Nanowire labelled direct-charge transfer biosensor for detecting *Bacillus* species. *Biosensors and Bioelectronics* 22:2329–2336.

151. Labbe, R. G. 2001. *Clostridium perfringens.* In *Compendium of Methods for the Microbiological Examination of Foods*, eds. F. P. Downes and K. Ito, pp. 325–330. Washington: American Public Health Association.

152. Hauschild, A. H. W. and Hilsheimer, R. H. 1974. Enumeration of foodborne *Clostridium perfringens* in egg yolk-free tryptose-sulfite cycloserine agar. *Applied and Environmental Microbiology* 27:521.

153. Baez, L. A. and Juneja, V. K. 1995. Nonradioactive colony hybridization assay for detection and enumeration of enterotoxigenic *Clostridium perfringens* in raw beef. *Applied and Environmental Microbiology* 61:807–810.

154. Garmory, H. S., Chanter, N., French, N. P., Bueschel, D., Songer, J. G., and Titball, R. W. 2000. Occurrence of *Clostridium perfringens* beta2-toxin amongst animals, determined using genotyping and subtyping PCR assays. *Epidemiology and Infection.* 124:61–67.

155. Wen, Q., Miyamoto, K., and McClane, B. A. 2004. Development of a duplex PCR genotyping assay for distinguishing *Clostridium perfringens* type A isolates carrying chromosomal enterotoxin (*cpe*) genes from those carrying plasmid borne enterotoxin (*cpe*) genes. *Journal of Clinical Microbiology* 41:1494–1498.

156. Solomon, E. B. and Hoover, D. G. 1999. *Campylobacter jejuni*: a bacterial paradox. *Journal of Food Safety* 19:121–136.

157. Rautelin, H. and Hanninen, M. L. 2000. Campylobacters: the most common bacterial enteropathogens in the Nordic countries. *Annals of Medicine* 32:440–445.

158. Fernandez, H. and Pison, V. 1996. Isolation of thermotolerant species of *Campylobacter* from commercial chicken livers. *International Journal of Food Microbiology* 29:75–80.

159. Stern, N. J., Line, J. E., and Chen, H. C. 2001. *Campylobacter.* In *Compendium of Methods for the Microbiological Examination of Foods*, eds. F. P. Downes and K. Ito, pp. 301–310. Washington: American Public Health Association.

160. ISO. 1995. International Standard, ISO 10272: *Microbiology of food and animal feeding stuffs— Horizontal method for detection and enumeration of Campylobacter spp.* Geneva: International Organization for Standardization.

161. George, H. A., Hoffman, P. S., Smibert, R. M., and Kreig, N. R. 1978. Improved media for growth and aerotolerance of *Campylobacter fetus. Journal of Clinical Microbiology* 8:36–41.

162. Giesendorf, B. A. J., Quint, W. G., Henkens, M. H., Stegeman, H., Huf, F. A., and Niesters, H. G. 1992. Rapid and sensitive detection of *Campylobacter* spp. in chicken products by using polymerase chain reaction. *Applied and Environmental Microbiology* 58:3804–3808.

163. Konkel, M. E., Gray, S. A., Kim, B. J., Garvis, S. G., and Yoon, J. 1999. Identification of enteropathogens *Campylobacter jejuni* and *Campylobacter coli* based on the *cad*F virulence gene and its product. *Journal of Clinical Microbiology* 37:510–517.

164. Denis, M., Refregier-Petton, J., Laisney, M. J., Ermel, G., and Salvat, G. 2001. *Campylobacter* contamination in French chicken production from farm to consumer, use of a PCR assay for detection and identification of *Campylobacter jejuni* and *Campylobacter coli. Journal of Applied Microbiology* 91:255–267.

165. Sails, A. D., Fox, J. A., Bolton, F. J., Wareing, D. R. A., and Greenway, D. L. A. 2003. A real-time PCR assay for the detection of *Campylobacter jejuni* in foods after enrichment culture. *Applied and Environmental Microbiology* 69:1383–1390.

166. Mateo, E., Carcamo, J., Urquijo, M., Perales, I., and Fernandez-Astorga, A. 2005. Evaluation of a PCR assay for the detection and identification of *Campylobacter jejuni* and *Campylobacter coli* in retail poultry products. *Research in Microbiology* 156:568–574.

167. Churruca, E., Girbau, C., Martínez, I., Mateo, E., Alonso, R., and Fernandez-Astorga, A. 2007. Detection of *Campylobacter jejuni* and *Campylobacter coli* in chicken meat samples by real-time nucleic acid sequence-based amplification with molecular beacons. *International Journal of Food Microbiology* 117:85–90.

Chapter 3

Parasites

Anu Näreaho

Contents

3.1 *Trichinella* spp. 60
    3.1.1 Direct Detection Methods 61
        3.1.1.1 Trichinoscopic Examination 62
        3.1.1.2 Methods for Digestion 62
        3.1.1.3 Histology 65
    3.1.2 Indirect Detection Methods 66
3.2 *Taenia* spp. 67
3.3 *Toxoplasma gondii* 68
3.4 *Sarcocystis* spp. 69
3.5 Some Other Parasites of Importance in Meat Inspection 69
    3.5.1 *Ascaris suum* 69
    3.5.2 *Fasciola hepatica* and Other Liver Flukes 70
    3.5.3 *Echinococcus* spp. 71
    3.5.4 *Parafilaria bovicola* 72
3.6 Future Visions 72
References 73

In the industrialized countries, people are relatively seldom bothered by parasites in meat, because of the high standards of meat hygiene and the advanced parasite control in animal production. However, when a primary toxoplasmosis is diagnosed in a pregnant woman, or a spot epidemic of trichinellosis is found in a village, it surely is devastating for the people involved. Also, the hospitalization costs for even one patient with a severe parasitic infection are considerable. Thus, meat parasites cannot be ignored.

To prevent meat-borne parasitic infections, it is good advice to cook the meat thoroughly. This is not, however, always the most desirable procedure from a culinary standpoint or the most tasteful way to prepare a meal. Tartar steak, dried ham, raw-marinated meat, and several kinds of smoked meat products are traditional and very much appreciated foods, but those treatments as such certainly are not adequate for destroying possible parasites. Also, in home cooking it is not always possible to follow the temperature of the meat very accurately. Color change of the meat is not a reliable indicator of sufficient temperature. In microwave cooking, in which the temperature is elevated unevenly, the risk is high even though the meal seems to be thoroughly cooked. Therefore, constant control work and research has to be done to maintain a high level of meat hygiene and safety.

In mammal muscles, both protozoan and helminth parasites can be found. Protozoans are microscopic organisms that can live intracellularly in a host. *Toxoplasma* and *Sarcocystis* are examples of zoonotic protozoans, which are infective to humans. Helminths are larger in size, and many of the species can be seen by naked eye. Of helminths, *Taenia* and *Trichinella* species can be transmitted to humans via the consumption of raw mammal meat or meat products [1].

Several parasite species live in the alimentary tract and may also be found in meat inspection, though not from the muscles. These parasites can affect the general welfare of the animal, cause production losses, and lower the quality of the meat. They can also have hygienic implications in food if parts of the alimentary tract are used as foodstuffs or if the meat is contaminated with the contents of the gut during the slaughtering process. Fecal examination for parasites or parasite eggs is not of importance in slaughterhouses, but could be a valuable diagnostic tool in animal husbandry. Intestinal parasites or external parasites either are not, however, discussed here. The focus in this chapter is on the parasites whose life cycle directly involves mammal muscles; some parasites affecting the liver are also briefly described.

In the meat inspection protocols at slaughterhouses, several checking, palpation, and incision steps are followed to verify parasites. Visual inspection and microscopic analysis of parasite morphology is useful and suitable for many species. Closer parasite species or strain differentiation, however, often requires molecular biological methods, such as polymerase chain reaction (PCR). Indirect methods that measure the immunological reaction of the host animal against parasite, including ELISA and immunoblot, are also useful in diagnostics of parasite infections, but also have their disadvantages, as described later.

Official meat inspection is regulated by law, and the analyses necessary are strictly stated. This chapter gives an overview of the diagnostic laboratory methods that are used or could be used to detect certain parasites in meat, and also introduces some methods that are designed for research purposes. The methods are introduced along with descriptions of the most important meat parasites in the industrialized countries.

3.1 *Trichinella* spp.

Trichinella nematodes are found worldwide, and can infect mammals, birds, and reptiles. They are well-known for their ability to cause illness to humans who eat undercooked infective meat; even deaths have been reported. Pork, horse, wild boar, and certain game meats are common sources for human infection. *Trichinella* larvae are freed from the muscle tissue in the stomach; molting and reproduction take place in the small intestine. The newly hatched larvae migrate through the circulatory system to the host's striated muscle, penetrate muscle cells, transform them into so-called "nurse cells," and settle in. Most species induce formation of a connective tissue capsule around them (Table 3.1). The larvae can survive inside the muscle cells in a dormant state for years, until the muscle is

Table 3.1 Summary of *Trichinella* Species and Genotypes

Isolates	Species/ Genotype	Capsule	Geographical Distribution	Freeze Resistance	Host Examples
T1	*T. spiralis*	Yes	Cosmopolitan	–	Pig, wild boar, rat
T2	*T. nativa*	Yes	Arctic, subarctic	+++	Bear, wolf, fox
T3	*T. britovi*	Yes	Europe, Asia	+	Wild boar, horse
T4	*T. pseudospiralis*	No	Cosmopolitan	–	Birds, marsupials
T5	*T. murrelli*	Yes	North America	–	Bear, raccoon
T6		Yes	North America	++	Bear, wolf
T7	*T. nelsoni*	Yes	Africa	–	Hyena, lion
T8		Yes	South Africa	–	Hyena, lion
T9		Yes	Japan	–	Bear, raccoon dog
T10	*T. papuae*	No	Papua New Guinea	–	Reptiles
T11	*T. zimbabwensis*	No	Africa	–	Reptiles

Sources: Dupoy-Camet, J. et al., in *Trichinellosis. Proceedings of the Eighth International Conference on Trichinellosis 1993*, F. Istituto Superiore de Sanità Press, Rome, Italy, 1994, 83; Pozio, E., *Vet. Parasitol.*, 93, 241, 2000; and Pozio, E. et al., *Parasitology*, 128, 333, 2004.

ingested by a new host. The life cycle is straightforward and involves only one host animal; *Trichinella* does not have separate definitive and intermediate hosts for different development stages [2–4].

In Europe, four species, *T. spiralis, T. nativa, T. britovi,* and *T. pseudospiralis,* are found. Owing to the risk of human infection, they are actively searched for with laboratory analysis in the meat inspection in the EU [5]. Swine and other potential *Trichinella* hosts (horse, wild boar, and certain game) are examined. An approved freezing protocol is an alternative for the *Trichinella* inspection of pork, because in pork, the infective species usually is *T. spiralis,* which is not resistant to below-zero temperatures. Some *Trichinella* species can, however, tolerate freezing. *T. britovi* has a moderate tolerance for low temperatures, and are infective to swine. Freezing alone is not recommended without *Trichinella* testing in the areas where *T. britovi* is endemic.

In horse and game meat, freeze-resistant *Trichinella* species can be more predominant than in pork. Freezing is never an alternative to testing in these animals. Large recent epidemics of human trichinellosis have unexpectedly been caused by the consumption of horse meat. These herbivore infections can be a consequence of rodents accidentally getting crushed in the feed or of feeding the horses on purpose with animal protein. Also, pig infections result from similar causes, but as omnivorous species, pigs are naturally more willing to ingest meat or even kill rodents by themselves.

3.1.1 Direct Detection Methods

In official meat inspection, only direct methods of detecting *Trichinella* larvae are used. Samples are taken as described in legislation (Commission regulation [EC] No. 2075/2005 [5], in EU countries). Sample sites are set according to the vulnerability of muscles of each animal species. If the predilection muscles of the animal are not known, diaphragm or tongue should be used. Sample size also varies depending on the animal species. Samples are taken at the slaughter line with a knife or with *Trichinella* forceps and arranged in a special tray in such a way that traceability is guaranteed. Pure muscle tissue, without fat or tendons, should be taken as a sample.

The purpose of *Trichinella* inspection is to prevent clinical trichinellosis in humans [6]. This means that absolutely *Trichinella*-free meat is not guaranteed with the methods used for testing at the moment, but the possible infection level is so low that people do not get sick. If more precise results are needed, the sensitivity of current methods can easily be improved by increasing the sample size.

3.1.1.1 Trichinoscopic Examination

Trichinoscopy, or compression technique, is the classic method of *Trichinella* inspection. Several (the number depending on the animal species) small, oat-kernel-size pieces of meat are pressed tightly between two glass plates, and paper-thin slices are then carefully scanned through with a microscope with 30–40 times magnification (Figure 3.1). The method is labor-intensive, slow to perform, and sensitivity is poor. Early infections, nonencapsulated species of *Trichinella* (*T. pseudospiralis, T. papuae, T. zimbabwensis*), or low infection levels are not easily recognized in trichinoscopy. It is currently used only in exceptional conditions, and is normally replaced by digestion methods.

3.1.1.2 Methods for Digestion

Artificial digestion is the method most commonly used in *Trichinella* examination. Digestion methods mimic the conditions in the stomach, with hydrochloride acid (HCl) and pepsin enzyme. The treatment enzymatically dissolves the muscle and the connective tissue of the capsule, enabling the count of the released, sedimented larvae (Figure 3.2). The method consists of three basic steps: digestion, sieving, and microscopic detection of the larvae.

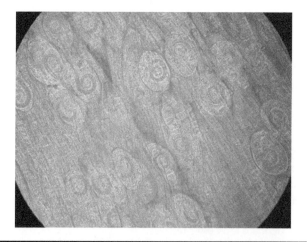

Figure 3.1 *Trichinella britovi* **encapsulated in experimentally infected mouse muscle. Light microscopic picture of compressorium plates. (Photo by Dr. J. Bien and Dr. K. Pastusiak, Witold Stefanski Institute of Parasitology, PAS.)**

Figure 3.2 *Trichinella spiralis* **muscle larvae after digestion. Light microscopic picture with 480× magnification. (Photo by Dr. J. Bien and Dr. K. Pastusiak, Witold Stefanski Institute of Parasitology, PAS.)**

In meat inspection, 1 g of muscle tissue from the diaphragm of 100 fattening pigs is pooled as one *Trichinella* digestion sample. For species other than pig, and also for sows and boars, larger samples per animal are required. In the event of a positive finding in the pooled sample, the potentially infected individuals are searched by repeating the examination with smaller subsets of the samples pooled together, until single sample digestions finally reveal the infected animal(s). In the EU, infected carcasses are condemned and removed from the food chain.

The theoretical sensitivity of 1 g muscle sample digestion is naturally one larva per gram (1 lpg). In practice, it is lower; a true sensitivity of 3–5 lpg is evaluated [7]. The sensitivity of the method can be improved by increasing the size of the sample. For example, 5 g samples are estimated to give a true sensitivity of 1 lpg [6].

All the digestion methods operate on the same principle, but the magnetic stirrer method (Table 3.2) is considered the gold standard [8]. Some digestion methods involve specialized laboratory apparatus for homogenization and warming of the samples. Certain filters with a pump can be used as well. The results of the different digestion methods may vary because of the different ways of handling the digestion fluids. The magnetic stirrer method is performed in a glass container, in which the freed larvae do not easily stick to the surfaces. In the Stomacher® apparatus (Seward Ltd., Worthing, U.K.) and similar methods, the plastic bags used in the liquid handling may offer tight corners or form capillary forces when the digestion fluid is poured out, and some of the larvae may get trapped. This could result in lower larvae per gram values or even false negative results if the infection level is low. These machines, however, enable several digestions at the same time.

Other steps in digestion may affect the results as well. Care must be taken not to inactivate the pepsin enzyme with concentrated HCl before digestion. The correct order to measure and add the regents for digestion is: water, HCl, and, when those are well mixed together, pepsin. The water temperature must be controlled carefully before adding pepsin. At temperatures over 50°C, the enzyme is inactivated. Below that temperature, the inactivation happens so slowly that it does not have an effect on the digestion.

After digestion, the mesh size used for sieving the fluid to remove excess undigested material is 180 μm [5]. Higher sensitivity (better larval recovery) with a larger mesh size, 355 μm, has been

Table 3.2 Magnetic Stirrer Digestion for *Trichinella* (100 g Pooled Sample)

• Add 16 mL of 25% hydrochloric acid into a beaker containing 2 L of water (46–48°C).
• Start magnetic stirring on a preheated plate. The digestion fluid must rotate at high speed but without splashing.
• Add 10 g of pepsin (1:10,000 U.S. National Formulary).
• Grind 100 g pool of samples with a kitchen blender, meat mincer, or scissors, and add to the beaker.
• Rinse all the equipment used for mincing with the digestion fluid to ensure that all the meat is included to the examination.
• Cover the beaker with aluminum foil to balance the temperature. Constant temperature of 44–46°C throughout the digestion must be mainteined. Overheating will inactivate the pepsin and interrupt the digestion.
• Continue stirring for 30 min, until the meat particles have disappeared. Longer digestion times may be necessary (not exceeding 60 min) for tongue, game meat, etc.
• Pour the digestion fluid through 180 µm mesh sieve into the sedimentation funnel. Not more than 5% of the starting sample weight should remain on the sieve. You may cool the sample with ice before pouring it into the funnel.
• Let the fluid stand in the funnel for 30 min to sediment the larvae. The sedimentation may be aided with periodic mechanical vibration.
• From the bottom tap of the funnel, quickly run 40 mL sediment sample of digestion fluid into a measuring cylinder or large centrifuge tube.
• Allow the 40 mL sample to stand for 10 min. Carefully remove with suction 30 mL of supernatant and leave a volume of not more than 10 mL.
• Pour the remaining 10 mL sample of sediment into a petri dish marked with 10 × 10 mm grid to ease the examination.
• Rinse the cylinder or centrifuge tube with not more than 10 mL of water, which has to be added to the sample for larval count.
• Examine the sample by trichinoscope or stereomicroscope at 15–40 times magnification. For suspect areas or parasite-like shapes, use higher magnifications of 60–100 times. Count the larvae.

Sources: EC (European Commission) Regulation No. 2075/2005, of 5 December 2005, laying down specific rules on official controls for *Trichinella* in meat. *Official Journal of the European Union* 22.12.2005; ICT recommendations: Gamble, H.R. et al., *Vet. Parasitol.*, 93, 393, 2000; and OIE standards: World Organisation for Animal Health. Health standards. Manual of diagnostic tests and vaccines of terrestrial animals. Trichinellosis (updated 21.11.2005). http://www.oie.int/eng/normes/mmanual/A_00048.htm.

reported [9], however. No more than 5% of the original pig sample weight should be retained in the sieve—the digestion should be considered inadequate if more material is found.

The meat sample itself could affect the digestion outcome as well. The structure of the muscle can slow the digestion. For certain muscle types (e.g., tongue) and animal species (horse, game), the time of digestion often has to be prolonged. Fat in the sample, besides decreasing the muscle mass and thus affecting the larvae per gram value, makes the digestion fluid hazy, and detecting the larvae with a microscope thus becomes more difficult. The separation of fat to the top of the digestion fluid, thereby diminishing the remaining lipids in the sample, can be made faster by cooling the fluid with ice before pouring it into the sedimentation funnel. Cooling also induces coiling and faster sedimentation of the larvae.

In meat inspection, the presence of bone pieces in the *Trichinella* sample is not a concern, but in research work various kinds of samples set limitations for the methods. In plastic bag digestion, sharp bones may break the plastic; the magnetic stirrer method is recommended. Bone, as well as fat and connective tissue, in the sample lowers the larvae per gram value—only muscle tissue should be used if infection intensity is analyzed.

The microscopic examination of the digestion fluid should be performed immediately after the digestion. If the examination is delayed, the fluid must be clarified [5]. Microscopic examination should not be postponed until the next day; digestion fluid containing acid and pepsin can cause degradation of the larvae.

All the containers, materials, and fluids should await the result of *Trichinella* examination before washing or disposing them; otherwise, they should be handled as would be done in the case of a positive sample. Digestion fluid and the instruments which have been in direct contact with a positive sample, should be sterilized to kill the *Trichinella* larvae. A few minutes in well-boiling water are enough. If potentially infective waste exists, it should be autoclaved or decontaminated in some other acceptable way. Of disinfectants, 1:1 mixtures of xylol and ethanol (95%), or xylol and phenol are lethal for infective larvae [10].

3.1.1.3 Histology

Trichinella can be found with a microscope in histological samples using common dyeing techniques for muscle and connective tissue—hematoxylin and eosin (HE), for example. If the section is not sufficiently representative, immunohistochemical confirmation can be done. Immunofluorescence techniques are also suitable. Histological methods are not used in slaughterhouses or in routine work, but in research, and sometimes in diagnostics, they can be useful. Biopsy is, however, an insensitive method because the size of the sample is so small. In the histological samples, the presence or absence of the capsule or differences in capsule formation and shape, as well as the cellular reaction around the parasite, can give hints about the infecting *Trichinella* species [11], but definitive differentiation is made using molecular biological methods.

3.1.1.3.1 PCR Methods

Identification of infecting *Trichinella* species is required in the EU when infection is found in meat inspection [5]. This analysis is performed in a national reference laboratory, or the examination may be ordered from some other qualified laboratory. Community Reference Laboratory for Parasites (Istituto Superiore di Sanita, Rome, Italy) offers services and detailed procedure descriptions for species identification [12].

Many methods for *Trichinella* species differentiation are published, but it is currently usually done with multiplex-PCR. In this method, several primer pairs are added to a single PCR reaction mixture, and thereby several DNA fragments can be amplified at the same time. *Trichinella* primers are generated according to sequence data from internal transcribed spacers ITS1 and ITS2 and from expansion segment V (ESV) of the ribosomal DNA repeat [13,14]. All the *Trichinella* species can be screened from one sample. Even one larva is enough for species analysis, but all the larvae found in the original digestion, plus some of the infected meat, should be sent as a sample to the laboratory. The laboratory analyzing for species gives detailed instructions for sending the samples, but usually digested, washed larvae in 90% ethanol are requested. Freeze-thaw cycles of the sample should be avoided before species analysis. Even more detailed diagnostics of the *Trichinella* isolates for epidemiological research purposes can be made with techniques based on sequence analysis.

Several methods are available to extract DNA from *Trichinella* larvae. When using commercial kits, usually the protocol application for tissue samples is the most suitable. A good yield of genomic DNA from even a single muscle larva can be achieved with a simple protocol in a small volume of buffer (10–20 μl) [15]. There are several modifications of this extraction method, and the yield can be enhanced with a longer incubation with proteinase K (overnight), by increasing the proteinase K concentration, adding detergents, such as sodium dodecyl sulfate (SDS), and combining the commercial Gene Releaser® (BioVentures Inc., Murfreesboro, TN) protocol after the extraction.

3.1.2 Indirect Detection Methods

In addition to the direct detection methods, *Trichinella* infection can be diagnosed by searching *Trichinella*-specific antibodies in serum, plasma, and whole blood or tissue fluid samples of the host. These methods are used in surveys and monitoring, not in the individual carcass testing. They are called indirect, since instead of detecting the parasite itself, they detect the immunological reaction of the host animal against the invader—antibodies show that the immune system has faced *Trichinella* antigens at some point. In addition to exposure to the parasite, the reaction is dependent on the host's immunological status and capability to produce antibodies. Production of antibodies to a detectable level takes time; early infections may be falsely diagnosed as negative reactions using these methods. Indirect methods are not used in meat inspection.

The antigen used in the test and the antibody type searched strongly affect the results. Validation of the antigens, dilutions, and cut-off levels to separate the positive samples from the negative ones should be done with a significant number of samples before accepting any method for routine use; otherwise, misdiagnosis may occur. Positive and negative control samples should always be included in the test, to ensure the proper performance of the test. There are several ways to count the cut-off level, which affects the interpretation of the results, that is, the sensitivity and specificity of the test. Serological methods can be very sensitive, with a detection level of one larva/100 g meat [16], but due to the disadvantages related to the indirect detection, they are not recommended for use in meat inspection to replace the conventional direct methods [17]. For surveillance studies and *Trichinella* diagnostics at herd or population levels, indirect methods are, however, very useful.

The immunological reaction of the host species should be studied before using any immunological test for diagnostics. In the horse, for example, the diagnostic value of serological *Trichinella* tests is questionable. False results are common. The circulating specific IgG antibodies are undetectable in horse already 4–5 months after the infection, even though the capsulated *Trichinella* larvae are alive and well, and ready to infect a new host [18]. Therefore, serological diagnostics of *Trichinella* infections should not be used in horses. Also, insufficient information about game animal and other wildlife testing is available; without decent validation of the method, serology should not be used even in surveillance studies.

Of serological laboratory techniques, bentonite flocculation, indirect immunofluorescence microscopy, latex agglutination, and enzyme-linked immunosorbent assay (EIA, ELISA) are commonly used for *Trichinella* diagnostics [19], ELISA being the most common method. Immunoblot is also often utilized.

In ELISA, antibodies from the sample are bound to antigens that are coated onto a microwell plate. The bound antibodies are detected with specific antibodies that are linked to an enzyme. The enzyme reacts with a substrate to form a color reaction, which is then spectrophotometrically measured. ELISA can be used for detection of antigens as well.

In immunoblot, antigens are first separated by their molecular weight in SDS polyacrylamide gel electrophoresis (PAGE) and transferred onto a nitrocellulose membrane. The membrane is exposed to the antibodies in the sample. Antigen–antibody binding is visualized, and the molecular weight of immunoreactive antigens can also be analyzed.

Noticeable effort in several laboratories has been put toward developing an optimal serological test for *Trichinella* infection. The test should recognize the infection as early as possible, with as low infection level as possible, but without cross-reactions with other parasites. Several antigens, crude, excretory–secretory (ES), and synthetic, have been used for detection. The replacement of crude parasite antigens with more specific antigens has lowered the risk of cross-reaction, and crude antigens are not recommended in immunological testing any more. To obtain more reliable results in immunological tests, several methods can be used for analysis of the samples.

Fast "patient-side" tests are developed for diagnostics, especially as preliminary tests, or when laboratory facilities are not available. A "dipstick assay" [20,21] has been reported to show useful sensitivity and specificity in analysis of human and swine *Trichinella* cases. In this method, the antigen is dotted onto a nitrocellulose membrane. The membrane is then dipped into a serum sample, and the antigen–antibody reaction is visualized. A commercial, so-called "lateral flow card" has been tested with blood, serum, and tissue fluids with satisfactory results [22]. It is recommended for use as a farm screening test or for preliminary screening of suspected pigs during slaughter, especially in countries with high *Trichinella* prevalence, to improve the food hygienic quality. In the muscle fluid there is about 10 times lower concentration of antibodies than in serum, but if serodiagnostics are not possible, muscle or body fluid samples can offer a good alternative [23,24].

3.2 *Taenia* spp.

Taenia tapeworms form larval cysts, cysticerci, in intermediate hosts. These are the infective forms for definitive hosts. There are three human *Taenia* species (*T. saginata, T. solium,* and *T. asiatica*), and several other species that can be found in the postmortem inspection of the intermediate hosts that are used for food.

Bovine cysticercosis is caused by *Cysticercus bovis*, the larval stage of *T. saginata*. The parasites spread through the bloodstream to skeletal muscles and the heart, forming 5–10 mm diameter, thick-walled, pearl-like infective cysts. The cyst remains viable for about 6 months, after which it starts to calcify. However, in light infections viable cysts can be found years after the onset of infection. The sites of predilection in cattle are the masticatory muscles, tongue, heart, and diaphragm. Incision of these muscles in the postmortem inspection reveal the possible infection. Infections can be classified as light, local, heavy, or generalized, and the judgment during meat inspection is made according to this analysis. In the case of a light or local infection, condemnation of the carcass is not necessary, though the meat should at minimum be heated or frozen before consumption.

T. solium is the infective agent in porcine cysticercosis, *Cysticercus cellulosae*. The species is significant, because humans can act as both definitive and intermediate hosts for the parasite. Thus, the infection can be manifested in humans not only as intestinal taeniosis, but also as cysticercosis or neurocysticercosis. Human cases are generally linked to inadequate sanitation, free-ranging swine, ineffective meat inspection, or ingestion of inadequately cooked pork [25]. Because pigs are slaughtered young, all cysts found in meat inspection in the industrialized countries should be considered as viable.

There are limitations with diagnosing cysticercosis by visual inspection only. Cases of low infection level, or early infections, are inevitably not diagnosed. On the other hand, overestimation of the disease may also be made in visual inspection: In a study of cysts diagnosed as cysticercosis in abattoirs, only 52.4% were confirmed positive in further studies with PCR [26]. In another study, 97% of viable *T. saginata* cysts were confirmed by PCR-restriction fragment polymorphism, while the percentage for dead cysts was approximately 73% [27].

The use of serological methods, such as ELISA and immunoblot, together with the visual inspection would improve the efficacy of meat inspection for the detection of cysticercosis [28]. Many serological methods are available for differentiation between viable and degenerated cysts [29–32].

3.3 *Toxoplasma gondii*

T. gondii is a common, worldwide, mammal- and bird-infecting, zoonotic, protozoan parasite. Its sexual reproduction takes place only in feline hosts, but asexual multiplication is possible in all host species. Infection can be acquired by ingestion of oocysts (for example, from soil or water contaminated with cat feces) or through tissue cysts ingested with infected meat. Transplacental infection from mother to fetus may also occur, and contact infection through mucous membranes has been reported. The infection is relatively common among humans but does not cause any severe symptoms for healthy adults; most of them are unaware that they have had *Toxoplasma* infection. The most severe threat in *Toxoplasma* infections is when a pregnant woman gets infected for the first time with no previous immunological protection. Damage caused to the fetus can be fatal. Also, immunocompromised people, such as AIDS patients or transplant recipients taking immunosuppressive medication, are at risk. In a study defining predisponating factors for pregnant women's *Toxoplasma* infections, eating undercooked meat, contact with soil, and traveling outside Europe and North America were risk factors, but contact with cats was not [33].

The majority of *Toxoplasma* isolates can be classified in three genetic lineages: type I, II, and III. There are differences in geographic distribution of the types. Their virulence, in addition to host genetic factors, has been shown to have an influence on the severity of the disease. Most human cases, especially in Europe and North America, are associated with type II, whereas lesions in the eye are often reported to be caused by type I [34,35].

Noticeable economic losses occur among farmed sheep due to *Toxoplasma* abortions and stillbirths. Goats and pigs are also susceptible; cattle and horses are more resistant to the disease. Several studies of prevalence among domestic animals over the world show that *Toxoplasma* infection is common. For example, in a German study, 19% of sows ($n = 2041$) were seropositive [36].

Toxoplasma is not searched during meat inspection. Instead, consumers at risk are advised not to eat undercooked meat. Nonspecific lesions in postmortem inspection, such as necrotic foci in organs, might be suggestive of *Toxoplasma* infection, but the tissue cysts are invisible to naked eye; thus, the diagnostics requires microscopic, immunological, or molecular biological analysis. Microscopic *Toxoplasma*-like findings (predilection in brain and placenta) can be confirmed with immunohistochemical or immunofluorescence methods. PCR identification can be performed from DNA isolated from tissue samples. For isolation and culturing the organism for research, infection of mice with placental or brain homogenate can be used. *In vitro* cultivation in cell cultures is commonly done to maintain the strains for research purposes.

In addition to conventional serology, analyzing immunoglobulin avidity has been successfully done to determine the onset of infection, which is of great importance in regard to infection in pregnant women [37]. The method measures the strength of the antigen–antibody binding: If the infection is new, the antibody is not yet mature and does not bind as strongly to the antigen as an older antibody would.

Genotyping of the isolates is done with multiplex-PCR combined with restriction fragment length polymorphism (RFLP) of the amplified loci. Closer analysis of the strains can be done with sequencing methods. Serotyping, based on the infected serum reaction in ELISA with polymorphic peptides derived from certain *Toxoplasma* antigens, is also used in human toxoplasmosis [38].

3.4 *Sarcocystis* spp.

Carnivores are the definitive hosts for *Sarcocystis* protozoan, whereas herbivores act as intermediate hosts. Birds and reptiles can be infected as well. Human infections occur, but they are usually asymptomatic. This parasite has worldwide distribution. There are several species of *Sarcocystis* with differing host preferences.

In meat inspection, light-colored *Sarcocystis* bradyzoite cysts between the muscle fibers of ruminants and pigs can be found. The size of the cysts varies according to the host species, and may be visible to the naked eye. Mild infections are asymptomatic, but in heavy infections various kinds of clinical signs may appear, depending on the location of the cysts. Death of the intermediate host has even been reported.

Diagnosis can be confirmed in histological samples; cysts stain basofilic in HE. *Sarcocystis* species can be ultrastructurally identified according to their cyst wall with transmission electron microscopy (TEM), but PCR with DNA isolated from the muscle with cycts reveals the species as well. This analysis can be completed with sequencing techniques. Common immunological methods, described earlier with other parasites, in addition to direct methods, can be used in diagnostics.

3.5 Some Other Parasites of Importance in Meat Inspection

3.5.1 *Ascaris suum*

Ascaris suum roundworms are macroscopic parasites of the pig's small intestine, but their life cycle also involves lungs and can affect abdominal organs as well. *Ascaris* eggs are secreted in the feces, and infection is feco-oral. Microscopic analysis of feces to find the eggs can be used in diagnostics. In meat inspection, common findings are so-called "milk spots" in the liver. These are light-colored fibrotic tissue areas in the tunnels made by the *Ascaris* larvae when migrating through the liver. They can easily be recognized by visual inspection of the liver surface. Economic losses result because the affected parts are condemned. *Ascaris* control should be done at the farm level. Besides pigs, other animals and humans might suffer from a condition called "visceral larva migrans" if *Ascaris* eggs are hatched in their intestines and the larvae begin to migrate, though the infection is not permanent in these coincidental hosts.

3.5.2 *Fasciola hepatica and Other Liver Flukes*

Liver trematodes are large and rather easily visible in meat inspection. Several species exist. *Fasciola hepatica* is 2–3 cm long and about 1 cm wide. It is predominantly a parasite of ruminants, but can also be found in horses, pigs, and humans. In heavy infections, bleeding and damage of the parenchymal liver tissue may even cause the death of the animal. Sheep are especially sensitive to acute fascioliasis. In milder infections, fairly small amounts of metacercaria forms can predisponate to secondary bacterial infections. The life cycle includes a development period in an intermediate host snail; wet grazing conditions are favorable for the infection, and seasonal variation occurs.

In meat inspection, in addition to the actual presence of the flukes, thickening or calcification of the bile ducts, darkish parasitic material in bile, and color changes on the liver surface or in the carcass (anemia, icterus) can be suggestive for fascioliasis (Figure 3.3). In addition, the lungs might be affected.

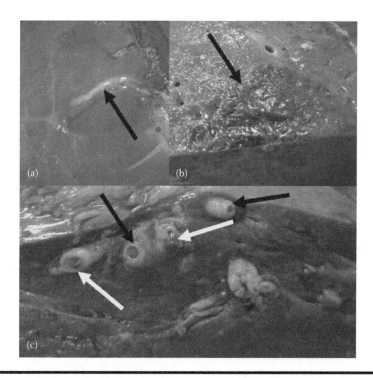

Figure 3.3 *Fasciola hepatica* lesions in bovine liver. (a) Thickened bile duct on the surface of the liver (arrow). (b) Hemorrhage area at the incision surface of the liver. (c) Thickened bile ducts are prominent at the incision surface of the liver (black arrows); the parasites are indicated with the white arrows. (Photo by Dr. M. Kozak, Witold Stefanski Institute of Parasitology, PAS.)

Dicrocoelium dendriticum, the lancet fluke, is smaller in size (5–10 mm long) than *Fasciola* but still visible in opened bile ducts. Cattle, sheep, and swine can be affected. The damages caused are small because this worm does not migrate in the liver parenchyme—usually no clinical signs or notable alterations are noticed in postmortem examination.

Liver fluke diagnosis is made in postmortem inspection of liver or by observing eggs in fecal samples. Immunological methods are available as well, and milk can also be conveniently used as a sample.

3.5.3   *Echinococcus spp.*

Echinococcus worms, belonging to the cestodes, cause lesions (cysts) mainly in the liver or lungs of livestock. These parasites are briefly discussed here because their cysts can be found in meat inspection, and the parasites are notable for their ability to cause the severe disease, hydatidosis, in humans.

To date, eight species (*E. multilocularis*, *E. shiquicus*, *E. vogeli*, *E. oligarthrus*, *E. granulosus*, *E. equinus*, *E. ortleppi*, and *E. canadensis*) are recognized, but the taxonomy is still incomplete. The distribution of *Echinococcus* is worldwide, with climate preferences varying among the species. Adult forms of these parasites reproduce in carnivores, but cattle, pigs, and sheep, for example, as well as humans, can act as intermediate hosts after ingestion of *Echinococcus* eggs, which are spread through feces of the definitive hosts. All the species except *E. equinus* are infective to humans. *E. multilocularis* is the most harmful because of its alveolar, cancer-like appearance in the liver. The main intermediate hosts for *E. multilocularis* are rodents. *E. granulosus* is important for its global distribution and common infections in humans. The main intermediate host for *E. granulosus* is sheep. Tissue cysts are not a source of infection for the intermediate hosts, such as humans, but it is necessary to control the *Echinococcus* life cycle at the meat inspection stage to prevent further carnivore infections.

Echinococcus worms are some millimeters long depending on species. The cysts in the intermediate hosts' organs can grow up to tens of centimeters in diameter, and are thus often easily noticed in meat inspection. The cysts are fluid-filled with concentrically calcified particles (hydatide sand) and protoscolices, preliminary heads of the worms.

In addition to the typical cyst findings in postmortem inspection, diagnosis can be confirmed histologically in uncertain cases. Formalin-fixed samples can be dyed with periodic acid Schiff (PAS) to observe the *Echinococcus* metacestode characteristic PAS-positive, acellular laminated layer.

Although PCR or coproantigen ELISA from fecal samples might be used to detect parasite eggs in the definitive host, molecular methods are not of great importance in intermediate host diagnostics of *Echinococcus*. For small and calcified lesions of *E. multilocularis*, PCR can be used for identification and confirmation of the diagnosis [39]. The analysis of the species of *Echinococcus* is also done with PCR-based methods. Immunological tests, which are successfully used in human diagnostics, are not sensitive or specific enough to replace careful conventional postmortem inspection of livestock. Furthermore, immunological methods do not distinguish current and past infections, and cross-reactions between *Taenia* species can occur as well. In a study of sheep echinococcosis, however, macroscopic diagnosis at the time of slaughter was found to have limitations, and histology or immunoblot were used with success [40]. In surveillance studies, immunological methods, mainly serology, are useful; suitable antigens are numerous [41]. Even ultrasonography has been used in mass screening for ovine hydatid cysts [42].

3.5.4 *Parafilaria bovicola*

The filarial nematode, *Parafilaria bovicola*, which is transmitted by a fly vector (*Musca*), can cause severe economic losses in the meat industry in areas where it is endemic. Filaria gets into a fly while it is feeding on damaged skin. After a 3-week development, the parasite is secreted in the fly's saliva and transmitted to a new host when the fly is again having a meal at a wound. A parafilaria female penetrates the skin of the host and deposits the new filaria into the surrounding tissue fluids. The length of the adult nematodes can be 3–6 cm. The lesions resemble a bruise and have a greenish color due to the presence of large numbers of eosinophilic granulocytes. The condition is therefore sometimes called "green meat." The subcutaneous tissue may appear swollen, and bleeding in the muscle might be found during inspection. Localized lesions can be trimmed, but in heavy, general infections the condemnation of the whole carcass should be considered. The infection can be diagnosed, in addition to a direct indication of the nematodes, with immunological methods such as ELISA.

3.6 Future Visions

Parasitological safety of muscle foods can be further improved in the future. Continuous research efforts will develop diagnostic tools, but legislation and administrative decisions can also have a powerful influence on safety: Control can be directed to the problematic areas based on risk analysis. If, for example, *Trichinella* epidemics due to private consumption of uninspected meat are a concern, inspection could be encouraged by removing inspection fees and by making the meat inspection easily achievable. Or if the meat is meant to be used in products that are consumed raw, the inspection could be stricter; *Toxoplasma* inspection or serological confirmation of *Trichinella* inspection, for example, could be demanded in addition. Even now, lower demands for *Trichinella* inspection are made in the areas that have officially recognized negligible risk of infection [5]. Careful monitoring of the infection pressure, however, must be performed with epidemiological studies on suitable indicator animals [43].

Modern molecular biological techniques are routinely used in current parasitological research. Meat production could benefit from these research tools as well, if supplemental parasite diagnostics is desired in addition to the official inspection. Serological laboratory procedures, for example, are easy, cheap, and can be almost fully automated. An antigen microarray for serodiagnostics [44,45], which detects several infections, including parasite infections, from a single sample at the same time, could be applied to slaughterhouse samples. Serology could be used together with conventional methods to complement the meat inspection; methods that actually replace parts of the parasite inspection should, however, be carefully evaluated—no elevated risk for the consumer can be allowed. On the other hand, consumers may considerably benefit from serological testing of animals, because parasites that might not be found in the official inspection can be diagnosed (*Toxoplasma*), or lower infection levels might be noticed (*Trichinella, Cysticercus*). Also, more precise diagnostics could be done in the future from serum samples, using, for example, a serological test differentiating the *Trichinella* species.

Modernization of meat inspection may change the practices at slaughterhouses. In the future, parasite control at the farm level may replace control at meat inspection, at least in the low prevalence areas. The tendency in animal production is to stress the overall welfare of the animals with health care programs and preventive actions. Part of the parasite inspection could be done at the farm as a normal part of the health care program. When farm testing is coordinated with serological testing at the time of slaughter, paired serum samples could be achieved, and the diagnostic value of the serological testing enhanced. In the endemic areas for severe parasitic diseases, however, only direct detection methods should be used for individual carcasses to prevent parasites in meat.

In the future, education on different levels of food production, the combination of good animal production practices with frequent control visits by authorities, and serological testing on the farms as well as in the slaughterhouses, together with risk-based, directed use of conventional methods for parasitic examination, could guarantee parasite-free meat.

References

1. Pozio, E. Foodborne and waterborne parasites. *Acta Microbiol. Pol.,* 52, 83, 2003.
2. Dupoy-Camet, J. et al. Genetic analysis of *Trichinella* isolates with random amplified polymorphic DNA markers, in: *Trichinellosis. Proceedings of the Eighth International Conference on Trichinellosis 1993.* Campbell, W.C., Pozio, E., and Bruschi, F. Istituto Superiore de Sanita Press, Rome, Italy. 1994, 83.
3. Pozio, E. Factors affecting the flow among domestic, synanthropic and sylvatic cycles of *Trichinella. Vet. Parasitol.,* 93, 241, 2000.
4. Pozio, E. et al. *Trichinella papuae* and *Trichinella zimbabwensis* induce infection in experimentally infected varans, caimans, pythons and turtles. *Parasitology,* 128, 333, 2004.
5. EC (European Commission) Regulation No. 2075/2005, of 5 December 2005, laying down specific rules on official controls for *Trichinella* in meat. *Official Journal of the European Union* 22.12.2005.
6. Gamble, H.R. et al. International Commission on Trichinellosis: recommendations on methods for the control of *Trichinella* in domestic and wild animals intended for human consumption. Review. *Vet. Parasitol.* 93, 393, 2000.
7. Webster, P. et al. Meat inspection for *Trichinella* in pork, horse meat and game within the EU: available technology and its implementation. *Eurosurveillance,* 11, 50, 2006.
8. World Organisation for Animal Health. Health standards. Manual of diagnostic tests and vaccines of terrestrial animals. Trichinellosis (updated 21.11.2005). http://www.oie.int/eng/normes/mmanual/A_00048.htm.
9. Gamble, H.R. Factors affecting the efficiency of pooled sample digestion for the recovery of *Trichinella spiralis* from muscle tissue. *Int. J. Food Microbiol.,* 48, 73, 1999.
10. Public Health Agency of Canada, Office of Laboratory Security. Material safety data sheet—infectious substances. *Trichinella* spp. http://www.phac-aspc.gc.ca/msds-ftss/index.html.
11. Sukura, A. et al. *Trichinella nativa* and *T. spiralis* induce distinguishable histopathologic and humoral responses in the raccoon dog (*Nyctereutes procyonoides*). *Vet. Pathol.,* 39, 257, 2002.
12. Community Reference Laboratory for Parasites. Istituto Superiore di Sanità. Identification of *Trichinella* muscle stage larvae at the species level by multiplex PCR. http://www.iss.it/binary/crlp/cont/PCR%20method%20WEB%20SITE.1177083731.pdf.
13. Zarlenga, D.S. et al. A multiplex PCR for unequivocal differentiation of all encapsulated and non-encapsulated genotypes of *Trichinella. Int. J. Parasitol.,* 29, 1859, 1999.
14. Zarlenga, D.S. et al. A single, multiplex PCR for differentiating all species of *Trichinella. Parasite,* 8, 24, 2001.
15. Rombout, Y.B., Bosch, S., and Van Der Giessen, J.W.B. Detection and identification of eight *Trichinella* genotypes by reverse line blot hybridization. *J. Clin. Microbiol.,* 39, 642, 2001.
16. Gamble, H.R. et al. Serodiagnosis of swine trichinosis using an excretory-secretory antigen. *Vet. Parasitol.,* 13, 349, 1983.
17. Gamble, H.R. et al. International Commission on Trichinellosis: Recommendations on the use of serological tests for the detection of *Trichinella* infection in animals and man. *Parasite,* 11, 3, 2004.
18. Pozio, E. et al. Evaluation of ELISA and Western blot analysis using three antigens to detect anti-*Trichinella* IgG in horses. *Vet. Parasitol.,* 108, 163, 2002.

19. Bruschi, F. and Murrel, K.D. New aspects of human trichinellosis: the impact of new *Trichinella* species. *Postgrad. Med. J.*, 78, 15, 2002.

20. Zhang, G.P. et al. Development and evaluation of an immunochromatographic strip for trichinellosis detection. *Vet. Parasitol.*, 137, 286, 2006.

21. Al-Sherbiny, M.M. et al. Application and assessment of a dipstick assay in the diagnosis of hydatidosis and trichinosis. *Parasitol. Res.*, 93, 87, 2004.

22. Patrascu, I. et al. The lateral flow card test: an alternative method for the detection of *Trichinella* infection in swine. *Parasite*, 8, 240, 2001.

23. Moller, L.N. et al. Comparison of two antigens for demonstration of *Trichinella* spp. antibodies in blood and muscle fluid of foxes, pigs and wild boars. *Vet. Parasitol.*, 132, 81, 2005.

24. Tryland, M. et al. Persistence of antibodies in blood and body fluids in decaying fox carcasses, as exemplified by antibodies against *Microsporum canis*. *Acta Vet. Scand.*, 48, 10, 2006.

25. Hoberg, E. *Taenia* tapeworms: their biology, evolution and socioeconomic significance. *Microbes Infect.*, 4, 859, 2002.

26. Abuseir, S. et al. Visual diagnosis of *Taenia saginata* cysticercosis during meat inspection: is it unequivocal? *Parasitol. Res.*, 99, 405, 2006.

27. Geysen, D. et al. Validation of meat inspection results for *Taenia saginata* cysticercosis by PCR-restriction fragment length polymorphism. *J. Food Prot.*, 70, 236, 2007.

28. European Commission, Health and Consumer Protection Directorate-General. Opinion of the scientific committee on veterinary measures relating to public health on: the control of taeniosis/cysticercosis in man and animals. (adopted on 27–28 September 2000).

29. Harrison, L.J. et al. Specific detection of circulating surface/secreted glycoproteins of viable cysticerci in *Taenia saginata* cysticercosis. *Parasite Immunol.*, 11, 351, 1989.

30. Brandt, J.R. et al. A monoclonal antibody-based ELISA for the detection of circulating excretory-secretory antigens in *Taenia saginata* cysticercosis. *Int. J. Parasitol.*, 22, 471, 1992.

31. Onyango-Abuje, J.A. et al. Diagnosis of *Taenia saginata* cysticercosis in Kenyan cattle by antibody and antigen ELISA. *Vet. Parasitol.*, 61, 221, 1996.

32. Wanzala, W. et al. Serodiagnosis of bovine cysticercosis by detecting live *Taenia saginata* cysts using a monoclonal antibody-based antigen-ELISA. *J. S. Afr. Vet. Assoc.*, 73, 201, 2002.

33. Cook, A.J. et al. Sources of *Toxoplasma* infection in pregnant women: European multicentre case-control study. European Research Network on Congenital Toxoplasmosis. *Br. Med. J.*, 15, 142, 2000.

34. Vallochi, A.L. et al. The genotype of *Toxoplasma gondii* strains causing ocular toxoplasmosis in humans in Brazil. *Am. J. Ophthalmol.*, 139, 350, 2005.

35. Switaj, K. et al. Association of ocular toxoplasmosis with type I *Toxoplasma gondii* strains: direct genotyping from peripheral blood samples. *J. Clin. Microbiol.*, 44, 4262, 2006.

36. Damriyasa, I.M. et al. Cross-sectional survey in pig breeding farms in Hesse, Germany: seroprevalence and risk factors of infections with *Toxoplasma gondii*, *Sarcocystis* spp. and *Neospora caninum* in sows. *Vet. Parasitol.*, 126, 271, 2004.

37. Hedman, K. et al. Recent primary *Toxoplasma* infection indicated by a low avidity of specific IgG. *J. Infect. Dis.*, 159, 736, 1989.

38. Kong, J.T. et al. Serotyping of *Toxoplasma gondii* infections in humans using synthetic peptides. *J. Infect. Dis.*, 187, 1484, 2003.

39. Deplazes, P., Dinkel, A., and Mathis, A. Molecular tools for studies on the transmission biology of *Echinococcus multilocularis*. Review. *Parasitology*, 127, 53, 2003.

40. Gatti, A. et al. Ovine echinococcosis I. Immunological diagnosis by enzyme immunoassay. *Vet. Parasitol.*, 143, 112, 2007.

41. Lorenzo, C. et al. Comparative analysis of the diagnostic performance of six major *Echinococcus granulosus* antigens assessed in a double-blind, randomized multicenter study. *J. Clin. Microbiol.*, 43, 2764, 2005.

42. Lahmar, S. et al. Ultrasonographic screening for cystic echinococcosis in sheep in Tunisia. *Vet. Parasitol.*, 143, 42, 2007.

43. Community Reference Laboratory for Parasites. Istituto Superiore di Sanità. Identification and development of sampling methods and design of suitable protocols for monitoring of *Trichinella* infection in indicator species. http://www.iss.it/binary/crlp/cont/Guideline%20sampling%20indicator%20 species.1166800931.pdf.
44. Mezzasoma, L. et al. Antigen microarrays for serodiagnosis of infectious diseases. *Clin. Chem.*, 48, 121, 2002.
45. Bacarese-Hamilton, T. et al. Serodiagnosis of infectious diseases with antigen microarrays. *J. Appl. Microbiol.*, 96, 10, 2004.

Chapter 4

Mycotoxin Analysis in Poultry and Processed Meats

Jean-Denis Bailly and Philippe Guerre

Contents

4.1 Introduction 78
4.2 Main Mycotoxins 81
    4.2.1 Trichothecenes 81
        4.2.1.1 Origin and Nature 81
        4.2.1.2 Structure and Physicochemical Properties 84
        4.2.1.3 Analytical Methods 84
    4.2.2 Zearalenone 86
        4.2.2.1 Origin and Nature 86
        4.2.2.2 Structure and Physicochemical Properties 86
        4.2.2.3 Analytical Methods 86
    4.2.3 Fumonisins 87
        4.2.3.1 Origin and Nature 87
        4.2.3.2 Physicochemical Properties 88
        4.2.3.3 Methods of Analysis 88
    4.2.4 Aflatoxins 89
        4.2.4.1 Origin and Nature 89
        4.2.4.2 Structure and Chemical Properties 89
        4.2.4.3 Analytical Methods 90
    4.2.5 Ochratoxin A 91
        4.2.5.1 Origin and Nature 91
        4.2.5.2 Physicochemical Properties 91
        4.2.5.3 Methods of Analysis 91

4.2.6 Other Toxins 92
    4.2.6.1 Citrinin 92
    4.2.6.2 Cyclopiazonic Acid 93
4.3 Mycotoxin Analysis and Prevalence in Poultry 94
    4.3.1 Trichothecenes 94
        4.3.1.1 Methods of Analysis 94
        4.3.1.2 Behavior and Residual Contamination of Poultry Tissues 94
    4.3.2 Zearalenone 97
        4.3.2.1 Methods of Analysis 97
        4.3.2.2 Behavior and Residual Contamination of Poultry Tissues 97
    4.3.3 Fumonisins 100
        4.3.3.1 Methods of Analysis 100
        4.3.3.2 Behavior and Prevalence in Poultry Tissues 100
    4.3.4 Aflatoxins 100
        4.3.4.1 Methods of Analysis 100
        4.3.4.2 Behavior and Prevalence in Poultry Tissues 100
    4.3.5 Ochratoxin A 101
        4.3.5.1 Methods of Analysis 101
        4.3.5.2 Behavior and Prevalence in Poultry Tissues 101
    4.3.6 Other Toxins 104
        4.3.6.1 Citrinin 104
        4.3.6.2 Cyclopiazonic Acid 104
4.4 Mycotoxin Analysis and Prevalence in Processed Meats 104
    4.4.1 Aflatoxin B1 105
    4.4.2 Ochratoxin A 105
    4.4.3 Citrinin 105
    4.4.4 Cyclopiazonic Acid 107
4.5 Conclusion 107
References 107

4.1 Introduction

Mycotoxins are a heterogeneous group of secondary metabolites elaborated by fungi during their development. About 30 molecules are of real concern for human and animal health [1]. They can be found as natural contaminants of many vegetal foods or feeds, mainly cereals, but also of fruits, nuts, grains, and forage, as well as of compound foods intended for human or animal consumption. The most important mycotoxins are produced by molds belonging to the *Aspergillus*, *Penicillium*, and *Fusarium* genera (Table 4.1) [2–4].

Mycotoxin toxicity is variable. Some have hepatotoxicity (aflatoxins), others have an estrogenic potential (zearalenone [ZEA]), or are immunotoxic (trichothecenes, fumonisins) (Table 4.1) [1]. Some mycotoxins are considered to be carcinogenic or are suspected to have carcinogenic properties [5]. Although some toxins display an important acute toxicity (after unique exposure to one high dose), chronic effects (observed after repeated exposure to weak doses) are probably more important in humans. Mycotoxins are suspected to be responsible for several pathological syndromes in humans, including ochratoxin A (OTA), which is associated with Balkan endemic

Table 4.1   Mycotoxins of Interest in Poultry and Processed Meat

Toxin	Main Producing Fungal Species	Toxicity	Structure
Deoxynivalenol	*Fusarium nivale*	Hematotoxicity	
	Fusarium crookwellense	Immunomodulation skin toxicity	
	Fusarium oxysporum		
	Fusarium avanaceum		
	Fusarium graminearum		
	Fusarium solani		
Zearalenone	*Fusarium graminearum*	Fertility and reproduction troubles	
	Fusarium culmorum		
	Fusarium crookwellense		
Fumonisin B1	*Fusarium verticillioides*	Lesion of central nervous system	
	Fusarium proliferatum	Hematotoxicity	
		Genotoxicity	
		Immunomodulation	

(continued)

Table 4.1 (continued) Mycotoxins of Interest in Poultry and Processed Meat

Toxin	Main Producing Fungal Species	Toxicity	Structure
Aflatoxin B1	Aspergillus flavus	Hepatotoxic	
	Aspergillus parasiticus	Genotoxic	
	Aspergillus nomius	Carcinogenic	
		Immunomodulation	
Ochratoxin A	Penicillium verrucosum	Nephrotoxic	
	Aspergillus ochraceus	Genotoxic	
	Aspergillus carbonarius	Immunomodulation	
Citrinin	Aspergillus terreus	Nephrotoxic	
	Aspergillus carneus		
	Aspergillus niveus		
	Penicillium verrucosum		
	Penicillium citrinum		
	Penicillium expansum		
Cyclopiazonic acid	Aspergillus flavus	Neurotoxicity	
	Aspergillus versicolor	Tremorgenic	
	Aspergillus tamarii		
	Penicillium camemberti		

nephropathy (BEN), and fumonisin B1, which is associated with esophageal cancer. Mycotoxin exposure of human consumers is usually directly linked with alimentary habits.

For human consumers, the main source of exposure to mycotoxins is represented by cereals and cereal-based products [6–8]. However, they may also be exposed to these toxic compounds after ingestion of animal-derived products. Indeed, foods prepared from animals that have been fed with contaminated feeds may contain residual contamination and represent a vector of mycotoxins. Depending on the mycotoxins, the residues may correspond to the native toxin or to metabolites that keep all or part of the toxic properties of the parental molecule.

Among farm animals, poultry species can be exposed to several different mycotoxins, due to their breeding and feeding conditions. Moreover, given the importance of poultry meat and poultry products in the diet of many people around the world, it is very important to characterize potential transfer within tissues of edible poultry products.

The exposure of human consumers may also result from mycotoxin synthesis during ripening of products. Indeed, ripened foods are favorable to mold development, because they often participate in organoleptic improvement of such products. Therefore, the contamination with a toxigenic strain may lead to mycotoxin synthesis and accumulation in the final product [9].

At the present time, few toxins are regulated in foods (Table 4.2) [10–11]. The risk management is mainly based on controlling the contamination of vegetal raw materials intended for both human and animal consumption and limiting animal exposure through feed ingestion. It may guarantee against the presence of residual contamination of mycotoxins in animal derived products. However, a high level of contamination may accidentally lead to a sporadic contamination of products coming from exposed animals. Moreover, some toxins, mainly from *Penicillium* species, may also appear later, particularly during ripening of dry-cured meat products.

The aim of this work is to present methodology described for mycotoxin quantification in poultry and processed meats. Owing to the important structural diversity of mycotoxins and to the variations in their metabolism, it is impossible to establish general rules; each toxin and each product has to be investigated as a particular case. Therefore, we will first present the main toxins with their most important characteristics. After that, their analysis and prevalence will be presented in poultry and processed meats.

4.2 Main Mycotoxins

Depending on the fungal species that produces them, mycotoxins can be classified as "field" or "storage" toxins. The former are mainly produced by *Fusarium* fungi that develop on living plants, because a high water activity is required for their growth [12]. The later are toxins from *Penicillium* that may grow on foods and feeds during storage when moisture and temperature are favorable [13]. Between these two groups, the toxins produced by *Aspergillus* may occur both in the field and during storage, depending on climatic conditions [14]. We will now focus on the most important toxins of these three groups, based on their toxicity or their prevalence in foods and feeds.

4.2.1 Trichothecenes

4.2.1.1 Origin and Nature

Trichothecenes constitute a large group of secondary metabolites produced by numerous species of *Fusarium*, such as *F. graminearum*, *F. culmorum*, *F. poae*, and *F. sporotrichioides*. More than

Table 4.2 EU Regulation for Mycotoxin Contamination (μg/kg)

Toxin	Destination	Matrix	Maximal Concentration (μg/kg)
Aflatoxins			
Aflatoxin B1		Groundnuts + grains + dry fruits	2, 5 or 8 depending on the product and the processing step
		Cereals	2 or 5 depending on the product and the processing step
		Spices	5
		Cereal based foods for young children	0, 1
Aflatoxins B1 + B2 + G1 + G2	Human food	Groundnuts + grains + dry fruits	4, 10 or 15 depending on the product and the processing step
		Cereals	4 or 10 depending on the product and the processing step
Aflatoxin M1		Spices	10
		Milk	0, 05
		Preparation for young children	0, 025
Aflatoxin B1	Animal feed	Raw material for animal feeds	20
		Compound feeds	5 to 20 depending on animal species
Ochratoxin A	Human food	Raw cereal grains	5
		All cereal products	3

		Dried vine fruits	10
		Coffee	3
Zearalenone	Human food	Raw cereals	100
		Cereal flours	75
		Bread, biscuits, corn flakes, snacks	50
		Baby food	20
Deoxynivalenol	Human food	Raw cereals	1250
		Durum wheat, maize	1750
		Cereal flours	750
		Bread, corn flakes, snacks, biscuits	750
		Pâtés	750
		Baby food	200
Fumonisins	Human food	Maize	2000
		Maize flour	1000
		Maize-based food	400
		Baby food	200

Source: European Union, Commission Regulation (EC) N° 466/2001 setting maximum levels for certain contaminants in foodstuffs, *Off. J. Eur. Un.,* L77, 1, 2001; and European Union, Commission Regulation n° 856/2005, toxins of *Fusarium, Off. J. Eur. Un.,* L143, 3, 2005.

160 trichothecenes have been identified, notably deoxynivalenol (DON), nivalenol (NIV), T-2 toxin, HT-2 toxin, diacetoxyscirpenol (DAS), and fusarenon X. DON is the most frequently found trichothecene. Trichothecenes are frequent worldwide contaminants of cereals, mainly wheat and maize, and cereal-based products [8,15–18].

Because trichothecenes are a large family grouping many compounds of variable structure and properties, their toxicity can be very different depending on the molecule, the animal species, the dose, and the exposure period. There are many reviews available on trichothecenes toxicity [19–22]; only the main features will be presented here.

Trichothecenes are potent inhibitors of eukaryotic protein synthesis, interfering with initiation, elongation, or termination stages.

Concerning their toxicity in animals, DAS, DON, and T-2 toxin are the most studied molecules. The symptoms include effects on almost all major systems of organisms; many of them are secondarily initiated by poorly understood metabolic processes connected with protein synthesis inhibition.

Among naturally occurring trichothecenes, DAS and T-2 toxin seem to be the most potent in animals. They have an immunosuppressive effect, decreasing resistance to microbial infections [21]. They also cause a wide range of gastrointestinal, dermatological, and neurological symptoms [23]. In humans, these molecules have been suspected to be associated with alimentary toxic aleukia. The disease, often reported in Russia during the nineteenth century, is characterized by inflammation of the skin, vomiting, and damage to haematopoietic tissues [24,25]. When ingested at high concentrations, DON causes nausea, vomiting, and diarrhea. At lower doses, pigs and other farm animals display weight loss and feed refusal [21]. For this reason, DON is often called vomitoxin or feed refusal factor.

4.2.1.2 Structure and Physicochemical Properties

Trichothecenes belong to the sesquiterpenoid group. They all contain a 12,13-epoxytrichothene skeleton and an olefinic bond with various side chain substitutions. Trichothecenes are classified as macrocyclic or nonmacrocyclic, depending on the presence of a macrocyclic ester or an ester–ester bridge between C-4 and C-15 [26]. The nonmacrocyclic trichothecene can be classified in two groups: type A, which does not have a ketone group on C-8 (T-2 toxin, HT-2 toxin, DAS), and type B, with a ketone group on C-8 (DON, NIV, fusarenon X) [27].

Trichothecenes have a molecular weight ranging from 154 to 697 Da, but it is often between 300 and 600 Da. They do not absorb ultraviolet (UV) or visible radiations, with the exception of type D, which absorbs UV light at 260 nm. They are neutral compounds, usually soluble in mildly polar solvents such as alcohols, chlorinated solvents, ethyl acetate, or ethyl ether. They are sometimes weakly soluble in water [27].

These molecules are very stable, even if stored for a long time at room temperature. They are not degraded by cooking or sterilization processes (15 min at 118°C) [28].

4.2.1.3 Analytical Methods

Methods reported mainly concern the most frequently found toxins in cereals, which are DON, NIV, T-2 toxin, and HT-2 toxin [29]. Validated methods are now available for DON [30], but this is not the case for type A trichothecenes, and reference material and interlaboratory studies are still required [31].

4.2.1.3.1 Type A Trichothecenes

Extraction from solid matrixes is usually done with binary mixtures associating water and acetonitrile, water and methanol, chloroform and methanol, or methanol alone.

Purification is done with solid-phase extraction (SPE) columns working in normal phase (silica, florisil) or inverse phase (C18). Another approach, employing ready-to-use Mycosep columns (Romer Labs Inc., Union, MO), may be applied. These columns are adsorbants (charcoal, celite, ion exchange resin) mixed in a plastic tube. These multifunctional columns are increasingly popular.

Immunoassays are the main method routinely used for T-2 and HT-2 determination in cereals. Detection limits are in accordance with the contamination levels that are observed for these contaminants, and range from 0.2 to 50 ng/g for T-2 toxin [32].

Other methods have also been described, but type A trichothecenes cannot be analyzed by high-pressure liquid chromatography (HPLC)-UV due to the absence of ketone group in C-8 position. That is why gas chromatography (GC) is the most popular approach for this family of compounds. The derivatization of the native compounds by silylation or fluoroacylation is necessary to increase the sensitivity of the measure. Detection can be performed with an electron capture detector or by mass spectrometry (MS). The limits of detection of these methods are a few tens ng/g [33]. Another method was reported using HPLC with fluorescence detection after immuno-affinity columns (IAC) purification of extract and derivatization of T-2 toxin with 1-anthroylnitrile. This procedure allowed a limit of detection of 5 ng/g [34].

4.2.1.3.2 Type B Trichothecenes

Extraction of type B trichothecenes is done with a mixture of acetonitrile–water or chloroform–methanol [35].

Many purification procedures have been reported for type B trichothecenes, such as liquid–liquid extraction (LLE), SPE, and IAC [36]. However, the use of mixed columns (charcoal–alumina–celite) is still widespread [37]. Once again, the Mycosep column is increasingly used for DON analysis.

Thin-layer chromatographic methods are still used for screening, particularly in countries where GC or HPLC are not easily available [38]. Since trichothecenes are not fluorescent, the detection of the molecules requires the use of revelators such as sulfuric acid, para-anisaldehyde, or aluminum chloride. Detection limits of thin-layer chromatography (TLC) range from 20 to 300 ng/g.

Enzyme-linked immunosorbent assay (ELISA) can also be of interest to get rapid and semiquantitative results with only minor purification of the extract. Many kits are commercially available for DON analysis in cereals [39,40].

GC coupled with an electron capture detector, a mass spectrometer, or in tandem (MS-MS) is regularly used after derivatization of the analyte [41–44]. Derivatization reactions are trimethylsilylation or perfluoroacylation. Fluoroacylation with anhydride perfluorated acid improves detection limits using an electron capture detector or MS. However, a European interlaboratory investigation of the official Association of Official Analytical Chemists (AOAC) method for DON measurement revealed that coefficient of variation between laboratories was very important (about 50%), despite the relatively high level of contamination of the material used (between 350 and 750 μg/kg). These observations increased interest in HPLC-MS methodology in trichothecene determination. This is progressively becoming the choice method [45].

4.2.2 Zearalenone

4.2.2.1 Origin and Nature

ZEA is a mycotoxin with estrogenic effect that is produced by *Fusarium* species such as *F. graminearum*, *F. proliferatum*, *F. culmorum*, and *F. oxysporum* [46,47]. Such molecules are suspected of reducing male fertility in human and wildlife populations, and is possibly involved in several types of cancer development [48]. This molecule is well known by farmers, often being responsible for reproduction perturbation, especially in pigs.

Acute toxicity of ZEA is usually considered as weak, with LD_{50} after oral ingestion ranging from 2,000 to more than 20,000 mg/kg body weight [49,50]. Subacute and chronic toxicity of the mycotoxin is more frequent and may be observed at the natural contamination levels of feeds. The effects are directly related to the fixation of ZEA and metabolites on estrogenic receptors [51]. Affinity with estrogenic receptors is, in decreasing order: α-zearalanol > α-zearalenol > β-zearalanol > ZEA > β-zearalenol. Pigs and sheep appear more sensitive than other animal species [49,50].

ZEA induces alteration in the reproductive tracts of both laboratory and farm animals. Variable estrogenic effects have been described, such as a decrease in fertility, a decrease in litter size, an increase in embryo-lethal resorptions, and change in adrenal, thyroid, and pituitary gland weight. In male pigs, ZEA can depress testosterone, weight of testes, and spermatogenesis while inducing feminization and suppressing libido [49,50,52]. Long-term exposure studies did not demonstrate any carcinogenic potential for this mycotoxin [5].

4.2.2.2 Structure and Physicochemical Properties

The structure of ZEA is shown in Table 4.1. α- and β-zearalenol, the natural metabolites of the native toxin, correspond to the reduction of the ketone function in C_6.

ZEA has a molecular weight of 318 g/mol. This compound is weakly soluble in water and in hexane. Its solubility increases with the polarity of solvents such as benzene, chloroform, ethyl acetate, acetonitrile, acetone, methanol, and ethanol [53]. The molecule has three maximal absorption wavelengths in UV light: 236, 274, and 314 nm. The 274-nm peak is the most characteristic and commonly used for UV detection of the toxin.

ZEA emits a blue fluorescence with maximal emission at 450 nm after excitation between 230 and 340 nm [54].

4.2.2.3 Analytical Methods

Owing to regulatory limits, methods for analysis of ZEA content in foods and feeds may allow the detection of several nanograms per gram. Reviews have been published detailing the analytical methods available [45,55,56]. ZEA is sensitive to light exposure, especially when in solution. Therefore, preventive measures have to be taken to avoid this photodegradation.

Solvents used for liquid extraction of ZEA and metabolites are mainly ethyl-acetate, methanol, acetonitrile, and chloroform, alone or mixed. The mixture acetonitrile–water is the most commonly used. For solid matrixes, more sophisticated and efficient methods may be applied: for example, ultrasounds or microwaves [57,58].

In biological matrixes (e.g., plasma, urine, feces), hydrolysis of phase II metabolites is necessary before the purification procedure. It can be achieved by an enzymatic or a chemical protocol [59].

In vegetal materials, the demonstrated presence of sulfate conjugates [60] or glucoside conjugates [61] is rarely taken into account in routine methods.

Purification may be achieved using LLE, SPE, or IAC procedures. For SPE, most stationary phases may be used: inverse phase (C18, C8, or C4), normal phase (florisil, SiOH, NH_2), or strong anion exchange (SAX) [62]. The ready-to-use Mycosep column allows a rapid purification of samples without any rinsing and with a selective retention of impurities [63].

IAC columns have also been developed for ZEA and are very popular [64–70]. Although purification is very selective and extraction yields usually high, several points have to be highlighted:

- Antibody may not have the same affinity for all metabolites, some not being accurately extracted.
- Fixation capacity of columns are limited; a great number of interfering substances may perturb the purification by saturation of the fixation sites [62].
- These columns may be reused, increasing the risk of cross-contamination of samples.

For quantification of ZEA and metabolites in cereals and other matrixes, several immunological methods have been set up, including radioimmunoassay and ELISA [71–75]. The limit of quantification of these methods is several tens of nanograms per gram. ELISA kits show a cross-reactivity with α- and β-zearalenol [76].

Physicochemical methods are also widely used. They mainly include HPLC and GC, TLC being nearly withdrawn [77,78]. Many methods using C18 as stationary phase and CH_3CN/H_2O as mobile phase have been described. More specific stationary phases have also been proposed, such as molecular printing (MIP) [79]. Detectors are often fluorimeters [64–67,69,80] or UV detectors [62,66]. Sensitivity of these methods varies, depending on the metabolites, and is less important for reduced metabolites (α- and β-zearalenol).

ZEA and metabolites can also be detected by GC. However, the usefulness of this method is limited due to the time-consuming need to derivatize phenolic hydroxy groups. Consequently, only GC-MS has been applied for confirmation of positive results [81,42].

Many liquid chromatography (LC)-MS methods have also been proposed for ZEA and metabolites detection [45]. The method of chemical ionization at atmospheric pressure is most often used followed by electrospray [70,82–85]. These methods allow the detection of ZEA and metabolites at levels below 1 ng/g [45].

In an international interlaboratory study, important variations were observed between results from the participant laboratories, probably related to differences in sample preparation (LLE, SPE, or IAC) and quantification (HPLC, GC, TLC, and ELISA) [73,86].

4.2.3 Fumonisins

4.2.3.1 Origin and Nature

Fumonisins were first described and characterized in 1988 from *F. verticillioides* (formerly *F. moniliforme*) culture material [87,88]. The most abundant and toxic member of the family is fumonisin B1. These molecules can be produced by several species of *Fusarium* fungi: *F. verticillioides*, *F. proliferatum*, and *F. nygamai* [89,90]. These fungal species are worldwide contaminants of maize, and represent the main source of fumonisins [91].

One major characteristic of fumonisins is that they induce very different syndromes depending on the animal species. FB1 is responsible for equine leukoencephalomalacia characterized by necrosis and liquefaction of cerebral tissues [92,93]. Horses appear to be the most sensitive species; clinical signs may appear after exposure to doses as low as 5 mg FB1/kg feed over a few weeks. Pigs are also sensitive to FB1 toxicity. In this species, fumonisins induce pulmonary edema after exposure to high doses (higher than 20 mg FB1/kg feed) of mycotoxins, and are hepatotoxic and immunotoxic at lower doses [94–96]. By contrast, poultry and ruminants are more resistant to this mycotoxin, and clinical signs appear only after exposure to doses higher than 100 mg FB1/kg, which may be encountered in natural conditions, but are quite rare [97–102]. In rodents, FB1 is hepatotoxic and carcinogenic, leading to the appearance of hepatocarcinoma in long-term feeding studies [103,104]. In humans, FB1 exposure has been correlated with a high prevalence of oesophageal cancer in some parts of the world, mainly South Africa, China, and Italy [105]. Finally, fumonisins can cause neural tube defects in experimental animals, and thus may also have a role in human cases [106–109]. At the cellular level, FB1 interacts with sphingolipid metabolism by inhibiting ceramide synthase [110]. This leads to the accumulation of free sphinganine (Sa) and, to a lesser extent, of free sphingosine (So). Therefore, the determination of the Sa/So ratio has been proposed as a biomarker of fumonisin exposure in all species in which it has been studied [111–114].

4.2.3.2 Physicochemical Properties

The structure of FB1 and related compounds is shown in Table 4.1. FB1 has a molecular weight of 722 g/mol. It is a polar compound, soluble in water and not soluble in apolar solvents. FB1 does not absorb UV light, nor is it fluorescent. Fumonisins are thermostable [115]. However, extrusion cooking may reduce fumonisin content in maize products [116].

4.2.3.3 Methods of Analysis

Because of their relatively recent discovery, analytical methodology for fumonisin analysis is still undergoing development. In most described methods, the food or foodstuff is corn. An HPLC method has been adopted by the AOAC and the European Committee for Standardization as a reference methodology for fumonisin B1 and B2 in maize [117–119].

An efficient extraction of fumonisins in solid matrix can be obtained with acetonitrile–water or methanol–water mixtures [120,121]. This was assessed by interlaboratory assay [122]. Increased contact time and solvent/sample ratio also increase yield of extraction step.

Purification of extracts is usually based on SPE with SAX, inverse phase (C18), or IAC [123,124].

Quantification of FB1 can be done by TLC, HPLC, or GC-MS. However, derivatization of the fumonisins is usually required. For TLC, this is usually done by spraying *p*-anisaldehyde on the plates after development in a chloroform-methanol-acetic acid mixture. It leads to the appearance of blue-violet spots that can be quantified by densitometry [115,125]. Quantification limits obtained with TLC methods often range from 0.1 to 3 mg/kg. That may be sufficient for rapid and costless screening of raw materials [126,127].

For HPLC analysis, fluorescent derivatives are formed with *o*-phtaldialdehyde (OPA), naphthalene-2,3-dicarboxaldehyde, or 4-fluoro-2,1,3-benzoxadiazole [128]. OPA derivatization offers the best response, and has been generally adopted, but the derivatization product is very unstable, and analysis of samples has to be quickly performed after derivatization [129]. HPLC with fluorescence detection (HPLC-FL) methods have detection limits usually ranging from 10 to 100 µg/kg [124,128,133].

GC has also been proposed for FB1 determination. It is based on partial hydrolysis of fumonisins before reesterification and GC-MS analysis. However, this structural change does not allow the distinction of different fumonisin molecules [130]. Another GC-MS method has been described, developing a derivatization step with trimethylsilylation coupled with detection by flame ionization [131].

The introduction of LC-MS with atmospheric pressure ionization has increased specificity and sensitivity of the detection. The majority of published fumonisin analysis with LC-MS was performed to the low parts per billion level in grains and maize-derived products. Furthermore, this methodology also appeared powerful in investigating for new fumonisin molecules, and elucidating structures and biosynthetic pathways and behavior during food processing [47].

ELISA kits are also commercially available for fumonisin quantification in vegetal matrix [132–134]. They usually offer detection limits around 500 µg/kg. However, the comparison with HPLC-FL shows that ELISA often overestimates the fumonisin content of samples. This may be due to cross-reactions between antibody and coextracted impurities [135]. This drawback could be overcome by purification of extracts before ELISA realization. This method can nevertheless be useful for rapid screening of maize and maize products. One ELISA kit has been validated by the AOAC for total fumonisin determination in corn [136].

4.2.4 Aflatoxins

4.2.4.1 Origin and Nature

Aflatoxins are probably the most studied and documented mycotoxins. They were discovered following a toxic accident in turkeys fed a groundnut oilcake supplemented diet (Turkey X disease) [137–139]. The four natural aflatoxins (B1, B2, G1, and G2) can be produced by strains of fungal species belonging to the *Aspergillus* genus, mainly *A. flavus* and *A. parasiticus* [14,140]. These are worldwide common contaminants of a wide variety of commodities, and therefore aflatoxins may be found in many vegetal products, including cereals, groundnuts, cotton seeds, dry fruits, and spices [141–146]. If these fungal species can grow and produce toxins in the field or during storage, climatic conditions required for their development are often associated with tropical areas (high humidity of the air, temperature ranging from 25 to 40°C) [147–151]. However, following extreme climatic conditions (an abnormally hot summer period), aflatoxins could be found in other parts of the world. For example, in 2003, controls on maize harvested in Europe were found contaminated by unusual AFB1 concentrations [152,153].

Aflatoxin B1 is a highly carcinogenic agent leading to primary hepatocarcinoma [154–157]. This property is directly linked to its metabolism and to the appearance of the highly reactive epoxide derivative. Formation of DNA adducts of AFB1-epoxide is well characterized [158]. Differences in AFB1 metabolism within animal species could explain the variability of the response in terms of carcinogenic potential of the mycotoxin [159,160].

AFM1, a hydroxyled metabolite of AFB1, can also be considered a genotoxic agent, but its carcinogenic potential is weaker than that of AFB1 [161]. Taking into account the toxicity of these molecules, the International Agency for Research on Cancer classified AFB1 in the group 1 of carcinogenic agents, and AFM1 in the 2B group of molecules that are carcinogenic in animals and possibly carcinogenic in humans [5].

4.2.4.2 Structure and Chemical Properties

The structures of aflatoxin B1 are presented in Table 4.1. Molecular weights of aflatoxins range from 312 to 320 g/mol. These toxins are weakly soluble into water, insoluble in nonpolar solvents, and very soluble in mildly polar organic solvents (i.e., chloroform and methanol). They are fluorescent under UV light (blue fluorescence for AF"B" and green for AF"G") [162].

4.2.4.3 Analytical Methods

Most common solvent systems used for extraction of aflatoxins are mixtures of chloroform–water [163–165] or methanol–water [166–170]. This latter mixture is mainly used for multiextraction of mycotoxins, and is not specific for aflatoxin extraction [171]. Whatever the solvent system used, the extract obtained still contains various impurities and requires further cleanup steps. The most commonly used extraction technique is SPE, which has replaced the traditional liquid–liquid partition for cleanup [165]. Stationary phase of the SPE columns used may be silica gel, C18 bonded-phase, and magnesium silicate (commercially available as Florisil) [163,172]. Antibody affinity SPE columns are also widely used.

IAC chromatography using antitoxin antibodies allowed the improvement of both specificity and sensitivity [173,174]. Indeed, methods were validated for grains [175], cattle feed [176,177], maize, groundnuts, and groundnut butter [178], pistachio, figs, and paprika [179], and baby food [180]. Analytical methods of the same kind were validated for quantification of AM1 in milk [181] and in powder milk [182,183], these methods show limits of quantification below the regulatory limit of 0.05 µg/L.

Aflatoxins are usually quantified by TLC, HPLC, or ELISA.

TLC was first developed in the early 1980s. Using strong fluorescence of the molecules, the characterization of signals with naked eyes or densitometric analysis could give semiquantitative to quantitative results (AOAC methods 980.20 and 993.17) [184]. Therefore, aflatoxin B1 could be measured in concentrations ranging from 5 to 10 µg/kg. A TLC method for quantification of AFM1 in milk was also validated by AOAC (980.21) [185] and normalized (International Standardization Organisation [ISO] 14675:2005) [186]. A method for semiquantitative analysis of AFB1 in cattle feed was also published (ISO 6651:2001) [187]. Confirmation of identity of aflatoxins B1 and M1 in foods and feeds is still classically done by TLC after bidimential migration and trifluoric acid–hexane (1:4) spraying of plates.

HPLC allowed the reduction of detection limits together with an improvement of the specificity of the dosage [188]. Therefore, new methods were validated for aflatoxin quantification in grains (AOAC 990.33), cattle feed (ISO 14718:1998), and AFM1 in milk (ISO/FDIS 14501) [189–191]. These methods are based on the use of a fluorescence detector allowing the quantification of low levels of aflatoxins. The sensitivity can be increased by the treatment of extracts with trifluoric acid to catalyze the hydratation of aflatoxins M1, B1, and G1 into their highly fluorescent M2a, B2a, and G2a derivatives.

ELISA has been developed for both total aflatoxins [192,193] and AFB1 detection in feeds and grains [194–197] and for AFM1 in milk [198]. These methods have limits of quantification in accordance with international regulations. Therefore, some commercially available kits have been validated by the AOAC, as for example the one referenced as AOAC 989.86, devoted to AFB1 dosage in animal feed. However, in spite of the development of ELISA methods for AFM1 detection [199], no ELISA kit has been validated following the harmonized protocol of ISO/AOAC/International Union of Pure and Applied Chemistry (IUPAC) for AFM1 quantification in milk. The AOAC has edited rules for characterization of antibodies used in immunochemical methods [200].

Detection limits in the low parts per trillion range can be achieved by these classical LC-fluorescence methods. Therefore, methods such as LC-MS may represent only a minor alternative or confirmation technique for already well-established methodologies [45]. It may however be useful to confirm positive results of TLC or ELISA-based screening analysis [201]. At the present time, few quantitative methods have been published for aflatoxin determination in food and milk [202–205].

4.2.5 Ochratoxin A

4.2.5.1 Origin and Nature

Ochratoxins A, B, and C are secondary metabolites produced by several *Aspergillus* and *Penicillium* species. According to its prevalence and toxicity, only OTA will be treated in this section. This molecule can be produced by *Aspergillus* species such as *A. ochraceus* [206], *A. carbonarius* [207,208], *A. alliaceus* [209], and *A. niger* [210], although the frequency of toxigenic strains in this species appears moderate [211–213]. OTA can also be synthesized by *Penicillium* species, mainly *P. verrucosum* (previously named *P. virridicatum*) [214–215].

The ability of both *Aspergillus* and *Penicillium* species to produce OTA makes it a worldwide contaminant of numerous foodstuffs. Indeed, *Aspergillus* is usually found in tropical or subtropical regions, whereas *Penicillium* is a very common contaminant in temperate and cold climate areas [216–219]. Many surveys revealed the contamination of a large variety of vegetal products such as cereals [220,221], grapefruit [222,223], and coffee [221,224]. For cereals, OTA contamination generally occurs during storage of raw materials, especially when moisture and temperature are abnormally high, whereas for coffee and wine, contamination occurs in the field or during the drying step [219,225–227]. When ingested by animals, OTA can be found at residue level in several edible organs (see 23.3.5). Therefore, the consumption of meat contaminated with OTA has also been suspected to represent a source of exposure for humans [228]. Recent surveys done in European countries demonstrated that the role of meat products in human exposure to OTA can be considered low [6,229].

Kidney is the primary target of OTA. This molecule is nephrotoxic in all animal species studied. For example, OTA is considered responsible for a porcine nephropathy that has been studied intensively in the Scandinavian countries [230,231]. This disease is endemic in Denmark, where rates of porcine nephropathy and ochratoxin contamination of pig feed are highly correlated [232]. Because the renal lesions observed in pig kidneys after exposure to OTA are quite similar to those observed in kidneys of patients suffering from BEN, OTA is suspected to play a role in this human syndrome [233–235]. BEN is a progressive chronic nephropathy that occurs in populations living in areas bordering the Danube River in Romania, Bulgaria, Serbia, and Croatia [236,237].

4.2.5.2 Physicochemical Properties

The structure of OTA is presented in Table 4.1. OTA has a molecular weight of 403.8 g/mol. It is a weak organic acid with a pKa of 7.1. At an acidic or neutral pH, it is soluble in polar organic solvents and weakly soluble in water. At a basic pH, it is soluble and stable in an aqueous solution of sodium bicarbonate (0.1 M; pH: 7.4), as well as in alkaline aqueous solutions in general.

OTA is fluorescent after excitation at 340 nm, and emits at 428 nm when nonionized and at 467 nm when ionized.

4.2.5.3 Methods of Analysis

Extraction of OTA is often achieved by using a mixture of acidified water and organic solvents. An IUPAC/AOC method validated for OTA determination in barley uses a chloroform–phosphoric acid mixture [238]. For coffee or wine, chloroform is successfully used [239,240]. Mixtures of methanol–water or acetonitrile–water have also been reported [241,242]. *Tert*-butylmethylether has been used for OTA extraction from baby food, and may represent an alternative to the use of chlorinated solvents [243].

Several efficient cleanup procedures based on IAC and SPE using C8, C18, and C-N stationary phases were developed to replace, when possible, conventional LLE [244]. Stationary phases based on the principle of MIP are emerging [245,246]. The specificity of such methods is comparable to that of IAC. Although their applicability in real matrixes has not been established, they may represent alternatives to IAC and SPE methods in the future.

Many methods have been developed for separation and detection of OTA. TLC methods have been published [247–249]. However, both specificity and sensitivity of TLC are limited, and interferences with the sample matrix often occur [250]. These drawbacks may be overcome by two-dimensional TLC [251]. However, HPLC is the most commonly used method for determination of OTA [244,252].

Most described HPLC methods use a reverse-phase C18 column and an acidic mobile phase composed of acetonitrile or methanol with acetic, formic, or phosphoric acid [242,253–255]. The property of OTA to form an ion-pair on addition of a counter ion to the mobile phase has been used [256]. This led to a shift in OTA fluorescence from 330 to 380 nm and allowed an improvement of the signal. Ion-pair chromatography was also used for detection of OTA in plasma and human and cows' milk, with detection levels of 0.02 and 10 ng/mL for plasma and milk, respectively [257,258]. The major limit of the method is that small changes in composition of mobile phase may change retention time of OTA.

HPLC methods using fluorescence detection are applicable to OTA detection in barley, wheat, and rye at concentrations of about 10 μg/kg [259]. For baby foods, a quantification limit of 8 ng/kg has been reached by postcolumn derivatization with ammoniac [240,243].

Today, several validated methods have been published for OTA detection in cereals and derived products [260], in barley and coffee [261–263], and in wine and beer [264].

Immunoassays such as ELISA and radioimmunoassays have been developed [265–268], and may be regarded as qualitative or semiquantitative methods, useful for rapid screening.

Owing to its toxicity and regulatory values, OTA analysis has to be performed down to the ppb range in foods and feeds. In addition, plasma and urine samples are analyzed to monitor OTA exposure in humans and animals. In this context, methods using LC-MS may be used to confirm OTA-positive results obtained by ELISA or HPLC-FL. They may also be powerful tools to elucidate structure of *in vivo* metabolites and OTA adducts in biological fluids. Many studies have described LC-MS methods for OTA determination [47].

4.2.6   Other Toxins

4.2.6.1   Citrinin

4.2.6.1.1   Origin and Nature

Citrinin is produced by different *Aspergillus* (*A. terreus, A. carneus, A. niveus*) and *Penicillium* species (*P. citrinum, P. verrucosum, P. expansum*) [269]. It may also be produced by fungi belonging to the *Monascus* genus [270]. It has been found at levels ranging from few micrograms per kilogram to several milligrams per kilogram in barley, wheat, and maize, and also in rice, nuts, dry fruits, and apple juice [1,271–273].

Citrinin is nephrotoxic in all animal species where it has been studied, leading to a time- and dose-dependent necrosis of renal tubules [274–276]. This is mainly due to citrinin-mediated oxidative stress [277].

4.2.6.1.2   Physicochemical Properties

Citrinin is an acidic phenolic benzopyrane with a molecular weight of 250 g/mol (Table 4.1). This molecule is insoluble in water but very soluble in most of organic solvents, such as methanol, ethanol, and acetonitrile [38]. Citrinin is heat labile in acidic or alkaline solution. It easily links to proteins.

4.2.6.1.3   Analytical Methods

Several methods have been used for citrinin determination in foods and feeds. A rapid TLC method allows the detection of 15–20 µg/kg in fruits [278]. Immunological methods such as ELISA have also been developed, and present good sensitivity [272]. HPLC allows the detection of citrinin in cereals, biological fluids (urine and bile), and fermentation media [272]. It has to be noted that efficiency of HPLC methods greatly depends on the extraction step, which must not degrade the toxin. Detection is made in UV at 254 or 366 nm [38]. The detection limits in cereals are usually about 10 µg/kg. A semiquantitiative fluorimetric method has also been set up to detect citrinin in fungal culture isolated from cheeses [279].

4.2.6.2   *Cyclopiazonic Acid*

4.2.6.2.1   Origin and Nature

Cylopiazonic acid (CPA) was first isolated from culture of *P. cyclopium*, but has also been shown to be produced by several species of *Aspergillus* and *Penicillium*, such as *A. flavus, A. tamarii*, or *P. camemberti* [280,281]. Therefore, CPA has been detected in many foods, especially cheeses [282], although few cases of intoxication have been described. However, retrospective analysis of "Turkey X disease" performed in 1986 by Cole suggested that clinical signs were not all typical of aflatoxicosis. He thus tried to demonstrate a possible role for cyclopiazonic acid in this affection. For instance, opisthotonos originally described in "Turkey X disease" can be reproduced by administration of a high dose of cyclopiazonic acid but not by ingestion of aflatoxin [283]. Cyclopiazonic acid is a specific inhibitor of the Ca2+ ATPase pump of the endoplasmic reticulum [284], which plays a key role in muscular contraction and relaxation. Principal target organs of cyclopiazonic acid in mammals are the gastrointestinal tract, liver, and kidneys [285,286]. Main symptoms observed after acute intoxication with CPA are nervous signs, including eyelid ptosis, ataxia with hypothermia, tremors, and convulsions [287].

4.2.6.2.2   Physicochemical Properties

CPA is a tetramic indole acid with a molecular weight of 336 g/mol (Table 4.1). It is produced by the amino acid pathway and derived from tryptophane, mevalonate, and two acetate molecules.

4.2.6.2.3 Methods of Analysis

TLC is still used to quantify CPA in cereals and milk products [280]. For milk products, several methods were developed: inverse phase LC [288] and LC-ion trap electrospray MS-MS [289] allowing a detection limit of 5 ng/mL. Methods using LC with UV detection were also developed for quantification in cheese [290] and cereals and derived products [291].

Immunoenzymatic methods allow detection of CPA in maize and animal organs (muscles and plasma) [292], and also in peanuts and mixed feed [293]. Detection limits of such methods range from 1 to 20 ng/g.

4.3 Mycotoxin Analysis and Prevalence in Poultry

If many methods have been developed and validated for vegetal matrix, due to the absence of regulation, few data are available on techniques that may be used for animal-derived foods. With the exception of the detection of Aflatoxin M1 in milk and milk products [294], no official method is available for such products.

Taking into account the great structural differences that exist between mycotoxins and their distinct metabolism after absorption in animal digestive tracts, no multidetection method can be carried out; methods have to be developed specifically for each toxin and metabolite.

In this section will be presented the analytical methods used for mycotoxin quantification in poultry organs, as well as the available data concerning the metabolism of these toxic compounds in avian species and the persistence of a residual contamination after dietary exposure. These data are helpful to evaluate the real risk of mycotoxin contamination of poultry products and the subsequent possible need for development of analytical methods.

4.3.1 Trichothecenes

4.3.1.1 Methods of Analysis

Few methods have been developed for trichothecenes analysis in poultry tissues. Indeed, first experiments on the pharmacokinetics and distribution of these mycotoxins were performed using radiolabeled toxins [295–299]. Because these experiments revealed that trichothecenes were rapidly excreted and carryover of the toxins in edible parts of poultries was minimal (see 24.3.1.2), few studies were carried out to evaluate trichothecene presence in muscle and other tissues of animals after exposure to unlabeled toxins. The methods used in these works are summarized in Table 4.3 [300–309].

4.3.1.2 Behavior and Residual Contamination of Poultry Tissues

Oral absorption of trichothecenes is limited (<10% at 6 h) in poultry, at least for DON and T-2 toxin. For example, in laying hens, after oral administration of 0.25 mg DON/kg BW, the mean plasmatic peak was reached after 2.25 h, and average bioavailability was 0.64%, with marked individual variations [297,310,311].

As is true for other animal species, distribution of trichothecenes is wide and rapid. Maximal tissue concentrations of DON, T-2 toxin, and their metabolites were observed after 3 h in liver and kidneys, and 4–6 h in muscle, fat, and the oviduct. Higher concentrations were found in

Table 4.3   Methods for Mycotoxin Analysis in Poultry Muscle and Tissues

Toxin	Organ (Species)	Extraction and Clean Up	Derivatisation– Quantification	LOD*	Reference
T-2	Liver, kidney, heart (chicken)	Acetonitrile	TFAA – tri-Sil TBT	—	300
		Amberlite XAD-2 resin column	GC-MS		
DON	Liver, kidney, muscle (hens)	Acetonitrile-water	Heptafluorobutyryl imidazole	10 ng/g	301
		Alumina-charcoal column	Gas–iquid chromatography		
ZEA	Muscle (laying hens)	Overnight treatment with 2/0.9U b-glucuronidase/ arylsulfatase	HPLC fluorescence	1 ng/g	302
		Ethyl acetate			
		IAC			
ZEA	Muscle (chicken)	Acetone-water	HPLC	4 ng/g	303
		Basic alumina and phosphate exchange AGMP-1 resin column	UV		
FB1	Muscle, kidney, liver (mule ducks)	Acetonitrile-mehanol	HPLC fluorescence	25 ng/g	304
		Fat removal with n-hexane			
		Immunoaffinity column			
AFB1	Liver, kidney, heart, muscle (Chicken)	Column chromatography	2D TLC fluorodensitometry	≤0.1 ng/g**	305

(continued)

Table 4.3 (continued)   Methods for Mycotoxin Analysis in Poultry Muscle and Tissues

Toxin	Organ (Species)	Extraction and Clean Up	Derivatisation– Quantification	LOD*	Reference
AFB1	Liver (Chicken)	Immunoaffinity columns	ELISA optical density	1 ng/g**	306
AFB1	Liver (Chicken)	Immunoaffinity columns	HPLC fluorescence	0.008 ng/g**	307
OTA	Muscle (Turkey, chicken)	Chloroform-orthophosphoric acid	HPLC fluorescence	0.04 ng/g**	308
		Immunoaffinity columns			
OTA	Muscle (broiler chicks)	0.1 M Phosphoric acid-chloroform	HPLC fluorescence	0.05 ng/g	309
		Diatomaceous hearth column			
OTA	Muscle (broiler chicks)	Dichloromethane-citric acid	ELISA optical density	0.042 ng/g	309

Note: "*", detection limit; and "**", quantification limit.

the anterior digestive tract, kidney, liver, gall bladder, and spleen. Plasmatic distribution profiles did not show a secondary peak correlated with the enterohepatic cycle [297,310,311]. When administration was prolonged, maximal DON values in tissues were reached rapidly and remained relatively constant throughout the exposure period. The highest concentrations were detected in the same organs as described above after a single administration [311]. Residual persistence of T-2 toxin and DON, as well as of their metabolites, in muscle, liver, and kidney, in the case of single or repeated administration, is summarized in Table 4.4 [296,297,311,312]. Detected levels of contamination were on the scale of micrograms per kilogram. Prolonged administration of trichothecenes led to a higher level of contamination than a single one, indicating an accumulation of toxins or metabolites. The decrease in the residual contamination was slower.

4.3.2 Zearalenone

4.3.2.1 Methods of Analysis

Owing to metabolism of the native molecule and the very weak carryover of ZEA in edible parts of farm animals (see Section 4.3.2.2), few "classical" physicochemical or immunological methods have been developed for ZEA detection in edible parts of poultry species (Table 4.3) [302,313–315]. HPLC-UV or HPLC-FL are used for quantification and display detection limits near 1ng/g.

4.3.2.2 Behavior and Residual Contamination of Poultry Tissues

Although metabolism is a key point of ZEA toxicity [316], few studies are available concerning poultry. An intracellular partitioning of reduction activity of ZEA in liver has been described, the extent varying depending on the species and on the isomer produced. *Ex vivo*, hens almost exclusively produced α-zearalenol with a microsomal fraction and β-zearalenol with a cytosolic fraction [317]. Hen hepatocytes are said to produce mainly β-zearalenol; only traces of α-zearalenol have been found [318]. These results are not in agreement with those obtained *in vivo*. In chickens, administration of a diet containing 100 mg/kg ZEA for 8 days, followed by exposure to 10^9 dpm/kg [^{3}H] ZEA, revealed that the kinetics of the toxin is rapid, with tissue half-life ranging from 24 to 48 h [319]. In addition to the digestive tracts and excreta (bile), most of the radioactivity was found in the liver and kidneys, and a concentration peak was reached 30 min after administration. The residue profile found in liver (GC-MS), in nanograms per gram, was the following: zearalenone 681, α-zearalenol 1200, β-zearalenol 662. After 24 h, total quantities found in liver, gizzard (without mucosa), muscle, plasma, skin, and fat were respectively 651, 297, 111, 91, 70, and 53 ng/g. These results are similar to those obtained by Maryamma et al. [320] after 20 days' administration of 10 mg/kg body weight (BW) of zearalenone to broilers. Hepatic and muscular concentrations of 207 and 170 ng/g were found 24 h after the last administration. Likewise, in turkeys, administration of feed containing 800 mg ZEA/kg for 2 weeks resulted in plasmatic concentrations of 66 ng/mL ZEA and 194 ng/mL α-zearalenol at the end of the experiment. Only traces of β-zearalenol were found [303]. All these studies were performed using very high doses of the toxin. A recent experiment in chickens using a 1.58 mg ZEA/kg feed for 16 weeks appears to confirm these results for low concentrations. Hepatic concentrations obtained at the end of the experiment were 2.1 ng/g ZEA and 3.7 ng/g α-zearalenol, mainly in conjugated forms, whereas β-zearalenol was below the detection limit (<3 ng/g) [321]. No trace of zearalenone or of its metabolites was found in muscles, fat, or eggs.

Table 4.4 Residues of Trichothecenes in Poultry Tissues (Expressed as Equivalent-Toxin)

Toxin	Species	Route	Dose (mg/kg b.w.)	Tissues	6 h	12 h	24 h	2 j	4 j	Half-life	Reference
							After a Single Administration — *Residues (µg/kg)*				
DON	Hen	VO	1.3–1.7	Muscle	8.46	6.6	4.3	2.1	ND		311
				Liver	74	56	30	13	ND	15.7 h	
				Kidney	165	123	44	19	2	8.2 h	
T-2	Chicken	VO	0.126–1.895	Muscle			17/220				296
				Liver			32/416				
				Kidney			24/327				
	Chicken/duck	VO	5	Muscle	30	30	<10	<10			297
				Liver	130/90	30/40	10/<10	<10			
				Kidney	30	20	<10	<10			

After a Repeated Administration

Toxin	Species	Route	Length	Dose	Tissue	Residues (µg/kg)							Reference
						2 j	4 j	6 j	8 j	10 j	12 j		
DON	Hen	VO	6 j	1.3–1.7 mg/ kg b.w.	Muscle	16	17	10	11	7	3		311
					Liver	37	41	39	25	15	9		
					Kidney	60	51	55	21	15	9		
	Chicken	VO	28–190 j	5 mg/kg feed	Muscle	<10							312
					Liver								
					Kidney								

Note: ND, not detectable.

4.3.3 Fumonisins

4.3.3.1 Methods of Analysis

Measurement of FB1 in poultry is poorly documented. Moreover, most of the data concern its toxicokinetic effect in animals and were obtained by using labeled molecules [322]. Finally, only one method was described concerning the determination of nonradiolabeled FB1 in duck tissues. It is based on the use of immunoaffinity columns for the extraction of the mycotoxin and quantification of derivatized FB1 by fluorescence detection after its separation by HPLC [304]. This method allowed fumonisin B1 detection in liver, kidney, and muscle, with a limit of quantification of 25 ng/g (Table 4.3).

4.3.3.2 Behavior and Prevalence in Poultry Tissues

No data are available on a possible metabolism of FB1 in poultry, and few data are available concerning its toxicokinetics.

Absorption after oral administration is reported to be very limited in laying hens (<1%), but higher in growing ducks (2.5–3.5%), close to values already described in rodents, pigs, and nonhuman primates [322,323]. Concentrations in the muscles were about 10-fold lower than in plasma, and no transfer to eggs has been reported.

4.3.4 Aflatoxins

4.3.4.1 Methods of Analysis

Techniques described for aflatoxin analysis in poultry tissues mainly use native fluorescence of these compounds after purification and separation of extract with chromatographic methods (TLC, HPLC) (Table 4.3). Since the 1980s, few studies and surveys have been carried out to characterize aflatoxin presence in poultry products [306,307]. Indeed, risk management is based on the control of animal feed quality, which may guarantee the absence of toxin residues in animal-derived products. These few surveys all demonstrated that muscle foods were not an important source of aflatoxin exposure in humans. It is, however, likely that recent alerts for unusual aflatoxin contamination of cereals produced in temperate climates and the possible consequent animal exposure may strengthen the interest of aflatoxin testing in animal-derived foods. That is why some authors investigated the possible use of ELISA for determination of aflatoxin residue in chicken livers [306].

4.3.4.2 Behavior and Prevalence in Poultry Tissues

Few data are available on aflatoxin behavior in poultry. Oral absorption seems to be comparable to that occurring in other monogastric species, and could represent 90% of the administrated dose [324]. This absorption could be decreased by several adsorbants [325]. Aluminosilicates and clays are among the most effective, and a protective effect has been demonstrated in numerous studies. These studies, the first of which was performed by Phillips in the 1980s, certainly help explain the interest in these kinds of compounds in animal feed [326]. Many studies are done each year to confirm the benefit of these molecules in the case of exposure to aflatoxin.

As is true in other animal species, in poultry metabolization and liver bioactivation in AFB1-8,9-epoxyde and in aflatoxicol could play a key role in the appearance of hepatic lesions.

Bioactivation could explain the greater sensitivity of ducks to aflatoxins, whereas quails could be more resistant due to their lower metabolic capacities [327].

Persistence of aflatoxin B1 and its metabolites at the residual level appears to vary depending on the species and the study. These differences cannot all be explained by differences in metabolization processes between species; differences in the procedures used for detection, extraction, and purification of the toxin and its metabolites from the tissues are more likely to be responsible. The most conclusive results are listed in Table 4.5 [328–333]. Liver and kidney contain more toxin and metabolites than muscles, with the exception of the gizzard, which is directly exposed. Quail appears to be a more important vector for residues than the other species. Hens could be a more important vector than chickens, because excretion in the eggs is also possible, at least after exposure to high concentrations of toxins.

4.3.5   Ochratoxin A

4.3.5.1   Methods of Analysis

All previously described methods were used to analyze OTA content of animal tissues and animal-derived products. The aim of such studies was to characterize the potential carryover of the mycotoxin in animal tissues and to assess human exposure. Most studies have been set up in pigs and pig tissues, because this species appears to be the most sensitive and exposed to OTA. For poultry meat samples, the solvent extraction step cannot be avoided, and precedes the purification step. Typical procedures include extraction with acidic chloroform or acidic ethyl acetate, followed by back extraction into $NaHCO_3$ before cleanup on IAC or C18 columns [308,309].

It appears that detection limits exhibited by HPLC-FL are sufficient to control meat products according to existing regulations. The use of IAC for cleanup allows the reduction of the limit of quantification (LOQ) below 1 ng/g [308].

By contrast, the use of HPLC-MS does not strongly increase the sensitivity of detection, but may be used as a confirmatory method in the case of a positive result.

ELISA tests usually display LOQ higher than other methods. Nevertheless, due to their simplicity and rapidity, these tests could be useful as screening methods in slaughterhouses [309].

4.3.5.2   Behavior and Prevalence in Poultry Tissues

In poultry, oral absorption of OTA appears to occur in the same way as in other monogastric species (passive diffusion of the nonionized lipophilic form), but absorption is apparently lower: about 40% in broilers and only 6.2% in quails. The concentration peak is more rapidly reached in broilers, after 0.33 h [334].

During circulation, OTA fixes to plasmatic proteins, its affinity constant for serum albumin being of about 5.1×10^4 mol/L, which is very close to the value observed in humans [335]. Distribution of OTA in chicken tissue appears to be higher than in other avian species (above 2 L/kg). The highest tissue concentrations were observed in the following organs: kidney > liver > muscles. No residue was found in fat or skin. Transfer to eggs is minimal or nil [334].

To our knowledge, no data are available on OTA metabolism in poultry. Plasmatic half-life of OTA after oral administration ranges from 4.1 h in chicken to 6.7 h in quail. This half-life is well below that reported in most mammalian species [336].

Table 4.5 Residues of Aflatoxin in Animal Tissues (Only the Most Demonstrative Studies are Reported)

Animal Species	Dose and Duration of Exposure	Tissues	Residues (µg/kg)	Metabolites	Reference
Poultry					
Turkey	50 and 150 µg/kg feed for 11 weeks	Liver	0.02–0.009 and 0.11–0.23	AFB1 + AFM1	328
		Kidney	0.02–0.04 and 0.11–0–21	AFB1 + AFM1	
		Gizzard	0.04–0.16 and 0.01–0.12	AFB1 (AFM1 <0.01)	
Turkey	50 and 150 µg/kg feed for 11 weeks and 1 week with toxin free feed	Liver	<0.01	AFB1 + AFM1	
		Kidney	<0.01	AFB1 + AFM1	
		Gizzard	0.04–1.9 and 0.09–0.24	AFB1 (AFM1 <0.01)	
Quail	3000 µg/kg feed for 8 days	Liver	7.83 ± 0.49 and 5.31 ± 0.22	Free and conjugated AFB1	329
			22.34 ± 2.4 and 10.54 ± 0.42	Free and conjugated metabolites	
		Muscle	0.38 ± 0.03 and <0.03	Free and conjugated AFB1	
			0.82 ± 0.05 and 0.32 ± 0.08	Free and conjugated metabolites	
Duck	3000 µg/kg feed for 8 days	Liver	0.52 ± 0.04 and 0.44 ± 0.16	Free and conjugated AFB1	
			2.74 ± 0.15 and 3.81 ± 0.25	Free and conjugated metabolites	
		Muscle	<0.03 and <0.03	Free and conjugated AFB1	
			0.21 ± 0.09 and 0.14 ± 0.05	Free and conjugated metabolites	

Animal	Feed	Tissue	Value	Analyte	Ref
Chicken	3000 µg/kg feed for 8 days	Liver	0.15 ± 0.09 and 0.10 ± 0.01	Free and conjugated AFB1	
			1.54 ± 0.36 and 0.93 ± 0.04	Free and conjugated metabolites	
		Muscle	<0.03 and <0.03	Free and conjugated AFB1	
			0.11 ± 0.02 and 0.08 ± 0.05	Free and conjugated metabolites	
Hen	3000 µg/kg feed for 8 days	Liver	0.34 ± 0.03 and 0.23 ± 0.08	Free and conjugated AFB1	
			2.38 ± 0.36 and 4.04 ± 0.1	Free and conjugated metabolites	
		Muscle	<0.03 and <0.03	Free and conjugated AFB1	
			0.14 ± 0.04 and 0.11 ± 0.04	Free and conjugated metabolites	
Laying hen	10000 µg/kg feed for 7 days	Eggs	0.28 ± 0.1 and 0.38 ± 0.11	AFB1 and total Aflatoxicol	330
Laying hen	8000 µg/kg feed for 7 days	Liver	0.49 ± 0.28 and 0.2 ± 0.09	AFB1 and total Aflatoxicol	331
		Kidney	0.32 ± 0.18 and 0.1 ± 0.04	AFB1 and total Aflatoxicol	
		Muscle	0.08 ± 0.03	Aflatoxicol	
		Eggs	0.24 ± 0.07 and 0.25 ± 0.09	AFB1 and total Aflatoxicol	
Laying hen	2500 µg/kg feed for 4 weeks	Liver	4.13 ± 1.95	AFB1	332
		Eggs	<0.5 and <0.01	AFB1 and AFM1	
Chicken	55 µg/kg feed for 9 days	Liver	0.26 and 0.02	AFB1 and AFM1	333
Chicken	4448 µg/kg feed for 9 days	Liver	1.52 and <0.1	AFB1 and AFM1	

4.3.6 Other Toxins

4.3.6.1 Citrinin

If several studies evaluating citrinin toxicity in avian species, no method was specially set up for the determination of residual contamination of edible organs with this toxin, although poultry appeared sensitive to citrinin toxicity. Indeed, administration of 125–250 ppm citrinin to young chicken leads to acute toxicity with diarrhea and increase in water consumption without any mortality [337,338]. Lesions were mainly digestive hemorrhages, lipidic infiltrations in liver, kidney, and pancreas, and an increase in kidney weight for birds treated with 250 ppm [338].

Administration of labeled toxin demonstrated that citrinin is only weakly absorbed after oral administration and quickly eliminated in urine and feces, at least in rodents [339]. In poultry, the administration of a contaminated diet containing 440 ppm of citrinin did not allow the detection of residual contamination in muscles, whereas only weak amounts of the toxin were found in liver of exposed animals. Lower doses (110–330 ppm) did not led to residual contamination of tissues [340]. Therefore, due to the natural contamination levels observed in poultry feeds [341], the risk of contamination of poultry tissues seems very low.

4.3.6.2 Cyclopiazonic Acid

Only one HPLC method was developed for CPA analysis in poultry tissues. Extraction is achieved with chloroform–methanol. Then partition into 0.1 N sodium hydroxide is done before acidification and dichloromethane extraction. The existence of an interfering compound requires cleanup with silica gel column. Mean recovery of CPA from meat samples spiked with pure toxin at levels ranging from 0.016 to 16.6 mg/kg is about 70% [342].

Tissue transfer in muscle was characterized after oral administration of 0.5, 5, and 10 mg/kg BW using this HPLC quantification. The highest levels of contamination were found in muscle 3 h after administration. For birds fed 0.5 and 5 mg/kg BW, the toxin was rapidly eliminated from meat in 24–48 h [343]. In laying hens, two studies on egg transfer were done after administration of cylopiazonic acid at 0, 2.5, 5, and 10 mg/kg BW/day for 9 days and 0, 1.25, and 2.5 mg/kg BW/day for 4 weeks. Whatever the group of animals concerned, all eggs contained cyclopiazonic acid from the first day of exposure. The concentration of toxin was higher in albumen than in yolk (average of 100 ng/g and 10 ng/g, respectively). All birds fed 10 mg/kg BW and four of the five treated with 5 mg/kg BW died after a decrease in feed intake, in body weight, and in egg production. Other authors have reported a reduction in egg production and shell quality [344,345].

4.4 Mycotoxin Analysis and Prevalence in Processed Meats

Several studies have shown that mold species belonging to the genera *Penicillium* and *Aspergillus* could be isolated from meat products such as ripened sausages or dry-cured ham [346–348]. This mycoflora actively participates in the acquisition or improvement of organoleptic qualities of these products. However, fungal development also raises the question of a possible mycotoxin synthesis in these products, leading to the contamination of final products. Usually, fungal ferments used are selected for their lack of toxigenic potential (*P. nalgiovenses* for instance). However, many studies have demonstrated that fungal mycoflora of dry-cured meat products is usually complex and made of many fungal species, from which several may be toxinogenic, at least *in vitro*. Indeed, some of these strains were found to be able to produce aflatoxins [349,350], ochratoxins [351], citrinin, or cyclopiazonic acid on culture medium [348,352]. Nevertheless, few studies have demonstrated the presence of mycotoxins in such processed meat. It can be linked to the lack of production of mycotoxins in this kind of substrate, to the rapid degradation of the toxins, or to both.

In this section, we will present the few available data on mycotoxin analysis in processed meat. We will focus on mycotoxins that may be produced during the ripening period. The analytical methods that may be used to evaluate the residual contamination of meat as a raw material in food making have already been presented elsewhere [353]. Therefore, fusariotoxins (trichothecenes, ZEA, and fumonisines) will not be presented. Indeed, production of these molecules cannot be observed in processed meats due to environmental conditions required for *Fusarium* development and toxinogenesis (mainly water activity) [12].

4.4.1 Aflatoxin B1

Several studies have indicated that processed meats can be contaminated with toxigenic *Aspergillus flavus* strains, especially when products are processed in countries with a hot climate [349,350, 354–356]. Moreover, it has been demonstrated that the processing conditions during aging of hams may allow aflatoxin synthesis [357]. Therefore, it is of public health importance to evaluate the possible production of aflatoxin B1 during meat processing and aging. Few studies have been carried out, but all demonstrated that the frequency of contamination of processed meat with aflatoxin B1 was low, and that the level of toxin within meat was usually below 10 ng/g [354,356]. However, it is not clear whether aflatoxin B1 was produced during meat processing or was present before at the residual level in muscles. It seems there is no relationship between the presence of toxigenic strains of *A. flavus* and aflatoxin contamination of meat samples [354]. Moreover, the frequent contamination of spices and additives used in such meat processing may also represent a source of mycotoxin [356,358]. All these studies were performed using classic methods for aflatoxin B1 analysis (see 4.2.4.3), and no special treatment was applied to samples according to their composition or process-induced changes.

4.4.2 Ochratoxin A

Many methods were devoted to the OTA analysis in processed meat; the most recent ones are summarized in Table 4.6 (Refs 359–363). They have essentially been set up in pig products because this species appears to be the most sensitive and exposed to OTA. It appears that detection limits exhibited by HPLC-FL are sufficient to control meat products according to existing regulations. The use of IAC for cleanup allows the reduction of the LOQ below 1 ng/g. However, a 10-fold OTA fluorescence enhancement obtained by using the alkaline eluent in HPLC permitted the determination of a very low level of OTA in muscle without any column purification or a concentration step [363].

However, all of these surveys essentially demonstrated the possible carryover of OTA in processed meat. Indeed, even if ochratoxigenic molds have been isolated from such foods [348,364,365], it appears that ripening and aging conditions are not favorable to toxin production [9,351].

4.4.3 Citrinin

Although citrinin-producing fungal strains have been isolated from dry-cured meat products [349,366], and it has been demonstrated that citrinin production may occur on dry-cured meat [9,367], no data are available on citrinin content in meat products, despite that this toxin has been

Table 4.6 Recent Methods for OTA Determination in Processed Meat

Quantification	Tissue	Extraction	Clean Up	LOQ (ng/g)	Reference
Fluorimetry HPLC-FL	Ham	Methanol–1% sodium bicarbonate (70:30)	IAC	0.7[a] 0.04[a]	359
HPLC-FL	Ham	Chloroform-orthophsphoric acid Centrifugation	Back extraction with $NaHCO_3$, pH 7.5 IAC	0.03	360
HPLC-FL	Salami	Ethyl acetate (0.5 mol NaCl)–phosphoric acid	Back extraction with $NaHCO_3$, pH 8.0 IAC	0.2	361
HPLC-FL	Pig liver derived pâté	Acidified acetonitrile-water	C8 columns	0.84	362
HPLC-FL[b]	Dry cured pork meat	Chloroform-phosphoric acid	Back extraction with Tris-HCl pH 8.5 Addition of chloroform until 90:10 ratio	0.06	363

Note: a, limits of detection; b, mobile phase: $NH_3/NH_4Cl:CH_3-CN$ (85:15), pH 9.8 instead of acetonitrile-water-acetic acid (99:99:2) in others.

suspected to play a role in BEN [368] and is mutagenic [369]. However, stability studies have demonstrated that this mycotoxin is only partially stable in cured ham, as already demonstrated in other animal derived foods [9,370]. Nevertheless, it may be of interest to develop methods able to quantify a possible contamination of processed meat with citrinin.

4.4.4 Cyclopiazonic Acid

As for citrinin, no survey is available concerning CPA contamination of meat products. It has been demonstrated that CPA-producing strains could be isolated from processed meats [348,352,371]. Moreover, it has been shown that toxigenic strains of *Penicillium* were able to produce the toxin on meat products, and that the toxin was stable on that substrate, with more than 80% of the initial contamination still recoverable after 8 days of incubation [9]. These results suggest that an accumulation of a relatively high level of CPA could be observed on cured meat after contamination and development of toxigenic strains. Owing to cyclopiazonic toxicity and its suspected role in "Kodua poisoning" in humans [372,373], fungal strains used in meat processing should be tested for their ability to produce cyclopiazonic acid before use in commercial products. This recommendation is in agreement with previous one concerning the use of fungal starters in cheese [374]. The development of micellar capillary electrophoresis for the detection of toxigenic mold strains may represent a useful alternative to classical analysis [375]. It has already been applied to fungal strains isolated from cured meat and allowed multidetection of mycotoxins such as CPA and also aflatoxin B1 [376]. It appears also important to develop or adapt existing analytical methods to allow the final control of processed meats.

4.5 Conclusion

Mycotoxins are widely found contaminants of cereals and other vegetal products. When contaminated feeds are distributed to farm animals, mycotoxin may be found as residues in edible parts of the animals. Owing to their breeding and feeding conditions, poultry may often be exposed to such contamination, which has consequences for the safety of edible organs. For the most important toxins, the available data on absorption, distribution within animal organism, and metabolism revealed that mainly aflatoxins and OTA may be found at significant levels in muscles and muscle foods. For these molecules, sensitive and specific methods are required to allow safety control of poultry and processed meats, because levels of contamination are usually in the low ppb range. Most commonly used methodologies are based on HPLC-FL detection of molecules. Mycotoxin contamination of meat may also result from toxigenic mold development during ripening and aging. It may lead to production and accumulation of toxins such as citrinin or cyclopiazonic acid, for which few if any methods have been established for meat control. Even if the toxicity of such molecules appears less important than the previous ones, their possible implications in human diseases or syndromes should lead to the implementation of methods able to control contamination of processed meat.

References

1. Bennett, J.W. and Klich, M., Mycotoxins, *Clin. Microbiol. Rev.*, 16, 497, 2003.
2. Bhatnagar, D., Yu, J., and Ehrlich, K.C., Toxins of filamentous fungi, *Chem. Immunol.*, 81, 167, 2002.
3. Pitt, J.I., Biology and ecology of toxigenic penicillium species, *Adv. Exp. Med. Biol.*, 504, 29, 2002.

4. Conkova, E. et al., Fusarial toxins and their role in animal diseases, *Vet. J.*, 165, 214, 2003.
5. IARC, some naturally occuring substances, food items and constituents, heterocyclic aromatic amines and mycotoxins, in *Monographs on the Evaluation of Carcinogenic Risks to Humans*, World health organization, Lyon, France, 1993, 56, 245.
6. Leblanc, J.C., Etude de l'alimentation totale en France: mycotoxines, mineraux et aliments traces, INRA (ed.) Paris, France, 2004.
7. SCOOP reports on Tasks 3.2.7. Assessment of dietary intake of Ochratoxin A by the population of EU members states, 2000.
8. Schothorst, R.C. and Van Egmond, H.P., Report from SCOOP task 3.2.10 "collection of *Fusarium* toxins in food and assessment of dietary intake by the population of EU member states"; subtask: trichothecenes, *Toxicol. Lett.*, 153, 133, 2004.
9. Bailly, J.D. et al., Production and stability of patulin, ochratoxin A, citrinin and cyclopiazonic acid on dry cured ham, *J. Food Prot.*, 68, 1516, 2005.
10. European Union, Commission Regulation (EC) N° 466/2001 setting maximum levels for certain contaminants in foodstuffs, *Off. J. Eur. Un.*, L77, 1, 2001.
11. European Union, Commission Regulation n° 856/2005, toxins of *Fusarium*, *Off. J. Eur. Un.*, L 143, 3, 2005.
12. Miller, J.D., Aspects of the ecology of *Fusarium* toxins in cereals. *Adv. Exp. Med. Biol.*, 504, 19, 2002.
13. Pitt, J.L., *Laboratory Guide to Common Penicillium Species*. Academic Press, London, 1988.
14. Rapper, K.B. and Fennell, D.I., *The Genus Aspergillus*. Williams & Wilkins, Baltimore, MD, 1965.
15. Pan, D. et al., Deoxynivalenol in barley samples from Uruguay, *Int. J. Food Microbiol.*, 114, 149, 2007.
16. Trucksess, M.W. et al., Determination and survey of deoxynivalenol in white flour, whole wheat flour, and bran, *J. AOAC Int.*, 79, 883, 1996.
17. Li, F.Q. et al., *Fusarium* toxins in wheat from an area in Henan Province, PR China, with a previous human red mould intoxication episode, *Food Addit. Contam.*, 19, 163, 2002.
18. Tanaka, T. et al., Worldwide contamination of cereals by the *Fusarium* mycotoxins nivalenol, deoxynivalenol, and zearalenone. I. Survey of 19 countries, *J. Agric. Food Chem.*, 36, 979, 1988.
19. Rocha, O., Ansari, K., and Doohan, F.M., Effects of trichothecene mycotoxins on eukaryotic cells: a review, *Food Add. Contam.*, 22, 369, 2005.
20. Pestka, J.J. and Smolinski, A.T., Deoxynivalenol: toxicology and potential effects on humans, *J. Toxicol. Environ. Health B Crit. Rev.*, 8, 39, 2005.
21. Rotter, B.A., Prelusky, D.B., and Pestka, J.J., Toxicology of deoxynivalenol (vomitoxin), *J. Toxicol. Environ. Health*, 48, 1, 1996.
22. Pieters, M.N. et al., Risk assessment of deoxynivalenol in food: concentration limits, exposure and effects, *Adv. Exp. Med. Biol.*, 504, 235, 2002.
23. Trenholm, H.L. et al., Lethal toxicity and nonspecific effects, in *Trichothecene Mycotoxicosis: Pathophysiologic Effects*, Beasley, V.L., Ed., Vol. 1, CRC press, Boca Raton, FL, 1989, p. 107.
24. Joffe, A.Z., *Fusarium poae* and *Fusarium sporotrichioides* as a principal causal agents of alimentary toxic aleukia, in *Mycotoxic Fungi, Mycotoxins, Mycotoxicoses*, Wyllie, T.D. and Morehouse, L.G., Eds., Vol. 3, Marcel Dekker, New York, 1978, p. 21.
25. Lutsky, I.N. et al., The role of T-2 toxin in experimental alimentary toxic aleukia: a toxicity study in cats, *Toxicol. Appl. Pharmacol.*, 43, 111, 1978.
26. Desjardin, A.E., Mc Cormick, S.P., and Appell, M., Struture activity of thrichothecene toxins in an *Arabidopsis thaliana* leaf assay. *J. Agric. Food Chem.*, 55, 6487, 2007.
27. Balzer, A. et al., Les trichothécènes: nature des toxines, présence dans les aliments et moyens de lutte, *Revue Med. Vet.*, 155, 299, 2004.
28. Hazel, C.M. and Patel, S., Influence of processing on trichothecene levels, *Toxicol. Lett.*, 153, 51, 2004.
29. Krska, R., Baumgartner, S., and Josephs, R., The state-of-the-art in the analysis of type-A and -B trichothecene mycotoxins in cereals, *Fresenius J. Anal. Chem.*, 371, 285, 2001.

30. AOAC. 986-18. Deoxynivalenol in wheat, 1995.

31. Josephs, R.D. et al., Trichothecenes: reference materials and method validation, *Toxicol. Lett.*, 153, 123, 2004.

32. Yoshizawa, T. et al., A practical method for measuring deoxynivalenol, nivalenol, and T-2 + HT-2 toxin in foods by an enzyme-linked immunosorbent assay using monoclonal antibodies, *Biosci. Biotechnol. Biochem.*, 68, 2076, 2004.

33. Koch, P., State of the art of trichothecenes analysis, *Toxicol. Lett.*, 153, 109, 2004.

34. Pascale, M., Haidukowski, M., and Visconti, A., Determination of T-2 toxin in cereal grains by liquid chromatography with fluorescence detection after immunoaffinity column clean-up and derivatization with 1-anthroylnitrile, *J. Chromatogr. A*, 989, 257, 2003.

35. Trenholm, H.L., Warner, R.M., and Prelusky, D.B., Assessment of extraction procedures in the analysis of naturally contamined grain products deoxynivalenol (vomitoxin), *J. Assoc. Off. Anal. Chem.*, 68, 645, 1985.

36. Scott, P.M. and Kanhere, S.R., Comparison of column phases for separation of derivatised trichothecenes by capillary gas chromatography, *J. Chromatogr.*, 368, 374, 1986.

37. Romer, T.R., Use of small charcoal/alumina cleanup columns in determination of trichothecene mycotoxins in foods and feeds, *J. Assoc. Off. Anal. Chem.*, 69, 699, 1986.

38. Betina, V., Thin layer chromatography of mycotoxins, *J. Chromatogr.*, 334, 211, 1985.

39. Morgan, M.R.A., Mycotoxin immunoassays: with special reference to ELISAs, *Tetrahedron*, 45, 2237, 1989.

40. Schneider, E. et al., Rapid methods for deoxynivalenol and other trichothecenes, *Toxicol. Lett.*, 153, 113, 2004.

41. Park, J.C., Zong, M.S., and Chang, I.M., Survey of the presence of the *Fusarium* myxotoxins nivalenol, deoxynivalenol an T-2 toxin in Korean cereals of the 1989 harvest, *Food Addit. Contam.*, 8, 447, 1991.

42. Ryu, J.C. et al., Survey of natural occurrence of trichothecene mycotoxins and zearalenone in Korean cereals harvested in 1992 using gas chromatography mass spectrometry, *Food Addit. Contam.*, 13, 333, 1996.

43. Langseth, W. and Rundberget, T., Instrumental methods for determination of non-macrocyclic trichothecenes in cereals, foodstuffs and cultures, *J. Chromatogr. A*, 815, 103, 1998.

44. Klötzel, M. et al., Determination of 12 type A and B trichothecenes in cereals by liquid chromatography-electrospray ionization tandem mass spectrometry, *J. Agric. Food Chem.*, 53, 8904, 2005.

45. Zölner, P. and Mayer-Helm, B., Trace mycotoxin analysis in complex biological and food matrices by liquid chromatography-atmospheric pressure ionisation mass spectrometry, *J. Chromatogr. A.*, 1136, 123, 2006.

46. Molto, G.A. et al., Production of trichothecenes and zearalenone by isolates of *Fusarium* spp. From Argentinian maize, *Food Addit. Contam.*, 14, 263, 1997.

47. Sydenham, E.W. et al., Production of mycotoxins by selected *Fusarium graminearum* and *F. crookwellense* isolates, *Food Addit. Contam.*, 8, 31, 1991.

48. Stopper, H., Schmitt, E., and Kobras, K., Genotoxicity of phytoestrogens, *Mutat. Res.*, 574, 139, 2005.

49. Kuiper-Goodman, T., Scott, P.M., and Watanabe, H., Risk assessment of the mycotoxin zearalenone, *Regul. Toxicol. Pharm.*, 7, 253, 1987.

50. JECFA, 53rd Report. Safety evaluation of certain food additives. WHO Food Additives Series 44, 2000.

51. Malekinejad, H., Colenbrander, B., and Fink-Gremmels, J., Hydroxysteroid dehydrogenases in bovine and porcine granulosa cells convert zearalenone into its hydroxylated metabolites alpha-zearalenol and beta-zearalenol, *Vet. Res. Commun.*, 30, 445, 2006.

52. Zinedine, A. et al., Review on the toxicity, occurrence, metabolism, detoxification, regulations and intake of zearalenone: an oestrogenic mycotoxin, *Food Chem. Toxicol.*, 45, 1, 2007.

53. Hidy, P.H. et al., Zearalenone and some derivates: production and biological activities, *Adv. Appl. Microbiol.*, 22, 59, 1977.

54. Gaumy, J.L. et al., Zearalénone: propriétés et toxicité expérimentale, *Revue Med. Vét.*, 152, 219, 2001.

55. Krska, R., Performance of modern sample preparation techniques in the analysis of *Fusarium* mycotoxins in cereals, *J. Chromatogr. A.*, 815, 49, 1998.

56. Krska, R. and Josephs, R., The state-of-the-art in the analysis of estrogenic mycotoxins in cereals, *Fresenius' J. Anal. Chem.*, 369, 469, 2001.

57. Pallaroni, L. et al., Microwave-assisted extraction of zearalenone from wheat and corn, *Anal. Bioanal. Chem.*, 374, 161, 2002.

58. Pallaroni, L. and Von Holst, C., Comparison of alternative and conventional extraction techniques for the determination of zearalenone in corn, *Anal. Bioanal. Chem.*, 376, 908, 2003.

59. Zöllner, P. et al., Concentration levels of zearalenone and its metabolites in urine, muscle tissue, and liver samples of pigs fed with mycotoxin-contamined oats, *J. Agric. Food Chem.*, 50, 2494, 2002.

60. Plasencia, J. and Mirocha, C.J., Isolation and characterization of zearalenone sulfate produced by *Fusarium* spp., *Appl. Environ. Microbiol.*, 57, 146, 1991.

61. Garels, M. et al., Cleavage of zearalenone-glycoside, a "masked" mycotoxin, during digestion in swine, *Zentralbl Veterinarmed B.*, 37, 236, 1990.

62. Llorens, A. et al., Comparison of extraction and clean-up procedures for analysis of zearalenone in corn, rice and wheat grains by high-performance liquid chromatography with photodiode array and fluorescence detection, *Food Addit. Contam.*, 19, 272, 2002.

63. Silva, C.M. and Vargas, E.A., A survey of zearalenone in corn using Romer Mycosep 224 column and high performance liquid chromatography, *Food Addit. Contam.*, 18, 39, 2001.

64. De Saeger, S., Sibanda, L., and Van Peteghem, C., Analysis of zearalenone and α-zearalenol in animal feed using high-performance liquid chromatography, *Anal. Chim. Acta*, 487, 137, 2003.

65. Eskola, M., Kokkonen, M., and Rizzo A., Application of manual and automated systems for purification of ochratoxin A and zearalenone in cereals with immunoaffinity columns, *J. Agric. Food Chem.*, 50, 41, 2002.

66. Fazekas, B. and Tar, A., Determination of zearalenone content in cereals and feedstuffs by immunoaffinity column coupled with liquid chromatography, *J. AOAC Int.*, 5, 1453, 2001.

67. Kruger, S.C. et al., Rapid immunoaffinity-based method for determination of zearalenone in corn by fluorometry and liquid chromatography, *J. AOAC Int.*, 82, 1364, 1999.

68. Zöllner, P., Jodlbauer, J., and Lindner, W., Determination of zearalenone in grains by high-performance liquid chromatography-tandem mass spectrometry after solid-phase extraction with RP-18 columns or immunoaffinity columns, *J. Chromatogr.*, 858, 167, 1999.

69. Visconti, A. and Pascale, M., Determination of zearalenone in corn by means of immunoaffinity clean-up and high-performance liquid chromatography with fluorescence detection, *J. Chromatogr.*, 815, 133, 1998.

70. Rosenberg, E. et al., High-performance liquid chromatography-atmospheric-pressure chemical ionization mass spectrometry as a new tool for the determination of the mycotoxin zearalenone in food and feed, *J. Chromatogr.*, 819, 277, 1998.

71. Meyer, K. et al., Zearalenone metabolites in bovine bile, *Archiv. Für Lebensmittelhygiene*, 53, 115, 2002.

72. Pichler, H. et al., An enzyme-immunoassay for the detection of the mycotoxin zearalenone by use of yolk antibodies, *Fresenius' J. Anal. Chem.*, 362, 176, 1998.

73. Josephs, R.D., Schuhmacher, R., and Krska, R., International interlaboratory study fort the *Fusarium* mycotoxins zearalenone and deoxynivalenol in agricultural commodities, *Food Addit. Contam.*, 18, 417, 2001.

74. Lee, M.G. et al., Enzyme-linked immunosorbent assays of zearalenone using polyclonal, monoclonal and recombinant antibodies, *Methods Mol. Biol.*, 157, 159, 2001.

75. Bennet, G.A., Nelsen, T.C., and Miller, B.M., Enzyme-linked immunosorbent assay for detection of zearalenone in corn, wheat, and pig feed: collaborative study, *J. AOAC Int.*, 77, 1500, 1994.

76. Maragos, C.M. and Kim, E.K., Detection of zearalenone and related metabolites by fluorescence polarization immunoassay, *J. Food Prot.*, 67, 1039, 2004.

77. Dawlatana, M. et al., An HPTLC method for the quantitative determination of zearalenone in maize, *Chromatographia*, 47, 217, 1998.

78. De Oliveira Santos Cazenave, S. and Flavio Midio, A., A simplified method for the determination of zearalenone in corn-flour, *Alimentaria*, 298, 27, 1998.

79. Weiss, R. et al., Improving methods of analysis for mycotoxins: molecularly imprinted polymers for deoxynivalenol and zearalenone, *Food Addit. Contam.*, 20, 386, 2003.

80. Ware, G.M. et al., Preparative method for isolating α-zearalenol and zearalenone using extracting disk, *J. AOAC Int.*, 82, 90, 1999.

81. Tanaka, T. et al., Simultaneous determination of trichothecene mycotoxins and zearalenone in cereals by gas chromatography-mass spectrometry, *J. Chromatogr.*, 882, 23, 2000.

82. Pallaroni, L., Björklund, E., and Von Holst, C., Optimization of atmospheric pressure chemical ionization interface parameters for the simultaneous determination of deoxynivalenol and zearalenone using HPLC/MS, *J. Liq Chromatogr. Relat. Technol.*, 25, 913, 2002.

83. Jodlbauer, J., Zöllner, P., and Lindner, W., Determination of zeranol, taleranol, zearalenone, α- and β-zearalenol in urine and tissue by high-performance liquid chromatography-tandem mass spectrometry, *Chromatographia*, 51, 681, 2000.

84. Zöllner, P. et al., Determination of zearalenone and its metabolites a- and b-zearalenol in beer samples by high-performance liquid chromatography-tandem mass spectrometry, *J. Chromatogr. B—Biomed. Sci. Appl.*, 738, 233, 2000.

85. Kleinova, M. et al., Metabolic profiles of the zearalenone and of the growth promoter zeranol in urine, liver, and muscle of heifers, *J. Agric. Food Chem.*, 50, 4769, 2002.

86. Schuhmacher, R. et al., Interlaboratory comparison study for the determination of the *Fusarium* mycotoxins deoxynivalenol in wheat and zearalenone in maize using different methods, *Fresenius' J. Anal. Chem.*, 359, 510, 1997.

87. Bezuidenhout, S.C. et al., Structure elucidation of the fumonisins, mycotoxins from *Fusarium moniliforme*, *J. Chem. Commun.*, 1988, 743, 1988.

88. Gelderblom, W.C. et al., Fumonisin-novel mycotoxins with cancer-promoting activity produced by *Fusarium moniliforme*, *Appl. Environ. Microbiol.*, 54, 1806, 1988.

89. Marasas, W.F.O. et al., Fumonisins-occurrence, toxicology, metabolism and risk assessment, in *Fusarium*, Summerell, B.A. et al., Eds., APS press, St Paul, MN, 2001, p. 332.

90. Marin, S. et al., Fumonisin-producing strains of *Fusarium*: a review of their ecophysiology, *J. Food Prot.*, 67, 1792, 2004.

91. Conseil Superieur d'Hygiène Publique de France (CSHPF), Les mycotoxines dans l'alimentation: évaluation et gestion du risque, Tec & Doc (ed.), Paris, France, 1999.

92. Bailly, J.D. et al., Leuco-encéphalomalacie des _tudes; cas rapportés au CNITV, *Rev. Méd. Vét.*, 147, 787, 1996.

93. Marasas, W. et al., Leucoencephalomalacia in a horse induced by fumonisine B1 isolated from *Fusarium moniliforme*, *Onderstepoort J. Vet. Res.*, 55, 197, 1988.

94. Harrison, L.R. et al., Pulmonary edema and hydrothorax in swine produced by fumonisin B1 a toxic metabolite of *Fusarium moniliforme*, *J. Vet. Diagn. Invest.*, 2, 217, 1990.

95. Harvey, R.B. et al., Effects of dietary fumonisin B1-containing culture material, deoxynivalenol-contaminated wheat, or their combination on growing barrows, *Am. J. Vet. Res.*, 57, 1790, 1996.

96. Oswald, I.P. et al., Mycotoxin fumonisin B1 increases intestinal colonization by pathogenic *Escherichia coli* in pigs, *Appl. Environ. Microbiol.*, 69, 5870, 2003.

97. Diaz, D.E. et al., Effect of fumonisin on lactating dairy cattle, *J. Dairy Sci.*, 83, 1171, 2000.

98. Osweiller, G.D. et al., Effects of fumonisin-contaminated corn screenings on growth and health of feeder calves, *J. Anim. Sci.*, 71, 459, 1993.

99. Bermudez, A.J., Ledoux, D.R., and Rottinghaus, G.E., Effects of *Fusarium moniliforme* culture material containing known levels of fumonisin B1 in ducklings, *Avian Dis.*, 39, 879, 1995.

100. Brown, T.P., Rottinghaus, G.E., and Williams, M.E., Fumonisin mycotoxicosis in broilers: performance and pathology, *Avian Dis.*, 36, 450. 1992.
101. Bailly, J.D. et al., Toxicity of *Fusarium moniliforme* culture material containing known levels of fumonisin B1 in ducks, *Toxicology*, 11, 22, 2001.
102. Ledoux, D. R. et al., Fumonisin toxicity in broiler chicks, *J. Vet. Diagn. Invest.*, 4, 330, 1992.
103. Gelderblom, W.C.A. et al., The cancer initiating potential of the fumonisin B1 mycotoxins, *Carcinogenesis*, 13, 433, 1992.
104. Gelderblom, W.C. et al., Fumonisin-induced hepatocarcinogenesis: mechanisms related to cancer initiation and promotion, *Environ. Health Perspect.*, 109, 291, 2001.
105. Marasas, W.F., Fumonisins: their implications for human and animal health, *Nat. Toxins*, 3, 193, 1995.
106. Hendricks, K., Fumonisins and neural tube defects in south Texas, *Epidiomology*, 10, 198, 1999.
107. Hendricks, K.A., Simpson, J.C., and Larsen, R.D., Neural tube defects along the Texas–Mexico border, 1993–1995, *Am. J. Epidemiol.*, 149L, 1119, 1999.
108. Missmer, S. et al., Fumonisins and neural tube defects, *Epidemiology*, 11, 183, 2000.
109. Marasas, W.F. et al., Fumonisins disrupt sphingolipid metabolism, folate transport, and neural tube development in embryo culture and in vivo: a potential risk factor for human neural tube defects among populations consuming fumonisin-contaminated maize, *J. Nutr.*, 134, 711, 2004.
110. Merrill, A.H. Jr. et al., Spingholipid metabolism: roles in signal tranduction and disruption by fumonisins, *Environ. Health Perspect.*, 109, 283, 2001.
111. Tran, S.T. et al., Sphinganine to sphingosine ratio and redicatieve biochemical markers of fumonisin B1 exposure in ducks, *Chem. Biol. Interact.*, 2003, 61, 2003.
112. Goel, S. et al., Effects of *Fusarium moniliforme* isolates on tissue and serum sphingolipid concentrations in horses, *Vet. Hum. Toxicol.*, 38, 265, 1996.
113. Garren, L. et al., The induction and persistence of altered sphingolipid biosynthesis in rats treated with fumonisin B1, *Food Addit. Contam.*, 18, 850, 2001.
114. Van der Westhuizen, L., Shephard, G.S., and Van Schalkwyk, D.J., The effect of repeated gavage doses of fumonisine B1 on the sphinganine and sphingosine levels in vervet monkeys, *Toxicon*, 39, 969, 2001.
115. Dupuy, J. et al., Thermostability of fumonisin B1 a mycotoxin from *Fusarium moniliforme*, in corn, *Appl. Environ. Microbiol.*, 59, 2864, 1993.
116. Castells, M. et al., Fate of mmycotoxins in cereals during extrusion cooking: a review, *Food Addit. Contam.*, 22, 150, 2005.
117. AOAC. 995.15. Fumonisins B1, B2 and B3 in corn, 2000
118. Sydenham, E.W. et al., Liquid chromatographic determination of fumonisins B1, B2 and B3 in corn: AOAC-IUPAC collaborative study, *J. AOAC Int.*, 79, 688, 1996.
119. AFNOR, NF. EN., 13585. Produits alimentaires: dosage des fumonisines B1 et B2 dans le maïs. Méthode CHLP avec purification par extraction en phase solide, 2002.
120. Rice, L.G. et al., Evaluation of a liquid chromatographic method for the determination of fumonisins in corn, poultry feed, and *Fusarium* culture material, *J. AOAC Int.*, 78, 1002, 1995.
121. De Girolamo, A. et al., Comparison of different extraction and clean-up for the determination of fumonisins in maize-based food products, *Food Addit. Contam.*, 18, 59, 2001.
122. Visconti, A. et al., European intercomparison study for the determination of the fumonisins content in two maize materials, *Food Addit. Contam.*, 13, 909, 1996.
123. Shephard, G.S., Chromatographic determination of the fumonisin mycotoxins, *J. Chromatogr. A*, 815, 31, 1998.
124. Dilkin, P. et al., Robotic automated clean-up for detection of fumonisins B1 and B2 in corn-based feed by high-performance liquid chromatography, *J. Chromatogr. A*, 925, 151, 2001.
125. Bailly, J.D. et al., Production and purification of fumonisins from a highly toxigenic *Fusarium verticilloides* strain, *Rev. Med. Vet.*, 156, 547, 2005.
126. Shephard, G.S. and Sewram, V., Determination of the mycotoxin fumonisin B1 in maize by reversed-phase thin layer chromatography: a collaborative study, *Food Addit. Contam.*, 21, 498, 2004.

127. Preis, R.A. and Vargas, E.A., A method for determining fumonisin B1 in corn using immunoaffinity column clean-up and thin layer chromatography/densitometry, *Food Addit. Contam.*, 17, 463, 2000.
128. Dorner, J.W., Mycotoxins in food: method of analysis, in *Handbook of Food Analysis*, Nollet, L.M., Ed., Vol. 2, CRC press, New York, 1996, p. 1089.
129. Shephard, G.S. et al., Quantitative determination of fumonisins B1 and B2 by high-performance liquid chromatography with fluorescence detection, *J. Liq. Chromatogr.*, 13, 2077, 1990.
130. Plattner, R. et al., Analysis of corn and cultured corn for fumonisin B1 by HPLC and GC/MS by four laboratories, *J. Vet. Diagn. Invest.*, 3, 357, 1991.
131. Jackson, M.A. and Bennett, G.A., Production of fumonisin B1 by *Fusarium moniliforme* NRRL 13616 in submerged culture, *Appl. Environ. Microbiol.*, 56, 2296, 1990.
132. Wang, S. et al., Rapid determination of fumonisin B1 in food samples by enzyme-linked immunosorbent assay and colloidal gold immunoassay, *J. Agric. Food Chem.*, 54, 2491, 2006.
133. Barna-Vetro, I. et al., Development of a sensitive ELISA for the determination of fumonisine B1 in cereals, *J. Agric. Food Chem.*, 48, 2821, 2000.
134. Bird, C.B. et al., Determination of total fumonisins in corn by competitive direct enzyme-linked immunosorbent assay: collaborative study, *J. AOAC Int.*, 85, 404, 2002.
135. Kulisek, E.S. and Hazebroek, J.P., Comparison of extraction buffers for the detection of fumonisin B1 in corn by immunoassay and high-performance liquid chromatography, *J. Agric. Food Chem.*, 48, 65, 2000.
136. AOAC. 2001.06. Determination of total fumonisins in corn, 2001.
137. Nesbitt, B.F. et al., *Aspergillus flavus* and turkey X disease. Toxic metabolites of *Aspergillus flavus*, *Nature*, 195, 1062, 1962.
138. De Iongh Berthuis, R.K. et al., Investigation of the factor in groundnut meal responsible for "turkey X disease", 65, 548, 1962.
139. Asao, T. et al., Aflatoxins B and G, *J. Am. Chem. Soc.*, 85, 1706, 1963.
140. Klich, M.A. and Pitt, J.I., Differentiation of *Aspergillus flavus* from *Aspergillus parasiticus* and other closely related species, *Trans. Brit. Mycol. Soc.*, 91, 99, 1968.
141. Detroy, R.W., Lillehoj, E.B., and Ciegler, A., Aflatoxin and related compounds, in *Microbial Toxins Vol VI: Fungal Toxins*, Ciegler, A., Kadis, S., and Ajl, S.J., Eds., Academic press, New York, 1971, p. 3.
142. Diener, U.L. et al., Epidemiology of aflatoxin formation by *Aspergillus flavus*, *Ann. Rev. Phytopathol.*, 25, 249, 1987.
143. Senyuva, H.Z., Gilbert, J., and Ulken, U., Aflatoxins in Turkish dried figs intended for export to the European Union, *J. Food Prot.*, 70, 1029, 2007.
144. Zinedine, A. et al., Limited survey for the occurrence of aflatoxins in cereal and poultry feeds from Rabat, Morocco, *Int. J. Food Microbiol.*, 115, 124, 2007.
145. Toteja, G.S. et al., Aflatoxin B1 contamination in wheat grain samples collected from different geographical regions of India: a multicenter study, *J. Food Prot.*, 69, 1463, 2006.
146. Fazekas, B., Tar, A., and Kovacs, M., Aflatoxin and ochratoxin A content of spices in Hungary, *Food Addit. Contam.*, 22, 856, 2005.
147. Kaaya, A.N. and Kyamuhangire, W., The effect of storage time and agroecological zone on mould incidence and aflatoxin contamination of maize from traders in Uganda, *Int. J. Food Microbiol.*, 110, 217, 2006.
148. Thompson, C. and Henke, S., Effect of climate and type of storage container on aflatoxin production in corn and its associated risks to wildlife species, *J. Wild. Dis.*, 36, 172, 2000.
149. Trenk, H.L. and Hartman, P.A., Effect of moisture content and temperature on aflatoxin production in corn, *Appl. Microbiol.*, 19, 781, 1970.
150. Northolt, M.D. and van Egmond, H.P., Limits of water activity and temperature for the production of some mycotoxins, *4th Meeting Mycotoxins in Animal Disease*, Weybridge, UR, 106, 1981.
151. Sanchis, V. and Magan, N., Environmental conditions affecting mycotoxins, in *Mycotoxins in Food: Detection and Control*, Magan, N. and Olsen, M., Eds., Woodhead Publishing Ltd, Oxford, 2004, p. 174.

152. Giorni, P. et al., Studies on *Aspergillus* section *Flavi* isolated from maize in northern Ital, *Int. J. Food Microbiol.*, 113, 330, 2007.
153. Battilani, P. et al., Monitoraggio della contaminazione da micotossine in mais, *Infor. Agro.*, 61, 47, 2005.
154. Newberne, P.M. and Butler, W.H., Acute and chronic effect of aflatoxin B1 on the liver of domestic animals: a review, *Cancer Res.*, 29, 236, 1969.
155. Peers, F.G. and Linsell, M.P., Dietary aflatoxins and human liver cancer—a population study based in Kenya, *Brit. J. Cancer*, 27, 473, 1973.
156. Shank, R.C. et al., Dietary aflatoxin and human liver cancer IV. Incidence of primary liver cancer in two municipal population in Thailand, *Food Cosmetol. Toxicol.*, 10, 171, 1982.
157. JECFA, Evaluation of certain food additives and contaminants. Forty-nine report. WHO Technical Report Series, Geneva, 40, 1999.
158. Cullen J.M. and Newberne, P.M., Acute hepatotoxicity of aflatoxins, in *Toxicol. Aflatoxins*, Eaton, D.L. and Groopman J.D., Eds., Academic press, San Diego, CA, 1994, p. 3.
159. Eaton, D.L. and Ramsdel, H.S., Species and related differences in aflatoxin biotransformation, in *Handbook of Applied Mycology, Vol 5: Mycotoxins in Ecological Systems*, Bhatnagar, D., Lillehoj, E.B., and Arora, D.K., Eds., Marcel Dekker, New York, 1992, p. 157.
160. Gallagher, E.P. and Eaton, D.L., In vitro biotransformation of aflatoxin B1 (AFB1) in channel catfish liver, *Toxicol. Appl. Pharmacol.*, 132, 82, 1995.
161. JECFA, Aflatoxin M1. Fifty-six report. WHO Technical Report Series, Geneva, 47, 2001.
162. Asao, T. et al., Structures of aflatoxins B_1 and G_1, *J. Am. Chem. Soc.*, 87, 822, 1965.
163. Van Egmond, H.P., Heisterkamp, S.H., and Paulsch, W.E., EC collaborative study on the determination of aflatoxin B1 in animal feeding stuffs, *Food Addit. Contam.*, 8, 17, 1991.
164. Otta, K.H. et al., Determination of aflatoxins in corn by use the personal OPLC basic system, *JPC—J. Planar Chromatogr—Modern TLC*, 11, 370, 1998.
165. Papp, E. et al., Liquid chromatographic determination of aflatoxins, *Microchem. J.*, 73, 39, 2002.
166. Vega, V.A., Rapid extraction of aflatoxin from creamy and crunchy peanut butter, *J. Assoc. Anal. Chem.*, 88, 1383, 2005.
167. Truckess, M.W., Brumley, W.C., and Nesheim, S., Rapid quantification and confirmation of aflatoxins in corn and peanut butter, using a disposable silica gel column, thin layer chromatography and gas chromatography/mass spectrometry, *J. Assoc. Off. Anal. Chem.*, 67, 973, 1984.
168. AOAC. Official methods of analysis, 17th edition, 49, 20, 1995.
169. Truckess, M.W. et al., Multifonctional column coupled with liquid chromatography for determination of aflatoxins B1, B2, G1 and G2 in corn, almonds, brazil nuts, peanuts and pistachio nuts: collaborative study, *J. Assoc. Off. Anal. Chem.*, 77, 1512, 1994.
170. Otta, K.H., Papp, E., and Bagocsi, B., Determination of aflatoxins in foods by overpressured-layer chromatography, *J. Chromatogr. A.*, 882, 11, 2000.
171. Takeda, Y. et al., Simultaneous extraction and fractionation and thin layer chromatographic determination of 14 mycotoxins, *J. Assoc. Off. Anal. Chem.*, 62, 573, 1979.
172. Van Egmond, H.P., Paulsch, W.E., and Sizoo, E.A., Comparison of six methods of analysis for the determination of aflatoxin B1 in feeding stuffs containing citrus pulp, *Food Addit. Contam.*, 5, 321, 1988.
173. Truckess, M.W. et al., Immunoaffinity column coupled with solution fluorimetry or liquid chromatography postcolumn derivatization for determination of aflatoxins in corn, peanuts and peanut butter: collaborative study, *J. Assoc. Anal. Chem.*, 74, 81, 1991.
174. Grosso, F. et al., Joint IDF-IUPAC-IAEA(FAO) interlaboratory validation for determining aflatoxin M1 in milk by using immunoaffinity clean up before thin layer chromatography, *Food Addit. Contam.*, 21, 348, 2004.
175. AOAC. 990.33. Aflatoxin determination in corn and peanut butter, 1995.
176. ISO. 14718. Aflatoxin B1 determination in compound feeds with HPLC, 1998.

177. AOAC. 2003.2. Determination of aflatoxin in animal feed by IAC clean up and HPLC with post column derivatization, 2003.

178. AOAC. 991.31. Aflatoxins determination in corn, raw peanut and peanut butter, 1994.

179. AOAC. 999.07. Aflatoxin B1 and total aflatoxins in peanut butter, pistachio paste, fig paste and paprika powder, 1999.

180. AOAC. 2000.16. Aflatoxin B1 in baby food, 2000.

181. AOAC. 2000.08. Aflatoxin M1 in milk, 2000.

182. ISO. 14501. Milk and powder milk. Determination of aflatoxin M1. IAC Purification and HPLC determination, 1998.

183. ISO. 14674. Milk and milk powder. Aflatoxin M1 determination. Clean up with immunoaffinity chromatography and determination by thin layer chromatography, 2005.

184. Park, D.L. et al., Solvent efficient thin layer chromatographic method for the determination of aflatoxins B1, B2, G1 and G2 in corn and peanut products: collaborative study, *J. Assoc. Anal. Chem.*, 77, 637, 1994.

185. AOAC. 980.21. Aflatoxin M1 in milk and cheese, 1990.

186. International Organisation for standardization. ISO 14675:2005 Norm. Milk and milk powder: determination of aflatoxin M1 content: clean-up by immunoaffinity chromatography and determination by thin layer chromatography, 2005.

187. International Organisation for standardization. ISO 6651:2001 norm. Animal feeding stuffs: semi quantitative determination of aflatoxin B1: thin layer chromatographic methods, 2001.

188. Park, D.L. et al., Liquid chromatographic method for the determination of aflatoxins B1, B2, G1 and G2 in corn and peanut products: collaborative study, *J. Assoc. Anal. Chem.*, 73, 260, 1990.

189. AOAC. 990.33. Aflatoxins determination in corn and peanut butter, 1995.

190. International Organisation for standardization. ISO 14718:1998 norm. Animal feeding stuffs: determination of aflatoxin B1 content of mixed feeding stuffs: method using high performance liquid chromatography, 1998.

191. International Organisation for standardization. ISO/FDIS 14501 norm. Milk and milk powder: determination of aflatoxin M1 content: clean up by immunoaffinity chromatography and determination by high performance liquid chromatography, 2007.

192. Truckess, M.W. et al., Enzyme linked immunosorbentt assay of aflatoxins B1, B2, G1 and G2 in corn, cottonseed, peanuts, peanut butter and poultry feed: collaborative study, *J. Assoc. Anal. Chem.*, 72, 957, 1989.

193. Zheng, Z. et al., Validation of an ELISA test kit for the detection of total aflatoxins in grain and grain products by comparison with HPLC, *Mycopathologia*, 159, 255, 2005.

194. Chu, F.S. et al., Improved enzyme linked immunosorbent assay for aflatoxin B1 in agricultural commodities, *J. Assoc. Anal. Chem.*, 70, 854, 1987.

195. Chu, F.S. et al., Evaluation of enzyme linked immunosorbent assay of clean up for thin layer chromatography of aflatoxin B1 in corn, peanuts and peanut butter, *J. Assoc. Anal. Chem.*, 71, 953, 1988.

196. Park, D.L. et al., Enzyme linked immunosorbent assay for screening aflatoxin B1 in cottonseed products and mixed feed: collaborative study, *J. Assoc. Anal. Chem.*, 72, 326, 1989.

197. Kolosova, A.Y. et al., Direct competitive ELISA based on monoclonal antibody for detection of aflatoxin B1; Stabilisation of ELISA kit component and application to grain samples, *Anal. Bioanal. Chem.*, 384, 286, 2006.

198. Thirumala-Devi, T. et al, Development and application of an indirect competitive enzyme-linked immunoassay for aflatoxin m(1) in milk and milk-based confectionery, *J. Agric. Food Chem.*, 50, 933, 2002.

199. Fremy, J.M. and Chu, F.S., Immunochemical methods of analysis for aflatoxin M1, in *Mycotoxin in Dairy Products*, Van Egmond, E., Ed., Elsevier Science, London, 1989, p. 97.

200. Fremy, J.M. and Usleber, E., Policy on the characterization of antibodies used in immunochemical methods of analysis for mycotoxins and phycotoxins, *J. AOAC*, 86 (4), 868, 2003.

201. Blesa, J. et al., Determination of aflatoxins in peanuts by matrix solid hase dispersion and liquid chromatography, *J. Chromatogr.*, A, 1011, 49, 2003.
202. Sorensen, L.K.and Elbaek, T.H., Determination of mycotoxins in bovine milk by liquid chromatography tandem mass spectrometry, *J. Chromatogr.*, B, 820, 183, 2005.
203. Kokkonen, M., Jestoi, M., and Rizzo, A., Determination of selected mycotoxins in mould cheeses with liquid chromatography coupled tandem with mass spectrometry, *Food Addit. Contam.*, 22, 449, 2005.
204. Takino, M. et al., Atmopheric pressure photoionisation liquid chromatography/mass spectrometric determination of aflatoxins in food, *Food Addit. Contam.*, 21, 76, 2004.
205. Cavaliere, C. et al., Liquid chromatography/mass spectrometric confirmatory method for determinig aflatoxin M1 in cow milk: comparison between electrospray and atmospheric pressure photoionization sources, *J. Chromatogr. A*, 1101, 69, 2006.
206. Van der Merwe, K.J. et al., Ochratoxin A, a toxic metabolite produced by *Aspergillus ochraceus* Wilh, *Nature*, 205, 1112, 1965.
207. Belli, N. et al., *Aspergillus carbonarius* growth and ochratoxin A production on a synthetic grape medium in relation to environmental factors, *J. Appl. Microbiol.*, 98, 839, 2005.
208. Abarca, M.L. et al., *Aspergillus carbonarius* as the main source of ochratoxin A contamination in dried vine fruits froom ther spanish market, *J. Food Prot.*, 66, 504, 2003.
209. Bayman, P. et al., Ochratoxin production by the *Aspergillus ochraceus* group and *Aspergillus alliaceus*, *Appl. Environ. Microbiol.*, 68, 2326, 2002.
210. Abarca, M.L. et al., Ochratoxin A production by strains of *Aspergillus niger* var. *niger*, *Appl. Environ. Microbiol.*, 60, 2650, 1994.
211. Teren, J. et al., Immunochemical detection of ochratoxin A in black *Aspergillus* strains, *Mycopathologia*, 134, 171, 1996.
212. Romero, S.M. et al., Toxigenic fungi isolated from dried vine fruits in Argentina, *Int. J. Food Microbiol.*, 104, 43, 2005.
213. Hajjaji, A. et al., Occurrence of mycotoxins (ochratoxin A, deoxynivalenol) and toxigenic fungi in Moroccan wheat grains: impact of ecological factors on the growth and ochratoxin A production, *Mol. Nutr. Food Res.*, 50, 494, 2006.
214. Pitt, J.I., *Penicillium viridicatum*, *Penicillium verrucosum*, and production of ochratoxin A, *Appl. Environ. Microbiol.*, 53, 266, 1987.
215. Pardo, E. et al., Ecophysiology of ochratoxigenic *Aspergillus ochraceus* and *Penicillium verrucosum* isolates; Predictive models for fungal spoilage prevention: a review, *Food Addit. Contam.*, 23, 398, 2006.
216. Pitt, J.I. and Hocking, A.D., Influence of solute and hydrogen ion concentration on the water relations of some xerophilic fungi, *J. Gen. Microbiol.*, 101, 35, 1977.
217. Pardo, E. et al., Prediction of fungal growth and ochratoxin A production by *Aspergillus ochraceus* on irradiated barley grain as influenced by temperature and water activity, *Int. J. Food Microbiol.*, 95, 79, 2004.
218. Pardo, E. et al., Effect of water activity and temperature on mycelial growth and ochratoxin A production by isolates of *Aspergillus ochraceus* on irradiated green coffee beans, *J. Food Prot.*, 68, 133, 2005.
219. Magan, N. and Aldred, D., Conditions of formation of ochratoxin A in drying, transport, and in different commodities, *Food Addit. Contam.*, 22, 10, 2005.
220. Sangare-Tigori, B. et al., Preliminary survey of ochratoxin A in millet maize, rice and peanuts in Cote d'Ivoire from 1998 to 2002, *Hum. Exp. Toxicol.*, 25, 211, 2006.
221. Jorgensen, K., Occurrence of ochratoxin A in commodities and processed food: a review of EU occurrence data, *Food Addit. Contam.*, 22, 26, 2005.
222. Battilani, P. et al., Black Aspergilli and ochratoxin A in grapes in Italy, *Int. J. Food Microbiol.*, 111, S53, 2006.
223. Battilani, P., Magan, N., and Logrieco, A., European research on ochratoxin A in grapes and wine, *Int. J. Food Microbiol.*, 111, S2, 2006.

224. Taniwaki, M.H., An update on ochratoxigenic fungi and ochratoxin A in coffee, *Adv. Exp. Med. Biol.*, 571, 189, 2006.

225. Cairns-Fuller, V., Aldred, D., and Magan, N., Water, temperature and gas composition interaction affect growth and ochratoxin A production isolates of *Penicillium verrucusum* on wheat grain, *J. Appl. Microbiol.*, 99, 1215, 2005.

226. MacDonald, S. et al., Survey of ochratoxin A and deoxynivalenol in stored grains from the 1999 harvest in UK, *Food Addit. Contam.*, 21, 172, 2004.

227. Bucheli, P. and Taniwaki, M.H., Research on the origin and on the impact of post harvest handling and manufacturing on the presence of ochratoxin A in coffee, *Food Addit. Contam.*, 19, 655, 2002.

228. JECFA. Safety evaluation of certain mycotoxins in food. Fifty-six report. WHO Technical Report Series, Geneva, 47, 2001.

229. Miraglia, M. and Brera, C., Assessment of dietary intake of Ochratoxin A by the population of EU Member States, 2002. Available at http://ec.europa.eu/food/fs/scoop/3.2.7–en.pdf.

230. Krogh, P., Ochratoxin A residues in tissues of slaughter pigs with nephropathy, *Nord. Vet. Med.*, 29, 402, 1977.

231. Elling, F., Feeding experiments with ochratoxin A contamined barley for bacon pigs. 4. Renal lesions, *Acta Agric. Scand.*, 33, 153, 1983.

232. Krogh, P., Porcine nephropathy associated with ochratoxin A, in *Mycotoxins and Animal Foods*, Smith, J.E. and Anderson, R.S., Eds., CRC press, Boca Raton, FL, 1991, p. 627.

233. Plestina, R. et al., Human exposure to ochratoxin A in areas of Yugoslavia with endemic nephropathy, *Environ. Pathol. Toxicol. Oncol.*, 10, 145, 1982.

234. Castegnaro, M. et al., Balkan endemic nephropathy: role of ochratoxin A through biomarkers, *Mol. Nutr. Food Res.*, 50, 519, 2006.

235. Fuchs, R. and Peraica, M., Ochratoxin A in human diseases, *Food Addit. Contam.*, 22, 53, 2005.

236. Abouzied, M.M. et al., Ochratoxin A concentration in food and feed from a region with Balkan endemic nephropathy, *Food Addit. Contam.*, 19, 755, 2002.

237. Vrabcheva, T. et al., Analysis of ochratoxin A in foods consummed by inhabitants from an area with Balkan endemic nephropathy: a 1 month follow up study, *J. Agric. Food Chem.*, 52, 2404, 2004.

238. Battaglia, R. et al., Fate of ochratoxin A during breadmaking, *Food Addit. Contam.*, 13, 25, 1996

239. AOAC. 973.37. Official mehtods of analysis, Ochratoxins in Barley. Part 2, 1207, 1990.

240. Zimmerli, B. and Dick, R., Determination of ochratoxin A at the ppt level in human blood, serum, milk and some foodstuffs by high-performance liquid chromatography with enhanced fluorescence detection and immunoaffinity column cleanup: methodology and Swiss data, *J. Chromatogr.*, B 666, 85, 1995.

241. Entwisle, A.C. et al., Liquid chromatographic method with immunoaffinity column cleanup for determination of ochratoxin A in barley: collaborative study, *J AOAC Int.*, 83, 1377, 2000.

242. Sharman, M., MacDonald, S., and Gilbert, J., Automated liquid chromatographic determination of ochratoxin A in cereals and in animal products using immunoaffinity column clean up, *J. Chromatogr. A*, 603, 285, 1992.

243. Burdaspal, P., Legarda, T.M., and Gilbert, J., Determination of ochratoxin A in baby food by immunoafinity column cleanup with liquid chromatography: interlaboratory study. *J. Assoc. Off. Anal. Chem. Int.*, 84, 1445, 2001.

244. Monaci, L., Tantillo, G., and Palmisano, F., Determination of ochratoxin A in pig tissues by liquid extraction and clean up and high performance liquid chromatography, *Anal. Bioanal. Chem.*, 378, 1777, 2004.

245. Baggiani, C., Giraudi, G., and Vanni, A., A molecular imprinted polymer with recognition properties towards the carcinogenic mycotoxin ochratoxin A, *Bioseparation*, 10, 389, 2001.

246. Jodibauer, J., Maier, N.M., and Lindner, W., Towards ochratoxin A selective molecularly imprinted polymers for solid phase extraction, *J. Chromatogr. A.*, 945, 45, 2002.

247. Le Tutour, B., Tantaoui-Elaraki, A., and Aboussalim, A., Simultaneous thin layer chromatographic determination of aflatoxin B1 and ochratoxin A in black olives, *J. Assoc. Anal. Chem.*, 67, 611, 1984.

248. Pittet, A. and Royer, D., Rapid low cost thin layer chromatographic screening method for the detection of ochratoxin A in green coffee at a control level of 10 μg/kg, *J. Agric. Food Chem.*, 50, 243, 2002.

249. Santos, E.A. and Vargas, E.A., Immunoaffinity column clean up and thin layer chromatography for determination of ochratoxin A in green coffee, *Food Addit. Contam.*, 19, 447, 2002.

250. Betina, V., Ed., *Chromatography of Mycotoxins*. Elsevier, Amsterdam, 1993.

251. Ventura, M. et al., Two dimentionnal thin layer chromatographic method for the analysis of ochratoxin A in green coffee, *J. Food Prot.*, 68, 1920, 2005.

252. Scott, P.M., Methods of analysis for ochratoxin A, *Adv. Exp. Med. Biol.*, 504, 117, 2002.

253. Levi, C.P., Collaborative study of a method for the determination of ochratoxin A in green coffee, *J. Assoc. Off. Anal. Chem.*, 58, 258, 1975.

254. Tangni, E.K. et al., Ochratoxin A in domestic and imported beers in Belgium: occurrence and exposure assessement, *Food Addit. Contam.*, 19, 1169, 2002.

255. Markaki, P. et al., Determination of ochratoxin A in red wine and vinegar by immunoaffinity high pressure liquid chromatography, *J. Food Prot.*, 64, 533, 2001.

256. Terada, H. et al., Liquid chromatographic determination of ochratoxin A in coffee beans and coffee products, *J. Assoc. Off. Anal. Biochem.*, 69, 960, 1986.

257. Breitholtz, A. et al., Plasma ochratoxin A levels in three Swedish populations surveyed using an ion-paired HPLC, *Food Addit. Contam.*, 8, 183, 1991.

258. Breitholtz-Emanuelsson, A. et al., Ochratoxin A in cow's milk and in human milk with corresponding human blood samples, *J. AOAC*, 76, 842, 1993.

259. Larsson K. and Möller T., Liquid chromatographic determination of ochratoxin A in barley, wheat bran and rye by the AOAC/IUPC/NMKL method: NMKL—collaborative study, *J. AOAC Int.*, 79(5), 1102, 1996.

260. EN-ISO 15141-1. Dosage of ochratoxin A in cereals and derived products: HPLC method, 1998.

261. EN 14132. Dosage of ochratoxin A in barley and roasted coffee: method with IAC purification and HPLC analysis, 2003.

262. AOAC. 2000.09. Ochratoxin A determination in roasted coffee, 2000.

263. AOAC. 2000.03. Ochratoxin A determination in barley, 2000.

264. EN 14133. Dosage of ochratoxin A in wine and beer. HPLC method after IAC purification, 2004.

265. Valenta, H. and Goll, M., Determination of ochratoxin A in regional samples of cow milk from Germany, *Food Addit. Contam.*, 13, 669, 1996.

266. Kuhn, I., Valenta, H., and Rohr, K., Determination of ochratoxin A in bile of swine by high performance liquid chromatography, *J. Chromatogr. B.*, 668, 333, 1995.

267. Rousseau, D.M. et al., Detection of ochratoxin A in porcine kidneys by a monoclonal antibody based radioimmunoassay, *Appl. Environ. Microbiol.*, 53, 514, 1987.

268. Solti, L. et al., Ochratoxin A content of human sera determined by sensitive ELISA, *J. Anal. Toxicol.*, 21, 44, 1997.

269. Sweeney, M.J. and Dobson, A.D., Mycotoxin production by *Aspergillus*, *Fusarium* and *Penicillium* species, *Int. J. Food Microbiol.*, 43, 141, 1998.

270. Blanc P.J., Loret, M.O., and Goma, G. Production of citrinin by various species of *Monascus*, *Biotechnol. Lett.*, 17, 291, 1995.

271. Vrabcheva, T. et al., Co-occurrence of ochratoxin A and citrinin in cereals from Bulgarian villages with a history of Balkan endemic nephropathy, *J. Agric. Food Chem.*, 48, 2483, 2000.

272. Abramson, D., Usleber, E., and Martlbauer, E., Rapid determination of citrinin in corn by fluorescence liquid chromatography and enzyme immunoassay, *J. AOAC Int.*, 82, 1353, 1999.

273. Abramson, D., Usleber, E., and Martlbauer, E., Immunochemical methods for citrinin, *Methods Mol. Biol.*, 157, 195, 2001.

274. Hanika, C., Carlton, W.W., and Tuite, J., Citrinin mycotoxicosis in the rabbit, *Food Chem. Toxicol.*, 21, 487, 1983.

275. Manning, R.O. et al., The individual and combined effects of citrinin and ochratoxin A in broiler chicks, *Avian Dis.*, 29, 986, 1985.

276. Kogika, M.M., Hagikawa, M.K., and Mirandola, R.M., Experimental citrinin nephrotoxicosis in dogs: renal function evaluation, *Vet. Human Toxicol.*, 35, 136, 1993.

277. Ribeiro, S.M. et al., Mechanism of citrinin induced dysfunction of mitochondria.V. Effect on the homeostasis of the reactive oxygen species, *Cell Biochem. Funct.*, 15, 203, 1997.

278. Martins, M.L. et al., Co-occurrence of patulin and citrinin in Portugese apples with roten spots, *Food Addit. Contam.*, 19, 568, 2002.

279. Vazquez, B.I. et al., Rapid semi-quantitative fluorimetric determination of citrinin in fungal cultures isolated from cheese and cheese factories, *Lett. Appl. Microbiol.*, 24, 397, 1997.

280. Martins, M.L. and Martins, H.M., Natural and in vitro coproduction of cyclopiazonic acid and aflatoxins, *J. Food Prot.*, 62, 292, 1999.

281. Le Bars, J., Cyclopiazonic acid production by *Penicillium camemberti* Thom and natural occurrence of this mycotoxin in cheese, *Appl. Environ. Microbiol.*, 38, 1052, 1979.

282. Le Bars, J., Detection and occurrence of cyclopiazonic acid in cheeses, *J. Environ. Pathol. Toxicol. Oncol.*, 10, 136, 1990.

283. Cole, R.J., Etiology of Turkey X disease in retrospect: a case for the involment of cyclopiazonic acid, *Mycotoxin Res.*, 2, 3, 1986.

284. Seidler, N.W. et al., Cyclopiazonic acid is a specific inhibitor of the Ca^{2+}-ATPase of sarcoplasmic reticulum, *J. Biol. Chem.*, 264, 17816, 1989.

285. Nishie, K., Cole R.J., and Dorner, J.W., Effects of cyclopiazonic acid on the contractility of organs with smooth muscles, and on frog ventricles, *Res. Commun. Chem. Pathol. Pharmacol.*, 53, 23, 1986.

286. Lomax, L.G., Cole, R.J., and Dorner, J.W., The toxicity of cyclopiazonic acid in weaned pigs, *Vet. Pathol.*, 21, 418, 1984.

287. Nishie, K., Cole, R.J., and Dorner, J.W., Toxicity and neuropharmacology of cyclopiazonic acid, *Food Chem. Toxicol.*, 23, 831, 1985.

288. Prasongsidh, B.C. et al., Behaviour of cyclopiazonic acid during manufacturing and storage of yogurt, *Aust. J. Dairy Technol.*, 53, 152, 1998.

289. Losito, I. et al., LC-ion trap electrospray MS-MS for the determination of cyclopiazonic acid in milk samples, *Analyst*, 127, 499, 2002.

290. Dorner, J.W., Recent advances in analytical methodology for cyclopiazonic acid, *Adv. Exp. Med. Biol.*, 50, 6148, 2002.

291. Aresta, A. et al., Simultaneous determination of ochratoxin A and cyclopiazonic, mycophenolic, and tenuazonic acid in cornflakes by solid phase microextraction coupled to high performance liquid chrtomatography, *J. Agric. Food Chem.*, 51, 5232, 2003.

292. Byrem, T.M. et al., Analysis and pharmacokinetics of cyclopiazonic acid in market weight pigs, *J. Anim. Sci.*, 77, 173, 1999.

293. Yu, W., Dorner, J.W., and Chu, F.S., Immunoaffinity column as clean up tool for a direct enzyme-linked immunosorbent assay of cyclopiazonic acid in corn, peanuts and mixed feed, *J. AOAC Int.*, 81, 1169, 1998.

294. AOAC. 2000.08. Determination of Aflatoxin M1 in liquid milk, 2000.

295. Prelusky, D.B. et al., Tissue distribution and excretion of radioactivity following administration of 14C-labeled deoxynivalenol to white leghom hens, *Fundam. Appl. Toxicol.*, 7, 635, 1986.

296. Chi, M.S. et al., Excretion and tissue distribution of radioactivity from tritium-labeled T-2 toxin in chicks, *Toxicol. Appl. Pharmacol.*, 45, 391, 1978.

297. Giroir, L.E., Ivie, G.W., and Huff, W.E., Comparative fate of the tritiated trichothecene mycotoxin, T-2 toxin, in chickens and ducks, *Poult. Sci.*, 70, 1138, 1991.

298. Yoshizawa, T., Swanson, S.P., and Mirocha, C.J., T-2 metabolites in the excreta of broiler chickens administered 3H-labeled T-2 toxin, *Appl. Environ. Microbiol.*, 39, 1172, 1980.

299. Prelusky, D.B. et al., Transmission of (14)deoxynivalenol to eggs following oral administration to laying hens, *J. Agric. Food Chem.*, 35, 182, 1987.

300. Visconti, A. and Mirocha, C.J., Identification of various T-2 toxin metabolites in chicken excreta and tissues, *Appl. Environ. Microbiol.*, 49, 1246, 1985.

301. Lun, A.K. et al., Effects of feeding hens a high level of vomitoxin-contamined corn on performance and tissue residues, *Poult. Sci.*, 65, 1095, 1986.

302. Dänicke, S. et al., On the interactions between *Fusarium* toxin-contamined wheat and non-starch-polysaccharide hydrolysing enzymes in turkey diets on performance, health and carry-over of deoxynivalenol and zearalenone, *Brit. Poult. Sci.*, 48, 39, 2007.

303. Olsen, M. et al., Metabolism of high concentrations of dietary zearalenone by young male turkey poults, *Poult. Sci.*, 65, 1905, 1986.

304. Tardieu, D. et al., Determination of fumonisin B1 in animal tissues with immunoaffinity purification, J. Chromatogr B., in press.

305. Stubblefield, R.D. and Shotwell, O.L., Determination of aflatoxins in animal tissues, *J. Assoc. Off. Anal. Chem.*, 64, 964, 1981.

306. Gathumbi, J.K. et al., Application of immunoaffinity chromatography and enzyme immunoassay in rapid detection of aflatoxin B1 in chicken liver tissues, *Poult. Sci.*, 82, 585, 2003.

307. Tavca-Kalcher, G. et al., Validation of the procedure for the determination of aflatoxin B1 in animal liver using immunoaffinity columns and liquid chromatography with postcolumn derivatisation and fluorescence detection, *Food Control*, 18, 333, 2007.

308. Moreno-Guillamont, E. et al., A comparative study of extraction apparatus in HPLC analysis of ochratoxin A in muscle, *Anal. Bioanal. Chem.*, 383, 570, 2005.

309. Biro, K. et al., Tissues distribution of ochratoxin A as determined by HPLC and ELISA and histopathological effects in chickens, *Avian Pathol.*, 31, 141, 2002.

310. Chi, M.S. et al., Transmission of radioactivity into eggs from laying hens (*Gallus domesticus*) administered tritium labelled T-2 toxin. *Poult. Sci.*, 57, 1234, 1978.

311. Prelusky, D.B. et al., Tissue distribution and excretion of radioactivity following administration of 14C-labeled deoxynivalenol to white leghorn hens. Fundam, *Appl. Toxicol.*, 7, 635, 1986.

312. El-Banna, A.A. et al., Nontransmission of deoxynivalenol (vomitoxin) to eggs and meat in chickens fed deoxynivalenol-contamined diets, *J. Agric. Food Chem.*, 31, 1381, 1983.

313. Dänicke, S. et al., Effect of addition of a detoxifying agent to laying hen diets containing uncontamined or *Fusarium* toxin-contamined maize on performance of hens and on carryover of zearalenone, *Poult. Sci.*, 81, 1671, 2002.

314. Medina, M.B. and Sherman, J.T., High performance liquid chromatography separation of anabolic oestrogens and ultraviolet detection of 17 beta-oestradiol, zeranol, diethylstilboestrol or zearalenone in avian muscle tissue extracts, *Food Addit. Contam.*, 3, 263, 1986.

315. Turner, G.V. et al., 1983. High pressure liquid chromatographic determination of zearalenone in chicken tissues, *J. Assoc. Off. Anal. Chem*, 66, 102.

316. Gaumy, J.L. et al., Zearalenone: origine et effets chez les animaux d'élevage, *Rev. Med. Vet.*, 152, 219, 2001.

317. Olsen, M. and Kiessling, K.H., Species differences in zearalenone-reducing activity in subcellular fractions of liver from female domestic animals, *Acta Pharmacol. Toxicol.*, 52, 287, 1983.

318. Pompa, G. et al., The metabolism of zearalenone in subcellular fractions from rabbit and hen hepatocytes and its estrogenic activity in rabbits, *Toxicology*, 42, 69, 1986.

319. Mirocha, C.J. et al., Distribution and residue determination of 3H zearalenone in broilers, *Tox. Appl. Pharmacol.*, 66, 77, 1982.

320. Maryamma, K.I. et al., Pathology of zearalenone toxicosis in chicken and evaluation of zearalenone residues in tissues, *Indian J. Anim Sci.*, 62, 105–107, 1992.

321. Danicke, S. et al., Excretion kinetics and metabolism of zearalenone in broilers in dependence of a detoxifying agent, *Arch. Tierernahr.*, 55, 299, 2001.

322. Vudathala, D.K. et al., Pharmacokinetic fate and pathological effect of 14C-fumonisin B1 in laying Hens, *Nat. Toxins*, 2, 81, 1994.

323. Bluteau C. 2005, toxicocinétique de la fumonsine B1 chez le canard mulard. Thèse de Doctorat Vétérinaire, Université Paul Sabatier, Toulouse.

324. Gregory, J.F. 3rd., Goldstein, S.L., and Edds, G.T., Metabolite distribution and rate of residue clearance in turkeys fed a diet containing aflatoxin B1, *Food Chem. Toxicol.*, 21, 463, 1983.

325. Guerre, P., Intérêt des traitements des matières premières et de l'usage d'adsorbants lors d'une contamination des aliments du bétail par des mycotoxines, *Rev. Med. Vet.*, 151, 1995, 2000.

326. Phillips, T.D. et al., Hydrated sodium calcium aluminosilicate: a high affinity sorbent for aflatoxin, *Poult. Sci.*, 67, 243, 1988.

327. Lozano, M.C. and Diaz, G.Z., Microsomal and cytosolic biotransformation of aflatoxin B1 in four poultry species, *Brit. Poult. Sci.*, 47, 734, 2006.

328. Richard, J.L. et al., Distribution and clearance of aflatoxins B1 and M1 in turkeys fed diets containing 50 or 150 ppb aflatoxin from naturally contamined corn, *Avian Dis.*, 30, 788, 1986.

329. Bintvihok, A. et al., Residues of aflatoxins in the liver, muscle and eggs of domestic fowls, *J. Vet. Med. Sci.*, 64, 1037, 2002.

330. Qureshi, M.A. et al., Dietary exposure of broiler breeders to aflatoxin results in immune dysfunction in progeny chicks, *Poult. Sci.*, 77, 812, 1998.

331. Trucksess, M.W. et al., Aflatoxicol and aflatoxins B1 and M1 in eggs and tissues of laying hens consuming aflatoxin-contamined feed, *Poult. Sci.*, 62, 2176, 1983.

332. Zaghini, A. et al., Mannanoligosaccharides and aflatoxin B1 in feed for laying hens: effects on eggs quality, aflatoxins B1 and M1 residues in eggs, and aflatoxin B1 levels in liver, *Poult. Sci.*, 84, 825, 2005.

333. Madden, U.A. and Stahr, H.M., Effect of soil on aflatoxin tissue retention in chicks added to aflatoxin-contaminated poultry rations, *Vet. Hum. Toxicol.*, 34, 521, 1992.

334. Galtier, P., Pharmacokinetics of ochratoxin A in animals, in *Mycotoxins, Endemic Nephropathy and Urinary Tract Tumors*, M. Castegnaro, R. Plestina, G. Dirheimer, Eds., IARC, Lyon, France, 1991, p. 187.

335. Galtier, P., Camguilhem, R., Bodin, G., Evidence for in vitro and in vivo interaction between ochratoxin A and three acidic drugs, *Food Cosmet. Toxicol.*, 18, 493, 1980.

336. Galtier, P., Alvinerie, M., Charpenteau, J.L., The pharmacokinetic profiles of ochratoxin A in pigs, rabbits and chicken, *Food Cosmet. Toxicol.*, 19, 735, 1981.

337. Kirby, L.K. et al., Citrinin toxicity in young chicks, *Poult. Sci.*, 66, 966, 1987.

338. Uma, M. and Reddy, M.V., Citrinin toxicity in broiler chicks: haematological and pathological studies, *Ind. J. Vet. Pathol.*, 19, 11, 1995.

339. Phillips, R.D., Berndt, W.O., and Hayes, A.W., Distribution and excretion of (14C)citrinin in rats, *Toxicology*, 12, 285, 1979.

340. Nelson, T.S., Beasley, J.N., and Kirby, L.K., Citrinin toxicity in chicks, *Poult. Sci.*, 59, 1643, 1980.

341. Abramson, D. et al., Mycotoxins in fungal contaminated samples of animal feed from western Canada, *Can. J. Vet. Res.*, 61, 49, 1997.

342. Norred, W.P. et al., Liquid chromatographic determination of cyclopiazonic acid in poultry meat, *J. Assoc. Off. Anal. Chem.*, 70, 121, 1987.

343. Norred, W.P. et al., Occurrence of the mycotoxin cyclopiazonic acid in meat after oral administration to chickens, *J. Agric. Food Chem.*, 36, 113, 1988

344. Cole R.J. et al., Cyclopiazonic acid toxicity in the lactating ewe and layinf hen, *Proc. Nutr., Soc. Aust.*, 13, 134, 1986.

345. Dorner, J.W. et al., Cyclopiazonic residue in milk and eggs, *J. Agric. Food Chem.*, 42, 1516, 1994.

346. Andersen, S.J., Compositional changes in surface mycoflora during ripening of naturally fermented sausages, *J. Food Prot.*, 58, 426, 1995.

347. Leistner, L., Mould-fermented foods: recent developments, *Food Biotechnol.*, 4, 433, 1990.

348. Tabuc, C. et al., Toxigenic potential of fungal mycoflora isolated from dry cured meat products: preliminary study, *Rev. Med. Vet.*, 156, 287, 2004.

349. El Kady, I., El Maraghy, S., and Zorhi, A.N., Mycotoxin producing potential of some isolates of *Aspergillus flavus* and Eurotium groups from meat products, *Microbiol. Res.*, 149, 297, 1994.

350. Rojas, F.J. et al., Mycoflora and toxigenic *Aspergillus flavus* in Spanish dry cured ham, *Int. J. Food Microbiol.*, 13, 249, 1991.

351. Escher, F.E., Koehler, P.E., and Ayres, J.C., Production of ochratoxins A and B on country cured ham, *Appl. Microbiol.*, 26, 27, 1973.

352. Lopez-Diaz, T.M. et al., Surface mycoflora of a spanish fermented meat sausage and toxigenicity of *Penicillium* isolates, *Int. J. Food Microbiol.*, 68, 69, 2001.

353. Bailly, J.D. and Guerre P., Mycotoxin analysis in muscle, in *Handbook of muscle Food Analysis*, Nollet, L., Ed., CRC press, chap. 36, in press.

354. Ismail, M.A. and Zaky, Z.M., Evaluation of the mycological status of luncheon meat with special reference to aflatoxigenic moulds and aflatoxin residues, *Mycopathologia*, 146, 147, 1999.

355. Cvetnik, Z., and Pepeljnjak, S., Aflatoxin producing potential of *Aspergillus flavus* and *Aspergillus parasiticus* isolated from samples of smoked-dried meat, *Nahrung*, 39, 302, 1995.

356. Aziz, N.H. and Youssef, Y.A., Occurrence of aflatoxins and aflatoxin-producing moulds in fresh and processed meat in Egypt, *Food Addit. Contam.*, 8, 321, 1991.

357. Bullerman, L.B., Hartman, P.A., and Ayres, J.C., Aflatoxin production in meats. II. Aged salamis and aged country cured hams, *Appl. Microbiol.*, 18, 718, 1969.

358. Refai, M.K. et al., Incidence of aflatoxin B1 in the Egyptian cured meat basterma and control by gamma-irradiation, *Nahrung*, 47, 377, 2003.

359. Chiavaro, E. et al., Ochratoxin A determination in ham by immoniaffinity clean up and a quick fluorometric method, *Food Addit. Contam.*, 19, 575, 2002.

360. Pietri, A. et al. Occurrence of ochratoxin A in raw mham muscle and in pork products from northern Italy, *J. Food Sci.*, 18, 1, 2006.

361. Monaci, L., Palmisano, F., Matrella, R., Tantillo, G., Determination of ochratoxin A at part per trillion level in italian salami by immunoaffinity clean up and high performance liquid chromatography with fluorescence detection, *J. Chromatogr. A.*, 1090, 184, 2005.

362. Jimenez, A.M. et al., Determination of ochratoxin A in pig derived pates by high performance liquid chromatography, *Food Addit. Contam.*, 18, 559, 2001.

363. Toscani, T. et al., Determination of ochratoxin A in dry cured meat products by a HPLC-FLD quantitative method, *J. Chromatogr. B.*, 855, 242, 2007.

364. Bogs, C., Battilani, P., and Geisen, R., Development of a molecular detection and differentiation system for ochratoxin A producing *Penicillium* species an dits application to analyse the occurrence of *Penicillium nordicum* in cured meats, *Int. J. Food Microbiol.*, 107, 39, 2006.

365. Battilani, P. et al., *Penicillium* populations in dry cured ham manufacturing palnts, *J. Food Prot.*, 70, 975, 2007.

366. Wu, M.T., Ayres, J.C., and Koehler, P.E., Toxigenic *Aspergilli* and *Penicillia* isolated from aged cured meat, *Appl. Microbiol.*, 28, 1094, 1974.

367. Wu, M.T., Ayres, J.C., and Koehler, P.E., Production of citrinin by *Penicillium viridicatum* on country cured ham, *Appl. Microbiol.*, 27, 427, 1974.

368. Pfohl-Lezkowicz, A. et al., Balkan endemic nephropathy and associated urinary tract tumors: a review on aetiological causes and the potential role of mycotoxins, *Food Addit. Contam.*, 19, 282, 2002.

369. Sabater-Vilar, M., Maas, R.F.M., and Fink-Gremmels, J., Mutagenicity of commercial monascus fermentation products and the role of citrinin contamination, *Mutation Res.*, 444, 7, 1999.

370. Bailly, J.D. et al., Citrinin production and stability in cheese, *J. Food Prot.*, 65, 1317, 2002.

371. Sosa, M.J. et al., Production of cyclopiazonic acid by *Penicillium commune* isolaed from dry cured ham on a meat extract based substrate, *J. Food. Prot.*, 65, 988, 2002.

372. Anthony, M., Janardhanan, K.K., and Shukla, Y., Potential risk of acute hepatotoxity of kodo poisoning due to exposure to cyclopiazonic acid, *J. Ethnopharmacol.*, 87, 211, 2003.

373. Rao, L.B. and Husain, A., Presence of cyclopiazonic acid in kodo millet causing "kodua poisoning" in man an dits production by associated fungi, *Mycopathologia*, 89, 177, 1985.
374. Le Bars, J. and Le Bars, P., Strategy for safe use of fungi and fungal derivatives in food processing, *Rev. Med. Vet.*, 149, 493, 1998.
375. Cantalon, P.F., Capillary electrophoresis: a useful technique for food analysis, *Food Technol.*, 49, 52, 1995.
376. Martin, A. et al., Characterization of molds from dry meat products and their metabolites by micellar electrokinetic capillary electrophoresis and random amplified polymorphic DNA PCR, *J. Food Prot.*, 67, 2234, 2004.

Chapter 5

Detection of Genetically Modified Organisms in Processed Meats and Poultry

Andrea Germini and Alessandro Tonelli

Contents

5.1 Introduction 126
    5.1.1 Genetically Modified Organism Production for Food and Feed 126
    5.1.2 Legislative Framework for Genetically Modified Organism Traceability 127
    5.1.3 Analytical Methods for Genetically Modified Organism Traceability 128
    5.1.4 Transgenic Material in Processed Meats and Poultry 129
5.2 Detection of Genetically Modified Organisms.. 130
    5.2.1 DNA–Based Methods 130
        5.2.1.1 DNA Extraction Methods 131
        5.2.1.2 PCR–Based Assay Formats 132
        5.2.1.3 Applications in Meat and Poultry Analysis 136
    5.2.2 Protein-Based Methods..143
        5.2.2.1 Antibody-Based Assay Formats...143
        5.2.2.2 Applications in Meat and Poultry Analysis..............................144
    5.2.3 Alternative Techniques for GMO Detection... 145
        5.2.3.1 DNA Microarray Technology..146
        5.2.3.2 Biosensors 147
References..148

5.1 Introduction

5.1.1 Genetically Modified Organism Production for Food and Feed

The advancement of biotechnologies applied to the agro-food industry has resulted, during the past few years, in an increasing number of genetically modified organisms (GMOs) being introduced into the food chain at various levels. Although the regulatory approach to this matter differs depending on the attitudes of different legislative bodies, to inform final consumers correctly and to be able to guarantee the safety of food production chains, the traceability of genetically modified products or ingredients coming from genetically modified products must be guaranteed.

GMOs can be defined as organisms in which the genetic material has been altered by recombinant deoxyribonucleic acid (DNA) technologies, in a way that does not occur naturally by mating or natural recombination. Recombinant DNA techniques allow the direct transfer of one or a few genes between either closely or distantly related organisms; in this way, only the desired characteristic should be safely transferred from one organism to another, speeding up the process of improving the characteristics of target organisms and facilitating the tracking of the genetic changes and of their effects.

The first transgenic plants obtained by recombinant DNA technologies were produced in 1984, and since then more than 100 plant varieties, many of which are economically important crop species, have been genetically modified. The majority of these GMOs have been approved, albeit with differences according to the various legislations worldwide, for use in livestock feed and human nutrition.[1]

Whereas only a few crops have been modified so far to improve their nutritional value, most of the first generation of genetically modified (GM) crops (i.e., those currently in, or close to, commercialization) aim to increase yields, and to facilitate crop management. This is achieved through the introduction of resistance to viral, fungal, and bacterial diseases, or insect pests, or through herbicide tolerance. So far the majority of GM crops can be clustered according to three main characteristics:

- *Insect-protected plants.* The majority of the commercialized products belonging to this category are engineered to express a gene derived from the soil bacterium *Bacillus thuringiensis* (Bt) that encodes for the production of a protein, the delta endotoxin, with insecticidal activity. Other genes that are used in developing this category of crops encode inhibitors of digestive enzymes of pest organisms, such as insect-specific proteinases and amylases, or direct chemically mediated plant defense by plant secondary metabolites.
- *Herbicide-tolerant plants.* A variety of products have been genetically engineered to create crops in which the synthesis of essential amino acids is not inhibited by the action of broad-spectrum herbicides like glufosinate, as happens for conventional plants.
- *Disease-resistant plants.* Using gene manipulation technology, specific disease resistance genes can be transferred from other plants that would not interbreed with the crops of interest, or from other organisms; this allows the transformed crops to express proteins or enzymes that interfere with bacterial or fungal growth. GM virus-resistant crops have also been developed using "pathogen-derived resistance," in which plants expressing genes for particular viral proteins are "immunized" to resist subsequent infection.

Other phenotypic characteristics, less common than those mentioned earlier, include: modified fatty acid composition, fertility restoration, male sterility, modified color, and delayed ripening.

According to the latest statistics available, GMO crop cultivation has been continuously growing, since its introduction in the agricultural practice, in both industrial and developing countries. "Although the first commercial GM crop (tomato) was planted in 1994, it has been in the last few years that a dramatic increase in planting has been observed, bringing the estimated global area of GM crops in 2007 to around 114, 3 million hectares, involving 12 million farmers in 23 countries worldwide, and with a global market value for biotech crops estimated to be around $6.9 billion. As for the kinds of cultivated crops, four GM crops represent at present almost 100% of the market: GM soybean accounts for the largest share, 51.3%, followed by maize with 30.8%, cotton, 13.1%, and canola, 4.8%. These figures confirm how globally widespread GM cultivation is and how important the numbers are becoming compared to traditional crops: in particular, in 2007, GM soybean accounted for 64% of total soybean-plantings worldwide, whereas maize, cotton, and canola represented 23, 43, and 20% of their respective global plantings.[2]

A new wave of genetically modified products, the second generation of GM-derived food and feed, is now at the end of its developing stage or already under evaluation from the competent authorities for approval. These products mainly respond with similar approaches to the same issues addressed by the first generation (herbicide resistance, pest protection, and disease resistance). However, an increasing number of products are trying to respond to various new problems, such as removing detrimental substances, enhancing health-promoting substances, enhancing vitamin and micronutrient content, altering fatty acids and starch composition, reducing susceptibility to adverse environmental conditions, and improving carbon and nitrogen utilization. This second generation of GMOs should constitute a new class of products in an attempt to respond to the needs of consumers and of industries in the near future.

5.1.2 Legislative Framework for Genetically Modified Organism Traceability

The need for monitoring the presence of GM plants in a wide variety of food and feed matrices has become an important issue both for countries with specific regulations on mandatory labeling of food products containing GM ingredients or products derived from GMOs, and for countries without mandatory labeling on food products but that are required to test for the presence of unapproved GM varieties in food products.

Among the countries with mandatory labeling, the European Union (EU) has devised an articulated regulatory framework on GMOs to guarantee an efficient control on food safety-related issues and to ensure correct information to European consumers; the use and commercialization of GM products and their derivatives have been strictly regulated in both food and feedstuffs, and compulsory labeling applies to all products containing more than 0.9% genetically modified ingredients (an adventitious presence threshold of 0.5% applies for GMOs that have already received a favorable risk evaluation but have not yet been approved). Other mandatory schemes for labeling are present worldwide in various countries, including Australia and New Zealand, Brazil, Cameroon, Chile, China, Costa Rica, Ecuador, India, Japan, Malaysia, Mali, Mauritius, Mexico, Norway, the Philippines, Russia, Saudi Arabia, South Africa, South Korea, Switzerland, Taiwan, Thailand, and Vietnam. Most of these countries have established mandatory labeling thresholds ranging from 0 to 5% of GMO content.[3,4] In other countries in which labeling is voluntary, such as the United States, Canada, and Argentina among the most important, being able to detect GM varieties is however of great importance, e.g., to prevent unauthorized transgenes from entering the food productions chains.

5.1.3 Analytical Methods for Genetically Modified Organism Traceability

One of the main challenges related to the use of GMOs is their traceability all along the food chain. In general, to be able to correctly identify the presence of transgenic material, a three-stage approach is needed:[5]

- *Detection.* A preliminary screening is performed to detect characteristic transgenic constructs used to develop GMOs (e.g., promoter and terminator sequences in the case of DNA analysis) and to gain initial insight into the composition of the sample analyzed.
- *Identification.* This stage allows researchers to gain information on the presence of specific transgenic events in the sample analyzed. According to the specific regulation framework in which the analysis is performed, the presence of authorized GMOs should then be quantified, and the presence of unauthorized GMOs should be reported to competent authorities and the product prevented from entering the food chain.
- *Quantitation.* Transgene-specific quantification methods should be used at this stage to determine the amount of one or more authorized GMOs in the sample, and to assess compliance with the labeling thresholds set in the context of the applicable regulative framework.

All along this analytical scheme for the detection of GMOs, particular attention should be paid to the evaluation of the degradation of the target DNA/protein during sampling and processing and to the robustness of the analytical methods. Thorough knowledge and understanding of the problems associated with both the sample to be analyzed and the method for the analysis are fundamental prerequisites to obtaining reliable results.

The first two stages of this scheme of analysis can essentially be accomplished by qualitative methods, whereas semiquantitative or quantitative methods need to be used to accomplish the third stage of analysis.

At present the two most important approaches for the detection of GMOs are (i) immunological assays based on the use of antibodies that bind to the novel proteins expressed, and (ii) polymerase chain reaction (PCR)-based methods using primer oligonucleotides that selectively recognize DNA sequences unique to the transgene.

The two most common immunological assays are enzyme-linked immunosorbent assay (ELISA) based methods and immunochromatographic assays (e.g., lateral flow strip tests). Whereas the former can produce qualitative, semiquantitative, and quantitative results according to the method employed, the latter, although fast and easy to perform, produces mainly qualitative results. However, both techniques require a sufficient protein concentration to be detected by specific antibodies, and thus their efficiency is strictly related to the plant environment, tissue-specific protein expression, and, not least, protein degradation during sampling and processing.

The most powerful and versatile methods for tracking transgenes are, however, based on the detection of specific DNA sequences by means of PCR methods. These methods are reported to be highly specific, and have detection limits close to a few copies of the target DNA sequence.

Qualitative and semiquantitative detection of GMOs can easily be achieved via end-point PCR combined with gel electrophoresis, whereas quantitative detection can only be obtained by applying specific real-time PCR protocols, which rely on the quantification of fluorescent reporter molecules that increase during the analysis with the amount of PCR product.

In addition to the aforementioned methods, other detection methods based on chromatography, mass spectrometry, and near-infrared (NIR) spectroscopy have been developed[5] and found to be suitable for specific applications, in particular when the genetic modifications create significant changes in the chemical composition of the host organism.

5.1.4 Transgenic Material in Processed Meats and Poultry

The significant increase of GM productions since the commercialization of the first genetically modified crop has generated interest and concern regarding the fate of transgenic material along the food chain. Questions have been posed both at public and at scientific levels about the potential appearance of novel proteins and recombinant DNA in products for human consumption, driven by animal products potentially containing GMOs. Considering the fact that livestock consume large amounts of plant material and that high-protein feeds are among the most common GM crops, it has become necessary to evaluate the fate of GMOs in the animals' diet and the possible consequences on human health. From a legislative point of view, however, countries that have implemented labeling regulation concerning GM feed have at present no mandatory regulations on products derived from livestock fed transgenic feed.

Although in the past few years several attempts to investigate the fate of transgenic proteins and DNA within the gastrointestinal tract of livestock fed GMOs and the incorporation of transgenic material into tissues have been reported,[3] to date very few results support the feasibility of detecting traces of transgenic material in animal tissues outside the gastrointestinal tract. Indeed several factors could influence the presence and hamper the detectability of DNA and protein targets in animal tissues as a result of GM crops feeding: (i) the kind of genetic modification and the type of plant tissue in which the protein is expressed, together with environmental conditions of growth of the GM crop, could cause the content of transgenic protein to vary greatly; (ii) postharvest feed processing, such as ensiling, steeping, wet-milling, and heating, often degrade DNA and protein to an undetectable level; (iii) the rapid degradation observed in the gastrointestinal tract dramatically reduces the absorption across the epithelial tissues of protein and DNA fragments suitable for analytical detection; and (iv) although the passage of dietary DNA fragments has been suggested by several researchers, currently available PCR techniques have only allowed detection of "high copy number genes" (e.g., plant endogenous genes such as rubisco and chloroplast-specific sequences), whereas transgenes are often the result of a single insertion event.

Considering the detectable presence of GM-derived materials outside the gastrointestinal tract in livestock as an extremely rare event, the main route for the presence of transgenic material in processed meat and poultry could be an external event, such as an adventitious contamination (e.g., during slaughtering, the gastrointestinal content could come in contact with other animals' parts) or the intentional addition of GM-derived additives intended to enhance meat products properties. In particular, apart from additives produced via the use of genetically modified microorganisms (GMMs) such as antioxidants (e.g., ascorbic acid), flavor enhancer (e.g., glutamate), and enzymes (e.g., proteases to be used as tenderizer), which do not require labeling because GMMs are not directly associated with the final purified product, several additives used during meat processing are produced from GMOs and mainly from GM soybean and maize. Soy proteins (in the form of soy flour, texturized vegetable protein [TVP], soy concentrates, and soy isolates) are by far the most commonly employed vegetal protein in the meat industry on account of their excellent water-binding properties, fat emulsification activity, and high biological value. Maize starches are often used on account of their water-binding properties, and the products obtained by their hydrolysis or thermal treatment, in the form of maltodextrin, are often used as filler or stabilizer. Soybean is also a source of lecithin and mono- and diglycerides commonly employed as emulsifiers in meat products to reduce the risk of fat and water separation, to lower cooking loss, and to improve the texture and firmness of the product.

5.2 Detection of Genetically Modified Organisms

Approved transgenes and detection methods are continuously updated, and official detection methods are validated and reported by the different national control agencies.[6] Online databases of protein and DNA-based methods that have been validated by different research agencies are also available for consultation.[7]

5.2.1 DNA–Based Methods

GMOs currently available are the result of transformation events that provide the stable insertion of an exogenous DNA fragment into a host's genome, by means of DNA recombinant technology. The insert contains at least three elements: the gene coding for a specific desired feature and the transcriptional regulatory elements, typically a promoter and a terminator. Several additional elements could be present, depending on the transformation system employed: selection markers such as antibiotic resistance, introns, or sequences coding for signaling peptides are commonly used.[8]

A wide spectrum of analytical methods based on PCR have been developed during the past decade, and PCR-based assays are generally considered the method of choice for regulatory compliance purposes. The general procedure for performing PCR analysis includes four subsequent phases: sample collection, DNA isolation, DNA amplification, and detection of products. The latter two steps may occur simultaneously in certain PCR applications, such as real-time PCR.

Sampling, DNA extraction, and purification are crucial steps in GMO detection. Sampling plans have to be carefully designed to meet important statistical requirements involving the level of heterogeneity, the type of material (raw material, ingredients, or processed food), and the threshold limit for acceptance.[9] DNA quality and purity are also parameters that dramatically affect the PCR efficiency.[10] DNA quality is strictly dependent on degradation caused by temperature, the presence of nucleases, and low pH, and determines the minimum length of DNA-amplifiable fragments. Moreover, the presence of contaminants from the food matrix or chemicals from the method used for DNA isolation can severely affect DNA purity and could cause the inhibition of PCR reactions.

The PCR scheme involves subsequent steps at different temperatures during which: (i) the DNA is heated to separate the two complementary strands of the DNA template (denaturation, 95°C), (ii) the oligonucleotide primers anneal to their complementary sequences on the single strand target DNA (annealing step, 50–60°C), and (iii) the double-strand DNA region formed by the annealing is extended by the enzymatic activity of a thermostable DNA polymerase (extension step, 72°C). All these cycles are automatically repeated in a thermal cycler for a certain number of cycles, and at the end of the process the original target sequence results in an exponential increase in the number of copies.

Several authors have classified PCR-based GMO assays according to a "level of specificity" criterion.[5,11]

1. *Methods for screening purposes* are usually focused on target sequences commonly present in several GMOs. The most commonly targeted sequences pursuing this strategy are two genetic control elements, the cauliflower mosaic virus (CaMV) 35S promoter (P-35S) and the nopaline synthase gene terminator (T-NOS) from *Agrobacterium tumefaciens*.
2. *Gene-specific methods* target a portion of DNA sequence of the inserted gene. These methods amplify a gene tract directly involved in the genetic modification event, typically structural genes such as Cry 1A(b) coding for endotoxin B_1 from Bt, or the 5-enolpyruvylshikimate-3-phosphate synthase (EPSPS) gene, coding for an enzyme conferring herbicide tolerance to the GM crop.

   Both the screening and the gene-specific approach are useful to investigate the presence of GMOs, but fail to reveal the GMO identity. Moreover, these methods are based on the detection of sequences naturally occurring in the environment, and this fact could lead to a significant increase of false-positive results.
3. Junction regions between two artificial construct elements such as the promoter and the functional gene are targeted by construct-specific methods; these reduce the risks of false-positive appearances and increase the chances of identifying the GM source of DNA. However, more than one GMO could share the same gene construct, preventing their unambiguous identification.
4. The highest level of specificity is obtained using event-specific methods that target the integration locus at the junction between the inserted DNA and the recipient genome.

An overview of validated PCR methods for the different strategies of GMO detection is provided in Section 5.2.1.2.

PCR assays can be followed by confirmation methods suitable to discriminate specific from unspecific amplicons. Gel electrophoresis is the simplest method to confirm the expected size of PCR products, but fails to identify the presence of unspecific amplicons having the same size of the expected PCR product. Sequencing the amplicons is the most reliable method of confirming the identity of PCR products, but it is an expensive approach and requires specific instrumentation not frequently available in control laboratories. Nested PCR is commonly used both in optimization steps and in routine analyses; it is based on a second PCR reaction in which a PCR product is reamplified using primers specifically designed for an inner region of the original target sequence. Since nested PCR consists of two PCR reactions in tandem, increased sensitivity is obtained. At the same time, however, it increases the risk of false positives by carryover or cross-contamination. Southern blot assays are another reliable confirmation method; after gel electrophoresis, DNA samples are fixed onto nitrocellulose or nylon membranes and hybridized to a specific DNA probe. Southern blot is time-consuming and quite labor-intensive, and its implementation in routine analysis is limited.

5.2.1.1 DNA Extraction Methods

Isolation of nucleic acids is one of the most crucial steps in genetic studies. The presence of a great variety of extraction and purification methods arises from the numerous parameters that analysts have to take into account (source organism, specific matrix to be analyzed, downstream application, etc.). Regardless of the specific extraction method, the overall aim of this part of the detection process is to obtain an adequate yield of recovered DNA of high quality and purity to be used in the subsequent steps of the PCR analysis. DNA quality essentially refers to the degree of degradation of the nucleic acids recovered; the presence of DNA fragments long enough to be amplifiable is a key factor to be taken into account when designing and performing a PCR test. DNA purity mainly refers to the possible presence of PCR inhibitors in the extracted solution; the presence of proteins, bivalent cations, polyphenols, polysaccharides, and other secondary metabolites can interfere with the enzyme activity and dramatically reduce the efficiency of PCR amplification.

The extraction of nucleic acids from biological material essentially requires the following basic steps: cell lysis/sample homogenization, inactivation of nucleases, separation of the nucleic acid from other matrix components, and recovery of the purified nucleic acids.[12]

Because food matrices in general and meat samples in particular can vary greatly in their physical and chemical properties, it is difficult to devise an all-purpose extraction procedure suitable for the different matrices and meeting all the necessary criteria. For this reason, customized DNA extraction methods need to be developed or adapted from more general methods, to respond to the particular problem of the specific matrix to be analyzed and to optimize the extraction efficiency. Common extraction and purification methods for the recovery of nucleic acids reported in the literature are fundamentally based in one of the following:

- Combination of phenol and chloroform for proteins removal followed by selective precipitation of nucleic acids with isopropanol or ethanol
- Use of the ionic detergent cetyltrimethylammonium bromide (CTAB) to lysate cells and selectively insolubilize nucleic acids in a low-salt environment, followed by solubilization and precipitation with isopropanol or ethanol
- Use of detergents and chaotropic agents followed by DNA binding on silica supports (e.g., spin column or magnetic silica particles) and elution in a low-salt buffer

Several commercial methods are currently available that employ combination of the strategies mentioned earlier to perform fast and reliable extractions for specific food and feed matrices.

An overview of customized DNA extraction procedures available in the literature, clustered according to the different meat and poultry samples to be analyzed and the different processing they underwent, is reported in Table 5.1, together with the corresponding bibliographic references.

5.2.1.2 PCR–Based Assay Formats

5.2.1.2.1 Qualitative PCR–Based Methods

Conventional end-point PCR has been extensively used as a qualitative method to detect the presence of transgenic plants as raw materials and in processed foods. PCR products are usually separated and visualized using agarose gel electrophoresis in combination with DNA staining.

The main advantages of this technique are the cost effectiveness and the simplicity. Conventional PCR is carried out using instrumentation commonly available in control laboratories. The amplification and the detection steps, occurring separately, extend the analysis time, increase the risk of contamination, and reduce the automation possibilities. Despite these potential limitations, several authors have developed methods for the sensitive detection of GM crops.

Table 5.1 Customized DNA Extraction Procedures for Different Meat Samples

Samples Type	Processing	Deoxyribonucleic Acid Extraction/Purification	Reference
Beef muscle	Unprocessed	CTAB extraction method followed by CTAB precipitation or chloroform extraction	13
Chicken muscle			
Pork muscle			
Broiler muscle	Unprocessed	In-house method based on ammonium acetate extraction followed by isopropanol precipitation	14
Pork muscle	Unprocessed	In-house method based on phenol/chloroform/ isoamyl alcohol and ammonium acetate extraction followed by isopropanol precipitation	15
Beef meat	Mincing	CTAB extraction method followed by purification through a silicon spin column (Qiagen)	16
Chicken meat	Freezing		
Lamb meat	Corned		
Pork meat	Steak Pie		
Turkey meat			
Beef meat	Curing	CTAB extraction method followed by QIAquick PCR Purification Kit (Qiagen)	17
Chicken meat	Cooking		
Pork meat	Smoking		
Sheep meat	Heating		
Turkey meat	Sterilization		
Beef meat	Canning under different conditions (home, industrial, tropical conditions, ultra high heat)	CTAB extraction method followed by QIAquick PCR Purification Kit (Qiagen)	18
Chicken meat			
Duck meat			
Goat meat			
Lamb meat			
Pork meat			
Turkey meat			

(*continued*)

Table 5.1 (continued)   Customized DNA Extraction Procedures for Different Meat Samples

Samples Type	Processing	Deoxyribonucleic Acid Extraction/Purification	Reference
Poultry meat	Light boiling	Wizard DNA extraction Kit (Promega)	19
	Heavy boiling		
	Light baking		
	Heavy baking		
	Canning		
	Autoclaving		
Turkey-based meat products (sausages, canned liver, ready-to-eat hamburgers)	Smoking	Wizard DNA clean-up system (Promega)	20
	Cooking		
	Sterilization		
	Frying		
	Roasting		

Conventional PCR assays have been improved, performing simultaneous amplification of several GMOs in the same reaction, and using more than one primer pair; this multiplex PCR format often requires longer optimization procedures, but results in more rapid and inexpensive assays. Several multiplex PCR methods have been developed that allow simultaneous screening of different GM events in the same reaction tube.[21–24]

5.2.1.2.2   Quantitative PCR–Based Methods

The threshold for compulsory labeling of products containing GMOs set in many countries greatly accelerated the development of quantitative PCR-based GMO assays to comply with legislative requirements. Usually, the efficiency of quantitative methods is described using at least two fundamental parameters: the limit of detection (LOD) and the limit of quantification (LOQ). One of the main drawbacks is that these values are usually determined using standard reference material with high-quality DNA, and their value dramatically decreases when faced with complex matrices or processed products. The availability of reference material containing known amounts of GMOs is another problematic aspect in calibrating and standardizing quantitative assays, because certified reference materials (CRMs) are commercially available only for a limited number of GMOs (e.g., JRC-IRMM in Europe[25]). To overcome problems related to CRMs, alternative strategies have been proposed, such as the use of plasmid constructs carrying the sequence to be quantified, which seems to represent a promising alternative strategy.[26,27]

5.2.1.2.2.1   Quantitative Competitive PCR   In quantitative competitive polymerase chain reaction (QC-PCR), the target amplification is coupled with coamplification of quantified internal controls that compete with target DNA for the same primers. The assay is carried out by amplifying samples with varying amounts of a previously calibrated competitor, finding the point that gives the same quantity of amplification products: the equivalence point. The end-point quantitation is then usually performed on agarose gel electrophoresis. QC-PCR methods for Roundup Ready (RR) soybean and Maximizer maize have been developed[28] and tested in an interlaboratory trial at the EU level.[29] A screening method targeting the 35S promoter and the NOS terminator has also been reported.[30] Even if the QC-PCR method potentially allows GMO detection with low limits of quantification, some drawbacks have limited the diffusion of this technique. The use of pipetting on a large scale increases the risk of cross-contamination and makes automation procedures difficult. Moreover, QC-PCR is time-consuming and often needs long optimization procedures.

5.2.1.2.2.2   Real-Time PCR   Real-time PCR-based methods have become more and more often recognized in the past few years as the method of choice for GMO quantitation. The most distinctive feature of this technique is that the amplicon can be monitored and quantified during each cycle of the PCR reaction: the increase in amplicon amount is indirectly measured as fluorescence signal variation during amplification. Quantitation by real-time PCR relies on the setting of two parameters: (i) the threshold fluorescence signal, defined as the value statistically significant above the noise; and (ii) the threshold cycle (C_t), which is the cycle number at which the fluorescence value is above the set threshold. Quantitation can be calculated directly comparing C_t values of the GM-specific targeted gene with a reference gene. To obtain reliable measures, it is essential to perform the reactions starting with the same concentration of DNA template. Moreover, this quantitation method relies on the assumption that both amplicons are amplified with the same efficiency. As an alternative to overcome this limitation, quantitation can be done building a standard curve with a series of PCR reactions using different known initial amounts of reference material. This method allows only C_t values of the same amplicons to be compared, reducing errors in measurements.

Several chemical strategies are currently available for real-time PCR analysis. Nonspecific methods use DNA intercalating agents such as SYBR Green, and others.[31] These assays have good sensitivity, but often require postanalysis confirmation methods to distinguish the amplicons' identity and avoid false positives. This purpose is achieved by some commercial instruments, which allow analysis of the thermal denaturation curve to define the amplicons' identity.[32]

Specific methods, however, allow the simultaneous detection and confirmation of target sequences using specific probes or primers labeled with fluorescent dyes. The most widely adopted technology in real-time PCR analysis of GMOs is the TaqMan approach: a DNA oligonucleotide probe containing both a fluorophore and a quencher conjugated at each side of the molecule. During the extension step, the probe is degraded by the 5•–3• exonuclease activity of the DNA polymerase, and the quenching molecule is consequently physically separated from the fluorophore reporter, allowing the reporter to emit a detectable fluorescence that increases at each amplification cycle. A further improvement compared to TaqMan assays has been achieved through the use of minor groove binding (MGB) probes, in which a minor groove binder group increases the melting temperature of the duplex, improving the probe's selectivity and sensitivity. Alternatives, based on the same principle of physical separation between fluorophore and quencher, have been developed in scorpion primers and in molecular beacons. In these approaches a conformational change induced by the specific annealing, instead of a degradation event, drives the mechanism of fluorescence emission (a passage from a hairpin-shaped structure in solution to an unfolded conformation upon target hybridization). Other alternative technologies such as fluorescence resonance energy transfer

(FRET) probes and light up probes could be promising tools also for the detection of GMOs.[31] Comparison of the different chemistries currently available for GMO detection has been recently reported.[33,34]

Compared to the other PCR-based methods, real-time PCR offers several advantages: (i) by performing both reaction and detection in a closed tube format, the risk of cross-contamination is greatly reduced; (ii) the high degree of automation makes real-time PCR less labor-intensive and time-consuming; and (iii) due to the possibility of setting multiplex assays and simultaneously performing several tests, the sample throughput result is increased compared with other PCR quantitation methods.

Real-time PCR has been successfully employed for quantitative analysis of genetically modified maize, soybean, rapeseed, cotton, potato, rice, tomato, and sugar beet (see Table 5.2). Several composite feed diets such as silage, commercial feed, and pellet mixed diet have been also investigated for their possible GMO content using real-time PCR.[35,36]

5.2.1.2.2.3 PCR Enzyme-Linked Immunosorbent Assay

An alternative method to perform end-point quantitation is coupling a conventional PCR with an enzymatic assay. In PCR-ELISA, a capture probe specific for the PCR amplicon is used to capture the amplicon in a well plate. PCR products, labeled during amplification, are then quantified by a conventional ELISA assay targeting the labeled amplicon. The main advantage of PCR-ELISA is that it offers a cheaper alternative to real-time PCR assays and requires less expensive instruments. Some PCR-ELISA applications have been developed for GMOs detection and quantitation.[37,38] However, this technique does not seem to be widely adopted for accurate GMO quantitation.

5.2.1.3 Applications in Meat and Poultry Analysis

Because of the recent interest in the fate of transgenic DNA after consumption by human and animals, several studies have attempted to detect DNA fragments, related to both endogenous genes and transgenes, using PCR-based technologies, in livestock and in the processed meat and poultry obtained.

The fate of chloroplast-specific gene fragments of different lengths (199 and 532 bp) and a Bt176–specific fragment has been evaluated in cattle and chicken fed a diet containing conventional or GM maize.[69] Only the short DNA amplicon from chloroplast was detected in blood lymphocytes of cows, but no plant DNA was detectable in muscle, liver, spleen, or kidney. In contrast, in all chicken tissues (muscle, liver, spleen, and kidney) the short maize chloroplast gene fragment was amplified. However, a Cry 1A(b)–specific sequence was not detectable in any of the analyzed sample.

An optimized DNA extraction protocol combined with PCR has been used to detect fed-derived plant DNA in muscle meat from chickens, swine, and beef steers fed MON 810 maize.[13] Short fragments (173 bp) amplified from the high copy number chloroplast-encoded maize rubisco gene (*rbcL*) were detected in 5, 15, and 53% of the muscle samples from beef steers, broiler chickens, and swine, respectively. Only one pork sample out of 118 tested positive for the screening of P-35S; however, further analysis performed with a specific MON 810 PCR method generated indeterminate results, suggesting that the number of target copies in the sample, where present, were below the detection limit of the method.

PCR has also been used to investigate the fate of feed-ingested foreign DNA in pigs fed Bt maize.[70] Fragments of transgenic DNA were detected in the gastrointestinal tract of pigs up to 48 h after the last feeding with transgenic maize. Chloroplast DNA was detected in blood, liver,

Table 5.2 Validated PCR Methods for the Different Strategies of GMO Detection

P	Target	Primer Sequences	TaqMan Probe if Real-Time (5'-FAM 3'-TAMRA)	Reference
1	Animal mtDNA 16S rRNA gene	5'-GGTTTACGACCTCGATGTT-3' 5' CCGGTCTGAACTCAGATCAC-3'		39
1	Myostatin gene of mammals and poultry species	5'-TTGTGCAAATCCTGAGACTCAT-3' 5'-ATACCAGTGCCTGGGTTCAT-3'	5'-CCCATGAAAGACGGTACAAGGTATACTG-3'	17
1	Cattle	5'-ACTCCTACCCATCATGCAGAT-3' 5'-TTTTTAAATATTTCAGCTAAGAAAAAAAG-3'	5'-AACATCAGGATTTTTGCTCGCATTTGC-3'	18
1	Chicken	5'-TGTTACCTGGGAGAAGTGGTTACT-3' 5'-TTTTCGATATTTTGAATAGCAGTTACAA-3'	5'-TGAAGAAAGAAACTGAAGATGACACT GAAATTAAAG-3'	18
1	Lamb	5'-ACCCGTCAAGCAGACTCTAACG-3' 5'-TAAATATTTCAGCTAAGGAAAAAAAGAAG-3'	5'-CAGGATTTTTGCCGCATTCGCTT-3'	18
1	Pig	5'-CCCCACCTCAAGTGCCT-3' 5'-CACAGACTTTATTTCTCCACTGC-3'	5'-CACAGCAAGCCCCTTAGCCC-3'	18
1	Turkey	5'-TGTATTTCAGTAGCACTGCTTATGACTACT-3' 5'-TTTATTAATGCTGGAAGAATTTCCAA-3'	5'-TTATGGAGCCATCGCTATCACCAGAAAA-3'	18
2	Chloroplast gene for vegetal species	5'-CGAAATCGGTAGACGCTACG-3' 5'-GGGGATAGAGGGACTTGAAC-3'		40
2	Canola	5'-GGCCAGGGTTTCCGTGAT-3' 5'-CCGGTCGTTGTAGAACCATTGG-3'	5'-AGTCCTTATGTGCTCCACTTTCTGGTGCA-3' (5'-VIC)	41

(continued)

Table 5.2 (continued) Validated PCR Methods for the Different Strategies of GMO Detection

P	Target	Primer Sequences	TaqMan Probe if Real-Time (5'-FAM 3'-TAMRA)	Reference
2	Cotton	5'-AGTTTGTAGGTTTGATGTTACATTGAG-3' 5'-GCATCTTTGAACCGCCTACTG-3'	5'-AAACATAAAATAATGGGAACAACCAT GACATGT-3'	42
2	Maize	5'-CTCCCAATCCTTTGACATCTGC-3' 5'-TCGATTTCTCTCTTGGTGACAGG-3'	5'-AGCAAAGTCAGAGGCGCTGCAATGCA-3'	43
2	Potato	5'-GGACATGTGAAGAGACGGAGC-3' 5'-CCTACCTCTACCCCTCCGC-3'	5'-CTACCACCATTACCTCGCACCTCCTCA-3'	44
2	Rice	5'-TGGTGAGCGTTTTGCAGTCT-3' 5'-CTGATCCACTAGCAGGAGGTCC-3'	5'-TGTTGTCGTGCCAATGTGGCCTG-3'	45
2	Soybean	5'-TCCACCCCCATCCACATTT-3' 5'-GGCATAGAAGGTGAAGTTGAAGGA-3'	5'-AACCGGTAGCGTTGCCAGCTTCG-3'	46
2	Sugarbeet	5'-GACCTCCATATTACTGAAAGGAAG-3' 5'-GAGTAATTGCTCCATCCTGTTCA-3'	5'-CTACGAAGTTTAAAGTATGTGCCGCTC-3'	47
2	Tomato	5'-GGATCCTTAGAAGCATCTAGT-3' 5'-CGTTGGTGCATCCCTGCATGG-3'		48
3	CaMV 35S promoter	5'-CCACGTCTTCAAAGCAAGTGG-3' 5'-TCCTCTCCAAATGAAATGAACTTCC-3'		49
3	Coat protein gene from potato potyvirus Y (PVY)	5'-GAATCAAGGCTATCACGTCC-3' 5'-CATCCGCACTGCCTCATACC-3'		50

	Target	Sequence(s)	Sequence	Ref.
3	CP4 EPSPS	5'-GCGTCGCCGATGAAGGTGCTGTC-3' 5'-CGGTCCTTCATGTTCGGCGGTCTC-3'		15
3	CrylA(b)	5'-CCGCACCCTGAGCAGCAC-3' 5'-GGTGGCACGTTGTTGTTCTGA-3'		46
3	Figwort mosaic virus (P-FMV) promoter	5'-GCCAAAAGCTACAGGAGATCAATG-3' 5'-GCTGCTCGATGTTGACAAGATTAC-3'		51
3	Hygromycin phosphotransferase (hph) gene	5'-CGCCGATGGTTTCTACAA-3' 5'-GGCGTCGGTTTCCACTAT-3'		52
3	Neomycin phosphotransferase II (nptII) gene	5'-GGATCTCCTGTCATCT-3' 5'-GATCATCCTGATCGAC-3'		53
3	Nopaline synthase (NOS) terminator	5'-GCATGACGTTATTTATGAGATGGG-3' 5'-GACACCCGCGCGCGATAATTTATCC-3'		49
4	Canola GT73	5'-CCATATTGACCATCATACTCATTGCT-3' 5'-GCTTATACGAAGGCAAGAAAAGGA-3'	5'-TTCCCGGACATGAAGATCATCCTCCT-3'	54
4	Canola Ms8	5'-GTTAGAAAAAGTAAACAATTAATATAGCCGG-3' 5'-GGAGGGTGTTTTTGGTTATC-3'	5'-AATATAATCGACGGATCCCCGGGAATTC-3'	55
4	Canola Rf3	5'-AGCATTTAGCATGTACCATCAGACA-3' 3'-CATAAAGGAAGATGGAGACTTGAG-3'	5'-CGCACGCGCTTATCGACCATAAGCCCA-3'	56
4	Canola T45 (HCN28)	5'-CAATGGACACATGAATTATGC-3' 5'-GACTCTGTATGAACTGTTCGC-3'	5'-TAGAGGACCTAACAGAACTCGCCGT-3'	41

(continued)

Table 5.2 (continued)   Validated PCR Methods for the Different Strategies of GMO Detection

P	Target	Primer Sequences	TaqMan Probe if Real-Time (5'-FAM 3'-TAMRA)	Reference
4	Cotton	5'-GGAGTAAGACGATTCAGATCAAACAC-3'	5'-ATCAGATTGTCGTTTCCCGCCTTCAGTTT-3'	57
	MON 1445	5'-ATCGACCTGCAGCCCAAGCT-3'		
4	Cotton	5'-CTCATTGCTGATCCATGTAGATTTC-3'	5'-TTGGGTTAATAAAGTCAGATTAGAGGG AGACAA-3'	42
	281-24-236	5'-GGACAATGCTGGGCTTTGTG-3'		
4	Cotton	5'-AAATATTAACAATGCATTGAGTATGATG-3'	5'-TACTCATTGCTGATCCATGTAGATTTCCCG-3'	42
	3006-210-23	5'-ACTCTTTCTTTTTCTCCATATTGACC-3'		
4	Cotton	5'-TCCCATTCGAGTTTCTCACGT-3'	5'-TTGTCCCCTCCACTTCTTCTC-3'	58
	MON 531	5'-AACCAATGCCACCCCACTGA-3'		
4	Cotton	5'-CAGATTTTGTGGGATTGGAATTC-3'	5'-CTTAACAGTACTCGGCCGTCGACCGC-3'	59
	LLCotton25	5'-CAAGGAACTATTCAACTGAG-3'		
4	Maize	5'-CACACAGGAGATTATTATAGGG-3'		60
	Bt10	5'-GGGAATAAGGGGCGACACGG-3'		
4	Maize	5'-AAAAGACCACAACAACAAGCCGC-3'	5'-CGACCATGGACAACAACCCAAACATCA-3'	43
	Bt11	5'-CAATGCGTTCTCCACCAAGTACT-3'		
4	Maize	5'-CCTTCGCAAGACCCTTCCTCTATA-3'		21
	CBH-351	5'-GTAGCTGTCGGTGTAGTCCTCGT-3'		

4	Maize	5'-GGGATAAGCAAGTAAAAGCGCTC-3'	5'-TTTAAACTGAAGGCGGGAAACGACAA-3'	61
	DAS-59122-7	5'-CCTTAATTCTCCGCTCATGATCAG-3'		
4	Maize	5'-TGTTCACCAGCAGCAACCAG-3'	5'-CCGACGTGACCGACTACCACATCGA-3'	43
	Event 176	5'-ACTCCACTTTGTGCAGAACAGATCT-3'		
4	Maize	5'-GAAGCCTCGGCAACGTCA-3'	5'-AAGGATCCGGTGCATGGCCG-3'	43
	GA21	5'-ATCCGGTTGGAAAGCGACTT-3'		
4	Maize	5'-GCGCACGCAATTCAACAG-3'	5'-AGGCGGGAAACGACAATCTGATCATG-3'	62
	MIR604	5'-GGTCATAACGTGACTCCCTTAATTCT-3'		
4	Maize	5'-GATGCCTTCTCCCTAGTGTTGA-3'	5'-AGATACCAAGCGGCCATGGACAACAA-3'	43
	MON 810	5'-GGATGCACTCGTTGATGTTTG-3'		
4	Maize	5'-GTAGGATCGGAAAGCTTGGTAC-3'	5'-TGAACACCCATCCGAACAAGTAGGGTCA-3'	63
	MON 863	5'-TGTTACGGCCTAAATGCTGAACT-3'		
4	Maize	5'-ATGAATGACCTCGAGTAAGCTTGTTAA-3'	5'-TGGTACCA CGCGACACACTTCCACTC-3'	64
	NK603	5'-AGAGATAAACAGGATCCACTCAAACACT-3'		
4	Maize	5'-GCCAGTTAGGCCAGTTACCCA-3'	5'-TGCAGGCATGCCCGCTGAAATC-3'	43
	T25	5'-TGAGCGAAACCCTATAAGAACCCT-3'		

(continued)

Table 5.2 (continued) Validated PCR Methods for the Different Strategies of GMO Detection

P	Target	Primer Sequences	TaqMan Probe if Real-Time (5'-FAM 3'-TAMRA)	Reference
4	Maize	5'-TAGTCTTCGGCCAGAATGG-3'	5'-TAACTCAAGGCCCTCACTCCG-3'	65
	TC1507	5'-CTTTGCCAAGATCAAGCG-3'		
4	Potato	5'-GTGTCAAAACACAATTTACAGCA-3'	5'-AGATTGTCGTTTCCCGCCTTCAGTT-3'	44
	EH92-527-1	5'-TCCCTTAATTCTCCGCTCATGA-3'		
4	Rice	5'-TCTAGGATCCGAAGCAGATCGT-3'	5'-CCACCTCCCAACAATAAAAGGCCTG-3'	66
	LLRICE601	5'-GGAGGGCGCGGGAGTGT-3'		
4	Rice	5'-AGCTGGCGTAATAGCGAAGAGG-3'	5'-CGCACCGATTATTTATACTTTTAGTCCACCT-3'	45
	LLRICE62	5'-TGCTAAACGGGTGCATCGTCTA-3'		
4	Soybean	5'-GCAAAAAGCCGGTTAGCTCCT-3'	5'-CGGTCCTCCGATCGCCCTTCC-3'	67
	A2704-12	5'-ATTCAGGCTGCGCAACTGTT-3'		
4	Soybean	5'-CCGGAAAGGCCAGAGGAT-3'	5'-CCGGCTGCTTGCACCGTGAAG-3'	68
	GTS 40-3-2	5'-GGATTTCAGCATCAGTGGCTACA-3'		
4	Sugarbeet	5'-TGGGATCTGGGTGGCTCTAACT-3'	5'-AAGGCGGGAAACGACAATCT-3'	47
	H7-1	5'-AATGCTGCTAAATCCTGAG-3'		
4	Tomato	5'-GGATCCTTAGAAGCATCTAGT-3'		48
	Nema 282F	5'-CATCCGCAAGACCGGCAACAG-3'		

Note: P, purpose of the analysis; 1, presence of animal amplifiable material/identification of animal species; 2, presence of vegetal amplifiable material/identification of vegetal species; 3, identification of transgenic constructs; 4, identification of transgenic events.

spleen, kidney, lymphatic glands, ovary, *musculus longissimus dorsi*, *musculus trapezius*, and *gluteus maximus*. In contrast, the Bt maize Cry 1A(b) gene was never detected in tissue samples.

The persistence of plant-derived recombinant DNA in sheep and pigs fed genetically modified (RR) canola has been assessed by PCR and Southern hybridization analysis of DNA extracted from digesta, gastrointestinal tract tissues, and visceral organs.[71] The study confirmed that feed-ingested DNA fragments (endogenous and transgenic) do survive to the terminal gastrointestinal tract, and that uptake into gut epithelial tissues does occur; furthermore, a very low frequency of transmittance to visceral tissue was confirmed in pigs, but not in sheep.

A study was performed to assess whether processing and thermal treatments influence the detection of genetically modified DNA in different kinds of processed meat products (sausages, canned liver, ready-to-eat hamburgers) prepared with soybean meal spiked with a known amount of RR soybean.[20] The products were tested for the presence of specific 35S promoter and NOS terminator sequences, at different stages of processing, by PCR. The lowest contamination level (0.5%) was successfully detected in all raw and processed meat products at the different degrees of processing evaluated.

In a recent work, the detection of transgenic soybean was performed using a nested PCR protocol applied to several meat additives (blends, spices, taste enhancers), soy protein–based ingredients for meat products (soy protein and texturized soy protein), and processed meat samples (chicken mortadella, hot dog, cooked ham, hamburger, chicken-fried steaks) present on the Brazilian market.[72] The reported results indicated that RR soybean was detectable in 3 out of 18 of the meat additives, 12 out of 14 of the soy protein ingredients, and 3 out of 8 processed meats tested.

5.2.2   Protein-Based Methods

Apart from transformation events bearing an antisense sequence, GM plants usually undergo the insertion of transgenes coding for novel proteins. These proteins represent in most cases suitable targets for GMO detection. A wide spectrum of immunoassay-based technologies has been developed in the past decades, covering an enormous range of purposes and scientific disciplines.

5.2.2.1   Antibody-Based Assay Formats

5.2.2.1.1   Enzyme-Linked Immunosorbent Assay

ELISA is the most commonly employed technique among immunoassay strategies. ELISA assays allow the detection, and often the quantitation, of several classes of molecules such as proteins, peptides, antibodies, hormones, and other small molecules able to elicit immune response (haptens). A standard 96-well (or 384-well) polystyrene plate is the most common format used to perform ELISAs. The first step of the assay usually involves the target protein (antigen) absorption to a solid surface (direct ELISA) or the bounding of the antigen to a specific antibody, fixed at the bottom of a plate well (sandwich ELISA). The antigen is then bound by an antibody coupled with an enzyme (typically horseradish peroxidase [HRP] or alkaline phosphatase [AP]). After the formation of the complex, a substrate that produces a detectable product is added. Several substrates and instruments (luminometers, spectrophotometers, fluorometers) are available to meet the different technical needs.

Variants of ELISA assay with improved sensitivity have been developed using signal amplification strategies. The most common approach is based on the addiction of a secondary enzyme-labeled antibody that binds a primary antibody specifically linked to the antigen. The binding of several secondary antibodies to a single primary immunoglobulin results in a strong signal enhancement. Another strategy consists of forming a biotin/streptavidin–derived complex linking more copy numbers of the enzyme to the same antibody.

Competitive ELISA formats have also been developed. These assays are particularly suitable for molecules that have only one epitope or when only one specific antibody is available. Several applications of this format are available. One of the most common uses an enzyme-conjugated antigen as standard: unlabeled antigen (from sample) competes with known amounts of labeled antigens for a limited number of specific binding sites of a capture antibody fixed on the well plate.

The main advantages of ELISA assay are that it provides quantitative information using an economical, high-throughput, and non-labor-intensive approach.

5.2.2.1.2 Lateral Flow Assays

Lateral flow assay technology commonly consists of a nitrocellulose strip containing specific antibodies conjugated to a color reactant. One end of the strip is placed in a tube containing the protein extract, which then starts to flow to the other end of the strip. When the target protein is present, a complex with color reagent–conjugated antibodies is formed and passes through two capture zones containing respectively a second antigen-specific antibody (test line) and an antibody for the labeled immunoglobulin excess (control line). When both lines give a positive signal, the test indicates a positive sample. When only the control line is positive, the test gives a negative sample. Lateral flow strip tests are very inexpensive, take a short time to analyze, and do not require a high degree of technical skills to be performed. All these reasons make this assay particularly suitable for field tests.

Several drawbacks have so far limited the application of antibody-based assay formats in GMO detection: (i) the presence of other substances in complex matrices (other proteins, phenolic compounds, surfactants, fatty acids) can interfere with the assay; (ii) GM protein can be expressed in a very low amount, and the amount of the target protein expressed could be highly variable in different plant tissues or development stages; and (iii) matrices that undergo industrial processing, e.g., heating, could change the conformational structure of active epitopes, resulting in nonreactive proteins. This problem should be carefully evaluated for each sample when choosing the appropriate assay format.

Although protein-based methods have not found wide application in GMO detection if compared to PCR, several works report their use in this field, and innovative applications have also been developed and tested.[9,73,74]

5.2.2.2 Applications in Meat and Poultry Analysis

The potential presence in food products of novel proteins as a consequence of GMOs entering the food chain has become, in the last few years, a relevant issue at national and international policy levels, also raising concern among citizens. On account of this, several attempts to investigate the fate of transgenic proteins have been performed on livestock and derived productions.

The possible transfer of the Cry 9C protein to blood, liver, and muscle in broiler chicks fed with StarLink corn has been investigated.[75] The determination of Cry 9C protein in the analytical materials was performed using a commercial GMO Bt9 maize test kit, and no positive samples were detected in the examined tissues.

A study was conducted to determine the content of GM protein from RR soybeans in tissues and eggs of laying hens.[76] A commercial double antibody sandwich incorporated in a lateral flow strip format, specific for the CP4 EPSPS protein, has been used. Whole egg, egg albumen, liver, and feces were all negative for GM protein.

The attempt to detect the Cry 1A(b) protein in chicken breast muscle samples from animals fed YieldGard Corn Borer Corn event MON 810 has been published.[14] Analyses were performed using an in-house developed competitive ELISA with an LOD of approximately 60 ng of protein per gram of chicken muscle. Neither the Cry 1A(b) protein nor the immunoreactive peptide fragments were detectable in the breast muscle samples.

Using a similar strategy, the same author also investigated the presence of CP4 EPSPS protein in the muscle of pigs fed a diet containing RR soybean.[15] A competitive immunoassay, with an LOD of approximately 94 ng of CP4 EPSPS protein per gram of pork muscle, was developed by the authors and used to test samples; neither the CP4 EPSPS protein nor immunoreactive peptide fragments were detected in any samples.

In another work, three different assays to detect Cry 1A(b) protein in the gastrointestinal contents of pigs fed genetically modified corn Bt11 were employed.[77] Two commercial kits (a conventional microplate-format ELISA and a test strip format immunochromatographic assay) and immunoblotting were used to test pig samples. The Cry 1A(b) protein was detected in the contents of stomach, duodenum, ileum, cecum, and rectum.

5.2.3 Alternative Techniques for GMO Detection

With the number of GMOs developed by biotech companies constantly increasing and expected to have an even higher impact on worldwide cultivations and markets in the coming years,[2] new technologies and instruments will be needed to face the challenges of high throughput and affordable detection of an increasing number of transgenes. For both qualitative and quantitative analysis, routine procedures such as PCR and immunodetection methods appear to be inadequate when confronted with the future demand to screen very large numbers of different GMOs. Several analytical approaches have been used to develop new detection systems able to implement the currently available methodologies in terms of sensitivity, specificity, robustness, and sample throughput.

Although most of the work on the development of new detection methods cited in the literature mainly focuses on analytical systems for the detection of GMOs in grains or plant products, several approaches also seem to be suitable for performing analysis on more complex matrices, such as meat products.

NIR spectroscopy, usually employed for the nondestructive analysis of grains for the prediction of moisture, protein, oil, fiber, and starch, has been described as a tool to discriminate between sample sets of RR soybean and nontransgenic soybeans.[78] More recently, visible/NIR (vis/NIR) spectroscopy combined with multivariate analysis was used to analyze tomato leaves and successfully discriminate between genetically modified and conventional tomatoes.[79] Although NIR techniques combine rapidity, ease of use, and cost effectiveness, their ability to resolve small quantities of GM varieties is assumed to be low: in fact the technique discriminates according to structural changes that are larger than those produced by single gene modifications. Further advancement in the development of the technique still needs to be accomplished before it could be evaluated for use in complex matrices.

Some authors have proposed chromatographic techniques for the detection of GMOs. Conventional chromatographic methods combined with efficient detection systems such as mass spectrometry could be applicable when significant changes occur in the composition of GM plants or derived products. This approach has been used to investigate the triglyceride patterns of oil derived from GM canola, showing that increased triacylglycerol content characterizes the transgenic canola variety.[80] Matrix-assisted laser desorption/ionization time-of-flight (MALDI-TOF) and nanoelectrospray ionization quadrupole time-of-flight (nano ESI-QTOF) were successfully applied to the detection of the transgenic protein CP4 EPSPS in 0.9% GM soybean after fractionation by gel filtration, anion-exchange chromatography, and sodium dodecyl sulfate polyacrylamide gel electrophoresis (SDS-PAGE).[81]

Again these methodologies, although very sensitive, appear at present only to be suitable for differentiating between GM and conventional varieties, but they lack the specificity needed for detection in composite food matrices.

A recent application has been described that uses anion exchange liquid chromatography coupled with a fluorescent detector in combination with peptide nucleic acid (PNA) probes to detect and univocally identify PCR amplicons of RR soybean or Bt176 maize both on CRM and in commercial samples.[82]

5.2.3.1 DNA Microarray Technology

With the number of genetic targets to be monitored constantly increasing, the detection of GMOs in the near future appears to be moving toward the need for higher throughput analysis that can simultaneously detect a high number of targets of interest and lower the cost of detecting an increased variety of genetic targets. In this context, one of the more promising technologies available appears to be microarray systems. In their general form, microarray systems are oligonucleotide probe–based platforms on which a high number of nucleic acid targets can be simultaneously detected with high specificity. This would imply, in the case of GMO detection, the potential for rapid and efficient screening of a large number of control, gene-specific, and transgene-specific nucleic acid targets.

The main advantages of DNA microarray technology are miniaturization, high sensitivity, and screening throughput. Its main limitation is at present the strict dependence on PCR or other amplification techniques to amplify and label DNA or mRNA target sequences before performing the microarray analysis of a sample. The presence of this PCR step, at present still not likely to be overcome, imposes on this technology all the limitations discussed in the previous PCR section. Moreover, the possibility of quantifying GMO content in the sample is lost, because amplification and labeling are performed using end-point PCR, which is strictly qualitative. Different DNA microarray approaches, at both the research and the commercial stage, have been described for the detection of GMOs in food and feed systems, and their approach could be valuable also for the specific analysis of meat products.

A recent paper describes the development of a method for screening GMOs using multiplex-PCR coupled with oligonucleotide microarray.[83] The authors developed an array of 20 oligonucleotide probes for the detection of the majority of the genetic construct, covering 95% of commercially available transgenes (soybean, maize, cotton, and canola), with a detection limit of 0.5 and 1.0% for transgenic soybean and maize, respectively.

A multiplex DNA microarray chip was developed for simultaneous identification of nine GMOs, five plant species, and three GMO screening elements.[84] The targets were labeled with biotin during amplification, and the arrays could be detected using a colorimetric analysis with a detection limit below 0.3%.

A commercial microarray system for the qualitative detection of EU-approved GMOs has been recently commercialized in Europe.[85] The system combines the identification of GMOs by characterization of their genetic elements with a colorimetric detection based on silver.

A multiplex quantitative DNA array–based PCR (MQDA-PCR) method has been described for the quantification of seven different transgenic maize types in food and feed samples.[86] The authors were able to correctly characterize the presence of transgenic maize in the range 0.1–2.0% using a two-step PCR, which used opportunely labeled primers, and a DNA array spotted on a nylon membrane.

Ligation detection reaction (LDR), in combination with multiplex PCR and a universal array, has been described as a sensitive tool for GMO detection.[87] The authors were able to detect trace amounts of five transgenic events (maize and soybean) in heterogeneous samples both in reference materials and in commercial samples.

A class of synthetic oligonucleotide analogs with increased hybridization sensitivity and specificity has been described in a recent paper,[88] in which the authors used PNAs as capture probes for the detection of five GM maize and soybean products amplified by a multiplex PCR with a LOD of 0.25%.

5.2.3.2 Biosensors

Although only at research stage, several biosensor-based methods have been developed and tested for the detection of GMOs. Their main advantage is the fact that detection is based on physical principles, resulting in the possibility of performing the analysis in a faster and more economical way than conventional techniques. Their major drawback is that, as do the previously described techniques, they rely on PCR, because their sensitivity is not high enough for standalone analysis. As research on biosensors has continuously improved over the past few years, innovative techniques and detection systems are likely to be developed, which could in the near future adequately fulfill the requirements of GMO detection.

A biosensor based on quartz crystal microbalance (QCM) has been described for the detection of sequences of the 35S promoter and NOS terminator.[89] PCR products obtained from CRM and real samples were correctly identified in a label-free hybridization reaction showing how this approach could be a sensitive and specific method for the detection of GMOs in food samples.

An electrochemical biosensor based on disposable screen-printed gold electrodes has been recently described for the detection of characteristic sequences of soybean and the 35S promoter.[90] The applied detection scheme, based on the enzymatic amplification of hybridization signals by a streptavidin-AP conjugate, led to a highly sensitive detection of the target sequences without the need for chemical or physical treatment of the electrode surfaces.

A biosensor based on surface plasmon resonance (SPR) has been reported to allow for the discrimination between samples containing 0.5 and 2.0% Bt176 maize reference material.[91] The PCR products amplified by multiplex PCR were immobilized on the surface of the sensor, and oligonucleotide probes were flowed through the cell and hybridized to their specific target, generating a quantifiable signal.

References

1. AGBIOS, http://www.agbios.com/, 2007.
2. Clive, J., Global Status of Commercialized Biotech/GM Crops: 2007, *ISAAA Briefs* 37, 2008, Executive Summary.
3. Alexander, T. W., Reuter, T., Aulrich, K., Sharma, R., Okine, E. K., Dixon, W. T., and McAllister, T. A., A review of the detection and fate of novel plant molecules derived from biotechnology in livestock production, *Animal Feed Science and Technology* 133 (1–2), 31–62, 2007.
4. The Food Center for Food Safety, http://www.centerforfoodsafety.org/, 2007.
5. Anklam, E., Gadani, F., Heinze, P., Pijnenburg, H., and Van Den Eede, G., Analytical methods for detection and determination of genetically modified organisms in agricultural crops and plant-derived food products, *European Food Research and Technology* 214 (1), 3–26, 2002.
6. JRC-IHCP, http://gmo-crl.jrc.it/statusofdoss.htm.
7. JRC-IHCP, http://biotech.jrc.it/home/ict/methodsdatabase.htm.
8. Garcia-Canas, V., Cifuentes, A., and Gonzalez, R., Detection of genetically modified organisms in foods by DNA amplification techniques, *Critical Reviews in Food Science and Nutrition* 44 (6), 425–436, 2004.
9. Emslie, K. R., Whaites, L., Griffiths, K. R., and Murby, E. J., Sampling plan and test protocol for the semiquantitative detection of genetically modified canola (*Brassica napus*) seed in bulk canola seed, *Journal of Agricultural and Food Chemistry* 55 (11), 4414–4421, 2007.
10. Cankar, K., Štebih, D., Dreo, T., Žel, J., and Gruden, K., Critical points of DNA quantification by real-time PCR—effects of DNA extraction method and sample matrix on quantification of genetically modified organisms, *BMC Biotechnology* 6, 2006.
11. Holst-Jensen, A., Ronning, S. B., Lovseth, A., and Berdal, K. G., PCR technology for screening and quantification of genetically modified organisms (GMOs), *Analytical and Bioanalytical Chemistry* 375 (8), 985–993, 2003.
12. Somma, M., The analysis of food samples for the presence of genetically modified organisms. Session 4: Extraction and purification of DNA., http://gmotraining.jrc.it/, 2007.
13. Nemeth, A., Wurz, A., Artim, L., Charlton, S., Dana, G., Glenn, K., Hunst, P., Jennings, J., Shilito, R., and Song, P., Sensitive PCR analysis of animal tissue samples for fragments of endogenous and transgenic plant DNA, *Journal of Agricultural and Food Chemistry* 52 (20), 6129–6135, 2004.
14. Jennings, J., Albee, L. D., Kolwyck, D. C., Surber, J. B., Taylor, M. L., Hartnell, G. F., Lirette, R. P., and Glenn, K. C., Attempts to detect transgenic and endogenous plant DNA and transgenic protein in muscle from broilers fed yieldgard1 corn borer corn, *Poultry Science* 82, 371–380, 2003.
15. Jennings, J. C., Kolwyck, D. C., Kays, S. B., Whetsell, A. J., Surber, J. B., Cromwell, G. L., Lirette, R. P., and Glenn, K. C., Determining whether transgenic and endogenous plant DNA and transgenic protein are detectable in muscle from swine fed Roundup Ready soybean meal, *Journal of Animal Science* 81 (6), 1447–1455, 2003.
16. Zhang, C. L., Fowler, M. R., Scott, N. W., Lawson, G., and Slater, A., A TaqMan real-time PCR system for the identification and quantification of bovine DNA in meats, milks and cheeses, *Food Control* 18 (9), 1149–1158, 2007.
17. Laube, I., Spiegelberg, A., Butschke, A., Zagon, J., Schauzu, M., Kroh, L., and Broll, H., Methods for the detection of beef and pork in foods using real-time polymerase chain reaction, *International Journal of Food Science and Technology* 38 (2), 111–118, 2003.

18. Laube, I., Zagon, J., Spiegelberg, A., Butschke, A., Kroh, L. W., and Broll, H., Development and design of a 'ready-to-use' reaction plate for a PCR-based simultaneous detection of animal species used in foods, *International Journal of Food Science and Technology* 42 (1), 9–17, 2007.

19. Hird, H., Chisholm, J., Sanchez, A., Hernandez, M., Goodier, R., Schneede, K., Boltz, C., and Popping, B., Effect of heat and pressure processing on DNA fragmentation and implications for the detection of meat using a real-time polymerase chain reaction, *Food Additives and Contaminants* 23 (7), 645–650, 2006.

20. Ujhelyi, G., Jánosi, A., and Gelencsér, É., Effects of different meat processing techniques on the detection of GM soy from model meat samples, *Acta Alimentaria* 36 (1), 39–48, 2007.

21. Matsuoka, T., Kuribara, H., Akiyama, H., Miura, H., Goda, Y., Kusakabe, Y., Isshiki, K., Toyoda, M., and Hino, A., A multiplex PCR method of detecting recombinant DNAs from five lines of genetically modified maize, *Journal of the Food Hygienic Society of Japan* 42 (1), 24–32, 2001.

22. Hernandez, M., Rodriguez-Lazaro, D., Zhang, D., Esteve, T., Pla, M., and Prat, S., Interlaboratory transfer of a PCR multiplex method for simultaneous detection of four genetically modified maize lines: Bt11, MON 810, T25, and GA21, *Journal of Agricultural and Food Chemistry* 53 (9), 3333–3337, 2005.

23. Germini, A., Zanetti, A., Salati, C., Rossi, S., Forre, C., Schmid, S., and Marchelli, R., Development of a seven-target multiplex PCR for the simultaneous detection of transgenic soybean and maize in feeds and foods, *Journal of Agricultural and Food Chemistry* 52 (11), 3275–3280, 2004.

24. James, D., Schmidt, A. M., Wall, E., Green, M., and Masri, S., Reliable detection and identification of genetically modified maize, soybean, and canola by multiplex PCR analysis, *Journal of Agricultural and Food Chemistry* 51 (20), 5829–5834, 2003.

25. JRC-IRMM, http://irmm.jrc.ec.europa.eu/

26. Lee, S. H., Kim, J. K., and Yi, B. Y., Detection methods for biotech cotton MON 15985 and MON 88913 by PCR, *Journal of Agricultural and Food Chemistry* 55 (9), 3351–3357, 2007.

27. Weighardt, F., Barbati, C., Paoletti, C., Querci, M., Kay, S., De Beuckeleer, M., and Van den Eede, G., Real-time polymerase chain reaction-based approach for quantification of the pat gene in the T25 Zea mays event, *Journal of AOAC International* 87 (6), 1342–1355, 2004.

28. Studer, E., Rhyner, C., Luthy, J., and Hubner, P., Quantitative competitive PCR for the detection of genetically modified soybean and maize, *Zeitschrift Fur Lebensmittel-Untersuchung Und-Forschung a-Food Research and Technology* 207 (3), 207–213, 1998.

29. Hubner, P., Studer, E., Hafliger, D., Stadler, M., Wolf, C., and Looser, M., Detection of genetically modified organisms in food: critical points for duality assurance, *Accreditation and Quality Assurance* 4 (7), 292–298, 1999.

30. Hardegger, M., Brodmann, P., and Herrmann, A., Quantitative detection of the 35S promoter and the NOS terminator using quantitative competitive PCR, *European Food Research and Technology* 209 (2), 83–87, 1999.

31. Kubista, M., Andrade, J. M., Bengtsson, M., Forootan, A., Jonak, J., Lind, K., Sindelka, R., Sjoback, R., Sjogreen, B., Strombom, L., Stahlberg, A., and Zoric, N., The real-time polymerase chain reaction, *Molecular Aspects of Medicine* 27 (2–3), 95–125, 2006.

32. Hernandez, M., Rodriguez-Lazaro, D., Esteve, T., Prat, S., and Pla, M., Development of melting temperature-based SYBR Green I polymerase chain reaction methods for multiplex genetically modified organism detection, *Analytical Biochemistry* 323 (2), 164–170, 2003.

33. LaPaz, J. L., Esteve, T., and Pla, M., Comparison of Real-time PCR detection chemistries and cycling modes using MON 810 event-specific assays as model, *Journal of Agricultural and Food Chemistry* 55 (11), 4312–4318, 2007.

34. Andersen, C. B., Holst-Jensen, A., Berdal, K. G., Thorstensen, T., and Tengs, T., Equal performance of TaqMan, MGB, Molecular Beacon, and SYBR green-based detection assays in detection and quantification of Roundup Ready soybean, *Journal of Agricultural and Food Chemistry* 54 (26), 9658–9663, 2006.

35. Alexander, T. W., Sharma, R., Deng, M. Y., Whetsell, A. J., Jennings, J. C., Wang, Y. X., Okine, E., Damgaard, D., and McAllister, T. A., Use of quantitative real-time and conventional PCR to assess the stability of the cp4 epsps transgene from Roundup Ready® canola in the intestinal, ruminal, and fecal contents of sheep, *Journal of Biotechnology* 112 (3), 255–266, 2004.

36. Novelli, E., Balzan, S., Segato, S., De Rigo, L., and Ferioli, M., Detection of genetically modified organisms (GMOs) in food and feedstuff, *Veterinary Research Communications* 27 Suppl 1, 699–701, 2003.

37. Petit, L., Baraige, F., Balois, A.-M., Bertheau, Y., and Fach, P., Screening of genetically modified organisms and specific detection of Bt176 maize in flours and starches by PCR-enzyme linked immunosorbent assay, *European Food Research and Technology* 217 (1), 83–89, 2003.

38. Brunnert, H. J., Spener, F., and Borchers, T., PCR-ELISA for the CaMV-35S promoter as a screening method for genetically modified Roundup Ready soybeans, *European Food Research and Technology* 213 (4–5), 366–371, 2001.

39. Sawyer, J., Wood, C., Shanahan, D., Gout, S., and McDowell, D., Real-time PCR for quantitative meat species testing, *Food Control* 14 (8), 579–583, 2003.

40. Taberlet, P., Gielly, L., Pautou, G., and Bouvet, J., Universal primers for amplification of three non-coding regions of chloroplast DNA, *Plant Molecular Biology* 17 (5), 1105–1109, 1991.

41. JRC-IHCP, Event-specific method for the quantification of oilseed rape line T45 using real-time PCR, http://gmo-crl.jrc.it/statusofdoss.htm.

42. JRC-IHCP, Event specific methods for the quantification of the hybrid cotton line 281-24-236/3006-210-23 using real-time PCR, http://gmo-crl.jrc.it/statusofdoss.htm.

43. Shindo, Y., Kuribara, H., Matsuoka, T., Futo, S., Sawada, C., Shono, J., Akiyama, H., Goda, Y., Toyoda, M., Hino, A., Asano, T., Hiramoto, M., Iwaya, A., Jeong, S. I., Kajiyama, N., Kato, H., Katsumoto, H., Kim, Y. M., Kwak, H. S., Ogawa, M., Onozuka, Y., Takubo, K., Yamakawa, H., Yamazaki, F., Yoshida, A., and Yoshimura, T., Validation of real-time PCR analyses for line-specific quantitation of genetically modified maize and soybean using new reference molecules, *Journal of AOAC International* 85 (5), 1119–1126, 2002.

44. JRC-IHCP, Event-specific method for the quantification of event EH92-527-1 potato using real-time PCR, http://gmo-crl.jrc.it/statusofdoss.htm.

45. JRC-IHCP, Event-specific method for the quantification of rice line LLRICE62 using real-time PCR, http://gmo-crl.jrc.it/statusofdoss.htm.

46. Pauli, U., Liniger, M., Schrott, M., Schouwey, B., Hübner, P., Brodmann, P., and Eugster, A., Quantitative detection of genetically modified soybean and maize: method evaluation in a Swiss ring trial, *Mitt. Lebensm. Hyg* 92, 145–158, 2001.

47. JRC-IHCP, Event-specific method for the quantitation of sugar beet line H7-1 using real-time PCR, http://gmo-crl.jrc.it/statusofdoss.htm.

48. LMBG, Food analysis, *Collection of Official Methods under Article 35 of the German Federal Foodstuffs Act.* L 25.03.01, 1999.

49. Lipp, M., Bluth, A., Eyquem, F., Kruse, L., Schimmel, H., Van den Eede, G., and Anklam, E., Validation of a method based on polymerase chain reaction for the detection of genetically modified organisms in various processed foodstuffs, *European Food Research and Technology* 212 (4), 497–504, 2001.

50. Akiyama, H., Sugimoto, K., Matsumoto, M., Isuzugawa, K., Shibuya, M., Goda, Y., and Toyoda, M., A detection method of recombinant DNA from genetically modified potato (NewLeaf Plus potato) and detection of NewLeaf Plus potato in snack, *J. Food Hyg. Soc. Japan* 43, 301–305, 2002.

51. Jaccaud, E., Hohne, M., and Meyer, R., Assessment of screening methods for the identification of genetically modified potatoes in raw materials and finished products, *Journal of Agricultural and Food Chemistry* 51 (3), 550–557, 2003.

52. LMBG, Food analysis, *Collection of Official Methods under Article 35 of the German Federal Foodstuffs Act* L 24.01-1, 1997.

53. Beck, E., Ludwig, G., Auerswald, E. A., Reiss, B., and Schaller, H., Nucleotide sequence and exact localization of the neomycin phosphotransferase gene from transposon Tn5, *Gene* 19 (3), 327–336, 1982.

54. JRC-IHCP, A recommended procedure for real-time quantitative TaqMan PCR for roundup ready canola RT73, http://gmo-crl.jrc.it/detectionmethods/MON-Art47-pcrGT73rapeseed.pdf.

55. JRC-IHCP, Event-specific method for the quantification of oilseed rape line Ms8 using real-time PCR, http://gmo-crl.jrc.it/summaries/Ms8_validated_Method.pdf.

56. JRC-IHCP, Event-specific method for the quantification of oilseed rape line Rf38 using real-time PCR, http://gmo-crl.jrc.it/summaries/Rf3_validated_Method.pdf.

57. JRC-IHCP, A recommended procedure for real-time quantitative TaqMan PCR for roundup ready cotton 1445, http://gmo-crl.jrc.it/detectionmethods/MON-Art47-pcr1445cotton.pdf.

58. JRC-IHCP, A recommended procedure for real-time quantitative TaqMan PCR for Bollgard cotton 531, http://gmo-crl.jrc.it/detectionmethods/MON-Art47-pcr531cotton.pdf.

59. JRC-IHCP, Event-specific method for the quantification of cotton line LLCotton25 using real-time PCR, http://gmo-crl.jrc.it/summaries/LLCotton25_validated_Method.pdf.

60. JRC-IHCP, PCR assay for the detection of maize transgenic event Bt10, http://gmo-crl.jrc.it/summaries/Bt10%20Detection%20Protocol.pdf.

61. JRC-IHCP, Event-specific method for the quantitation of maize line DAS-59122-7 using real-time PCR, http://gmo-crl.jrc.it/statusofdoss.htm.

62. JRC-IHCP, Event-specific method for the quantification of maize line MIR604 using real-time PCR, http://gmo-crl.jrc.it/summaries/MIR604_validated_Method.pdf.

63. JRC-IHCP, Event-specific method for the quantitation of maize line MON 863 using real-time PCR, http://gmo-crl.jrc.it/statusofdoss.htm.

64. JRC-IHCP, Event-specific method for the quantitation of maize line NK603 using real-time PCR, http://gmo-crl.jrc.it/statusofdoss.htm.

65. JRC-IHCP, Event-specific method for the quantitation of maize line TC1507 using real-time PCR, http://gmo-crl.jrc.it/statusofdoss.htm.

66. JRC-IHCP, Report on the verification of an event-specific detection method for identification of rice GM-event LLRICE601 using a real-time PCR assay, http://gmo-crl.jrc.it/LLRice601update.htm.

67. JRC-IHCP, Event-specific method for the quantification of soybean line A2704-12 using real-time PCR, http://gmo-crl.jrc.it/summaries/A2704-12_soybean_validated_Method.pdf.

68. Hird, H., Powell, J., Johnson, M. L., and Oehlschlager, S., Determination of percentage of Roundup Ready® Soya in soya flour using real-time polymerase chain reaction: interlaboratory study, *Journal of AOAC International* 86 (1), 66–71, 2003.

69. Einspanier, R., Klotz, A., Kraft, J., Aulrich, K., Poser, R., Schwagele, F., Jahreis, G., and Flachowsky, G., The fate of forage plant DNA in farm animals: a collaborative case-study investigating cattle and chicken fed recombinant plant material, *European Food Research and Technology* 212 (2), 129–134, 2001.

70. Reuter, T. and Aulrich, K., Investigations on genetically modified maize (Bt-maize) in pig nutrition: fate of feed-ingested foreign DNA in pig bodies, *European Food Research and Technology* 216 (3), 185–192, 2003.

71. Sharma, R., Damgaard, D., Alexander, T. W., Dugan, M. E. R., Aalhus, J. L., Stanford, K., and McAllister, T. A., Detection of transgenic and endogenous plant DNA in digesta and tissues of sheep and pigs fed roundup ready canola meal, *Journal of Agricultural and Food Chemistry* 54 (5), 1699–1709, 2006.

72. Ca Brod, F. and Arisi, A. C. M., Recombinant DNA in meat additives: specific detection of Roundup Ready Soybean by nested PCR, *Journal of the Science of Food and Agriculture* 87 (10), 1980–1984, 2007.

73. Fantozzi, A., Ermolli, M., Marini, M., Scotti, D., Balla, B., Querci, M., Langrell, S. R. H., and VandenEede, G., First application of a microsphere-based immunoassay to the detection of genetically modified organisms (GMOs): quantification of Cry1Ab protein in genetically modified maize, *Journal of Agricultural and Food Chemistry* 55 (4), 1071–1076, 2007.

74. Ermolli, M., Prospero, A., Balla, B., Querci, M., Mazzeo, A., and Eede, G. V. D., Development of an innovative immunoassay for CP4EPSPS and Cry1AB genetically modified protein detection and quantification, *Food Additives & Contaminants* 23 (9), 876–882, 2006.

75. Yonemochi, C., Fujisaki, H., Harada, C., Kusama, T., and Hanazumi, M., Evaluation of transgenic event CBH 351 (StarLink) corn in broiler chicks, *Animal Science Journal* 73 (3), 221–228, 2002.

76. Ash, J., Novak, C., and Scheideler, S. E., The fate of genetically modified protein from roundup ready soybeans in laying hens, *The Journal of Applied Poultry Research* 12 (2), 242–245, 2003.

77. Chowdhury, E. H., Kuribara, H., Hino, A., Sultana, P., Mikami, O., Shimada, N., Guruge, K. S., Saito, M., and Nakajima, Y., Detection of corn intrinsic and recombinant DNA fragments and Cry1Ab protein in the gastrointestinal contents of pigs fed genetically modified corn Bt11, *Journal of Animal Science* 81 (10), 2546–2551, 2003.

78. Roussel, S. A., Hardy, C. L., Hurburgh, C. R., and Rippke, G. R., Detection of Roundup Ready$^{(TM)}$ Soybeans by near-infrared spectroscopy, *Applied Spectroscopy* 55 (10), 1425–1430, 2001.

79. Xie, L., Ying, Y., and Ying, T., Quantification of chlorophyll content and classification of non-transgenic and transgenic tomato leaves using visible/near-infrared diffuse reflectance spectroscopy, *Journal of Agricultural and Food Chemistry*, 55 (12), 4645–4650, 2007.

80. Byrdwell, W. C. and Neff, W. E., Analysis of genetically modified canola varieties by atmospheric pressure chemical ionization mass spectrometric and flame ionization detection, *Journal of Liquid Chromatography & Related Technologies* 19 (14), 2203–2225, 1996.

81. Ocana, M. F., Fraser, P. D., Patel, R. K. P., Halket, J. M., and Bramley, P. M., Mass spectrometric detection of CP4 EPSPS in genetically modified soya and maize, *Rapid Communications in Mass Spectrometry* 21 (3), 319–328, 2007.

82. Rossi, S., Lesignoli, F., Germini, A., Faccini, A., Sforza, S., Corradini, R., and Marchelli, R., Identification of PCR-amplified genetically modified organisms (GMOs) DNA by peptide nucleic acid (PNA) probes in anion-exchange chromatographic analysis, *Journal of Agricultural and Food Chemistry* 55 (7), 2509–2516, 2007.

83. Xu, J., Miao, H. Z., Wu, H. F., Huang, W. S., Tang, R., Qiu, M. Y., Wen, J. G., Zhu, S. F., and Li, Y., Screening genetically modified organisms using multiplex-PCR coupled with oligonucleotide microarray, *Biosensors & Bioelectronics* 22 (1), 71–77, 2006.

84. Leimanis, S., Hernández, M., Fernández, S., Boyer, F., Burns, M., Bruderer, S., Glouden, T., Harris, N., Kaeppeli, O., Philipp, P., Pla, M., Puigdomènech, P., Vaitilingom, M., Bertheau, Y., and Remacle, J., A Microarray-based detection system for genetically modified (GM) food ingredients, *Plant Molecular Biology* 61 (1), 123–139, 2006.

85. Eppendorf, DualChip® GMO kit, http://www.eppendorf.com, 2007.

86. Rudi, K., Rud, I., and Holck, A., A novel multiplex quantitative DNA array based PCR (MQDA-PCR) for quantification of transgenic maize in food and feed, *Nucleic Acids Research* 31 (11), e62, 2003.

87. Bordoni, R., Germini, A., Mezzelani, A., Marchelli, R., and De Bellis, G., A microarray platform for parallel detection of five transgenic events in foods: a combined polymerase chain reaction—ligation detection reaction—universal array method, *Journal of Agricultural and Food Chemistry* 53 (4), 912–918, 2005.

88. Germini, A., Rossi, S., Zanetti, A., Corradini, R., Fogher, C., and Marchelli, R., Development of a peptide nucleic acid array platform for the detection of genetically modified organisms in food, *Journal of Agricultural and Food Chemistry* 53 (10), 3958–3962, 2005.

89. Mannelli, I., Minunni, M., Tombelli, S., and Mascini, M., Quartz crystal microbalance (QCM) affinity biosensor for genetically modified organisms (GMOs) detection, *Biosensors & Bioelectronics* 18 (2–3), 129–140, 2003.
90. Carpini, G., Lucarelli, F., Marrazza, G., and Mascini, M., Oligonucleotide-modified screen-printed gold electrodes for enzyme-amplified sensing of nucleic acids, *Biosensors & Bioelectronics* 20 (2), 167–175, 2004.
91. Feriotto, G., Gardenghi, S., Bianchi, N., and Gambari, R., Quantitation of Bt-176 maize genomic sequences by surface plasmon resonance-based biospecific interaction analysis of multiplex polymerase chain reaction (PCR), *Journal of Agricultural and Food Chemistry* 51 (16), 4640–4646, 2003.

Chapter 6

Detection of Adulterations: Addition of Foreign Proteins

María Concepción García López
and María Luisa Marina Alegre

Contents

6.1   Reasons for the Addition of Foreign Proteins in Processed Meats.................................156
    6.1.1   Stabilization and Sensory Improvement of Processed Meats...........................156
    6.1.2   Reduction of Meat Fat Content...156
    6.1.3   Exploitation of Low-Quality Meats...157
    6.1.4   Health Benefits..157
6.2   Kinds of Foreign Proteins Added to Processed Meats...157
6.3   Methods Used for the Detection of Foreign Proteins in Processed Meats.......................158
6.4   Electrophoretic Methods for the Detection of Foreign Proteins in
    Processed Meats 158
6.5   Immunological Methods for the Detection of Foreign Proteins in
    Processed Meats 164
    6.5.1   Serology..164
    6.5.2   Immunodiffusion 164
    6.5.3   Indirect Hemagglutination..168
    6.5.4   Immunological Methods Comprising Electrophoretical Separations...................169
    6.5.5   Immunoassays...170
6.6   Chromatographic Methods for the Detection of Foreign Proteins in
    Processed Meats 172
    6.6.1   Analysis of Amino Acids 172
    6.6.2   Analysis of Peptides 174
    6.6.3   Analysis of Whole Proteins..174

6.7   Other Methods for the Detection of Foreign Proteins in Processed Meats 176
Acknowledgment 177
References 177

6.1   Reasons for the Addition of Foreign Proteins in Processed Meats

The addition of foreign proteins to processed meats is a very common practice. The main aims of such addition are to assist in the management and production of these products, especially to improve the water-binding capacity of meat, resulting in less water exudation upon sterilization, and, in the case of comminuted meats, to assist in the emulsion of fat particles. Other reasons are to obtain less-fatty meat products, to exploit low-quality meat pieces, and, in the case of soybean proteins, to obtain health benefits.

6.1.1   Stabilization and Sensory Improvement of Processed Meats

Comminuted meat products are complex food systems in which water absorption, gelation, and emulsion formation influence stability and sensory characteristics of the cooked product. During comminution of fine sausage emulsions, a relatively large amount of small fat or oil droplets are liberated from the fat cells. All this fat needs sufficient protein coating to prevent it from flowing back together during heating. This task is performed by the soluble myofibrillar proteins present in the meat, which also act to bind meat water. Nevertheless, frequently the meat protein content in processed meats is insufficient to support an emulsion, and foreign proteins are usually added to stabilize it. Different sources of foreign proteins have been added to meat emulsions and numerous studies have reported the benefits of these additions.[1–4]

Foreign proteins are also added for the improvement of organoleptic characteristics such as texture,[5–9] color,[4,10] flavor,[11] and, in general, the quality of the final product.[12,13] Fermented sausages are another kind of processed meats (not heat treated) to which the addition of foreign proteins is standard. The reason for such addition is to improve water-binding and textural properties that are damaged during vacuum packaging. For example, the addition of 2.5% of soybean protein isolate (SPI) prevents drip loss without introducing any change in the flavor, aroma, or juiciness characteristics of the product.[14,15]

6.1.2   Reduction of Meat Fat Content

Processed meats normally contain higher fat content than whole-muscle products. Fat provides flavor, texture, juiciness, and water entrapment. Therefore, lowering the fat content in emulsified products has been reported to increase toughness and significantly alter the texture, flavor, and color of the resulting low-fat product.[16,17]

The replacement of fat by water is an alternative, but resulting products have been reported to increase cooking and purge losses. Another challenge is the formulation of low-salt meat products, since the use of low sodium chloride content affects the water-holding capacity and emulsifying properties of meat. The addition of foreign proteins, especially soybean and milk proteins, to comminuted meats can balance these negative effects.[18–23] In fact, added proteins are capable of forming gels upon heating entrapping liquid and moisture. This gelling action in a low-fat/high-added-water formulation has the potential to return some of the texture often lost when levels of water addition are high.[24]

Another approach to the reduction of fat content is the direct addition of foreign proteins as fat replacers (protein-based fat replacers or substitutes).[17,25–27]

6.1.3 Exploitation of Low-Quality Meats

The meat industry is constantly looking for ways to enable the efficient utilization of meat from spent or aged animals. Spent animal meat is tougher and less juicy due to high collagen content and a high degree of crosslinkages. Quality attributes of spent animal meat can be improved by the addition of foreign proteins, especially milk proteins.[13,28–31]

Another approach to the exploitation of low-quality meat is the manufacture of *restructured meats*. Restructuration of meat uses less-valuable meat pieces to produce palatable meat products at reduced cost. Binding of these meat pieces and texture of the final product are the main characteristics that influence the acceptability of these products. Cohesion among meat pieces in structured meat products is accomplished by the formation of a protein matrix after extraction of muscle proteins, which requires the addition of salts and tumbling. The process brings salt-soluble meat proteins to the meat surface, forming a tacky exudate that coagulates upon cooking to bond the meat pieces into a continuous body. Nevertheless, due to damage to muscle texture produced during the tumbling and to the increasing concern of consumers over the sodium content of food, nonmeat proteins have been in demand as binders in restructured meats.[32–38]

6.1.4 Health Benefits

The consumption of soybean protein is related to health benefits. New food-based recommendations issued by the American Heart Association with the objective of reducing risk for cardiovascular disease promoted the inclusion in the diet of specific foods with cardioprotective effects, including soybean. The available evidence indicates that the daily consumption of 25 g of soybean protein could decrease total and low-density lipoprotein (LDL)-cholesterol levels in hypercholesterolemic individuals.[39–41]

6.2 Kinds of Foreign Proteins Added to Processed Meats

The foreign proteins most frequently added to processed meats are soybean proteins, wheat gluten, and milk proteins. Other proteins used to a lesser extent are corn gluten, blood plasma, pea proteins, and egg proteins.[24]

Soybean proteins can be added to meat products as textured soybean (50% protein), soybean protein concentrate (70% proteins), or SPI (90% proteins).[42] Water solubility of soybean proteins significantly contributes to improve functional properties of soybean-containing products, including water-holding capacity, foaming properties, appearance, and texture. Moreover, modification of soybean proteins by ultracentrifugation, low-dose irradiation, or treatment with various chemicals (e.g., proteolytic enzymes) contributes to the improvement of soybean protein functionality.[43,44] An even greater improvement of soybean protein functionality can be achieved by the heating of these proteins before their addition to meat. In fact, the high denaturation temperatures of the major soybean proteins (75–90°C) prevent the protein from undergoing sufficient structural changes under common meat heating conditions (65–73°C), thereby limiting their interaction with meat proteins and not contributing to meat gelling properties.[45]

Other vegetable proteins have also been added to processed meats (especially sausages), but their use is far less common than the industry applications of soybean proteins. For example, the addition of wheat gluten is advantageous due to its functionality and low cost but is limited due to its poor solubility. Chemical (acid deamidation), enzymatical, or physical modification of wheat gluten can result in a product with enhanced functional properties.[46,47] The case of corn gluten meal is similar, since it is not suitable for use in the food industry due to its low functionality, poor solubility, etc. Nevertheless, a simple hydrolysis of native corn gluten meal or increasing the pH of the native corn gluten meal results in an improvement of functional properties.[48]

Various milk products (nonfat dry milk, whey proteins, sodium caseinate, etc.) have been added to meat products. Skim milk powder (35% protein), which is widely used as filler in comminuted meat products, has good water-binding properties, but lactose may cause discoloration of meat products because of Maillard reactions. Whey proteins act as binders and extenders, gelling when they are heated. Sodium caseinate (90% protein) is completely soluble in water and in solutions with pH lower than 9, emulsifying up to 188 mL of oil/g of protein.[49] Nevertheless, in comparison with SPI, the incorporation of sodium caseinate results in high moisture loss.[50]

6.3 Methods Used for the Detection of Foreign Proteins in Processed Meats

There is an extensive literature dealing with the detection of foreign proteins, especially soybean proteins, in processed meat products. Methods can be divided in two groups—methods determining soybean proteins based on the presence of substances accompanying these proteins and methods based on the determination of proteins themselves.

Chemical methods have been employed for the determination of certain compounds or tracers that could reveal the presence of certain foreign proteins. The compounds analyzed were oligosaccharides, amino acids, phytate or phytic acid, metals, etc. The main drawback of these methods is their low specificity.[51] Microscopic methods enable the visualization of characteristic structural forms of the soybean such as palisade and hourglass cells present in the bean hull and calcium oxalate crystals from the cotyledon cells. In the case of soybeans, histological methods based on the selective stain of certain compounds present in the bean, normally carbohydrate-containing cells, have also been employed. These methods proved useful when soybean flour and textured soybean were added, but their application was limited when soybean protein concentrates or isolates were employed.[52–54]

Currently, the most common methods employed for the determination of foreign proteins in meat products are based on electrophoresis, immunological reactions, and chromatography.

6.4 Electrophoretic Methods for the Detection of Foreign Proteins in Processed Meats

The use of electrophoretic techniques for the determination of foreign proteins in meat products requires the prior solubilization of these proteins. Protein solubilization is more difficult, with the most severely heated samples necessitating the use of detergents or concentrated solutions of urea containing mercaptoethanol to disrupt disulfide crosslinks. Regarding the support material, most electrophoretic methods use polyacrylamide gels (polyacrylamide gel electrophoresis [PAGE]), although starch gels and cellulose acetate membranes have also been employed. Most PAGE methods employ sodium dodecyl sulfate (SDS). SDS not only solubilizes the proteins but also confers a negative charge in proportion to their mass. Since the mass-to-charge ratio is uniform for most proteins, all proteins migrating to the cathode will cross the gel matrix and will separate as a function of their molecular weights. Table 6.1 groups the electrophoretic methods developed for the determination of foreign proteins in processed meats, most of them devoted to soybean proteins.

Olsman[55] and Thorson et al.[56] reported the first methods using electrophoresis for the detection of soybean proteins and caseins in heated meats. In both cases, urea was employed for the solubilization of foreign proteins, although Olsman mixed it with mercaptoethanol. The main difference between them was the supporting material and the electrophoretic mode employed, a starch gel in slab in the case of Olsman and a polyacrylamide gel in tube in the case of Thorson and coworkers. Olsman obtained significantly better detection limits than Thorson and coworkers, who, in addition, had difficulties in the detection of soybean proteins due to co-elution of meat bands with the main soybean protein bands. Nevertheless, Olsman's method was not adequate for routine analysis due to the lengthy time required for a single analysis (24 h). Detection of soybean proteins by PAGE was improved by Freimuth and Krause[57] (in the slab mode) and by Fischer and Belitz[58] (in tube). While Freimuth and Krause extracted soybean proteins with a urea-lactate buffer and separated them at pH 3.1, Fischer and Belitz employed a tris-glycine buffer and the separation was carried out at basic pH. Fischer and Belitz's method was valid for highly cooked sausages, yielding results within 12 h.

Hofmann and Penny[59,60] developed another approach based on the use of SDS-PAGE in slab and a tris-boric acid buffer for the extraction of proteins. The method enabled the detection of soybean proteins in meat products heated up to 100°C, whereas those heated to higher temperatures (121°C) showed less clearly defined bands. Hofmann[61,62] also applied the method to the identification of foreign proteins other than soybean (egg white, egg yolk, milk, and wheat proteins) in meat products. Every protein showed a characteristic pattern that enabled its identification, with the exceptions of egg yolk proteins and wheat proteins, which could not be identified because their protein pattern was very complex (in the case of egg yolk) or was not stained properly (in the case of wheat proteins).[61,62] Other authors tried to improve Hofmann and Penny's method. Mattey[85] and Smith[86] used a 6% acrylamide gel and Bergen and Bosch[87] employed 10% instead of the 8% used by Hofmann and Penny. Moreover, Smith[86] and Endean[88] also cooled the front of the gel to avoid band distortions. Parsons and Lawrie[63] also applied an electrophoretic method similar to Hofmann and Penny's. In this case, proteins were extracted with a buffered solution containing 10 M urea and the acrylamide concentration was varied from 3 to 8%. The method enabled the quantification of soybean proteins in meat products heated up to 100°C, while at sterility temperatures (127°C for 24 min) only qualitative identification was possible, with no interference observed from field beans or egg albumin.[63] A further investigation on the reliability of this method was performed by Tateo.[89]

Spell[64] and Frouin et al.[65] focused their efforts on the improvement of sensitivity in the determination of soybean and milk proteins by PAGE in sterilized meats. Frouin et al.[65] proposed a first fractionation of proteins to eliminate those high molecular-weight interfering proteins. Detection limits obtained by this method were better than those yielded by the PAGE method of Spell.[64] Lee et al.[66] proposed the use of a preconcentration technique based on SDS-PAGE to detect soybean proteins in cooked meat-soybean blends. This preconcentration step yielded high-resolution separations and accurate determinations of soybean proteins in the presence of milk proteins and egg white proteins.

Table 6.1 Electrophoretic Methods for the Determination of Foreign Proteins in Processed Meats

Sample	Foreign Proteins	Technique	Detection Limit	References
Luncheon meat heated at 115°C and liver paste heated at 105°C	Soybean proteins and caseins	Urea-starch gel electrophoresis	0.50% for soybean proteins and 0.25% for caseins	55
Heated meat products (110°C)	Soybean proteins and caseins	PAGE (in tube)	3% for caseins	56
Cooked sausages	Soybean proteins	PAGE	—	57
Sausages (116°C)	Soybean proteins	PAGE (in tube)	—	58
Heated meats (pork and beef) (100°C) and sausages	Soybean, egg, milk, and wheat proteins	SDS-PAGE	5% for soybean proteins	59–62
Heated meats (sausages, pies, and beefburgers) (100°C)	Soybean proteins, field bean proteins, and egg albumin	SDS-PAGE	—	63
Heated meats (sausages) (120°C)	Soybean and milk proteins	PAGE	2% for soybean and milk proteins	64
Meat products (pate, ham, and sausages) sterilized at 117°C for 1 h 15 min	Soybean and milk proteins	SDS-PAGE (in tube)	1% for soybean and milk proteins	65
Cooked model meats	Soybean, milk, and egg white proteins	Stacking SDS-PAGE	—	66
Beef burgers, sausages, pies, and canned meat autoclaved at 110–115°C	Soybean proteins	PAGE	—	67,68
Cooked meats (paté, corned beef, bolognaise sauce, ravioli, and sausages)	Soybean proteins, wheat gluten, and milk proteins	Urea-PAGE (in tube)	1% for soybean proteins	69
Sausages and beefburgers	Soybean proteins	Isoelectric-focusing PAGE	—	70–72

Sample	Proteins	Method	Detection limit	Ref.
Heated pork and beef (74°C for 150 min)	Soybean proteins and caseins	PAGE (in tube)	1% for soybean proteins and caseins	73
Cooked sausages	Soybean proteins and caseins	PAGE (in tube)	—	74
Pasteurized meats (sausages and ham; 70°C)	Soybean proteins	SDS-PAGE	0.5% for soybean proteins	75
Model products consisting of roe and deer meats heated up to 70°C	Soybean proteins, caseins, and egg white proteins	PAGE (in tube)	—	76
Model cooked beef and pork meats	Soybean proteins	SDS-PAGE	—	77
Frankfurters	Soybean proteins	PoroPAGE	—	78
Frankfurters	Soybean proteins	SDS-PAGE	3% for soybean proteins	79
Model pork and beef meats autoclaved to 118°C for 20 min	Soybean proteins, sunflower proteins, and field bean proteins	SDS-PAGE	—	80
Model beef frankfurters	Soybean proteins	SDS-PAGE	<1% for soybean proteins	81
Cooked pork meat products (ham)	Soybean proteins, whey proteins, and caseins	SDS-PAGE	0.5% for soybean proteins and caseins and 1% for whey proteins	82
Model meat samples cooked to 100°C for 15 min, hamburgers, and sausages	Soybean proteins, wheat gluten, milk proteins, and egg proteins	SDS-PAGE	—	83,84

A comparative study of two different methods (extraction of proteins with a solution containing 8 M urea and 1% 2-mercaptoethanol at 18–20°C for 16 h (method 1, based on Olsman's[55] approach) and extraction of proteins with 10 M urea and 4% 2-mercaptoethanol at 100°C for 30 min (method 2, based on Parsons and Lawrie's[63] approach)) for the extraction of soybean proteins in meat products) was published by Guy et al.[67] Figure 6.1 shows the densitograms obtained for different soybean protein sources and for meatloaf with and without 5% SPI using both methods. The three soybean protein sources showed characteristic peaks that could be observed in the pattern corresponding to the meatloaf containing soybean proteins (peaks 1, 2, and 3 by method 1 and peaks 4 and 5 by method 2). From these results, the authors concluded that method 1 provided a better separation of soybean proteins in cooked meats than method 2. Moreover, this method was reproducible and free from interference from other nonmeat proteins (milk proteins, egg proteins, and wheat gluten).[68]

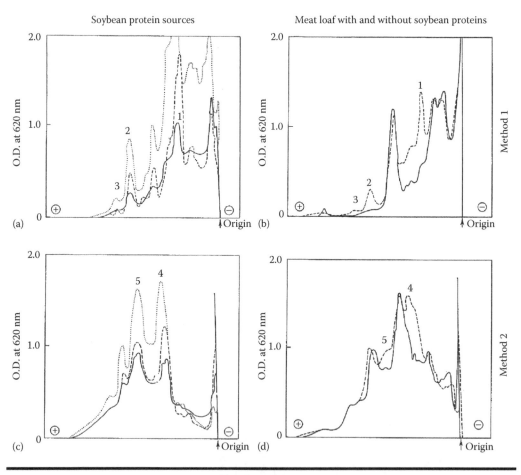

Figure 6.1   Densitograms corresponding to the SDS-PAGE separation of different soybean protein sources (textured soybean [—], SPI [- - -], and soybean flour [. . .]) and a meat loaf with (- - -) and without (—) SPI by method 1 (based on Olsman's[55] approach) (a,b) and method 2 (based on Parsons and Lawrie's[67] approach) (c,d). Labeled bands 1, 2, 3, 4, and 5 were from soybean proteins. (From Guy, R.C.E. et al., *J. Sci. Food. Agric.*, 24, 1551, 1973. With permission.)

Homayounfar[69] took up again Freimuth and Krause's[57] idea of developing the protein separation in an acid environment instead of the neutral or slightly alkaline conditions normally used. Four bands pertaining to soybean proteins were observed when the concentration of soybean proteins in the meat was higher than 5%, two of them disappearing at lower proportions. This method was used by Baylac et al. for the determination of various foreign proteins (soybean proteins, caseins, whey proteins, egg white proteins, wheat gluten, and blood plasma) in model fresh, pasteurized, or canned meat products.[90]

Isoelectric focusing in polyacrylamide gels enables higher resolution than conventional electrophoresis and has been applied to the determination of soybean proteins in cooked meats.[70–72] Although patterns observed by isoelectric focusing were more complex than those obtained by conventional electrophoresis, they could be simplified since meat bands disappeared when applying limited heating. The technique proved adequate for raw meats but failed with severely heated meat products since denaturation of soybean proteins made them insoluble in the extracting solution (urea-mercaptoethanol). A similar conclusion was drawn by Vállas-Gellei,[73] who observed that samples heated to 74°C for 150 min yielded weaker meat bands due to the high sensitivity of meat proteins to thermal denaturation, while soybean protein and casein bands remained unchanged or even stronger.

Other approaches have been developed to improve different aspects of the application of electrophoresis to the determination of foreign proteins in processed meats. Richardson[74] reduced the whole analysis time from the 2–3 days usually required in the slab mode to a single working day. Armstrong et al.[75] proposed the use of an internal standard protein (hemocyanin) to compensate for variations in the meat pattern and obtained accurate determination of soybean proteins in meats. Ring et al.[76] developed a unique separation method enabling the simultaneous differentiation between closely related meat species and the identification of added nonmeat proteins (caseins, egg white albumin, and soybean proteins) in cooked meat products. Molander[77] compared standard curves obtained by SDS-PAGE for the determination of soybean proteins in meat products subjected to different degrees of heat treatment. Although the method was accurate for raw or slightly heated meats, it failed with severely heated meats. In any case, the presence of other ingredients (milk powder, potato flour, bread-crumbs, caseins, and whole blood) did not seem to affect the determination of soybean proteins. Heinert and Baumann[78] proposed the use of a porosity gradient in PAGE in the presence of SDS and urea to obtain two soybean protein bands separated from those of meat proteins, which proved adequate for the detection of soybean proteins in sausages. Feigl[79] proposed an SDS-PAGE method using commercially available gel plates for its application as a routine procedure for the determination of soybean proteins in meat products.

Lacourt et al.,[80] Woychik et al.,[81] and López et al.[82] applied essentially the Laemmli[91] SDS-PAGE procedure using a tris-glycine buffer for the detection of soybean proteins in heated meats. This stacked buffer system provided a resolution above that obtained without stacking. Lacourt et al.[80] studied model beef and pork meats sterilized at 118°C for 20 min that contained soybean, sunflower, or field bean proteins. Despite the high resolution power of the method, they observed that, especially at low concentrations, differentiation among these three foreign proteins was not feasible. Woychik et al.[81] applied the Laemmli procedure to quantitate soybean proteins in pasteurized frankfurters based on the α-conglycinin/actin peak height ratios. López et al.[82] applied the method to the determination of soybean proteins in cooked ham, and was able to quantitate down to 0.5% of soybean proteins and caseins and 1% of whey proteins.

Olivera Carrión and Valencia[83] developed a PAGE method in the slab mode enabling the identification of soybean proteins in various model and commercial processed meats heated to 100°C. Quantification was performed from the area ratio corresponding to the bands appearing at 19,500 and 52,000 Da. No interferences from meat proteins or other extrinsic proteins (egg proteins, wheat gluten, milk casein, and whey) were observed.[84]

6.5 Immunological Methods for the Detection of Foreign Proteins in Processed Meats

Many immunological methods have been developed for the determination of foreign proteins in processed meat, soybean proteins being those most extensively determined. The determination of soybean proteins, and foreign proteins in general, in meats is limited by the low extraction efficiency observed, a result of the mild extracting conditions used to avoid the loss of protein antigenicity. In fact, the use of extracting solutions containing urea or SDS, despite being very efficient, could destroy the immunogenic properties of proteins. In this respect, Hyslop[92] suggested the possibility of using a 2% SDS solution for the extraction of soybean proteins with the posterior removal of SDS to regain protein immunogenicity. Moreover, in the case of processed meats there is an additional limitation related to the structural changes occurring in foreign proteins due to the processing. The subjection of soybean proteins to heat treatment improves their nutritional value by denaturing various antinutritional factors. Nevertheless, the susceptibility of the major soybean proteins to heat processing has been well documented.[93] Moreover, in the case of soybean proteins, their antigenic properties depend on the source of the added soybean protein (soybean flour, textured soybean, soybean protein concentrate, or SPI).[51,52]

Table 6.2 groups the immunological methods that have been developed and applied to the analysis of foreign proteins in processed meat products. Immunological methods have been grouped in five categories: serology, immunodiffusion, indirect hemagglutination, methods involving an electrophoretic separation and an immunological reaction, and immunoassays.

6.5.1 Serology

Early immunological methods consisted of serological reactions applied to the determination of soybean proteins. Serological methods are based on the specific interaction between an antigen and an antibody. Major limitations were observed in their application to meats heated to extremes.[142] In 1939, Glynn published a serological method enabling the detection of soybean flour in sausages.[94] Other research refined this method (by the optimization of the time and temperature of incubation of the serum with the soybean proteins) for application in quantitative analysis.[95,96] Degenkolb and Hingerle[97,98] developed a screening method for the detection of foreign proteins in meats. Samples yielding a positive precipitation reaction were later subjected to a volumetric assay. This assay proved useful with products heated up to 110–115°C, using antibodies different from those employed with products heated up to 70°C. Krüger and Grossklaus,[99] using this method for the determination of soybean proteins in canned meats heated at 100°C, obtained a detection limit of 0.2%. Moreover, quantitative determination of added soybean proteins was possible in scalded meat products (heated to 75°C).

6.5.2 Immunodiffusion

In immunodiffusion, antigen–antibody reactions take place in an agar or agarose gel medium. Single immunodiffusion involves the antigen diffusing into a gel containing the corresponding

Table 6.2 Immunological Methods for the Determination of Foreign Proteins in Processed Meats

Sample	Foreign Proteins	Detection Limit	References
Serology			
Sausages	Soybean proteins	—	94–96
Sausages and canned meats heated to 120°C	Soybean proteins	0.2%	97–99
Immunodiffusion			
Canned meat	Soybean proteins	—	100
Model sausages	Soybean proteins	—	101
Sausages	Soybean proteins, hydrolyzed milk proteins, and ovalbumin	0.3 mg/mL for soybean proteins and 0.5 mg/mL for hydrolyzed milk proteins and ovalbumin	102,103
Heated meats	Soybean proteins	—	104
Heated and unheated model meats (60 min at 121°C) and commercial beefburgers, meat balls, sausages, and canned stewed steak	Soybean proteins	1%	105
Canned meat heated to 120°C for 50 min	Soybean proteins	—	106
Indirect Hemagglutination			
Sausages	Soybean proteins, milk proteins, and ovalbumin	—	107,108
Model frankfurters (75–120°C)	Soybean proteins	—	109
Sausages	Soybean proteins, hydrolyzed milk proteins, and ovalbumin	1.0 mg/mL for soybean and hydrolyzed milk proteins and 5.0 mg/mL for ovalbumin	102
Heated meats (121°C)	Soybean proteins	—	110

(continued)

Table 6.2 (continued) Immunological Methods for the Determination of Foreign Proteins in Processed Meats

Sample	Foreign Proteins	Detection Limit	References
Electrophoresis + Immunological Methods			
Heated meats	Milk proteins	—	111
Luncheon meat	Soybean proteins and caseins	—	112
Model heated meats (65–125°C)	Soybean proteins	—	113
Model frankfurters (75–120°C)	Soybean proteins		109
Model cooked sausages (78 and 114°C) and commercial meat products (sausages, luncheon meat, meatballs, ham, and roast turkey)	Caseins	—	114,115
Model sausages	Soybean and mustard proteins	—	116
Model cooked meats (71°C)	Soybean proteins	—	117
Model cooked meats (60–125°C)	Soybean proteins	2.5%	118
Model heated sausages (121°C for 45 min)	Soybean proteins	0.1%	119
Model heated meats (100°C)	Soybean proteins	0.02%	120
Model heated meats (100°C)	Soybean proteins, caseins, whey proteins, ovalbumin, and wheat gluten (modified and nonmodified)	0.1% for each protein	121,122
Model heated meats (60–100°C)	Soybean proteins	0.5%	123

Sausages	Egg proteins	—	124
Sausages	Milk proteins	—	125
Immunoassays			
Model heated meats	Soybean proteins	—	126,127
Pasteurized hamburger and canned luncheon meats sterilized at 120°C for 30 min	Soybean proteins	0.1%	128
Meat balls, beef croquettes, fried chicken, and hamburger	Soybean proteins	—	129
Commercial hamburger	Soybean proteins	2 ppm (0.0002%)	130
Commercial hamburger	Soybean proteins	—	131
Autoclaved model meats (121°C for 20 min), sausages, ham, paté, and hamburger	Soybean proteins	—	132,133
Model and commercial sausages	Soybean proteins	—	134
Model pork sausages (80°C for 20 min)	Soybean proteins	—	135
Fermented sausage (chorizo)	Soybean proteins	1%	136
Model heated meats	Wheat gluten	—	137,138
Model heated meats (100°C for 5 min) and commercial sausages	Wheat gluten	0.2%	139,140
Sausages	Soybean proteins, pea proteins, and wheat gluten	0.05–0.1% for soybean and pea proteins and 0.025–0.5% for wheat gluten	141

antibodies. Peter[100] found this technique to be adequate for the screening of soybean proteins in meats. In this respect, Hauser et al.[101] prepared ready-to-use agar layers for the routine application of this technique to the determination of soybean proteins in meats. They concluded that the successful application of this technique required knowing the soybean protein source added.

Double immunodiffusion, or Ouchterlony immunodiffusion, involves both antigen and antiserum to diffuse from different wells in an agarose or agar gel. This technique has been applied for the screening of various foreign proteins (soybean proteins, hydrolyzed milk proteins, and ovalbumin) in sausages. After 2 h incubation the method enabled the detection of up to 0.3 mg/mL of soybean proteins and 0.5 mg/mL of hydrolyzed milk proteins and ovalbumin.[102] Appelqvist et al.[103] also applied this method to the determination of soybean and milk proteins in meat products. Günther and Baudner[104] found that the use of cellulose acetate membranes was also suitable for the qualitative detection of soybean proteins in processed meats, though agar gels were more adequate for quantitation.

Several approaches have been developed to improve the antisera performance. The use of a commercial soybean protein antiserum proved useful with raw meats but did not solve the problem of decreasing sensitivity observed when meats are severely heated.[52,143] Hammond et al.[105] prepared an antiserum against both heated (121°C) and unheated SPI. Nevertheless, the lack of specificity due to cross-reactivity with certain spices, onion, and hydrolyzed vegetable proteins, combined with the inability of the method to respond to severely processed products, limited its application. Another proposal was suggested by Baudner et al.,[106] who proved the suitability of an antiserum against a soybean protein fragment stable at 120°C and conjugated with a carrier for the detection of soybean proteins in meats.

The double immunodiffusion method proposed by Ouchterlony and the starch gel electrophoretic method proposed by Olsman[55] were evaluated in a collaborative study for the detection of caseins and soybean proteins in meat products. In general, results were more successful by immunodiffusion, since electrophoretic patterns were difficult to interpret. Nevertheless, and as expected, soybean proteins could not be detected in meats heated to temperatures higher than 100°C. In the case of caseins, false positives were obtained due to the presence of undenatured bovine blood proteins with similar immunogenic properties.[144]

6.5.3 Indirect Hemagglutination

Indirect hemagglutination uses erythrocytes coated with antigenic molecules. When these aggregates are added to a solution containing the corresponding antibodies, the cells agglutinate and, due to their large size, their detection is possible even in low concentrations. Kotter et al.[107,108] applied this technique to the determination of different foreign proteins in meats, concluding that the high labor intensity and time requirements limited its application. Regarding feasibility and reliability, conclusions published by various authors have been contradictory.[97,98,145] Krüger and Grossklaus[109] obtained quantitative results for products heated at 75°C, but the technique failed with more severely heated products, even when using antiserum against soybean proteins heated at 110°C. Kraack[102] used this technique for the confirmation of results obtained by a screening serological test. He observed detection limits much higher than that obtained by immunodiffusion. Herrmann and Wagenstaller[110] could quantify soybean proteins in meat products heated up to 115°C, and found it possible to detect soybean proteins in products heated up to 121°C.

6.5.4 Immunological Methods Comprising Electrophoretical Separations

In this section, all immunological methods consisting of a first electrophoretical separation have been grouped. Among these methods are immunoelectrophoresis and Western blot (immunoblotting).

Immunoelectrophoretical methods combine electrophoresis and immunodiffusion. Proteins separated by electrophoresis are transferred onto a membrane and detected by radio- or enzyme-labeled antibodies.[146] The development of electroimmunodiffusion, also known as Laurell immunoelectrophoresis, constituted a significant advance. Since the electrophoretic separation takes place on a gel containing a uniform concentration of the antiserum, no transference of proteins is required. The antigenic proteins present in the sample form complexes with antibodies, which migrate as well, resulting in rocket-shaped precipitation lines (rocket electrophoresis). The length of these lines is proportional to the concentration of antigen in the sample. Laurell immunoelectrophoresis is rapid compared to immunodiffusion methods and can be applied for quantitative analysis.

Early applications of immunoelectrophoresis were devoted to the qualitative analysis of soybean and milk proteins in meat products.[111,112] Kamm[113] was the first to propose the immunochemical quantitation of soybean proteins in cooked meats by immunoelectrophoresis. He prepared an antiserum against crude soybean globulin that contained three antigenic species. One of these species disappeared after heating at 65°C, others after heating at 100°C, and the most stable one was removed at commercial sterility temperature (125°C for 25–30 min), meaning the method was not adequate for severely cooked products. Krüger and Grossklaus[109] studied the effect of temperature on the immunoelectrophoretical signal. They applied the method to model canned frankfurters heated to temperatures ranging from 75 to 120°C and containing from 0.1 to 0.4% of soybean proteins. The method yielded quantitative results for products subjected to scalding temperatures (75°C), but inadequate when products were subjected to higher temperatures. Sinell and Mentz[114,115] used Laurell's technique to quantitate milk proteins in sausages with antibodies against α- and β-caseins. The quantitative determination of this part of milk proteins enabled the measurement of the whole.

Various efforts have been made for the improvement of these results. Merkl[116] avoided cross-reactivity in the determination of soybean proteins in meat products containing mustard by pH adjustment of the agarose gel. Koh[117] prepared antibodies against renatured soybean proteins by extracting soybean proteins under denaturing conditions with urea and mercaptoethanol and removing them by dialyzing. The renatured proteins surprisingly kept their antigenic properties, making the method suitable for the identification and quantification of soybean proteins in heated (71°C) beef mixtures. Poli et al.[118] developed a rapid and sensitive method combining electrophoretic separation with an indirect immunofluorescence detection. The method enabled the detection down to 2.5% of soybean proteins in meat products, even when they were sterilized. A further reduction of detection limits (0.1% of soybean proteins) was obtained by Heitmann,[119] who also used immunofluorescence detection. Janssen et al.[120] proposed the use of a Western blot method for the sensitive determination of soybean proteins in processed meats. In this case, proteins separated by SDS-PAGE are transferred to a nitrocellulose membrane and immunostained with peroxidase. Under these conditions meat proteins did not stain and soybean proteins were detected at a level of 0.02%. The method was also valid for the detection of other nonmeat proteins (ovalbumin, wheat gluten, caseins, and whey proteins) added at a level down to 0.1% in meats heated up to 100°C.[121] Moreover, the elimination of the separation step enabled the rapid screening of samples by a dot blot procedure.[127] This rapid method using immunoperoxidase staining was compared with an immunoglod-silver staining method. Though the immunoglod-silver procedure proved to be more sensitive, it was much more expensive than the immunoperoxidase method.[147] In any case, it was recommended that positive samples be re-examined using the whole procedure, including the electrophoretic separation. Körs[123] could improve the proposed method by the substitution of SDS by a less denaturing detergent (CTAB, *N*-cetyl-*N*,*N*,*N*-trimethylammonium bromide). He concluded that the intensity of the soybean protein bands depended only on the heating temperature at low additions (0.5–1%), independent of temperature at higher proportions.

Although the use of wheat gluten as meat extender has not been as extensive as the use of soybean proteins, the modification of wheat gluten to obtain a more readily soluble product has opened new possibilities for its application in the meat industry. Janssen et al.[122] proved that their proposed Western blot method[121] was capable of detecting this modified gluten and could also discriminate between modified and nonmodified wheat gluten.

Brehmer et al. focused their efforts on the determination of foreign proteins present in cooked meats other than soybean proteins and wheat gluten. They developed immunoelectrophoretical methods sensitive to egg proteins[124] and milk proteins (based on the detection of the α-casein fraction)[125] in cooked meats.

6.5.5 Immunoassays

A number of immunoassays have been developed for the detection of foreign proteins, especially soybean proteins, in cooked meats. The most commonly used immunoassay, the enzyme-linked immunosorbent assay (ELISA), has shown certain advantages compared to previous immunological techniques, such as their suitability for routine analysis and easy semi-automation. Unlike the classical immunochemical methods, ELISA does not rely on the precipitation of the antigen–antibody complex since the presence of the complex is monitored by colorimetric measurement of an enzyme linked to it.

Based on the idea of Koh[117] for the extraction of proteins, Hitchcock et al.[126] developed an ELISA method working with sterilized meat products for the detection of soybean proteins. The sample extract, prepared in a hot concentrated solution of urea, was cooled, diluted for the renaturation of soybean proteins, and treated with a known excess of soybean protein antiserum. The soybean protein in the sample (the antigen) interacted with the antibody while the unreacted antibody was trapped on an immunosorbent that contained an immobilized standard of soybean protein antigen. The captured antibody was determined after adding a second antibody to which an enzyme had been covalently attached (conjugate). The captured enzyme (alkaline phosphatase) was determined by adding *p*-nitrophenyl phosphate as a chromogenic substrate. Finally, the optical density after incubation was measured at 405–410 nm. Olsman et al.[127] organized a collaborative trial in which various meat products heated at 80°C, containing soybean proteins from different sources, were analyzed using an SDS-PAGE method[76] and the ELISA method of Hitchcock et al.[126] Both methods were suitable for qualitative purposes, with SDS-PAGE being more precise and ELISA more accurate. In 1985 this method was adopted as the AOAC official first action.

Although this method was considered one of the best methods for high specificity and sensitivity, reliable quantitative analysis could be obtained only if the source of soybean proteins was known and when meats were not subjected to severe heating processing. Moreover, the long time needed for completion of an analysis (several days were needed to prepare samples) also limited its routine application. Several approaches have been developed to overcome these limitations:

1. *Improvement of antibody performance:* Menzel and Hagemeister[148] reported that antibodies against formaldehyde-treated soybean proteins reacted with both native and heated soybean proteins (125°C). The author suggested the applicability of these antibodies for soybean protein determination in processed meats, but no corroborative data demonstrated it. Ravestein and Driedonks[128] prepared antibodies against soybean proteins denatured with SDS instead of the urea used by Hitchcock. This modification made the method feasible for heated meats, independent of the soybean variety and soybean protein source. Moreover, this method was demonstrated to have no interference from meat proteins and other non-soybean vegetable proteins, making possible the quantitation and detection down to 0.5% and 0.1%, respectively, of soybean proteins. Monoclonal and polyclonal antibodies against different fractions of soybean proteins, rather than against all soybean proteins, have been proposed and used in ELISA systems for the detection of soybean proteins in processed meat products. Tsuji et al.[129] prepared two monoclonal antibodies against the major soybean allergen (Gly m Bd 30K) and used them in an ELISA method for the measurement of this allergen in different meat products. Yeung and Collins[130] developed polyclonal antibodies specific to soybean proteins with no demonstrated cross-reactivity with any nuts, legumes, or other ingredients in hamburgers. Macedo-Silva et al.[131] proposed the use of the 7S fraction of soybean proteins to prepare a polyclonal antibody since it yielded higher immunogenicity than the 11S fraction.

2. *Reduction of analysis time:* Griffiths et al.[149] modified the ELISA method of Hitchock, using commercial immunoreagents (antisera and labeled antiglobulin) and commercial microtiter plates. This method was subjected to a collaborative trial involving 23 U.K. laboratories.[150] Rittenburg et al.[132] developed a ready-to-use kit containing standardized reagents that enabled the complete analysis of a meat sample in a working day. The performance of this kit was evaluated in another collaborative trial, which concluded that using a single arbitrary soybean standard as a reference enabled the reliable estimation of the level of soybean proteins in a pasteurized meat product of entirely unknown composition. Moreover, suitable repeatability and reproducibility (RSD values of 1 and 2%, respectively) and recoveries ranging from 80 to 100% were obtained.[133] Another improvement was introduced by Medina,[134] who reduced the analysis time and complexity of the ELISA procedure by the use of a simple and rapid sample preparation based on the direct extraction of soybean proteins in a carbonate buffer. Results reported for the analysis of various model and commercial sausages demonstrated the validity of the proposed method. On the other hand, Koppelman[151] demonstrated that the use of an extremely high pH (pH 12) for the extraction of soybean proteins yielded higher recoveries than those observed in other (native) conditions (Tris buffer, pH 8.2) or commercially available test conditions (urea and dithiothreitol); it was possible to detect down to 1 ppm of soybean proteins. Although this extraction procedure was suggested as a solid alternative to other preparation procedures used for the determination of soybean proteins by ELISA in meats, no corroborative data in meats was shown.

3. *Improvement of sensitivity and accuracy:* The denaturation of soybean proteins by heating made their determination by immunological methods limited in sensitivity and accuracy. Since protein denaturation rarely affects its primary structure, Yasumoto et al.[135] proposed the detection of the presence of soybean proteins by the identification of characteristic peptides. For that purpose, they prepared antibodies against a peptide fragment of the 11S soybean globulin, the major soybean protein exhibiting the most heat-stable antigenicity. Quantitative results obtained in model sausages demonstrated agreement between the added and the determined soybean protein content.

The application of the ELISA procedure has extended to the determination of soybean proteins in meat products with processing other than heating. González-Córdova et al.[136] developed an ELISA method for use in the determination of soybean proteins in fermented sausages (chorizo). The method proved specific and accurate, and the total time needed for the completion of an analysis was just 4 hours.

Although most methods were focused on the analysis of soybean proteins, there are some examples in which other nonmeat proteins have been analyzed in processed meats. Skerritt and Hill[137] developed an immunological method based on the detection of ω-gliadins for the determination of wheat gluten. Since ω-gliadins are heat-stable proteins, this test seemed suitable for the detection of wheat gluten in heat-processed meats. The main limitation of this test was the dependence of the response on the wheat gluten standard used. This method was subjected to a collaborative study in 15 laboratories. In the case of processed meats, the method proved semi-quantitative.[138] The use of antibodies allowing the recognition of total gliadins instead of only a part of them yielded more accurate determinations they were more affected by heating.[152] Marcin et al.[139,140] proposed a dot EIA (enzyme immunoassay) test that enabled the detection despite down to 0.2% of wheat gluten in sausages.

Finally, Brehmer et al.[141] have applied the ELISA method to quantitate various foreign proteins (soybean proteins, pea proteins, and wheat gluten) in sausages, observing very low detection limits.

6.6 Chromatographic Methods for the Detection of Foreign Proteins in Processed Meats

The analysis of amino acids, peptides, or whole proteins by chromatography has been an alternative to electrophoretic and immunological methods for the detection of foreign proteins in processed meats. This section is devoted to a discussion of chromatographic methods, grouped in Table 6.3, applied to the determination of foreign proteins in processed meats.

6.6.1 Analysis of Amino Acids

The chromatographic analysis of amino acids consists of three steps: hydrolysis of the sample, chromatographic analysis of the hydrolyzed sample, and comparison of the amino acid pattern with a collection of amino acid patterns from different proteins. This comparison is assisted by a computer program based on a regression method, which can determine the types of proteins present in a sample. The main advantage is that this strategy works equally well for mixtures of native or denatured proteins since the amino acids are less prone to undergoing changes during processing than are proteins. The principal difficulties observed are due to the fact that all proteins contain all the major 17 amino acids, though in varying amounts. An additional problem in the case of soybean and meat proteins is that soybean and muscle proteins present a similar amino acid composition.[51]

Lindqvist et al.[165] published the first application of this mathematical approach to the determination of proteins in mixtures. They used a stepwise multiregression analysis adapted to perform the comparison of the amino acid pattern corresponding to a composite sample with those of simple substances arranged in a data bank. This program selected from the bank those proteins whose amino acid patterns best matched that of the sample and calculated the proportion of

Table 6.3 Chromatographic Methods for the Determination of Foreign Proteins in Processed Meats

Sample	Foreign Proteins	Chromatographic Mode	Detection Limit	References
Analysis of Amino Acids				
Pasteurized meat samples	Soybean proteins, egg white proteins, wheat proteins, caseins, potato proteins, and sinew proteins	Ion exchange	—	51
Model heated meats	Soybean proteins, caseins, and whey proteins	RP	—	153
Model heated meats	Soybean and wheat proteins	RP	—	154
Analysis of Peptides				
Model heated meats (120°C for 3 h)	Soybean proteins	Ion exchange	5–10%	155–157
Model heated meats (100°C for 30 min)	Soybean proteins	Ion exchange	—	158
Model heated meats (120°C for 3 h)	Soybean proteins	Ion exchange	2%	72,159
Analysis of Whole Proteins				
Commercial loaf meats	Soybean proteins	RP	0.19%	160
Model heated meats (pork, turkey, chicken, and beef), sausages, and meatloaf	Soybean proteins, caseins, and whey proteins	RP (perfusion)	0.07% for soybean proteins	161–163
Commercial cured meats (dry-fermented (Spanish chorizo) and to spread)	Soybean proteins	RP (perfusion)	0.04%	164

every protein in the mixture. They applied the method to two model mixtures containing soybean and milk proteins but in no case used meat proteins.

Olsman[51] applied, for the first time, a similar multiregression procedure to identify foreign proteins (soybean proteins, egg white protein, wheat proteins, caseins, potato proteins, and sinew proteins) in pasteurized meat products. Lindberg et al.[153] applied partial least-squares regression analysis to determine various proteins in model heated meat products containing ground beef mixed with some common meat extenders (collagen, soybean proteins, and milk proteins). Samples were totally hydrolyzed, derivatized with dansyl chloride, and analyzed by reversed-phase (RP-) high-performance liquid chromatography (HPLC). Separation was carried out with a binary gradient acetonitrile-phosphate buffer water in 25 min. The method seemed to be very little affected by heating, with observing accuracies of 94% for heated meats. Zhi-Ling et al.[154] employed a similar procedure to determine muscle, collagen, shrimp, wheat, and soybean proteins in heated simulated mixtures. Chromatographic separation was performed in a column similar to that used previously, with an analysis time of 12 min, using a binary gradient acetonitrile–acetate buffer water. The accuracy in the determination of soybean proteins was not as good as that observed for wheat gluten and collagen since soybean proteins presented a similar amino acid profile to shrimp and muscle proteins.

6.6.2 Analysis of Peptides

Another approach to the determination of foreign proteins in meat samples has been the determination of characteristic peptides of the searched proteins. This proposal involves the partial hydrolysis, normally by enzymatic digestion, of proteins and the separation of soluble peptides by HPLC. Special care is needed in the case of heated samples in order to avoid aggregation of individual proteins, which could be difficult to dissolve. The studies published using this idea were focused on the analysis of soybean proteins and used ion-exchange chromatography for the separation of characteristic peptides.

Bailey et al. applied this approach for the first time to the determination of soybean proteins in heated meats.[155–157] They isolated a characteristic peptide from soybean proteins (*Ser-Gln-Gln-Ala-Arg* from 11S globulin) by ion-exchange chromatography of the extracts obtained by trypsin digestion. The method was valid for heated samples but was not as sensitive as other methods. Moreover, the analysis time was extremely long (180 min), and this characteristic peak was badly resolved from meat. Llewellyn et al.[158] improved Bailey's method by the introduction of a filtration step before separation, the use of a larger column, and the reduction of flow rate by half.[158] Two characteristic peptides from soybean were selected for the determination of soybean proteins. Despite these efforts, the method continued to be inaccurate since these two soybean peptides proved to overlap with some minor peaks from meat. A further development of the method using an even longer column could improve the resolution of the target peaks and yield lower detection limits for soybean proteins. Nevertheless, the method presented limitations for quantitative purposes and it was not adequate for routine analysis since the total time required for a single analysis was 5–6 days.[72,159]

6.6.3 Analysis of Whole Proteins

The determination of whole soybean proteins in cooked meats has also been approached by HPLC. Various methods enabling the determination of soybean proteins in raw meats have appeared, the group of Marina et al. being the first to focus its efforts on the determination of soybean proteins in heat-processed meats by the analysis of whole proteins by HPLC. They developed conventional and perfusion HPLC methods in the RP mode, applying them to the

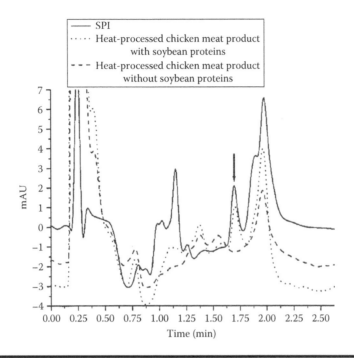

Figure 6.2 **Chromatograms obtained by perfusion HPLC from a heat-processed chicken meat with and without SPI and an SPI. The arrow shows the selected soybean protein peak. (From Castro, F. et al.,** *Food Chem.,* **100, 468, 2007. With permission.)**

determination of soybean proteins in commercial heated meats. Moreover, they could also identify additions of caseins and whey proteins in the meats by perfusion HPLC. As examples, Figure 6.2 shows the chromatograms corresponding to a heat-processed chicken product with and without soybean proteins and an SPI using the perfusion method; the separations obtained by conventional HPLC for a commercial heat-processed meat (containing turkey and pork) and an SPI are presented in Figure 6.3. As expected, perfusion chromatography enabled a much shorter separation than conventional HPLC. Nevertheless, in both cases it was possible to obtain a soybean protein peak totally isolated from meat bands, which was used for quantitation. Both methods enabled detection limits significantly lower than those obtained with any previous technique. Quantitative results obtained by both methods were very similar, with the soybean protein content in commercial meats between 0.60 and 1.54%. Moreover, the results obtained by perfusion HPLC were compared with those observed applying the official ELISA method, with the conclusion that the proposed method could be a serious alternative to the official ELISA method, enabling a significant reduction of analysis time, price, and the complexity of the method itself.[160–163]

The same group has also extended its interest to the analysis of other processed meats, such as cured meat products also containing soybean proteins. They have proposed a new perfusion HPLC method that enabled the isolation of a soybean protein peak that proved adequate for the detection and determination of soybean proteins. Figure 6.4 shows the separations obtained for an SPI and for various cured meat products with and without soybean proteins.[164]

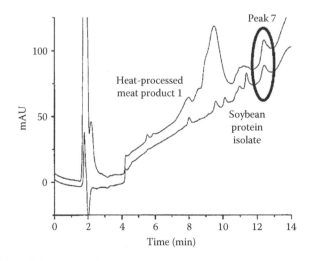

Figure 6.3 Chromatograms obtained from a heat-processed meat product and an SPI by conventional HPLC. (From García, M.C. et al., *Anal. Chim. Acta*, 559, 215, 2006. With permission.)

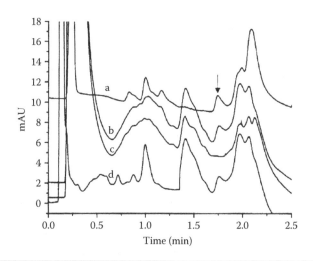

Figure 6.4 Separations corresponding to an SPI (a), a cured meat product to spread spiked with SPI (b), a cured meat product to spread without soybean proteins (c), and a cured meat product (dry-fermented sausage) with soybean proteins (d) in its composition obtained by perfusion HPLC. The arrow shows the selected soybean protein peak. (From Criado, M. et al., *J. Sep. Sci.*, 28, 987, 2005. With permission.)

6.7 Other Methods for the Detection of Foreign Proteins in Processed Meats

Deoxyribonucleic acid (DNA) analysis has also been applied to the detection of foreign food constituents. The stability of DNA made these methods appropriate for the analysis of heated products where antibody based methods fail. Moreover, the unique specificity of the target in these methods ensures the discrimination and avoids cross-reactivity. The main disadvantage of these methods is that they are qualitative or semiquantitative (by the incorporation of internal standards). Superior quantification could be achieved by using real-time polymerase-chain reaction (PCR) or a PCR-ELISA.[115,166] Meyer et al.[167] designed a PCR protocol for the amplification of 414 and 118 bp fragments of the Lectin gene *Le*1 and compared its performance with the commercial ELISA test (based on polyclonal antibodies against renatured soybean proteins) for the detection of soybean proteins in both fresh and processed meats (hamburgers, frankfurters, and heat processed mixtures of soybean and beef meat). The ELISA kit yielded higher recoveries and could quantify soybean proteins in meat products. However, sample preparation using a denaturation–renaturation step was very time consuming. In contrast, the oligonucleotides used in PCR were synthesized rapidly and could be stored for several years. They concluded that PCR could be an interesting method to confirm ELISA results.

Boutten et al.[168] combined immunohistochemistry and video image analysis and applied the method to the detection of soybean proteins in processed meats. They used the visual images provided by histochemical techniques and the specificity of antibodies. Polyclonal antibodies against both raw and heated SPI and soybean protein concentrate were employed. No interference was observed when other proteins were added. Moreover, the labeled soybean surface was proportional to the percentage of soybean proteins added, making this method adequate for the estimation of soybean proteins.

Acknowledgment

The authors thank the Comunidad Autónoma de Madrid (Spain) for project S-0505/AGR/0312.

References

1. Ozimek, G. and Poznánski, S., Influence of an addition of textured milk proteins upon physicochemical properties of meat mixtures, *J. Food Sci.*, 47, 234, 1981.
2. Zorba, Ö., Kurt, S., and Genccelep, H., The effects of different levels of skim milk powder and whey powder on apparent yield stress and density of different meat emulsions, *Food Hydrocoll.*, 19, 149, 2005.
3. Gujral, H.S., Kaur, A., Singh, N., and Sodhi, N.S., Effect of liquid whole egg, fat and textured soy protein on the textural and cooking properties of raw and baked patties from goat meat, *J. Food Eng.*, 53, 377, 2002.
4. Schilling, M.W., Marriott, N.G., Acton, J.C., Anderson-Cook, C., Alvarado, C.Z. and Wang, H., Utilization of response surface modeling to evaluate the effects of non-meat adjuncts and combinations of PSE and RFN pork on water holding capacity and cooked color in the production of boneless cured pork, *Meat Sci.*, 66, 371, 2004.
5. Ulu, H., Effect of wheat flour, whey protein concentrate and soya protein isolate on oxidative processes and textural properties of cooked meatballs, *Food Chem.*, 87, 523, 2004.
6. Gnanasambandam, R. and Zayas, J.F., Microstructure of frankfurters extended with wheat germ proteins, *J. Food Sci.*, 59, 474, 1994.
7. Yusof, S.C.M. and Babji, A.S., Effect of non-meat proteins, soy protein isolate and sodium caseinate, on the textural properties of chicken bologna, *Int. J. Food Sci. Nutr.*, 47, 323, 1996.
8. Li, R.R., Carpenter, J.A., and Cheney, R., Sensory and instrumental properties of smoked sausage made with mechanically separated poultry (MSP) meat and wheat protein, *J. Food Sci.*, 63, 923, 1998.

9. Barbut, S., Effects of caseinate, whey and milk powders on the texture and microstructure of emulsified meat batters, *Lwt-Food Sci. Technol.*, 39, 660, 2006.

10. Slesinski, A.J., Claus, J.R., Anderson-Cook, C.M., Eigel, W.E., Graham, P.P., Lenz, G.E., and Noble, R.B., Response surface methodology for reduction of pinking in cooked turkey breast mince by various dairy protein combinations, *J. Food Sci.*, 65, 421, 2000.

11. Rhee, K.S., Oilseed food ingredients used to minimize oxidative flavour deterioration in meat-products, *ACS Symposium Series*, 506, 223, 1992.

12. Gnanasambandam, R. and Zayas, J.F., Quality characteristics of meat batters and frankfurters containing wheat germ protein flour, *J. Food Qual.*, 17, 129, 1994.

13. Girish, P.S., Sanyal, M.K., Anjaneyulu, A.S.R., Keshari, R.C., and Naganath, M., Quality of chicken patties incorporated with different milk proteins, *J. Food Sci. Technol.*, 41, 511, 2004.

14. Stiebing, A., Influence of proteins on the ripening of fermented sausages, *Fleischwirtschaft*, 78, 1140, 1998.

15. Porcella, M.I., Sánchez, G., Vaudagna, S.R., Zanelli, M.L., Descalzo, A.M., Meichtri, L.H., Gallinger, M.M., and Lasta, J.A., Soy protein isolate added to vacuum-packaged *chorizos*: effect on drip loss, quality characteristics and stability during refrigerated storage, *Meat Sci.*, 57, 437, 2001.

16. Shand, P.J., Mimetic and synthetic fat replacers for the meat industry, in *Production and Processing of Healthy Meat, Poultry and Fish Products*, Pearson, A.M. and Dutson T.R., Eds., Blackie Academic & Professional, London, 1997, chap. 9.

17. Keeton, J.T., Low-fat meat products—technological problems with processing, *Meat Sci.*, 36, 261, 1994.

18. Pietrasik, Z., Effect of content of protein, fat and modified starch on binding textural characteristics, and colour of comminuted scalded sausages, *Meat Sci.*, 51, 17, 1999.

19. Pietrasik, Z. and Duda, Z., Effect of fat content and soy protein/carrageenan mix on the quality characteristics of comminuted, scalded sausages, *Meat Sci.*, 56, 181, 2000.

20. Lin, K.W. and Mei, M.Y., Influences of gums, soy protein isolate, and heating temperatures on reduced-fat meat batters in a model system, *J. Food Sci.*, 65, 48, 2000.

21. Su, Y.K., Bowers, J.A., and Zayas, J.F., Physical characteristics and microstructure of reduced-fat frankfurters as affected by salt and emulsified fats stabilized with nonmeat proteins, *J. Food Sci.*, 65, 123, 2000.

22. Grochalska, D. and Mroczek, J., Influence of soya bean preparations and reduced salt content on the quality of poultry sausages, *Medycyna Wet.*, 57, 54, 2001.

23. Cengiz, E. and Gokoglu, N., Changes in energy and cholesterol contents of frankfurter-type sausages with fat reduction and far replacer addition, *Food Chem.*, 91, 443, 2005.

24. Mandigo, R.W. and Eilert, S.J., Strategies for reduced-fat processed meats, in *Low Fat Meats*, Hafs, H. and Zimbelman, F.K., Eds., Academic Press, Orlando, FL, 1994, chap. 9.

25. Eilert, S.J. and Mandigo, R.W., Use of additives from plant and animal sources in production of low fat meat and poultry products, in *Production and Processing of Healthy Meat, Poultry and Fish Products*, Pearson, A.M. and Dutson T.R., Eds., Blackie Academic & Professional, London, 1997, chap. 10.

26. Chin, K.B., Keeton, J.T., Longnecker, M.T., and Lamkey, J.W., Utilization of soy protein isolate and konjac blends in a low-fat bologna (model system), *Meat Sci.*, 53, 45, 1999.

27. Sheshata, H.A., Attia, E.S.A., and Attia, A.A., Effect of some additives on the quality of low fat cooked emulsion sausage of beef, *J. Food Sci. Technol.*, 35, 447, 1998.

28. Aimiuwu, O.C. and Lilburn, M.S., Protein quality of poultry by-product meal manufactured from whole fowl co-extruded with corn or wheat, *Poultry Sci.*, 85, 1193, 2006.

29. Singh, R.R.B., Rao, K.H., Anjaneyulu, A.S.R., Rao, K.V.S.S., Dubey, P.C., and Yadav, P.L., Effect of caseinates on physico-chemical, textural and sensory properties of chicken nuggets from spent hens, *J. Food Sci. Technol.*, 34, 316, 1997.

30. Hoogenkamp, H.M., Formulated chicken foods—mastering further processing variables, *Fleischwirtschaft*, 78, 190, 1998.

31. Gennadios, A., Hanna, M.A., and Kollengode, A.N.R., Extruded mixtures of spent hens and soybean meat, *Transactions of the ASAE*, 43, 375, 2000.

32. Tsai, S.J., Unklesbay, N., Unklesbay, K., and Clarke, A., Textural properties of restructured beef products with five binders at four isothermal temperatures, *J. Food. Qual.*, 21, 397, 1998.

33. Kumar, S. and Sharma, B.D., Effect of skimmed milk powder incorporation on the physico-chemical and sensory characteristics of restructured buffalo meat blocks, *J. Appl. Anim. Res.*, 23, 217, 2003.

34. Kumar, S., Sharma, B.D., and Biswas, A.K., Influence of milk co-precipitates on the quality of restructured buffalo meat blocks, *Asian-Aust. J. Anim. Sci.*, 17, 564, 2004.

35. Means, W.J. and Schmidt, G.R., Restructuring fresh meat without the use of salt or phosphate, in *Advances in Meat Research*, AVI, B., Person, A.M., and Dutson, T. R., Eds., New York, 1987, chap. 14.

36. Pietrasik, Z. and Li-Chan, E.C.Y., Binding and textural properties of beef gels as affected by protein, κ-carrageenan and microbial transglutaminase addition, *Food Res. Int.*, 35, 91, 2002.

37. Hand, L.W., Crenwelge, C.H., and Terrell, R.N., Effects of wheat gluten, soy isolate and flavorings on properties of restructured beef steaks, *J. Food Sci.*, 49, 1004, 1981.

38. Siegel, D.G., Tuley, W.B., Norton, H.W., and Schmidt, G.R., Sensory, textural and yield properties of a combination ham extended with isolated soy protein, *J. Food Sci.*, 44, 1049, 1979.

39. Endres, J.G., *Soy Protein Products. Characteristics, Nutritional Aspects, and Utilization*, AOAC Press, Champaign, IL, 2001.

40. Kris-Etherton, P.M., Etherton, T.D., Carlson, J., and Gardner, C., Recent discoveries in inclusive-based approaches and dietary patterns for reduction in risk for cardiovascular disease, *Curr. Opin. Lipidol.*, 13, 397, 2002.

41. Wang, Y., Jones, P.J.H., Ausman, L.M., and Lichtenstein, A.H., Soy protein reduces triglyceride levels and triglyceride fatty acid fractional synthesis rate in hypercholesterolemic subjects, *Atherosclerosis*, 173, 269, 2004.

42. Rakosky, J., Soy products for the meat industry, *J. Agric. Food Chem.*, 18, 1005, 1970.

43. Ramezani, R., Aminlari, M., and Fallahi, H., Effect of chemically modified soy proteins and ficin-tenderized meat on the quality attributes of sausage, *J. Food Sci.*, 68, 85, 2003.

44. Feng, J., Xiong, Y.L., and Mikel, W.B., Textural properties of pork frankfurters containing thermally/enzymatically modified soy proteins, *J. Food Sci.*, 68, 1220, 2003.

45. Feng, J. and Xiong, Y.L., Interaction of myofibrillar and preheated soy proteins, *J. Food Sci.*, 67, 2851, 2002.

46. Comfort, S. and Howell, N.K., Gelation properties of salt soluble meat protein and soluble wheat protein mixtures, *Food Hydrocoll.*, 17, 149, 2003.

47. Kong, X.Z., Zhou, H.M., and Qian, H.F., Enzymatic preparation and functional properties of wheat gluten hydrolysates, *Food Chem.*, 101, 615, 2007.

48. Homco-Ryan, C.L., Ryan, K.J., and Brewer, M.S., Comparison of functional characteristics of modified corn gluten meal in vitro and in an emulsified meat model system, *J. Food Sci.*, 68, 2638, 2003.

49. Ellekjaer, M.R., Naes, T., and Baardseth, R., Milk proteins affect yield and sensory quality of cooked sausages, *J. Food Sci.*, 61, 660, 1996.

50. Mittal, G.S. and Usborne, W.R., Meat emulsion extenders, *Food Technol.*, 39, 121, 1985.

51. Olsman, W.J., Methods for detection and determination of vegetable proteins in meat products, *J. Am. Oil Chem. Soc.*, 56, 285, 1979.

52. Olsman, W.J. and Hitchcock, C., Detection and determination of vegetable proteins in meat products, in *Developments in Food Analysis Techniques—2*, King, R.D., Ed., Applied Science Publishers, London, 1980, chap. 6.

53. Eldridge, A.C., Determination of soya protein in processed foods, *JAOCS*, 58, 483, 1981.

54. Belloque, J., García, M.C., Torre, M., and Marina, M.L., Analysis of soyabean proteins in meat products: a review, *Crit. Rev. Food Sci. Nutr.*, 42, 507, 2002.

55. Olsman, W.J., Detection of non-meat proteins in meat products by electrophoresis, *Z. Lebensm. Unters. Forsch.*, 141, 253, 1969.

56. Thorson, B., Skaare, K., and Höyem, T., Identification of caseinate and soy protein by means of polyacrylamide gel electrophoresis, *Nord. Vet. Med.*, 21, 436, 1969.

57. Freimuth, U. and Krause, W., Detection of foreign protein in meat products. Part II. Detection and determination of soy protein in hot sausages with the aid of electrophoresis in polyacrylamide gel, *Mitt. Nahrung*, 14, 19, 1970.

58. Fischer, K.H. and Belitz, H.D., Detection of soybean-protein in meat products by electrophoresis, *Z. Lebensm. Unters. Forsch.*, 145, 271, 1971.

59. Hofmann, K. and Penny, I.F., Identifizierung von soja- und fleischeiweiß mittels dodecylsulfat-poly-acrylamidgel-elektrophorese, *Fleischwirtschaft*, 4, 577, 1971.

60. Hofmann, K. and Penny, I.F., A method for the identification and quantitative determination of meat and foreign proteins using sodium-dodecylsulphate-polyacrylamide gel electrophoresis, *Fleischwirtschaft*, 53, 252, 1973.

61. Hofmann, K., Identification and determination of meat and foreign protein by means of dodecyl-sulphate polyacrylamide gel electrophoresis, *Z. Anal. Chem.*, 267, 355, 1973.

62. Hofmann, K., Identification and determination of meat and foreign proteins by means of dodecyl sulphate polyacrylamide gel electrophoresis, *Ann. Nutr. Alim.*, 31, 207, 1977.

63. Parsons, A.L. and Lawrie, R.A., Quantitative identification of soya protein in fresh and heated meat products, *J. Food Technol.*, 7, 455, 1972.

64. Spell, E., Detecting milk protein and soya protein in meat products by means of vertical disc electrophoresis, *Fleischwirtschaft*, 11, 1451, 1972.

65. Frouin, A., Barraud, C., and Jondeau, D., Detection des proteines de soja ou de lait dans les pro-duits de viande sterilises ou non, *Ann. Fals. Exp. Chim.*, 66, 214, 1973.

66. Lee, Y.B., Rickansrud, D.A, Hagberg, E.C., Briskey, E.J., and Greaser, M.L., Quantitative determi-nation of soybean protein in fresh and cooked meat-soy blends, *J. Food Sci.*, 40, 380, 1975.

67. Guy, R.C.E., Jayaram, R., and Willcox, C.J., Analysis of commercial soya additives in meat products, *J. Sci. Food Agric.*, 24, 1551, 1973.

68. Guy, R.C.E. and Willcox, C.J., Analysis of soya proteins in commercial meat products by polyacryla-mide gel electrophoresis of the proteins extracted in 8 M urea and 1% 2-mercaptoethanol, *Ann. Nutr. Alim.*, 31, 193, 1977.

69. Homayounfar, H., Identification électrophorétique des protéines étrangéres et en particulier des protéines de soja dans les produits alimentaires carnés frais ou en conserve, *Ann. Nutr. Alim.*, 31, 187, 1977.

70. Flaherty, B., Progress in the identification of non-meat food proteins, *Chem. Ind.*, 12, 495, 1975.

71. Llewellyn, J.W. and Flaherty, B., The detection and estimation of soya protein in food products by isoelectric focusing, *J. Food Technol.*, 11, 555, 1976.

72. Llewellyn, J.W. and Sawyer, R., Application and limitation of isoelectric focusing and high per-formance chromatography in the estimation of soya proteins in meat products, *Ann. Nutr. Alim.*, 31, 231, 1977.

73. Vállas-Gellei, A., Detection of soya and milk proteins in the presence of meat proteins, *Acta Alimen-taria*, 6, 215, 1977.

74. Richardson, F.M., Separating meat and nonmeat proteins by using disc electrophoresis, *J. Assoc. Off. Anal. Chem.*, 61, 986, 1978.

75. Armstrong, D.J., Richert, S.H., and Riemann, S.M., The determination of isolated soybean protein in raw and pasteurized meat products, *J. Food Technol.*, 17, 327, 1982.

76. Ring, C., Weigert, P., and Hellmannsberger, L., Determination of non-meat and meat protein with the discontinuous acrylamide gel electrophoresis system, *Fleischwirtschaft*, 62, 648, 1982.

77. Molander, E., Determination of soya protein in meat products by standard curves obtained from SDS gel electrophoresis, *Z. Lebensm. Unters. Forsch.*, 174, 278, 1982.

78. Heinert, H.H. and Baumann, H.J., Detecting the presence of soya protein in frankfurter-type sausages PoroPAGE with added SDS and urea, *Fleischwirtschaft*, 64, 89, 1984.

79. Feigl, E., Electrophoretic soja determination in frankfurter-type sausage using commercially available SDS-containing gel plates and an apparatus for isoelectric focussing, *Fleischwirtschaft*, 70, 702, 1990.

80. Lacourt, A., Malicrot, M.T., and Dauphant, J., Détection des protéines étrangéres á la viande par électrophorése en gel de polyacrylamide en présence de SDS, *Ann. Nutr. Alim.*, 31, 217, 1977.

81. Woychik, J.H., Happich, M.C., Trinh, H., and Seilers, R., Quantitation of soy protein in frankfurters by gel electrophoresis, *J. Food Sci.*, 52, 1532, 1987.

82. López, L.B., Greco, C.B., Ronayne de Ferrer, P., and Valencia, M.E., Identification of extrinsic proteins in boneless cooked ham by SDS-PAGE: detection level in model systems, *Archivos Latinoamericanos de Nutrición*, 56, 282, 2006.

83. Olivera Carrion, M. and Valencia, M.E., Detección y cuantificación de soja en productos cárnicos por electroforesis. I. Estudio en sistema modelo, *Rev. Agroquím. Tecnol. Alim.*, 30, 509, 1990.

84. Olivera-Carrión, M. and Valencia, M.E., Detección y cuantificación de soja en productos cárnicos por electroforesis. II. Identificación e interferencias de otras proteínas diferentes de soja, *Rev. Agroquín. Tecnol. Alim.*, 30, 518, 1990.

85. Mattey, M.E., B.F.M.I.R.A. Technical Circular N° 518, 1972, Leatherhead, Surrey.

86. Smith, P.S., Detection and estimation of vegetable protein in the presence of meat proteins, *IFST Proc.*, 8, 154, 1975.

87. Bergen, J. and Bosch, G., Analyse van eiwitten met natriumdodecyl-acrylamidevlakgel-elektroforese, *De Ware(n)-Chemicus*, 6, 193, 1976.

88. Endean, M.E., Leatherhead Food International Technical Circular N° 565, 1974, Leatherhead, Surrey.

89. Tateo, F., Determination of soybean proteins in meat-base products. A report on results obtained by Parson-Lawrie's electrophoretic method, *La Rivista Italiana delle Sostanze Grasse*, 11, 155, 1974.

90. Baylac, P., Luigi, R., Lanteaume, M., Pailler, F.M., Lajon, A., Bergeron, M., and Durand, P., Vérification des limites de l'identification de différents liants protéiques dans un produit carné soumis a des valeurs cuisatrices connues, par utilisation de méthodes électrophorétiques immunologiques et histologiques, *Ann. Fals. Exp. Chim.*, 81, 333, 1988.

91. Laemmli, U.K., Cleavage of structural proteins during the assembly of the head of bacteriophage T4, *Nature*, 227, 680, 1970.

92. Hyslop, N.S.G., Extraction methods and test techniques for detection of vegetable proteins in meat-products. 1. Qualitative detection of soya derivatives, *J. Hyg.*, 76, 329, 1976.

93. Koie, B. and Djurtoft, R., Changes in the immunochemical response of soybean proteins as a result of heat treatment, *Ann. Nutr. Alim.*, 31, 183, 1977.

94. Glynn, J.H., The quantitative determination of soy-bean protein in sausage or other protein mixtures, *Science*, 89, 444, 1939.

95. Ferguson, C.S., Racicot, P.A., and Rane, L., Study of use of precipitin test for determination of soybean flour in sausages, *J. Assoc. Off. Analyt. Chem.*, 25, 533, 1942.

96. Hale, C.M.W., Determination of soybean flour in meat products, *Food Res.*, 10, 60, 1945.

97. Degenkolb, E. and Hingerle, M., Untersuchungen über den serologischen nachweis von fremdeiweißzusätzen in hitzedenaturierten fleischerzeugnissen, *Arch. Lebensm. Hyg.*, 18, 241, 1967.

98. Degenkolb, E. and Hingerle, M., Eine serologische schnellmethode zum quantitativen nachweis von fremdeiweißzusätzen in hitzedenaturierten fleischerzeugnissen, *Arch. Lebensm. Hyg.*, 20, 73, 1969.

99. Krüger, H. and Grossklaus, D., Serological detection of soya protein in heated meat products. II. Experiments using indirect haemagglutination and immune electrophoresis, *Fleischwirtschaft*, 2, 181, 1971.

100. Peter, M., Die vertikale immunodiffusion—ein hilfsmittel zur quantitativen beurteilung von fremdeiweißzusätzen in fleischwaren, *Arch. Lebensm. Hyg.*, 10, 220, 1970.

101. Hauser, E., Bicanova, J., and Künzler, W., Erfassung fleischfremder eiweiße in hitzebehandelten fleischwaren durch ein standardisiertes immunodiffusionsverfahren, *Mitt. Gebiete Lebensm. Hyg.*, 65, 82, 1974.

102. Kraack, J., Determination of foreign protein in sausage products, *Fleischwirtschaft*, 5, 697, 1973.

103. Appelqvist, L.A., Persson, B., and Wallin, B., The determination of non-meat protein in meat products, *Näringsforskning. Arg.*, 19, 217, 1975.
104. Günther, H.O. and Baudner, S., Nachweis von zusatzeiweißstoffen in lebensmitteln, *Lebensmittelchemie u. Gerichtl. Chemie*, 32, 105, 1978.
105. Hammond, J.C., Cohen, I.C., Everard, J., and Flaherty, B., A critical assessment of Ouchterlony's immunodiffusion technique as a screening test for soya protein in meat products, *J. Assoc. Pub. Anal.*, 14, 119, 1976.
106. Baudner, S., Schweiger, A., and Günther, H.O., Nachweis von sojaeiweiß aus auf 120°C erhitzten fleischkonserven, *Mitt. Gebiete Lebensm. Hyg.*, 68, 183, 1977.
107. Kotter, L. and Herrmann, C., Die indirekte hämagglutination als nackweisreaktion für fremdeiweiße, *Arch. Lebensm. Hyg.*, 19, 267, 1968.
108. Kotter, L., Herrmann, C., and Corsico, G., Zum quantitativen nachweis von aufgeschlossenem milcheiweiß in hocherhitzten fleischwaren mit hilfe der indirekten hämagglutination, *Z. Lebensm. Unters. Forsch.*, 133, 15, 1966.
109. Krüger, H. and Grossklaus, D., Serological detection of soya protein in heated meat products. III. Using qualitative precipitate measurements, *Fleischwirtschaft*, 3, 315, 1971.
110. Herrmann, C. and Wagenstaller, G., Über den nachweis von sojaprotein in fleischerzcugnissen mittels der indirekten hämagglutination, *Arch. Lebensm. Hyg.*, 24, 131, 1973.
111. Sinell, H.J., Probleme bei der spezies-identifizierung von proteinen in lebensmitteln, *Arch. Lebensm. Hyg.*, 19, 121, 1968.
112. Günther, H., Bestimmung von Fremdeiweiß in fleischwaren, *Arch. Lebensm. Hyg.*, 6, 128, 1969.
113. Kamm, L., Immunochemical quantitation of soybean protein in raw and cooked meat products, *JAOAC*, 53, 1248, 1970.
114. Sinell, H.J. and Mentz, I., Electro immunodiffusion assay for the quantitative detection of milk proteins in meat products, *Arch. Lebensm. Hyg.*, 26, 41, 1975.
115. Sinell, H.J. and Mentz, I., Use of electroimmunodiffusion for quantitative determination of non meat proteins added to meat products, *Folia Vet. Lat.*, 7, 41, 1977.
116. Merkl, H., Differenzierungsmöglichkeiten von sojaprotein und senfmehl, *Fleischwirtschaft*, 10, 1458, 1976.
117. Koh, T.Y., Immunochemical method for the identification and quantitation of cooked or uncooked beef and soya proteins in mixtures, *J. Inst. Can. Sci. Technol. Alim.*, 11, 124, 1978.
118. Poli, G., Balsari, A., Ponti, W., Cantoni, C., and Massaro, L., Crossover electrophoresis with indirect immunofluorescence in the detection of soy protein in heated meat products, *J. Food Technol.*, 14, 483, 1979.
119. Heitmann, J., Demonstration of soya protein in heated meat products with an indirect immunofluorescence test, *Fleischwirtschaft*, 67, 621, 1987.
120. Janssen, F.W., Voortman, G., and Baaij, J.A., Detection of soya proteins in heated meat products by "blotting" and "dot-blot," *Z. Lebensm. Unters. Forsch.*, 182, 479, 1986.
121. Janssen, F.W., Voortman, G., and Baaij, J.A., Detection of wheat gluten, whey protein, casein, ovalbumin, and soy protein in heated meat products by eletrophoresis, blotting, and immunoperoxidase staining, *J. Agric. Food Chem.*, 35, 563, 1987.
122. Janssen, F.W., Baaij, J.A., and Hägele, G.H., Heat-treated meat products. Detection of modified gluten by SDS-electrophoresis, western-blotting and immunochemical staining, *Fleischwirtschaft*, 74, 168, 1994.
123. Körs, M. and Steinhart, H., CTAB electrophoresis and immunoblotting: a new method for the determination of soy protein in meat products, *Z. Lebensm. Unters. Forsch. A*, 205, 224, 1997.
124. Brehmer, H., Kothe, E., and Baumann, H.J., Semiquantitative determination of whole egg additions to frankfurter-type sausage products using electro-immune diffusion, *Fleischwirtschaft*, 70, 700, 1990.
125. Brehmer, H., Borowski, U., and Fessel, P., Quantitative determination of milk protein in food by means of an electroimmunodiffusion (Laurell-technique), *Deutsche Lebensmittel-Rundschau*, 96, 167, 2000.

126. Hitchcock, C.H.S., Bailey, F.J., Crimes, A.A., Dean, D.A.G., and Davis, P.J., Determination of soya protein in food using an enzyme-linked immunosorbent assay procedure, *J. Sci. Food Agric.*, 32, 157, 1981.

127. Olsman, W.J., Dobbelaere, S., and Hitchcock, C.H.S., The performance of an SDS-PAGE and an ELISA method for the quantitative analysis of soya protein in meat products: an international collaborative study, *J. Sci. Food Agric.*, 36, 499, 1985.

128. Ravestein, P. and Driedonks, R.A., Quantitative immunoassay for soya protein in raw and sterilized meat products, *J. Food Technol.*, 21, 19, 1986.

129. Tsuji, H., Okada, N., Yamanishi, R., Bando, N., Kimoto, M., and Ogawa, T., Measurement of *Gly m* Bd 30K, a major soybean allergen, in soybean products by a sandwich enzyme-linked immunosorbent assay, *Biosci. Biotech. Biochem.*, 59, 150, 1995.

130. Yeung, J.M. and Collins, P.G., Determination of soy proteins in food products by enzyme immunoassay, *Food Technol. Biotechnol.*, 35, 209, 1997.

131. Macedo-Silva, A., Shimokomaki, M., Vaz, A.J., Yamamoto, Y.Y., and Tenuta-Filho, A., Textured soy protein quantification in commercial hamburger, *J. Food Comp. Anal.*, 14, 469, 2001.

132. Rittenburg, J.H., Adams, A., Palmer, J., and Allen, J.C., Improved enzyme-linked immunosorbent assay for determination of soy protein in meat products, *J. Assoc. Off. Anal. Chem.*, 70, 582, 1987.

133. Hall, C.C., Hitchcock, C.H.S., and Wood, R., Determination of soya protein in meat products by a commercial enzyme immunoassay procedure: collaborative trial, *J. Assoc. Publ. Anal.*, 25, 1, 1987.

134. Medina, M.B., Extraction and quantitation of soy protein in sausages by ELISA, *J. Agric. Food Chem.*, 36, 766, 1988.

135. Yasumoto, K., Sudo, M., and Suzuki, T., Quantitation of soya protein by enzyme linked immunosorbent assay of its characteristic peptide, *J. Sci. Food Agric.*, 50, 377, 1990.

136. González-Córdova, A.F., Calderón de la Barca, A.M., Cota, M., and Vallejo-Córdoba, B., Immunochemical detection of fraudulent adulteration of pork chorizo (sausage) with soy protein, *Food Sci. Technol. Int.*, 4, 257, 1998.

137. Skerritt, J.H. and Hill, A.S., Monoclonal antibody sandwich enzyme immunassays for determination of gluten in foods, *J. Agric. Food Chem.*, 38, 1771, 1990.

138. Skerritt, J.H. and Hill, A.S., Enzyme immunoassay for the determination of gluten in foods: collaborative study, *J. Assoc. Off. Anal. Chem.*, 74, 257, 1991.

139. Marcin, A., Fencik, R., Belickova, E., and Siklenka, P., DOT-BLOT technique for the quantitative detection of the wheat protein in the sausages, *Vet. Med. Czech.*, 40, 227, 1995.

140. Marcin, A., Fencik, R., Belickova, E., and Siklenka, P., Detection of wheat protein in sausages by DOT-EIA technique, *J. Food Sci. Technol.*, 33, 421, 1996.

141. Brehmer, H., Schleiser, S., and Borowski, U., Determination of soya protein, pea protein and gluten in frankfurter-type sausages by means of an enzyme-linked-immunosorbent assay (ELISA), *Fleischwirtschaft*, 8, 74, 1999.

142. Herrmann, C., Merkle, C., and Kotter, L., The problem of serological reactions for detecting foreign proteins in heated meat products, *Fleischwirtschaft*, 1, 97, 1973.

143. Hargreaves, L.L., Jarvis, B., and Wood, J.M., Leatherhead Food International Research Report N° 206, Leatherhead, Surrey, 1974.

144. Beljaars, P.R. and Olsman, W.J., Collaborative study on the detection of caseinate and soya protein isolate in meat products, *Ann. Nutr. Alim.*, 31, 233, 1977.

145. Herrmann, C. and Merkle, C. Zur standardisierung und konservierung der sensibilisierten hammelerythrozyten für die indirekte hämagglutination zum nachweis von fremdeiweiß, *Arch. Lebensm. Hyg.*, 22, 189, 1971.

146. Poms, R.E., Klein, C.L., and Anklam, E., Methods for allergen analysis in food: a review, *Food Add. Contam.*, 21, 1, 2004.

147. Jansen, F.W., Voortman, G., and Baaij, J.A., Rapid detection of soya protein, casein, whey protein, ovalbumin and wheat gluten in heat-treated meat products by means of dot blotting. A comparison between immunoperoxidase and immunoglod enhancing systems, *Fleischwirtschaft*, 67, 577, 1987.

148. Menzel, J.E. and Hagemeister, H., Solid-phase radioimmunoassay for native and formaldehyde-treated soya protein, *Z. Lebensm. Unters. Forsch.*, 175, 211, 1982.

149. Griffiths, N.M., Billington, M.J., Crimes, A.A., and Hitchcock, C.H.S., An assessment of commercially available reagents for an enzyme-linked immunosorbent assay (ELISA) of soya protein in meat products, *J. Sci. Food Agric.*, 35, 1255, 1984.

150. Crimes, A.A., Hitchcock, C.H.S., and Wood, R., Determination of soya protein in meat products by an enzyme-linked immunosorbent assay procedure: collaborative study, *J. Assoc. Publ. Anal.*, 22, 59, 1984.

151. Koppelman, S.J., Lakemond, C.M.M., Vlooswijk, R., and Hefle, S.L., Detection of soy proteins in processed foods: literature overview and new experimental work, *J. AOAC Int.*, 87, 1398, 2004.

152. Denery-Papini, S., Nicolas, Y., and Popineau, Y., Efficiency and limitations of immunochemical assays for the testing of gluten-free foods, *J. Cereal Sci.*, 30, 121, 1999.

153. Lindberg, W., Ohman, J., Wold, S., and Martens, H., Simultaneous determination of five different food proteins by high-performance liquid chromatography and partial least-squares multivariate calibration, *Anal. Chim. Acta*, 174, 41, 1985.

154. Zhi-Ling, M., Yan-Ping, W., Chun-Xu, W., and Fen-Zhi, M., HPLC determination of muscle, collagen, wheat, shrimp, and soy proteins in mixed food with the aid of chemometrics, *Am. Lab.*, 29, 27, 1997.

155. Bailey, F.J., A novel approach to the determination of soya proteins in meat products using peptide analysis, *J. Sci. Food Agric.*, 27, 827, 1976.

156. Bailey, F.J. and Hitchcock, C., A novel approach to the determination of soya protein in meat products using peptide analysis, *Ann. Nutr. Alim.*, 31, 259, 1977.

157. Bailey, F.J., Llewellyn, J.W., Hitchcock, C.H.S., and Dean, A.C., The determination of soya protein in meat products using peptide analysis and the characterisation of the specific soya peptide used in calculations, *Chem. Ind.*, 13, 477, 1978.

158. Llewellyn, J.W., Dean, A.C., Sawyer, R., Bailey, F.J., and Hitchcock, C.H.S., Technical note: the determination of meat and soya proteins in meat products by peptide analysis, *J. Food Technol.*, 13, 249, 1978.

159. Agater, I.B., Briant, K.J., Llewellyn, J.W., Sawyer, R., Bailey, F.J., and Hitchcock, C.H.S., The determination of soya and meat protein in raw and processed meat products by specific peptide analysis. An evaluation, *J. Sci. Food Agric.*, 37, 317, 1986.

160. García, M.C., Domínguez, M., García-Ruiz, C., and Marina, M.L., Reversed-phase high-performance applied to the determination of soybean proteins in commercial heat-processed meat products, *Anal. Chim. Acta*, 559, 215, 2006.

161. Castro-Rubio, F., García, M.C., Rodríguez, R., and Marina, M.L., Simple and inexpensive method for the reliable determination of additions of soybean proteins in heat-processed meat products: an alternative to the AOAC official method, *J. Agric. Food Chem.*, 53, 220, 2005.

162. Castro, F., Marina, M.L., Rodríguez, J., and García, M.C., Easy determination of the addition of soybean proteins to heat-processed meat products prepared with turkey meat or pork-turkey meat blends that could also contain milk proteins, *Food Add. Contam.*, 22, 1209, 2005.

163. Castro, F., García, M.C., Rodríguez, R., Rodríguez, J., and Marina, M.L., Determination of soybean proteins in commercial heat-processed meat products prepared with chicken, beef or complex mixtures of meats from different species, *Food Chem.*, 100, 468, 2007.

164. Criado, M., Castro-Rubio, F., García-Ruiz, C., and García, M.C., and Marina, M.L., Detection and quantitation of additions of soybean proteins in cured-meat products by perfusion revered-phase high-performance liquid chromatography, *J. Sep. Sci.*, 28, 987, 2005.

165. Lindqvist, B., Ostgren, J., and Lindberg, I., A method for the identification and quantitative investigation of denatured proteins in mixtures based on computer comparison of amino-acid patterns, *Z. Lebensm. Unters. Forsch.*, 159, 15, 1975.

166. Malmheden Yman, I., Detection of inadequate labelling and contamination as causes of allergic reactions to food, *Acta Alimentaria*, 33, 347, 2004.

167. Meyer, R., Chardonnens, F., Hübner, P., and Lüthy, J., Polymerase chain reaction (PCR) in the quality and safety assurance of food: detection of soya in processed meat products, *Z. Lebensm. Unters. Forsch.*, 203, 339, 1996.
168. Boutten, B., Humbert, C., Chelbi, M., Durand, P., and Peyraud, D., Quantification of soy proteins by association of immunohistochemistry and video image analysis, *Food Agric. Immunol.*, 11, 51, 1999.

Chapter 7

Detection of Adulterations: Identification of Animal Species

Johannes Arjen Lenstra

Contents

7.1 Introduction 188
7.2 Alternatives to Polymerase Chain Reaction 188
7.3 Deoxyribonucleic Acid Methods 188
    7.3.1 Deoxyribonucleic Acid Extraction 188
    7.3.2 Polymerase Chain Reaction 189
        7.3.2.1 Design of Polymerase Chain Reaction 189
        7.3.2.2 Universal Primers 189
        7.3.2.3 Determination of Polymerase Chain Reaction Products 190
        7.3.2.4 Species-Specific Amplification 192
        7.3.2.5 Multiplex Polymerase Chain Reaction 194
        7.3.2.6 Fingerprinting 194
        7.3.2.7 Real-Time Polymerase Chain Reaction 194
7.4 Conclusion 198
Acknowledgments 198
References 198

7.1 Introduction

Species substitution during food production results from economic fraud or negligence. It may not only lead to unwanted disrespect of religious rules, but can also have harmful health effects. For these reasons several methods have been developed for the identification of the species origin of meat samples. In addition, the same methodology can be applied to the control of poaching and illegal trade in animal products.

Earlier reviews described the state of the art in species identification in 2001 [1] and 2003 [2], more general aspects of food forensics [3,4], or traceability at the level of the subspecies or breed [5]. In this chapter we review the considerable progress during recent years. The almost complete dominance of deoxyribonucleic acid (DNA)-based methods has not led to the abandonment of other techniques. However, most reports describe wider applications or refinement of polymerase chain reaction (PCR)-based species identification. There is now a growing emphasis on convenient real-time PCR assays, which allow a quantitative interpretation of the results.

In addition to the published work, the Web site www.molspec.org offers a detailed description of the detection of several food species.

7.2 Alternatives to Polymerase Chain Reaction

Immunochemical methods require no expensive equipment or elaborate protocols and are still in use. Species-specific proteins, or epitopes, have been developed for most animals used for meat production, including pig, cattle, sheep, and poultry, but threshold values have yet to be determined empirically [6]. Although heating decreases the sensitivity and specificity of the antisera, adequate performance of a species-specific enzyme-linked immunosorbent assay (ELISA) with commercial antisera [6] and of a pork-specific indirect ELISA [7] has been reported. However, ELISA procedures are not yet adequate for a sensitive detection of ruminant material in feed [8].

Capillary electrophoresis has been described as a flexible tool for the analysis of species-specific proteins in unheated meat product [9].

7.3 Deoxyribonucleic Acid Methods

7.3.1 Deoxyribonucleic Acid Extraction

For most applications, DNA is now purified by using one of several commercially available kits, which are based on the adsorption of DNA to special resins. Apart from convenience and speed, the major advantage of these procedures is the effective removal of various inhibitors of the PCR reaction that often are present in food samples. However, the relative performance of the kits depends on the food commodity [10,11], and for large-scale applications different kits should be compared.

Heating for prolonged periods destroys DNA, which especially hinders the DNA-based species identification of extremely heated meat and bone meal. However, bovine DNA could be amplified from meat subjected to the most common cooking procedures with the exception of panfrying for 80 min [12].

A promising approach is the binding and subsequent sequence analysis of highly fragmented DNA to beads, followed by emulsion PCR and high-throughput sequencing. This advanced technology has been used for the partial sequence analysis of Neanderthal DNA extracted from fossil remains [13].

7.3.2 Polymerase Chain Reaction

7.3.2.1 Design of Polymerase Chain Reaction

Any PCR reaction critically depends on the design of the primers. With only a few exceptions, primers for animal species identification target variable regions in the mitochondrial DNA (mtDNA). Mitochondrial DNA is more variable than nuclear DNA, but its high copy number increases the sensitivity relative to the PCR of single-copy nuclear sequences. However, because of its maternal origin, mtDNA may not be representative if samples originate from hybrids between species [2].

Remarkably, earlier species identification methods [1,2] were based on hybridization to species-specific repetitive elements, which combine a high copy number with often absolute specificity for a species, suborder, order, or higher taxon. In general, a centromeric satellite DNA sequence is confined to one species, whereas homologous satellites from related species can be differentiated by a restriction fragment length polymorphism (RFLP) assay [1,2]. Further, the dispersed short interspersed nuclear elements (SINE) are specific for mammalian order or suborder, which is useful, for instance, in detection of ruminant DNA [14,15]. However, repetitive elements must be characterized for each species, which is not practical for exotic animals. Furthermore, standardization of the PCR across species with several nonhomologous repetitive elements will be more difficult than for mitochondrial DNA.

Several different strategies for the PCR-based species detection are being adopted.

1. One strategy relies on the design of universal primers in conserved regions that amplify a DNA fragment from all species to be detected (Section 7.3.2.2). Subsequent analysis of the PCR product then allows the determination of the species origin (Section 7.3.2.3).
2. In another strategy, PCR primers match the sequence of a single species. Species identification follows from the presence or absence of an amplification product (Section 7.3.2.4). If different components have to be detected, primers can be combined in a multiplex reaction, often with one common forward primer and for each species a specific reverse primer (Section 7.3.2.5).
3. Several methods are available for the generation of fingerprints by PCR. The resulting patterns depend on the species and thus allow their detection (Section 7.3.2.6).
4. The latest development is real-time quantitative PCR, which often is able to differentiate low levels of target DNA from insignificant background signals (Section 7.3.2.7).

7.3.2.2 Universal Primers

A seminal paper in 1989 [16] described a number of universal mtDNA primers. These or similar primers often allow the sequencing or detection of various mtDNA segments from known or unknown species [2,17]. However, with species not previously tested, the matching of the primers and the amplification should be checked. Further, even for the most common meat species [16], matching to the mtDNA target sequences is incomplete [2]. This may necessitate a low annealing temperature, but then invites nonspecific amplification of, for example, nuclear mtDNA copies. In addition, it is likely to cause uneven amplification of different targets with samples of mixed-species composition.

For purposes of detecting all animal DNA in foodstuffs, primers specific for the 16S mtDNA gene were designed that (with two ambiguities) matched completely to species from all mammalian orders [18,19]. In the same gene, other primers were designed to generate a short amplicon from mammalian and avian species for real-time PCR [20]. Primers in the mtDNA *ATP8* gene were designed to be specific for nonhuman mammalian DNA [21] or ruminant DNA. However, as can be checked by a Genbank search, these mammalian primers match completely only to bovine DNA, whereas the ruminant primers match only four ruminant species. As a consequence, it is not likely that DNA from all targeted species will be amplified with the same efficiency, if at all. Similarly, other universal primers [22,23] match most completely to ruminant DNA and indeed appeared to amplify only ruminants and horse [22].

Different primer pairs in the mtDNA *12S rRNA* gene designed to match the cattle, sheep, and pig sequences [24] also matched to several other mammals. Curiously, the 3′ end of the reverse primer [24] does not match any mammalian mtDNA sequence, including the Genbank sequences used for the design of the primer. Other primers with cross-species specificity in the same gene were used for PCR-RFLP of several ruminants [25,26] or for quantitation of mammalian DNA [27]. However, the amplicon of 425 bp [27] is rather long for this purpose.

Trading sensitivity for broad specificity, universal primers may be derived from nuclear genes. Primers specific for an intron in an actin gene were found to be suitable for species identification by sequence analysis in a wide range of species [17]. Mammalian primers have also been based on the myostatin [28] or growth hormone [29] genes. Truly universal eukaryotic primers have been derived from the nuclear *18S rRNA* gene to serve as positive control of species-specific PCR reactions [30,31].

The nucleotide database now contains mitochondrial and genomic DNA sequences of most, if not all, species that are used for meat production. However, more often than not, allegedly universal primers have not been aligned with all relevant homologous sequences to check their taxonomic range. Further, the implicit assumption that in a sample of mixed-species origin the primers target the different components with the same efficiency has in most cases not been validated.

7.3.2.3 Determination of Polymerase Chain Reaction Products

For samples with single-species origin, sequencing of the PCR product is the most straightforward way of species identification. It is especially useful if it is not known beforehand which species is to be expected, for instance with game species. For this, the mtDNA cytochrome *b* gene is the most popular target [32–35], since this gene has been used frequently for phylogenetic studies. If the sample is derived from an exotic species for which no sequence data are available, a basic local alignment search tool (BLAST) search in the nucleotide database will turn up a number of related species. Other genes suitable for species identification are the mtDNA *12S rRNA* [36], *16S rRNA*, and *ND4* genes [17], or the nuclear actin genes [17].

A simple way to determine the species origin of PCR products is digestion by a restriction enzyme that cleaves at a species-specific (diagnostic) site. Although RFLP for restriction enzyme length polymorphism formally refers to a genetic polymorphism within a species, the term "PCR-RFLP" is now commonly used to denote the procedure to detect the species-specific restriction sites. The method requires only simple equipment and is most practical if few samples have to be tested. In general, admixtures of 1% can be detected. Table 7.1 summarizes a number of PCR-RFLP assays, most of which use the original universal primers [16]. Most of these reports confirm or add other species to the report of Meyer et al. [37]. Maede [38] gives the most complete list of species and restriction patterns and also describes a number of species-specific primers.

Apart from preferential amplification by the use of the original universal primers, another caveat is that the diagnostic site can be polymorphic with the consequence that the assay does not

Table 7.1   PCR-RFLP Systems for Species Identification

References	Target Gene	Primers	Detected Species	Remarks
39	mt *cyt*B	Universal [16]	Cattle, sheep, goat, roe-, red deer	
40	mt *cyt*B	Universal [16]	Cattle, fallow, roe-, red deer, pig, chicken, turkey, quail, Muscovy duck	
41,42	*cyt*B	Universal [16]	Pig	
38	mt *cyt*B	Universal [16]	24 mammalian and avian species	Several enzymes
		Horse-specific	Horse, donkey	
		Poultry-specific	Chicken turkey, mallard duck, Muscovy duck, goose	
		Deer-specific	Red-, roe-, fallow deer, elk	
38	Growth hormone	Cattle-specific	Cattle, water buffalo, etc.	Amplification of other related species not excluded
		Sheep/goat specific	Sheep, goat, etc.	
43	mt *cyt*B, *CO2*	Bovine-specific	Cattle, zebu, gayal, banteng	Several enzymes
	Satellite IV			
	Satellite 1.711b			
26	mt *12S rRNA*	Ruminants	Cattle, sheep, goat, red-, roe-, fallow deer	
44	mt *12S rRNA*	Ruminants	Chamois, ibex, mouflon	
	mt D-loop	Sheep, mouflon	Sheep, mouflon	
45	mt *12S rRNA*	Universal [16]	Cattle, water buffalo, sheep, goat	
46	mt *cyt*B	Dog-, cat-specific	Dog, cat	
47	mt *cyt*B	Universal [16]	Two ostrich species, chicken, turkey	

(*continued*)

Table 7.1 (continued)   PCR-RFLP Systems for Species Identification

References	Target Gene	Primers	Detected Species	Remarks
48	mt *12S rRNA*	Universal [16]	Chicken, mallard duck, turkey, guinea fowl, quail	
49	mt *12S rRNA*	Universal [16]	Peacock, chicken, turkey	Two enzymes
50	mt *cytB*	Turtle-specific	Ten turtle species	
51	mt *12S rRNA, 16S rRNA*	Snail-specific	Two snail species	

Note: *ATPase6,* gene for ATPase subunit 6; *ATPase8,* gene for ATPase subunit 8; *CO1,* gene for cytochrome oxidase subunit I; *CO2,* gene for cytochrome oxidase subunit II; *cytB,* cytochrome *b* gene; mt, mitochondrial; *ND5,* gene for NADH dehydrogenase subunit 5; *t-Glu,* tRNA^{Glu} gene; *t-Lys,* tRNA^{Lys} gene; *t-Phe,* tRNA^{Phe} gene; and *t-Val,* tRNA^{Val} gene.

detect all individuals from a species [2]. This can be circumvented by testing for more than one diagnostic site. Further, the taxonomic range of the diagnostic site should be checked in alignment with homologous sequences of closely related species. For instance, it is relevant to know if a bovine pattern is the same in zebu, bison, and water buffalo and which of the several deer species share a diagnostic site.

Alternatively, species can be detected by hybridization of PCR products to immobilized species-specific probes. For analysis of feed, mtDNA cytochrome *b* fragments generated by ruminant-specific primers were spotted on polyester cloth and hybridized to probes specific for cattle, sheep, goat, elk, and deer [23]. Using newly developed cytochrome *b* primers, PCR products were hybridized to microarrays containing probes for cattle, sheep, goat, pig, chicken, and turkey [52]. The commercially available kit CarnoCheck (http://www.jainbiologicals.com/PDF/carno_cryo.pdf) has been developed for use with the original universal cytochrome *b* primers [16]. Hybridization of amplicons to an array of probes targeted to the detection of cattle, sheep, goat, pig, horse, donkey, chicken, and turkey allows the detection of admixtures of 1% or less.

7.3.2.4   Species-Specific Amplification

Although most universal primers are a compromise of specificity and taxonomic range, primers targeted at a single species potentially offer better selectivity, that is, a more sensitive and specific detection in the presence of a complex and dominating background of other components in the sample. Several of these methods have been developed for the detection of bovine or ruminant material in feed to prevent a further spread of transmissible spongiform encephalopathy, but are equally applicable for analysis of processed meat products.

Specific primers have been described in several publications (Table 7.2). Although the design of these primers for any species-variable sequence on the basis of an alignment of homologous sequences is straightforward, published data will lend credibility to test results in the event of prosecution.

Table 7.2 Species-Specific PCR Amplifications

References	Target Gene	Detected Species	Detection Limit (w/w)
53	Lactoferrin	Cattle	0.02% in foodstuff
54	mt *CO1*	Cattle	0.5% in water buffalo cheese
55	mt *cytB*	Cattle	0.025%
56	mt *ATPase8* [57]	Cattle, sheep, pig	0.1% in animal feed, ring trial
58	mt *ATPase8* [56]	Cattle	0.006–0.03% in feed
59	mt *12S rRNA*	Cattle	0.1% in sheep or goat cheese
60	mt *12S rRNA*	Goat	1% goat milk in sheep milk
21	mt *ATPase8*	Ruminants, cattle, sheep, goat	0.1–0.01% meat and bone meal in vegetable meal
61	mt *cytB*	Pig	
	mt *APTase8* [57]	Cattle	0.1%
62	mt *12S rRNA*	Ruminants, pig, poultry	0.125–0.5% in fish meal
63	mt *12S rRNA*	Cattle, sheep, goat	0.1% in feedstuff
64	mt *t-Lys, ATPase8, ATPase6*	Cattle, sheep, pig, chicken	0.01% meat and bone meal in grain concentrates
65	mt *12S rRNA, 16S rRNA*	Cattle, sheep, goat, deer, ruminant	0.05% in vegetable meal
66	mt D-loop	Chamois, ibex, mouflon	0.1% in pork after sterilization
25	mt *12S rRNA*	Red-, roe-, fallow deer	
67	mt *cytB* [68]	Pig	
69	mt D-loop	Dog	0.05%
70	*cytB*	Tiger	
71	*cytB*	Chicken, turkey	
72	mt *12S rRNA*	Chicken, turkey, mule duck, goose	0.1% in oats
73	mt *12S rRNA*	Four duck species	0.1–1% in goose meat
		Muscovy duck	0.1–1% in goose meat
74	α-Actin	Mule duck, goose	1% duck in goose *foie gras*
75	mt *cytB*	Goose	
76	mt *cytB*	Ostrich, emu	
77	mt *cytB*	Chinese alligator	
78,79	mt *cytB*	Basking shark	

Note: For abbreviations, see Table 7.1. Different primers were developed for each species or taxon listed in the third column.

7.3.2.5 Multiplex Polymerase Chain Reaction

Often, only a limited number of species is to be expected in a sample. This obviously applies to dairy products, but also to meat products if possible adulterations likely originate from the available livestock species. Further, with game species, the number of species that can be present in a sample is in practice limited by their geographical distribution.

To detect these species, species-specific primers can be combined in one multiplex reaction (Table 7.3). However, increasing the number of primers also increases the chance of nonspecific amplification. This can be reduced by combining one common forward primer with a specific primer for each species to be detected [80]. Amplification products can be differentiated either by gel electrophoresis (see the various references in Table 7.3) or by their melting temperature [81,82].

7.3.2.6 Fingerprinting

PCR amplification with random primers [17,94], or primers specific for an ancient mammalian repetitive element [95], generate a fingerprint pattern that is specific for the species. Although this would allow the detection of several different species with one protocol, these methods suffer the disadvantages of problematic reproducibility and exchange of patterns between institutes. Further, the methods are not very well suited for the detection of a species against a background of other species. However, a qualitative PCR with species- or taxon-specific primers will not target all DNA components in a mixture and will not always differentiate trace amounts or contamination from a complete species substitution. In this case, a species-specific pattern would yield additional evidence for the species origin of a sample.

7.3.2.7 Real-Time Polymerase Chain Reaction

Quantification of species composition is mainly relevant if low but significant levels of a species must be differentiated from an insignificant signal, which, for instance, may originate from nonspecific side reactions or from contaminations of the reagents. For instance, qualitative PCR reactions described earlier would not be suitable for a sensitive yet specific detection of potentially pathogenic ruminant material in animal feed or for traces of porcine material in food for Jewish or Islamic consumers. For these applications, quantification has already been accomplished by competitive PCR [96–98]. However, much more accurate and convenient is real-time PCR that can be based on the binding of the fluorescent reporter SYBR Green to double-stranded probe, on relieving the quenching of fluorescence by the 5′ nuclease degradation of an internal probe (the TaqMan procedure), or on fluorescence resonance energy transfer (FRET) between two internal probes (often performed in a Lightcycler apparatus). In fact, because of its closed-tube format without post-PCR steps, real-time PCR is now becoming the method of choice for species identification.

As for the qualitative PCR methods, most published real-time PCR protocols (Table 7.4) exploit the high copy number of mtDNA or DNA repetitive elements [14,15,99,100]. Short amplicons (150 bp or shorter) are most suitable [101,102]. Hird et al. [103] give a few hints for deriving species-specific real-time PCR primers from alignments of homologous sequences. However, for most assays of common livestock species, no information is available about results with closely related species, either in the wild or kept locally as domesticates.

Detection limits (Table 7.4) are variable, but most assays appear adequate to detect significant adulterations or potentially harmful trace amounts.

Table 7.3   Multiplex PCR Amplifications for Species Identification

Reference	Target Gene	Detected Species	Detection Limit (w/w)
Two Primers per Species			
62	mt *12S rRNA*	Ruminants, pig, poultry	0.25%
83	mt *cytB*	Cattle, water buffalo	
84	mt *16S rRNA*	Cattle	0.002–0.004% in maize
	mt *12S rRNA—t-Val*	Pig	
	mt *12S rRNA*	Fish, poultry	
81	mt *t-Glu—cytB*	Cattle, horse	1% cattle, 5% horse by melting temperature analysis
		Cattle, wallaroo	5% cattle, 5% wallaroo by melting temperature analysis
		Pig, horse	5% pig, 1% horse by melting temperature analysis
		Pig, wallaroo	60% pig, 1% wallaroo by melting temperature analysis
One Primer per Species and One Common Primer			
80	mt *cytB*	Cattle, sheep, goat, pig, horse, chicken	ca. 10%
85	mt *12S rRNA, 16S rRNA*	Cattle, sheep	0.1% bovine milk in ovine cheese
86	mt *12S rRNA, 16S rRNA*	Cattle, goat	0.1% bovine milk in goat cheese
87	mt *12S rRNA, 16S rRNA*	Cattle, sheep, goat	0.5% in cheese
88	mt *cytB*	Cattle, water buffalo	1% in cheese
89	mt *cytB* [80]	Pig, horse	
90	mt *12S rRNA*	Cattle, sheep, goat, pig	1% for monoplex reactions
68	α-Actin	Chicken, pork	0.1% in goose and mule duck *foie gras*
91	5S rDNA	Mule duck, goose	

(continued)

Table 7.3 (continued)   Multiplex PCR Amplifications for Species Identification

Reference	Target Gene	Detected Species	Detection Limit (w/w)
92	mt *12S rRNA*	Pig, chicken, turkey, mule duck, goose	1% in *foie gras*
93		Pig, goose	
82	mt *t-Phe—12S rRNA*	Six Tasmanian carnivores	

Note: For abbreviations, see Table 7.1. One or two primers were developed for each species or taxon listed in the third column.

Table 7.4   Real-Time PCR Amplifications for Species Identification

References	Target Gene	Detected Species	Detection Limit (w/w)
SYBR Green Detection			
15	Bov-B SINE	Ruminants	0.1% ruminant material in processed chicken feed samples
81	mt *t-Glu—cytB*	Cattle, pig, horse, wallaroo	0.04 pg pig, wallaroo DNA, 0.4 pg cattle, horse DNA
99	Satellite DNA	Cattle	0.005%
	PRE-1 SINE	Pig	0.0005%
	Bov-tA2 SINE	Ruminants	
	Cr1 SINE	Chicken	0.05%
100	SINE and LINE elements	Birds, rodents, horse, dog, cat, rat, hamster, guinea pig, rabbit	0.1–100 pg
104	mt *cytB*	Tiger	0.5%
TaqMan Detection			
105	mt *ATPase8*	Cattle	0.0001% bovine material in meat and bone meal
106	mt *12S rRNA*	Goat	0.6% goat milk in sheep milk
107	mt *12S rRNA*	Cattle	0.6% cow milk in sheep milk
14	Bov-A2 SINE	Ruminants	10 fg bovine DNA
108	mt *16S rRNA*	Ruminants	
29	Growth hormone	Cattle	
		Mammals	

Table 7.4 (continued)   Real-Time PCR Amplifications for Species Identification

References	Target Gene	Detected Species	Detection Limit (w/w)
101	mt *t-Lys—ATPase8*	Cattle, pig	0.1% in compound feeds
109	mt *cyt*B	Cattle, sheep, chicken	35 pg bovine DNA
110	mt *t-Lys, ATPase8, ATPase6*	Cattle, sheep, pig, chicken	0.01% in grain concentrates
111	mt *cyt*B	Cattle, sheep, pig, chicken, turkey	0.5%
112	Prion protein	Cattle + sheep + goat, pig, chicken	10 pg DNA after heating
30	mt *t-Glu—cyt*B	Cattle, sheep, pig	1% pig, 5% cattle, lamb in binary mixtures
	mt *ND5*	Chicken, ostrich, turkey	1% chicken, turkey
	18S rRNA	Eukaryotes	
28,113,114	Phosphodiesterase	Cattle, sheep, goat	0.1% in processed food
	Ryanodin	Pig	
	Interleukin-2 precursor	Chicken, turkey, duck	
	Myostatin	Several mammals and birds	
27	mt *12S rRNA*	Pig, mammals	0.5% pig in beef
115	mt *cyt*B	Horse, donkey	1 pg donkey DNA, 25 pg horse DNA
116	*MC1R*	Dog	
117	mt *cyt*B	Mallard duck, Muscovy duck	
118	mt *12S rRNA*	Mule duck, mule duck + goose	1% duck in goose *foie gras*
FRET (Lightcycler)			
119	mt *ATPase8* [21]	Cattle	0.001% bovine gelatin in gelatin
120	mt *cyt*B	Cattle	0.001% bovine material in cattle feed

(continued)

Table 7.4 (continued) Real-Time PCR Amplifications for Species Identification

References	Target Gene	Detected Species	Detection Limit (w/w)
121	mt *cytB*	Cattle, sheep	0.05% cattle MBM, 0.1% sheep MBM in feed
	Chloroplast *rpoβ*	Plants (positive control)	
20	mt *16S rRNA*	Mammals + birds	
	mt D-loop	Cattle (scorpion reverse primer)	0.1%

Note: For abbreviations, see Table 7.1. Separate assays were developed for each species, combination of species, or taxon listed in the third column.

7.4 Conclusion

The technical progress of the methodology of species identification mirrors the fast and continuing progress in DNA technology. As a consequence, several methods have been replaced before being put in practice and validated by routine testing. Quite often, the same authors successively publish various methods for the detection of the same species without an explicit evaluation of the relative merits of the different approaches.

Quantitative real-time PCR is now accessible to most laboratories and is likely to dominate the field during the coming years. Future progress is likely to come from bead-based technologies, which are now being established in single nucleotide polymorphism (SNP) typing, microbial typing, and high-throughput sequencing.

Acknowledgments

We acknowledge support from the Framework V project MolSpecID, funded by the European Union (www.molspec.org). The contents of this publication do not represent the views of the Commission or its services. We thank Hermann Broll, Jutta Zagon, Andreas Butschke (BfR—Federal Institute for Risk Assessment, Berlin), and Ies Nijman (Utrecht) for useful discussions.

References

1. Lenstra, J.A., Buntjer, J.B., and Janssen, F.W., On the origin of meat. DNA techniques for species identification in meat products. *Vet. Sci. Tomorrow*, http://www.vetscite.org/, 2001.
2. Lenstra, J.A., DNA methods for identifying plant and animals species in foods, in *Food Authenticity and Traceability*, Lees, M., Ed., Woodhead Publishing, Cambridge, 2003, chap. 2.
3. Woolfe, M. and Primrose, S., Food forensics: using DNA technology to combat misdescription and fraud, *Trends Biotechnol.*, 22, 222, 2004.
4. Teletchea, F., Maudet, C., and Hanni, C., Food and forensic molecular identification: update and challenges. *Trends Biotechnol.*, 23, 359, 2005.
5. Lenstra, J.A., Primary identification: DNA markers for animal and plant traceability, in *Improving Traceability in Food Processing and Distribution*, Smith, I. and Furness, T. Eds., Woodhead Publishing, Cambridge, 2005, chap. 8.

6. Giovannacci, I., Guizard, C., Carlier, M., Duval, V., Martin, J.-L., and Demeulemester, C., Species identification of meat products by ELISA, *Int. J. Food Sci. Tech.*, 39, 863–867, 2004.

7. Jha, V.K., Kumar, A., and Mandokhot, U.V., Indirect enzyme-linked immunosorbent assay in detection and differentiation of cooked and raw pork from meats of other species, *J. Food Sci. Tech.*, 40, 254–256, 2003.

8. Myers, M.J., Yancy, H.F., Farrell, D.E., Washington, J.D., Deaver, C.M., and Frobish, R.A., Assessment of two enzyme-linked immunosorbent assay tests marketed for detection of ruminant proteins in finished feed, *J. Food Protect.*, 70, 692–699, 2007.

9. Vallejo-Cordoba, B., González-Córdoba, A., Mazorra-Manzano, M.A., and Rodríguez-Ramírez, R., Capillary eletrophoresis for the analysis of meat authenticity, *J. Sep.Sci.*, 28, 826, 2005.

10. Di Bernardo, G., Del Gaudio, S., Galderisi, U., Cascino, A., and Cipollaro, M., Comparative evaluation of different DNA extraction procedures from food samples, *Biotech. Prog.*, 23, 297–301, 2007.

11. Di Pinto, A.D., Forte, V., Guastadisegni, M.C., Martino, C., Schena, F.P., and Tantillo, G., A comparison of DNA extraction methods for food analysis, *Food Control*, 18, 76–80, 2007.

12. Arslan, A., Ilhak, O.I., and Calicioglu, M., Effect of method of cooking on identification of heat processed beef using polymerase chain reaction (PCR) technique, *Meat Sci.*, 72, 326–330, 2006.

13. Green, R.E. et al., Analysis of one million base pairs of Neanderthal DNA, *Nature*, 444, 330–336, 2006.

14. Romero-Mendoza, L., Verkaar, E.L.C., Savelkoul, P.H.M., Catsburg, A., Aarts, H.J.M., Buntjer, J.B., and Lenstra, J.A., 2004, Real-time PCR detection of ruminant DNA, *J. Food Protect.*, 67, 550, 2004.

15. Aarts, H.J.M., Bouw, E.M., Buntjer, J.B., Lenstra, J.A., and Van Raamsdonk, L.W.D., Detection of bovine meat and bone meal in animal feed at a level of 0.1%, *J. AOAC Int.*, 89, 1443, 2006.

16. Kocher, T.D., Thomas, W.K., Meyer, A., Edwards, S.V., Pääbo, S., Villablanca, F.X., and Wilson, A.C., Dynamics of mitochondrial DNA evolution in animals: amplification and sequencing with conserved primers, *P. Natl. Acad. Sci. USA*, 86, 6196, 1989.

17. Rastogi, G., Dharne, M.S., Walujkar, S., Kumar, A., Patole, M.S., and Shouche, Y.S., Species identification and authentication of tissues of animal origin using mitochondrial and nuclear markers, *Meat Sci.*, 76, 666, 2007.

18. Bottero, M.T., Civera, T., Nucera, D., and Turi, R.M., Design of universal primers for the detection of animal tissues in feedstuff, *Vet. Res. Commun.*, 27(Suppl. 1), 667, 2003.

19. Bottero, M.T., Dalmasso, I.A., Nucera, D., Turi, R.M., Rosatim, S., Squadrone, S., Goria, M., and Civera, T., Development of a PCR assay for the detection of animal tissues in ruminant feeds, *J. Food Protect.*, 66, 2307, 2003.

20. Sawyer, J., Wood, C., Shanahan, D., Gout, S., and McDowell, D., Real-time PCR for quantitative meat species testing, *Food Control*, 14, 579, 2003.

21. Kusama, T., Nomura, T., and Kadowaki, K., Development of primers for detection of meat and bone meal in ruminant feed and identification of the animal of origin, *J. Food. Protect.*, 67, 1289, 2004.

22. Myers, M.J., Yancy, H.F., and Farrell, D.E., Characterization of a polymerase chain reaction-based approach for the simultaneous detection of multiple animal-derived materials in animal feed, *J. Food Protect.*, 66, 1085, 2003.

23. Armour, J. and Blais, B.W., Cloth-based hybridization array system for the detection and identification of ruminant species in animal feed, *J. Food Protect.*, 69, 453, 2006.

24. Sun, Y.-L. and Lin, C.-S., Establishment and application of a fluorescent polymerase chain reaction-restriction fragment length polymorphism (PCR-RFLP) method for identifying porcine, caprine, and bovine meats, *J. Agr. Food Chem.*, 51, 1771, 2003.

25. Fajardo, V., González, I., López-Calleja, I., Martín, I., Hernández, P.E., García, T., and Martín, R., PCR-RFLP authentication of meats from red deer (*Cervus elaphus*), fallow deer (*Dama dama*), roe deer (*Capreolus capreolus*), cattle (*Bos taurus*), sheep (*Ovis aries*), and goat (*Capra hircus*), *J. Agr. Food Chem.*, 54, 1144, 2006.

26. Fajardo, V., González, I., López-Calleja, I., Martín, I., Rojas, M., Hernández, P.E., García, T., and Martín, R., Identification of meats from red deer (*Cervus elaphus*), fallow deer (*Dama dama*), and roe deer (*Capreolus capreolus*) using polymerase chain reaction targeting specific sequences from the mitochondrial 12S rRNA gene, *Meat Sci.*, 76, 234, 2007.

27. Rodríguez, M.A., García, T., González, I., Hernández, P.E., and Martín, R., TaqMan real-time PCR for the detection and quantitation of pork in meat mixtures, *Meat Sci.*, 70, 113, 2005.

28. Laube, I., Zagon, J., Spiegelberg, A., Butschke, A., Kroh, L.W., and Broll, H., Development and design of a "ready-to-use" reaction plate for a PCR-based simultaneous detection of animal species used in foods, *Int. J. Food Sci. Tech.*, 42, 9, 2007.

29. Brodmann, P.D. and Moor, D., Sensitive and semi-quantitative TaqMan™ real-time polymerase chain reaction systems for the detection of beef (*Bos taurus*) and the detection of the family *Mammalia* in food and feed, *Meat Sci.*, 65, 599, 2003.

30. López-Andreo, M., Lugo, L., Garrido-Pertierra, A., Prieto, M.I., and Puyet, A., Identification and quantitation of species in complex DNA mixtures by real-time polymerase chain reaction, *Anal. Biochem.*, 339, 73, 2005.

31. Martin, I., Garcia, T., Fajardo, V., Rojas, M., Hernandez, P.E., Gonzalez, I., and Martin, R., Technical Note, Detection of cat, dog, and mouse/rat tissues on food and animal feed using species-specific polymerase chain reaction, *J. Anim. Sci.*, 85, 2734, 2007.

32. Brodmann, P.D., Nicolas, G., Schaltenband, P., and Ilg, E.C., Identifying unknown game species: experience with nucleotide sequencing of the mitochondrial cytochrome b gene and a subsequent basic local alignment too search, *Eur. Food Res. Technol.*, 212, 491, 2001.

33. Hsieh, H.-M., Huang, L.-H., Tsai, L.-C., Kuo, Y.-C., Meng, H.-H., Linacre, A., and Lee, J.C.-I., Species identification of rhinoceros horns using the cytochrome b gene, *Forensic Sci. Int.*, 136, 1, 2003.

34. Colombo, F., Cardia, A., Renon, P., and Cantoni, C., A note on the identification of *Rupicapra rupicapra* species by polymerase chain reaction product sequencing, *Meat Sci.*, 66, 753, 2004.

35. Kyle, C.J. and Wilson, C.C., Mitochondrial DNA identification of game and harvested freshwater fish species, *Forensic Sci. Int.*, 166, 68, 2007.

36. Girish, P.S., Anjaneyulu, A.S.R., Viswas, K.N., Anand, M., Rajkumar, N., Shivakumar, B.M., and Bhaskar, S., Sequence analysis of mitochondrial 12S rRNA gene can identify meat species, *Meat Sci.*, 66, 551, 2004.

37. Meyer, R., Höfelein, C., Lüthy, J., and Candrian, U., Polymerase chain reaction-restriction fragment length polymorphism analysis: a simple method for species identification in food, *J. AOAC Int.*, 78, 1542, 1995.

38. Maede, D., A strategy for molecular species detection in meat and meat products by PCR-RFLP and DNA sequencing using mitochondrial and chromosomal genetic sequences, *Eur. Food Res. Technol.*, 224, 209, 2006.

39. Pfeiffer, I., Burger, J., and Brenig, B., Diagnostic polymorphisms in the mitochondrial cytochrome b gene allow discrimination between cattle, sheep, goat, roe buck and deer by PCR-RFLP, *BMC Genet.*, 5, 2004.

40. Pascoal, A., Prado, M., Castro, J., Cepeda, A., and Barros-Velázquez, J., Survey of authenticity of meat species in food products subjected to different technological processes, by means of PCR-RFLP analysis, *Eur. Food Res. Technol.*, 218, 306, 2004.

41. Aida, A.A., Man, Y.B.C., Wong, C.M.V.L., Raha, A.R., and Son, R., Analysis of raw meats and fats of pigs using polymerase chain reaction for Halal authentication, *Meat Sci.*, 69, 47–52, 2005.

42. Aida, A.A., Che Man, Y.B., Raha, A.R., and Son, R., Detection of pig derivatives in food products for halal authentication by polymerase chain reaction-restriction fragment length polymorphism, *J. Sci. Food Agr.*, 87, 569, 2007.

43. Verkaar, E.L.C., Boutaga, K., Nijman, I.J., and Lenstra, J.A., Differentiation of bovine species in beef by PCR-RFLP of mitochondrial and satellite DNA, *Meat Sci.*, 60, 365, 2002.

44. Fajardo, V., González, I., López-Calleja, I., Martín, I., Rojas, M., Pavón, M.Á., Hernández, P.E., García, T., and Martín, R., Analysis of mitochondrial DNA for authentication of meats from chamois (*Rupicapra rupicapra*), pyrenean ibex (*Capra pyrenaica*), and mouflon (*Ovis ammon*) by polymerase chain reaction-restriction fragment length polymorphism, *J. AOAC Int.*, 90, 179, 2007.

45. Girish, P.S., Anjaneyulu, A.S.R., Viswas, K.N., Shivakumar, B.M., Anand, M., Patel, M., and Sharma, B., Meat species identification by polymerase chain reaction-restriction fragment length polymorphism (PCR-RFLP) of mitochondrial 12S rRNA gene, *Meat Sci.*, 70, 107, 2005.

46. Abdulmawjood, A., Schönenbrücher, H., and Bülte, M., Development of a polymerase chain reaction system for the detection of dog and cat meat in meat mixtures and animal feed, *J. Food Sci.*, 68, 1757, 2003.

47. Abdulmawjood, A. and Bülte, M., Identification of ostrich meat by restriction fragment length polymorphism (RFLP) analysis of cytochrome b gene, *J. Food Sci.*, 1688, 2002.

48. Girish, P.S., Anjaneyulu, A.S.R., Viswas, K.N., Santhosh, F.H., Bhilegaonkar, K.N., Agarwal, R.K., Kondaiah, N., and Nagappa, K., Polymerase chain reaction-restriction fragment length polymorphism of mitochondrial 12S rRNA gene: a simple method for identification of poultry meat species, *Vet. Res. Commun.*, 31, 447, 2007.

49. Saini, M., Das, D.K., Dhara, A., Swarup, D., Yadav, M.P., and Gupta, P.K., Characterisation of peacock (Pavo cristatus) mitochondrial 12S rRNA sequence and its use in differentiation from closely related poultry species, *Brit. Poultry Sci.*, 48, 162, 2007.

50. Moore, M.K., Bemiss, J.A., Rice, S.M., Quattro, J.M., and Woodley, C.M., Use of restriction fragment length polymorphisms to identify sea turtle eggs and cooked meats to species, *Conserv. Genet.*, 4, 95, 2003.

51. Abdulmawjood, A. and Bulte, M., Snail species identification by RFLP-PCR and designing of species-specific oligonucleotide primers, *J. Food Sci.*, 1287, 2001.

52. Peter, C., Brünen-Nieweler, C., Cammann, K., and Börchers, T., Differentiation of animal species in food by oligonucleotide microarray hybridization, *Eur. Food Res. Technol.*, 219, 286, 2004.

53. Gao, H.-W., Zhang, D.-B., Pan, A.-H., Liang, W.-Q., and Liang, C.-Z., Multiplex polymerase chain reaction method for detection of bovine materials in foodstuffs, *J. AOAC Int.*, 86, 764, 2003.

54. Feligini, M., Bonizzi, I., Curik, V.C., Parma, P., Greppi, G.F., and Enne, G., Detection of adulteration in Italian mozzarella cheese using mitochondrial DNA templates as biomarkers, *Food Technol. Biotech.*, 43, 91, 2005.

55. Pascoal, A., Prado, M., Calo, P., Cepeda, A., and Barros-Velázquez, J., Detection of bovine DNA in raw and heat-processed foodstuffs, commercial foods and specific risk materials by a novel specific polymerase chain reaction method, *Eur. Food Res. Technol.*, 220, 444, 2005.

56. Myers, M.J., Yancy, H.F., Araneta, M., Armour, J., Derr, J., Hoostelaere, L.A.D., Farmer, D., Jackson, F., Kiessling, W.M., Koch, H., Lin, H., Liu, Y., Mowlds, G., Pinero, D., Riter, K.L., Sedwick, J., Shen, Y., Wetherington, J., and Younkins, R., Validation of a PCR-based method for the detection of various rendered materials in feedstuffs using a forensic DNA extraction kit, *J. Food Protect.*, 69, 205, 2006.

57. Lahiff, S., Glennon, M., Obrien, L., Lyng, J., Smith, T., Maher, M., and Shilton, N., Species-specific PCR for the identification of ovine, porcine and chicken species in meat and bone meal (MBM), *Mol. Cell. Probe.*, 15, 27, 2001.

58. Yancy, H.F., Mohla, A., Farrell, D.E., and Myers, M.J., Evaluation of a rapid PCR-based method for the detection of animal material, *J. Food Protect.*, 68, 2651, 2005.

59. López-Calleja, I., González, I., Fajardo, V., Rodríguez, M.A., Hernández, P.E., García, T., and Martín, R., Rapid detection of cows' milk in sheeps' and goats' milk by a species-specific polymerase chain reaction technique, *J. Dairy Sci.*, 87, 2839, 2004.

60. López-Calleja Díaz, I., González Alonso, I., Fajardo, V., Martín, I., Hernández, P., García Lacarra, T., and Martin De Santos, R., Application of a polymerase chain reaction to detect adulteration of ovine cheeses with caprine milk, *Eur. Food Res. Technol.*, 225, 345, 2007.

61. Castello, A., Francino, O., Cabrera, B., Polo, J., and Sanchez, A., Identification of bovine material in porcine spray-dried blood derivatives using the polymerase chain reation technique, *Biotech. Agr. Soc. Environ.*, 8, 267, 2004.

62. Bellagamba, F., Valfrè, F., Panseri, S., and Moretti, V.M., Polymerase chain reaction-based analysis to detect terrestrial animal protein in fish meal, *J. Food Protect.*, 66, 682, 2003.

63. Martín, I., García, T., Fajardo, V., López-Calleja, I., Hernández, P.E., González, I., and Martín, R., Species-specific PCR for the identification of ruminant species in feedstuffs, *Meat Sci.*, 120, 2007.

64. Krcmar, P. and Rencova, E., Identification of species-specific DNA in feedstuffs, *J. Agr. Food Chem.*, 51, 7655, 2003.

65. Ha, J.C., Jung, W.T., Nam, Y.S., and Moon, T.W., PCR identification of ruminant tissue in raw and heat-treated meat meals, *J. Food Protect.*, 69, 2241, 2006.

66. Fajardo, V., González, I., López-Calleja, I., Martín, I., Rojas, M., García, T., Hernández, P.E., and Martín, R., PCR identification of meats from chamois (*Rupicapra rupicapra*), Pyrenean ibex (*Capra pyrenaica*), and mouflon (*Ovis ammon*) targeting specific sequences from the mitochondrial D-loop region, *Meat Sci.*, 76, 644, 2007.

67. Che Man, Y.B., Aida, A.A., Raha, A.R., and Son, R., Identification of pork derivatives in food products by species-specific polymerase chain reaction (PCR) for halal verification, *Food Control*, 885, 2007.

68. Rodríguez, M.A., García, T., González, I., Asensio, L., Hernández, P.E., and Martin, R., Qualitative PCR for the detection of chicken and pork adulteration in goose and mule duck *foie gras*, *J. Sci. Food Agr.*, 83, 1176, 2003.

69. Gao, H.-W., Xu, B.-L., Liang, C.-Z., Zhang, Y.-B., and Zhu, L.-H., Polymerase chain reaction method to detect canis materials by amplification of species-specific DNA fragment, *J. AOAC Int.*, 87, 1195–1211, 2004.

70. Wan, Q.-H. and Fang, S.-G., Application of species-specific polymerase chain reaction in the forensic identification of tiger species, *Forensic Sci. Int.*, 131, 75, 2003.

71. Hird, H., Goodier, R., and Hill, M., Rapid detection of chicken and turkey in heated meat products using the polymerase chain reaction followed by amplicon visualisation with vistra, *Meat Sci.*, 65, 1117, 2003.

72. Martín, I., García, T., Fajardo, V., López-Calleja, I., Rojas, M., Pavón, M.A., Hernández, P.E., González, I., and Martín, R., Technical note: detection of chicken, turkey, duck, and goose tissues in feedstuffs using species-specific polymerase chain reaction, *J. Anim. Sci.*, 85, 452–458, 2006.

73. Martín, I., García, T., Fajardo, V., López-Calleja, I., Rojas, M., Hernández, P.E., González, I., and Martín, R., Mitochondrial markers for the detection of four duck species and the specific identification of Muscovy duck in meat mixtures using the polymerase chain reaction, *Meat Sci.*, 76, 721–729, 2007.

74. Rodríguez, M.A., García, T., González, I., Asensio, L., Mayoral, B., López-Calleja, I., Hernández, P.E., and Martín, R., Development of a polymerase chain reaction assay for species identification of goose and mule duck in *foie gras* products, *Meat Sci.*, 65, 1257, 2003.

75. Colombo, F., Marchisio, E., Pizzini, A., and Cantoni, C., Identification of the goose species (*Anser anser*) in Italian "Mortara" salami by DNA sequencing and a polymerase chain reaction with an original primer pair, *Meat Sci.*, 61, 291, 2002.

76. Colombo, F., Viacava, R., and Giaretti, M., Differentiation of the species ostrich (*Struthio camelus*) and emu (*Dromaius novaehollandiae*) by polymerase chain reaction using an ostrich-specific primer pair, *Meat Sci.*, 56, 15, 2000.

77. Yan, P., Wu, X.-B., Shi, Y., Gu, C.-M., Wang, R.-P., and Wang, C.-L., Identification of Chinese alligators (*Alligator sinensis*) meat by diagnostic PCR of the mitochondrial cytochrome b gene, *Biol. Conserv.*, 121, 45, 2005.

78. Hoelzel, A.R., Shark fishing in fin soup, *Conserv. Genet.*, 2, 69, 2001.

79. Magnussen, J.E., Pikitch, E.K., Clarke, S.C., Nicholson, C., Hoelzel, A.R., and Shivji, M.S., Genetic tracking of basking shark products in international trade, *Anim. Conserv.*, 10, 199, 2007.

80. Matsunaga, T., Chikuni, K., Tanabe, R., Muroya, S., Shibata, K., Yamada, J., and Shinmura, Y., A quick and simple method for the identification of meat species and meat products by PCR assay, *Meat Sci.*, 143, 1999.

81. López-Andreo, M., Garrido-Pertierra, A., and Puyet, A., Evaluation of post-polymerase chain reaction melting temperature analysis for meat species identification in mixed DNA samples, *J. Sci. Food Agr.*, 54, 7973, 2006.

82. Berry, O. and Sarre, S.D., Gel-free species identification using melt-curve analysis, *Mol. Ecol. Notes*, 7, 1, 2007.

83. Bottero, M.T., Civera, T., Anastasio, A., Turi, R.M., and Rosati, S., Identification of cow's milk in "buffalo" cheese by duplex polymerase chain reaction, *J. Food Protect.*, 65, 362, 2002.

84. Dalmasso, A., Fontanella, E., Piatti, P., Civera, T., Rosati, S., and Bottero, M.T., A multiplex PCR assay for the identification of animal species in feedstuffs, *Mol. Cell. Probe.*, 81, 2004.

85. Mafra, I., Ferreira, I.M., Faria, M.A., and Oliveira, B.P.P., A novel approach to the quantification of bovine milk in ovine cheeses using a duplex polymerase chain reaction method, *J. Agr. Food Chem.*, 52, 4943, 2004.

86. Mafra, I., Roxo, A., Ferreira, I.M., and Oliveira, M.B.P.P., A duplex polymerase chain reaction for the quantitative detection of cows' milk in goats' milk cheese, *Int. Dairy J.*, 17, 1132, 2007.

87. Bottero, M.T., Civera, T., Nucera, D., Rosati, S., Sacchi, P., and Turi, R.M., A multiplex polymerase chain reaction for the identification of cows', goats' and sheep's milk in dairy products, *Int. Dairy J.*, 13, 277, 2003.

88. Rea, S., Chikuni, K., Branciari, R., Sangamayya, R.S., Ranucci, D., and Avellini, P., Use of duplex polymerase chain reaction (duplex-PCR) technique to identify bovine and water buffalo milk used in making mozzarella cheese, *J. Dairy Res.*, 68, 689, 2001.

89. Di Pinto, A., Forte, V.T., Conversano, M.C., and Tantillo, G.M., Duplex polymerase chain reaction for detection of pork meat in horse meat fresh sausages from Italian retail sources, *Food Control*, 16, 391, 2005.

90. Rodríguez, M.A., García, T., González, I., Asensio, L., Hernández, P.E., and Martín, R., PCR Identification of beef, sheep, goat, and pork in raw and heat-treated meat mixtures, *J. Food Protect.*, 67, 172, 2004.

91. Rodríguez, M.A., García, T., González, I., Asensio, L., Fernández, A., Lobo, E., Hernández, P.E., and Martín, R., Identification of goose (*Anser anser*) and mule duck (*Anas platyrhynchos* × *Cairina moschata*) foie gras by multiplex polymerase chain reaction amplification of the 5S rDNA gene, *J. Agr. Food Chem.*, 49, 2717, 2001.

92. Rodríguez, M.A., García, T., González, I., Asensio, L., Mayoral, B., López-Calleja, I., Hernández, P.E., and Martin, R., Identification of goose, mule duck, chicken, turkey, and swine in *foie gras* by species-specific polymerase chain reaction, *J. Agr. Food Chem.*, 51, 1524, 2003.

93. Colombo, F., Marchisio, E., and Cantoni, C., Use of polymerase chain reaction (PCR) and electrophoretic gel computer-assisted statistical analysis to semi-quantitatively determine pig/goose DNA ratio, *Ital. J. Food Sci.*, 14, 71, 2002.

94. Saez, R., Sanz, Y., and Toldrá, F., PCR-based fingerprinting techniques for rapid detection of animal species in meat products, *Meat Sci.*, 66, 659, 2004.

95. Buntjer, J.B. and Lenstra, J.A., Mammalian species identification by interspersed repeat PCR fingerprinting, *J. Ind. Microbiol. Biot.*, 21, 121, 1998.

96. Wolf, C. and Luthy, J., Quantitative competitive (QC) PCR for quantification of porcine DNA, *Meat Sci.*, 57, 161, 2001.

97. Calvo, J.H., Osta, R., and Zaragoza, P., Quantitative PCR detection of pork in raw and heated ground beef and pâté, *J. Agr. Food Chem.*, 50, 5265, 2002.

98. Frezza, D., Favaro, M., Vaccari, G., Von-Holst, C., Giambra, V., Anklam, E., Dove, D., Battaglia, P.A., Agrimi, U., Brambilla, G., Ajmone-Marsan, P., and Tartaglia, M., A competitive polymerase chain reaction-based approach for the identification and semiquantification of mitochondrial DNA in differently heat-treated bovine meat and bone meal, *J. Food Protect.*, 66, 103, 2003.

99. Walker, J.A., Hughes, D.A., Anders, B.A., Shewale, J., Sinha, S.K., and Batzer, M.A., Quantitative intra-short interspersed element PCR for species-specific DNA identification, *Anal. Biochem.*, 316, 259, 2003.

100. Walker, J.A., Hughes, D.A., Hedges, D.J., Anders, B.A., Laborde, M.E., Shewale, J., Sinha, S.K., and Batzer, M.A., Quantitative PCR for DNA identification based on genome-specific interspersed repetitive elements, *Genomics*, 83, 518, 2004.

101. Fumière, O., Dubois, M., Baeten, V., Von Holst, C., and Berben, G., Effective PCR detection of animal species in highly processed animal byproducts and compound feeds, *Anal. Bioanal. Chem.*, 385, 1045, 2006.

102. Hird, H., Chisholm, J., Sanchez, A., Hernandez, M., Goodier, R., Schneede, K., Boltz, C., and Popping, B., Effect of heat and pressure processing on DNA fragmentation and implications for the detection of meat using a real-time polymerase chain reaction, *Food Addit. Contam.*, 23, 645, 2006.

103. Hird, H., Goodier, R., Schneede, K., Boltz, C., Chisholm, J., Lloyd, J., and Popping, B., Truncation of oligonucleotide primers confers specificity on real-time polymerase chain reaction assays for food authentication, *Food Addit. Contam.*, 21, 1035, 2004.

104. Wetton, J.H., Tsang, C.S.F., Roney, C.A., and Spriggs, A.C., An extremely sensitive species-specific ARMs PCR test for the presence of tiger bone DNA, *Forensic Sci. Int.*, 140, 139, 2004.

105. Lahiff, S., Glennon, M., Lyng, J., Smith, T., Shilton, N., and Maher, M., Real-time polymerase chain reaction detection of bovine DNA in meat and bone meal samples, *J. Food Protect.*, 65, 1158, 2002.

106. López-Calleja, I., González, I., Fajardo, V., Martín, I., Hernández, P.E., García, T., and Martín, R., Quantitative detection of goats' milk in sheep's milk by real-time PCR, *Food Control*, 18, 1466, 2007.

107. López-Calleja, I., González, I., Fajardo, V., Martín, I., Hernández, P.E., García, T., and Martín, R., Real-time TaqMan PCR for quantitative detection of cows' milk in ewes' milk mixtures, *Int. Dairy J.*, 17, 729, 2007.

108. Chiappini, B., Brambilla, G., Agrimi, U., Vaccari, G., Aarts, H.J.M., Berben, G., Frezza, D., and Giambra, V., Real-time polymerase chain reaction approach for quantitation of ruminant-specific DNA to indicate a correlation between DNA amount and meat and bone meal heat treatments, *J. AOAC Int.*, 88, 1399, 2005.

109. Zhang, C.-L., Fowler, M.R., Scott, N.W., Lawson, G., and Slater, A., A TaqMan real-time PCR system for the identification and quantification of bovine DNA in meats, milks and cheeses, *Food Control.*, 18, 1149, 2007.

110. Krcmar, P. and Rencova, E., Quantitative detection of species-specific DNA in feedstuffs and fish meals, *J. Food Protect.*, 68, 1217, 2005.

111. Dooley, J.J., Paine, K.E., Garrett, S.D., and Brown, H.M., Detection of meat species using TaqMan real-time PCR assays, *Meat Sci.*, 68, 431, 2004.

112. Bellagamba, F., Comincini, S., Ferretti, L., Valfrè, F., and Moretti, V.M., Application of quantitative real-time PCR in the detection of prion-protein gene species-specific DNA sequences in animal meals and feedstuffs, *J Food Protect.*, 69, 891, 2006.

113. Laube, I., Spiegelberg, A., Butschke, A., Zagon, J., Schauzu, M., Kroh, L., and Broll, H., Methods for the detection of beef and pork in foods using real-time polymerase chain reaction, *Int. J. Food Sci. Tech.*, 38, 111, 2003.

114. Laube, I., Zagon, J., and Broll, H., Quantitative determination of commercially relevant species in foods by real-time PCR, *Int. J. Food Sci.Tech.*, 42, 336, 2007.

115. Chisholm, J., Conyers, C., Booth, C., Lawley, W., and Hird, H., The detection of horse and donkey using real-time PCR, *Meat Sci.*, 70, 727, 2005.

116. Evans, J.J., Wictum, E.J., Penedo, M.C.T., and Kanthaswamy, S., Real-time polymerase chain reaction quantification of canine DNA, *J. Forensic Sci.*, 52, 93, 2007.

117. Hird, H., Chisholm, J., and Brown, J., The detection of commercial duck species in food using a single probe-multiple species-specific primer real-time PCR assay, *Eur. Food Res. Technol.*, 221, 559, 2005.

118. Rodríguez, M.A., García, T., González, I., Asensio, L., Hernández, P.E., and Martín, R., Quantitation of mule duck in goose *foie gras* using Taqman real-time polymerase chain reaction, *J. Agr. Food Chem.*, 52, 1478, 2004.

119. Tasara, T., Schumacher, S., and Stephan, R., Conventional and real-time PCR-based approaches for molecular detection and quantitation of bovine species material in edible gelatin, *J. Food Protect.*, 68, 2420, 2005.

120. Rensen, G.J., Smith, W.L., Jaravata, C.V., Osburn, B., and Cullor, J.S., Development and evaluation of a real-time FRET probe based multiplex PCR assay for the detection of prohibited meat and bone meal in cattle feed and feed ingredients, *Foodborne Pathogens Disease*, 3, 337, 2006.

121. Rensen, G., Smith, W., Ruzante, J., Sawyer, M., Osburn, B., and Cullor, J., Development and evaluation of a real-time fluorescent polymerase chain reaction assay for the detection of bovine contaminates in cattle feed, *Foodborne Pathogens Disease*, 2, 152, 2005.

Chapter 8

Detection of Irradiated Ingredients

Eileen M. Stewart

Contents

8.1　Introduction to Food Irradiation 208
8.2　Can Irradiated Foodstuffs Be Identified in the Marketplace? 209
　　8.2.1　Gas Chromatographic Analysis of Hydrocarbons (EN1784) 210
　　8.2.2　Gas Chromatography: Mass Spectrometric Analysis
　　　　　of 2-Alkylcyclobutanones (EN1785) 212
　　8.2.3　Electron Spin Resonance Spectroscopy...213
　　　　　8.2.3.1　Detection of Irradiated Food Containing Bone
　　　　　　　　　by Electron Spin Resonance Spectroscopy (EN1786)..............214
　　　　　8.2.3.2　Detection of Irradiated Food Containing Cellulose
　　　　　　　　　by Electron Spin Resonance Spectroscopy (EN1787)............215
　　　　　8.2.3.3　Detection of Irradiated Food Containing Crystalline Sugar
　　　　　　　　　by Electron Spin Resonance Spectroscopy (EN13708)..........216
　　8.2.4　Luminescence Methods: Detection of Irradiated Food from
　　　　　Which Silicate Minerals Can Be Isolated 216
　　　　　8.2.4.1　Thermoluminescence Detection of Irradiated Food from
　　　　　　　　　Which Silicate Minerals Can Be Isolated (EN1788) 217
　　　　　8.2.4.2　Detection of Irradiated Food Using Photostimulated
　　　　　　　　　Luminescence (EN13751) 217
　　8.2.5　DNA Comet Assay...218

8.2.6 Measurement of Microbiological Changes...219
    8.2.6.1 Direct Epifluorescent Filter Technique/Aerobic Plate
    Count (DEFT/APC) (EN13783) 219
    8.2.6.2 *Limulus* Amebocyte Lysate/Gram-Negative Bacteria
    Test (EN14569) 220
8.2.7 Other Methods Explored..220
8.2.8 Application of Detection Methods in the Marketplace.......................221
8.3 Conclusions...223
References...223

8.1 Introduction to Food Irradiation

Food irradiation is a process by which food is exposed to ionizing radiation in a controlled manner, either using gamma rays (produced mostly from cobalt 60) or by electron beams or x-rays (generated electrically). These are high-energy sources, which act in the same way to bring about changes to the foodstuff. When food is irradiated, energy is absorbed, and it is this absorbed energy that leads to the ionization or excitation of the atoms and molecules of the food, which in turn results in chemical changes. These changes may result from "direct" or "indirect" action. In "direct" action, a sensitive target such as the deoxyribonucleic acid (DNA) of a living organism is damaged directly by an ionizing particle or ray, whereas "indirect" action is caused mostly by the products of water radiolysis, which disappear quickly by reacting with each other or with other food components [1].

The use of ionizing radiation as a preservation method for foodstuffs is not new. In 1896 H. Minsch, Germany, published a proposal to use ionizing radiation for the preservation of food by destroying spoilage microorganisms. Thus, there is a long history of research on the radiation processing of foodstuffs, including extensive safety studies on irradiated food [2]. In 1980, the Joint FAO/IAEA/WHO Expert Committee on the Wholesomeness of Irradiated Food (JECFI) met in Geneva, and their landmark report published in 1981 concluded that the "irradiation of any food commodity up to an overall average dose of 10 kGy presents no toxicological hazard." The Committee also concluded that irradiation up to 10 kGy "introduces no special nutritional or microbiological problems" [3].

As a result of the JECFI report [3], in 1983 the Codex Alimentarius Commission (CAC) adopted the Codex General Standard for Irradiated Foods and the Recommended Code of Practice for the Operation of Radiation Facilities Used for the Treatment of Foods. Irradiated food in international trade should therefore conform to the provisions of the Codex General Standard and recommended Code of Practice. In an effort to harmonize the law of the Member States on food irradiation, the European Union (EU) adopted framework Directive 1999/2/EC and implementing Directive 1999/3/EC [4]. The framework directive sets out the general and technical aspects for carrying out food irradiation, labeling of irradiated foods, and the conditions for authorizing the process, whereas the implementing directive established an initial "positive list" specifying food categories that may be irradiated and freely traded in the EU. The list is still under discussion and currently includes only dried aromatic herbs, spices, and vegetable seasonings. Until this list is complete, EU Member States may continue to apply their own existing national authorizations of irradiated foodstuffs not included in the initial "positive list."

The two main drivers for treating foods with ionizing radiation are the enhancement of food safety and of trade in agricultural products [5]. The process should not be used as a substitute for good manufacturing practices, but rather as a means of reducing risk. As food poisoning bacteria are highly sensitive to ionizing radiation, food irradiation has a proven efficacy for destroying microorganisms of public health importance, for example, *Escherichia coli* O157:H7 and *Salmonella* spp., as well as controlling parasitic organisms, such as *Trichinella spiralis*. According to Molins et al. [6], irradiation could be a critical control point in ensuring the microbiological safety of raw foods such as poultry, meat, meat products, fish, seafood, fruits, and vegetables.

Food irradiation can be used to extend the shelf life of perishable foods such as fruits, vegetables, meat, and meat products. As an example, spoilage bacteria such as *Pseudomonas putida* found in poultry meat are highly sensitive to irradiation, thus treatment with doses of 2–3 kGy can extend shelf life by as much as 2 weeks when combined with refrigeration.

Another beneficial use of the process is the prevention of food losses by inhibition of sprouting in bulb and tuber crops. Irradiation of potatoes to prevent sprout inhibition is carried out in Japan with approximately 16,000 t of irradiated potato per annum being distributed on the domestic market [7].

Irradiation is a "cold process," and thus is suitable for reducing the microbial load in herbs, spices, and seasonings. It is an effective alternative to using chemical fumigants such as ethylene oxide, which are now banned for use in Europe and the United States. One of the benefits of using ionizing radiation is that it does not cause any adverse changes to the important quality characteristics of herbs and spices such as color, aroma, or flavor.

Quarantine security is required to protect the ecology and agriculture of importing regions from pests that may be present on imported goods, while facilitating trade between different regions [5]. Research has demonstrated the suitability of ionizing radiation for the disinfection of cereals, grains, and certain fruits, such as mango and papaya [8], thus the process could play a significant role in fulfilling quarantine needs.

The use of ionizing radiation is, however, not suitable for all food products. Its use for the treatment of foods with a high fat content may lead to off-odors and tastes, as ionizing radiation is known to accelerate rancidity, and food with a high amount of protein can have changes in flavor and odor. It is therefore important that the suitability of a foodstuff is rigorously assessed before treatment and the irradiation conditions optimized to ensure a product of highest quality.

8.2 Can Irradiated Foodstuffs Be Identified in the Marketplace?

Irradiated food on sale in the marketplace should be clearly labeled so that consumers can choose whether or not to buy it. Under EU regulations, and those of other countries, irradiated food must be clearly labeled as "irradiated" or "treated with ionizing radiation." Such labeling should allow consumers to make informed choices about their food purchases. Thus, if a food is being marketed as irradiated or if irradiated goods are being sold without the appropriate labeling, then detection tests should be able to prove the authenticity of the product.

The reasons for the development of detection methods for irradiated foods can be summarized as follows:

- To control any legislative prohibitions regarding irradiation of specific foods, for example, reirradiation
- To control limitations imposed on the irradiation process
- To control the labeling of irradiated foodstuffs

■ To enhance consumer confidence in the correct application of the radiation process and its proper control by the inspection authorities
■ To protect the consumers' freedom of choice between irradiated and nonirradiated food products [9,10]

Before the 1980s, little progress was made in the development of detection methods for irradiated foods. The lack of emphasis was partly due to the fact that detection methods were considered unnecessary, because it was believed that food products would be irradiated in licensed facilities and that appropriate documentation would accompany the irradiated food throughout the food chain. However, because of the individual efforts of research teams in many countries and the noteworthy international cooperation in this field, between the years 1985 and 1995 considerable progress was made in the development of reliable methods to identify irradiated foods. The European Community (EC), through its Community Bureau of Reference (BCR), set up a collaborative program to develop methods to identify irradiated food while, on a worldwide basis, the Joint FAO/IAEA Division of Nuclear Techniques in Food and Agriculture set up a co-ordination program on Analytical Detection Methods in Irradiation Treatment of food (ADMIT), which promoted cooperation in this area.

Although it would have been ideal to have developed one method to detect all irradiated foodstuffs, this was not feasible, mainly due to differences in the nature of the foodstuffs being irradiated and the diverse range of changes produced in foods by ionizing radiation. The development of these methods also proved difficult due to the fact that the radiolytic changes that occur in food upon irradiation are minimal and often similar to those produced by other food-processing technologies, such as cooking. The methods that were developed are in fact based on particular physical, chemical, biological, and microbiological changes induced in foods during the irradiation process.

Under EU legislation it also states that Member States shall ensure that the analytical methods used to detect irradiated foods are validated or standardized. In 1993, the European Commission (EC) gave a mandate to the European Committee for Standardization (CEN) to standardize these methods. Consequently, CEN created within its Technical Committee 275 "Food Analysis– Horizontal Methods" (CEN/TC 275) Working Group 8 "Irradiated Foodstuffs" (CEN/TC275/ WG8), which had its first meeting in November 1993. As a result of the efforts of this Working Group, 10 European Standards are now available from national standardization institutes [11]. These European Standards have also been adopted by the CAC as General Methods and are referred toin the Codex General Standard for Irradiated Foods in Section 6.4 on "Postirradiation Verification." Table 8.1 lists the 10 methods that are now available and used worldwide for the detection of irradiated foodstuffs. The rest of this chapter will outline these methods and demonstrate how they have been used to detect irradiated foodstuffs on sale in the marketplace and not labeled correctly.

8.2.1 Gas Chromatographic Analysis of Hydrocarbons (EN1784)

European Standard EN1784 was developed for the identification of irradiated food containing fat. As for all the standard methods, EN1784 was validated by a series of interlaboratory trials as a reliable test for the detection of irradiated products such as chicken meat, pork, beef, camembert, papaya, and mango [12]. It is based on the gas chromatography (GC) detection of radiation-induced hydrocarbons.

Table 8.1 European Standards for the Detection of Irradiated Foodstuffs [11]

EN1784:2003	Foodstuffs—detection of irradiated food containing fat—gas chromatographic analysis of hydrocarbons
	Validated with raw chicken, pork, liquid whole egg, salmon, Camembert
EN1785:2003	Foodstuffs—detection of irradiated food containing fat—gas chromatographic/mass spectrometric analysis of 2-alkylcyclobutanones
	Validated with raw meat, Camembert, fresh avocado, papaya, mango
EN1786:1996	Foodstuffs—detection of irradiated food containing bone—method by ESR spectroscopy
	Validated with beef bones, trout bones, chicken bones—expected that method can be applied to all meat and fish species containing bone
EN1787:2000	Foodstuffs—detection of irradiated food containing cellulose, method by ESR spectroscopy
	Validated with pistachio nut shells, paprika powder, fresh strawberries
EN1788:2001	Foodstuffs—detection of irradiated food from which silicate minerals can be isolated, method by thermoluminescence
	Validated with herbs and spices as well as their mixtures, shellfish including shrimps and prawns, both fresh and dehydrated fruits and vegetables, potatoes
EN13708:2001	Foodstuffs—detection of irradiated food containing crystalline sugar by ESR spectroscopy
	Validated with dried figs, dried mangoes, dried papayas, raisins
EN13751:2002	Detection of irradiated food using photostimulated luminescence
	Validated with shellfish, herbs, spices, seasonings
EN13783:2001	Detection of irradiated food using Direct Epifluorescent Filter Technique/Aerobic Plate Count (DEFT/APC)—Screening method
	Validated with herbs and spices
EN13784:2001	DNA comet assay for the detection of irradiated foodstuffs—Screening method
	Validated with chicken bone marrow, chicken muscle, pork muscle, almonds, figs, lentils, linseed, rosé pepper, sesame seeds, soyabeans, sunflower seeds
EN14569:2004	Microbiological screening for irradiated foodstuffs—Screening method (LAL/GNB)
	Validated for chilled or frozen chicken fillets (boneless) with or without skin

Source: European Commission, Food irradiation—analytical methods. http://ec.europa.eu/food/food/biosafety/irradiation/ anal_methods_en.htm.

As most of the volatile products formed in food by irradiation originate from the fat or lipid content, in 1988 Nawar [13] proposed that measurement of radiolytic products from food lipids could form the basis for a method to identify irradiated foods. Research showed that both the quantitative and qualitative patterns of the radiolytic products depend largely on the fatty acid composition of the fat. Thus, if the fatty acid composition of the fat is known, the composition of the products formed by irradiation of a fat, or fat-containing food, can be predicted to a certain degree [14].

Upon irradiation of foods containing fat, two hydrocarbons are formed in relatively large quantities [15]. In the fatty acid moieties of triglycerides, breaks in chemical bonds occur mainly in the alpha and beta positions with respect to the carbonyl groups. Thus, one hydrocarbon has a carbon atom less than the parent fatty acid, resulting from cleavage at the carbon–carbon bond alpha to the carbonyl group (C_{n-1}), whereas the other has two carbons less and one extra double bond resulting from cleavage beta to the carbonyl ($C_{n-2:1}$).

In 1970, Nawar and Balboni [15] reported on the feasibility of detecting irradiation in pork meat at doses between 1 and 60 kGy by analysis of the six "key hydrocarbons." Tetradecene ($C_{14:1}$) and pentadecane ($C_{15:0}$) are produced from palmitic acid ($C_{16:0}$) upon irradiation, hexadecene ($C_{16:1}$) and heptadecane ($C_{17:0}$) from stearic acid ($C_{18:0}$), whereas hexadecadiene ($C_{16:2}$) and heptadecene ($C_{17:1}$) are typically produced from oleic acid ($C_{18:1}$). Nawar and Balboni [15] demonstrated a linear relationship between irradiation dose and each of these compounds, with neither of them, nor water, having a significant effect on the quantitative pattern. Work on irradiated chicken reported by Nawar et al. [16] in 1990 considered tetradecene, hexadecadiene, and heptadecene to be the most promising hydrocarbons for reliable detection of irradiation treatment in meat, because they were found in the highest concentrations and were absent or present at a low level in nonirradiated samples.

For detection of irradiated hydrocarbons, the fat is isolated from the sample by melting it out or by solvent extraction. The hydrocarbon fraction is obtained by adsorption chromatography before separation using GC and detection with a flame ionization detector or a mass spectrometer (MS) [12].

Alternatively, the hydrocarbons may be detected using liquid chromatography-GC (LC-GC) coupling [17]. Horvatovich et al. [18] showed how supercritical carbon dioxide can be used to carry out a selective and fast extraction (30 min) of volatile hydrocarbons and 2-alkylcyclobutanones contained in irradiated foods. The supercritical fluid extraction (SFE) method was successfully applied to freeze-dried samples (1 g or less) of cheese, chicken, avocados, and various ingredients (chocolate, liquid whole eggs) included in nonirradiated cookies. The method proved to be 4–5 h faster than the standardized hydrocarbon (EN1784) [12] and 2-alkylcyclobutanone (EN1785) [19] methods, which take 1.5 days each to determine if a food has been irradiated. In addition, the minimal dose detectable by this method was slightly lower than those of the standardized methods.

8.2.2 Gas Chromatography: Mass Spectrometric Analysis of 2-Alkylcyclobutanones (EN1785)

European Standard EN1785, along with EN1784, can be used for the identification of irradiated food containing fat. This method is based on the mass spectrometric detection of 2-alkylcyclobutanones after gas chromatographic separation [19]. It has been proposed that the formation of the 2-alkylcyclobutanones in irradiated foods results from cleavage at the acyl–oxygen bond in triglycerides, with the pathway involving a six-membered ring intermediate. The cyclobutanones so formed contain the same number of carbon atoms as the parent fatty acid, and the alkyl group is located in ring position 2 [10,14]. To date, the cyclobutanones are the only cyclic compounds reported in the radiolytic products of saturated triglycerides. As for the hydrocarbons, if the fatty acid composition of a lipid is known, then the products formed upon irradiation can be predicted to a certain degree. Thus, for example, if the fatty acids palmitic, stearic, oleic, and linoleic acid are exposed to ionizing radiation, then the respective 2-dodecyl-, 2-tetradecyl-, 2-tetradecenyl-, and 2-tetradecadienyl-cyclobutanones will be formed [20].

The method is based on the detection of 2-dodecylcyclobutanone (2-DCB) and 2-tetradecyl-cyclobutanone (2-TCB), these being the two markers most commonly used for identification purposes. These cyclobutanones have been identified in irradiated foods treated with irradiation doses as low as 0.1 kGy, and to date have not been detected in nonirradiated foods or microbiologically spoiled products. The specificity of the compounds as irradiation markers has been demonstrated in extensive experimental work, which has shown that they are not produced by cooking, by packaging in air, vacuum, or carbon dioxide, or during storage [21].

The 2-alkylcyclobutanones are extracted from the sample using either hexane or pentane along with the fat. The extract is then fractionated using adsorption chromatography before separation by GC and detection using a mass spectrometer [19]. As most foods contain some fat, the method is applicable to a wide range of products, and interlaboratory trials have successfully validated EN1785 for the identification of irradiated raw chicken, pork, liquid whole egg, salmon, and camembert. 2-DCB and 2-TCB have been detected postcooking in such products as irradiated meat, poultry, and egg [22,23]. Detection of irradiated ingredients such as irradiated liquid whole egg in cakes is also possible [10,22,23].

2-Tetradecenylcyclobutanone (2-TDCB) has been detected in irradiated chicken meat, papaya, and mango [24,25]. However, as this cyclobutanone is more difficult to detect and quantify in comparison with 2-DCB and 2-TCB, it is not used routinely for detection of irradiation treatment.

Since the initial development of the 2-alkylcyclobutanone method, alternative procedures have been developed for the extraction and purification of these radiation markers. Studies published by Stewart et al. [23], Gadgil et al. [26], and Horvatovich et al. [27] demonstrated that SFE could be used for the selective and rapid extraction of the cyclobutanones from irradiated foodstuffs without prior extraction of the fat. Obana et al. [22] used an accelerated solvent extraction (ASE) system for extraction of the cyclobutanones. Work by Ndiaye et al. [28] showed that inclusion of a purification step by silver ion chromatography in the EN1785 protocol considerably improved the quality of the chromatograms obtained, thereby allowing the detection of food samples irradiated at doses as low as 0.1 kGy. In addition, Horvatovich et al. [29] used a column containing 60 g silica gel for cleanup and the use of isobutane as a reactant for chemical ionization–mass spectrometric analysis of saturated and monounsaturated alkyl side-chains of 2-alkylcyclobutanones to improve both the sensitivity and selectivity of the method. However, it should be noted that these procedures have not been validated by interlaboratory trials.

8.2.3 Electron Spin Resonance Spectroscopy

Three of the European Standards for detection of irradiated foodstuffs use the technique of electron spin resonance (ESR) spectroscopy, also known as electron paramagnetic resonance (EPR) spectroscopy. ESR spectroscopy is a physical technique that detects species with unpaired electrons. Electrons are almost invariably paired. However, some molecules do contain an odd number of electrons, and the one that is unpaired is referred to as a free radical. Free radicals are highly reactive and consequently are short-lived. Some do exist in a stable state for some time, and it is these that are examined by ESR spectroscopy. Ionizing radiation produces free radicals in food, and because ESR spectroscopy detects free radicals, it can be used to determine whether certain foods have been irradiated. In foodstuffs with a relatively high moisture content, such as vegetables and meat, the induced radicals disappear rapidly. On the other hand, if food contains components with a relatively large proportion of dry matter, such as bones, seeds, or shells, the radicals may be trapped and be sufficiently stable to be detected by ESR [30]. The three ESR methods standardized by CEN are used for the detection of irradiated food containing bone (EN1786) [31], cellulose (EN1787) [32], and crystalline sugar (EN13708) [33].

8.2.3.1 Detection of Irradiated Food Containing Bone by Electron Spin Resonance Spectroscopy (EN1786)

When bone is subjected to ionizing radiation, free radicals are trapped in the crystal lattice of the bone, and these can be detected by ESR spectroscopy. The use of ESR to detect the presence of radiation-induced free radicals in bone dates back to the mid-1950s, being used to date archeological specimens, and also as an *in vivo* dosimeter for human to assess their exposure to radiation [34]. Nonirradiated bone gives a weak, broad ESR signal that increases in magnitude if the bone is ground into a powder. The signal derived from irradiated bone (Figure 8.1) is a large axially asymmetrical singlet, and can easily be distinguished from the endogenous signal [10,34]. Two prevailing types of paramagnetic species have been observed after the irradiation of bone tissue. One species is derived from bone collagen, and the other is attributed to the mineral constituent of bone, the hydroxyapatite. It is surmised that the characteristic signal produced on irradiation of the bone is due to either the CO^{2-} or the CO_3^{3-} radical trapped in the hydroxyapatite matrix.

Significant work has been carried out on chicken bone [30,35,36], with the bones from duck, turkey, goose, beef, pork, lamb, and frog legs also being studied to a more limited extent [37–40]. The signal produced from all sources of bone is essentially the same, thus it is evident that ESR can be used for the qualitative detection of irradiation in a wide range of meats containing bone. Interlaboratory trials have validated the method for beef bones, trout bones, and chicken bones [31].

Gray and Stevenson [41] also demonstrated that the method could be used for the identification of irradiated mechanically recovered meat (MRM), a secondary food product from

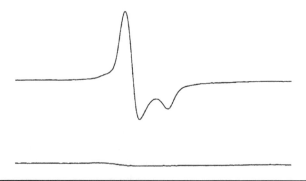

Figure 8.1 ESR spectra derived from irradiated (top spectrum) and nonirradiated (bottom spectrum) bone from frog legs.

which small bone fragments can be extracted. It has also been shown by Stevenson et al. [42] that ESR could be used to detect irradiated MRM as an ingredient in a food product, for example, burgers, at inclusion levels as low as 3 g/100 g. Work published by Marchioni et al. [43,44] also proved that ESR can be used for the detection of irradiated mechanically recovered poultry meat at very low inclusion levels in tertiary food products such as poultry quenelles and precooked meals.

An ESR signal similar to that of bone has also been derived from irradiated eggshell, as demonstrated by Onori and Pantaloni [45]. When tested by an interlaboratory trial [46], samples of irradiated eggshell were identified with a 100% success rate, even when treated at doses as low as 0.3 kGy.

8.2.3.2 Detection of Irradiated Food Containing Cellulose by Electron Spin Resonance Spectroscopy (EN1787)

European Standard EN1787 specifies a method for the detection of foods containing cellulose that have been treated with ionizing radiation [32]. It was Raffi [47] who first examined the ESR signal derived from the seeds of strawberries and derived a multicomponent signal that is typical of that from foodstuffs containing cellulose. A central single line is present in both irradiated and nonirradiated samples (Figure 8.2) that is thought to arise from a semiquinone radical. This single line increases with increasing irradiation dose, but will vary to a large extent with the water content of the sample. For irradiated samples (Figure 8.2), a pair of outlying lines occurs to the left and right of the central signal, the left one of which is most easily detected. It was proposed that these lines originate from cellulose and, as they are not present in nonirradiated samples, they can be used to detect irradiation treatment.

The method has been validated by interlaboratory trials for pistachio nut shells, paprika powder, and fresh strawberries [32].

This method could be used for a wide range of fruits, and has been employed for the detection of irradiated nuts, some aromatic herbs and spices, and for certain packaging materials, containing a high percentage of cellulose [48–50].

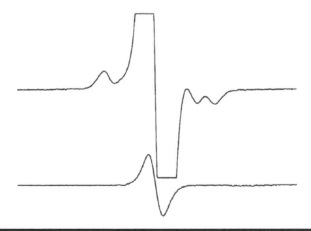

Figure 8.2 ESR spectra derived from irradiated (top spectrum) and nonirradiated (bottom spectrum) paper containing cellulose.

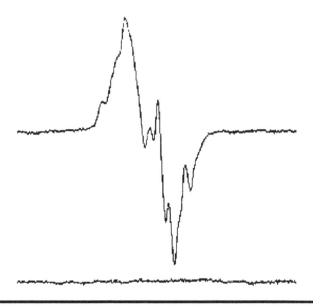

Figure 8.3 ESR spectra derived from irradiated (top spectrum) and nonirradiated (bottom spectrum) samples of dried fruits containing crystalline sugars.

8.2.3.3 Detection of Irradiated Food Containing Crystalline Sugar by Electron Spin Resonance Spectroscopy (EN13708)

EN13708 [33] uses ESR spectroscopy for the detection of irradiated food containing crystalline sugar. A multicomponent ESR signal is derived from irradiated dried fruits such as dates, grapes, mango, papaya, and pineapple, being easily distinguishable from the single line obtained from nonirradiated samples (Figure 8.3). It was proposed that the complex signal induced by ionizing radiation arises from sugar radicals [51], as the overall sugar content of fruits is high, varying from 60 to 75%, the main components being D-fructose, D-glucose, and D-saccharose. These radiation-induced signals are, in general, sufficiently stable for the identification of irradiated samples, even when they are stored for several months.

Interlaboratory trials have successfully demonstrated that the method can be used to identify irradiated dried figs, dried mangoes, dried papayas, and raisins [33]. The lower detection limit will mainly depend on the crystallinity of the sugar in the sample. The presence of sufficient amounts of crystalline sugar in the sample at all stages of handling between irradiation and testing will determine the applicability of the method.

8.2.4 Luminescence Methods: Detection of Irradiated Food from Which Silicate Minerals Can Be Isolated

The luminescence methods are probably the most sensitive means by which irradiated products such as herbs, spices, and seasonings can be identified. The methods involve either the thermoluminescence (TL) or photostimulated luminescence (PSL) analysis of contaminating silicate minerals. Mineral debris, typically silicates or bioinorganic materials such as calcite that originate from shells or exoskeletons, or hydroxyapatite from bones or teeth, can be found on most foods [52].

These materials store energy in charge carriers trapped at structural, interstitial, or impurity sites, when exposed to ionizing radiation. Luminescence is the emission of light when this trapped energy is liberated by the addition of either heat (TL) or light (PSL). Two European Standards have been developed based on the use of TL (EN1788) and PSL (EN13751) for the detection of irradiated foodstuffs containing silicate minerals.

8.2.4.1 Thermoluminescence Detection of Irradiated Food from Which Silicate Minerals Can Be Isolated (EN1788)

European Standard EN1788 is applicable to those foodstuffs from which silicate minerals can be isolated [53]. The energy stored within the silicate minerals is released by controlled heating of isolated silicate minerals so that light is emitted, the intensity of the emitted light being measured as a function of temperature, resulting in a so-called glow curve.

It was first thought that the TL arose from the organic component of the samples, but research [54,55] has clearly shown that the signals from herbs and spices actually originated from adhering mineral grains, although they accounted for less than 1% of the sample weight. In this method, the silicate minerals are separated from the food matrix, mostly by a density separation step. The isolated minerals should be as free from organic constituents as possible, so as not to obscure the TL. A first glow of the separated mineral extracts is recorded (glow 1). However, as various amounts and types of minerals exhibit variable integrated TL intensities, a second glow (glow 2) of the sample is measured after exposure to a fixed dose of ionizing radiation. The latter step is necessary to normalize the TL response. Thus, a ratio of glow 1 to glow 2 is obtained and used to indicate irradiation treatment of the food, as irradiated samples normally yield higher TL glow ratios than nonirradiated samples. Glow shape parameters can also be used as additional evidence for the identification of irradiated foods. As the method relies solely on the separated silicate minerals, it is not on principle influenced by the kind of food product.

Interlaboratory trials have validated the TL method for a wide range of herbs and spices as well as their mixtures, shellfish including shrimps and prawns, fresh fruits and vegetables (strawberries, avocados, mushrooms, papayas, mangoes, potatoes), dehydrated fruits and vegetables (sliced apples, carrots, leeks, onions, powdered asparagus). In the case of shrimps and prawns, the mineral grains present in the intestinal gut are isolated and analyzed [53].

8.2.4.2 Detection of Irradiated Food Using Photostimulated Luminescence (EN13751)

The PSL standard method (EN13751) uses excitation spectroscopy for optical stimulation of minerals to release stored energy [56]. It has been shown that the same spectra can be obtained from whole herbs and spices and other foods using photostimulation. PSL measurements do not destroy the sample, thus whole samples, or other mixtures of organic and inorganic material, can be measured repeatedly. The PSL signals obtained do, however, decrease if the same sample is measured repeatedly.

The method has overcome the need for full mineral separation, and a low-cost instrument is now commercially available for high-sensitivity PSL measurements from food samples using the highly radiation-specific ultraviolet–visible (UV–Vis) luminescence signals, which can be stimulated using infrared sources [57,58]. The SURRC pulsed photostimulated luminescence system (SURRC Pulsed PSL System) was designed and developed at the Scottish Universities Research and Reactor Centre (SURRC). The system is commercially available from the Scottish Universities Environmental Research Centre (SUERC), and has been supplied to more than 80 laboratories in the United Kingdom, Europe, and United States for routine commercial quality testing, and in support of labeling requirements. Originally developed for rapid screening of irradiated herbs, spices, and seasonings, it has been validated for a wider range of foodstuffs, and is finding other scientific applications in assessment of fire-damaged structures and in environmental dosimetry [59].

Two modes of operation can be employed; the screening mode, where the luminescence intensity detected from the samples is used for preliminary classification into negative, intermediate, or positive bands, and calibrated PSL (CalPSL), which can distinguish between low- and high-sensitivity samples, thus resolving ambiguous or low-sensitivity cases. It is necessary to confirm a positive screening result using CalPSL or another standardized method such as EN1788.

The method has been validated by interlaboratory trials [56] for shellfish, herbs, spices, and seasonings. For shellfish, the signals from intestinally trapped silicates can be stimulated through the membranes of dissected guts, and in some cases through the whole body of the creature. From the results of other studies, it has been concluded that PSL is applicable to a large variety of foods [60,61].

8.2.5 DNA Comet Assay

The DNA Comet Assay EN13784 [62] is a rapid and inexpensive screening test to identify irradiated food [63]. As the DNA molecule is an easy target for ionizing radiation, it was logical to investigate whether radiation damage to DNA in food could be used as a means of detecting irradiation treatment. The irradiation of DNA has been shown to induce three major classes of lesions—double-strand breaks, single-strand breaks, and base damage [64]. A sensitive technique to detect this fragmentation is microgel electrophoresis. The technique analyzes the leakage of DNA from single cells or nuclei extracted from food material and embedded in agarose gel on microscopic slides. In irradiated samples (Figure 8.4), the fragmented DNA leaks from the nuclei during electrophoresis, forming a tail in the direction of the anode and giving the appearance of a "comet" when the gel is stained with a fluorescent dye and viewed with a microscope. The head of the comet is formed by the remaining nucleus, whereas the tail is dominated by the fragments. The extension of the tail is closely related to the damage intensity. Cells from nonirradiated samples will appear as nuclei with no or only slight tails (Figure 8.4). The method is restricted to foods that have not been subjected to heat or other treatments, which would induce DNA fragmentation, resulting in comets similar to those of samples treated with ionizing radiation [65]. It is also necessary to establish background DNA damage in nonirradiated samples for each new type of food under investigation.

(a)　　　　　　　　　　　　　　　　　　(b)

Figure 8.4 Typical DNA comets from (a) irradiated (at 7.5 kGy) and (b) nonirradiated tissues. (Haine, H., Cerda, H., and Jones, L., *Food Sci. Technol. Today*, 9(3), 139, 1995. Copyright IFST.)

As the DNA Comet Assay is not radiation-specific, positive results must be confirmed using specific standardized methods such as EN1784 or EN1785. The method has been validated by interlaboratory trials for identification of irradiated chicken bone marrow, chicken, and pork muscle tissue given irradiation doses of 1, 3, or 5 kGy and plant foods (almonds, figs, lentils, linseed, rosé pepper, sesame seeds, soybeans, and sunflower seeds) given 0.2, 1, or 5 kGy [62]. Research has shown that the method can be applied to a wide range of products, but the limitations outlined previously apply [63,66–68], with further development of the method also being reported to allow for more rapid detection and dose estimation [69].

8.2.6 Measurement of Microbiological Changes

Any kind of processing will destroy the microbial flora in food or change the flora present so that the vegetative cells are killed off, whereas the bacterial spores survive. Such microbial reduction and change is to be expected in all kinds of food processing, including irradiation. Thus, it was presumed that simple detection tests for foods could be developed comparing the microbiological quality of nonirradiated and irradiated foods to determine if irradiation treatment has been applied [70]. Consequently, two screening methods were successfully developed, validated, and standardized for the identification of irradiated foods based on modification of the microbiological flora of samples.

8.2.6.1 Direct Epifluorescent Filter Technique/Aerobic Plate Count (DEFT/APC) (EN13783)

One microbiological method that has been developed, validated, and standardized as a screening method for irradiated foods is the DEFT/APC test (EN13783) [71]. The DEFT/APC method can be used for the detection of irradiation treatment of herbs and spices, using the combined direct epifluorescent filter technique (DEFT) and aerobic plate count (APC). The method is based on comparison of the APC with the count obtained using the DEFT. The APC gives the number of viable microorganisms in the sample after irradiation, whereas the DEFT count determines the total number of microorganisms present in the sample, including cells rendered nonviable by irradiation. For a nonirradiated sample, the counts by DEFT are in close agreement with those by APC, because nearly all the cells present are alive. However, when the APC of an irradiated sample is compared with the DEFT count on the same sample, the APC is found to be considerably less than that obtained by DEFT, and the difference indicates that the samples could have been irradiated [72].

The difference between the DEFT and the APC counts in spices treated with doses of 5–10 kGy is generally about or above 3–4 log units. Similar differences between DEFT and APC counts can be induced by other treatments of the foods that lead to death of microorganisms, for example, heat or fumigation treatment. Thus, as the method is not radiation-specific, positive results should be confirmed by another suitable standardized method, such as TL (EN1788) or PSL (EN13751). It has been shown that some spices such as cloves, cinnamon, garlic, and mustards can contain inhibitory components with antimicrobial activity, which may lead to decreasing APC, thereby giving false-positive results.

The DEFT/APC method has been successfully validated for herbs and spices (including whole allspice, whole and powdered black pepper, whole white pepper, paprika powder, cut basil, cut marjoram, and crushed cardamom) by interlaboratory trials [71].

8.2.6.2 Limulus *Amebocyte Lysate/Gram-Negative Bacteria Test (EN14569)*

The *Limulus* amebocyte lysate/Gram-negative bacteria (LAL/GNB) test, European Standard EN14569 [73], is another microbiological screening method comprising two procedures carried out in parallel to detect an abnormal microbiological profile of foods typically contaminated with predominantly Gram-negative bacteria. It is based on the principle that relatively low doses of irradiation can render large numbers of bacteria nonviable.

The two procedures to be carried out are (i) enumeration of total resuscitated GNB in the test samples and (ii) determination of lipopolysaccharide (bacterial endotoxin) concentration in the test sample using the LAL test. The level of endotoxin (measured in endotoxin units) is directly related to the number of GNB, although it is not species-specific. Thus the test determines the number of viable GNB present in a sample, and the concentration of bacterial endotoxin serves as a measure for the estimation of the amount of total GNB, both viable and dead. If a high LAL value is obtained in the absence of significant numbers of viable GNB, this indicates the presence of a large population of dead bacteria. In the absence of any visible processing of the sample, for example cooking, this profile is indicative of some other processing, such as treatment with ionizing radiation [73,74].

This method is not radiation-specific, as a high amount of dead bacteria in comparison with numbers of viable microorganisms can be due to other reasons, such as cooking or some form of chemical preservation. Freezing after irradiation can also influence the ratio of GNB to endotoxin units due to loss of the viability of microorganisms. On the other hand, regrowth of bacterial flora can occur in irradiated samples that are stored unfrozen.

This screening method was validated by interlaboratory trials [73,74] using boneless chicken breasts with skin and boneless chicken breast fillets. The method is generally applicable to whole parts of poultry, such as breast, legs, and wings of fresh, chilled, or frozen carcasses with or without skin. In addition, it can also provide useful information about the microbiological quality of a product before irradiation.

8.2.7 Other Methods Explored

The methods presented up to this point are those that have been validated and standardized. However, it is worthy of note that other methods have been explored, but for one reason or another have not been standardized. For example, the use of ESR spectroscopy was investigated for the identification of irradiated crustacea. It was found that the ESR signal derived from the shell of prawns or shrimp is species-dependent, with the geographical origin also being shown to influence ESR signal shape. Thus, while detection of irradiation treatment is possible, it is not without its problems, as demonstrated by a number of interlaboratory blind trials [75,76], where the identification rate of certain species was extremely poor. More research would certainly need to be undertaken before the method could be standardized. ESR can also be employed to detect irradiation treatment of shellfish such as mussels, oysters, and scallops [77] and other crustaceans such as crab [78].

Other physical methods investigated included measurement of changes in the viscosity of products, such as suspensions of herbs, spices, and seasonings [79,80], and the electrical impedance of potatoes [81,82]. Studies on chemical methods also explored the potential use of orthotyrosine, formed from phenylalanine, as a radiation marker [83,84]. However, studies showed that this compound can also be found in nonirradiated products, thus it is not radiation-specific. But it was concluded that if the difference in the amounts present in nonirradiated and irradiated samples was sufficiently large, the compound could still have potential as a radiation marker. Significant work on using gas evolution to detect irradiated foods was undertaken by workers such as Furuta et al. [85], Delincée [86], and Hitchcock [87]. The method was based on the detection of evolved gases such as carbon monoxide, hydrogen, hydrogen sulfide, and ammonia.

The use of agarose electrophoresis of mitochondrial DNA (mtDNA) for identification of irradiated foods was studied by Marchioni et al. [88,89]. This method is potentially applicable to foods, particularly meat products, treated with ionizing radiation at doses of 1 kGy or greater, as long as mtDNA can be extracted. The use of immunoassays for the detection of irradiated products has also been explored. Work published by Tyreman et al. [90] described the development of a competitive enzyme-linked immunoassay (ELISA) to detect irradiated prawns. The ELISA described uses a monoclonal antibody against dihydrothymidine, a modified DNA base. It has been successfully applied for the detection of irradiated North Atlantic prawn (*Pandalus borealis*) and Tiger prawn (*Penaeus monodon*), having a working range of 0.5–2 kGy, with detection of irradiation treatment being possible for prawns stored up to 12 months at –20°C. Potentially this method could be applied to a range of foodstuffs, as most food contains DNA, and it is also simple and inexpensive to carry out.

The half-embryo test to measure inhibition of seed germination was also studied as a simple detection method for products such as irradiated apples, cherries, grapefruits, lemons, and oranges [91–93]. The embryos are taken out of the seed shells for germination so that irradiation treatment can be detected within 2–4 days at dose levels as low as 0.15 kGy. The test is simple and inexpensive to perform, not requiring any specialized equipment.

8.2.8 Application of Detection Methods in the Marketplace

Currently with the EU, 10 Member States have facilities approved in accordance with Article 7(2) of Directive 1999/2/EC for the irradiation of food. In 2005, as only eight Member States forwarded to the Commission the results of checks carried out in irradiation facilities, the precise amount of foodstuffs irradiated in the Union could not be determined [94]. During 2005 the main products treated by ionizing radiation within the EU were dried herbs and spices, frog legs, poultry, and dried vegetables.

Within the EU, to ensure that current labeling regulations are being complied with, analytical checks are carried out on foods placed on the market. In 2005 a total of 16 Member States reported checks on foods placed on the market, with a total of 7011 food samples being tested. About 4% of products tested from the marketplace were found to be illegally irradiated or not labeled [94]. Table 8.2 is a summary of the numbers of samples analyzed and the results obtained for the EU as a whole in 2005.

It was found that the infringements were unevenly distributed over product categories. Products from Asia, especially Asian-type noodles and food supplements, represented a significant proportion of the samples that were irradiated and not labeled as such. Only six of the 287 samples found to be irradiated complied with the regulations. It was noted that in 2005, there were no irradiation facilities in Asia approved by the EC. Such incorrectly labeled Asian products were found in Germany, the Republic of Ireland, and the United Kingdom. Incorrectly labeled food supplements were also detected in the same countries as well as in Finland and the Netherlands. In Germany, 47 samples out of 96 soups and sauces tested were found to be treated with ionizing radiation, with irradiation being unauthorized or samples not being correctly labeled. Other products found to be irradiated within the EU and not labeled correctly included dried herbs, spices, vegetable seasonings, fish and fisheries products, frogs legs, dried mushrooms, and tea

Table 8.2 Summary of Samples Analyzed for Irradiation Treatment and Results Obtained for the EU as a Whole in 2005

Member State	No. of Samples Nonirradiated	No. of Samples Irradiated	Percentage of Samples Irradiated, Not Labeled Correctly
Austria	115	0	0
Belgium	148	0	0
Cyprus	NAC	NAC	NAC
Czech Republic	70	8	10
Germany	3798	143[a]	3.6
Denmark	NAC	NAC	NAC
Estonia	NAC	NAC	NAC
Greece	54	0	0
Spain	NI	NI	NI
Finland	264	13	5
France	80	6	7
Hungary	134	7[a]	2
Ireland (Republic)	439	20	4
Italy	107	5	5
Latvia	NAC	NAC	NAC
Lithuania	12	0	0
Luxembourg	40	0	0
Malta	NAC	NAC	NAC
The Netherlands	761	31	4
Poland	116	6	4
Portugal	NAC	NAC	NAC
Sweden	6	0	0
Slovakia	56	0	0
Slovenia	10	0	0
The United Kingdom	514[b]	42	6
Total	6724	281	4.0

Source: European Union, *Off. J. Eur. Union*, 2007/C122/03, 2 June 2007.

Note: NI = no information forwarded by the Member State, NAC = no analytical checks performed in 2005.

[a] Germany and Hungary found respectively 2 and 4 samples that were legally irradiated and correctly labeled.
[b] The United Kingdom classified 101 samples as inconclusive.

and tealike products. TL (EN1788) and PSL (EN13751) were the most commonly used methods within the Member States for detection purposes, with PSL being used for screening purposes, and confirmation of positive results being undertaken using TL. The results of these tests within the EU is indicative of the successful detection of irradiated products using standardized analytical methods.

8.3   Conclusions

This chapter has briefly summarized the main methods currently available for the detection of irradiated foodstuffs, whether they are whole products or ingredients within a foodstuff. As noted, the methods have been successfully applied for the detection of irradiated foodstuffs in the marketplace, thereby giving assurance to retailers and consumers alike that irradiated foods on sale and incorrectly labeled can be identified. The availability and regular use of these methods could even help to facilitate international trade in irradiated food [95]. A number of reviews have been written on methods for the detection of irradiated foods; for further reference the author suggests reading McMurray et al. [96], which contains the proceedings of an International Meeting on Analytical Detection Methods for Irradiation Treatments of Foods held in June 1994, as well as reviews by Delincée [95,97], Stewart [10], and Marchioni [98].

References

1. Stewart, E.M., Food irradiation chemistry, in *Food Irradiation—Principles and Applications*, Molins, R., Ed., Wiley, 2001, chap. 3.
2. Diehl, J.F., Chemical effects of radiation, in *Safety of Irradiated Foods*, 2nd Ed., Marcel Dekker, New York, 1995.
3. WHO, *Wholesomeness of Irradiated Food: A Report of a Joint FAO/IAEA/WHO Expert Committee on Food Irradiation*, WHO Technical Report Series, 659, World Health Organization, Geneva, 1981.
4. European Commission, Food irradiation—community legislation. http://ec.europa.eu/food/food/biosafety/irradiation/ comm_legisl_en.htm.
5. Borsa, J., Outlook for food irradiation in the 21st century, in *Irradiation of Food and Packaging: Recent Developments*, Komolprasert, V. and Morehouse, K.M., Eds., ACS Symposium Series 875, Oxford University Press, Washington, DC, 2004, p. 326.
6. Molins, R.A., Motarjemi, Y., and Käferstein, F.K., Irradiation: A critical control point in ensuring the microbiological safety of raw foods, *Food Control*, 12, 347, 2001.
7. Furuta, M., Current status of information transfer activity on food irradiation and consumer attitudes in Japan, *Radiat. Phys. Chem.*, 71, 499, 2004.
8. Moy, J.H. and Wong, L., The efficacy and progress of using radiation as a quarantine treatment of tropical fruits—a case study in Hawaii, *Radiat. Phys. Chem.*, 63, 397, 2002.
9. Delincée, H., International cooperation in the field of detection of irradiated food, *Z. Lebensm. Unters. Forsch.*, 197, 217, 1993.
10. Stewart, E.M., Detection methods for irradiated foods, in *Food Irradiation—Principles and Applications*, Molins, R., Ed., Wiley, 2001, chap. 14.
11. European Commission, Food irradiation—analytical methods. http://ec.europa.eu/food/food/biosafety/irradiation/anal_methods_en.htm.
12. Anonymous, Foodstuffs—detection of irradiated food containing fat—gas chromatographic analysis of hydrocarbons, *European Standard EN1784:2003, European Committee for Standardization*, Brussels, 2003.

13. Nawar, W.W., Analysis of volatiles as a method for the identification of irradiated foods, in *Health Impact, Identification and Dosimetry of Irradiated Food, Report of a WHO Working Group (Neuherberg/ Munich, 17–21 November 1986)*, Bögl, K.W., Regulla, D.F. and Suess, M.J., Eds., WHO, Copenhagen, Denmark, 1988.

14. LeTellier, P.R. and Nawar, W.W., 2-Alkylcyclobutanones from radiolysis of triglycerides, *Lipids*, 7, 75, 1972.

15. Nawar, W.W. and Balboni, J.J., Detection of irradiation treatment in foods, *J. Assoc. Off. Anal. Chem.*, 53, 726, 1970.

16. Nawar, W.W., Zhu, R., and Yoo, Y.J., Radiolytic products of lipids as markers for the detection of irradiated meats, in *Food Irradiation and the Chemist*, Johnston, D.E. and Stevenson, M.H., Eds., Royal Society of Chemistry, Special Publication no. 86, Cambridge, U.K., 1990, p. 13.

17. Schulzki, G., Spiegelberg, A., and Bögl, K.W., Detection of radiation-induced hydrocarbons in irradiated fish and prawns by on-line coupled liquid chromatography-gas chromatography, *J. Agric. Food Chem.*, 45, 3921, 1997.

18. Horvatovich, P. et al., Supercritical fluid extraction of hydrocarbons and 2-alkylcyclobutanones for the detection of irradiated foodstuffs, *J. Chromatogr. A*, 897, 259, 2000.

19. Anonymous, Foodstuffs – detection of irradiated food containing fat – gas chromatographic/mass spectrometric analysis of 2-alkylcyclobutanones, EN1785:2003, *European Committee for Standardization*, Brussels, 2003.

20. Elliott, C.T. et al., Detection of irradiated chicken meat by analysis of lipid extracts for 2-substituted cyclobutanones using an enzyme linked immunosorbent assay, *Analyst*, 120, 2337, 1995.

21. Stevenson, M.H., Identification of irradiated foods, *Food Technol.*, 48, 141, 1994.

22. Obana, H., Furuta, M., and Tanaka, Y., Detection of 2-alkylcyclobutanones in irradiated meat, poultry and egg after cooking, *J. Health Sci.*, 52, 375, 2006.

23. Stewart, E.M. et al., Isolation of lipid and 2-alkylcyclobutanones from irradiated food by supercritical fluid extraction, *J. AOAC Int.*, 84, 976, 2001.

24. Hamilton, L. et al., Detection of 2-substituted cyclobutanones as irradiation products of lipid-containing foods: Synthesis and applications of cis- and trans-2-(tetradec-5′-enyl)cyclobutanones and 11-(2′-oxocyclobutyl)-undecanoic acid, *J. Chem. Soc. Perkin. Trans.*, 1, 139, 1995.

25. Stewart, E.M. et al., 2-Alkylcyclobutanones as markers for the detection of irradiated mango, papaya, Camembert cheese and salmon meat, *J. Sci. Food Agric.*, 80, 121, 2000.

26. Gadgil, P. et al., 2-Alkylcyclobutanones as irradiation dose indicators in irradiated ground beef patties, *J. Agric. Food Chem.*, 50, 5746, 2002.

27. Horvatovich, P. et al., Supercritical fluid extraction for the detection of 2-dodecylcyclobutanone in low dose irradiated plant foods, *J. Chromatogr. A*, 968, 251, 2002.

28. Ndiaye, B. et al., 2-Alkylcyclobutanones as markers for irradiated foodstuffs III. Improvement of the field of application on the EN 1785 method by silver ion chromatography, *J. Chromatogr. A*, 858, 109, 1999.

29. Horvatovich, P. et al., Determination of 2-alkylcyclobutanones with electronic impact and chemical ionization gas chromatography/mass spectrometry (GC/MS) in irradiated foods, *J. Agric. Food Chem.*, 54, 1990, 2006.

30. Desrosiers, M.F. and Simic, M.G., Post-irradiation dosimetry of meat by electron spin resonance spectroscopy of bones, *J. Agric. Food Chem.*, 36, 601, 1988.

31. Anonymous, Foodstuffs—detection of irradiated food containing bone—method by ESR spectroscopy, *European Standard EN 1787:2000, European Committee for Standardization*, Brussels, 2000.

32. Anonymous, Foodstuffs—detection of irradiated food containing cellulose—method by ESR spectroscopy, *European Standard EN 1786:1996, European Committee for Standardization*, Brussels, 2000.

33. Anonymous, Foodstuffs—detection of irradiated food containing crystalline sugar by ESR spectroscopy, *European Standard EN 13708:2001, European Committee for Standardization*, Brussels, 2001.

34. Stevenson, M.H. and Gray, R., Can ESR spectroscopy be used to detect irradiated food? in *Food Irradiation and the Chemist*, Johnston, D.E. and Stevenson, M.H., Eds., Royal Society of Chemistry, Special Publication no. 86, Cambridge, U.K., 1990, p. 80.

35. Stevenson, M.H. and Gray, R., An investigation into the effect of sample preparation methods on the resulting ESR signal from irradiated chicken bone, *J. Sci. Food Agric.*, 48, 261, 1989.

36. Stevenson, M.H. and Gray, R., Effect of irradiation dose and temperature on the ESR signal in irradiated chicken bone, *J. Sci. Food Agric.*, 48, 269, 1989.

37. Goodman, B.A., McPhail, D.B., and Duthie, D.M.L., Electron spin resonance spectroscopy of some irradiated foodstuffs, *J. Sci. Food Agric.*, 47, 101, 1989.

38. Raffi, J.J. et al., ESR analysis of irradiated frog legs and fishes, *Appl. Radiat. Isotopes*, 40, 1215, 1989.

39. Chawla, S.P., Detection of irradiated lamb meat with bone: Effect of chilled storage and cooking on ESR signal strength, *Int. J. Food Sci. Technol.*, 34, 41, 1999.

40. Sin, D.W.M., Identification and stability study of irradiated chicken, pork, beef, lamb, fish and mollusk shells by electron paramagnetic resonance (EPR) spectroscopy, *Eur. Food Res. Technol.*, 221, 84, 2005.

41. Gray, R. and Stevenson, M.H., Detection of irradiated deboned turkey meat using ESR spectroscopy, *Radiat. Phys. Chem.*, 34, 899, 1989.

42. Stevenson, M.H. et al., The use of ESR spectroscopy for the detection of irradiated mechanically recovered meat (MRM) in tertiary food products, in *Detection Methods for Irradiated Foods—Current Status*, McMurray, C.H., Stewart, E.M., Gray, R. and Pearce, J., Eds., Royal Society of Chemistry, Special Publication no. 171, Cambridge, 1996, p. 53.

43. Marchioni, E. et al., Detection of irradiated ingredients included in low quantity in non-irradiated food matrix. 1. Extraction and ESR analysis of bones from mechanically recovered poultry meat, *J. Agric. Food Chem.*, 53, 3769, 2005.

44. Marchioni, E. et al., Detection of irradiated ingredients included in low quantity in non-irradiated food matrix. 2. ESR analysis of mechanically recovered poultry meat and TL analysis of spices, *J. Agric. Food Chem.*, 53, 3773, 2005.

45. Onori, S. and Pantaloni, M., Electron spin resonance technique identification and dosimetry of irradiated chicken eggs, *Int. J. Food Sci. Technol.*, 29, 671, 1995.

46. Desrosiers, M.F. et al., Interlaboratory trials of the EPR method for the detection of irradiated spices, nutshell and eggshell, in *Detection Methods for Irradiated Foods—Current Status*, McMurray, C.H., Stewart, E.M., Gray, R. and Pearce, J., Eds., Royal Society of Chemistry, Special Publication no. 171, Cambridge, U.K., 1996, p. 108.

47. Raffi, J.J., Electron spin resonance identification of irradiated strawberries, *J. Chem. Soc. Fara. Trans. 1.*, 84(10), 3359, 1988.

48. Delincée, H. and Soika, C., Improvement of the ESR detection of irradiated food containing cellulose employing a simple extraction method, *Radiat. Phys. Chem.*, 63, 437, 2002.

49. Tabner, B.J. and Tabner, V.A., Electron spin resonance spectra of γ-irradiated citrus fruits skins, skin components and stalks, *Int. J. Food Sci. Technol.*, 29, 143, 1994.

50. Stevenson, M.H. and Gray, R., The use of ESR spectroscopy for the identification of irradiated food, *Ann. Rep. NMR Spect.*, 31, 123, 1995.

51. Raffi, J., Agnel, J.-P., and Ahmed, S.H., Electron spin resonance identification of irradiated dates, *Food Technol.*, 34, 26, 1991.

52. Sanderson, D.C.W., Carmichael, L.A., and Naylor, J.D., Photostimulated luminescence and thermoluminescence techniques for the detection of irradiated food, *Food Sci. Technol. Today*, 9(3), 150, 1995.

53. Anonymous, Detection of irradiated food using photostimulated luminescence, *European Standard EN 13751:2002, European Committee for Standardization*, Brussels, 2002.

54. Sanderson, D.C.W., Slater, C., and Cairns, K.J., Detection of irradiated food, *Nature*, 340, 23, 1989.

55. Sanderson, D.C.W., Slater, C., and Cairns, K.J., Thermoluminescence of foods: Origins and implications for detecting irradiation, *Radiat. Phys. Chem.*, 34(6), 915, 1989.

56. Anonymous, Thermoluminescence detection of irradiated food from which silicate minerals can be isolated, *European Standard EN 1788:2001, European Committee for Standardization*, Brussels, 2001.

57. Sanderson, D.C.W. et al., Luminescence studies to identify irradiated food, *Food Sci. Technol. Today*, 8(2), 93, 1994.

58. Sanderson, D.C.W., Carmichael, L.A., and Naylor, J.D., Recent advances in thermoluminescence and photostimulated luminescence detection methods for irradiated foods, in *Detection Methods for Irradiated Foods—Current Status*, McMurray, C.H., Stewart, E.M., Gray, R. and Pearce, J., Eds., Royal Society of Chemistry, Special Publication no. 171, Cambridge, U.K., 1996, p. 124.

59. SUERC, SUERC Luminescence Systems—SURRC Pulsed PSL System. http://www.gla.ac.uk/suerc/luminescence/psl.html.

60. Sanderson, D.C.W., Carmichael, L.A., and Fisk, S., Photostimulated luminescence detection of irradiated herbs, spices and seasonings, International interlaboratory trial, *J. AOAC Int.*, 86(5), 990, 2003.

61. Sanderson, D.C.W., Carmichael, L.A., and Fisk, S., Photostimulated luminescence detection of irradiated shellfish: International interlaboratory trial, *J AOAC Int.*, 86(5), 983, 2003.

62. Anonymous, DNA comet assay for the detection of irradiated foodstuffs—screening method, *European Standard EN 1788:2001, European Committee for Standardization*, Brussels, 2001.

63. Cerda, H. et al., The DNA "comet assay" as a rapid screening technique to control irradiated food, *Mutation Res.*, 375, 167, 1997.

64. von Sonntag, C., *The Chemical Basis of Radiation Biology*, Taylor & Francis, London, U.K., 1987.

65. Haine, H., Cerda, H., and Jones, L., Microgel electrophoresis of DNA to detect irradiation of foods, *Food Sci. Technol. Today*, 9(3), 139, 1995.

66. Delincée, H., Introduction to DNA methods for identification of irradiated foods, in *Detection Methods for Irradiated Foods—Current Status*, McMurray, C.H., Stewart, E.M., Gray, R. and Pearce, J., Eds., Royal Society of Chemistry, Special Publication no. 171, Cambridge, U.K., 1996, p. 345.

67. Jo, D. and Kwon, J.H., Detection of radiation-induced markers from parts of irradiated kiwifruits, *Food Control*, 17(8), 617, 2006.

68. Marin-Huachacea, N., Use of the DNA Comet Assay to detect beef treated by ionizing radiation, *Meat Sci.*, 71, 446, 2005.

69. Verbeek, F., Automated detection of irradiated food with the comet assay. *Radiat. Prot. Dos.*, Published on-line 6 October 2007. http://rpd.oxfordjournals.org/cgi/content/abstract/ncm433v1.

70. Gibbs, P.A. and Wilkinson, V.M., Feasibility of detecting irradiated foods by reference to the endogenous microflora: A literature review, Scientific and Technical Surveys, no. 149, Leatherhead Food Research Association, U.K., 1985.

71. Anonymous, Detection of irradiated food using Direct Epifluorescent Filter Technique/Aerobic Plate Count (DEFT/APC), EN 13783:2001, *European Committee for Standardization*, Brussels, 2001.

72. Betts, R.P., The detection of irradiated foods using the direct epifluorescent filter technique, *J. Appl. Bacteriol*, 64, 329, 1988.

73. Anonymous, Microbiological screening for irradiated food using LAL/GNB procedures. EN 13784:2004, *European Committee for Standardization*, Brussels, 2004.

74. Scotter, S.L., Beardwood, K., and Wood, R., *Limulus* amoebocyte lysate test/gram negative bacteria count method for the detection of irradiated poultry: Results of two inter-laboratory studies, *Food Sci. Technol. Today*, 8(2), 106, 1994.

75. Schreiber, G.A. et al., Interlaboratory tests to identify irradiation treatment of various foods via gas chromatographic detection of hydrocarbons, ESR spectroscopy and TL analysis, in *Detection Methods for Irradiated Foods—Current Status*, McMurray, C.H., Stewart, E.M., Gray, R. and Pearce, J., Eds., Royal Society of Chemistry, Special Publication no. 171, Cambridge, U.K., 1996, p. 98.

76. Stewart, E.M. and Kilpatrick, D.J., An international collaborative blind trial on electron spin resonance (ESR) identification of irradiated Crustacea, *J. Sci. Food Agric.*, 74, 473, 1997.
77. Raffi, J. et al., ESR detection of irradiated seashells, *Appl. Radiat. Isotopes*, 47(11/12), 1633, 1996.
78. Maghraby, A., Identification of irradiated crab using EPR, *Radiat. Meas.*, 42, 220, 2007.
79. Farkas, J., Sharif, M.M., and Koncz, A., Detection of some irradiated spices on the basis of radiation induced damage of starch, *Radiat. Phys. Chem.*, 36, 621, 1990.
80. Hayashi, T., Todoriki, S., and Koyhama, K., Applicability of viscosity measurements to the detection of irradiated peppers, in *Detection Methods for Irradiated Foods—Current Status*, McMurray, C.H., Stewart, E.M., Gray, R. and Pearce, J., Eds., Royal Society of Chemistry, Special Publication no. 171, Cambridge, U.K., 1996, p. 215.
81. Hayashi, T., Iwamato, M., and Kawashima, K., Identification of Irradiated Potatoes by Means of Electrical Conductivity, *Agric. Biol. Chem.*, 46, 905, 1982.
82. Hayashi, T. et al., Detection of irradiated potatoes by imedance measurement, in *Detection Methods for Irradiated Foods—Current Status*, McMurray, C.H., Stewart, E.M., Gray, R. and Pearce, J., Eds., Royal Society of Chemistry, Special Publication no. 171, Cambridge, U.K., 1996, p. 202.
83. Karam, L.R. and Simic, M.G., Ortho-tyrosine as a marker in post-irradiation dosimetry (PID) of chicken, in *Health Impact, Identification and Dosimetry of Irradiated Food, Report of a WHO Working Group, Neuherberg/Munich, 17–21 November 1986*, K.W. Bögl, K.W., Regulla, D.F. and Suess, M.J., Eds., WHO, Copenhagen, Denmark, 1988, p. 297.
84. Meier, W., Hediger, H., and Artho, A., Determination of o-tyrosine in shrimps, mussels, frog legs and egg-white, in *Detection Methods for Irradiated Foods—Current Status*, McMurray, C.H., Stewart, E.M., Gray, R. and Pearce, J., Eds., Royal Society of Chemistry, Special Publication no. 171, Cambridge, U.K., 1996, p. 303.
85. Furuta, M. et al., Detection of irradiated frozen meat and poultry using carbon monoxide gas as a probe, *J. Agric. Food Chem.*, 40, 1099, 1992.
86. Delincée, H., A rapid and simple screening test to identify irradiated food using multiple gas sensors, in *Detection Methods for Irradiated Foods—Current Status*, McMurray, C., Stewart, C.H., Gray, R. E.M. and Pearce, J., Eds., Royal Society of Chemistry, Special Publication no. 171, Cambridge, U.K., 1996, p. 326.
87. Hitchcock, C.H., Hydrogen as a marker for irradiated food, *Food Sci. Technol. Today*, 12(2), 112, 1998.
88. Marchioni, E. et al., Alterations of mitochondrial DNA: A method for the detection of irradiated beef liver, *Radiat. Phys. Chem.*, 40, 485, 1992.
89. Marchioni, E. et al., Detection of irradiated fresh, chilled and frozen foods by the mitochondrial DNA method, in *Detection Methods for Irradiated Foods—Current Status*, McMurray, C.H., Stewart, E.M., Gray, R. and Pearce, J., Eds., Royal Society of Chemistry, Special Publication no. 171, Cambridge, U.K., 1996, p. 355.
90. Tyreman, A. et al., Detection of irradiated food by immunoassay—development and optimization of an ELISA for dihyrothumidine in irradiated prawns, *Int. J. Food Sci. Technol.*, 39, 533, 2004.
91. Kawamura, Y., Uchiyama, S., and Saito, Y., Improvement of the half-embryo test for detection of gamma-irradiated grapefruit and its application to irradiated oranges and lemons, *J. Food Sci.*, 54, 1501, 1989.
92. Kawamura, Y., Uchiyama, S., and Saito, Y., A half-embryo test for identification of gamma-irradiated grapefruit, *J. Food Sci.*, 46, 371, 1989.
93. Kawamura, Y. et al., Half-embryo test for identification of irradiated citrus fruit: Collaborative study, *Radiat. Phys. Chem.*, 48(5), 665, 1996.
94. European Union, Report from the commission on food irradiation for the year 2005, *Off. J. Eur. Union*, 2007/C122/03, 2 June 2007.
95. Delincée, H., Analytical methods to identify irradiated food—a review, *Radiat. Phys. Chem.*, 63, 455, 2002.

96. McMurray, C.H., Stewart, E.M., Gray, R., and Pearce, J., Eds., *Detection Methods for Irradiated Foods—Current Status*, Royal Society of Chemistry, Special Publication no. 171, Cambridge, U.K., 1996.
97. Delincée, H., Detection of food treated with ionizing radiation, *Food Sci. Technol.*, 9, 73, 1998.
98. Marchioni, E., Detection of irradiated food, in *Food Irradiation Research and Technology*, Sommers, C.H., Fan, X., Eds., IFT Press, Iowa, USA, 2006, chap. 6.

Chapter 9

Growth Promoters

Milagro Reig and Fidel Toldrá

Contents

9.1 Introduction 230
9.2 Control of Growth Promoters 231
9.3 Sampling and Sample Preparation 233
    9.3.1 Samples from Animal Farms 233
        9.3.1.1 Water 233
        9.3.1.2 Feed 233
        9.3.1.3 Urine 233
        9.3.1.4 Hair 233
    9.3.2 Meat Samples 234
9.4 Methods for Cleanup of Growth Promoters and Their Residues 234
    9.4.1 Extraction Procedures 234
    9.4.2 Immunoaffinity Chromatography 234
    9.4.3 Molecular Recognition 235
9.5 Screening Methods 235
    9.5.1 Immunological Techniques 235
    9.5.2 Biosensors 235
    9.5.3 Chromatographic Techniques 236
9.6 Confirmatory Analytical Methods 236
References 243

9.1 Introduction

Growth promoters include a wide range of substances that are generally used in farm animals for therapeutic and prophylactic purposes. These substances can be administered through the feed or the drinking water. In some cases, the residues may proceed from contaminated animal feedstuffs.[1] Anabolic promoters have been administered in the United States to meat-producing animals where estradiol, progesterone, and testosterone are some of the allowed substances. The regulations in 21 Code of Federal Regulations (CFR) Part 556 provide the acceptable concentrations of residues of approved new animal drugs that may remain in edible tissues of treated animals.[2] Other countries allowing the use of certain growth promoters are Canada, Mexico, Australia, and New Zealand. However, the use of growth promoters is officially banned in the European Union since 1988 due to concerns about harmful effects on consumers.[3]

Growth promoters increase growth rate and improve efficiency of feed utilization and thus, contribute to the increase in protein deposition that is usually linked to fat utilization, which means a reduction in the fat content in the carcass and an increase in meat leanness.[4] In addition, some fraudulent practices consist in the use of low amounts of several substances such as β-agonists (clenbuterol) and corticosteroids (dexamethasone) and anabolic steroids, mixtures known as "cocktails," that have a synergistic effect and exert growth promotion but make their analytical detection more difficult.

The presence of residues of growth promoters or their metabolites in meat and their associated harmful health effects on humans make necessary the continuous improvement of analytical methodologies to guarantee consumer protection. The use of veterinary drugs in food animal species is strictly regulated in the European Union and, in fact, only some of them can be permitted for specific therapeutic purposes under strict control and administration by a veterinarian.[5]

Sanitary authorities in different countries are concerned about the presence of residues of veterinary drugs or their metabolites in meat because they may exert some adverse toxic effects on consumers' health. The European Food Safety Authority has recently issued an opinion about substances with hormonal activity, specifically testosterone and progesterone, as well as trenbolone acetate, zeranol, and melengestrol acetate. The exposure to residues of the hormones used as growth promoters could not be quantified. Although epidemiological data in the literature provided evidence for an association between some forms of hormone-dependent cancers and red meat consumption, the contribution of residues of hormones in meat could not be assessed.[6] Other substances such as β-agonists have shown adverse effects on consumers. This was evident in the case of intoxications in Italy, with symptoms described as gross tremors of the extremities, tachycardia, nausea, headaches, and dizziness, after consumption of lamb and bovine meat containing residues of clenbuterol.[7]

Meat quality is also affected by the use of substances used as growth promoters.[4] The connective tissue production is increased and collagen cross links at a higher rate giving a tougher meat,[8–10] whereas muscle proteases responsible for protein breakdown in postmortem meat are inhibited.[8,11] The lipolysis rate and the breakdown of triacylglycerols are accelerated.[12,13] The use of anabolic steroids reduces marbling and tenderness and may have a negative effect on palatability.[14] The "aggressive" use of anabolic implants may compromise the quality grades of beef carcasses and increase the incidence of dark cutting carcasses.[14] When cocktails of clenbuterol and dexamethasone are used, the meat quality is also affected; but it has been reported to be less tough than when using clenbuterol alone.[15]

Meat must be monitored for the presence of residues of veterinary drugs. Control strategies also include sampling at farm, which helps in prevention before animals reach the slaughterhouse. The samples include hair and urine as well as feed and water. This chapter reports the important strategies for the control of growth promoters as part of the wide range of residues of veterinary drugs in meat. The analysis of antibiotic residues is discussed in Chapter 10.

9.2 Control of Growth Promoters

The monitoring of residues of substances having hormonal or thyreostatic action as well as β-agonists is regulated in the European Union through the Council Directive 96/23/EC[16] on measures to monitor certain substances and residues in live animals and animal products. The European Union Member States have set up national monitoring programs and sampling procedures following this directive.

The major veterinary drugs and substances with anabolic effect are listed in Table 9.1, where group A includes unauthorized substances that have anabolic effect and group B includes veterinary drugs some of which have established maximum residue limits (MRLs). Commission Decisions 93/256/EC[17] and 93/257/EC[18] gave criteria for the analytical methodology regarding the

Table 9.1   Lists of Substances Having Anabolic Effect Belonging to Groups A and B According to Council Directive 96/23/EC[16]

Group A: Substances having anabolic effect
1. Stilbenes
2. Antithyroid agents
3. Steroids
Androgens
Gestagens
Estrogens
4. Resorcyclic acid lactones
5. Beta-agonists
6. Other compounds
Group B: Veterinary drugs
1. Antibacterial substances
Sulfonamides and quinolones
2. Other veterinary drugs
a. Antihelmintics
b. Anticoccidials, including nitroimidazoles
c. Carbamates and pyrethroids
d. Sedatives
e. Nonsteroidal antiinflammatory drugs
f. Other pharmacologically active substances

screening, identification, and confirmation of these residues. Council Directive 96/23/EC[16] was implemented by the Commission Decision 2002/657/EC,[19] which is in force since September 1, 2002. This directive provides rules for the analytical methods to be used in testing of official samples and specific criteria for the interpretation of analytical results of official control laboratories for such samples. This means that when using mass spectrometric detection, substances in group A would require four identification points whereas those in group B would only require a minimum of three. The relative retention of the analyte must correspond to that of the calibration solution at a tolerance of ±0.5% for gas chromatography (GC) and ±2.5% for liquid chromatography (LC). The guidelines given in this new directive also imply new concepts such as the decision limit (CCα) or the detection capability (CCβ) that are briefly defined in Table 9.2. Both limits permit the daily control of the performance of a specific method qualified when used with a specific

Table 9.2 Definitions of Main Performance Criteria and Other Requirements for Analytical Methods[19]

Term	Definition
Decision limit (CCα)	It is defined as the limit at and above which it can be concluded with an error probability of α that a sample is noncompliant
Detection capability (CCβ)	It is the smallest content of the substance that may be detected, identified, and quantified in a sample with an error probability of β
Minimum required performance limit (MRPL)	It means the minimum content of an analyte in a sample, which at least has to be detected and confirmed
Precision	The closeness of agreement between independent test results obtained under stipulated conditions
Recovery	The percentage of the true concentration of a substance recovered during the analytical procedure
Reproducibility	Conditions where test results are obtained within the same method on identical test items in different laboratories with different operators using different equipment
Specificity	Ability of a method to distinguish between the analyte being measured and other substances
Ruggedness	Susceptibility of an analytical method to changes in experimental conditions that can be expressed as a list of the sample materials, analytes, storage conditions, environmental, and sample preparation conditions under which the method can be applied as presented or with specified minor conditions
Interlaboratory study	Organization, performance, and evaluation of tests on the same sample by two or more laboratories in accordance with predetermined conditions to determine testing performance
Within-laboratory reproducibility	Precision obtained in the same laboratory under stipulated conditions

instrument and under specific laboratory conditions, and thus contribute to the determination of the level of confidence in the routine analytical result.

9.3 Sampling and Sample Preparation

9.3.1 Samples from Animal Farms

9.3.1.1 Water

It is the easiest sample because it requires no general treatment. It does not require homogenization, just mild centrifugation to remove any suspended particle before further analysis.

9.3.1.2 Feed

The sample must be representative, especially taking into account the heterogeneous nature of most feeds. Thus, feeds must be milled and well homogenized before sampling. Adequate liquid extraction and solid-phase extraction (SPE) are performed for sample cleanup and concentration.

9.3.1.3 Urine

As a fluid, it does not require homogenization and, after centrifugation, an aliquot is diluted with the buffer and pH adjusted to correct values. Many analytes form conjugates such as sulfates and glucuronides and must be hydrolyzed to release the free analyte. Enzymatic hydrolysis with the juice of *Helix pomatia*, which has sulfatase and β-glucuronidase, is a milder treatment that usually gives good results. Special caution must be taken if other types of hydrolysis are performed (i.e., an acidic or alkaline hydrolysis) because they might affect and degrade the analyte.

9.3.1.4 Hair

The possibility of using hair to detect the illegal addition of clenbuterol even after 3 weeks of withdrawal, which is undetectable in urine and tissues, has been reported.[20] The amount of clenbuterol increased up to 20 days in the washout period and then slightly decreased even though still detectable after 40 days.[21] It must be pointed out that black hairs accumulate more clenbuterol and steroids than colored hairs;[22,23] the residue has been detected at 23 weeks after treatment with clenbuterol.[24] The hair, previously cleaned with detergents (sodium dodecyl sulfate [SDS]), is extracted with methanol and evaporated to dryness. The residue is dissolved in phosphate buffer and immunoextracted by affinity chromatography using monoclonal antisalbutamol immunoglobulin (IgG) that displays cross-reactivity (75%) with clenbuterol.[25] The residue is purified with SPE and silyl derivatized for its analysis by GC-MS (mass spectrometry). Other residues such as 17β-estradiol-3-benzoate have also been detected in hair up to 2 weeks after administration. However, 17α-methyltestosterone and medroxyprogesterone acetate could not be detected in hair.[23] The confirmation was possible above 5 ng/g by using liquid chromatography with mass spectrometry (LC-MS/MS) detection.[26] The analysis of a wide range of steroid residues such as estrogens, resorcylic acid lactones, and stilbens has been recently reported. The hair was extracted with methanol before the acid hydrolysis followed by specific liquid–liquid extraction and SPE to get four different fractions that were analyzed separately.[27]

9.3.2 Meat Samples

Preparation procedures and handling of meat samples are very important to improve the sensitivity of the screening tests.[28] Typical procedures include cutting, blending, and homogenization of the meat in an appropriate buffer. Enzymatic digestion with proteases such as subtilisin may be alternatively performed. The homogenate is extracted with an organic solvent usually followed by an SPE for sample cleanup and concentration. Previously, the residues may be bound or conjugated (i.e., as sulfates or glucuronides) and need further cleavage by treatment with the juice of the snail *H. pomatia*, which has sulfatase and β-glucuronidase, to release the free analytes. Some authors prefer enzymatic digestion with subtilisin to release steroids as they state that using enzymatic hydrolysis with the juice of *H. pomatia* may not reflect the conjugated fraction of steroids.[29]

9.4 Methods for Cleanup of Growth Promoters and Their Residues

9.4.1 Extraction Procedures

Extraction is primarily performed to remove interfering substances while retaining most of the analyte. Extraction solvents must be carefully chosen for each analyte as determined by pH, polarity, and solubility in different solvents. For instance, polar extraction methods for the determination of anabolic steroids of beef are used because they avoid some cleanup problems when following nonpolar extraction, but they are insufficient. It has been reported that polar extraction followed by nonpolar extraction gives better results.[30] Supercritical fluid extraction of meat with unmodified supercritical CO_2 has also been used for certain residues such as steroids.[31]

Matrix solid-phase dispersion consists in the mechanical blending of the sample with a solid sorbent that progressively retains the analyte by hydrophobic and hydrophilic interactions. The solid matrix is then packed into a column and eluted with an adequate solvent.

SPE is extensively used for the isolation of a group or class of analytes. The type of extractant and cartridge depends on the target analyte.[32] Small cartridges (C18, C8, NH_2) are commercially available at reasonable prices and have low affinity and specificity but have high capacity. Furthermore, they can be performed in parallel and thus, they allow the simultaneous extraction of a large number of samples.

9.4.2 Immunoaffinity Chromatography

This type of chromatography is based on the antigen–antibody interaction, which is very specific for a particular residue. The columns are packaged with a specific antibody that is bound to the solid matrix, usually a gel. When the extract is injected, the analyte (antigen) is retained. These chromatographic columns are highly specific and are only limited by potential interferences (i.e., substances that may cross-react with the antibody) that must be checked. These columns are rather expensive and can only be reused a certain number of times. In any case, due to the nature of the specific antibody when preparing the immunosorbent material, an in-depth assessment is necessary before considering its use in a routine analytical method.[33]

9.4.3   Molecular Recognition

There are several methods based on molecular recognition mechanisms for cleanup. Molecular imprinted polymers (MIPs) have shown promising results for the isolation of low amounts of residues such as those found in meat. These are cross-linked polymers prepared in the presence of a template molecule like a β-agonist. When this template is removed, the polymer offers a binding site complementary to the template structure. MIPs have better stability than antibodies because they can support high temperatures, larger pH ranges, and a wide range of organic solvents. The choice of the appropriate molecule as template is the critical factor for a reliable analysis.[34] The extracted residues are then analyzed by LC-MS and have shown good quantitative results for cimaterol, ractopamine, clenproperol, clenbuterol, brombuterol, mabuterol, mapenterol, and isoxsurine but not for salbutamol and terbutaline.[35]

9.5   Screening Methods

The wide range of veterinary drugs and residues potentially present in a meat sample necessitates the use of screening procedures for routine monitoring. Screening methods are used to detect the presence of the suspect analyte in the sample at the level of interest. If the searched residue has a MRL, then the screening method can detect the residue below this limit. These controls are based on the screening of a large number of samples and thus must have a large throughput, low cost, and enough sensitivity to detect the analyte with a minimum of false negatives.[36,37] Compliant samples are accepted whereas the suspected noncompliant samples would be further analyzed using confirmatory methods. According to the Commission Decision 2002/657/EC,[19] the screening methods must be validated and have a detection capability (CCβ) with an error probability (β) less than 5%.

9.5.1   Immunological Techniques

Immunological techniques are specific for a given residue because they are based on the antigen–antibody interaction. The most well-known and extensively used technique is the enzyme-linked immunosorbent assay (ELISA). A wide range of assay kits with measurement based on color development are commercially available. The possibility of interferences by cross-reactions with other substances must be taken into account. Other immunological techniques are radioimmunoassay (RIA), based on the measurement of the radioactivity of the immunological complex;[38] dipsticks, based on membrane strips with the receptor ligands and measurement of the developed color;[39] or the use of luminiscence or fluorescence detectors.[40]

9.5.2   Biosensors

The need to screen a large number of meat samples in relatively short time has prompted the development of biosensors that are based on an immobilized antibody that interacts with the analyte in the sample and the optical or electronic detection of the resulting signal.[41,42] Biosensors can simultaneously detect residues of multiple veterinary drugs in a sample at a time[43,44] without the need for sample cleanup.[45] There are different types of biosensors, such as the surface plasmon resonance (SPR) that measures variations in the refractive index of the solution close to the sensor[46] and has been successfully applied to the detection of residues of different veterinary drugs,[47,48] or the biosensors based on the use of biochip arrays that are specific for a certain number of residues[49] and are also applied to the detection of residues.[50]

9.5.3 Chromatographic Techniques

High-performance thin-layer chromatography (HPTLC) has been successfully used for multi-residue screening purposes in meat. Samples are injected onto the plates and the residues eluted from the plate with the appropriate eluent. Once eluted, residues can be viewed under UV or fluorescent lights or visualized by spraying with a chromogenic reagent. HPTLC has been applied to meat to screen different residues in meat such as agonists,[51,52] nitroimidazol,[53] and thyreostatic drugs.[54,55]

GC and high-performance liquid chromatography (HPLC) are powerful separation techniques capable of separating the analyte from most of the interfering substances by varying the type of column and elution conditions.[37] In some cases, the analyte can be detected after appropriate derivatization.[56] In addition, these techniques can be used for multiresidue screening. The recent development of ultraperformance liquid chromatography systems and new types of columns with packagings of reduced size offer valuable improvements for residue detection such as a considerable reduction in elution times and the possibility of a larger number of samples per day.[37,57] This procedure has been applied to meat for detection of residues of a wide range of veterinary drugs,[58–62] anabolic steroids,[63,64] quinolone,[65] and corticosteroids.[66–69] Additional advantages of GC and HPLC are automation and the possibility to couple the chromatograph to mass spectrometry detectors for further confirmatory analysis. Recently, a rapid, specific, and highly sensitive multiresidue method has been reported for the determination of anabolic steroid residues in bovine, pork, and poultry meat.[29] The methodology involves enzymatic digestion, methanol extraction, and SPE for final purification. The detection is carried out with LC-MS/MS in both ESI$^+$ and ESI$^-$ with a CCα and CCβ below 0.5 ng/g, but the method shows good performance for qualitative screening but not for quantitation.[29]

9.6 Confirmatory Analytical Methods

Confirmatory methods are preferentially based on mass spectrometry because they provide direct information on the molecular structure of the suspect compound and thus an unambiguous identification and confirmation of the residue in meat. However, these methods are costly in terms of time, equipment, and chemicals. When the target analyte is clearly identified and quantified above the decision limit for a forbidden substance (i.e., substances of group A) or exceeding the MRL in the case of substances having an MRL, the sample is considered as noncompliant (unfit for human consumption). A suitable internal standard must be added to the test portion at the beginning of the extraction procedure. If no suitable internal standard is available, the identification of the analyte can be done by cochromatography. This consists in dividing the sample extract into two parts. The first part is injected into the chromatograph as such. The second part is mixed with the standard analyte to be detected and injected into the chromatograph. The amount of added standard analyte must be similar to the estimated amount of the analyte in the extract. Identification is easier for a limited number of target analytes and matrices of constant composition.[70]

GC with mass spectrometry detection has been used for many years even though derivatization (i.e., silyl or boronate derivatives) was required for some nonvolatile residues such as agonists. An example of some agonists such as boronate derivatives giving good identification ions is shown in Figure 9.1. But derivatization entails a serious limitation adding to time and cost of the analysis.

In recent years, the rapid development of mass spectrometry coupled to LC has expanded its applications in this field, especially for nonvolatile or thermolabile compounds. Tandem mass spectrometry (MS/MS) has shown high selectivity and sensitivity and thus allows the analysis of more complex matrices such as meat with easier sample preparation procedures. LC-MS/MS allows the selection of a precursor *m/z* that is performed first. This eliminates any uncertainty on the origin of the observed fragment ions, eliminates potential interferences from the meat sample or from the mobile phase, reduces the chemical noise, and increases the sensitivity.[71]

The interface technology has been rapidly developed. Electrospray ionization (ESI) and atmospheric pressure chemical ionization (APCI) interfaces are preferred depending on the polarity and molecular mass of analytes.[71] ESI ionization technique facilitates the analysis of small to relatively large and hydrophobic to hydrophilic molecules.[56,72,73] An important limitation of LC-MS/MS quantitative analysis is its susceptibility to matrix effect that is dependent on the ionization type, type of sample, and sample preparation. APCI ionization has been reported to be less sensitive than ESI to matrix effects.[74–77] ESI is preferred for the MS analysis of nonsteroidal antiinflammatory drugs (NSAIDs) due to their polar nature; however, some interfering substances of the matrix such as fat may lead to ion suppression problems.[78] The extraction of the analyte must be more selective and further purified and cleaned up.

A rapid qualitative method using online column-switching LC-MS/MS has been developed and validated for screening 13 target veterinary drugs in different animal muscles.[79] This system may reduce the cost and time for confirmatory analysis. A list of recent performance reports of the analysis of veterinary drug residues is shown in Table 9.3.

The ion suppression phenomenon in LC-MS must be taken into account because of matrix effect problems and the presence of interfering compounds that affect the analyte detection. A number of reviews about ion suppression phenomenon and its consequences for residue analysis has been recently published.[84] The major mechanism for ion suppression corresponds to the presence of matrix-interfering compounds that reduce the evaporation efficiency leading to reduced detection capability and repeatability. The ion ratios, linearity, and quantification are also affected. It could even lead to the lack of detection of an analyte or the underestimation of its concentration or the nonfulfilment of the identification criteria.[84] The prevention of this phenomenon includes an improved purification and cleanup of the sample as well as the use of an appropriate internal standard. Another strategy is to modify the elution conditions for the analytes to elute in an area nonaffected by ion suppression.[84]

According to the Commission Decision 2002/657/EC,[19] a system of identification points is used for confirmatory purposes with a minimum of 4 points required for the substances of group A and a minimum of 3 for group B substances. So, 1 identification point can be earned for the precursor ion with a triple quadrupole spectrometer and 1.5 points for each product ion. A high-resolution mass spectrometer acquires 2 identification points for the precursor ion and 2.5 for each product ion. Variable window ranges for MS peak abundances are also established in the new decision (EC 2002). So, the relative ion intensities must be >50, >20–50, >10–20, and ≤10%. In the case of electron impact-GC-MS (EI-GC-MS), the maximum permitted tolerances are ±10, ±15, ±20, and ±50%, respectively, whereas in the case of collision-induced GC-MS (CI-GC-MS), GC-MS[n], LC-MS, and LC-MS[n] are ±20, ±25, ±30, and ±50%, respectively.

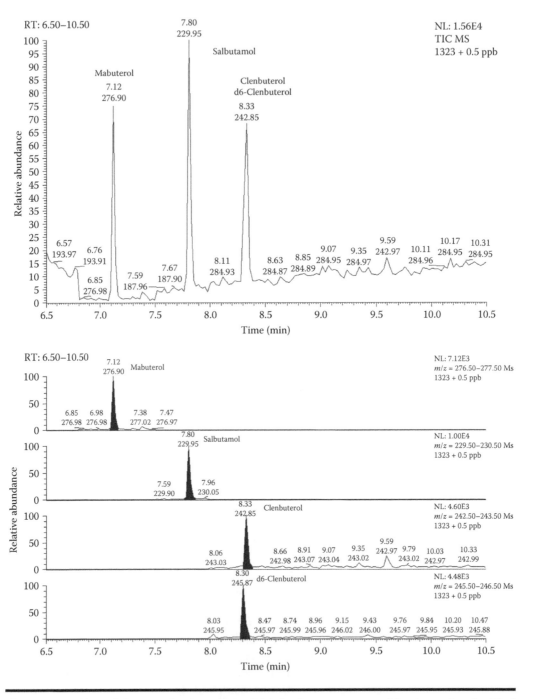

Figure 9.1 Selected ion monitoring (SIM) GC-MS chromatogram of bovine urine fortified with 0.5 ng/mL for clenbuterol, 0.5 ng/mL for mabuterol, 0.75 ng/mL for salbutamol, and 1 ng/mL for d6-clenbuterol as IS. (Reproduced from Reig, M. et al., *Anal. Chim. Acta*, 529, 293, 2005. With permission.)

Table 9.3 Performance of Some Recent Methods of Analysis of Growth Promoters in Antemortem (Farm Samples) and Postmortem (Meat) Samples

Analyte	Matrix	Extraction	Column	System/ Detector	CCα (ng/g)	CCβ (ng/g)	Recovery (%)	Reference
17β-Estradiol-3-benzoate	Bovine hair	Methanol extraction, SPE NH$_2$	Nucleosil C18AB, 5 μm	LC-MS/MS, ESI$^+$	LOD 4.1	LOI 5.0	—	23
Zeranol	Bovine hair	Methanol extraction, acid hydrolysis, SPE	OV-1, 0.25 μm	GC-MS/MS	LOD 2.66	LOI 4.48	—	27
17α-Trenbolone	Bovine hair	Methanol extraction, acid hydrolysis, SPE	OV-1, 0.25 μm	GC-MS/MS	LOD 0.76	LOI 1.99	—	27
Methyltestosterone	Bovine hair	Methanol extraction, acid hydrolysis, SPE	OV-1, 0.25 μm	GC-MS/MS	LOD 1.02	LOI 1.74	—	27
17-Estradiol	Bovine hair	Methanol extraction, acid hydrolysis, SPE	OV-1, 0.25 μm	GC-MS/MS	LOD 0.12	LOI 0.19	—	27
17α-Testosterone	Bovine hair	Methanol extraction, acid hydrolysis, SPE	OV-1, 0.25 μm	GC-MS/MS	LOD 0.29	LOI 0.85	—	27
Melengestrol	Bovine hair	Methanol extraction, acid hydrolysis, SPE	OV-1, 0.25 μm	GC-MS/MS	LOD 1.12	LOI 1.97	—	27
Hexestrol	Pork meat	Liquid extraction, SPE C18	DB-5, 30 m, 0.25 μm	GC-MS/MS	LOD 0.2	—	84.6	80

(continued)

Table 9.3 (continued) Performance of Some Recent Methods of Analysis of Growth Promoters in Antemortem (Farm Samples) and Postmortem (Meat) Samples

Analyte	Matrix	Extraction	Column	System/ Detector	CCα (ng/g)	CCβ (ng/g)	Recovery (%)	Reference
Diethylestilbestrol	Pork meat	Liquid extraction, SPE C18	DB-5, 30 m, 0.25 μm	GC-MS/MS	LOD 0.1	—	80.1	80
Androsterone	Pork meat	Liquid extraction, SPE C18	DB-5, 30 m, 0.25 μm	GC-MS/MS	LOD 0.2	—	91.0	80
Estradiol	Pork meat	Liquid extraction, SPE C18	DB-5, 30 m, 0.25 μm	GC-MS/MS	LOD 0.1	—	95.8	80
Zeranol	Pork meat	Liquid extraction, SPE C18	DB-5, 30 m, 0.25 μm	GC-MS/MS	LOD 0.1	—	95.3	80
α-Zearalenol	Pork meat	Liquid extraction, SPE C18	DB-5, 30 m, 0.25 μm	GC-MS/MS	LOD 0.1	—	94.6	80
17α-Hydroxyl-progesterone	Pork meat	Liquid extraction, SPE C18	DB-5, 30 m, 0.25 μm	GC-MS/MS	LOD 0.4	—	102.4	80
Diethylestilbestrol	Beef muscle	Solvent extraction, freezing-lipid filtration, SPE C8	DB-1MS, 30 m, 0.25 μm	GC-MS/MS	LOD 0.3	—	91.0	81
17β-Estradiol	Beef muscle	Solvent extraction, freezing-lipid filtration, SPE C8	DB-1MS, 30 m, 0.25 μm	GC-MS/MS	LOD 0.2	—	85.0	81
Testosterone	Beef muscle	Solvent extraction, freezing-lipid filtration, SPE C8	DB-1MS, 30 m, 0.25 μm	GC-MS/MS	LOD 0.1	—	100.0	81

Zeranol	Beef muscle	Solvent extraction, freezing-lipid filtration, SPE C8	DB-1MS, 30 m, 0.25 μm	GC-MS/MS	LOD 0.2	—	83.0	81
Progesterone	Beef muscle	Solvent extraction, freezing-lipid filtration, SPE C8	DB-1MS, 30 m, 0.25 μm	GC-MS/MS	LOD 0.3	—	80.0	81
Dexamethasone	Feed	Liquid extraction, SPE NH$_2$	Sinergy MAX-RP 80A, 4 μm	LC-DAD	190	217	108.9	69
Dexamethasone	Drinking water	Centrifugation			26	30	105.1	69
β-Boldenone glucuronide	Urine	Extraction, SPE	Nucleosil C18, 5 μm	LC-MS/MS, API$^-$	0.40	0.55	75.0	82
β-Boldenone sulfates	Urine	Extraction, SPE	Nucleosil C18, 5 μm	LC-MS/MS, API$^-$	0.75	0.99	72.0	82
β-Boldenone	Urine	Extraction, SPE	Nucleosil C18, 5 μm	LC-MS/MS, API$^-$	0.52	0.70	76.0	82
α-Boldenone	Urine	Extraction, SPE	Nucleosil C18, 5 μm	LC-MS/MS, API$^-$	0.70	0.93	71.0	82
5β-Androst-1en-17ol-3one	Urine	Extraction, SPE	Nucleosil C18, 5 μm	LC-MS/MS, API$^-$	0.42	0.56	79.0	82
Trenbolone	Poultry muscle	Matrix solid-phase dispersion	Alltima C18, 5 μm	LC-MS/MS, APCI$^+$		0.13	99	83
Testosterone	Poultry muscle	Matrix solid-phase dispersion	Alltima C18, 5 μm	LC-MS/MS, APCI$^+$	0.03	0.21	97	83

(continued)

Table 9.3 (continued) Performance of Some Recent Methods of Analysis of Growth Promoters in Antemortem (Farm Samples) and Postmortem (Meat) Samples

Analyte	Matrix	Extraction	Column	System/Detector	CCα (ng/g)	CCβ (ng/g)	Recovery (%)	Reference
Melengestrol acetate	Poultry muscle	Matrix solid-phase dispersion	Alltima C18, 5 μm	LC-MS/MS, APCI+	0.03	0.26	90	83
Progesterone	Poultry muscle	Matrix solid-phase dispersion	Alltima C18, 5 μm	LC-MS/MS, APCI+	0.21	0.16	96	83
α-Zeranol	Poultry muscle	Matrix solid-phase dispersion	Alltima C18, 5 μm	LC-MS/ MS, TIS−	0.08	0.87	90	83
α-Estradiol	Poultry muscle	Matrix solid-phase dispersion	Alltima C18, 5 μm	LC-MS/ MS, TIS−	0.11	0.85	100	83
Diethylestilbestrol	Poultry muscle	Matrix solid-phase dispersion	Alltima C18, 5 μm	LC-MS/ MS, TIS−	0.04	0.33	80	83

Note: ESI, electrospray ionization; LOD, limit of detection; LOI, limit of identification; APCI, atmospheric pressure chemical ionization; TIS, turbo ion spray.

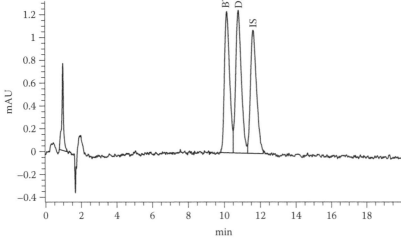

Figure 9.2 Example of detection of dexamethasone, a substance in group B 2f, and closely related substances in feed through LC-DAD. Retention times 10.14, 10.78, and 11.60 min corresponded to betamethasone (BTM), dexamethasone (DXM), and flumethasone (IS), respectively. (Reproduced from Reig, M. et al., *Meat Sci.*, 74, 676, 2006. With permission.)

Other methods are allowed for group B substances.[19] So, liquid chromatography-full scan diode array detection (LC-DAD) can be used as a confirmatory method if specific requirements for absorption in UV spectrometry are met. This means that the absorption maxima of the spectrum of the analyte shall be at the same wavelengths of the calibration standard within a margin of ±2 nm for diode array detection. Furthermore, the spectrum of the analyte above 220 nm will not be visibly different (at no point greater than 10%) from the spectrum of the calibration standard. An example of identification of dexamethasone, a substance in group B 2f, in feed through LC-DAD is shown in Figure 9.2.

References

1. McEvoy, J.D.G. Contamination of animal feedstuffs as a cause of residues in food: a review of regulatory aspects, incidence and control. *Anal. Chim. Acta* 473: 3–26, 2002.
2. Brynes, S.D. Demystifying 21 CFR Part 556—Tolerances for residues of new animal drugs in food. *Regul. Toxicol. Pharmacol.* 42: 324–327, 2005.
3. Council Directive 88/146/EEC of March 7, 1988, prohibiting the use in livestock farming of certain substances having a hormonal action. *Off. J. Eur. Commun.* L070: 16, 1988.
4. Lone, K.P. Natural sex steroids and their xenobiotic analogs in animal production: growth, carcass quality, pharmacokinetics, metabolism, mode of action, residues, methods, and epidemiology. *Crit. Rev. Food Sci. Nutr.* 37: 93–209, 1997.
5. Van Peteguem, C. and Daeselaire, E. Residues of growth promoters. In: L.M.L. Nollet (Ed.), *Handbook of Food Analysis*, 2nd ed. New York, NY: Marcel Dekker, 1037–1063, 2004.

6. EFSA. Opinion of the scientific panel on contaminants in the food chain on a request from the European Commission related to hormone residues in bovine meat and meat products. *EFSA J.* 510: 1–62, 2007.

7. Barbosa, J. et al. Food poisoning by clenbuterol in Portugal. *Food Addit. Contam.* 22: 563–566, 2005.

8. Moloney, A. et al. Influence of beta-adrenergic agonists and similar compounds on growth. In: A.M. Pearson and T.R. Dutson (Eds.), *Growth Regulation in Farm Animals.* London, UK: Elsevier Applied Science, 455–513, 1991.

9. Miller, L.F. et al. Relationships among intramuscular collagen, serum hydroxyproline and serum testosterone in growing rams and wethers. *J. Anim. Sci.* 67: 698, 1989.

10. Miller, L.F., Judge, M.D., and Schanbacher, B.D. Intramuscular collagen and serum hydroxyproline as related to implanted testosterone and estradiol 17β in growing wethers. *J. Anim. Sci.* 68: 1044, 1990.

11. Fiems, L.O. et al. Effect of a β-agonist on meat quality and myofibrillar protein fragmentation in bulls. *Meat Sci.* 27: 29–35, 1990.

12. Duquette, P.F. and Muir, L.A. Effect of the β-adrenergic agonists isoproterenol, clenbuterol, L-640,033 and BRL 35135 on lipolysis and lipogenesis in rat adipose tissue *in vitro. J. Anim. Sci.* 61(Suppl. 1): 265, 1985.

13. Brockman, R.P. and Laarveld, R. Hormonal regulation of metabolism in ruminants. Review. *Livest. Prod. Sci.* 14: 313–317, 1986.

14. Dikeman, M.E. Effects of metabolic modifiers on carcass traits and meat quality. *Meat Sci.* 77: 121–135, 2007.

15. Monsón, F. et al. Carcass and meat quality of yearling bulls as affected by the use of clenbuterol and steroid hormones combined with dexamethasone. *J. Muscle Foods* 18: 173–185, 2007.

16. Council Directive 96/23/EC of April 29, 1996, on measures to monitor certain substances and residues thereof in live animals and animal products. *Off. J. Eur. Commun.* L125: 10, 1996.

17. Commission Decision 93/256/EC of May 14, 1993, laying down the methods to be used for detecting residues of substances having hormonal or a thyreostatic action. *Off. J. Eur. Commun.* L118: 64, 1993.

18. Commission Decision 93/257/EC of April 15, 1993, laying down the reference methods and the list of the national reference laboratories for detecting residues. *Off. J. Eur. Commun.* L118: 73, 1993.

19. Commission Decision 2002/657/EC of August 17, 2002, implementing Council Directive 96/23/EC concerning the performance of the analytical methods and the interpretation of results. *Off. J. Eur. Commun.* L221: 8, 2002.

20. Appelgren, L.-E. et al. Analysis of hair samples for clenbuterol in calves. *Fleischwirtsch.* 76: 398–399, 1996.

21. Panoyan, A. et al. Immunodetection of clenbuterol in the hair of calves. *J. Agric. Food Chem.* 43: 2716–2718, 1995.

22. Gleixner, A. and Meyer, H.H.D. Hair analysis for monitoring the use of growth promoters in meat production. *Fleischswirtsch.* 76: 637–638, 1996.

23. Rambaud, L. et al. Study of 17-estradiol-3-benzoate, 17β-methyltestosterone and medroxyprogesterone acetate fixation in bovine hair. *Anal. Chim. Acta* 532: 165–176, 2005.

24. Haasnoot, W. et al. A fast immunoassay for the screening of β-agonists in hair. *Analyst* 123: 2707–2710, 1998.

25. Adam, A. et al. Detection of clenbuterol residues in hair. *Analyst* 119: 2663–2666, 1994.

26. Hooijerink, H. et al. Liquid chromatography-electrospray ionisation-mass spectrometry based method for the determination of estradiol benzoate in hair of cattle. *Anal. Chim. Acta* 529: 167–172, 2005.

27. Rambaud, L. et al. Development and validation of a multi-residue method for the detection of a wide range of hormonal anabolic compounds in hair using gas chromatography–tandem mass spectrometry. *Anal. Chim. Acta* 586: 93–104, 2007.

28. McCracken, R.J., Spence, D.E., and Kennedy, D.G. Comparison of extraction techniques for the recovery of veterinary drug residues from animal tissues. *Food Addit. Contam.* 17: 907–914, 2000.

29. Blasco, C., Poucke, C.V., and Peteghem, C.V. Analysis of meat samples for anabolic steroids residues by liquid chromatography/tandem mass spectrometry. *J. Chromatogr. A* 1154: 230–239, 2007.

30. Schmidt, G. and Steinhart, H. Impact of extraction solvents on steroid contents determined in beef. *Food Chem.* 76: 83–88, 2002.

31. Stolker, A.A.M., Zoonties, P.W., and Van Ginkel, L.A. The use of supercritical fluid extraction for the determination of steroids in animal tissues. *Analyst* 123, 2671–2676, 1998.

32. Stubbings, G. et al. A multi-residue cation-exchange clean up procedure for basic drugs in produce of animal origin. *Anal. Chim. Acta* 547: 262–268, 2005.

33. Godfrey, M.A.J. Immunoaffinity extraction in veterinary residue analysis: a regulatory viewpoint. *Analyst* 123: 2501–2506, 1998.

34. Wistrand, C. et al. Evaluation of MISPE for the multi-residue extraction of β-agonists from calves urine. *J. Chromatogr. B* 804: 85–91, 2004.

35. Kootstra, P.R. et al. The analysis of beta-agonists in bovine muscle using molecular imprinted polymers with ion trap LCMS screening. *Anal. Chim. Acta* 529: 75–81, 2005.

36. Toldrá, F. and Reig, M. Methods for rapid detection of chemical and veterinary drug residues in animal foods. *Trends Food Sci. Technol.* 17: 482–489, 2006.

37. Reig, M. and Toldrá, F. Veterinary drug residues in meat: concerns and rapid methods for detection. *Meat Sci.* 78: 60–67, 2008.

38. Samarajeewa, U. et al. Application of immunoassay in the food industry. *Crit. Rev. Food Sci. Nutr.* 29: 403–434, 1991.

39. Link, N., Weber, W., and Fussenegger, M. A novel generic dipstick-based technology for rapid and precise detection of tetracycline, streptogramin and macrolide antibiotics in food samples. *J. Biotechnol.* 128: 668–680, 2007.

40. Roda, A. et al. A rapid and sensitive 384-well microtitre format chemiluminescent enzyme immunoassay for 19 nortestosterone. *Luminescence* 18: 72–78, 2003.

41. Patel, P.D. Biosensors for measurement of analytes implicated in food safety: a review. *Trends Food Sci. Technol.* 21: 96–115, 2002.

42. White, S. Biosensors for food analysis. In: L.M.L. Nollet (Ed.), *Handbook of Food Analysis*, 2nd ed. New York, NY: Marcel Dekker, 2133–2148, 2004.

43. Gründig, B. and Renneberg, R. Chemical and biochemical sensors. In: A. Katerkamp, B. Gründig, and R. Renneberg (Eds.), *Ullmann's Encyclopedia of Industrial Chemistry*. Verlag Wiley-VCH, Berlin, Germany, 87–98, 2002.

44. Franek, M. and Hruska, K. Antibody based methods for environmental and food analysis: a review. *Vet. Med. Czech.* 50: 1–10, 2005.

45. Elliott, C.T. et al. Use of biosensors for rapid drug residue analysis without sample deconjugation or clean-up: a possible way forward. *Analyst* 123: 2469–2473, 1998.

46. Gillis, E.H., Gosling, J.P., Sreenan, J.M., and Kane, M. Development and validation of a biosensor-based immunoassay for progesterone in bovine milk. *J. Immunol. Methods* 267: 131–138, 2002.

47. Bergweff, A.A. Rapid assays for detection of residues of veterinary drugs. In: A. van Amerongen, D. Barug, and M. Lauwars (Eds.), *Rapid Methods for Biological and Chemical Contaminants in Food and Feed*. Wageningen Academic Publishers, Wageningen, The Netherlands, 259–292, 2005.

48. Haughey, S.A. and Baxter, C.A. Biosensor screening for veterinary drug residues in foodstuffs. *J. AOAC Int.* 89: 862–867, 2006.

49. Johansson, M.A. and Hellenas, K.E. Sensor chip preparation and assay construction for immunobiosensor determination of beta-agonists and hormones. *Analyst* 126: 1721–1727, 2001.

50. Zuo, P. and Ye, B.C. Small molecule microarrays for drug residue detection in foodstuffs. *J. Agric. Food Chem.* 54: 6978–6983, 2006.

51. Degroodt, J.-M. et al. Clenbuterol residue analysis by HPLC–HPTLC in urine and animal tissues. *Z. Lebens. Unters. Forsch.* 189: 128–131, 1989.

52. Degroodt, J.-M. et al. Cimaterol and clenbuterol residue analysis by HPLC–HPTLC. *Z. Lebens. Unters. Forsch.* 192: 430–432, 1991.

53. Gaugain, M. and Abjean, J.P. High-performance thin-layer chromatographic method for the fluorescence detection of three nitroimidazole residues in pork and poultry tissue. *J. Chromatogr. A* 737: 343–346, 1996.

54. De Brabender, H.F., Batjoens, P., and Van Hoof, V. Determination of thyreostatic drugs by HPTLC with confirmation by GC-MS. *J. Planar Chromatogr.* 5: 124–130, 1992.

55. De Wasch, K. et al. Confirmation of residues of thyreostatic drugs in thyroid glands by multiple mass spectrometry after thin-layer chromatography. *J. Chromatogr. A* 819: 99–111, 1998.

56. Bergweff, A.A. and Schloesser, J. Residue determination. In: B. Caballero, L. Trugo, and P. Finglas (Eds.) *Encyclopedia of Food Sciences and Nutrition*, 2nd ed. London, UK: Elsevier, 254–261, 2003.

57. Aristoy, M.C., Reig, M., and Toldrá, F. Rapid liquid chromatography techniques for detection of key (bio)chemical markers. In: L.M.L. Nollet and F. Toldrá (Eds.), *Advances in Food Diagnostics*. Ames, IA: Blackwell Publishing, 229–251, 2007.

58. Cooper, A.D. et al. Development of multi-residue methodology for the HPLC determination of veterinary drugs in animal-tissues. *Food Addit. Contam.* 12: 167–176, 1995.

59. Aerts, M.M.L., Hogenboom, A.C., and Brinkman, U.A.T. Analytical strategies for the screening of veterinary drugs and their residues in edible products. *J. Chromatogr. B Biomed. Appl.* 667: 1–40, 1995.

60. Horie, M. et al. Rapid screening method for residual veterinary drugs in meat and fish by HPLC. *J. Food Hyg. Soc. Jpn.* 39: 383–389, 1998.

61. Kao, Y.M. et al. Multiresidue determination of veterinary drugs in chicken and swine muscles by high performance liquid chromatography. *J. Food Drug Anal.* 9: 84–95, 2001.

62. Reig, M. et al. Stability of β-agonist methyl boronic derivatives before GC-MS analysis. *Anal. Chim. Acta* 529: 293–297, 2005.

63. Gonzalo-Lumbrearas, R. and Izquierdo-Hornillos, R. High-performance liquid chromatography optimization study for the separation of natural and synthetic anabolic steroids. Application to urine and pharmaceutical samples. *J. Chromatogr. B* 742: 1–11, 2000.

64. De Cock, K.J.S. et al. Detection and determination of anabolic steroids in nutritional supplements. *J. Pharm. Biomed. Anal.* 25: 843–852, 2001.

65. Verdon, E., Hurtaud-Pessel, D., and Sanders, P. Evaluation of the limit of performance of an analytical method based on a statistical calculation of its critical concentrations according to ISO standard 11843: application to routine control of banned veterinary drug residues in food according to European Decision 657/2002/EC. *Accred. Qual. Assur.* 11: 58–62, 2006.

66. Shearan, P., O'Keefe, M., and Smyth, M. Reversed-phase high-performance liquid chromatographic determination of dexamethasone in bovine tissues. *Analyst* 116, 1365–1368, 1991.

67. Mallinson, E.T. et al. Determination of dexamethasone in liver and muscle by liquid chromatography and gas chromatography/mass spectrometry. *J. Agric. Food Chem.* 43: 140–145, 1995.

68. Stolker, A.A.M. et al. Comparison of different liquid chromatography methods for the determination of corticosteroids in biological matrices. *J. Chromatogr. A* 893, 55–67, 2000.

69. Reig, M. et al. A chromatography method for the screening and confirmatory detection of dexamethasone. *Meat Sci.* 74: 676–680, 2006.

70. Milman, B.L. Identification of chemical compounds. *Trends Anal. Chem.* 24: 493–508, 2005.

71. Gentili, A., Perret, D., and Marchese, S. Liquid chromatography-tandem mass spectrometry for performing confirmatory analysis of veterinary drugs in animal-food products. *Trends Anal. Chem.*, 24: 704–733, 2005.

72. Hewitt, S.A. et al. Screening and confirmatory strategies for the surveillance of anabolic steroid abuse within Northern Ireland. *Anal. Chim. Acta* 473: 99–109, 2002.

73. Thevis, M., Opfermann, G., and Schänzer, W. Liquid chromatography/electrospray ionization tandem mass spectrometric screening and confirmation methods for β_2-agonists in human or equine urine. *J. Mass Spectrom.* 38, 1197–1206, 2003.

74. Dams, R. et al. Matrix effects in bio-analysis of illicit drugs with LC-MS/MS: influence of ionization type, sample preparation and biofluid. *J. Am. Soc. Mass Spectrom.* 14: 1290–1294, 2003.

75. Puente, M.L. Highly sensitive and rapid normal-phase chiral screen using high-performance liquid chromatography-atmospheric pressure ionization tandem mass spectrometry (HPLC/MS). *J. Chromatogr.* 1055: 55–62, 2004.

76. Maurer, H.H. et al. Screening for library-assisted identification and fully validated quantification of 22 beta-blockers in blood plasma by liquid chromatography-mass spectrometry with atmospheric pressure chemical ionization. *J. Chromatogr.* 1058: 169–181, 2004.

77. Turnipseed, S.B. et al. Analysis of avermectin and moxidectin residues in milk by liquid chromatography-tandem mass spectrometry using an atmospheric pressure chemical ionization/atmospheric pressure photoionization source. *Anal. Chim. Acta* 529: 159–165, 2005.

78. Gentili, A. LC–MS methods for analyzing anti-inflammatory drugs in animal-food products. *Trends Anal. Chem.* 26: 595–608, 2007.

79. Tang, H.P., Ho, C., and Lai, S.S. High-throughput screening for multi-class veterinary drug residues in animal muscle using liquid chromatography/tandem mass spectrometry with on-line solid-phase extraction. *Rapid Commun. Mass Spectrom.* 20: 2565–2572, 2006.

80. Fuh, M.-R. et al. Determination of residual anabolic steroid in meat by gas chromatography–ion trap–mass spectrometer. *Talanta* 64: 408–414, 2004.

81. Seo, J. et al. Simultaneous determination of anabolic steroids and synthetic hormones in meat by freezing-lipid filtration, solid-phase extraction and gas chromatography–mass spectrometry. *J. Chromatogr. A* 1067: 303–309, 2005.

82. Buiarelli, F. et al. Detection of boldenone and its major metabolites by liquid chromatography—tandem mass spectrometry in urine samples. *Anal. Chim. Acta* 552: 116–126, 2005.

83. Gentili, A. et al. High- and low-resolution mass spectrometry coupled to liquid chromatography as confirmatory methods of anabolic residues in crude meat and infant foods. *Rapid Commun. Mass Spectrom.* 20: 1845–1854, 2006.

84. Antignac, J.P. et al. The ion suppression phenomenon in liquid chromatography-mass spectrometry and its consequences in the field of residue analysis. *Anal. Chim. Acta* 529: 129–136, 2005.

Chapter 10

Antibiotic Residues in Muscle Tissues of Edible Animal Products

Eric Verdon

Contents

10.1 Introduction 250
    10.1.1 Antibiotics and Antibacterials in Veterinary Practice 250
    10.1.2 Veterinary Drug Residue Regulatory Control for Food Safety 251
    10.1.3 Strategies for Screening and Confirmation of Antimicrobial Residues in Meat.....252
    10.1.4 Analytical Methods for Control of Antimicrobials in Meat 253
10.2 Screening Analysis by Means of Biological Methods 253
    10.2.1 Microbiological Methods for Antimicrobial Residues 253
    10.2.2 Other Biological Methods 254
10.3 Confirmatory Analysis by Means of Chromatographic Methods 256
    10.3.1 Chromatographic Methodologies for Antimicrobial Residues 256
    10.3.2 Other Modes of Chemical Analysis 257
    10.3.3 Sample Preparation for Liquid Chromatography 257
    10.3.4 Modes of Separation in Liquid Chromatography 258
    10.3.5 Modes of Detection in Liquid Chromatography 259
10.4 Applications of Chromatographic Methods to Antimicrobial Residues 260
    10.4.1 Aminoglycosides 260
    10.4.2 Amphenicols 262
    10.4.3 Beta-Lactams: Penicillins and Cephalosporins 266

10.4.4   Macrolides and Lincosamides 272
10.4.5   Nitrofurans 280
10.4.6   Nitroimidazoles 285
10.4.7   Quinolones 290
10.4.8   Sulfonamide Antiinfectives 301
10.4.9   Tetracyclines 308
10.4.10 Polypeptidic Antibiotics 309
10.4.11 Polyether Antibiotics 316
10.4.12 Other Antibiotics (Novobiocin, Tiamulin) 316
10.4.13 Other Antibacterials (Carbadox and Olaquindox) 322
10.5 Conclusion 322
Abbreviations 325
References 327

10.1   Introduction

10.1.1   *Antibiotics and Antibacterials in Veterinary Practice*

Antibiotics are considered as the most important class of drugs. They play a key role in control-ling bacterial infections in both human and animals. The need of food for human consumption is growing and expansion of intensive livestock farming is of major concern. Antibiotics contribute, for a large part, to this industrialization of the farming practice (treatment and prevention of animal diseases and growth-promoting feed additives, even though the latter is being reassessed in some countries such as in the European Union [EU]). Their use in animal husbandry requires them to be on the top of the veterinary drug production for the pharmaceutical industry. Both the rational usage of these substances and the monitoring of their residual concentrations in animal products for human consumption would contribute to prevent their excessive content in human food, reducing the risks for human health. Analytical methods dedicated to monitor these sub-stances in food-producing animal products for human consumption are one of the tools for food safety control. However, it is the mutual concern of all the actors involved in the human food supply—farmers, veterinarians, feed manufacturers, food industry, and regulatory agencies. They should create the conditions congenial to human food safety.

For a better understanding of the terms "antibiotics" and "antibacterials," it is essential to clarify their meaning and usage. Today, the term "antibiotic" is often wrongly used in the place of "anti-bacterial" or "antimicrobial" because not only antibiotics possess an antibacterial activity. In fact, according to an internationally recognized classification, the term "antibiotic" should strictly apply to a range of compounds that are of biological origin; some are produced metabolically from molds of filamentous fungi such as from *Penicillium* species (i.e., benzylpenicillin, 6-aminopenicillenic acid [6-APA], cephalosporin C, 7-aminocephalosporanic acid [7-ACA]), others are extracted from cultures of specific bacteria such as from several *Streptomyces* species (i.e., streptomycin, gentami-cin, tetracycline, spiramycin), and many others are semisynthetic substances that are additionally modified by chemical synthesis (i.e., florfenicol, amoxycillin, cephalexin). The terms "antibacterial" or "antimicrobial" apply to a broader set of compounds including not only the natural and semi-synthetic antibiotics, but also several other classes of molecules having an antibacterial property, besides those that are produced by chemical synthesis; quinolones, nitrofurans, nitroimidazoles, sul-fonamides belong to this category. The compounds covered by the term "antibiotics" fall into seven categories: aminoglycosides, amphenicols, cephalosporins, macrolides, penicillins, polypeptides, and tetracyclines. The penicillins and the cephalosporins are frequently merged into the wider family called beta-lactams. According to the recognized classification, the synthetic compounds from the four families of nitrofurans, nitroimidazoles, quinolones, and sulfonamides can only be considered as antibacterials or antimicrobials but do not belong to the antibiotic class. Finally, there are several other compounds that are included in this chapter dedicated to antibiotics. Some of them are considered as subfamilies or quasifamilies to those described in the preceding discussion: lincosamides and cephamycins fall into this category. Other compounds are from less important veterinary drug families that also present some antibacterial activities: carbadox, dapsone, malachite green, olaquindox, novobiocin, and virginiamycin. In chemical terms, the collections of substances that exhibit antibiotic properties feature diverse groups characterized by very different molecular structures and bearing widely divergent functionalities and mode of operation. This diversity poses a tremendous challenge to the analyst when subtle structural variations in closely related antibiotic compounds can lead to large variations in the chemical toxicity and biological activity of the antibiotic.

10.1.2 Veterinary Drug Residue Regulatory Control for Food Safety

The administration of licensed veterinary antimicrobials to food-producing animals may lead to the occurrence of residues in the food, primarily in the meat produced for human consumption. With the increasing concern for the safety of human food supply, monitoring for animal drug residues has become an important regulatory issue. To safeguard human health, safe tolerance levels or maximum residue limits (MRLs) in food products from animal origin have been established in various countries around the world. In the EU, the establishment of MRLs is governed by Council Regulation 2377/90/EEC [1]. These limits account as part of the regulation for controlling the safety of food with regard to residues of veterinary drugs in tissues and fluids of animals entering the human food chain. To ensure that human food is entirely free from potentially harmful concentrations of residues, MRLs are calculated from toxicological data and with a safety margin ranging from a factor of 10 to 100, depending on the drug considered. The Regulation 2377/90/EEC establishes the lists of compounds that have a fixed MRL (Annex 1) or that need no MRL (Annex 2). Provisional MRL can be supported for a limited period in certain cases (Annex 3); other substances including some antibiotics (chloramphenicol) and some antibacterials (nitrofurans, nitroimidazoles), which are excluded from Annexes 1, 2, or 3, are enlisted in the Annex 4 of the Regulation. This enlisting in Annex 4 has the consequence of prohibiting their use in livestock production. The Council Regulation 2377/90/EEC is amended continuously since 1990 for the implementation of new MRLs while authorizing new veterinary substances. Therefore, the surveillance of veterinary drug residues in food products is an issue for each country subjected to the EU legislation. In essence, there are two types of regulatory residue programs. One deals with the direct-targeted control where the animal or the product is under consignment pending the result of the analysis. The other is built through the implementation of a National Residue Monitoring Plan (NRMP) that is used to monitor the residue status of food from animal origin, without systematic rejection of the specific product from the food market. The regulatory NRMP is established under EU Council Directive 96/23/EC [2], and more recently, also included into Regulation 882/2004/EC [3]. This Directive describes the quantity of samples to be tested for each species of food-producing animals (i.e., bovine species, etc.) or of animal products (milk, etc.) and the different groups of residual compounds to be monitored (i.e., antimicrobials, anabolics, antiparasitics, etc.). In both cases, suspected samples should be efficiently separated from the bulk of negative samples.

The criteria establishing the performance expected from the analytical methods for the screening and for the confirmatory control of residues have been established in the EU by Commission Decision 2002/657/EC [4] replacing in 2002 the former EU Commission Decision 93/256/EEC [5]. Efforts have been made to develop analytical tools capable of supporting the surveillance of the residues in food products from animal origin according to this set of laws.

10.1.3 Strategies for Screening and Confirmation of Antimicrobial Residues in Meat

There is a need to develop rapid analytical methods for controlling drug residues in food products of animal origin, particularly antimicrobials. While developing or selecting analytical procedures for residue control programs, certain aspects have to be taken into account, some of which are governed by external economical/political factors or internal organizational constraints. These factors/constraints are important to build the strategy that must be, at the same time, in line with the national, the European community, and the international food safety legislations as it is described in a relevant paper dedicated to the control of chloramphenicol [6]. Various methodological options for the screening and confirmation of veterinary drug residues, particularly the antimicrobials, can be implemented. Traditionally, the microbiological assays involving bacterial inhibition of a probe microorganism on a medium containing the antibiotic are the basis of antimicrobial residue control in food. But these methods are time-consuming and labor-intensive. In addition, microbiological assays often cannot differentiate univocally among the various forms and derivatives of a given antibiotic family. The quantitative information offered by such an approach reflects a lack of selectivity with the total amount of all forms of a given antibiotic, rather than providing distinct information related to quantitation and identification on the different analogs. These drawbacks are counterbalanced by the cost-effective and high sample throughput implementation of these techniques. In contrast to microbiological methods, chemical chromatographic approaches such as gas chromatography (GC) and liquid chromatography (LC) with various detectors (gas chromatography–electron capture detector [GC-ECD], gas chromatography–flame ionization detector [GC-FID], gas chromatography–nitrogen-phosphorus detector [GC-NPD], liquid chromatography–ultraviolet detector [LC-UV], liquid chromatography–visible detector [LC-Vis], liquid chromatography–fluorescence detector [LC-FLD]) can provide a more selective response with both high sensitivity and good separation efficiencies for most of them. Thus, chromatographic methods hold a real potential to display many of the characteristics necessary for systematic screening of antimicrobial residues in food products. However, the extremely diverse chemical nature of antibiotic substances requires that a variety of separation modes, detection strategies, and sample preparation procedures be used to achieve the goals outlined previously as necessary for rapid and sensitive screening. Moreover, the changes in the regulations, enforced during the past 20 years, highlighted the need to monitor drug residues in food, starting from a single rapid cost-effective screening to now achieving a univocal confirmation of the residual substance(s) primarily suspected in the food products. The analytical strategies, applied then by the networks of control laboratories involved in the food safety legislation, now require at least a two-step analytical monitoring; sometimes, to adjust the cover of regulatory needs, a strategy involving three or more steps may be required. Fortunately, in the time legislation strengthened during the past 20 years, the introduction of versatile and highly sensitive detectors built from different modes of mass spectrometric analyzers (single-quadrupole, triple-quadrupole, also called tandem-quadrupole, ion-trap, time-of-flight [ToF], quadrupole ToF, and quadrupole ion-trap) improved and sometimes simplified the strategies in the veterinary residue control. These rather expensive instruments introduced into the residue control laboratories gave the opportunity to readily modify the strategies, improving the quality and enhancing the efficiency of the control.

10.1.4 Analytical Methods for Control of Antimicrobials in Meat

Trends in analytical method development for drug residue control, particularly for antimicrobial residue control in meat products, changed significantly during the 1990s and early 2000s, with the increasing reliability of high-performance liquid chromatographic (HPLC) instruments. During the past 20 years, there have been a number of reviews covering the analysis of residual antimicrobials in food [7–21], which indicate that comparatively few analytical methods capable of measuring residual concentrations of many antimicrobials, at or near their MRL, existed in the 1980s. For example, developing procedures to extract and concentrate their residues from biological matrices became difficult due to low solubility of some antimicrobials in organic solvents. The other antimicrobials are either insufficiently volatile or thermally unstable (or both) to permit their analysis using GC or GC coupled to mass spectrometry (GC-MS). As a consequence, many methods for measuring antimicrobial residues have been developed by using HPLC. Liquid chromatographic technologies received more and more attention in the late 1980s, and much innovations and reliability occurred in the 1990s [22–25]. However, HPLC with UV detection is not considered sufficiently specific for use as a reliable confirmatory technique, at least in the present 2000's EU legislation. A spectral recognition of the compound (by means of multiwavelength UV-visible detectors such as diode array detector [DAD] and photodiode array [PDA]) or a more specific signal such as with fluorescence detection is mandatory. The development of LC coupled to mass spectrometry (LC-MS) instruments has significantly increased the range of antimicrobials for which reliable and identificative assays based on molecular spectrometry can be developed.

Today, the strategy of surveillance for the presence of antimicrobial residues in meat can be divided into two main categories of analytical methods. The biological methods (inhibitory plate tests, receptor test kits, immunoenzymological kits, and immunochemical biosensors) are generally aimed at wide range of antimicrobial screening, sometimes proposing a reduced monitoring to only one antimicrobial family or even to a single substance. The chromatographic methods (GC, GC-MS, LC, and LC-MS) often bring a higher degree of selectivity, sensitivity, and chemical structure recognition. These physicochemical methods are dedicated to a more specialized control of single substances with monoresidue methods. However, they are also able to cover the control of a family or of a set of substances and hence considered as multiresidue methods. The trends with the brand-new versatile technologies provided by LC-tandem-MS and LC-hybrid-MS lead to even wider ranges of antimicrobials and families of antimicrobials potentially analyzed altogether, and in the near future, even possibly together with other veterinary drugs (antiparasitics, anticoccidials, and antiinflammatories).

10.2 Screening Analysis by Means of Biological Methods

10.2.1 Microbiological Methods for Antimicrobial Residues

Ideally, a screening method should allow to establish the presence or absence of veterinary drugs by detecting all suspect samples (avoiding false negatives), preferably using a simple, routinely applicable procedure. Microbiological methods able to control the inhibitory activity of a majority of antimicrobials are of premium interest in this regard. Yet, it may sometimes be rather complex in analyzing inhibitory data that result due to the variety of antimicrobials of interest or also due to the desired limit of detection. In certain cases of inhibitory tests, postscreening orientative information is needed before confirming with adequate identification/quantification. Most of the developed microbiological tests dedicated to the control of meat products focus on muscle or kidney as a target tissue. The obvious advantage of analyzing muscle tissue lies in the fact that this is the edible part of the animal for which MRLs have primarily been established. Another advantage is that false positives due to naturally inhibiting substances are not likely to occur or are at least considerably reduced compared to other potential target matrices (kidney, liver). A disadvantage is that a variety of microorganisms have to be used to meet the MRLs for the commonly used antimicrobials. The major advantage of using kidney is the highest factor of concentration of many antimicrobial veterinary drugs in that kind of tissue compared to muscle [26]. This is an advantage of first interest when the analytical technologies for residue testing seriously lack sensitivity. The major drawback is the false-positive samples that are likely to occur due, in part, to the inhomogeneity of this offal and also due to the lack of stability of this tissue and to the difficulties in extracting several of the protein-bound antimicrobial compounds.

Most of the microbiological methods for tissues detect inhibitory substances diffusing from a piece of tissue [27–37], or from a paper disk soaked with tissue fluid [26,38–40], into an agar layer seeded with a susceptible bacterial strain. These tests are multiresidue screening methods and use either only one plate [41,42] or different plates, different combinations of pH, media, and different test microorganisms to try to improve the detection of different families of drugs [29, 33–35,37,43]. Some of these methods have also been modified to perform a postscreening analysis [34,35,37] often proposed in antibiotic residue control strategy to orientate toward the appropriate antimicrobial family before more sophisticated chemical identification and quantification.

10.2.2 Other Biological Methods

Several other biologically derived analytical methods were recently or are still in use in some control laboratories: the radioimmunological Charm II Test® applied to muscle tissue and derived from milk control [44–46], the high voltage gel electrophoresis [47–50], or the TLC-bioautography [51–56].

Owing to the fact that the microbiological methods with wide-range antimicrobial residue screening are considered time-consuming incubation procedures with regard to the fast processing in agrifood industrial practice, several other microbiological or immunological receptor test technologies have been developed for residue control. The analytical strategy for meat control is very similar and can be compared to that built for antimicrobial residue control in milk. The most common tools used for time-saving strategies are rapid microbiological tube test assays such as the Charm Farm Test® [46,57,58], the more recently developed Premi Test® [59–61], rapid receptor tests such as the Tetrasensor® for tetracyclines [62], or solid-phase fluorescence immunoassays (SPFIA) for gentamicin, several antibiotics, and sulfonamides [63–65], or enzyme-linked immunosorbent assays (ELISA kits) based on monoclonal or polyclonal antibodies. These tools generally focus on an immunoenzymatic action for one or two specific antimicrobial families or, even in some cases, on only one very specific antimicrobial compound. For example, ELISA kits were developed not only for chloramphenicol, enrofloxacin, gentamicin, halofuginone, nicarbazin, specific nitrofuran metabolites, streptomycin, tylosin but also for beta-lactams, fluoroquinolones, macrolides, nitroimidazoles, and tetracyclines [6,36,66–84]. A new immunological technology developed in the 1990s, the surface plasmon resonance-based biosensor immunoassay (SPR-BIA), based on both immunological receptors and signal reading by a specific light-scattering property, has been of great interest in the 2000s. It has been applied first as screening/postscreening strategy in milk products for residues of sulfonamide compounds such as sulfamethazine and sulfadiazine [85–90] and then also extended in milk and in tissues to several other antimicrobial compounds or families of antimicrobials such as streptomycin, dihydrostreptomycin, nicarbazin metabolite dinitrocarbanilamide (DNC), and a range of penicillins and fluoroquinolones [83,91–96]. However, lack of wide-range screening for antimicrobials generally put these methodologies in the position to be prescribed for prescreening strategies or for very specific and selective control. In certain conditions, they might be useful when the regulatory residue control enforces a ban on antimicrobial substances such as chloramphenicol, nitrofurans, and nitroimidazoles. They are also available for some screening strategies when the microbiological wide-range screening methods lack sensitivity with regard to the requested MRL for specific families or substances (sulfonamides, aminoglocosides). A survey of screening methodologies implemented in the EU by the reference laboratories in the period 2000–2003 for monitoring authorized antimicrobials in meat is shown in Figure 10.1. The same survey is displayed in Figure 10.2 but for the screening in meat of the residues of a prohibited drug: chloramphenicol.

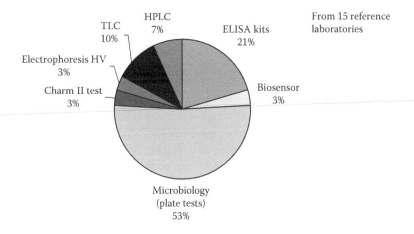

Figure 10.1 **Screening strategy for meat control: methodologies used for screening in meat products considering 15 national reference laboratories from the European Union Member States (2003).**

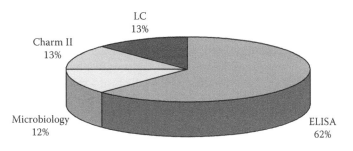

Figure 10.2 **Screening strategy for meat control—chloramphenicol: methodologies used for screening in meat products for chloramphenicol residues considering 15 national reference laboratories from the European Union Member States (2003).**

10.3 Confirmatory Analysis by Means of Chromatographic Methods

10.3.1 Chromatographic Methodologies for Antimicrobial Residues

The widely used methods in residue analysis are those based on chromatographic procedures. Their efficiency comes, for a large part, from their separative properties making them selective to the compounds to be analyzed with regard to the complex endogenous interfering substances extracted from the biological tissues. A part of their efficiency also derives from the sensitivity of their detectors, often enabling them to monitor traces of antimicrobials at low parts per billion equivalent to μg of residue/kg of tissue (ppb) level. Today they are the preferred methodologies for the confirmatory step in most of the analytical strategies. Ideally, a method set for drug residue confirmation should unequivocally establish the identity of the residue. During the regulatory control of nonprohibited drug residues, reliable quantification should be additionally carried out at an appropriate stage. Quantification is a mandatory procedure of the analytical residue method when the control has to reliably establish whether the residue concentration exceeds the MRL. On the contrary, for the regulatory control of residues of prohibited drugs, the unequivocal identification of the drug is necessary; its reliable quantification lies only at the second level even though, for chemical analytical methods, the quality of the identification is often correlated to the quality of the quantification. Among the chromatographic procedures, one can quote TLC as an efficient method when used for screening/identifying within a single assay several compounds from the same family of antimicrobials [97–104]. In its use of screening, this technique displays an acceptable resolutive property compared to microbiological methods, especially when these nonselective wide-range inhibitory methods significantly lack sensitivity as it is for the sulfonamides [103]. GC [105] has also been used for a long time since the 1970s for analyzing a large number of compounds in a variety of matrices but with complicated development and variable success for antimicrobial substances [106–110]. Antibiotics are, for most of them, nonvolatile, polar-to-very polar, and thermally unstable compounds precluding their correct analysis by GC. Chloramphenicol is the most often cited antibiotic to be analyzed by GC or GC-MS techniques [6,78,111–117].

HPLC is by far the most suitable and widely used technique for antimicrobial analysis [22–25]. It can be applied to the measurement of almost all the antimicrobial compounds. Considering its high resolutive properties, it is an appropriate technology for both the separation and the simultaneous quantification of closely related residues (parent drug and its possible metabolites). High precision in the measurement can easily be achieved with relative standard deviation (RSD) lower than any other technologies ever used for analyzing traces of antimicrobials in muscle tissue (1% < RSD < 15%). A wide choice in acidic–basic aqueous solvents combined with several organic ones (generally acetonitrile [ACN] or methanol [MeOH]) leads to numerous mobile phases enabling the control of the elution of antimicrobials on an extended library of column packings (normal phase, reverse phase, ion exchange phase, and mixed phases) and column geometries (conventional, narrow-bore, and micro-bore) [118]. Multiantimicrobial residue methods are then to be considered readily achievable in such a choice of analytical parameters. Still, some disadvantages need to be mentioned. The low sample throughput as compared to microbiological methods is probably the major drawback. The limited number of compounds to be separated within the same run should also be considered. Additionally, not all antimicrobials bear chromophore or fluorophore properties, making them undetectable by conventional HPLC detectors. Derivatization procedures before detection are generally employed at precolumn or postcolumn stage to recover the detectability by UV-visible or fluorescence detectors [119–122]. Another limitation derives from the implementation of mass spectrometers for which not all compositions of mobile phase can be applied to LC-MS instruments (phosphate buffers, etc.). Nevertheless, the LC-MS technology is readily becoming one of the most powerful LC techniques in the field of drug residue analysis. Research in new technologies in MS, notably LC-MS, with new quadrupolar MS detectors in the 1990s (LC-tandem MS, also called LC-triple quadrupolar MS), with ion-trap and hybrid-trap technologies (LC-ion trap MS and LC-quadrupolar ion trap MS), or with ToF and hybrid ToF technologies in the 2000s (LC-ToF-MS and LC-quadrupolar ToF-MS), has greatly benefited from the international need of protecting food quality. Monitoring nonvolatile and polar antimicrobials with a high degree of sensitivity and specificity is now possible, thanks to the large variety of interfaces developed for mass analyzers during the past 20 years: thermospray ionization (TSP), particle beam (PB), electrospray ionization (ESI), atmospheric pressure chemical ionization (APCI), and atmospheric pressure photochemical ionization (APPI). Several interesting review articles have been published recently, some of them dedicated to the advances in mass spectrometry analysis coupled to LC and GC separative systems [123–134], and others dealing with LC-MS analysis of drug residues in food [8,11,13,14,135–137].

10.3.2 Other Modes of Chemical Analysis

The other technologies potentially dedicated to polar nonvolatile compounds, such as many antimicrobials, were also investigated in the past 15 years. At the end of the 1990s, supercritical fluid chromatography (SFC) was reported to be a potentially innovative concept of separation evaluated on sulfonamide antimicrobials extracted from swine tissues [138–141]. Electrokinetic technologies with their major component, the capillary zone electrophoresis (CZE or CE), have focused on the high separation efficiencies possible with this separative mode [142–154]. Its three major variants are the micellar electrokinetic capillary chromatography (MEKC), the capillary isotachophoresis, and the electrochromatography (CEC) [155–161]. But none of these innovating analytical technologies reached a sufficient degree of reliability to trigger their implementation in the field of drug residue control in food. The capillary zone electrophoresis coupled to a mass spectrometer (CZE-MS or CE-MS) can become a potentially interesting technique as soon as the major technical problems related to the hyphenation between the electrophoretic nanocapillary tube and the atmospheric source of the mass spectrometer are reliably stabilized [162–169].

10.3.3 Sample Preparation for Liquid Chromatography

The sample preparation procedures developed for the antimicrobial residue analysis in muscle tissues by means of chromatographic instruments are all described with a similar process. The very first step always involves the extraction of the compounds of interest from the tissue by deproteinizing the sample using organic solvents such as ACN acidic or buffer aqueous solutions ethyl acetate (EtOAc), or through such as hydrochloric acid (HCl) or trichloroacetic acid (TCA) or phosphate buffer solutions (PBS) mixed with miscible organic solvents such as ACN or dichloromethane (DCM). The efficiency of the deproteinization depends on the degree and strength of the binding of the residues to the tissue proteins and on the adequateness of the deproteinizing solvent or mixture of solvents to be used. A centrifugation is often applied at this stage to separate the liquid phase containing the residues from the solid phase principally made of the precipitated proteins and other remaining substrates. The following step in sample preparation is the proper extraction from the biological liquid into a suitable solvent by partitioning process or liquid–liquid extraction (LLE). The choice of the solvent(s) depends on the polarity of the residue(s) of interest and must be adjusted (pH, volume, ionic strength, salt saturation, homogenization process, salting-out process, breaking emulsions) to give a maximum recovery of the residue(s). An acidic antimicrobial compound must be extracted by nonpolar solvents at low pH values where its acidic function is suppressed. A basic antimicrobial substance must be extracted by nonpolar solvents at high pH values where its basic function is neutralized. A neutral antimicrobial analyte is extracted whatever the pH of the sample/solvent is. An amphoteric antimicrobial molecule is extracted by nonpolar solvents at its specific pH of neutralization. Protein-bound antimicrobials can be released more efficiently by a stronger solvent at high acidic pH such as sulfuric acid (H_2SO_4), HCl, phosphoric acid (H_3PO_4) or at basic pH such as trisaminomethane (TRIS) and dithioerythreitol (DTE), enabling to cut the bind between the attached residue and the protein. Release of the residue is either as a neutral substance or more generally as an ionic species that can be neutralized by using an ion-pair extraction mode. The counterionic species neutralizes the ionicity of the residue and forms a neutral ion pair easily transferred into the organic phase. Further to this extraction step is carried out another possible centrifugation where separate liquid phases can be obtained leading to discard one phase and to retain of the other for further sample extract purification. Rather than neutralizing the antimicrobial residue by the ion pairing mode, it is also possible to make it react with a labeling substance to enhance the detectability of the residual antimicrobial by UV-visible or fluorescence detectors, in case it lacks chromophoric or fluorogenic properties. The last step in the sample preparation is generally the cleanup process. Purification is often necessary in biological matrices when too much endogenous substances still remain in the extract, leading to problems of chromatographic separation (column plugging or retention time variation) or interference in the detection. The liquid–liquid partition between immiscible solvents was the most common procedure. But now liquid–solid extraction with pouring the sample extract through sorbing material packed in a short column, also called the solid-phase extraction (SPE), is very often performed in residue analysis and the preferred cleanup for routine control. Most of the sorbents employed are derived from the different classes of chromatographic packings such as pure silica normal-phase sorbents, or alkyl-bonded silica such as 8-carbon-alkyl bonded silica stationary phase (C8), 18-carbon-alkyl bonded silica stationary phase (C18), and phenyl reverse-phase sorbents or ion-exchange material such as strong anionic or cationic exchange sorbents (SAX, SCX) or now mixed hydrophilic–lipophilic balanced cartridge (HLB), medium anion exchange cartridge (MAX), or medium cation exchange cartridge (MCX) mixed sorbents balanced with reverse-phase and ionic exchange properties. A number of reviews deal with this interesting SPE approach [170–175]. Further to the cleanup procedure is sometimes added another step to remove the last fat content from the extract by a liquid–liquid partition with apolar organic solvent (*n*-hexane or *iso*-octane). The very last sample preparation step is the transfer of the purified extract into the correct mixture of solvents, which should be closely similar to the mobile phase employed for the separative chromatography. Very often, this step requires a partial evaporation by rotary evaporator or a complete evaporation to dryness under gentle heating (40–50°C) combined with a gentle nitrogen flow. Direct extract is finally reconstituted in the mobile phase or in a closely related mixture of solvents before injection in the chromatograph.

10.3.4 Modes of Separation in Liquid Chromatography

Several modes of separation are applicable in liquid chromatography; the three major ones are the normal-phase mode (adsorption on silica stationary phases and elution driven by a mobile phase composed of mixtures of organic solvents), the reverse-phase mode (liquid partitioning between carbon-bonded stationary phases and a mobile phase mixing an aqueous buffer with an organic modifier), and the ion-exchange mode (cationic or anionic exchange with specific ion-bonded stationary phases). Because the heterogeneity in the chemical behavior of the antimicrobial substances is governed by their polarity or their weaker or stronger ionizability, different modes of separation can be considered depending on the compounds to be analyzed. Polar and highly ionizable antimicrobials can be successfully and specifically separated in the ion-exchange mode. Less polar antimicrobials can be separated in the normal-phase mode. Nonpolar antimicrobials or neutralized ionic antimicrobials are generally separated in the reverse-phase mode. Because it is easier to neutralize ionic compounds to chromatograph them together with the nonpolar ones, the reverse-phase mode has often been considered the preferred mode of separation for antimicrobial residues in food. Different techniques of neutralization have been investigated from the buffered pH displacement of the mobile phase for weak acidic and alkaline antimicrobials (penicillins, quinolones) to the use of ion-pairing agents for the more polar and ionic antimicrobials (aminoglycosides). All these strategies are generally employed to adjust the resolution and to improve the efficiency of the separation. It is convenient for chromatographing several compounds not only selected in the same family of antimicrobials, such as different neutral and amphoteric penicillins [176], but also for several families within the same runs, such as penicillins, cephalosporins, sulfonamides, and macrolides or aminoglycosides, quinolones, and tetracyclines [177]. The analytical parameters for the analyst to control the LC in the reverse-phase mode are numerous ranging from the chemistry of the mobile phase (pH, buffering, organic modifiers) to the chemistry of the stationary phase, including controlling the physical parameters of the pumping device by working in isocratic or gradient mode. The choice in the reverse stationary phase (C4, C8, C18, phenyl, etc.) can be relevant to resolve certain analytes even though C18-bonded silica stationary phase is actually the most common one. Polymeric reverse stationary phases (PLRP-S) are also sometimes useful when the ionic or highly polarizable antimicrobials may react with the silica of the column packing or with impurities contained in it. Some new brands of stationary phases appeared in the recent years such as the mixing of reverse-phase mode and weak ion-exchange mode or as the HLB stationary phase. A promising stationary phase for chromatographing highly polar antimicrobials is the hydrophilic liquid chromatography (HILIC) pseudonormal phase. Depending on the quality of detection mode, the separation can receive more or less attention. The science of chromatographic separation recently received less interest as soon as highly selective and sensitive detectors, such as new-generation mass spectrometers, have replaced the conventional UV-visible or diode array or fluorescence detectors.

10.3.5 Modes of Detection in Liquid Chromatography

The three commonly used detectors for analyzing antimicrobials are UV-visible detectors in a monowavelength version or now, since the 1990s, in a multiwavelength version, such as the DAD or the PDA detector; fluorescence detectors with a higher degree of specificity compared to the UV-visible ones while considering possible coextractive substances from the food matrix; and mass spectrometers with a large variety of combinations from single quadrupolar (SQ-MS) or triple quadrupolar (TQ-MS or tandem MS) instruments to ion-trap (IT-MS) devices and now to ToF-MS instruments, the most attractive and reliable instrument being the LC-TQ-MS with its high degree of reliability in identifying and quantifying drug residues in food matrix. Several hybrid instruments are now also commercially available, such as LC-Q/IT-MS or LC-Q/ToF-MS or LC-Q/Trap/Orbitrap-MS. To bring some antimicrobials to a more selective and sensitive analysis, derivatization by tagging the compounds of interest either with a chromogenic or with a fluorogenic labeling agent is also a useful alternative to extend the field of detection for some undetectable compounds such as for the aminoglycosides. Derivatization can also be proposed for some antimicrobials to enhance their signal in mass spectrometry by adjusting the antimicrobial(s) of interest to a correctly detectable range of mass and/or by improving its (their) capacity of ionization in the source of the mass spectrometer.

10.4 Applications of Chromatographic Methods to Antimicrobial Residues

Considering the wide range of antimicrobials potentially found as their residues in meat and particularly in muscle tissues of food-producing animals, a large number of methods have been developed during the last decades to monitor these compounds in accordance with the regulations enforced in various countries interested in food safety and food-producing animal husbandry.

The following paragraphs present some relevant methodological developments proposed in the field of antimicrobial residue analysis in the past 20 years. They are sorted according to the different families in an alphabetic order, although the more recent selective LC methods now open the field of multiresidue antimicrobial analysis, including more than 50 compounds and up to 100 compounds at once in some cases of LC-MS approaches.

10.4.1 Aminoglycosides

The aminoglycosides are broad-spectrum antibiotics produced by members of two types of bacterial genii, either *Streptomyces* sp. (streptomycin, neomycin) or *Micromonospora* sp. (gentamicin, amikacin). With the first streptomycin compound reported in 1944, this family of antibiotics is the second one discovered following the penicillins. Structurally, they belong to the chemical family of carbohydrates with two or more amino-sugars linked via a glycosidic bond to an aglycone moiety called the "aminocyclitol" ring (Figure 10.3). The aminoglycosides are divided into two subgroups: one small subgroup containing a streptamine moiety and the other larger one containing a 2-deoxystreptamine moiety. Streptomycin and dihydrostreptomycin belong to the streptamine subgroup. Neomycins, paromomycins, gentamicins, kanamycins, and apramycin belong to the deoxystreptamine subgroup. The deoxystreptamine subgroup is further structured in subclasses depending on the substituents attached to the deoxystreptamine moiety, leading to classes such as neomycins (neomycin A, neomycin B, neomycin C), or gentamicins (gentamicin C1, gentamicin C2, gentamicin C1a, gentamicin C2a). Useful reviews have been published on this family of antibiotics [18,19].

In food animal production the most commonly used aminoglycosides are gentamicin, neomycin, dihydrostreptomycin, and streptomycin. The aminoglycosides are not metabolized in the body but rather are bound to proteins (generally <30%) and excreted as the parent compound. Residues concentrate in the kidney (cortex) for more than 40% and in cochlear tissues, making these drugs both nephrotoxic and ototoxic. Muscle tissues are less subjected to aminoglycoside residue concentrations—more than 150 times lesser than in kidney tissues. Aminoglycosides are water-soluble, polar, and weakly basic compounds. They are stable at both high and low pH levels, and are heat resistant. A number of methods can be quoted for aminoglycoside residue analysis in tissues. For a screening strategy based on the antibacterial property of aminoglycosides, several bioassays have been developed such as microbiological inhibitory plate tests with different strains (*Bacillus subtilis*, *B. stearothermophilus*, *B. megaterium*, and *B. aureus*) [27–30,

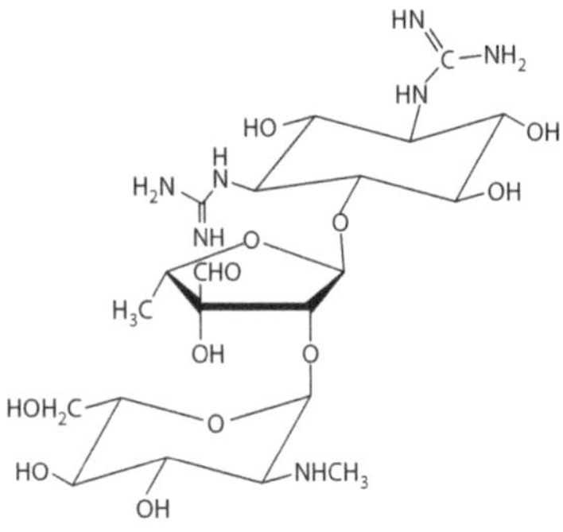

Streptomycin MW 581.57

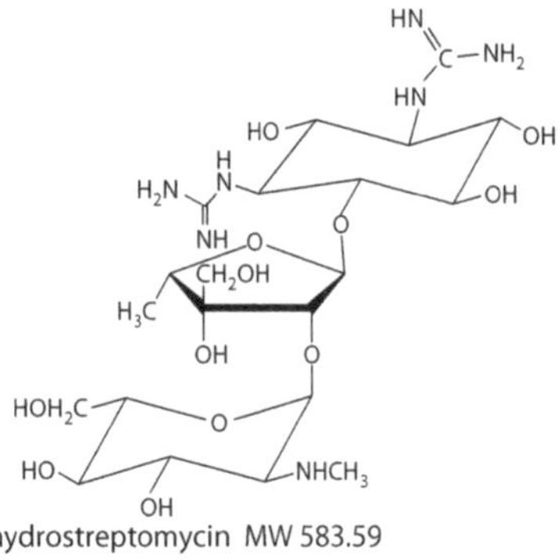

Dihydrostreptomycin MW 583.59

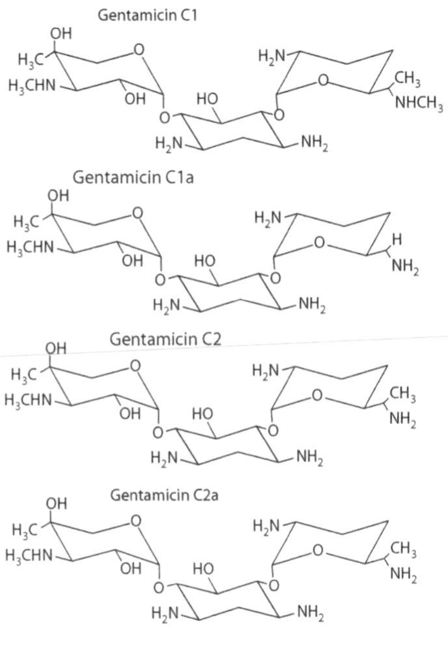

Gentamicins MW 694–723

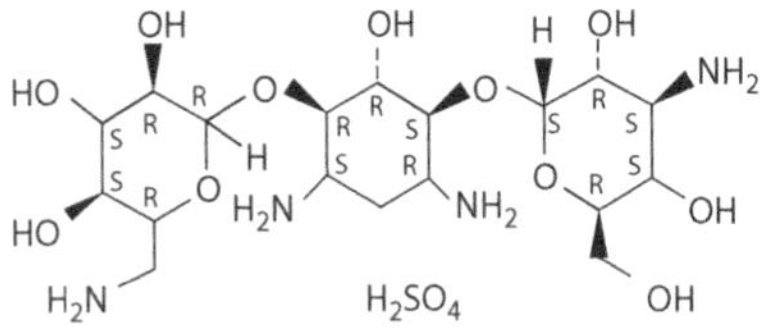

Kanamycin sulfate MW 582.57

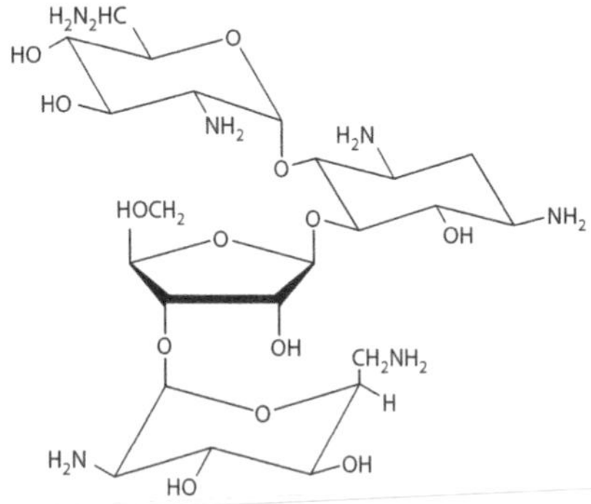

Neomycin B or Framycetin MW 614.64

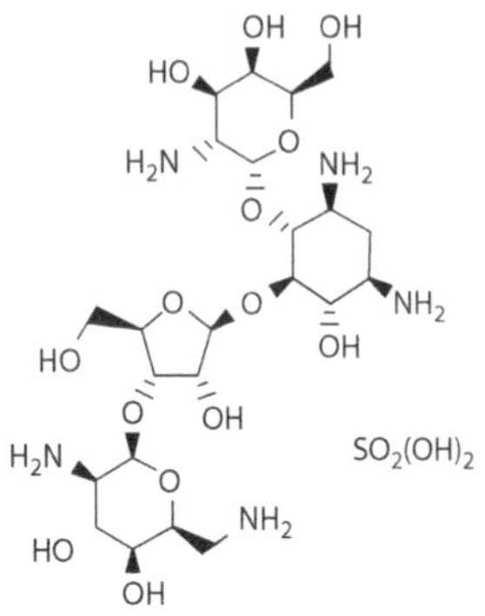

Paromomycin sulfate MW 713.71

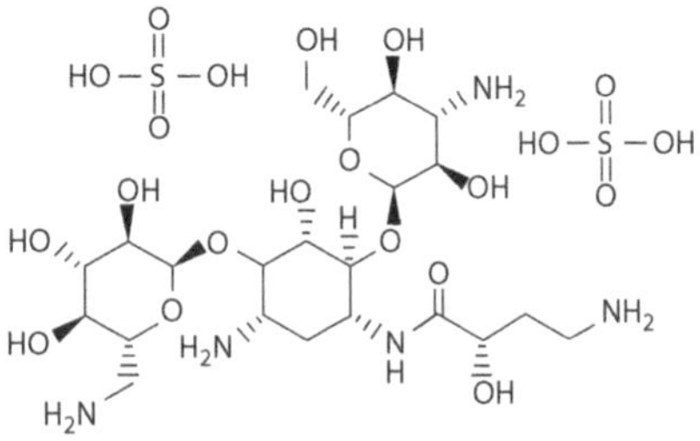

Amikacin sulfate MW 781.76

Figure 10.3 Structures of some aminoglycosides.

32–35,37,39,40,42,43]. Several microbiological tube test kits such as the Charm Farm Test Kit [57,58] or the Premi Test Kit [60,61] are also available. Some immunological bioassays are also dedicated to specific aminoglycosides: a fluorescence polarization immunoassay test (FPIA) for gentamicin [63]; and for streptomycin and dihydrostreptomycin, either the radioimmunological Charm II test [45–46], an ELISA test kit [71], or a SPR-BIA immunoassay [93]. TLC methods have also been proposed for some aminoglycosides in tissues and urines of swine and calves [178] or in the form of TLC-bioautography [46,55,56]. For their detection by analytical physicochemistry (Table 10.1), aminoglycosides lack chromophores and fluorophores; thus, derivatization is usually required. Because of their nonvolatility, HPLC procedures with fluorescent labeling are preferred; several of them were developed in the 1980s and 1990s [179–192]. Very few HPLC-UV methods have been proposed [193]. In addition, the ionicity of the aminoglycosides induces active use of the ion-pairing technique to achieve sufficient separation in the reverse-phase mode. Recently, in the 2000s, mass spectrometry (LC-MS techniques) has also become a convenient alternative to detecting and identifying these antibiotics [194–196]. But efficient and reliable extraction from animal tissue matrix still remains an issue in the analytical chemistry of aminoglycoside residues.

10.4.2 Amphenicols

The amphenicols are composed of three substances, namely chloramphenicol, thiamphenicol, and florfenicol. They are broad-spectrum bacteriostatic antibiotics. Chloramphenicol was the first of the family to be produced in 1947 from cultures of *Streptomyces venezuelae* and, due to its frequent use in veterinary and human medicine, was further produced synthetically starting from dichloro-acetic acid. This natural compound is rather unique in that it contains a nitrobenzene moiety. But, following extensive reports of adverse reactions, primarily aplastic anemia, in humans and other side effects after chloramphenicol treatment, the drug was found toxic enough to be banned world-wide in the end of the 1980s and early 1990s in veterinary practice and also in most human medicines. Thiamphenicol and, more recently, florfenicol, which have chemical structures similar to that of chloramphenicol (Figure 10.4), have been permitted as substitutes but have been licensed with effective tolerance limits at the 50–500 ppb level such as for the EU-MRL ones [1]. Because they are chemically and thermally rather stable [74,197–199], they can be readily analyzed by GC techniques and residues can be found in frozen stored meat for several months. The chromatographic methods for the analysis of chloramphenicol in edible animal products were reviewed thoroughly by Allen in 1985 [200]. But the ban of chloramphenicol changed significantly the strategy for the analysis of this family of compounds. In the 1990s, chloramphenicol became an important issue in residue monitoring because of the particular public health concern due to considerable levels of drug residues that may occur in edible animal products from illegally treated animals. To replace the conventional screening methods inefficient in detecting chloramphenicol residues at very low ppb levels, such as microbiological inhibitory plate tests [201] and microbiological tube tests [58] or TLC analysis [66,103,202] or TLC/bioautography [55], several sensitive and rapid immuno-enzymological test kits (ELISA) were developed specifically for the detection of chloramphenicol residues at <1.0 μg/kg in muscle tissues and other edible animal products [6,26,74,203,204].

At the confirmatory stage (Table 10.2) of the residue control strategy for chloramphenicol, it became mandatory in the 1990s to change from conventional GC [107,112,114,115] and HPLC-UV methods [6,77,197,199,205–208] to sensitive and unequivocally identifying techniques such as GC-MS(MS) [78,111,116] and LC-MS(MS) ones [78, 209–212]. Several mass spectrometric detection methods were also developed for the three amphenicols within a single method [113,195,213,214] and sometimes also including the major metabolite of florfenicol, the florfenicol amine [117,215].

Table 10.1 Summary of Literature Methods for Determination of Aminoglycoside Residues in Tissues Using Liquid Chromatography

Tissue	Analytes	Sample Treatment	Derivatization	LC Column Technique	Detection	Limit Range (ppm)	Year (Reference)
Kidney (cortex)	Gentamicin, kanamycin	Buffer extn, cell lysis, ion exchange SPE	Postcolumn: OPA	RPLC C18, HPSA ion pairing	Fluoresc. Ex:340 nm, Em:440 nm	—[a]	1983 (179)
Kidney, muscle	Neomycin, paromomycin	Buffer extn, heat deprot.	Postcolumn: OPA	RPLC C18, PSA ion pairing	Fluoresc. Ex:340 nm, Em:455 nm	0.5	1985 (180)
Kidney, muscle	Kanamycin	TCA deprot. ether defat, ion exchange SPE	Precolumn: OPA	RPLC C18, ion pairing	Fluoresc. Ex:335 nm, Em:440 nm	0.04	1986 (181)
Kidney, muscle	Streptomycin	PCA deprot., C8 SPE	Postcolumn: nihydrin	RPLC C18, OSA ion pairing	Fluoresc. Ex:400 nm, Em:495 nm	0.5	1988 (182)
Muscle	Gentamicin	Buffer extn, H_2SO_4 deprot., ion exchange, silica SPE	Precolumn: OPA	RPLC C18, HPSA ion pairing	Fluoresc. Ex:340 nm, Em:418 nm	0.2	1989 (183)
Muscle, liver, kidney, fat	Gentamicin	TCA deprot., ion exchange SPE	Postcolumn: OPA	RPLC C18, CPS ion pairing	Fluoresc. Ex:340 nm, Em:418 nm	0.5	1993 (184)
Muscle, liver, kidney	Streptomycin, DHS	PCA deprot., cation exchange SPE	Postcolumn: NQS	RPLC C18, HSA ion pairing	Fluoresc. Ex:347 nm, Em:418 nm	0.02	1994 (185)

(continued)

Table 10.1 (continued) Summary of Literature Methods for Determination of Aminoglycoside Residues in Tissues Using Liquid Chromatography

Tissue	Analytes	Sample Treatment	Derivatization	LC Column Technique	Detection	Limit Range (ppm)	Year (Reference)
Kidney, liver	Neomycin	TCA deprot., cation exchange SPE	Postcolumn: OPA	RPLC C18, CPS ion pairing	Fluoresc. Ex:340 nm, Em:440 nm	0.05	1995 (186)
Kidney, muscle	Streptomycin, DHS	TCA deprot., cation exchange SPE	Postcolumn: NQS	RPLC C18, OSA ion pairing	Fluoresc. Ex:375 nm, Em:420 nm	0.02	1997 (187)
Kidney, muscle	DHS	Acid deprot., ion exchange SPE, C8 SPE	Postcolumn: NQS	RPLC C18, HSA ion pairing	Fluoresc. Ex:375 nm, Em:420 nm	0.015	1998 (188)
Muscle, liver, kidney	Spectinomycin	TCA deprot., WCX SPE	Postcolumn: NQS	RPLC C18, ion pairing	Fluoresc. Ex:340 nm, Em:460 nm	0.05	1998 (189)
Feed	Amikacin, kanamycin, gentamicin, neomycin	HCl deprot.	Postcolumn: OPA	RPLC C18, ion pairing	Fluoresc. Ex:355 nm, Em:415 nm	0.2	1998 (190)
Muscle, kidney	Neomycin	PCA deprot., WCX SPE	Postcolumn: FMOCl	RPLC C4, ion pairing	Fluoresc. Ex:260 nm, Em:315 nm	0.12	1999 (191)
Muscle, liver, kidney	Gentamicin, neomycin	Buffer extn, TCA deprot., C18 SPE	Postcolumn: FMOCl	RPLC C18, HSA ion pairing	Fluoresc. Ex:260 nm, Em:315 nm	0.05, 0.10	2001 (192)

Tissue	Analytes	Extraction	Postcolumn: OPA	Column	Detection	LOD	Year
Muscle, liver, kidney, milk	Gentamicin netilmicin[IS]	Buffer extn, C18 SPE	Postcolumn: OPA	RPLC C18, HSA ion pairing	UV:330 nm	1.0	1995 (193)
Muscle, injection site	Gentamicin, neomycin, spectinomycin tobramycin[IS]	Buffer extn, TCA deprot, C18 SPE	—	RPLC C18	LC-MS^n esi+	—[a]	2000 (194)
Muscle, liver, kidney, fat	Gentamicin, tobramycin[IS]	Deprot, WCX SPE	—	RPLC C18, PFPA ion pairing	LC-tandem MS esi+	0.025	2003 (195)
Muscle, liver	Spectinomycin, apramycin, streptomycin, DHS, neomycin, gentamicin, kanamycin, paromomycin, amikacin, tobramycin, sisomycin	TCA extn, SAX SPE, HLB SPE	—	RPLC C18, HFBA ion pairing	LC-tandem MS esi+	0.015–0.040	2005 (196)

Note: ppm: parts per million equivalent to mg of residue/kg of tissue; OPA: ortho-phthalaldehyde; RPLC: reverse-phase liquid chromatography; HPSA: 1-heptanesulfonic acid; Fluoresc.: Fluorescence; Ex: excitation; Em: emission; PSA: 1-pentane sulfonic acid; TCA: trichloroacetic acid; SPE: solid-phase extraction; PCA: perchloric acid; OSA: octane sulfonic acid; CPS: dl-camphor-10-sulfonate; DHS: dihydrostreptomycin; extn: extraction; NQS: beta-naphtoquinone-4-sulfonate; HSA: hexane-1-sulfonic acid; WCX: weak cation exchange; FMOCl: 9-fluorenylmethylchloroformate; MS^n: ion trap; esi+: electrospray source in positive mode; PFPA: pentafluoropropionic acid; MSMS: triple quadrupole; MS: single quadrupole; SAX: strong anion exchange; HLB: hydrophilic lipophilic balance; HFBA: heptafluorobutyric acid.

[a] Not reported.
[IS] Internal standard.

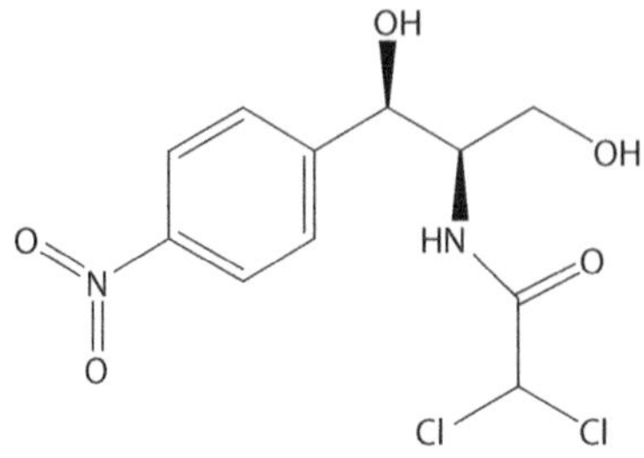

Chloramphenicol MW 323.13

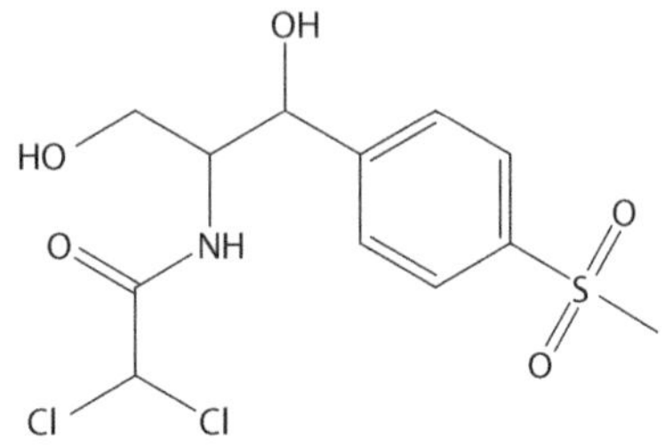

Thiamphenicol MW 356.22

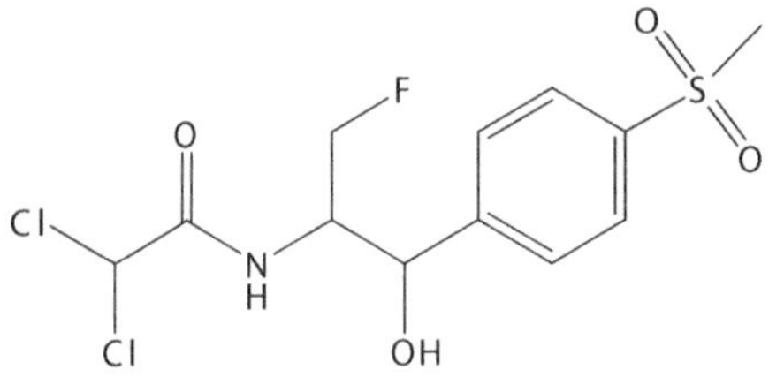

Florfenicol MW 358.21

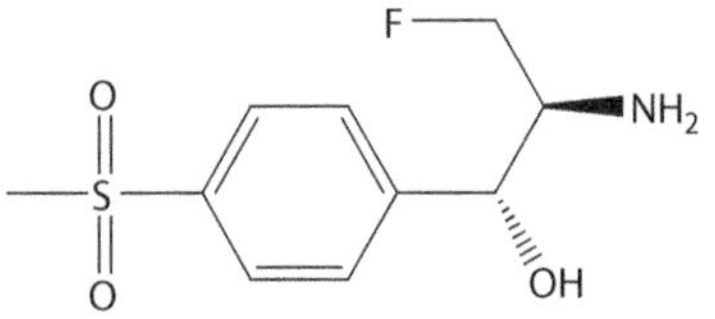

Florfenicol amine MW 247.08

Figure 10.4 Structures of amphenicols.

10.4.3 Beta-Lactams: Penicillins and Cephalosporins

The beta-lactams are antibiotics active against gram-positive bacteria. They consist, basically, of two classes of thermally labile compounds: penicillins and cephalosporins. Most of the commonly used beta-lactam antibiotics are produced semisynthetically either from the 6-APA for semisynthetic penicillins or from the 7-ACA for semisynthetic cephalosporins. Only the benzylpenicillin (or penicillin G), the 6-APA, and the phenoxymethylpenicillin (or penicillin V) are three naturally occurring penicillins extracted directly from molds of *Penicillium chrysogenum* and *Penicillium notatum*. Cephalosporin C and 7-ACA are the naturally occurring cephalosporins extracted from the molds of *Cephalosporium*. All other beta-lactam compounds are derived semisynthetically from these natural precursors. Penicillin G was the first among all antibiotics discovered accidentally in 1928 by Alexander Fleming, revealed by inhibiting the growth of *Staphylococcus aureus* strains.

Table 10.2 Summary of Literature Methods for Determination of Amphenicol Residues in Tissues Using Gas or Liquid Chromatography

Tissue	Analytes	Sample Treatment	Derivatization	Separation Technique	Detection	Limit Range (ppb)	Year (Reference)
Muscle	CAP, TAP	TCA extn, ion exchange SPE	Pyr, TMSA, TMSI, TMCS	GC	FID	2.5–2.5	1992 (107)
Muscle, kidney, liver	CAP	ACN/NaCl extn, hexane defat., EtOAc purif., IAC cleanup	Pyr, TMCS, HDMDS	GC	ECD	0.2	1995 (112)
Muscle	CAP, metaCAP[IS]	Water extn, hexane defat., silica SPE	Pyr, HDMDS, TMCS	GC	ECD	1.0	2002 (114)
Fish flesh, shrimp flesh	CAP	EtOAc extn, hexane defat.	TMSA,TMCS	GC	Micro-ECD	0.1	2005 (115)
Muscle	CAP	H_2O extn, silica SPE, toluene defat.	—	RPLC	UV-285 nm	1.5	1989 (205)
Muscle, kidney, liver	CAP	EtOAc extn, hexane/$CHCl_3$ defat.	—	RPLC	UV-278 nm	1.0	1991 (197)
Muscle	CAP PCB[IS]	H_2O extn, silica SPE, toluene defat.	—	RPLC	DAD-240–320 nm	1.0	1992 (6)
Muscle	CAP	MSPD, hexane defat.	—	RPLC	UV-290 nm	6.0	1997 (206)
Muscle, kidney, liver	CAP	Various extn tested (glucuronidase digestion for kidney), silica SPE, toluene defat.	—	RPLC	UV-285 nm	—[a]	1998 (207)
Muscle, kidney	CAP	EtOAc/ACN extn, acidic purif., heat, neutral., C_{18} SPE	—	RPLC	UV-270 nm	0.5	2003 (77)

(continued)

Table 10.2 (continued)   Summary of Literature Methods for Determination of Amphenicol Residues in Tissues Using Gas or Liquid Chromatography

Tissue	Analytes	Sample Treatment	Derivatization	Separation Technique	Detection	Limit Range (ppb)	Year (Reference)
Chicken muscle	CAP	EtOAc/anh. Na_2SO_4 extn, silica SPE, DCM defat., ACN/EtOAc partition., hexane/$CHCl_3$ defat.	—	RPLC	UV-278 nm	2.0	2003 (199)
Fish feed	CAP	ACN/water extn, C_{18} SPE	—	RPLC	UV-225 nm	200	2005 (208)
Muscle	CAP, metaCAP[IS]	EtOAc extn, NaCl partition., C_{18} SPE	Pyr, HDMDS, TMCS	GC	MS nci	0.5	1994 (111)
Shrimp flesh	CAP, CAP-d_5 [IS]	EtOAc extn, hexane defat., C_{18} SPE	MSTFA	GC	MSMS nci	0.1	2003 (78)
Shrimp flesh, crayfish flesh	CAP, metaCAP[IS], CAP-d_5 [IS]	ACN/NaCl extn, hexane defat., EtOAc purif., C_{18} SPE	BSA, n-heptane	GC	MS nci	0.07	2006 (116)
Muscle	CAP	ACN extn, $CHCl_3$ defat., C_{18} SPE	—	RPLC	MS esi–	0.5	2001 (209)
Shrimp flesh	CAP, CAP-d_5 [IS]	EtOAc extn, hexane defat., C_{18} SPE	—	RPLC	MS^n esi	0.1	2003 (78)
Muscle	CAP	EtOAc extn, NaCl partition., C_{18} SPE	—	RPLC	MSMS apci–	0.02	2003 (210)
Crab meat	CAP	EtOAc extn, MeOH/NaCl partition., heptane defat., EtOAc purif.	—	RPLC	MSMS esi–	0.25	2005 (211)

Tissue	Analyte	Extraction/purification		Chromatography	Detection	LOD	Year (ref.)
Muscle, liver, kidney	CAP	ACN extn, hexane defat.	—	RPLC	MS esi−	0.2–0.6	2005 (212)
Fish flesh	CAP, TAP, FLF	EtOAc extn, hexane defat., EtOAc purif., silica SPE	BSA	GC	MS esi	5	1996 (113)
Muscle, injection sites	CAP, FLF	—[a]	—	RPLC	MS^n esi+/esi−	—[a]	2003 (195)
Kidney	CAP, TAP, FLF	—[a]	—	RPLC	MSMS esi−	0.3	2006 (213)
Muscle	CAP, FLF	ACN/MeOH extn, hexane defat.	—	RPLC	MSMS esi+	3 and 2	2006 (214)
Shrimp flesh	CAP, TAP, FLF, FFA	EtOAc/ACN extn, hexane defat., C_{18} SPE, cation exchange SPE	—	RPLC	MS^n esi+	0.5	2003 (215)
Fish flesh, shrimp flesh muscle	FLF, FFA	EtOAc extn, hexane defat., EtOAc purif., MCX SPE	—	GC	micro-ECD	0.5 and 1.0	2006 (117)

Note: ppb: parts per billion equivalent to μg of residue/kg of tissue; CAP: chloramphenicol; TCA: trichloroacetic acid; Extn: extraction; SPE: solid-phase extraction; Pyr: pyridine; TMSA: *N*,O-bis(trimethylsilyl)acetamide; TMSI: *N*-trimethylsilylimidazole; TMCS: trimethylchlorosilane; GC: gas chromatography; FID: flamme ionization detector; ACN: acetonitrile; defat.: defattening; EtOAc: ethyl acetate; purif.: purification; IAC: immunoaffinity column; ECD: electron capture detector; HDMDS: hexadimethyldisilazane; RPLC: reverse-phase liquid chromatography; MSPD: matrix solid-phase dispersion; Neutral.: NaOH neutralization; DCM: dichloromethane; MS:single quadrupole; nci: negative ion chemical ionization; MSTFA: *N*-methyl-*N*-(trimethylsilyl)trifluoroacetamide; MSMS: triple quad-rupole; BSA: *N*,O-Bis(trimethylsilyl)acetamide; esi−: electrospray source in negative mode;. MS^n: ion trap; MeOH: methanol; partition.: aqueous partitioning; FLF: florfenicol; esi+: electrospray source in positive mode; FFA: florfenicol amine; MCX: medium cation exchange; HLB: hydrophilic–lipophilic balance.

[a] Not reported.
[IS]Internal standard.

Both classes of beta-lactams contain bulky side chain attached, respectively, to 6-APA or 7-ACA nuclei (Figure 10.5). The penicillin nucleus is a reactive unstable beta-lactam (four-membered) ring coupled to a thiazolidine (five-membered) ring to form the penam ring system. The cephalosporin nucleus differs from the penicillin nucleus by having a dihydrothiazine (six-membered) ring coupled to the beta-lactam ring to form the 3-cephem ring system and conferring a better stability to the molecule compared to the penam ring.

Several other beta-lactam antibiotic subfamilies have been discovered and synthetically modified, such as the cephamycins (extracted from *Actinomycetes*) or the clavulanic acid (a beta-lactamase inhibitor). The beta-lactam antibiotics act on bacteria by binding to peptidoglycan transpeptidase and inhibiting its normal cross-linking role in completing the cell wall synthesis in bacteria, causing the bacterial cell to undergo lysis and death. Beta-lactams are used at therapeutic levels in veterinary practice primarily to treat disease and prevent infection. Used at sub-therapeutic levels, they increase feed efficiency and promote growth of food-producing animals besides preventing spread of disease from the herd or flock of animals kept at a level of optimum productivity.

Penicillins are medium acidic polar drugs ($pK_{a\ (-COOH)}$ ranging from 2.4 to 2.7) and are relatively unstable in aqueous solutions. Their degradation is catalyzed by both acids and bases. They are also extremely susceptible to nucleophilic reactions in aqueous solutions. The optimum stability for the amphoteric penicillins (ampicillin, amoxicillin) with their second $pK_{a(NH_2)}$ ranging from 7.2 to 7.4, occurs at a pH that coincides with their respective isoelectric point; for the monobasic penicillins (penicillin G, cloxacillin, etc.), all of which have measured acid dissociation constants $pK_{a\ (-COOH)}$ less than 3, this generally occurs at a pH between 6 and 7. The presence of an unstable four-term ring in the beta-lactam moiety makes these compounds prone to degradation by heat and/or in the presence of alcohols. Penicillins are also readily isomerized in an acidic solution. The chemical instability of the beta-lactams, particularly of the penicillins, has important consequences for extraction and chromatography conditions. The relative stability of incurred penicillin G, amoxicillin and ampicillin residues in animal muscle tissues under various storage conditions has also been studied [216–223], demonstrating fairly good stability incurred in muscle tissues stored at –70 to –80°C (more than one year) compared to a lower stability at –20 to –30°C (less than 3 months [216,222]). Cephalosporins are somewhat more resistant to breakdown than many other beta-lactam compounds. But, as with the penicillins, they can be subjected to enzymatic degradation by specific beta-lactamases (cephalosporinases). New highly stable third generation of cephalosporins such as cefquinome and ceftiofur are available in veterinary practice since the 1990s.

Several useful reviews including methods for residue analysis in meat have been published on this family of antibiotics [9,11,13,14,16,20]. For screening the beta-lactams as residues in meat, many biological methods have been proposed. Most of them are based on the inhibitory properties of these antibiotics related to bacterial growth. They are readily detected in the presence of highly reactive specific strains such as *B. subtilis* [38,42,56,198], *B. stearothermophilus* [58,59,70,222,224], *B. megaterium* [39,40], or *Escherichia coli* [36]. The comparison of different screening methodologies has been published [32,42,60,61,225,226]. Often, these inhibitory screening methods, developed for beta-lactam at first, have been extended to large-scale multifamily methods, often called the "plate test methods." The first of these large screening tests, still officially in use in many countries in the 2000s, was the four-plate test method first described by Bogaerts and Wolf in 1980 [27] and revisited by Currie in 1998 [31]. Several other three-plate [227], five-plate [33,35], six-plate [34,37], seven-plate [29], and other multicombination/multistrain plate test [216,43] inhibitory methods have additionally been proposed either to extend the number of antimicrobials or to

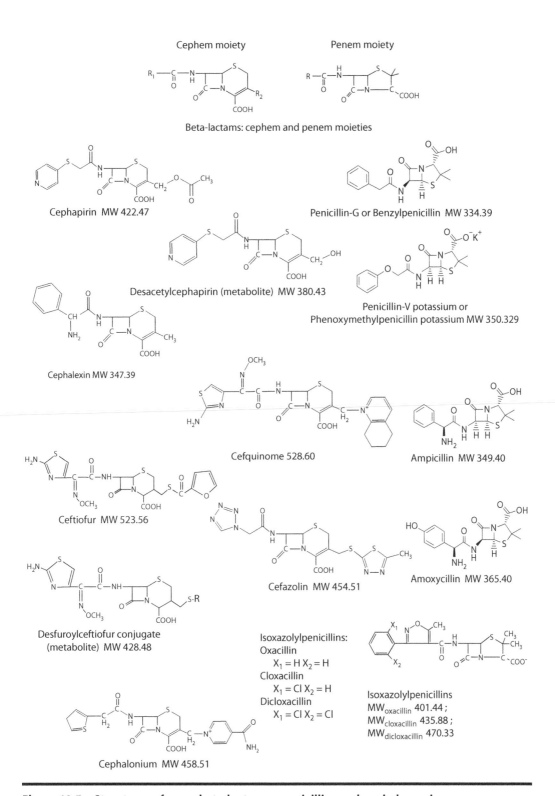

Figure 10.5   Structures of some beta-lactams—penicillins and cephalosporins.

improve the sensitivity for some of them. TLC-bioautography was a method employed also for penicillins in the 1980s and 1990s [46,51,53–56]. A TLC method with fluorescence detection was published for ampicillin in milk and muscle tissue [104]. Beta-lactam antibiotics are also easily detected by means of rapid biological test kits such as the Charm Farm Test [46, 57,58] or the more recently developed European Premi Test [59–61]. Very few immunological ELISA test kits have been developed for beta-lactams [70,76,228]. The radioimmunological Charm II Test kit is also adapted to screen beta-lactam residues in animal tissues [46]. Finally, a SPR-BIA immunological method for penicillins was proposed in 2001, but for detection in milk matrix [20, 229].

For the confirmation of the presence of beta-lactams in meat, several chromatographic methodologies have been investigated (Table 10.3). But, due to the unstable behavior of the beta-lactam ring and the thermal lability of these compounds, very few GC methods were developed in the 1990s [106,107], and fewer GC-MS methods [109]. Most of the developments required a liquid chromatographic separation. The UV detection was also quite problematic when considering extraction of the nonfluorescent and low chromophoric beta-lactams from biological matrices. Two key options were proposed: the first one was to improve the sample preparation by purifying as much as possible the biological extracts before detection at low UV wavelength (200–230 nm), where most of the beta-lactams display a good UV absorption [230–240]; the other option was to shift the UV detection by means of derivatizing reagents to higher UV wavelengths (>300 nm), where very few biological endogenous substances absorbed UV photons [176, 218,221,241–249], or, when possible, to move to fluorescent conditions by means of appropriate reagents as it was proposed for the two amphoteric penicillins: amoxycillin and ampicillin [236,250–252]. In the 2000s, most of the developments for beta-lactam residues in meat received the benefit of the mass spectrometry advances. From the methods with LC-MS thermospray (TSP) or PB sources and SQ detector employed in the 1990s [253–256], to those, in 2000s, with LC coupled to a tandem mass spectrometer (LC-MSMS) atmospheric pressure ionization (API) sources such as electrospray (ESI) or APCI and TQ detector (LC-tandem MS) [223,257–261] or ion-trap detector (LC-MS^n) [195,262], the confirmation of penicillin and cephalosporin residues became more easily achievable at the ppb levels requested by the regulations. However, because of the chemical behavior of beta-lactams, analytical chemistry poses a challenge for these substances present as residues in food from animal origin.

10.4.4 Macrolides and Lincosamides

The macrolide antibiotics are mostly produced by *Streptomyces* genera, characterized by a large macrocyclic lactone ring structure of 12–16-carbon-lactone ring to which several amino groups and neutral sugars are bound (Figure 10.6). Erythromycin was the first macrolide to be isolated in 1952 from *Saccharopolyspora erythraea*. It is active against gram-positive bacteria and mycoplasmas and is widely available in animal veterinary practice to treat respiratory diseases. Some macrolides (tylosin, spiramycin) have also been used as feed additives to promote growth. Other examples of macrolides include oleandomycin, tilmicosin, josamycin, and more recently, isovaleryltylosin. They are easily absorbed after oral administration and distribute extensively to tissues, especially the lungs, liver, and kidneys. They are weak bases slightly soluble in water but readily soluble in common organic solvents. Most of them are multicomponent systems containing lesser amounts of related compounds. For example, tylosin A, which is a commercialized compound, also contains small amounts of desmycosin (tylosin B), macrocin (tylosin C), and relomycin (tylosin D) [263]. In following the same principle, erythromycin A is commercialized with small but quantifiable

Table 10.3 Summary of Literature Methods for Determination of Beta-Lactam Residues in Tissues Using Gas or Liquid Chromatography

Tissue	Analytes	Sample Treatment	Derivatization	GC and LC Columns Technique	Detection	Limit Range (ppb)	Year (Reference)
Muscle	PenG, PenV, MTH, OXA, CLX, DCX, NAF	PBS/ACN extn, water removal, ether/PBS partition,, buffer/DEE partition,, buffer/ CH$_2$Cl$_2$ partition,, H$_3$PO$_4$/CH$_2$Cl$_2$ partition,, SCX SPE, buffer/ CH$_2$Cl$_2$ partition,, cyclohexane drying	Diazomethane	GC	NPD	3.0	1991 (106)
Muscle	PenG, PenV, CLX, DCX, AMP, AMOX	TCA extn, ion exchange SPE	Pyr, TMSA, TMSI, TMCS	GC	FID	5.0– 10.0	1992 (107)
Muscle	PenG, [13] C$_2$PenG[(IS)], PenV[(IS)]	PBS/ACN extn, water removal, ether/PBS partition,, buffer/ DEE partition,, buffer/CH$_2$Cl$_2$ partition,, H$_3$PO$_4$/CH$_2$Cl$_2$ partition,, SCX SPE, buffer/ CH$_2$Cl$_2$ partition,, cyclohexane drying	Diazomethane	GC	MS ei	3.0	1998 (109)
Muscle, liver, kidney	PenG	NaWO$_4$/H$_2$SO$_4$ extn-deprot,, Al$_2$O$_3$ SPE, C18 SPE	—	RPLC-C18	UV, 210 nm	5.0	1985 (230)
Muscle	PenG, CEP AMP[(IS)]	MSPD/MeOH extn, PBS extn	—	RPLC-C18	UV, 230 nm	5.0	1989 (231)
Muscle	PenG, PenV, CLX	ACN extn, H$_3$PO$_4$/CH$_2$Cl$_2$ partition,, ACN extn, hexane defat,, PBS extn	—	PLRP-S	UV, 210 nm	5.0	1992 (232)

(continued)

Table 10.3 (continued) Summary of Literature Methods for Determination of Beta-Lactam Residues in Tissues Using Gas or Liquid Chromatography

Tissue	Analytes	Sample Treatment	Derivatization	GC and LC Columns Technique	Detection	Limit Range (ppb)	Year (Reference)
Muscle	CEFT, DFCC	DTE extn, C18 SPE, SAX SPE, SCX SPE	Iodoacetamide	RPLC-C18	UV, 266 nm	100	1995 (233)
Muscle, kidney	PenG	Et₄NCI/ACN extn, LC fract. cleanup	—	RPLC-C18	UV, 215 nm	5.0	1998 (234)
Muscle, liver, kidney	DFCC	Et₄NCI/ACN extn, LC fract. cleanup	—	RPLC-C18, ion pairing DSF, DDSF	UV, 270 nm	—ᵃ	1998 (235)
Muscle, kidney, liver	AMOX, AMP, PenG, CLX, DFCC, DACEP	PBS extn, LC fract. cleanup	—	RPLC-C18	UV 210 nm, 270 nm	5.0	1998 (236)
Muscle, kidney, liver	PenG	PBS extn, ultrafiltration	—	RPLC-C4	DAD, 211 nm	40	2001 (237)
Muscle	DACEP, CEP, CEFQ, CEPX	Buffer extn, isooctane defat., C18 SPE	—	RPLC-C18	UV, 252 nm	12, 12, 9, 45	2002 (238)
Muscle, kidney	CEFT, DFCA	DTE extn, C18 SPE, SAX SPE, SCX SPE	Iodoacetamide	RPLC-C18	UV, 266 nm	100	2003 (239)
Muscle	PenG, PenV, CLX, DCX, OXA, NAF, AMP, AMOX, CEP, CEFT	ACN extn, hexane defat., C18 SPE	—	RPLC-C18	DAD, 21 nm and 228 nm and 270 nm	40	2004 (240)
Muscle, kidney, liver	PenG, PenV⁽ᴵˢ⁾	PBS extn, C18 SPE	Triazole-HgCl₂ and acetic anhydride	RPLC-C18, ion pairing THS	UV, 325 nm	5.0	1991 (241)

Matrix	Analytes	Extraction/Cleanup	Derivatization	HPLC	Detection	LOD	Year (Ref.)
Muscle, kidney, liver	PenG, PenV[(IS)]	PBS extn, C18 SPE	Triazole-HgCl$_2$ and acetic anhydride	RPLC-C18, ion pairing THS	UV, 325 nm	5.0	1992 (218)
Muscle	AMOX, CFDX	PBS extn, online dialysis, ion pair C18 SPE	Postcolumn: NaOH	RPLC-C18, ion pairing HTACl	UV, 260 nm	50, 200	1994 (242)
Muscle	AMP	PBS extn, hexane defat., C18 SPE	Triazole-HgCl$_2$ and acetic anhydride	RPLC-C8, ion pairing THS/TBA	UV, 325 nm	5.0	1996 (243)
Muscle, liver	AMOX, AMP	NaWO$_4$/H$_2$SO$_4$ extn-deprot., SCX SPE, PGC SPE	Triazole-HgCl$_2$ and acetic anhydride	RPLC-C8, ion pairing THS	UV, 325 nm	5.0	1997 (244)
Muscle, liver	PenG, PenV[(IS)]	PBS extn, C18 SPE	Triazole-HgCl$_2$ and acetic anhydride	RPLC-C18, ion pairing THS	UV, 325 nm	5.0	1997 (221)
Muscle	PenG, CLX, AMP, AMOX, PenV[(IS)]	PBS extn, t-C18 SPE	Triazole-HgCl$_2$ and acetic anhydride	RPLC-C18, ion pairing THS	UV, 325 nm, 340 nm	5.0	1998 (245)
Muscle	PenG, CLX, AMP, AMOX, PenV[(IS)]	PBS extn, t-C18 SPE	Triazole-HgCl$_2$ and acetic anhydride	RPLC-C18, ion pairing THS	UV, 325 nm, 340 nm	5.0	1998 (246)
Muscle	PenG, CLX, DCX, AMP, AMOX	MSPD extn, hexane defat, C18 SPE	Triazole-HgCl$_2$ and acetic anhydride	RPLC-C18	UV, 325 nm, 340 nm	20.0	1998 (247)
Muscle	PenG, NAF, CLX, DCX, OXA, AMP, AMOX	PBS extn, isooctane defat, HLB SPE	Triazole-HgCl$_2$ and benzoic anhydride	RPLC-C18	UV, 325 nm, 340 nm	8.0–11.0	1999 (248)

(continued)

Table 10.3 (continued) Summary of Literature Methods for Determination of Beta-Lactam Residues in Tissues Using Gas or Liquid Chromatography

Tissue	Analytes	Sample Treatment	Derivatization	GC and LC Columns Technique	Detection	Limit Range (ppb)	Year (Reference)
Muscle	PenG, PenV, CLX, DCX, OXA, NAF, AMP, AMOX	PBS extn, isooctane defat., C18 SPE	Triazole-HgCl$_2$ and benzoic anhydride	RPLC-C8, ion pairing THS/TBA	UV, 325 nm, 340 nm	3.0–10.0	1999 (176)
Muscle	AMP, AMOX, PenG, PenV, CLX, DCX, OXA, NAF	PBS extn, isooctane defat., C18 SPE	Triazole-HgCl$_2$ and benzoic anhydride	RPLC-C8, ion pairing THS/TBA	UV, 325 nm, 340 nm	3.0–10.0	2002 (249)
Muscle, kidney, liver	AMOX, AMP, PenG, CLX, DFCC, DACEP	PBS extn, LC fract. cleanup	Formaldehyde	RPLC-C18	FLD, Exc 358 nm, Em 440 nm	5.0	1998 (236)
Muscle	AMOX	PBS extn, C18 SPE	Formaldehyde	RPLC-C18	FLD, Exc 358 nm, Em 440 nm	5.0	2000 (250)
Muscle, kidney, liver	AMOX	TCA extn/ deprot.	Salicylaldehyde	RPLC-C18	FLD, Exc 358 nm, Em 440 nm	6.0–16	2000 (251)
Feed	AMOX, AMP	Water/ACN extn,	Formaldehyde	RPLC-C18	FLD, Exc 358 nm, Em 440 nm	5000	2003 (252)

Muscle	PenG	—[a]	—	RPLC-C18	MS pb-nci	—[a]	1990 (253)
Muscle, kidney	PenG / Nafcillin[(IS)]	Water/ACN extn, o-H_3PO_4 neutral., DCM purif.	—	RPLC-C18; ion pairing	MS esi–	25	1994 (254)
Muscle	OXA, CLX, DCX	EtOAc acidic extn, C18 SPE	—	TBARPLC-C18	MS pb-nci	40–50	1994 (255)
Muscle, kidney, liver	PenG, PenV, CLX, DCX	TCA/Acetone extn, C18 SPE	—	RPLC-C18	MS esi–	15	1998 (256)
Muscle, kidney, liver	PenG, PenV, DCX, OXA, NAF	Extn, C18 SPE, QMA ion exchange SPE	—	RPLC-C18; ion pairing DBAA	MSMS esi–	20	2001 (257)
Muscle, kidney, liver	PenG, PenV, CLX, DCX, AMP / Pheneticillin[(IS)]	Aqueous extn, C18 SPE	—	RPLC-C18	MSMS esi–	6.0–15.0	2003 (258)
Muscle, kidney, liver	AMOX	—[a]	—	RPLC-C18	MSMS esi–	—[a]	2003 (223)
Muscle, kidney, liver	PenG, PenV, CLX, DCX, OXA, NAF	Extn, C18 SPE, QMA ion exchange SPE	—	RPLC-C18	MSMS esi–	20	2003 (259)
Muscle, kidney, liver	PenG, PenV, CLX, DCX, OXA, NAF, PenG-d5[(IS)] / Nafcillin-d6[(IS)]	NaCl aqueous extn, and $NaWO_4$/H_2SO_4 deprot. for liver and kidney	—	RPLC-C18	MSMS esi–	2.0–10.0	2004 (260)

(continued)

Table 10.3 (continued) Summary of Literature Methods for Determination of Beta-Lactam Residues in Tissues Using Gas or Liquid Chromatography

Tissue	Analytes	Sample Treatment	Derivatization	GC and LC Columns Technique	Detection	Limit Range (ppb)	Year (Reference)
Kidney	PenG, AMOX, CEP, DACEP, DCCD, AMP, CFZ, OXA, CLX, NAF, DCX, / PenV[(IS)]	ACN extn, MSPD	–	RPLC-C18	MSMS esi+	1.0	2005 (261)
Muscle, injection sites	PenG	MeOH extn, evap., water dilution, C18 SPE, Evap., reconst.	–	RPLC-C18	MS^{n}esi+/esi–	–[a]	2003 (195)
Fish flesh	PenG, AMOX, CEPX, AMP, OXA, CLX, DCX	ACN Extn; hexane defat.; water/ ACN; hexane defat.	–	RPLC-phenyl	MS^{n}esi+/esi–	100–1000	2005 (262)

Note: ppb: parts per billion equivalent to µg of residue/kg of tissue; PenG: benzylpenicillin or Penicillin G; PenV: phenoxymethylpenicillin or Penicillin V; MTH: methicillin; OXA: oxacillin; CLX: cloxacillin, DCX: dicloxacillin, NAF: nafcillin; PBS: phosphate buffer solution; ACN: acetonitrile; SCX: strong cation exchange; SPE: solid-phase extraction; NPD: nitrogen specific detector; TCA: trichloroacetic acid; TMSA: N,O-bis(trimethylsilyl)acetamide; TMSI: N-trimethylsilylimidazole; TMCS: trimethylchlorosilane; UV: ultraviolet detection; AMP: ampicillin, RPLC: reverse phase liquid chromatography; defat.: defattening; DSF: decanesulfonate; DDSF: dodecylsulfate; DFCC: desfuroylceftiofur cysteine; DACEP: desacetylcephapirin; CEP: cephapirin; CEFQ: cefquinome; CEPX: cephalexin; AMX: amoxycillin; DCCD: desfuroylceftiofur cysteine disulfide; DFCA: desfuroylceftiofuracetamide; CEFT: ceftiofur; CFZ: cefazolin; CFDX: cefadroxil; Extn: extraction; MeOH: methanol; Et$_4$NCl: tetraethylammonium chloride; DTE: dithioerythritol; deprot.: deproteinisation; SAX: strong anion exchange; PGC: porous graphitic carbon; HLB: hydrophilic Lipophilic Balance; MCX: medium cation exchange; LC Fract.: LC fractionation; MSPD: matrix solid phase dispersion; EtOAc: ethyl acetate; purif.: purification; partition.: aqueous partitioning; PLRP-S: copolymeric reverse phase column; HTACl: hexadecyltrimethylammonium chloride; DCM: dichloromethane; DBAA: di-n-butylamine acetate; Pyr.: pyridine; FID: flamme ionization detector; FLD: fluorescence detection; DAD: UV diode array detection; MS:single quadrupole; ei: electron impact source; pb-nci: particle beam source in negative ion chemical ionization; MSMS: triple quadrupole; MS^n: ion trap; nci: negative ion chemical ionization; esi+: electrospray source in positive mode; esi–: electrospray source in negative mode.

a Not reported.

[(IS)]Internal standard.

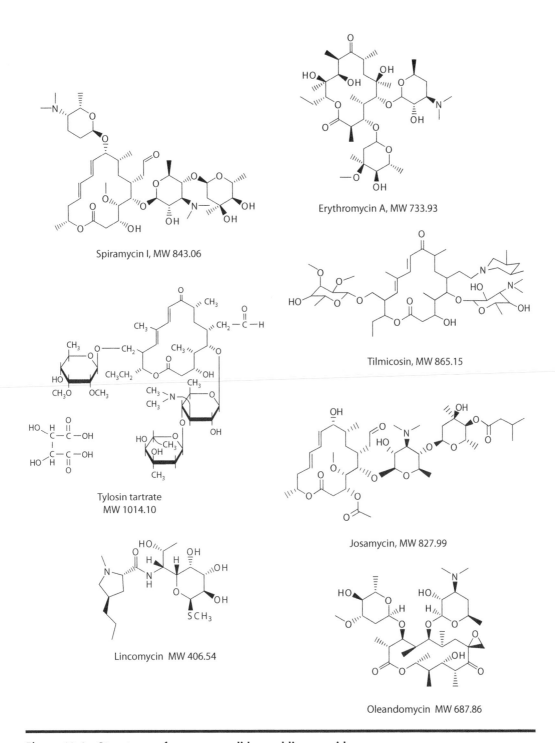

Figure 10.6 Structures of some macrolides and lincosamides.

amounts of erythromycin B and C [264]. Spiramycin I is generally found with spiramycins II and III [265]. Moreover, spiramycin can easily be degraded to neospiramycin under acidic conditions and the sum of neospiramycin and of spiramycin should be taken into account for spiramycin residue control in food as of regulation 1442/95/EC [266].

No comprehensive review specifically dedicated to macrolide residue analysis in food has been addressed in the recent years but several reviews dealing with antibiotic residue analysis attributed a part of their content to macrolide antibiotics [9,11,13–15,267].

For screening the macrolide antibiotics as their residues in meat, several biological methods have been proposed. Most of them are based on the inhibitory properties of these antibiotics related to bacterial growth. As for beta-lactam antibiotics, some of them (erythromycin, tylosin, tilmicosin, spiramycin, and lincomycin) are readily detected by microbiological inhibitory tube tests in the presence of specific strains such as *B. stearothermophilus*. It is the case of (the Charm Farm Test [58], the Premi Test [59]), or by microbiological inhibitory plate tests in multiplate configurations in the presence of specific strains such as *Micrococcus luteus* [29,37,43] or *B. megaterium* [40]. The comparison of different screening methodologies has been published to evaluate among several other antimicrobial families the response and detectability of macrolide residues in regard to regulated limits as for EU-MRLs [32,42,60,61]. TLC methods have also been proposed for some macrolides in the form of TLC-bioautography with *B. subtilis* as the revealing strain [46,55]. An ELISA test kit for macrolide was also investigated by Draisci et al. [72]. Because of the complex structure and composition of macrolides and of their relatively weak UV absorption, the development of chromatographic methods for their determination in foods and muscle tissues has been rather limited in the 1980s [268]. Nevertheless, several HPLC-UV methods (Table 10.4), which were able to cover at least two, tylosin and tilmicosin, and often more macrolides within the same multiresidue run of analysis, were proposed in the past 15 years [240,269–272]. Few but potentially interesting methods proposed HPLC with fluorescence detection after fluorescent labeling derivatization of some macrolide compounds such as josamycin, erythromycin, and oleandomycin [273,274]. Considering the mass spectrometric detector, it opens widely the field of macrolide detection and identification. From the methods with LC-MS thermospray (TSP) or PB sources and SQ detector employed in the 1990s [275,276], to those in 2000s with LC-MS API sources such as electrospray (ESI), APCI, TQ detector (LC-tandemMS) [214,223,262,277–281], or LC-MS^n [195], the confirmation of macrolide residues became more easily achievable at the ppb levels requested by the regulations.

10.4.5 Nitrofurans

Nitrofurans are synthetic compounds adapted from the 5-nitrofuran nucleus, all displaying a broad-spectrum activity. They are bacteriostatic antimicrobials acting by inhibition of some microbial enzymes involved in carbohydrate metabolism. They were widely used in veterinary medicine against gastrointestinal infections in cattle, pigs, and poultry. The major members of this antibacterial family are furazolidone, furaltadone, nitrofurazone, and nitrofurantoin (Figure 10.7). Following evidence of mutagenicity and genotoxicity of furazolidone in the late 1980s and early 1990s, legislation changed regarding the 5-nitrofuran nucleus compounds, which were all prohibited for use in food-producing animals in many countries. Listed in the Annex IV of EU Council Regulation 2377/90/EC [1,266,282], a minimum required performance limit (MRPL) for analytical methods developed for the residue control of nitrofurans in food has been set in the EU at 1.0 µg/kg [283]. Besides, the detection of nitrofurans is difficult in tissue matrices

Table 10.4 Summary of Literature Methods for Determination of Macrolide and Lincosamide Residues in Tissues Using Gas or Liquid Chromatography

Tissue	Analytes	Sample Treatment	Derivatization	LC Column Technique	Detection	Limit Range (ppb)	Year (Reference)
Muscle, kidney	Tylo	ACN extn, CH$_2$Cl$_2$ partition., ACN/ether defat.	—	RPLC-C18	UV, 278 nm	200	1982 (268)
Muscle, kidney	Tylo, Tilmico	ACN extn, C18 SPE	—	RPLC-C18	UV, 287 nm	20, 10	1994 (269)
Muscle, kidney, liver	Spira, Tylo, Josa, Kitasa, Mirosa	MPA/MeOH deprot., SCX SPE	—	RPLC-C18	UV, 232 nm and 287 nm	50	1998 (270)
Muscle	Spira, Neospira, Tylo, Tilmico	ACN extn, C18 SPE	—	RPLC-C18	UV, 232 nm (spira, neospir) and 287 nm (tylo, tilmico)	30, 25, 15, 15	1999 (271)
Muscle	Spira, Tilmico, Tylo, Josa, Kitasa, Erythro, Oleando	MPA/MeOH deprot., SCX SPE	—	RPLC-C18	UV, 232 nm and 287 nm	6–33 (S,T,J,K) and 400 (E,O)	2001 (271)
Muscle	Tylo, Spira, Neospira, Tilmico, Josa, Kitasa, Mirosa, Roxithro	ACN extn, hexane purif., HLB-SCX SPE	—	RPLC-C18	UV, 210 nm, 228 nm, and 287 nm	40	2004 (240)
Muscle, liver, kidney	Josa	PBS/ACN	Cyclohexane-1, 3-dione	RPLC-C18	FLD, Exc:375 nm, Em:450 nm	100	1994 (273)
Muscle, liver, kidney	Erythro, Oleando Roxithro[IS]	ACN extn, hexane purif., SCX SPE,	FMOC	RPLC-C18	FLD, Exc:260 nm, Em:305 nm	50–100	2002 (274)
Muscle	Tylo, Spira, Erythro	CHCl$_3$ extn, Diol SPE	—	RPLC-C18	MS pb-nci	20	1994 (275)
Muscle	Tylo, Spira, Erythro, Josa, Tilmico	CHCl$_3$ extn, Diol SPE	—	RPLC-C18	MS pb-pci and pb-nci	50	1996 (276)

(continued)

Table 10.4 (continued) Summary of Literature Methods for Determination of Macrolide and Lincosamide Residues in Tissues Using Gas or Liquid Chromatography

Tissue	Analytes	Sample Treatment	Derivatization	LC Column Technique	Detection	Limit Range (ppb)	Year (Reference)
Muscle	Erythro, Tylo, Tilmico	CHCl$_3$ extn, Diol SPE	—	RPLC-C18	MSMS esi+	25–40	2001 (277)
Muscle, liver, kidney	Spira, Tylo, Erythro, Timico, Josa Roxithro[(IS)]	Tris buffer extn, acetic acid/NaWO$_4$ deprot., HLB SPE	—	RPLC-C18	MSMS esi+	25–35	2001 (278)
Muscle	Erythro, Roxithro, Tylo, Tiamul	—	—	RPLC-C18	MS esi+	1.0–10	2002 (279)
Muscle	Tylo A Spiramycin[(IS)]	—[a]	—	RPLC-C18	MSMS esi	—[a]	2003 (223)
Feed	Spira, Tylo	MeOH/H$_2$O extn, HLB SPE, Dilut ACN/H$_2$O,	—	RPLC-C18	MSMS esi+	<1000	2003 (280)
Fish flesh	Erythro, ^{13}C$_2$-Erythro[(IS)]	ACN extn, hexane defat.	—	RPLC-C18	MSMS esi+	20	2005 (281)

Fish flesh	Tylo, Tilmico, Erythro, Linco	ACN extn, hexane defat., water/ACN, hexane defat.,	—	RPLC-phenyl	MS^n esi+/esi–	10	2005 (262)
Muscle	Erythro, Tylo, Tilmico, Josa, Kitasa, Linco, Clinda, Oleando	ACN/MeOH extn, hexane/ACN defat.	—	RPLC-C18	MSMS esi+	0.2–2.0	2006 (214)
Muscle, injection site	Tilmico, Linco	MeOH extn, Evap., Water dilution, C18 SPE, Evap., Reconst.	—	RPLC-C18	MS^n esi+/esi–	—ᵃ	2003 (195)

Note: ppb: parts per billion equivalent to µg of residue/kg of tissue; ACN: acetonitrile; Extn: extraction; defat.: defattening; RPLC: reverse phase liquid chromatography; UV: ultraviolet detection; Tylo: tylosin; Tilmico: tilmicosin; MPA: metaphosphoric adic; MeOH: methanol; deprot.: deproteinisation; SCX: strong cation exchange; HLB: hydrophilic Lipophilic Balance; SPE: solid-phase extraction; Spira: spiramycin; Josa: josamycin; Kitasa; kitasamycin; Mirosa: mirosamycin; Neospira: neospiramycin; Roxithro: roxithromycin; Linco: lincomycin; Oleando: oleandomycin; Tiamul: tiamulin (not a macrolide but a pleuromutiline); Clinda: clindamycin; purif.: purification; partition: aqueous partitioning; Neutral.: NaOH neutralization; Tris: tris(hydroxymethyl)aminomethane; EtOAc: ethyl acetate; PBS: phosphate buffer solution; FMOC: 9-fluoromethylchloroformate; MSPD: matrix solid phase dispersion; PLRP-S: copolymeric reverse phase column; FLD: fluorescence detection; MS:single quadrupole; pb: particle beam source; pci: positive ion chemical ionization; nci: negative ion chemical ionization; MSMS: triple quadrupole; MS^n: ion trap; esi⁺: electrospray source in positive mode; esi⁻: electrospray source in negative mode.

ᵃ Not reported.
(IS)Internal standard.

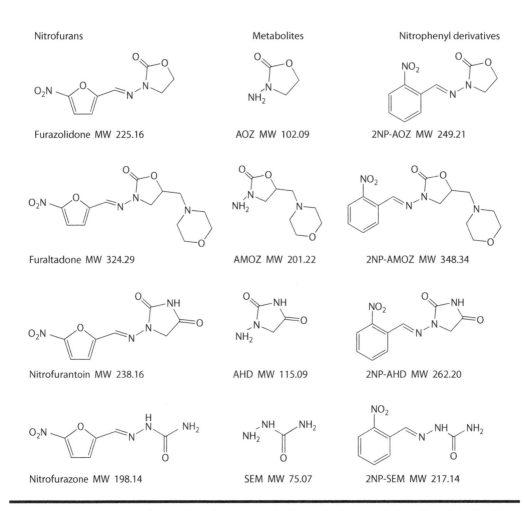

Figure 10.7 Structures of some nitrofurans and their metabolites and nitrophenyl derivatives.

because of extremely rapid metabolization *in vivo* (<1 day). But it was found in the late 1980s that furazolidone was able to bind extensively to proteinaceous tissues, and that acidic treatment could be efficiently applied to release a metabolized compound directly related to furazolidone, that is, the 3-amino-2-oxazolidinone (AOZ) [284]. Following this finding, the corresponding metabolites for the other nitrofurans were also extracted from tissues, confirming that parent compound monitoring was actually ineffective in food-producing animal tissues. The 3-amino-5-morpholino methyl-2-oxazolidinone (AMOZ) metabolite was extracted from protein-bound furaltadone, the 1-aminohydantoin (AHD) from nitrofurantoin, and the semicarbazide (SEM) from nitrofurazone (Figure 10.7).

Several reviews over the past 15 years covered the analysis of nitrofurans and their metabolites [11,13–15,267].

Owing to the aforesaid regulations and to the problematic issue in residue analysis of the rapidly metabolized nitrofurans, their monitoring in edible animal products has been, since the 2000, focused essentially on the LC-MS technology, avoiding the screening by microbiological methods or TLC [103, 202], which were found unadapted for nitrofuran compounds, and also disregarding the conventional GC or LC techniques that had been developed in the 1980s and 1990s [107, 206, 285–288]. Yet, in relation to the screening step, it can be mentioned an immunological screening of nitrofurans by means of two recent ELISA kits able to detect AOZ and AMOZ at very low ppb level in different matrices including muscle tissues (<1 μg/kg) [79]. ELISA screening for SEM and AHD is still under development. HPLC-UV methods have also been investigated in the early 2000 (Table 10.5), considering the four major nitrofuran metabolites as target residues in muscle tissue [289,290]. But, most of the recent developments on nitrofuran residue control in meat has been extensively supported by the LC-MS technology, including seldom methods with SQ detectors [291] and frequently with tandem MS or ion-trap MS detectors [214,284,292–297]. Apart from the four major nitrofurans, a fifth one, the nifursol, was employed as a feed additive against histomoniosis poultry infections. As a consequence of its ban in 2002 under regulation 1756/2002/EC [298], LC-MSMS methods have been proposed to monitor the nifursol metabolite, the 3,5-dinitrosalicylic acid hydrazide (DNSH), either specifically as a single metabolite monitored in poultry muscle tissues [299,300] or as the fifth residue monitored in a multinitrofuran metabolite analysis [301]. In addition, the problematic analysis of semicarbazide, which can be found as a protein-bound, but also as unbound, compound in meat and in many other food products (e.g., baby food), should be cited. It is not generated by nitrofurazone metabolization after a veterinary treatment, but produced by several external contaminations of meat products, one of which being the flour coating of poultry meat. Cereal flours can be treated with legal concentrations of azodicarbonamide (ADC)—a chemical substance easily transformed to biurea and finally to free semicarbazide [302–303].

10.4.6 Nitroimidazoles

Nitroimidazoles are antiprotozoals and bactericidal antimicrobials active against gram-negative and also many gram-positive bacteria. They are obtained synthetically and their structure is based on a 5-nitroimidazole ring while two protonic positions in N1 and C2 can be substituted by several groups to give different members of the family. Two methyl substitutions lead to the dimetridazole compound. A methyl and an ethanolic substitution give the metronidazole (Figure 10.8). The destructive action of nitroimidazoles takes place in the bacterial cell when the 5-nitro group is reduced by nitro-reductase bacterial proteins of the anaerobic bacteria, leading to free radicals or intermediate products, most of them cytotoxic for the bacteria.

Four major nitroimidazoles were commonly used in veterinary medicine or employed as feed additives in poultry prophylactic treatments against histomoniosis and coccidiosis: the dimetridazole, the metronidazole, the ronidazole, and the ipronidazole. The metronidazole is known to be easily transformed *in vivo* into both its alcoholic metabolite, the hydroxymetronidazole that is even more active against anaerobic bacteria and into its acidic metabolite, the acetylmetronidazole that bears no bactericidal activity anymore.

Dimetridazole can also experience an *in vivo* metabolization to give the hydroxydimetridazole metabolite, that is, the 2-hydroxymethyl-1-methyl-5-nitroimidazole (HMMNI). Ronidazole is extensively and quite exclusively metabolized to compounds without the intact nitroimidazole ring structure, also generating small amounts of the HMMNI too. Ipronidazole is also metabolized *in vivo* to its specific hydroxylated counterpart. Owing to their mutagenic, carcinogenic, and toxic properties toward eukaryotic cells, the nitroimidazoles have been prohibited for use in food-producing animals in the mid-1990s as enforced in the EU by three regulations—3426/93/EC, 1798/95/EC, and 613/98/EC [304–306].

Table 10.5 Summary of Literature Methods for Determination of Nitrofuran Residues in Tissues Using Gas or Liquid Chromatography

Tissue	Analytes	Sample Treatment	Derivati-zation	LC Column Technique	Detection	Limit Range (ppb)	Year (Reference)
Muscle, liver	Nitrofurazone, furazolidone, nitromide, sulfanitran	CHCl$_3$/EtOAc/DMSO Extn, Alumina SPE	—	RPLC-C18	Electrochemical, reductive mode	2.0–6.0	1989 (285)
Muscle, liver	Furazolidone, nitrofurazone, furaltadone, nitrofurantoin	ACN/EtOAc/DCM Extn, hexane defat., PBS dilution	—	RPLC-CN	UV, 365 nm	1.0	1989 (286)
Muscle	Furazolidone, nitrofurazone	TCA extn, ion exchange SPE	Pyr, TMSA, TMSI, TMCS	GC	FID	100, 50	1992 (107)
Kidney	Furazolidone, furaltadone, nitrofurantoin	ACN extn, C18 SPE, Silica SPE	—	RPLC-C18	UV-DAD, 359 nm, 370 nm	2.0–3.0	1995 (287)
Muscle	Furazolidone, nitrofurazone[IS]	ACN extn, Partition., Reconst.	—	RPLC-C18	UV-DAD, 365 nm	3.0	1995 (288)
Muscle	Furazolidone	MSPD	—	RPLC-C18	UV-DAD, 365 nm	3.5	1997 (206)
Liver	AOZ	PED, HCl hydrol., 2-NBA deriv, neutral., MAX SPE, HLB SPE	2-NBA	RPLC-C18	UV, 275 nm	—[a]	2002 (289)
Liver	Protein-bound AOZ	H$_2$O,MeOH,EtOH, EtOAc washings, HCl hydrol. extn, 2-NBA deriv, Neutral/PBS dilution, EtOAc partition., Tris dilution; MAX SPE; HLB SPE	2-NBA	RPLC-C18	UV, 275 nm / MSMS esi+	2.0, <1.0	2003 (290)

Matrix	Analytes	Sample preparation	Derivative	LC	Detection	LOD	Year (ref)
Muscle	AOZ, AMOZ, AHD, SEM	TCA extn, hexane defat, 2-NBA deriv, neutral/PBS dilut., C18 SPE, CHCl$_3$ extn, H$_2$O dilut.	2-NBA	RPLC-C18	MS esi+	0.5	2004 (291)
Muscle, liver	Protein-bound AOZ	MeOH, EtOH, EtOAc washings, HCl hydrol. extn, 2-NBA deriv, neutral/PBS dilution, EtOAc partition., ACN/H$_2$O reconst.	2-NBA	RPLC-C18	MS TSP+	0.5	1997 (284)
Muscle	AOZ, AMOZ, AHD, SEM, 4NBA-SEM[IS]	HCl hydrol. extn, 2-NBA deriv, neutral, PBS dilution, C18 SPE	2-NBA, 2-NBA-d4	RPLC-C18	MSMS esi+	0.5–5.0	2001 (292)
Muscle	AOZ, AMOZ, AHD, SEM	HCl hydrol. extn, 2-NBA deriv, neutral/PBS dilution, hexane defat., HLB SPE	2-NBA	RPLC-C18	MSMS esi+	0.2–0.5	2003 (293)
Muscle	AOZ, AMOZ, AHD, SEM, AOZd4[IS], AMOZd5[IS]	HCl hydrol. extn, 2-NBA deriv, Neutral/ PBS dilution, EtOAc partition., reconst. ACN/Acetic acid	2-NBA	RPLC-C18	MSMS esi+	0.1–0.5	2004 (294)
Muscle	AOZ, AMOZ, AHD, SEM, AOZd4[IS], AMOZd4[IS] / AHDd4[IS], SEMd4[IS]	HCl hydrol. extn, 2-NBA deriv, neutral/ PBS dilution, EtOAc partition., reconst. H$_2$O, hexane defat, C18 SPE, reconst. H$_2$O/ACN	2-NBA, 2-NBA-d4	RPLC-C18	MSMS esi+	0.2	2005 (295)
Muscle, Egg	AOZ, AMOZ, AHD, SEM, AOZd4[IS], AMOZd5[IS] / ^{13}C^{15}N$_2$SEM[IS]	HCl hydrol. extn, 2-NBA deriv, Neutral/PBS dilution, EtOAc partition., reconst. H$_2$O/ACN	2-NBA	RPLC-C18	MSMS esi+	0.3	2005 (296)

(continued)

Table 10.5 (continued) Summary of Literature Methods for Determination of Nitrofuran Residues in Tissues Using Gas or Liquid Chromatography

Tissue	Analytes	Sample Treatment	Derivatization	LC Column Technique	Detection	Limit Range (ppb)	Year (Reference)
Shrimp flesh	AOZ, AMOZ, AHD, SEM, AOZd4[IS], AMOZd5[IS]	HCl hydrol. extn, 2-NBA deriv, neutral/ PBS dilution, EtOAc partition., reconst. acetic acid	2-NBA	RPLC-C18	MSMS esi+	<0.5	2006 (297)
Muscle	Nifuroxazide	ACN/MeOH extn, Hexane/ACN defat.	—	RPLC-C18	MSMS esi+	0.2	2006 (214)
Muscle, liver	DNSH SH[IS]	HCl hydrol. extn, 2-NBA deriv, ammonia neutral., reconst. ACN	2-NBA	RPLC-C18	MSMS esi−	0.05	2005 (299)
Muscle, liver	DNSH, HBH[IS]	HCl hydrol. extn, 2-NBA deriv, neutral/ PBS dilution, EtOAc partition., reconst. ACN/NH$_4$OH	2-NBA	RPLC-C18	MSMS esi+	0.10, 0.06	2005 (300)
Muscle	AOZ, AMOZ, AHD, SEM, DNSAH, AOZd4[IS], AMOZd5[IS], 13C3AHD[IS], 13C15N$_2$SEM[IS], SH[IS]	HCl hydrol. extn, 2-NBA deriv, Neutral/ PBS dilution, EtOAc partition., reconst. MeOH/CH$_3$ONH$_4$	2-NBA	RPLC-C18	MSMS esi+	0.08–0.20	2007 (301)
Flour-coated meat	AOZ, AMOZ, AHD, SEM, AOZd4[IS], AMOZd5[IS] ADC, Biurea	HCl hydrol. extn, 2-NBA deriv, EtOAc partition., reconst. ACN/H$_2$O/acetic acid	2-NBA	RPLC-C18	MSMS esi+	—[a]	2004 (302)

Muscle, liver, hen eye	Nitrofurazone, $^{13}C^{15}N_2$NFZ[IS], total SEM, bound SEM $^{13}C^{15}N_2$SEM[IS]	For muscle, liver: Solvent washings, HCl hydrol. extn, 2-NBA deriv, Neutral/ PBS dilution, EtOAc partition., reconst. MeOH/H_2O	For muscle: 2-NBA	MSMS esi+	<0.5	2005 (303)
		For eye: EtOAc extn, reconst. ACN, hexane defat., reconst. MeOH/H_2O	For eye tissue: –		–ᵃ	

Note: ppb: parts per billion equivalent to μg of residue/kg of tissue; EtOAc: ethyl acetate; DMSO: dimethyl sulfoxyde; Extn: extraction; RPLC: reverse phase liquid chromatography; SPE: solid-phase extraction; ACN: acetonitrile; DCM: dichloromethane; UV: ultraviolet detection; Pyr.: pyridine; TMSA: N,O-bis(trimethylsilyl)acetamide; TMSI: N-trimethylsilylimidazole; TMCS: trimethylchlorosilane; FID: flame ionization detection; MSPD: matrix solid phase dispersion; AOZ: 3-amino-2-oxazolidinone; PED: protease enzyme digestion; 2-NBA: 2-nitrobenzaldehyde; Neutral.: NaOH neutralization; MAX: medium anion exchange; HLB: hydrophilic Lipophilic Balance; MeOH: methanol; EtOH: ethanol; PBS: phosphate buffer solution; MSMS: triple quadrupole; esi+: electrospray source in positive mode; AMOZ: 3-amino-5-morpholinomethyl-2-oxazolidinone; AHD: 1-aminohydantoin; SEM: semicarbazide; AOZd4: 4 times deuterated AOZ; AMOZd5: 5 times deuterated AMOZ; $^{13}C^{15}N_2$SEMd4[IS]; $^{13}C^{15}N$ isotopic SEM; partition.: aqueous partitioning; Reconst.: aqueous solvent reconstitution; DNSH, DNSAH: 3,5-dinitrosalicylic acid hydrazine; esi–: electrospray source in negative mode; SH: salicylic acid; HBH: 4-hydroxy-3 5-dinitrobenzoic acid hydrazide; NFZ: nitrofurazone; ADC: azodicarbonamide; DAD: diode array detection; defat.: defattening; deprot.: deproteinisation; Hydrol.: hydrolytic extraction; MS: single quadrupole; MSⁿ: ion trap; nci: negative ion chemical ionization; pci: positive ion chemical ionization; purif.: purification; TSP: thermospray.

ᵃ Not reported.
[IS]Internal standard.

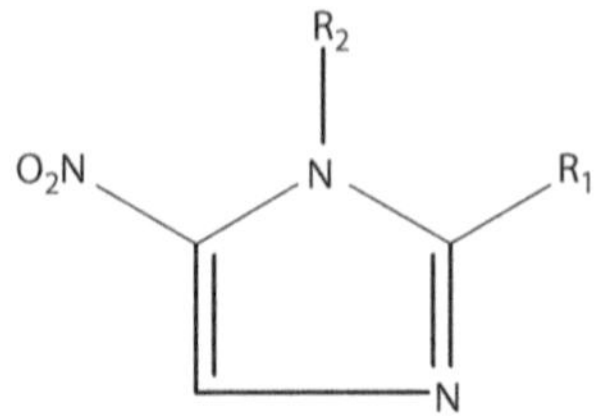

Compound	R$_1$	R$_2$	MW
Dimetridazole	CH$_3$	CH$_3$	141.13
Ronidazole	CH$_2$OOCNH$_2$	CH$_3$	200.15
Ipronidazole	CH(CH$_3$)$_2$	CH$_3$	169.18
Metronidazole	CH$_3$	C$_2$H$_4$OH	171.15
Hydrolylated metabolites			
HMMNI	CH$_2$OH	H	158.14
MNZOH	CH$_2$OH	C$_2$H$_4$OH	202.19
IPZOH	C(CH$_3$)$_2$OH	H	186.19

Figure 10.8   Structures of some nitroimidazoles and their hydroxylated metabolites.

Several reviews have been published on nitroimidazole residues over the past 10 years [13–15]. Recently, an immunological method has been investigated with production of polyclonal antibodies against a range of nitroimidazoles: metronidazole, ronidazole, dimetridazole, and ipronidazole, and their hydroxy metabolites [75]. An ELISA test kit for the screening of several anticoccidials including the nitroimidazoles in chicken muscle and eggs was further developed with analytical limits ranging from 2 μg/kg (dimetridazole) to 40 μg/kg (ipronidazole) [84]. Not many HPLC methods were published during the late 1990s and early 2000s for nitroimidazole monitoring in animal food matrices (Table 10.6) [307–312] together with rare LC-MS methods [313–314]. More recently, several up-to-date GC-MS [110,315] and LC-MS(MS) methods [316–320] were proposed following the ban of nitroimidazoles and the request for higher level of sensitivity and for unequivocal identification for the confirmatory methods employed in the residue control for food safety. This request is legally stated in the EU Decision 657/2002/EC regarding the criteria for performance of the official analytical methods for residue monitoring in food from animal origin [4].

10.4.7   Quinolones

Quinolones are broad-spectrum synthetic antimicrobial compounds used in the treatment of livestock and in aquaculture. They act against bacteria by inhibiting the DNA gyrase—a key component in DNA replication. They are a relatively new family of antibacterials synthesized from 3-quinolonecarboxylic acid, the carboxylic group at position 3 providing them with acidic properties (Figure 10.9). Nalidixic acid, oxolinic acid, and flumequine represent the oldest subgroup of compounds from the first generation of acidic quinolones, generally called the pyridonecarboxylic acid (PCA) antibacterials. Oxolinic acid is more restricted to treating fish diseases such

Table 10.6 Summary of Literature Methods for Determination of Nitroimidazole Residues in Tissues Using Gas or Liquid Chromatography

Tissue	Analytes	Sample Treatment	Derivatization	LC Column/ Technique	Detection	Limit Range (ppb)	Year (Reference)
Muscle, liver	DMZ, HMMNI, MNZ, MNZOH	EtOAc extn, evap., HCl/EtOAc dilution, hexane defat., aqueous neutral., C18 SPE, evap., reconst.	—	RPLC-C18	UV-DAD	—	1992 (307)
Muscle, liver	DMZ, MNZ	ACN extn, SPE cleanup	—	RPLC-C18	UV-DAD 450 nm	2.0, 5.0	1995 (308)
Muscle, liver	DMZ	ACN extn, SPE cleanup	—	RPLC-C18	UV-DAD 450 nm	2.0	1996 (309)
Muscle, liver	DMZ, RNZ, HMMNI	ACN extn, NaSO₄ deprot., acetic acid dilution, SCX SPE, reconst. PBS	—	RPLC-C18	UV 315 nm MS apci+	0.5 0.1–0.5	1998 (310)
Muscle	DMZ, HMMNI, RNZ	ACN extn, EtOAc/hexane partition., silica SPE	—	RPLC-C18	UV 315 nm	0.5	1999 (311)
Fish flesh	MNZ, MNZOH TNZ[IS]	ACN extn, C18 SPE	—	RPLC-C18	UV 325 nm	1.5–2.0	2000 (312)
Muscle, liver	DMZ, HMMNI, IPZ, IPZOH	EtOAc extn (DMZ-DMZOH), C18 SPE, reconst. benzene extn (IPZ-IPZOH), reconst.	—	RPLC-C18	MS TSP+	2.0	1992 (313)

(continued)

Table 10.6 (continued) Summary of Literature Methods for Determination of Nitroimidazole Residues in Tissues Using Gas or Liquid Chromatography

Tissue	Analytes	Sample Treatment	Derivatization	LC Column/ Technique	Detection	Limit Range (ppb)	Year (Reference)
Muscle, egg	DMZ, DMZ-d3(IS)	DCM extn, silica SPE, reconst. MeOH/H$_2$O, hexane defat.	—	RPLC-C18	MS TSP+	<5.0	1997 (314)
Muscle, liver	DMZ, MNZ, RNZ, HMMNI, MNZOH, IPZOH, TNZ, IPZ-d3(IS), IPZOH-d3(IS), HMMNI-d3(IS), RNZ-d3(IS), DMZ-d3(IS)	Protease+PBS hydrolysis extn, PBS partition., silica SPE, deriv	BSA-50	GC	MS nci	0.6–2.8, 5.2(IPZOH)	2001 (110)
Retina, plasma	DMZ, MNZ, RNZ, HMMNI, MNZOH, IPZOH	Protease+PBS hydrolysis extn, PBS partition., hexane defat., silica SPE, deriv	BSA-50	GC	MS nci	0.5–4.0	2002 (315)
Muscle	DMZ, MNZ, RNZ, HMMNI	EtOAc extn, Hexane/CCl$_4$/ Formic acid partition.,	—	RPLC-C18	MS esi+	—[a]	2000 (316)
Muscle, liver	DMZ, RNZ, MNZ	ACN extn, NaCl/DCM partition., H$_2$O dilut., hexane defat., silica SPE	—	RPLC-C18	MS esi+	2.0–4.0	2001 (317)

Muscle, liver	DMZ, RNZ, MNZ, IPZ, HMMNI, TNZ, RNZ-d3[(IS)], DMZ-d3[(IS)]	ACN extn, NaSO$_4$ deprot., acetic acid dilution, SCX SPE, reconst. PBS/ACN	—	RPLC-C18	MSMS esi+	2.0–5.0	2004 (318)
Muscle, liver	DMZ, RNZ, MNZ, IPZ, HMMNI, MNZOH, IPZOH, IPZ-d3[(IS)], IPZOH-d3[(IS)], HMMNI-d3[(IS)], RNZ-d3[(IS)], DMZ-d3[(IS)]	—[a]	—	RPLC-C18	MSMS apci+	—[a]	2004 (319)
Muscle, liver	RNZ, DMZ, MNZ, HMMNI	2xACN extn with NaCl partition., partial ACN evap., filtration	—	RPLC-C8	MSMS esi+	0.1–0.3	2006 (320)

Note: DMZ: dimetridazole; MNZ: metronidazole; EtOAc: ethyl acetate; Extn: extraction; RPLC: reverse phase liquid chromatography; MNZOH: hydroxymetronidazole; Reconst.: aqueous solvent reconstitution; Reconst.: reconstitution prior to injection; ACN: acetonitrile; SPE: solid-phase extraction; RNZ: ronidazole; deprot.: deproteinisation; SCX: strong cation exchange; PBS: phosphate buffer solution; partition.: aqueous partitioning; TNZ: tinidazole; IPZ: ipronidazole; IPZOH: hydroxyipronidazole; BSA-50: N,O-bis(trimethylsilyl)acetamid; esi+: electrospray source in positive mode; MSMS: triple quadrupole; DAD: diode array detection; DCM: dichloromethane; defat.: defattening; DMSO: dimethyl sulfoxyde; DMZOH: hydroxydimetridazole; esi–: electrospray source in negative mode; HLB: hydrophilic Lipophilic Balance; MeOH: methanol; MS:single quadrupole; MS^n: ion trap; MSPD: matrix solid phase dispersion; nci: negative ion chemical ionization; Neutral.: NaOH neutralization; ppb: parts per billion equivalent to µg of residue/kg of tissue; purif.: purification; TSP: thermospray; UV: ultraviolet detection.

[a] Not reported.
[(IS)] Internal standard.

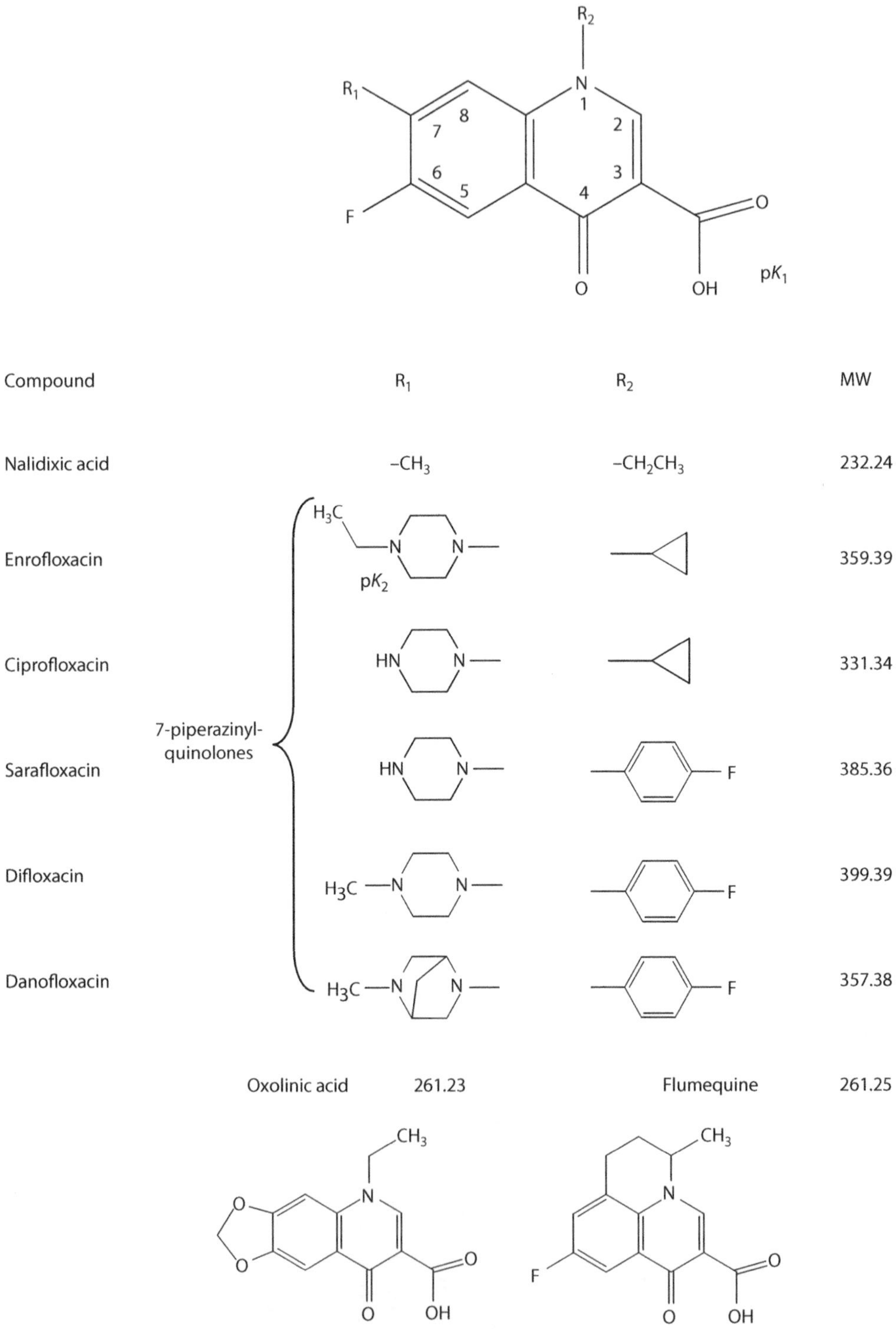

Compound	R_1	R_2	MW
Nalidixic acid	$-CH_3$	$-CH_2CH_3$	232.24
Enrofloxacin			359.39
Ciprofloxacin			331.34
Sarafloxacin			385.36
Difloxacin			399.39
Danofloxacin			357.38

7-piperazinyl-quinolones

Oxolinic acid	261.23	Flumequine	261.25

Figure 10.9 Structures of some quinolones and fluoroquinolones.

as killing the bacteria causing furunculosis in salmon. Nalidixic acid is not used in veterinary medicine. Representatives of the second generation are the fluoroquinolones, such as enrofloxacin, ciprofloxacin, or sarafloxacin, with higher potency in regard to the first generation. They bear a piperazinyl moiety in the C-7 position which gives these amino-quinolones some additional basic properties, and depending on the chemical environment leads to zwitterionic, cationic, or anionic behaviors in aqueous solution. At pH of 6–8, they hold poor water solubility due to their amphoteric characteristics. However, they are readily soluble in polar organic solvents and also in acidic or basic aqueous/organic solutions. Their extraction from biological matrices needs to be considered taking into account their incell intranuclear accumulation. They all display a native fluorescence, which is of particular interest when ppb residual quantities are detected from biological tissues and fluids of food-producing animals. Other fluorescent fluoroquinolones employed in veterinary medicine are the danofloxacin, the difloxacin, and the marbofloxacin. Several reviews on quinolone residue analysis have been published over the past 10 years, most of them dealing with LC methods [13–15]. Microbiological methods aimed at screening quinolones in meat have also been reported. Several papers present relevant inhibitory methods using strains of *B. subtilis* or *E. coli* [32,34,37,49,321]. Specific applications of microbiological inhibitory testing have also been reported for quinolones using particular strains of bacteria such as *Klebsellia pneumoniae* [322] or *Yersinia ruckeri* [323].

ELISA have also been recently developed [73,80,83]. A TLC method was proposed about 30 years ago for a very specific quinolonic substance, the decoquinate, for its monitoring in chicken muscle with a fluorescent detection [324], and was followed by a HPLC-FLD method [325]. In the 1990s, many papers were published on quinolone residue analysis (Table 10.7), especially for poultry meat monitoring, and a large part of them focused on RPLC-FLD technique [73,326–335]. In regard to potential multiresidue testing including quinolones from first and second generations, some of these papers present comparative studies of different LC methods, either HPLC-FLD to HPLC-UV/DAD [336] or HPLC-FLD to LC-MS methodologies [337–338] or even LC-UV to LC-MS [339].

Recently, a multifamily analysis of both quinolones and tetracyclines in chicken muscle has been proposed within the same HPLC-FLD method [340]. Few recent articles discuss the HPLC-UV analysis of quinolones in muscle tissues [341]. This is probably due to the lack of sensitivity of this type of detection with a reduced UV absorbance efficiency for most of the quinolones. Considering the chromatographic separation on reverse-phase mode, it is worth noting the ability of the ampholytic quinolones to interact with silanols and metal impurities of the stationary phases, causing peak tailing and reducing drastically the quality of the quantification in multiquinolone analysis. To cope with this problem, polymeric phases (PLRP-S) have sometimes been preferred to sustain reliable separative performance in LC methods [326,331]. Another alternative is to utilize the ion-pairing properties of sulfonic acids [333] or to use phenyl stationary phases instead of conventional C8 or C18 ones [337]. In the 2000s, investigations have been carried out for several quinolones in chicken muscle tissues with the capillary zone electrophoretic techniques (CZE or CE) with UV detector [342–343] or with laser-induced fluorescence detection (CZE-LIF) [344] or even more recently, with MS detection [344]. Following first articles published in the 1990s dealing with MS detection of quinolones in different tissues [346], the 2000s have been the period for exploring the use of LC-MSMS techniques in the analysis of very large multiquinolone residues in muscle and in other matrices (kidney, etc.) from animal origin [262,338,347–350]. Table 10.7 displays different techniques published for LC analysis of quinolones in muscle tissues of different food-producing animal species.

Table 10.7 Summary of Literature Methods for Determination of Quinolone Residues in Tissues Using Gas or Liquid Chromatography

Tissue	Analytes	Sample Treatment	Derivatization	LC Column Technique	Detection	Limit Range (ppb)	Year (Reference)
Muscle, liver, kidney	Decoquinate	MeOH/CHCl$_3$ Extn, Acid/CHCl$_3$ Partition.	—	RPLC-Florisil	FLD: Ex290 nm Em370 nm	<100	1973 (325)
Muscle, milk	ENR, SAR[IS]	ACN/NH$_4$OH Extn, EtOAc/Hexane/NaCl partition., H$_3$PO$_4$ acidif.	—	PLRP-S	FLD: Ex278 nm Em440 nm	5	1994 (326)
Muscle, liver, egg, honey	OXA, NLA, FLU, DAN, ENR, CIP, SAR, MAR, NOR, ENO, LOM, OFL	Fluoroquinos: ACN/Acetic acid/Na$_2$SO$_4$ extn, SCX SPE, Drying, Reconst. Acidic quinos: ACN/Na$_2$SO$_4$ extn	—	RPLC-C8	FLD: Ex278 nm Em445 nm / UV (MAR): 302 nm	5; 10; 50	1998 (327)
Muscle	MAR, DAN, ENR, CIP, DIF, SAR, NOR	Drying, PBS dilution, SAX SPE, Evap., Reconst.	—	RPLC-C18	FLD: Ex278 nm Em440 nm	—[a]	2000 (328)
Muscle	FLU, OXA	DCM Extn, NaOH partition.	—	RPLC-C8	FLD: Ex328 nm Em365 nm	<1	2000 (329)
Fish muscle	7OH-NLA, NLA, OXA, CIN	A: CHCl$_3$ Extn, Evap., Reconst. / B: NaOH Extn, CHCl$_3$ partition., ChloroAcetic acid partition., CHCl$_3$ extn, Na$_2$SO$_4$ drying, Evap., Reconst.	—	RPLC-C18	FLD: Ex260 nm Em360 nm / FLD: Ex270 nm Em440 nm	<20	2000 (330)

Sample	Analytes	Extraction/Cleanup		HPLC	Detection	LOD	Year (Ref.)
Muscle	CIP,n Extn, Evap., Tris buffer Reconst., Hexane defat.		—	PLRP-S	FLD: Ex280 nm Em450 nm / FLD: Ex294 nm Em514 nm / FLD: Ex312 nm Em366 nm	0.5–35	2000 (331)
Fish muscle	OXA, FLU	ACN/NH$_4$OH Extn, EtOAc/Hexane/NaCl partition., H$_3$PO$_4$ acidif./Acetone defat., H$_2$O dilution	—	PLRP-S	FLD: Ex325 nm Em360 nm	20, 30	2001 (332)
Muscle, liver	ENR, CIP, SAR, DIF	TCA/ACN Extn, comparison: C8,C18,NH$_2$, BSA, SDB SPE cleanups	—	RPLC-C8; ion-pairing HSA	FLD: Ex278 nm Em440 nm	—[a]	2001 (333)
Muscle, liver	ENR	ACN Extn, Hexane defat., Evap., Reconst., Filtration	—	RPLC-C18	FLD: Ex290 nm Em455 nm	< 50	2002 (73)
Muscle	ENR, CIP, SAR, OXA, FLU	PBS Extn, C18 SPE	—	RPLC-C18	FLD: Ex280 nm Em450 nm / FLD: Ex312 nm Em366 nm	5.0 / 10 (SAR)	2003 (334)
Muscle, liver, kidney, fish flesh, egg, milk	MAR, NOR, ENR, CIP, DAN, SAR, DIF, OXA, NLA, FLU	TCA extn, Filtration	—	RPLC-C18	FLD: Ex294 nm Em514 nm / FLD: Ex328 nm Em425 nm / FLD: Ex312 nm Em366 nm	4–36	2005 (335)

(continued)

Table 10.7 (continued) Summary of Literature Methods for Determination of Quinolone Residues in Tissues Using Gas or Liquid Chromatography

Tissue	Analytes	Sample Treatment	Derivatization	LC Column Technique	Detection	Limit Range (ppb)	Year (Reference)
Feeds	CIP, ENR, DAN, OXA, NLA, FLU, DIF, NOR, OFL, ENO, RUF, PIP, CIN	ASE extn ACN/MPA, HLB SPE, Evap. Reconst.	—	RPLC-C5	DAD: 278 nm; FLD: Ex278 nm Em446 nm / FLD: Ex324 nm Em366 nm	500–1500	2003 (336)
Muscle, liver	DESCIP, NOR, CIP, DAN, ENR, ORB, SAR, DIF	PBS extn, ACN/NaOH extn, Hexane/EtOAc/NaCl cleanup, Evap., Reconst.	—	RPLC-Phenyl	FLD: Ex278 nm Em440 nm	0.1–0.5	2002 (337)
Shrimp flesh	DESCIP, NOR, CIP, DAN, ENR, ORB, SAR, DIF	PBS extn, ACN/ NH_4OH partition., Hexane/EtOAc/ NaCl cleanup, Evap., Reconst.	—	RPLC-Phenyl	FLD: Ex278 nm Em465 nm	0.1–1.0	2005 (338)
Muscle	ENR, CIP, DIF, DAN, MAR, AR, OXA, FLU, NOR[IS]	ACN/ H_3PO_4 Extn, ENV+isolute SPE	—	RPLC-C8	UV: 250 nm, 280 nm, / 290 nm	7–13	2006 (339)
Muscle	DAN, CIP, ENR, DIF, SAR	ACN/Citrate buffer/$MgCl_2$ extn, Evap., Reconst. in Malonate/$MgCl_2$	—	RPLC-Phenyl	FLD: Ex275 nm Em425 nm	0.5–5.0	2007 (340)
Muscle	CIP, ENR, DAN, SAR, DIF, OXA, FLU	DCM extn, NaOH partition., comparison HLB-MAX-SDB SPEs	—	RPLC-C8	UV-DAD: 250 nm, 280 nm	16–30	2004 (341)
Muscle	ENR, CIP MAR[IS]	DCM extn, C18 SPE	—	CZE	UV-DAD: 270 nm	<25	2001 (342)

Sample	Analytes	Extraction		Separation	Detection	LOD	Year (Ref)
Muscle	CIP, ENR, DAN, DIF, MAR, OXA, FLU, PIR(IS)	DCM extn, NaOH partition., comparison C18-SCX-SAX-HLB-MAX-SDB SPEs	—	CZE	UV-DAD: 260 nm	7–30	2004 (343)
Muscle	ENR, CIP DIF(IS)	H$_2$O homogen., Buffer extn, DCM partition., H$_3$PO$_4$ Dilution, Evap., Hexane defat., Filtration	—	CZE	LIF: HeCd Exc 325 nm	5; 20	2002 (344)
Muscle, fish flesh	DAN, ENR, FLU, OFL, PIP		—	CZE	MS^n esi+	20	2006 (345)
Muscle	ENR, CIP, DAN, MAR, SAR, DIF		—	RPLC-C18	MS apci+	7.5	1998 (346)
Muscle, milk, prawn flesh, Eel flesh	ENR, CIP, DAN, SAR, LOM, ENO, OFL	ACN/Formic acid extn, C18 SPE, Dilution IPCC-MS3	—	RPLC-C18	MSMS esi+	1.0–2.0	2004 (347)
Muscle	ENR, CIP, DAN, SAR, DIF, OXA, FLU	ACN extn, Hexane defat., Evap. H$_2$O/ACN extn, Hexane defat.	—	RPLC-C18	MSMS esi+/esi−	10	2005 (262)
Kidney	NOR, MAR, ENR, CIP, DAN, OXA, NLA, FLU, CIN, OFL, ENO, LOM(IS), CIN(IS)	ACN extn, Evap., Reconct., SDB SPE, Dilution	—	RPLC-C8	MSMS esi+	0.3–2.0	2005 (348)
Kidney	NOR, MAR, ENR, CIP, DAN, OXA, NLA, FLU, CIN, OFL, ENO, LOM(IS), CIN(IS)	ACN extn, Evap., Reconct., SDB SPE, Dilution	—	RPLC-C8	MSMS esi+	0.3–2.0	2005 (349)
Muscle	DESCIP, NOR, CIP, DAN, ENR, ORB, SAR, DIF	ACN/NH$_4$OH Extn, EtOAc/Hexane/NaCl partition., Evap., PBS dilution	—	RPLC-Phenyl	MSMS apci+	0.1–1.0	2005 (338)

(continued)

Table 10.7 (continued) Summary of Literature Methods for Determination of Quinolone Residues in Tissues Using Gas or Liquid Chromatography

Tissue	Analytes	Sample Treatment	Derivatization	LC Column Technique	Detection	Limit Range (ppb)	Year (Reference)
Muscle	ENR, CIP, DIF, NOR, OFL, ORB	ACN/MeOH extn, Hexane/ACN defat., Evap., Reconst.	—	RPLC-C18	MSMS esi+	0.3–3.0	2006 (214)
Muscle	ENR, CIP, DIF, DAN, MAR, SAR, OXA, FLU, NOR[(IS)]	ACN/ H$_3$PO$_4$ Extn, ENV+isolute SPE	—	RPLC-C8	MS esi+	0.3–1.8	2006 (339)
					MSMS esi+	<0.2	
Muscle	ENR, CIP, DIF, DAN, DESCIP, SAR, OXA, NLA, FLU, OFL, PIR, NOR, ENO, CIN, LOM[(IS)], PIP[(IS)]	EtOH/ Acetic acid, HCl dilution, hexane defat., SCX SPE	—	RPLC-C18	MSMS esi+	0.1–0.4	2007 (350)

Note: ppb: parts per billion equivalent to μg of residue/kg of tissue; MeOH: methanol; partition:: aqueous partitioning; RPLC: reverse phase liquid chromatography; ENR: enrofloxacin; SAR: sarafloxacin; ACN: acetonitrile; EtOAc: ethyl acetate; PLRP-S: polymeric stationary phase; OXA: oxolinic acid; NLA: nalidixic acid; FLU: flumequine; DAN: danofloxacin; CIP: ciprofloxacin; NOR: norfloxacin; ENO: enoxacin; LOM: lomefloxacin; OFL: ofloxacin; SCX: strong cation exchange; SPE: solid-phase extraction; PBS: phosphate buffer solution; SAX: strong anion exchange; Evap.: evaporation before reconstitution; defat.: defattening; UV: ultraviolet detection; DIF: difloxacin; DCM: dichloromethane; Neutral.: neutralization; CIN: cincophen; TCA: trichloroacetic acid; BSA: benzene sulfonic acid; HSA: heptane sulfonic acid; SDB: styrenedivinylbenzene; RUF: rufoxacin; PIP: pipemidic acid; ASE: accelerated solvent extractor; MPA: metaphosphoric acid; HLB: hydrophilic Lipophilic Balance; DAD: diode array detection; DESCIP: desethylciprofloxacin; ORB: orbifloxacin; MAX: medium anion exchange; PIR: piromidic acid; LIF: laser-induced fluorescence; MSn: ion trap; esi+: electrospray source in positive mode; esi–: electrospray source in negative mode; MS: single quadrupole; MSMS: triple quadrupole; Extn: extraction; EtOH: ethanol; CIN: cinoxacin; PIR: piromidic acid; deprot.: deproteinisation; purif.: purification; Reconst. reconstitution prior to injection.

a Not reported.

[(IS)]Internal standard.

10.4.8 Sulfonamide Antiinfectives

Sulfonamides are an important antiinfective family of drugs with bacteriostatic properties. Owing to their broad-spectrum activity against a range of bacterial species, both gram-positive and gram-negative, they are widely used in veterinary medicine with more than 12 licensed compounds. Synthetically prepared from *para*-aminobenzenesulfonic acid, they act by competing with *para*-aminobenzoic acid in the enzymatic synthesis of dihydrofolic acid, leading to a decreased availability of the reduced folates that are essential molecules in the synthesis of nucleic acids. In practice, they are usually combined with synthetic diaminopyrimidine, trimethoprime, to enhance synergistic action against bacterial DNA synthesis even though the synergy was never really demonstrated. Most of the sulfonamide drugs are readily soluble in polar solvents such as ethanol, ACN, and chloroform but relatively insoluble in nonpolar ones. They are considered as weak acids but behave as amphoteric compounds due to the interaction between an acidic N–H link in the vicinity of a sulfonyl group (pK_a 4.6) and an alkaline character at the *para*-NH$_2$ group (pK_a 11.5), leading to a particular behavior in extraction and cleanup process in the 7–9 pH range. Sulfamethazine (also called sulfadimidine and sulfadimerazine) is probably the most widely used sulfa drug. But, several other sulfonamides are also employed in food-producing animal treatments such as sulfadiazine, sulfadoxine, sulfaquinoxaline, sulfapyridine, sulfapyridazine, sulfadimethoxine, sulfamerazine, sulfathiazole, sulfachloropyridazine, sulfamonomethoxine, and sulfamethoxazole (Figure 10.10). Extensive and updated reviews of analytical methods for sulfonamide analysis in food from animal origin have been published over the past 15 years [9,13–15,351].

Most of these reviews relate to chromatographic methods. Yet, several microbiological screening techniques have also been tentatively applied to these compounds even though sulfonamides do not react with much sensitivity to bacteria such as *B. subtilis* [29,37,42,43,60,66,198], *B. stearothermophilus* [32,58,59,61], or even *B. megaterium* [39–40]. It is of great concern that they are not easily detected in food products (muscle, milk, egg, honey, etc.) in the 10–100 ppb range where they are generally regulated even after enhanced sensitivity brought by the addition of trimethoprim. Therefore, TLC technique is one alternative for screening sulfa drugs, which is sometimes proposed in regulatory control achieving good sensitivity in the 100 ppb range [103,202,352]. TLC-bioautography was employed in the 1990s [46] and ELISA kits have also been proposed especially for sulfamethazine screening in urine and plasma with predictive concentration in porcine muscle tissues [67,68]. More recently, biosensor-based immunochemical screening assays for the detection at 10 ppb level of sulfamethazine and sulfadiazine in bile and in muscle extracts from pigs and in chicken serum were developed [86,87,89]. High cross-reactivities (50–150%) in chicken serum were found with several other sulfa drugs such as sulfamerazine, sulfathiazole, sulfachloropyrazine, sulfachloropyridazine, and sulfisoxazole [90]. On the confirmatory quantitative stage of sulfa drug strategy of control, many relevant chromatographic techniques have been developed (Table 10.8). From GC with flame ionization detection in the early 1990s [107] to GC-MS [108] for sulfamethazine-specific detection, it is the LC methods that took the leadership in an even wider scale and not only for sulfamethazine monitoring with multisulfa drug analysis by LC-UV or LC-PDA instruments [206,287,353–363], and by LC-Fluo detections after derivatization of the sulfonamides [206,354,364–366]. The investigation of LC-MS and LC-MSMS methods with a wide range of sulfonamides started in the mid-1990s [367–369]; they were still improved with enhanced mass detectors in the 2000s [195,214,262,370–375]. A representative overview of LC-UV, LC-Fluo, and LC-MS methods used to monitor sulfa drugs in the early 2000s is presented in a recent paper related to European proficiency testing studies on sulfonamide residue in muscle and milk [376].

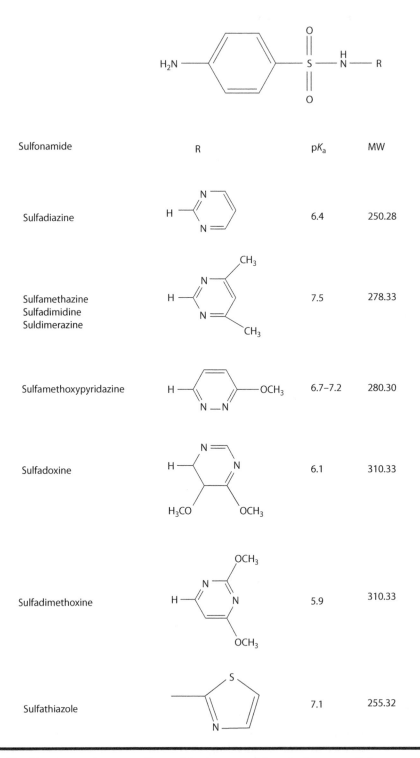

Sulfonamide	R	pKₐ	MW
Sulfadiazine		6.4	250.28
Sulfamethazine Sulfadimidine Suldimerazine		7.5	278.33
Sulfamethoxypyridazine		6.7–7.2	280.30
Sulfadoxine		6.1	310.33
Sulfadimethoxine		5.9	310.33
Sulfathiazole		7.1	255.32

Figure 10.10 Structures of some sulfonamide drugs used in veterinary medicine.

Table 10.8 Summary of Literature Methods for Determination of Sulfonamide Residues in Tissues Using Gas or Liquid Chromatography

Tissue	Analytes	Sample Treatment	Derivatization	LC Column Technique	Detection	Limit Range (ppb)	Year (Reference)
Muscle	SMM, SDM, SMT, SMX, SQX	TCA extn, ion exchange SPE	Pyr, TMSA, TMSI, TMCS	GC	FID	100	1992 (107)
Muscle	SMT	CH$_2$Cl$_2$/acetone extn, silica SPE, SCX SPE, PBS/MTBE partition., evap., deriv.	methylation, silylation	GC	MS	10–20	1996 (108)
Muscle	SMT SEPDZ[IS]	CHCl$_3$ extn, alkaline NaCl partition., C18 SPE	—	RPLC-C18	UV: 265 nm	2	1994 (353)
Kidney	SMT,SQX, SDZ	ACN extn, drying evap., buffer dilution, C18 SPE, drying evap., DCM dilution, silica SPE, drying evap., buffer reconst.	—	RPLC-C8	UV-DAD: 220–400 nm; 246 nm (SQX), 251 nm (SDZ), 299 nm (SMT)	2–18	1995 (287)
Muscle, liver, kidney	SMT, N^4-metabolites	EtOAc/acetic acid /Na$_2$WO$_4$, NH$_2$-SCX SPE, drying evap., HCl reconst.	—	RPLC-C18	UV: 270 nm	—^{a}	1995 (354)
Muscle, liver, kidney	STZ, SMR, SCP, SMT, SMPDZ, SMX, SQX, SDM	MSPD extn, partition.n 1 with CH$_2$Cl$_2$ and partition.n 2 with EtOAc	—	RPLC-C18	UV: 270 nm	1–66	1997 (206)
Muscle	SMT	MAE	—	RPLC-C18	UV: 450 nm	—^{a}	1998 (355)
Muscle	SMT, SDZ, SPD, SMR, SDX	—^{a}	—	RPLC-C18	UV: 270 nm	—^{a}	1998 (356)
Muscle, liver, kidney	SDZ, STZ, SPD, SMR, SMZ, SMT, SMPDZ	Saline extn, dialysis, C18 SPE	—	RPLC-C18	UV: 280 nm	40	1999 (357)

(continued)

Table 10.8 (continued) Summary of Literature Methods for Determination of Sulfonamide Residues in Tissues Using Gas or Liquid Chromatography

Tissue	Analytes	Sample Treatment	Derivatization	LC Column Technique	Detection	Limit Range (ppb)	Year (Reference)
Muscle	SMM, SDM, SQX	MeOH/H$_2$O extn, IAC cleanup	—	RPLC-C18	UV: 370 nm	1–2	2000 (358)
Muscle, liver, kidney	SMT	EtOH/H$_2$O extn, ultrafiltration	—	RPLC-C4	UV-DAD: 263 nm	24–27	2001 (359)
Shrimp flesh	SDZ, STZ, SQX, SDM, SMR	EtOAc extn, Drying evap., SEC cleanup, Drying evap., ACN/acetic acid reconst.	—	RPLC-phenyl	UV: 270 nm	10	2003 (360)
Muscle, liver, kidney	SMT	PCA ultrasonic extn	—	RPLC-C4	UV: 267 nm	<90	2003 (361)
Muscle	SDZ, STZ, SPD, SMR, SMT, SMM, SCP, SMX, SQX, SDM	EtOAc/Na$_2$SO$_4$ extn, drying evap., EtOAc dilution, SCX SPE, drying evap., reconst. acetate buffer	—	RPLC-C8	UV-DAD: 270 nm	30–70	2004 (362)
Muscle	SDZ, STZ, SMT, SMR, SDX, SMM, SCP, SMX, SQX, SMZ	Acetone/CHCl$_3$ extn, SCX SPE, drying evap., MeOH reconst.	—	RPLC-C18	UV-DAD: 270 nm		2007 (363)
Muscle, liver, kidney, serum	SMT, SMM, SMX, SDM SDZ[(IS)]	ACNdeprot. Extn, Evap., H$_2$O/ACN dilution, evap., TCA dilution, hexane defat.	Fluoresc-amine	RPLC-C18	FLD: Ex390 nm Em475 nm	0.1	1995 (364)
Muscle, liver, kidney	SMT, N^4-metabolites	EtOAc/acetic acid /Na$_2$WO$_4$, NH$_2$-SCX SPE, drying evap., HCl reconst.	Fluoresc-amine	RPLC-C18	FLD: Ex405 nm Em495 nm	—[a]	1995 (354)

Matrix	Analytes	Extraction	Derivatization	Separation	Detection	LOD	Year (Ref)
Muscle, liver, kidney	STZ, SMR, SCP, SMT, SMPDZ, SMX, SQX, SDM	MAE	DMABA	RPLC-C18	FLD: Ex405 nm Em495 nm	2.5	1997 (206)
Muscle	SCP, SDZ, SDM, SDX, SMT, SQX, STZ; SPD[IS]	EtOAc extn, glycine/PBS/HCl purif., hexane defat., CH$_2$Cl$_2$/Na$_2$SO$_4$ purif., DEA dilution, drying evap., reconst. PBS/ACN	Fluoresc-amine	RPLC-C18	FLD: Ex405 nm Em495 nm	15	2004 (365)
Muscle	SDZ, SMR, SMT, SMPDZ, SMX, SDM	ACN extn, C18 MSPD, drying evap., reconst. acetate buffer	Fluoresc-amine	RPLC-C18	FLD: Ex405 nm Em495 nm	1–5	2005 (366)
Kidney	SMT, SMR, SDZ, SQX	Acidic EtOAc extn, NH$_2$ + SCX SPE, drying evap., acetone storage, drying evap., reconst.	–	RPLC-C18	MS esi+ / MSMS esi+	–[a]	1994 (367)
Muscle	SMT, SDM, SEPDZ[IS]	CHCl$_3$ extn, alkaline NaCl partition., C18 SPE, drying evap., reconst.	–	RPLC-C18	UV: 265 nm / MS TSP+	<10	1995 (368)
Muscle	TMP	CHCl$_3$/acetone extn, drying evap., MeOH/H$_2$O/acetic acid dilution, hexane defat.	–	RPLC-C18	MS TSP+	4	1997 (369)
Kidney	Sulfa drugs	On-line extn, sample cleanup	–	RPLC-C18	MSMS esi+	–[a]	2000 (370)
Muscle	SDX	MeOH extn, evap., water dilution, C18 SPE, evap., reconst.	–	RPLC-C18	MSMS esi+	–[a]	2003 (195)

(continued)

Table 10.8 (continued) Summary of Literature Methods for Determination of Sulfonamide Residues in Tissues Using Gas or Liquid Chromatography

Tissue	Analytes	Sample Treatment	Derivatization	LC Column Technique	Detection	Limit Range (ppb)	Year (Reference)
Muscle	SDZ, STZ, SPD, SMR, SMT, SMZ, SMPDZ, SCP, SMX, SMM, SDM, SQX, SME[IS]	MSPD with 80°C water extn	—	RPLC-C18	MS esi+	3–15	2003 (371)
Muscle	SDZ, STZ, SMT, SMR, SDM	ACN/Na$_2$PO$_4$ extn, C18 SPE, drying evap., reconst.	—	RPLC-C8	MS apci+ + MS esi+	—[a]	2003 (372)
Raw meat, infant food	SIM, SDZ, SPD, SMR, SMO, SMT, SMTZ, SMPDZ, SCP, SMM, SMX, SQX, SDM	C18 ASE with 160°C/100 atm water extn	—	RPLC-C18	MSMS esi+	0.25	2004 (373)
Fish muscle	SDZ, SMT, SDM, TMP, OMP	ACN extn, hexane defat., evap. H$_2$O/ACN extn, hexane defat.	—	RPLC-C18	MSMSesi+/esi–	10	2005 (262)
Muscle	SMT, SCP, SBZ, SDZ, SMZ, SDX, SMR, SMX, SMM, SMPDZ, STZ, SPZ, SDM, SQX, SSZ, SOZ, TMP, SNT, SPD	ACN/MeOH extn, hexane/ACN defat., evap., reconst.	—	RPLC-C18	MSMS esi+	0.1–0.6	2006 (214)

Muscle	SMT, SDZ, SMM, SMX, SDM, SQX	Alumina MSPD extn, drying evap., reconst.	–	RPLC-C4	MS apci+	<50	2007 (374)
Muscle	SMT, SAA, SGN, SNL, SPD, SDZ, STZ, SMR, SMX, SMO, SOZ, SMPDZ, SMM, SDM, SQX, SCP	ACN/H$_2$O extn, hexane defat., CHCl$_3$ partition., drying evap. reconst.	–	RPLC-C18	MSMS esi+	0.1–0.9	2007 (375)
Muscle	SMT, SDZ, SMM, SMX, SDM, SQX, STZ, SGN, SMPDZ	ACN extn, hexane defat., drying evap. reconst.	–	RPLC-C18	MSMS esi+	<2	2005 (376)

Note: ppb: parts per billion equivalent to µg of residue/kg of tissue; SMM: sulfamonomethoxine; SDM: sulfadimethoxine; SMT: sulfamethazine 'also called sulfadimidine or sulfadimerazine); SMX: sulfamethoxazole; SQX: sulfaquinoxaline; TCA: trichloroacetic acid; SPE: solid-phase extraction; Pyr.: pyridine; TMSA: N,O-bis(trimethylsilyl)acetamide; TMSI: N-trimethylsilylimidazole; TMCS: trimethylchlorosilane; SCX: strong cation exchange; PBS: phosphate buffer solution; MTBE: methyl-tert-butyl-ether; partition.: aqueous partitioning; Evap.: evaporation before reconstitution; derivat.: derivatization; MS: single quadrupole; SEPDZ: sulfaethoxypyridazine; Extn: extraction; SDZ: sulfadiazine; ACN: acetonitrile; DCM: dichloromethane; RPLC: reverse phase liquid chromatography; UV: ultraviolet detection; DAD: diode array detection; EtOAc: ethyl acetate; SMR: sulfamerazine; SMPDZ: sulfamethoxypyridazine; MSPD: matrix solid phase dispersion; MAE: microwave assisted extraction; SDX: sulfadoxine; SMZ: sulfamethizol; MeOH: methanol; SEC: size-exclusion chromatography; PCA: perchloric acid; defat. defattening; purif.: purification; DMABA: dimethylaminobenzaldehyde; DEA: diethylamine; esi+: electrospray source in positive mode; MSMS: triple quadrupole; TSP: thermospray; TMP: trimethoprim; SME: sulfameter; SIM: sulfisomidine; esi–: electrospray source in negative mode; ASE: accelerated solvent extraction; SMO: sulfamoxole; SMTZ: sulfamethizole; OMP: ormethoprim; SBZ: sulfabenzamide; SPZ: sulfaphenazole; SSZ: sulfasalazine; SOZ: sulfisoxazol; SNT: sulfanitran; SAA: sulfacetamide; SGN: sulfaguanidine; SNL: sulphanilamide; SCPZ: sulfachloropyridazine; Reconst: reconstitution prior to injection; PLRP-S: polymeric stationary phase; HSA: heptane sulfonic acid; MSn: ion trap.

a Not reported.
(IS)Internal standard.

10.4.9 Tetracyclines

Tetracyclines are broad-spectrum bacteriostatic antibiotics, some of which are produced by bacteria of the genus *Streptomyces* and others obtained as semisynthetic products. They act by inhibiting protein biosynthesis through their binding to the 30S ribosome. Owing to their high degree of activity against both gram-positive and gram-negative bacteria, they are commonly used in veterinary medicine to treat respiratory diseases in cattle, sheep, pig, and chicken. They may be employed prophylactically as additives in feed or in drinking water. Oxytetracycline, tetracycline, and chlortetracycline are the three major compounds licensed in veterinary medicine. Doxycycline is also, to some extent, a veterinary drug candidate to monitoring in tissues. Oxytetracycline can be found after treatment of various bacterial diseases in fish farming. The basic structure of tetracyclines is derived from the polycyclic naphthacenecarboxamide and contains four fused rings (Figure 10.11). They are polar compounds due to the different functional groups attached to the four fused rings. Particularly active are an acidic hydroxyl group in position 3 (pK_a 3.3), a dimethylamino group in position 4 (pK_a 7.5), and a basic hydroxyl group in position 12 (pK_a 9.4). Tetracyclines are photosensitive, nonvolatile compounds, existing as bipolar ions in aqueous solution in the pH range 4–7, able to lose their dimethylamino group in the pH range 8–9, and capable of reversible epimerization in the pH range 2–6. This chemical behavior leads to difficulties in extracting them from biological matrices where they can easily bind to proteins to form macromolecules. Acidic extraction is often utilized, but further purification by liquid–liquid partitioning and SPE cleanup through organic solvents remain a critical issue. Ion pairing and chelation process are also used to achieve acceptable recoveries. The analysis of tetracyclines by reverse-phase LC is also of concern. Silanol-encapped, metal-purified, alkyl-bonded silica stationary phases are

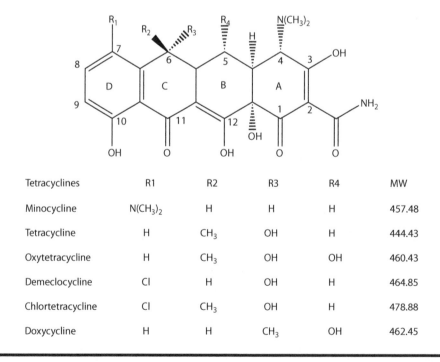

Tetracyclines	R1	R2	R3	R4	MW
Minocycline	N(CH$_3$)$_2$	H	H	H	457.48
Tetracycline	H	CH$_3$	OH	H	444.43
Oxytetracycline	H	CH$_3$	OH	OH	460.43
Demeclocycline	Cl	H	OH	H	464.85
Chlortetracycline	Cl	CH$_3$	OH	H	478.88
Doxycycline	H	H	CH$_3$	OH	462.45

Figure 10.11 Structures of tetracyclines used in veterinary medicine.

required for their satisfactory separation and elution. Polymeric phases have also been successfully investigated. New mixed polymeric/alkyl-bonded silica stationary phases are now promising separative instruments.

An extensive development over the past years has been dedicated to analytical methods for monitoring tetracycline residues at the ppb level in meat products (MRL in muscle tissue is 100 μg/kg). Several reviews are reported on this subject [9,11,13–15]. On the part dealing with residue screening methods, several papers describe microbiological bioassays using the inhibitory properties regarding bacterial growth. Strains such as *B. subtilis*, *B. stearothermophilus*, *Bacillus cereus*, and *E. coli* have been employed to attempt developing inhibitory plate tests capable of detecting, with more or less success, the tetracyclines at the 100–500 ppb level in muscle or kidney tissues [29,32,34,37,40,42,43,58,69,198,216,321]. In the consideration to find the best strategy to screen in meat products antibiotic residues and thus tetracyclines as a part of it, some comparative studies of the performance between inhibitory plate tests and rapid test kits such as Tetrasensor have also been evaluated recently [59–62,377]. TLC with or without bioautography has also been an alternative to detect tetracycline residues in meat tissues but some 20 years ago [55,202].

Regarding the confirmatory methods (Table 10.9), GC was investigated in the previous years but with a limited extent [107], and due to the polar nonvolatile chemical properties of tetracyclines, HPLC was largely preferred to GC. First applied with UV or DAD detection [378–391], and also with fluorescence detection in regard to the high capacity of tetracycline to form fluorophoric metal complexes [340,370, 392–401], it was more recently coupled to different mass spectrometric detectors, with SQ detectors and now with TQ detectors (LC-MS/MS) or ion-trap mass spectrometric detectors (LC-IT/MS or LC-MS[n]) or even time-of-flight mass spectrometric detectors (LC-TOF-MS) [195,223,262,402–405]. Capillary electrophoretic techniques have also been tested for tetracyclines [388]. One of the challenges in modern analysis of tetracycline residues in muscle or other food products is to separate and analyze simultaneously all the four tetracyclines along with the existing 4-epimers and some possible degradation compounds [223,382,384,387,390,396,402–405].

10.4.10 Polypeptidic Antibiotics

Polypeptide antibiotics include flavomycin (also named bambermycin or flavophospholipol), avoparcin, virginiamycin, and among polymyxin polypeptides, bacitracin and colistin (Figure 10.12). They are all derived from fungi or bacteria (*Streptomyces bambergiensis*, *Streptomyces candidus*, *Streptomyces virginiae*, *Streptomyces orientalis*, *B. subtilis*, *Bacillus polymyxa*) and exist as complexes of several related macromolecules. Avilamycin, a polysaccharide antibiotic, obtained from *Streptomyces viridochromogenes* can also be added to this group of substances. Most of these antibiotics were used as growth promoters and efficient feed converters except for colistin. Formerly regulated under the feed additive legislation in the EU by Directive 70/524/EC [406], they are now extensively proposed to be prohibited. The risk that resistance to antibiotics might be transferred through them to pathogenic bacteria was assessed at the end of the 1990s and beginning of the 2000s. It led to food safety recommendations from the antimicrobial resistance research program [407]. Avoparcin in 1997, and bacitracin along with virginiamycin in 1999 were immediately banned as feed additives in the EU by Directive 97/6/EC [408] and by Regulation 2821/98/EC [409]. Flavomycin is still on the market but on its way to be banned too. Under Regulation 2562/99/EC, a period of 5–10 years is granted from 2004 to 2014 for reevaluation of the drug by the supporting pharmaceutical stakeholders [410]. The polypeptidic antibiotics are macromolecular compounds. They often feature

Table 10.9 Summary of Literature Methods for Determination of Tetracycline Residues in Tissues Using Gas or Liquid Chromatography

Tissue	Analytes	Sample Treatment	Derivatization	LC Column Technique	Detection	Limit Range (ppb)	Year (Reference)
Muscle	TTC, OTC, CTC	TCA extn, ion exchange SPE	Pyr, TMSA, TMSI, TMCS	GC	FID	50	1992 (107)
Muscle	CTC, isoCTC	HCl/glycine extn, cyclohexyl SPE	pH 12 isoCTC conversion	PLRP-S	FLD: Ex:340 nm Em:420 nm	20–50	1994 (377)
Muscle	TTC, OTC, CTC	EDTAMIB extn, hexane/ DCM purif., TCA deprot., C18 SPE	—	RPLC-C18	UV-DAD: 360 nm	50	1994 (379)
Muscle	TTC, OTC, CTC, DMCTC	SEPSA extn, C8 SPE or XAD2resin SPE, Cu²⁺gel chelate purif.	—	MCAC + PLRP-S	UV: 350 nm	10–20	1996 (380)
Muscle	TTC, OTC, CTC	Liquid–Liquid extn, C18 SPE cleanup	—	RPLC-C18	UV: 360 nm	100	1996 (381)
Muscle	OTC, 4-epiOTC, alpha-apoOTC, beta-apoOTC	Oxalic acid extn, C18 SPE cleanup	—	RPLC-C18	UV-DAD	—ᵃ	1996 (382)
Kidney	OTC	Citrate buffer/ EtOAc extn, Na₂SO₄ drying, filtration	—	MCAC + PLRP-S	UV: 350 nm	—ᵃ	1998 (383)
Muscle	OTC, 4-epiOTC	LiqLiq extn, C18 SPE cleanup	—	RPLC-C18	UV: 350 nm	5–10	1998 (384)
Muscle, kidney	TTC, OTC, CTC	Oxalic acid/ACN extn-dechelation- deprot., SDB SPE	—	PLRP-S	UV: 360 nm	10–40	1999 (385)

Muscle, kidney	TTC, OTC, CTC	ACN/H$_3$PO$_4$ extn, hexane/DCM defat., limited evap., filtration	—	RPLC-C18 + ion pairing DSA	UV: 370 nm	50–100	2000 (386)
Kidney	TTC, OTC, CTC, DC	LiqLiq extn, C18 SPE cleanup	—	RPLC-C18	UV: 350 nm	50–100	2001 (387)
Muscle, kidney, liver	OTC	EDTAMIB extn, C18 or HLB SPE	—	RPLC-C8 + CZE	LC-UV: 350 nm CZE-UV: 365 nm	80–160	2001 (388)
Muscle, milk	TTC, OTC, CTC, DC	TCA/EDTAMIB extn, HLB SPE	—	RPLC-C18	UV-DAD: 365 nm	10–30	2003 (389)
Kidney	CTC+4-epiCTC	Oxalic acid/TCA extn, SDB SPE	—	RPLC-C8	UV-DAD: 365 nm	70–90	2005 (390)
Plasma	OTC	MeOH/EDTAMIB extn, C18 SPE	—	RPLC-C18	UV: 360 nm	3.5–12	2006 (391)
Muscle	CTC, isoCTC	HCl/Glycine extn, cyclohexyl SPE	pH 12 isoCTC conversion	PLRP-S	FLD: Ex:340 nm Em:420 nm	20–50	1989 (392)
Muscle, liver, fish, milk, egg	OTC, TTC$^{(IS)}$	ASTED dialysis, online enrichment SDB cartridge	NaOH + irradiation 366 nm	PLRP-S + HSA ion-pairing +	FLD: Ex:358 nm Em:460 nm	3–4	1992 (393)
Muscle, kidney, liver	TTC, OTC, CTC	HCl/glycine extn, cyclohexyl SPE	Al^{3+} postcol deriv	RPLC-C18	FLD: Ex:390 nm Em:490 nm	20–230	1995 (394)
Muscle, kidney	OTC	ACN/EDTAMgIB extn, hexane defat., ultrafiltration	Mg^{2+} deriv	RPLC-C18	FLD: Ex:380 nm Em:520 nm	40–50	1996 (395)

(continued)

Table 10.9 (continued) Summary of Literature Methods for Determination of Tetracycline Residues in Tissues Using Gas or Liquid Chromatography

Tissue	Analytes	Sample Treatment	Derivatization	LC Column Technique	Detection	Limit Range (ppb)	Year (Reference)
Muscle, kidney, liver *fresh and lyophilized*	CTC+4-epiCTC	HCl/glycine extn, cyclohexyl SPE	pH 12 isoCTC conversion	PLRP-S	FLD: Ex:340 nm Em:420 nm	20–50	1998 (396)
Muscle, liver	DC+4-epiDC DMCTC[IS]	Succinate buffer extn, MeOH dilut, MCAC cleanup, SDBRPS SPE cleanup	Postcol Zr^{2+}deriv	RPLC-C18	FLD: Ex:406 nm Em:515 nm	1.0	1998 (397)
Muscle, liver	DC+4-epiDC DMCTC[IS]	Succinate buffer extn, MeOH dilut, MCAC cleanup, SDBRPS SPE cleanup	Postcol Zr^{2+}deriv	RPLC-C18	FLD: Ex:406 nm Em:515 nm	1.0	2000 (398)
Fish muscle	TTC, OTC DMCTC[IS]	EDTAMIB extn, hexane defat., TCA deprot., HLB SPE	—	RPLC-C18	FLD: Ex:385 nm Em:500 nm	50	2003 (399)
Chicken muscle	TTC, OTC, CTC	EDTAMIB extn, HLB SPE	Tris/Eu^{3+}/ CTAC deriv	—	TRL: Ex:388 nm Em:615 nm	3–20	2004 (400)
Fish muscle	OTC, 4-epiOTC, anhydroOTC, alpha-apoOTC, beta-apoOTC	EDTAMIB extn, C18 SPE, NH$_2$ SPE	Tris/Mg^{2+} deriv	RPLC-phenyl	FLD: Ex:378 nm Em:500 nm	100	2005 (401)

Muscle	TTC, OTC, CTC	ACN/citrate buffer/ MgCl$_2$ extn, evap., reconst. in malonate/ MgCl$_2$	—	RPLC-phenyl	FLD: Ex375 nm Em535 nm	1.0–2.0	2007 (340)
Muscle, kidney	TTC, OTC, CTC+isomers	HCl/glycine extn, cyclohexyl SPE	—	RPLC-C8 + ion pairing HFBA/EDTA	MS apci+ammoniac	10–20	1997 (402)
Kidney	TTC, OTC, CTC	On-line extn, sample cleanup	—	RPLC-C18	MSMS esi+	—[a]	2000 (370)
Kidney	OTC	MeOH extn, Evap., water dilution, C18 SPE, evap., reconst.	—	RPLC-C18	MSn esi−	—[a]	2003 (195)
Muscle, liver, kidney	OTC+ 4-epiOTC / DMCTC[IS]	Succinate buffer extn, TCA deprot., HLB SPE	—	PLRP-S	MSn esi+	1–48	2003 (403)
Muscle	OTC+ 4-epiOTC, selected[IS]	—[a]	—	PLRP-S	MSMS esi+/esi−	—[a]	2003 (223)
Fish muscle	TTC, OTC, CTC, DC	ACN extn, hexane defat., evap. H$_2$O/ACN extn, hexane defat.	—	RPLC-C18	MSMS esi+/esi−	10–100	2005 (262)

(continued)

Table 10.9 (continued) Summary of Literature Methods for Determination of Tetracycline Residues in Tissues Using Gas or Liquid Chromatography

Tissue	Analytes	Sample Treatment	Derivatization	LC Column Technique	Detection	Limit Range (ppb)	Year (Reference)
Muscle	TTC, OTC, CTC, DC + / 4-epimers, DMCTC[IS]	EDTAMIB extn, HLB SPE	—	RPLC-C18	MSMS esi+	10	2006 (404)
Muscle	TTC, OTC, CTC, DC, 4-epiOTC, 4-epiTTC, 4-epiCTC, DMCTC[IS]	EDTA/SW extn	—	RPLC-C8	MSMS esi+	1–10	2006 (405)

Note: TTC: tetracycline; OTC: oxytetracycline; CTC: chlortetracycline; TCA: trichloroacetic acid; SPE: solid-phase extraction; PLRP-S: polymeric stationary phase; EDTAMIB: Ethylenediaminetetraacetic acid and McIlvaine Buffer pH 4; DCM: dichloromethane; RPLC: reverse phase liquid chromatography; UV: ultraviolet detection; DAD: diode array detection; DMCTC: minocycline, demeclocycline or demethylchlortetracycline; SEPSA: succinate/EDTA/pentane sulfonic acid buffer; MCAC: metal-chelate affinity chromatographic precolumn; EtOAc: ethyl acetate; ACN: acetonitrile; SDB: styrene divylnilbenzene cartridge; DSA: decane sulfonic acid; DC: doxycycline; Extn: extraction; HLB: hydrophilic Lipophilic Balance; MeOH: methanol; HSA: heptane sulfonic acid; EDTAMgIB: Ethylenediaminetetraacetic acid and Mg²⁺ in imidazole buffer pH 7.2; 4-epiDC: 4-epimer of doxycycline; SDBRPS: polystyrene-divinylbenzene-reverse phase sulfonated cartridge; CTAC: cetyltrimethylammonium chloride; HFBA: heptafluorobutyric acid; EDTA: Ethylenediaminetetraacetic acid; MSMS: triple quadrupole; Evap.: evaporation before reconstitution; Reconst: reconstitution prior to injection; MSn: ion trap; esi+: electrospray source in positive mode; esi−: electrospray source in negative mode; SW: subcritical water; 4-epiOTC: 4-epimer of oxytetracylcine; 4-epiCTC: 4-epimer of chlortetracycline; deprot.: deproteinisation; defat. defattening; purif.: purification; Dilut.: dilution; PBS: phosphate buffer solution; Tris: tris(hydroxymethyl)-aminomethane pH 9; PLRP-S: polymeric stationary phase; TRL: time resolved luminescene; MS: single quadrupole; ppb: parts per billion equivalent to μg of residue/kg of tissue.

ᵃ Not reported.

[IS]Internal standard.

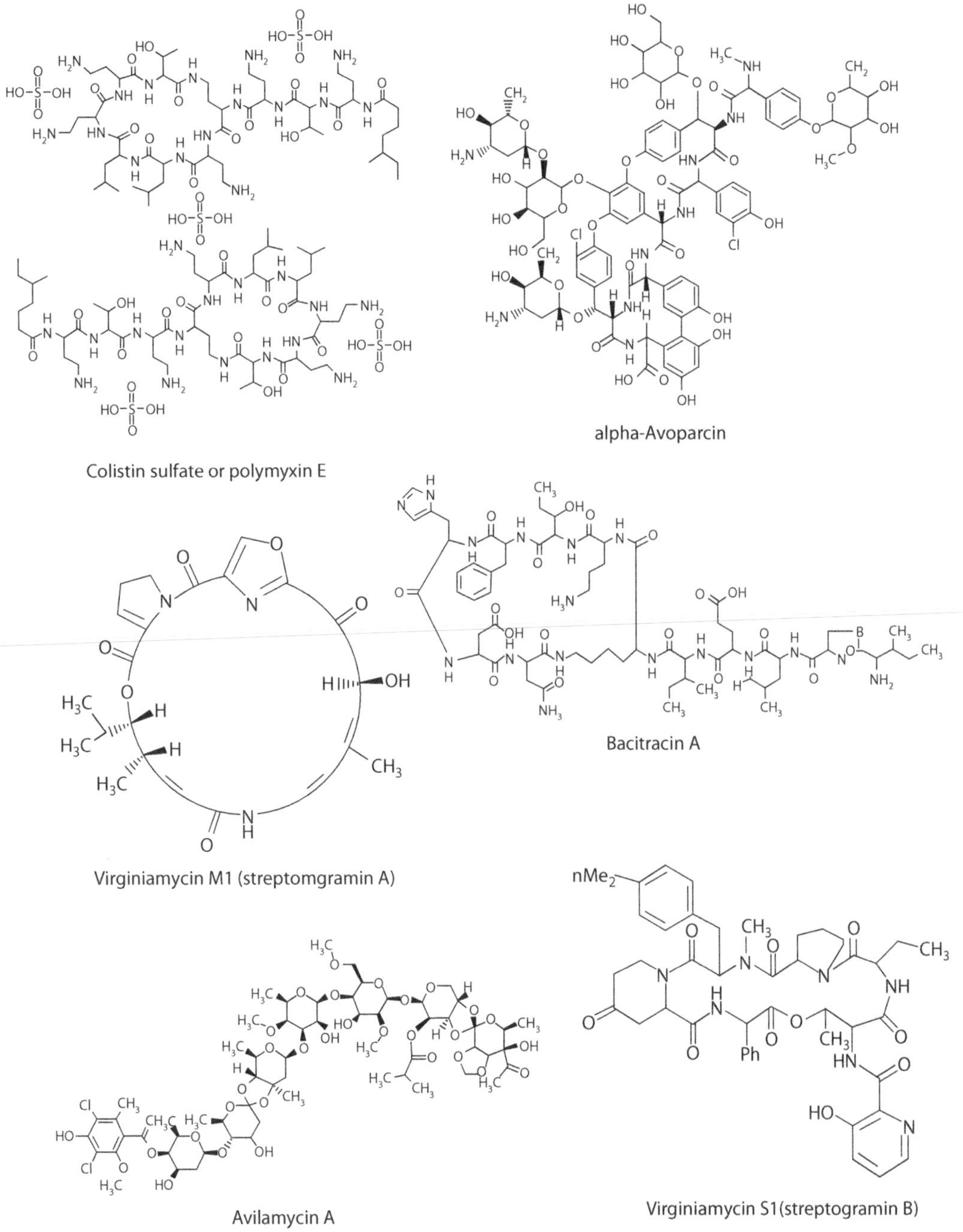

Colistin sulfate or polymyxin E

alpha-Avoparcin

Virginiamycin M1 (streptomgramin A)

Bacitracin A

Avilamycin A

Virginiamycin S1(streptogramin B)

Figure 10.12 Structures of some polypeptidic antibiotics.

a mixture of several molecules, for example, factor M1 and factor S1 principal components for virginiamycin or compound A and compound F principal components for bacitracin or even alpha and beta major components for avoparcin. The macromolecular structure makes them difficult to selectively be extracted among and purified from the naturally occurring polypeptidic molecules found in food products from animal origin. Few attempts have been undertaken during the past 10 years for developing selective analytical methods aimed at monitoring polypeptides in meat (muscle and kidney tissues) in the ppb range in line with their illegal use. Table 10.10 displays some of these methods for the glycopeptidic antibiotic avoparcin [411], for the streptograminic antibiotic virginiamycin [206,214,412–415], for the polypeptidic antibiotic bacitracin, and for the cyclic polypeptidic antibiotic colistin [280,416]. In line with their use in feedingstuffs, most of the methods developed for their monitoring analyze polypeptides as additives in the feedingstuff matrices instead of residues in meat tissues. Two examples are presented in Table 10.10, one for colistin [190] and one for bacitracin and virginiamycin [280]. Polypeptidic antibiotics are still a challenge in drug residue analysis from biological matrix.

10.4.11 Polyether Antibiotics

The ionophores are polyether antibiotics obtained mostly by fermentation of several *Streptomyces*. They hold the specificity to be licensed essentially for use against protozoal coccidial infections in poultry instead of being directed against bacteria. They are, therefore, more generally considered as anticoccidials or coccidiostats even though formerly employed as feed additives for promoting growth in cattle and sheep. They are regulated as feed additives and growth-promoting agents under the feed additive legislation in the EU by Directive 70/524/EC [406]. As for the polypeptidic antibiotics, under regulation 2562/99/EC, a period of 5–10 years is granted from 2004 to 2014 for reevaluation of these drugs by the supporting pharmaceutical stakeholders [410]. Their principal compounds are lasalocid, maduramicin, monensin, narasin, salinomycin, and semduramicin. In terms of chemistry, the basis of their structure is a sequence of tetrahydrofuran and tetrahydropyran units linked together in the form of spiroketal moieties (Figure 10.13). In spite of hydroxylic and carboxylic functions at both ends of these macromolecules, they are rather poor soluble antibiotics in aqueous solutions due to their macrocyclic conformation with polar groups oriented inward and nonpolar groups oriented outward. As a consequence, organic solvent extraction is the preferred one. But, on the counterbalance, purification by liquid–liquid partitioning is difficult to achieve due to similar solubility properties of the ionophores in their free acid and salt forms and due to instability in acidic media. The term "ionophore" is attributed to these macromolecules in relation to their ability to stabilize by complexing with such alkaline cations as Ca^{2+} or Mg^{2+}. Four interesting reviews were published on ionophore polyethers: one by Weiss and MacDonald in 1985 essentially dedicated to their chemistry [417] and three other more recent ones by Asukabe and Harada in 1995 [418], by Botsoglou and Kufidis in 1996 [11], and by Elliott et al. in 1998 [419] and dedicated to their chemical analysis in food products from animal origin. Since then, the analysis of polyether ionophores in muscle tissues relied essentially on the confirmatory LC-MS technique as described in Table 10.11 [214,420–425]. Only few attempts are reported of the screening with fluoroimmunoassays [426].

10.4.12 Other Antibiotics (Novobiocin, Tiamulin)

Novobiocin is an antibiotic produced by *Streptomyces spheroides* and *Streptomyces niveus* with a narrow-spectrum activity against some gram-positive bacteria. It is soluble in polar organic solvents such as alcohols, acetone, and EtOAc but rather insoluble in aqueous solution below pH 7.5 and

Table 10.10 Summary of Literature Methods for Determination of Polypeptidic Antibiotic Residues in Tissues Using Liquid Chromatography

Tissue	Analytes	Sample Treatment	Derivatization	LC Column Technique	Detection	Limit (ppb)	Year (Reference)
Muscle	Virginiamycin M1 Virginiamycin S1	ACN extn, Evap., MeOH dilution, CHCl₃ partition., H₂O purif., drying evap.	–	RPLC-C18	FLD: Ex311 nm Em427 nm	100, 10	1987 (412)
Muscle, kidney, liver, serum	Virginiamycin M1	PBS/MeOH extn, PE defat.	–	RPLC-C18	UV: 254 nm	10	1988 (413)
Muscle	Virginiamycin M1	MeOH/PTA extn, CHCl₃ partition., silica SPE	–	RPLC-C18	UV: 235 nm	50	1989 (414)
Muscle	Virginiamycin M1	C18 MSPD, EtOAc extn, evap.	–	RPLC-C18	UV-DAD: 254 nm	2–7	1997 (206)
Feed	Colistin A, Colistin B	HCl extn	Postcol. OPA	RPLC-C18	FLD: Ex355 nm Em415 nm	100	1998 (190)
Kidney	Avoparcin	Hot water/EtOH ASE, XAD-7 MSPD, HILIC SPE	–	HILIC-LC	UV: 225 nm	500	2002 (411)
Feed	Virginiamycin M1, bacitracin A	MeOH/Water extn, HLB SPE	–	RPLC-C18	MSMS esi+	200–600	2003 (280)
Standards	Polymyxins, Bacitracin A	–	–	RPLC-C18	MS^n	–[a]	2003 (415)
Muscle	Virginiamycin	ACN/MeOH extn, hexane/ACN defat, evap., reconst.	–	RPLC-C18	MSMS esi+	2–8	2006 (214)
Muscle	Bacitracin A, Colistin A, Colistin B	Acid extn, strata-X SPE	–	RPLC-C18	MSMS esi+	14–47	2006 (416)

Note: ACN: acetonitrile; Evap.: evaporation of volatile solvent; MeOH: methanol; RPLC: reverse phase liquid chromatography; PBS: phosphate buffer solution; PE: petroleum ether; UV: ultraviolet detection; PTA: phosphotungstic acid; SPE: solid-phase extraction; MSPD: matrix solid pagse dispersion; EtOAc: ethyl acetate; OPA: ortho-phthalaldehyde; EtOH: ethanol; ASE: accelerated solvent extractor; XAD-7 HP: acrylic polymer resin; HILIC: hydrophilic interaction chromatography; HLB: hydrophilic Lipophilic Balance; MSMS: triple quadrupole; esi+: electrospray source in positive mode; MSⁿ: ion trap; Extn: extraction; defat. defattening; purif.: purification; partition.: aqueous partitioning; DCM: dichloromethane; ppb: parts per billion equivalent to µg of residue/kg of tissue.

ᵃ Not reported.
⁽ᴵˢ⁾Internal standard.

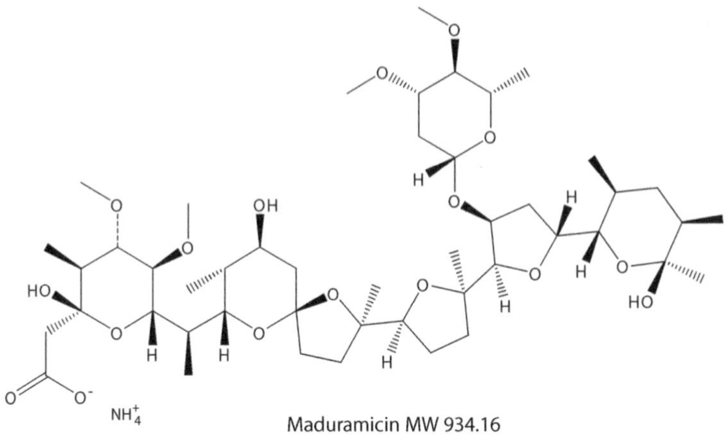

Maduramicin MW 934.16

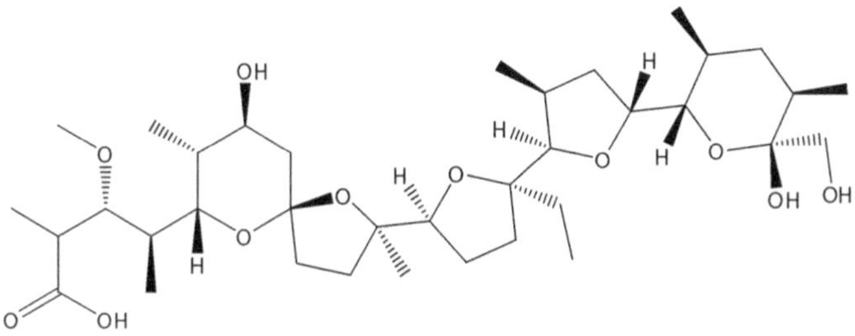

Monensin MW 670.88

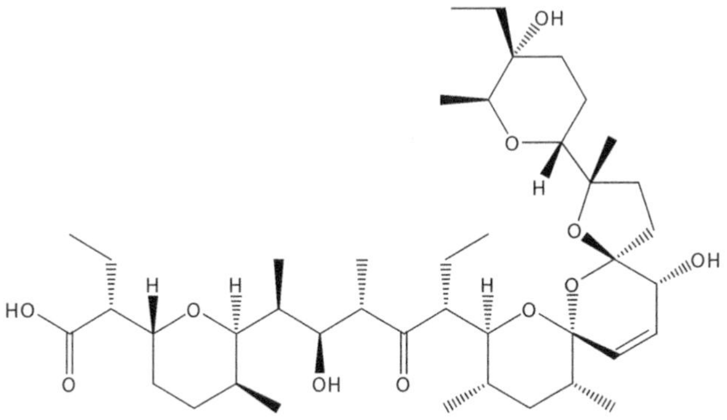

Salinomycin MW 751.01

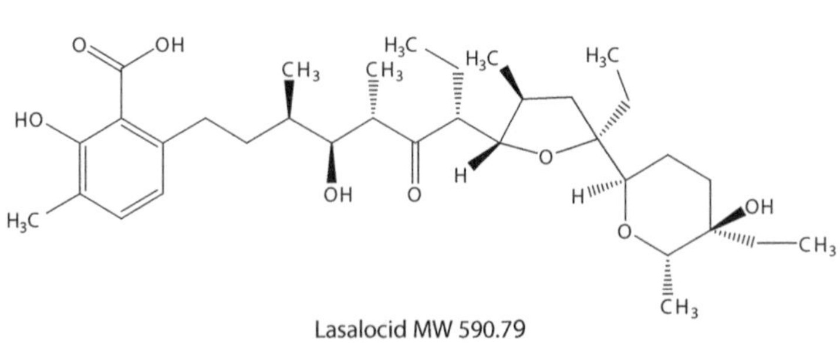

Lasalocid MW 590.79

Figure 10.13 Structures of ionophore polyethers used in food-producing animal feeding.

Table 10.11 Summary of Literature Methods for Determination of Polyether Antibiotic Residues in Tissues Using Liquid Chromatography

Tissue	Analytes	Sample Treatment	Derivati-zation	LC Column Technique	Detection	Limit (ppb)	Year (Reference)
Standards	Lasalocid, salinomycin, narasin, monensin	—	—	RPLC-C18	MS esi+	—[a]	1998 (420)
Standards	Salinomycin, narasin, lasalocid	—	—	RPLC-C18	MSMS esi+	—[a]	1999 (421)
Muscle	Lasalocid	—	—	RPLC-C18	MS esi+	—[a]	2002 (422)
Liver, eggs	Narasin, monensin, salinomycin, lasalocid	MeOH extn	—	RPLC-C18	MSMS esi+	0.5	2002 (423)
Muscle, eggs	Narasin, monensin, salinomycin, lasalocid, maduramycin, nigericin[(IS)]	ACN extn, silica SPE	—	RPLC-C18	MSMS esi+	0.1–0.5	2004 (424)
Eggs	Lasalocid, salinomycin, narasin, monensin, nigericin[(IS)]	Organic extn	—	RPLC-C18	MSMS esi+	1	2005 (425)
Muscle	Lasalocid, salinomycin, narasin, monensin	ACN/MeOH extn, hexane/ACN defat., evap., reconst.	—	RPLC-C18	MSMS esi+	0.6–2	2006 (214)

Note: RPLC: reverse phase liquid chromatography; MS: single quadrupole; esi+: electrospray source in positive mode; MSMS: triple quadrupole; MeOH: methanol; ACN: acetonitrile; SPE: solid-phase extraction; Extn: extraction; defat.: defattening; Evap.: evaporation before reconstitution; Reconst: reconstitution prior to injection; purif.: purification; EtOAc: ethyl acetate; SDB: styrene divinylbenzene; HLB: hydrophilic Lipophilic Balance; PLRP-S: polymeric stationary phase; UV: ultraviolet detection; DAD: diode array detection; LIF: laser-induced fluorescence; MS^n: ion trap; esi⁻: electrospray source in negative mode; ppb: parts per billion equivalent to µg of residue/kg of tissue.

[a] Not reported.
[(IS)] Internal standard.

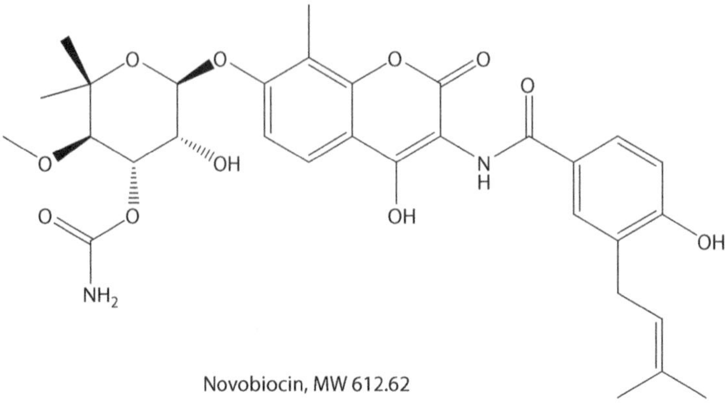

Novobiocin, MW 612.62

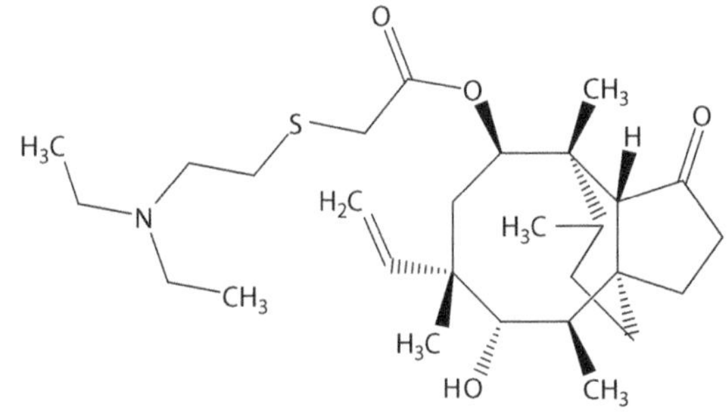

Tiamulin, MW 493.74

Figure 10.14 Structures of novobiocin and tiamulin.

in chloroform. It bears both an enolic (pK_a 4.3) and a phenolic (pK_a 9.2) character (Figure 10.14), leading to a weak dibasic acid behavior. As a consequence, liquid–liquid partitioning is not an efficient process for extraction-purification from biological matrices. Veterinary treatments for lactating cows may lead to novobiocin residues in milk, and feed additive practice in poultry may give novobiocin residues in chicken muscle. Microbiological methods have been tentatively applied to novobiocin residue detection in meat [29]. TLC-bioautography was a formerly screening method applied to novobiocin [55]. Although very few LC methods for monitoring novobiocin in meat are reported [427], recent articles relate to residues in egg [428] or in milk [429] as displayed in Table 10.12.

Tiamulin is a diterpene antibiotic with a pleuromutilin chemical structure similar to that of valnemulin with an eight-membered carbocyclic ring at the center of the structure. Pleuromutilines are biosynthetically produced by *Pleurotus mutilus*. Tiamulin's activity is largely confined to gram-positive microorganisms. This antibiotic acts by inhibiting bacterial protein synthesis at the ribosomal level. Its usage in veterinary medicine applies for treatment and prophylaxis of dysentery, pneumonia, and mycoplasmal infections in pigs and poultry. The principal residue to be monitored in muscle and in other meat tissues is the metabolite 8-alpha-hydroxymutilin. A bioscreening assay was reported in 2000 to monitor tiamulin activity [430]. Two LC methods (Table 10.12) are reported for tiamulin with one in meat by UV detection [431] and one in honey by MS detection [432].

Table 10.12 Summary of Literature Methods for Determination of Several Antibiotic Residues of Lower Interest in Tissues Using Gas or Liquid Chromatography

Tissue	Analytes	Sample Treatment	Derivatization	LC Column Technique	Detection	Limit (ppb)	Year (Reference)
Muscle, milk	Novobiocin	MeOH extn deprot., filtering, online SPE	—	RPLC-C18	UV: 340 nm	50	1988 (427)
Eggs	Novobiocin	MeOH extn, silica SPE, hexane defat.	—	RPLC-C18	MS^n esi+	3	2004 (428)
Milk	Novobiocin	Buffer dilution, MeOH deprot., filtering	—	RPLC-C18	UV: 340 nm	50	2005 (429)
Muscle	Tiamulin	ACN extn, Evap., hexane purif., C18 SPE	—	RPLC-C18	UV: 210 nm	25	2006 (431)
Honey	Tiamulin	Organic extn, SDB SPE	—	RPLC-C18	MS esi+	0.5–1.2	2006 (432)

Note: MeOH: methanol; SPE: solid-phase extraction; RPLC: reverse phase liquid chromatography; UV: ultraviolet detection; MS^n: ion trap; esi+: electrospray source in positive mode; ACN: acetonitrile; Extn: extraction; Purif.: purification; deprot.: deproteinisation; defat. defattening; DAD: diode array detection; MS: single quadrupole; MSMS: triple quadrupole; ppb: parts per billion equivalent to μg of residue/kg of tissue.

[a] Not reported.
(IS)Internal standard.

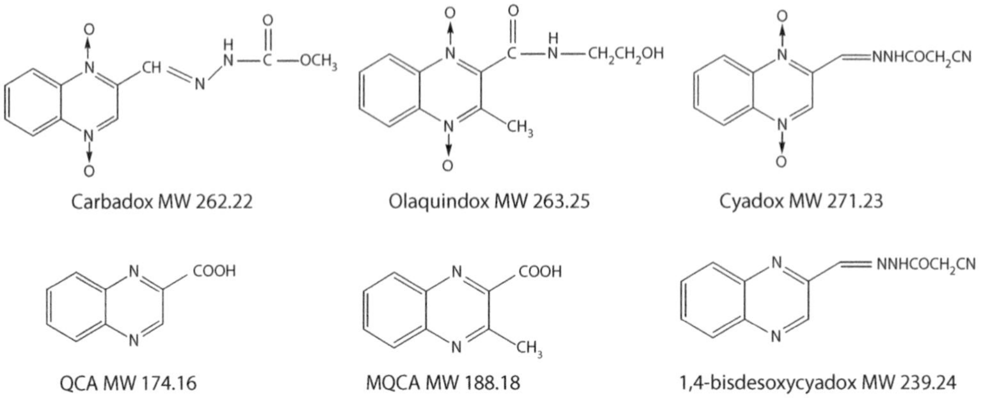

Carbadox MW 262.22   Olaquindox MW 263.25   Cyadox MW 271.23

QCA MW 174.16   MQCA MW 188.18   1,4-bisdesoxycyadox MW 239.24

Figure 10.15   Structures of 3 *N,N'*-di-*N*-oxide quinoxalines and their major metabolites—QCA, MQCA, and 1,4-bisdesoxycyadox.

10.4.13   Other Antibacterials (Carbadox and Olaquindox)

Carbadox and olaquindox are two widely available antibacterial synthetic *N,N'*-di-*N*-oxide qui-noxaline compounds used as growth promoters (Figure 10.15). Possible mutagenicity and carcino-genicity have been demonstrated for these active molecules, leading to their ban in many countries including EU under regulation 2788/98/EC [433]. As metabolism studies have shown the rapid conversion of these compounds into their monooxy- and desoxy-metabolites which are also pos-sible mutagenic and carcinogenic entities, it was important to monitor either the intermediate desoxycarbadox or final metabolite quinoxaline-2-carboxylic acid (QCA) as residual targets in muscle or liver tissues and also the methyl-3-quinoxaline-2-carboxylic acid (MQCA) as the stable metabolite for olaquindox. As described in Table 10.13, several LC-UV and LC-MS methods were proposed recently to cover the control of the ban of these two quinoxaline compounds [434–441]. Another quinoxaline 1,4-dioxide with antimicrobial properties, the cyadox (CYX), might be of interest for monitoring as it is possibly used as a growth promoter. It metabolizes as carbadox and olaquindox in animal tissues to give, among other intermediate compounds, the 1,4-bisdesoxy-cyadox (BDCYX). A paper related to the analysis of its major metabolites by LC-UV in chicken muscle tissues is reported in the literature [442].

10.5   Conclusion

Antimicrobials are one of the largest families of pharmaceuticals used in veterinary medicine either to treat animal diseases or to prophylactically prevent their occurrence. Also used to pro-mote the growth of food-producing animals, this practice is prone to drastic reduction in the near future. The control of veterinary drug residues in meat and other food from animal origin is one of the concerns for food safety regulation. It is important to prevent risks for human health. Farmers, veterinarians, feed manufacturers, food industry, and regulatory agencies together have to create the conditions of the food safety for the consumers. The control of anti-microbial resistance of certain bacteria is also another challenge for human health. Regulating

Table 10.13 Summary of Literature Methods for Determination of Carbadox and Olaquindox Residues in Tissues Using Gas or Liquid Chromatography

Tissue	Analytes	Sample Treatment	Derivatization	LC Column Technique	Detection	Limit (ppb)	Year (Reference)
Liver	QCA	Tris buffer/Subtilisin A enzymat. digestion heat extn; acetic acid neutral.; DEE LLE; EtOAc LLE; HCl dilution; EtOAc extn; evap.; H2SO4/propanol esterification; hexane defat.; evap. reconst.; HPLC fraction purif.; Evap.; hexane extn; evap.; EtOH reconst	H2SO4/propanol esterification	GC	MS	10	1990 (434)
Liver, muscle	QCA	Alkaline hydrolysis, neutral., EtOAc LLE, citric acid dilution, SCX SPE	Methylation for GC	RPLC-C18 GC	UV: 320 nm MS nci	3–5	1996 (435)
Liver	QCA, QCA-d4(IS)	Alkaline hydrolysis, neutral., LLE	—	RPLC-C18	MSMS esi+	0.2	2002 (436)
Muscle, liver	QCA, MQCA, QCA-d4(IS), MQCA-d4(IS)	Alkaline hydrolysis, HCl acidif., SCX SPE, HCl acidif, EtOAc LLE, evap., reconst.		RPLC-C18	MSMS esi+	1–4	2005 (437)
Feeds	Carbadox, olaquindox	ACN/CHCl3 extn, evap., reconst.	—	RPLC-C18	MSMS esi+	500	2005 (438)
Liver	QCA, MQCA	Protease digestion, LLE, SPE cleanup, LLE	—	RPLC-C18	MSMS esi+	1–3	2005 (439)
Liver	QCA, QCA-d4(IS)	MPA/MeOH deprot., PBS/EtOAc extn, MAX SPE	MTBSTFA, or TMSDM	GC	MS nci	0.7	2007 (440)
Liver, muscle	QCA, MQCA	Acid hydrolysis, LLE, MAX SPE	—	RPLC-C18	UV	1–5	2007 (441)

(continued)

Table 10.13 (continued) Summary of Literature Methods for Determination of Carbadox and Olaquindox Residues in Tissues Using Gas or Liquid Chromatography

Tissue	Analytes	Sample Treatment	Derivatization	LC Column Technique	Detection	Limit (ppb)	Year (Reference)
Muscle	QCA, BDCYX	QCA: alkaline hydrolysis, neutral., EtOAc LLE, citric acid dilution, IEC purif., HCl/CHCl$_3$ partition., evap., MeOH	—	RPLC-C18	UV	20–25	2005 (442)
		BDCYX: ACN extn, evap., ACN/hexane defat., evap., MeOH reconst.					

Note: QCA: quinoxaline-2-carboxylic acid; EtOAc: ethyl acetate; Evap.: evaporation before reconstitution; MS: single quadrupole; LLE: liquid-liquid extraction/partitioning; RPLC: reverse phase liquid chromatography; UV: ultraviolet detection; SCX: strong cation exchange; SPE: solid-phase extraction; MSMS: triple quadrupole; esi+: electrospray source in positive mode; MQCA: methyl-3-quinoxaline-2-carboxylic acid; ACN: acetonitrile; Reconst: reconstitution prior to injection; MPA: metaphosphoric acid; MeOH: methanol; PBS: phosphate buffer solution; MAX: medium anion exchange; MTBSTFA: N-methyl-N-tert-butyldimethylsilyltrifluoroacetamide; TMSDM: , trimethylsilyldiazomethane; BDCYX: 1,4-bisdesoxycyadox; IEC: ion exchange cartridge; Extn: extraction; DEE: diethyl ether; HLB: hydrophilic Lipophilic Balance; PLRP-S: polymeric stationary phase; MS^n: ion trap; esi-: electrospray source in negative mode; nci: negative chemical ionization; ppb: parts per billion equivalent to μg of residue/kg of tissue.

a Not reported.

(IS)Internal standard.

antibacterials in animal husbandry is the first step to contain bacterial resistance and control human diseases and their potential cure.

Analytical methods developed for monitoring antimicrobial residues in meat and in other food products from animal origin can be ranged in two different stages. At first stage, there is the strategy of screening to be adapted for evaluating in a reduced period of time the presence or absence of antimicrobials. The screening should generally be as large as possible in terms of the residues tested. This is the concept applied in the microbiological inhibitory methods (plate tests, swab tests, or receptor tests). But, screening methods reduced to one family or even one compound (immunological tests) can also be proposed in some specific cases, for specific antimicrobial monitoring, or with particular food products. At the second stage, there is an increasing interest in the strategy of confirmation of the residues with unequivocal identification of the analyte(s). During the past 15 years, physicochemical technologies have been developed and implemented for that specific purpose. From TLC to HPLC, from HPLC to LC-MS, and from LC-MS to LC-MSMS or LC-MS^n systems, it is obvious that nowadays it is the innovative mass spectrometric technology that is in application at the confirmatory step. One methodology that has been increasingly disregarded because of the use of innovative mass spectrometric technologies is the chemical extraction/purification process. However, this is often a bad consideration because optimizing the extraction is still of great importance in antimicrobial residue testing. Following the same idea, a thorough purification before injecting into the analytical instruments, including the LC-MS ones, can be of great help to improve the reliability of the analysis. In fact, chemical diversity of antimicrobials always requires the setting up of different approaches for their extraction from food matrices. Extraction methods have considerably changed in the past 20 years. LLE has been miniaturized or largely replaced by solid-phase extraction. Matrix solid-phase extraction (MSPD) and accelerated solvent extraction (ASE) are also emerging techniques. Extracting and purifying residues of antimicrobials from meat and other food matrices still need to be satisfactorily undertaken for quality and reliability of the analyses. On the instrumental part, a chromatographic separation is always necessary to optimize the selective control of an analyte with regard to the others and to the interfering substances in the purified extract. At the detection level of the analytical instrument, the mass spectrometer is now considered the optimal detector for controlling reliable identification and sufficient quantification of antibiotic residues in the ppb (μg/kg) range of concentrations. Several food safety legislations including the EU one have now enforced this concept, in particular for the prohibited substances [4].

Abbreviations

6-APA	6-aminopenicillenic acid
7-ACA	7-aminocephalosporanic acid
ACN	acetonitrile
ADC	azodicarbonamide
AHD	1-aminohydantoin
AMOZ	3-amino-5-morpholinomethyl-2-oxazolidinone
AOZ	3-amino-2-oxazolidinone
APCI	atmospheric pressure chemical ionization source
API	atmospheric pressure ionization source
APPI	atmospheric pressure photochemical ionization source
ASE	accelerated solvent extraction
BDCYX	1,4-bisdesoxycyadox

C8	8-carbon alkyl-bonded silica stationary phase
C18	18-carbon alkyl-bonded silica stationary phase
CEC	electrochromatography
CE-MS	capillary zone electrophoresis coupled to mass spectrometry
CYX	cyadox
CZE or CE	capillary zone electrophoresis
DAD	diode array detector
DCM	dichloromethane
DNA	desoxyribonucleic acid
DNC	dinitrocarbanilamide
DNSH	3,5-dinitrosalicylic acid hydrazide
DTE	dithioerythritol
ECD	electron capture detector
ELISA	enzyme-linked immunosorbent assay
ESI	electrospray source of ionization
EtOAc	ethyl acetate
EU	European Union
FID	flame ionization detector
FLD	fluorescence detector
Fluo	fluorescence detector
FPIA	fluorescence polarization immunoassay test
GC	gas chromatography
GC-MS	gas chromatography coupled to mass spectrometry
HCl	hydrochloric acid
HILIC	hydrophilic liquid chromatography
HLB	mixed hydrophilic–lipophilic balanced cartridge
HMMNI	2-hydroxymethyl-1-methyl-5-nitroimidazole
HPLC	high-performance liquid chromatography
HPLC-UV	high performance liquid chromatography connected to UV detector
IT	ion trap
LC	liquid chromatography
LC-MS	liquid chromatography coupled to mass spectrometry
LC-MSMS	liquid chromatography coupled to a tandem mass spectrometer also called triple quadrupolar mass spectrometer
LC-MS^n	liquid chromatography coupled to a ion trap mass spectrometer
LC-tandemMS	liquid chromatography coupled to a tandem mass spectrometer
LIF	laser-induced fluorescence detection
LLE	liquid–liquid extraction
MAX	medium anion exchange cartridge
MCX	medium cation exchange cartridge
MEKC	micellar electrokinetic capillary chromatography
MeOH	methanol
MQCA	methyl-3-quinoxaline-2-carboxylic acid
MRL	maximum residue limit
MRPL	minimum required performance limit
NPD	nitrogen-phosphorus detector
NRMP	national residue monitoring program (or plan)

MSPD	matrix solid-phase extraction
PB	particle beam source of ionization
PBS	phosphate buffer solution
PCA	pyridonecarboxylic acid
PDA	photodiode array
PLRP-S	polymeric reverse stationary phase
ppb	parts per billion equivalent to μg of residue/kg of tissue
ppm	parts per million equivalent to μg of residue/kg of tissue
QCA	quinoxaline-2-carboxylic acid
RPLC	reverse-phase liquid chromatography
RSD	relative standard deviation
SAX	strong anion exchange cartridge
SCX	strong cation exchange cartridge
SEM	semicarbazide
SFC	supercritical fluid chromatography
SPE	solid-phase extraction
SPFIA	solid-phase fluorescence immunoassay
SPR-BIA	surface plasmon resonance-based biosensor immunoassay
SQ	single quadrupole
TCA	trichloroacetic acid
TLC	thin layer chromatography
ToF	time-of-flight
TQ	triple quadrupole
TRIS	tris(hydroxymethyl)-aminomethane
TSP	thermospray source of ionization
UV	ultra-violet

References

1. Council Regulation (EEC) No. 2377/90 of 26th June 1990, *Off. J. Eur. Commun.*, L224 (1990) 1.
2. Council Directive 96/23/EC of 29th April 1996, *Off. J. Eur. Commun.*, L125 (1996) 10.
3. Council Regulation (EEC) No. 882/2004 of 29th April, 29th, 2004, *Off. J. Eur. Commun.*, L165 (2004) 1 and its Corrigendum of 28th May, 28th, 2004, *Off. J. Eur. Commun.*, L191 (2004) 1.
4. Commission Decision 2002/657/EC of 12th August 2002, *Off. J. Eur. Commun.*, L221 (2002) 8.
5. Commission Decision 93/256/EEC of 14th April 1993, *Off. J. Eur. Commun.*, L118 (1993) 64.
6. Keukens, H.J. et al., Analytical strategy for the regulatory control of residues of chloramphenicol in meat: preliminary studies in milk, *J. Assoc. Off. Anal. Chem.*, 75, 245, 1992.
7. Moats, W.A., Liquid chromatography approaches to antibiotic residue analysis, *J. Assoc. Off. Anal. Chem.*, 73, 343, 1990.
8. Bobbitt, D.R. and Ng, K.W., Chromatographic analysis of antibiotic materials in food, *J. Chromatogr.*, 624, 153, 1992.
9. Shaikh, B. and Moats, W.A., Liquid chromatographic analysis of antibacterial drug residues in food products of animal origin, *J. Chromatogr.*, 643, 369, 1993.
10. Aerts, M.M.L., Hogenboom, A.C., Brinkman, U.A.Th., Analytical strategies for the screening of veterinary drugs and their residues in edible products, *J. Chromatogr. B*, 667, 1, 1995.
11. Botsoglou, N.A. and Kufidis, D.C., Determination of antimicrobial residues in edible animal products by HPLC, in *Handbook of Food Analysis*, Vol. 2, Nollet, L.M.L., Ed., Marcel Dekker, New York, 1996.

12. Korsrud, G.O. et al., Bacterial inhibition tests used to screen for antimicrobial veterinary drug residues in slaughtered animals, *J. Assoc. Off. Anal. Chem. Int.*, 81, 21, 1998.

13. Di Corcia, A. and Nazzari, M., Liquid chromatographic-mass spectrometric methods for analyzing antibiotic and antibacterial agents in animal food products, *J. Chromatogr. A*, 974, 53, 2002.

14. Kennedy, D.G. et al., Use of liquid chromatography-mass spectrometry in the analysis of residues of antibiotics in meat and milk, *J. Chromatogr. A*, 812, 77, 1998.

15. Stolker, A.A.M. and Brinkman, U.A.Th., Analytical strategies for residue analysis of veterinary drugs and growth-promoting agents in food-producing animals—a review, *J. Chromatogr. A*, 1067, 15, 2005.

16. Moats, W.A., Liquid chromagraphic approaches to determination of beta-lactam antibiotic residues in milk and tissues, in *Analysis of Antibiotic Drug Residues in Food Products of Animal Origin*, Agarwal, V.K., Ed., Plenum Press, New York, NY, 1992, 133.

17. Boison, J.O., Review—Chromatographic methods of analysis for penicillins in food-animal tissues and their significance in regulatory programs for residue reduction and avoidance, *J. Chromatogr.*, 624, 171, 1992.

18. Shaikh, B. and Allen, E.H., Overview of physico-chemical methods for determining aminoglycoside antibiotics in tissues and fluids of food-producing animals, *J. Assoc. Off. Anal. Chem.*, 68, 1007, 1985.

19. Salisbury, C.D.C., Chemical analysis of aminoglycoside antibiotics, in *Chemical Analysis for Antibiotics Used in Agriculture*, Oka, H. et al., Eds., AOAC International, Arlington, VA, 1995.

20. Boison, J.O., Chemical analysis of beta-lactam antibiotics, in *Chemical Analysis for Antibiotics Used in Agriculture*, Oka, H. et al., Eds., AOAC International, Arlington, VA, 1995.

21. Wang, S. et al., Analysis of sulphonamide residues in edible animal products: a review, *Food Addit. Contam.*, 23, 362, 2006.

22. Dorsey, J.G. et al., Liquid chromatography: theory and methodology, *Anal. Chem.*, 66, 1R, 1994.

23. Bruckner, C.A. et al., Column liquid chromatography: equipment and instrumentation, *Anal. Chem.*, 66, 1R, 1994.

24. La Course, W.R., Column liquid chromatography: equipment and instrumentation, *Anal. Chem.*, 72, 37R, 2000.

25. La Course, W.R., Column liquid chromatography: equipment and instrumentation, *Anal. Chem.*, 74, 2813, 2002.

26. Nouws, J.F.M. et al., The new Dutch kidney test, *Arch. Lebensmittelhygiene*, 39, 133, 1988.

27. Bogaerts, R. and Wolf, F., A standardized method for the detection of residues of antibacterial substances in fresh meat, *Die Fleischwirtschaft*, 60, 672, 1980.

28. Ellerbroek, L., The microbiological determination of the quinolone carbonic acid derivatives enrofloxacin, ciprofloxacin and flumequine, *Die Fleischwirtschaft*, 71, 187, 1991.

29. Calderon, V. et al., Evaluation of a multiple bioassay technique for determination of antibiotic residues in meat with standard solutions of antimicrobials, *Food Addit. Contam.*, 13, 13, 1996.

30. Korsrud, G.O. et al., Bacterial inhibition tests used to screen for antimicrobial veterinary drug residues in slaughtered animals, *J. Assoc. Off. Anal. Chem. Internat. Internat.*, 81, 21, 1998.

31. Currie, D., Evaluation of a modified EC four plate method to detect antimicrobial drugs, *Food Addit. Contam.*, 15, 651, 1998.

32. Okerman, L., De Wasch, K., and Van Hoof, J., Detection of antibiotics in muscle tissue with microbiological inhibition tests: effects of the matrix, *Analyst*, 123, 2361, 1998.

33. Fuselier, R., Cadieu, N., and Maris, P., S.T.A.R.: screening test for antibiotic residues in muscle—results of a European collaborative study, in *Proceedings Euroresidue IV Conference on Residues of Veterinary Drugs in Food*, Veldhoven, Netherlands, May 8–10, Van Ginkel, L.A. and Ruiter, A., Eds., Faculty of Medicine, Utrecht, Netherlands, 2000, 444.

34. Myllyniemi, A.-L. et al., A microbiological six-plate method for the identification of certain antibiotic groups in incurred kidney and muscle samples, *Analyst*, 126, 641, 2001.

35. Gaudin, V. et al., Validation of a microbiological method: the STAR protocol, a five-plate test, for the screening of antibiotic residues in milk, *Food Addit. Contam.*, 21, 422, 2004.

36. Pena, A. et al., Antibiotic residues in edible tissues and antibiotic resistance of faecal *Escherichia coli* in pigs from Portugal, *Food Addit. Contam.*, 21, 749, 2004.

37. Ferrini, A.-M., Mannoni, V., and Aureli, P., Combined plate microbial assay (CPMA): a 6-plate-method for simultaneous first and second level screening of antibacterial residues, *Food Addit. Contam.*, 23, 16, 2006.

38. Johnston, R.W., A new screening method for the detection of antibiotic residues in meat and poultry tissues, *J. Food Prot.*, 44, 828, 1981.

39. Dey, B.P. et al., Calf antibiotic and sulfonamide test (CAST) for screening antibiotic and sulfonamide residues in calf carcasses, *J. Assoc. Off. Anal. Chem. Internat. Internat.*, 88, 440, 2005.

40. Dey, B.P. et al., Fast antimicrobial screen test (FAST): improved screen test for detecting antimicrobial residues in meat tissue, *J. Assoc. Off. Anal. Chem. Internat. Internat.*, 88, 447, 2005.

41. Koenen-Dierick, K. et al., A one-plate microbiological screening test for antibiotic residue testing in kidney tissue and meat: an alternative to the EEC four-plate method ?, *Food Addit. Contam.*, 12, 77, 1995.

42. Cornet, V. et al., Interlaboratory study based on a one-plate screening method for the detection of antibiotic residues in bovine kidney tissue, *Food Addit. Contam.*, 22, 415, 2005.

43. Myllyniemi, A.-L. et al., Microbiological and chemical identification of antimicrobial drugs in kidney and muscle samples of bovine cattle and pigs, *Food Addit. Contam.*, 16, 339, 1999.

44. Charm, S.E. and Chi, R., Microbial receptor assay for rapid detection and identification of seven families of antimicrobial drugs in milk: collaborative study, *J. Assoc. Off. Anal. Chem.*, 71, 304, 1988.

45. Korsrud, G.O. et al., Evaluation and testing of Charm test II receptor assays for the detection of antimicrobial residues in meat, in *Analysis of Antibiotic Drug Residues in Food Products of Animal Origin*, Agarwal, V.K., Ed., Plenum Press, New York, NY, 1992, 75.

46. Korsrud, G.O. et al., Investigation of Charm test II receptor assays for the detection of antimicrobial residues in suspect meat samples, *Analyst*, 119, 2737, 1994.

47. Smither, R. and Vaughan, D.R., An improved electrophoretic method for identifying antibiotics with special reference to animal tissues and animal feeding stuffs, *J. Appl. Bacteriol.*, 44, 421, 1978.

48. Stadhouders, J., Hassing, F., and Galesloot, T.E., An electrophoretic method for the identification of antibiotic residues in small volumes of milk, *Neth. Milk Dairy J.*, 35, 23, 1981.

49. Lott, A.F., Smither, R., and Vaughan, D.R., Antibiotic identification by high voltage electrophoresis bioautography, *J. Assoc. Off. Anal. Chem.*, 68, 1018, 1985.

50. Tao, S.H. and Poumeyrol, M., Detection of antibiotic residues in animal tissues by electrophoresis, *Rec. Med. Vet.*, 161, 457, 1985.

51. Bossuyt, R., Van Renterghem, R., and Waes, G., Identification of antibiotic residues in milk by thin-layer chromatography, *J. Chromatogr.*, 124, 37, 1976.

52. Herbst, D.V., Applications of TLC and bioautography to detect contaminated antibiotic residues: tetracycline identification scheme, *J. Pharm. Sci.*, 69, 616, 1980.

53. Yoshimura, H., Itoh, O., and Yonezawa, S., Microbiological and thin-layer chromatographic identification of benzylpenicillin and ampicillin in animal body, *Jpn. J. Vet. Sci.*, 43, 833, 1981.

54. Kondo, F., A simple method for the characteristic differentiation of antibiotics by TLC-bioautography in graded concentration of ammonium chloride, *J. Food Prot.*, 51, 786, 1988.

55. Salisbury, C.D.C., Rigby, C.E., and Chan, W., Determination of antibiotic residues in Canadian slaughter animals by thin-layer chromatography-bioautography, *J. Agric. Food Chem.*, 37, 105, 1989.

56. Lin, S.-Y. and Kondo, F., Simple bacteriological and thin-layer chromatographic methods for determination of individual drug concentrations treated with penicillin-G in combination with one of the aminoglycosides, *Microbios*, 77, 223, 1994.

57. MacNeil, J.D., Current laboratory testing strategy for the identification and confirmation of antibiotic residues in fresh meat, in *Proceedings Euroresidue II Conference on Residues of Veterinary Drugs in Food*, Veldhoven, Netherlands, Haagsma, N., Ruiter, A., and Czedik-Eysenberg, P.B., Eds., Faculty of Medicine, Utrecht, Netherlands, 1993, 469.

58. Korsrud, G.O. et al., Laboratory evaluation of the Charm Farm test for antimicrobial residues in meat, *J. Food Prot.*, 58, 1129, 1995.
59. Stead, S. et al., Meeting maximum residue limits: an improved screening technique for the rapid detection of antimicrobial residues in animal food products, *Food Addit. Contam.*, 21, 216, 2004.
60. Fabre, J.-M. et al., Résidus d'antibiotiques dans la viande de porc et de volaille en France: situation actuelle et évaluation d'un nouveau test de détection, *Bull. des G.T.V.*, 23, 305, 2004.
61. Cantwell, H. and O'Keeffe, M., Evaluation of the Premi® test and comparison with the one plate test for the detection of antimicrobials in kidney, *Food Addit. Contam.*, 23, 120, 2006.
62. Okerman, L. et al., Evaluation and establishing the performance of different screening tests for tetracycline residues in animal tissues, *Food Addit. Contam.*, 21, 145, 2004.
63. Brown, S.A. et al., Extraction methods for quantitation of gentamicin residues from tissues using fluorescence polarization immunoassay, *J. Assoc. Off. Anal. Chem.*, 73, 479, 1990.
64. Okerman, L. et al., Simultaneous determination of different antibiotic residues in bovine and porcine kidneys by solid-phase fluorescence immunoassay, *J. Assoc. Off. Anal. Chem. Internat.*, 86, 236, 2003.
65. Korpimäki, T. et al., Generic lanthanide fluoroimmunoassay for the simultaneous screening of 18 sulfonamides using an engineered antibody, *Anal. Chem.*, 76, 3091, 2004.
66. Corrégé, I. et al., Comparaison des méthodes rapides de détection des résidus d'antibactériens dans la viande de porc, *Techni-Porc*, 17/04/94, 29, 1994.
67. Crooks, S.R.H. et al., The production of pig tissue sulphadimidine reference material, *Food Addit. Contam.*, 13, 211, 1996.
68. Haasnoot, W. et al., Application of an enzyme immunoassay for the determination of sulphamethazine (sulphadimidine) residues in swine urine and plasma and their use as predictors of the level in edible tissue, *Food Addit. Contam.*, 13, 811, 1996.
69. De Wasch, K. et al., Detection of residues of tetracycline antibiotics in pork and chicken meat: correlation between results of screening and confirmatory tests, *Analyst*, 123, 2737, 1998.
70. Medina, M.B., Poole, D.J., and Anderson, M.R., A screening method for beta-lactams in tissues hydrolyzed with penicillinase I and lactamase II, *J. Assoc. Off. Anal. Chem. Internat.*, 81, 963, 1998.
71. Edder, P., Cominoli, A., and Corvi, C., Determination of streptomycin residues in food by SPE and LC with post-column derivatization and fluorimetric detection, *J. Chromatogr. A*, 830, 345, 1999.
72. Draisci, R. et al., A new electrochemical enzyme-linked immunosorbent assay for the screening of macrolide antibiotic residues in bovine meat, *Analyst*, 126, 1942, 2001.
73. Watanabe, H. et al., Monoclonal-based enzyme-linked immunosorbent assay and immunochromatographic assay for enrofloxacin in biological matrices, *Analyst*, 127, 98, 2002.
74. Gaudin, V., Cadieu, N., and Maris, P., Inter-laboratory studies for the evaluation of ELISA kits for the detection of chloramphenicol residues in milk and muscle, *Food Agric. Immunol.*, 15, 143, 2003.
75. Fodey, T.L. et al., Production and characterisation of polyclonal antibodies to a range of nitroimidazoles, *Anal. Chim. Acta*, 483, 193, 2003.
76. Grunwald, L. and Petz, M., Food processing effects on residues: penicillins in milk and yoghurt, *Anal. Chim. Acta*, 483, 73, 2003.
77. Posyniak, A., Zmudski, J., and Niedzielska, J., Evaluation of sample preparation for control of chloramphenicol residues in porcine tissues by enzyme-linked immunosorbent assay and liquid chromatography, *Anal. Chim. Acta*, 483, 307, 2003.
78. Impens, S. et al., Screening and confirmation of chloramphenicol in shrimp tissue using ELISA in combination with GC-MSMS and LC-MSMS, *Anal. Chim. Acta*, 483, 153, 2003.
79. Gaudin, V., Cadieu, N., and Sanders, P., Validation of commercial ELISA kits for banned substances: chloramphenicol and nitrofurans, communication presented at *Workshop on Validation of Screening Methods*, Fougeres, France, June 3–4, 2005 (unpublished data).
80. Kuhlhoff, S. and Diehl, Y., Schneller Nachweis von Fluoroquinolonen bei Geflügelfleischproben aus dem Handel, *Deutsche Lebensmittel-Rundschau*, 101, 384, 2005.

81. Norgaard, A., Validation of a screening method for detection of chloramphenicol in muscle, casings and urine, sing previous obtained results, communication presented at *Workshop on Validation of Screening Methods*, Fougeres, France, June 3-4, 2005 (unpublished data).

82. Huet, A.-C. et al., Simultaneous determination of (fluoro)quinolone antibiotics in kidney, marine products, eggs and muscle by ELISA, communication presented at the *5th International Symposium on Hormone and Veterinary Drug Residue Analysis*, Antwerp, Belgium, May 16–19, 2006.

83. Huet, A.-C. et al., Simultaneous determination of (fluoro)quinolone antibiotics in kidney, marine products, eggs, and muscle by enzyme-linked immunosorbent assay (ELISA), *J. Agric. Food Chem.*, 54, 2822, 2006.

84. Huet, A.-C. et al., Development of an ELISA screening test for halofuginone, nicarbazin and nitro-imidazoles in egg and chicken muscle, communication presented at the *5th International Symposium on Hormone and Veterinary Drug Residue Analysis*, Antwerp, Belgium, May 16–19, 2006.

85. Sternesjö, A., Mellgren, C., and Björck, L., Determination of sulfamethazine residues in milk by a surface plasmon resonance-based biosensor assay, *Anal. Biochem.*, 226, 175, 1995.

86. Crooks, S.R.H. et al., Immunobiosensor—an alternative to enzyme immunoassay screening for residues of two sulfonamides in pigs, *Analyst*, 123, 2755, 1998.

87. Baxter, G.A. et al., Evaluation of an immunobiosensor for the on-site testing of veterinary drug residues at an abattoir. Screening for sulfamethazine in pigs, *Analyst*, 124, 1315, 1999.

88. Gaudin, V. and Pavy, M.L., Determination of sulfamethazine in milk by biosensor immunoassay, *J. Assoc. Off. Anal. Chem. Internat.*, 82, 1316, 1999.

89. Bjurling, P. et al., Biosensor assay of sulfadiazine and sulfamethazine residues in pork, *Analyst*, 125, 1771, 2000.

90. Haasnoot, W., Bienenmann-Ploum, M., and Kohen, F., Biosensor immunoassay for the detection of eight sulfonamides in chicken serum, *Anal. Chim. Acta*, 483, 171, 2003.

91. Gaudin, V., Fontaine, J., and Maris, P., Screening of penicillin residues in milk by a surface plasmon resonance based biosensor assay: comparison of chemical and enzymatic sample treatment, *Anal. Chim. Acta*, 436, 191, 2000.

92. Gaudin, V. and Maris, P., Development of a biosensor based immunoassay for screening of chloramphenicol residues in milk, *Food Agric. Immunol.*, 13, 77, 2000.

93. Ferguson, J.P. et al., Detection of streptomycin and dihydrostreptomycin residues in milk, honey and meat samples using an optical biosensor, *Analyst*, 127, 951, 2002.

94. McCarney, B. et al., Surface plasmon resonance biosensor screening of poultry liver and eggs for nicarbazin residues, *Anal. Chim. Acta*, 483, 165, 2003.

95. Gustavsson, E., SPR biosensor analysis of beta-lactam antibiotics in milk. Development and use of assays based on a beta-lactam receptor protein, Doctoral thesis, Swedish University of Agricultural Sciences, Uppsala, 2003.

96. Mellgren, C. and Sternesjö, A., Optical immunobiosensor assay for determining enrofloxacin and ciprofloxacin in bovine milk, *J. Assoc. Off. Anal. Chem. Internat.*, 81, 394, 1998.

97. Langner, H.J. and Teufel, U., Chemical and microbiological detection of antibiotics 1. Thin-layer chromatographic separation of various antibiotics, *Die Fleischwirtschaft*, 52, 1610, 1972.

98. Parks, O.W., Screening test for sulfamethazine and sulfathiazole in swine liver, *J. Assoc. Off. Anal. Chem.*, 65, 632, 1982.

99. Thomas, M.H., Soroka, K.E., and Thomas, S.H., Quantitative thin-layer chromatographic multi-sulfonamide screening procedure, *J. Assoc. Off. Anal. Chem.*, 66, 881, 1983.

100. Herbst, D.V., Identification and determination of four beta-lactam antibiotics in milk, *J. Food Prot.*, 45, 450, 1982.

101. Moats, W.A., Detection and semiquantitative estimation of penicillin-G and cloxacillin in milk by thin-layer chromatography, *J. Agric. Food Chem.*, 31, 1348, 1983.

102. Moats, W.A., Chromatographic methods for determination of macrolide antibiotic residues in tissues and milk of food-producing animals, *J. Assoc. Off. Anal. Chem.*, 68, 980, 1985.

103. Abjean, J.P., Planar chromatography for the multiclass, multiresidue screening of chloramphenicol, nitro-furan, and sulphonamide residues in pork and beef, *J. Assoc. Off. Anal. Chem. Internat.*, 80, 737, 1997.

104. Abjean, J.P. and Lahogue V., Planar chromatography for quantitative determination of ampicillin residues in milk and muscle, *J. Assoc. Off. Anal. Chem. Internat.*, 80, 1171, 1997.

105. Eiceman, G.A., Instrumentation of gas chromatography, in *Encyclopedia of Analytical Chemistry*, Meyers, R.A., Ed., Wiley, Chichester, 1994.

106. Meetschen, U. and Petz, M., Gas chromatographic method for the determination of residues of seven penicillins in foodstuffs of animal origin, *Z. Lebensm. Unters. Forsch.*, 193, 337, 1991.

107. Mineo, H. et al, An analytical study of antibacterial residues in meat: the simultaneous determination of 23 antibiotics and 13 drugs using gas chromatography, *Vet. Hum. Toxicol.*, 34, 393, 1992.

108. Kennedy, D.G. et al., Gas-chromatographic-mass spectrometric determination of sulfamethazine in animal tissues using a methyl/trimethylsilyl derivative, *Analyst*, 121, 1457, 1996.

109. Preu, M. and Petz, M., Isotope dilution GC-MS of benzylpenicillin residues in bovine muscle, *Analyst*, 123, 2785, 1998.

110. Polzer, J. and Gowik, P., Validation of a method for the detection and confirmation of nitroimidazoles and corresponding hydroxy metabolites in turkey and swine muscle by means of gas chromatography-negative ion chemical ionization mass spectrometry, *J. Chromatogr. B*, 761, 47, 2001.

111. Epstein, R.L. et al., International validation study for the determination of chloramphenicol in bovine muscle, *J. Assoc. Off. Anal. Chem. Internat.*, 77, 570, 1994.

112. Gude, Th., Preiss, A., and Rubach, K., Determination of chloramphenicol in muscle, liver, kidney and urine of pigs by means of immunoaffinity chromatography and gas chromatography with electron-capture detection, *J. Chromatogr. B*, 673, 197, 1995.

113. Nagata, T. and Oka, H., Detection of residual chloramphenicol, florfenicol, and thiamphenicol in yellowtail fish muscles by capillary gas chromatography-mass spectrometry, *J. Agric. Food Chem.*, 44, 1280, 1996.

114. Cerkvenik, V., Analysis and monitoring of chloramphenicol residues in food of animal origin in Slovenia from 1991 to 2000, *Food Addit. Contam.*, 19, 357, 2002.

115. Ding, S. et al., Determination of chloramphenicol residue in fish and shrimp tissues by gas chromatography with a microcell electron capture detector, *J. Assoc. Off. Anal. Chem. Internat.*, 88, 57, 2005.

116. Polzer, J. et al., Determination of chloramphenicol residues in crustaceans: preparation and evaluation of a proficiency test in Germany, *Food Addit. Contam.*, 23, 1132, 2006.

117. Zhang, S. et al., Simultaneous determination of florfenicol and florfenicol amine in fish, shrimp, and swine muscle by gas chromatography with a microcell electron capture detector, *J. Assoc. Off. Anal. Chem. Internat.*, 89, 1437, 2006.

118. Vissers, J.P.C., Claessens, H.A., and Cramers, C.A., Microcolumn liquid chromatography: instrumentation, detection and applications, *J. Chromatogr. A* 770, 1, 1997.

119. Krull, I.S. and Lankmayr, E.P., Derivatization reaction detector in HPLC, *Am. Lab.*, 14, 18, 1982.

120. Frei, R.W., Jansen, H., and Brinkman, U.A.Th., Postcolumn reaction detectors for HPLC, *Anal. Chem.*, 57, 1529, 1985.

121. Krull, I.S., Deyl, Z., and Lingeman, H., General strategies and selection of derivatization reactions for liquid chromatography and capillary electrophoresis, *J. Chromatogr. B Biomed. Appl.*, 659, 1, 1994.

122. Decolin, D. et al., Analyse de traces d'antibiotiques dans les tissus animaux par chromatographie en phase liquide couplée à la fluorescence, *Ann. Fals. Exp. Chim.*, 89, 11, 1996.

123. Yergey, A.L. et al., *Liquid Chromatography-Mass Spectrometry, Techniques and Applications*, Plenum Press, New York, NY, 1990, 316 pp.

124. Niessen, W.M.A. and Van der Greef, J., in *Liquid Chromatography-Mass Spectrometry: Principles and Applications*, *Chromatographic Science Series*, Vol. 58, Marcel Dekker, New York, 1992, 479 pp.

125. Kinter, M., Mass spectrometry, *Anal. Chem.*, 67, 493, 1994.

126. March, R.E., An introduction to quadrupole ion trap mass spectrometry, *J. Mass Spectrom.*, 32, 351,1997.

127. Cole, R.B., *Electrospray Ionization Mass Spectrometry—Fundamentals, Instrumentation and Applications*, Cole, R.B., Ed., Wiley, New York, NY, 1997.

128. Gaskell, S.J., Electrospray: principles and practice, *J. Mass Spectrom.*, 32, 677, 1997.
129. Burlingame, A.L., Boyd, R.K., and Gaskell, S.J., Mass spectrometry, *Anal. Chem.*, 70, 467, 1998.
130. Abian, J., The coupling of gas and liquid chromatography with mass spectrometry, *J. Mass Spectrom.*, 34, 157, 1999.
131. Niessen, W.M.A., Liquid chromatography-mass spectrometry, in *Chromatographic Science Series*, 2nd ed., Vol. 79, Niessen, W.M.A., Ed., Marcel Dekker, New York, NY, 1999.
132. Amad, M.H. et al., Importance of gas-phase proton affinities in determining the electrospray ionization response for analytes and solvents, *J. Mass Spectrom.*, 35, 784, 2000.
133. Shukla, A.K. and Futrell, J.H., Tandem mass spectrometry: dissociation of ions by collisional activation, *J. Mass Spectrom.*, 35, 1069, 2000.
134. Chernushevich, I.V., Loboda, A.V., and Thomson, B.A., An introduction to quadrupole-time-of-flight mass spectrometry, *J. Mass Spectrom.*, 36, 849, 2001.
135. Brown, M.A., Liquid chromatography-mass spectrometry, in *ACS Symposium Series*, Vol. 420, Brown, M.A., Ed., American Chemical Society, Washington, DC, 1990.
136. Strege, M.A., High-performance liquid chromatographic-electrospray ionization mass spectrometric analyses for the integration of natural products with modern high-throughput screening, *J. Chromatogr. B Biomed. Appl.*, 725, 67, 1999.
137. Careri, M., Bianchi, F., and Corradini, C., Recent advances in the application of mass spectrometry in food-related analysis, *J. Chromatogr. A*, 970, 3, 2002.
138. Perkins, J.R. et al., Analysis of sulphonamides using supercritical fluid chromatography and supercritical fluid chromatography-mass spectrometry, *J. Chromatogr.*, 540, 239, 1991.
139. Chester, T.L., and Pinkston, J.D., and Raynie, D.E., Supercritical fluid chromatography and extraction, *Anal. Chem.*, 66, 106, 1994.
140. Verillon, F., Dossier: supercritical-fluid chromatography, *Analusis*, 27, 671, 1999.
141. Chester, T.L. and Pinkston, J.D., Supercritical fluid and unified chromatography, *Anal. Chem.*, 72, 129, 2000.
142. Lloyd, D.K., Capillary electrophoretic analyses of drugs in body fluids: sample pretreatment and methods for direct injection of biofluids, *J. Chromatogr. A*, 735, 29, 1996.
143. Pesek, J.J. and Matyska, M.T., Column technology in capillary electrophoresis and capillary electro chromatography, *Electrophoresis*, 18, 2228, 1997.
144. El Rassi, Z., Capillary electrophoresis reviews, *Electrophoresis*, 18, 2121, 1997.
145. Robson, M.M. et al., Capillary electrochromatography: a review, *J. Microcol. Sep.*, 9, 357, 1997.
146. Thormann, W. and Caslavska, J., Capillary electrophoresis in drug analysis, *Electrophoresis*, 19, 2691, 1998.
147. Altria, K.D., Smith, N.W., and Turnbull, C.H., Analysis of acidic compounds using capillary electrochromatography, *J. Chromatogr. B Biomed. Appl.*, 717, 341, 1998.
148. Dermaux, A. and Sandra, P., Applications of capillary electrochromatography, *Electrophoresis*, 20, 3027, 1999.
149. Frazier, R.A., Ames, J.M., and Nursten, H.E., The development and application of capillary electrophoresis methods for food analysis, *Electrophoresis*, 20, 3156, 1999.
150. Colon, L.A. et al., Recent progress in capillary electrochromatography, *Electrophoresis*, 21, 3965, 2000.
151. Osbourn, D.M., Weiss, D.J., and Lunte, C.E., On-line preconcentration methods for capillary electrophoresis, *Electrophoresis*, 21, 2768, 2000.
152. Waterval, J.C.M. et al., Derivatization trends in capillary electrophoresis, *Electrophoresis*, 21, 4029, 2000.
153. Colon, L.A., Maloney, T.D., and Fermier, A.M., Packing columns for capillary electrochromatography, *J. Chromatogr. A*, 887, 43, 2000.
154. Pursch, M. and Sander, L.C., Stationary phases for capillary electrochromatography, *J. Chromatogr. A*, 887, 313, 2000.
155. Nishi, H. and Tsumagari, N., Effect of tetraalkylammonium salts on micellar electrokinetic chromato-graphy of ionic substances, *Anal. Chem.*, 61, 2434, 1989.

156. Nishi, H. et al., Separation of beta-lactam antibiotics by micellar electrokinetic chromatography, *J. Chromatogr.*, 477, 259, 1989.

157. Tsikas, D., Hofrichter, A., and Brunner, G., Capillary isotachophoretic analysis of beta-lactam antibiotics and their precursors, *Chromatographia*, 30, 657, 1990.

158. Zhu, Y.X. et al., Micellar electrokinetic capillary chromatography for the separation of phenoxymethylpenicillin and related substances, *J. Chromatogr. A*, 781, 417, 1997.

159. De Boer, T. et al., Selectivity in capillary electrokinetic separations, *Electrophoresis*, 20, 2989, 1999.

160. Smith, N.W. and Carterfinch, A.S., Electrochromatography, *J. Chromatogr. A*, 892, 219, 2000.

161. Legido-Quigley, C. et al., Advances in capillary electrochromatography and micro-high performance liquid chromatography monolithic columns for separation science, *Electrophoresis*, 24, 917, 2003.

162. Banks, J.F., Recent advances in capillary electrophoresis/electrospray/mass spectrometry, *Electrophoresis*, 18, 2255, 1997.

163. Bateman, K.P., Locke, S.J., and Volmer, D.A., Characterization of isomeric sulfonamides using capillary zone electrophoresis coupled with nano-electrospray quasi-MS/MS/MS, *J. Mass Spectrom.*, 32, 297, 1997.

164. Chen, Y.C. and Lin, X., Migration behavior and separation of tetracycline antibiotics by micellar electrokinetic chromatography, *J. Chromatogr. A*, 802, 95, 1998.

165. Berzas-Nevado, J.J., Castaneda-Penalvo, G., and Guzman-Bernardo, F.J., Micellar electrokinetic capillary chromatography as an alternative method for the determination of sulfonamides and their associated compounds, *J. Liq. Chromatogr Relat. Technol.*, 22, 975, 1999.

166. Spikmans, V. et al., Hyphenation of capillary electrochromatography with mass spectrometry: the technique and its application, *LC-GC Int.*, 13, 486, 2000.

167. Desiderio, C. and Fanali, S., Capillary electrochromatography and CE-electrospray MS for the separation of non-steroidal anti-inflammatory drugs, *J. Chromatogr. A*, 895, 123, 2000.

168. von Broke, A., Nicholson, G., and Bayer, E., Recent advances in capillary electrophoresis/electrospray-mass spectrometry, *Electrophoresis*, 22, 1251, 2001.

169. von Broke, A. et al., On-line coupling of packed capillary electrochromatography with coordination ion spray-mass spectrometry for the separation of enantiomers, *Electrophoresis*, 23, 2963, 2002.

170. Berrueta, L.A., Gallo, B., and Vicente, F., A review of solid-phase extraction: basic principles and new developments, *Chromatographia*, 40, 474, 1995.

171. Thurman, E.M. and Mills, M.S., *Solid Phase Extraction: Principles and Practice*, Wiley, New York, NY, 1998.

172. Franke, J.P. and De Zeeuw, R.A., Solid-phase extraction procedures in systematic toxicological analysis, *J. Chromatogr. B Biomed. Appl.*, 713, 51, 1998.

173. Hubert, P. et al., Préparation des échantillons d'origine biologique préalable à leur analyse chromatographique, *S.T.P. Pharma pratiques*, 9, 160, 1999.

174. Huck, C.W., and Bonn, G.K., Recent developments in polymer-based sorbents for solid-phase extraction, *J. Chromatogr. A*, 885, 51, 2000.

175. Buldini, P.L., Ricci, M.C., and Sharma, J.L., Recent applications of sample preparation techniques in food analysis, *J. Chromatogr. A*, 975, 47, 2002.

176. Verdon, E. and Couëdor, P., Multiresidue analytical method for the determination of eight penicillin antibiotics in muscle tissue by ion-pair reversed-phase HPLC after precolumn derivatization, *J. Assoc. Off. Anal. Chem. Internat.*, 82, 1083, 1999.

177. Delepine, B. and Hurtaud-Pessel D., Liquid chromatography/tandem mass spectrometry: screening method for the identification of residues of antibiotics in meat, communication presented at the *4th International Symposium on Hormone and Veterinary Drug Residue Analysis*, Antwerp, Belgium, June 3–7, 2002 (unpublished data).

178. Yoshimura, H., Itoh, O., and Yonezawa, S., Microbiological and TLC identification of aminoglycoside antibiotics in animal body, *Jpn. J. Vet. Sci.*, 44, 233, 1982.

179. Lachatre, G. et al., Séparation des aminosides et de leurs fractions par chromatographie en phase liquide—application à leur dosage dans le plasma, l'urine et les tissus, *Analusis*, 11, 168, 1983.

180. Shaikh, B., Allen, E.H., and Gridley, J.C., Determination of neomycin in animal tissues using ion-pair liquid chromatography with fluorometric detection, *J. Assoc. Off. Anal. Chem.*, 68, 29, 1985.
181. Nakaya, K.-I., Sugitani, A., and Yamada, F., Determination of kanamycin in beef and kidney of cattle by HPLC, *J. Food Hyg. Soc. Jpn*, 27, 258, 1986.
182. Okayama, A. et al., Fluorescence HPLC determination of streptomycin in meat using ninhydrin as a postcolumn labelling agent, *Bunseki Kagaku*, 37, 221, 1988.
183. Agarwal, V.K., HPLC determination of gentamicin in animal tissue, *J. Liq. Chromatogr.*, 12, 613, 1989.
184. Sar, F. et al., Development and optimization of a liquid chromatographic method for the determination of gentamicin in calf tissues, *Anal. Chim. Acta*, 275, 285, 1993.
185. Gerhardt, G.C., Salisbury, C.D.C., and MacNeil, J.D., Determination of streptomycin and dihydrostreptomycin in animal tissue by on-line sample enrichment liquid chromatography, *J. Assoc. Off. Anal. Chem. Internat.*, 77, 334, 1994.
186. Guggisberg, D. and Koch, H., Methode zur fluorimetrischen bestimmung von neomycin in der niere und in der leber mit HPLC und nachsaülenderivatisation, *Mitt. Gebiete Lebensmittel Hygiene*, 86, 449, 1995.
187. Hormazabal, V. and Yndestad, M., H.P.L.C. determination of dihydrostreptomycin sulfate in kidney and meat using post column derivatization, *J. Liq. Chromatogr. Relat. Technol.*, 20, 2259, 1997.
188. Abbasi, H. and Hellenäs, K.E., Modified determination of DHS in kidney, muscle and milk by HPLC, *Analyst*, 123, 2725, 1998.
189. Bergwerff, A.A., Scherpenisse, P., and Haagsma, N., HPLC determination of residues of spectinomycin in various tissue types from husbandry animals, *Analyst*, 123, 2139, 1998.
190. Morovjan, G., Csokan, P.P., and Nemeth-Konda, L., HPLC determination of colistin and aminoglycoside antibiotics in feeds by post-column derivatization and fluorescence detection, *Chromatographia*, 48, 32, 1998.
191. Reid, J.-A. and MacNeil, J.D., Determination of neomycin in animal tissue by liquid chromatography, *J. Assoc. Off. Anal. Chem. Internat.*, 82, 61, 1999.
192. Posyniak, A., Zmudski, J., and Niedzielska, J., Sample preparation for residue determination of gentamicin and neomycin by liquid chromatography, *J. Chromatogr. A*, 914, 59, 2001.
193. Fennell, M.A. et al., Gentamicin in tissue and whole milk: an improved method for extraction and cleanup of samples for quantitation on HPLC, *J. Agric. Food Chem.*, 43, 1849, 1995.
194. Cherlet, M., De Baere, S., and De Backer, P., Determination of gentamicin in swine and calf tissues by HPLC combined with electrospray ionization mass spectrometry, *J. Mass Spectrom.*, 35, 1342, 2000.
195. de Wasch, K. et al., Identification of "unknown analytes" in injection sites: a semi-quantitative interpretation, *Anal. Chim. Acta*, 483, 387, 2003.
196. Kaufmann, A. and Maden, K., Determination of 11 aminoglycosides in meat and liver by liquid chromatography with tandem mass spectrometry, *J. Assoc. Off. Anal. Chem. Internat.*, 88, 1118, 2005.
197. Sanders, P., LC determination of chloramphenicol in calf tissue: studies of stability in muscle, kidney and liver, *J. Assoc. Off. Anal. Chem.*, 74, 483, 1991.
198. O'Brien, J.J., Campbell, N., and Conaghan, T., Effect of cooking and cold storage on biologically active antibiotic residues in meat, *J. Hyg. Camb.*, 87, 511, 1981.
199. Ramos, M. et al., Chloramphenicol residues in food samples: their analysis and stability during storage, *J. Liq. Chromatogr. Relat. Technol.*, 26, 2535, 2003.
200. Allen, E.H., Review of chromatographic methods for chloramphenicol residues in milk, eggs, and tissues from food-producing animal, *J. Assoc. Off. Anal. Chem.*, 68, 5, 1985.
201. McCracken, A., O'Brien, J.J., and Campbell, N., Antibiotic residues and their recovery from animal tissues, *J. Appl. Bacteriol.*, 41, 129, 1976.
202. Kruzik, P. et al., Über den Nachweis und die Bestimmung antibiotsch wirksamer Substanzen in Lebensmitteln tierischer Herkunft: Sulfonamide, Nitrofurane, Nicarbazin, Tetracycline, Tylosin und Chloramphenicol, *Wiener Tierärztliche Monatsschrift*, Vol. 77, p. 141, 1989.

203. Cazemier, G., Haasnoot, W., and Stouten, P., Screening of chloramphenicol in urine, tissue, milk and eggs in consequence of the prohibitive regulation, in *Proceedings Euroresidue III Conference on Residues of Veterinary Drugs in Food*, Veldhoven, Netherlands, May 6–8, Haagsma, N. and Ruiter, A., Eds., Faculty of Medicine, Utrecht, Netherlands, 1996, 315.

204. Lynas, L. et al., Screening for chloramphenicol residues in the tissues and fluids of treated cattle by the four plate test, Charm II radioimmunoassay and Ridascreen CAP-glucuronid enzyme immunoassay, *Analyst*, 123, 2773, 1998.

205. Aerts, R.M.L., Keukens, H.J., and Werdmuller, G.A., Liquid chromatographic determination of chloramphenicol residues in meat: interlaboratory study, *J. Assoc. Off. Anal. Chem.*, 72, 570, 1989.

206. Le Boulaire, S., Bauduret, J.C., and André, F., Veterinary drug residues survey in meat: an HPLC method with a matrix solid phase dispersion extraction, *J. Agric. Food Chem.*, 45, 2134, 1997.

207. Cooper, A.D. et al., Aspects of extraction, spiking and distribution in the determination of incurred residues of chloramphenicol in animal tissues, *Food Addit. Contam.*, 15, 637, 1998.

208. Hayes, J.M., Determination of florfenicol in fish feed by liquid chromatography, *J. Assoc. Off. Anal. Chem. Internat.*, 88, 1777, 2005.

209. Hormazabal, V. and Yndestad, M., Simultaneous determination of chloramphenicol and ketoprofen in meat and milk and chloramphenicol in egg, honey and urne using LC-MS, *J. Liq. Chromatogr. Relat. Technol.*, 24, 2477, 2001.

210. Gantveg, A., Shishani, I., and Hoffman, M., Determination of chloramphenicol in animal tissues and urine. LC-tandem MS versus GC-MS, *Anal. Chim. Acta*, 483, 125, 2003.

211. Rupp, H.S., Stuart, J.S., and Hurlbut, J.A., Liquid chromatography/tandem mass spectrometry analysis of chloramphenicol in cooked crab meat, *J. Assoc. Off. Anal. Chem. Internat.*, 88, 1155, 2005.

212. Penney, L. et al., Determination of chloramphenicol residues in milk, eggs, and tissues by liquid chromatography/mass spectrometry, *J. Assoc. Off. Anal. Chem. Internat.*, 88, 645, 2005.

213. Batas, V. et al., Determination of fenicols (chloramphenicol, thiamphenicol and florfenicol) residues in live animals and primary animal products by liquid chromatography-tandem mass spectrometry, communication presented at the *5th International Symposium on Hormone and Veterinary Drug Residue Analysis*, Antwerp, Belgium, May 16–19, 2006 (unpublished data).

214. Yamada, R. et al., Simultaneous determination of residual veterinary drugs in bovine, porcine, and chicken muscle using liquid chromatography coupled with electrospray tandem mass spectrometry, *Biosci. Biotechnol. Biochem.*, 70, 54, 2006.

215. Turnipseed, S.B. et al., Use of ion-trap liquid chromatography-mass spectrometry to screen and confirm drug residues in aquacultured products, *Anal. Chim. Acta*, 483, 373, 2003.

216. Nouws, J.F.M. and Ziv, G., The effect of storage at 4°C on antibiotic residues in kidney and meat tissues of dairy cows, *Tijdschrift voor diergeneeskunde (Neth. J. Vet. Sci.)*, 101, 1145, 1976.

217. Wiese, B. and Martin, K., Determination of benzylpenicillin in milk at the pg/ml level by reversed-phase liquid chromatography in combination with digital subtraction chromatography technique part I, *J. Pharm. Biomed. Anal.*, 7, 95, 1989.

218. Boison, J.O., Effect of cold-temperature storage on stability of benzylpenicillin residues in plasma and tissues of food-producing animals, *J. Assoc. Off. Anal. Chem.*, 76, 974, 1992.

219. Boison, J.O., Keng, L., and MacNeil, J.D., Analysis of penicillin-G in milk by liquid chromatography, *J. Assoc. Off. Anal. Chem. Internat.*, 77, 565, 1994.

220. Gee, H.-E., Ho, K.-B., and Toothill, J., Liquid chromatographic determination of benzylpenicillin and cloxacillin in animal tissues and its application to a study of the stability at –20°C of spiked and incurred residues of benzylpenicillin in ovine liver, *J. Assoc. Off. Anal. Chem. Internat.*, 79, 640, 1995.

221. Rose, M.D. et al., The effect of cooking on veterinary drug residues in food. Part 8: benzylpenicillin, *Analyst*, 122, 1095, 1997.

222. Verdon, E. et al., Stability of penicillin antibiotic residues in meat during storage—Ampicillin, *J. Chromatogr. A*, 882, 135, 2000.

223. Croubels, S., De Baere, S., and De Backer, P., Practical approach for the stability testing of veterinary drugs in solutions and in biological matrices during storage, *Anal. Chim. Acta*, 483, 419, 2003.
224. Vilim, A.B. and Larocque, L., Determination of penicillin-G, ampicillin and cephapirin residues in tissues, *J. Assoc. Off. Anal. Chem.*, 66, 176, 1983.
225. McNeil, J.D. et al., Performance of five screening tests for the detection of penicillin-G residues in experimentally injected calves, *J. Food Prot.*, 54, 37, 1991.
226. Suhren, G. and Knappstein, K., Detection of cefquinome in milk by LC and screening methods, *Food Addit. Contam.*, 483, 363, 2003.
227. Ellerbroek, L. et al., Zur mikrobiologischen erfassung von rückstanden antimikrobiell wirksamer stoffe beim fisch, *Archiv für Lebensmittelhygiene*, 48, 1, 1997.
228. Fitzgerald, S.P. et al., Stable competitive enzyme-linked immunosorbent assay kit for rapid measurement of 11 active beta-lactams in milk, tissue, urine, and serum, *J. Assoc. Off. Anal. Chem. Internat.*, 90, 334, 2007.
229. Gustavsson, E. et al., Determination of beta-lactams in milk using a surface plasmon resonance-based biosensor, *J. Agric. Food Chem.*, 52, 2791, 2004.
230. Terada, H., Sakabe, Y., and Asanoma, M., Studies on residual antibacterials in foods. III. High performance liquid chromatographic determination of penicillin-G in animal tissues using an on-line pre-column concentration and purification system, *J. Chromatogr.*, 318, 299, 1985.
231. Barker, S.A., Long, A.R., and Short, C.R., Isolation of drug residues from tissues by solid phase dispersion, *J. Chromatogr.*, 475, 355, 1989.
232. Moats, W.A., High performance liquid chromatographic determination of penicillin-G, penicillin V and cloxacillin in beef and pork tissues, *J. Chromatogr.*, 593, 15, 1992.
233. Beconi-Barker, M.G. et al., Determination of ceftiofur and its desfuroylceftiofur-related metabolites in swine tissues by high-performance liquid chromatography, *J. Chromatogr. B*, 673, 231, 1995.
234. Moats, W.A., and Romanovski, R.D., Determination of penicillin-G in beef and pork tissues using an automated LC cleanup, *J. Agric. Food Chem.*, 46, 1410, 1998.
235. Moats, W.A. and Buckley, S.A., Determination of free metabolites of ceftiofur in animal tissues with an automated liquid chromatographic cleanup, *J. Assoc. Off. Anal. Chem. Internat.*, 81, 709, 1998.
236. Moats, W.A., Romanovski, R.D., and Medina, M.B., Identification of beta-lactam antibiotics in issue samples containing unknown microbial inhibitors, *J. Assoc. Off. Anal. Chem. Internat.*, 81, 1135, 1998.
237. Furusawa, N., Liquid chromatographic determination/identification of residual penicillin G in food-producing animal tissues, *J. Liq. Chromatogr. Relat. Technol*, 24, 161, 2001.
238. Verdon, E. and Couëdor, P., HPLC/UV determination of residues of cephalosporins in pork muscle tissue, communication presented at the *116th AOAC International Meeting*, Los Angeles, CA, September 22–26, 2002 (unpublished data).
239. Hornish, R.E., Hamlow, P.J., and Brown, S.A., Multilaboratory trial for determination of ceftiofur residues in bovine and swine kidney and muscle, and bovine milk, *J. Assoc. Off. Anal. Chem. Internat.*, 86, 30, 2003.
240. Nagata, T., Ashizawa, E., and Hashimoto, H., Simultaneous determination of residual fourteen kinds of beta-lactam and macrolide antibiotics in bovine muscles by high-performance liquid chromatography with a diode array detector, *J. Food Hyg. Soc. Jpn*, 45, 161, 2004.
241. Boison, J.O. et al., Determination of penicillin-G residues in edible animal tissues by liquid chromatography, *J. Assoc. Off. Anal. Chem.*, 74, 497, 1991.
242. Snippe, N. et al., Automated column LC determination of amoxicillin and cefadroxil in bovine serum and muscle tissue using on-line dialysis for sample preparation, *J. Chromatogr. B*, 662, 61, 1994.
243. Verdon, E. and Couëdor, P., Determination of ampicillin residues in bovine muscle at the residue level (25 to 100 μg/kg) by HPLC with precolumn derivatization, in *Proceedings Euroresidue III Conference on Residues of Veterinary Drugs in Food*, Veldhoven, Netherlands, May 6–8, Vol. 2, Haagsma, N. and Ruiter, A., Eds., Faculty of Medicine, Utrecht, Netherlands, 1996, 963.

244. Rose, M.D. et al., Determination of penicillins in animal tissues at trace residue concentrations: II. Determination of amoxicillin and ampicillin in liver and muscle using cation exchange and porous graphitic carbon solid phase extraction and high-performance liquid chromatography, *Food Addit. Contam.*, 14, 127, 1997.

245. Boison, J.O. and Keng, J.-Y., Multiresidue liquid chromatographic method for determining residues of mono- and dibasic penicillins in bovine muscle tissues, *J. Assoc. Off. Anal. Chem. Internat.*, 81, 111, 1998.

246. Boison, J.O. and Keng, J.-Y., Improvement in the multiresidue LC analysis of residues of mono- and dibasic penicillins in bovine muscle tissues, *J. Assoc. Off. Anal. Chem. Internat.*, 81, 1257, 1998.

247. McGrane, M., O'Keeffe, M., and Smyth, M.R., Multi-residue analysis of penicillin residues in porcine tissue using matrix solid phase dispersion, *Analyst*, 123, 2779, 1998.

248. Sorensen, L.K. et al., Simultaneous determination of seven penicillins in muscle, liver and kidney tissues from cattle and pigs by a multiresidue HPLC method, *J. Chromatogr. B*, 734, 307, 1999.

249. Verdon, E. et al., Liquid chromatographic determination of ampicillin residues in porcine muscle tissue by a multipenicillin analytical method: European collaborative study, *J. Assoc. Off. Anal. Chem. Internat.*, 85, 889, 2002.

250. Luo, W. and Ang, C.W.Y., Determination of amoxicillin residues in animal tissues by solid-phase extraction and liquid chromatography with fluorescence detection, *J. Assoc. Off. Anal. Chem. Internat.*, 83, 20, 2000.

251. Csokan, P. and Bernath, S., A simple and fast HPLC assay for amoxicillin residues in swine and cattle tissues, in *Proceedings of the 8th International Congress of the European Association for Veterinary Pharmacology and Toxicology (EAVPT)*, Jerusalem, Israel, July 30–August 10, Soback, S. and McKellar, Q.A., Eds., *J. Vet. Pharmacol. Toxicol. Ther.*, 29, 1, 2000.

252. Gamba, V. and Dusi, G., LC with fluorescence detection of amoxicillin and ampicillin in feeds using pre-column derivatization, *Anal. Chim. Acta*, 483, 69, 2003.

253. Voyksner, R.D., Smith, C.S., and Knox, P.C., Optimization and application of particle beam high-performance liquid chromatography/mass spectrometry to compounds of pharmaceutical interest, *Biomed. Environ. Mass Spectrom.*, 19, 523, 1990.

254. Blanchflower, W.J., Hewitt, S.A., and Kennedy, D.G., Confirmatory assay for the simultaneous detection of five penicillins in muscle, kidney and milk using liquid chromatography-electrospray mass spectrometry, *Analyst*, 119, 2595, 1994.

255. Hurtaud, D., Delepine, B., and Sanders, P., Particle beam liquid chromatography-mass spectrometry method with negative ion chemical ionization for the confirmation of oxacillin, cloxacillin and dicloxacillin residues in bovine muscle, *Analyst*, 119, 2731, 1994.

256. Hormazabal, V. and Yndestad, M., Determination of benzylpenicillin and other beta-lactam antibiotics in plasma and tissues using liquid chromatography-mass spectrometry for residual and pharmacokinetic studies, *J. Liq. Chromatogr. Relat. Technol.*, 21, 3099, 1998.

257. Ito, Y. et al., Application of ion-exchange cartridge clean-up in food analysis. IV. Confirmatory assay of benzylpenicillin, phenoxymethylpenicillin, oxacillin, cloxacillin, nafcillin and dicloxacillin in bovine tissues by liquid chromatography-electrospray tandem mass spectrometry, *J. Chromatogr. A*, 911, 217, 2001.

258. Hatano, K., Simultaneous determination of five penicillins in muscle, liver and kidney from slaughtered animals using liquid chromatography coupled to electrospray ionization tandem mass spectrometry, *J. Food Hyg. Soc. Jpn*, 44, 1, 2004.

259. Ito, Y., Development of analytical methods for residual antibiotics and antibacterials in livestock products, *Yakugaku Zasshi*, 123, 19, 2003.

260. Ito, Y. et al., Application of ion-exchange cartridge clean-up in food analysis. VI. Determination of six penicillins in bovine tissues by liquid chromatography-electrospray ionization tandem mass spectrometry, *J. Chromatogr. A*, 1042, 107, 2004.

261. Fagerquist, C.K., Lightfield, A.R., and Lehotay, S.J., Confirmatory and quantitative analysis of beta-lactam antibiotics in bovine kidney tissue by dispersive solid-phase extraction and liquid chromatography-tandem mass spectrometry, *Anal. Chem.*, 77, 1473, 2005.

262. Smith, S. et al., Multiclass confirmation of veterinary drug residues in fish by LC-MS[n], communication presented at the *119th AOAC International Meeting*, Orlando, FL, USA, September 11–15, 2005 (unpublished data).

263. Zuzulova, M. et al., In vivo activity of tylosin and its derivatives against ureaplasma urealyticum, *Arzneimittel-Forschung*, 45, 1222, 1995.

264. Chepkwony, H.K., Liquid chromatographic determination of erythromycins in fermentation broth, *Chromatographia*, 53, 89, 2001.

265. Chepkwony, H.K., Development and validation of an reversed-phase liquid chromatographic method for analysis of spiramycin and related substances, *Chromatographia*, 54, 51, 2001.

266. Council Regulation (EEC) No. 2377/90 of 26th June 1990, *Off. J. Eur. Commun.*, L224 (1990) 1 as amended by regulation No. 1442/95 of 26th June 1995, *Off. J. Eur. Commun.*, L143 (1995) 26.

267. Woodward, K.N. and Shearer, G., Antibiotic use in animal production in the European Union—regulation and current methods for residue detection, in *Chemical Analysis for Antibiotics used in Agriculture*, Oka, H. et al., Eds., AOAC International, Arlington, VA, 1995.

268. Moats, W.A., Determination of tylosin in tissues, milk and blood serum by reversed phase high performance liquid chromatography, in *Instrumental Analysis of Foods*, Vol. 1, Charalambous, G. and Inglett, G., Eds., Academic Press, Orlando, 1983, 357.

269. Chan, W., Gerhardt, G.C., and Salisbury, C.D.C., Determination of tylosin and tilmicosin residues in animal tissues by reversed-phase liquid chromatography, *J. Assoc. Off. Anal. Chem. Internat.*, 77, 331, 1994.

270. Horie, M. et al., Simultaneous determination of five macrolide antibiotics in meat by high performance liquid chromatography, *J. Chromatogr. A*, 812, 295, 1988.

271. Gaugain-Juhel, M., Anger, B., and Laurentie, M., Multiresidue chromatographic method for the determination of macrolide residues in muscle by high performance liquid chromatography with UV detection, *J. Assoc. Off. Anal. Chem. Internat.*, 82, 1046, 1999.

272. Leal, C. et al., Determination of macrolide antibiotics by liquid chromatography, *J. Chromatogr. A*, 910, 285, 2001.

273. Leroy, P., Decolin, D., and Nicolas, A., Determination of josamycin residues in porcine tissues using high performance liquid chromatography with pre-column derivatization and spectrofluorimetric detection, *Analyst*, 119, 2743, 1994.

274. Edder, P. et al., Analysis of erythromycin and oleandomycin residues in food by high-performance liquid chromatography with fluorometric detection, *Food Addit. Contam.*, 19, 232, 2002.

275. Delepine, B., Hurtaud, D., and Sanders, P., Identification of tylosin in bovine muscle at the maximum residue limit level by liquid chromatography-mass spectrometry, using a particle beam interface, *Analyst*, 119, 2717, 1994.

276. Delepine, B., Hurtaud-Pessel, D., and Sanders, P., Multiresidue method for confirmation of macrolide antibiotics in bovine muscle by liquid chromatography/mass spectrometry, *J. Assoc. Off. Anal. Chem. Internat.*, 79, 397, 1996.

277. Draisci, R. et al., Confirmatory method for macrolide residues in bovine tissues by micro-liquid chromatography-tandem mass spectrometry, *J. Chromatogr. A*, 926, 97, 2001.

278. Dubois, M. et al., Identification and quantification of five macrolide antibiotics in several tissues, eggs and milk by liquid chromatography-electrospray tandem mass spectrometry, *J. Chromatogr. B*, 753, 189, 2001.

279. Hwang, Y.-H. et al., Simultaneous determination of various macrolides by liquid chromatography/mass spectrometry, *J. Vet. Sci.*, 3, 103, 2002.

280. Van Poucke, C. et al., Liquid chromatographic-tandem mass spectrometric detection of banned antibacterial growth promoters in animal feed, *Anal. Chim. Acta*, 483, 99, 2003.

281. Lucchetti, D. et al., Simple confirmatory method for the determination of erythromycin residues in trout: a fast liquid–liquid extraction followed by liquid chromatography-tandem mass spectrometry, *J. Agric. Food Chem.*, 53, 9689, 2005.

282. Council Regulation (EEC) No. 2377/90 of 26th June 1990, *Off. J. Eur. Commun.*, L224 (1990) 1 as amended by regulation No. 2901/93 of 18th October 1993, *Off. J. Eur. Commun.*, L264 (1993) 1.

283. Commission Decision No. 2003181/EC of 13th March 2003, *Off. J. Eur. Commun.*, L71 (2003) 17 and amending Decision 2002/657/EC.

284. McCracken, R.J. and Kennedy, D.G., Determination of the furazolidone metabolite, 3-amino-2-oxazolidinone, in porcine tissues using LC-thermospray MS and the occurrence of residues in pigs produced in Northern Ireland, *J. Chromatogr. B*, 691, 87, 1997.

285. Parks, O.W., Liquid chromatographic electrochemical detection screening procedure for six nitro-containing drugs in chicken tissues at low ppb level, *J. Assoc. Off. Anal. Chem.*, 72, 567, 1989.

286. Laurensen, J.J. and Nouws, J.F.M., Simultaneous determination of nitrofuran derivatives in various animal substrates by HPLC, *J. Chromatogr.*, 472, 321, 1989.

287. Cooper, A.D. et al., Development of multi-residue methodology for the high-performance liquid chromatography determination of veterinary drugs in animal tissues, *Food Addit. Contam.*, 12, 167, 1995.

288. Hormazabal, V. and Yndestad, M., Simple and rapid method of analysis for furazolidone in meat tissues by HPLC, *J. Liq. Chromatogr.*, 18, 1871, 1995.

289. Conneely, A., Nugent, A., and O'Keeffe, M., Use of solid phase extraction for the isolation and clean-up of a derivatised furazolidone metabolite from animal tissues, *Analyst*, 127, 705, 2002.

290. Conneely, A. et al., Isolation of bound residues of nitrofuran drugs from tissue by solid-phase extraction with determination by liquid chromatography with UV and tandem mass spectrometric detection, *Anal. Chim. Acta*, 483, 91, 2003.

291. Hormazabal, V. and Norman Asp, T., Determination of the metabolites of nitrofuran antibiotics in meat by liquid chromatography-mass spectrometry, *J. Liq. Chromatogr. Relat. Technol.*, 27, 2759, 2004.

292. Leitner, A., Zöllner, P., and Lindner, W., Determination of the metabolites of nitrofuran antibiotics in animal tissue by high-performance liquid chromatography-tandem mass spectrometry, *J. Chromatogr. A*, 939, 49, 2001.

293. Edder, P. et al., Analysis of nitrofuran metabolites in food by high-performance liquid chromatography with tandem mass spectrometry detection, *Clin. Chem. Lab. Med.*, 41, 1608, 2003.

294. O'Keeffe, M. et al., Nitrofuran antibiotic residues in pork - The FoodBRAND retail survey, *Anal. Chim. Acta*, 520, 125, 2004.

295. Mottier, P. et al., Quantitative determination of four nitrofuran metabolites in meat by isotope dilution liquid chromatography-electrospray ionisation-tandem mass spectrometry, *J. Chromatogr. A*, 1067, 85, 2005.

296. Finzi, J.K. et al., Determination of nitrofuran metabolites in poultry muscle and eggs by liquid chromatography-tandem mass spectrometry, *J. Chromatogr. B*, 824, 30, 2005.

297. Hurtaud-Pessel, D., Verdon, E., and Sanders, P., Proficiency study for the determination of nitrofuran metabolites in shrimps, *Food Addit. Contam.*, 23, 569, 2006.

298. Council Regulation (EEC) No. 2377/90 of 26th June 1990, *Off. J. Eur. Commun.*, L224 (1990) 1 as amended by regulation No. 1756/2002/EC of 23rd September 2002 *Off. J. Eur. Commun.*, L265 (2002) 1.

299. Vahl, M., Analysis of nifursol residues in turkey and chicken meat using liquid chromatography-tandem mass spectrometry, *Food Addit. Contam.*, 22, 120, 2005.

300. Mulder, P.P.J. et al., Determination of nifursol metabolites in poultry muscle and liver tissue. Development and validation of a confirmatory method, *Analyst*, 130, 763, 2005.

301. Verdon, E., Couëdor, P., and Sanders, P., Multi-residue monitoring for the simultaneous determination of five nitrofurans (furazolidone, furaltadone, nitrofurazone, nitrofurantoine, nifursol) in poultry muscle tissue through the detection of their five major metabolites (AOZ, AMOZ, SEM, AHD, DNSAH), by liquid chromatography coupled to electrospray tandem mass spectrometry—in-house validation in line with Commission Decision 657/2002/EC, *Anal. Chim. Acta*, 586, 336, 2007.

302. Pereira, A.S., Donato, J.L., and De Nucci, G., Implications of the use of semicarbazide as a metabolic target of nitrofurazone contamination in coated products, *Food Addit. Contam.*, 21, 63, 2004.

303. Cooper, A.D., McCracken, R.J., and Kennedy, D.G., Nitrofurazone accumulates in avian eyes–a replacement for semicarbazide as a marker of abuse, *Analyst*, 130, 824, 2005.

304. Council Regulation (EEC) No. 2377/90 of 26th June 1990, *Off. J. Eur. Commun.*, L224 (1990) 1 as amended by regulation No. 3426/93/EC, *Off. J. Eur. Commun.*, L312 (1993) 15.

305. Council Regulation (EEC) No. 2377/90 of 26th June 1990, *Off. J. Eur. Commun.*, L224 (1990) 1 as amended by regulation No. 1798/95/EC, *Off. J. Eur. Commun.*, L174 (1995) 20.

306. Council Regulation (EEC) No. 2377/90 of 26th June 1990, *Off. J. Eur. Commun.*, L224 (1990) 1 as amended by regulation No. 613/98/EC, *Off. J. Eur. Commun.*, L82 (1998) 14.

307. Mallinson, E.T. and Henry, A.C., Determination of nitroimidazole metabolites in swine and turkey muscle by liquid chromatography, *J. Assoc. Off. Anal. Chem.*, 75, 790, 1992.

308. Semeniuk, S. et al., Determination of nitroimidazole residues in poultry tissues, serum and eggs by high-performance liquid chromatography, *Biomed. Chromatogr.*, 9, 238, 1995.

309. Posyniak, A. et al., Tissue concentration of dimetridazole in laying hens, *Food Addit. Contam.*, 13, 871, 1996.

310. Sams, M.J. et al., Determination of dimetridazole, ronidazole and their common metabolite in poultry muscle and eggs by HPLC with UV detection and confirmatory analysis by apci-MS, *Analyst*, 123, 2545, 1998.

311. Rose, M.D., Bygrave, J., and Sharman, M., Effect of cooking on veterinary drug residues in food. Part 9. Nitroimidazoles, *Analyst*, 124, 289, 1999.

312. Sorensen, L.K. and Hansen, H., Determination of metronidazole and hydroxymetronidazole in trout by a high-performance liquid chromatography method, *Food Addit. Contam.*, 17, 197, 2000.

313. Matusik, J.E. et al., Identification of dimetridazole, ipronidazole, and their alcohol metabolites in turkey tissues by thermospray tandem mass spectrometry, *J. Agric. Food Chem.*, 40, 439, 1992.

314. Cannavan, A. et al., Gas chromatographic-mass spectrometric determination of sulfamethazine in animal tissues using a methyl/trimethylsilyl derivative, *Analyst*, 121, 1457, 1997.

315. Polzer, J. et al., Validation of a method for the determination of nitroimidazoles in plasma and retina of turkeys, communication presented at the *4th International Symposium on Hormone and Veterinary Drug Residue Analysis*, Antwerp, Belgium, June 4–7, 2002 (unpublished data).

316. Hurtaud-Pessel, D., Delepine, B., and Laurentie, M., Determination of four nitroimidazole residues in poultry meat by liquid chromatography-mass spectrometry, *J. Chromatogr. A*, 16, 89, 2000.

317. Hormazabal, V. and Yndestad, M., Determination of nitroimidazole residues in meat using mass spectrometry, *J. Liq. Chromatogr. Relat. Technol.*, 24, 2487, 2001.

318. Govaerts, Y., Degroodt, J.M., and Srebrnik, S., Nitroimidazole drug residues in poultry and pig tissues, in *Proceedings Euroresidue IV Conference on Residues of Veterinary Drugs in Food*, Veldhoven, Netherlands, Van Ginkel, L.A. and Ruiter, A., Eds., Faculty of Medicine, Utrecht, Netherlands, 2000, 470.

319. Radeck, W., Determination of nitroimidazoles using apci+ mass spectrometry, *Proceedings Euroresidue IV Conference on Residues of Veterinary Drugs in Food*, Veldhoven, Netherlands, May 8–10, Van Ginkel, L.A. and Ruiter, A., Eds., Faculty of Medicine, Utrecht, Netherlands, 2000.

320. Xia, X. et al., Determination of four nitroimidazoles in poultry and swine muscle and eggs by liquid chromatography/tandem mass spectrometry, *J. Assoc. Off. Anal. Chem. Internat.*, 89, 94, 2006.

321. Calderon, V. et al., Screening of antibiotic residues in Spain, communication presented at the *Workshop on Validation of Screening Methods*, Fougeres, France, June 3–4, 2005.

322. Schneider, M.J. and Donoghue, D.J., Comparison of a bioassay and a LC-Fluorescence-MS[n] Method for the detection of incurred enrofloxacin residues in chicken tissues, *Poult. Sci.*, 83, 830, 2004.

323. Pikkemaat, M.G. et al., Improved microbial screening assay for the detection of quinolone residues in poultry and eggs, *Food Addit. Contam.*, 24, 842, 2007.

324. Laurent, M.R., Terlain, B.L., and Caude, M.C., Estimation of decoquinate residues in chicken tissues by TLC and measurement of fluorescence on plates, *J. Agric. Food Chem.*, 19, 55, 1971.

325. Stone, L.R., Fluorometric determination of decoquinate in chicken tissues, *J. Assoc. Off. Anal. Chem.*, 56, 71, 1973.

326. Hormazabal, V. and Yndestad, M., Rapid assay for monitoring residues of enrofloxacin in milk and meat tissues by HPLC, *J. Liq. Chromatogr. Relat. Technol.*, 17, 3775, 1994.

327. Rose, M.D., Bygrave, J., and Stubbings, G.W.F., Extension of multi-residue methodology to include the determination of quinolones in food, *Analyst*, 123, 2789, 1998.

328. Prat, M.D. et al., Liquid chromatographic separation of fluoroquinolone antibacterials used as veterinary drugs, *Chromatographia*, 52, 295, 2000.

329. Hernandez-Arteseros, J.A., Compano, R., and Prat, M.D., Analysis of flumequine and oxolinic acid in edible animal tissues by LC with fluorimetric detection, *Chromatographia*, 52, 58, 2000.

330. Duran-Meras, I. et al., Comparison of different methods for the determination of several quinolonic and cinolonic antibiotics in trout muscle tissue by HPLC with fluorescence detection, *Chromatographia*, 51, 163, 2000.

331. Yorke, J.C. and Froc, P., Quantitation of nine quinolones in chicken tissues by high-performance liquid chromatography with fluorescence detection, *J. Chromatogr. A*, 882, 63, 2000.

332. Hormazabal, V. and Yndestad, M., A simple assay for the determination of flumequine and oxolinic acid in fish muscle and skin by HPLC, *J. Liq. Chromatogr. Relat. Technol.*, 24, 109, 2001.

333. Posyniak, A., Zmudski, J., and Semeniuk, S., Effects of the matrix and sample preparation on the determination of fluoroquinolone residues in animal tissues, *J. Chromatogr. A*, 914, 89, 2001.

334. Ramos, M. et al., Simple and sensitive determination of five quinolones in food by liquid chromatography with fluorescence detection, *J. Chromatogr. B Analyt. Technol. Biomed. Life Sci.*, 789, 373, 2003.

335. Verdon, E. et al., Multiresidue method for simultaneous determination of ten quinolone antibacterial residues in multimatrix/multispecies animal tissues by liquid chromatography with fluorescence detection: single laboratory validation study, *J. Assoc. Off. Anal. Chem.*, 88, 1179, 2005.

336. Pecorelli, I. et al., Simultaneous determination of 13 quinolones from feeds using accelerated solvent extraction and liquid chromatography, *Anal. Chim. Acta*, 483, 81, 2003.

337. Schneider, M.J. and Donoghue, D.J., Multiresidue analysis of fluoroquinolone antibiotics in chicken tissue using liquid chromatography-fluorescence-multiple mass spectrometry, *J. Chromatogr. B*, 780, 83, 2002.

338. Schneider, M.J., Vazquez-Moreno, L., and Del Carmen Bermudez-Almada, M., Multiresidue determination of fluoroquinolones in shrimp by liquid chromatography-fluorescence-mass spectrometry, *J. Assoc. Off. Anal. Chem. Internat.*, 88, 1160, 2005.

339. Hermo, M.P., Barron, D., and Barbosa, J., Development of analytical methods for multiresidue determination of quinolones in pig muscle samples by liquid chromatography with ultraviolet detection, liquid chromatography-mass spectrometry and liquid chromatography-tandem mass spectrometry, *J. Chromatogr. A*, 1104, 132, 2006.

340. Schneider, M.J. et al., Simultaneous determination of fluoroquinolones and tetracyclines in chicken muscle using HPLC with fluorescence detection, *J. Chromatogr. B*, 846, 8, 2007.

341. Bailac, S. et al., Determination of quinolones in chicken tissues by liquid chromatography with ultraviolet absorbance detection, *J. Chromatogr. A*, 1129, 145, 2004.

342. Barron, D. et al., Determination of residues of enrofloxacin and its metabolite ciprofloxacin on biological materials by capillary electrophoresis, *J. Chromatogr. B*, 759, 73, 2001.

343. Jimenez-Lozano, E. et al., Effective sorbents for solid-phase extraction in the analysis of quinolones in animal tissues by capillary electrophoresis, *Electrophoresis*, 25, 65, 2004.

344. Horstkötter, C. et al., Determination of residues of enrofloxacin and its metabolite ciprofloxacin in chicken muscle by capillary electrophoresis using laser-induced fluorescence detection, *Electrophoresis*, 23, 3078, 2002.

345. Juan-Garcia, A., Font, G., and Pico, Y., Determination of quinolone residues in chicken and fish by capillary electrophoresis-mass spectrometry, *Electrophoresis*, 27, 2240, 2006.

346. Delepine, B., Hurtaud-Pessel, D., and Sanders, P., Simultaneous determination of six quinolones in pig muscle by liquid chromatography-atmospheric pressure chemical ionisation mass spectrometry, *Analyst*, 123, 2743, 1998.

347. Hatano, K., Simultaneous determination of quinolones in foods by LC/MS/MS, *J. Food Hyg. Soc. Jpn*, 45, 239, 2004.

348. Toussaint, B. et al., Determination of (fluoro)quinolone antibiotic residues in pig kidney using liquid chromatography-tandem mass spectrometry. Part I. Laboratory-validated method, *J. Chromatogr. A*, 1088, 32, 2005.

349. Toussaint, B. et al., Determination of (fluoro)quinolone antibiotic residues in pig kidney using liquid chromatography-tandem mass spectrometry. Part II: Intercomparison exercise, *J. Chromatogr. A*, 1088, 40, 2005.

350. Dufresne, G. and Fouquet, A., Multiresidue determination of quinolone and fluoroquinolone antibiotics in fish and shrimp by liquid chromatography-tandem mass spectrometry, *J. Assoc. Off. Anal. Chem. Internat.*, 90, 604, 2007.

351. Agarwal, V.K., High-performance liquid chromatographic methods for the determination of sulfonamides in tissue, milk and eggs, *J. Chromatogr. A*, 624, 411, 1992.

352. Diserens, J.-M., Renaud-Bezot, C., and Savoy-Perroud, M.-C., Simplified determination of sulfonamides residues in milk, meat and eggs, *Deutsche Lebensmittel Rundschau*, 87, 205, 1991.

353. Boison, J.O. and Keng, J.-Y., Determination of sulfamethazine in bovine and porcine tissues by HPLC, *J. Assoc. Off. Anal. Chem. Internat.*, 77, 558, 1994.

354. Rose, M.D., Farrington, W.H.H., and Shearer, G., The effect of cooking on veterinary drug residues in food: 3.sulfamethazine (sulphadimidine), *Food Addit. Contam.*, 12, 739, 1995.

355. Humayoun-Akhtar, M. et al., Extraction of incurred sulphamethazine in swine tissue by microwave assisted extraction and quantification without clean up by HPLC following derivatization with dimethylaminobenzaldehyde, *Food Addit. Contam.*, 15, 542, 1998.

356. Alfredsson, G. and Ohlsson, A., Stability of sulphonamide drugs in meat during storage, *Food Addit. Contam.*, 15, 302, 1998.

357. McGrane, M., O'Keeffe, M., and Smyth, M.R., The analysis of sulphonamide drug residues in pork muscle using automated dialysis, *Anal. Lett.*, 32, 481, 1999.

358. Li, J.S. et al., Determination of sulfonamides in swine meat by immunoaffinity chromatography, *J. Assoc. Off. Anal. Chem. Internat.*, 83, 830, 2000.

359. Furusawa, N., Determining the procedure for routine residue monitoring of sulfamethazine in edible animal tissues, *Biomed. Chromatogr.*, 15, 235, 2001.

360. Roybal, J.E. et al., Application of size-exclusion chromatography to the analysis of shrimp for sulphonamide residues, *Anal. Chim. Acta*, 483, 147, 2003.

361. Furusawa, N., A clean and rapid LC technique for sulfamethazine monitoring in pork tissues without using organic solvents, *J. Chromatogr. Sci.*, 41, 377, 2003.

362. Pecorelli, I. et al., Validation of a confirmatory method for the determination of sulphonamides in muscle according to the European Union regulation 2002/657/EC, *J. Chromatogr. A*, 1032, 23, 2004.

363. Di Sabatino, M. et al., Determination of ten sulphonamide residues in meat samples by liquid chromatography, *J. Assoc. Off. Anal. Chem. Internat.*, 90, 598, 2007.

364. Tsai, C.-E. and Kondo, F., A sensitive high-performance liquid chromatography method for detecting sulfonamide residues in swine serum and tissues after fluorescamine derivatization, *J. Liq. Chromatogr.*, 18, 965, 1995.

365. Salisbury, C.D.C., Sweet, J.C., and Munro, R., Determination of sulfonamide residues in the tissues of food animals using automated precolumn derivatization and liquid chromatography with fluorescence detection, *J. Assoc. Off. Anal. Chem. Internat.*, 87, 1264, 2004.

366. Posyniak, A., Zmudski, J., and Mitrowska, K., Dispersive solid-phase extraction for the determination of sulfonamides in chicken by muscle liquid chromatography, *J. Chromatogr. A*, 1087, 259, 2005.

367. Porter, S., Confirmation of sulphonamide residues in kidney tissue by liquid chromatography-mass spectrometry, *Analyst*, 119, 2753, 1994.

368. Boison, J.O. and Keng, L.J.-Y., Determination of sulfadimethoxine residues in animal tissues by liquid chromatography and thermospray mass spectrometry, *J. Assoc. Off. Anal. Chem. Internat.*, 78, 651, 1995.

369. Cannavan, A. et al., Determination of trimethoprim in tissues using liquid chromatography-thermospray mass spectrometry, *Analyst*, 122, 1379, 1997.

370. Van Eekhout, N. and Van Peteghem, C., The use of LC/MS/MS in the determination of residues of veterinary drugs and hormones in foodstuffs, in *Proceedings of the 8th International Congress of the European Association for Veterinary Pharmacology and Toxicology (EAVPT)*, Jerusalem, Israel, July 30–August 10, Soback, S. and McKellar, Q.A., Eds., 2000.

371. Bogialli, S. et al., A liquid chromatography-mass spectrometry assay for analyzing sulfonamide antibacterials in cattle and fish muscle tissues, *Anal. Chem.*, 75, 1798, 2003.

372. Kim, D.-H. and Lee, D.W., Comparison of separation conditions and ionization methods for the LC-MS determination of sulfonamides, *J. Chromatogr. A*, 984, 153, 2003.

373. Gentili, A. et al., Accelerated solvent extraction and confirmatory analysis of sulfonamide residues in raw meat and infant foods by liquid chromatography electrospray tandem mass spectrometry, *J. Agric. Food Chem.*, 52, 4614, 2004.

374. Kishida, K., Quantitative and confirmation of six sulphonamides in meat by liquid chromatography-mass spectrometry with photodiode array detection, *Food Control*, 18, 301, 2007.

375. Potter, R.A. et al., Simultaneous determination of 17 sulfonamides and the potentiators ormetoprim and trimethoprim in salmon muscle by liquid chromatography with tandem mass spectrometry detection, *J. Assoc. Off. Anal. Chem. Internat.*, 90, 343, 2007.

376. Juhel-Gaugain, M. et al., European proficiency testing of national reference laboratories for the confirmation of sulfonamide residues in muscle and milk, *Food Addit. Contam.*, 22, 221, 2005.

377. Stead, S.L. et al., New method for the rapid identification of tetracycline residues in foods of animal origin using Premi'test in combination with a metal ion chelation assay, *Food Addit. Contam.*, 24, 583, 2007.

378. Mc Evoy, J.D.G. et al., Origin of chlortetracycline in pig tissue, *Analyst*, 119, 2603, 1994.

379. Sokol, J. and Matisova, E., Determination of tetracycline antibiotics in animal tissues of food-producing animals by high-performance liquid chromatography using solid-phase extraction, *J. Chromatogr. A*, 669, 75, 1994.

380. Stubbings, G., Tarbin, J.A., and Shearer, G., On-line metal chelate affinity chromatography clean-up for the HPLC determination of tetracycline antibiotics in animal tissues, *J. Chromatogr. B Biomed. Appl.*, 679, 137, 1996.

381. MacNeil, J.D. et al., Chlortetracycline, oxytetracycline and tetracycline in edible animal tissues, liquid chromatographic method: collaborative study, *J. Assoc. Off. Anal. Chem. Internat.*, 79, 405, 1996.

382. Rose, M.D. et al., The effect of cooking on veterinary drug residues in food: 4. Oxytetracycline, *Food Addit. Contam.*, 13, 275, 1996.

383. Cooper, A.D., Effects of extraction and spiking procedures on the determination of incurred residues of tetracycline in cattle kidney, *Food Addit. Contam.*, 15, 645, 1998.

384. Juhel-Gaugain, M. et al., Results of a European interlaboratory study for the determination of oxytetracycline in pig muscle by HPLC, *Analyst*, 123, 2767, 1998.

385. Posyniak, A. et al., Validation study for the determination of tetracycline residues in animal tissues, *J. Assoc. Off. Anal. Chem. Internat.*, 82, 862, 1999.

386. Moats, W.A., Determination of tetracycline antibiotics in beef and pork tissues using ion-paired liquid chromatography, *J. Agric. Food Chem.*, 48, 2244, 2000.

387. Oka, H. et al., Survey of residual tetracyclines in kidneys of diseased animals in Aichi Prefecture, Japan (1985–1997), *J. Assoc. Off. Anal. Chem. Internat.*, 84, 350, 2001.

388. Hernandez, M., Borrull, F., and Calull, M., Capillary zone electrophoresis determination of oxytetracycline in pig tissue samples at maximum residue limits, *Chromatographia*, 54, 355, 2001.

389. Cinquina, A.L. et al., Validation of a HPLC method for the determination of oxytetracycline, tetracycline, chlortetracycline and doxycycline in bovine milk and muscle, *J. Chromatogr. A*, 987, 227, 2003.

390. Posyniak, A. et al., Analytical procedure for the determination of chlortetracycline and 4-epi-chlortetracycline in pig kidneys, *J. Chromatogr. A*, 1088, 169, 2005.

391. Kowalski, C., Pomorska, M., and Slavik, T., Development of HPLC with UV-VIS detection for the determination of the level of oxytetracycline in the biological matrix, *J. Liq. Chromatogr. Relat. Technol.*, 29, 2721, 2006.

392. Blanchflower, W.J., McCracken, R.J., and Rice, D.A., Determination of chlortetracycline residues in tissues using high-performance liquid chromatography with fluorescence detection, *Analyst*, 114, 421, 1989.

393. Agasoster, T., Automated determination of oxytetracycline residues in muscle, liver, milk and egg by on-line dialysis and post-column reaction detection HPLC, *Food Addit. Contam.*, 9, 615, 1992.

394. McCracken, R.J. et al., Simultaneous determination of oxytetracycline, tetracycline and chlortetracycline in animal tissues using liquid chromatography, post-column derivatization with aluminium and fluorescence detection, *Analyst*, 120, 1763, 1995.

395. Kawata, S. et al., LC determination of oxytetracycline in swine tissue, *J. Assoc. Off. Anal. Chem. Internat.*, 79, 1463, 1996.

396. McEvoy, J.D.G. et al., Production of CTC-containing porcine reference materials, *Analyst*, 123, 2535, 1998.

397. Croubels, S. et al., Liquid chromatography separation of doxycycline and 4-epidoxycycline in a tissue depletion study of doxycycline in turkeys, *J. Chromatogr. B*, 708, 145.

398. Croubels, S., De Baere, S., and De Backer, P., The proposed MRL for doxycycline: controversy about the inclusion of the 4-epimer as marker, in *Proceedings of the 8th International Congress of the European Association for Veterinary Pharmacology and Toxicology (EAVPT)*, Jerusalem, Israel, July 30–August 10, Soback, S. and McKellar, Q.A., Eds., 2000.

399. Pena, A.L., Lino, C.M., and Silveira, M.I.N., Determination of tetracycline antibiotics in salmon muscle by liquid chromatography using post-column derivatization with fluorescence detection, *J. Assoc. Off. Anal. Chem. Internat.*, 86, 925, 2003.

400. Schneider, M.J. and Chen, G., Time-resolved luminescence screening assay for tetracyclines in chicken muscle, *Anal. Lett.*, 37, 2067, 2004.

401. Rupp., H.S. and Anderson, C.R., Determination of oxytetracycline in salmon by liquid chromatography with metal-chelate fluorescence detection, *J. Assoc. Off. Anal. Chem. Internat.*, 88, 505, 2005.

402. Blanchflower, W.J. et al., Confirmatory assay for the determination of tetracycline, oxytetracycline, chlortetracycline and its isomers in muscle and kidney using liquid chromatography-mass spectrometry, *J. Chromatogr. B Biomed. Appl.*, 692, 351, 1997.

403. Cherlet, M., De Baere, S., and De Backer, P., Quantitative analysis of oxytetracycline and its 4-epimer in calf tissues by HPLC combined with positive electrospray ionization mass spectrometry, *Analyst*, 128, 871, 2003.

404. Berendsen, B.J.A. and Van Rhijn, J.A., Residue analysis of tetracyclines in poultry muscle: shortcomings revealed by a proficiency test, *Food Addit. Contam.*, 23, 1141, 2006.

405. Bogialli, S. et al., A rapid confirmatory method for analyzing tetracycline antibiotics in bovine, swine, and poultry muscle tissues: matrix solid-phase dispersion with heated water as extractant followed by liquid chromatography-tandem mass spectrometry, *J. Agric. Food Chem.*, 54, 1564, 2006.

406. Council Directive 70/524/EEC of 23rd November 1970, *Off. J. Eur. Commun.*, L270 (1970) 1.

407. Lönnroth, A., *Antimicrobial Resistance Research 1999–2002*, Revised and extended edition, European Commission-DG for Research, Directorate F: Life sciences, genomics and biotechnology for Health, Office for Official Publications of the European Communities, Luxembourg, 2003.

408. Council Directive 97/6/EC of 30th January 1997, *Off. J. Eur. Commun.*, L35 (1997) 11.

409. Council Regulation (EC) No. 2821/98 of 17th December 1998, *Off. J. Eur. Commun.*, L351 (1998) 4.

410. Council Regulation (EC) No. 2562/99 of 3rd December 1999, *Off. J. Eur. Commun.*, L310 (1999) 11.

411. Curren, M.S.S. and King, J.W., New sample preparation technique for the determination of avoparcin in pressurized hot water extracts from kidney samples, *J. Chromatogr. A*, 954, 41, 2002.

412. Nagasen, M. and Fukamachi, K., Determination of virginiamycin in swine, cattle and chicken muscles by HPLC with fluorescence detection, *Jpn Analyst*, 36, 297, 1987.

413. Moats, W.A. and Leskinen, L., Determination of virginiamycin residues in swine tissue using high performance liquid chromatography, *J. Agric. Food Chem.*, 36, 1297, 1988.

414. Saito, K. et al., Determination of virginiamycin in chicken or swine tissues by high performance liquid, *Eisei Kagaku*, 35, 63, 1989.

415. Govaerts, C. et al., Hyphenation of liquid chromatography to ion trap mass spectrometry to identify minor components in polypeptide antibiotics, *Anal. Bioanal. Chem.*, 377, 909, 2003.

416. Wan, E.C.-H. et al., Detection of residual bacitracin A, colistin A, and colistin B in milk and animal tissues by liquid chromatography tandem mass spectrometry, *Anal. Bioanal. Chem.*, 385, 181, 2006.

417. Weiss, G. and MacDonald, A., Methods for determination of ionophore-type antibiotic residues in animal tissues, *J. Assoc. Off. Anal. Chem.*, 68, 972, 1985.

418. Asukabe, H. and Harada, K.I., Chemical analysis of polyether antibiotics, in *Chemical Analysis for Antibiotics Used in Agriculture*, Oka, H. et al., Eds, AOAC International, Arlington, VA, 1995.

419. Elliott, C.T., Kennedy, D.G., and McCaughey, W.J., Methods for the detection of polyether ionophore residues in poultry, *Analyst*, 123, 45R, 1998.

420. Harris, J.A., Russell, C.A.L., and Wilkins, J.P.G., The characterisation of polyether ionophore veterinary drugs by HPLC-electrospray MS, *Analyst*, 123, 2625, 1998.

421. Davis, A.L. et al., Investigations by HPLC-electrospray mass spectrometry and NMR spectroscopy into the isomerisation of salinomycin, *Analyst*, 124, 251, 1999.

422. Lopes, N.P. et al., Fragmentation studies on lasalocid acid by accurate mass electrospray mass spectrometry, *Analyst*, 127, 1224, 2002.

423. Rosen, J., Efficient and sensitive screening and confirmation of residues of selected ionophore antibiotics in liver and eggs by liquid chromatography-electrospray mass spectrometry, *Analyst*, 126, 1990, 2002.

424. Dubois, M., Pierret, G., and Delahaut, Ph., Efficient and sensitive detection of residues of nine coccidiostats in egg and muscle by liquid chromatography-electrospray mass spectrometry, *J. Chromatogr. B*, 813, 181, 2004.

425. Mortier, L., Daeseleire, E., and Van Peteghem, C., Determination of the ionophoric coccidiostats narasin, monensin, lasalocid and salinomycin in eggs by liquid chromatography/tandem mass spectrometry, *Rapid Commun. Mass Spectrom.*, 19, 533, 2005.

426. Peippo, P. et al., Rapid time-resolved fluoroimmunoassay for the screening of narasin and salinomycin residues in poultry and eggs, *J. Agric. Food Chem.*, 52, 1828, 2004.

427. Moats, W.A. and Leskinen, L., Determination of novobiocin residues in milk, blood, and tissues by liquid chromatography, *J. Assoc. Off. Anal. Chem.*, 71, 776, 1988.

428. Heller, D.N. and Nochetto, C.R., Development of multiclass methods for drug residues in eggs: silica SPE cleanup and LC-MS/MS analysis of ionophore and macrolide residues, *J. Agric. Food Chem.*, 52, 6848, 2004.

429. Reeves, V.B., Liquid chromatographic procedure for the determination of novobiocin residues in bovine milk: interlaboratory study, *J. Assoc. Off. Anal. Chem. Internat.*, 78, 55, 2005.

430. Beechinor, J.G. and Bloomfield, J., Effect of cooking on residues of tiamulin in beef, in *Proceedings of the 8th International Congress of the European Association for Veterinary Pharmacology and Toxicology (EAVPT)*, Jerusalem, Israel, July 30–August 10, Soback, S. and McKellar, Q.A., Eds., 2000.

431. Chen, H.-C. et al., Determination of tiamulin residue in pork and chicken by solid phase extraction and HPLC, *J. Food Drug Anal.*, 14, 80, 2006.

432. Nozal, M.J. et al., Trace analysis of tiamulin in honey by liquid chromatography–diode array–electrospray ionization mass spectrometry detection, *J. Chromatogr. A*, 1116, 102, 2006.
433. Council Regulation 2788/98/EC of 22nd December 1998. *Off. J. Eur. Commun.*, L347 (1998) 31.
434. Van Ginkel, J.A. et al., The detection and identification of quinoxaline-2-carboxylic acid, a major metabolite of carbadox, in swine tissue, in *Proceedings Euroresidue Conference on Residues of Veterinary Drugs in Food*, Noordwijkerhout, Netherlands, Haagsma, N., Ruiter, A., and Czedik-Eysenberg, P.B., Eds., Faculty of Medicine, Utrecht, Netherlands, 1990, 189.
435. Rutalj, M. et al., Quinoxaline-2-carboxylic acid (QCA) in swine liver and muscle, *Food Addit. Contam.*, 13, 879, 1996.
436. Hutchinson, M.J. et al., Development and validation of an improved method for confirmation of the carbadox metabolite, quinoxaline-2-carboxylic acid, in porcine liver using LC-electrospray MS-MS according to revised EU criteria for veterinary drug residue analysis, *Analyst*, 127, 342, 2002.
437. Hurtaud-Pessel, D. et al., An LC/MS-MS method for the determination of QCA and MQCA, the metabolites of carbadox and olaquindox in porcine liver and muscle, communication presented at the *5th International Symposium on Hormone and Veterinary Drug Residue Analysis*, Antwerp, Belgium, May 16–19, 2006.
438. Hutchinson, M.J., Young, P.B., and Kennedy, D.G., Confirmatory method for the analysis of carbadox and olaquindox in porcine feedingstuffs using liquid chromatography–electrospray mass spectrometry, *Food Addit. Contam.*, 22, 113, 2005.
439. Hutchinson, M.J., Young, P.B., and Kennedy, D.G., Confirmation of carbadox and olaquindox metabolites in porcine liver using liquid chromatography–electrospray-tandem mass spectrometry, *J. Chromatogr. B*, 816, 15, 2005.
440. Sin, D.W.M. et al., Determination of quinoxaline-2-carboxylic acid, the major metabolite of carbadox, in porcine liver by isotope dilution gas chromatography–electron capture negative ionization mass spectrometry, *Anal. Chem. Acta*, 508, 147, 2004.
441. Wu, Y. et al., Development of a high-performance liquid chromatography method for the simultaneous quantification of quinoxaline-2-carboxylic acid and methyl-3-quinoxaline-2-carboxylic acid in animal tissues, *J. Chromatogr. A*, 1146, 1, 2007.
442. Zhang, Y. et al., Effects of cooking and storage on residues of cyadox in chicken muscle, *J. Agric. Food Chem.*, 53, 9737, 2005.

Chapter 11

Determination of Persistent Organic Pollutants in Meat

Anna Laura Iamiceli, Igor Fochi, Gianfranco Brambilla, and Alessandro di Domenico

Contents

11.1 Introduction 350
    11.1.1 General Remarks 350
    11.1.2 Laboratory Safety 350
    11.1.3 The Stockholm Convention on POPs 351
    11.1.4 Organochlorine Pesticides 360
    11.1.5 Polychlorodibenzo-*p*-Dioxins and Polychlorodibenzofurans 360
    11.1.6 Polychlorobiphenyls 361
    11.1.7 Polybrominated Diphenyl Ethers 361
    11.1.8 Polyfluorinated Alkylated Substances 362
11.2 Chemical Methods 363
    11.2.1 Organochlorine Pesticides 363
        11.2.1.1 Analytical Methods 363
    11.2.2 Polychlorodibenzo-*p*-Dioxins, Polychlorodibenzofurans, and
        Dioxin-Like Polychlorobiphenyls 369
        11.2.2.1 Analytical Methods 369
    11.2.3 Non-Dioxin-Like Polychlorobiphenyls and Polybrominated Diphenyl Ethers 380
        11.2.3.1 Analytical Methods 380
    11.2.4 Polyfluorinated Alkylated Substances 384
        11.2.4.1 Sampling and Sample Storage 384
        11.2.4.2 Extraction and Cleanup 384

11.2.4.3 Instrumental Identification and Determination 385
11.2.4.4 Observations.. 386
11.3 Bioassays to Screen in Meat Polychlorodibenzo-*p*-Dioxins,
Polychlorodibenzofurans, and Dioxin-Like Polychlorobiphenyls................................ 387
11.3.1 Introduction.. 387
11.3.2 Cell-Based Bioassays... 387
11.3.3 Bioassay Based on Polymerase Chain Reaction................................. 387
11.3.4 Bioassay Reliability and Applicability.. 387
Abbreviations... 388
References.. 390

11.1   Introduction

11.1.1   General Remarks

As defined by the United Nations Environment Programme (UNEP), persistent organic pollutants (POPs) are "chemical substances that persist in the environment, bioaccumulate through the food web, and pose a risk of causing adverse effects to human health and the environment" [1]. Although many POPs are already strictly regulated or are no longer in production, they are found in the environment and can enter the food chain mainly through the intake of animal fats (meat, fish, and milk) [2]. The measurement of POPs in food and, in particular, in products of animal origin is particularly important for the protection of human health. Maximum residue limits (MRLs) for some POPs (organochlorine pesticides) in a variety of food commodities were established by the European Union (EU), thus making necessary the development of sensitive methods to analyze these pollutants in food.

This chapter evaluates the methods in use for the determination of POPs in meat. The chapter is divided into three parts. The first part provides an overview of the POPs under the Stockholm Convention [3] and of those proposed for inclusion. The second part deals with the chemical methods available for the determination of some selected classes of POPs. The similarities among these methods often result in the simultaneous determination of several families of pollutants—for example, polychlorinated dibenzo-*p*-dioxins (PCDDs), polychlorinated dibenzofurans (PCDFs), and polychlorinated biphenyls (PCBs), and polybrominated diphenyl ethers (PBDEs)—after a common preparative procedure of the sample. Nevertheless, for practical reasons separate sections will be devoted to the methods used in the analysis of each class of pollutants: organochlorine pesticides (Section 11.2.1); PCDDs, PCDFs, and dioxin-like PCBs (DL-PCBs) (Section 11.2.2); non-dioxin-like PCBs (NDL-PCBs) and PBDEs (Section 11.2.3); and polyfluorinated alkylated substances (PFAS) (Section 11.2.4). The third part of the chapter provides some information about the use of bioassays for PCDDs, PCDFs, and DL-PCBs for screening.

11.1.2   Laboratory Safety

The POPs under investigation should be treated as a potential health hazard. A strict safety program for handling these substances and the chemicals used for their determination should be developed by the laboratory.

Following are suggested readings taken from the vast literature available: Organochlorine pesticides were evaluated for their risk to human health and the environment within the International Programme on Chemical Safety (IPCS) [4–14]. Evaluation of carcinogenic risk for humans from PCDDs and PCDFs was performed by IARC in 1997 [15]. More recently, the Scientific Committee on Food (SCF) of the European Commission (EC) adopted an opinion on PCDDs, PCDFs, and DL-PCBs in food [16], updating its opinion of 2000 [17]. As regards PBDEs, a draft risk profile on pentabromodiphenyl ether (penta-BDE) (commercial mixture) was adopted by the POPs Review Committee [18]; a risk assessment report on octabromodiphenyl ether (octa-BDE) (commercial mixture) was prepared on behalf of the EU [19]. A draft risk profile on perfluorooctane sulfonate (PFOS) was adopted by the POPs Review Committee [20], while a risk assessment opinion is being recently finalized at the European Food Safety Authority (EFSA).

Waste handling and decontamination of glassware, towels, laboratory coats, etc., are described in Refs. 21–25.

11.1.3 The Stockholm Convention on POPs

POPs are by definition persistent, thus representing a risk of a long-time exposure. Their persistence is generally correlated to their chemical stability, which makes these substances highly resistant to biological and chemical degradation. POPs can be found worldwide, even in areas where human activities are almost completely absent, that is, in the Antarctic and Arctic regions. POPs are also characterized by their ability to bioaccumulate. Bioaccumulation magnitude depends on several factors, one among which is the solubility of the substance in lipids [2]. Highly lipophilic substances are substantially insoluble in water, as commonly shown by the high values of *n*-octanol–water partition coefficient (K_{OW}). This strengthens the tendency of these substances to be concentrated in the fatty tissue of a living organism. As a result of bioaccumulation, several organic persistent substances are subjected to biomagnification process and are found at higher concentrations in animals at higher levels of the food chain [26].

Potential adverse effects on the environment and human health caused by exposure to POPs are of considerable concerns for governments, nongovernmental organizations, and the scientific community. The persistence of POPs in the environment and the capacity of covering long distances away from the point of their release have required that concerted international measures were adopted to efficiently control release. To this end, the global Stockholm Convention on POPs, opened for signatures in 2001 and entered into force in May 2004 [27], provides an international framework, based on the precautionary principle that seeks to guarantee the elimination of POPs or the reduction of their production and use. Since the beginning, the Convention has concerned 12 chlorinated chemicals, but every country for which the Convention is in force may submit a proposal for listing new POPs in Annex A (substances to be eliminated), Annex B (substances whose production and use is restricted), or Annex C (substances unintentionally produced whose releases have to be reduced and finally eliminated). The substances actually under the Convention are listed in Table 11.1 and include eight individual organochlorine pesticides, hexachlorobenzene, PCBs, PCDDs, and PCDFs. At present, a second group of chemicals (candidate POPs) is under consideration for inclusion in the Convention on the basis of their risk profile prepared by the POP Review Committee (Table 11.2); a third group (proposed POPs) has been proposed for risk evaluation to the Review Committee (Table 11.3). The inclusion of organic chemicals in the frame of the Convention presupposes that some requirements be met:

a. Persistence, measured as half-life, greater than 2 months in water, or 6 months in soil or sediments, or any evidence of sufficient persistence to justify the consideration of a substance within the Convention;

Table 11.1 Some Physical–Chemical Properties of the POPs under the Stockholm Convention

Compound	Molecular Structure	Molecular Weight	Water Solubility	$logK_{OW}$	Vapor Pressure (mmHg)	Half-life in Soil (years)	Reference
Aldrin (CAS 309-00-2)		365	27 µg/L (25°C)	5.17–7.4	2.3×10^{-5} (20°C)	<1.6	28
Chlordane (CAS 57-74-9)	Cis Trans	410	56 µg/L (25°C)	4.58–5.57	0.98×10^{-5} (20°C)	4	28
pp'-DDT[a] (CAS 50-29-3)		355	1.2–5.5 µg/L (25°C)	6.19	0.2×10^{-6} (20°C)	15	28

Dieldrin (CAS 60-57-1)		381	140 μg/L (20°C)	3.69–6.2	1.78×10^{-7} (20°C)	3–4	28
PCDDs[b,c]		322–460	19.3– 0.074 ng/L (25°C)	6.80– 8.20	1.5×10^{-9}– 8.25×10^{-13} (25°C)	10–12[d]	28, 29
PCDFs[e,f]		306–444	419–1.16 ng/L (25°C)	6.53–8.7	1.5×10^{-8}– 3.75×10^{-13} (25°C)	—	28, 29
Endrin (CAS 72-20-8)		381	220–260 μg/L (25°C)	3.21– 5.34	2.7×10^{-7} (25°C)	12	28

(continued)

Table 11.1 (continued) Some Physical–Chemical Properties of the POPs under the Stockholm Convention

Compound	Molecular Structure	Molecular Weight	Water Solubility	$logK_{OW}$	Vapor Pressure (mmHg)	Half-life in Soil (years)	Reference
Hexachlorobenzene (CAS 118-74-1)		285	50 µg/L (20°C)	3.93–6.42	1.09×10^{-5} (20°C)	2.7–5.7	28
Heptachlor (CAS 76-44-8)		373	180 ng/mL (25°C)	4.4–5.5	3×10^{-4} (20°C)	0.75–2	28
Mirex (CAS 2385-85-5)		546	0.07 µg/L (25°C)	5.28	3×10^{-7} (25°C)	10	28

PCBs[g]	189–499	0.0001–0.01 µg/L (25°C)	4.3–8.26	$0.003–1.6 \times 10^{-6}$ (25°C)	>6	28
Toxaphene (CAS 8001-35-2)	414	550 µg/mL (20°C)	—	0.2–0.4 (25°C)	0.3–12	28

[a] 1,1,1-trichloro-2,2-*bis*(4-chlorophenyl)ethane.
[b] Polychlorinated dibenzo-*p*-dioxins.
[c] Data refer to the seven 2,3,7,8-chlorosubstituted toxic congeners only.
[d] Data refer to 2,3,7,8-T4CDD.
[e] Polychlorinated dibenzofurans.
[f] Data refer to the ten 2,3,7,8-chlorosubstituted toxic congeners only.
[g] Polychlorinated biphenyls.

Table 11.2 Some Physical–Chemical Properties of POPs Candidate for Inclusion in Annex A, B, or C of the Stockholm Convention

Compound	Molecular Structure	Molecular Weight	Water Solubility	$logK_{OW}$	Vapor Pressure (mmHg)	Half-life in Soil (years)	References
Chlordecone (CAS 143-50-0)		490.6	1–3 mg/L	4.50–5.41	2.25×10^{-5} -3×10^{-5} (25°C)	1–2	28,30
Hexabromobiphenyl[a] (CAS 6355-01-8)		627.58	3–11 µg/L	6.39	5.2×10^{-8} -5.6×10^{-6} (25°C)	—	31
Lindane (gamma-hexachloro-cyclohexane) (CAS 58-89-9)		290.83	7 mg/L (20°C)	3.8	4.2×10^{-5} (20°C)	>1	28,32

PFOS[b] (CAS 2795-39-3)		538	519–680 mg/L (20–25°C)	Not measurable	2.45×10^{-6}	—	20
Penta-BDE[c] (commercial mixture)		485.8–564.7	13 µg/L (25°C)	5.9–7.0	7.2×10^{-10} –3.5×10^{-7} (20–25°C)	—	33

[a] Only one isomeric structure is shown.
[b] Perfluorooctane sulfonate (the potassium salt is shown).
[c] Pentabromodiphenyl ether. The commercial mixture contains penta- through heptabromo-substituted homologs.

Table 11.3 Some Physical–Chemical Properties of the POPs Proposed for Risk Evaluation as Prescribed for Inclusion in the Stockholm Convention

Compound	Molecular Structure	Molecular Weight	Water Solubility	$\log K_{OW}$	Vapor Pressure (mmHg)	Half-life in Soil (days)	References
Pentachloro-benzene (CAS 608-93-5)		250.32	0.56 mg/L (25°C)	4.8–5.18	1.65×10^{-2}	194–345	34
Octa-BDF[a] (commercial mixture) (CAS 32536-52-0)		801.38	0.0005 mg/L	6.29	4.94×10^{-8}	—	35
SCCPs[b] (CAS 85535-84-8)		320–500	0.0224–0.994 mg/L	4.39–8.69	2.1×10^{-9} –1.9 × 10^{-2}	>365	36

alpha-HCH[c] (CAS 319-84-6)	 (+)-alpha-HCH (−)-alpha-HCH	290.83	10 mg/L (28°C)	3.46–3.85	2×10^{-2} (20°C)	48–125	33,37
beta-HCH[d] (CAS 319-85-7)		290.83	5 mg/L (20°C)	3.78–4.50	5×10^{-3} (20°C)	91–122	33,38

[a] Octabromodiphenyl ether. The commercial mixture contains penta- through decabromo-substituted homologs.
[b] Short-chained chlorinated paraffins.
[c] alpha-Hexachlorocyclohexane.
[d] beta-Hexachlorocyclohexane.

b. Bioaccumulation, measured as bioconcentration factor (BCF) or bioaccumulation factor (BAF), greater than 5000, or as $\log K_{OW}$, greater than 5, or any evidence of bioaccumulation for consideration within the Convention;

c. Long-range environmental transport, evidenced by the measured levels of the chemical far from the source of release, or by modeled data demonstrating the potentiality for the substance to be transported through air, water, or migratory species;

d. Adverse effects to human health and the environment.

Owing to their physical–chemical properties, bioaccumulative behavior in lipid tissues, and possible toxicological effects, POPs represent a relevant and growing interest for human beings, with the food of animal origin representing the main source of exposure. For most of them (i.e., organochlorine pesticides), regulatory limits have been already set on meat commodities at European level and in non-European countries, with possible different maximum levels (MLs) of acceptance according to the animal species [39].

11.1.4 Organochlorine Pesticides

The term organochlorine pesticides refers to a wide range of organic chemicals containing chlorine atoms and used in agriculture and public health activity to effectively control pest. Although most of them have been banned during the 1970s and 1980s, they are still found in the environment [40–42] and in biological matrices [43,44]. In fact, the intrinsic characteristics of these substances (i.e., highly lipophilic, low chemical and biological degradation rate) have led to their accumulation in the biosphere where they magnify in concentrations progressing through the food web. Organochlorine pesticides under the Stockholm Convention and their principal chemical–physical properties are listed in Table 11.1.

Food is considered to represent a constant source of exposure. For this reason, regulations concerning pesticides in food have become more and more severe in the past decades. To harmonize registration of pesticides and tolerances throughout the community, the EU Directive 91/414/EEC [45] lays down the basic rules with respect to plant protection products. Regulation 396/2005/EC [46] indicates that temporary maximum residue limits (TMRLs) have to be established at the EU level for all the active substances for which harmonized MRLs are not yet set, that is, pesticides currently not covered by the EU MRL Directives 86/362/EEC [47], 86/363/EEC [39], and 90/642/EEC [48]. Current EU MRLs established for the organochlorine pesticides of interest in animal products are set between 0.02 and 1 mg/kg on fat basis. In the United States, legislation was enacted in 1996 with the Food Quality Protection Act, including stricter safety standards, especially for infants and children, and a complete reassessment of all existing pesticide tolerances. For the pesticides of our concern, U.S. residue limits are established between 0.1 and 7 mg/kg fat. At the international level, MRLs in meat and meat products recommended by Food and Agriculture Organization (FAO)/World Health Organization (WHO) vary from 0.05 to 3 mg/kg fat [49].

11.1.5 Polychlorodibenzo-p-Dioxins and Polychlorodibenzofurans

PCDDs and PCDFs (altogether also commonly known as "dioxins") are two groups of tricyclic aromatic compounds containing between one and eight chlorine atoms, thus resulting in 210 congeners (75 PCDDs and 135 PCDFs), different in the number and position of chlorine atoms. As shown in Table 11.1, PCDDs and PCDFs are insoluble in water, exhibit a strong lipophilic character, and are very persistent. Neither PCDDs nor PCDFs are produced intentionally. In fact, their formation and release into the environment occur primarily in thermal or combustion processes or as unwanted by-products of industrial processes involving chlorine. Of the 210 positional isomers, only the 17 congeners with chlorines at positions 2, 3, 7, and 8 are of toxicological interest. To facilitate risk assessment/management, a toxicity equivalency factor (TEF) relative to 2,3,7,8-T_4CDD was assigned to each of the toxic congeners: for food and feeding stuffs, the WHO-TEFs adopted in 1997 by the WHO [50] are presently used. An update of WHO-TEFs was carried out in 2005 [51]. The "international" system (I-TEFs) [52] is still used for environmental samples.

Humans are exposed to PCDDs and PCDFs through the diet. The contribution of meat and meat products and fish and fishery products together may be higher than 90% of the total exposure to PCDDs, PCDFs, and DL-PCBs [53–57]. To reduce human exposure and protect consumer health, the EU has progressively issued regulatory measures setting MLs for PCDDs, PCDFs, and DL-PCBs in food and feeding stuffs. For example, an ML of 3.0 pg WHO-TE/g fat was established for PCDDs and PCDFs in bovine and sheep meat corresponding to an ML of 4.5 pg WHO-TE/g fat when DL-PCBs are considered [58]; in pork meat, the corresponding ML values are 1.0 and 1.5 pg WHO-TE/g fat.

11.1.6 Polychlorobiphenyls

PCBs are a family of 209 chlorinated compounds produced commercially under various trade names by direct chlorination of biphenyl. As PCDDs and PCDFs, PCBs are substantially insoluble in water, strongly lipophilic, and very persistent (Table 11.1). For their chemical–physical stability and dielectric properties, they were used worldwide as or in transformer and capacitor oils, hydraulic and heat exchange fluids, and lubricating and cutting oils. PCBs are divided into two groups according to their toxicological mode of action: the DL-PCBs consist of 12 congeners with toxicological properties similar to PCDDs and PCDFs, and NDL-PCBs, with a different toxicological profile. 2,3,7,8-T_4CDD WHO-TEFs have also been assigned to DL-PCBs [50,51]. Both NDL-PCBs and DL-PCBs bioaccumulate in animals and humans and biomagnify in the food chain. No MLs for NDL-PCBs in food and feeding stuffs have been set at the community level as yet.

11.1.7 Polybrominated Diphenyl Ethers

PBDEs are a group of 210 congeners, differing in the number of bromine atoms and in their position on two phenyl rings linked by oxygen. Their nomenclature is identical to that of PCBs. These chemicals are persistent and lipophilic, which results in bioaccumulation in fatty tissues of organisms and enrichment through the food chain [59]. PBDEs were first introduced into the market in the 1960s and used as flame retardants to improve fire safety in various consumer products and in electronics. There are three types of commercial PBDE products—penta-BDE, octa-BDE, and decabromodiphenyl ether (deca-BDE)—each product being a mixture of various PBDE congeners [60]. The EU has prohibited the uses of penta- and octa-BDE [61], but these substances are still on the market in many regions of the world. In any case, a substantial reservoir of PBDEs exists in products that could release them to the environment.

Despite the fact that dietary intake is probably the main route of exposure to PBDEs for the general population [62,63], no MLs for PBDEs in food have yet been set by the EU.

11.1.8 Polyfluorinated Alkylated Substances

The PFAS are compounds consisting of a hydrophobic alkyl chain of variable length (typically C_4 to C_{16}) and a hydrophilic end group. The hydrophobic part may be fully or partially fluorinated: for instance, the "6:2" formula (Table 11.4) indicates that, in the C_8-chain, six carbons are fully fluorinated whereas the remaining two bear hydrogen atoms. When fully fluorinated, the molecules are called perfluorinated alkylated substances, whereas the partially fluorinated ones, because of the telomerization production process, are named telomers. The hydrophilic end group can be neutral or positively or negatively charged. The resulting compounds are nonionic, cationic, or anionic surface-active agents: due to their amphiphilic features, most of the perfluorinated compounds will not accumulate in fatty tissues as is usually the case with other persistent halogenated compounds.

Table 11.4 Examples of PFAS of Environmental Interest

Compound	Molecular Structure	Acronym	Molecular Weight
Perfluorobutyl sulfonate (CAS 29420-49-3)		PFBS	299.21
Perfluorooctanoic acid (CAS 335-67-1)		PFOA	414.07
6:2 Fluorotelomer sulfonate (CAS 29420-49-3)		6:2 FTS	427.16
Perfluorooctane sulfonamide (CAS 754-91-6)		PFOSA	499.14
Perfluorooctyl sulfonate (CAS 2795-39-3; 1763-23-1 (acid))		PFOS	499.23
N-Methyl perfluorooctane sulfonamidoethanol (CAS 24448-09-7)		N-MeFOSE	557.23
Perfluorotetradecanoic acid (CAS 376-06-7)		PFTeDA	714.12

PFAS can be widely found in the environment, primarily resulting from anthropogenic sources as a consequence of industrial and consumer applications, including stain-resistant coatings for fabrics and carpets, oil-resistant coatings for paper products, fire-fighting foams, mining and oil well surfactants, floor polishes, and insecticide formulations [64].

At present, PFOS and perfluorooctanoic acid (PFOA) are the most investigated molecules in the environment and humans, due to their widespread occurrence, bioaccumulation, and persistence. The latter is determined by the strong covalent C–F bond [65–69]. Many of the neutral PFAS—such as perfluorooctane sulfonamide (PFOSA) and *n*-ethyl perfluorooctane sulfonamidoethanol (*n*-EtFOSE)—are considered to be potential precursors of PFOS. In addition, PFOA could be generated from the 8:2 fluorotelomer alcohol (8:2 FTOH), PFOSA, and *n*-EtFOSE. Some selected PFAS of environmental interest are reported in Table 11.4.

11.2 Chemical Methods

11.2.1 *Organochlorine Pesticides*

11.2.1.1 *Analytical Methods*

The EU MRLs set for organochlorine pesticides in products of animal origin require the development of highly sensitive methods to analyze these pesticides in different sample matrices. The most suitable and efficient approach involves the use of multiresidue procedures whose properties were recently reported by Hercegovà et al. [70] and are summarized as follows: (a) possibility to determine a number of pesticides as high as possible in a single analysis, (b) high recoveries, (c) high selectivity obtained by means of effective removal of potential interferences from the sample, (d) high sensitivity, (e) high precision, (f) good ruggedness, (g) low cost, (h) high speed, and (i) use of less harmful solvents and in low amounts.

Multiresidue methods developed for organochlorine pesticides follow the general scheme shown in Figure 11.1. After a pretreatment step aimed at obtaining a homogenized and dried sample, the analytes and fat are extracted together from the test matrix and the extract is purified to obtain a suitable sample for instrumental determination. The lipids coextracted with the analytes are separated using different nondisruptive procedures, which include liquid–liquid partitioning and gel permeation chromatography (GPC). Pesticides are further cleaned up by adsorption chromatography with Florisil® (U.S. Silica Company, Berkeley Springs), alumina, or silica, used as adsorbent phases. Instrumental determination is performed by high-resolution gas chromatography (HRGC) coupled to electron capture detection (ECD) or mass spectrometry (MS). The principal procedures currently used for the analysis of organochlorine pesticides in meat samples (Table 11.5) are examined in the following sections. Most of them were reviewed by the Codex Committee on Pesticide Residues [71] and included in the Official Methods of Analysis of the Association of Official Analytical Chemists (AOAC International, 2005) [72] and in the Pesticide Analytical Manual of the Food and Drug Administration (FDA) [73].

11.2.1.1.1 Pretreatment

A good preparation of sample matrix is an essential step to enhance extraction efficiency. The ideal sample for extraction is a dry, finely divided solid. In fact, a high surface area of the test matrix is recommended to improve the contact of the solvent with test molecules and, finally, to obtain

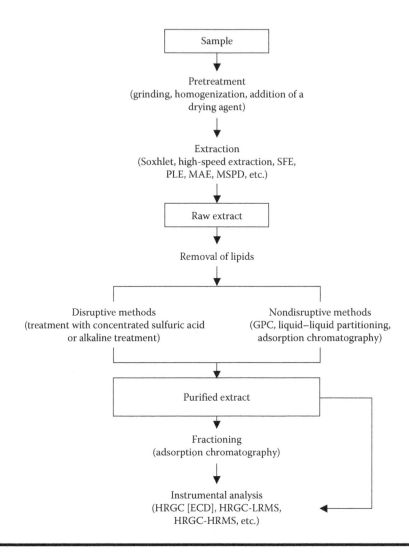

Figure 11.1 General scheme adopted for the analysis of organochlorine pesticides, PCDDs, PCDFs, PCBs, and PBDEs.

quantitative recoveries. As reported in the general procedure related to preparation of test samples for meat and meat products [80], samples of animal origin are minced and homogenized. Anhydrous sodium sulfate or Hydromatrix™ (Varian Associates, Inc.) are added until a friable mixture is obtained.

11.2.1.1.2 Extraction

Extraction techniques include classical procedures (i.e., Soxhlet extraction, column extraction, partitioning extraction, high-speed extraction, etc.) and innovative methods, such as supercritical fluid extraction (SFE), pressurized liquid extraction (PLE)—also known as accelerated solvent extraction (ASE)—microwave-assisted extraction (MAE), and matrix solid-phase dispersion (MSPD) extraction.

Table 11.5 Examples of Organochlorine Pesticide Determination in Samples of Animal Origin

Sample	Pretreatment	Extraction	Cleanup	Instrumental Analysis	Method Performance	Reference
Cattle fat, swine internal organ tissues 5 g	Homogenization, addition of Hydromatrix™	PLE, 1:1 dichloromethane–acetone, 1500 psi, two cycles™	Fat removal with GPC, SX-3 BioBeads column; fractioning into two fractions on silica gel SPE cartridges (PCBs, PBDEs, and nonpolar chlorinated pesticides in fraction I, polar chlorinated pesticides in fraction II)	HRGC-LRMS, DB-5 ms GC column; MS operating in NCI mode with methane as a reagent gas	Mean recoveries, 24–111%	74
Fatty samples 4–5 g	Addition of Hydromatrix	SFE, CO_2 modified with 3% acetonitrile at 27.58 MPa, 60°C; fat removal on a C_1-bonded phase at 95°C	Adsorption chromatography on Florisil column	HRGC coupled to an electrolytic conductivity detector; DB-1 GC column	Mean recoveries, 85–115%	75
Meat (chicken, pork, and lamb) 5 g	Homogenization, addition of anhydrous Na_2SO_4	Extraction at high speed with ethyl acetate; alternatively, Soxhlet extraction with ethyl acetate	GPC, SX-3 BioBeads column	HRGC-MS/MS (triple quadrupole)	Mean recoveries, 75–96% (extraction at high speed); mean recoveries, 67–86% (Soxhlet); LOQ, 0.8–2.7 µg/kg	76

(continued)

Table 11.5 (continued)   Examples of Organochlorine Pesticide Determination in Samples of Animal Origin

Sample	Pretreatment	Extraction	Cleanup	Instrumental Analysis	Method Performance	Reference
Meat (chicken, pork, and lamb) 5 g	Homogenization, freeze-drying, addition of Hydromatrix™	PLE, ethyl acetate, 120°C, 1800 psi, static extraction time 5 min	GPC, SX-3 BioBeads column		Mean recoveries, 64–87%; LOQ, 0.8–2.7 µg/kg	76
Liver of chicken, pork, and lamb 5 g	Homogenization	Extraction at high speed with ethyl acetate	GPC, SX-3 BioBeads column	HRGC-MS/MS (triple quadrupole)	Mean recoveries, 65–111%	77
Liver of chicken, pork, and lamb 0.5 g	Homogenization, addition of C_{18}	MSPD, sample/C_{18} transferred in a cartridge containing 2 g Florisil; elution with ethyl acetate	No additional cleanup steps	HRGC-MS/MS (triple quadrupole)	Mean recoveries, 69–86%, except for lindane and endrin <30 % (MSPD); LOQ, 3.5–4.9 µg/kg	77
Pork fat 1.25 g		Blending with 1:1 ethyl acetate–cyclohexane	GPC, two Environsep-ABC columns, elution with 1:1 cyclohexane–ethyl acetate	HRGC-MS/MS (triple quadrupole); VF-5 ms GC column	Mean recoveries, 66–101%; LOD, 0.1–2 µg/kg	78
Meat 50–100 g		Extraction at high speed with 2:1 petroleum ether–acetone	GPC, Environgel™ column; adsorption chromatography on Florisil column	GC-ECD; DB-5 GC column	Mean recoveries, 101–103%[a]; LOQ, 0.002–0.05 µg/kg[a]	79

[a] Data refer to chlordane only.

Soxhlet extraction has been widely used in the organochlorine pesticide analysis [81,82]. The continuous contact of the sample with freshly distilled solvent ensures high extraction efficiency, usually higher than 70% for the pesticides of interest [73], so that it allows its employment as reference method.

Besides Soxhlet, other classical techniques are widely used for the extraction of meat sample. Among these, extraction with solvent at high speed is largely used, in which the sample is transferred into a blender cup or into a homogenizer and extracted with organic solvent [73,76,82]. The extract is decanted and separated from the matrix by filtration or centrifugation. Alternatively, extraction of meat sample (in the form of a friable product) is carried out directly in centrifuge in presence of organic solvent [82].

Fast extraction of organochlorine pesticides is also performed by the column extraction technique. This is carried out in a glass column where the sample, dried and homogenized, is transferred and eluted with organic solvent [82].

The need for determining a high number of residues in a time as short as possible had led to the development of innovative techniques for the extraction of pesticides from fatty food. Among these techniques, SFE with CO_2 and PLE with pressurized solvents appear to be equivalent to liquid-base techniques in the extraction of pesticides from fatty matrices. In the past decades, SFE has received wide attention for its low solvent consumption and its high degree of selectivity [72,83]. The use of supercritical fluids, characterized by densities close to those of the liquid solvents but with lower viscosity and higher diffusion capability, results in extraction agents that are more penetrative and with a higher solvating power. In addition, the combination with solid sorbent traps allows to obtain more purified extracts, eventually resulting in a single-step extraction and cleanup. Applications of SFE to the analysis of pesticides in fatty matrices are reported by Snyder et al. [84] and Hopper [72].

PLE is an innovative method for the rapid extraction of analytes. It is based on the use of solvents at temperatures (from 60 to 200 °C) higher than their boiling points at atmospheric pressure, and at high pressure (from 5×10^5 to 2×10^7 Pa) to maintain the solvent in a liquid state. PLE has found wide applications, especially in the field of environmental samples (soil, sediments, sludges, dust, etc.). More recently, the extraction efficiency of PLE as well as its application to the analysis of pesticides in fatty matrices was investigated in the isolation of lipids from biological tissues [71,85]. This technique appears to be effective in the quantitative extraction of organochlorine compounds from tissue samples (liver, heart, kidney, and adipose tissue) and from muscle of chicken, pork, and lamb (recoveries in the range from 70 to 93%) [73].

In the MAE, a sample is suspended in a suitable solvent and the mixture irradiated in a microwave oven. The irradiation step is generally repeated until the maximum yield of extraction is obtained. In the field of organochlorine pesticides, this technique has been applied especially for the extraction of environmental samples, such as soil, sediments, and vegetables [86,87], whereas applications to the extraction of fatty samples are limited [88,89].

The MSPD extraction was introduced by Barker in 1989 [90] and has proven to be an efficient procedure for the extraction of a wide range of drugs, pesticides, and naturally occurring constituents in samples of vegetable and animal origin [91,92]. It involves the use of octadecyl-bonded silica (C_{18}), octyl-bonded silica (C_8), or other sorbents obtained by chemical modification of silica surface, blended with the sample by means of a mortar and pestle. The material is successively transferred to a syringe, compressed by a syringe plunger, and eluted with a suitable organic solvent. Applications of the MSPD technique to the extraction of organochlorine pesticides in animal fat and animal tissue are reported in the studies by Long et al. [93], Furusawa [94,95], and Frenich et al. [77].

11.2.1.1.3 Cleanup

Owing to the lipophilic properties of organochlorine pesticides and their tendency to accumulate in fat, their extraction from the matrix is always accompanied by coextraction of lipidic material, which makes the instrumental analysis difficult without a preliminary purification of the extract. In the determination of some chlorinated environmental contaminants (e.g., PCDDs, PCDFs, PCBs), fat is often efficiently removed by treatment with sulfuric acid. However, pesticides such as dieldrin, endrin, and DDT (different isomers and metabolites) are not sufficiently stable and are decomposed by this method [71,96]. Therefore, nondestructive procedures, such as GPC and liquid–liquid partitioning, are widely applied for the elimination of the lipidic fraction in the analysis of organochlorine residues in fatty samples.

GPC is an automated procedure that is highly effective in removing high-molecular-weight substances (i.e., lipids, proteins, and pigments) due to the difference in molecular size between interferences and the target analytes. In the analysis of organochlorine pesticides, the divinyl-benzene-linked polystyrene gel (BioBeads SX-3) is the most commonly used sorbent [97]; several solvent mixtures have been recommended as eluents [71]. Owing to separation principle, which is not selective with respect to interferences with low molecular weight, the application of additional cleanup steps is generally necessary. In exceptional cases when highly selective instrumental detectors are used (e.g., MS/MS detector), samples can be directly injected after GPC cleanup [73].

Liquid–liquid partitioning uses the differences in polarity between the analytes and the interfering species to separate pesticides from the lipidic fraction. Partitioning between petroleum ether (or light petroleum) and acetonitrile is one of the most traditional procedures to separate pesticides from fat [98,99]. Owing to the low solubility of lipidic compounds in acetonitrile, fat is retained in petroleum ether whereas partition of organochlorine compounds into acetonitrile is a function of their partitioning coefficients. In a subsequent step, residues in acetonitrile are partitioned back into petroleum ether when acetonitrile is diluted with excess water, which is added to reduce pesticide solubility in acetonitrile. As observed for GPC, liquid–liquid partitioning is not effective for the separation of organochlorine pesticides and other coextractive species. This is generally accomplished by various cleanup procedures, based on the use of different adsorbent phases (commonly Florisil, alumina, or silica) employed either in traditional column chromatography or in solid-phase extraction (SPE) cartridges [98–100]. Recently, this last approach has gained wide acceptance because of its simplicity and low time and solvent consumption. Today several commercial SPE cartridges are available for cleanup of organochlorine pesticides in fatty samples. The most widely used SPE cartridges include octadecyl (C_{18})-bonded porous silica, silica gel, Florisil, and alumina [101].

11.2.1.1.4 Instrumental Analysis

Common methods for the quantification of organochlorine pesticides involve HRGC (ECD) and HRGC-MS.

11.2.1.1.4.1 HRGC (ECD) The high efficiency of capillary columns allows the separation of a large number of organochlorine pesticides. In this field, the most widely used phases are the nonpolar 100% dimethylpolysiloxane and (5% phenyl) methylpolysiloxane columns. Operative conditions and relative retention data obtained on these two GC columns can be found in Refs. 81, 102, and 103. ECD is the most common detector used for the detection of organochlorine pesticides. It presents high sensitivity but, according to the guidelines proposed by EC DG Health and Consumer Protection [104] for pesticide residue analysis, does not provide enough selectivity. To overcome this problem, the use of two columns of different polarity is mandatory to obtain unambiguous identification when HRGC (ECD) is used as a confirmatory method in the determination of residues of organic contaminants in live animals and their products [105]. In two-dimensional GC, two columns of different selectivity are serially coupled via a modulation device that cuts small portions of the effluent from the first column and refocuses them onto the second column, thus obtaining an improvement of the overall resolution [106]. Applications of two-dimensional GC to pesticide analysis were reported by Focant et al. [107], Korytár et al. [108], Seemamahannop et al. [109], and Chen et al. [110].

11.2.1.1.4.2 HRGC-MS As observed, the determination of organochlorine pesticides in samples with high fat content requires selective techniques to unambiguously confirm their presence, since interfering species often mask the analytical signal of the target compounds. With respect to ECD, MS detectors coupled to HRGC provide greater identification and confirmation power, thus generally avoiding false positive and false negative errors due to matrix interferences [111]. Nowadays, the most widely used MS technique in the field of organochlorine pesticides relies on low-resolution (LR) apparatuses such as the quadrupole analyzer, operating with an electron impact (EI) ion source, an electron energy of 70 eV, and in the selected ion monitoring (SIM) mode. It allows the determination of the target analytes at levels in compliance with the regulations established for a wide range of pesticides and food commodities, including those of animal origin. However, confidence in the confirmation of identity could be reduced if one or more of the selected ions are affected by matrix interferences [78]. A remarkable enhancement in terms of selectivity and sensitivity is observed with tandem MS (MS/MS, ion trap, or triple quadrupole analyzers) where a single ion is subjected to a second fragmentation to confirm the identity of the parent compound. Application of triple quadrupole MS in the analysis of organochlorine pesticides in samples of animal origin has recently been reported by Patel et al. [78] and Frenich et al. [76,77].

11.2.2 Polychlorodibenzo-p-Dioxins, Polychlorodibenzofurans, and Dioxin-Like Polychlorobiphenyls

11.2.2.1 Analytical Methods

The standard methods developed to determine these contaminants in food samples generally include the assessment of the 17 2,3,7,8-chlorosubstituted PCDDs and PCDFs, and the 12 DL-PCBs. The latter include four non-*ortho* congener PCBs (PCBs 77, 81, 126, and 169), and eight mono-*ortho* congener PCBs (PCBs 105, 114, 118, 123, 156, 157, 167, and 189). The analysis of these classes of substances in food is complicated by their low contamination levels (in the order of pg/g) and by the complexity of the matrix. This generally contains large amounts of interfering species (i.e., lipids) whose appropriate removal requires laborious cleanup procedures. To give an overall view of the analytical methods in use for the quantification of PCDDs, PCDFs, and DL-PCBs in meat samples, the most topical literature has been reviewed (Table 11.6). Recent advances in determination of PCDDs, PCDFs, and DL-PCBs are in particular reported by Reiner et al. [112]; the importance of matrix pretreatment, sample extraction, cleanup, and fractionation of PCBs from food matrices are exhaustively described by Ahmed [97]; and a critical review of the

Table 11.6 Examples of PCDD/F, DL-PCB, NDL-PCB, and PBDE Determinations in Samples of Animal Origin

Analytes	Sample	Pretreatment	Extraction	Cleanup	Instrumental Analysis	Method Performances[a]	Found Levels[a,b]	Reference
PCDDs, PCDFs	Meat (beef, pork, chicken, lamb) 5–10 g	Homogenization, freeze-drying, addition of anhydrous Na_2SO_4	Soxhlet, toluene, 24 h	Fat removal with adsorption chromatography on a multilayer column; fractioning with an alumina column	HRGC-HRMS, DB-5 GC column; MS operating in the EI mode at resolution of 10,000	LOD, 0.02–0.2 ng/kg dw	Beef: 0.7 pg TEQ/g fat; Pork: 0.3 pg TEQ/g fat; Chicken: 0.8 pg TEQ/g fat; Lamb: 0.7 pg TEQ/g fat	115
PCDDs, PCDFs, DL-PCB	Meat (beef, chicken) 25 g	Homogenization	PLE, toluene, 135°C, 1500 psi, static extraction time 10 min, three cycles	Fat removal with 20 g Extrelut impregnated with 40 g H_2SO_4; fractioning with disposable prepacked columns containing multilayer silica, alumina, and carbon	HRGC-HRMS, DB-5 GC column; MS operating in the EI mode at 35 eV and a resolution of 10,000		Beef: PCDD/Fs 2.95 pg TEQ/g fat; DL-PCBs 3.44 pg TEQ/g fat; Chicken: PCDD/Fs 1.37 pg TEQ/g fat; DL-PCB 1.12 pg TEQ/g fat	116
PCDDs, PCDFs	Meat (beef, pork, poultry)	Homogenization, freeze-drying, crushing with blender, addition of anhydrous Na_2SO_4	Blending with 2:1 n-hexane–acetone	Adsorption chromatography on multilayer silica, charcoal, and Florisil columns	HRGC-HRMS, DB-5 and SP2331 GC column; MS operating in the EI mode and a resolution of 10,000		Beef: 0.72 pg I-TEQ/g fat; Pork: 0.27 pg I-TEQ/g fat; Poultry: 0.46 pg I-TEQ/g fat	117

Analytes	Sample	Preparation	Extraction/Cleanup	Fractionation/Cleanup	Instrumentation	Results	Concentrations	Ref.
PCDDs, PCDFs, DL-PCBs, PBDE	Fatty samples (aliquots tested corresponding to 10 g of fat)	Homogenization	Blending with *n*-hexane and acidified silica gel (1:1.5 H_2SO_4:silica); extraction through a multilayer column	Fractioning of the extract on a carbon column into two fractions (mono- to tetra-*ortho*-PCBs, and PBDEs in fraction I, non-*ortho*-PCBs, PCDDs, and PCDFs in fraction II); purification of fraction II by treatment with H_2SO_4 and elution through silica gel and alumina columns	HRGC-HRMS, (non-*ortho*-PCBs, PCDD/Fs), DB-5 GC column; MS operating in the EI mode and a resolution of 10,000; HRGC-MS, (mono- to tetra-*ortho*-PCBs), DB-5 GC column; MS operating in the EI mode	LOD, 0.01–0.05 ng/kg; precision, 5–11%; recoveries, >50%		118
PCDDs, PCDFs, DL-PCB	Meat (poultry)	Homogenization, freezing under liquid nitrogen, addition of anhydrous Na_2SO_4, freeze-drying, grounding	PLE, *n*-hexane, 1500 psi, static extraction time 5 min, two cycles	Fat removal with GPC (4 g fat loaded), SX-3 BioBeads column; fractioning with disposable prepacked columns containing multilayer silica, alumina, and carbon	HRGC-HRMS, RTX-5SIL-MS GC column; MS operating in the EI mode at 60 eV and a resolution of 10,000	Recoveries, >75 %	PCDD/Fs 1.6 pg I-TEQ/g fat; non-*ortho*-PCBs 2.2 pg I-TEQ/g fat	119
NDL-PCBs	Animal fat (pork, poultry) 1–5 g	Homogenization, melting	Blending with *n*-hexane–acetone	Adsorption chromatography on acidified silica column (1:1 concentrated H_2SO_4-silica)	HRGC-ECD, HT-8 GC column; HRGC-LRMS, DB-5 ms GC column	Recoveries, 72–78%	Poultry: 171–3753 ng/g fat[c] Pork: 591–2855 ng/g fat[c]	120

(continued)

Table 11.6 (continued) Examples of PCDD/F, DL-PCB, NDL-PCB, and PBDE Determinations in Samples of Animal Origin

Analytes	Sample	Pretreatment	Extraction	Cleanup	Instrumental Analysis	Method Performances[a]	Found Levels[a,b]	Reference
PCDDs, PCDFs, DL-PCBs	Animal fat (pork, poultry) 1–5 g	Homogenization	Blending with n-hexane–acetone	Fat removal with concentrated H_2SO_4 adsorbed on and Florisil; fractioning on activated carbon	HRGC-HRMS	LOD, <0.2 pg/g (tetra- to hexa-PCDD/Fs)	Poultry: PCDD/Fs 3–118 pg TEQ/g fat; non-ortho-PCBs 3–6 pg TEQ/g fat[d]; mono-ortho-PCBs 8–187 pg TEQ/g fat Pork: PCDD/Fs 3–118 pg TEQ/g fat; non-ortho-PCBs 1–1.5 pg TEQ/g fat[d] mono-ortho-PCBs 13–63 pg TEQ/g fat	120
PCDDs, PCDFs, DL-PCB	Meat (chicken, pork)	Homogenization, addition of anhydrous Na_2SO_4 (Na_2SO_4/sample 1.5–2.0)	PLE with fat retainer, n-heptane, 100°C, static extraction time 5 min, two cycles	Fractioning of extract on activated carbon column AX-21 with Celite® (Celite Corporation) into three fractions (bulk PCBs in fraction I, mono-ortho-PCBs in fraction II, non-ortho-PCBs, PCDD/Fs in fraction III)	HRGC-HRMS, DB-5 GC column; MS operating in the EI mode at 65 eV and a resolution of 10,000	Recoveries, 74–92%		121
NDL-PCBs	Meat (beef, pork, poultry, horse)	Homogenization, freezing under liquid nitrogen, freeze-drying	PLE, n-hexane, 1500 psi, static extraction time 5 min, two cycles	GPC (4 g fat loaded), SX-3 BioBeads column; fractioning with disposable prepacked columns containing multilayer silica, alumina, and carbon	HRGC-ITMS/MS, RTX-5Sil-MS GC column; MS operating in the EI mode	Recoveries, 60–101%	Beef: 5910 pg/g fat[c]; Pork: 8828 pg/g fat[c]; Poultry: 4770 pg/g fat[c]; Horse: 21,588 pg/g fat[c]	122

| PCDDs, PCDFs, DL-PCBs | Fast food samples containing meat 10–15 g dw | Homogenization, freezing under liquid nitrogen, freeze-drying, addition of anhydrous Na₂SO₄ | PLE, *n*-hexane, 1500 psi, static extraction time 5 min, two cycles | Fractioning (4 g fat) with disposable prepacked columns containing multilayer silica, alumina, and carbon | HRGC-HRMS (non-*ortho*-PCBs, PCDD/Fs), GC column RTX-5SIL-MS; MS operating in the EI mode at 60 eV and a resolution of 10,000; HRGC-ITMS/MS, (mono-*ortho*-PCBs), RTX-5SIL-MS GC column; MS operating in the EI mode | McDonald's Big Mac® | PCDD/Fs ND–1.07 pg TEQ/g fat; DL-PCBs ND–2.31 pg TEQ/g fat | 123 |
| PBDEs | Meat 0.2–10 g | Homogenization, addition of anhydrous Na₂SO₄ | Soxhlet with 3:1 *n*-hexane–acetone | Column chromatography on silica impregnated with concentrated H_2SO_4 | GC-NCI/MS, HT-8 GC column (for tri- to hepta-BDE congeners) and AT-5 GC column (for 209 congener) | Beef steak: 31 pg/g ww[e]

Minced meat: 110 pg/g ww[e]

Hamburger: 120 pg/g ww[e]

McDonald's Big Mac: 160 pg/g ww[e] | | 124 |

(continued)

Table 11.6 (continued) Examples of PCDD/F, DL-PCB, NDL-PCB, and PBDE Determinations in Samples of Animal Origin

Analytes	Sample	Pretreatment	Extraction	Cleanup	Instrumental Analysis	Method Performances[a]	Found Levels[a,b]	Reference
PCDDs, PCDFs, DL-PCBs, PBDEs	Meat (hamburger, fat of chicken, pork, and beef) 5 g	Homogenization in dichloromethane and drying with anhydrous Na$_2$SO$_4$ (fat samples); mixing with Celite (hamburger samples)	PLE, 35:30:35 2-propanol-*n*-hexane–dichloromethane, 125°C, 1500 psi (hamburger samples)	Cleanup and fractioning with disposable prepacked jumbo columns containing multilayer silica, alumina, and carbon	HRGC-HRMS, (non-*ortho*-PCBs, PCDD/Fs), GC column DB-5; MS operating in the EI mode at 35 eV and a resolution of 10,000; HRGC-MS, (PBDEs), DB-5 ms GC column; MS operating in EI mode at a resolution of 2500	Recoveries, 35–150% (except for PBDE 209 occasionally <20%); accuracy and precision better than 20%	Hamburger: PCDD/Fs 1.3 pg I-TEQ/g fat; non-*ortho*-PCBs 0.2 pg I-TEQ/g fat; PBDEs 648 pg/g fat[f]	125
							Chicken fat: PCDD/Fs 0.3 pg I-TEQ/g fat; non-*ortho*-PCBs 0.1 pg I-TEQ/g fat; PBDEs 2911 pg/g fat[f]	
							Pork fat: PCDD/Fs 0.2 pg I-TEQ/g fat; non-*ortho*-PCBs 0.01 pg I-TEQ/g fat; PBDEs 2588 pg/g[f]	
							Beef fat: PCDD/Fs 0.6 pg I-TEQ/g fat; non-*ortho*-PCBs 0.1 pg I-TEQ/g fat; PBDEs 244 pg/g fat[f]	

Analytes	Sample	Pretreatment	Extraction	Cleanup	Instrumental	Recoveries, LOQ	Results	Ref.
PCDDs, PCDFs, DL-PCBs, PBDEs	Meat and eggs (processed meat products, beef, pork, poultry, eggs)	Freeze-drying	Soxhlet, toluene, 24 h	Fat removal by elution through a silica gel column containing acidic and neutral silica layers; fractioning on an activated carbon column into two fractions (PCBs and PBDEs in fraction I, and PCDD/Fs in fraction II); purification of the two fractions on an activated alumina column; further fractioning of fraction I into two subfractions on activated alumina column (sub-fraction IA containing non-*ortho*-PCBs, and sub-fraction IB containing other PCBs and PBDEs)	HRGC-HRMS, DB-Dioxin GC column; MS operating in the EI mode at a resolution of 10,000	Recoveries, >50%; LOQ, 0.0007–0.63 pg/g ww (PCDD/Fs); LOQ, 0.0007–0.13 pg/g ww (non-*ortho*-PCBs); LOQ, 0.048–3.2 pg/g ww (mono-*ortho*-PCBs and NDL-PCBs); LOQ, 0.035–13 pg/g ww (PBDEs)	Meat and eggs (market basket): PCDD/Fs 0.0082 pg I-TEQ/g ww; non-*ortho*-PCBs 0.59 pg I-TEQ/g fw[d]; mono-*ortho*-PCBs 52 pg I-TEQ/g ww; NDL-PCBs 410 pg/g ww[g]; PBDEs 13 pg/g ww[h]	126
PBDEs	Chicken fat 1 g	Homogenization, filtration through anhydrous Na_2SO_4		Stirring with 10 g of 40% acid silica, purification with prepacked disposable columns containing multilayer silica, and alumina	HRGC-MS, DB-5 ms GC column; MS operating in EI mode at 70 eV	Recoveries, >75%	1.76–39.43 ng/g[i]	127

(continued)

Table 11.6 (continued)   Examples of PCDD/F, DL-PCB, NDL-PCB, and PBDE Determinations in Samples of Animal Origin

Analytes	Sample	Pretreatment	Extraction	Cleanup	Instrumental Analysis	Method Performances[a]	Found Levels[a,b]	Reference
PBDEs	Meat (pork, beef, chicken) 20 g	Homogenization	Saponification with 1 M KOH/ EtOH containing 10% H_2O and extraction with n-hexane	Purification with multilayer column chromatography	HRGC-LRMS, SPB-5 GC column; MS operating at 70 eV; HRGC-MRMS, SPB-5 GC column; MS operating at 38 eV and a resolution of 5000	LOD, 9.0–27 pg/g (HRGC-LRMS) LOD, 2.0–8.0 pg/g (HRGC-MRMS)	Pork: 63.6 pg/g ww[i] Beef: 16.2 pg/g fw[i] Chicken: 6.25 pg/g fw[i]	128
PBDEs	Meat (pork, beef, chicken) 5–200 g	Homogenization, addition of anhydrous Na_2SO_4	Column extraction, cyclohexane–dichloromethane	Fat removal with acid treatment; adsorption chromatography on activated silica gel and alumina columns	HRGC-HRMS, DB-5 GC column; MS operating at a resolution of 10,000		Pork: 41 pg/g ww[k] Ground beef: 78.3 pg/g ww[k] Chicken breast: 283 pg/g ww[k]	62
PBDEs	Cattle fat, swine internal organ tissues 5 g	Homogenization, addition of Hydromatrix	PLE, 1:1 dichloromethane–acetone, 1500 psi, two cycles	Fat removal with GPC, SX-3 BioBeads column; fractioning into two fractions on silica gel SPE cartridges (PCBs, PBDEs, and nonpolar chlorinated pesticides in fraction I, polar chlorinated pesticides in fraction II)	HRGC-LRMS, DB-5 ms GC column; MS operating in NCI mode with methane as a reagent gas	Recoveries, >68%		74

| DL-PCBs, NDL-PCBs PBDEs | Animal fat (beef, chicken) 0.5 g | Homogenization, addition of anhydrous Na$_2$SO$_4$ and Florisil | MSPD, sample/Florisil transferred to a cartridge containing 5 g acidic silica; elution with n-hexane | Fractioning into two fractions on silica gel SPE cartridges (PCBs in fraction I, PBDEs in fraction II) | HRGC-ECD, HP-5 GC column; HRGC-MS/MS (ion trap), HP-5 GC column | Mean recoveries, 74–99 % LOQ, 0.4–3 ng/g | 129 |

a dw, dry weight; ww, wet weight.

b Where not specified, TEQs were obtained by the WHO-TEFs system of 1997.

c Sum of NDL-PCBs 28, 52, 101, 138, 153, and 180.

d Sum of non-*ortho*-PCBs 77, 126, and 169.

e Sum of PBDEs 28, 47, 99, 100, 153, 154, 183, and 209.

f Sum of PBDEs 28, 47, 99, 100, 153, 154, and 183.

g Sum of NDL-PCBs 18, 28, 33, 49, 52, 60, 66, 74, 99, 101, 110, 122, 128, 138, 141, 153, 170, 180, 183, 187, 194, 206, and 209.

h Sum of PBDEs 47, 99, 100, 153, and 154.

i Sum of PBDEs 47, 99, 100, 153, 154, 183, and 209.

j Sum of PBDEs 28, 47, 99, 100, 153, and 154.

k Sum of PBDEs 17, 28, 47, 66, 77, 85, 99, 100, 138, 153, 154, 183, and 209.

various methods used in the analysis of DL-PCBs is given by Hess et al. [113]. Reference is also made to the methods elaborated by the U.S. Environmental Protection Agency (EPA) for determination of the tetra- through octachloro-substituted PCDD and PCDF toxic congeners [23] and for PCB congeners [24] by HRGC-HRMS (HRMS—high-resolution mass spectrometry). Basic requirements for analytical methods used in the EU for official controls of PCDD, PCDF, and DL-PCB levels in foodstuffs are reported in EU Regulation 1883/2006 [114].

Many analytical methods follow the general scheme of Figure 11.1. As observed for organochlorine pesticide analysis, the test sample has to be homogenized and dehydrated with anhydrous sodium sulfate to obtain a friable product. Because some amounts of the chemicals under investigation may be lost during the complex preparative procedures, the internal standard (IS) technique is generally adopted to provide proper correction for analyte losses [23,24]. To this aim, known quantities of isotopically labeled analytes are added to the samples at the earliest possible stage of extraction. The analytes of interest are extracted with a suitable organic solvent and the extract is purified to remove interfering compounds and prepare the sample for instrumental determination. Many of the purification procedures are based on the use of sulfuric acid, generally adsorbed on an inert support such as Extrelut® (Merck KGaA, Darmstadt, Germany). The rationale is that all the analytes of interest are resistant to acid treatment and this property is exploited to selectively destroy most of the interfering species coextracted with the target compounds. Owing to the difference in concentrations between planar (PCDDs, PCDFs, and non-*ortho* DL-PCBs) and nonplanar analytes (mono-*ortho* DL-PCBs) and the presence of other coextractive compounds resistant to cleanup procedure (i.e., chlorinated pesticides), fractioning steps are generally included during purification before instrumental analysis by HRGC-HRMS.

11.2.2.1.1 Pretreatment

Tissue samples are dissected into small pieces and preserved by fast freezing in liquid nitrogen [97] or, else, normal deep-freeze. Before analysis, samples are grinded to rupture cell membranes and homogenized. Addition of anhydrous sodium sulfate in the ratio sodium sulfate-to-sample 1.5–2.0 (w/w) is carried out to dry sample [121]. Alternatively, a freeze-drying procedure is sometimes adopted.

11.2.2.1.2 Extraction

Extraction techniques for meat samples are generally based on the principle that lipophilic organic compounds such as PCDDs, PCDFs, and PCBs are predominantly associated with the fat fraction of the matrix. Therefore, the extraction methods used for removal of these compounds are based on general methods employed for the isolation of the lipidic fraction. As observed for organochlorine pesticides (Section 11.2.1.1.2), a number of well-established techniques, including Soxhlet extraction or sonication with solvent, are available for the extraction of PCDDs, PCDFs, and DL-PCBs in fatty samples. These procedures are shown to be highly efficient (Soxhlet is the extraction method indicated by the U.S. EPA Methods 1613 and 1668A in the case of tissue samples) and do not require expensive instrumentation. For these reasons they are still in use in several routine laboratories. However, the main disadvantages presented by these procedures, that is, the large solvent consumption and the long time required for extraction, are determining their gradual replacement with more sophisticated extraction techniques such as PLE, MAE, and SFE. As previously observed, the possibility of working at elevated temperatures and pressures drastically improves the speed of the extraction process. These innovative techniques were object of evaluation within the DIFFERENCE research project [130], requested by the EU as a result of the "Belgian dioxin crisis" to develop fast and cheap analytical procedures for determination of PCDDs, PCDFs, and PCBs. In this context, a promising new procedure is the inclusion of a fat retainer (sulfuric acid impregnated silica) in PLE extraction cells [131,132]. As demonstrated by Sporring and Björklund [133], the presence of a fat retainer efficiently removes lipidic substances by oxidizing them and hindering their coeluting compounds.

11.2.2.1.3 Cleanup and Fractionation

The nonselective nature of the exhaustive extraction procedures results in complex extracts that contain the analytes of interest together with lipidic material and other organohalogen compounds (e.g., organochlorine pesticides, polychlorinated naphthalenes, polychlorinated camphenes, toxaphene). Therefore, the purification methodology used for PCDD, PCDF, and PCB analysis requires first lipid elimination, then fractionation to separate the groups of analytes from other coextractive species.

For the removal of lipids, two approaches are generally employed: destructive and nondestructive methods. The nondestructive lipid removal principally includes the use of GPC with SX-3 BioBeads columns, and adsorption chromatography with alumina, silica, and Florisil. Destructive methods comprise oxidative dehydration by concentrated sulfuric acid mixed with the lipid extract [134] or adsorbed on solid support through which the extract is eluted [135].

Fractionation of the extract into groups of analytes is normally required before instrumental analysis. In fact, with the exception of DL-PCB 118, and to a minor extent DL-PCB 105, all mono-*ortho* and non-*ortho* DL-PCBs, PCDDs, and PCDFs are present at substantially lower concentrations with respect to the remaining NDL-PCBs [97,113]. Therefore, the range of concentrations of target compounds is normally too large to measure all congeners without additional dilution or concentration. The methods available for the isolation of the analytes of interest into separate fractions utilize spatial planarity of these molecules to selectively adsorb them on the surface of carbonaceous material such as activated or graphitized carbon. Recently, an automated cleanup system (Power-Prep™, Fluid Management Systems, Inc.) has been developed, which is capable of rapidly separating planar and nonplanar organochlorine molecules [119]. This system uses high-capacity disposable multilayer silica columns, basic alumina columns, and PX-21 carbon columns. Fractionation allows isolation of two fractions, one containing NDL-PCBs and the eight mono-*ortho* DL-PCBs, the other containing the 17 PCDDs and PCDFs and the four coplanar non-*ortho* DL-PCBs [123,136].

11.2.2.1.4 HRGC-HRMS Instrumental Analysis

In the determination of PCDDs, PCDFs, and DL-PCBs, NDL-PCB interferences can be eliminated by fractioning the extract into analyte groups or by analyzing the final extract on multiple column [24]. In the attempt to reduce the need for multicolumn analysis, a number of analyte-specific columns have been developed. The low-polarity 5% phenyl columns exhibit multiple coelution for PCDDs, PCDFs, and DL-PCBs. However, they are generally considered sufficiently selective for biological samples, containing a smaller number of congeners in comparison with environmental samples [137].

The HRMS based on magnetic sector instruments is the reference method for the determination of PCDDs, PCDFs, and DL-PCBs [23,24] at 10^{-12}–10^{-15} g/g levels in complex matrices. EI ion sources are normally used in the HRMS determination of these compounds, with conventional electron energies of 30–35 eV. SIM mode is canonically employed to improve specificity and sensitivity. MS/MS with triple quadrupole and ion trap detectors has also been investigated for the analysis of PCDDs, PCDFs, and dioxin-like compounds [138,139]. For PCDDs and PCDFs, the selectivity of MS/MS is usually higher, due to the specific loss of the COCl fragment, never observed in any other halogenated organic compounds [140]. In the case of DL-PCBs, this enhanced selectivity is not observed, because the loss of Cl_2 from the parent molecule is not uniquely related to PCB molecules [112]. The sensitivity of MS/MS instruments is generally lower than HRMS [141], but it can be compensated by adjustments to sample size and final extract volume.

11.2.3 Non-Dioxin-Like Polychlorobiphenyls and Polybrominated Diphenyl Ethers

11.2.3.1 Analytical Methods

Most analytical studies on NDL-PCBs and PBDEs are limited to the determination of a small number of congeners as indicators of the presence of NDL-PCBs and PBDEs, respectively. In the case of NDL-PCBs, data on their occurrence in food are generally reported as the sum of the six congeners—PCBs 28, 52, 101, 138, 153, and 180—often termed as "indicator PCBs" or "marker PCBs," that represent some 50% of the total NDL-PCBs in food [142]. For PBDEs, the EFSA Scientific Panel on Contaminants in the Food Chain has recently recommended the inclusion of the following congeners in a European monitoring program for feed and food: PBDEs 28, 47, 99, 100, 153, 154, 183, and 209 [143]. As for NDL-PCBs, this "core group" reflects the most frequently found PBDEs in food and biological samples.

The analytical procedures for NDL-PCBs and PBDEs are reviewed here (Table 11.7) on the basis of the recent literature. Particular attention is given to the articles by Ahmed [97] on PCB analysis in food and by Covaci et al. [144,145] on the advances in the analysis of brominated flame retardants.

The analytical methods developed for NDL-PCBs and PBDEs are based on the same protocols used for PCDDs, PCDFs, and DL-PCBs. Differences may be observed in the chromatographic and detection systems used for instrumental determination. The general scheme adopted for the analysis of these pollutants is reported in Figure 11.1. NDL-PCBs and PBDEs are extracted with an organic solvent, most frequently by Soxhlet or PLE. Lipids are removed by GPC or treatment with sulfuric acid and coextracted substances are eliminated by adsorption chromatography. The final determination is performed by HRGC (ECD) or, preferably, by HRGC-MS. As a function of the detection system used for the analysis, the IS technique is generally adopted in accord with the U.S. EPA Methods 1668 and 1614 (draft) [24,25].

11.2.3.1.1 Pretreatment

The pretreatment step for NDL-PCBs and PBDEs is similar to that adopted for PCDD, PCDF, and DL-PCB analysis. Specific information and references can be found in Section 11.2.2.1.1.

11.2.3.1.2 Extraction

Given the similarity with the extraction methods used for PCDDs, PCDFs, and DL-PCBs, more detailed information of the extractive procedures applied in the case of NDL-PCBs can be found in Section 11.2.2.1.2. For PBDEs, the use of Soxhlet [25,152], elution through multilayer column

Table 11.7 Examples of PFAS Determinations in Samples of Animal Origin

Analytes	Sample	Pretreatment	Extraction	Cleanup	Instrumental Analysis	Method Performances[a]	Found Levels[a]	Reference
Acid compounds, PFOSA	Rabbit liver 0.20 g	Homogenization	IPE with tetrabutyl ammonium hydrogen sulfate; extraction with methyl tert-butyl ether (pH 10)	Centrifuging (speed not specified); filtration with 0.2 μm nylon mesh filter	HPLC-MS/MS, C$_{18}$ column 50 × 2 mm (5 μm), water (2 mM ammonium acetate)–methanol mobile phase	Recovery, 87% (PFOA), 100% (PFOS); LOD, 5 ng/g ww (PFOA), LOD, 8.5 ng/g ww (PFOS)		146
N-EtFOSA, N,N-Et$_2$FOSA, PFOSA	Hamburger 10 g	Homogenization	Extraction at high speed with 2:1 n-hexane–acetone	Centrifuging and adsorption chromatography on silica gel column impregnated with concentrated sulfuric acid	GC-PCI/MS, DB-1701 GC column, reagent gas, methane	Recovery, 74–101%; MDL 0.10–0.25 ng/g ww	0.23–0.70 ng/g ww	147
Acid compounds, PFOSA	Cod, gull liver 1 g	Homogenization	Mixing with Vortex; ultrasonic extraction with methanol–water (50:50; 2 mM ammonium acetate)	Filtration at high speed with YM-3 centrifugal filter	HPLC-HRMS (mass tolerance, 0.06 u), C$_{18}$ column 150 × 2.1 mm (3 μm); methanol and water (2 mM ammonium acetate each phase) mobile phase	Recovery, 83–84% (PFOA), 79–90% (PFOS); MDL, 1.25–1.28 ng/g ww (PFOA), MDL, 0.23–0.30 ng/g ww (PFOS)	Gull liver: PFOA <1.28 ng/g ww; PFOS 183 ng/g ww	148

(continued)

Table 11.7 (continued)   Examples of PFAS Determinations in Samples of Animal Origin

Analytes	Sample	Pretreatment	Extraction	Cleanup	Instrumental Analysis	Method Performances[a]	Found Levels[a]	Reference
Acid compounds, N-EtFOSA, alcoholic telomer	Beaver liver 1 g	Homogenization	Blending with 0.01 N KOH methanolic solution	Fractioning on weak anionic exchange solid-phase column into two fractions (nonacid compounds in fraction I, acid compounds in fraction II)	HPLC-MS/MS, C$_{18}$ column 50 × 2.1 mm (5 μm), water (2 mM ammonium acetate)/methanol mobile phase	Mean recovery, 85%; MDL, 0.03–3 ng/g ww	PFOA 0.29 ng/g ww; PFOS 133 ng/g ww	149
Acid compounds	Fish 0.01 g	Homogenization	IPE with tetrabutyl ammonium hydrogen sulfate; extraction solvent methyl tert-butyl ether (pH 10)	Centrifuging (speed not specified)	HPLC-ITD, C$_{18}$ column 50 × 2.1 mm (5 μm), water (1 mM ammonium acetate)/methanol mobile phase	Recovery, 80–81% (PFOA), 99–102% (PFOS) LOD, 10 ng/g ww (PFOA); LOD, 2.5 ng/g ww (PFOS)	PFOA 100 ng/g ww, PFOS 200 ng/g ww	150
Acid compounds	Meat (composite samples) 2 g	Homogenization	Blending with methanol	Centrifuging at high speed	HPLC-MS/MS, C$_{18}$ column 50 × 2.1 mm, water (5 mM ammonium formate)/acetonitrile–methanol (2:1) mobile phase	Recovery 91–116% (PFOA), 85–108% (PFOS); LOD 0.5–1 ng/g ww	PFOA <0.5–2.6 ng/g ww, PFOS <0.6–2.7 ng/g ww	151

[a] ww, wet weight.

[121], extraction at high speed [153], MSPD [114], PLE [154], and MAE [155] are reported for the extraction of these contaminants in samples of animal tissue.

11.2.3.1.3 Cleanup and Fractionation

As observed for PCDDs, PCDFs, and DL-PCBs, the nonselective nature of exhaustive extraction procedures and the complexity of sample matrices result in a complex extract that requires efficient purification. Lipid elimination, performed by destructive or nondestructive methods (see Section 11.2.2.1.3), is generally followed by isolation of the target analytes from other organohalogenated compounds. Fractionation of NDL-PCBs and PBDEs from coextractive species with similar chemical–physical properties (i.e., organochlorine pesticides, DL-PCBs, PCDDs, and PCDFs) is based on the different polarity of NDL-PCBs and PBDEs in comparison with other chlorinated compounds, and the attitude of NDL-PCBs and PBDEs to be easily eluted from activated carbon with respect to other molecules with planar structure (i.e., PCDDs, PCDFs, and non-*ortho* DL-PCBs). With regard to this, the use of silica, alumina, Florisil, and activated carbon is widely described in the literature [63,156,157]. Recently, the PowerPrep automated cleanup procedure used for separation of PCDDs, PCDFs, and PCBs [122] has been extended also to include PBDEs, after optimization of the type and volume of the solvent necessary to isolate the different chemical families [158].

11.2.3.1.4 Instrumental Analysis

HRGC combined with ECD or MS detectors is the method of choice for the analysis of NDL-PCBs. A comprehensive review on developments in the HRGC of PCBs is given by Cochran and Frame [159], who evaluated a variety of stationary phases commonly used for PCB analysis. The 5%-phenyl type column has substantially become the standard for PCB analysis. Although alternative phases, such as phenyl carborane and that present in DB-XLB columns, have been an attempt to overcome the problem of coelution of the most significant congeners, no column phase can resolve all PCBs in a single injection. More complete separation can be achieved with a different column configuration based on the use of a single injection split coupled to two columns in parallel that end in two ECD detectors [97,160]. ECD is the most utilized detection method for PCBs for its high sensitivity, low cost, and ease in use and maintenance. As observed for organochlorine pesticides (Section 11.2.1.1.4.1), the main disadvantages are its poor selectivity and, as observed by Cochran and Frame [159], nonlinear response over a relative narrow concentration range. The application of low-resolution mass spectrometry (LRMS) operating either in the EI mode or with negative chemical ionization (NCI) [161] provides higher specificity than ECD and allows to obtain qualitative information for analyte identification along with HRGC retention time. Recently, the use of ion trap tandem MS systems has been evaluated for the analysis of PCBs in environmental samples and biota [162,163].

PBDEs are generally quantified by HRGC-MS. Given the degradation problems sometimes experienced for certain congeners (i.e., PBDE 209), the characteristics of the GC system have to be properly selected. In fact, as observed by Björklund et al. [164], the column brand, type of retention gap, press-fit connector, stationary phase, column length, and injection technique strongly influence the accuracy and precision of nona- and deca-PBDE analysis. Determination of PBDEs can be relatively easily performed on nonpolar or semipolar columns such as 100% methyl-polysiloxane (DB-1) and 5% phenyl-dimethyl-polysiloxane (DB-5, CP-Sil 8, or AT-5) [144]. A selection of the most suitable GC columns for PBDE congener-specific analysis can be easily done on the base of the work of Korytár et al. [165] who reported the elution order of 126 PBDEs on seven different GC stationary phases; a DB-XLB column was found to be the most efficient for the separation of PBDE congeners, with a DB-1 column as runner-up. The most commonly used detectors for PBDE analysis is MS operating in the NCI or EI mode [166]. Although NCI presents a higher sensitivity than EI, it is less selective, since only bromine can be monitored, and less accurate since IS method with ^{13}C-labeled PBDEs cannot be utilized. HRMS with EI ionization is preferred in principle over LRMS for its higher sensitivity and selectivity. Nevertheless, due to the complexity of the analysis and cost of HRMS, the LRMS is the most widely used. Recently the use of ion trap MS or quadrupole MS has been evaluated for PBDE analysis. Application of these analytical approaches to the determination of PBDE in abiotic, biotic, and food samples is reported by Wang et al. [167], Gómara et al. [168], Petinal et al. [169], and Yusà et al. [170].

11.2.4 Polyfluorinated Alkylated Substances

In the recent scientific literature, only few works deal with specific analytical methods for PFAS in meat (Table 11.7). More information can be drawn from analyses carried out on biota of environmental interest. In this section, when not further specified, the assay of PFOS and PFOA is mainly dealt with.

Tittlemier et al. [151] describe a liquid chromatography in tandem with mass spectrometry (LC-MS/MS) multiresidue method to analyze PFOS, PFOA, and related compounds in composite samples of several foods (e.g., chicken, lamb, beef, pork) with a limit of determination (LD) ranging from 0.5 to 1 ng/g fresh weight. Nevertheless, the following possible problems in the analysis of PFAS have been reported by Martin et al. [171]:

a. Ion suppression in electrospray interface (ESI), a widely reported phenomenon using such interface
b. Presence of teflonated materials in the analytical tools, which can release PFAS
c. Capability of glassware to sequestrate PFAS when in aqueous solutions
d. Ambiguous quantification of branched and linear isomers, possibly due to poor resolution in liquid chromatography or insufficient purity of the standards
e. Limited availability of ^{13}C-labeled PFAS to be used as ISs for quantification
f. Nonavailability of reference materials

11.2.4.1 Sampling and Sample Storage

Contamination may occur during sampling if the gloves, dresses, or tools that the operators use are made of material releasing PFAS (e.g., Gore-Tex®, W. L. Gore & Associates, Inc.; Teflon®, Dupont). Polypropylene sample bottles should be precleaned by rinsing with polar solvents, such as methanol. Sample storage at −20°C seems to be generally appropriate to preserve the analytes [172].

11.2.4.2 Extraction and Cleanup

Owing to their tensioactive properties, PFAS tend to interact with materials that are used in an undedicated analytical laboratory. Therefore, it is recommended to limit the extraction and cleanup procedures to the essential steps, capable of guaranteeing quantitative recoveries and a selectivity that can minimize the ion-suppression phenomenon during the instrumental acquisition of data [172,173]. The following examples describe some reference methods for PFAS analysis according to their evolution in time.

Ylinen et al. [174] proposed an ion-pair extraction (IPE) using tetra-*n*-butylammonium hydrogen sulfate as a counterion; the approach was subsequently modified by Hansen et al. [146]. This extraction technique was also applied to several biological and environmental matrices [150,175,176]. More recently, a sample extraction with alkalinized methanolic solution followed by a weak anionic exchange solid-phase cleanup was applied by Taniyasu et al. [149] on fish samples (Figure 11.2), with an optimization of recovery rates. A selective elution from an SPE column of the analytes related to their polarity was achieved, thereby obtaining two fractions with neutral and ionic analytes, respectively, with a low matrix overload. Powley et al. [177,178] proposed a dispersive solid phase with graphitized carbon as a cleanup step. Carbon was directly added to the extract and mixed thoroughly by vortexing, thus allowing to sequestrate the hydrophobic substances that were further removed by centrifugation. This procedure yielded a decreased ion-suppression phenomenon and the possible release of PFAS from SPE column cartridges was avoided. Berger and Haukas [148] proposed to use a polar solvent extraction followed by a clarification of the extract via centrifugation and a selective filtration at 3000 nominal molecular weight limit cutoff, before instrumental analysis.

11.2.4.3 Instrumental Identification and Determination

The performance of analytical instruments may condition the choice of extraction and cleanup procedures, allowing the injection of extracts more or less diluted, which possibly did not undergo a cleanup. Moreover, according to the geometry of LC-MS interfaces, the ion-suppression

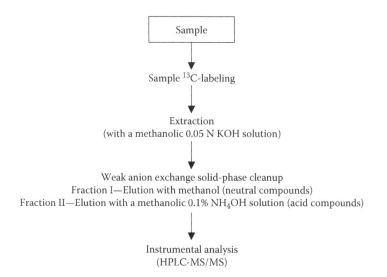

Figure 11.2 Analytical method employing solid-phase cleanup. (From Taniyasu, S. et al., *J. Chromatogr. A*, 1093, 89, 2005.)

Table 11.8   Principal MS/MS Transition of Some PFAS

Compound/Acronym	Precursor Ion $(M – H)^-$ m/z	Transition	Nature of Product Ion
PFBS	299	$299 \rightarrow 99$	FSO_3^-
PFOA	413	$413 \rightarrow 369$	$(M–COOH)^-$
6:2 FTS	427	$427 \rightarrow 81$	HSO_3^-
PFOSA	498	$498 \rightarrow 78$	SNO_2^-
PFOS	499	$499 \rightarrow 80$	SO_3^-
N-MeFOSE	556	$556 \rightarrow 526$	$(M–CH_2OH)^-$
PFTeDA	713	$713 \rightarrow 669$	$(M–COOH)^-$

phenomena can be reduced. LC coupled to MS/MS detectors with an ESI is the instrumental technique of choice to identify and determine PFAS. Data can be acquired in the selected reaction monitoring (SRM) mode as reported in Table 11.8.

A particular attention is required when using ion trap mass spectrometers. Owing to cutoff limitations, very wide transitions—such as SO_3^- produced by the molecular ion of PFOS—cannot be achieved in a quantitative way [150,179]. Berger and Haukas [148] analyzed carboxylic acid, sulfonate compounds, and PFOSA with LC coupled to HRMS such as a time of flight (TOF), as an alternative to the MS/MS technique. Chromatographic separation is generally achieved on reverse phase C_8 or C_{18} columns, using methanol and ammonium acetate or formic acid or acetic acid aqueous solutions as mobile phases. Possible background contributions, originating from teflonated parts in the LC system, should be carefully evaluated. LC-grade water should be decontaminated through Amberlite XAD-7 resin to remove any possible perfluorinated compound [151].

It is generally agreed to prepare calibration curves using the real matrix to account for ion-suppression phenomena as well, especially when the ^{13}C-labeled ISs are not available for all the analytes. A volumetric standard is also employed.

Nonpolar fluorinated compounds, such as PFOSA, can be directly determined with HRGC-MS with positive chemical ionization (PCI) [180]. An inventory of MS-based techniques is reported in Table 11.7.

11.2.4.4   Observations

As the first two international intercalibration studies yielded unsatisfactory results [181,182], the organizers of the third round (2007) decided to meet all the participating laboratories to define the "best" analytical method(s) to be adopted for the determination of PFAS. This concern provides an indication that the aforesaid analytical methods still need to be consolidated to be adequately reliable.

11.3 Bioassays to Screen in Meat Polychlorodibenzo-*p*-Dioxins, Polychlorodibenzofurans, and Dioxin-Like Polychlorobiphenyls

11.3.1 Introduction

The use of bioassay as a screening tool aims at dosing the biological activity of contaminants, by comparing their effects with those of a standard preparation or a reference material, on a culture of living cells [183]. Within this frame, rather than the amount of contaminant(s) bound to a biological macromolecule, as in the case of immunoassay determinations [184], bioassays allow to dose the response elicited as a result of the interaction between the analyte(s) and a specific receptor. As a consequence, the signal measured on the selected biological substrate results from the cumulative effects of the different substances present in the extract to be analyzed that share the same mode of action, according to their concentration and their relative potency (REP) [185]. This can be the case of PCDD, PCDF, and DL-PCB congeners [58], and of other categories of pesticides and contaminants, for which a cumulative assessment on a toxicological basis has been suggested by regulatory agencies [186].

11.3.2 Cell-Based Bioassays

In the literature, the most consolidated applications of bioassays on food samples are based on the use of chemically activated fluorescence or luminescence gene expression in engineered cell lines [187]. Briefly, the contaminants present in the extract interact with the specific aryl hydrocarbon receptor (AhR) expressed on cell membranes. The complex is transported to the nucleus where it activates the deoxyribonucleic acid (DNA) sequence for the synthesis of a specific enzyme (i.e., luciferase). After extraction and cleanup, and incubation of the extract on the cell culture, the addition of luciferine as substrate to the supernatant from cell lysis—containing the induced luciferase—produces a chemoluminescence or fluorescence signal whose intensity is related to both the amount(s) and REP(s) of the contaminant(s) present.

11.3.3 Bioassay Based on Polymerase Chain Reaction

Another bioassay, which has been only preliminarily applied to food matrices, is based on DNA real-time amplification and fluorescence detection [188]. This technique has the advantage that no cell lines and related laboratory facilities are needed to perform the test. The target compounds activate the AhR to a form that binds to DNA. The activated complex is then trapped onto a micro-well; the receptor-bound DNA is amplified through the polymerase chain reaction (PCR) and read in real-time mode.

11.3.4 Bioassay Reliability and Applicability

The EU legislation has recently established some specific requirements that should be fulfilled in the cell-based bioassay screening of PCDDs, PCDFs, and DL-PCBs [114]. A series of reference concentrations of 2,3,7,8-T_4CDD or a PCDD, PCDF, and DL-PCB mixture should be tested to obtain a significant full dose–response curve; it is recommended to use reference materials and build appropriate quality control charts to ensure that the relative standard deviation shall not be above 15% in a triplicate determination for each sample (repeatability) and not above 30% between three independent experiments (reproducibility). For quantitative calculations, the induction of the sample dilution used must fall within the linear portion of the response curve, with an LD sixfold the standard deviation of the solvent blank or the background. Information on the correspondence between bioassay and HRGC-HRMS results should also be provided. For official use, positive results from screening must always be confirmed; false negative rates must be below 1%.

The following critical points can be identified as the main causes of possible inconsistencies between bioassay screening and confirmatory (HRGC-HRMS) analysis outputs:

a. The samples should be appropriately processed, allowing an exhaustive fat extraction and the removal of other possible AhR ligands—such as polycyclic aromatic hydrocarbons—capable of eliciting a bioassay response, if present in large quantities.

b. The congener REPs may differ from the consensus-based TEFs used for conversion of HRGC-HRMS data into toxicology-based WHO-TEQs: this may cause deviations from HRGC-HRMS results (the magnitude of deviations is affected by the contamination profile) [189].

c. Possible deviations from simple additivity of the bioassay measured effects may be expected in the presence of PCDD, PCDF, and PCB mixtures (e.g., Aroclors 1242, 1254, 1260) [190].

Abbreviations

AhR	aryl hydrocarbon receptor
AOAC	Association of Official Analytical Chemists
ASE	accelerated solvent extraction
pp'-DDT	1,1,1-trichloro-2,2-*bis*(4-chlorophenyl)ethane
deca-BDE	decabromodiphenyl ether
DG	direction-general
DL-PCB	dioxin-like polychlorinated biphenyls
DNA	deoxyribonucleic acid
EC	European Commission
ECD	electron capture detector
EFSA	European Food Safety Authority
EI	electron impact
EPA	Environmental Protection Agency
ESI	electrospray interface
n-EtFOSE	*n*-ethyl perfluorooctane sulfonamidoethanol
EU	European Union
FAO	Food and Agriculture Organization
FDA	Food and Drug Administration
8:2 FTOH	8:2 fluorotelomer alcohol
6:2 FTS	6:2 fluorotelomer sulfonate
GPC	gel permeation chromatography
alpha-HCH	alpha-hexachlorocyclohexane

beta-HCH	beta-hexachlorocyclohexane
HPLC	high-performance liquid chromatography
HRGC	high-resolution gas chromatography
HRMS	high-resolution mass spectrometry
IPCS	International Programme on Chemical Safety
IPE	ion-pair extraction
I-TEF	international toxicity equivalency factor
IS	internal standard
LC	liquid chromatography
LD	limit of determination
LOD	limit of detection
LOQ	limit of quantification
LRMS	low-resolution mass spectrometry
MAE	microwave-assisted extraction
MDL	minimum detection level
N-MeFOSE	*N*-methyl perfluorooctane sulfonamidoethanol
ML	maximum level
MRL	maximum residue limit
MSPD	matrix solid-phase dispersion
NCI	negative chemical ionization
NDL-PCB	non-dioxin-like polychlorinated biphenyl
octa-BDE	octabromodiphenyl ether
PBDE	polybrominated diphenyl ether
PCDD	polychlorinated dibenzo-*p*-dioxin
PCDF	polychlorinated dibenzofuran
PCI	positive chemical ionization
penta-BDE	pentabromodiphenyl ether
PFAS	polyfluorinated alkylated substances
PFBS	perfluorobutyl sulfonate
PFOA	perfluorooctanoic acid
PFOS	perfluorooctane sulfonate
PFOSA	perfluorooctane sulfonamide
PFTeDA	perfluorotetradecanoic acid
PLE	pressurized liquid extraction
POP	persistent organic pollutant
REP	relative potency
SCCP	short-chained chlorinated paraffin
SCF	Scientific Committee on Food
SFE	supercritical fluid extraction
SIM	single (or selected) ion monitoring
SRM	selected reaction monitoring
SPE	solid-phase extraction
TMRL	temporary maximum residue limit
TOF	time of flight
UNEP	United Nations Environment Programme
WHO	World Health Organization
WHO-TEF	WHO toxicity equivalency factor

References

1. Available at http://www.chem.unep.ch/pops/default.html.
2. Bernes, C., Where do persistent pollutants come from? in *Persistent Organic Pollutants: A Swedish View of an International Problem*, Swedish Environmental Protection Agency, Stockholm, Sweden, 1998, chap. 2.
3. Available at http://www.pops.int/.
4. IPCS, Aldrin and Dieldrin, *Environmental Health Criteria* 91, WHO, Geneva, 1989.
5. IPCS, Camphechlor, *Environmental Health Criteria* 45, WHO, Geneva, 1984.
6. IPCS, Chlordane, *Environmental Health Criteria* 34, WHO, Geneva, 1984.
7. IPCS, Chlordecone, *Environmental Health Criteria* 43, WHO, Geneva, 1984.
8. IPCS, DDT and its derivatives, Environmental Health Criteria 9, WHO, Geneva, 1979.
9. IPCS, Endrin, *Environmental Health Criteria* 130, WHO, Geneva, 1992.
10. IPCS, Heptachlor, *Environmental Health Criteria* 38, WHO, Geneva, 1984.
11. IPCS, Hexachlorobenzene, *Environmental Health Criteria* 195, WHO, Geneva, 1997.
12. IPCS, *Alpha-* and *beta-*hexachlorocyclohexane, *Environmental Health Criteria* 123, WHO, Geneva, 1992.
13. IPCS, Lindane, *Environmental Health Criteria* 124, WHO, Geneva, 1991.
14. IPCS, Mirex, *Environmental Health Criteria* 44, WHO, Geneva, 1984.
15. IARC, Polychlorinated dibenzo-para-dioxins and dibenzofurans. *IARC Monographs on the Evaluation of Carcinogenic Risks to Humans.* Vol. 69, IARC, Lyon, France, 1997.
16. SCF, Opinion of the SCF on the Risk Assessment of Dioxins and Dioxin-like PCBs in food. Update based on new scientific information available since the adoption of the SCF opinion of 22nd November 2000 (adopted on 30 May 2001), http://ec.europa.eu/food/fs/sc/scf/out90_en.pdf.
17. SCF, Opinion of the SCF on the Risk Assessment of Dioxins and Dioxin-like PCBs in food (adopted on 22 November 2000), http://ec.europa.eu/food/fs/sc/scf/out78_en.pdf.
18. UNEP, Draft risk profile: pentabromodiphenyl ether, 2006, available at http://www.pops.int/documents/meetings/poprc_2/meeting_docs.htm.
19. European Commission, Diphenyl ether, octabromo derivative. Summary Risk Assessment Report, Joint Research Centre, 2003.
20. UNEP, Draft risk profile: perfluorooctane sulfonate (PFOS), 2006, available at http://www.pops.int/documents/meetings/poprc_2/meeting_docs.htm.
21. FDA, General Analytical Operations and Information, in *Pesticide Analytical Manual, Volume I*, FDA. 3rd Edition, 1994, chap. 2.
22. U.S. EPA, Method 1618A, *Organo-halide Pesticides, Organo-phosphorus Pesticides, and Phenoxy-acid Herbicides by Wide Bore Capillary Column Gas Chromatography with Selective Detectors*, Industrial Technology Division, Office of Water, US Environmental Protection Agency (Washington), 1989.
23. U.S. EPA, Method 1613, Tetra- through octachlorinated dioxins and furans by isotope dilution HRGC-HRMS, Engineering and Analysis Division (4303), Office of Water, US Environmental Protection Agency (Washington), 1994.
24. U.S. EPA Method 1668, Revision A. Chlorinated biphenyl congeners in water, soil, sediment, bio-solids and tissue by HRGC/HRMS, Engineering and Analysis Division (4303), Office of Water, US Environmental Protection Agency (Washington), 1999.
25. U.S. EPA Method 1614 Draft, Brominated diphenyl ethers in water, soil, sediment, and tissue by HRGC/HRMS. Engineering and Analysis Division (4303), Office of Water, US Environmental Protection Agency (Washington), 2003.
26. Moriarty, F., Prediction of ecological effects, in *Ecotoxicology: The Study of Pollutants in Ecosystems*, 2nd Edition, Academic Press, San Diego, CA, 1988, chap. 6.
27. EU Council Decision 2006/507/EC of 14 October 2004 concerning the conclusion, on behalf of the European Community, of the Stockholm Convention on Persistent Organic Pollutants. OJ L 209, 31.7.2006, 1–2.

28. United Nation/UNEP, Global Report 2003. Regionally Based Assessment of Persistent Toxic Substances, 2003.

29. Iamiceli, A.L., Turrio-Baldassarri, L., and di Domenico, A., Determination of PCDDs and PCDFs in water, in *Handbook of Water Analysis*, Nollet, L.M.L., ed., Marcel Dekker, 2000, chap. 31.

30. UNEP, Draft risk profile: chlordecone, 2006, available at http://www.pops.int/documents/meetings/poprc_2/meeting_docs.htm.

31. UNEP, Draft risk profile: hexabromobiphenyl, 2006, available at http://www.pops.int/documents/meetings/poprc_2/meeting_docs.htm.

32. UNEP, Draft risk profile: lindane, 2006, available at http://www.pops.int/documents/meetings/poprc_2/meeting_docs.htm.

33. Joint WHO/Convention task force on the health aspects of air pollution, Health risks of persistent organic pollutants from long-range transboundary air pollution, WHO, 2003.

34. UNEP, Summary of pentachlorobenzene proposal, 2006, available at http://www.pops.int/documents/meetings/poprc_2/meeting_docs.htm.

35. UNEP, Summary of octabromodiphenyl ether proposal, 2006, available at http://www.pops.int/documents/meetings/poprc_2/meeting_docs.htm.

36. UNEP, Summary of short-chained chlorinated paraffins proposal, 2006, available at http://www.pops.int/documents/meetings/poprc_2/meeting_docs.htm.

37. UNEP, Summary of *alpha*-hexachlorocyclohexane proposal, 2006, available at http://www.pops.int/documents/meetings/poprc_2/meeting_docs.htm.

38. UNEP, Summary of *beta*-hexachlorocyclohexane proposal, 2006, available at http://www.pops.int/documents/meetings/poprc_2/meeting_docs.htm.

39. EU Council Directive 86/363/EEC of 24 July 1986 on the fixing of maximum levels for pesticide residues in and on foodstuffs of animal origin. *OJ L 221, 7.8.1986, 43–47.*

40. Kostantinou, I.K. et al., The status of pesticide pollution in surface waters (rivers and lakes) of Greece. Part I. Review on occurrence and levels, *Environ. Pollut.*, 141, 555, 2006.

41. Li, J. et al., Organochlorine pesticides in the atmosphere of Guangzhou and Hong Kong: regional sources and long-range atmospheric transport, *Atmos. Environ.*, 41, 3889, 2007.

42. Vagi, M.C. et al., Determination of organochlorine pesticides in marine sediments samples using ultrasonic solvent extraction followed by GC/ECD, *Desalination*, 210, 146, 2007.

43. Torres, M.J. et al., Organochlorine pesticides in serum and adipose tissue of pregnant women in Southern Spain giving birth by cesarean section, *Sci. Total Environ*, 372, 32, 2006.

44. Meeker, J.D., Altshul, L., and Hauser, R., Serum PCBs, p,p'-DDE and HCB predict thyroid hormone levels in men, *Environ. Res.*, 104, 296, 2007.

45. EU Council Directive 91/414/EEC of 15 July 1991 concerning the placing of plant protection products on the market. *OJ L 230, 19.8.1991, 1–32.*

46. EU Regulation 396/2005/EC of the European Parliament and of the Council of 23 February 2005 on maximum residue levels of pesticides in or on food and feed of plant and animal origin and amending Council Directive 91/414/EECText with EEA relevance. *OJ L 70, 16.3.2005, 1–16.*

47. EU Council Directive 86/362/EEC of 24 July 1986 on the fixing of maximum levels for pesticide residues in and on cereals. *OJ L 221, 7.8.1986, 37–42.*

48. EU Council Directive 90/642/EEC of 27 November 1990 on the fixing of maximum levels for pesticide residues in and on certain products of plant origin, including fruit and vegetables. *OJ L 350, 14.12.1990, 71–79.*

49. FAO/WHO Codex Alimentarius Commission. Maximum Residue Limits for pesticides, FAO/WHO, Rome, Italy, 2001.

50. Van den Berg, M. et al., Toxic equivalency factors (TEFs) for PCBs, PCDDs, PCDFs for humans and wildlife, *Environ. Health Perspect.*, 106, 775, 1998.

51. Van den Berg, M. et al., The 2005 World Health Organization reevaluation of human and mammalian toxic equivalency factors for dioxins and dioxin-like compounds, *Toxicol. Sci.*, 93, 223, 2006.

52. NATO/CCMS, International toxicity equivalency factor (I-TEF) method of risk assessment for complex mixtures of dioxins and related compounds, Report No. 176, Committee on the Challenges of Modern Society, North Atlantic Treaty Organization, 1988.

53. Galliani, B. et al., Occurrence of ndl-PCBs in food and feed in Europe, *Organohalogen Compd.*, 66, 3561, 2004.

54. Fattore, E. et al., Current dietary exposure to polychlorodibenzo-*p*-dioxins, polychlorodibenzofurans, and dioxin-like polychlorobiphenyls in Italy, *Mol. Nutr. Food Res.*, 50, 915, 2006.

55. Fattore, E. et al., Assessment of the dietary exposure to non-dioxin-like PCBs of the Italian general population, *Chemosphere*, submitted.

56. Brambilla, G. et al., Persistent organic pollutants in meat: a growing concern, *Meat Sci.*, 78, 25–33, 2008.

57. Domingo, J.L. and Bocio, A., Levels of PCDD/PCDFs and PCBs in edible marine species and human intake: a literature review, *Environ. Int.*, 33, 397, 2007.

58. EU Commission Regulation 1881/2006/EC of 19 December 2006 setting maximum levels for certain contaminants in foodstuffs. OJ L 364, 20.12.2006.

59. Law, R.J. et al., Levels and trends of polybrominated diphenylethers and other brominated flame retardants in wildlife, *Environ. Int.*, 29, 757, 2003.

60. Alaee, M. et al., An overview of commercially used brominated flame retardants, their applications, their use patterns in different countries/regions and possible modes of release, *Environ. Int.*, 29, 683, 2003.

61. EU Directive 2002/95/EC of the European Parliament and of the Council of 27 January 2003 on the restriction of the use of certain hazardous substances in electrical and electronic equipment. OJ L 37, 13.2.2003, 19–23.

62. Schecter, A. et al., Polybrominated diphenyl ethers contamination of United States Food, *Environ. Sci. Technol.*, 5306, 2004.

63. Schuhmacher, M. et al., Concentrations of polychlorinated biphenyls (PCBs) and polybrominated diphenyl ethers (PBDEs) in milk of women from Catalonia, Spain, *Chemosphere*, 67, S295, 2007.

64. Renner, R., Growing concern over perfluorinated chemicals, *Environ. Sci. Technol.*, 35, 154A, 2001.

65. Hoff, P. et al., Perfluorooctane sulfonic acid in bib (trisopterus luscus) and plaice (pleuronectes platessa) from the Belgian North Sea: distribution and biochemical effects, *Environ. Toxicol. Chem.*, 22, 608, 2003.

66. Kannan, K. et al., Perfluorooctanesulfonate and related fluorochemicals in human blood from several countries, *Environ. Sci. Technol.*, 38, 4489, 2004.

67. Kannan, K. et al., Perfluorinated compounds in aquatic organisms at various trophic levels in a great lakes food chain, *Arch. Environ. Contam. Toxicol.*, 48, 559, 2005.

68. Yamashita, N. et al., A global survey of perfluorinated acids in oceans, *Mar. Pollut. Bull.*, 51, 658, 2005.

69. Haukas, M. et al., Bioaccumulation of per- and polyfluorinated alkyl substances (PFAS) in selected species from the Barents Sea food web, *Environ. Pollut.*, 148, 360, 2007.

70. Hercegová, A., Dömötörová, M., and Matisová, E., Sample preparation methods in the analysis of pesticide residues in baby food with subsequent chromatographic determination, *J. Chromatogr.*, doi:10.1016/j.chroma.2007.01.008.

71. Codex Alimentarius Commission, Analysis of Pesticide Residues: Recommended Methods, CODEX STAN 229-1993, REV.1-2003.

72. AOAC International, Official Methods of Analysis of AOAC International, 18th Edition, 2005.

73. FDA, Multiresidue methods, in *Pesticide Analytical Manual, Volume I*, FDA, 3rd Edition, 1994, chap. 3.

74. Saito, K. et al., Development of a accelerated solvent extraction and gel permeation chromatography analytical method for measuring persistent organohalogen compounds in adipose and organ tissue analysis, *Chemosphere*, 57, 373, 2004.

75. Hopper M.L., Automated one-step supercritical fluid extraction and clean-up system for the analysis of pesticide residues in fatty matrices, *J. Chromatogr. A,* 840, 93, 1999.

76. Frenich, A.G. et al., Multiresidue analysis of organochlorine and organophosphorus pesticides in muscle of chicken, pork and lamb by gas chromatography–triple quadrupole mass spectrometry, *Anal. Chim. Acta*, 558, 42, 2006.

77. Frenich, A.G., Bolaños, P.P., and Vidal, J.L.M., Multiresidue analysis of pesticides in animal liver by gas chromatography using triple quadrupole tandem mass spectrometry, *J. Chromatogr. A*, doi: 10.1016/j.chroma.2007.01.066.

78. Patel, K. et al., Evaluation of gas chromatography–tandem quadrupole mass spectrometry for the determination of organochlorine pesticides in fats and oils, *J. Chromatogr. A*, 1068, 289, 2005.

79. Janouskova, E. et al., Determination of chlordane in foods by gas chromatography, *Food Chem.*, 93, 161, 2005.

80. AOAC Official method 983.18, Meat and Meat Products. Preparation of Test Sample Procedure.

81. RIVM, Pesticides amenable to gas chromatography: multiresidue method 1. RIVM report No. 638817014, 1996.

82. UNI EN 1528-2, Fatty food. Determination of pesticides and polychlorinated biphenyl (PCBs). Part 2: extraction of fat, pesticides and PCBs, and determination of fat, 1996.

83. Valcarel, M. and Tena, M.T., Applications of supercritical fluid extraction in food analysis, *Fresenius J. Anal. Chem.*, 35, 561, 1997.

84. Snyder , J.M. et al., Supercritical fluid extraction of poultry tissues containing incurred pesticide residues, *J. AOAC Int.* 76, 888, 1993.

85. Gallina Toschi, T. et al., Pressurized solvent extraction of total lipids in poultry meat, *Food Chem.*, 83, 551, 2003.

86. Eskilsson, C.S. and Björklund, E., Analytical-scale microwave-assisted extraction, *J. Chromatogr. A*, 902, 227, 2000.

87. Pereira, M.B. et al., Comparison of pressurized liquid extraction and microwave assisted extraction for the determination of organochlorine pesticides in vegetables, *Talanta*, 71, 1345, 2007.

88. Hummert, K., Vetter, W., and Luckas, B., Fast and effective sample preparation for determination of organochlorine compounds in fatty tissue of marine mammals using microwave extraction, *Chromatographia*, 42, 300, 1996.

89. Hummert, K., Vetter, W., and Luckas, B., Combined microwave assisted extraction and gel permeation chromatography for the determination of organochlorine compounds in fatty tissue of marine mammals, *Organohalogen Compd.*, 27, 360, 1996.

90. Barker, S.A., Long, A.R., and Short, C.R., Isolation of drug residues from tissues by solid phase dispersion, *J. Chromatogr. A*, 475, 353, 1989.

91. Barker, S.A., Matrix solid phase dispersion (MSPD), *J. Biochem. Biophys. Methods*, 70, 151, 2007.

92. Bogialli, S. and Di Corcia, A., Matrix solid-phase dispersion as valuable tool for extracting contaminants from foodstuffs, *J. Biochem. Biophys. Methods*, 70, 163, 2007.

93. Long, A.R., Soliman, M.M., and Barker, S.A., Matrix solid phase dispersion (MSPD) extraction and gas chromatographic screening of nine chlorinated pesticides in beef fat, *J. AOAC Int.*, 74, 493, 1991.

94. Furusawa, N., A toxic reagent-free method for normal-phase matrix solid-phase dispersion extraction and reversed-phase liquid chromatographic determination of aldrin, dieldrin, and DDTs in animal fats, *Anal. Bioanal. Chem.*, 378, 2004, 2004.

95. Furusawa, N., Determination of DDT in animal fats after matrix solid-phase dispersion extraction using an activated carbon filter, *Chromatographia*, 62, 315, 2005.

96. van der Hoff, G.R. and van Zoonen, P., Trace analysis of pesticides by gas chromatography, *J. Chromatogr. A*, 843, 301, 1999.

97. Ahmed, F.E., Analysis of pesticides and their metabolites in foods and drinks, *Trends Anal. Chem.*, 20, 649, 2001.

98. AOAC Official Method 970.52, Organochlorine and Organophosphorus Pesticides Residues. General Multiresidue Method, in *Official Methods of Analysis of AOAC International*, 18th Edition, AOAC International, Gaithersburg, MD, 2005.

99. UNI EN 1528-3, Fatty food. Determination of pesticides and polychlorinated biphenyl (PCBs). Part 3: Clean-up methods, 1996.

100. EPA Method 3620C, Florisil clean-up. Revision 3. US Environmental Protection Agency, Washington, DC, November 2000.

101. Burke, E.R., Holden, A.J., and Shaw, I.C., A method to determine residue levels of persistent organo-chlorine pesticides in human milk from Indonesian women, *Chemosphere*, 50, 529, 2003.

102. UNI EN 1528-4, Fatty food. Determination of pesticides and polychlorinated biphenyl (PCBs). Part 4: Determination, confirmatory tests, miscellaneous, 1996.

103. U.S. EPA Method 8081A, Organochlorine pesticides by gas chromatography. Revision 1. US Environmental Protection Agency, Washington, DC, December 1996.

104. European Commission, Quality Control Procedures for Pesticide Residue Analyses, SANCO/10232/2006, 24 March 2006, available at http://ec.europa.eu/food/plant/protection/resources/qualcontrol_en.pdf.

105. EU Decision/2005/657/EC implementing Council Directive 96/23/EC concerning the performance of analytical methods and the interpretation of results, OJ L 221 of 17.8.2002.

106. Zrostlíková, J., Hajšlová, J., and Cajka, T., Evaluation of two-dimensional gas chromatography–time-of-flight mass spectrometry for the determination of multiple pesticide residues in fruit, *J. Chromatogr. A*, 1019, 173, 2003.

107. Focant, J.F., Sjödin, A., and Patterson, D.G., Qualitative evaluation of thermal desorption-program-mable temperature vaporization-comprehensive two-dimensional gas chromatography–time-of-flight mass spectrometry for the analysis of selected halogenated contaminants, *J. Chromatogr. A*, 1019, 143, 2003.

108. Korytár, P. et al., Group separation of organohalogenated compounds by means of comprehensive two-dimensional gas chromatography, *J. Chromatogr. A*, 1086, 29, 2005.

109. Seemamahannop, R. et al., Uptake and enantioselective elimination of chlordane compounds by common carp (*Cyprinus carpio*, L.), *Chemosphere*, 59, 493, 2005.

110. Chen, S. et al., Determination of organochlorine pesticide residues in rice and human and fish fat by simplified two-dimensional gas chromatography, *Food Chem.*, doi: 10.10167j.foodchem.2006.10.032, 2006.

111. Reyes, J.F.G. et al., Determination of pesticide residues in olive oil and olives, *Trends Anal. Chem.*, 26, 239, 2007.

112. Reiner, E.J. et al., Advances in analytical techniques for polychlorinated dibenzo-*p*-dioxins, polychlorinated dibenzofurans and dioxin-like PCBs, *Anal. Bioanal. Chem.*, 386, 797, 2006.

113. Hess, P. et al., Critical review of the analysis of non- and mono-*ortho*-chlorobiphenyls, *J. Chromatogr. A*, 703, 417, 1995.

114. EU Commission Regulation 1883/2006/EC of 19 December 2006 laying down methods of sampling and analysis for the official control of levels of dioxins and dioxin-like PCBs in certain foodstuffs, OJ L 364, 20.12.2006, 32–43.

115. Bocio, A. and Domingo, J.L., Daily intake of polychlorinated dibenzo-*p*-dioxins/polychlorinated dibenzofurans (PCDD/PCDFs) in foodstuffs consumed in Tarragona, Spain: a review of recent studies (2001–2003) on human PCDD/PCDF exposure through the diet, *Environ. Res.*, 97, 1, 2005.

116. Loutfy, N. et al., Monitoring of polychlorinated dibenzo-*p*-dioxins and dibenzofurans, dioxin-like PCBs and polycyclic aromatic hydrocarbons in food and feed samples from Ismailia city, Egypt, *Chemosphere*, 66, 1962, 2007.

117. Mayer, R., PCDD/F levels in food and canteen meals from Southern Germany, *Chemosphere*, 43, 857, 2001.

118. Fernandes, A. et al., Simultaneous determination of PCDDs, PCDFs, PCBs and PBDEs in food, *Talanta*, 63, 1147, 2004.

119. Focant, J.F. et al., Fast clean-up for polychlorinated dibenzo-*p*-dioxins, dibenzofurans and coplanar polychlorinated biphenyls analysis of high-fat-content biological samples, *J. Chromatogr. A*, 925, 207, 2001.

120. Covaci, A., Ryan, J.J., and Schepens, P., Patterns of PCBs and PCDD/PCDFs in chicken and pork fat following a Belgian food contamination incident, *Chemosphere*, 47, 207, 2002.

121. Wiberg, K. et al., Selective pressurized liquid extraction of polychlorinated dibenzo-*p*-dioxins, dibenzofurans and dioxin-like polychlorinated biphenyls from food and feed samples, *J. Chromatogr. A*, 1138, 55, 2007.

122. Pirard, C., Focant, J.F., and De Pauw, E., An improved clean-up strategy for simultaneous analysis of polychlorinated dibenzo-*p*-dioxins (PCDD), polychlorinated dibenzofurans (PCDF), and polychlorinated biphenyls (PCB) in fatty food samples, *Anal. Bioanal. Chem.*, 372, 373, 2002.

123. Focant, J.F., Pirard, C., and De Pauw, E., Levels of PCDDs, PCDFs and PCBs in Belgian and international fast food samples, *Chemosphere*, 54, 137, 2004.

124. Voorspoels, S. et al., Dietary PBDE intake: a market-basket study in Belgium, *Environ. Int.*, 33, 93, 2007.

125. Huwe, J.K. and Larsen, G.D., Polychlorinated dioxins, furans, and biphenyls, and polybrominated diphenyl ethers in a U.S. meat market basket and estimates of dietary intake, *Environ. Sci. Technol.*, 39, 5606, 2005.

126. Kiviranta, H., Ovaskainen, M.J., and Vartiainen, T., Market basket study on dietary intake of PCDD/Fs, PCBs, and PBDEs in Finland, *Environ. Int.*, 30, 923, 2004.

127. Huwe, J.K. et al., Analysis of mono- to deca-brominated diphenyl ethers in chickens at the part per billion level, *Chemosphere*, 46(5), 635, 2002.

128. Ohta, S. et al., Comparison of polybrominated diphenyl ethers in fish, vegetables, and meats and levels in human milk of nursing women in Japan, *Chemosphere*, 46, 689, 2002.

129. Martínez, A. et al., Development of a matrix solid-phase dispersion method for the screening of polybrominated diphenyl ethers and polychlorinated biphenyls in biota samples using gas chromatography with electron-capture detection, *J. Chromatogr. A*, 1072, 83, 2005.

130. European Commission, DIFFERENCE Project G6RD-CT-2001-00623, available at www.dioxins nl.

131. Sporring, S., Holst, C., and Björklund, E., Selective pressurized liquid extraction of PCBs from food and feed samples: effects of high lipid amounts and lipid type on fat retention, *Chromatographia*, 64, 553, 2006.

132. Björklund, E. et al., New strategies for extraction and clean-up of persistent organic pollutants from food and feed samples using selective pressurized liquid extraction, *Trends Anal. Chem.*, 25, 318, 2006.

133. Sporring, S. and Björklund, E., Selective pressurized liquid extraction of polychlorinated biphenyls from fat-containing food and feed samples. Influence of cell dimensions, solvent type, temperature and flush volume, *J. Chromatogr. A*, 1040, 155, 2004.

134. Harrad, S.J. et al., A method for the determination of PCB congeners 77, 126 and 169 in biotic and abiotic matrices, *Chemosphere*, 24, 1147, 1992.

135. Berdié, L. and Grimalt, J.O., Assessment of the sample handling procedures in a labor-saving method for the analysis of organochlorine compounds in a large number of fish samples, *J. Chromatogr. A*, 823, 373, 1998.

136. Pirard, C., Focant, J.F., and De Pauw, E., An improved clean-up strategy for simultaneous analysis of polychlorinated dibenzo-*p*-dioxin (PCDD), polychlorinated dibenzofurans (PCDF), and polychlorinated biphenyls (PCB) in fatty samples, *Anal. Bioanal. Chem.*, 372, 373, 2001.

137. Maier, E.A., Griepink, B., and Fortunati, U., Round table discussions. Outcome and recommendations, *Fresenius J. Anal. Chem.*, 348, 171, 1994.

138. March, R.E. et al., A comparison of three mass spectrometric methods for the determination of dioxins/furans, *Int. J. Mass Spectrom.*, 194, 235, 2000.

139. Lorán, S. et al., Evaluation of GC-ion trap-MS/MS methodology for monitoring PCDD/Fs in infant formula, *Chemosphere*, 67, 513, 2007.

140. Focant, J.F. et al., Recent advances in mass spectrometric measurement of dioxins, *J. Chromatogr. A*, 1067, 265, 2005.

141. Petrovic, M. et al., Recent advances in the mass spectrometric analysis related to endocrine disrupting compounds in aquatic environmental samples, *J. Chromatogr. A*, 974, 23, 2002.

142. EFSA, Opinion of the Scientific Panel on contaminants in the food chain on a requested from the Commission related to the presence of non dioxin-like polychlorinated biphenyls (PCB) in feed and food. Question N° EFSA-Q-2003-114. Adopted on 8 November 2005, *EFSA J.*, 284, 1, 2005.

143. EFSA, Advice of the scientific panel on contaminants in the food chain on a request from the commission related to relevant chemical compounds in the group of brominated flame retardants for monitoring in feed and food. Question N° EFSA-Q-2005-244. Adopted on 24 February 2006, *EFSA J.*, 328, 1, 2006.

144. Covaci, A., Voorspoels, S., and de Boer, J., Determination of brominated flame retardants, with emphasis on polybrominated diphenyl ethers (PBDEs) in environmental and human samples—a review, *Environ. Int.*, 29, 735, 2003.

145. Covaci, A. et al., Recent developments in the analysis of brominated flame retardants and brominated natural compounds, *J. Chromatogr. A,* 1153, 145, 2007.

146. Hansen, K.J. et al., Compound-specific, quantitative characterization of organic fluorochemicals in biological matrices, *Environ. Sci. Technol.*, 35, 766, 2001.

147. Tittlemier, S.A. et al., Development and characterization of a solvent extraction-gas chromatographic/mass spectrometric method for the analysis of perfluorooctane sulfonamide compounds in solid matrices, *J. Chromatogr. A*, 1066, 189, 2005.

148. Berger, U. and Haukas, M., Validation of a screening method based on liquid chromatography coupled to high-resolution mass spectrometry for analysis of perfluoroalkylated substances in biota, *J. Chromatogr. A*, 1081, 210, 2005.

149. Taniyasu, S. et al., Analysis of fluorotelomer alcohols, fluorotelomer acids, and short- and long-chain perfluorinated acids in water and biota, *J. Chromatogr. A*, 1093, 89, 2005.

150. Tseng, C.L. et al., Analysis of perfluorooctanesulfonate and related fluorochemicals in water and biological tissue samples by liquid chromatography-ion trap mass spectrometry, *J. Chromatogr. A*, 1105, 119, 2006.

151. Tittlemier, S.A. et al., Dietary exposure of Canadians to perfluorinated carboxylates and perfluorooctane sulfonate via consumption of meat, fish, fast foods, and food items prepared in their packaging, *J. Agric. Food Chem.*, 55, 3203, 2007.

152. Morris, S. et al., Determination of the brominated flame retardant, hexabromocyclododane, in sediments and biota by liquid chromatography-electrospray ionisation mass spectrometry, *Trends Anal. Chem.*, 25, 343, 2006.

153. de Boer, J. et al., Method for the analysis of polybrominated diphenylethers in sediments and biota, *Trends Anal. Chem.*, 20, 591, 2001.

154. Eljarrat, E. et al., Occurrence and bioavailability of polybrominated diphenyl ethers and hexabromocyclododecane in sediment and fish from the Cinca River, a tributary of the Ebro River (Spain), *Environ. Sci. Technol.*, 38, 2603, 2004.

155. Bayen, S., Lee, H.K., and Obbard, J.P., Determination of polybrominated diphenyl ethers in marine biological tissues using microwave-assisted extraction, *J. Chromatogr. A*, 1035, 291, 2004.

156. Basu, N., Scheuhammer, A.M., and O'Brien, M., Polychlorinated biphenyls, organochlorinated pesticides, and polybrominated diphenyl ethers in the cerebral cortex of wild river otters (*Lontra canadensis*), *Environ. Pollut.*, 2007, doi:10.1016/j.envpol.2006.12.026.

157. Ingelido, A.M. et al., Polychlorinated biphenyls (PCBs) and polybrominated diphenyl ethers (PBDEs) in milk from Italian women living in Rome and Venice, *Chemosphere*, 67, S301, 2007.

158. Pirard, C., De Pauw, E., and Focant, J.F., New strategy for comprehensive analysis of polybrominated diphenyl ethers, polychlorinated dibenzo-*p*-dioxins, polychlorinated dibenzofurans and polychlorinated biphenyls by gas chromatography coupled with mass spectrometry, *J. Chromatogr. A*, 998, 169, 2003.

159. Cochran, J.W. and Frame, G.M., Recent developments in the high-resolution gas chromatography of polychlorinated biphenyls, *J. Chromatogr. A*, 843, 323, 1999.

160. Galceran, M.T., Santos, F.J., Barceló, D., and Sanchez, J., Improvements in the separation of polychlorinated biphenyl congeners by high-resolution gas chromatography: application to the analysis of two mineral oils and powdered milk, *J. Chromatogr. A*, 655, 275, 1993.

161. Chernetsova, E.S. et al., Determination of polychlorinated dibenzo-p-dioxins, dibenzofurans, and biphenyls by gas chromatography/mass spectrometry in the negative chemical ionization mode with different reagent gases, *Mass. Spectrom. Rev.*, 21, 373, 2002.

162. Verenitch, S.S. et al., Ion-trap tandem mass spectrometry-based analytical methodology for the determination of polychlorinated biphenyls in fish and shellfish: performance comparison against electron-capture detection and high-resolution mass spectrometry detection, *J. Chromatogr. A*, 1142, 199, 2007.

163. Gómara, B. et al., Feasibility of gas chromatography– ion trap tandem mass spectrometry for the determination of polychlorinated biphenyls in food, *J. Sep. Sci.*, 123, 2006.

164. Björklund, J. et al., Influence of the injection technique and the column system on gas chromatographic determination of polybrominated diphenyl ethers, *J. Chromatogr. A*, 1041, 201, 2004.

165. Korytár, P. et al., Retention-time database of 126 polybrominated diphenyl ether congeners and two Bromkal technical mixtures on seven capillary gas chromatographic columns, *J. Chromatogr. A*, 1065, 239, 2005.

166. Thomsen, C. et al., Comparing electron ionization high-resolution and electron capture low-resolution mass spectrometric determination of polybrominated diphenyl ethers in plasma, serum and milk, *Chemosphere*, 46, 641, 2002.

167. Wang, D. et al., Gas chromatography/ion trap mass spectrometry applied for the determination of polybrominated diphenyl ethers in soil, *Rapid. Commun. Mass Spectrom.*, 19, 83, 2005.

168. Gómara, B. et al., Quantitative analysis of polybrominated diphenyl ethers in adipose tissue, human serum and foodstuff samples by gas chromatography with ion trap tandem mass spectrometry and isotope dilution, *Rapid Commun. Mass Spectrom.*, 20, 69, 2006.

169. Petinal, C.S. et al., Headspace solid-phase microextraction gas chromatography tandem mass spectrometry for the determination of brominated flame retardants in environmental solid samples, *Anal. Bioanal. Chem.*, 385, 637, 2006.

170. Yusà, V. et al., Optimization of a microwave-assisted extraction large-volume injection and gas chromatography–ion trap mass spectrometry procedure for the determination of polybrominated diphenyl ethers, polybrominated biphenyls and polychlorinated naphthalenes in sediments, *Anal. Chim. Acta*, 557, 304, 2006.

171. Martin, J.W. et al., Analytical challenges hamper perfluoroalkyl research, *Environ. Sci. Technol.*, 38, 248A, 2004.

172. van Leeuwen, S.P.J. and de Boer, J., Extraction and clean-up strategies for the analysis of poly- and perfluoroalkyl substances in environmental and human matrices, *J. Chromatogr. A*, 1153, 172, 2007.

173. de Voogt, P. and Saez, M., Analytical chemistry of perfluoroalkylated substances, *Trends Anal. Chem.*, 25, 326, 2006.

174. Ylinen, M. et al., Quantitative gas chromatographic determination of perfluorooctanoic acid as the benzyl ester in plasma and urine, *Arch. Environ. Contam. Toxicol.*, 14, 713, 1985.

175. Giesy, J.P. and Kannan, K., Global distribution of perfluorooctane sulfonate in wildlife, *Environ. Sci. Technol.*, 35, 1339, 2001.

176. De Silva, A. and Mabury, S., Isolating isomers of perfluorocarboxylates in polar bears (ursus maritimus) from two geographical locations, *Environ. Sci. Technol.*, 38, 6538, 2004.

177. Powley, C.R., Ryan, T.W., and George, S.W., Matrix-effect free analytical methods for determination of perfluorinated carboxylic acids in environmental and biological samples, Proceeding in "Fourth SETAC World Congress, 25th Annual Meeting in North America", Portland, Oregon, USA, 2004.

178. Powley, C.R. et al., Matrix effect-free analytical methods for determination of perfluorinated carboxylic acids in environmental matrixes, *Anal. Chem.*, 77, 6353, 2005.

179. Langlois, I. and Oehme, M., Structural identification of isomers present in technical perfluorooctane sulfonate by tandem mass spectrometry, *Rapid Commun. Mass Spectrom.*, 20, 844, 2006.

180. Tittlemier, S.A. et al., Development and characterization of a solvent extraction-gas chromatographic/mass spectrometric method for the analysis of perfluorooctane sulfonamide compounds in solid matrices, *J. Chromatogr. A*, 1066, 189, 2005.

181. Van Leeuwen, S. et al., First worldwide interlaboratory study on perfluorinated compounds in human and environmental matrices, Joint report of the Netherlands Institute for Fisheries Research (ASG-RIVO) (The Netherlands), Man-Technology-Environment (MTM) Research Centre, Sweden, and Institute of Water Technology Laboratory, Malta, 2005.

182. Fluoros Report 2006, 2nd Worldwide Interlaboratory Study on PCFs. December 2006.

183. Behnisch, P.A., Hosoea, K., and Sakai, S., Bioanalytical screening methods for dioxins and dioxin-like compounds—a review of bioassay/biomarker technology, *Environ. Int.*, 27, 413, 2001.

184. Płaza, G., Ulfig, K., and Tien, A.J., Immunoassays and environmental studies, *Polish J. Environ. Studies*, 9, 231, 2000.

185. Scippo, M.L. et al., DR-CALUX® screening of food sample: evaluation of the quantitative approach to measure dioxin, furans and dioxin-like PCBs, *Talanta*, 63, 1193, 2004.

186. U.S. EPA, Guidance on Cumulative Risk Assessment of Pesticide Chemicals That Have a Common Mechanism of Toxicity, Office of Pesticide Programs U.S. Environmental Protection Agency, Washington, D.C. 20460 January 14, 2002, available at http://www.epa.gov/pesticides/trac/science/cumulative_guidance.pdf.

187. Hoogenboom, L. et al., The CALUX bioassay: current status of its application to screening food and feed, *Trends Anal. Chem.*, 25, 410, 2006.

188. U.S. EPA, Interim Report on the Evolution and Performance of the Eichrom Technologies Procept® Rapid Dioxin Assay for Soil and Sediment Samples, available at http://costperformance.org/monitoring/pdf/epa_eichrom_dioxin_assay.pdf.

189. Fochi, I. et al., Modeling of DR CALUX® bioassay response to screen PCDDs, PCDFs, and dioxin-like PCBs in milk from dairy herds, *Chemosphere*, submitted.

190. Schroijen, C. et al., Study of the interference problems of dioxin-like chemicals with the bio-analytical method CALUX, *Talanta*, 63, 1261, 2004.

Chapter 12

Biogenic Amines

M. Carmen Vidal-Carou, M. Luz Latorre-Moratalla,
and Sara Bover-Cid

Contents

12.1 Introduction 400
    12.1.1 Biogenic Amines: Origin and Classification 400
    12.1.2 Relevance of Biogenic Amines in Food... 400
12.2 Biogenic Amines in Meat and Meat Products 402
    12.2.1 Aminogenic Microorganisms Associated with Meat
         and Meat Products 402
    12.2.2 Occurrence of Biogenic Amines in Meat and Meat Products 402
         12.2.2.1 Fresh Meat and Fresh Meat Products.................................. 402
         12.2.2.2 Cooked Meat Products 403
         12.2.2.3 Cured Meat Products 403
         12.2.2.4 Fermented Meat Products 403
    12.2.3 Biogenic Amine Index 404
         12.2.3.1 Biogenic Amines to Evaluate the Loss of Meat Freshness.... 404
         12.2.3.2 Biogenic Amines to Monitor the Hygienic Quality
             of Raw Materials in Meat Products.................................... 405
12.3 Determination of Biogenic Amines in Meat and Meat Products 406
    12.3.1 Biogenic Amine Extraction and Cleanup 406
    12.3.2 Analytical Procedures to Detect and Quantify Biogenic Amines 407
         12.3.2.1 Chromatographic Quantification Procedures..................... 407
         12.3.2.2 Rapid Screening Procedures.. 409
References 414

12.1 Introduction

12.1.1 Biogenic Amines: Origin and Classification

Biologically active amines, also known as biogenic amines, are nitrogenous compounds of basic nature that show biological activity. They are synthesized and degraded by animal, plant, and microbial metabolisms, and consequently are found in a wide variety of food products [1–3]. On the basis of their chemical structure, the biogenic amines most commonly found in food are grouped as

- Aromatic monoamines—tyramine and phenylethylamine
- Heterocyclic amines—histamine and tryptamine
- Aliphatic diamines—cadaverine and putrescine
- Aliphatic polyamines—agmatine, spermidine, and spermine

Classically, biogenic amines are defined as "biogenic" or "endogenous/natural," depending on their synthesis. However, sometimes there is no clear division between these two categories [4]. The former result from the activity of decarboxylase enzymes against precursor amino acids. Within this group, tyramine, phenylethylamine, histamine, tryptamine, cadaverine, putrescine, and agmatine originate from the decarboxylation of tyrosine, phenylalanine, histidine, tryptophan, lysine, ornithine, and arginine, respectively. The decarboxylase enzymes responsible for the synthesis of these biogenic amines in food are mainly of bacterial origin and usually inducible by certain environmental conditions (e.g., unfavorable acidic pH). Although bacterial decarboxylases are generally specific for one amino acid, in some cases they may have activity, although with a lower affinity, against other amino acids of a similar chemical structure, such as tyrosine and phenylethylamine [5] or ornithine and lysine [4].

The so-called endogenous or natural amines are formed as a result of the intracellular metabolic processes of animals, plants, and microorganisms. The aliphatic polyamines spermine and spermidine are the most relevant amines within this category, the synthesis of which follows other reactions apart from the decarboxylation of arginine during the early stages of the biosynthetic pathway. Small amounts of putrescine, as a precursor of polyamines, can also be considered of endogenous origin [4]. In addition to putrescine, several other biogenic amines, such as cadaverine and agmatine, may occur in certain foods both endogenously and as microbial metabolic products. When these biogenic amines are present in low concentrations, it is difficult to differentiate their true origin, and it is difficult to know the significance of their occurrence in food products.

12.1.2 Relevance of Biogenic Amines in Food

Interest in biogenic amines is related to both food safety and food quality issues. Traditionally, these compounds have been regarded as undesirable toxic components of food. Tyramine, histamine, and to lesser extent phenylethylamine, are the main dietary biogenic amines associated with several acute adverse reactions in consumers. Interaction with monoamine-oxidase-inhibitor (MAOI) drugs, histaminic intoxication, food intolerance related to enteral histaminosis, and food-induced migraines may occur following the ingestion of biologically active amines [4,6–9]. These compounds trigger vasoactive and psychoactive reactions. The vasoconstrictive properties of tyramine and phenylethylamine have been reported to be directly responsible for increases in blood pressure, and may also cause headaches, sweating, vomiting, and pupil dilatation, among other effects. Histamine causes vasodilatation and subsequent hypotension as well as other dermal (flushing and pruritus), gastrointestinal (diarrhea, cramps, vomiting), and neurological (headache, dizziness) effects [7,8,10]. The severity of the disorders associated with biogenic amines varies depending on individual sensitivity, but, in general, reactions are mild and medical attention is rarely required [11]. It is precisely the mild nature of the symptoms together with misdiagnosis and the lack of a mandatory or adequate system for reporting these food diseases that explain the poor statistics on the incidence of intoxications caused by dietary amines.

In spite of compelling evidence that biogenic amines are the causative agents of adverse food reactions, the toxic dose is difficult to estimate. Not only do the concentrations of biogenic amines vary greatly among food products, but the amounts ingested also vary greatly among consumers, who in turn show a wide range of inter- and intraindividual clinical responses to a given amount of these dietary compounds [7,8,12]. Moreover, there are numerous potentiating factors of dietetic-, physiological-, and pharmacological nature that contribute to the variability of the response to biogenic amines in food. The toxicity of tyramine may be of special concern for individuals taking MAOI drugs, which may increase the vasoconstrictive effects of this dietary amine. Nevertheless, according to literature, amounts from 50 to 150 mg of tyramine are well tolerated by patients under a new generation MAOI treatment [13–15]. According to a review by Shalaby [8], ingestion of 8–40 mg of histamine causes slight toxicity, over 40 mg moderate toxicity, and over 100 mg severe poisoning. Although these doses are repetitively cited in the literature, no toxicological studies supporting them are available. In fact, histamine food poisoning incidents are related to fish containing high concentrations (usually above 600 mg/kg) of this biogenic amine [12,16]. Therefore, if an average fish portion weighs 200–300 g, the toxicological effects of histamine would appear after ingestion of more than 120–180 mg of this biogenic amine. However, histamine intolerance by sensitive individuals has been described after the intake of variable amounts of this biogenic amine, ranging from 50 μg accompanied by wine to 75 mg of pure histamine [17].

The diamines putrescine and cadaverine, although not considered toxic individually, may enhance the absorption of vasoactive amines as a result of the saturation of intestinal barriers through competition for mucin attachment sites, and detoxification enzymes [6,18].

Biologically active amines present in food products can also act as precursors of nitroso compounds with potential carcinogenic activity, thereby constituting an indirect additional risk. Nitrosamines result from the action of nitrite on secondary amines, which in turn may be formed from primary amines (such as the aliphatic diamines and polyamines) by a cyclization reaction under certain circumstances [1,8,19]. Some aromatic amines, such as tyramine, have also been proposed as possible precursors of diazotyramine, which shows mutagenic activity [20]. The occurrence of nitrosating agents (i.e., nitrites and nitrates), mild acidic pH, and high temperatures during food manufacture favor nitrosamine formation. Cured and cooked or smoked meat products (such as cooked and fried bacon) are sources of nitrosamines.

Given the potential effects of biogenic amines on health and their microbial origin, the occurrence of these substances in food is relevant from the technological and food quality standpoints. Indeed, the accumulation of biogenic amines can be associated with fermentation processes but also with spoilage. In this regard, dietary biogenic amines are of particular interest, because they can be used as chemical indicators or monitors of the hygienic quality of raw materials and manufacturing conditions.

12.2 Biogenic Amines in Meat and Meat Products

In general, all protein-rich food subjected to conditions that allow bacterial development and activity (e.g., storage, maturation, fermentation) is expected to accumulate certain amounts of biogenic amines, in addition to those present naturally. Meat and meat products contain moderate or high amounts of these compounds. Apart from spermine and spermidine, the main origin of notable amounts of biologically active amines in food in general, and in meat products in particular, is widely attributed to the action of bacterial decarboxylase enzymes [1,2]. However, there is no common origin for all biogenic amines, and the final type and content will depend on the conditions of manipulation, treatment, and storage as well as microorganism activity.

12.2.1 Aminogenic Microorganisms Associated with Meat and Meat Products

Several bacterial groups associated with meat and meat products can generate biogenic amines. The capacity to decarboxylate certain amino acids has generally been attributed to specific bacterial families or genera. For instance, enterobacteria are frequently histamine and diamine (cadaverine and putrescine) producers, and although fewer studies have addressed *Pseudomonas*, they have also been reported as notably aminogenic [2,21,22]. Among Gram-positive bacteria, lactic acid bacteria, especially enterococci and certain lactobacilli such as *Lactobacillus curvatus*, are usually associated with tyramine production. In contrast, staphylococci are much less frequently reported as powerful aminogenic organisms [21,23–25]. Despite these general rules of thumb, the capacity to produce one or more biogenic amines simultaneously is strain-dependent [21], thus explaining why the biogenic amine content of a given product cannot always be statistically correlated with the global counts of specific bacterial groups in the same product.

12.2.2 Occurrence of Biogenic Amines in Meat and Meat Products

12.2.2.1 Fresh Meat and Fresh Meat Products

In freshly slaughtered meat, spermine, and spermidine are the main biogenic amines [26]. Apart from small amounts of putrescine, the other amines are usually undetectable and appear only under conditions that allow bacterial activity. The contents of spermine and spermidine may vary widely in meat. In contrast to vegetable products, meat and products of animal origin contain higher amounts of spermine than spermidine, with a ratio of approximately 10:1 [27]. Concentrations of 15–50 mg/kg of spermine and 1–5 mg/kg of spermidine are commonly reported [2,27–31]. The animal species does appear to be a determinant of this variability, because the differences between pork, beef, and poultry products, for example, are not as wide as between organs or parts of the same animal or another animal of the same species. One of the factors influencing the cellular levels of polyamines is the metabolic activity of the tissue. The synthesis *de novo* and the accumulation of polyamines are particularly stimulated in tissues and organs that show rapid growth or in phases with a considerable cellular regeneration rate [32]. This observation could explain, at least in part, the range of polyamine concentrations detected in meat from distinct animals or even from different parts of the same animal.

12.2.2.2   Cooked Meat Products

In heat-treated meat products (cooked ham, cooked meat sausages, etc.), spermine and spermidine are the only biogenic amines usually detected. The levels of these polyamines in these products are in general slightly lower than in fresh meat. This fact is attributed to a dilution effect produced when lean meat is mixed with fat and other ingredients included in the product formula [30]. Although polyamines are considered heat-resistant, a small reduction of these compounds has also been reported during thermal treatments of products [33–35].

The contents of other biogenic amines in cooked products are much more variable than those of polyamines. In general, concentrations of tyramine, histamine, and diamines are quite low, with some punctual exceptions. Rarely are phenylethylamine and tryptamine detected. In some particular cooked meat products, a short maturation/fermentation step is applied before cooking, for instance for bologna sausage, Catalan sausage (*butifarra*). In this case, the activity of aminogenic organisms can be notable, and may result in a significant accumulation of biogenic amines [36].

12.2.2.3   Cured Meat Products

The manufacture of cured products involves large pieces or whole muscle parts without mincing or mixing. Common salt is an essential ingredient not only for product safety but also for the development of the organoleptic characteristics during ripening at relatively low temperatures [37]. Although the pH does not drop, under these conditions microbial growth is strongly limited and only halophile bacteria, such as Gram-positive catalase-positive cocci (staphylococci, micrococci, and kocuria) grow, with counts ranging from 10 to 10^6 colony-forming units (cfu)/g. Yeast and some lactic acid bacteria may also develop to a lesser extent. Consequently, the contents of biogenic amines, such as tyramine, histamine, cadaverine, and putrescine, in this type of product are quite low (with median values from 2 to 80 mg/kg), with only particular exceptions [30,36,38–40]. The occurrence of significant amounts of phenylethylamine and tryptamine has not been described in cured meat products.

The length of ripening is a critical factor that determines the extent of biogenic amine accumulation, especially tyramine [40]. In contrast, a considerable formation of diamines, especially cadaverine, during dry-cured ham manufacture has been reported to depend on the type of ripening [41]. Short (rapid) ripening allows greater accumulation of biogenic amines in comparison to a long (slow) ripening process. These findings are attributed to the higher temperatures applied during drying in the former. The proteolytic phenomena occurring during ripening increase the concentration of precursor amino acids and correlate with biogenic amine formation [40,42].

12.2.2.4   Fermented Meat Products

Fermented sausages and cheese are foods that register the highest biogenic amine contents. However, fewer studies have addressed the former. According to the literature [30,36,43–47], biogenic amine levels vary greatly between fermented sausages of diverse types, between manufacturers, and also between samples from distinct batches of the same kind of product and from the same producer. In retail fermented sausages, tyramine is usually the most frequent and most abundant biogenic amine. The literature describes an average tyramine content of 140 mg/kg (relative standard deviation [RSD] of 89%) in these meat products. The diamines, putrescine and cadaverine, are also quite common, though with a higher variability (RSD of 145% for putrescine and 187% for cadaverine). Most samples of fermented sausages show relatively low amounts of diamines; however, some may accumulate large amounts, which may exceed the tyramine content. As a consequence of this variability, the mean values of 89 mg/kg for putrescine and 44 mg/kg for cadaverine are much higher than the corresponding median values (36 and 8 mg/kg, respectively). This variability is even more pronounced for histamine (median value of 4 mg/kg and RSD of 222%), which is not detected in most retail fermented sausages, but in some particular samples may reach quite high levels, usually accompanied by high amounts of other biogenic amines. Similarly, the contents of phenylethylamine (RSD of 206%) and tryptamine (RSD of 170%) are relatively low (median of 2 and 4 mg/kg, respectively) in these meat products. These two amines could be considered minor amines in fermented sausages and their accumulation appears to depend on the occurrence of high concentrations of tyramine.

Fermented sausages are significant sources of physiological polyamines, although these amines have received less attention [3,31]. Polyamines are found naturally in raw meat, and therefore their levels are much less variable than those of biogenic amines of microbial origin. According to data in the literature, the average content of spermine in fermented meat products is 23 mg/kg, and that of spermidine is 7 mg/kg. Occasionally a decrease in polyamine content during meat fermentation has been reported [48–50], which is attributed to uptake by microorganisms as a nitrogenous source [4] or to deamination reactions [2].

12.2.3 Biogenic Amine Index

As a result of their microbiological origin, biogenic amines have been used as criteria to evaluate the hygienic quality and freshness of certain foods, especially fish, but also meat and a number of meat products.

12.2.3.1 Biogenic Amines to Evaluate the Loss of Meat Freshness

Biogenic amines in fresh meat and fresh meat products (such as hamburgers, raw sausages, and packaged fresh meat) are usually below the detection limit, except for the physiological polyamines spermine and spermidine. When monitoring aminogenesis during the storage of meat under aerobic conditions, several biogenic amines, such as cadaverine, putrescine, tyramine, and histamine, progressively increase to variable extents. The higher the storage temperature, the faster the accumulation of these compounds [36,51]. Significant accumulation of biogenic amines generally occurs before the appearance of sensorial signs of spoilage, when counts of aerobic mesophile bacteria reach 10^5–10^7 cfu/g [52,53]. In contrast, polyamines usually remain constant or may even decrease [53–55]. These observations have been attributed to consumption by microorganisms [4].

Therefore, biogenic amines individually or in combination have been proposed as objective chemical indexes to evaluate meat freshness. The biogenic amine index (BAI) put forward by Mietz and Karmas [56] (cadaverine + putrescine + histamine/1 + spermine + spermidine) to evaluate fish freshness was applied by Sayem El Daher et al. [34] to assess the hygienic quality of beef. A highly significant correlation between BAI values and microbial counts was detected in this meat. Maijala et al. [28] also used this index to compare the effect of pH on aminogenesis during meat spoilage. Several authors defend the use of only one biogenic amine for evaluation purposes, for instance putrescine [51,57], cadaverine [58–60], or both [52,61] for aerobically stored meat, mainly pork or beef. However, although the use of one biogenic amine for evaluation is more straightforward, the application of a multiple amine index may increase specificity and selectivity. Tyramine increases considerably during meat storage; therefore, this biogenic amine should also be included, together with cadaverine, putrescine, and histamine, in a BAI. This is the case of the BAI of tyramine + cadaverine + putrescine + histamine, proposed by Wortberg and Woller [27] and Hernández-Jover et al. [53]. Wortberg and Woller [27] established a spoilage limit at 500 mg/kg, but, according to other findings on pork and beef meat, Hernández-Jover et al. [53] reported that spoilage is evident at 10-fold lower values.

In particular for poultry meat, cadaverine concentrations have been proposed for the monitoring of chicken meat spoilage by Vinci and Antonelli [60], whereas Patsias et al. [62] suggested tyramine and putrescine limits for precooked chicken meat. Other authors consider the sum of tyramine, cadaverine, and putrescine to be the most promising indicator for both storage time and temperature, as well as for the microbiological quality of modified atmosphere and aerobically packaged chicken meat [63,64].

In vacuum-packaged meat, bacterial flora varies with the environment in the package, and thus the pattern of biogenic amine formation in meat packed in this way differs from that packaged aerobically. Lactic acid bacteria become dominant in the microflora of vacuum-packaged meats from early storage. As a result, tyramine may be a better indicator of spoilage/acceptability of vacuum-packaged meat stored at chilled temperatures [65,66].

12.2.3.2 Biogenic Amines to Monitor the Hygienic Quality of Raw Materials in Meat Products

The heat treatments commonly applied by the meat industry inactivate microorganisms but do not reduce the contents of biogenic amines, because these compounds are thermoresistant. Moreover, cooking does not favor aminogenesis. Consequently, cooked meat products should contain only the physiological amines spermine and spermidine. The occurrence of other biogenic amines in these products would indicate the decarboxylation of amino acids by undesirable contaminant microorganisms before, during, or even after manufacture of the product. Although meat products made of blood or liver may contain certain amounts of histamine of endogenous origin, the concentrations from this source are much lower than those formed by bacterial activity.

Therefore, because biogenic amines are thermoresistant, BAIs have been considered useful to evaluate the quality of the raw material used and the hygienic conditions prevalent during the manufacturing processes, and contribute valuable information relevant to quality control processes [67]. Indeed, most retail samples of cooked meat products contain low levels of biogenic amines (optimally <5 mg/kg). Only occasionally do some show considerable amounts of tyramine, cadaverine, putrescine, and histamine, which allow producers to monitor the hygienic quality of raw materials used during manufacturing.

Cured meat products, such as cured ham or cured loin, are subjected to the action of brine and maturation, and their manufacture does not include a fermentation step or a cooking process. In this case, the halophilic microorganisms surviving high salt concentrations are not usually related to notable decarboxylase activity [21]. In general, no significant formation or degradation of biogenic amines is observed in cured meat products when these are manufactured following proper hygienic practices. Therefore, BAIs could also be applied to evaluate the hygienic quality of raw meat materials as well as conditions during maturation.

On the whole, the application of BAIs as criteria for quality evaluation of fermented meat products is more difficult because the formation of biogenic amines cannot be directly and exclusively associated with the quality of raw materials [36,55]. A number of microorganisms that produce biogenic amines, especially tyramine, have been reported in these meat products (such as salami, *salchichón*, and other dry sausages). It has been demonstrated that fermented sausages practically free from tyramine and other biogenic amines can be produced, for instance, through scrupulously hygienic conditions and the inoculation of selected starter cultures [46,68,69]. Abundant data are available on biogenic amine contents in retail fermented sausages as well as on biogenic amine accumulation during the manufacture of this type of product using raw materials with optimal hygienic quality [30,36,43,47]. The consequences of using raw materials of poor hygienic quality [70] and also the contribution of contaminant enterobacteria and lactic acid bacteria [71] to overall aminogenesis during sausage fermentation have been reported. On the basis of the results from these two studies, it could be inferred that meat fermentation leads to the accumulation of certain amounts of biogenic amines. In particular, tyramine is the most important amine associated with fermented sausages, registering average concentrations from 100 to 200 mg/kg. Putrescine and cadaverine can also be accumulated at concentrations below 50 mg/kg, but histamine is rarely found in fermented sausages manufactured under proper hygienic and manufacturing conditions. Therefore, biogenic amine accumulation above the levels described earlier could be considered the result of poor hygienic practices, and therefore biogenic amines could also be used to monitor the hygienic quality of fermented meat products.

12.3   Determination of Biogenic Amines in Meat and Meat Products

Several procedures have been developed and improved for the detection and determination of various biogenic amines in meat and meat products. From an analytical perspective, the measurement of biogenic amines and polyamines in food in general, and in meat and meat products in particular, is not a simple procedure, mainly because of (a) the diverse chemical structures of biogenic amines (aromatic, heterocyclic, and aliphatic); (b) the wide range of concentrations at which each biogenic amine can be present in the product; and (c) the complexity of the sample matrix (high protein content and often high fat content).

Analytical study of biogenic amines in meat products involves two well-differentiated phases: (1) extraction of amines from the solid food matrix, in some cases including a further purification or cleanup of the raw extract; and (2) the analytical determination of these amines, which can be carried out by means of a variety of approaches including enzymatic, spectrofluorometric, and chromatographic procedures.

12.3.1   Biogenic Amine Extraction and Cleanup

In solid samples, biogenic amines are extracted to a liquid phase and separated from potentially interfering compounds. This separation step is crucial for the accuracy of the methodology, because it is probably the most decisive factor for the analytical recovery of each amine. Although some authors have extracted amines from solid matrixes with water at room or higher temperatures, the most common extracting solvents used for this purpose include acid solutions, such as hydrochloric acid (e.g., 0.1 M), trichloroacetic acid (e.g., 5–10%), and perchloric acid (e.g., 0.4–0.6 M), as well as organic solvents, such as methanol, acetone, acetonitrile–perchloric acid, or dichloromethane–perchloric acid [80].

The selectivity and recovery of the extraction is influenced by the type of acid used. Although several studies have compared the extraction capacity of distinct acids on amines, the results obtained are not always concordant or conclusive. In the case of meat products, because of sample turbidity and the occurrence of interfering substances, hydrochloric acid is not a suitable choice [72]. However, perchloric acid [29,73–75] or trichloroacetic acid, which show a high capacity to precipitate proteins, are recommended [80,76].

A cleanup of the extract before analysis of biogenic amines is required, depending on the final analytical technique applied. A number of approaches have been proposed to purify raw extracts, including column chromatography with alumina or ion-exchange resins [61,77] and solid-phase extraction [78,79]. Liquid–liquid extraction with organic solvents is also applied [78,72]. In this procedure, the raw extract is saturated with a salt, adjusted to an alkaline pH, and partitioned with an organic solvent (butanol, butanol/chloroform) that can selectively extract free amines and leave free amino acids in the aqueous layer. Because the pH optimum for extraction varies between amines, a strict control of this parameter is required to ensure satisfactory recovery and reproducibility. A pH of 11.5 is considered a suitable compromise for all biogenic amines [80]. A cleanup step increases the time required for the analysis and introduces a factor of uncertainty and variability as a result of sample handling. The use of an internal standard may help to address this limitation. However, several procedures for the extraction of biogenic amines from meat and meat products that do not include this step have been reported [29,60,81].

12.3.2 Analytical Procedures to Detect and Quantify Biogenic Amines

12.3.2.1 Chromatographic Quantification Procedures

The analytical methodologies to determine biogenic amines in meat and meat products are usually based on a chromatographic separation coupled with distinct detection techniques. Chromatographic procedures are the most extensively used methods because they provide high resolution, sensitivity, and versatility, and sample treatment is simple. Thin-layer chromatography [82,83], gas chromatography [84], and micellar liquid chromatography [85] have been applied for the analysis of biogenic amines in meat products. However, high-performance liquid chromatography (HPLC) with ion-exchange columns [39,86,87] or reverse-phase columns using ion pairs to separate biogenic amines as neutral [29,75] or nonneutral [76,81,88,89] compounds are the most frequently reported methods in the literature. Recent studies have addressed capillary (zone) electrophoresis [90,91].

Most biogenic amines, especially those of an aliphatic nature, have low absorption coefficients or quantum yields and require derivatization when the methods involve ultraviolet (UV)-visible (Vis) absorption or fluorescence detection. Chemical derivatization of these compounds can be carried out with a variety of reagents. The most often used are 5-dimethylamino-1-naphtalene-sulfonyl chloride (dansyl chloride [DnCl]), which forms stable compounds after reaction with both primary and secondary amino groups, and *o*-phthaldialdehyde (OPA), which reacts rapidly (i.e., 30 sec) with primary amines in the presence of a reducing agent such as 2-mercaptoethanol (ME) or *N*-acetylcyteine [92]. Figure 12.1 shows the representative derivatization reactions for

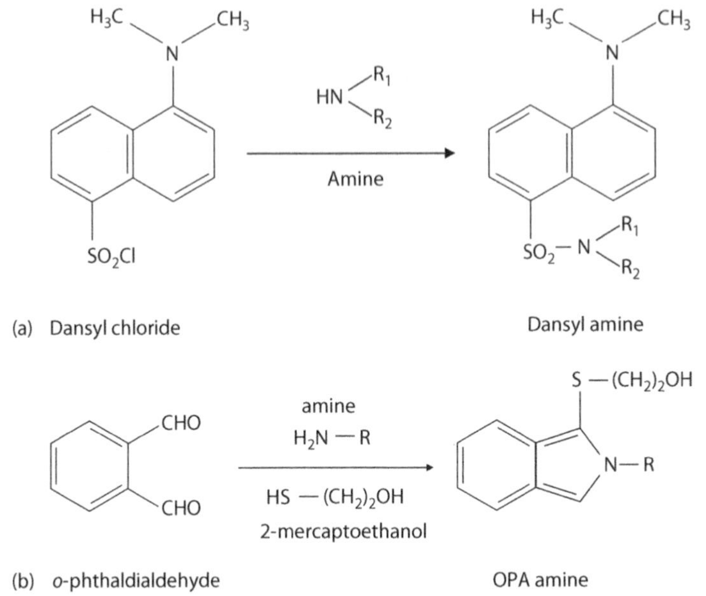

Figure 12.1 Representative derivatization reactions of biogenic amines with dansyl chloride (a) and *o*-phthaldialdehyde (b).

biogenic amines with these reagents. Other alternatives for the formation of detectable amine derivatives include fluorescamine, fluorescein isothiocyanate (FITC), phenylisothiocyanate (PITC), 6-aminoquinoyl-*N*-hydroxysuccinimidyl-carbamate (ACCQ), 2-naphthyloxycarbonyl chloride (NOC-Cl), benzoyl chloride, and ninhydrin [93,94].

Amine derivatives can be formed before (precolumn), during (on-column), or after (postcolumn) the chromatographic separation. Prederivatization comprises a series of time-consuming manual steps and may introduce imprecision to the overall analytical procedure. The use of an internal standard is critical to guarantee precision and accuracy (e.g., 1,7-diaminoheptane, 1,8-diaminooctane, or benzylamine have been described for the DnCl precolumn methodologies). Postcolumn derivatization has the advantage that it is automatically performed online, thereby avoiding sample manipulation and shortening the time required for the analysis. Moreover, changing the pH (to alkaline as required for derivatization reaction) is simple, easy, and quick with a postcolumn system. Nevertheless, it adds complexity to the instrumentation, because an extra pump is required. However, although postcolumn reactions have been criticized because of the occurrence of peak widening, this problem can be easily addressed using capillary connections and tubes.

Measurements of biogenic amines in meat and meat products have been taken by means of several techniques, such as fluorimetry [29,75,89], UV absorption [76,81], diode array-UV multichannel [75,78], and mass spectrometry [39]. Most of these methods are related to pre-, post- or on-column derivatization.

Conductometry, as applied by Kvasnicka and Voldrich [95], does not involve a derivatization step, but uses chemical suppression of the eluent conductivity, which also leads to a loss of some analytes. This technique also detects common alkaline and alkaline-earth cations found in food matrices [87]. Pulsed amperometric detection, with dedicated wave-form [87] for complex matrixes such as meat and meat products, is less affected by the already mentioned drawbacks, although electrode damage effects may arise.

Of the extraction alternatives for biogenic amines described, the most used for meat and meat products, as deduced from the literature, involve acid extraction of these compounds, followed either by (a) DnCl precolumn derivatization, reverse-phase HPLC separation coupled with UV detection (Figure 12.2) [80,81], or by (b) ion-pair reverse-phase HPLC with OPA-ME (post- or precolumn) derivatization coupled with fluorescence detection (Figure 12.3) [29,75]. The main conditions of these techniques are summarized in Table 12.1. The use of OPA instead of DnCl or fluorescamine is advantageous because of its greater selectivity for primary amines and the increase in method sensitivity as a result of fluorometric rather than spectrophotometric detection [76,96]. Moreover, DnCl and fluorescamine reactions result in several interfering by-products [80,93] that must be removed before chromatographic analysis to prevent coelution with biogenic amines. The addition of ammonia or proline [97] has been proposed for this purpose. The stability of OPA-amine derivatives is low, and postcolumn derivatization or an automated precolumn derivatization immediately before HPLC analysis is recommended. It has been reported that the natural polyamines spermidine and spermine can be analyzed only by means of DnCl derivatization, but not with OPA because the latter reacts only with primary amines [76]. However, several authors have described OPA-based methodologies that allow accurate measurement of these two polyamines (Table 12.1) [29,75,98]. In fact, spermidine and spermine bear primary amino groups and thus react with OPA-ME reagent as other biogenic amines do.

Few studies have compared the performance and the concordance between analytical methods for biogenic amines in meat and meat products. In an examination of Czech dry fermented sausages, HPLC procedures after precolumn derivatization of DnCl and OPA gave similar results in terms of detection limit, repeatability, recovery, and accuracy [76]. However, these authors reported that OPA derivatization was faster and much simpler in terms of sample pretreatment, which can be fully automated by the autosampler. In another study [25], the application of modifications of DnCl-based methodologies by three laboratories significantly affected the results obtained on biogenic amine accumulation in European fermented sausages. Two laboratories used 0.4 M perchloric acid as the extractant and 1,7-diaminoheptane as the internal standard, the derivatization was carried out for 40 min, after which the sample was dissolved in acetonitrile [81]. The third laboratory used acetone and 5% trichloroacetic acid as extractant solvent and 1,8-diaminooctane as internal standard. The derivatization was performed for a longer period (4 h), followed by a further extraction of the amines with diethyl ether before the sample was dissolved in acetonitrile. The amines most affected by the method of analysis were spermine and spermidine, for which this factor accounted for 43 and 83% of the total variance [25].

12.3.2.2 Rapid Screening Procedures

Alternatives to the instrumental procedures described earlier for the meat industry include the application of less expensive, less time-consuming, and simpler analytical techniques, especially for routine screening or controls.

An automated OPA derivatization and flow injection analysis for rapid (<1 min) histamine determination has been developed to screen fish and seafood products (though not tested for meat products) that does not include a sample cleanup other than extraction and crude filtration [99].

An enzymatic method has been described specifically for histamine determination. The procedure involves the use of amine-specific enzymes that recognize and rapidly transform the substrate into another measurable product. In the presence of oxygen, diaminooxidase (DAO) deaminates

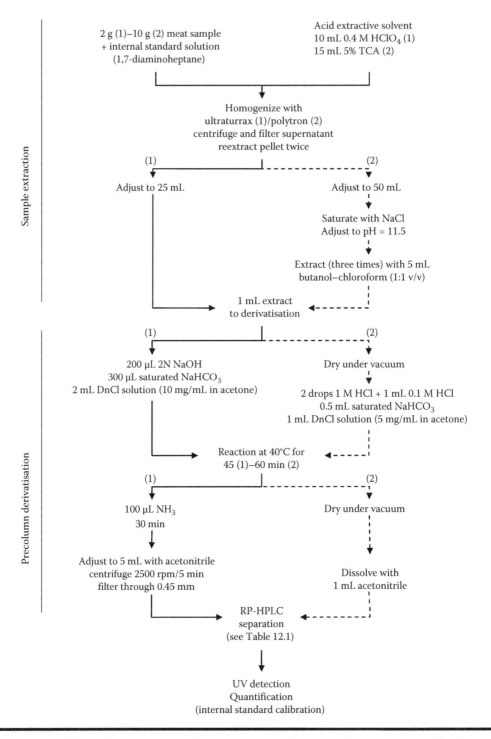

Figure 12.2 Schematic protocol for biogenic amine determination by dansyl chloride pre-column derivatization as described in (1) (Eerola, S., Hinkkanen, R., Lindfors, E. and Hirvi, T., *J. AOAC Int.*, 76(3), 575–577, 1993) and (2) (Moret, S., Conte, L., and Callegarin, F., *Ind. Aliment.*, 35(349), 650–657, 1996).

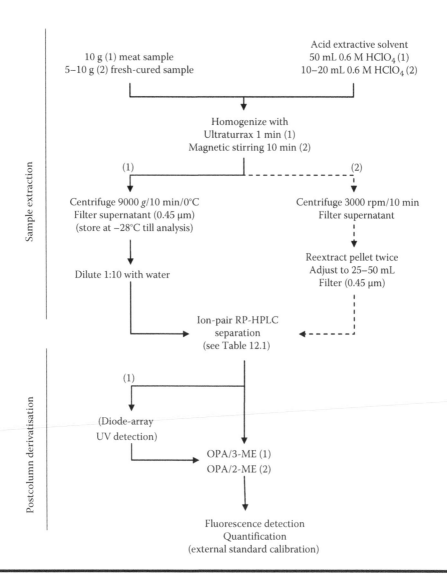

Figure 12.3 Schematic protocol for biogenic amine determination by *o*-phthaldialdehyde postcolumn derivatization as described in (1) (Straub, B., Schollenberger, M., Kicherer, M., Luckas, B., and Hammes, W.P., *Z. Lebensm. Unters. For.*, **197**(3), 230–232, 1993) and (2) (Hernández-Jover, T., Izquierdo-Pulido, M., Veciana-Nogués, M.T. and Vidal-Carou, M.C., *J. Agr. Food Chem.*, **44**(9), 2710–2715, 1996).

histamine, thereby forming hydrogen peroxide, which, coupled with horseradish peroxidase (HRP), converts a reduced dye (leucocrystal violet) to its oxidized form (crystal violet). The accompanying color development allows colorimetric quantification. This methodology was initially developed for detecting histamine in fish from a neutralized extract [100], and was reported to be suitable for routine analysis, providing simplicity, and speed. However, it tended to overestimate histamine concentrations below 10 mg/kg [101]. A number of limitations have been reported for this technique. Although DAO can also act on other biogenic amines, only little absorbance is developed by tyramine, and no change in the absorbance value for histamine is observed when

Table 12.1 Conditions of Some of the Most Used Chromatographic Procedures to Determine Biogenic Amines in Meat and Meat Products

	Eerola et al. [81]	Moret and Conte [80]	Straub et al. [75]	Hernández-Jover et al. [29]
BA determined	TY, HI, PHE, TR, SE, PU, CA, SD, SM 1,7-Diaminoheptane (IS)	TY, HI, PHE, TR, PU, CA, SD, SM[b] 1,7-Diaminoheptane (IS)	TY, HI, PHE, PU, CA	TY, HI, PHE, TR, SE, OC, DO, PU, CA, AG, SD, SM
Meat sample assayed	Dry sausages	Salami	Minced meat (mixed lean pork, lean beef, and pork fat)	Fresh meat Cooked product Ripened (fermented) product
Sample preparation	Acid extraction	Acid extraction plus clean-up (liquid–liquid partitioning)	Acid extract	Acid extract
Precolumn derivatization	Dansyl chloride	Dansyl chloride, dry, and redisolve	—	—
Column	Reverse phase (Spherisorb ODS2)	Reverse phase (Spherisorb 3s TG)	Reverse-phase (Nucleosil 100 7C18)	Reverse phase (NovaPak C18)
Mobile phase	Solvent A (0.1 M ammonium acetate) Solvent B (acetonitrile)	Solvent A (water) Solvent B (acetonitrile)	Solvent A (0.05 M hexanesulfonic acid [ion pair] and 0.1 M KH_2PO_4, pH 3.5) Solvent B (solvent A/acetonitrile [3/1])	Solvent A (0.01 M sodium octanesulfonate [ion pair] and 0.1 M sodium acetate, pH 5.20) Solvent B (solvent C [0.2 M sodium acetate and 0.01 M sodium octanesulfonate, pH 4.5] and acetonitrile [6.6/3.4])
Elution	Gradient (19 min + 10 min equilibration)	Gradient (12 min + equilibration)	Gradient (total time 49 min) (ca. 100 min to separate and resolve amino acids)	Gradient (54 min + 10 min equilibration)

	At 1.0 mL/min flow	At 0.8 mL/min flow	Flow not specified	At 1.0 mL/min
Postcolumn derivatizing reagent	—	—	0.5 M borate buffer OPA/3-ME Brij 35	Borate buffer OPA/2-ME Brij 35
Detection	UV 254 nm	UV 254 nm	UV diodearray detector (before derivatization) Fluorescence after derivatization Ex 340 nm; Em 455 nm	Fluorescence after derivatization Ex 340 nm; Em 445 nm
Accuracy (percentage recovery)	TY, 98; HI, 96; PHE, 104; TR, 84; SE, 56; PU, 90; CA, 101; SD, 91; SM, 90	TY, 85; HI, 76; PHE, 87; TR, nqª; PU, 2.9; CA, 82; SD, 67; SM, 72	TY, 102; HI, 113; PHE, 96; PU, 111; CA, 108	TY, 96; HI, 98; PHE, 95; TR, 93; SE, 93; OC, 97; DO, 98; PU, 99; CA, 97; AG, 99; SD, 99; SM, 100
Precision (percentage [RSD], for Straub et al. is SD)	TY, 4; HI, 4; PHE, 6; TR, 4; SE, 5; PU, 18; CA, 7; SD, 9; SM, 5	TY, 2.5; HI, 3.2; PH, 2.6; TR, nq PU, 4.0; CA, 2.6; SD, 4.2; SM, 6.0	TY, 9.9; HI, 10.8; PHE, 9.1; PU, 9.5; CA, 10.1	TY, 3.22; HI, 3.9; PHE, 3.1; TR, 4.8; SE, 4.2; OC, 3.7; DO, 6.1; PU, 4.1; CA, 4.7; AG, 5.1; SD, 3.4; SM, 3.1
Sensitivity	Determination limit 1 mg/kg (TY, CA, SD, SM) 2 mg/kg (PHE, TR, PU) 5 mg/kg (SE)	Not studied	Detection limit (0.5 mg/kg)	Determination limit <1.00 mg/kg (TY, HI, PHE, TR, OC, DO, PU, CA, AG) <1.50 mg/kg (SE, SM)

Note: TY, tyramine; HI, histamine; PHE, phenylethylamine; TR, tryptamine; SE, serotonine; OC, octopamine; DO, dopamine; PU, putrescine; CA, cadaverine; AG, agmatine; SD, spermidine; SM, spermine; IS, Internal Standard; Ex, Excitation; and En, Emission.

ª Not quantificable due to interfering peaks.

equimolar solutions of diamines and histamine are determined [102]. These limitations imply that although other biogenic amines give little interference, this technique is not useful to detect amines other than histamine. Alternative specific amine–oxidase enzymes have also been applied in rapid tests to screen for the presence of other biogenic amines. A specific biosensor for tyramine was constructed either with a monoamine oxidase (MAO), from *Aspergillus niger* and beef plasma immobilized in a collagen membrane [103], or with a tyramine–oxidase (from *Micrococcus luteus*) immobilized on porous microglass beads [58,66]. As a result, tyramine is oxidized to aldehydes, and the oxygen consumption is monitored amperometrically with an oxygen electrode detector. These tyramine biosensors have been used to estimate bacterial spoilage during meat storage.

DAO from porcine kidney immobilized onto a porous nylon membrane attached to an amperometric electrode has been used to estimate the total concentrations of histamine, cadaverine, and putrescine accumulated in fish fillets during storage [104]. DAO from peas (*Cicer arietinum*) seem to be more selective to the diamines putrescine and cadaverine [105] and could be used together with other sources of DAO and MAO to distinguish the spoilage pattern [106].

An enzyme sensor array has been developed to simultaneously determine histamine, tyramine, and cadaverine, with a combination of specific amine oxidases of distinct origin [90]. The cross-reactivities of these enzymes against several biogenic amines were characterized, and data were included in an artificial neural network for pattern recognition. The best discrimination was obtained for samples containing tyramine (91%), followed by histamine (75%) and putrescine (57%). The use of enzymes exhibiting higher specific activities would improve the biosensor system.

An alternative potentiometric (nonenzymatic) sensor to measure putrescine has been proposed for monitoring pork freshness [107]. Sample pretreatment is required before analysis of biogenic amines, and therefore this system cannot be used as an online sensor.

An immunological approach has also been developed for the analysis of histamine. This method is simple, rapid, and relatively low-cost in comparison with HPLC. Commercial kits are already available to analyze histamine from aqueous food extracts (e.g., fish, cheese, or sausage) through an enzyme immunoassay. However, the antibodies used in these tests require chemical derivatization of histamine before analysis (propionic acid esters), or require toxic reagents (*p*-benzoquinone), both of which are time-consuming. Alternatively, polyclonal antihistamine antibodies recognizing intact histamine have been included in commercial competitive direct enzyme-linked immunosorbent (CD-ELISA) test kits [108] (e.g., R-Biopharm GmbH, Darmstadt, Germany, or Veratox® histamine test from NEOGEN Corporation, Lausing, MI, USA). Because a number of nonpurified aqueous extracts can be analyzed simultaneously in a microtiter plate, the maximum daily throughput of the CD-ELISA method is much higher than by HPLC. The CD-ELISA for the detection of histamine in fish [108,109], cheese [110], and other dairy products [108] as well as in wine [111] is a suitable alternative method that provides results comparable with the official fluorometric or HPLC methods. However, no studies have addressed the application of CD-ELISA in meat and meat products.

References

1. Ten Brink, B., Damink, C., Joosten, H.M.L.J. and Huis in 't Veld, J.H.J. 1990. Occurrence and formation of biologically active amines in foods. *International Journal of Food Microbiology* 11(1), 73–84.
2. Halász, A., Baráth, A., Simon-Sarkadi, L. and Holzapfel, W. 1994. Biogenic amines and their production by microorganisms in food. *Trends in Food Science and Technology* 5(2), 42–49.

3. Kalac, P. and Krausova, P. 2005. A review of dietary polyamines: formation, implications for growth and health and occurrence in foods. *Food Chemistry* 90(1–2), 219–230.

4. Bardócz, S. 1995. Polyamines in food and their consequences for food quality and human health. *Trends in Food Science and Technology* 6(10), 341–346.

5. Marcobal, A., de las Rivas, B. and Muñoz, R. 2006. First genetic characterization of a bacterial β-phenylethylamine biosynthetic enzyme in *Enterococcus faecium* RM58. *FEMS Microbiology Letters* 258(1), 144–149.

6. Taylor, S.L. 1986. Histamine food poisoning: toxicology and clinical aspects. *CRC Critical Reviews in Toxicology* 17(2), 91–128.

7. Mariné-Font, A., Vidal-Carou, M.C., Izquierdo-Pulido, M., Veciana-Nogués, M.T. and Hernández-Jover, T. 1995. Les amines biògenes dans les aliments: leur signification, leur analyse. *Annales des Falsifications, de l'Expertise Chimique et Toxicologique* 88, 119–140.

8. Shalaby, A.R. 1996. Significance of biogenic amines to food safety and human health. *Food Research International* 29(7), 675–690.

9. Amon, U., Bangha, E., Kuster, T., Menne, A., Vollrath, I.B. and Gibbs, B.F. 1999. Enteral histaminosis: clinical implications. *Inflammation Research* 48(6), 291–195.

10. Stratton, J.E., Hutkins, R.W. and Taylor, S.L. 1991. Biogenic amines in cheese and other fermented foods. *Journal of Food Protection* 54, 460–470.

11. Sumner, J., Ross, T. and Ababouch, L. 2004. Application of risk assessment in the fish industry. In FAO Fisheries Technical Paper, Vol. 442, p. 78. Rome: FAO.

12. Lehane, L. and Olley, J. 2000. Histamine fish poisoning revisited. *International Journal of Food Microbiology* 58(1–2), 1–37.

13. Korn, A., Da Prada, M., Raffesberg, W., Allen, S. and Gasic, S. 1988. Tyramine pressor effect in man: studies with moclobemide, a novel, reversible monoamine oxidase inhibitor. *Journal of Neural Transmission Supplement* 26, 57–71.

14. Patat, A., Berlin, I., Durrieu, G., Armand, P., Fitoussi, S., Molinier, P. and Caille, P. 1995. Pressor effect of oral tyramine during treatment with befloxatone, a new reversible monoamine oxidase-A inhibitor, in healthy subjects. *Journal of Clinical Pharmacology* 35(6), 633–643.

15. Dingemanse, J., Wood, N., Guentert, T., Oie, S., Ouwekerk, M. and Amrein, R. 1998. Clinical pharmacology of moclobemide during chronic administration of high doses to healthy subjects. *Psychopharmacology* 140(2), 164–172.

16. Taylor, S.L., Stratton, J.E. and Nordlee, J.A. 1989. Histamine poisoning (scombroid fish poisoning): an allergy-like intoxication. *Journal of Toxicology Clinical Toxicology*, 27(5–7), 225–240.

17. Veciana-Nogués, M.T. and Vidal-Carou, M.C. 2006. Dietas bajas en histamina. In *Nutrición y dietética clínica*, J. Salas-Salvadó, A. Bomada, R. Trallero and M.E. Saló, Eds., Elsevier-Masson, Barcelona, Chapter 46, pp. 405–408.

18. Hui, J.Y. and Taylor, S.L. 1985. Inhibition of invivo histamine-metabolism in rats by foodborne and pharmacologic inhibitors of diamine oxidase, histamine n-methyltransferase, and monoamine-oxidase. *Toxicology and Applied Pharmacology* 81(2), 241–249.

19. Warthesen, J.J., Scanlan, R.A., Bills, D.D. and Libbey, L.M. 1975. Formation of heterocyclic N-nitrosamines from the reaction of nitrite and selected primary diamines and amino acids. *Journal of Agricultural and Food Chemistry* 23(5), 898–902.

20. Ochiai, M., Wakabayashi, K., Nagao, M. and Sigimura, T. 1984. Tyramine is a major mutagen precursor in soy sauce, being convertible to a mutagen by nitrite. *Gann* 75(1), 1–3.

21. Bover-Cid, S., Hugas, M., Izquierdo-Pulido, M. and Vidal-Carou, M.C. 2001. Amino acid-decarboxylase activity of bacteria isolated from fermented pork sausages. *International Journal of Food Microbiology* 66(3), 185–189.

22. Durlu-Ozkaya, F., Ayhan, K. and Vural, N. 2001. Biogenic amines produced by *Enterobacteriaceae* isolated from meat products. *Meat Science* 58(2), 163–166.

23. Maijala, R. and Eerola, S. 1993. Contaminant lactic acid bacteria of dry sausages produce histamine and tyramine. *Meat Science* 35(3), 387–395.

24. Masson, F., Montel, M.C. and Talon, R. 1996. Histamine and tyramine production by bacteria from meat products. *International Journal of Food Microbiology* 32(1–2), 199–207.

25. Ansorena, D., Montel, M.C., Rokka, M., Talon, R., Eerola, S., Rizzo, A., Raemaekers, M. and Demeyer, D. 2002. Analysis of biogenic amines in northern and southern European sausages and role of flora in amine production. *Meat Science* 61(2), 141–147.

26. Silva, C.M.G. and Gloria, M.B.A. 2002. Bioactive amines in chicken breast and thigh after slaughter and during storage at 4+/–1°C and in chicken-based meat products. *Food Chemistry* 78(2), 241–248.

27. Wortberg, B. and Woller, R. 1982. Qualität und Frische von Fleisch und Fleischwaren im Hinblick auf ihren Gehalt an biogenen Aminen. *Fleischwirtschaft* 62(11), 1457–1460, 1463.

28. Maijala, R., Eerola, S., Aho, M. and Hirn, J.A. 1993. The effect of GDL-induced pH decrease on the formation of biogenic amiens in meat. *Journal of Food Protection* 56(2), 125–129.

29. Hernández-Jover, T., Izquierdo-Pulido, M., Veciana-Nogués, M.T. and Vidal-Carou, M.C. 1996. Ion-pair high-performance liquid chromatographic determination of biogenic amines in meat and meat products. *Journal of Agricultural and Food Chemistry* 44(9), 2710–2715.

30. Hernández-Jover, T., Izquierdo-Pulido, M., Veciana-Nogués, M.T., Mariné-Font, A. and Vidal-Carou, M.C. 1997. Biogenic amine and polyamine contents in meat and meat products. *Journal of Agricultural and Food Chemistry* 45(6), 2098–2102.

31. Kalac, P. 2006. Biologically active polyamines in beef, pork and meat products: a review. *Meat Science* 73(1), 1–11.

32. Bardócz, S., Grant, G., Brown, D.S., Ralph, A. and Pusztai, A. 1993. Polyamines in food - implications for growth and health. *Journal of Nutritional Biochemistry* 4(2), 66–71.

33. Lakritz, L., Spinelli, A.M. and Wassermann, A.E. 1975. Determination of amines in fresh and processed pork. *Journal of Agricultural and Food Chemistry* 23(2), 344–346.

34. Sayem El Daher, N., Simard, R.E., Fillion, J. and Roberge, A.G. 1984. Extraction and determination of biogenic amines in ground beef and their relation to microbial quality. *Lebensmittel-Wissenschaft&Technologie* 17(1), 20–23.

35. Paulsen, P. and Bauer, F. 2007. Spermine and spermidine concentrations in pork loin as affected by storage, curing and thermal processing. *European Food Research and Technology* 225(5), 921–924.

36. Vidal-Carou, M.C., Izquierdo-Pulido, M.L., Martín-Morro, M.C. and Mariné-Font, A. 1990. Histamine and tyramine in meat products: relationship with meat spoilage. *Food Chemistry* 37(4), 239–249.

37. Arnau, J. 1998. Tecnología del jamón curado en distintos países. Eurocarne, Eds., Estrategias Alimentarias, S.L., Madrid Simposio especial 44th ICoMST, pp. 10–21.

38. Alfaia, C.M., Castro, M.F., Reis, V.A., Prates, J.M., de Almeida, I.T., Correia, A.D. and Dias, M.A. 2004. Changes in the profile of free amino acids and biogenic amines during the extended short ripening of portuguese dry-cured ham. *Food Science and Technology International* 10(5), 297–304.

39. Saccani, G., Tanzi, E., Pastore, P., Cavalli, S. and Rey, A. 2005. Determination of biogenic amines in fresh and processed meat by suppressed ion chromatography-mass spectrometry using a cation-exchange column. *Journal of Chromatography A* 1082(1), 43–50.

40. Virgili, R., Saccani, G., Gabba, L., Tanzi, E. and Soresi Bordini, C. 2007. Changes of free amino acids and biogenic amines during extended ageing of Italian dry-cured ham. *Lebensmittel-Wissenschaft Und-Technologie–Food Science and Technology* 40(5), 871–878.

41. Hortós, M. and García-Regueiro, J.A. 1998. Biogenic amines in two processes of Spanish dry-cured ham. In Proceeding of *44th International Congress of Meat Science and Technology (ICoMST)*, Vol. 1, pp. 396–397.

42. Córdoba, J., Antequera-Rojas, T., García-González, C., Ventanas-Barroso, J., López-Bote, C. and Asensio, M. 1994. Evolution of free amino acids and amines during ripening of Iberian cured ham. *Journal of Agricultural and Food Chemistry* 42, 2296–2301.

43. Bover-Cid, S., Schoppen, S., Izquierdo-Pulido, M. and Vidal-Carou, M.C. 1999. Relationship between biogenic amine contents and the size of dry fermented sausages. *Meat Science* 51(4), 305–311.

44. Suzzi, G. and Gardini, F. 2003. Biogenic amines in dry fermented sausages: a review. *International Journal of Food Microbiology* 88(1), 41–54.

45. Miguélez-Arrizado, M.J., Bover-Cid, S., Latorre-Moratalla, M.L. and Vidal-Carou, M.C. 2006. Biogenic amines in Spanish fermented sausages as a function of diameter and artisanal or industrial origin. *Journal of the Science of Food and Agriculture*, 86(4), 549–557.

46. Vidal-Carou, M.C., Latorre-Moratalla, M.L., Veciana-Nogués, M.T. and Bover-Cid, S. 2007. Biogenic amines: risks and control. In *Handbook of Fermented Meat and Poultry*, F. Todrà, Y.H. Hui, I. Astiasarán, Wai-Kit Nip, J.G. Sebranek, E.T.F. Silveira, L.H. Stahnke and R. Talon, Eds., Blackwell Publishing, Oxford, Chapter 43, pp. 455–468.

47. Latorre-Moratalla, M.L., Veciana-Nogues, T., Bover-Cid, S., Garriga, M., Aymerich, T., Zanardi, E., Ianieri, A., Fraqueza, M.J., Patarata, L., Drosinos, E.H., Laukova, A., Talon, R. and Vidal-Carou, M.C. 2008. Biogenic amines in traditional fermented sausages produced in selected European countries. *Food Chemistry* 107(2), 912–921.

48. Hernández-Jover, T., Izquierdo-Pulido, M., Veciana-Nogués, M.T., Mariné-Font, A. and Vidal-Carou, M.C. 1997. Effect of starter cultures on biogenic amine formation during fermented sausage production. *Journal of Food Protection* 60(7), 825–830.

49. Bover-Cid, S., Izquierdo-Pulido, M. and Vidal-Carou, M.C. 2001. Changes in biogenic amine and polyamine contents in slightly fermented sausages manufactured with and without sugar. *Meat Science* 57(2), 215–221.

50. Bover-Cid, S., Miguelez-Arrizado, J. and Vidal-Carou, M.C. 2001. Biogenic amine accumulation in ripened sausages affected by the addition of sodium sulphite. *Meat Science* 59(4), 391–396.

51. Sayem El Daher, N. and Simard, R.E. 1985. Putrefactive amine changes in relation to microbial counts of ground beef during storage. *Journal of Food Protection* 48(4), 54–58.

52. Slemr, J. and Beyermann, K. 1985. Concentration profiles of diamines in fresh and aerobically stored pork and beef. *Journal of Agricultural and Food Chemistry* 33(3), 336–339.

53. Hernández-Jover, T., Izquierdo-Pulido, M., Veciana-Nogués, M.T. and Vidal-Carou, M.C. 1996. Biogenic amine sources in cooked cured shoulder pork. *Journal of Agricultural and Food Chemistry* 44(10), 3097–3101.

54. Nakamura, M., Wada, Y., Sawaya, H. and Kawabata, T. 1979. Polyamine content in fresh and processed pork. *Journal of Food Science* 44, 515–523.

55. Maijala, R., Nurmi, E. and Fischer, A. 1995. Influence of processing temperature on the formation of biogenic amines in dry sausages. *Meat Science* 39(1), 9–22.

56. Mietz, J.L. and Karmas, E. 1977. Chemical quality index of canned tuna as determined by high-pressure liquid chromatography. *Journal of Food Science* 42(1), 155–158.

57. Yamanaka, H., Matsumoto, M. and Yano, Y. 1989. Polyamines as potential indexes for decomposition of pork, beef and chicken. *Journal of Food Hygiene and Society of Japan* 30(5), 401–405.

58. Yano, Y., Murayama, F., Kataho, N., Tachibana, M. and Nakamura, T. 1992. Evaluation of the quality of meat and fermented dairy and meat products by tyramine sensor. *Animal Science and Technology (Japan)* 63, 970–977.

59. Chen, C.M., Lin, L.C. and Yen, G.G. 1994. Relationship between changes in biogenic amine contents and freshness of pork during storage at different temperatures. *Journal of the Chinese Agricultural Chemical Society* 32, 47–60.

60. Vinci, G. and Antonelli, M.L. 2002. Biogenic amines: quality index of freshness in red and white meat. *Food Control* 13(8), 519–524.

61. Schmitt, R.E., Haas, J. and Amado, R. 1988. Bestimmung von biogenen Aminen mit RP-HPLC zur Erfassung des mikrobiellen Verderbs von Schlachtgefluegel. *Zeitschrift für Lebensmittel-Untersuchung und -Forschung* 187(2), 121–124.

62. Patsias, A., Chouliara, I., Paleologos, E.K., Savvaidis, I. and Kontominas, M.G. 2006. Relation of biogenic amines to microbial and sensory changes of precooked chicken meat stored aerobically and under modified atmosphere packaging at 4°C. *European Food Research and Technology* 223(5), 683–689.

63. Rokka, M., Eerola, S., Smolander, M., Alakomi, H.L. and Ahvenainen, R. 2004. Monitoring of the quality of modified atmosphere packaged broiler chicken cuts stored in different temperature conditions B. Biogenic amines as quality-indicating metabolites. *Food Control* 15(8), 601–607.

64. Balamatsia, C.C., Paleologos, E.K., Kontominas, M.G. and Savvaidis, I. 2006. Correlation between microbial flora, sensory changes and biogenic amine formation in fresh chicken meat stroed aerobically or under modified atmosphere packaging at 4°C: possible role of iogenic amines as spoilage indicators. *Antonie van Leeuwenhoek* 89, 9–17.

65. Edwards, R.A., Dainty, R.H., Hibard, C.M. and Ramantanis, S.V. 1987. Amines in fresh beef of normal pH and the role of bacteria in changes in concentration observed during storage in vacuum packs at chill temperatures. *Journal of Applied Bacteriology* 63, 427–434.

66. Yano, Y., Kataho, N., Watanabe, M., Nakamura, T. and Asano, Y. 1995. Changes in the concentration of biogenic amines and application of tyramine sensor during storage of beef. *Food Chemistry* 54(2), 155–159.

67. Izquierdo-Pulido, M., Veciana-Nogués, M.T., Mariné-Font, A. and Vidal-Carou, M.C. 1999. Polyamine and biogenic amine evolution during food processing. In *Polyamines in Health and Nutrition*, S. Bardócz and A. White, Eds., Kluwer Academic Publishers, London, Chapter 12, pp. 139–159.

68. Bover-Cid, S., Hugas, M., Izquierdo-Pulido, M. and Vidal-Carou, M.C. 2000. Reduction of biogenic amine formation using a negative amino acid-decarboxylase starter culture for fermentation of Fuet sausages. *Journal of Food Protection* 63(2), 237–243.

69. Bover-Cid, S., Izquierdo-Pulido, M. and Vidal-Carou, M.C. 2001. Effectiveness of a *Lactobacillus sakei* starter culture in the reduction of biogenic amine accumulation as a function of the raw material quality. *Journal of Food Protection* 64(3), 367–373.

70. Bover-Cid, S., Izquierdo-Pulido, M. and Vidal-Carou, M.C. 2000. Influence of hygienic quality of raw materials on biogenic amine production during ripening and storage of dry fermented sausages. *Journal of Food Protection* 63(11), 1544–1550.

71. Bover-Cid, S., Hernandez-Jover, T., Miguelez-Arrizado, M.J. and Vidal-Carou, M.C. 2003. Contribution of contaminant enterobacteria and lactic acid bacteria to biogenic amine accumulation in spontaneous fermentation of pork sausages. *European Food Research and Technology* 216(6), 477–482.

72. Moret, S., Conte, L. and Callegarin, F. 1996. Determination of biogenic amines in fish and meat foods. *Industrie Alimentari* 35(349), 650–657.

73. Zee, J.A., Simard, R.E. and L'Heureux, L. 1983. Evaluation of analytical methods for determination of biogenic amines in fresh and processed meat. *Journal of Food Protection* 46(12), 1044–1049.

74. Hurst, W.J. 1990. A review of HPLC methods for the determination of selected biogenic amines in foods. *Journal of Liquid Chromatography* 13(1), 1–23.

75. Straub, B., Schollenberger, M., Kicherer, M., Luckas, B. and Hammes, W.P. 1993. Extraction and determination of biogenic amines in fermented sausages and other meat products using reversed-phase-HPLC. *Zeitschrift für Lebensmittel-Untersuchung und-Forschung* 197(3), 230–232.

76. Smela, D., Pechova, P., Komprda, T., Klejdus, B. and Kuban, V. 2004. Chromatographic determination of biogenic amines in meat products during fermentation and long-term storage. *Chemicke Listy* 98(7), 432–437.

77. Treviño, E., Beil, D. and Steinhart, H. 1997. Determination of biogenic amines in mini-salami during long-term storage. *Food Chemistry* 58(4), 385–390.

78. Lange, J., Thomas, K. and Wittmann, C. 2002. Comparison of a capillary electrophoresis method with high-performance liquid chromatography for the determination of biogenic amines in various food samples. *Journal of Chromatography B* 779(2), 229–239.

79. Molins-Legua, C. and Campins-Falco, P. 2005. Solid phase extraction of amines. *Analytica Chimica Acta* 546(2), 206–220.

80. Moret, S. and Conte, L.S. 1996. High-performance liquid chromatographic evaluation of biogenic amines in foods an analysis of different methods of sample preparation in relation to food characteristics. *Journal of Chromatography A* 729(1–2), 363–369.

81. Eerola, S., Hinkkanen, R., Lindfors, E. and Hirvi, T. 1993. Liquid chromatographic determination of biogenic amines in dry sausages. *Journal of AOAC International* 76(3), 575–577.

82. Santos-Buelga, C., Nogales-Alarcon, A. and Mariné-Font, A. 1981. A method for the analysis of tyramine in meat products: its content in some Spanish samples. *Journal of Food Science* 46(6), 1794–1795.

83. Shalaby, A.R. 1999. Simple rapid and valid thin layer chromatographic method for determining biogenic amines in foods. *Food Chemistry* 65, 117–121.

84. Slemr, J. and Beyermann, K. 1984. Determination of biogenic amines in meat by combined ion-exchange and capillary gas chromatography. *Journal of Chromatography A* 283, 241–250.

85. Paleologos, E.K. and Kontominas, M.G. 2004. On-line solid-phase extraction with surfactant accelerated on-column derivatization and micellar liquid chromatographic separation as a tool for the determination of biogenic amines in various food substrates. *Analytical Chemistry* 76(5), 1289–1294.

86. Draisci, R., Giannetti, L., Boria, P., Lucentini, L., Palleschi, L. and Cavalli, S. 1998. Improved ion chromatography-integrated pulsed amperometric detection method for the evaluation of biogenic amines in food of vegetable or animal origin and in fermented foods. *Journal of Chromatography A* 798(1–2), 109–116.

87. Favaro, G., Pastore, P., Saccani, G. and Cavalli, S. 2007. Determination of biogenic amines in fresh and processed meat by ion chromatography and integrated pulsed amperometric detection on Au electrode. *Food Chemistry* 105(4), 1652–1658.

88. Alberto, M.R., Arena, M.E. and Manca de Nadra, M.C. 2002. A comparative survey of two analytical methods for identification and quantification of biogenic amines. *Food Control* 13(2), 125–129.

89. Tamim, N.M., Bennett, L.W., Shellem, T.A. and Doerr, J.A. 2002. High-performance liquid chromatographic determination of biogenic amines in poultry carcasses. *Journal of Agricultural and Food Chemistry* 50(18), 5012–5015.

90. Lange, J. and Wittmann, C. 2002. Enzyme sensor array for the determination of biogenic amines in food samples. *Analytical and Bioanalytical Chemistry* 372(2), 276–283.

91. Ruiz-Jiménez, J. and Luque De Castro, M.D. 2006. Pervaporation as interface between solid samples and capillary electrophoresis: determination of biogenic amines in food. *Journal of Chromatography A* 1110(1–2), 245–253.

92. Onal, A. 2007. A review: current analytical methods for the determination of biogenic amines in foods. *Food Chemistry* 103(4), 1475–1486.

93. Molnár-Perl, I. 2003. Quantitation of amino acids and amines in the same matrix by high-performance liquid chromatography, either simultaneously or separately. *Journal of Chromatography A* 987(1–2), 291–309.

94. Chiu, T.-C., Lin, Y.-W., Huang, Y.-F. and Chang, H.-T. 2006. Analysis of biologically active amines by CE. *Electrophoresis* 27(23), 4792–4807.

95. Kvasnicka, F. and Voldrich, M. 2006. Determination of biogenic amines by capillary zone electrophoresis with conductometric detection. *Journal of Chromatography A* 1103(1), 145–149.

96. IzquierdoPulido, M., Vidal-Carou, M.C. and Mariné-Font, A. 1993. Determination of biogenic amines in beers and their raw materials by ion-pair liquid chromatography with post-column derivatisation. *Journal of AOAC International* 76(5), 1027–1032.

97. Vallé, M., Malle, P. and Bouquelet, S. 1997. Optimization of a liquid chromatographic method for determination of amines in fish. *Journal of AOAC International* 80(1), 49–56.

98. Salazar, M.T., Smith, T.K. and Harris, A. 2000. High-performance liquid chromatographic method for determination of biogenic amines in feedstuffs, complete feeds, and animal tissues. *Journal of Agricultural and Food Chemistry* 48(5), 1708–1712.

99. Hungerford, J.M., Hollingworth, T.A. and Wekell, M.M. 2001. Automated kinetics-enhanced flow-injection method for histamine in regulatory laboratories: rapid screening and suitability requirements. *Analytica Chimica Acta* 438(1–2), 123–129.

100. Lerke, P.A., Porcuna, M.N. and Chin, H.B. 1983. Screening test for histamine in fish. *Journal of Food Science* 48(1), 155–157.

101. Ben-Gigirey, B., Craven, C. and An, H. 1998. Histamine formation in albacore muscle analyzed by AOAC and enzymatic methods. *Journal of Food Science* 63(2), 210–214.

102. Sumner, S. and Taylor, S. 1989. Detection method for histamina-producing, dairy-related bacteria using diamine oxidase and leucocrystal violet. *Journal of Food Protection* 52, 105–108.

103. Karube, I., Satoh, I., Araki, Y., Suzuki, S. and Yamada, H. 1980. Monoamine oxidase electrode in freshness testing of meat. *Enzyme and Microbial Technology* 2(2), 117–120.

104. Male, K.B., Bouvrette, P., Luong, J.H.T. and Gibbs, B.F. 1996. Amperometric biosensor for total histamine, putrescine and cadaverine using diamine oxidase. *Journal of Food Science* 61(5), 1012–1016.

105. Hall, M., Sykes, P.A., Fairclough, D.L., Lucchese, L.J., Rogers, P., Staruszkiewicz, W.F. and Bateman, R.C. 1999. A test strip for diamines in tuna. *Journal of AOAC International* 82(5), 1102–1108.

106. Tombelli, S. and Mascini, M. 1998. Electrochemical biosensors for biogenic amines: a comparison between different approaches. *Analytica Chimica Acta* 358(3), 277–284.

107. Kaneki, N., Miura, T., Shimada, K., Tanaka, H., Ito, S., Hotori, K., Akasaka, C., Ohkubo, S. and Asano, Y. 2004. Measurement of pork freshness using potentiometric sensor. *Talanta* 62(1), 215–219.

108. Leszczynska, J., Wiedlocha, M. and Pytasz, U. 2004. The histamine content in some samples of food products. *Czech Journal of Food Science* 22, 81–86.

109. Rogers, P.L. and Staruszkiewicz, W.F. 2000. Histamine test kit comparison. *Journal of Aquatic Food Product Technology* 9(2), 5–17.

110. Aygün, O., Schneider, E., Scheuer, R., Usleber, E., Gareis, M. and Martlbauer, E. 1999. Comparison of ELISA and HPLC for the determination of histamine in cheese. *Journal of Agricultural and Food Chemistry* 47(5), 1961–1964.

111. Marcobal, A., Polo, M.C., Martín-Alvarez, P.J. and Moreno-Arribas, M.V. 2005. Biogenic amine content of red Spanish wines: comparison of a direct ELISA and an HPLC method for the determination of histamine in wines. *Food Research International* 38(4), 387–394.

Chapter 13

Nitrosamines

Susanne Rath and Felix Guillermo Reyes Reyes

Contents

13.1 Introduction 421
13.2 Chemistry 422
13.3 Formation and Occurrence in Meat and Meat Products 424
13.4 Toxicological Aspects 425
13.5 Regulatory Aspects.. 425
13.6 Analytical Aspects... 426
     13.6.1 Sample Preparation... 427
          13.6.1.1 Distillation and Clean-Up Procedures 427
          13.6.1.2 Solvent Extraction Followed by Cleanup Using Liquid–Solid
                   Extraction 432
          13.6.1.3 Matrix Solid-Phase Dispersion and Liquid–Liquid Extraction............ 432
          13.6.1.4 Solid-Phase Microextraction 433
          13.6.1.5 Supercritical Fluid Extraction 434
          13.6.1.6 Quantitation Methods 435
13.7 Conclusions.. 437
References 437

13.1 Introduction

Nitrosamines are N-nitroso compounds that have received considerable attention worldwide during the past half century, since Barnes and Magee[1] first reported in 1954 the association between dimethylnitrosamine (NDMA) and liver damage in rats. Two years later, the same British scientists confirmed the induction of liver tumors in rats by feeding them NDMA.[2]

During the period of 1957–1962, liver disorders, including cancer, in various farm animals in Norway were attributed to herring meal that had been preserved by the addition of large amounts of sodium nitrite.[3] Further investigations showed that the fishmeal was contaminated with NDMA, which was formed as a result of a chemical reaction between dimethylamine, a commonly occurring amine in this meal, and a nitrosating agent formed from sodium nitrite. This finding led to the idea that nitrosamines might also occur in human food through the interaction between naturally occurring or added precursor compounds. This was the beginning of a world-wide investigation of the presence of nitrosamines in several matrices, including foodstuffs. As a result, NDMA was detected by European scientists in beer.[4] Since then, nitrosamines have been found in a large variety of products such as foods (in particular, cured meat products), alcoholic beverages, water, soil, air, tobacco, rubber products, pesticides, cosmetics, and drugs. Nowadays, it is well established that nitrosamines are potential carcinogenic compounds.[5]

Although the occurrence of nitrosamines in food products was reported before 1970, some of these early results are untrustworthy, due to the lack of a reliable analytical method available at that time that could identify and determine nitrosamines at the low concentration level required, because many of the methods then available had limits of detection above the levels of nitrosamines now known to be present in foods. This situation was overcome with the development of analytical methodologies for the determination of volatile nitrosamines by gas chromatography (GC) associated with thermal energy analyzer (TEA) or mass spectrometric (MS) detection devices. The number of scientific papers reporting the presence of volatile nitrosamines in meat products peaked in the 1980s. It is worth emphasizing that most of these studies were conducted in the United States, Canada, Germany, and Japan.

This chapter will provide some insight on the chemistry, formation, and occurrence of nitrosamines in meat products, as well as toxicological information, the main focus being analytical aspects.

13.2 Chemistry

N-nitrosamines are aliphatic or aromatic compounds, which have a nitroso functional group attached to nitrogen. The chemical and physical properties depend on the substituents (R_1 and R_2) on the amine nitrogen. Whereas the low molar mass dialkylnitrosamines are water-soluble liquids, the high molar mass nitrosamines are soluble in organic solvents and food lipids. The chemical structures and physicochemical parameters of some nitrosamines commonly found in meat products are presented in Table 13.1.[6,7]

In general, nitrosamines are stable compounds in neutral and strongly alkaline solutions, and are difficult to destroy once they are formed. Under ultraviolet (UV) radiation or strongly acidic conditions nitrosamines decompose with cleavage of the nitroso group.[8]

Nitrosamine formation in food generally is related to the nitrosation of secondary amines, where the main nitrosating agent is nitrous anhydride produced from nitrite (Equations 13.1 through 13.3).

$$NO_2^- + H_2O \rightleftharpoons HNO_2 + OH^- \tag{13.1}$$

$$2HNO_2 \rightleftharpoons N_2O_3 + H_2O \tag{13.2}$$

$$R_1R_2NH + N_2O_3 \rightarrow R_1R_2NN{=}O + HNO_2 \tag{13.3}$$

Table 13.1 Chemical Structures and Physicochemical Parameters of Some *N*-Nitrosamines Commonly Found in Meat Products

N-Nitrosamine	Chemical Structure	CAS	MM	sp gr (g/cm³)	bp (°C)	vp (mm Hg)	Solubility (mg/mL)	General Description
N-nitrosodimethylamine (NDMA)	CH₃—N-NO / CH₃	62-75-9	74.08	1.0048	151–153	2.7	>100 / 1000[a]	Yellow oil
N-nitrosodiethylamine (NDEA)	C₂H₅—N-NO / C₂H₅	55-18-5	102.14	0.9422	175–177	0.86	>100 / 106[a]	Yellow liquid
N-nitrosopiperidine (NPIP)	N-NO	100-75-4	114.2	1.0631	217–219	0.14	10–50 / 76.5[a]	Yellow oil
N-nitrosopyrrolidine (NPYR)	N-NO	930-55-2	100.14	1.085	214–216	0.06	1000[a]	Yellow liquid
N-nitrosomorpholine (NMOR)	N-NO	59-89-2	116.14	N/A	225–227	0.036	>100 / 861.5[a]	Liquid/ yellow crystals

Source: Adapted from CAMEO Chemicals, http://cameochemicals.noaa.gov/, accessed June 2007.

Note: CAS: CAS registry number; MM: molar mass; sp gr: specific gravity; bp: boiling point; vp: vapor pressure; N/A: not available.

[a] NIOSH Manual of Analytical Method (NIMAM), Nitrosamines, Method 2522, Fourth edition, 1994.

The nitrosation rate is first and second order in terms of the amine (R_1R_2NH) and nitrite concentrations, respectively.[9] The kinetics of the nitrosating reaction depends on the pH of the medium and the basicity of the amine. The optimum pH value lies between 2.5 and 3.5, conditions where the formation of nitrous acid (pKa 3.35) is favored while molecules of amine still exist in their nonprotonated forms. This explains the fact that the reaction rate increases as the basicity of the amine decreases.[10]

Several conditions can contribute to an increase of rate or inhibition of the nitrosation reactions in food. It is well documented that the nitrosation of secondary amines is catalyzed by nucleophilic anions (thiocyanate, bromide, chloride), because the concentration of the available nitrosating agent is increased. The effectiveness of the catalysis is related to the nucleophilic strength of the anion. On the other hand, several compounds, such as ascorbic acid (vitamin C), erythorbic acid, and α-tocopherol (vitamin E), are well recognized as nitrite scavengers and, in consequence, act as inhibitors of the nitrosation reaction.[11]

13.3 Formation and Occurrence in Meat and Meat Products

Several authors have reviewed the formation and occurrence of N-nitrosamine in meat products.[12–15] The formation of nitrosamines in meat and meat products is a complex process, and several factors and substances could influence nitrosation reaction. The nitrosamine concentration in meat products depends on the residual nitrite concentration, presence of nitrosation catalysts and inhibitors, cooking method, cooking temperature and time, storage conditions, and presence of microorganisms, which are able to reduce nitrate to nitrite and promote degradation of proteins to amines and amino acids.

The food matrices that have received most attention are cured and smoked meats, because sodium nitrite is used as a food additive in the manufacturing process. Several model-system studies have been carried out to explain nitrosamine formation in meat products. The effect of the cooking process on nitrosamine formation in cured and smoked meat products was also extensively investigated. Accordingly, it has been postulated that NDMA is derived from creatine, a muscle constituent, through its breakdown to sarcosine, followed by the decarboxylation of its N-nitroso derivative. In the same manner, proline and lysine are considered to be the precursors of NPYR and NPIP in meat products, respectively.[16]

Pensabene and Fiddler were the first to associate the presence of N-nitrosothiazolidine (NTHZ) in bacon with smokehouse processing, indicating the nitrogen oxides generated during the smoking process and the residual nitrite in the bacon as the nitrosating elements.[17] The nitrosable amine is formed by the condensation of cysteine with formaldehyde, a component of the wood smoke. In fried meat the nitrosating agent was identified as N_2O_3, which could be formed during the heating of nitrite in meat, or to NO radical formed by dissociation of N_2O_3 at high temperature.[18,19]

Byun et al.[20] verified that gamma irradiation (>10 kGy) reduced the content of volatile nitrosamines (NDMA and NPYR) in pepperoni and salami sausages during storage, and Rywotycki[21] evaluated the nitrosamine content (NDMA and NDEA) in raw meat (gilts, saws, hogs, boars, heifers, cows, bullocks, bulls, calves, horses, rams, and goats) and verified that the nitrosamine level depended on the animal species, breeding factors, and the season of the year.

In general, the concentration of nitrosamines in meat products currently lies at levels lower than 30 μg/kg, which demonstrates the efficacy of actions taken by the meat industry, such as the use of nitrosation reaction inhibitors and a decrease in the nitrite concentration used for the curing process.

13.4 Toxicological Aspects

The great majority of the over 300 *N*-nitroso compounds tested in laboratories, including nitrosamines, were found to be carcinogenic in a wide variety of experimental animals. In addition, they also present mutagenic and teratogenic activity.[5]

N-nitrosamines are readily absorbed from the gastrointestinal tract,[5,22] do not undergo bioaccumulation, and require metabolic activation to exhibit their mutagenic and carcinogenic action. The initial step of the biotransformation involves hydroxylation of the α-carbon, which is catalyzed by the cytochrome P450 system, mainly CYP2E1[23,24] and the cytochrome P450 isoform, CYP2A6.[24,25] The resultant α-hydroxyalkylnitrosamine breaks down to an alkyldiazonium ion and the corresponding carbonyl compound. The diazonium ion could alkylate a variety of nucleophilic sites such as deoxyribonucleic acid (DNA) and ribonucleic acid (RNA). This biotransformation is considered a fundamental step in cancer initiation.[26] The liver is the main organ of nitrosamine biotransformation, but other human tissues also have this capacity.[27]

Carcinogenic effects induced by the nitrosamines have been reported in all the mammalian species tested, including monkeys, and *in vitro* studies suggest that *N*-nitrosamines present similar biologic activity in humans and experimental animal tissues.[26] Consequently, it is assumed that humans are susceptible to the toxic action of these compounds. In fact, the International Agency for Research on Cancer (IARC) concluded, for those *N*-nitrosamines evaluated by the agency, that although no epidemiological data were available, nitrosamines should be regarded for practical purposes as if they were carcinogenic to humans.[28]

Volatile nitrosamines induce tumors in several organs including liver, lung, kidney, bladder, pancreas, esophagus, and tongue, depending on the animal species.[29,30] Among the nitrosamines, the volatile nitrosamines show higher carcinogenic potential and, of those found in foods, NDEA is the one that shows the higher carcinogenic activity.[27] Tumor induction could occur in different organs, according to the chemical structure of the nitrosamine, the dose, the route of exposure, and the animal species, which makes difficult the extrapolation of the data obtained from experimental animals to humans.

Nitrosamines are more effective as carcinogenic agents to the experimental animals when administered at low repeated doses than in a higher single dose. This is the situation of human low-dose exposures (traces) to nitrosamines present in foods.[16] Consequently, the presence of nitrosamines in foods, and particularly in meat, should be a matter of concern from the toxicological and public health standpoint.

13.5 Regulatory Aspects

Only a few countries have reported data related to the formation and to the presence of nitrosamines in foods, which would allow control of the nitrosamines to negligible levels to reduce exposure to levels that may not represent a higher risk to consumers.[31] Moreover, only a few countries have specific legislation for the presence of nitrosamines in foods. Table 13.2 shows the maximum levels established in some countries for the presence of nitrosamines in foods.

It is worth emphasizing that the regulatory levels provide guidelines for the minimum required limit of determination of the analytical method to be used by the governmental agencies to conduct action on food surveillance.

Table 13.2   Maximum Levels Permitted in Some Countries for the Presence of N-Nitrosamines in Foods

Country	Level (μg/kg)	N-Nitrosamine	Food	Reference
United States	10	Total volatile N-nitrosamines	Cured meat products	22
Canada	10	NDMA, NDEA, NDBA, NPIP, NMOR	Meat products	32
	15	NPYR		
Chile	30	NDMA	Meat products	33
Russia	2	N-nitrosamines	Raw foods	34
	4	N-nitrosamines	Smoked foods	34
Estonia	3	NDMA, NDEA	Raw and smoked fish	19

13.6   Analytical Aspects

Traditionally, for analytical purposes the nitrosamines have been divided into nonvolatile and volatile compounds, the latter ones being considered a group of relatively nonpolar, low-molar mass nitrosamines, which present sufficiently high vapor pressure to be removed from the food matrix by distillation. Whereas long chain dialkylnitrosamines, nitrosopeptides, and nitroso-amino acids possess lower vapor pressure and are considered nonvolatile compounds, short-chain dialkylnitrosamines, such as NDMA, NDEA, and low molar mass cyclic compounds, such as NPYR and NTHZ, are considered volatile nitrosamines. The differences in their physicochemical properties hinder the establishment of analytical methods of general application.

During the 1970s intense research efforts were carried out toward development of analytical methodologies for the determination of volatile nitrosamines. As a consequence, there are a great number of scientific papers reported in the literature for the period of 1970–1990 on the presence of volatile nitrosamines in food matrices. In general, these methods recommend the extraction of nitrosamines from the food matrix by vacuum, steam, or mineral oil distillation with subsequent quantitation by GC-TEA. TEA was developed as a specific nitrosamine detector[35] and has been widely employed in the past half century for the determination of volatile nitrosamines in food. Usually, these methods are simpler and receive more attention than those required for the determination of nonvolatile nitrosamines, because they do not require sophisticated sample preparation before the quantitation step. Nowadays, it is well known that in foods, and in particular in meat products, among the nitrosamines the volatiles are certainly the compounds of main relevance, and for this reason in this chapter more attention will be devoted to these compounds.

In the past 10 years, novel analytical methodologies and techniques have been proposed, improving selectivity, detectability, analysis time, and cost. In addition, several analytical methodologies have been subjected to collaborative studies carried out under the auspices of the Association of Official Analytical Chemists (AOAC).[36]

The extraction of the nitrosamine from the complex food matrices and the cleanup of the extract have been the critical points of the sample preparation step, and several approaches are documented in the literature, including distillation (steam, vacuum, or atmospheric), solvent extraction, solid-phase extraction, solid-phase microextraction (SPME), and supercritical fluid extraction. Thus, in this chapter, the analytical aspects of the determination of nitrosamines in meat products will be presented in terms of sample preparation procedures and quantitation steps. In addition, it should be mentioned that a worldwide single analytical method is not available; most of the methods comprise two or more clean-up steps, depending on the nitrosamine, the food matrix, and the detection device. Nonetheless, most of the methods recommend that artifactual nitrosamine formation during sample preparation should be inhibited by adding sulfamic acid, ascorbate, or other nitrosation inhibitors.

An overview of the possible steps in the analytical procedure for the determination of nitrosamines in foods is presented in Figure 13.1 and Table 13.3.

13.6.1 *Sample Preparation*

13.6.1.1 *Distillation and Clean-Up Procedures*

Distillation was extensively used in the past as the primary stage for the extraction of the volatile nitrosamines from food matrices, including steam distillation and mineral oil vacuum distillation (MOVD). The clean-up procedures that follow the extraction have included liquid–liquid extraction (LLE), liquid–solid extraction (LSE), and SPME. The concentration of the separated nitrosamines to a small volume before quantitation has generally been carried out using a Kuderna-Danish (K-D) evaporator.

The MOVD became the AOAC Official Method for the determination of volatile nitrosamines in fried bacon. For this purpose, 25.0 g of sample is added to 2 mL of 0.2 mol/L NaOH and 25 mL mineral oil. The mixture is introduced into a pumping and distillation assembly, vacuum (<2 torr) is applied, and the temperature is increased from ambient temperature to 120°C in 55–60 min. The distillate is collected in a vapor trap inserted in a Dewar flask containing liquid nitrogen. The nitrosamines are removed from the distillate by LLE using dichloromethane. The final extracted volume is reduced to 1.0 mL in a K-D flask, and the quantitation is carried out by GC-TEA.[54]

Although the volatile nitrosamines are efficiently extracted from foods by vacuum distillation, this sample extraction procedure presents limitations, such as long analysis time, being work-intensive, possibility of contamination, loss of the analyte during the concentration process, formation of emulsions during LLE, and environmental problems related to discarding solvents.

A combination of vacuum steam distillation and solid-phase extraction for the determination of NDMA, NDEA, NMOR, NPIP, and NPYR in sausages was proposed by Sanches Filho et al.[50] For this purpose, 150 g of sample was added to 100 mL of water, and the nitrosamines were separated by vacuum steam distillation using a rotary evaporator (65°C for 80 min). To the distillate active carbon powder (100–400 mesh) was added, and the mixture was shaken for 45 min. The sorbent was removed by filtration, and the nitrosamines were eluted from it with acetone and dichloromethane. After concentration under a nitrogen stream, nitrosamine quantitation and identity confirmation were performed by micellar electrokinetic chromatography (MEKC) and GC-MS, respectively. Powdered activated carbon for the cleanup and concentration of NDMA and NDEA from aqueous solutions (water and beer samples) was also employed by Ayügin et al.[55]

Sen et al.[48] described the use of SPME for the clean-up step in the determination of *N*-nitrosodibutylamine (NDBA) and *N*-nitrosodibenzylamine (NDBZA) in smoked hams. The method consists of the isolation of the volatile nitrosamines by steam distillation. A polyacrylate coated silica fiber was introduced into the headspace of the distillate. Quantitation

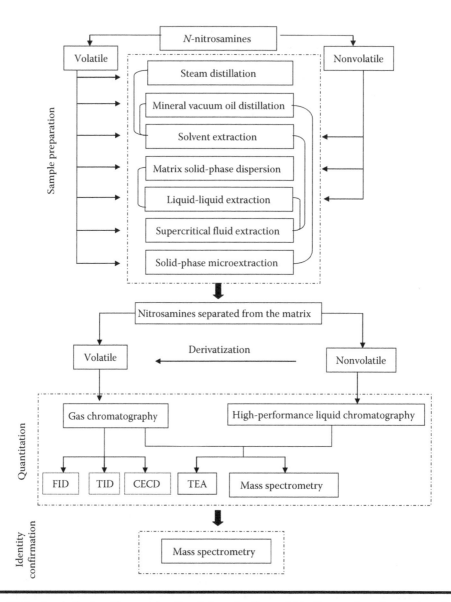

Figure 13.1 Analytical pathways for the determination of *N*-nitrosamines in meat and meat products. (FID, Flame ionization detector; TID, Thermoionic detector; and CECD, Coulson electrolytic conductivity detector.)

was conducted by GC-TEA, and the identity confirmation was done by GC-MS. The authors stated that the SPME extraction efficiency, using an extraction time of 60 min at room temperature, was too low for most of the nitrosamines evaluated (NDMA: 0.08%; NDEA: 0.17%; *N*-nitrosodipropylamine [NDPA]: 2.04%; NDBA: 19.3%; NPIP: 0.07%; NPYR: 0.07%; NMOR: 0.02%; NDBZA: 1.9%; and lower than 1% for *N*-nitrosodioctylamine [NDOA]). Using a temperature and a time of extraction of 80°C and 60 min, respectively, and by addition of alkali (3 mol/L KOH) and salt saturation (NaCl), better efficiencies were achieved only for NDBZA and NDBA.

Table 13.3 Some Analytical Methods Reported for Nitrosamine Determination in Meat Products

Food	Nitrosamine	Sample Preparation	Analytical Technique	LOD (μg/kg)	Publ. Year	Reference
Cooked bacon, cooked bacon fat	NDMA, NDEA, NPIP, NPYR	VD, LLE, C, LSE (alumina), K-D	GC-CECD, GC-MS, GC-TEA (Q)	N/A	1976	37
Bacon	NPRO	SE-LSE (anion exchange column), LLE, K-D	HPLC with phtotohydrolysis (Q), GC-TEA (D, Q)	N/A	1977	38
Meat loaf, liver loaf, and bologna	NDMA, NPYR, NPIP	MOVD, LLE, K-D	GC-TEA (Q)	N/A	1978	39
			GC-MS (IC)			
Bacon, boiled ham, bologna	NHPYR	Celite, C, LLE, LSE (alumina +acidic cellulose), C	GC-TEA (D)	0.2 ng	1978	39
Ham, frankfurters, pork shoulder, canned meats	NDMA, NDEA, NDPA, NPYR, NPIP	Digestion in methanolic KOH, LLE, distillation, LLE, LSE (silica gel)	GLC-TEA (Q)	N/A	1972	40
Fried bacon, fried pork	NDMA, NDEA, NDPA, NDBA, NPIP, NPYR, NMOR	MOVD, LLE, K-D	GC-TEA (Q)	N/A	1980	41
Cooked bacon	NTHZ	MOVD, LLE, K-D or MSPD (Celite), + LSE (alumina column)	GC-TEA (Q)	N/A	1982	42
			GC-MS (IC)			
Fried bacon	NPYR	MSPD Celite (dry column), K-D	GC-TEA (Q)	N/A	1982	43

(continued)

Table 13.3 (continued)   Some Analytical Methods Reported for Nitrosamine Determination in Meat Products

Food	Nitrosamine	Sample Preparation	Analytical Technique	LOD (μg/kg)	Publ. Year	Reference
Fried bacon	NTHZ, NPYR	MSPD Celite (dry column), K-D, LSE (alumina column), K-D	GC-MS (Q)	N/A	1982	44
Smoked bacon	HMNTHZ	LLE, C, + SLE (alumina column), C	HPLC-TEA (Q) GC-TEA (Q) (D) GC-MS (IC)	1–2	1989	45
Bacon, smoked poultry	HMNTCA, HMNTHZ	SE, LLE, C, LLE, C, LSE (alumina column), LLE, C, deriv	HPLC-TEA (D, Q) GC-MS, deriv. (IC)	N/A	1992	46
Fried bacon	NPYR NDMA	SFE, SPE (silica), C	GC-TEA (Q)	0.2–0.5	1996	47
Ham	NDBA	Steam distillation, HS-SPME	GC-TEA (Q) GC-MS (IC)	1	1997	48
Sausages	NDMA, NEMA, NDEA, NDPA, NMOR, NPYR, NPIP, NDBA	MSPD (Extrelut), K-D, SPE (Florisil), K-D	GC-TEA (Q)	0.3	1997	49

Sample	Compounds	Extraction	Analysis	Value	Year	Ref
Sausages	NDMA, NDEA, NMOR, NPIP, NPYR	VSD and SPE (activated carbon), C	MEKC (Q) GC-MS (IC)	22.5–36.0[a]	2003	50
Fermented sausages	NDMA, NPYR	Steam distillation on a steam generator, LLE, K-D	GC-TEA	N/A	2004	20
Sausages	NDMA, NDEA, NPIP, NPYR	HS-SPME (DVB/PDMS)	GC-TEA (Q)	3	2005	51
Meat patties, gelatine (food model)	NDMA, NMEA, NDEA, NPYR, NMOR, NDPA, NPIP, NDBA, NDPheA	SPME (DVB/PDMS)-DED	GC-TEA (Q/IC)	0.142–9.539 (Gelatine)	2006	52
Sausages	NDMA, NDEA, NMOR, NPIP, NPYR	SFE (CO$_2$) + florisil trap	MEKC (Q) GC-MS (IC)	Range mg/kg	2007	53
Meat (raw, fried, smoked, grilled, pickled, and canned)	NDMA, NDEA, NPIP, NPYR, NDBA	SPE (Extrelut), C, SPE (Florisil), C	GC-MS	0.09	2007	19

Note: N/A: not available; MOVD: mineral oil vacuum distillation; VSD: vacuum steam distillation; VD: vacuum distillation; MSPD: matrix solid-phase dispersion; LLE: liquid–liquid extraction; SE: solvent extraction; LSE: liquid–solid extraction; SFE: supercritical fluid extraction; SPME: solid-phase microextraction; HS: head space sampling; SPE: solid-phase extraction; K-D: Kuderna Danish concentrator; C: concentration; GC: gas chromatography; TEA: thermal energy analyzer; MS: mass spectrometry; MEKC: micellar electrokinetic chromatography; IC: identity confirmation; Q: quantitation; D: derivatization.

[a] μg/L.

13.6.1.2 Solvent Extraction Followed by Cleanup Using Liquid–Solid Extraction

Solvent extraction has been widely used as the clean-up step of the aqueous distillate obtained by the extraction of volatile nitrosamines from the food matrices. In a few circumstances, solvent extraction was employed as the extraction step of nonvolatile nitrosamines from meat samples. Nevertheless, Sen et al.[46] described a solvent extraction procedure for the determination of 2-hydroxymethyl-*N*-nitrosothiazolidine (HMNTHZ) and 2-hydroxymethyl-*N*-nitrosothiazolidine-4-carboxilic acid (HMNTCA) in smoked meats. For this purpose, 10–20 g of the food sample are mixed with sulfuric acid and sulfamic acid, and extracted with 100 mL acetonitrile (for processed meat) or methanol (bacon). After the first filtration, the residue was further extracted with two 60-mL portions of the solvent. The combined filtrates were washed with 80 mL of isooctane to remove fats and lipids. NH_4OH was added to the remaining extract, and the mixture was evaporated to 10 mL in a rotary evaporator. Water was added to the evaporated residue, the pH adjusted to 2–2.3, and the solution saturated with NaCl before extraction with three portions of 50 mL ethyl acetate. The combined extract was concentrated to 1 mL in a rotary evaporator and cleaned up by LSE on an acidic alumina cartridge.

Another LSE procedure for extraction of nitrosoamino acids *N*-nitrosoproline (NPRO) from bacon was reported by Hansen et al.[38] The raw bacon (100 g) was added to water and then homogenized, centrifuged, and stored at 0°C until the fat had been solidified. The supernatant was removed, and the procedure was repeated two more times. The combined supernatants were filtered and cleaned up on an anion-exchange column (Dowex 2X8-100 strongly basic anion-exchange resin). After LSE, the nitrosoamino acid was extracted with dichloromethane from the eluate and concentrated in a K-D flask. NPRO was quantified by reverse-phase HPLC using a photohydrolysis system.

13.6.1.3 Matrix Solid-Phase Dispersion and Liquid–Liquid Extraction

Pensabene et al.[43] introduced a rapid method for the determination of NPYR in fried bacon using a dry column of acid—Celite. The ground food sample (10 g) was mixed thoroughly with 25 g anhydrous sodium sulfate and 20 g Celite and added to the chromatographic column containing 10 g Celite previously washed with phosphoric acid. At the top of the column, 30 g of anhydrous sodium sulfate was added. The column was rinsed with 100 mL pentane–dichloromethane (95 + 5 v/v) and 125 mL dichloromethane. Only the last 40 mL of the eluate was collected and concentrated in a K-D flask to a final volume of 1 mL before GC-TEA quantitation. In this sample preparation technique, the solid food sample matrix is dispersed into the adsorbent material (diatomaceous earth), which is subsequently packed into a column from which the nitrosamines are eluted. In this manner, the sample becomes dispersed throughout the column and is part of the overall chromatographic character of the system. Interactions involve the stationary phase, the solid support, the mobile or eluting phase, and all of the sample matrix components as well.

Pensabene and Fiddler[44] also reported a method using a dual-column chromatographic procedure (Celite + alumina columns) for the determination of NTHZ (nonvolatile nitrosamine) and NPYR (volatile nitrosamine) in fried bacon, and compared this procedure with the MOVD. The extraction procedure reported is as described previously, with the modification that the first 85 mL of dichloromethane eluted from the Celite column was collected and concentrated to a final volume of 6 mL in a K-D flask. The concentrate was added to 2 mL hexane and quantitatively transferred to an alumina column containing anhydrous sodium sulfate at the top. An initial volume of 25 mL hexane was added to the column, and the NTHZ was eluted with 125 mL dichloromethane. The eluate was concentrated, and NTHZ was quantified by GC-TEA. The authors verified that the MOVD extraction procedure introduces artifacts and observed *in situ* nitrosamine formation during this analytical step, which thus requires the addition of nitrosating inhibitors. Sulfamic acid and ascorbic acid were shown to be effective for this purpose, as sulfamic acid reduces the pH and thereby removes any nitrite present in the sample, as well as prevents the bacterial reduction of nitrate.[56]

Raoul et al.[49] presented a rapid, time- and solvent-sparing MSPD plus SPE method to determine NDMA, *N*-nitrosoethylmethylamine (NEMA), NDEA, NDPA, NMOR, NPYR, NPIP, and NDBA in thermally processed sausages. The food sample (6 g) homogenized in 6 mL 0.1 mol/L NaOH was dispersed in Extrelut (6 g) and packed into a column. The nitrosamines were eluted with 40 mL of hexane:dichloromethane (60:40 v/v), and the eluate was concentrated in a K-D flask. The extract was cleaned up on a commercial Florisil cartridge. In comparison to the vacuum distillation technique, this sample preparation approach requires less food sample and solvents without affecting the detectability of the method, and could also be applied to determine the less volatile nitrosamines NDBZA and NTHZ, which have been found in smoked meat products. This sample preparation procedure, using Extrelut and Florisil, was also employed by Yurchenko and Mölder[19] for the determination of volatile nitrosamines in several meat matrices.

13.6.1.4 Solid-Phase Microextraction

SPME was first described by Pawliszyn, and since then this technique have been extensively used for several analytical purposes in substitution of traditional solvent extraction, including the evaluation of the volatile compounds present in the vapor or in the liquid phase of solid and liquid foods. The advantages of the SPME method over other methods of extraction are numerous. SPME can be significantly faster and easier than solvent extraction methods, it is easily automated, and it does not require the use of potentially toxic and expensive solvents.[57] SPME has gained widespread acceptance in many areas in recent years, and has been applied to a wide spectrum of analytes, including the determination of nitrosamines in food. Commercially available fused-silica fibers coated with polydimethylsiloxane (PDMS), carboxen–polydimethylsiloxane (CAR/PDMS) polyacrylate (PA), divinylbenzene–carboxen–polydimethylsiloxane (DVB/CAR/PDMS), carbowax–divinylbenzene (CW/DVB), and carbowax-templated resin (CW/TPR) are available.[58]

Andrade et al.[51] described a simple method using headspace sampling by SPME with GC-TEA detection (HS-SPME-GC-TEA) for the determination of NDMA, NDEA, NPIP, and NPYR in sausages. Two fused-silica fibers, one coated with PDMS/DVB and another with PA, were evaluated, and the experimental conditions (equilibrium time, salt addition, extraction time, and temperature) were optimized using an experimental design. The PDMS/DVB-coated fiber showed better recoveries for the extraction of NDMA and NDEA in sausages in comparison with the PA-coated fiber, which presented higher efficiency for NPIP and NPYR. The optimum recoveries were obtained with the following experimental conditions: PDMS/DVB (equilibrium time: 10 min; salt addition: 36% w/v NaCl; temperature: 30°C; and extraction time: 30 min) and PA (equilibrium time: 10 min; salt addition: 36% w/v NaCl; temperature: 50°C; and extraction time: 20 min).

The outstanding advantage of HS using the SPME technique in food analysis is the prevention of direct contact of the fiber with the food matrix; therefore, the fiber has a longer lifetime, and the selectivity of the method could be enhanced. On the other hand, HS-SPME is limited to volatile nitrosamines, which present high vapor pressure. The extraction efficiency onto the fiber depends on the polarity and the thickness of the stationary phase, extraction time, and concentration of the nitrosamine in the sample. Extraction efficiency could be improved by agitation, addition of salt, pH, and temperature.[58]

Ventanas et al.[59] employed SPME coupled to a direct extraction device (DED) for extracting nine volatile nitrosamines (NDMA, *N*-nitrosomethylethylamine [NMEA], NDEA, NPYR, NMOR, NDPA, NPIP, NDBA, and *N*-nitrosodiphenylamine [NDPheA]) from a solid food model system (gelatin) at refrigeration and at room temperature. The DED enables the introduction of the SPME fiber in the core of the solid matrices, with the advantage of determining volatile compounds from solid foods without deterioration of the product. In a subsequent work, Ventanas and Ruiz[52] studied the feasibility of using SPME-DED for extraction of nitrosamines from solid matrices mimicking solid foodstuffs, and compared the efficiency of different fiber-coatings for extraction (CAR/PDMS, DVB/CAR/PDMS, and DVB/PDMS). Meat patties spiked with nitrosamines were also analyzed using a PDMS/DVB-coated fused-silica fiber. The authors reported that with the patties instead the gelatin matrix, lower reproducibility and poorer linearity were obtained, and concluded that quantitation of nitrosamines in solid meat samples using SPME-DED was not fully reliable. However, the proposed technique is promising for qualitative assessment.

13.6.1.5 Supercritical Fluid Extraction

Supercritical fluids have been successfully used to extract a wide variety of analytes from several matrices, including food, with the advantages of providing fairly clean extracts, minimizing sampling handling, reducing the use of toxic solvents, and expediting sample preparation. Fiddler and Pensabene[47] reported a method using supercritical extraction (SFE) of NPYR and NDMA from fried bacon. Fried bacon (5 g) was added to 250 mg propyl gallate and 5.0 g Hydromatrix. The homogenized mixture was transferred to the extraction vessel of the SFE system attached to an SPE cartridge (silica). The extraction was carried out at 10,000 psi with a flow rate of expanded CO_2 of 2.8 mL/min for a total of 50 L. The SPE cartridge was washed with pentane–dichloromethane (72:25 v/v), and the nitrosamines were eluted with dichloromethane:ether 70:30 v/v. The quantitation was performed by GC-TEA. The authors compared the SFE method to SPE, mineral oil distillation, and low-temperature vacuum distillation, and concluded that SFE was superior in relation to recovery, repeatability, rapidity of analysis, and lower solvent consumption, and that the method is not susceptible to artifactual nitrosamine formation.

Recently, Sanches Filho et al.[53] reported a procedure for the extraction of NDMA, NDEA, NMOR, NPIP, and NPYR from sausages, using CO_2 as extraction fluid. Several parameters were evaluated and optimized such as density, temperature at constant pressure of 200 bar (40°C), dynamic extraction time (20 min), organic modifier, flow rate (3 mL/min), and trap adsorbent (Florisil). The quantitation was done by MEKC. The nitrosamine recoveries from spiked sausages (0.2 g sample) at three concentration levels (0.4, 1.0, and 10 mg/kg) ranged from 20.9 to 81.6%. The authors attributed the low recovery values to the presence of lipids in the matrix and losses during the evaporation and change of solvents. The method, due to instrumental limitations, was developed for the quantitation of nitrosamines at the milligram per kilogram level and needs to be improved in relation to sample amount and concentrations steps to allow the determination of nitrosamines in food at the microgram per kilogram level.

13.6.1.6   Quantitation Methods

Several analytical methods have been employed in the past for the semiquantitative and quantitative determination of nitrosamines in food, including thin-layer chromatography,[60,61] spectrophotometry, colorimetry, and polarography.[62,63] In general, these methods lack selectivity and do not allow nitrosamine determination at the microgram per kilogram level required for foodstuffs. Only after the development of chromatographic methods with adequate sample preparation procedures, including clean-up and concentration steps, and the use of selective detector devices, did it become possible to establish reliable methods for the determination of volatile and nonvolatile nitrosamines in food.

13.6.1.6.1   Gas Chromatography

GC has been the method of choice for the determination of volatile nitrosamines around the world. Furthermore, some nonvolatile nitrosamines, such as hydroxylated nitrosamines and nitrosoamino acids, were determined by GC after derivatization by acylation or trimethylsilylation.[64]

Several stationary phases of moderate to strong polarity in packed, megabore, and capillary columns have been employed for the separation of the nitrosamines using GC-TEA, including 15% Carbowax 20 M/terephthalic acid on 100/120 mesh Gaschrom Q,[39] glass capillary column coated with UCON 5100,[42] silica capillary column coated with Supelcowax 10,[48] 88% methyl, 7% phenyl, 5% cyanopropyl capillary column,[49] 11% Carbowax 20M on 60/120 Chromosorb W packed column,[41] and HP-INOWAX megabore column.[51] For GC-MS analysis, silica capillary columns coated with DB-5,[48] 5% phenyl-methyl silicone (HP-5),[59] 14% cyanopropyl–86% methyl polysiloxane (HP 1701),[19] and HP-1[50] have been employed.

In the past, flame ionization detectors, thermionic detectors, Coulson electrolytic conductivity detectors, and electron capture detectors were employed for volatile nitrosamines quantitation. A comparison of the performance of those different detection devices was reported by Fine and coworkers.[37] Later, the selective TEA became the internationally recognized standard detector for quantitation purposes. Despite the high selectivity characterizing the TEA detector for *N*-nitroso compounds, which allows reduced clean-up procedures in the sample preparation step, identity confirmation by mass spectrometry is mandatory. Basically, the TEA is composed of a catalytic pyrolyzer, a trap, a reaction chamber, and a photomultiplier tube. The principle of operation of the TEA consists of the cleavage of the N–NO bond of the nitrosamine in the catalytic pyrolyzer chamber, forming the nitrosyl radical (NO•). The by-products of the pyrolysis are removed in the trap. The nitrosyl radicals are conducted by vacuum to the reaction chamber, where they are oxidized with ozone, forming electronically excited nitrogen dioxide (NO_2^* ["*" electronically excited state]). When the excited molecule decays to its ground state, it emits near infrared radiation (600 nm). At the last stage the radiation is detected by a sensitive photomultiplier tube, where the intensity of the radiation is proportional to the nitrosamine content present in the sample. Detectability is at the picogram level.

Nowadays, the mass spectrometer is the most recommended detector for volatile and nonvolatile nitrosamine determination, due to the fact that the technique allows accurate quantitation, as well as confirmation in one run. In this regard, the use of the GC-MS technique for the determination of volatile nitrosamines in meat products has been reported.[19,48,50,52] Nonetheless, although GC-MS/MS and LC-MS/MS have become a routinely applicable technique for the quantitation of a large number of toxic compounds in several matrices, no scientific publications were found in the literature for the determination of nitrosamines in meat and meat products using these instruments.

13.6.1.6.2   High-Performance Liquid Chromatography

HPLC coupled with the TEA detector (HPLC-TEA) was first employed by Fine et al.[37] Afterward, this technique was employed for the determination of nonvolatile nitrosamines in foods, including hydroxyl nitrosamines and *N*-nitrosoamino acids. Early on, the HPLC technique presented several drawbacks, such as the incompatibility of the TEA system with components of the mobile phase from the HPLC. Furthermore, *N*-nitrosamines do not show relevant absorption in the UV region of the spectra, and derivatization reactions are required to improve the detectability with a UV detector.

Owing to the different physical and chemical properties of the nonvolatile *N*-nitrosamines, a general sample preparation procedure before HPLC quantitation is not possible, and the extension of the clean-up step is related to the selectivity of the detection device used.

Considering the more polar characteristics of the nonvolatile nitrosamines, normal phase HPLC has been, in general, the method of choice, using silica or cyano stationary phases.[46,65] Only a few papers report performing the nitrosamine separation on reversed-phase octadecyl columns.[38]

Among the *N*-nitrosoamino acids, NPRO has been the most studied, the reason being that it could originate from the amino acid proline, which is present in all proteins. Wolfram et al.[65] reported a method for the determination of NPRO using fluorimetric detection (HPLC-FL). The fluorescent derivative was formed by NPRO denitrosation, followed by the derivatization of proline with 7-chloro-4-nitro-benzo-2-oxa-1,3-diazole. The HPLC conditions comprised a LiChrosorb Si stationary phase and a mobile phase composed of *n*-hexane:ethyl acetate:acetic acid (50:50:0.5 v/v/v).[65] An HPLC-method was reported for the determination of *N*-nitrosobenzylphenylamine (NBPHA) in cooked bacon, luncheon meat, and dried beef.[37] For the chromatographic separation, a μ-Porasil column and acetone:2,2,4-trimethylpentane (5:95 v/v) as column and mobile phase, respectively, were used. Sen et al.[46] used a LiChrosorb Si 100 column for the determination of HMNTHZ and HMNTCA in meat products. Whereas HMNTHZ did not require derivatization before HPLC-TEA quantitation, the HMNTCA was derivatized with diazomethane. For the identity confirmation, both compounds were derivatized: HMNTCA with heptafluorobutyric anhydride, whereas HMNTHZ was converted to its *O*-methyl ester derivative.

Sen et al.[45] described a method employing HPLC-TEA for the determination of HMNTHZ in fried bacon, as its *O*-methyl ester derivative, using a Lichrosorb Si 100 column (5 μm) and a mobile phase composed of acetone and *n*-hexane with linear gradient elution. The detection limit is about 1–2 μg/kg.

The detectability for the determination of nonvolatile *N*-nitroso compounds can be improved, in relation to precolumn and postcolumn derivatization, by denitrosation of the nitrosamines and derivatization of the liberated secondary amines with fluorescent agents, such as dansyl chloride. In this regard, Cárdenes et al.[66] described a microwave-assisted method (radiation power 378 W, maximum pressure 1.4 bar, reaction time 5 min) for dansylation of NMOR, NDMA, NPYR, NDEA, and NPIP with subsequent quantitation by HPLC with fluorimetric detection. The denitrosation was achieved using hydrobromic acid–acetic acid. The method was employed to study the recoveries of *N*-nitrosamines from beer.

13.6.1.6.3 Electrophoresis

Although capillary electrophoresis has been increasingly used in the separation of a large variety of compounds in several matrices, only one paper using MEKC reports the determination of volatile nitrosamines (NDMA, NMOR, NPYR, NDEA, NPIP), employing a fused-silica capillary and a diode array detector; sodium dodecyl sulfate was used as the pseudo-stationary phase. The limit of quantitation was between 520 and 820 μg/L, and the authors pointed out that the method is simple, and has a short analysis time and high efficiency.[50]

13.7 Conclusions

Reliable analytical methods are available for determination of volatile nitrosamines at concentration levels lower than 10 μg/kg in meat and meat products. Although a large number of sample preparation procedures for the volatile nitrosamines are reported in the literature, most of them are time-consuming and labor-intensive, and require large volumes of solvents. Artifactual nitrosamine formation during analysis should be considered in all analytical procedures, and should be evaluated for each sample preparation to allow the acquisition of reliable results. There is a need for development of simple, low-cost, and environmentally friendly sample preparation procedures for the quality control of meat and meat products in relation to the content of nitrosamines to avoid or minimize human exposure to these toxic compounds through consumption of foods.

Undoubtedly, the use of the specific TEA detector coupled to chromatographic systems has simplified sample preparation without minimizing selectivity and detectability, and represents a great contribution to the quantitation of nitroso compounds in food—in particular, volatile nitrosamines in meat products. Nevertheless, this detector lacks versatility in comparison to the mass spectrometer. In addition, in view of the fact that the confirmatory evidence for an analyte is indispensable in the quality control of any toxic compound in food, the use of the mass spectrometer coupled to the chromatographic becomes the technique of choice. As a consequence, the TEA has been replaced in many laboratories dealing with the quantitation of toxic compounds in food.

The newer generation of mass spectrometers, including tandem mass spectrometers, coupled to gas or liquid chromatographic systems, due to their higher detectability and selectivity capacities, as well as simplified sample preparation procedure requirement, have been shown to be a potential technique for the determination of nitrosamines in meat and meat products.

References

1. Barnes, J.M. and Magee, P.N., Some toxic properties of dimethylnitrosamine, *Brit. J. Ind. Med.*, 11, 167, 1954.
2. Magee, P.N. and Barnes, J.M., The production of malignant primary hepatic tumors in the rat by feeding dimethylnitrosamine, *Brit. J. Cancer*, 10, 114, 1956.
3. Ender, F. et al., Isolation and identification of a hepatotoxic factor in herring meal produced from sodium nitrite preserved herring, *Naturwissenschaften*, 51, 637, 1964.
4. Spiegelhalder, B., Eisenbrand, G., and Preussmann, R., Contamination of beer with trace quantities of N-nitrosodimethylamine, *Food Cosmet. Toxicol.*, 17, 29, 1979.
5. IARC. Monographs on the evaluation of the carcinogenic risk of chemicals to humans. Some N-nitrosocompounds. IARC, Lyon 17, 1978.
6. CAMEO Chemicals, http://cameochemicals.noaa.gov/, accessed June 2007.

7. National Institute for Occupational Safety and Health (NIOSH), Manual of Analytical Method (NINAM), Nitrosamines, Method 2522, Fourth edition, 1994. (http://www.cdc.gov/niosh/nmam/pdfs/2522.pdf, accessed, June, 22, 2008.)

8. Douglass, M.L. et al., The chemistry of nitrosamine formation, inhibition and destruction, *J. Soc. Cosmet. Chem.*, 29, 581, 1978.

9. Mirvish, S.S., Kinetics of dimethylamine nitrosation in relation to nitrosamine carcinogenesis, *J. Natl. Cancer Inst.*, 44, 663, 1970.

10. Mirvish, S.S., Formation of nitroso compounds: chemistry, kinetics, and in vivo occurrence, *Toxicol. Appl. Pharmacol.*, 31, 325, 1975.

11. Archer, M.C., Catalysis and inhibition of N-nitrosation reactions, *IARC Sci. Publ.*, 57, 263, 1984.

12. Gough, T.A. et al., An examination of some foodstuffs for the presence of volatile nitrosamines, *J. Sci. Food Agric.*, 28, 345, 1977.

13. Scanlan, R.A., N-nitrosamines in food, *Crit. Rev. Food Technol.*, 5, 357, 1975.

14. Reyes, F.G.R. and Scanlan, R.A., N-nitrosaminas: formação e ocorrência em alimentos, *Bol. SBCTA*, 18, 299, 1984.

15. Walker, R., Nitrates, nitrites and N-nitrosocompounds: a review of the occurrence in food and diet and the toxicological implications, *Food Addit. Contam.*, 7, 717, 1990.

16. Walters, C.L., Reactions of nitrate and nitrite in foods with special reference to the determination of N-nitroso compounds, *Food Addit. Contam.*, 9, 441, 1992.

17. Pensabene, J.W. and Fiddler, W., N-nitrosothiazolidine in cured meat products, *J. Food Sci.*, 48, 1870, 1983.

18. Sen, N.P., Formation and occurrence of nitrosamines in food, in *Diet, Nutrition and Cancer: A Critical Evaluation. Micro Nutrients, Nonnutritive Dietary Factors, and Cancer*, Cohen, L.A. and Reddy, B.S., Eds., CRC Press, Boca Raton, FL, 1986, Vol. 2, pp. 135–160.

19. Yurchenko, S. and Mölder, U., The occurrence of volatile N-nitrosamines in Estonian meat products, *Food Chem.*, 100, 1713, 2007.

20. Byun, M.W. et al., Determination of volatile N-nitrosamines in irradiated fermented sausage by gas-chromatography coupled to a thermal energy analyzer, *J. Chromatogr. A*, 1054, 403, 2004.

21. Rywotycki, R., Meat nitrosamine contamination level depending on animal breeding factors, *Meat Sci.*, 65, 669, 2003.

22. USDA. US Code of Federal Regulations. Food Safety and Inspection Service, USDA § 424.22. Certain other permitted uses. 9 CFR Ch. III (1-1-03 Edition). (http://www.usda.gov., accessed February 4, 2004).

23. Lin, H.L. et al., N-nitrosodimethylamine- mediated cytotoxicity in a cell line expressing P450 2E1 evidence for apoptotic cell death, *Toxicol. Appl. Pharmacol.*, 157, 117, 1999.

24. Rossini, A. et al., CYP2A6 and CYP2E1 polymorphisms in a Brazilian population living in Rio de Janeiro, *Braz. J. Med. Biol. Res.*, 39, 195, 2006.

25. Kamataki, T. et al., Genetic polymorphism of CYP2A6 in relationship to cancer, *Mutat. Res.*, 428, 125, 1999.

26. Tricker, A.R. and Preussmann, R., Carcinogenic N-nitrosamines in diet: occurrence, formation, mechanism and carcinogenic potential, *Mutat. Res.*, 259, 277, 1991.

27. Järgestad, M. and Skog, K., Review: Genotoxicity of heat-processed foods, *Mutat. Res.*, 574, 156, 2005.

28. IARC. International Agency for Research on Cancer. (http://monographs.iarc.fr/ENG/Classification/index.php, accessed, June, 22, 2008.)

29. Biaudet, H., Mavelle, T., and Debry, G., Mean daily intake of N-nitrosodimethylamine from foods and beverages in France in 1987–1992, *Food Chem. Toxicol.*, 32, 417, 1994.

30. Lijinsky, W., N-nitroso compounds in the diet, *Mutat. Res.*, 443, 129, 1999.

31. Villegas, R., Mendoza, N., and Marisol, L.N., Seguridad alimentaria, necesidad de implementar técnicas modernas de análisis de xenobióticos em alimentos. *Anales de la Universidad de Chile*, Santiago, n.11, 2000.

32. Canada, Canadian Food Inspection Agency. Animal products food of animal origin. Livestock and meat processing-nitrosamines. Date modified: 2003-02-07. (http://www.inspection.gc.ca/english/anima/meavia/meaviae.shtml, accessed, June, 22, 2008.)

33. Chile. Reglamento sanitario de los alimentos, n.977. Santiago, seis de agosto de 1997. Diário Oficial de la Republica de Chile, treze de maio de 1997.

34. Komarova, N.V. and Velikanov, A.A., Determination of volatile N-nitrosamines by high-performance liquid chromatography with fluorescence detection, *J. Anal. Chem.* (*Moscow*), 56, 359, 2001.

35. Fine, D.H., Lieb, F.R., and Rounbehler, D.P., Description of the thermal energy analyzer (TEA) for trace determination of volatile and non-volatile N-nitroso compounds, *Anal. Chem.*, 47, 1188, 1975.

36. Greenfield, E.L., Smith, W.J., and Malanoski, A.J., Mineral oil vacuum method for nitrosamines in fried bacon, with thermal energy analyzer: collaborative study, *J. Assoc. Off. Anal. Chem.*, 65, 1319, 1982.

37. Fine, D.H., Rounbehler, D.P., and Sen, N.P., A comparison of some chromatographic detectors for the analysis of volatile N-nitrosamines, *J. Agric. Food Chem.*, 24, 980, 1976.

38. Hansen, T. et al., Analysis of N-nitrosoproline in raw bacon. Further evidence that nitroproline is not a major precursor of nitrosopyrrolidine, *J. Agric. Food Chem.*, 25, 1423, 1977.

39. Eisenbrand, G. et al., Volatile and non-volatile N-nitroso compounds in foods and other environmental media, *IARC Sci. Publ.*, 19, 311, 1978.

40. Fazio, T., Howard, J.W., and White, R., Multidetection method for analysis of volatile nitrosamines in foods. N-nitroso compounds: analysis and formation, *IARC Sci. Publ.*, 16, 16–24, 1972.

41. Hotchkiss, J.H. et al., Combination of a GC-TEA and a GC-MS-Data system for the μg/kg estimation and confirmation of volatile N-nitrosamines in foods, *IARC Sci. Publ.*, 31, 361, 1980.

42. Kimoto, W.I. and Fiddler, W., Confirmatory method for N-nitrosodimethylaimne and N-nitrosopyrrolidine in food by multiple ion analysis with gas chromatography-low resolution mass spectrometry before and after ultraviolet photolysis, *J. Assoc. Off. Anal. Chem.*, 65, 1162, 1982.

43. Pensabene, J.W. et al., Rapid dry column method for determination N-nitrosopyrrolidine in fried bacon, *J. Assoc. Off. Anal. Chem.*, 65,151, 1982.

44. Pensabene, J.W. and Fiddler, W., Dual column chromatographic method for determination of N-nitrosothiazolidine in fried bacon, *J. Assoc. Off. Anal. Chem.*, 65, 1346, 1982.

45. Sen, N.P. et al., Determination and occurrence of 2-(hydroxymethyl)-N-nitrosothiazolidine in fried bacon and other cured meat products, *J. Agric. Food Chem.*, 37, 717, 1989.

46. Sen, N.P. et al., Simultaneous determination of 2-hydroxymethyl-N-nitrosothiazolidine-4-carboxylic acid and 2-hydroxymethyl-N-nitrosothiazolidine in smoked meats and cheese, *J. Agric. Food Chem.*, 40, 221, 1992.

47. Fiddler, W. and Pensabene, J.W., Supercritical fluid extraction of volatile N-nitrosamines in fried bacon and its drippings: Method comparison, *J. AOAC Int.*, 79, 895, 1996.

48. Sen, N.P., Seaman, S.W., and Page, B.D., Rapid semi-quantitative estimation of N-nitrosodibutylamine and N-nitrosodibenzylamine in smoked hams by solid-phase microextraction followed by gas chromatography-thermal energy analysis, *J. Chromatogr. A*, 788, 131, 1997.

49. Raoul, S. et al., Rapid solid-phase extraction method for the detection of volatile nitrosamines in food, *J. Agric. Food Chem.,* 45, 4706, 1997.

50. Sanches Filho, P.J. et al., Determination of nitrosamines in preserved sausages by solid-phase extraction-micellar electrokinetic chromatography, *J. Chromatogr. A*, 985, 503, 2003.

51. Andrade, R., Reyes, F.G.R., and Rath, S., A method for the determination of volatile N-nitrosamines in food by HS-SPME-GC-TEA, *Food Chem.*, 91, 173, 2005.

52. Ventanas, S. and Ruiz, J., On-site analysis of volatile nitrosamines in food model systems by solid-phase microextraction coupled to a direct extraction device using SPME-DED at different temperatures and times of extraction, *Talanta*, 70, 1017, 2006.

53. Sanches Filho, P.J. et al., Method of determination of nitrosamines in sausages by CO₂ supercritical fluid extraction (SFE) and micellar electrokinetic chromatography (MEKC), *J. Agric. Food Chem.*, 5, 603, 2007.

54. AOAC Official Methods of Analysis of the Association of Official Analytical Chemists, Method 982.22 (16th ed.) Gaithersburg, 1997.

55. Ayügin, S.F., Uyanik, A., and Bati, B., Adsorption of N-nitrosodiethylamine on activated carbon: a pre-concentration procedure for gas chromatographic analysis, *Mikrochim. Acta.*, 146, 279, 2004.

56. Walters, C.L., Smith, P.L., and Reed, P.I., Pitfalls to avoid in determining N-nitroso compounds as a group, in *N-nitroso Compounds: Occurrence, Biological Effects and Relevance to Human Cancer*, O'Neill, I.K., Von Borstel, R.C., Miller, C.T., Long, J. and Bartsch, H., Eds., *IARC Sci. Publ.*, International Agency for Research on Cancer, Lyon, 1984, Vol. 57, pp. 113–119.

57. Arthur, C.L. and Pawliszyn, J., Solid-phase microextraction with thermal-desorption using fused-silica optical fibers, *Anal. Chem.*, 62, 2145, 1990.

58. Kataoka, H., Lord, H.L., and Pawliszyn, J., Applications of solid-phase microextraction in food analysis, *J. Chromatogr. A*, 880, 35, 2000.

59. Ventanas, S., Martin, D., and Ruiz, E.J., Analysis of volatile nitrosamines from a model system using SPME–DED at different temperatures and times of extraction, *Food Chem.*, 99, 842, 2006.

60. Young, J.C., Detection and determination of N-nitrosoamines by thin-layer chromatography using fluorescamine, *J. Chromatogr.*, 124, 17, 1976.

61. Young, J.C., Detection and determination of N-nitrosoaminoacids by thin-layer chromatography using fluorescamine, *J. Chromatogr.*, 151, 215, 1978.

62. Walters, C.L., Johnson, E.M., and Ray, N., Separation and detection of volatile and non-volatile N-nitrosamines, *Analyst*, 95, 485, 1970.

63. Hasebe, K. and Osteryoung, J., Differential pulse polarographic determination of some carcinogenic nitrosamines, *Anal. Chem.*, 47, 2412, 1975.

64. Ohshima, H. and Kawabata, T., Gas chromatographic separation of hydroxylated N-nitrosamines, *J. Chromatogr.*, 169, 279, 1979.

65. Wolfram, J.H., Feinberg, R.C., and Fiddler, W., Determination of N-nitrosoproline at the nanogram level, *J. Chromatogr.*, 132, 37, 1977.

66. Cárdenes, L., Ayala, J.H., Gonzáles, V., and Afonso, A.M., Fast microwave-assisted dansylation of N-nitrosamine analysis by high-performance liquid chromatography with fluorescence detection, *J. Chromatogr. A*, 946, 133, 2002.

Chapter 14

Polycyclic Aromatic Hydrocarbons

Peter Šimko

Contents

14.1 Introduction 442
    14.1.1 Principles of Smoking 442
        14.1.1.1 Traditional Procedures of Smoking 442
        14.1.1.2 Alternatives to Traditional Smoking Procedures 443
    14.1.2 Polycyclic Aromatic Hydrocarbons 443
        14.1.2.1 Behavior of Polycyclic Aromatic Hydrocarbons in an Organism 443
        14.1.2.2 Legislative Aspects and International Normalization of Polycyclic
                Aromatic Hydrocarbons in Smoked Meat and Liquid Smoke Flavor 444
14.2 Analysis of Polycyclic Aromatic Hydrocarbons 444
    14.2.1 Sample Preparation 445
        14.2.1.1 Sample Treatment of Smoked Meat 445
        14.2.1.2 Sample Treatment of Liquid Smoke Flavors 446
    14.2.2 Preseparation Procedures 447
        14.2.2.1 Thin-Layer Chromatography 447
        14.2.2.2 Gas Chromatography 448
        14.2.2.3 High-Pressure Liquid Chromatography 448
    14.2.3 Comparison of Gas Chromatography and High-Pressure
        Liquid Chromatography 456
    14.2.4 Occurrence of Polycyclic Aromatic Hydrocarbons 457
References 457

14.1 Introduction

Meat smoking is one of the oldest food technologies, having been used by mankind for a minimum of 10,000 years. Probably as a protection against canines a man might hang a catch over the fire, and from this smoking came to be widely used, not only for the production of smoked products with a special organoleptic profile, but also for its inactivating effects on enzymes and microorganisms. The techniques of smoking have gradually improved and various procedures have been developed in different regions for treating meat and fish. Currently, the technology is used mainly for enrichment of foods with specific taste, odor, and appearance that are in wide demand on the market. On the other hand, the role of the preservative effects is gradually diminishing in importance as a result of more recent trends in alternative preservation procedures. Today it is supposed that the technology is used, in many forms, to treat 40–60% of the total amount of meat products [1] and 15% of fish [2].

14.1.1 Principles of Smoking

In general, smoke is a polydispersed mixture of liquid and solid components with diameters of 0.08–0.15 μm in gaseous phase of air, carbon oxide, carbon dioxide, water vapor, methane, and other gases. Smoke has a variable composition depending on various conditions including procedure and temperature of smoke generation, origin and composition of wood, water content in wood, etc. [1]. To date, up to 1100 various chemical compounds have been identified and published in the literature [3]. The smoking treatment itself is based on successive deposition of compounds such as phenol derivates, carbonyls, organic acids and their esters, lactones, pyrazines, pyrols, and furan derivates [4] on a food surface and their subsequent migration into the food bulk. Smoke is generated during thermal combustion of wood, consisting roughly of 50% cellulose, 25% hemicellulose, and 25% lignin, with limited access to oxygen. The thermal combustion of hemicelluloses, cellulose, and lignin occurs at 180–300, 260–350, and 300–500°C, respectively. However, the decomposition of the wood components also proceeds at temperatures reaching up to 900°C and, in the presence of an excess of oxygen, even 1200°C. The smoke produced at 650–700°C is richest in components able to impart desirable organoleptic properties to treated products. The temperature of generation of smoke can be decreased by increasing the humidity of the wood [5]. The quantitative composition of smoke depends not only on the kind of wood used, on the temperature of the generation, and the excess of oxygen, but also on cleaning procedures applied immediately after smoke generation [1].

14.1.1.1 Traditional Procedures of Smoking

After generation, smoke is driven into a kiln, during which time its temperature is going down, which is accomplished by partial condensation of smoke components (especially compounds with high boiling point) in pipes, walls, or on foods. The rate of smoke deposition depends on the temperature, humidity, volatility, and velocity of a smoke stream. When the smoke comes into contact with a food surface, there are three modes of smoke treatment procedures, related to the temperature of smoke, as follows:

1. *Cold smoking.* Temperature of the smoke between 15 and 25°C (used for aromatization of uncooked sausage, raw hams, and fermented—not thermally treated—salami)
2. *Warm smoking.* Temperature between 25 and 50°C (used for aromatization and mild pasteurization of frankfurters, sausages, meat pieces, and gammon)
3. *Hot smoking.* Temperature between 50 and 85°C (used for both aromatization and thermal treatment of hams, salami, sausages, etc.)

To achieve a rich, deep brown coloring on the surface and very strong aroma profile formation, the time of smoking must be considerably prolonged. Such products are frequently termed "black-smoked" or "farmhouse-smoked." These products contain far higher contents of polycyclic aromatic hydrocarbons (PAH) [3,6]. "Wild" smoking occurs under uncontrolled technological conditions and without legislative regulation, which is typical for households and developing countries; this can lead to very high PAH content in smoked foods [7–9].

14.1.1.2 *Alternatives to Traditional Smoking Procedures*

A Kansas pharmacist named Wright developed and patented the first liquid smoke flavor (LSF) to be prepared from primary smoke condensate in the late nineteenth century. The use of LSF has important advantages: It reduces considerably the time necessary to reach the required organoleptic profile of flavored foods and makes it possible to control more effectively the "addition" of contaminants, including PAH, into aromatized products. Currently, LSF is used in the following forms:

- Liquids for spraying, nebulization, immersion, or showering
- Emulsions incorporated into foods by injection or mixing
- Water-mixable emulsions for showering or curing brine
- Powders such as maltodextrins, salt, saccharides, starch, proteins, and seasonings
- Solutions in vegetable oils [10].

14.1.2 *Polycyclic Aromatic Hydrocarbons*

Apart from the compounds mentioned earlier, there are also conditions suitable for formation of other compounds during smoke production. One of the most important groups that are actually harmful to human health are PAH. These are formed during the thermal decomposition of wood, especially under limited oxygen access, in the range of 500–900°C [11]. PAH are characterized by two or more condensed aromatic rings in a molecular structure and have a strong lipophilic character. The temperature of smoke generation plays a decisive role, because the amounts of PAH contained in smoke (which are formed during a pyrolysis) increase linearly with the temperature of smoke generation in the interval of 400–1000°C [12]. Apart from the formation of the compounds, the temperature also affects the structure and number of PAH. The number of PAH present in smoked fish can reach up to 100 different compounds [13] that have various effects on living organisms.

14.1.2.1 *Behavior of Polycyclic Aromatic Hydrocarbons in an Organism*

According to current knowledge, some PAH are able to interact in organisms with enzymes (such as aryl hydrocarbon hydroxylases) to form PAH dihydrodiol derivatives. These reactive products (so-called "bay region" dihydrodiol epoxides) are believed to be ultimate carcinogens that are able to form covalently bounded adducts with proteins and nucleic acids. In general, deoxyribonucleic acid (DNA) adducts are thought to initiate cell mutation, resulting in a malignancy [11]. A direct mutagenic potential of 14 PAH and PAH, containing fractions isolated from smoked and charcoal-broiled samples, was studied for strains TA 98 and TA 100 using the Ames test. The greatest potential mutagenicity was observed with PAH fractions isolated from smoked fish treated before smoking with nitrites in an acid solution [14]. To simplify an interpretation of the real risk of PAH to human health, there have been attempts to express objectively the risk using toxic equivalency factors (TEF) [15]. However, this approach does not reflect wider aspects of the potential toxicity of oxidized PAH products due to the effect of ultraviolet (UV) light or other environmental factors [16]. Moreover, PAH content in smoked foods can be affected not only by environmental factors, but also by diffusion processes from plastic packaging materials [17].

14.1.2.2 Legislative Aspects and International Normalization of Polycyclic Aromatic Hydrocarbons in Smoked Meat and Liquid Smoke Flavor

With regard to the harmful effects of PAH on living organisms, some European countries have enacted maximum limits for these compounds in smoked meat products. To simplify problems associated with the variability of PAH composition, benzo[*a*]pyrene (BaP) has been accepted as the indicator of total PAH presence in smoked foods, although BaP constitutes only between 1 and 20% of the total carcinogenic PAH [18]. At present, the situation in the European Union (EU) has been resolved by adoption of the European Commission (EC) Regulation 208/2005 limiting BaP content to a level of 5 μg kg^{-1} in smoked meats, smoked meat products, muscle meat of smoked fish, and smoked fish products. The regulation entered into a force as of February 28, 2005, to be applied from April 1, 2005. The EC has also adopted Directive 2005/10/EC, describing sampling methods and methods of analysis for the official control of BaP levels in foodstuffs and the recommendation 2005/108/EC on the further investigation into the levels of PAH in certain foods, such as benzo[*a*]anthracene (BaA), benzo[*b*]fluoranthene (BbF), benzo[*j*]fluoranthene (BjF), benzo[*k*]fluoranthene (BkF), benzo[*g,h,i*]perylene (BghiP), chrysene (Chr), BaP, cyclopenta[*c,d*] pyrene (CcdP), dibenzo[*a,h*]anthracene (DahA), dibenzo[*a,e*]pyrene (DaeP), dibenzo[*a,h*]pyrene (DahP), dibenzo[*a,i*]pyrene (DaiP), dibenzo[*a,l*]pyrene (DalP), indeno[*1,2,3-cd*]pyrene (IcdP), and 5-methylchrysene. The Joint Expert Committee on Food Additives (JECFA) of FAO and WHO (JECFA) has defined another compound benzo[*c*]fluorene (BcF), which should also be monitored with regard to its effects on living organisms. Concerning LSF, the EC has adopted Regulation 2065/2003, relating to the production of smoke flavorings intended to be used for food flavoring. This regulation limited the maximum acceptable concentrations of BaP to 10 μg kg^{-1} and BaA to 20 μg kg^{-1} in these products. Finally, the Directive 88/388/EEC limited the maximum residual levels of BaP to 0.03 μg kg^{-1} in foodstuffs flavored by LSF. For international trade purposes, JECFA has adopted a specification that tolerates the concentration in LSF at the levels of 10 μg kg^{-1} for BaP, and 20 μg kg^{-1} for BaA [19].

14.2 Analysis of Polycyclic Aromatic Hydrocarbons

Owing to the fact that PAH are present in food at the micrograms per kilogram levels, analysis usually consists of such steps as extraction/hydrolysis of food matrix, liquid/liquid partition, cleanup procedures, concentration, chromatographic separation, and, of course, determination. Although all steps are very important, chromatographic separation is the most important for correct evaluation of real risk assessment; for example, while BaP is a very strong carcinogenic agent, the carcinogenic activity of its isomer benzo[*e*]pyrene (BeP) is quite low. The methodology of PAH analysis has been strongly affected by levels of development of chromatographic methods. In the middle of the last century, a separation of BaP isomers by paper and column chromatography was practically impossible [20]. With regard to complex mixtures of PAH, the presence of a variety of interfering substances and the need to assess correctly the concentrations of the most dangerous compounds made it necessary to overcome problems regarding resolution of so-called "benzopyrene fraction," which consisted of BaP and its isomer BeP, BkF, BbF, and perylene (Per). In 1968, at a joint meeting of Indiana University Cancer Center and the International Agency for Research on Cancer, it had been specified that any acceptable analytical method should be capable of separating at least BaA, BaP, BeP, BghiP, pyrene (Py), BkF, and Cor [21]. Collaborative studies of a method specific for BaP and a general procedure for PAH were conducted under the auspices of the Association of Official Analytical Methods (AOAC) and the International Union of Pure and Applied Chemistry (IUPAC). Procedures consisted of an initial saponification of the sample in ethanolic potassium hydroxide solution, followed by a partition step involving dimethylsulfoxide (DMSO) and an aliphatic solvent, followed by column chromatography on pretreated Florisil. For determination of individual PAH, a cellulose reverse-phase technique in conjunction with cellulose acetate multiphase technique was used. This method was adopted as an AOAC official first action method in 1973 and accepted as a recommended method by IUPAC. Statistical evaluation of the data obtained by interlaboratory tests, in which ham samples were fortified with BaP, BeP, BaA, and BghiP at a level of 10 $\mu g\ kg^{-1}$ and analyzed by the aforementioned method, showed standard deviation between 7.4 and 12.7%. On this basis, the method has been adopted as official method of the AOAC [22].

14.2.1 Sample Preparation

Smoked meat and LSF represent two different matrices, which have in common the organoleptic profile and compounds to be determined. For this, various procedures for sample pretreatment are taken to reach the highest recoveries of analytes possible.

14.2.1.1 Sample Treatment of Smoked Meat

From an analytical point of view, meat and its products belong to problematic matrices with regard to the presence of various interfering compounds. Moreover, PAH, as lipophile compounds, have a tendency to diffuse not only into the nonpolar part of the sample but also inside tissue cells depending on the existing concentration gradient. For this reason a simple solvent extraction with nonpolar solvent seems to be insufficient to reach high recovery. Grimmer and Böhnke [13] isolated PAH from smoked fish and smoked-dried cobra with boiling methanol prior sample hydrolysis with methanolic KOH. It was found that only about 30% of BaP and other PAH was extractable from the samples, whereas an additional alkaline hydrolysis of meat protein yielded another 60% of PAH. It was concluded that PAH were linked adsorptively to high molecular-weight structures not destroyed with boiling methanol. Although more than 80% of the methanol used could be recovered, this contained only one-third of the PAH contained in sample. As postulated, alkaline hydrolysis with aqueous methanolic KOH is an absolute necessity to isolate PAH quantitatively from such samples. Alkaline hydrolysis usually takes 2–4 h of time, depending on the character of the sample. Lean tissues take less time than adipose and collagen containing tissues. This sample treatment has been adopted in many experimental works [23–26]. On the other hand, in a study by Vassilaros et al. [27], the use of an alcohol is superfluous and contributes to interference problems because of methyl esters formed from fatty acids and methanol, which are than difficult to remove from the PAH fraction. Takatsuki et al. [28] found that during alkaline hydrolysis BaP may be partially decomposed by the coexistence of alkaline conditions, light oxygen, and peroxides in aged ethyl ether. They proposed to use amber glass, the addition of Na_2S as an antioxidant, distillation with ethyl ether just before use, and prevention of air from contact with adsorbents. To protect PAH from light decomposition, Karl and Leinemann [29] used brown glassware carefully rinsed with acetone before using an alkaline hydrolysis. Some authors also recommended direct extraction with organic solvents. Potthast and Eigner [30] proposed a procedure based on mixing of preground sample with chloroform and anhydrous Na_2SO_4 to remove water from the extract. After adding Celite, the portion became uniformly distributed over the surface of the adsorbent. Although the authors achieved a recovery 95–100% of added BaP at a level of 10 μg, there is an assumption that they recovered only "free" PAH accessible with solvent. This procedure was also used in the work of Alonge [8]. Cejpek et al. [31] tested the efficiency of several organic solvents to obtain fat from meat samples. The most efficient solvent was a mixture of chloroform:methanol (2:1); less effective was chloroform; and the worst yields were achieved with methanol. This confirms observations of Grimmer and Böhnke [13] regarding the inability of methanol to extract quantitatively PAH from meat samples. The chloroform–methanol mixture, called the Folch agent, is widely used in food analysis for the extraction of lipids, while methanol makes possible the extraction of lipids from inside cells by denaturation of the cell wall proteins. Joe et al. [32] digested samples of smoked food with KOH, with PAH extracted with Freon 113 (1,1,2-trichloro-1,2,2-trifluoroethane). Chen et al. [33] compared the efficiency of extraction from freeze-dried sample using sonication and Soxhlet procedures. Recovery studies showed that Soxhlet extraction was more suitable than the sonication method. An accelerated procedure of extraction was tested by Wang et al. [34]. Samples were extracted in a Dionex extractor as well as a Soxhlet apparatus. Advanced solvent extraction (ASE) technique was found to be comparable to or even better than the reference Soxhlet method, and significant reductions in time of extraction and solvent consumption were achieved. García-Falcón et al. [35] accelerated extraction of PAH from freeze-dried samples into hexane with microwave treatment and hexane extract, then saponified with ethanolic KOH.

14.2.1.2 Sample Treatment of Liquid Smoke Flavors

Sample treatment of LSF matrix is different from the treatment of processed meats due to easy access of organic solvent "inside" a liquid matrix. For this, there is not usually any reason to treat samples by time-consuming hydrolysis under reflux. Other situations could arise when LSF are in solid state (e.g., applied on starch, gelatine, or encapsulated). Despite this, some authors preferred alkaline hydrolysis of liquid LSF under reflux. However, addition of KOH is strongly recommended to transform phenols to polar, nonextractable phenolates prior the PAH extraction with a nonpolar solvent. White et al. [36] alkalized water-soluble LSF (and also resinous condensates that settled out of LSF after storage) with KOH solution and extracted PAH into isooctane. Silvester [37] extracted PAH from alkalized liquid SFA with hexane. Radecki et al. [38] alkalized LSF with ethanolic KOH solution and maintained it at 60°C for 30 min prior to extraction into cyclohexane. After alkalization, a direct extraction of PAH with cyclohexane was used by Šimko et al. [39]. On the other hand, Gomaa et al. [40] saponified liquid LSF with methanolic KOH for 3 h and than extracted PAH into cyclohexane. Laffon Lage et al. [41] used a solid-phase extraction (SPE) technique on Sep Pak C18 for PAH isolation and compared it to the supercritical fluid extraction (SFE) procedure, in which the sample for SFE was mixed with alumina and extracted PAH were concentrated in an octadecylsilane (ODS) trap. In both cases, 91% recoveries of BaP spiked at 15 ng were found and no statistically significant differences were observed. Taking into account the expensive SFE extractor, they recommended the use of the simple SPE procedure. Guillén et al. [42,43] alkalized LSF with methanolic KOH and heated under reflux for 3 h, following with extraction of PAH into dichlormethane or cyclohexane.

14.2.2 Preseparation Procedures

At this time, both procedures are more or less equivalent for processed meats and LSF. But sometimes, mainly after adipose tissue hydrolysis, a presence of lipoproteins in nonpolar solvent requires removal prior to preseparation with a one-step liquid–liquid partition between nonpolar and polar solvent (e.g., hexane–water/dimethylforamid [13], methanol/water, or DMSO/water–cyclohexane [26,29]), a two-step liquid–liquid partition (e.g., NaCl/water and dimethylformamide/water [44]), or precipitation of lipoproteins with Na_2WO_4 [6,45–47]. For preseparation, deactivated Florisil [6,26,34,40,43,47–49], silica gel [25,28,48], alumina [44], and Celite [36,37] are used frequently. Only one study [37] reported that elution of BaP from Florisil and silica gel with hexane was impossible, and for this reason alumina was recommended for preseparation of concentrated PAH extracts. Guillén et al. [44] preferred elution of silica with cyclohexane prior to Florisil dichlormethane elution to obtain higher recoveries, with reduced amounts of interfering substances, which were eluted from Florisil with dichlormethane. Another preseparation procedure is gel permeation chromatography (GPC) on Sephadex LH 20 [28] or BioBeads S-X3 [31]. Mottier et al. [48] cleaned concentrated cyclohexane extracts by SPE, using conditioned isolute aminopropyl and C_{18} columns. Also, the use of two different cleaning techniques is possible, with cyclohexane extract first cleaned with GPC on Sephadex LH 20, then cleaned on silica gel [44]. The last procedure can also be carried out in reverse mode [9]. In all cases, removal of organic solvents by vacuum evaporation to concentrate PAH is an unavoidable operation. This may be a critical step, especially if there is a presumption of the presence of light PAH such as fluorene (Flu), anthracene (Ant), or phenanthrene (Phe) in the extracts. In this case, organic solvents should not be evaporated to dryness because these PAH could be lost due their volatility. This cautious manipulation is not necessary if only PAH with boiling points above 370°C are determined [13].

14.2.2.1 Thin-Layer Chromatography

Thin-layer chromatography (TLC) is one of the older analytical methods used for determination of PAH in various matrices. Haenni [50] discussed the development of analytical tools for control of PAH in food additives and in food by the use of UV specification within specific wavelength ranges. Schaad [20] reviewed various chromatographic separation procedures, including TLC. White et al. [36] used two systems for PAH separation. The first consisted of 20% *N,N*-dimethylformamide in ethyl ether as the stationary phase and isooctane as the mobile phase. Fluorescent spots were scraped out from cellulose layer and eluted with hot methanol. After concentration, the sample was developed in the second system, using ethanol–toluene–water (17:4:4) as developer. Fluorescent spots were eluted again from the cellulose acetate layer and a UV spectrum was recorded against isooctane in a reference cell. The observed maxima were compared with those in the spectra of known PAH obtained under the same instrumental conditions. Estimation of the quantity of the identified compounds was made by the baseline technique in conjunction with spectra of these PAH and the identification was confirmed by spectrophotofluorometry. This method has become a base of AOAC Official Method 973.30, adopted in 1974 [22].

14.2.2.2 Gas Chromatography

Currently, gas chromatography (GC) is widely used for determination of PAH in food analysis. The determination of the large number of PAH in samples requires columns with high efficiency. To separate some critical pairs as well as isomers of methyl derivatives of certain PAH, capillary columns (50 m × 0.3–0.5 mm) which can achieve 50,000–70,000 high equivalent theoretical plate (HETP) are especially convenient. However, packed columns used for determination of PAH [13] had lower HETP, ranging between 20,000 and 30,000, and for this reason were not suitable for quantity determination. Two stationary phases, OV-17 and OV-101, were used for separation of BaP from BeP, DajA from DahA, and Phe from Ant. Successful separation of Chr from BaA was achieved using the OV-17 stationary phase, but separation of BbF, BjF, and BkF isomers on packed columns was not possible [13]. Radecki et al. [38] tested various stationary phases (GE SE 30, OV-1, SE-52, OV-7, OV-101, BMBT, BBBT) on Chromosorb W, Chromosorb W HP, Gas Chrom, and Diatomite CQ supports in packed columns to develop a precise GC method for assaying BaP in LSF. Separation of BaP from BeP and Per was not possible using SE 30, OV-1, SE-52, OV-7, or OV-101 stationary phases. Nematic phases gave a good separation of BaP from its isomers, but they were not suitable for analysis due to their poor thermal stability. Detection of PAH is not a serious problem, because the response of a flame ionization detector (FID) is practically equal for all compounds and is linear over a large concentration range (about $1–1.10^6$), according to the carbon content. However, the use of FID is sometimes hampered by the need for very thorough cleanup procedures with the accompanying risk of severe losses and possible misidentification [51]. A mass spectrometry detector (MSD) has also successfully been used for PAH analysis in many cases [52]. In particular, the use of MSD operating in selected ion monitoring mode makes it possible to simplify the time-consuming cleanup procedure [51], and it is recommended especially for quantitative analysis. The ion trap detector (ITD) has some advantages over traditional MSD. The ITD utilizes electric fields to hold ions within the ion storage regions. The ITD is then scanned through the mass range, causing the ions to be ejected from this region sequentially, from low to high mass. The ejected ions are detected by a conventional electron multiplier. Thus, the characteristic of the ITD is that ionization and mass analysis take place in the same space. This contrasts with a conventional MSD, which requires a separated ionization source, focusing lenses and analyzer [53]. Sometimes, separation of isomers is quite a serious problem even when capillary columns are used. Dennis et al. were not able [54] to separate BjF from BkF. Speer et al. [55] were not able to separate Chr from triphenylene (Tph); BbF, BjF, and BkF from each other; or DahA from DacA. Problems associated with separation of Chr from Tph are also reported in works of Guillén et al. [42,43]. Wise et al. [56] discussed difficulties in separating isomers BbF and BkF. On the other hand, Chen and Chen [57] separated BbF and BkF sufficiently on a DB-1 fused silica capillary column. Review of preseparation procedures as well as GC conditions to be used for determination of PAH in smoked meat products and LSF are summarized in Table 14.1.

14.2.2.3 High-Pressure Liquid Chromatography

In recent years, the high-pressure liquid chromatography (HPLC) method has been used intensively for determination of PAH in food, as reported in review works [11,58,59]. Formerly used stationary phases such as alumina and silica gel were later replaced with chemically bonded phases, particularly reverse phases such as ODS, widely used at the time. For determination of PAH in food, Hunt et al. [60] developed a pthalimidopropylsilane (PPS) stationary phase and compared

Table 14.1 Preparation Procedures as Well as GC Conditions to Be Used for Determination of PAH in Smoked Meat Products and LSFs

Sample	Sample Treatment and Preseparation	Column/Stationary Phase	Temperature Program	Detection	Reference
Barbecued sausages	Saponification with mixture of ethanol, water, and KOH, extraction with cyclohexane, preseparation by SPE on isolute aminopropyl and C_{18} columns	25 m × 0.2 mm capillary column/SPB-5	80°C for 0.5 min → 230°C at 8°C min^{-1} → 300°C at 5°C min^{-1}	MSD	48
Smoked fish	Extraction with pentane, precleaning on silica gel and Sephadex LH-20	25 m × 0.2 mm quartz capillary column/SE-54	100 → 260°C, 3°C min^{-1}	MSD	9
Smoked fish	Saponification in methanolic KOH, liquid–liquid extraction (methanol–water–cyclohexane and DMF–water–cyclohexane), and GPC on Sephadex LH 20	10 m × 2 mm packed columns/5% OV-101 and OV-17 on sorbent Gas Chrom.	120 → 250°C, 1°C min^{-1} 250°C isothermal	FID, MSD	13
Smoked sausages	Saponification in methanolic KOH, liquid–liquid extraction (methanol–water–cyclohexane and DMF–water–cyclohexane), precleaning on silica gel, and GPC on Sephadex LH 20	10 m × 2 mm packed column/5% OV-101 on sorbent Gas Chrom.	260°C isothermal	FID	23
Smoked meat products	Saponification with mixture of methanol, water, and KOH, partition with DMF, precleaning on Kiesel gel 60	25 m × 0.28 mm capillary column/SE-54	240°C isothermal	MSD	24
Smoked fish and fish products	Saponification in methanolic KOH, liquid–liquid extraction (methanol–water–cyclohexane and DMF–water–cyclohexane), precleaning by CC on silica gel, and GPC on Sephadex LH 20	55 m × 0.3 mm glass capillary column/SE-54	165°C for 6 min, 165 → 255°C, at 4°C min^{-1}	FID	25

(continued)

Table 14.1 (continued)   Preseparation Procedures as Well as GC Conditions to Be Used for Determination of PAH in Smoked Meat Products and LSFs

Sample	Sample Treatment and Preseparation	Column/Stationary Phase	Temperature Program	Detection	Reference
Smoked fish, smoked meat spreads	Saponification with mixture of methanol, water, and KOH, extraction with cyclohexane, cleanup on Florisil, partitioning with DMSO/hexane	30 m × 0.25 mm capillary column/DB-5	$25 \rightarrow 180°C$ rapidly $\rightarrow$ 320°C at 8°C min^{-1}	FID, MSD	26
Smoked fish	Saponification with methanol–water–KOH mixture under reflux, extraction into cyclohexane, extraction of PAHs with caffeine/formic acid, washing with NaCl solution, extraction into cyclohexane, preseparation on silica gel	30 m × 0.25 mm capillary fused silica column/DB-5	110°C isothermal for 1.5 min $\rightarrow$ 210°C at 30°C min^{-1} $\rightarrow$290°C at 3°C min^{-1} $\rightarrow$ 300°C at 10°C min^{-1}	MSD	29
Smoked salmon, sausages, pork	Direct solvent extraction (ASE), cleanup on Florisil	30 m × 0.25 mm capillary column/cross-linked 5% phenyl methyl siloxane HP-5MS	40°C isothermal for 1 min $\rightarrow$ 250°C at 12°C/min $\rightarrow$ 310°C at 5°C/min	MSD	34
LSF	Alkalization with KOH solution, extraction with cyclohexane, cleanup on silica	25 m × 0.2 mm fused silica capillary column/HP-5 cross linked with 5% henylmethylsilicone	50°C isothermal for 0.5 min $\rightarrow$ 180°C at 30°C min^{-1} $\rightarrow$ 300°C at 7°C min^{-1}	MSD	39

LSF	Heating with methanolic KOH under reflux, extraction with cyclohexane, cleaningup by SPE technique on Florisil	60 m × 0.25 mm fused silica capillary column/HP-5MS, 5% phenyl methyl siloxane	50°C isothermal for 0.5 min → 130°C at 8°C min⁻¹ → 290°C at 5°C min⁻¹	MSD	42
LSF	Heating with methanolic KOH under reflux, extraction with cyclohexane, cleaningup by SPE technique on LC silica	60 m × 0.25 mm fused silica capillary column/HP-5MS, 5% phenyl methyl siloxane	50°C isothermal for 0.5 min → 130°C at 8°C/ min → 290°C at 5°C/ min	MSD	43
Smoked meats	Saponification with methanolic KOH, extraction with cyclohexane, partition with DMF/water, cleanup on silica gel, and with GPC on Bio Beads S-X3	50 capillary column/DB-5	70 → 280°C at 5°C min⁻¹	MSD	55
Smoked chicken	Extraction with methanol in Soxhlet app., + KOH, extraction into *n*-hexane, cleanup on Sep-Pak Florisil cartridge	30 m × 0.32 mm/DB-5	70°C isothermal for 1 min → 150°C at 10°C min⁻¹ → 280°C at 4°C min⁻¹ hold for 14 min	ITD	63

Source: Reprinted with permission from Šimko, P., *J. Chromatogr. B*, 770, pp. 3–18, 2002. Copyright © Elsevier Science B.V. 2002.

Table 14.2 Preseparation Procedures as Well as HPLC Conditions to Be Used for Determination of PAH in Smoked Meat Products and LSFs

Sample	Sample Treatment and Preseparation	Column/Stationary Phase	Mobile Phase	Detection	Reference
LSF, smoked meats	Saponification with ethanolic KOH, extraction into cyclohexane, washing with saturate NaCl solution, cleanup on silica gel	25 cm × 4 mm Lichrosorb RP 18	Acetonitrile/water 8:2, isocratic, 1.5 mL min^{-1}	FLD Ex: 305, 381 nm Em: 389, 430, 520 nm	7
Smoked meat products	Saponification with mixture of methanol, water, and KOH, extraction with cyclohexane, washing with Na_2WO_4 solution, cleanup on Florisil	30 cm × 3 mm, Separon SGX C$_{18}$ RP, 5 μm	Acetonitrile/water 3:1, isocratic, 1.5 mL min^{-1}	FLD Ex/Em 310/410 nm	6,45–47
Smoked fish, smoked meat spreads	Saponification with mixture of methanol, water, and KOH, extraction with cyclohexane, cleanup on Florisil, partitioning with DMSO/hexane	25 cm × 4.6 mm, RP–18, 5 μm	Acetonitrile/water 7:3, isocratic, 3 mL min^{-1}	UVD 254 nm FLD Ex/Em 250/370 nm	26
Fish, shellfish	Saponification with methanol–water–KOH mixture under reflux, extraction into n-hexane, cleanup on silica gel	Radial-Pak PAH	Acetonitrile/water 8:2, isocratic, 1 mL min^{-1}	FLD Ex/Em 370/410 nm	28
Smoked fish	Saponification with methanol–water–KOH mixture under reflux, extraction into cyclohexane, extraction of PAHs with caffeine/formic acid, washing with NaCl solution, extraction into cyclohexane, preseparation on silica gel	ET 15 cm × 4 mm, Nucleosil 5 C$_{10}$ PAH	Acetonitrile/water 7:3 for 1 min, then gradient linearly up to 9:1 in 19th min, then to 100% acetonitrile from 20 to 40 min, then isocratic till 55 min	UVD 240, 254, 260 nm FLD Ex/Em 300/408 and 280/395 nm	29

Sample	Extraction/Cleanup	LC Column	Mobile Phase	Detection	Ref.
Smoked sausage, smoked meat	Extraction with chloroform/methanol mixture, preseparation by GPC on Bio Beads S-X3	15 cm × 4.6 mm Supelcosil LC PAH, 5 µm	A: methanol/acetonitrile/water 50:25:25 B: acetonitrile; 1 min 100% A, 25th min 100% B	FLD Variable Ex (240–293) Em (340–498) nm	31
Smoked frankfurters, smoked meats	Extraction with methanol in Soxhlet app. + KOH, extraction into *n*-hexane, cleanup on Pep-Pak Florisil	12.5 cm × 4.6 mm Envirosep-pp C_{18} 5 µm	I. Acetonitrile/water 7:3, isocratic, 2 mL min^{-1} II. Acetonitrile/water 40:60, gradient to 100% acetonitrile within 25 min III. Acetonitrile/water 55:45, gradient to 100% acetonitrile within 23 min	UVD 230–360 nm FLD Variable Ex (232–302) Em (330–484) nm	33
LSF	Alkalization with NaOH solution, extraction with hexane, cleanup on alumina	25 cm × 4.6 mm Partisil 10 ODS	Methanol/acetonitrile/water 35:35:30, isocratic	FLD Ex/Em 280/390 nm	37
LSF	Alkalization with ethanolic and aqueous NaOH, extraction into cyclohexane, partitioning with DMSO/water, extraction into cyclohexane	30 cm × 4 mm, µBondapak C_{18}/Corasil	Methanol/water 7:3, 2 mL min^{-1}	UVD 280 nm	38
LSF, smoked food products	LSF: Saponification with methanolic KOH, extraction into cyclohexane, purification on Florisil Meat products: digestion with KOH solution, extraction with Freon 113, purification on Florisil	25 cm × 4.6 mm, Supelcosil LC-PAH	Acetonitrile/water 60:40 for 5 min, then 100% of acetonitrile in 15 min hold for 15 min, then decrease to 60% over 10 min	FLD Ex/Em 254/375 nm	40

(continued)

Table 14.2 (continued)   Preseparation Procedures as Well as HPLC Conditions to Be Used for Determination of PAH in Smoked Meat Products and LSFs

Sample	Sample Treatment and Preseparation	Column/Stationary Phase	Mobile Phase	Detection	Reference
Smoked fish	Direct extraction with chloroform, preseparation on preparation silica column	Preparation column: 25 cm × 4.6 mm, silica 5 μ	Preparation column: pentane/5% DCM, 0.8 mL min^{-1}	FLD	63
		Analytical column: 15 × 4.6 mm 5 μ particle, Supelcosil LC-PAH	Analytical column: water/ acetonitrile 6:4 for 5 min, then to 100% acetonitrile over 40 min, 1.5 mL min^{-1}	Variable nm	
Smoked fish, ham	Saponification with mixture of methanol, water, and KOH, extraction with cyclohexane, partitioning with DMSO/ hexane	Spherisorb ODS 5 μm precolumn and 5 μm VydacODS analytical column	Acetonitrile/water 6:4, linearly to 9:1 over 35 min	FLD Ex/Em 290/430 nm	54
Smoked meats products	Saponification with methanolic KOH, extraction with n-hexane, preseparation by SPE on CN bonded silica	Nucleosil 100–5 C 18 PAK	Acetonitrile/water 8:2, isocratic, 0.5 mL min^{-1}	FLD Ex/Em 290/430 nm	64
Smoked fish	Saponification with methanol–water– KOH mixture under reflux, extraction into n-hexane, cleanup on silica gel	15 cm × 6 mm, ODS, 5 μm particles, 1 mL min^{-1}	Acetonitrile/water 8:2, isocratic	FLD Ex/Em 370/410 nm	65

LSF, smoked foods	LSF: Saponification with methanolic KOH, extraction into cyclohexane, purification on Florisil	12.5 cm × 4 mm Lichrosphere 100 RP−18	A: water; B: methanol/acetonitrile 1:1 I. segment: 1:80–100% B for 20 min II. segment: 100% B for 5 min	FLD	66
	Smoked products: digestion with KOH solution, extraction with Freon 113, purification on Florisil		III. segment: 100–80 B for 5 min	Ex/Em 365/418 nm	
Smoked meat products	Saponification with methanolic KOH, extraction into cyclohexane, preseparation by SPE on Kiesel gel	12.5 cm × 4 mm, Chrompack PAH-Säule	Acetonitrile/water 9:1, isocratic, 0.5 mL min^{-1}	FLD Ex/Em 290/430 nm	67
Smoked chicken	Extraction with methanol in Soxhlet app. + KOH, extraction into n-hexane, cleanup on Sep-Pak Florisil	12.5 cm ×4.6 mm Envirosep-pp 5 μm C$_{18}$	Acetonitrile/water 55:45, gradient to 100% acetonitrile within 23 min 1.2 mL min^{-1}	FLD Variable nm	68

Source: Reprinted with permission from Šimko, P., *J. Chromatogr.* B, 770, 2002, pp. 3–18. Copyright © Elsevier Science B.V. 2002.

Note: Ex, Excitation and Em, Emission.

it with ODS. The PPS column was able to separate BkF from Per, which was impossible by ODS column. HPLC has some advantages in PAH analysis, as follows [58]:

- Separation of isomers shows very good resolution
- Sufficient sensitivity and specificity of ultraviolet detection (UVD) and fluorescence detection (FLD)
- Molecular sizes of PAH can be estimated on the base of retention time using a reversed-phase (RP) column
- Ability to determine compounds with high molecular weight
- Analysis is usually carried out at ambient temperature; there is no risk of thermal decomposition of analytes

HPLC equipped with MSD is an effective tool for characterization of high molecular-weight, thermally unstable compounds; for example, BaP metabolites were identified and determined by this method in microbore mode [61]. Owing to a high absorption of light in the UV part of spectrum and intensive fluorescence (FL), both types of detectors are able to detect reliable concentrations at the micrograms per kilogram levels. On the other hand, measurements by nonspecific detection systems, particularly optical detectors, though often precise, can be much less accurate due to possible chemical interferences not having been chromatographically resolved or otherwise avoided prior to the measurement. The major impurities in the PAH fractions appear to be alkylated PAH, which have responses in optical detection systems very similar to their unsubstituted analogs [62]. Regarding diode array detector (DAD), confirmation of peak purity and identification is possible, but due to the broad absorption bands in UV spectra it is highly probable that there will be some interference if one particular wavelength is chosen for quantification. In any case, identification must be based on retention time. The FL detector provides very high selectivity and sensitivity, particularly those with excitation and emission wavelengths that can be varied throughout the analysis. However, FL suffers from not being able to provide "broad-spectrum" analyses (i.e., a wide variety of compounds) because of the presence of alkylated PAH compounds. Review of preseparation procedures as well as HPLC conditions to be used for determination of PAH in smoked meat products and SFA are summarized in Table 14.2.

14.2.3 Comparison of Gas Chromatography and High-Pressure Liquid Chromatography

In many works, authors studied advantages and drawbacks of both methods, with studies aimed especially at recovery procedures, quality of separation processes, time of analysis, price of equipment, etc. Dennis et al. [54] compared results of analysis of some food (two smoked) obtained by GC and HPLC. Thirty-five pairs of analyses were tested using statistical procedure (student *t*-test). Of these, 25 were not significantly different within the 95% confidence limits employed. But data for BkF/benzofluorantenes and DahA/dibenzoanthracenes were not compared because different analytes were measured. Standard deviations indicated that repeatability of both methods was very good, usually within 10%, and provided comparable data throughout a wide range (0.2–1000 μg kg^{-1}). In the conclusion of this study it was stressed that capillary GC possessed a much greater resolving power, in terms of plate number, so that many more PAH can be separated and determined. On the other hand, HPLC was able to separate individual isomers (BbF and BkF; Chr and Tph); that is, it had greater selectivity. Chiu et al. [63] compared separation and detection conditions of both methods analyzing smoked chicken. As found, 16 priority PAH pollutants defined by the Environmental Protection Agency (EPA) can be separated simultaneously by HPLC using a gradient solvent system and detection by FLD at variable wavelength settings due to different excitation/FL spectra. The same mixture can also be separated successfully by GC using an appropriate temperature program. The presence of impurities in smoked meat products can interfere with the identification and quantification of PAH by HPLC. With ITD, the PAH can be identified even in the presence of fat- or PAH-like impurities. The retention times by HPLC were shorter than those by GC, while HPLC had better separation for most compounds than GC. Sim et al. [62] compared GC and HPLC methods analyzing 16 PAH pollutants. Chromatographic resolution involves a combination of column capacity, column efficiency, and separation selectivity. GC has a higher column efficiency and thus has an advantage for complex mixture analysis, but HPLC can often have a higher column selectivity, which is more suitable for separation of isomeric compounds. Thus, the two methods should be viewed as complementary in the analysis of PAH, and they are essential for precise and reliable analysis.

14.2.4 Occurrence of Polycyclic Aromatic Hydrocarbons

After gleaning information regarding carcinogenic effect, research workers started to find real situations of PAH content in smoked meat products. These data prove that technologically correct smoking process contaminate meat products with only small levels of PAH content—usually bellow 1 $\mu g\ kg^{-1}$. Far more dangerous is the smoking process under uncontrolled conditions, typical of home "wild" smoking in the preparation of heavily smoked "farm" products, as well as smoking being done in developing countries, without any technological knowledge or hygienic control. These products bring a serious real risk to consumer in terms of cancer, especially after a long period of consumption due to BaP content reaching even up to 100 $\mu g\ kg^{-1}$ [69].

References

1. Sikorski, Z.E. Traditional smoking, in *Encyclopedia of Meat Sciences*, Jensen, W.K., Devine, C., and Dikeman, K., Eds., Elsevier, London, 2004, pp. 1265–1277.
2. Stołyhwo, A. and Sikorski, Z.E. Polycyclic aromatic hydrocarbons in smoked fish—a critical review. *Food Chem.*, 91, 303, 2005.
3. Wilms, M. The developing of modern smokehouses—ecological and economical aspects. *Fleischwirtschaft Int.*, 4, 8, 2000.
4. Maga, J.A. The flavor chemistry of wood smoke. *Food Rev. Int.*, 3, 139, 1987.
5. Tóth, L. and Potthast, K. Chemical aspects of the smoking of meat and meat products, in *Advances in Food Research*, Chichester, C.O., Mrak, E.M.K., and Schweigert, B.S., Eds., Academic Press, New York, 1984, pp. 87–158.
6. Šimko, P., Gombita, M., and Karovičová, J. Determination and occurrence of benzo(a)pyrene in smoked meat products. *Nahrung/Food*, 35, 103, 1991.
7. Alonge, D.O. Carcinogenic polycyclic aromatic hydrocarbons (PAH) determined in Nigerian kundi (smoke-dried meat). *J. Sci. Food Agric.*, 43, 167, 1988.
8. Alonge, D.O. Factors affecting the quality of smoke-dried meats in Nigeria. *Acta Aliment.*, 16, 263, 1987.
9. Afolabi, A.O., Adesulu, E.A., and Oke, O.L. Polynuclear aromatic hydrocarbons in some Nigerian preserved freshwater fish species. *J. Agric. Food Chem.*, 31, 1083, 1983.

10. Borys, A. Liquid smoke application, in *Encyclopedia of Meat Sciences*, Jensen, W.K., Devine, C., and Dikeman, K., Eds., Elsevier, London, 2004, pp. 1272–1277.

11. Bartle, K.D. Analysis and occurrence of polycyclic aromatic hydrocarbons in food, in *Food Contaminants, Sources and Surveillance*, Creaser, C. and Purchase, R., Eds., The Royal Society of Chemistry, Cambridge, 1991, chap. 3.

12. Tóth, L. and Blaas, W. Einfluss der Räeuchertechnologie auf den Gehalt von geraeucherten Fleischwaren an cancerogenen Kohlenwasserstoffen. II. Einfluss der Glimmtemperatur des Holzes sowie der Kuehlung, Waesche und Filtration des Räeucherrauches. *Fleischwirtschaft*, 52, 1419, 1972.

13. Grimmer, G. and Böhnke, H. Polycyclic aromatic hydrocarbon profile analysis of high protein foods, oils and fats by gas chromatography. *J. Assoc. Off. Anal. Chem.*, 58, 725, 1975.

14. Kangsadalampai, K., Butryee, C., and Manoonphol, K. Direct mutagenicity of the polycyclic aromatic hydrocarbon-containing fraction of smoked and charcoal-broiled foods treated with nitrite in acid solution. *Food Chem. Toxicol.*, 35, 213, 1997.

15. Nisbet, I.C.T. and La Goy, P.K. Toxic equivalency factors (TEFS) for polycyclic aromatic hydrocarbons. *Regul. Toxicol. Pharmacol.*, 16, 290, 1992.

16. Law, R.J. et al. Toxic equivalency factors for PAH and their applicability in shellfish pollution monitoring studies. *J. Environ. Monit.*, 4, 383, 2002.

17. Šimko, P. Factors affecting elimination of polycyclic aromatic hydrocarbons in smoked meat foods and liquid smoke flavours. *Mol. Nutr. Food Res.*, 49, 637, 2005.

18. Andelman, J.B. and Suess, M.J. PAH in the water environment. *Bulletin WHO*, 43, 479, 1970.

19. Report of the Join FAO/WHO Expert Commission on Food Additives No. 31, 759, 1987.

20. Schaad, R. Chromatographie (karzinogener) polyzyclicer aromatischer Kohlenwasserstoffe. *Chromatogr. Rev.*, 13, 61, 1970.

21. Howard, J.W. and Fazio, T. PAH in foods. *J. Off. Anal. Chem.*, 63, 1077, 1980.

22. AOAC Official Method 973.30. 16th ed. AOAC International, Arlington, 48-11995.

23. Fretheim, K. Carcinogenic PAH in Norwegian smoked meat sausages. *J. Agric. Food Chem.*, 24, 976, 1976.

24. Binnemann, P.H. Benz(a)pyrene in Fleischerzeugnissen. *Z.Lebensm. Unters. Forsch.*, 169, 447, 1979.

25. Larsson, B.K. Polycyclic aromatic hydrocarbons in smoked fish. *Z. Lebensm. Unters. Forsch.*, 174, 101, 1982.

26. Lawrence, J.F. and Weber, D.F. Determination of polycyclic aromatic hydrocarbons in Canadian samples of processed vegetable and dairy products by liquid chromatography. *J. Agric. Food Chem.*, 32, 795, 1984.

27. Vassilaros, D.L. et al. Capillary gas chromatographic determination of polycyclic aromatic compounds in vertebrate fish tissue. *Anal. Chem.*, 54, 106, 1982.

28. Takatsuki, K. et al. Liquid chromatographic determination of polycyclic aromatic hydrocarbons in fish and shellfish. *J. Assoc. Off. Anal. Chem.*, 68, 945, 1985.

29. Karl, H. and Leinemann, M. Determination of polycyclic aromatic hydrocarbons in smoked fishery products from different smoking kilns. *Z. Lebensm. Unters. Forsch.*, 202, 458, 1996.

30. Potthast, K. and Eigner, G. A new method for the rapid isolation of polycyclic aromatic hydrocarbons from smoked meat products. *J. Chromatogr.*, 103, 173, 1975.

31. Cejpek, K. et al. Simplified extraction and cleanup procedure for the determination of PAHs in fatty and protein rich matrices. *J. Int. J. Environ. Anal. Chem.*, 61, 65, 1995.

32. Joe, F.L., Salemme, J., and Fazio, T. Liquid chromatographic determination of trace residues of polynuclear aromatic hydrocarbons in smoked foods. *J. Assoc. Off. Anal. Chem.*, 67, 1076, 1984.

33. Chen, B.H., Wang, C.Y., and Chiu, C.P. Evaluation of analysis of polycyclic aromatic hydrocarbons in meat products by liquid chromatography. *J. Agric. Food Chem.*, 44, 2244, 1996.

34. Wang, G. et al. Accelerated solvent extraction and gas chromatography/mass spectrometry for determination of polycyclic aromatic hydrocarbons in smoked food samples. *J. Agric. Food Chem.*, 47, 1062, 1999.

35. García-Falcón, M.S. Simal-Gandara, J., and Carril-Gonzalez-Barros, S.T. Analysis of benzo(a)pyrene in spiked fatty foods by second derivative synchronous spectrofluorimetry after microwave-assisted treatment of samples. *Food Addit. Contam.*, 17, 957, 2000.

36. White, R.H., Howard, J.W., and Barnes, C.J. Determination of polycyclic aromatic hydrocarbons in liquid smoke flavours. *J. Agric. Food Chem.*, 19, 143, 1971.

37. Silvester, D.S. Determination of 3,4-benzpyrene and benzanthracene in phenolic smoke concentrates. *J. Food Technol.*, 15, 413, 1980.

38. Radecki, A. et al. Separation of polycyclic aromatic hydrocarbons and determination of benzo(a)pyrene in liquid smoke preparations. *J. Chromatogr.*, 150, 527, 1978.

39. Šimko, P., Petrík, J., and Karovičová, J. Determination of benzo[a]pyrene in liquid smoke preparations by high pressure liquid chromatography and confirmation by gas chromatography-mass spectrometry. *Acta Aliment.*, 21, 107, 1992.

40. Gomaa, E.A. et al. Polycyclic aromatic hydrocarbons in smoked food products and commercial liquid smoke flavourings. *Food Addit. Contam.*, 10, 503, 1993.

41. Laffon Lage, B. et al. Comparison of supercritical fluid extraction and conventional liquid–solid extraction for the determination of benzo(a)pyrene in water-soluble smoke. *Food Addit. Contam.*, 14, 469, 1997.

42. Guillén, M.D., Sopelana, P., and Partearroyo, A. Determination of polycyclic aromatic hydrocarbons in commercial liquid smoke flavourings of different composition by gas chromatography–mass spectrometry. *J. Agric. Food Chem.*, 48, 126, 2000.

43. Guillén, M.D., Sopelana, P., and Partearroyo, A. Polycyclic aromatic hydrocarbons in liquid smoke flavorings obtained from different types of wood. Effect of storage in polyethylene flasks on their concentrations. *J. Agric. Food Chem.*, 48, 5083, 2000.

44. Vaessen, G.M.A.H., Jekel, A.A., and Wilbers, M.M.A.A. Dietary intake of polycyclic aromatic hydrocarbons. *Toxic. Environm. Chem.*, 16, 281, 1988.

45. Šimko, P. Changes of benzo[a]pyrene content in smoked fish during storage. *Food Chem.*, 40, 293, 1991.

46. Šimko, P., Karovičová, J., and Kubincová, M. Changes in benzo[a]pyrene content in fermented salami. *Z. Lebensm. Unters Forsch.*, 193, 538, 1991.

47. Šimko, P. et al. Influence of cooking on benzo[a]pyrene content in smoked sausages. *Meat Sci.*, 34, 301, 1993.

48. Mottier, P., Parisod, V., and Turesky, R.J. Quantitative determination of polycyclic aromatic hydrocarbons in barbecued meat sausages by gas chromatography coupled to mass spectrometry. *J. Agric. Food Chem.*, 48. 1160, 2000.

49. Stijve, T. and Hischenhuber, C. Simplified determination of benzo(a)pyrene and other aromatic hydrocarbons in various food materials by HPLC and TLC. *Dtsch. Lebensm. Rundsch.*, 83, 276, 1987.

50. Haenni, E.O. Analytical control of PAH in food and food additives. *Residue Rev.*, 24, 42, 1968.

51. Tuominen, J., Wickström, K., and Pyysalo, H. Determination of polycyclic aromatic hydrocarbons by GC-selected ion monitoring technique. *J. High Resol. Chrom. Chrom. Comm.*, 9, 469, 1986.

52. Lee, M.L., Novotny, M.V., and Bartle, K.D. *Analytical Chemistry of Polycyclic Aromatic Hydrocarbons.* 1st ed., Academic Press, New York, 1981, p. 242

53. Williams, P.T. et al. Analysis of the polycyclic aromatic compounds of diesel fuel by gas chromatography with ion trap detection. *Biomed. Environ. Mass Spectrom.*, 15, 517, 1988.

54. Dennis, M.J. et al. Comparison of a capillary gas chromatographic and a high-performance liquid chromatographic method of analysis for polycyclic aromatic hydrocarbons in food. *J. Chromatogr.*, 285, 127, 1984.

55. Speer, K. et al. Determination and distribution of polycyclic aromatic hydrocarbons in native vegetable oils, smoked fish products, mussels and oysters, and bream from the river Elbe. *J. High Resol. Chrom. Chrom. Comm.*, 13, 104, 1990.

56. Wise, A.S., Sander, L.C., and May, W.E. Determination of polycyclic aromatic hydrocarbons by liquid chromatography. *J. Chromatogr.*, 642, 329, 1993.

57. Chen, J. and Chen, S. Removal of polycyclic aromatic hydrocarbons by low density polyethylene from liquid model and roasted meat. *Food Chem.*, 90, 461, 2005.

58. Tamakawa, K. Polycyclic aromatic hydrocarbons in food, in *Handbook of Food Analysis*, Nollet, L.M.L., Ed., Marcel Dekker, New York, 2004, chap. 38.

59. Stahl, W. and Eisenbrand, G. Determination of polynuclear aromatic hydrocarbons and nitrosamines, in *HPLC in Food Analysis*, Macrae, R., Ed., Academic Press, London, 1988, chap. 10.

60. Hunt, D., Wild, P., and Crosby, N.T. A new chemically bonded stationary phase for the determination of polynuclear aromatic hydrocarbons by high pressure liquid chromatography. *J. Chromatogr.*, 130, 320, 1977.

61. Bieri, R.H. and Greaves, J. Characterization of benzo(a)pyrene metabolites by high performance liquid chromatography-mass spectrometry with a direct liquid introduction interface and using negative chemical ionization. *Biom. Environ. Mass Spectr.*, 14, 555, 1987.

62. Sim, P.G. et al. A comparison of chromatographic and chromatographic/mass spectrometric techniques for determination of polycyclic aromatic hydrocarbons in marine sediments. *J. Biom. Environ. Mass Spectr.*, 14, 375, 1987.

63. Moret, S., Conte, L., and Dean, D. Assessment of polycyclic aromatic hydrocarbon content of smoked fish by means of a fast HPLC/HPLC method. *J. Agric. Food Chem.*, 47, 1367, 1999.

64. Hartmann, K. Benzo[a]pyren Bestimmung bei mit räucharomageräucherten Fleischerzeugnissen. *Deutsch. Lebensm. Rundsch.*, 96, 163, 2000.

65. Ova, G. and Onaran, S. Polycyclic aromatic hydrocarbons contamination in salmon-trout and eel smoked by two different methods. *Adv. Food Sci.*, 20, 168, 1998.

66. Yabiku, H.Y. Martins, M.S., and Takahashi, M.Y. Levels of benzo(a)pyrene and other polycyclic aromatic hydrocarbons in liquid smoke flavour and some smoked foods. *Food Addit. Contam.*, 10, 399, 1993.

67. Rauter, W. Content of benzo(a)pyrene in smoked foods. *Ernährung*, 21, 447, 1997.

68. Chiu, C.P., Lin, Y.S., and Chen, B.H. Comparison of GC–MS and HPLC for overcoming matrix interferences in the analysis of PAHs in smoked food. *Chromatographia*, 44, 497, 1997.

69. Šimko, P. Determination of polycyclic aromatic hydrocarbons in smoked meat products and liquid smoke flavourings by gas chromatography and high pressure liquid chromatography. *J. Chrom. B*, 770, 3, 2002.

FISH AND SEAFOODS

Chapter 15

Assessment of Seafood Spoilage and the Microorganisms Involved

Robert E. Levin

Contents

15.1 Introduction 464
15.2 Chemical Causes of Seafood Spoilage 465
15.3 Assays for Assessing the Quality of Seafood 465
    15.3.1 Assay for TMA 465
    15.3.2 Assessment of Fish Quality Based on the Refractive Index of Eye Fluid 466
    15.3.3 Vacuum Distillation Procedure for Determination of Volatile Acids and
            Volatile Bases in Fish Tissue 466
15.4 Taxonomy of Psychrotrophic Intense Spoilage and Nonspoilage Bacteria
    on Seafood 466
    15.4.1 The Genus *Pseudomonas* 466
            15.4.1.1 *Pseudomonas fragi* 469
            15.4.1.2 *Pseudomonas perolens* 469
    15.4.2 The Genus *Alteromonas* 469
            15.4.2.1 *Alteromonas nigrifaciens* 469
    15.4.3 The Genus *Shewanella* 469
            15.4.3.1 *Shewanella putrefaciens* 470
    15.4.4 The Genera *Moraxella* and *Acinetobacter* 470
    15.4.5 The Genera *Flavobacterium* and *Cytophaga* 471
    15.4.6 The Genus *Brochothrix* 472

15.4.7 The Genus *Photobacterium* 472
15.4.8 The Genus *Lactobacillus*...472
15.4.9 The Genus *Vibrio*..473
15.4.10 The Genus *Aeromonas*...473
15.5 The Microbiology of Modified Atmosphere Stored Seafood 473
15.6 The Microbiology of Gamma-Irradiated Seafood 474
15.7 Determination of Varius Bacterial Counts from Seafood 474
    15.7.1 Determination of Total Aerobic Plate Counts 474
        15.7.1.1 Selective Enumeration of Members of the Genus *Pseudomonas* 475
        15.7.1.2 Enumeration of Fluorescent Pseudomonads 475
15.8 Useful Tests for Confirming the Identity of Seafood Spoilage Bacteria 476
    15.8.1 Genetic Transformation Assay for Confirming the Identity of *Psychrobacter*
        immobilis Isolates...476
        15.8.1.1 Motility..476
        15.8.1.2 The Oxidase Test 478
    15.8.2 Litmus Milk 478
    15.8.3 Proteolysis 478
    15.8.4 Detection of H_2S Production 479
    15.8.5 DNase Activity..479
        15.8.5.1 Molecular Techniques for Detection and Enumeration of Seafood
        Spoilage Bacteria..480
References 481

15.1 Introduction

Seafood can be expected to harbor a wide variety of bacterial species and genera. However, among the psychrotrophic seafood spoilage bacteria one finds relatively few genera and species that can be considered intense spoilage organisms. An early study undertaken to identify the major intense fish spoilage bacterial genera and species was by Castell and Anderson [1]. Their study involved the use of an autoclaved fish medium composed of equal weights of macerated fresh cod muscle and water in addition 0.05% agar that was poured into petri dishes. Known numbers of organisms from pure bacterial cultures were inoculated into this fish tissue medium followed by incubation at 3°C with daily assessment of spoilage odors. Three categories of pure cultures were described. The first group of organisms represented by enteric bacteria, *Bacilli*, and micrococci yielded no off odors at 3°C because they were unable to grow at this low temperature. The second group produced musty, sour, or sweetish odors at 3°C and consisted of flavobacteria, *Achromobacter*, and micrococci. The third group consisted of organisms that produced offensive odors rapidly at 3°C and consisted of *Pseudomonas* spp. *Achromobacter* sp., *Serratia marcescens*, and *Proteus vulgaris*. *P. vulgaris* is never found on seafood and the isolate used was from the American type culture collection (ATCC) (C. Castell, personal communication). This reduces the intense spoilage organisms to members of the genera *Pseudomonas* and *Achromobacter* as recognized in 1948. A later study, undertaken to identify the major psychrotrophic bacterial genera on freshly caught cod was by Georgala [2]. Among a total of 727 isolates, the following were identified: 51.5% *Pseudomonas*, 41.8% *Achromobacter*, 3.3% *Vibrio*, 1.5% *Flavobacterium*, 0.7% *Micrococcus*, and 0.7% miscellaneous. Since these early studies, various attempts have been made to further elucidate the major intense fish spoilage bacterial genera and species. This has resulted in a certain expansion and taxonomic alterations of the two originally recognized intense spoilage genera, *Pseudomonas* and *Achromobacter*, which this chapter elucidates.

15.2 Chemical Causes of Seafood Spoilage

A number of chemical agents that are products of microbial metabolism have been found to be associated with seafood spoilage. Included most notably among these is trimethylamine (TMA), the cause of the characteristic odor associated with spoiled seafood. Trimethylamine oxide (TMAO) is considered a compatible osmolyte in the muscle tissue of marine fish and is reduced by the bacterial enzyme TMAO-reductase to TMA. TMAO is present primarily in pelagic fish. In addition, other marines are encountered, such as putrescine and cadaverine derived from the bacterial decarboxylation of the amino acids arginine and lysine, respectively. Ammonia is usually derived from the oxidative deamination of amino acids by spoilage bacteria and is most readily sensed in the later stages of spoilage. Mercaptans are noxious sulfur-containing compounds derived from the activity of bacteria on cysteine in addition to the bacterial release of hydrogen sulfide from sulfur-containing amino acids. Volatile spoilage compounds fall into several categories: volatile amines, volatile acids (formic and acetic), and volatile reducing substances. All volatile compounds can readily be quantified presently by gas chromatography. However, the original description of a vacuum distillation procedure for the determination of volatile acids and bases in fish tissue by Tomiyama et al. [3] may be of value to some in that it requires no extraction and low cost equipment (see the following text).

15.3 Assays for Assessing the Quality of Seafood

15.3.1 Assay for TMA

The quantitative presence of TMA in pelagic fish still remains a major chemical criterion of quality since most of the TMA is produced by bacterial reduction of the TMAO and therefore reflects the numbers of spoilage bacteria on fish tissue and the days of refrigeration or iced time. The chemical assay for quantitation of TMA in fish as described by Dyer [4] is presented. Fish tissue (100 g) is blended with 200 mL of 7.5% trichloroacetic acid (TCA) and filter clarified. One mL is then transferred to a tube and 3.0 mL of water added. One mL of 4.0% formaldehyde is added in addition to 10 mL of toluene and 3.0 mL of 50% potassium carbonate. The tube is capped and vortexed for 30 s. Five mL of the toluene (top) layer is transferred to a tube containing about 0.3 g of anhydrous sodium sulfate and is shaken for 10 s to remove trace amounts of water. The 5.0 mL of toluene is decanted into a dry tube and mixed with 5.0 mL of 0.02% picric acid–toluene solution. The intensity of the resulting yellow color is determined at 410 nm. For a standard curve use 0, 0.2, 0.4, 0.6, 0.8, and 1.0 mL of a stock TMA solution (0.682% TMA·HCl in water). Results are reported in mg of TMA nitrogen (TMA-N) per 100 g of muscle tissue. Fresh cod and haddock have been found to contain no more than 0.2 mg of TMA nitrogen per 100 g of tissue. In contrast, at the time of spoilage the TMA content has been found to be 6–8 mg of TMA nitrogen per 100 g of tissue [5]. Castell et al. [6] found that cod and haddock could be graded into three categories based on sensory and TMA analysis. Group I consisted of fresh fish having TMA values (mg of TMA nitrogen per 100 g of tissue) of 0.52–0.89 with an average of 0.59. Group II consisted of stale fish having TMA values of 1.3–4.5 with an average of 3.1. Group II consisted of spoiled fish with TMA values of 2.6–8.3 with an average of 6.8. However, Hillig et al. [7] found a lack of correlation between sensory evaluation and TMA values for pollock and whiting.

15.3.2 Assessment of Fish Quality Based on the Refractive Index of Eye Fluid

Proctor et al. [8] found a linear relationship between the refractive index of eye fluid from haddock and organoleptic score. They placed the quality of haddock into four refractive index categories: very good (1.3347–1.3366), fair to good (1.3367–1.3380), poor (1.3381–1.3393), and not marketable (1.3394 or higher). The procedure involves removal of the eyes from the head, making a slit in the posterior portion of the eyes, and allowing the fluids from each eye to drain into the same beaker. The fluids are then centrifuged (speed not specified) and then they are passed through glass wool. An Abbé refractometer or similar refractive index monitor is then used to determine the refractive index to the fourth decimal point using two or three drops of the eye fluid.

15.3.3 Vacuum Distillation Procedure for Determination of Volatile Acids and Volatile Bases in Fish Tissue

The procedure of Tomiyama et al. [3] involves the blending of 85 g of fish tissue with 200 mL of water. Four hundred mL of a $MgSO_4$ solution (600 g made to 1 L, followed by the addition of 20 mL of 6 N H_2SO_4) is then added, and thoroughly agitated, and then filtered through a filter paper. The filtrate should be adjusted to pH 2.0. Fifty mL of the filtrate (equivalent to 5.0 g of tissue) are then added to a 500 mL round bottom three-neck flask in a temperature controlled water bath (75°C). If the volatile acid number is to be determined 10 of 0.01 N NaOH are placed in the receiving vessel; if the volatile base content is to be determined 10 mL of N/28 H_2SO_4 are placed in the receiving vessel. For volatile acid determination, vacuum distillation is then initiated. If volatile bases are to be determined, 10 mL of 10% NaOH are added to the sample flask. In the determination of volatile acid number neutral red is used as the indicator and titration is with 0.01 N NaOH; in the determination of the volatile bases content methyl red is used as the indicator and titration is with 0.005 N H_2SO_4. An illustration of the vacuum distillation apparatus and a detailed description of its use is presented by Ref. [55].

15.4 Taxonomy of Psychrotrophic Intense Spoilage and Nonspoilage Bacteria on Seafood

15.4.1 The Genus Pseudomonas

The genus *Pseudomonas* is characterized as consisting of obligately aerobic gram-negative rods with polar flagella. The molar G + C (guanine + cytosine) content for members of this genus is recognized as being from 58% to 70%. Any organism outside this range is not considered a member of the genus *Pseudomonas*. The intense fish spoilage psychrotrophic species of the genus *Pseudomonas* can be divided into two convenient major groups consisting of fluorescent and nonfluorescent isolates. Among the fluorescent pseudomonads, we find that isolates of *P. fluorescens* are protease positive while isolates of *P. putida* are protease negative which constitute the major distinction between these two intense fish spoilage fluorescent species. Stanier et al. [9]

established seven biotypes for isolates of *Pseudomonas fluorescence* (A–G) and two biotypes (A and B) for isolates of *P. putida* based on metabolic characteristics which Gennari and Fragotto [10] made use of for distinguishing fluorescent isolates from seafood and other food products. However, when it comes to the nonfluorescent fish spoilage pseudomonads, little is known about their species designations.

Hugh and Leifson [11] developed a convenient culture method for distinguishing between oxidative and fermentative Gram-negative bacteria. It is known as the "Hugh–Leifson test" or the "O/F" test. The medium involved consists of: peptone, 0.2%; NaCl, 0.5%; KH_2PO_4, 0.03%; glucose, 1.0%; bromthymol blue, 0.03%; and agar, 0.3% pH 7.1. The low-level phosphate is used to slightly stabilize the pH and to promote fermentation. The low level of agar is to prevent convection currents. Glucose is added after autoclaving from a sterile 10% solution. Two tubes are inoculated with each culture and one is sealed with sterile mineral oil to a depth of about 1 cm to exclude oxygen. The initial pH of 7.1 results in a green color. An oxidative organism will produce a yellow acid reaction in the open tube starting at the top and with time proceeding downward and no reaction in the sealed tube. Some nonoxidizers and nonfermentors produce no change in the covered tube and only an alkaline reaction in the open tube. Other nonoxidizers and nonfermentors produce no reaction in either tube. Fermentative organisms will produce an acid reaction throughout both tubes.

Shewan et al. [12] established a broad grouping of Gram-negative organisms found in fish and in other habitats that was based on 10 phenotypic characteristics. They then applied the Hugh–Leifson test to distinguish the various Gram-negative organisms prevailing on fish, which yielded four distinguishable metabolic groups of *Pseudomonas* from fish (Figure 15.1). This grouping of

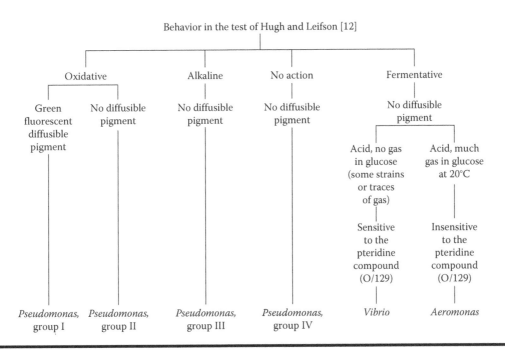

Figure 15.1 **A grouping of the Gram-negative asporogenous rods, polar-flagellate, oxidase positive, and not sensitive to 2.5 i.u. of penicillin, on the results of four other tests. (Redrawn from Shewan, J. et al.,** *J. Appl. Bacteriol.*, **23, 379, 1960. With permission.)**

Pseudomonas isolates from seafood is still used because many such isolates do not adhere to recognized species of *Pseudomonas*.

Shewan et al. [13] presented a dichotomous key for the screening of cultures from seafood involving all of the major genera. This diagrammatic outline made use of the gram stain, pigmentation, flagellation, the cytochrome oxidase test, and the medium of Hugh and Leifson for the determination of oxidative versus fermentative metabolism and is presented in Figure 15.2.

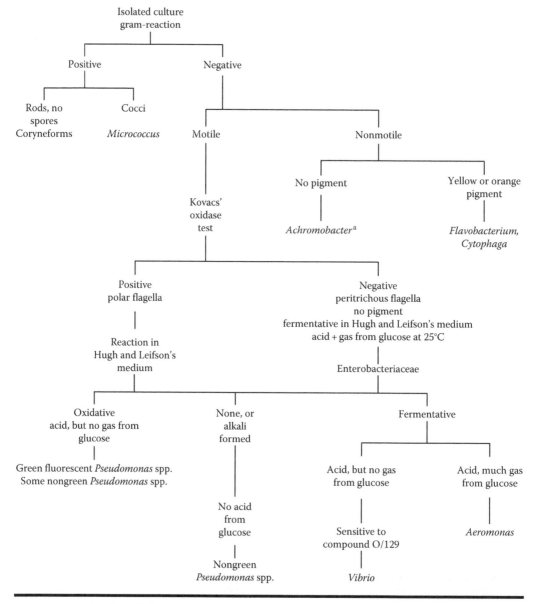

Figure 15.2 Outline of the sequence of tests used in the screening of bacterial cultures from seafood. [a]Members of the genus *Achromobacter* presently allocated to the genera *Moraxella* and *Acinetobacter*. (Redrawn from Shewan, J. et al., *J. Appl. Bacteriol.*, 23, 463, 1960. With permission.)

15.4.1.1   *Pseudomonas fragi*

Among the off-odors that frequently develop during the early stages in the spoilage of refrigerated fillets are those that have been described as "sweet" and "fruity." The responsible organism was found to be nonproteolytic and identified as *Pseudomonas fragi* which has been characterized as producing a "sweet, ester-like odor resembling that of the flower of the May apple" [14]. Isolates of *P. fragi* from seafood are characterized as being nonfluorescent, nonproteolytic, do not produce TMA, or H_2S, but are capable of producing ammonia from amino acids, are lipolytic and are isolated from both fresh and spoiled fillets [14].

15.4.1.2   *Pseudomonas perolens*

During the early stages of fish spoilage, a "musty" odor is sometimes noted. When the implicated organisms are in pure culture, they give rise to a "stored potato" odor. The responsible organism has been found to be *P. perolens*. Isolates of this organism are neither proteolytic nor lipolytic, do not produce TMA, and produce little or no change in milk, but do produce ammonia from amino acids, and H_2S [15].

15.4.2   The Genus *Alteromonas*

The genus *Alteromonas* was originally created to accommodate organisms having typical phenotypic characteristics of the genus *Pseudomonas* but which have a molar G + C content of less than 58%, thereby excluding them from the genus *Pseudomonas*.

15.4.2.1   *Alteromonas nigrifaciens*

A. nigrifaciens was originally known as *Pseudomonas nigrifaciens* in the older literature. This is an extremely intense fish spoilage organism characterized by producing an intense black melanin type of pigment. The organism is often overlooked in that maximum pigment production occurs with 1.5%–2.5% NaCl added to the culture medium with incubation from 4°C to 15°C. The presence of tyrosine (0.1%) has been found to be essential for pigment production [16]. In the absence of pigment production, this organism appears as a typical pseudomonad. The cells of this organism are motile by means of a single polar flagellum. Cultures are obligately aerobic, cytochrome oxidase positive, gelatinase positive, lipase positive, amylase positive, and produce putrescine, cadaverine, and spermidine. Sodium ions are required for growth. The molar G + C content is 39%–41%.

15.4.3   The Genus *Shewanella*

Members of the genus *Shewanella* were formerly considered pseudomonads. Venkateswaran [17] has reviewed the taxonomy of this genus at length. All isolates are Gram-negative, nonsporeforming rods, motile by means of a single polar flagellum, and are 2–3 μm in length. There are presently 12 recognized species, some of which produce salmon or pink-colored colonies. All species are cytochrome oxidase and catalase positive and negative for the production of amylase. Most species are gelatinase positive and lipase has been reported to be produced by several species. All species reduce TMAO to TMA and reduce nitrate to nitrite and the majority produce H_2S from thiosulfate. Several species reduce elemental sulfur.

15.4.3.1 Shewanella putrefaciens

The organism presently known as *Shewanella putrefaciens* was first isolated from tainted butter and classified as a member of the genus *Achromobacter* by Derby and Hammer [18]. It was transferred to the genus *Pseudomonas* in 1941 by Long and Hammer [19]. In 1972, it was allocated to the genus *Alteromonas* by Lee et al. [20] on the basis of its much lower mol% $G + C$ DNA content than the acceptable range of 58–70 mol% $G + C$ for members of the genus *Pseudomonas* [21]. In 1985, it was transferred to the newly established genus *Shewanella* under the family *Vibrionaceae* due to its perceived closer relationship with the genus *Vibrio* [22]. The type species of *S. putrefaciens*, ATCC strain 8071 has a molar $G + C$ content of 46% and strains in this species vary from 43% to 48% $G + C$ [23–25].

Isolates of *S. putrefaciens* are intense psychrotrophic fish spoilage organisms as the species designation implies. They are visually characterized as producing salmon pigmented colonies, particularly on the surface of Peptone Iron Agar (PIA) where crowded surface colonies produce uniformly black colonies while well-isolated colonies usually produce salmon pigmented colonies with intense black centers (Figure 15.3). With pour plates of PIA, intensely black pinpoint size subsurface colonies develop. All such isolates produce an extracellular DNAse [26] in addition to an extracellular protease and lipase. Isolates of *S. putrefaciens* have been found on occasion to dominate at the time of intense fish spoilage [27]. Mg ions are a critical requirement for maintaining the integrity of the cell membrane [28]. If one prepares decimal dilutions of fish tissue in saline for plate counts the organism will rupture unless at least 0.001 M Mg^{++} ions are added to the saline by way of $MgCl_2$. This requirement is not widely recognized. Phosphate buffer enhances the lytic phenomenon by presumably pulling Mg^{++} ions out of the membrane or sacculus.

15.4.4 The Genera Moraxella and Acinetobacter

The genera *Moraxella* and *Acinetobacter* were originally the Gram-negative nonpigmented, nonflagellated, obligately aerobic coccobacilli that were allocated to the former genus *Achromobacter*

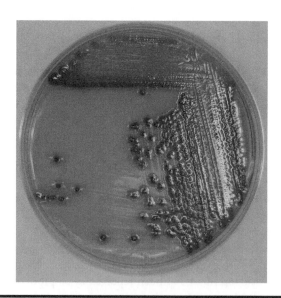

Figure 15.3 Typical colonies with black centers of S. *putrefaciens* on PIA.

in the older literature. The genus *Achromobacter* was eventually eliminated so that now all such Gram-negative coccobacilli from seafood are placed into the genera *Moraxella* or *Acinetobacter*. Without species differentiation, such bacterial isolates have often been placed into the "*Moraxella–Acinetobacter*" group. The molar G + C value for isolates of *Moraxella* varies from 40% to 46% and for *Acinetobacter* varies from 40% to 47%. These two genera are distinguished primarily on the basis that the *Moraxella* are sensitive to penicillin (1 i.u. disc) and are cytochrome oxidase positive while members of the genus *Acinetobacter* are resistant to penicillin and are cytochrome oxidase negative. Juni and Hyme [29] however found that fishery isolates of both *Acinetobacter* and *Moraxella* were cytochrome oxidase positive. A major metabolic distinction between these two genera results from the ability of *Moraxella* isolates to produce significant amounts of phenylethanol from the amino acid phenylalanine [30] while *Acinetobacter* isolates produce little or no phenylethanol. Members of both genera however appear to be closely related genetically. Juni and Hyme [29] developed a genetic transformation assay whereby the DNA from members of both of these psychrotrophic genera isolated from fish and meat products is able to transform a mutant recipient unable to synthesize hypoxanthine-to-hypoxanthine synthesis. In a subsequent report by Juni and Hyme [31] the designation *Psychrobacter immobilis* was proposed for all Gram-negative, aerobic, and cytochrome oxidase positive coccobacilli on the basis of genetic compatibility (genetic transformation). Such a designation eliminates the use of the genera *Moraxella* and *Acinetobacter* for such psychrotrophic food isolates and to that extent may be more convenient for individuals working in the area of seafood microbiology.

In a later study, Rossau et al. [32] found that on the basis of DNA–rRNA hybridization, members of the genera *Moraxella*, *Psychrobacter*, and *Acinetobacter* constitute a separate genotype cluster and proposed the family *Moraxellaceae* to accommodate these organisms. González et al. [33] isolated 979 Gram-negative, nonmobile, aerobic coccobacilli from fresh water fish stored in ice. A total of 106 randomly selected isolates were found to consist of *Psychrobacter* (64 strains), *Acinetobacter* (24 strains), *Moraxella* (6 strains), *Chryseobacterium* (5 strains), *Myroides odoratus* (2 strains), *Flavobacterium* (1 strain), *Empedobacter* (1 strain), and three unidentified strains. Among the 64 *Psychrobacter*, 14 isolates were *P. phenylpyruvica*. These authors considered *Acinetobacter* to be the only oxidase negative genus within the family *Moraxellaceae*. These authors concluded that identification at the genomic species level is complex, time consuming, and in some cases impossible. In contrast to earlier studies, these authors concluded that there is considerable confusion of the role of *Moraxellaceae* in the spoilage of presumably raw refrigerated proteinaceous foods. In their study, this genus contributed only 9.0% of the total flora, and the majority failed to produce typical spoilage compounds (TMA and H_2S). However, among the 106 randomly selected isolates, 50 were found to be mesophiles, and 56 psychrotrophs indicating that a disproportionate number of isolates were derived from fish prior to prolonged iced storage. Their observation that the *Moraxellaceae* decreased as spoilage progressed contradicts earlier studies on the spoilage of marine fish. A key factor in this discrepancy may be that the predominant species at the time of spoilage of fresh water fish may differ notably from that of marine fish.

15.4.5   The Genera *Flavobacterium* and *Cytophaga*

Members of both these genera are characterized as producing yellow, orange, or red carotenoid pigments. The *Flavobacteria* may be motile by peritrichous flagella or nonmotile. The *Cytophaga*, if motile exhibit gliding motility and lack flagella. Both genera are characterized as being obligately aerobic and weakly active on carbohydrates. Not all isolates of both genera are capable of utilizing glucose. Many isolates of both genera have little or no effect on litmus milk. Some isolates have been found to be proteolytic and some are able to produce H$_2$S, moreover none are lipolytic. McMeekin [34] subjected 59 yellow-pigmented bacterial isolates from various foods to 78 phenotypic properties and with the aid of adansonian taxonomy distinguished six groups. One of the problems encountered in attempting to allocate individual strains to one or the other of these pigmented genera involves atypical strains. In an earlier study, Castell and Maplebeck [35] examined 245 isolates of *Flavobacterium* (132 yellow and 113 orange) from fish for the ability to exhibit fish spoilage activities. Seventy-eight percent of the yellow isolates and 92% of the orange isolates grew at 2°C–3°C. Thirty-six percent of the yellow isolates and only 4% of the orange isolates produced TMA. Forty percent of the yellow isolates and 84% of the orange isolates were proteolytic. When isolates were inoculated onto sterile fish tissue incubated at 3°C the orange cultures of *Flavobacterium* began to develop disagreeable odors after 5–8 days and many became quite putrid by the 10th or 11th day. Sterile fish tissue inoculated with the yellow cultures yielded no perceptible spoilage odors, even after 15 days but did discolor the fish tissue yellow, indicative of growth. In contrast, fish tissue inoculated with *Pseudomonas* isolates became offensive after 48 and 72 h. It is a widely recognized observation, that members of the brightly pigmented genera *Flavobacterium* and *Cytophaga* are frequently encountered on fresh fish where they may constitute 10%–30% of the initial flora and are rarely among the dominant flora of stale fish. Although isolates of these pigmented organism have been found under noncompetitive conditions to eventually spoil fish tissue, on a practical basis, under commercial conditions they are generally outgrown by the more intense spoilage pseudomonads that grow more rapidly under refrigerated conditions than members of the these two pigmented genera. As a result, many workers group such isolates into the "flavobacterium–cytophaga" group rather than attempting to determine clearly and arduously which pigmented genus they belong.

15.4.6 The Genus Brochothrix

These are Gram-positive nonsporeforming rods closely related to the genus *Lactobacillus* and are considered heterofermentative with regard to lactic acid production. Log-phase cells are typically rods, while older cells are coccoids, a feature common to coryneforms. Only two species are recognized: *B. thermosphacta* and *B. campestris*. These organisms are important in the spoilage of modified atmosphere (MA) stored seafood [36]. In contrast to *B. thermosphacta*, *B. campestris* is rhamnose and hippurate positive. Both species have a molar G + C content of 36%.

15.4.7 The Genus Photobacterium

These are Gram-positive nonsporeforming, peritrichously flagellated rods possessing fermentative metabolism with sugars and are therefore facultative anaerobes. Isolates are luminescent (glow in the dark). The molar G + C for the genus is 39%–42%. Because of their facultatively anaerobic metabolism, they have been frequently found to be among the major spoilage organisms in MA storage when oxygen is excluded. The species associated with spoiled seafood is *P. phosphoreum* [37]. Sivertsvik et al. [38] have reviewed the relationship between this organism and MA storage.

15.4.8 The Genus Lactobacillus

These organisms are Gram-positive nonsporeforming rods 2–9 μm long. All species of lactobacilli produce at least 1.0% lactic acid from glucose, are nutritionally fastidious, and are catalase negative. Members of the genus *Lactobacillus* do not predominate when seafood is stored under normal iced or refrigerated conditions. However, under conditions of MA storage, lactobacilli can dominate at the termination of storage. They are most readily enumerated from seafood products as dominant members of the prevailing flora with the use of Lactobacilli MRS Agar, which is designed to support luxuriant growth of all lactobacilli and is not a selective medium. Therefore, identification is based on the phenotypic properties of isolated colonies.

15.4.9   The Genus Vibrio

Members of the genus *Vibrio* are Gram-negative short asporogenous curved or straight rods, which are motile by means of polar flagella. They are all facultative anaerobes exhibiting fermentative metabolism in the absence of oxygen producing acid but no gas (H_2 or CO_2). They are cytochrome oxidase positive and are usually nonpigmented. All members of the genus are considered sensitive to the vibriostatic agent 2,4-diamino-6,7-diisopropyl pteridine (O/129) which is considered a diagnostic criterion for the genus. The molar G + C value for the genus ranges from 40% to 50%. Members of the genus are not considered spoilage organisms and are usually not found among the dominant flora of stale fish. The genus does have several species that are notable human pathogens such as *V. cholerae*, *V. vulnificus*, and *V. parahaemolyticus* that are associated with the consumption of raw seafood.

15.4.10   The Genus Aeromonas

Members of the genus *Aeromonas* are straight rod shaped Gram-negative polarly flagellated cells. They are facultative anaerobes exhibiting fermentative metabolism in the absence of oxygen with the production of acid and gas ($H_2 + CO_2$). Isolates are proteolytic, produce extracellular DNase, are cytochrome oxidase positive, and insensitive to the vibriostatic agent O/129. The molar G + C content ranges from 57% to 63%. The genus contains several species pathogenic to fish such as *A. hydrophila* and *A. salmonicida*. Isolates of *Aeromonas* have on occasion been implicated in gastroenteritis.

15.5   The Microbiology of Modified Atmosphere Stored Seafood

MA storage of seafood results in a dramatic change in the bacterial flora that develop under refrigerated temperatures compared to the normal atmosphere of 79% nitrogen and 21% oxygen. An atmosphere of 20% CO_2 and 80% air is commonly used, however studies involving the use of 50% CO_2 and 50% nitrogen have indicated that the complete absence of oxygen in the storage atmosphere greatly influences the resulting dominant flora. Another factor that may influence the dominant flora at spoilage is fresh water farm raised fish versus ocean caught fish. The presence of 20% air is preferred to ensure that *Clostridium botulinum* does not develop. Mokhele et al. [39] found that with rock cod fillets stored in such a MA at 4°C for 21 days, the total bacterial population increased by only 2 log cycles, the fillets were not spoiled, and only *Aeromonas*-like (31%) and *Lactobacillus* (69%) isolates were recovered. In contrast, control fillets stored under conditions of a normal atmosphere at 4°C were spoiled after 7 days and had undergone an increase of

3–4 log cycles in total bacterial counts. Johnson and Ogrydziak [40] using an identical MA for the storage of rock cod at 4°C found that after 21 days only *Lactobacillus* (71%–87%) and tan-colored *Pseudomonas*-like isolates were recovered. Their observations suggested that the tan isolates underwent mutation to enhanced tolerance to the MA. Some of the tan isolates produced H_2S, strongly suggesting that they were *S. putrefaciens*. These authors also indicated that the tan isolates grew slowly under anaerobic conditions. *S. putrefaciens* does not have a fermentative metabolism but isolates are able to couple anaerobic growth to iron and manganese reduction, and are able to utilize an array of other electron acceptors anaerobically such as NO_3^{-1}, NO_2^{-1}, $S_2O_3^{-2}$, S^0, and fumarate; TMAO with most strains was able to utilize lactate, pyruvate, and some amino acids anaerobically [41].

Hovda et al. [36] found that when fresh water farm raised halibut were stored in a MA of 50%CO_2:50%N_2 that 16 days at 4°C were required for a 5 log increase in total counts to occur, whereas in a CO_2:50%O_2 MA no more than a 3 log increase occurred after 23 days, and that storage in air resulted in a 7 log increase in 16 days. 16S rDNA sequencing following denaturing gradient gel electrophoresis (16S rDNA-DGGE) indicated that in air *Pseudomonas* spp., *P. putida*, *P. phosphoreum*, and *B. thermosphacta* were dominant at the time of spoilage. With a MA of 50% CO_2:50%N_2 the predominant organisms at the time of spoilage were *P. phosphoreum*, and *B. thermosphacta*. *S. putrefaciens* was found only sporadically and in low numbers, while pseudomonads were not detected. In a MA of 50% CO_2:50%O_2 *B. thermosphacta* and *Staphylococcus* spp. dominated after 23 days.

Rudi et al. [42] stored salmon and coalfish in a MA of 60% CO_2:40%N_2 for up to 18 days at 1°C and 5°C. The predominant organisms after 18 days on coalfish identified by 16S rDNA analysis were found to be *Photobactrium* spp., while those on salmon were found to be *Carnobacterium* spp. and *Brochothrix* spp. These authors made the interesting observation that *Photobacterium* spp. were underrepresented when identification was based on colony isolation compared to 16S rDNA sequencing following cloning which was presumably due to technical difficulties in culturing photobacterium. Their study also detected the presence of *C. botulinum* at the end of storage.

15.6 The Microbiology of Gamma-Irradiated Seafood

The organisms predominating at the time of spoilage on petrale sole subjected to gamma irradiation (4 kGy) and then stored in air at 0.5°C have been found to be members of the bacteria genera *Moraxella* and *Acinetobacter* (formerly *Achromobacter*) and the yeast *Trichosporon* [43]. When stored under conditions of vacuum packaging after irradiation (4 kGy) lactobacilli dominated [44]. These observations show that these organisms are more resistant to gamma irradiation than the other members of the bacterial flora such as the pseudomonads. Pseudomonads are seldom encountered with irradiation over 1 kGy.

15.7 Determination of Varius Bacterial Counts from Seafood

15.7.1 Determination of Total Aerobic Plate Counts

The conventional plate count method described in the following text for examining frozen and chilled foods conforms to the AOAC Official Methods of Analysis sec. 966.23 with procedural changes as indicated. The suitable counting colony range is 25–250. The frequently recommended diluent for dilution blanks consists of Butterfield's phosphate-buffered dilution water (KH_2PO_4, 3.4%, pH 7.2 adjusted with NaOH) is not appropriate. The author has found that with seafoods, higher counts will result with dilution blanks prepared with 0.85% NaCl and 0.001 M $MgCl_2$ in that organisms such as *S. putrefaciens* will readily undergo autolysis if placed into phosphate buffer [28]. Although standard methods recommend stomaching or blending 50 g of tissue with 450 mL of diluent, total aerobic counts from seafood tissue are as a rule sufficiently high so that 10 g of tissue can be added to 90 mL of diluent without compromising results. Dilutions should be prepared by transferring 10–90 mL of diluent and shaken 30 times, avoiding foaming. Reshake if dilutions stand for more than 3 min. One mL of each dilution is then transferred to duplicate petri dishes and 12–15 mL of Plate Count Agar (tryptone, 0.5%; yeast extract, 0.25%; glucose, 0.1%; agar, 1.5%; supplemented with 0.5% NaCl, pH 7.0) or Tryptic Soy Agar (tryptone, 1.7%; soytone, 0.3%; NaCl, 0.5%; K_2HPO_4, 0.15%, agar, 1.5%, pH 7.3) at 45°C is added and the plates agitated on a flat surface by traversing a figure eight motion at least six times. The plates are then allowed to solidify before stacking. Alternatively, one can smear-plate 0.1 mL of dilutions onto prepoured agar plates, the surface of which have been allowed to dry for 1–2 days. Plates should be incubated at 20°C for 72 h. An incubation temperature of 20°C will yield approximately 10-fold higher counts than at 35°C [45] and will usually yield total counts that are within 70% of the true counts. However, if absolute maximum counts from stale fish are to be obtained, then it is necessary to incubate the plates at 3°C for 6 days, where with stale fish, counts have been found to be 30% higher than at 20°C [46]. This is due to the development of obligately psychrotrophic organisms (mostly vibrios) which do not grow above 15°C [46]. It is important to keep in mind that obligately psychrotrophic bacteria are extremely sensitive to elevated temperatures of 25°C–35°C [47]. Hence, for truly maximum counts one should use prepoured and solidified agar plates whose temperature is not allowed to rise above 10°C once in contact with the sample, and which are smear plated with 0.1 mL of decimal dilutions. Pipets, pipet tips, and dilution blanks must also be chilled. In addition, blending or stomaching must also be done under refrigerated conditions. Such low temperature plate counts are best achieved by performing all procedures in a refrigerated room or laboratory at about 2°C–4°C so that samples and supplies are never allowed to increase above this refrigerated temperature range [46].

15.7.1.1 Selective Enumeration of Members of the Genus Pseudomonas

The selective enumeration of essentially all pseudomonads from seafood tissue can be achieved with the use of *Pseudomonas* Isolation Agar. This medium contains: peptone, 20 g; $MgCl_2$, 1.4 g; K_2SO_4 1.5 g; Irgasan, 0.025 g; agar, 13.6 g; and deionized water, pH 7.0.

15.7.1.2 Enumeration of Fluorescent Pseudomonads

All culture media designed for the detection of fluorescent pseudomonads are high in magnesium and notably low in iron; the latter greatly suppressing fluorescence. The fluorescent pseudomonad agar medium of Sands and Rovira [48] is ideal for the selective isolation of only fluorescent pseudomonads. The basal medium consists of: proteose peptone no. 3, 20 g; Oxoid Ionagar no. 1, 12 g; glycerol, 8 mL; K_2SO_4 1.5 g; $MgSO_4 7H_2O$, 1.5 g; distilled water, 940 mL, pH 7.2. Penicillin G (75,000 units, novobiocin, 45 mg; and cycloheximide, 75 mg are mixed together in 3 mL of 95% ethanol, and are then diluted with 50 mL of sterile distilled water, and added to the 940 mL of the melted basal medium at 45°C. The surface of the plates is allowed to dry overnight. Antibiotic activity will decrease significantly with prolonged storage of plates. Alternatively, the nonselective medium Pseudomonas Agar F can be used for distinguishing fluorescent *Pseudomonas* colonies from nonfluorescent colonies from seafood issue. However, total counts are frequently lower on this medium than on plate count or Tryptic Soy Agar. Pseudomonas Agar F contains: tryptone, 10 g; proteose peptone no. 3, 10 g; K_2HPO_4, 1.5 g; $MgSO_4$, Agar 15 g; and distilled water, 1000 mL, pH 7.0. Plates are viewed under a black UV lamp for detection of fluorescent colonies.

15.8 Useful Tests for Confirming the Identity of Seafood Spoilage Bacteria

15.8.1 Genetic Transformation Assay for Confirming the Identity of Psychrobacter immobilis Isolates (Juni and Hyme [29])

1. The auxotrophic mutant A351-Hyx-7 (ATCC 43117) recipient culture requiring hypoxanthine is cultured onto a slant of brain heart infusion agar (BHI-A) or heart infusion agar (HI-A). Do not culture sequentially to prevent spontaneous reversion to prototrophy.

2. Colorless colonies of unknown organisms are picked to a slant of Nutrient Agar or Tryptic Soy Agar.

3. Several large loops of cell growth from an unknown slant are transferred to a vial of sterile lysing solution (0.05% sodium dodecyl sulfate in 0.15 M NaCl and 0.015 M trisodium citrate) and the cells are dispersed by vigorous agitation. The vials are then held at 65°C for 1 h to lyse the cells and achieve sterility.

4. Divide a BHI-A or an HI-A plate into four sections and aseptically transfer a full loop of crude DNA from each sample to a separate quadrant and smear a circular area of about 1 in. in diameter. Prepare a duplicate control plate for determining the sterility of the DNA samples.

5. To one set of BHI-A plates apply a loop of the recipient 135-Hyx-7 culture to each area smeared with DNA and leave the other plate as a sterility control. In addition, inoculate one quadrant of the BHI-A plate with just the recipient to detect spontaneous revertants. Incubate the plate overnight at 20°C.

6. Transfer a loop of cell growth from each area of the BHI-A plate that has grown up, to a quadrant of an M9A Agar plate by streaking the quadrant and incubate the plate at 20°C for 3 days and observe for the development of isolated colonies (genetic transformants) at the end of the streaks (see Figure 15.4). Discount solid areas of growth at the initial areas of the streaks, which are due to hypoxanthine carry over from the BHI-A plate. M9A Agar plates contain 0.8% vitamin free casein hydrolysate, 0.28% Na_2HPO_4, 0.1% KH_2PO_4, 0.5% NaCl, 0.045% $MgSO_4 \cdot 7H_2O$, 5.0 mL of 60% sodium lactate, 4.0 g of glucose sterilized separately, and 15 g of agar in a total volume of 1000 mL prepared with deionized water.

15.8.1.1 Motility

The experience of the author has indicated that the most reliable assay for motility is to observe a young broth culture under the microscope. An additional advantage to microscopic observation is that with experience, the observer can distinguish motility of a peritrichously flagellated

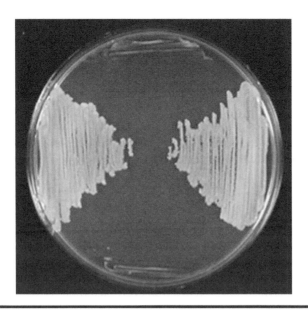

Figure 15.4 Genetic transformation of *P. immobilis*. Growth of recipient A351-Hyx-7 genetically transformed by DNA from a fishery isolate of *P. immobilis* on an M9A plate lacking hypoxanthine. Quadrants without growth have been streaked with the recipient A351-Hyx culture without prior contact with exogenous DNA and serve as controls. Slight growth at the initiation of the control streaks can be observed because of nutrient carryover from the BHI plate.

organism versus a polarly flagellated organism. Polarly flagellated organisms will frequently move at a much higher rate of speed than peritrichously flagellated organisms. In addition, polarly flagellated organisms tend to lift the flagellated end of the cell due to a rapid rotary swirling of the polar flagella. However, motility test agar may still be found useful by some and consists of tubes filled with 10 mL of: 1.0% peptone, 0.5% NaCl, and 3.5% agar. Inoculation is by stabbing. Motile cultures will exhibit turbid lateral growth, usually just below the surface. The flagella stain is notably time consuming and unreliable unless performed by an individual with extensive experience and technical expertise. However, it is still the only reliable method for determining the cellular location of flagella outside of electron microscopy. The flagella stain in the following text [49] has been found to be superior to several others and yields results amenable to photomicroscopy.

1. Cells from 18 h. Nutrient Agar slants are suspended in distilled water to yield a cloudy suspension.
2. A glass slide is scrupulously cleaned by scrubbing with a household diatomaceous earth cleaning powder and paper towel moistened with water to achieve a thick slurry. Blot the slide dry. Care should be taken not to handle the surface of the slide with bare fingers.
3. Several drops of the cell suspension are placed on the end of a cleaned glass slide and the excess fluid is allowed to run down and off the slide to facilitate orienting polar flagella in one direction. After air-drying, the slide is immersed in reagent I for 15 min. This step serves as both fixative and mordant. Reagent I consists of: deionized water, 100 mL; tannic acid, 5.0 g; FeSO$_4$ (saturated solution), 1.5 mL; formaldehyde, 2.0 mL; and 1.0% NaOH, 1.5 mL.

4. Without rinsing the slide, it is then covered with reagent II until a brown color appears and is then washed with distilled water and air-dried. Reagent II contains: $AgNO_3$, 5.0 g; distilled water, 100 mL; and NH_4OH (several drops added to 10 mL of $AgNO_3$ solution; when the precipitate clears add to 90 mL of remaining solution).

15.8.1.2 The Oxidase Test

The test for cytochrome oxidase is highly definitive for the distinction of various taxonomic groups of seafood spoilage bacteria. All members of the genera *Pseudomonas*, *Vibrio*, and *Aeromonas* in addition to *P. immobilis* are presently considered positive for cytochrome oxidase. Enteric organisms are all negative. The methods of Kovács [50] and Bovre and Henriksen [51] are both recommended to ensure against false negative reactions. The cytochrome oxidase test is performed as follows: moisten a sheet of filter paper with a 1% aqueous solution of tetramethyl-*p*-phenylenediamine [50], or dimethyl-*p*-phenylenediamine [51] and smear onto the surface bacterial cells from a slant or colony using a platinum loop (not Nichrome). A positive test is indicated by the development of a dark purple color within 20 s. The use of a metal loop with an iron content can result in weak false positive reactions derived from corroded iron debri shed from the loop.

15.8.2 Litmus Milk

Tubes of litmus milk inoculated with bacterial isolates from seafood afford the observer a variety of reactions for distinguishing cultures and facilitating identification. Reactions to be observed are: (1) peptonization or proteolysis resulting in the initially opaque tube becoming translucent, (2) an alkaline reaction derived from the deamination of amino acids and resulting in a blue coloration of the medium, (3) an acid reaction, derived from the utilization of lactose resulting in a pink coloration, (4) an acid clot resulting in precipitation of the casein, and (5) reduction of the litmus dye resulting in the tube turning white. Combinations of several reactions are also frequently encountered such as (a) reduction and proteolysis, and (b) proteolysis with an alkaline reaction. Litmus milk consists of 10% rehydrated dry skimmed milk and 0.5% litmus.

15.8.3 Proteolysis

There are several approaches for determining if cultures are proteolytic. Fish press juice offers a natural substrate for determining proteolysis of bacterial isolates. Fillets are placed into a press and the liquid expressed is collected. Alternatively, the fillets are first ground and then pressed. It is important to maintain the temperature of the fish juice as close to 0°C as possible to prevent thermal denaturation and precipitation of initially soluble proteins at or near room temperature. The juice can then be centrifuged at 4°C in a precooled rotor for 10 min at $10,000 \times g$ to sediment tissue particles. The juice is then sterilized by passage through a 0.25 m porosity filter membrane, maintaining the temperature as close to 0°C as possible. Sterile agar is then mixed with 9%–24% fish juice (final concentration 1.5%) and plates are poured. At room temperature, a significant amount of initially soluble proteins will precipitate out and impart a cloudy appearance to the agar. Inoculation of circular zones onto the agar surface and incubation at 20°C for 48 h will yield areas of growth surrounded by large zones of clearing by proteolytic organisms [52]. Alternatively, 15 mL of rehydrated skimmed milk (autoclaved separately) is added to 85 mL of sterile Nutrient Agar and plates are poured [52].

Gelatin deeps are used frequently for the detection of proteolysis. It should be kept in mind that gelatin is not as readily hydrolyzed by some proteases as is casein in skimmed milk so that an organism exhibiting definite proteolysis of casein may not yield a positive proteolysis result with gelatin. Nutrient gelatin deeps (10 mL) in tubes are prepared and contain peptone, 5%; beef extract, 3%; and gelatin, 12%; in distilled water, pH 7.0. Nutrient gelatin will liquefy at temperatures above 20°C. It is therefore important to incubate the tubes at 20°C and on removal from the incubator, it is necessary to place them in an ice bath to prevent temperature-induced liquefaction before they are read. Tubes of Nutrient gelatin are stab inoculated to the bottom of the tubes with an inoculating needle from a broth tube. For greatest sensitivity, the surface of a soft-agar-gelatin overlay plate can be inoculated [53]. This medium consists of bottom agar prepared with Nutrient broth to which is added $MnSO_4$, 0.005%; NaCl, 0.8%; and agar, 1.5%, pH 7.0. Plates are poured and allowed to solidify. A soft-agar-gelatin overlay (3 mL) containing Nutrient broth, $MnSO_4$, 0.005%; NaCl, 0.8%; and agar 0.8%, and 1.5% gelatin, pH 7.0 is poured onto the bottom agar. The surface is then carefully inoculated. After incubation, 5% acetic acid is applied to the plate to precipitate intact gelatin and clear zones surrounding areas of growth are noted. This procedure is ideally suited for direct enumeration of proteolytic organisms from seafood where the surface of a series of bottom agar plates is smear inoculated with 0.1 mL of decimal dilutions of the homogenized tissue. A soft-agar-overlay is then applied to each plate. If one desires to isolate the proteolytic organisms, a soft-agar-overlay prepared with skimmed milk can be used which does not require the application of acetic acid as with soft-agar-gelatin plates.

15.8.4 Detection of H_2S Production

Detection of H_2S production by seafood isolates is most readily observed by stab inoculating tubes containing 10 mL of peptone-Fe-agar (peptone, 1.5%; proteose peptone, 0.5%; ferric ammonium citrate, 0.5%; sodium glycerophosphate, 0.1%; sodium thiosulfate, 0.008%; and agar, 1.5%, pH 6.7). Positive tubes will yield an intense black area of growth within 72 h along the stab resulting from FeS formation. It should be kept in mind that some isolates will be slow to form H_2S, and may be weak H_2S producers and that prolonged incubation, usually beyond 7 days, results in the fading of the intense black FeS due to slow oxidation by the penetration of atmospheric oxygen.

15.8.5 DNase Activity

The production of extracellular DNase activity by seafood isolates is most conveniently determined with the use of DNase test Agar w/methyl Green, which consists of tryptose, 2.0%; deoxyribonucleic acid, 0.2%; NaCl, 0.5%; methyl green, 0.005%; and agar, 1.5%, pH 7.3. Methyl green combines with high molecular weight DNA to impart a green coloration to the agar. When a DNase positive organism is streaked onto the surface, the extracellular DNase hydrolyzes the DNA immediately surrounding the area of growth yielding a colorless zone. Applying cells from an agar slant or broth tube to a circular area of about 0.5 cm allows a single plate to be used for five or six cultures. Detection of extracellular DNase is particularly useful in confirming the identity of *S. putrefaciens* isolates.

15.8.5.1 Molecular Techniques for Detection and Enumeration of Seafood Spoilage Bacteria

The minimum number of colony forming units (CFU) per gram of raw seafood tissue is about 1×10^4 unless the tissue is excised aseptically. On a more practical basis, commercially processed fresh fish fillets will usually have an initial CFU count of about 1×10^5 CFU per gram of tissue. Counts in the range of 10^7 to 10^8 per gram are usually associated with some degree of spoilage and poor quality. Universal primers have been successfully applied to the quantification of the total bacterial population on fish tissue [54–56]. The universal forward primer DG74 5'-AGG-AGG-TGA-TCC-AAC-CGA-A-3' and the universal reverse primer RW01: 5'-ACC-TGG-AGG-AAG-AAG-GTG-GGG-AT-3' primer [57] amplify a 370-bp sequence of the 16S rRNA gene derived from all bacteria. An extremely close linear relationship was found between the number of CFU per gram determined from plate counts and the total number of genomic targets determined by conventional and real-time polymerase chain reaction (PCR). In addition, this methodology has been extended to the use of the PCR for distinguishing the total number of dead and viable bacteria on fish tissue with the use of the selectively permeable DNA binding dye ethidium bromide monoazide [58].

Venkitanarayanan et al. [59] developed a pair of primers for the amplification of a 207-bp amplicon derived from the 23S rDNA sequence of meat spoilage bacteria. The assay was designed for detection of the following typical meat spoilage bacteria: *P. fluorescence*, *P. putida*, *P. fragi*, *P. areofaciens*, *Acinetobacter calcoaceticus*, *Enterobacter liquefaciens*, *Flavobacerium breve*, *Moraxella osloensis*, and *Brochotrix teremosphacta*. The assay should be equally applicable for the collective quantitative PCR enumeration of most of these spoilage organisms on seafood. However, these authors did not indicate the specificity of the assay. The sequence of the forward primer PF is 5'-AAG-CTT-GCT-GGA-GGT-ATC-AGA-AGT-GC and that of the reverse primer PR is CTC-CGC-CCC-TCC-ATC-GCA-GT.

Seafood isolates of *S. putrefaciens* can be confirmed as such by the PCR with the use of the primers SP-1: 5'-TTC-GTC-GAT-TAT-TTG-AAC-AGT AND SP-2r: 5'-TTC-TCC-AGC-AGA-TAA-TCG-TTC which amplify a 422-bp sequence of the *Gyr B* sequence [17].

The development of a primer pair specific for members of the genus *Pseudomonas* [60] has allowed the use of the PCR for the identification of *Pseudomonas* isolates. This PCR assay is ideally suited for the confirmation of presumptive isolates of *Pseudomonas* from seafood and has the potential to be used to quantify the total number of pseudomonads per gram of seafood tissue numerically. The assay is based on the presence of two *Pseudomonas* specific and conserved sequences, one at the middle of the 16S rDNA sequence and the other at the beginning of the 23S rDNA sequence. As a result, the amplified region includes the 3'-half of the 16S rDNA with the whole 16S–23S rDNA Internal Transcribed Spacer (ITS1) sequence in addition to the first 25 nucleotides of the 23S rDNA sequence from the 5'-end. The *Pseudomonas* specific primers generated amplicons of 1300-bp. The sequence of the forward primer fPs16S is 5'-ACT-GAC-ACT-GAG-GTG-CGA-AAG-CG that of the reverse primer rPs23S is 5'-ACC-GTA-TGC-GCT-TCT-TCA-CTT-GAC-C. All 33 *Pseudomonas* strains representing 14 species yielded amplicons while none of the 13 Gram-negative on *Pseudomonas* species with the exception of *Azotobacter chroococcum*. In addition, several of the *Pseudomonas* yielded two or three bands varying from 1100 to 1300-bp. The multiple bands are thought to reflect the number and variation in length of ITS1 sequences in a given species.

References

1. Castell, C. and Anderson, G. Bacteria associated with spoilage of cod fillets. *J. Fish. Res. Board Can.* 7, 370–377, 1948.
2. Georgala, D. The bacterial flora of the skin of North Sea Cod. *J. Gen. Microbiol.* 18, 84–91, 1958.
3. Tomiyama, T., da Sota, A., and Stern, J. A rapid vacuum distillation procedure for the determination of volatile acids and volatile bases in fish flesh. *Food Technol.* 10, 614–617, 1956.
4. Dyer, W. Report on trimethylamine in fish. *JAOAC* 42, 292–294, 1959.
5. Dyer, W. and Mounsy, T. Amines in fish muscle. II. Development of trimethylamine and other amines. *J. Fish. Res. Board Can.* 6, 351–358, 1945.
6. Castell, C., Greenough, M., Rodgers, R., and MacFarlane, A. Grading of fish for quality. 1. Trimethylamine values of fillets cut from graded fish. *J. Fish. Res. Board Can.* 15, 701–716, 1958.
7. Hillig, F., Selton, L. Jr., Loughrey, J., and Bethia, S. Chemical indexes of decomposition in pollock and whiting. *JAOAC* 44, 499–507, 1961.
8. Proctor, B., Nickerson, J., Fazino, T., Ronsivalli, L., Smith, R., and Stern, Rapid determination of quality of whole eviscerated haddock. *J. Food Technol.* 13, 224–228, 1958.
9. Stanier, R., Palleroni, N., and Doudoroff, M. The aerobic pseudomonads: A taxonomic study. *J. Gen. Microbiol.* 43, 159–271, 1966.
10. Gennari, M. and Fragotto, F. A study of the incidence of different fluorescent *Pseudomonas* species and biovars in the microflora of fresh and spoiled meat and fish, raw milk, cheese, soil and water. *J. Appl. Bacteriol.* 72, 281–288, 1992.
11. Hugh, R. and Leifson, E. The taxonomic significance of fermentative versus oxidative metabolism of carbohydrates by various gram negative bacteria. *J. Bacteriol.* 66, 24–26, 1953.
12. Shewan, J., Hobbs, G., and Hodgkiss, W. A determinative scheme for the identification of certain genera of gram-negative bacteria, with special reference to the Pseudomonadaceae. *J. Appl. Bacteriol.* 23, 379–390, 1960.
13. Shewan, J., Hobbs, G., and Hodgkiss, W. The *Pseudomonas* and *Achromobacter* groups of bacteria in the spoilage of marine white fish. *J. Appl. Bacteriol.* 23, 463–468, 1960.
14. Castell, C., Greenough, M., and Dale, J. The action of Pseudomonas on fish muscle. 3. Identification of organisms producing fruity and oniony odours. *J. Fish. Res. Board Can.* 16, 13–19, 1959.
15. Castell, C., Greenough, M., and Jenkin, N. The action of Pseudomonas on fish muscle. 2. Musty and potato-like odours. *J. Fish. Res. Board Can.* 14, 775–782, 1957.
16. Ivanova, E., Kiprianova, E., Valery, M., Levanova, G., Garagulya, A., Gorshkova, N., Yumoto, N., and Yoshikawa, S. Characterization and identification of marine *Alteromonas nigrifaciens* strains and emendation of the description. *Int. J. Syst. Bacteriol.* 46, 223–228, 1996.
17. Venkateswaran, K., Moser, D., Dollhopf, M., Lies, D., Saffarini, D., MacGregor, B., Ringelberg, D., White, D., Nishijima, M., Sano, H., Burghardt, J., Stackebrandt, E., and Nealson, K. Polyphasic taxonomy of the genus *Shewanella* and description of *Shewanella oneidensis* sp. nov. *Int. J. Syst. Bacteriol.* 49, 705–724, 1999.
18. Derby, H. and Hammer, B. Bacteriology of butter. IV. Bacteriological studies on surface taint butter. *Iowa Agric. Exp. Stn. Res. Bull.* 145, 387–416, 1931.
19. Long, H. and Hammer, B. Classification of organisms important in dairy products. III. *Pseudmonas putrefaciens. Iowa Agric. Exp. Stn. Res. Bull.* 285, 176–195, 1941.
20. Lee, J., Gibson, D., and Shewan, J. A numerical taxonomic study of *Pseudomonas*-like marine bacteria. *J. Gen. Microbiol.* 98, 439–451, 1977.
21. Baumann, L., Baumann, P., Mandel, M., and Allen, R.D. Taxonomy of aerobic marine eubacteria. *J. Bacteriol.* 110, 02–429, 1972.
22. MacDonell, M. and Colwell, R. Phylogeny of the Vibrionaceae and recommendation for two new genera, *Listonella* and *Shewanella. Syst. Appl. Microbiol.* 6, 171–182, 1985.

23. Levin, R.E. Correlation of DNA base composition and metabolism of *Pseudomonas putrefaciens* isolates from food, human clinical specimens, and other sources. *Antonie Van Leeuwenhoek* 38, 121–127, 1972.

24. Nozue, H., Hayashi, T., Hashimoto, Y., Ezaki, T., Hamasaki, K., Ohwada, K., and Terawaki, Y. Isolation and characterization of *Shewanella alga* from human clinical specimens and emendation of the description of *S. alga* Simidu et al., 1990, 335. *Int. J. Syst. Bacteriol.* 42, 628–634, 1992.

25. Vogel, B.F., Jørgensen, K., Christensen, H., Olsen, J.E., and Gram, L. Differentiation of *Shewanella putrefaciens* and *Shewanella alga* on the basis of whole-cell protein profiles, ribotyping, phenotypic characterization, and 16S rRNA gene sequence analysis. *Appl. Environ. Microbiol.* 63, 2189–2199, 1997.

26. Sadovski, A. and Levin, R. Extracellular nuclease activity of fish spoilage bacteria, fish pathogens and related species. *Appl. Microbiol.* 17, 787–789, 1969.

27. Chai, T., Chen, C., Rosen, A., and Levin, R. Detection and incidence of specific spoilage bacteria on fish. II. Relative incidence of *P. putrefaciens* and fluorescent pseudomonads on haddock fillets. *J. Appl. Microbiol.* 16, 1738–1741, 1968.

28. Van Sickle, C. and Levin, R. Relative rates of autolysis of high and low GC isolates of *Pseudomonas putrefaciens* and other gram negative bacteria. *Microbios Lett.* 6, 85–94, 1978.

29. Juni, E. and Hyme, G. Transformation assay for identification of psychrotrophic achromobacters. *Appl. Environ. Microbiol.* 40, 1106–1114, 1980.

30. Chen, T. and Levin, R. Taxonomic significance of phenethyl alcohol production by *Achromobacter* isolates from fishery sources. *Appl. Microbiol.* 28, 681–687, 1974.

31. Juni, E. and Hyme, G. *Psychrobacter immobilis* gen. nov., sp. Non.: Genospecies composed of gram-negative, aerobic, oxidase-positive coccobacilli. *Int. J. Syst. Bacteriol.* 36, 388–391, 1986.

32. Rossau, R., Van Landschot, A., Gillis, M., and De Ley, J. Taxonomy of Moraxellaceae fam. Nov., a new bacterial family to accommodate the genera *Moraxella, Acinetobacter*, and *Psychrobacter* and related organisms. *Int. J. Syst. Bacteriol.* 41, 310–319, 1991.

33. González, C., Santos, A., García-López, M., and Otero, A. Psychrobacters and related bacteria in freshwater fish. *J. Food Protect.* 63, 315–321, 2000.

34. McMeekin, T. The adensonian taxonomy and the deoxyribonucleic acid base composition of some gram negative, yellow pigmented rods. *J. Appl. Bacteriol.* 35, 129–137, 1972.

35. Castell, C. and Mapplebeck, E. The importance of *Flavobacterium* in fish spoilage. *J. Fish. Res. Board Can.* 9, 148–156, 1952.

36. Hovda, M., Sivertsvik, M., Lunestad, B., Lorentzen, G., and Rosnes, J. Characterization of the dominant bacterial population in modified atmosphere packaged farmed halibut (*Hippoglossus hippoglossus*) based on 16S rDNA-0DGGE. *Food Microbiol.* 24, 362–371, 2007.

37. Hovda, M., Lunestad, M., Sivertsvik, M., and Rosnes, J. Characterization of the bacterial flora of modified atmosphere packaged farmed atlantic cod (*Gadus morhua*) by PCr-DGGE of conserved 16S rrNA gene regions. *Int. J. Food Microbiol.* 117, 68–75, 2007.

38. Sivertsvik, M., Jeksrud, W., and Rosnes, T. A review of modified atmosphere packaging of fish and fishery products—Significance of microbial growth, activities and safety. *Int. J. Food Sci. Technol.* 37, 107–127, 2002.

39. Mokhele, K., Johnson, A., Barrette, E., and Ogrydziak, D. Microbiological analysis of rock cod (*Sebasts* spp.) stored under elevated carbon dioxide atmosphere. *Appl. Environ. Microbiol.* 45, 878–883, 1983.

40. Johnson, A. and Ogrydziak, D. Genetic adaptation to elevated carbon dioxide atmospheres by *Pseudomononas*-like bacteria isolated from rock cod (*Sebastes* spp.). *Appl. Environ. Microbiol.* 48, 486–490, 1984.

41. Nealson, K. and Saffarini, D. Iron and manganese in anaerobic respiration: Environmental significance, physiology, and regulation. *Ann. Rev. Microbiol.* 48, 311–343, 1994.

42. Rudi, K., Maudesten, T., Hannevik, S., and Nissen, H. Explorative multivariate analysis of 16S rRNA gene data from microbial communities in modified-atmosphere packed salmon and coalfish. *Appl. Environ. Microbiol.* 70, 5010–5018, 2004.

43. Pelroy, G., Seman, J. Jr., and Eklund, M. Changes in the microflora of irradiated petrale sole (*Eopsetta jordani*) fillets stored aerobically at 0.5 C. *Appl. Microbiol.* 15, 92–96, 1967.

44. Pelroy, G. and Eklund, M. Changes in the microflora of vacuum-packaged irradiated petrale sole (*s*) fillets stored at 0.5 C. *Appl. Microbiol.* 14, 921–927, 1966.

45. Silverrio, R. and Levin, R. Evaluation of methods for determining the bacterial population of fresh fillets. *J. Milk Food Technol.* 30, 242–246, 1967.

46. Makarios-Laham, I. and Levin, R. Isolation from haddock tissue of psychrophilic bacteria with maximum growth temperatures below 20°C. *Appl. Environ. Microbiol.* 48, 439–440, 1984.

47. Haight, R. and Morita, R. Thermally induced leakage from *Vibrio marinus*, an obligately psychrophilic marine bacterium. *J. Bacteriol.* 92, 1388–1393, 1966.

48. Sands, D. and Rovira, A. Isolation of fluorescent pseudomonads with a selective medium. *Appl. Microbiol.* 20, 513–514, 1970.

49. Rosen, A. and Levin, R. Vibrios from fish pen slime which mimic *Escherichia coli* on Violet Red Bile Agar. *Appl. Microbiol.* 20, 107–112, 1970.

50. Kovács, N. Identification of *Pseudomonas pyocyanea* by the oxidase reaction. *Nature* (Lond.) 178, 703, 1956.

51. Bovre, K. and Hendriksen, S. Minimal standards for description of new taxa within the genera *Moraxella* and *Acinetobacter*: Proposal by the subcommittee on *Moraxella* and allied bacteria. *Int. J. Syst. Bacteriol.* 26, 92–96, 1976.

52. Kazanas, N. Proteolytic activity of microorganisms isolated from freshwater fish. *Appl. Microbiol.* 16, 128–132, 1968.

53. Levin, R. Detection and incidence of specific spoilage bacteria on fish. I. Methodology. *Appl. Microbiol.* 16, 1734–1737, 1968.

54. Lee, J. and Levin, R. Selection of universal primers for PCR quantification of total bacteria associated with fish fillets. *Food Biotechnol.* 20, 275–286, 2006.

55. Lee, J. and Levin, R. Direct application of the polymerase chain reaction for quantification of total bacteria on fish fillets. *Food Biotechnol.* 20, 287–298, 2006.

56. Lee, J. and Levin, R. Rapid quantification of total bacteria on cod fillets by using real-time PCR. *J. Fisheries Sci.* 1, 58–67, 2007.

57. Greisen, K., Loeffelholz, M., Purohit, A., and Leong, D. PCR primers and probes for the 16S rRNA gene of most species of pathogenic bacteria, including bacteria found in cerebrospinal fluid. *J. Clin. Microbiol.* 32, 335–351, 1994.

58. Lee, J. and Levin, R. Use of ethidium bromide monoazide for quantification of viable and dead mixed bacterial flora from fish fillets by polymerase chain reaction. *J. Microbiol. Methods* 67, 456–462, 2006.

59. Venkitanarayanan, K., Khan, M., Faustman, C., and Berry, B. Detection of meat spoilage bacteria by using the polymerase chain reaction. *J. Food Protect.* 59, 845–848, 1996.

60. Locatelli, L., Tarnawsi, S., Hamelin, J., Rossi, P., Aragno, M., and Fromin, N. Specific PCR amplification for the genus *Pseudomonas* targeting the 3′ half of 16S rDNA and the whole 16S–23S rDNAS spacer. *Syst. Appl. Microbiol.* 25, 220–227, 2002.

Chapter 16

Detection of the Principal Foodborne Pathogens in Seafoods and Seafood-Related Environments

David Rodríguez-Lázaro and Marta Hernández

Contents

16.1 Introduction 486
16.2 Detection of the Principal Seafoodborne Pathogens 488
    16.2.1 Detection of Pathogenic *Vibrio* Species in Seafoods and Seafood-Related
        Environments 488
        16.2.1.1 *V. parahaemolyticus* 488
        16.2.1.2 *V. vulnificus* 493
        16.2.1.3 *V. cholerae* 495
    16.2.2 Detection of *L. monocytogenes* in Seafoods and Seafood-Related
        Environments 496
    16.2.3 Detection of *Salmonella* spp. in Seafoods and Seafood-Related
        Environments 498
References 500

16.1 Introduction

The importance of foodborne pathogens in public health is substantial. They cause more than 14 million illnesses, 60,000 hospitalizations, and 1,800 deaths per year in the United States [88] with annual medical and productivity losses above 6,500 million dollars [23]. In England and Wales, the figures are similar, and they cause 1.3 million illnesses, 20,759 hospitalizations and 480 deaths each year [1]. The number of bacterial gastroenteritis associated to seafood products has been increased considerably during the last decades by the rapid globalization of the food market, the increase of personal and food transportation, and profound changes in the food consumption habits [66,88]. Among the bacterial pathogen that can produce gastroenteritis associated to seafood products, three can be considered as a primary threat: the enteropathogenic *Vibrio*, *Listeria monocytogenes*, and *Salmonella* spp.

Three *Vibrio* species, *Vibrio parahaemolyticus*, *Vibrio vulnificus*, and *Vibrio cholerae*, are well-documented human pathogens, specially associated to the consumption of raw or undercooked seafood products [67,87,103]. *V. parahaemolyticus* is an important seafoodborne pathogen worldwide [71]. It was first identified as a cause of foodborne illness in Japan in 1950 [35], and it has been reported to account for 20%–30% of foodborne illnesses in Japan [3] and a common cause of seafoodborne gastroenteritis in Asian countries [26,130]. In contrast, infections are occasional in Europe, and only sporadic outbreaks have been reported in Spain and France [113]. In the United States, *V. parahaemolyticus* is the leading cause of gastroenteritis associated with seafood consumption, and between 1973 and 1998 approximately 40 outbreaks were reported [24]. Consumption of raw or undercooked seafood, particularly shellfish, contaminated with *V. parahaemolyticus* may produce a self-limiting gastroenteritis involving symptoms such as vomiting, nausea, diarrhea with abdominal cramps, headache, and low-grade fever. *V. parahaemolyticus* is disseminated worldwide in estuarine, marine, and coastal water environments [65]. Some environmental factors such as the water temperature, salinity, zooplankton blooms, tidal flushing, and dissolved oxygen modulate its spatial and temporal distribution [95]. The increase of the prevalence of *V. parahaemolyticus* in raw shellfish is also correlated to the warm seawaters. The *V. parahaemolyticus* loads in oysters is usually lower than 10^3 cfu g^{-1} [70], but it can increase notably when the shellfish is cultivated in warmer seawater [28].

V. vulnificus produces one of the most severe foodborne infections, with a case-fatality rate greater than 50% [92]. It can cause fatal septicemia, wound infections, and gastroenteritis especially in immunocompromised individuals [11]. It was first isolated by the Center for Disease Control (CDC) in 1964 [112]. This organism is also disseminated worldwide in waters of different temperatures and salinities [131]. Environmental conditions such as water temperature and salinity modulate the variation in its prevalence [44]. Most of the outbreaks in United States have been reported during the summer generally associated to the consumption of raw seafoods [22,46,75,93].

V. cholerae is the causative agent of the cholera outbreaks and epidemics. There is a direct relationship between the consumption of raw, undercooked, contaminated, or recontaminated seafood and outbreaks produced by *V. cholerae* [30,34,67]. Foodstuff can be contaminated by this pathogen through contaminated irrigation water or human origin-fertilizer [30,91]. The O1 serogroup is the group predominantly isolated in cholera epidemics [34], and a new pathogenic serogroup, O139, has been also identified [4]. However, non-O1/O139 serogroups are sporadically involved in cholera-like diarrheal episodes, but infrequently in outbreaks [85,110]. Toxigenic *V. cholerae* O1 is rarely isolated and no isolations of serogroup O139 have been reported in western countries. In contrast, non-O1/O139 isolates are commonly found in estuarine water and shellfish [6]. Various O1 strains have become endemic in many regions in the world, including Australia and the U.S. Gulf Coast [21,123].

L. monocytogenes is an important foodborne pathogen, which usually (20%–50% of the cases) produces a fatal infection. It has been isolated from a wide range of sources, and seafood and seafood-related environments have been reported as important niches for this bacterium [105]. Cao et al. [17] reported the recurrent presence of this pathogen in shrimp samples and a frozen shrimp-processing line environment, without a positive correlation between its presence and the accompanying environmental microbiota. Farber [32] reported a low incidence of *L. monocytogenes* in imported seafood products between 1996 and 1998 (below 1%), and a complete absence in Canadian seafood products. Van Coillie et al. [118] studied the prevalence of *L. monocytogenes* in different ready-to-eat (RTE) seafood products on the Belgian market. The occurrence of *L. monocytogenes* was 23.9%, and the contamination levels were low in most cases (84% below 100 cfu g^{-1}). The most prevalent serotype was 1/2a and serotypes 1/2b, 1/2c, and 4b were also present. In a longitudinal study in seafoods between 2001 and 2005 in France, Midelet-Bourdin et al. [90] observed similar findings (a prevalence of 28% with a low level of contamination). The presence of *L. monocytogenes* in tropical fish and shellfish in Mangalore, India was 17% and 12%, respectively [63]. Similar results were obtained by Nakamura et al. [96,97] in RTE seafood products commercially available or in a cold-smoked fish-processing plant in Osaka, Japan (13% and 7%, respectively). Its incidence was mainly in the summer and autumn, and it was only isolated in cold-smoked fish samples and in low numbers (below 100 cfu g^{-1}). The serotype 1/2a was the most prevalent in both studies, and serotypes 1/2b, 3b, 4b, and 3a were also present.

The consumption of seafoods and outbreaks of listeriosis is well documented [105]. For example, in a small human outbreak occurred in Ontario, Canada, the relationship between the presence of *L. monocytogenes* in seafood products (imitation crab meat) and the outbreak was clearly established [33]. Although all the foodstuffs obtained from the refrigerator of the two patients contained *L. monocytogenes*, three of them were heavily contaminated: imitation crab meat, olives, and salad. Molecular typing of the isolates by randomly amplified polymorphic DNA (RAPD) and pulsed-field gel electrophoresis (PFGE) typing demonstrated that the imitation crab meat and the clinical strains were indistinguishable. In addition, challenge studies performed with a pool of *L. monocytogenes* strains showed that imitation crab meat, but not olives, supported growth of this pathogen.

Salmonella spp. is a major public health problem because of its large and varied animal reservoir, the existence of human and animal carrier states, and the lack of a concerted nationwide program to its control [42]. Furthermore, *Salmonella* is the main cause of documented foodborne human illnesses in most developed countries [18,117,124]. Of the outbreaks of foodborne illness recorded in the World Health Organization (WHO) report for 1993–1998, Salmonellae were most often reported as causative agent (54.6% of cases) [108]. Food items with a greater hazard include raw meat and some products intended to be eaten raw, raw or undercooked products, such as seafood and seafood products [31]. The presence of *Salmonella* spp. in tropical seafood products collected from different landing centers and open markets in Mangalore, India was studied by Kumar et al. [78]. The overall incidence of *Salmonella* spp. was 17%, suggesting that the contamination of seafoods with *Salmonella* may be occurring during postprocess handling and processing. A similar study was conducted in fish, shellfish, ice, and water obtained from the market and fish-landing center in Mangalore, India [109]. Twenty percent of the samples were positive using conventional methods, but the number of positives increased up to 52% when PCR was used, indicating the prevalence of *Salmonella* in seafood may be much more than that reported by conventional isolation techniques. The most prevalent serotype was *Salmonella enterica* serotype Weltevreden, and *S. enterica* serotype Worthington and *S. enterica* serotype Newport were also present.

16.2   Detection of the Principal Seafoodborne Pathogens

As a consequence of the potential hazards described above, microbiological quality control programs are being increasingly applied throughout the seafood production chain in order to minimize the risk of infection for the consumer. Classical microbiological methods to detect the presence of those microorganisms involve enrichment and isolation of presumptive colonies of bacteria on solid media, and final confirmation by biochemical and/or serological identification. It is laborious and time consuming, and usually more than 3–5 days are needed for definitive results. Although remaining the approach of choice in routine analytical laboratories, the adoption of alternative techniques such as molecular-based methods in microbial diagnostics has become an alternative approach, as they possess inherent advantages such as shorter time to results, excellent detection limits, specificity, and potential for automation.

16.2.1   Detection of Pathogenic Vibrio Species in Seafoods and Seafood-Related Environments

16.2.1.1   V. parahaemolyticus

The most widely used methods for the detection of *V. parahaemolyticus* in foods are the International Organization for Standardization (ISO) standard 8914:1990 [51] and the most probable number (MPN) method described in the U.S. Food and Drug Administration (FDA) *Bacterial Analytical Manual* (BAM) [72]. In the International Standard ISO 8914:1990, food samples are incubated at 35°C for 7–8 h in parallel in two enrichment broths (salt polymyxin B broth and alkaline saline peptone water or saline glucose culture medium with sodium dodecyl sulfate), and then streaked on two selective media (thiosulfate–citrate–bile salts–sucrose agar [TCBS] and triphenyltetrazolium chloride soya tryptone agar [TSAT]). After incubation for 18 h on TCBS or 20–24 h on TSAT, colonies being 2–3 mm, smooth, and green on TCBS or 2–3 mm, smooth, flat, and dark red on TSAT can be considered presumptive colonies of *V. parahaemolyticus*, and they must be confirmed by biochemical tests. Recently a new ISO standard (ISO/TS 21872-1:2007) has been published describing a horizontal method in food for detection of *V. parahaemolyticus* and *V. cholerae* [58]. In the FDA BAM method, after the MPN analysis, the tubes must be plated on TCBS selective medium and several presumptive isolates must be confirmed by biochemical testing. In both cases, these methods are cumbersome and laborious, and definitive results can be only obtained after more than 4–5 days. To overcome those disadvantages, different PCR methods have been developed for detection of *V. parahaemolyticus* in seafood products and seafood-related environments (Table 16.1).

Some authors have reported PCR methods for the detection *V. parahaemolyticus* independently of the pathogenic capacity of the strains detected. For this purpose, different PCR targets and DNA protocols have used. Lee et al. [80] developed a PCR method based on a specific fragment, *pR72H*, cloned and sequencedv in that laboratory. To determine its selectivity, 124 *V. parahaemolyticus* and 50 non-*V. parahaemolyticus* isolates were assayed. The PCR assay was 100% selective. Finally, the applicability of the method was evaluated in oysters. Ten milliliters of oyster homogenate was inoculated with decreasing amounts of *V. parahaemolyticus*, and 1 mL of each homogenate was then mixed with 9 mL of tryptose soy broth (TSB) containing 2.5% NaCl and incubated at 35°C. After enrichment, the DNA was extracted following three different protocols (by heating; by addition of 10% Triton X-100 and heating; and by enzymatic digestion with lysozyme and proteinase followed by boiling). The limit of detection after 3 h enrichment, using enzymatic digestion and boiling was as few as 9.3 cfu g^{-1}.

Table 16.1   PCR-Based Method for the Detection of Pathogenic *Vibrio* Species in Seafood Products

Organism	Method	Target Sequence	Food Matrix	Reference
	PCR	pR72H	Oyster	[80]
	PCR	tdh	Oyster	[68]
	PCR	gyrB	Shrimp	[120]
	PCR	toxR	—	[73]
	PCR	tdh	Oyster	[43]
	PCR	orf8	—	[94]
	PCR	vmp	—	[83]
	Multiplex PCR	tdh, trh	—	[114]
	Multiplex PCR	tlh, th, trh	Oyster	[8]
V. parahaemolyticus	Multiplex PCR	tlh, tdh, trh	Seafoods	[84]
	Real-time PCR	tdh	Oyster	[9]
	Real-time PCR	tlh	—	[126]
	Real-time PCR	tlh	Oyster	[69]
	Real-time PCR	toxR	Clams	[115]
	Real-time PCR	gyrB	Oyster	[15]
	Multiplex real-time PCR	tlh, tdh, trh	Mussels	[25]
	Multiplex real-time PCR	tlh, orf8	Oyster	[104]
	Multiplex real-time PCR	tlh, tdh, Trh, orf8	Oyster	[129]
	Multiplex real-time PCR	tlh, tdh, trh	Oyster	[98]
	PCR	vvhA	—	[45]
	PCR	vvhA	—	[13]
	PCR	gyrB	Oyster	[79]
	Nested PCR	23 S rDNA	Fish	[1]
	Multiplex PCR	vvhA	Oysters, shrimp	[128]
	Multiplex PCR	vvhA	Oysters	[12]
V. vulnificus	Multiplex PCR	vvhA	Oyster	[82]

(continued)

Table 16.1 (continued)   PCR-Based Method for the Detection of Pathogenic *Vibrio* Species in Seafood Products

Organism	Method	Target Sequence	Food Matrix	Reference
	RT PCR	vvhA	Octopus	[81]
	Real-time PCR	vvhA	Oyster	[16]
	Real-time PCR	vvhA	Stools	[36]
	Real-time PCR	vvhA	Oyster	[100]
	Real-time PCR	vvhA	Clam	[126]
	Real-time PCR	16 S rDNA	—	[121]
	Real-time PCR	16 S rDNA	Oyster	[41]
	PCR	ctxAB	Oyster, crab	[77]
V. cholerae	PCR	ctxA	Oyster	[27]
	PCR	Ctx	Oyster	[10]

Other gene marker used for *V. parahaemolyticus*-specific detection is the thermolabile hemolysin (*tlh*) gene. Wang and Levin [125] observed a linear relationship between the fluorescent intensity of the *tlh* PCR products in the agarose gel and the bacterial populations. Kaufman et al. [69] devised an alternative strategy for detection of *V. parahaemolyticus* in oyster. They used mantle fluids as food matrix instead of homogenized oyster tissues, since they observed that the levels of natural contamination of *V. parahaemolyticus* were similar in mantle fluids and oyster tissues. They developed a *tlh*-specific real-time PCR, which was 100% selective as determined using 37 *V. parahaemolyticus*, 27 other *Vibrio*, and 37 non-*Vibrio* isolates. A strong linear correlation between the PCR results and the concentration of cells inoculated into mantle fluids was observed, and the mantle fluid exhibited less PCR inhibition than the homogenized oyster tissue.

Kim et al. [73] reported a PCR method based on the toxin transcriptional activator (*toxR*) gene. After testing 373 *V. parahaemolyticus* isolates and 290 isolates of other bacterial species, they concluded that the method was 100% selective. Similarly, Takahashi et al. [115] developed a *toxR*-based real-time PCR method. It was fully selective as tested 25 *V. parahaemolyticus* and 30 non-*V. parahaemolyticus* isolates. They also evaluated its applicability in shellfish. Twenty-five grams of short-neck clams was homogenized with phosphate-buffered saline (PBS), artificially contaminated with decreasing amounts of *V. parahaemolyticus*, and the DNA was extracted with the MagExtractor-Genome Kit (Toyobo). The real-time PCR detected as few as 100 cfu g^{-1}.

Venkateswaran et al. [120] reported a PCR method based on the B subunit of DNA gyrase (*gyrB*) gene. The selectivity of the method was evaluated using 117 strains of *V. parahaemolyticus* isolated from various environments, food, and clinical sources, and 150 isolates of other species. Twenty-five gram samples of shrimp were homogenized in 225 mL of alkaline peptone water (APW) and artificially contaminated with decreasing amounts of *V. parahaemolyticus* and *V. alginolyticus*, and incubated at 37°C. The homogenates were centrifuged and resuspended in 1 mL of sterile PBS. Ten microliters was used for PCR without extraction of DNA. The analytical sensitivity was as few as 1.5 *V. parahaemolyticus* cfu g^{-1} of homogenate. Similarly, Cai et al. [15] designed a *gyrB*-based real-time PCR. The selectivity was confirmed using 27 *V. parahaemolyticus* and 10 non-*V. parahaemolyticus* isolates. One gram oyster meat homogenate was artificially contaminated and 1 mL aliquot was used for the DNA extraction using the Wizard genomic DNA purification (Promega). The limit of detection of the method was 100 cfu mL^{-1} of oyster homogenates. When 300 seafood samples collected from local supermarkets in eastern China were tested, 32% of the samples were positive using the method. However, only 26% of the samples were positive using the conventional culture method. Interestingly, all culture-positive were also real-time PCR positive, indicating that the real-time PCR method was more sensitive that the conventional culture method.

PCR methods have been also developed for the only specific detection of pathogenic strains of *V. parahaemolyticus*. Tada et al. [114] developed a PCR method based on the thermostable hemolysin (*tdh*) gene and *tdh*-related hemolysin (*trh*) gene. The selectivity was demonstrated using 263 *V. parahaemolyticus* and 133 isolates of other species. Karunasagar et al. [68] reported a PCR method for the detection of Kanagawa-positive strains in seafoods. The primers targeted the *tdh* gene. It was fully selective as tested in 4 Kanagawa-positive *V. parahaemolyticus*, 20 Kanagawa-negative *V. parahaemolyticus*, and 31 other *Vibrio* isolates. For the detection in seafoods, 50 g of samples was homogenized with 450 mL APW. One milliliter of homogenate was centrifuged at 100 × *g*, and the supernatant was again centrifuged, resuspended, and lysed by heating. The analytical sensitivity was less than 10 cells of *V. parahaemolyticus* after 8 h enrichment. A real-time PCR method was also developed using the same molecular marker, *tdh* [9]. The sensitivity was demonstrated using 42 *tdh*+ *V. parahaemolyticus* isolates, 12 *tdh*- *V. parahaemolyticus* isolates, and 103 nontarget isolates. For detection of the pathogenic strains in oyster samples, a 50 mL aliquot of 1:1 oyster homogenate was added to 200 mL of APW and enriched overnight at 35°C. After the enrichment, 1 mL was boiled and 2.5 µL of the supernatant was used for PCR. The real-time PCR detected as few as 1 cfu per reaction. Finally, 131 natural oyster samples collected from Alabama, United States were analyzed by both conventional microbiological methods and real-time PCR. Forty-two percent of negative samples for the microbiological method were positive for the real-time PCR indicating a significantly higher detection rate ($p < 0.05$) and only a 20% of the samples positive for the microbiological method were negative for the real-time PCR method.

Hara-Kudo et al. [43] optimized a PCR method using different DNA extraction procedures for the detection of the pathogenic *V. parahaemolyticus* in seafoods. The primers targeted the *tdh* gene, whose PCR selectivity had been tested previously [114]. Three different DNA extraction methods were evaluated: a silica membrane method using the NucleoSpin Tissue Kit (Macherey-Nagel), a glass fibber method using the High Pure PCR Template Precipitation Kit (Roche), or a magnetic separation method using the MagExtractor-Genome Kit (Toyobo). The use of the silica membrane and the glass fibber methods increased notably the analytical sensitivity.

Taking in consideration the importance for public health of this pathogen, distinguishing between potentially pathogenic and nonpathogenic *V. parahaemolyticus* isolates is of critical importance. Bej et al. [8] reported a multiplex PCR method for the detection of total and hemolysin-producing *V. parahaemolyticus* in shellfish. The method targeted the *tlh* gene for the detection of all *V. parahaemolyticus* strains and the *tdh* and *trh* genes for the specific detection of the pathogenic strains. The selectivity of the method was evaluated using 111 *V. parahaemolyticus* isolates from different origins and 19 non-*V. parahaemolyticus* isolates. The *tlh* primers were 100% selective. Fifty-four percent of the *V. parahaemolyticus* isolates showed positive PCR amplification for the *tdh* primers and 39% showed amplification of the *trh* primers. Interestingly, three isolates showed no *tdh*- and *trh*-PCR amplification but were

Kanagawa positive, and three other isolates were *tdh*-PCR positive, but produced a negative Kanagawa reaction. Finally, 10 g of oyster homogenate was artificially contaminated with decreasing amounts of *V. vulnificus* strains with different *tll/tdh/trh* profiles, diluted in 350 mL of APW and incubated at 35°C for 6 h. DNA was extracted following a previously described method [38]. The limit of detection for all the three PCR primers was 100 cells for the *tdh*-primers, and 10 cells for *tlh*- and *trh*-primers. Using the same set of primers, Luan et al. [84] used a rapid MPN–PCR method for quantification of this pathogen in seafood samples purchased at local retail markets in Qingdao, China. Seventy-three percent of the samples were *V. parahaemolyticus* (*tlh*) positive with values higher than 719 MPN g^{-1}, and 41.5% of samples were positive for *tdh* gene-possessing cells, indicating the presence of pathogenic strains.

Nordstrom et al. [98] developed a multiplex real-time PCR method for detection of the total and pathogenic strains of this organism in oysters using the same targets: *tlh*, *tdh*, and *trh* genes, but this method included an internal amplification control (IAC). The IAC is a nontarget nucleic acid sequence present in every reaction, which is amplified simultaneously with the target sequence [106]. In PCR diagnostics, IACs are essential to identify false negative results [49] as in a reaction with an IAC, a control signal will always be produced when there is no target sequence present. The selectivity was evaluated using 117 *V. parahaemolyticus* isolates with different *tlh/tdh/trh* profiles and 36 isolates of other species of the genus *Vibrio*. A perfect correlation was shown between the results obtained for the *V. parahaemolyticus* isolates and the *tlh/tdh/trh* profiles, however 75% of the *Vibrio hollisae* strains gave a low positive signal for *tdh*. Twenty-seven natural oyster samples were collected at Alaska, and 1 g of homogenate was added to 10 mL of APW and incubated overnight at 35°C. After the enrichment, 1 mL aliquots were boiled and 2 μL of supernatant was used for PCR. Forty-four percent, 44% and 52% of the oyster samples were positive for *tlh*, *tdh*, and *trh*, respectively. However, only 33%, 19%, and 26% were positive for *tlh*, *tdh*, and *trh* using conventional culture methods. Davis et al. [25] used a similar strategy to evaluate *V. parahaemolyticus* strains isolated from mussels and associated with a foodborne outbreak happening in 2002, in Florida, United States. The selectivity of the assay was confirmed using 20 *V. parahaemolyticus* isolates. The mussels were the only food sample with positive results. More than 21% of the mussels samples were positive for *tlh* indicating the presence of the *V. parahaemolyticus* in the samples, and almost 17% of the samples were positive for *tdh*, indicating the presence of pathogenic variants in those samples.

The emergence of the O3:K6 serotype and its widespread distribution have fostered the development of detection methods to detect such pathogenic variants. Myers et al. [94] developed a PCR method for the specific detection of this serotype. The PCR target was the open reading frame 8 of phage f237 (*orf8*). They tested 37 *V. parahaemolyticus* O3:K6 serotype, 123 *V. parahaemolyticus* non-O3:K6 serotype, 114 isolates from other species, and they observed that the method was 100% selective. The method could detect down to 10^4 cells per 100 mL of water samples after the DNA purification using the FastDNA SPIN kit (Bio 101). Rizvi et al. [104] designed *orf8* primers coupled with *tlh* primers for the simultaneous detection of total *V. parahaemolyticus* and pandemic O3:K6 serovar using a multiplex real-time PCR. The selectivity of the assay was evaluated using 37 *V. parahaemolyticus* O3:K6, 26 *V. parahaemolyticus*, 7 non-*parahaemolyticus Vibrio*, and 9 non-*Vibrio* isolates. All the *V. parahaemolyticus* and all the *V. parahaemolyticus* O3:K6 isolates were positive for the *tlh*- and *orf8*-PCRs, respectively, and none of the nontarget isolates was positive. One gram oyster tissue homogenates and Gulf water were artificially contaminated with *V. parahaemolyticus* O3:K6, and incubated at 37°C. After the enrichment, DNA extraction was performed using the Instagene matrix (Bio-Rad). The limit of detection of the real-time PCR method was 1 cfu of pandemic *V. parahaemolyticus* O3:K6 serovar per mL of Gulf water or 1 g of oyster tissue homogenate after 8 h enrichment. Ward and Bej [129] developed a multiplex real-time PCR assay for the simultaneous detection of *V. parahaemolyticus* using the *tlh* gene, pathogenic strains using the *tdh* and *trh* genes, and the pandemic O3:K6 serotype using the *orf8*. Detection of 1 cfu g^{-1} of oyster tissue homogenate was possible after overnight enrichment. Finally the method was applied to 33 natural samples from the Gulf of Mexico, Alabama (United States). Fifty-two percent of the samples were positive for *tlh* indicating the presence of *V. parahaemolyticus* in these samples, and 12% were positive for *tdh* indicating the samples contained pathogenic *V. parahaemolyticus* strains.

Luan et al. [83] compared the performance of four PCR assays for the detection of *V. parahaemolyticus*. The PCR assays targeted the *toxR* [13], *tlh*, *tdh*, and *trh* [8], *gyrB* [120] and the *V. parahaemolyticus* metalloprotease (*vpm*) gene. Eighty-six *V. parahaemolyticus* and 16 non-*V. parahaemolyticus* isolates were tested with the four set of primers. All the four PCR assays were 100% selective. However the analytical sensitivity varied: the *vpm*-PCR assay detected as few as 4 pg of genomic *V. parahaemolyticus* DNA, whereas the *toxR*-PCR, *tlh*-PCR, and *gyrB*-PCR detected a minimum of 375, 100, and 800 pg, respectively.

16.2.1.2 V. vulnificus

The current guidelines recommended by the ISSC indicates that less than 30 cfu g^{-1} in postharvest-treated oysters is the threshold to consider a food item as safe for consumption [60]. The detection protocol approved by the FDA BAM method is based on the MPN enrichment series in APW coupled with isolation in selective medium and biochemical or molecular confirmation of *V. vulnificus* and on the direct isolation on minimally selective media followed by identification of *V. vulnificus* by colony blot DNA–DNA hybridization [72]. Recently the ISO/TS 21872-1:2007 standard has been published describing a horizontal method in food for detection of other potentially entero-pathogenic *Vibrio* species than *parahaemolyticus* and *V. cholerae* [59], which is based in similar principles. In Table 16.2 are summarized the currently available selective media for *V. vulnificus*.

As for *V. parahaemolyticus*, a battery of PCR-based methods have been devised to overcome the disadvantages of the microbiological culture methods (Table 16.1). Hill et al. [45] reported a PCR method based on the cytolysin gene (*vvhA*). The selectivity of the primers was evaluated by testing 5 *V. vulnificus*, 12 non-*vulnificus Vibrio*, and 10 non-*Vibrio* strains. The PCR method was fully selective. Using the *vvhA* gene as PCR target, Brauns et al. [13] confirmed the selectivity of the PCR assay testing one *V. vulnificus*, five non-*vulnificus Vibrio*, and nine non-*Vibrio* isolates. Campbell and Wright [16] developed a real-time PCR method based on the same gene. The selectivity of the assay was evaluated with 28 *V. Vulnificus* and 22 non-*V. vulnificus* isolates, showing to be 100%. Detection of *V. vulnificus* in pure cultures was possible down to 10^2 cfu mL^{-1}. The applicability of this method for detection of *V. vulnificus* in oysters was evaluated using natural and artificially contaminated oysters. Thirty grams of oyster meat was 1:10 diluted in ASW and homogenized for 90 s. Ten milliliters of oyster homogenates was artificially contaminated with decreasing amounts of *V. vulnificus*. DNA was extracted using the QIAamp DNA minikit and concentrated with precipitation with ethanol. The results obtained by real-time PCR correlated well with plate counts based on colony blot hybridization enumeration. Similarly, another real-time PCR method using SYBR Green was developed targeting the *vvhA* gene [100]. The method was fully selective as 80 *V. vulnificus* isolates produced PCR signals and 47 isolates from other species did not produce any PCR amplification. One gram aliquots of oyster tissue homogenate were 10-fold serially diluted in sterile GWP-16 and artificially contaminated with *V. vulnificus* and incubated for 5 h at 37°C. After the enrichment, 5 mL-aliquots were used for DNA extraction using the Instagene matrix (Bio-Rad). The real-time PCR method detected as few as 1 cfu of

Table 16.2 Selective Culture Media for Isolation and Identification of *V. vulnificus*

Medium	Abbreviation	Incubation Temperature (°C)	Carbon Source	Colony Color	Reference
Thiosulfate citrate bile salt agar	TCBS	37	Sucrose	Green	[76]
V. vulnificus agar	VV		Salicin	Grey, dark center	[14]
SDS polymyxin sucrose agar	SPS		Sucrose	Blue with halo	[74]
Cellobiose polymyxin B colistin agar	COC	40	Cellobiose	Yellow	[86]
Modified cellobiose polymyxin B colistin agar	mCPC	40	Cellobiose	Yellow	[116]
V. vulnificus enumeration agar	VVE	37	Cellobiose, lactose, X-Gal	Blue green	[89]
Cellobiose colistin agar	CC	40	Cellobiose	Yellow	[48]
V. vulnificus medium	VVM	37	Cellobiose	Yellow	[19]
V. vulnificus medium + colistin	VVMc	37	Cellobiose	Yellow	[20]

Source: Adapted from Harwood, V.J. et al., *J. Microbiol. Methods*, 59, 301, 2004.

V. vulnificus in 1 g of oyster homogenate. Using the same SYBR Green real-time PCR assay, Wang and Levin [126] optimized a DNA extraction protocol for clam samples. One gram homogenates were artificially contaminated with decreasing amounts of *V. vulnificus*. The aliquots were centrifuged at 1000×*g* for 5 min, and the supernatants were washed twice and lysed with TZ lysis. The DNA was purified using Micropure EZ minicolumns. The real-time PCR detected as few as 100 cfu g^{-1} of clam tissue and 1 cfu g^{-1} after an enrichment step for 5 h at 37°C. Panicker and Bej [99] compared three previously reported sets of primers targeting the *vvhA* gene [16,36,100]. A TaqMan probe was developed for the first two sets of primers, and the probe previously described was used for the former [16]. The selectivity was evaluated using 81 *V. vulnificus* and 37 isolates from other species. The first two PCR systems were 100% selective, however the former was not fully selective as detected more than 32% of non-*V. vulnificus* isolates. Both PCR systems were used for detection of *V. vulnificus* in naturally and artificially contaminated oysters. For artificially contaminated oysters, 1 g aliquots homogenized samples were added to 50 mL of GWP-18 and the solution was artificially contaminated with decreasing amounts of *V. vulnificus*, and incubated at

37°C for 5 h. One milliliter aliquots were used for the DNA extraction using the Instagene matrix (Bio-Rad). The PCR methods detected as few as 1 cfu g^{-1}.

Other PCR targets have been used for the detection of *V. vulnificus*. Kumar et al. [79] developed a PCR method based on the *gyrB* gene. The PCR assay was 100% selective as tested with 45 *V. vulnificus* and 49 other *Vibrio* isolates. The analytical sensitivity was evaluated using *V. vulnificus* pure cultures and artificially contaminated oyster meat. For artificially contaminated samples, 1 g of fresh homogenates was spiked with decreasing amounts of *V. vulnificus*, and lysed by heating. The PCR method detected as low as 3 *V. vulnificus* cfu mL^{-1} of pure cultures, and 300 cfu g^{-1} in artificially contaminated oyster homogenate without enrichment or 30 cfu g^{-1} after 18 h enrichment in APW. The method was also evaluated in 79 natural oyster samples collected from four different estuaries along the Mangalore coast, India. The homogenates were incubated for 0, 6, and 18 h. The best results were obtained after 18 h enrichment, where *V. vulnificus* was detected in 75% of natural oyster samples, while the conventional microbiological method (isolation on mCPC agar plates after 18 h enrichment) only detected *V. vulnificus* in 45.5% of samples.

Vickery et al. [121] reported a real-time PCR method for the classification of *V. vulnificus* based on 16 S rRNA genotype (type A or B). A re-evaluation of the 67 U.S. isolates demonstrated that 45.5% of the isolates originally identified as 16 S rRNA type A were actually type AB, and 76% of clinical isolates tested were type B, 9% type A, and 15% type AB, and in contrast, 91% of nonclinical isolates were found to be of either type A or type AB, and only 9% type B. Other additional 18 strains were also examined, and all of the isolates were classified as type A, all the Biotype 3 strains isolated from an outbreak in Israel were type AB. Using a similar approach, Gordon et al. [41] distinguished *V. vulnificus* strains form environmental and clinical sources. In addition, no amplification was observed with any of the non-*V. vulnificus* isolates tested. Tissues from single oysters collected, in United States were 1:10 diluted in APW, artificially contaminated with *V. vulnificus* and incubated at 37°C for 4 and 24 h. After enrichment, the homogenates were 10-fold diluted. Two milliliters was boiled and 2 μL was used for PCR. The limits of detection were 10^3 and 10^2 cfu per reaction for type A and type B, respectively. Using this method, the authors described that the type A/B ratio of Florida clinical isolates was 19:17. The ratio in oysters harvested from restricted sites in Florida with poor water quality was 5:8, but it was 10:1 in oysters from permitted sites with good water quality. A substantial percentage of isolates from oysters (19.4%) were type AB.

16.2.1.3   V. cholerae

The FDA BAM method for detection of *V. cholerae* in foods relies on the overnight enrichment in APW of 25 g of food samples at 42°C, the isolation on selective medium and final confirmation for biochemical and molecular tests [72]. Similarly the ISO Committee has developed a reference method for this pathogen, the ISO/TS 21872-1:2007 [58].

Another analytical approach is the screening of the samples for toxigenic *V. cholerae* with PCR assays targeting a portion of the *ctx* operon without or after enrichment (Table 16.1). Koch et al. [77] developed a PCR method, which targeted the cholera toxin operon, *ctxAB*. The selectivity was tested using 3 *V. cholerae* and 10 non-*V. cholerae* isolates, showing to be 100%. Analytical sensitivity was tested in artificially contaminated crab or oysters with *V. cholerae* before homogenization in APW. Ten percent APW homogenates were prepared and 1 mL aliquots were taken immediately and again after the 37°C incubation, boiled and 2–5 μL of supernatants was used for

PCR. Crabmeat homogenates inoculated with as few as 4×10^4 *V. cholerae* cfu g^{-1} without further enrichment (equivalent to 10 cells in the reaction) and oysters homogenates artificially contaminated with as few as 10 *V. cholerae* cfu g^{-1} after 8 h enrichment produced positive amplification. DePaola and Hwang [27] evaluated the effects of dilutions, incubation times, and incubation temperatures on detection of *V. cholerae* by a *ctxA*-based PCR method. PCR detection of *V. cholerae* was significantly improved using oyster homogenates diluted 1:100 in APW and incubated at 42°C for 18–21 h.

Blackstone et al. [10] developed a real-time PCR method for detection of toxigenic *V. cholearae* in seafood and seafood-related environments. The system targeted the cholera toxin (*ctxA*) gene, found in toxigenic *V. cholerae* strains. The real-time PCR assay was 100% selective as tested with 32 toxigenic *V. cholerae* and 59 non-*V. cholerae* isolates as well as DNA from different environments and eukaryotic organisms. The limit of detection of the method was less than 1 cfu per reaction in oyster. Finally, 6 shellfish and 10 related environmental samples collected in Mobile Bay, United States were evaluated. Twenty-five grams of oyster homogenate was added to 2475 mL of APW and incubated overnight at 42°C. A 1 mL aliquot of enrichment was boiled and 2–2.5 μL of the boiled aliquot was used for PCR. For environmental samples, 25 g of sediment and ballast water was added to 225 mL of APW and incubated overnight at 42°C. None of the seafood and environmental samples showed a positive signal for toxigenic *V. cholerae*.

16.2.2 Detection of L. monocytogenes in Seafoods and Seafood-Related Environments

ISO has developed reference methods for detection and enumeration of *L. monocytogenes*: ISO 11290-1 and 11290-2, respectively [52,53,56,57]. In the ISO 11290-1, 25 g of food sample is homogenized in a primary enrichment medium (Half Fraser broth) and incubated at 30°C for 24 h. Subsequently, primary culture is plated on Agar Listeria according to Ottaviani and Agosti (ALOA) and in other selective medium (e.g., Oxford or PALCAM media) and incubated at 37°C for 24 h, and in parallel 0.1 mL primary enrichment aliquot is also transferred into a tube with 10 mL of the secondary enrichment medium, and incubated at 35°C or 37°C for 48 h. Afterwards, the secondary enrichment is also streaked on ALOA and other selective medium (e.g., Oxford or PALCAM media), and incubated at 37°C for 24 h. Finally, the typical *L. monocytogenes* colonies (green-blue colonies surrounded by an opaque halo in ALOA plates) are confirmed by biochemical tests. In the protocol for detection of *L. monocytogenes* recommended by the FDA [50], 25 g of seafoods is homogenized in 225 mL of buffered *Listeria* enrichment broth base containing sodium pyruvate without selective agents (BLEB), and incubated at 30°C for 4 h, and then the selective agents are added and incubated for 44 h more at 30°C. At 24 and 48 h, BLEB culture are plated onto one selective isolation medium such as Oxford agar, PALCAM agar, modified Oxford agar (MOX), and Lithium chloride–phenylethanol–moxalactam (LPM) agar fortified with esculin and Fe^{3+}, and incubated at 35°C for 24–48 h for Oxford, PALCAM, or MOX plates or at 30°C for 24–48 h for fortified LPM plates. In addition primary cultures must be plated onto one *L. monocytogenes–L. ivanovii* differential selective agar (e.g., BCM, ALOA, RapidL'mono, or CHROMagar Listeria) after 48 h of enrichment (optionally at 24 h, too). Finally the typical *L. monocytogenes* colonies are confirmed by biochemical tests.

In the ISO 11290-2, 10-fold dilutions of the seafood product homogenate are prepared and plated on ALOA, and incubated at 37°C for 24 h for the enumeration of *L. monocytogenes*. After the enrichment, the typical *L. monocytogenes* colonies are confirmed by biochemical tests. However, in the FDA protocol for enumeration of *L. monocytogenes,* only the positive food samples for presence of *L. monocytogenes* are tested by colony count on *L. monocytogenes* differential selective agar in conjunction with MPN enumeration using selective enrichment in BLEB with subsequent plating on ALOA or BCM differential selective agar.

A study compared the reference ISO methods (ISO 11290-1 and 11290-2) with an in-house method in 543 seafood product samples collected from 21 different companies between 2001 and 2005 in France [90]. For the in-house method, 25 g of seafood product was homogenized with 225 mL of Listeria repair broth (LRB) [40,107], and left at room temperature up to 60 min. To enumerate *L. monocytogenes*, homogenates were spread over Listeria selective agar (LA) plates [64] and incubated at 37°C for 48 h. To detect *L. monocytogenes*, 0.90 mL of selective supplement LRB (Oxoid, U.K.) was added to the homogenate, and incubated at 30°C for 24 h, and subsequently streaked on ALOA and *L. monocytogenes* blood agar (LMBA) plates [64] and incubated 37°C for 48 h. For the second enrichment step, 0.1 mL of the 24-h culture was transferred to a tube with 10 mL of the Fraser broth, and the mixture was incubated at 37°C for 48 h. This second enrichment culture was streaked on ALOA and on LMBA plates and incubated at 37°C for 48 h. For each plate with suspect *L. monocytogenes* colonies, several colonies were spread on LA plates and incubated at 37°C for 48 h, and subsequently respread on Trypticase Soy Agar supplemented with yeast extract (TSAYE). Isolated colonies were taken into microcentrifuge tube containing 100 μL of sterile distilled water, and lysed by heating at 95°C for 25 min, then centrifuged and 3 μL of the supernatant was used for confirmation by PCR. Four sets of primers were used; one for the identification of *Listeria* spp. targeting the 16 S rRNA gene [47], and three specific for the identification of *L. monocytogenes* targeting the *hly* [7,101], and *iap* [47] genes. Twenty eight percent of the samples were positive by at least one of the methods and 16% were positive by both methods. The sensitivity of the methods was higher than 78%, being slightly higher than 79.5% in the case of the in-house method, and the efficiency of isolation was different depending on the nature of the seafood product. The international standard methods confirmed as positive more samples in smoked salmon and herb-flavored slices of smoked salmon, but the in-house method in carpaccio-like salmon, herb-flavored slices of raw salmon, and smoked trout.

Agersborg et al. [2] were the first to develop a specific PCR method for the detection of *L. monocytogenes* in seafood products. They artificially contaminated 5 g of fish cakes, fish pudding, peeled frozen shrimps, salted herring, and marinated and sliced coalfish in oil with 500, 10, 5, and 1 *L. monocytogenes* cells. The seafood samples were homogenized in 20 mL of Tryptone Soy Broth or universal pre-enrichment broth (UPB) and incubated for 24 h. Afterwards, 0.5 mL aliquots were inoculated to 5 mL of UPB and incubated for other 24 h, and subsequently 1.5 mL aliquots were centrifuged for 10 min at 16,000×*g*, and submitted to bacterial DNA extraction. Three different protocols were used by the DNA isolation: the bacterial pellets were resupended (1) in 500 μL of double-distilled (dd-)water and treated by heating; (2) in 750 μL of dd-water and treated with lysozyme and proteinase K; (3) in 400 μL of dd-water and 400 μL of 2% Triton X-100 was added. In all the cases, the DNA solutions were centrifuged, and 10 μL of the supernatants was used by the PCR. The PCR systems targeted different regions of the *hly* [37,39] and *iap* genes [62]. Lysis by Triton X-100 was the most reliable DNA extraction procedure. After 48 h of incubation, samples inoculated with one to five *L. monocytogenes* cells were clearly positive for the three different set of primers.

Isonhood et al. [61] developed an upstream processing method to facilitate the detection by PCR of *L. monocytogenes* in RTE (ready to eat) seafood salads. Eleven grams of the salads was diluted in 99 mL of sterile saline, and artificially contaminated with decreasing amounts of *L. monocytogenes*. After homogenizing, 80 mL of the filtrate was removed for a two-steps centrifugation, consisting of one centrifugation step (119×g for 15 min at 5°C) to remove large food particulates and a second centrifugation step (11,950×g for 10 min at 5°C) to concentrate the bacterial cells in the supernatant that was recovered after the first centrifugation. DNA extraction was done on the 1 g bacterial pellets using DNAzol (Invitrogen). The DNA was serially diluted and subjected to dilution series PCR amplification using a set of primers targeting the 16 S rDNA gene [111] and confirmed by chemiluminescent Southern blot hybridization. The mean recovery after the two-step method was 49.0%, and consistent PCR detection of *L. monocytogenes* was possible down to 103 cfu g^{-1}.

Destro et al. [29] combined RAPD and PFGE analysis to trace *L. monocytogenes* contamination in a shrimp-processing plant in Brazil, over a 5 month period (May to September 1993). Two random primers were used for the RAPD analysis, generating more than 10 different RAPD profiles, a lower number than reported previously. PFGE was performed using *Sma*I and *Apa*I restriction endonucleases, obtaining more than 12 restriction endonuclease digestion profiles (REDP), a number similar to previous studies. The combined profile generated when the two RAPD primers and the two PFGE enzymes were used, increased the discriminatory ability to detect differences among isolates of *L. monocytogenes* within serogroups. The combination of these two typing methods allowed tracking the origin of the isolates; i.e., natural isolates from inside the processing plant, and isolates introduced from outside the plant and restricted to the receiving area.

16.2.3 Detection of Salmonella spp. in Seafoods and Seafood-Related Environments

The International reference method for detection of *Salmonella* is the ISO 6579 [54,55]. In this standard, 25 g of food sample is homogenized with buffered peptone water (BPW), and incubated at 37°C for 18 h. Subsequently, a 0.1 mL pre-enrichment aliquot is transferred into 10 mL Rappaport-Vassiliadis (RV) medium with soya (RVS broth) and incubated for 24 h at 41.5°C and in parallel another 1 mL aliquot is transferred into 10 mL Muller–Kauffmann tetrathionate novobiocin (MKTTn) broth and incubated for 24 h are incubated at 37°C. After the 24 h-incubation, a loop of the RVS and MKTTn broths are streaked onto xylose lysine desoxycholate (XLD) agar and other selective medium, and incubate the plates at 37°C for 24 h. Afterwards, typical *Salmonella* colonies (pink colonies with or without black centers in XLD agar) are confirmed by biochemical (TSI agar test, urea agar test, L-lysine decarboxylation medium test, detection of β-galactosidase, Voges-Proskauer reaction, indole reaction), and serological tests. In the FDA protocol for detection of *Salmonella* [5] small differences can be noted. Twenty-five grams of food sample is homogenized in 225 mL sterile lactose broth. After 1 h at room temperature, 2.25 mL steamed Tergitol Anionic 7 or Triton X-100 are used, and the seafood homogenate is incubated for 24 h at 35°C. Subsequently, a 0.1 mL pre-enrichment aliquot is transferred into 10 mL RV medium and incubated for 24 h at 42°C and in parallel another 1 mL aliquot is transferred into 10 mL tetrathionate (TT) broth and incubated for 24 h at 35°C. Afterwards, the RV and TT enrichments are streaked on bismute sulfite (BS) agar, XLD agar, and Hektoen enteric (HE) agar, and the plates are incubated for 24 h at 35°C. Finally, typical *Salmonella* colonies (brown, grey, or black colonies; sometimes with a metallic sheen in BS agar, pink colonies with or without black centers in XLD agar; and blue-green to blue colonies with or without black centers in HE agar) are confirmed by biochemical or alternative tests.

As for pathogenic *Vibrio* and *L. monocytogenes* rapid alternatives based on molecular methods have been also devised. The research group led by Bej at the University of Alabama developed a multiplex PCR method for the simultaneous detection of *Escherichia coli*, *S. enterica* serotype Typhimurium, *V. vulnificus*, *V. cholerae*, and *V. parahaemolyticus* [12]. The PCR primers targeted the *E. coli uidA*, *S. typhimurium invA*, *V. vulnificus cth*, *V. cholerae ctx*, and *V. parahaemolyticus tl* genes. The multiplex PCR was totally selective as each specific primer only detected the corresponding target. One gram of sterilized shellstocks from oysters obtained from local seafood restaurants was artificially contaminated with decreasing loads of these organisms. The sample was diluted in 30 mL of APW and incubated at 35°C for 6 h. After the enrichment, the oyster homogenates were centrifuged and the DNA was extracted using the Chelex 100 resin (Biorad). To achieve maximum sensitivity, a 5 µL aliquot of the initial multiplex PCR-amplified products was subjected to a reamplification by a second PCR. The minimum level of detection of each target in a single multiplex PCR was 100 cfu g^{-1}. However, the detection limit was improved to 10 cells cfu g^{-1} using the second PCR round. The same research group improved the detection of *S. enterica* serotype Typhimurium, *V. vulnificus*, *Vibrio cholerae*, and *Vibrio parahaemolyticus* using a multiplex PCR followed by DNA–DNA sandwich hybridization [82]. The target genes were the *Salmonella hns* and *spvB*, *V. vulnificus vvh*, *V. cholerae ctx*, and *V. parahaemolyticus tlh* genes. Oyster samples were processed according to standard methods and 1 g of oyster homogenates was diluted in 5 mL of APW and artificially contaminated with 10-fold dilutions of those four bacterial pathogens. The homogenates were enriched for 3 h at 37°C. The bacterial DNA extraction was performed as described above. The multiplex PCR allowed the detection of all four bacterial pathogens, and it was further confirmed by the nonradioactive and colorimetric CovaLinkk NH microtiter plate hybridization assay. The analytical sensitivity was down to 10^2 cells g^{-1} of oyster tissue homogenate.

Vantarakis et al. [119] devised a multiplex PCR method for the simultaneous detection of *Salmonella* spp. and *Shigella* spp. in mussels. The multiplex PCR primers targeted specific nucleotide sequences of the *Salmonella invA* (215 bp) [122] and *Shigella virA* (275 bp) [102] genes. The PCR method was 100% selective as evaluated with six different Enterobacteriaceae genera. For the mussels analysis, 25 g of mussel meat was diluted in 90 mL of BPW. Decreasing amounts of *Salmonella* spp. and *Shigella* spp. were added to 1 mL of mussel homogenates and submitted to DNA extraction. Guanidine isothiocyanate was added to 1 mL homogenates and incubated at 65°C for 90 min, diluted and boiled for 5 min. The samples were cooled to room temperature, then sodium acetate was added to the samples, and centrifuged at 14,000×g for 10 min. The supernatants were transferred to new tubes and extracted twice with an equal volume of chloroform. Finally the DNA was precipitated with 95% ethanol and the DNA was resuspended in sterile distilled water. The PCR method detected less than 10 *Salmonella* cells mL^{-1} of homogenate. However the authors introduced a pre-enrichment step to increase the analytical sensitivity as well as to guarantee the only detection of viable cells. After a 22 h pre-enrichment in BPW, 10–100 cells of *Salmonella* spp. and *Shigella* per milliliter of homogenate were detected by the multiplex PCR.

Wang and Yeh [127] developed a novel PCR method for the detection of *Salmonella enteritidis*, and evaluated its performance in different food samples, including seafoods. The PCR system targeted the *Salmonella IE* gene. All of the 24 *Salmonella enteritidis* strains generated positive PCR signals. Ninety-six non-*enteritidis Salmonella* and 40 non-*Salmonella* isolates including strains of the family Enterobacteriaceae such as *E. coli*, *Shigella*, and *Citrobacter*, did not produce any amplification signal, therefore, the PCR assay was 100% selective. The detection limit of the PCR assay was 10^2 cfu mL^{-1} of cell extracts prepared by heat lysis. For the analysis of seafood samples, the authors followed the FDA procedure, and 10 μL of the final enrichment was lysed by heating, and used for the PCR detection. None of the 15 samples were detected by either completed BAM method or by PCR.

References

1. Adak, G.K., Long, S.M., and O'Brien, S.J. Trends in indigenous foodborne disease and deaths, England and Wales: 1992–2000, *Gut*, 51, 832–841, 2002.
2. Agersborg, A., Dahl, R., and Martinez, I. Sample preparation and DNA extraction procedures for polymerase chain reaction identification of *Listeria monocytogenes* in seafoods, *Int. J. Food Microbiol.*, 35, 275–280, 1997.
3. Alam, M.J. et al. Environmental investigation of potentially pathogenic *Vibrio parahaemolyticus* in the Seto-Inland Sea, Japan, *FEMS Microbiol. Lett.*, 208, 83–87, 2002.
4. Albert, M.J. *Vibrio cholerae* O139 Bengal, *J. Clin. Microbiol.*, 32, 2345–2349, 1994.
5. Andrews, W.H. and Hammack, T. Chapter 5: *Salmonella*, in: *Bacteriological Analytical Manual*, 8th edn., Gaithersburg, MD: U.S. Food and Drug Administration, 2007.
6. Arias, C.R., Garay, E., and Aznar, R. Nested PCR method for rapid and sensitive detection of *Vibrio vulnificus* in fish, sediments, and water, *Appl. Environ. Microbiol.*, 61, 3476–3478, 1995.
7. Bansal, N.S. et al. Multiplex PCR assay for the routine detection of *Listeria* in food, *Int. J. Food Microbiol.*, 33, 293–300, 1996.
8. Bej, A.K. et al. Detection of total and hemolysin-producing *Vibrio parahaemolyticus* in shellfish using multiplex PCR amplification of *tl*, *tdh* and *trh*, *J. Microbiol. Methods*, 36, 215–225, 1999.
9. Blackstone, G.M. et al. Detection of pathogenic *Vibrio parahaemolyticus* in oyster enrichments by real time PCR, *J. Microbiol. Methods*, 53, 149–155, 2003.
10. Blackstone, G.M. et al. Use of a real time PCR assay for detection of the *ctxA* gene of *Vibrio cholerae* in an environmental survey of Mobile Bay, *J. Microbiol. Methods*, 68, 254–259, 2007.
11. Blake, P.A. et al. Disease caused by a marine *Vibrio*. Clinical characteristics and epidemiology, *N. Engl. J. Med.* 300, 1–5, 1979.
12. Brasher, C.W. et al. Detection of microbial pathogens in shellfish with multiplex PCR, *Cur. Microbiol.*, 37, 101–107, 1998.
13. Brauns, L.A., Hudson, M.C., and Oliver, J.D. Use of the polymerase chain reaction in detection of culturable and non-culturable *Vibrio vulnificus* cells, *Appl. Environ. Microbiol.*, 57, 2651–2655, 1991.
14. Brayton, P.R. et al. New selective plating medium for isolation of *Vibrio vulnificus* biogroup 1, *J. Clin. Microbiol.*, 17, 1039–1044, 1983.
15. Cai, T. et al. Application of real-time PCR for quantitative detection of *Vibrio parahaemolyticus* from seafood in eastern China, *FEMS Immunol. Med. Microbiol.*, 46, 180–186, 2006.
16. Campbell, M.S. and Wright, A.C. Real-time PCR analysis of *Vibrio vulnificus* in oysters, *Appl. Environ. Microbiol.*, 69, 7137–7144, 2003.
17. Cao, J. et al. Concentrations and tracking of *Listeria monocytogenes* strains in a seafood-processing environment using a most-probable-number enrichment procedure and randomly amplified polymorphic DNA analysis, *J. Food Prot.*, 69, 489–494, 2006.
18. CAST. CAST Report: Foodborne Pathogens: Risks and Consequences. Task Force Report No. 122, Washington, DC: Council for Agricultural Science and Technology, 1994.
19. Cerda-Cuellar, M., Jofre, J., and Blanch, A.R. A selective medium and a specific probe for detection of *Vibrio vulnificus*, *Appl. Environ. Microbiol.*, 66, 855–859, 2000.
20. Cerda-Cuellar, M. et al. Comparison of selective media for the detection of *Vibrio vulnificus* in environmental samples, *J. Appl. Microbiol.*, 91, 322–327, 2001.
21. Colwell, R.R. et al. Occurrence of *Vibrio cholerae* O1 in Maryland and Louisiana estuaries, *Appl. Environ. Microbiol.*, 41, 555–558, 1981.

22. Cook, D.W. et al. *Vibrio vulnificus* and *Vibrio parahaemolyticus* in U.S. retail shell oysters: A national survey from June 1998 to July 1999, *J. Food Prot.*, 65, 79–87, 2002.
23. Crutchfield, S. and Roberts, T. Food safety efforts accelerate in 1990's, *USDA Economic Res. Service Food Rev.*, 23, 44–49, 2000.
24. Daniels, N.A. et al. *Vibrio parahaemolyticus* infections in the United States, 1973–1998, *J. Infect. Dis.*, 181, 1661–1666, 2000.
25. Davis, C.R. et al. Real-time PCR detection of the thermostable direct hemolysin and thermolabile hemolysin genes in a *Vibrio parahaemolyticus* cultured from mussels and mussel homogenate associated with a foodborne outbreak, *J. Food Prot.*, 67, 1005–1008, 2004.
26. Deepanjali, A. et al. Seasonal variation in abundance of total and pathogenic *Vibrio parahaemolyticus* bacteria in oysters along the southwest coast of India, *Appl. Environ. Microbiol.*, 71, 3575–3580, 2005.
27. DePaola, A. and Hwang, G.C. Effect of dilution, incubation time, and temperature of enrichment on cultural and PCR detection of *Vibrio cholerae* obtained from the oyster *Crassostrea virginica*, *Mol. Cell. Probes*, 9, 75–81, 1995.
28. DePaola, A. et al. Environmental investigations of *Vibrio parahaemolyticus* in oysters after outbreaks in Washington, Texas, and New York (1997 and 1998), *Appl. Environ. Microbiol.*, 66, 4649–4654, 2000.
29. Destro, M.T., Leitao, M.F.F., and Farber, J.M. Use of molecular typing methods to trace the dissemination of *Listeria monocytogenes* in a shrimp processing plant, *Appl. Environ. Microbiol.*, 62, 705–711, 1996.
30. Dobosh, D., Gomez-Zavaglia, A., and Kuljich, A. The role of food in cholera transmission, *Medicina*, 55, 28–32, 1995.
31. EC (European Commission). Trends and Sources of Zoonotic Agents in Animals, Feedstuffs, Food and Man in the European Union and Norway to the European Commission in Accordance with Article 5 of the Directive 92/117/EEC. Working document SANCO/927/2002, Part 1, 2002, pp. 45–122.
32. Farber, J.M. Present situation in Canada regarding *Listeria monocytogenes* and ready-to-eat seafood products, *Int. J. Food Microbiol.*, 62, 247–251, 2000.
33. Farber, J.M. et al. A small outbreak of listeriosis potentially linked to the consumption of imitation crab meat, *Lett. Appl. Microbiol.*, 31, 100–104, 2000.
34. Faruque, S.M., Albert, M. J., and Mekalanos, J.J. Epidemiology, genetics, and ecology of toxigenic *Vibrio cholerae*, *Microbiol. Molec. Biol. Rev.*, 62, 1301–1314, 1998.
35. Fujino, T. et al. On the bacteriological examination of Shirasu food poisoning, *Med. J. Osaka Univ.*, 4, 299–304, 1953.
36. Fukushima, H., Tsunomori, Y., and Seki, R. Duplex real-time SYBR green PCR assays for detection of 17 species of food- or waterborne pathogens in stools, *J. Clin. Microbiol.*, 41, 5134–5146, 2003.
37. Furrer, B. et al. Detection and identification of *Listeria monocytogenes* in cooked sausage products and in milk by in vitro amplification of haemolysin gene fragments, *J. Appl. Bacteriol.*, 70, 372–379, 1991.
38. Gannon, V.P. et al. Rapid and sensitive method for detection of shiga-like toxin-producing *Escherichia coli* in ground beef using the polymerase chain reaction, *Appl. Environ. Microbiol.*, 58, 3809–3815, 1992.
39. Golsteyn-Thomas, E.J. et al. Sensitive and specific detection of *Listeria monocytogenes* in milk and ground beef with the polymerase chain reaction, *Appl. Environ. Microbiol.*, 57, 2576–2580,1991.
40. Gombas, D.E. et al. Survey of *Listeria monocytogenes* in ready-to-eat foods, *J. Food Prot.*, 66, 559–569, 2003.
41. Gordon, K.V. et al. Real-time PCR assays for quantification and differentiation of *Vibrio vulnificus* strains in oysters and water, *Appl. Environ. Microbiol.*, 74, 1704–1709, 2008.
42. Humphrey, T. Public-health aspects of *Salmonella* infection, in: *Salmonella in Domestic Animals*, Way, C. and Way, A., Eds., Oxon, U.K.: CABI Publishing, 2000, pp. 245–263.

43. Hara-Kudo, Y. et al. Increased sensitivity in PCR detection of *tdh*-positive *Vibrio parahaemolyticus* in seafood with purified template DNA, *J. Food Prot.*, 66, 1675–1680, 2003.

44. Harwood, V.J., Gandhi, J.P., and Wright, A.C. Methods for isolation and confirmation of *Vibrio vulnificus* from oysters and environmental sources: A review, *J. Microbiol. Methods*, 59, 301–316, 2004.

45. Hill, W.E. et al. Polymerase chain reaction identification of *Vibrio vulnificus* in artificially contaminated oysters, *Appl. Environ. Microbiol.*, 57, 707–711, 1991.

46. Hlady, W.G. and Klontz, K.C. The epidemiology of *Vibrio* infections in Florida, 1981–1993, *J. Infect. Dis.*, 173, 1176–1183, 1996.

47. Herman, L.M.F., de Ridder, H.F.M., and Vlaemynck, G.M.M. A multiplex PCR method for the identification of *Listeria* spp. and *Listeria monocytogenes* in dairy samples, *J. Food Prot.*, 58, 867–872, 1995.

48. Høi, L., Dalsgaard, I., and Dalsgaard, A. Improved isolation of *Vibrio vulnificus* from seawater and sediment with cellobiosecolistin agar, *Appl. Environ. Microbiol.*, 64, 1721–2174, 1998.

49. Hoorfar, J. et al. Practical considerations in design of internal amplification controls for diagnostic PCR assays, *J. Clin. Microbiol.*, 42, 1863–1868, 2004.

50. Hitchins, A.D. Detection and enumeration of *Listeria monocytogenes* in foods, in: *Bacteriological Analytical Manual*, 8th edn., Chapter 10: Gaithersburg, MD: U.S. Food and Drug Administration, 2003.

51. International Organization for Standardization (ISO). *ISO 8914 General Guidance for the Detection of Vibrio parahaemolyticus*, Geneva, Switzerland, 1990.

52. International Organization for Standardization (ISO). *ISO 1190-1 Microbiology of Food and Animal Feeding Stuffs—Horizontal Method for the Detection and Enumeration of Listeria monocytogenes. Part 1: Detection Method*, Geneva, Switzerland, 1996.

53. International Organization for Standardization (ISO). *ISO 1190-2 Microbiology of Food and Animal Feeding Stuffs—Horizontal Method for the Detection and Enumeration of Listeria monocytogenes. Part 2: Enumeration Method*, Geneva, Switzerland, 2000.

54. International Organization for Standardization (ISO). *ISO 6579 Microbiology of Food and Animal Feeding Stuffs—Horizontal Method for the Detection of Salmonella spp.*, Geneva, Switzerland, 2002.

55. International Organization for Standardization (ISO). *ISO 6579 Microbiology of Food and Animal Feeding Stuffs—Horizontal Method for the Detection of Salmonella spp.*, Technical Corrigendum 1, Geneva, Switzerland, 2004.

56. International Organization for Standardization (ISO). *ISO 1190-1 Microbiology of Food and Animal Feeding Stuffs—Horizontal Method for the Detection and Enumeration of Listeria monocytogenes. Part 1: Detection Method. AMENDMENT 1: Modification of the Isolation Media and the Haemolysis Test, and Inclusion of Precision Data*, Geneva, Switzerland, 2004.

57. International Organization for Standardization (ISO). *ISO 1190-2 Microbiology of Food and Animal Feeding Stuffs—Horizontal Method for the Detection and Enumeration of Listeria monocytogenes. Part 2: Enumeration Method, AMENDMENT 1: Modification of the Enumeration Medium*, Geneva, Switzerland, 2005.

58. International Organization for Standardization (ISO). *ISO/TS 21872-1:2007 Microbiology of Food and Animal Feeding Stuffs. Horizontal Method for the Detection of Potentially Enteropathogenic Vibrio spp. Part 1: Detection of Vibrio parahaemolyticus and Vibrio cholerae*, Geneva, Switzerland, 2007a.

59. International Organization for Standardization (ISO). *ISO/TS 21872-2:2007 Microbiology of Food and Animal Feeding Stuffs. Horizontal Method for the Detection of Potentially Enteropathogenic Vibrio spp. Part 2: Detection of Species Other Than Vibrio parahaemolyticus and Vibrio cholerae*, Geneva, Switzerland, 2007b.

60. Interstate Shellfish Sanitation Conference. (ISSC) Issue relating to a *Vibrio vulnificus* risk management plan for oysters. *Proceedings of the Interstate Shellfish Sanitation Conference*, Columbia, SC, 2003.

61. Isonhood, J., Drake, M.A., and Jaykus, L.A. Upstream sample processing facilitates PCR detection of *Listeria monocytogenes* in mayonnaise-based ready-to-eat (RTE) salads, *Food Microbiol.*, 23, 584–590, 2006.

62. Jaton, K., Sahli, R., and Bille, J. Development of polymerase chain reaction assays for detection of *Listeria monocytogenes* in clinical cerebrospinal fluid samples, *J. Clin. Microbiol.*, 30, 1931–1936, 1992.

63. Jeyasekaran, G., Karunasagar, I., and Karunasagar, I. Incidence of *Listeria* spp. in tropical fish, *Int. J. Food Microbiol.*, 31, 333–340, 1996.

64. Johansson, T. Enhanced detection and enumeration of *Listeria monocytogenes* from foodstuffs and food-processing environments, *Int. J. Food Microbiol.*, 40, 77–85, 1998.

65. Joseph, S.W., Colwell, R.R., and Kaper, J.B. *Vibrio parahaemolyticus* and related halophilic Vibrios. *Crit. Rev. Microbiol.* 10, 77–124, 1982.

66. Käferstein, F.K., Motarjemi, Y., and Bettcher, D.W. Foodborne disease control: A transnational challenge, *Emerg. Infect. Dis.* 3, 503–510, 1997.

67. Kaper, J.B., Morris, J.G., and Levine. M.M. Cholera, *Clin. Microbiol. Rev.*, 8, 48–86, 1995.

68. Karunasagar, I. et al. Rapid polymerase chain reaction method for detection of Kanagawa positive *Vibrio parahaemolyticus* in seafoods, *Int. J. Food Microbiol.*, 31, 317–323, 1996.

69. Kaufman, G.E. et al. Real-time PCR quantification of *Vibrio parahaemolyticus* in oysters using an alternative matrix, *J. Food Prot.*, 67, 2424–2429, 2004.

70. Kaysner, C.A. and DePaola, A. Outbreaks of *Vibrio parahaemolyticus* gastroenteritis from raw oyster consumption: Assessing the risk of consumption and genetic methods for detection of pathogenic strains, *J. Shellfish Res.*, 19, 657, 2000.

71. Kaysner, C.A. and DePaola, A. *Vibrio* In: *Compendium of Methods for the Microbiological Examination of Foods*, 4th edn., Downes, F.P. and Ito, K., Eds., Washington, DC: American Public Health Association, 2001, pp. 405–420.

72. Kaysner, C.A. and DePaola, A. *Vibrio cholerae, V. parahaemolyticus, V. vulnificus,* and other *Vibrio* spp., In *Bacteriological Analytical Manual*, 8th edn., Chapter 9: Revision A, Gaithersburg, MD: U.S. Food and Drug Administration, 2004.

73. Kim, Y.B. et al. Identification of *Vibrio parahaemolyticus* strains at the species level by PCR targeted to the toxR gene, *J. Clin. Microbiol.*, 37, 1173–1177, 1999.

74. Kitaura, T. et al. Halo production by sulfatase activity in *V. vulnificus* and *V. cholerae* O1 on a new selective sodium dodecyl sulfate containing agar medium: A screening marker in environmental surveillance, *FEMS Microbiol. Lett.*, 17, 205–209, 1983.

75. Klontz, K.C. et al. Raw oyster-associated *Vibrio* infections: Linking epidemiologic data with laboratory testing of oysters obtained from a retail outlet, *J. Food Prot.* 56, 977–979, 1994.

76. Kobayashi, T. et al. A new selective isolation medium for vibrio group on a modified Nakanishi's medium (TCBS agar medium), *Jpn. J. Bacteriol.*, 18, 387–392, 1963.

77. Koch, W.H. et al. Rapid polymerase chain reaction method for detection of *Vibrio cholerae* in foods, *Appl. Environ. Microbiol.*, February, 556–560, 1993.

78. Kumar, H.S. et al. Detection of *Salmonella* spp. in tropical seafood by polymerase chain reaction, *Int. J. Food Microbiol.*, 88, 91–95, 2003.

79. Kumar, H.S. et al. A gyrB-based PCR for the detection of *Vibrio vulnificus* and its application for direct detection of this pathogen in oyster enrichment broths, *Int. J. Food Microbiol.*, 111, 216–220, 2006.

80. Lee, C.Y., Pan, S.F., and Chen, C.H. Sequence of a cloned pR72H fragment and its use for detection of *Vibrio parahaemolyticus* in shellfish with the PCR, *Appl. Environ. Microbiol.*, 61, 1311–1317, 1995.

81. Lee, J.Y., Eun, J.B., and Cho, S.N. Improving detection of *Vibrio vulnificus* in *Octopus variabilis* by PCR, *J. Food Sci.* 62, 179–182, 1997.

82. Lee, C.Y., Panicker, G., and Bej, A.K. Detection of pathogenic bacteria in shellfish using multiplex PCR followed by CovaLink NH microwell plate sandwich hybridization, *J. Microbiol. Methods*, 53, 199–209, 2003.

83. Luan, X. et al. Comparison of different primers for rapid detection of *Vibrio parahaemolyticus* using the polymerase chain reaction, *Lett. Appl. Microbiol.*, 44, 242–247, 2007.

84. Luan, X. et al. Rapid quantitative detection of *Vibrio parahaemolyticus* in seafood by MPN-PCR, *Curr. Microbiol.*, 57, 218–221, 2008.

85. Madden, J.M. et al. Virulence of three clinical isolates of *Vibrio cholerae* non O-1 serogroup in experimental enteric infections in rabbits, *Infect. Immun.*, 33, 616–619, 1981.

86. Massad, G. and Oliver, J.D. New selective and differential medium for *Vibrio cholerae* and *Vibrio vulnificus*, *Appl. Environ. Microbiol.*, 53, 2262–2264, 1987.

87. McLaughlin, J.C. *Vibrio*, In *Manual of Clinical Microbiology*, 6th edn., Murray, P.R., Baron, E.J., Pfaller, M.A., Tenover, F.C., and Yolken, R.H., Eds., Washington, DC: ASM Press, 1995, pp. 465–474.

88. Mead, P.S., Slutsker, L., Griffin, P.M., and Tauxe, R.V. Food-related illness and death in the United States, *Emerging Infect. Dis.*, 5, 607–625, 1999.

89. Miceli, G.A., Watkins, W.D., and Rippey, S.R. Direct plating procedure for enumerating *Vibrio vulnificus* in oysters (*Crassostrea virginica*), *Appl. Environ. Microbiol.*, 59, 3519–3524, 1993.

90. Midelet-Bourdin, G., Leleu, G., and Malle, P. Evaluation of the International Reference Methods NF EN ISO 11290-1 and 11290-2 and an in-house method for the isolation of *Listeria monocytogenes* from retail seafood products in France, *J. Food Prot.*, 70, 891–900, 2007.

91. Mintz, E.D., Popovic, T., and Blake, P.A. *Transmission of Vibrio cholerae O1 in Vibrio cholerae and Cholera: Molecular to Global Perspectives*, Wachsmuth, I.K., Blake, P.A., and Olsvik, O., Eds., Washington, DC: ASM press, 1994.

92. MMWR. *Vibrio vulnificus* infections associated with raw oyster consumption—Florida, 1981–1992, *Morb. Mortal. Wkly. Rep.*, 42, 405–407, 1993.

93. Motes, M.L. et al. Influence of water temperature and salinity on *Vibrio vulnificus* in Northern Gulf and Atlantic Coast oysters (*Crassostrea virginica*), *Appl. Environ. Microbiol.*, 64, 1459–1465, 1998.

94. Myers, M.L., Panicker, G., and Bej, A.K. PCR detection of a newly emerged pandemic *Vibrio parahaemolyticus* O3:K6 pathogen in pure cultures and seeded waters from the Gulf of Mexico, *Appl. Environ. Microbiol.*, 69, 2194–2200, 2003.

95. Nair, G.B. et al. Global dissemination of *Vibrio parahaemolyticus* serotype O3:K6 and its serovariants, *Clin. Microbiol. Rev.*, 20, 39–48, 2007.

96. Nakamura, H. et al. *Listeria monocytogenes* isolated from cold-smoked fish products in Osaka City, Japan, *Int. J. Food Microbiol.*, 94, 323–328, 2004.

97. Nakamura, H. et al. Molecular typing to trace *Listeria monocytogenes* isolated from cold-smoked fish to a contamination source in a processing plant, *J. Food Prot.*, 69, 835–841, 2006.

98. Nordstrom, J.L. et al. Development of a multiplex real-time PCR assay with an internal amplification control for the detection of total and pathogenic *Vibrio parahaemolyticus* bacteria in oysters, *Appl. Environ. Microbiol.*, 73, 5840–5847, 2007.

99. Panicker, G. and Bej, A.K. Real-time PCR detection of *Vibrio vulnificus* in oysters: Comparison of oligonucleotide primers and probes targeting *vvhA*, *Appl. Environ. Microbiol.*, 71, 5702–5709, 2005.

100. Panicker, G., Myers, M.L., and Bej, A.K. Rapid detection of *Vibrio vulnificus* in shellfish and Gulf of Mexico water by real-time PCR, *Appl. Environ. Microbiol.*, 70, 498–507, 2004.

101. Paziak-Domanska, B. et al. Evaluation of the API test, phosphatidyl-inositol specific phospholipase C activity and PCR method in identification of *Listeria monocytogenes* in meat foods, *FEMS Microbiol. Lett.*, 171, 209–214, 1999.

102. Rahn, K.J., De Grandis, S.A., and Clarke, R.C. Amplification of an *invA* gene sequence of *Salmonella typhimurium* by polymerase chain reaction as a specific method of detection of *Salmonella* spp., *Mol. Cell. Probes*, 6, 271–279, 1992.

103. Rippey, S.R. Infectious diseases associated with molluscan shellfish consumption, *Clin. Microbiol. Rev.*, 7, 419–425, 1994.

104. Rizvi, A.V. et al. Detection of pandemic *Vibrio parahaemolyticus* O3:K6 serovar in Gulf of Mexico water and shellfish using real-time PCR with Taqman fluorescent probes, *FEMS Microbiol. Lett.*, 262, 185–192, 2006.

105. Rocourt, J., Jacquet, Ch., and Reilly, A. Epidemiology of human listeriosis and seafoods, *Int. J. Food Microbiol.*, 62, 197–209, 2000.

106. Rodríguez-Lázaro, D. et al. Trends in analytical methodology in food safety and quality: Monitoring microorganisms and genetically modified organisms, *Trends Food Sci Technol.*, 18, 306–319, 2007.

107. Ryser, E.T. et al. Recovery of different *Listeria* ribotypes from naturally contaminated, raw refrigerated meat and poultry products with two primary enrichment media, *Appl. Environ. Microbiol.*, 62, 1781–1787, 1996.

108. Schmidt, K. and Tirado, C. WHO Surveillance Programme for Control of Foodborne Infections and Intoxications in Europe, 7th Report 1993–1998, Berlin, Germany: Federal Institute for Health Protection of Consumers and Veterinary Medicine (BgVV), 2001.

109. Shabarinath, S. et al. Detection and characterization of *Salmonella* associated with tropical seafood, *Int. J. Food Microbiol.*, 114, 227–233, 2007.

110. Sharma, C. et al. Molecular analysis of non-O1 non-O139 *Vibrio cholerae* associated with an unusual upsurge in the incidence of cholera-like disease in Calcutta, India, *J. Clin. Microbiol.*, 36, 756–763, 1998.

111. Somer, L. and Kashi, Y. A PCR method based on 16 S rRNA sequence for simultaneous detection of the genus *Listeria* and the species *Listeria monocytogenes* in food products, *J. Food Prot.*, 66, 1658–1665, 2003.

112. Strom, M.S. and Paranjpye, R.N., Epidemiology and pathogenesis of *Vibrio vulnificus*, *Microbes Infect.*, 2, 177–188, 2000.

113. Su, Y.G. and Liu, C. *Vibrio parahaemolyticus*: A concern of seafood safety, *Food Microbiol.*, 24, 549–558, 2007.

114. Tada, J. et al. Detection of the thermostable direct hemolysin gene (*tdh*) and the thermostable direct hemolysin-related hemolysin gene (*trh*) of *Vibrio parahaemolyticus* by polymerase chain reaction, *Mol. Cell. Probes*, 6, 477–487, 1992.

115. Takahashi, H. et al. Development of a quantitative real-time PCR method for estimation of the total number of *Vibrio parahaemolyticus* in contaminated shellfish and seawater, *J. Food Prot.*, 68, 1083–1088, 2005.

116. Tamplin, M.L. et al. Enzyme immuno assay for identification of *Vibrio vulnificus* in seawater, sediment, and oysters, *Appl. Environ. Microbiol.*, 57, 1235–1240, 1991.

117. Tirado, C., and Schmidt, K. WHO surveillance programme for control of foodborne infections and intoxications: Preliminary results and trends across greater Europe. World Health Organization, *J. Infect.*, 43, 80–84, 2001.

118. Van Coillie, E. et al. Prevalence and typing of *Listeria monocytogenes* in ready-to-eat food products on the Belgian market, *J. Food Prot.*, 67, 2480–2487, 2004.

119. Vantarakis, A. et al. Development of a multiplex PCR detection of *Salmonella* spp. and *Shigella* spp. in mussels, *Lett. Appl. Microbiol.*, 31, 105–109, 2000.

120. Venkateswaran, K., Dohmoto, N., and Harayama, S. Cloning and nucleotide sequence of the *gyrB* gene of *Vibrio parahaemolyticus* and its application in detection of this pathogen in shrimp, *Appl. Environ. Microbiol.*, 64, 681–687, 1998.

121. Vickery, M.C.L. et al. A real-time PCR assay for the rapid determination of 16 S rRNA genotype in *Vibrio vulnificus*, *J. Microbiol. Methods*, 68, 376–384, 2007.

122. Villalobo, E. and Torres, A. PCR for detection of *Shigella* spp. in mayonnaise, *Appl. Environ. Microbiol.*, 64, 1242–1245, 1998.

123. Wachsmuth, K. et al. Molecular Epidemiology of Cholera in *Vibrio cholerae* and Cholera: Molecular to Global Perspectives, Wachsmuth, I.K., Blake, P.A., and Olsvik, O., Eds., Washington, DC: ASM Press, 1994, pp. 357–370.

124. Wallace, D.J. et al. Incidence of foodborne illnesses reported by the foodborne diseases active surveillance network (FoodNet)-1997, *J. Food Prot.*, 63, 807–809, 2000.

125. Wang, S. and Levin, R.E. Quantitative determination of *Vibrio parahaemolyticus* by polymerase chain reaction, *Food Biotechnol.*, 18, 279–287, 2004.
126. Wang, S. and Levin, R.E. Rapid quantification of *Vibrio vulnificus* in clams (*Protochaca staminea*) using real-time PCR, *Food Microbiol.*, 23, 757–761, 2006.
127. Wang, S.J. and Yeh, D.B. Designing of polymerase chain reaction primers for the detection of *Salmonella enteritidis* in foods and faecal samples, *Lett. Appl. Microbiol.*, 34, 422–427, 2002.
128. Wang, R.F., Cao, W.W., and Cerniglia, C.E. A universal protocol for PCR detection of 13 species of foodborne pathogens in foods, *J. Appl. Microbiol.*, 83, 727–736, 1997.
129. Ward, L.N. and Bej, A.K. Detection of *Vibrio parahaemolyticus* in shellfish by use of multiplexed real-time PCR with TaqMan fluorescent probes, *Appl. Environ. Microbiol.*, 72, 2031–2042, 2006.
130. Wong, H.C. et al. Characterization of *Vibrio parahaemolyticus* isolates obtained from foodborne illness outbreaks during 1992 through 1995 in Taiwan, *J. Food Prot.*, 63, 900–906, 2000.
131. Wright, A.C. et al. Distribution of *Vibrio vulnificus* in the Chesapeake Bay. *Appl. Environ. Microbiol.*, 62, 717–724, 1996.

Chapter 17

Parasites

Juan Antonio Balbuena and Juan Antonio Raga

Contents

17.1 Protozoa 509
17.2 Trematodes 510
17.3 Cestodes 515
17.4 Anisakid Nematodes 515
17.5 Other Nematodes 523
17.6 Acanthocephalans 523
17.7 Seafood Safety 523
    17.7.1 Primary Production and Handling 523
    17.7.2 Thermal Processing 524
    17.7.3 Recommendations for Consumers and Restaurateurs 525
    17.7.4 Recommendations for Allergic and Immunosuppressed Persons 525
17.8 Further Developments 525
Acknowledgments 526
References 526

The consumption of seafood has increased steadily over the last years. The yearly per capita consumption has augmented by 77% (from 9.14 kg in 1961 to 16.1 kg in 2003) [1]. As any other food, fish and shellfish are carriers of a wide range of parasites, but only a few have zoonotic significance. Nevertheless, their incidence on human health should not be neglected. For instance, of the ~41 million people infected with foodborne trematodes, it is reckoned that ~18 million cases correspond to fish trematodes [2,3].

Parasite infections in humans resulting from the consumption of fish and shellfish have been known for centuries. Indeed, the occurrence of fish helminths has been documented in ancient human remains in China [4] and Korea [5], and in pre-Columbian civilizations of both North [6] and South America [7,8]. Humans become infected mostly by eating raw, marinated, smoked, or undercooked seafood (fish, squid, oysters, shrimps, crabs, etc.) carrying larval stages of the parasites (Figure 17.1). These may or may not develop into adults in humans, but can lead to disorders whose severity varies depending of the species involved.

Most seafoodborne parasites are metazoan helminths, particularly cestodes and, above all, trematodes. Infections with anisakid nematodes represent one of the most relevant emerging zoonoses worldwide. By contrast the significance of protozoan infections resulting from consumption of seafood is still poorly understood. Some helminth species can be detected visually by their size, color, and texture, which allow differentiation from fish tissues. Helminths located within the muscle or under the skin, can downgrade the product resulting in economic loss, but their detection is simpler, especially by candling of fish fillets. Other helminths, in contrast, occur in body cavities, viscera, or digestive tract. So detection by visual inspection under sanitary controls is more difficult and, thus, molecular or immunological assays are usually needed to reveal their occurrence.

Diseases caused by parasites of fish and shellfish have been traditionally regarded as typical in communities with high-risk culinary traditions or in developing countries, where food hygiene and processing are limited. This is still so, but from the 1970s on, fish and shellfishborne zoonoses have increased progressively, in terms of both number of people and world regions affected. Demographic changes, market globalization, and improvement of transportation, which facilitate both food exports to almost every part of the world, and people movements to and from endemic areas, are factors accounting for this increment.

The present chapter will review the main parasites with significance for human health of fish and shellfish from fresh, brackish, and marine waters, paying particular attention to the

Figure 17.1 Ceviche stand in Peru illustrates traditional consumption of dishes made with raw fish.

worldwide emerging anisakidoses. (Parasitic diseases mentioned in this chapter are named after the Standardized Nomenclature of Parasitic Diseases [9].)

17.1 Protozoa

Information about protozoa of zoonotic relevance in fish and seafood is scanty. Most case reports concern species occurring in fecal water, which can contaminate all types of food. The pathogenic protozoa *Cryptosporidium parvum* and *Giardia duodenalis* (=*G. intestinalis*, *G. lamblia*) are well known for their capacity to produce waterborne disease outbreaks. Part of the life cycle of *G. duodenalis* occurs in the intestine of humans, as trophozoites and cysts. Although both forms are passed in stool (Figure 17.2), only the cysts can survive outside the host and are infectious to humans. Infections occur by ingestion of cysts contaminating drinking water, food, hands, or fomites (Figure 17.3). Fishborne transmission of *G. duodenalis* has been documented in the United States, via home-canned salmon, and in China, by means of koipla, a soup prepared with uncooked freshwater fish [10]. In addition, oocysts of *Crystosporidium* and cysts of *Giardia* spp. have been reported in different species of commercial marine bivalve mollusks [11–13]. Microsporidian spores have also been reported in fish and crustaceans. Moreover, it has been shown that spores of human-infectious microsporidians can accumulate in the Asian oyster, *Crassostrea ariakensis* [14,15]. Therefore, consumption of raw or undercooked bivalves can lead to protozoan infections in humans.

Given that detection of *G. duodenalis* by stool analysis is difficult, enzyme-linked immunosorbent assay (ELISA) is the most usual approach for diagnosis in humans, whereas ELISA and immunofluorescent antibody analysis (IFA) are employed for detection in food [16]. The presence of *Crystosporidium* species in patients can be confirmed by both serological methods and stool analyses. The detection of oocysts in shellfish is relatively easy: the gills are removed, washed by vortexing and centrifugation, and oocysts present are examined and quantified by IFA. In addition, a polymerase chain reaction (PCR) protocol has been developed to genotype *Crystosporidium* oocysts in shellfish [17].

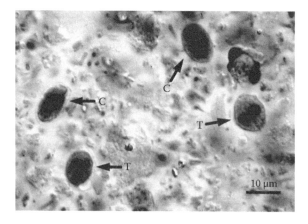

Figure 17.2 *G. duodenalis*: trophozoites (T) and cysts (C) from human stool.

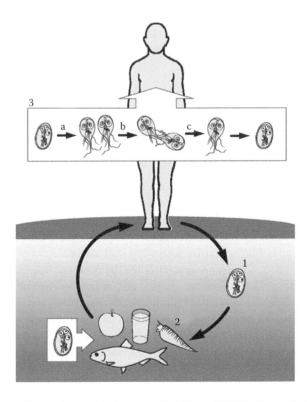

Figure 17.3 Life cycle of *G. duodenalis*, a waterborne protozoan. (1) Cysts are shed in water along with feces. (2) Cysts contaminate water, food, hands, or fomites. (3) Humans become infected with cysts by consumption of contaminated water or food, or by contact with hands or fomites. (3a) Excystation occurs in the small intestine; two trophozoites emerge from each cyst. (3b) Trophozoites multiply asexually by binary fission. (3c) As they transit toward the colon, trophozoites form cysts, which are eventually passed in stool.

17.2 Trematodes

Trematodes represent the most diverse group, in terms of number of species, of freshwater and seafood parasites infecting humans (Table 17.1). Most often, humans are accidental hosts of these trematodes, but in some species, such as *Clonorchis sinensis* or *Paragonimus westermani*, the parasites can attain sexual maturity and complete the life cycle in humans (Figure 17.4).

The Paragonimidae represent a family of lung flukes, whose most pathogenic species belong to the genus *Paragonimus* and *Pagumogonimus*. Paragonimidosis is a severe lung disorder. The common symptoms are fever, cough, chest pain and, occasionally, subcutaneous nodules. Secondarily the disease can affect other organs, reaching the central nervous system and leading to meningitis. Exceptionally, paragonimidosis can be fatal. The most important species causing paragonimidosis are *P. westermani* and *Paragonimus africanus*, and *Pagumogonimus skrjabini*.

China, Korea, Japan, and Thailand, in Asia, Cameroon and Liberia, in Africa and Venezuela, in South America, are countries where paragonimidosis is endemic [18]. Transmission to humans occurs by consumption of undercooked, marinated or raw crayfish or crab. It has been shown that metacercariae can survive outside crayfish, remaining viable for weeks on contaminated

Table 17.1   Main Fish and Shellfish Trematode Species Reported from Humans

Species	Site	Intermediate and Paratenic Host	Other Definitive Hosts	Geographic Distribution
Acanthoparyphium tyosenense	Intestine	Estuarine bivalves and snails	Ducks	Korea
C. sinensis	Liver	Freshwater snails and fish	Carnivores, pigs, rats, buffaloes	Southeast and East Asia, Russia
Cryptocotyle lingua	Intestine	Marine fish	Piscivorous birds and mammals	Alaska, Greenland
Echinostoma hortense	Stomach, intestine	Freshwater snails and fish	Carnivores, rats, mice	Eastern Asia
Echinochasmus japonicus	Intestine	Freshwater snails and fish	Ducks, chickens	Eastern Asia
Echinochasmus perfoliatus	Intestine	Freshwater snails and fish	Carnivores, rats	Eastern Asia, Hungary, Italy, Rumania, Russia
Echinochasmus liliputanus	Intestine	Freshwater snails and fish	Carnivores	Middle East, China
Echinochasmus fujianensis	Intestine	Freshwater snails and fish	Carnivores, pigs, rats	China
Gymnophalloides seoi	Pancreas	Oysters	Wading birds	Korea
Haplorchis taichui	Intestine	Freshwater snails and fish	Carnivores, egret	Middle East, East and South Asia, North East Africa
Haplorchis pumilio	Intestinal	Freshwater snails and fish	Carnivores, pelicans	Thailand, Laos, China
Haplorchis yokogawai	Intestinal	Freshwater snails and fish	Carnivores, egret	Middle East and East and South Asia, Egypt
H. heterophyes	Intestine	Brackish water snails and fish	Carnivores, pelicans	Middle East and East Asia, North East Africa, Spain, Russia
Heterophyes nocens	Intestinal	Brackish water snails and fish	Cats	Eastern Asia

(continued)

Table 17.1 (continued)   Main Fish and Shellfish Trematode Species Reported from Humans

Species	Site	Intermediate and Paratenic Host	Other Definitive Hosts	Geographic Distribution
Heterophyopsis continua	Intestine	Marine and brackish water fish	Cats, ducks, fish-eating birds	Eastern Asia
Metagonimus miyatai	Intestine	Freshwater snails and fish	Dogs, mice, rats, hamsters	Eastern Asia
Metagonimus takahashii	Intestine	Freshwater snails and fish	Dogs, mice	Eastern Asia
M. yokogawai	Intestine	Freshwater snails and fish	Carnivores, rats	Middle East, East and South Asia, Russia, Israel, Spain
Metorchis conjunctus	Bile ducts	Freshwater snails and fish	Carnivores	North America
Nanophyetus salmincola	Intestine	Freshwater and marine fish and snails	Carnivores	Northwest America, Eastern Siberia
Opisthorchis felineus	Liver	Freshwater snails and fish	Carnivore, pigs, rats, rabbits, martens, wolverines, seals	Eastern and South Europe, Russia, Caucasus
O. viverrini	Liver	Freshwater snails and fish	Carnivores, pigs, rats	South East Asia
P. skrjabini (=Paragonimus skrjabini)	Lungs	Freshwater snails and crabs	Carnivores	China, India
P. africanus	Lungs	Freshwater snails and crabs	Primates	Cameroon, Liberia
P. westermani	Lungs	Freshwater snails and crabs	Carnivores	East and South Asia
Paragonimus spp.	Lungs	Freshwater snails and crabs	Carnivores, pigs, rodents	South America
Pygidiopsis genata	Intestine	Brackish and freshwater fish	Domestic carnivores, piscivorous birds	Egypt

Table 17.1 (continued)   Main Fish and Shellfish Trematode Species Reported from Humans

Species	Site	Intermediate and Paratenic Host	Other Definitive Hosts	Geographic Distribution
Pygidiopsis summa	Intestine	Brackish and freshwater fish	Domestic carnivores, piscivorous birds	Japan, Korea
Stellantchasmus falcatus	Intestine	Brackish water fish	Piscivorous birds	South East Asia, Hawaii
Stictodora fuscata	Intestine	Brackish water fish	Piscivorous birds	South East Asia
Stictodora lari	Intestine	Brackish water fish	Seagulls	Korea

Source: Based on Blair, D., in *Marine Parasitology*, Rhode, K., Ed., CSIRO Publishing, Collingwood, Victoria, Australia, 427, 2005; Chai, J.Y. et al., *Int. J. Parasitol.*, 35, 1233, 2005.

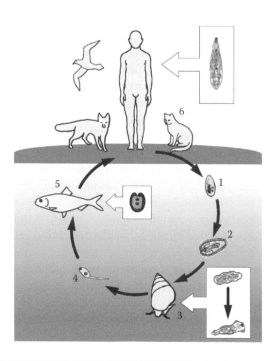

Figure 17.4   The life cycle of *C. sinensis* illustrates the transmission of fishborne trematodes to humans. (1) Eggs are shed with feces in water. (2) Miracidia emerge from the eggs and swim freely to reach and penetrate into snails (first intermediate hosts). (3) Miracidia give rise to sporocysts and rediae, which multiply asexually in the snail. (4) Cercariae leave the snail and swim actively in search of the second intermediate host (fish and, more rarely, crustaceans). (5) Cercariae penetrate the host's tissues and encyst as metacercariae. (6) Metacercariae develop into adults in the bile ducts of fish-eating birds or mammals (including humans).

kitchen utensils [19]. Human infections can be diagnosed by chest x-ray, computer tomography, magnetic resonance imaging, sputum, and stool analyses (to reveal the occurrence of eggs), and by serological assays [18,20]. Apart from visual inspection of opened specimens, there are currently no other detection methods of metacercariae in crayfish.

The species of intestinal flukes transmitted by fish belong to the families Heterophyidae and Echinostomidae. The heterophyids are fairly small and are not considered as highly pathogenic, although certain species can produce important damage in the heart and central nervous system. Of the >30 heterophyid species that can infect humans, the most significant ones are *Metagonimus yokogawai* and *Heterophyes heterophyes*. The former is endemic in Korea and Japan, due to consumption of sweetfish, *Plecoglossus altivelis*. Human infections are also known in China, Taiwan, Siberia, and Europe. *H. heterophyes* is typical in Egypt, especially in the Nile Delta, where it is transmitted by consumption of salted or insufficiently baked grey mullets, *Mugil cephalus*. Human infections of this parasite have also been reported in Sudan and Saudi Arabia, and more rarely in Korea and Japan.

The echinostomids usually occur in the digestive tract of their definitive hosts, birds and mammals, including humans, causing stomach, and duodenal ulcers. *Echinostoma japonicus* and *Echinochasmus hortense* are the most important of the numerous echinostomids reported in humans. *Echinostoma japonicus* is widely distributed in Korea and China, particularly in the Anhui, Fujian, Guangdong, Guangxi, and Jiangsu provinces. *Echinochasmus hortense* occurs mostly in Japan, Korea, and China, and it is transmitted by eating raw loach, *Misgurnus anguillicaudatus*, in Northeast China. Stool analyses often indicate mixed infections, making identification of the species involved more difficult [21].

Most liver flukes transmitted by fish and shellfish correspond to the family Opisthorchiidae. In addition, the pancreatic fluke *Gymnophalloides seoi* (family Gymnophallidae) is known to infect humans in Korea by consumption of oysters, *Crassostrea gigas* [22]. Opisthorchiosis can lead to severe inflammation of hepatic ducts and pancreatitis; chronic conditions are known in endemic regions. The most representative species are *C. sinensis* (Figure 17.5), especially abundant in China, with ~12.5 million people infected [23], but also in Korea and Vietnam, *Metorchis conjunctus*, in North America (particularly among aboriginal people from Northern Canada and Greenland), *Opisthorchis felineus*, distributed from Southeastern Europe to Russia, and *Opisthorchis viverrini*, endemic in Thailand, Laos, Cambodia, and Vietnam. It has been pointed out that transmission of the latter species varies seasonally, increasing during the monsoon, because floods propitiate fecal contamination of water. A typical dish from northeastern Thailand and Laos, known as koi-pla, based on raw fish with garlic and vegetables, is the main source of infection of *O. viverrini*. Traditional eating habits, such as the morning congee with slices of raw freshwater fish in Southern China or slices of raw freshwater fish with red pepper sauce in Korea, are the main infection routes of *C. sinensis*. Detection of liver flukes in humans is based on stool analysis by cellophane tic smear or Kato-Katz techniques. ELISA assays are also used, and recently a PCR technique has been developed to detect *O. viverrini* in snails and fish [21].

1 mm

Figure 17.5 *C. sinensis*, a liver fluke infecting fish-eating birds and mammals; whole mount of specimen extracted from the bile duct of a patient in Vietnam.

Sometimes, mixed infections of liver flukes are reported. For instance, a 69-year-old man in Korea harbored 69,125 specimens of *Gymnophalloides seoi*, 328 of *Heterphyes nocens*, and 1 of *Stictodora lari*. The first mentioned species is the most pathogenic one, since it invades pancreatic ducts and leads to pancreatitis [24].

Migrations have contributed a great deal to the extension of liver fluke infections. So, numerous reports of *C. sinensis* and *Opisthorchis* spp. in the United States and Canada can be linked to Asian immigrants. Tourism has also contributed to the expansion of infections. For instance, an outbreak of gastroenteritis produced by trematodes was reported in a group of American tourists returning from a trip to Kenya and Tanzania in 1983, and several similar cases have been reported from Canada [10].

17.3   Cestodes

The main cestodes infecting humans transmitted by fish and seafood belong to the family Diphyllobotriidae, particularly to the genera *Diphyllobothrium* and *Diplogonoporus* (Table 17.2). Although only the life cycle of some species is known in detail, transmission occurs through aquatic food webs (Figure 17.6).

Diphyllobothriosis can be asymptomatic, but usual manifestations are abdominal pain, diarrhea, nausea, anorexia, and fatigue. Sometimes infections lead to pernicious anemia by depletion of vitamin B_{12}. *Diphyllobothrium latum* and *Diphyllobothrium pacificum* (Figure 17.7) are significant representatives of this group. The former is typical in continental waters of the Holarctic region, and the latter occurs in marine waters along the Pacific coast of South America, where its abundance is influenced by El Niño event.

Diphyllobothriosis occurs in communities where consumption of raw or little cooked fish is common. Dishes related to diphyllobothriosis include sushi and sashimi in Japan, gravlax in Scandinavia, strogonina in Eurasia, and ceviche, tiradito, and chinguirito in Peru, Ecuador, and Chile. Diphyllobothriosis is apparently declining worldwide, particularly in North America and Europe, as a result of effective public health policies. However, infections still persist in some endemic regions, such as the Russian Far East and Japan. In the last years, new cases have been reported in Chile, due to the introduction for angling of exotic freshwater fishes, such as rainbow trout, *Oncorhynchus mykiss*, and in Western Europe, owing to consumption of imported North Pacific salmon, *Oncorhynchus keta* [25,26].

Detection in humans is based on standard stool analyses to reveal the occurrence of eggs or proglottids in feces. It is difficult to physically detect plerocercoids, in fish requiring meticulous analysis by specialized personnel. However, the presence of plerocercoids in fish elicits an immune response that can be detected by immunofluorescence techniques and ELISA [27].

17.4   Anisakid Nematodes

Anisakid nematodes are probably the most common parasites associated with seafood worldwide. Their larvae occur in fish and squid and their incidental ingestion by humans can cause anisakidosis and allergic reactions [28,29]. The most commonly reported anisakids causing disease in humans are *Anisakis simplex* and, to a lesser extent, *Pseudoterranova decipiens* [10,21,30], so that the more specific terms anisakiosis and pseudoterranovosis are used to designate infections with these species. Other larval anisakids, *A. physeteris*, *Contracaecum osculatum*, and *Hysterothylacium aduncum* (see Table 17.3), have been reported very rarely in humans [30].

Table 17.2 Main Fish and Shellfish Tapeworm Species Reported from Humans

Species	Site	Intermediate and Paratenic Hosts	Other Definitive Hosts	Geographic Distribution
Diphyllobothrium alascense	Intestine	Burbot, smelt	Dog	Alaska
Diphyllobothrium cameroni	Intestine	Marine fish	Seals	Pacific
Diphyllobothrium cordatum	Intestine	Marine fish	Dog, seals, walrus, sea lions	North Pacific, Arctic
Diphyllobothrium dalliae	Intestine	Freshwater fish	Dog, gulls	Alaska, Siberia
Diphyllobothrium dendriticum	Intestine	Freshwater fish	Fish-eating birds and mammals	Circumpolar, Switzerland
Diphyllobothrium elegans	Intestine	Marine fish	Seals, sea lions	North Sea, Greenland
Diphyllobothrium hians	Intestine	Marine fish	Seals	North Atlantic, Pacific Siberia
Diphyllobothrium klebanovski	Intestine	Salmonids	Unknown	Eastern Eurasia, Sea of Japan, Sea of Okhostsk
Diphyllobothrium lanceolatum	Intestine	Whitefishes	Dog, seals, porpoises	North Atlantic, North Pacific
Diphyllobothrium latum	Intestine	Burbot, pike, percids	Dog, bears	North and South America, Europe, Russia, Korea
Diphyllobothrium nihonkaiense	Intestine	Pacific salmon	Unknown	Japan, Korea, Canada, France, Switzerland
Diphyllobothrium orcini	Intestine	Marine fish	Killer whale	Japan
D. pacificum	Intestine	Marine fish	Sea lions, fur seals	Alaska, Japan, South Eastern Pacific
Diphyllobothrium scoticum	Intestine	Marine fish	Sea lions, seals	South Atlantic Ocean
Diphyllobothrium stemmacephalum	Intestine	Marine fish	Toothed whales	North Atlantic, North Sea, Eastern Asia
Diphyllobothrium ursi	Intestine	Red salmon	Bears	North Eastern Pacific
Diphyllobothrium yonagoensis	Intestine	Salmon	Unknown	Japan, Eastern Siberia
Diplogonoporus balaenopterae (=D. grandis)	Intestine	Japanese anchovy	Baleen whales, sea lions, seals	Circumboreal, Antarctic, Spain

Source: Based on Blair, D., in *Marine Parasitology,* Rhode, K., Ed., CSIRO Publishing, Collingwood, Victoria, Australia, 427, 2005; Chai, J.Y. et al., *Int. J. Parasitol.,* 35, 1233, 2005.

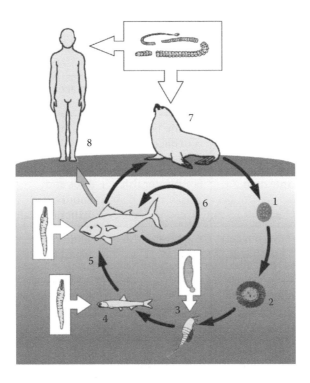

Figure 17.6 Life cycle of *D. pacificum*, a marine diphyllobothrid tapeworm infecting humans. (1) Eggs are shed in the feces of the definitive host (fur seals and sea lions). (2) Hatching occurs in water; emerging coracidia swim actively. (3) Coracidia are ingested by copepods (first intermediate hosts), where they develop into procercoids. (4) Copepods are preyed by fish (second intermediate hosts) and the procercoids become plerocercoids. (5) Second intermediate hosts can be preyed by larger fish (paratenic hosts). (6) Predation can occur several times, but plerocercoids undergo no further change. (7) Fish is eaten by the definitive hosts; plerocercoids develop then into adult worms. (8) Incidental infections occur by ingestion of plerocercoids in raw or undercooked fish.

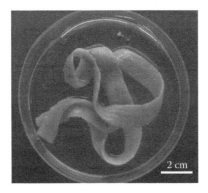

Figure 17.7 Proglottids of *D. pacificum* extracted from the intestine of a patient in Peru.

Table 17.3 Main Fish and Shellfish Nematode Species Reported from Humans

Species	Site	Intermediate and Paratenic Hosts	Other Definitive Hosts	Geographic Distribution
A. simplex	Stomach and intestine	Marine fish, squid	Cetaceans	Worldwide
Anisakis physeteris	Stomach and intestine	Marine fish, crustaceans, squid	Sperm whales	Worldwide
C. philippinensis	Intestine	Freshwater fish	Birds and monkeys experimentally	Southeast Asia
Gnathostoma spp.	Stomach and esophagus	Freshwater fish	Felids, pigs, weasels	Southeast and East Asia, India, Middle-East
Pseudoterranova decipiens	Stomach and intestine	Marine fish, crustaceans, squid	Seals, cetaceans	Cold waters worldwide
Contracaecum osculatum	Stomach and intestine	Marine fish, crustaceans	Seals, fish-eating birds	Worldwide
Hysterothylacium aduncum	Stomach and intestine	Marine fish, crustaceans, squid	Marine fish	Worldwide

Source: Based on Ko, R.C., in *Fish Diseases and Disorders. Vol. I. Protozoan and Metazoan Infections*, Woo, P.T.K., Ed., CAB International, Oxon, 631, 1995; Nagasawa, K., in *Marine Parasitology*, Rhode, K., Ed., CSIRO Publishing, Collingwood, Victoria, Australia, 430, 2005.

Different genetic studies over the last 20 years have shown that *A. simplex* and *P. decipiens* in fact represent two respective complexes of sibling species, which exhibit some degree of geographic and/or definitive host differentiation [31,32]. The epidemiology and pathogenic manifestations in humans seem similar within each complex (although no formal study has been conducted to date in order to analyze potential interspecific differences) and, for convenience, in the present chapter the two species complexes will be referred to collectively as *A. simplex* and *P. decipiens*.

A. simplex and *P. decipiens* utilize food webs for transmission to marine mammals (whales and seals, respectively), which act as definitive hosts. The life cycle of *A. simplex* has long been considered as pelagic, but recent studies have revealed differences between species within the complex, having pelagic, demersal, or benthic cycles [33]. In addition, the number and type of hosts used by each species varies depending on their availability and abundance in each geographic area [34]. However, a common trait is that the life cycle of *A. simplex* occurs offshore and, thus, *A. simplex* is virtually absent from estuarine and other brackish environments. Despite differences within the species complex, a generalized life cycle of *A. simplex* can be outlined (Figure 17.8). The life cycle of *P. decipiens* is similar to that of *A. simplex*, but the food web used is benthic or benthopelagic in order to target seals, instead of whales, as definitive hosts. The eggs sink to the

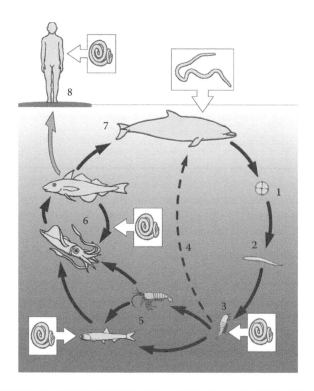

Figure 17.8 Generalized life cycle of *A. simplex*. (1) Adult worms reside in the stomach of cetaceans (definitive hosts), where gravid females shed eggs that are passed in the host's feces. (2) Free swimming larvae emerge from eggs. (3) Planktonic crustaceans (first intermediate hosts) ingest the larvae, which penetrate through the intestine into the crustacean hemocoel. (4) Third-stage larvae (L3s) occurring in planktonic crustaceans are infective and can already reach definitive hosts, such as baleen whales feeding on zooplankton. (5) Most often, infected crustaceans are preyed by fish, squid, or larger crustaceans, which in turn are consumed by larger predators. (6) Predation can occur several times and, at each instance, the ingested third-stage larvae bore the intestinal wall to encapsulate in the visceral cavity of the new host. (7) When prey are eaten by cetaceans, the L3s molt three times to become fourth- and fifth-stage larvae and eventually adult worms. (8) Incidental infections occur by consumption of raw or undercooked seafood.

bottom and the hatched larvae adhere to the substrate by their tails awaiting consumption by benthic crustaceans [35].

The bottom-up exploitation of marine food webs by anisakids (as illustrated in Figure 17.8) is very efficient to reach marine mammals because of their position as top predators. In particular, transmission is enhanced because the encapsulated third-stage larvae can use a wide range of hosts where they remain viable for long and, thus, accumulate over time in individual hosts [34,36]. So, in areas where marine mammals are abundant, anisakid larvae can be widespread, and common among fish and squid of commercial size [33,37] (Figure 17.9), which accounts for the numerous incidental infections in humans.

Incidental infections occur when the anisakid third-stage larvae are eaten with either raw or lightly cooked fish or squid. Given the physiological and anatomical similarity between humans

Figure 17.9 Third-stage larvae of *A. simplex* in the visceral cavity of blue whiting, *Micromesistius poutassou*. Note the characteristic larval print on the liver surface (arrow), which can prove the presence of the parasite even if larvae are inadvertently lost during examination.

and marine mammals, the larvae can survive and occasionally molt to fourth-stage larvae [38], but cannot mature and reach the adult stage. Some infections are asymptomatic; the larvae remain in the gastrointestinal tract without penetrating the tissues and can only be discovered when expelled by coughing, vomiting, or defecating. In other instances, the larvae release histolytic enzymes allowing them to penetrate into or through the stomach or intestinal mucosa (leading to gastric or intestinal anisakidosis, respectively) (Figure 17.10). More rarely, the larvae can invade other sites, such as the lung, liver, throat, and subcutaneous tissues [10,21,30].

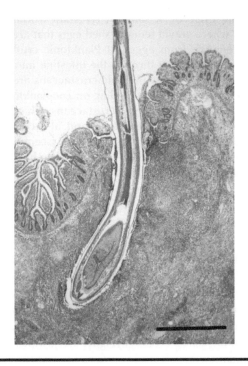

Figure 17.10 *A. simplex* larva penetrating the small intestine wall, surrounded by a thick cuff of acute inflammatory cells (bar 1 mm). (From Takei, H. and Powell, S.Z., *Ann. Diagn. Pathol.*, 11, 350, 2007. With permission.)

The clinical manifestations of anisakidosis are varied and unspecific. The disease is characterized by a sudden onset of epigastric pain, sometimes accompanied by nausea and vomiting. Additional reported disorders include urticaria, pulmonary complications, allergic edema, hypersialorrhea, and polyarthritis. When the condition is chronic, histopathological examination reveals the larvae, with distinctive Y-shaped lateral chords, embedded in the gastrointestinal wall, accompanied by inflammatory infiltrate, forming eosinophilic abscesses or eosinophilic granulomas (Figure 17.11) [21,30]. The gastric condition can be confirmed by endoscopy, but clinical diagnosis of intestinal anisakidosis is extremely difficult because the symptoms can be easily attributed to common disorders, such as appendicitis, intestinal obstruction, or peritonitis [21].

Detection of anisakid larvae in fish is mostly made by visual inspection. Fillets and napes are usually inspected by candling on a light table and the larvae spotted are removed manually [35]. This procedure is less efficient with *A. simplex* (smaller and whitish larvae) than with *P. decipiens* [39]. Candling under ultraviolet (UV) light causes the larvae to fluoresce, but in *A. simplex*, it has been observed that fluorescent emission is only clearly seen in previously frozen fish [40]. In addition, worms embedded >0.5 mm deep in fish tissue are not visible [35]. Recently, PCR and ELISA methods have been developed to detect and quantify *A. simplex* larvae in seafood [41,42].

Some patients of anisakiosis show symptoms of urticaria and/or other allergic reactions, usually after manifestation of the digestive disorders, a condition known as gastroallergic anisakiosis. It has also become clear in recent years that the ingestion of *A. simplex* larvae can cause an immediate allergic response without showing further digestive symptoms [29]. The clinical symptoms range from urticaria, angioedema, and even arthralgia, to life-threatening anaphylactic shock [21,43]. Exposure by contact or inhalation of *A. simplex* allergens can also elicit allergic responses [29,43]. The strength of allergy to other anisakids is not yet known [21].

The diagnosis of allergy to *A. simplex* is based on a compatible medical history (such as allergic symptoms after ingestion of fish) and immunological tests. Patients show positive skin tests and specific IgE against *A. simplex*, with a marked increase in total IgE and lack of reaction to

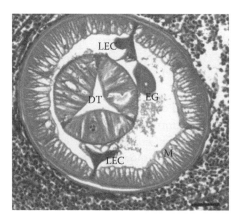

Figure 17.11 **Cross section through an *A. simplex* larva embedded in the small intestine wall. A thick cuff of acute inflammatory cells with numerous eosinophils surround the larva (bar 100 μm). M, muscle layer; LEC, epidermal chord; EG, excretory gland; DT, digestive tract. Note the characteristic Y-shaped LEC of *A. simplex*. (From Takei, H. and Powell, S.Z., *Ann. Diagn. Pathol.*, 11, 350, 2007. With permission.)**

proteins of fish [28,43]. However, specific IgE antibodies against *A. simplex* cross-react with those of other invertebrates, including dust mites, cockroaches, shrimps, and other helminths [43,44]. So a patient with, for instance, shrimp allergy might be misdiagnosed as allergic to *A. simplex*. An additional problem is that a considerable proportion of the population is sensitized to *A. simplex*, resulting from being exposed to the larvae without having developed allergy symptoms [45,46]. Therefore, the issue of distinguishing patients with clinical and subclinical sensitization has become highly relevant.

In order to better understand the molecular bases of cross-reactivity and improve serological diagnosis of allergic anisakidosis, many studies have focused on molecular identification and characterization of *A. simplex* allergens [43–45,47]. To date, nine allergens of *A. simplex* have been molecularly characterized (www.allergen.org) [29,47], but the total number is probably much higher [48,49].

Somatic antigens (Ani s 2 and Ani s 3), obtained by homogenization of the whole larvae, seem to account for most cross-reactivity between *A. simplex* and other organisms [44]. So, whole-larva extracts used to detect specific IgE lack enough specificity for diagnosis of allergy to *A. simplex*.

Excretory–secretory antigens, i.e., histolytic enzymes secreted by the parasite to penetrate the gastric mucosa, seem better suited for diagnosis [43]. In particular, Ani s 1 shows no or little cross-reactivity with other allergens [43,50]. Ani s 7 is another major excretory–secretory antigen. Due to its sugar epitopes, it can cross-react with homologous glycoproteins of other organisms, but after *o*-deglycosylation, it is recognized by monoclonal antibody UA3 without false positive results [43]. Moreover, molecular characterization of *A. simplex* allergens and isolation of their encoding cDNAs is currently a very active field of research that is expected to facilitate diagnosis in the near future [29].

Anisakidosis is a health problem particularly important in countries and communities in which consumption of raw or undercooked fish is widespread. *A. simplex* accounts for the vast majority (~97%) of human infections; *P. decipiens* represents only ~2.7% of the total, whereas the remaining 0.3% corresponds to either unidentified anisakids or to other species [30].

Of the ~20,000 human infections reported to date, ~90% have been reported in Japan, where ~2,000 new infections are diagnosed each year [21]. Anisakidosis in Europe is fairly common, with ~600 cases reported mostly in France, Germany, the Netherlands, and Spain [39]. In Korea, 107 cases were reported between 1989 and 1992 [10], and ~50 infections were known from the United States until the 1990s [51]. Other countries where anisakidosis has been reported sporadically include Canada, Chile, Egypt, Iceland, New Zealand, Oman, and Peru [21,52–54], and possibly China [23].

Evidence gathered over the last two decades shows that *A. simplex* is a frequent agent of food-related allergy. A serological survey performed in Japan on ~2000 patients initially diagnosed with urticaria or food allergy revealed specific IgE against *A. simplex* in 29.8% of the subjects [55]. Another Japanese study has shown IgE responses to excretory–secretory antigens in 87.5% of patients with gastric anisakiosis, 75% of subjects initially diagnosed with fish-induced urticaria, 8.3% of individuals with idiopathic urticaria, and 10% of healthy controls [56]. Investigations carried out in Spain also point to a high prevalence of allergy to *A. simplex*. In a survey involving 868 subjects from three geographic areas, the prevalence among patients of urticaria or angioedema was 19.2%, and 13.1% of individuals without allergic symptoms were sensitized to the parasite [57].

Note that the above prevalence estimations might be inflated by cross-reactivity with other allergens and the large number of people manifesting subclinical sensitization [29]. However, a recent study using the more specific UA3-based ELISA assay revealed that the prevalence of sensitized individuals in Madrid was 12.4% [58], which is only slightly lower (15.7%) than that previously reported in the same city using a classical test [57]. Furthermore, current data from the Basque Country (Northern Spain) indicates that *A. simplex* accounts for 10% of anaphylaxis cases in adult patients, which is a figure similar to that of other food allergies combined [29]. Therefore, allergy to *A. simplex* should be a public-health concern in countries with high per capita consumption of fish. An interesting finding is that three HLA class II alleles are overrepresented in patients allergic to *A. simplex* [59]. Note also that sensitized subjects with repeated exposure to *A. simplex* will be at risk of developing acute anisakiosis with severe symptoms [46].

17.5 Other Nematodes

Nematodes of freshwater fishes (Table 17.3), such as *Capillaria philippinensis* and *Gnathostoma* spp., can even be fatal for humans. *C. philippinensis* is typical in the Philippines, although it has extended to other countries, and it is currently considered an emerging zoonosis. Infections are common among farmers via ingestion of raw fish as lunch, during fishing activities in lagoons, lakes, and rivers. Another route of transmission is drinking basi (an alcoholic beverage) with raw food.

Gnathostomiosis is mostly caused by four species: *Gnathostoma hispidum*, *G. spinigerum*, *G. doloresi*, and *G. nipponicum*. Thailand is one of the most affected countries, where *G. spinigerum* is widespread in the population of the country's central regions by means of consumption of raw fish dishes, such as hu-sae, som-fak, and pla-som [60].

17.6 Acanthocephalans

Human infections with acanthocephalans are extremely rare. However, several species use fish as paratenic hosts, thereby making incidental infections possible. *Acanthocephalus rauschi* and *A. bufonis* from the peritoneum of an Alaskan Eskimo and small intestine of an Indonesian man, respectively, represent two isolated case reports [61]. Other sporadic episodes concern members of the genera *Bolbosoma* and *Corynosoma*, which use cetaceans and seals as definitive hosts. They have been detected in an Eskimo from Alaska, and Japanese fishermen after consumption of sashimi [61,62].

17.7 Seafood Safety

As any other food-related hazard, prophylaxis of parasitoses transmitted with seafood involves preventive steps at every level of the production–consumption chain [10].

17.7.1 Primary Production and Handling

The chances of human infections can be reduced by avoiding fishing and harvesting in particular areas, avoiding certain species or given sizes. In addition, control of feed in farmed species is highly relevant, since parasite infections are drastically reduced in species fed exclusively with artificial feed [39]. The occurrence of infective larvae in fish muscle is due to migration from the visceral cavity. Thus, evisceration on board reduces, but does not completely eliminate, the risk of human infections [37]. Note, however, that evisceration is not feasible in small species, such as sardines or anchovies. In addition, it has been suggested that viscera discarded over board can enhance transmission of parasites [33]. Trimming away the belly flaps of fish, candling and physically removing parasites detected visually can reduce the number of parasites. However, they do not completely eliminate the hazard, nor do they minimize it to an acceptable level [37,39,63].

Depuration is a standard safety procedure in which bivalve mollusks are placed in clean water under controlled conditions for varying periods of time in order to eliminate or minimize infective agents. The method is effective at reducing bacterial pathogens, but its efficacy with protozoa is currently unclear. Oocysts of *Cryptosporidim* sp. have been reported in bivalve samples with depuration times >72 h [11]. Industrial UV depuration protocols applied to spiked Pacific Oysters, *Crassotrea* gigas, resulted in a 13-fold increase in inactivation of *C. parvum* oocysts. However, low numbers of human-infectious oocysts still occurred in oysters after depuration [64]. In addition, it has been shown that depuration times of *Cryptosporidim* sp. oocysts and microsporidium spores in the Asian oyster are quite long (>33 days) [15]. Therefore, current depuration protocols are not enough to completely eliminate the risk of human protozoan infections.

17.7.2 Thermal Processing

Thermal processing (either cooking or freezing) is the most effective way for eliminating the risk of parasitic disease from seafood. Other procedures (radiation, high pressure, acidification, salting, etc.) either are in experimental phase, affect the organoleptic properties of the product, or do not reduce risk to acceptable level [39,63].

Parasites are inactivated by heating until the inner part of the product reaches at least 63°C for 15 s or longer [10]. So, fully cooked, hot-smoked, and pasteurized seafood are safe from a parasitological point of view. When using conventional cooking, heating the food to 65°C for 10 min is recommended. Microwaving can be problematic, because microwave heating only acts on the outermost part of the product, so that heat is only transmitted further inside by conduction. Therefore, microwave cooking requires a higher safety temperature (74°C) [65].

In the European Union, products to be consumed raw, cold-smoked, marinated, or salted must be frozen at a temperature of not more than –20°C, in all parts of the product for not less than 24 h [66]. This regulation is open to some misinterpretation, since the time needed to reach –20°C depends greatly on the type of freezer. For instance, a recent study [67] has shown that that it can take ~10 h to attain –20°C inside fish placed in household freezers. Moreover, the same investigation showed that 50.5 h were required to kill all *A. simplex* larvae in fish kept at an internal temperature of –20°C [67]. This result suggests that the European regulation may not be stringent enough. The U.S. Food and Drug Administration recommendations [63], namely, freezing and storing at –20°C or below for 7 days or –35°C or below for 15 h, seem more in line with the above scientific evidence.

The efficacy of proper freezing at preventing parasite infections in humans is unquestionable. Indeed, freezing is the only critical control point considered reliable in Hazard Analysis and Critical Control Points (HACCP) protocols for production of seafood [39]. In recent years, however, concern has been raised about the efficacy of thermal processing at preventing allergy to *A. simplex*. This issue is currently unsettled. Several studies suggest that allergic responses are only elicited by living larvae [39], at least in most patients [45]. However, other investigations have shown that *A. simplex* allergens are thermostable and, thus, cooking and freezing may not be protective enough [29,49]. The latter opinion is currently subscribed by the European Food Safety Authority [68].

17.7.3 Recommendations for Consumers and Restaurateurs

Information to consumers is of outmost importance to minimize the risks associated with parasites found in seafood. In order to avoid excessive public alarm, consumers should be aware that only a small portion of parasite species found in seafood are hazardous to public health. The following recommendations are largely based on information provided by the European Food Safety Authority [68], the Spanish Agency of Food Safety and Nutrition [39], and the U.S. Food and Drug Administration [51,63,65].

Fish and seafood should be purchased as fresh as possible. Medium and large fish should be bought eviscerated or be eviscerated immediately after purchase. The abdominal cavity should be washed and examined for parasites. Cook or freeze without delay.

Raw, cold-smoked, marinated, or lightly cooked seafood should not be consumed, unless proof is given that the product has been kept frozen at –20°C or less, for at least 1 week.

Seafood dishes should be properly cooked. Baking, boiling, and frying are safer than broiling and microwaving. In the latter case, setting the oven 14°C higher than the safety temperature (i.e., 74°C + 14°C) is recommended. The use of cooking thermometers placed at the thickest part of the food is encouraged to guarantee that safety temperatures are attained. In addition, consumers and restaurateurs should always check that the dish is cooked throughout before serving or eating.

17.7.4 Recommendations for Allergic and Immunosuppressed Persons

In addition to the above, persons suffering from immunological deficiency or from allergy to *A. simplex* should consider to the following recommendations:

Infections with waterborne protozoa can be fatal for immunosuppressed patients. Thus people with this ailment should by all means avoid consumption of raw or undercooked seafood [15].

Evidence suggests that patients allergic to *A. simplex* react differently after ingestion of frozen fish [45,49]. So only a doctor can provide adequate dietary advice to each individual. However, most patients (between 80% and 90% [49]) show good tolerance to frozen fish. Fish eviscerated and blast-frozen on factory vessels are to be preferred, since the risk of larval emigration from the visceral cavity to the muscle is greatly reduced. Fish tails and marine farmed and freshwater fish are safer. Allergic patients should always avoid eating the hypaxial musculature of fish, as well as small whole fish [39].

17.8 Further Developments

Seafood has played a major role in doubling animal protein consumption in developing countries over the last 30 years. In China, for instance, per capita fish consumption has quadrupled between 1970 and 2003 [1]. Although less dramatically, consumption has also increased in developed countries, driven by both globalization of culinary tastes and promotion of healthy food to prevent cardiovascular disease and obesity. Thus seafoodborne parasites are likely to remain a global issue in the years to come.

Although some public health plans have proven successful to control and reduce some parasitic diseases (such as diphyllobothriosis in Scandinavia), worldwide the incidence of fishborne parasites is on the rise. One reason for this phenomenon is increased pressure on food sources to meet world population growth. Increasing poverty and malnutrition, coupled to limited public health services in many countries, propitiates the extension of seafoodborne parasitoses, particularly those caused by trematodes.

In this context, steady growth of aquaculture as a source of inexpensive animal proteins (particularly in continental waters of Asia) will imply higher risk of seafoodborne zoonoses, if food habits and control policies remain unchanged [69]. Economy and trade globalization have fostered exports of seafood. The effects of food safety regulations imposed by developed countries to imports from developing countries, such as HACCP processes and technical barriers to trade, have already introduced high costs that tend to exclude small producers and processors from the export supply chain. Thus, future efforts should be directed at assisting and supporting fishermen, fish farmers, and processors in developing countries to adopt technology for efficient HACCP processes, and thereby realize benefit from global trade [70].

Low risk perception among consumers of developed countries of imported seafoodborne parasites favors uncommon human infections. For instance, case reports involving *Diplogonoporus* sp. and *A. simplex* in Zaragoza, inland in Spain, have been reported [71,72]. Many of these new human infections stem from cultural globalization by the adoption of cooking traditions based on dishes made with raw or little cooked seafood. The increment of human infections with *H. heterophyes* in North America [73] illustrates this aspect. Moreover, the increase of travel from and to endemic areas has facilitated the expansion of these parasitic zoonoses [21,74].

Therefore, the development of appropriate food safety programs poses new challenges to public health managers. In addition, improved diagnosis techniques and increased awareness among health personnel will result in detection of more cases that otherwise could go unnoticed. As in any other health hazard, better consumer information and education is of paramount importance for prevention. Providing the correct message and format to the different at-risk groups is a constant challenge to educational programs. However, providing alternative sustainable, affordable solutions to elicit behavioral change has also been a limitation of control programs [74].

Acknowledgments

The authors thank Drs. Jorge Manuel Cárdenas (Asociación Peruana de Helmintología e Invertebrados Afines), Angélica Terashima (Instituto de Medicina Tropical Alexander von Humboldt, Universidad Peruana Cayetano Heredia), Hidehiro Takei (Baylor College of Medicine, Houston), and Suzanne Z. Powell (The Methodist Hospital, Houston) for kindly contributing with photographs illustrating this review.

References

1. Food and Agricultural Organization of the United Nations. Food Balance Sheets. http://faostat. fao.org.
2. WHO. Control of Foodborne Trematode Infections. WHO Technical Report Series, 849. World Health Organization, Geneva, 1995.
3. Murrell, K.D. and Bernard F. *Food-Borne Parasitic Zoonoses. Fish and Plant-Borne Parasites*. Springer, New York, 2008.
4. Yang, W.Y. et al. Parasitologische Untersuchung einer alten Leiche aus der Chu-Dynastie der Streitenden Reiche aus dem Mazhuan-Grab Nr. 1, Kreis Jiangling, Provinz Hubei. *Acta Acad. Med. Wuhan*, 4, 23, 1984.
5. Seo, M. et al. *Gymnophalloides seoi* eggs from the stool of a 17th century female mummy found in Hadong, Republic of Korea. *J. Parasitol.*, 94, 467, 2008.

6. Bathurst, R.R. Archaeological evidence of intestinal parasites from coastal shell middens. *J. Archaeol. Sci.*, 32, 115, 2005.

7. Ferreira, L.F. et al. The finding of eggs of *Diphyllobothrium* in human coprolites (4,100–1,950 B.C.) from Northern Chile. *Mem. Inst. Oswaldo Cruz*, 79, 175, 1984.

8. Reinhard, K. and Urban, O. Diagnosing ancient Diphyllobothriasis from Chinchorro mummies. *Mem. Inst. Oswaldo Cruz*, 98, 191, 2003.

9. Kassai, T. Nomenclature for parasitic diseases: Cohabitation with inconsistency for how long and why? *Vet. Parasitol.*, 138, 169, 2006.

10. Butt, A.A. et al. Infections related to the ingestion of seafood. Part II: Parasitic infections and food safety. *Lancet Infect. Dis.*, 4, 294, 2004.

11. Freire-Santos, A.M. et al. Detection of *Cryptosporidium* oocysts in bivalbe molluscs destined for human consumption. *J. Parasitol.*, 86: 853, 2000.

12. Gómez-Couso, H. et al. Contamination of bivalve molluscs by *Crystosporidium* oocysts: The need for new quality control standards. *Int. J. Food Microbiol.*, 87, 97, 2003.

13. Gómez-Couso, H. et al. Detection of *Crystosporidium* and *Giardia* in molluscan shellfish by multiplexed nested-PCR. *Int. J. Food Microbiol.*, 91, 279, 2004.

14. Slifko, T.R. et al. Emerging parasite zoonoses associated with water and food. *Int. J. Parasitol.*, 30, 1379, 2000.

15. Graczyk, T.K. et al. Recovery, bioaccumulation and inactivation of human waterborne pathogens by the Chesapeake Bay nonnative oyster, *Crassotrea ariakensis*. *Appl. Environ. Microbiol.*, 72, 3390, 2006.

16. Sulaiman, I.M., and Cama, V. The biology of *Giardia* parasites, in *Foodborne Parasites*. Ortega, Y.R., Ed., Springer, New York, 2006, Chapter 2.

17. Xiao, L. and Cama, V. *Cryptosporidium* and cryptosporidiosis, in *Foodborne Parasites*. Ortega, Y.R., Ed., Springer, New York, 2006, Chapter 4.

18. Liu, Q. et al. Paragonimiasis: An important food-borne zoonosis in China. *Trends Parasitol.*, 24, 318, 2008.

19. Cross, J.H. Fish- and invertebrate-borne helminths, in *Foodborne Disease Handbook*, 2. Hui, Y.H., Sattar, S.A., Murrell, K.D., Nip, W.K., and Stanfield, P.D., Eds., Marcel Dekker Inc., New York, 2001, Chapter 12.

20. Adams, A.M. Foodborne trematodes, in *Foodborne Parasites*. Ortega, Y.R., Ed., Springer, New York, 2006, Chapter 7.

21. Chai J.Y. et al. Fish-borne parasitic zoonoses: Status and issues. *Int. J. Parasitol.*, 35, 1233, 2005.

22. Chai J.Y. et al. *Gymnophalloides seoi*: A new human intestinal trematode. *Trends Parasitol.*, 19, 109, 2003.

23. Zhou, P. et al. Food-borne parasitic zoonoses in China: Perspective for control. *Trends Parasitol.*, 24, 190, 2008.

24. Chai J.Y. et al. *Stictodora lari* (Digenea: Heterophyidae): The discovery of the first human infections. *J. Parasitol.*, 88, 627, 2002.

25. Yeraa, H. et al. Putative *Diphyllobothrium nihonkaiense* acquired from a Pacific salmon (*Oncorhynchus keta*) eaten in France; Genomic identification and case report. *Parasitol. Int.*, 55, 45, 2006.

26. Wichta, B., de Marvalb, F., and Peduzzia, R. *Diphyllobothrium nihonkaiense* (Yamane et al., 1986) in Switzerland: First molecular evidence and case reports. *Parasitol. Int.*, 56, 195, 2007.

27. Sharp, G.J.E., Pike, A.W., and Secombes, C.J. The immune response of wild rainbow trout, *Salmo gairdneri* Richardson, to naturally acquired plerocercoid infections of *Diphyllobothrium dendriticum* (Nitzsch, 1824) and *D. ditremum* (Creplin, 1825). *J. Fish Biol.*, 35, 781, 1989.

28. Audicana, M.T. et al. *Anisakis simplex*: Dangerous—Dead and alive? *Trends Parasitol.*, 18, 20, 2002.

29. Audicana, M.T. and Kennedy, M.W. *Anisakis simplex*: From obscure infectious worm to inducer of immune hypersensitivity. *Clin. Microbiol. Rev.*, 21, 360, 2008.

30. Nagasawa, K. Anisakiasis, in *Marine Parasitology*. Rhode, K., Ed., CSIRO Publishing, Collingwood, Victoria, Australia, 2005, 430.

31. Mattiucci, S. et al. Genetic and ecological data on the *Anisakis simplex* complex with evidence for a new species (Nematoda, Ascaridoidea, Anisakidae). *J. Parasitol.*, 83, 401, 1997.

32. Zhu, X.Q. et al. SSCP-based identification of members within the *Pseudoterranova decipiens* complex (Nematoda:Ascaridoidea:Anisakidae) using genetic markers in the internal transcribed spacers of ribosomal DNA. *Parasitology*, 124, 615, 2002.

33. Abollo, E., Gestal, C., and Pascual, S. *Anisakis* in marine fish and cephalopods from Galician waters: An updated perspective. *Parasitol. Res.*, 87, 492, 2001.

34. Klimpel S. et al. Life cycle of *Anisakis simplex* in the Norwegian Deep (northern North Sea). *Parasitol. Res.*, 94, 1, 2004.

35. McClelland, G. The trouble with sealworms (*Pseudoterranova decipiens* species complex, Nematoda): A review. *Parasitology*, 124, S183, 2002.

36. Herreras, M.V. et al. Anisakid larvae in the musculature of the Argentinean Hake, *Merluccius hubbsi*. *J. Food Prot.*, 63, 1141, 2000.

37. Adroher, F.J. et al. Larval anisakids (Nematoda:Ascaridoidea) in horse mackerel (*Trachurus trachurus*) from the fish market in Granada (Spain). *Parasitol. Res.*, 82, 253, 1996.

38. Rosales, J. et al. Acute intestinal anisakiasis in Spain: A fourth-stage *Anisakis simplex* larva. *Mem. Inst. Oswaldo Cruz*, 94, 823, 1999.

39. Scientific Committee, Spanish Agency of Food Security and Nutrition. La alergia por *Anisakis* y medidas de prevención. *Rev. Com. Cient. AESAN*, 1, 19, 2005. http://www.informacionconsumidor. org/Documentacioacuten/tabid/57/Default.aspx?xspc = anisakis.

40. Tejada, M. et al. Scanning electron microscopy of *Anisakis* larvae following different treatments. *J. Food Prot.*, 69, 1379, 2006.

41. Santos, A.T. et al. A method to detect the parasitic nematodes from the family Anisakidae in *Sardina pilchardus*, using specific primers of 18 S DNA gene. *Eur. Food Res. Technol.*, 222, 71, 2006.

42. Arilla, M.C. et al. An antibody-based ELISA for quantification of Ani s 1, a major allergen from *Anisakis simplex*. *Parasitology*, 135, 735, 2008.

43. Valls, A. et al. *Anisakis* allergy: An update. *Rev. Fr. Allergol. Immunol. Clin.*, 45, 108, 2005.

44. Guarneri, F., Guarneri, C., and Benvenga, S. Cross-reactivity of *Anisakis simplex*: Possible role of Ani s 2 and Ani s 3. *Int. J. Dermatol.*, 46, 146, 2007.

45. Baeza, M.L. et al. Characterization of allergens secreted by *Anisakis simplex* parasite: Clinical relevance in comparison with somatic allergens. *Clin. Exp. Allergy*, 34, 296, 2004.

46. Toro, C. et al. Seropositivity to a major allergen of *Anisakis simplex*, Ani s 1 in dyspeptic patients with *Helicobacter pylori* infection: Histological and laboratory findings and clinical significance. *Clin. Microbiol. Infect.*, 12, 453, 2006.

47. Rodríguez-Pérez, R. et al. Cloning and expression of Ani s 9, a new *Anisakis simplex* allergen. *Mol. Biochem. Parasitol.*, 159, 92, 2008.

48. Arlian, L.G. et al. Characterization of allergens of *Anisakis simplex*. Allergy, 58, 1299, 2003.

49. Moneo, I. et al. Sensitization of the fish parasite *Anisakis simplex*: Clinical and laboratory aspects. *Parasitol. Res.*, 101, 1051, 2007.

50. Ibarrola, I. et al. Expression of a recombinant protein immunochemically equivalent to the major *Anisakis simplex* allergen Ani s 1. *J. Investig. Allergol. Clin. Immunol.*, 18, 78, 2008.

51. Center for Food Safety and Applied Nutrition, U.S. Food and Drug Administration. *The Bad Bug Book: Foodborne Pathogenic Microorganisms and Natural Toxins Handbook*. International Medical Publishing, McLean, VA, 2004, Washington, D.C., 1992, Chapter 25. http://www.cfsan.fda.gov/~mow/chap25.html.

52. Bhargava, D. et al. Anisakiasis of the tonsils. *J. Laryngol. Otol.*, 110, 387, 1996.

53. Cabrera, R. and Trillo-Altamirano, M.P. Anisakidosis: ¿Una zoonosis parasitaria marina desconocida o emergente en el Perú? *Rev. Gastroenterol. Perú*, 24, 335, 2004.

54. Skírnisson, K. Hringormar berast í folk á Íslandi við neyslu á lítið elduðum fiski. *Læknablaðið*, 92, 21, 2006.

55. Kimura, S. et al. IgE response to *Anisakis simplex* compared to seafood. *Allergy*, 54, 1224, 1999.

56. Kasuya, S. and Koga, K. Significance of detection of specific IgE in *Anisakis*-related diseases. *Arerugi*, 41, 106, 1992 [in Japanese].

57. Fernández de Corres, L. et al. Prevalencia de la sensibilización a *Anisakis simplex* en tres áreas españolas, en relación a las diferentes tasas de consumo de pescado. Relevancia de la alergia a *Anisakis simplex*. *Alergol. Immunol. Clín.*, 16, 337, 2001.

58. Puente, P. et al. *Anisakis simplex*: The high prevalence in Madrid (Spain) and its relation with fish consumption. *Exp. Parasitol.*, 118, 271, 2008.

59. Sánchez-Velasco, P. et al. Association of hypersensitivity to the nematode *Anisakis simplex* with HLA class II DRB1*1502-DQB1*0601 haplotype. *Hum. Immunol.*, 61, 314, 2000.

60. Ko, R.C. Fish-borne parasitic zoonoses, in *Fish Diseases and Disorders. Vol. I. Protozoan and Metazoan Infections*. Woo, P.T.K., Ed., CAB International, Oxon, 1995, 631.

61. Schmidt, G.D. Acanthocephalan infection of man, with two new records. *J. Parasitol.*, 57, 582, 1971.

62. Williams, H. and Jones, A. *Parasitic Worms of Fish*, Taylor & Francis, London, 1994.

63. Center for Food Safety and Applied Nutrition. *Fish and Fisheries Products Hazards and Controls Guidance*, 3rd edn., U.S. Food and Drug Administration, Washington, D.C., 2001, Chapter 5. http://www.cfsan.fda.gov/~comm/haccp4.html.

64. Sunnotel, O. et al. Effectiveness of standard UV depuration at inactivating *Cryptosporidium parvum* recovered from spiked Pacific Oysters (*Crassostrea gigas*). *Appl. Environ. Microbiol.*, 73, 5083, 2007.

65. Center for Food Safety and Applied Nutrition. *Food Code* Report PB 2005-102200. U.S. Food and Drug Administration, College Park, MD, 2005, 75. http://www.cfsan.fda.gov/~dms/fc05-toc.html.

66. European Parliament and Council of the European Union. Regulation (EC) No. 853/2004 of 29 April 2004 laying down specific hygiene rules for on the hygiene of foodstuffs. *Official Journal of the European Union*, L 139/55, 30 April 2004, section VIII. http://eur-lex.europa.eu/LexUriServ/LexUriServ.do?uri=OJ:L:2004:139:0055:0205:EN:PDF.

67. Adams, A.M. et al. Survival of *Anisakis simplex* in arrowtooth flounder (*Atheresthes stomias*) during frozen storage. *J. Food Prot.*, 68, 1441, 2005.

68. European Food Safety Authority. Opinion of the Scientific Committee on veterinary measures relating to public health—Allergic reactions to ingested *Anisakis simplex* antigens and evaluation of the possible risk to human health. April 27 1998. http://ec.europa.eu/food/fs/sc/scv/out05_en.html.

69. Thu, N.D. et al. Survey for zoonotic liver and intestinal trematode metacercariae in cultured and wild fish in An Giang Province, Vietnam. *Kor. J. Parasitol.*, 45, 45, 2007.

70. Ahmed, M. Outlook for fish to 2020: A win-win-win for the oceans, fisheries and the poor? in *Fish Aquaculture and Food Security, Sustaining Fish as a Food Supply, Record of a conference conducted by the ATSE Crawford Fund*. Canberra, Australian Capital Territory, Australia, 11 August, 2004, 66.

71. Clavel A. et al. A live *Anisakis physeteris* larva found in the abdominal cavity of a woman in Zaragoza, Spain. *Jpn. J. Parasitol.*, 42, 445, 1993.

72. Clavel A. et al. Diplogonoporiasis presumably introduced into Spain: First confirmed case of human infection acquired outside the Far East. *Am. J. Trop. Med. Hyg.*, 57, 317, 1997.

73. Dixon B.R. and Flohr R.B. Fish- and shellfish-borne trematode infections in Canada. *Southeast Asian J. Trop. Med. Publ. Health*, 28, 58, 1997.

74. Macpherson, C.N.L. Human behaviour and the epidemiology of parasitic zoonoses. *Int. J. Parasitol.*, 35, 1319, 2005.

Chapter 18

Techniques of Diagnosis of Fish and Shellfish Virus and Viral Diseases

Carlos Pereira Dopazo and Isabel Bandín

Contents

18.1 Introduction: The Need for Diagnosis 532
18.2 Diagnosis: Its Definition 532
18.3 Validation of Diagnostic Tests..533
18.4 Factors Affecting the Accuracy of Diagnosis: Sample Processing....................................534
18.5 Methods of Diagnosis for Aquatic Animal Diseases 536
18.6 Clinical, Histological, and Microscopical Techniques 537
    18.6.1 Gross Signs 537
    18.6.2 Histopathology 537
    18.6.3 Immunohistochemistry...537
    18.6.4 Electron Microscopy 541
18.7 Isolation in Cell Culture 541
    18.7.1 Cell Lines and Cell Culture..541
    18.7.2 Selection of a Cell Line for Diagnosis 542
    18.7.3 Viral Isolation in Cell Culture 542
    18.7.4 Performance of the Diagnostic Procedure 543
18.8 Serological and Immunological Techniques of Diagnosis 544
    18.8.1 Scientific Basis 544
    18.8.2 Advantages and Disadvantages of the Immunological Diagnostic Tools 544

18.8.3 Description of Immune Diagnostic Procedures 547

18.8.4 Antibody Detection Diagnosis 549

18.9 Molecular Diagnosis..550

18.9.1 Scientific Basis: An Overview 550

18.9.2 Performance of the Molecular Techniques of Diagnosis: Critical
Steps and Critical Factors 555

18.9.3 Molecular Methods of Diagnosis: Brief Description of Protocols 559

18.10 Nonlethal Methods of Diagnosis.. 562

References 563

18.1 Introduction: The Need for Diagnosis

Viral diseases cause important loses in fish and shellfish aquaculture. They are especially a worrying issue for fish farmers since they can cause either high mortality short after first symptoms are discovered in a stock, and quickly spread within the farm, or low but continuous deaths that end with high cumulative mortalities. In addition, the survivors of a viral disease will, in many cases, become asymptomatic carriers and spread the disease for a long time before they are detected. Therefore, although, at least to present knowledge, they do not represent a threat to human health, viral diseases in aquaculture can compromise the otherwise unstoppable worldwide development of this industry.

On the other hand, those tools for controlling diseases caused by other agents are not available, poorly developed, or of low efficiency for viral disease. For instance, chemotherapy treatment, though available, is not affordable (for expensive) and its efficiency questionable. In addition, in spite of the important efforts in improving existing vaccines, and in the designing of new strategies for vaccination, this method for controlling viral diseases, though promising, is far from being effective.

Therefore, control of fish and shellfish diseases caused by viral agents mostly relies on access to highly sensitive, rapid, and reliable diagnostic procedures. The importance of diagnostics is unquestionable. It is focused to two main roles in aquatic animal health: (1) to determine the cause of a disease previously detected in a culture facility or fish stock and (2) to be applied in specifically designed surveillance and monitoring programs. In the first case, diagnostic is demanded by the industry to identify the cause of mortality in the farm, and to provide the appropriate tools to reduce its effect on the production. The second role can be aimed to perform epidemiology studies to determine the origin of, and/or to eradicate, a certain infection, or to demonstrate freedom from a disease or infection in a certain population or geographic zone. In this former case, diagnostics is part of a strategy to reduce risk of spreading pathogens due to national and international trade of live fish, which is well described and widely employed by different National Administrations and International Organizations [1–5].

18.2 Diagnosis: Its Definition

Under a strict point of view, a diagnostic test is applied to determine the nature of a disease. Thus, the term *diagnosis* should only be employed if the test is applied to clinical diseased individuals, whereas when applied to asymptomatic fish those tests must be considered as for *screening* instead [6,7]. Considering that in most cases the same types of tests are applied for both purposes, such definition seems to be too strict. In fact, under a wider point of view, a diagnostic test might be defined as a method, procedure, or technique that is employed for the detection and the identification of a certain agent (a virus, for instance) in a given sample (e.g., a fish tissue) and/or to determine the health status of the corresponding fish.

Definitions apart, as the final product of a diagnostic tool is the knowledge of the agent causing a disease in a fish or shellfish population, which will be the bases for important decisions regarding animal health control, that diagnostic tool must be reliable. This obvious remark (together with its applicability) is one of the most important aspects to be considered when selecting a diagnostic technique for any specific case. Therefore, parameters determining the performance of any diagnostic procedure must be defined and quantified.

18.3 Validation of Diagnostic Tests

Any test of diagnosis should be validated before it is applied for a specific purpose, and under well-defined conditions. Several parameters can be used to quantify the accuracy and the reliability of a diagnostic test.

For instance, *sensitivity* is one of the parameters most frequently employed. There are two ways of understanding this parameter [7]. Analytical sensitivity refers to the minimum amount of analyte that the test is capable to detect in a sample, and under the specific conditions assayed. It is equivalent to the *detection limit* (DL) mostly shown in reports on the design and the optimization of diagnostic methods. It provides information on the viral load threshold, independently from the level of infection of the animal. Diagnostic sensitivity or clinical sensitivity is defined as the percentage of diseased individuals that the test is capable to currently detect in a population. This kind of information is useful for field application and for surveillance programs statistics, and is therefore demanded by epidemiologists. However, its use has been scarce in the literature.

As for sensitivity, there are two ways of defining *specificity*. For the pathologists and the designers of the analytical methods of diagnosis the term specificity deals with the premise that a diagnostic method must not yield false positive results, meaning that the test must detect the agent (i.e., the specific virus for which it has been designed) only if it is actually present in the animal [8]. This is known as analytical specificity, and is the reason why in most reports (if not all) on the design or the evaluation of diagnostic procedures negative controls are included to rule out false positives due to endogenous reactivity with other analytes or chemicals, and/or unexpected reactivity with other phylogenetically related or unrelated viruses. For an epidemiologist, however, specificity is defined as the probability to correctly detect healthy individuals in a population (diagnostic specificity). In this case, it deals with a second way of understanding specificity for a pathologist. Thus, if the intention is to detect diseased fish, the causative virus must be detected independently from the viral type. Therefore, the diagnostic method must be validated against the different types (serotypes or genotypes) known for that virus.

Repeatability and reproducibility (R&R) are two parameters frequently misjudged. They are crucial to define the performance of a procedure since they are the only parameters quantifying its precision. Both deal with the probability to always obtain the same result (or the uncertainty of the obtained results). However, repeatability is defined as the precision determined under conditions where the same method and equipment are used by the same operator on a sample (equivalent to comparing results obtained from different replicas), reproducibility is the precision determined under conditions where the same method (with the same protocol and materials) but different equipment, in different days or laboratories are used by different operators. Although different statistical tests can be employed to quantify R&R [9], the most simple one is the use of the coefficient of variation between results when numeric values are yielded (e.g., viral load, or limit of detection (LD)), or the percentage of identical results when presence/absence of the virus is determined. In spite of its importance, as we will further show, R&R values are rarely calculated and provided in reports on the evaluation of diagnostic tests, mainly due to the special effort that required repetitions represent for the laboratory.

Other parameters frequently provided are *dynamic range* (range of viral concentration accurately detectable), analytical time required, cost, or applicability (which is a subjective, nonquantifiable parameter). On the other hand, other parameters (as predictive values or likelihood ratios), which are important for epidemiologist but rarely employed (or not at all) in the literature on the development of diagnostic procedures, will not be employed in the present chapter for the comparison of methods.

18.4 Factors Affecting the Accuracy of Diagnosis: Sample Processing

The reliability of the result obtained from a diagnosis does not only depend upon the performance of the diagnostic test itself. There are previous steps that strongly influence the result: sampling procedure and type of sample, conditions of transportation and conservation of samples, and sample processing and concentration. These steps will be the subject of this chapter. In addition, after the application of the analytical test two final steps such as the confirmation and the interpretation of the results must be taken into consideration. Their description and influence on the result will be tackled for each specific diagnostic procedure.

This chapter is on viral diagnosis, not on surveillance programs. Therefore, the description of the sampling methodology is not the scope of this chapter. The sampling procedure may influence the statistics of the health situation of a population under study, but not really, at least directly, the result of a test. For those really interested in sampling procedures, further reading is recommended [6,7,9,10]. We must remark that, no matter what is the kind of sample to be employed, it must be obtained from alive or moribund animals not from dead animals. On the other hand, if the viral load in a sample is critical for the efficacy of the diagnostic produce, symptomatic individuals must be chosen.

For general purposes, the Diagnostic Manual of the Office International des Épizooties (OIE, World Organization for Animal Health; [7]) recommends the sampling of whole fish larvae, or head, kidney, spleen, and encephalon from fish or hemolymph and hepatopancreas from shellfish. However, at least theoretically, depending on its target organ, for each specific virus there should be an optimum tissue or organ to be sampled for diagnosis (i.e., the one with the higher viral load). Therefore, in the literature we can find the use of almost any kind of tissue with different levels of efficiency.

The selection of a wrong organ or tissue to detect a specific virus in a specific fish or shellfish obviously influences the final efficiency of the diagnosis. For instance, using ovarian fluid to detect the viral hemorrhagic septicemia virus (VHSV, a virus which has not demonstrated vertical transmission) in an adult trout strongly reduces the chance (if any) to detect the virus, but seems to work properly for other viruses as infectious pancreatic necrosis virus (IPNV) [12] or infectious hematopoietic necrosis virus (IHNV; [11]). As another example, the use of certain organs may affect the performance of specific diagnostic tests. It is the case of using undiluted homogenates of liver or pyloric caeca to infect cell cultures—the toxicity of the homogenate might produce a false cytopathic effect (CPE) [12]—or the use of blood samples for viral detection by the polymerase chain reaction (PCR) or PCR-based methods (incorrect viral genome extraction may produce false negatives due to contamination with enzyme inhibitors).

As published elsewhere, to avoid viral inactivation and drop viral load in the tissues, samples must be transported from their original location to the diagnostic laboratory in optimal conditions. The OIE [7] stipulates that they must be stored at 4°C for no longer than 24 h after sampling (though 48 h are also acceptable). In an interesting study, Hostnik et al. [13] demonstrated the effect of the temperature of conservation of fish tissues on the detection of IHNV in cell culture (CC) and by reverse transcription (RT)-PCR. They observed that at 4°C the virus could be detected at a maximum of 3 d and 35 d, respectively, whereas maximum periods were reduced to 1 d and 8 d, respectively, at room temperature (rT).

Processing of samples is an important step with the objective of exposing the analyte to the detection system of a diagnostic test. In the case of solid samples, i.e., tissues and organs, processing begins with the homogenization of the tissue on a buffer specifically designed for cell culture. A variety of options are available but the most frequently used are Hanks' balanced salt solution (HBSS) and Earles' salt solution (ESS), which must be supplemented with antibiotics to eliminate microbial contamination (1000 μg/mL gentamicin, or 800 iu/mL penicillin plus 800 μg/mL streptomycin, and 400 iu/mL mycostatin or fungizone) [7].

For virus extraction, different methods of homogenization have been employed and published. The most frequently employed is the mortar and pestle [7,12,14–16]. Others have employed freezing and thawing [17] but its efficiency has proven not to be too high. In an old study, Agius et al. [14] demonstrated that the sonication of tissues or tissue homogenates favored the isolation of the virus in cell culture in comparison with the simple use of mortar, and more recently other authors reported similar results [12,16]. However, the application of sonication on large numbers of samples can be uncomfortable (or even harmful) for the operator, and additionally needs special equipment for protection. Other authors reported the use of trypsinization with high efficiencies of viral recovery, as demonstrated by isolation in cell culture, but the procedure can occasionally produce toxicity for the cell monolayer [12,18]. Finally, several devices have been designed with a performance as least as good as for regular homogenization with mortar [14,15], as Omnitron, Polytron, or Stomacher, which facilitate homogenization and reduce processing time for diagnosis.

After homogenization, cell debris must be removed by centrifugation and the supernatant incubated for antibiotic treatment [7] before using the viral suspension for diagnosis. Other authors, however, have employed the filtration of supernatants (instead of antibiotic treatment), though it can sometimes retain part of the viruses and thus reduce the overall performance of the diagnosis [18]. In addition, for the application of molecular diagnostic tests, the homogenate pallets can also be employed. Other kind of samples that can be employed must be processed in different ways. For instance, to detect virus or viral components in blood, sera [19–22] or different cell fractions can be chosen [21,23–25]; mucus may be simply diluted for cell culture inoculation [24] as well as ovarian fluid [7] unless it includes cavity cells [11].

Finally, if a molecular technique of diagnosis is to be applied, nucleic acid extraction must be performed. Several methods have been published as proteinase K or pronase treatments, followed by phenol–chloroform extraction and ethanol precipitation [26–28]. Commercial methods that applied the old known system of lysis of tissues by guanidine-phenol and ethanol precipitation are available, such as RNAzol, Trizol, or similar products. In addition, other methods based on nucleic acid filter capture devices are available that considerably reduce the time required for the procedure. All these methods have been used and reported in the literature. However, their performance (i.e., reliability to recover viral nucleic acid) has never been compared. In an ongoing study (unpublished data), the authors have observed that some of these methods of nucleic acid extraction show really low R&R, being strongly influenced by the equipment or the operator. Therefore, to ensure the accuracy of the extraction and hence of the overall diagnosis, the diagnostic laboratory must perform previous validation of the method of extraction to be employed.

18.5 Methods of Diagnosis for Aquatic Animal Diseases

Many analytical techniques can be applied for diagnosis in aquaculture. Some are based on the effect that the virus produces in the fish tissues or on cells; some are based on the detection of a viral component (protein or genome), and others depend on the host response to viral infection. In this chapter, we describe some of the techniques most employed in diagnosis of viral diseases of aquatic animals. They are described from a general point of view, but with references to their application to particular viruses.

How to Select a Diagnosis Method?: The selection of a technique of diagnosis for a specific purpose should be based mainly on a deep knowledge (theoretical and practical) of the procedure and factors affecting its performance. In the literature, there is a large list of reports on optimized, modified, and even on the new methods of diagnosis. However, before introducing any in the routine of a diagnostic laboratory, previous evaluation and quantification of its reliability and accuracy should be carried out under different conditions, and on different fish or shellfish species and tissues. The quantitative knowledge of the performance of all methods available in a laboratory would allow us to choose the best method for any particular case based on objective criteria. Unfortunately, in most cases, such previous validation has not been performed, and only in few reports some data are provided.

Frequently, the diagnostic techniques are grouped into "traditional" or "molecular," the former includes histopathology (HP) and microscopy, CC isolation, and sero/immune techniques. This is perhaps because they have a longer history of application and therefore experts in diagnosis feel more confident on their performance. To a certain extent this is partially true, because although the so-named traditional methods are quite standardized and have been included for long time in the recommendations of international organizations such as the OIE [7], EU [3], FDA [29], or the Australian Administration [30], only few reports on their validation are available, or they are even absent in some cases. On the other hand, the molecular methods have a relatively shorter history (of around two decades), and thus they still need an important effort for standardization. However, in this case many reports do provide quantitative data that make comparisons with other techniques quite easy. Nevertheless, the knowledge of all the factors affecting their performance is still in an ongoing process, and this must be a serious consideration for any diagnostic laboratory. In this sense, as we will show in this chapter, these kind of techniques can be extremely sensitive, at least theoretically, but the risk of false positive and/or negative results from inexperienced hands is a real threat.

In conclusion, in the absence of own objective criteria (based on experience), the best criteria is the use of the official recommendations of organisms as the OIE [7] in order to decide what technique to use, and how to apply it, for each specific case. Therefore, we will frequently reference to the OIE diagnostic manual, mainly for the traditional methods.

18.6 Clinical, Histological, and Microscopical Techniques

18.6.1 Gross Signs

The analysis of the clinical signs of the diseases (internal and external, including behavior of the affected individuals), and the consideration of the clinical history of the population, is the first step to be followed in diagnosis. Abundant information on clinical signs of each viral disease is available elsewhere [7,31–35], which may help the expert to decide what agents can be putatively affecting the population and thus representing an important support to decide what kind of analytical method must be selected. In general, this first step cannot be considered a diagnostic procedure by itself due to its low sensitivity (only disease situations are detected) and specificity: In most cases, different viruses can share similar symptoms. However, in some cases, the specificity of this method can be high because a specific virus can exclusively produce certain signs. This is the case of the white spots that appear in the body of the shrimp affected by the white spot syndrome virus (WSSV) [31,32]; the skin nodules that appear in the fish affected by the lymphocystis virus, or the spiral swimming of IPNV-infected salmonid fry [34,35].

18.6.2 Histopathology

Viral replication in the host may provoke lesions in some tissues. There is a certain relationship between the viral group and the type of alteration of the affected tissues (Table 18.1). For some virus, the detection of the specific lesions may be determinant for their diagnosis [7]. However, in most cases its sensitivity does not reach acceptable values, and specificity might be excessively low [7,36–39]. Additionally, in some cases similar histopathological signs can be the consequence of noninfectious factors, thus yielding wrong diagnostic results. Therefore, the first condition for a laboratory to introduce HP-based tests in its routine diagnosis is to demonstrate sufficient skills and experience.

The procedure for light microscope demonstration of tissue lesions is quite simple, and only requires two specific equipment: a dark field light microscope (available in any diagnostic laboratory) and a microtome (for those working with shrimp virus, this can even be avoided by employing squash mount preparations). The first step is the fixation of the tissues. Most frequently used fixative is 10% neutral buffered formalin, followed by 10% ethanol wash. For shrimp tissues, two fixatives can be employed: Davidson's AFA (alcohol, formalin, acetic) or nonacidic R-F (RNA-friendly) [7]. The fixed fish tissues must be paraffin embedded and 5 μm sections hematoxylin-eosin stained.

18.6.3 Immunohistochemistry

To confirm the histopathological analysis, the immunodetection of the agent can be applied on the tissues. For this purpose, different types of immune labeling are reported as fluorescein iso-thiocyanate (FITC) in an immunofluorescence antibody test (IFAT), or enzymatic labels as horse-radish peroxidase (HRP) or alkaline phosphatase (AP). The use of immunohistochemistry (IHC) procedures requires a previous reduction of the background due to endogenous activity [40]. Independently from the type of label employed, the most important factor is the use of specific anti-sera. Best results are obtained from the use of monoclonal antibodies (MAbs) [40–42]. However, in addition this kind of antibodies is not available for all laboratories (but we must remark that

Table 18.1   Fish and Shellfish Viruses: Host Species, Gross Signs, and Diagnostic Tools

Virus	Susceptible Species	Symptoms and Histopathology	Diagnostic Procedures
Channel catfish virus	Catfish (*Ictalurus punctatus*)	Renal tubules necrosis, exophthalmia, ascites accumulation, hemorrhages in muscle and fins.	CC, NT, IFAT, ELISA, PCR
Epizootic hematopoietic necrosis virus (EHNV)	Perch (*Perca fluviatilis*), rainbow trout (*Oncorhynchus mykiss*), sheatfish (*Silurus glanis*), catfish (*Ictalurus melas*)	Necrosis in liver, spleen, and kidney.	CC, IFAT, ELISA, EM, PCR
Infectious hematopoietic necrosis virus (IHNV)	Rainbow trout (*O. mykiss*), Atlantic salmon (*Salmo salar*), Pacific salmon (*Oncorhynchus* spp.)	Skin darkening, pale gills, edema, ascites and distended abdomen, exophthalmia, internal and external petechial hemorrhages, pseudolethargy. White color trailing fecal casts. Spinal deformities.	CC, NT, ELISA, NAH, IFAT (on imprints and on CC), RT-PCR
Infectious pancreatic necrosis virus (IPNV)	Susceptible species: Salmonid fish; host species: practically any.	Sudden and increasing mortality. Skin darkening, distended abdomen, black color trailing fecal casts, spiral swimming.	CC, NT, IFAT, ELISA, NAH, RT-PCR
Infectious salmon anemia virus (ISAV)	Atlantic salmon (*S. salar*)	Anemia, accumulation of ascites. Hepatic and kidney necrosis. Abnormally large and dark liver. Petechia in peritoneo.	CC, IHC, IFAT, RT-PCR
Lymphocystis disease virus (LDV)	Sea bass, grouper, sturgeon	Skin nodules. Cell hyperplasia.	EM, ISH, PCR
Oncorhynchus masou virus (OMV)	Pacific salmon and rainbow trout	Epithelioma around mouth and body surface. Skin ulcers. White spots in liver. Lethargy.	CC, NT, IFAT, ELISA.
Red sea bream iridovirus (RSIV)	Red sea bream (*Pagrus major*) and many other species of *Perciformes* and *Pleuronectiformes*	Lethargy, severe anemia, petechias in gills, spleen abnormally large.	CC, IFAT, PCR

Table 18.1 (continued)   Fish and Shellfish Viruses: Host Species, Gross Signs, and Diagnostic Tools

Virus	Susceptible Species	Symptoms and Histopathology	Diagnostic Procedures
Salmonid alphavirus (SAV) (including sleeping disease (SD) and pancreas disease (PD) viruses)	Atlantic salmon, common trout, sea trout (*Salmo trutta*), rainbow trout	Lesions in pancreas and muscle. Anorexia, lethargy, reduced growth. Yellow-white trailing casts. Peripheral swimming.	HP, IFAT, PCR
Spring viremia of carp virus (SVCV)	Common carp (*Cyprinus carpio carpio*), koi carp (*C. carpio koi*), silver carp (*Hypophthalmichthys molitrix*), bighead carp (*Aristichthys nobilis*), grass carp (*Ctenopharyngodon idella*), goldfish (*Carassius auratus*), tench (*Tinca tinca*), Northern pike (*Esox lucius*), sheatfish	Degeneration of the gill lamellae, ascitic fluid containing blood, inflammation of the intestines. Hemorrhagic visceral organs. Petechia in swim bladder, muscle, and fat tissue.	CC, NT, IFAT, ELISA
Viral encephalopathy and retinopathy (VER) or nervous necrosis virus (NNV)	Sea bass (*Lates calcarifer*, and *Dicentrarchus labrax*), grouper (*Epinephelus* spp.), jack (*Pseudocaranx dentex*), parrotfish (*Oplegnathus fasciatus*), puffer (*Takifugu rubripes*), and flatfish (halibut, *Hippoglossus hippoglossus*; Japanese flounder, turbot)	Retina and brain cells vacuolization. Neuronal necrosis. Abnormal swimming behavior (spiral whirling or upside down swimming).	CC, IFAT, RT-PCR
Viral hemorrhagic septicemia virus (VHSV)	Rainbow trout, pike, Japanese flounder (*Paralychthys olivaceus*), turbot (*Scophthalmus maximus*)	Skin darkening, exophthalmia, anemia (pale gills), skin, hemorrhages in fins and gills, distended abdomen, abnormal swimming. Lethargy. Rapid onset of mortality.	CC, IFAT, RT-PCR

(*continued*)

Table 18.1 (continued)   Fish and Shellfish Viruses: Host Species, Gross Signs, and Diagnostic Tools

Virus	Susceptible Species	Symptoms and Histopathology	Diagnostic Procedures
White sturgeon iridovirus (WSIV)	White sturgeon (*Acipenses transmontanus*)	Anorexia. Diffuse hyperplasia of skin. Abdominal hemorrhages.	CC, NT, IFAT
Infectious hypodermal and hematopoietic necrosis virus (IHHNV)	Shrimp (*Penaeus stylirostris*)	Irregular growth. Cuticular deformities. Anemia. Weakness, roll over movement. Motted appearance. Nuclear inclusion bodies, margination of chromatine.	Dot-blot, ISH
Taura syndrome virus (TSV)	White shrimp (*P. vannaemei*)	Lesions and necrosis in the epithelium of different body parts. Pale reddish coloration. Red tail. Soft shell.	HP, RT-PCR
White spot syndrome virus (WSSV)	Penaeid shrimp	White spots, anorexia, surface swimming. Hypertrophied nuclei.	HP, ISH, PCR
Yellow head disease virus (YHDV)	Penaeid shrimp	Yellowing of encephalothorax; clarified body. Systemic necrosis of ectodermal and mesodermal cells.	HP, RT-PCR

CC, Cell culture; ELISA, enzyme-linked immunosorbent assay; EM, electron microscope; HP, histopathology; IFAT, immunofluorescence antibody test; IHC, immunohistochemistry; ISH, in situ hybridization; NAH, nucleic acid hybridization; NT, neutralization; PCR, polymerase chain reaction; RT-PCR, reverse transcription-PCR.

for certain viruses, MAbs can be purchased from some companies), and they can yield excessive specificity. Therefore, the use of polyclonal antibodies is not rejected (though in this case certain background can be produced).

FITC, AP, and HRP protocols are quite similar, and can be applied on frozen [43] and dewaxed sections [44–46], imprints [40], squash tissues [7], or even directly on larvae [7,44]. The procedure begins in the treatment with a blocking agent, constituted by a solution of unspecific protein (normally skimmed milk), and followed by washes with buffer and incubation with the virus-specific antibody. New washes precede the treatment with the antispecific-labeled conjugate. Some authors have reported the use of biotinylated conjugates, employing in those cases, biotin-avidin AP [41], or streptavidin-FITC or HRP [46]. The final detection is performed by the addition of the corresponding substrate for AP [43,46] or HRP [40,41,45,47,48], or under UV light in a fluorescence microscope with the corresponding filters [41,43,46,48].

18.6.4 Electron Microscopy

The application of this technology on fish tissues is not too frequent due to the need of special equipment (including ultramicrotome) and special skills for processing and interpretation of results. Few reports can be found that employ staining of ultra thin section for electron microscope (EM) diagnosis [49–52]. In most cases, the use of EM is focused to the preliminary identification of a virus isolated in cell culture.

18.7 Isolation in Cell Culture

18.7.1 Cell Lines and Cell Culture

Since viruses are intracellular parasites, the unique procedure to detect them and simultaneously demonstrate that they are active and infective is to propagate them in an alive system. The old procedure (and still the only one for certain viruses) was the inoculation in experimental healthy animals (or avian eggs) to develop the disease. Later on, the use of primary cells from disaggregated tissues represented an important advantage in the study of viruses, and is still a useful tool in some cases. However, the revolution in virology (and fish virology) came from the production of CC of continuous line, or cell lines.

The use of cell cultures simplifies the propagation of virus and hence their isolation from infected samples. The only requisite is that the cell line selected for the isolation of a specific virus must be susceptible to its replication, yielding alteration and/or cell lysis in the culture monolayers, ending with the development of a specific CPE easily detected under light microscopy.

For fish viruses there is a large list of susceptible fish cell lines that have been described by many laboratories in the literature, and in most cases are available from the international culture type collection as the American Type Culture Collection (ATCC) or the European Collection of Cell Culture (ECACC). The cells can be bought in a ready-to-use monolayer, or in a frozen format (indications of the seller must be carefully followed to prepare the cell monolayers from the frozen vial). Working with CC is quite simple, and precise instructions can be found in specific manuals [53,54], but brief and useful indications can be found in the OIE diagnostic manual [7] and reported elsewhere. To culture cells, a laboratory needs (besides skills and experience) specific equipment and materials, as well as culture media and supplements. The equipment includes a sterile flow chamber and an inverted light microscope. Plastic flasks and plates specially treated to favor the adherence of cells are available from different companies with similar qualities; the best advice is to test different brands for different lines. There is a variety of culture media that can be chosen, all them sharing high concentration of basic nutrients. Most frequently employed media are the traditional Eagle's minimum essential medium (EMEM) with Earle's salt solution (ESS), and the Leibovitz L-15 medium. The OIE advises the use of amino acid and vitamin-enriched media, as the Stoker medium; however, other authors have reported the use of other media with good results. The media must be buffered with 0.16 M Tris or 0.02 M Hepes, and/or sodium bicarbonate (if closed flasks are employed). In all cases, the media must be additionally enriched with sera, normally fetal bovine sera (FBS), and supplemented with antibiotics to reduce the risk of bacterial (100 iu penicillin and 100 µg/mL streptomycin) and fungal (2 µg/mL fungizone or 50 iu/mL mycostatin) contamination.

18.7.2 Selection of a Cell Line for Diagnosis

The most important issue in this method of diagnosis is the correct selection of the susceptible cell line or lines for the target. For each virus, the OIE recommends a short number of susceptible cells, but data from other authors might help us in the selection of the most appropriate one. For its recommendation, the OIE has selected the lines with demonstrated susceptibility to any strain or type of the specific virus. In this sense, in many cases, certain cells employed by some authors show low susceptibility for some viral types, which should be avoided to reduce the risk of false negatives. For instance, in spite of some authors reported the use of Chinook salmon embryo (CHSE-214) cells for the isolation of IHNV [55], others reported that those cells may fail in the isolation of some strains [56]. As another example, in a recent study, Ogut and Reno [57] reported that fathead minnow (FHM) and epithelioma papillosum cyprini (EPC) were suitable for the isolation of American types of IPNV; however, they have demonstrated failure to develop CPE with other strains.

It is surprising the large variety of cell lines that the scientists use in the diagnosis of a specific virus in spite of the "officially" recommended one, or even in some cases using cells with demonstrated lower sensitivity than others. For example, although EPC, FHM, and CHSE-214 have been demonstrated to be of lower sensitivity to VHSV than the recommended BF-2 [7,58], some authors still report their use in diagnosis. In this sense, although they can be employed to propagate specific strains, their use is not advised in blind diagnosis, precisely to avoid false negatives due to excessive specificity.

This does not mean that research on testing new and established cell lines for the isolation of virus under different condition is not advised. On the contrary, much effort must be focused on the validation of each available cell line for the detection of any viral type of each group. As a requisite, the introduction of a new cell line in the diagnostic routine of a virus must be preceded by its testing (preferable with all the corresponding types) to determine its optimal temperature and the range of permissiveness, characteristic of the CPE and time for its development at each temperature and for all strains, and range of viral titters yielded in each case. Unfortunately, in spite of the large number of reports based on this method of diagnosis, many fail to provide important data.

18.7.3 Viral Isolation in Cell Culture

For its isolation in a cell monolayer, the virus must be previously extracted, in a suspension, from the fish tissues by any of the procedures of homogenization described earlier. However, this is not actually a strict requisite, because some authors have reported the isolation of virus from monolayers cocultivated with fractioned, disaggregated, or trypsinized fish tissues [14,23,24,59,60] with different efficiencies. In addition, fluid samples as for crude virus (viral suspensions from infected monolayers) or sera can be directly inoculated.

The cells must be inoculated before confluence. For this purpose, the culture medium must be removed, and the viral suspension incubated on the monolayer, at the corresponding optimum temperature, for an adsorption period of around 1 h. The remaining inoculum must be removed and the monolayer covered with the same culture medium but supplemented with lower percentages of FBS (normally 2%) to reduce advance of the cell growth (favoring the replication of virus in low loads or with slow replication). Afterward, the infected monolayers must be incubated at the selected optimum temperature, and daily visualized for the detection of characteristic CPE.

Although there are some CPEs quite specific for a virus, in most cases different viruses can share the cytopathic alteration. Due to that specificity failure, this diagnostic procedure must be considered exclusively for viral detection and should be followed by a method of identification.

This is also the general rule in the official diagnostic procedure in most of the cases. Thus, the OIE stipulates, for each virus, an initial protocol for viral isolation followed by a group of recommended techniques of identification (Table 18.1). A variety of techniques can be employed which will be further described. In some cases, the identification procedure can be directly applied onto the infected cells, even before a clear CPE is visualized. Those are the IHC-type techniques, as immunoperoxidase (IP) or immunofluorescence [43,44,61–63], and in situ hybridization (ISH) [59,61]. Frequently, the identification is performed on the isolated virus, using crude, concentrated, or, less frequently, purified virus. The most simple, though not conclusive, identification method is the visualization of the size and the morphology of the isolated and concentrated virus under electron microscopy [44,51,63]. However, the most frequently employed are viral neutralization [59,61,63–65] or the enzyme-linked immunosorbent assay (ELISA) in liquid or solid (immunodot-blot) phase [42,44,62,63,66–70], using specific polyclonal or monoclonal antisera.

Molecular techniques such as nucleic acid hybridization (NAH) and, in the last decade, the PCR and PCR-based procedures have been introduced as a complement of the CC isolation for identification of the isolate. [27,71–74].

18.7.4 Performance of the Diagnostic Procedure

In fish virus diagnosis, the isolation in CC is still considered, after many decades, a method of reference, not indeed due to its sensitivity/specificity but because it is the only technique that simultaneously detects the virus and confirms its infectivity. In this sense, although it is the gold standard for official organizations as the OIE [7], the EU [3,75], and the American Fisheries Society [29], and is theoretically considered to have a limit of detection (LD) of one viral particle, much is still to be known on its real sensitivity (in quantitative terms), specificity, and R&R. Regarding the first one, sensitivity of this procedure is really a "still to know" parameter. In fact, there are few reports with a real quantification of the sensitivity in term of LD, and in most cases, the supported data actually apply to the identification method: CC plus ELISA (10^1 or $10^{3.5}$ TCID$_{50}$/mL; [69,76]), CC plus electropherotyping (10^5 TCID$_{50}$/mL; [77]), CC plus ISH (0.5–1×10^3 TCID$_{50}$/mL; [17,61]) or CC plus PCR (100 TCID$_{50}$/mL; [78]).

Several authors have reported the improvement of the sensitivity of viral isolation by the use of certain substances as polyethylene glycol (PEG; [79]) or certain patented proteins [80]. However, their use is not extended and therefore more data are needed before being introduced into a routine diagnosis.

In viral isolation, all those parameters are in fact closely related. In a Delphi panel study, Bruneau et al. [81] interviewed a set of experts in diagnosis from reference laboratories on the sensitivity and specificity (in probabilistic terms) of the method for diagnosis of IPNV and IHNV, and the result was really worrying. Thus, not only the authors concluded that the sensitivity of the method is far from perfect but also most remarkable is the list of factors that the experts believe that can strongly influence sensitivity. Among them not only are included, as expected, the sampling procedure and sample processing, or the level of infection of the sample, but other factors that will also affect R&R, as the cell line and cell line age, materials, the interpretation of the results, and the staff involved in the procedure. Similar inconveniences of the method have been reported by other authors [8,82,83]. Additional factors that may strongly affect the specificity of the technique have been published as the development of defective interference particles, presence of neutralizing factors in tissue homogenates, or tissue toxicity that can yield false negative or positive results [18,84,85].

18.8 Serological and Immunological Techniques of Diagnosis

18.8.1 Scientific Basis

In the present section, we will describe the methods of diagnosis that are based on a specific antigen–antibody reaction, allowing the detection (and identification) of any of both: the antibody, by means of the sero-diagnostic techniques, or the antigen, by the immune-diagnostic techniques. Nevertheless, the procedures described here are similar independently from their application to sera- and immune-diagnosis. The techniques included in this chapter are the neutralization test (NT), the IFAT, and the immunoenzymatic assays (IEA) (including the immunodot (ID), AP and IP, and the ELISA). Although all the methods share similar scientific basis (all of them depend upon a specific reaction between the virus and its specific antibody) they show a big difference in the way used to detect such specific reaction. In this sense, in NT the antibody binds to the cell-attachment specific epitopes of the virus, therefore blocking its capacity to infect a susceptible cell (i.e., being neutralized) (Figure 18.1). This is an important difference with the remaining techniques which, to detect the specific reaction, use a label linked to the specific antibody (direct methods) or to an anti-antibody known as conjugate (indirect procedures) (Figure 18.2).

This difference is one of the causes for the different levels of sensitivity and specificity between both types of techniques, because the number of putative binding sites for fish antibodies is broader for the label-linked immune techniques in comparison with the few neutralizing epitopes generally present in a virus [86,87].

The second difference relies on the substrate employed to perform the diagnostic procedure. Whereas IFAT, IP, and AP-IEA are applied onto infected cells or tissues, ID uses nitrocellulose membranes, and ELISA microwell plastic plates. Therefore, although similar, the procedures of these label-linked-based techniques and their applicability obviously differ. In Section 18.8.2, all these techniques are approached as a unique group to analyze their advantages and disadvantages, and their application in sera- and immune-diagnosis. Section 18.8.3 focuses on the description of each method.

18.8.2 Advantages and Disadvantages of the Immunological Diagnostic Tools

The detection of a specific antigen is the objective of the immunological diagnostic tools, which use, for such purpose, a homologous antibody. The better specificity of those antibodies implies the highest specificity of the reaction, and thus the best reliability of the diagnosis. Therefore, the first critical factor in these methods is the type of antisera and the procedure to obtain it.

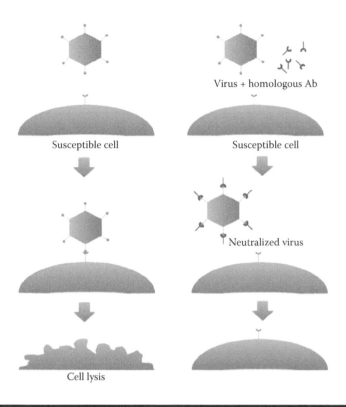

Figure 18.1 Viral neutralization. After mixing the virus with a homologous antiserum, the viral receptors are blocked by the antibodies, preventing those from attaching to the cell receptors. Therefore, the cell monolayer will not show the cytopathic effect developed in the absence of the specific antibodies.

There are two types of antisera. Polyclonal antisera are produced by means of inoculation of the virus in an animal. In fish virology, this is normally a rabbit (young New Zealand type). In using this system to obtain the antisera, the most important concern is the use of purified virus to reduce the production of nonspecific antibodies against any kind of contaminants that can be present in crude viral supernatants (i.e., cellular antigens and proteins from bovine sera). In addition, for the highest production and the quality of the polyclonal antisera, the inoculation schedule must be carefully chosen for each viral type. The description of those procedures is not the scope of this chapter. Therefore, the readers must refer to the extensive literature available. A polyclonal antiserum is actually constituted by a pool of antibodies specific for all antigenic epitopes present in the viral particles, including the neutralizing antibodies. In addition, even if the sera have been obtained from purified virus, before use they must preferably be absorbed on cell monolayers to remove unspecific antibodies [88].

On the other hand, MAbs (obtained by means of the hybridoma technology) are epitope specific. Therefore, many scientists opine that they are superior to polyclonal antisera for many applications [89] since they improve the sensitivity and the specificity of the immune-assays. However, this assumption must be carefully considered for each technique and case. In fact, some authors have reported lower sensitivity of some MAbs, or even cross-reactivity with close-related virus [90,91]. In addition, the high specificity of MAbs can actually represent a handicap because

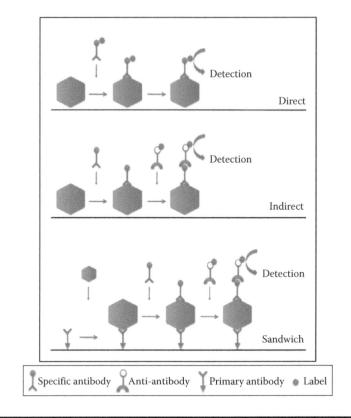

Figure 18.2 Scientific bases of the immunolabel diagnostic techniques.

certain strains of a viral type might be miss-detected [82]. This is the case reported by Ariel and Olesen [92], from the Community Reference Laboratory for Fish Diseases, who tested a set of commercial kits (based on ELISA or IFAT using MAbs) for the detection of IHNV, IPNV, Spring viremia of carp virus (SVCV), and VHSV, in comparison with their own reference methods. They observed that the only reliable one was the kit for IHNV detection, the remaining being nonspecific or too specific for some strains. However, Dixon and Longshaw [93], who tested two commercial kits for the detection of a series of fish rhabdovirus: SVCV, grass carp rhabdovirus (GCRV), pike fry rhabdovirus (PFRV), and tenca rhabdovirus, observed that, whereas the IFAT-based kit (employing a monoclonal antibody) specifically detected SVCV, the ELISA kit, that used a rabbit polyclonal antiserum, did not discriminate among the four viruses.

Many authors consider that the immunological techniques of diagnosis exhibit a level of sensitivity at least as high as the isolation in cell culture. This is especially true for those techniques that are strictly employed to identify a previously isolated virus (e.g., the neutralization test). However, in some cases these techniques can be applied on the infected cell, and detect the virus before CPE is visualized [89,90,94,95], which itself represents a clear advantage. In terms of LD, in general those techniques have shown LD values between 10^3 and 10^5 $TCID_{50}/mL$ for IPNV and VHSV, or 10^4–10^5 for IHNV [66,87,88,95–101]. For other viruses the values do not differ: $10^{3.5}$ $TCID_{50}/mL$ for epizootic hematopoietic necrosis virus (EHNV) [69] or 10^3–10^4 $TCID_{50}$ for betanodavirus [102], although some authors have reported really low DLs of 10^2 $TCID_{50}/mL$ for yellow head disease virus (YHDV) by ID [103] or ever 10–50 $TCID_{50}/mL$ for VHSV by ELISA [83]. Nevertheless, for Davis et al. [66], those values are understandable from the physical limitations of binding assays, and thus, those reported DLs of 10^2/mL should be treated with caution.

18.8.3 Description of Immune Diagnostic Procedures

Although all these procedures are well documented in the OIE manual [7], in this section we present a summary from scientific reports employing the techniques.

Neutralization: This technique is widely employed in fish virology and thus there are many reports describing the procedure for both, identification and typing of viruses [104,105], with minor differences. Briefly, 0.1 mL of the isolated virus, diluted to around 10^3–10^4 TCID$_{50}$/mL must be mixed with the serial dilutions of the specific neutralizing antisera (normally polyclonal antisera are employed) and incubated for 1 h at rT to 37°C. The dilutions are then incubated in replicate monolayers and daily visualized for the detection of CPE. Positive control reactions (reference virus assayed with the same antisera) are necessary to confirm the neutralizing activity of the sera on the isolated virus and to calculate the neutralization ratios [104]. Negative controls (with heterologous virus) are employed to ensure specificity. The technique can also be employed with a unique precalculated antiserum dilution, yielding only a qualitative result (positive or negative identification).

Immunodot: This method can be applied for identification and typing of viral isolates, and has been employed for both fish [101,104,106,107] and crustacean [103,108] viruses. Although time consuming, due to the high number of steps required, this procedure is quite simple, and no special apparatus is required. Most frequently, the "indirect" version is employed, using an enzyme-labeled anti-immunoglobulin (conjugate) specific for the virus-specific antibodies. Some authors, however, have employed the "direct" version, which reduces one-step but requires the virus-specific antibodies to be previously linked to the labeling enzyme [107]. In addition, for best performance virus-specific MAbs should be used. However, polyclonal rabbit antisera obtained from purified virus and adsorbed onto cell monolayers have also been employed with good results [104]. For the procedure, activated nitrocellulose membranes are loaded with the virus (either by pipetting or by using a vacuum filtration multiwell device). After drying, the membrane is immersed for 1 h at rT to 25°C in a blocking solution containing nonspecific protein in buffer. This solution may be 2%–5% skim milk or 1%–1.5% bovine seroalbumin (BSA) in PBS or Tris buffer. After 3–5 washes with 0.05% Tween 20 in the same buffer, the membrane is immersed for 1 h (rT to 25°C) in the virus-specific antisera, and then washed again. The conjugate (normally with HRP) is added and incubated as earlier (this step is avoided if the "direct" version is chosen), the membrane washed again and developed by the corresponding chromogenic substrate (for HRP-conjugate: 4-Cl-α-nafol plus H_2O_2, or carbazol plus H_2O_2; for substrate preparation uses the referenced reports). The results are visualized as defined dark spots in the membrane. The method can be quantitative if reference concentrations are assayed and special software and hardware are available.

Immunohistochemistry of infected cells: Labeling of the viral antigen directly on the infected cells opens the opportunity of detecting and identifying the virus in situ. The procedure is quite simple on infected monolayers [45–46,109–111], but has also been applied on infected lymphoid primary cells [112], imprints [40], and frozen [43] and deparaffinized tissue sections [45,105,108]. The initial step for tissue sections is the deparaffinization and rehydration. Infected cell cultures must be fixed using acetone (–20°C) for 10 min or Canay's solution (acetic acid:methanol; 1:3). After washing, the elimination of endogenous enzyme activity is recommended mainly if peroxidase conjugate will be further employed (treat 30 min with blocking solution: 0.3% H_2O_2 in methanol). Rinse with buffer (phosphate buffer saline, PBS, or Tris buffer saline, TBS, can be employed) and immerse the cells or tissue in serum blocking solution (10% normal goat serum, 5% skim powdered milk, or 1.5 BSA) for 20 min to 1 h, at rT to 37°C. After washing with 0.05% Tween 20 in PBS or TBS, the primary antibody is added (most frequently, the "indirect" version is employed, though the "direct" version, using HRP-labeled specific antibody can be chosen; [111]). For best performance, mouse MAbs must be employed, though polyclonal rabbit antisera can also be used. The only requisite is the right selection of the corresponding secondary antibody (antimouse or antirabbit IgG conjugates, respectively). After the incubation of the primary specific antibody for 1 h at rT to 37°C, and three to five washes, the secondary antibody (conjugate) must be added and incubated for 1 h. The cells are washed again and the chromogenic/ substrate supplemented. Different systems have been reported. Perhaps the more convenient way is using the commercial systems available, as Vector Red (Vector) or True Blue (Kirkegaard and Perry laboratories, KPL), which use AP, HRP, or biotinylated conjugates followed by the specific chromogen/substrate provided by he manufacturer. However, the traditional procedure (using AP or HRP conjugate and the corresponding substrate) can also be employed.

Immunofluorescense antibody test: This is a technique well standardized [7] and widely employed for most laboratories. It can be applied on infected CC [47,71,103,114,115], purified cells [24], imprints, [110,116], or fixed tissues [19,113], and allow the detection of the virus in the infected cells time before any CPE is visualized [94], and with a sensitivity at least as high as that of the isolation in CC [115]. In addition, it has also been designed and employed for viral quantification by the fluorescent foci counting method [117]. The procedure begins with the fixation of cell monolayers or tissues by cold acetone and rinsing with PBS or TBS buffer. Although some authors skip the blocking step, it is recommended. Therefore, the cells are immersed in a solution of unspecific protein (normally 5% skim milk) for 1 h at rT. After several rinses, the primary antibody (preferably a MAb, though polyclonal Ab can also be employed) is added for 1 h at rT to 37°C. After rinsing, the secondary Ab (FITC-conjugate anti-Ig) is added. The cells are then rinsed, mounted with glycerol, and visualized in a UV light microscope for brilliant green light foci specifically localized in the infected cells on a black or dark background.

Enzyme-linked immunosorbent assay: This is a well-recognized, standardized, and in many cases a validated method of diagnosis. It is well described for most viruses in the OIE manual [7], and has been widely described for fish and shellfish viruses in the literature [67,69,99–102,118–126]. Therefore, we recommend the reader to consider the following description of the procedure as a summary, not as a ready-to-follow protocol. Most frequently, the procedure employed is the antigen capture or sandwich ELISA (swELISA), which provides higher sensitivities and specificities than the indirect ELISA (iELISA). Moreover, frequent is the employment of MAbs as primary antisera, as it reinforces specificities.

Microwell ELISA plates are employed. In the swELISA, the wells are coated with antivirus-specific antiserum (normally polyclonal, though MAbs have also been applied by some authors) for 2 h at 37°C, or overnight (o/n) at 4°C. Then, the wells are rinsed with any of the following ELISA buffer: TNE (0.05 M Tris, 0.15 NaCl, 0.001 M EDTA), 0.1 M carbonate–bicarbonate buffer (pH 9.6), or most frequently PBS (supplemented with 0.05 Tween 20). Afterward, and before applying the viral suspension (crude virus or fish tissue extracts), the wells are blocked with ELISA buffer supplemented with unspecific protein (1.2% BSA or 3%–5% skim milk) for 1–2 h at rT to 37°C. After washing, the viral sample is added and incubated as earlier. In the iELISA version, the wells are coated with the viral sample, diluted in buffer with BSA or skim milk, by incubation o/n at 4°C. Then, the wells are rinsed and blocked as mentioned earlier. In both versions, the protocol continues with the rinsing of the wells and incubation with the specific antivirus antiserum, for 1–2 h at rT to 37°C. The wells are rinsed again and incubated with the conjugate (1–2 h/rT to

37°C), rinsed again and covered with the corresponding substrate. Two types of conjugate can be employed, which use different chromogenic substrates: AP, which uses *p*NPP (*p*-nitrophenyl phosphate) as chromogen substrate, or peroxidase, which can be complemented with different chromogens (orthophenylene diamine; tetramethylenediamine; 5-aminosalicylic acid or ABTS [2,2′-azino-di-3-ethylbenzthiazoline-6-sulphonic acid]) supplemented with H_2O_2 as the substrate for the enzyme. After incubation for 15 min to 1 h at rT, the reaction is stopped and optical density (OD) measured in a ELISA plaque reader at 450 nm (for AP) or 498 nm (for peroxidase).

18.8.4 Antibody Detection Diagnosis

In general, the capacity of antigen detection by the immune-techniques is over the viral loads characteristic of asymptomatic carrier fish [66,95]. The immediate solution relies on the previous amplification in cell culture. However, another approach is the application of sero-diagnostic techniques to detect specific antibodies in the fish sera. Crustaceans do not produce humoral response to infection, thus making these methods useless.

In fish, infection with most viruses yield a humoral immune response that, in spite of the lower complexity than in higher vertebrates, can be easily detected by means of the same techniques employed for immune diagnosis [95,97], such as plaque NT (PNT) [22,37,86], IFAT, and ELISA [36,127–131]. Nevertheless the level of antibodies in the fish depends on many factors, including the general status of the fish, stress situation, and time post-infection, which can really compromise the relative sensitivity of these methods of diagnosis [88]. On the other hand, although recognizing that these methods can be helpful in certain occasions, the OIE does not accept the use of direct diagnostic methods arguing that the serological methodology is insufficiently developed or validated to ensure the detection of specific antibodies in fish sera. In the following text, the most frequently used methods for antibody detection will be described. Since IFAT for antibody detection has been poorly reported, only PNT and ELISA will be approached here.

Plaque neutralization test: This procedure for antibody detection has been described for IHNV and VHSV [86,132,133]. As an initial step, the reference virus (against which specific antibodies are to be diagnosed in fish sera) must be freshly titrated by the plaque assay method. In addition, the heat inactivated fish sera must be mixed with fish complement for 30 min at 15°C–18°C. Then, the reference virus is added to a final concentration of 2 to 8×10^3 pfu/mL and incubated for 30 min (to o/n). After quantifying the nonneutralized virus by the plaque assay counting method, the 50% PNT is calculated as the reciprocal value of the highest serum dilution yielding 50% reduction of the average pfu with respect to the control titration.

ELISA for fish antibody detection: Two versions of the procedure have been employed [86,128,130,131,134,135]: iELISA and antigen-capture ELISA or sandwich ELISA (swELISA). In both cases, immunosorbent 96 wells plastic plates are employed. In the swELISA, the plate is initially coated with antivirus immunoglobulin (normally from rabbit) diluted in carbonate [86] or borate [132] buffer, incubated for 1–2 h, and washed with 0.05% Tween 20 in ELISA buffer (EBT) then, the procedure follows same steps as in the iELISA. Different ELISA buffer have been reported, but the most frequent is PBS.

For iELISA, the procedure starts with the coating of the well's bottom with the reference-specific virus, normally o/n at 4°C or at rT. Exclusively for iELISA, some authors have developed recombinant viral coat protein that can be used for this first step [134,135]. After washing three times with EBT, the remaining binding sites are blocked with unspecific protein for 1–2 h between rT and 37°C. Different blocking solutions have been reported with similar performance: 1% (w/v) gelatine, 1.0%–1.5% BSA, or 2%–5% skim milk. After washes, the test fish serum (previously treated 30 min at 45°C) is added (diluted in ELISA buffer) to each well and incubated rT to 37°C 1–2 h (some authors have reported shorter incubation). The wells are washed and covered with anti fish Ig (preferable monoclonal, although polyclonal antiserum has also been employed with good results). After washing, the conjugate must be added and incubated as before. For instance, if the former anti fish Ig has been obtained in mouse, a rabbit-anti-mouse-enzyme labeled conjugate must be employed. Two types of enzyme label have been employed: HRP and AP. The wells are then washed and covered with the corresponding chromogen/substrate (e.g., orthophenylenedi-amine and H_2O_2 x for HRP conjugate, or *p*NPP for AP-conjugate). After incubation at rT to 37°C (incubation time varies from 15 min to 1 h, depending on the chromogen), the reaction must be stopped (e.g., with 1 M H_2SO_4 for HRP, or 3 M NaOH for AP) and absorbance readings measured as OD at 405–492 nm in an ELISA reader.

18.9 Molecular Diagnosis

The techniques based on the detection of a specific sequence of the viral genome or mRNA (those known as molecular techniques of diagnosis) were first applied in aquaculture in the eighteens of the twentieth century. They were initially thought to substitute the traditional and sero-/immune-diagnostic tests since they were expected to yield better performance. Nevertheless, nowadays, more than two decades later, they are far from being the reference methods in fish and shellfish disease diagnosis. They are still theoretically of higher sensitivity and specificity, but other parameters such as R&R must be improved throughout the standardization of the procedures.

In this chapter, we approach to the scientific basis of this kind of techniques, their advantages and disadvantages, the parameters defining its performance, and the critical steps affecting it. Finally, a description of the different procedures is also included.

18.9.1 Scientific Basis: An Overview

In broad terms, molecular diagnostic techniques can be classified into two groups: those strictly based on the detection of a specific sequence, and those that additionally include the amplification of the target and/or signal. In the first group are included the NAH and NAH-derived procedures (e.g., ISH) and the hybridization based arrays (DNA chips). The second group is constituted by the PCR and PCR-based procedures, the nucleic acid sequence based amplification (NASBA), and the loop-mediated isothermal amplification (LAMP). There are other molecular techniques that will not be included in this chapter since they are more devoted to viral typing than to diagnosis: electropherotyping, T1 ribonuclease fingerprinting, ribonuclease protection assay (RPA), restriction fragment length polymorphism (RFLP), or genome sequencing are among them [136].

Nucleic acid hybridization (NAH): The NAH-derived techniques are based on two single-stranded nucleic acid molecules hybridize just if they exhibit a minimum sequence homology. In its use for diagnosis, the procedure includes a specific probe, which will hybridize with a homol-ogous viral sequence if present in a sample. The probe is constituted by a nucleotide genome sequence complementary to a target genome or mRNA, and is linked to a reporter molecule (biotine, fluorophor, or isotope). The probe can be obtained by cloning procedures or by PCR. The sample under examination must be processed to extract the viral genome and/or mRNA, which will be subjected to hybridization with a specific probe. At the end of the process, the presence of the reporter (demonstrated from a colorimetric reaction, fluorescent emission, or radioactivity detection) allows simultaneous detection and identification. The procedure can be performed on a nylon membrane (dot-blot hybridization; DB) or on infected cells or tissues (ISH).

The ISH is a useful technique to detect viruses in tissue imprints and squashes, and is therefore extensively applied to those viruses that cannot be grown and isolated in cell culture, as the crustacean viruses [7,137–141]. It has also been applied to fish viruses as aquatic birnaviruses [142,143], infectious salmon anemia virus (ISAV; [144]), iridovirus [145], or herpesvirus [146], in fish tissues, and also directly on infected monolayers. Regarding its application on cell culture, although the procedure is time consuming in comparison with other techniques [145], it has been demonstrated to yield less background than immunohistochemistry [142], and to let the detection of the virus in infected monolayers at shorter times p.i. than with immunofluorescence [143].

A more frequent procedure among the hybridization-derived techniques is the dot blot. In this case, the technique can be applied just for the identification of previously isolated virus [27,55,71,72,147,148] or to detect viral genomes directly from infected fish tissues [27,149–152]. In any of both cases, the extracted target nucleic acid is denatured and blotted onto a positively charged nylon membrane, which is immersed in the following solutions. Finally, the visualization of a dot is interpreted as a positive detection and identification.

PCR and PCR-based diagnostic techniques: PCR technology was developed in 1983 by the Cetus Corporation [153] and soon introduced in diagnostic laboratories. The procedure is based on the amplification of a sequence (of the viral genome or mRNA) in between a selected pair of primes which hybridize in specific positions to yield a fragment of known size. The complete process is constituted by a series of cycles consisting of a sequence of three steps corresponding to denaturing, re-annealing, and DNA polymerization. In each cycle, two size-classes of amplified products are produced: the specific fragment of the expected size, and a larger intermediate (Figure 18.3). Whereas the number of intermediate molecules increases arithmetically ($I_{c+1} = I_c + 1$, where I is the number of intermediate molecules and c the cycle number), the number of double-stranded molecules of the specific fragment increases in an exponential mode ($F_{c+1} = 2F_c + I_c$, where F is the number of specific fragments amplified). In this way, as shown in Table 18.2, amplification from a unique target molecule would yield around 10^{10} specific fragments (amplicons), all the process in about 1–3 h. Such a quantity of molecules is easily visualized, by means of a simple intercalating dye, in agarose gels. This is, together with the confirmation of the size of the fragment (by comparison with molecular size standards) the simplest way to develop the result, and for some authors it is enough for diagnosis in most cases [154]. However, to avoid the failure of the specificity due to false positives, an additional confirmatory final step is advisable. Several approaches have been reported: *Nested PCR* (Nt-PCR) is the most frequently employed since it simultaneously increases sensitivity [18,21,23,62,149,155–160]. It consists in a second round of PCR applied to the amplicon yielded in the first PCR, but using a second set of primers hybridizing in the internal positions of the first specific fragment. The detection of the second amplicon in a gel is considered as confirmatory of the diagnosis, which comes from a secondary amplification from a nonspecific primary amplification is of extremely low probability.

A second approach also commonly employed is based on the use of specific probes. They are usually employed for confirmation of the specificity of the band in agarose gel by *Southern blot* (SB; blotting of the band in a nylon membrane, and application of NAH with a labeled probe; [21,23,26,158,161–164]), but have also been employed to avoid gel electrophoresis by the detection of the PCR product by dot blot hybridization [149], or even by using a *miniarray* system with colorimetric detection of the amplicon [165]. Others have reported the use of ELISA detection of DIG-labeled amplicon (labeled during PCR amplification; [166]).

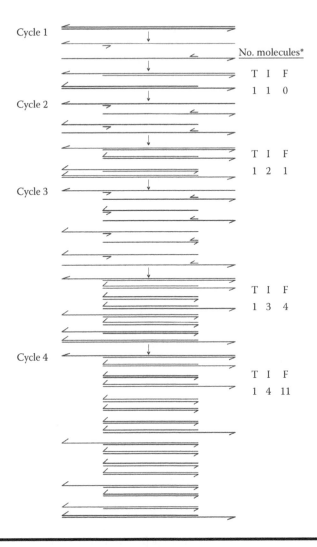

Figure 18.3 Specific sequence amplification by polymerase chain reaction. In the first cycle, extension from the primers produce two intermediate (I) chains of longer length than the specific fragment. In the following cycles, more intermediate chains are produced from the target (T) DNA in an arithmetical kinetics. From the second cycle, the specific fragments (F) will be produced in a geometrical-type kinetics. *: number of expected molecules of each type (T, I, and F) after each cycle.

Another alternative is the *sequencing* of the amplified fragment, which additionally provides phylogenetic information [120,136,139,167]. Finally, some authors have reported the use of RFLPs, for confirmatory purpose of PCR results [138,154]. However, the procedure is time consuming and cumbersome.

If the selected target is RNA, the procedure requires an initial reserve transcription step to produce the cDNA to be subjected to amplification by PCR. This can be performed in a separate reaction (2-step RT-PCR) or in a single tube (1-step RT-PCR). The performance of the diagnostic will not be necessarily affected by the procedure chosen. In addition, each version has its own advantages

Table 18.2 Number of Fragments Amplified by PCR after *n* Cycles from a Unique Target DNA

Cycle	Number of Amplified Fragments
1	0
2	1
3	4
4	11
5	26
6	57
7	120
8	247
9	502
10	1013
15	1.7×10^4
20	3.8×10^5
25	1.2×10^7
30	3.9×10^8
35	1.2×10^{10}
40	3.9×10^{11}

and thus, it is a matter of personal choice (based on experience). For instance, the 1-step RT-PCR has the advantage of reducing the risk of contamination due to manipulation. On the other hand, the 2-step version allows the use of the same RT product for PCR application for different virus.

Real-time PCR (Rt-PCR) and quantitative PCR (Qt-PCR): Although this should be included in the former section since it is also a PCR-based method of diagnosis, its peculiarities makes it to deserve a specific section. Although of quite recent development, this technology is being increasingly employed in fish and shellfish diagnosis. It is based on the labeling of the amplicon with a fluorescent reporter. The detection of the fluorescent signal (and its intensity) therefore provides information (in real time) on the level of production of specific fragments. In addition, the level of monitored signals depends on the original quantity of the target DNA, which allows to use the method not only for diagnosis, but also for quantitative purposes. The disadvantage of this method of diagnosis is the need of special equipment (not affordable for any laboratory) and skills. However, it is being increasingly employed in shellfish [20,168–170] and fish virus diagnosis including RNA viruses such as IHNV and VHSV [171–173], ISAV [143], salmonid alphavirus (SAV) [174–176], betanodavirus [163,177,178], or DNA viruses such as herpesvirus [179], and iridovirus [180,181].

Two types of procedures, depending on the method of labeling, are available. The simplest one is SYBR-Green Rt-PCR, which uses a dye (SYBR-Green) that unspecifically binds dsDNA-molecule. Therefore, the amplification must be complemented by determining the melting temperature of the amplicon to confirm if it coincides with that of the expected fragment. This procedure is easy to introduce for PCR-expert laboratories, and has produced good results in fish virus diagnosis [143,182–184]. Another method uses internal probes labeled with the fluorofor, and which is only activated throughout the hybridization of the probe to the specific amplified fragment. In this case, two alternatives exist, depending on the kind of probe employed: Molecular Beacon or TaqMan probe, though the latter one has been most frequently applied for fish and shellfish diagnosis [169,171,172,175,176,181,183,184]. The advantage of using probes is that it eliminates the need of any additional test for confirmation of the detection, thus reducing manipulation and time.

Nucleic acid sequence based amplification (NASBA): This technology was first described by Compton in the early 1990s [185]. It is also based on amplification, which ensures low DLs as for PCR, but has the advantage that the procedure is performed at constant temperature and thus the process for diagnosis is shorter because it does not waste time in raising/lowering reaction temperatures. However, it has not been frequently employed in fish viral diagnosis [186,187]. As for PCR, NASBA uses a pair of primers though one of them (secondary primers) is complemented with a T7 promoter. The amplification is based on the coordinate activity of three enzymes: reserve transcriptase, RNase H, and T7 RNA polymerase. The first step is the synthesis of a cDNA chain, and throughout the formation of the intermediate hybrid DNA–RNA, the original target RNA chain is eliminated by means of the RNase H activity. The enzyme responsible of the cycling amplification is the T7 RNA-polymerase, which synthesizes a single-strand RNA (using the secondary primer) of opposite polarity to that of the original target. The process is performed in the presence of a molecular beacon as specific internal probe. Similarly to its use for PCR, the probe is linked to two molecules, a reporter and a quencher. In its hybridized form, the probe shows a stem-loop structure and thus the fluorescence of the reporter is quenched due to this close proximity. The open form allows the emission of the fluorescence, which allow the detection of the signal and thus of the production of the specific RNA chain.

Loop-mediated isothermal amplification (LAMP): Recently a very novel technique has been developed and described [188], which allows amplification of a target DNA from a few copies to 10^9 in just 1 h. It is also based on amplification but has three main advantages: (1) it is performed under isothermal condition (at 60°C–65°C, depending on the selected primers), and therefore special reaction equipments are not needed, (2) it uses a set of four specific primers that recognize a total of six sites on the target DNA, which increases specificity, and (3) do not need secondary test or special equipment to interpret the result, because it just depends on the appearance of a white precipitate corresponding to magnesium pyrophosphate [189]. It is based on a sequence of synthesis and the displacement of strands (http://loopamp.eiken.co.jp/e/lamp/principle.html) generating different classes of stem-looped DNA molecules that can be separated in multiple bands of different sizes in agarose gel electrophoresis. The main requirement is the right selection of the set of four primers using specific software (http//primerexaplorer.jp/lamp3o.ol): A forward inner primer (fip), with sequences of the sense and antisense strand; a forward outer primer (fop); a backward inner primer (bip), with sequences of the sense and antisense strands; an outer backward primer (bop) (Figure 18.4). The process is based on the principle of autocycling strand displacement DNA synthesis, and is performed by a special DNA-polymerase, the *Bst* polymerase, which exhibits a strong strand displacement activity [189]. In its short history, this technology has been developed for diagnosis of a few fish [180,190–193] and shellfish [194,197] viruses, of both DNA [180,190,194] and RNA [191,193–195] genomes.

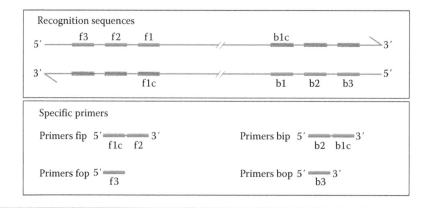

Figure 18.4 **Specific primers and recognition sites in the isothermal loop-mediated amplification (LAMP) diagnostic technique.**

18.9.2 Performance of the Molecular Techniques of Diagnosis: Critical Steps and Critical Factors

The molecular methods of diagnosis share theoretical high sensitivities and specificities. However, they are considered to yield different results among laboratories [196], which seem to indicate low reproducibility. For all these techniques, a number of critical steps and factors can compromise one or several performance parameters.

Extraction of viral nucleic acid: Independently from the method selected, all molecular diagnostic techniques (except perhaps ISH) have in common the first step. In this sense the viral genome or mRNA must be extracted from the infected cell or fresh tissues. Even for those techniques with a theoretical DL of one viral genome, if only one molecule is actually present in the sample, the method of extraction should have an absolute efficiency to facilitate its detection. There are different types of methods. The traditional ones were based on lysis by guanidine thiocyanate [197] or by proteinase K treatment [27,198] followed by phenol–chloroform extraction and ethanol precipitation. These procedures are constituted by many steps and are time consuming; therefore, nowadays they have been substituted by commercial kits of different types [69,82]. The most economical are those based on the adaptation of the traditional methods, such as the RNAzol and TriReagen LS from SIGMA, or the Trizol and Trizol LS Reagent from Invitrogen. A second group is based on lysis by guanidimicin thiocyanante and precipitation by Cs-triofluoroacetate-LiCl solutions. The third type is constituted by those methods that use lysis with guanidium thiocyanate followed by ethanol precipitation and resin-capture by silico membranes.

In a long study performed in the author's laboratory (unpublished data) to validate the methods of extraction, they have studied the recovery, the purity, and the R&R of a selection of methods and kits under different conditions to analyze the effect of (1) the storage conditions of the kit components, (2) the experience of the technician, and (3) the time of the day (at the beginning and at the end of the day), and the day of performance. From the results the authors can only conclude that there is not a unique method to advice. The traditional-based methods give the best recoveries in terms of quantity of nucleic acid, as spectrophotometrically measured. However in some cases, as when applied by inexperienced technicians, the purity of the extracted nucleic acid is quite different, and the failure of techniques as PCR is probable. On the other hand, the resin-based methods exhibit, probably due to their significant reduction of steps, higher purities and better R&R, although the recoveries are significantly lower.

Critical steps and factors in the diagnostic procedures: Once the optimum extraction method has been selected, the detection method must be also subjected to optimization. Regarding hybridization, the first parameter to optimize is the probe. In this sense, oligonucleotide probes are usually employed [55,199] because they are easy to obtain and usually provide good sensitivities. However, other authors prefer the use of longer probes, to ensure specificity though reducing sensitivity [27,72]. A critical step in HAN is the hybridization solution; therefore, special care must be evinced in testing the different reported solutions to introduce in routine diagnosis. Finally, the selection of the probe label is also important. The radioactive probes were originally used as they provided higher sensitivities than biotinylated ones [27]. However, the new chemiluminescence labeling provide high sensitivities using x-ray plaques, or even higher with special reader devices.

The number of critical steps is even higher in PCR-based techniques. The choice of primers is the first concern. The selection of primers should not be performed on a unique reference sequence, but on multi-alignments to search for conserved regions. In addition, the selected sets should be tested against a broad variety of strains from different origins and hosts, and against not related viruses. Both approaches will ensure the specificity of the primers by reducing the risk of false positives and false negatives. In the case of Rt-PCR, the selection of the set of primers can even be complicated if an internal probe must be used. Special software must be employed that does not allow the use of multi-alignments. Therefore, different sets of primers and probes should be tested. Nevertheless, to ensure detection of any strain of a virus multiplex method might be required.

There are several commercial kits available for PCR and RT-PCR, and it is not possible to select one by its theoretical characteristics because most of them have been demonstrated to perform appropriately. The best advice is the laboratory to test them, and select the one showing best performance. Once a kit has been validated, it should not be substituted for a different one before its efficiency is also confirmed. To quantify the performance of the procedure [82,154,200] and to facilitate quantitative epidemiology [200], the RT and PCR conditions, including the temperature and the time of RT and extension steps, the concentration of target nucleic acid, primers and Mg^{2+}, the number of cycles, and the length and the temperature of the steps in each cycle should be tested for each kit and virus.

Another important step is the development and the confirmation of the result. Although for some scientists visualization of the right size amplicon in a gel should be enough for giving a positive diagnosis, this approach is a serious risk for specificity and therefore any of the additional tests previously described must be included. This can be avoided by the use of Rt-PCR with internal probes. However, even in the best conditions, failure of any step, or even errors, can occur. Therefore, the use of correct controls is crucial to detect false results. For PCR or RT-PCR positive and negative controls, corresponding to nucleic acid from strains of homologous and heterologous viruses, should be subjected to the same protocol. However, this is not enough because failure could occur during the extraction step. Therefore, infected and uninfected tissues should be employed as positive and negative control, respectively.

The use of positive controls can compromise the result of simultaneously processed samples. Therefore, the ideal positive control would be that producing a PCR product amplified with the same set of primers, but giving a different size. This has been the approach by Cunningham and Hoffs [201], who designed a plasmid with an insert to be used for positive control, yielding an amplified fragment of 200 bp, easily differentiated from the 155 bp specific amplicon obtained from infected material. However, this control does not detect failures in the extraction step. Therefore, an additional approach is the amplification of a gene from the fish [202]. To avoid false results due to carryover of contaminants, a correct organization of the working areas is also important, separating the different steps and tools [82,154,203].

Limit of detection of molecular techniques: Table 18.3 gives a summary of the DLs reported for different fish and shellfish viruses. Comparing with other techniques, the DL values obtained by HAN are relatively high, perhaps due to the absolute limit of sensitivity of $10^4–10^5$ molecules by direct hybridization reported by Desselberger [203]. The values given in the table are the lowest ones; other authors have reported different values, which demonstrate that the performance of a technique depends on many factors and therefore standardization is a requirement before being able to compare protocols. For instance, from the data shown by Arakawa et al. [26] and López-Vázquez et al. [204], it seems that the base line for RT-PCR detection of IHNV (1 pg) is 10^4 times over that for VHSV (0.1 fg) in terms of the minimum detectable quantity of purified genome. However, the value of 0.1 fg reported by López-Vázquez et al. [206] was obtained after re-amplification by nested PCR. In this same study, DL observed by plain RT-PCR was 10 fg,

Table 18.3 Detection Limits of Fish and Shellfish Viruses by Different Molecular Techniques of Diagnosis

	NAH[a]		PCR/RT-PCR[b]		Rt-PCR[c]	
	na[d]	vt[e]	na	vt	na	vt
IPNV	15 ng [27]	10^4 TCID$_{50}$ [27]	15 fg [206]	100 TCID$_{50}$/mL [205]	—	—
IHNV	1 pg [55]	—	1 pg [26]	1 TCID$_{50}$ [205]	—	100 copies[h] [172]
VHSV	—	—	10 fg (1 fg w/Nt[f]) [204]	32 TCID50 [205] 1–100 inf. Units[g] [228]	—	0.5 ffu[i] [171]
SVCV	—	10^5 TCID$_{50}$/g [151]	—	10^{-1} TCID$_{50}$ [167]	—	—
ISAV	—	—	37 fg [202]	0.01–0.1 TCID$_{50}$ [229]	—	—
VERV	—	—	25 fg [207]	100 copies [159,230]	—	10 TCID$_{50}$ [178] 10 copies [176]
SAV	—	—	—	—	—	0.1 TCID$_{50}$ [177]
TRV	50 ng [72]	—	0.1 pg [231]	—	—	—
EHNV	—	—	—	1–10 pfu [162]	—	—
OMV	—	10 copies [232]	—	—	—	—
RSIV	—	—	10 fg [227]	$10^{3.5}$ pfu [227]	1 pg [182]	—
GIV	—	—	50 fg [156]	—	—	—
LDV	—	—	2.5 ng [161]	—	—	—

(continued)

Table 18.3 (continued) Detection Limits of Fish and Shellfish Viruses by Different Molecular Techniques of Diagnosis

	NAH[a]		PCR/RT-PCR[b]		Rt-PCR[c]	
	na[d]	vt[e]	na	vt	na	vt
LMBIV	—	—	—	2.7 pfu [233]	—	—
KHV	—	—	1 pg [209]	—	—	10 copies [179]
CCV	—	—	1 fg [155]	—	—	—
WSSV	—	—	100 g [212]	—	—	2 copies [169]
IHHNV	—	—	99 fg [212]	—	—	—

CCV, Channel catfish virus; EHNV, Epizootic hematopoietic necrosis virus; GIV, Grouper iridovirus; IHNV, Infectious hematopoietic necrosis virus; IHHNV, Infectious hypodermal and hematopoietic necrosis virus; IPNV, Infectious pancreatic necrosis virus; ISAV, Infectious salmon anemia virus; KHV, Koi herpesvirus; LDV, Lymphocystis disease virus; LMBIV, Largemouth bass iridovirus; OMV, *Oncorhynchus masou* virus; RSIV, Red sea bream iridovirus; SAV, Salmonid alphavirus; SVCV, Spring viremia of carp virus; TRV, Turbot aquareovirus; VERV, Viral encephalopathy and retinopathy virus; VHSV, Viral hemorrhagic septicemia virus; WSSV, White spot syndrome virus.

a Nucleic acid hybridization (NAH).
b Polymerase chain reaction or reverse transcription-PCR.
c Real-time PCR.
d Extracted nucleic acid.
e Viral titer (or na copies).
f With nested PCR.
g Infectious units.
h Copies of plasmid or of in vitro synthetized RNA.
i Fluorescent foci units.

which still is 100 times lower than that from Arakawa et al. [26]. On the other hand, in a study by Williams et al. [205] the DL for IHNV was 30 times lower. Similar levels to those for IHNV were reported by López-Lastra et al. [198] and Rimstad et al. [148] for IPNV, though Wang et al. [206] obtained 1000 times more sensitivity. Similar DL values have been reported for other RNA viruses such as ISAV (37 fg [202]) or betanodavirus (25 fg [207] or 100 fg [208]). For DNA viruses, low DL have also been reported (Table 18.3). Except for two cases of koi herpesvirus (KHV), with a DL of 1 fg by PCR [209] or, even more remarkable, the DL reported by PCR for lymphocystis disease virus (LDV) (2.5 ng [161]). For channel catfish virus (CCV), first reports gave DL values in the level of pg (0.1 pg [210]), though it has been lowered to 10 fg by RT-PCR plus SB [157], and even to 1 fg by Nt-PCR [155]. Finally, for shellfish viruses the limits seem to be between 0.1 and 1 pg [150,211–213]. In terms of minimum detectable viral titers, the best result was reported for SVCV by RT-PCR [168] and SRV by Rt-PCR [175], with a DL of 0.1 $TCID_{50}$, or WSSV [169] with a DL of two genome copies.

18.9.3 Molecular Methods of Diagnosis: Brief Description of Protocols

The aim of this chapter is not to constitute a manual or a list of protocols for each technique, but to show a general view of the procedure. For each step, different materials (enzymes, buffers, primes…) and different conditions (time and temperature of incubation) have been reported by many authors. As shown in the previous section, it is not easy to decide which protocol (if any) is actually the best one. Even, it might depend on so many factors, including the type of virus, the best advice to the diagnostician is to test several options before using one in routine diagnosis. In addition, only the techniques most frequently employed will be approached, NAH, PCR/RT-PCR, and Rt/Qt-PCR. Those interested in other techniques described earlier may consult the specific references.

Nucleic acid hybridization: Table 18.4 gives a summary of the steps of a basic protocol for dot blot hybridization. The first step is necessarily the preparation of the probe. For this purpose, two alternatives were originally available. The most common is the use of oligonucleotides [55,147,148], which have the convenience that they are easily selected with available software, and can be purchased already labeled. However as already cited, some authors believe they can be occasionally responsible of false positives [82]. The second alternative was the cloning technology [27,72] to obtain larger fragments of 200–600 bp. Fortunately, this cumbersome procedure to obtain this type of probe has been substituted by PCR, which allows the simultaneous labeling of the specific fragment, producing large quantities of probe in a short time.

The most convenient choices of reporters for probe labeling are biotin and digoxigenin, though better sensitivities are obtained with chemiluminescence. Any method for target extraction can be chosen. The unique condition is the purity of the extracted nucleic acid, which must be free of contaminated salts. Then, the sample nucleic acid must be blotted onto a nylon membrane previously soaked in 2× SSC (1× SSC: 0.15 M NaCl, 0.015 M sodium citrate). If the target is dsRNA, it must be pre-denatured by treatment at 100°C for 5 min, immediately transferred to an ice bath, and methylmercury (10 mM) added [72,82]. In the case of ssRNA, denaturing is not needed [55]. DNA targets may be pre-denatured by boiling, though it can be denatured by NaOH treatment after blotting [215]. Finally, the samples are blotted onto the membrane using a 96 wells dot-blotting minifold. The nucleic acid can be fixed by baking 1–2 h at 80°C under vacuum, or by UV light using a crosslinker.

Table 18.4 Steps in the Protocol of Nucleic Acid Hybridization

I. Probe preparation
I.1. Designing of the probe
I.2. Labeling
II. Sample preparation
II.1. The extraction of target nucleic acid
II.2. Blotting onto membrane, and denaturing
III. Hybridization
III.1. Prehybridization solution and selection of stringency
III.2. Prehybridization
III.3. Hybridization
III.4. Posthybridization washes
IV. Development of results

For hybridization, the important concerns are the stringency conditions [214] and the prehybridization solution. Stringence must be selected based on previous tests but in their absence medium level stringency can be chosen. For the prehybridization solution, Denhart's [55,71] and Hybrisol (Oncor), in combination with sonicated salmon sperm DNA or thyme DNA demonstrated good results. The membrane is subjected to prehybridization for 1–2 h at the temperature corresponding to the chosen stringency (40°C–55°C). The probe is then denatured by boiling and added to the prehybridization solution. After incubating from 2 to 24 h (usually o/n is enough) at the same temperature, the membrane is rinsed and subjected to several washes in salt concentrations specific for the selected stringency, using SSC buffer supplemented with 0.1% sodium dodecyl sulfate. Finally, the membrane is developed following the instruction of the manufacturer of the labeling kit employed.

PCR and RT-PCR: Table 18.5 enumerates the steps and parameters to consider in the designing of a PCR protocol for the diagnosis of DNA viruses. The first parameter is the quantity of DNA subjected to amplification, which normally must be around 100 ng to 1 µg, though much lower quantities (2.5 ng) have been reported [156,209]. The primers, usually designed to amplify fragments of 200–600 bp, must be supplied in a final concentration of 0.2–0.5 µM. Higher concentrations are usually responsible for unspecific bands at the bottom of the lane, which can be confusing if small specific fragments have been chosen. Lower concentration, on the other hand, may get exhausted during the cycles and cause poor amplifications. Taq DNA polymerase is added to a final concentration of 0.04 U/mL in a buffer (supplied by the manufacturer) supplemented with 1.5 M $MgCl_2$ (previous tests are advised to optimize the concentration of Mg^{2+}), and dNTPs to a final concentration normally between 0.4 and 0.6 mM, although lower values have been published.

The amplification normally starts with a previous denaturation step at 95°C for 5 min, though some laboratories skip this step [156,215], followed by a number in between 30 and 40 cycles consisting of denaturation (normally 94°C, 30 s), annealing, and extension (normally 72°C, 30 s). The

Table 18.5 Steps and Parameters to Consider in DNA Amplification

I. PCR Mix
I.1. Target DNA
I.2. Primers set
I.3. Enzyme
I.4. Buffer
I.5. MgCl$_2$/MgSO$_4$
I.6. dNTP
II. Initial denaturation
III. Cycle
III.1. Number of cycles
III.2. Denaturation
III.3. Annealing
III.4. Extension
IV. Final extension
V. The development of results

annealing temperature depends on the characteristics of primers and target DNA and, therefore must be assayed for each case. In some cases, low (46°C [166]) or high (68°C–72°C [209,216]) temperatures have been employed, but most frequently annealing is performed at 56°C–63°C for 30 s. After the cycles, a final extension period (72°C, 7 min) is advised to end incomplete synthesis.

For RNA viruses, a previous step of cDNA synthesis is required. In the case of ssRNA viruses, as for the amplification of viral mRNA, all buffers must be nuclease-free, and a supplement of RNase inhibitors is recommended [161,208,215,217]. An important parameter is the quantity of template RNA to use in the RT reaction. It is usually set at 0.5–1 µg, but 2 µg [218] or even higher values (5 µg [219]) have been reported. The author's laboratory has performed a validation of the optimum quantity of RNA (extracted by resin-based methods) for RT synthesis of IPNV, VHSV, and betanodavirus RNA, and observed that 100–200 ng of RNA (or around 10 ng/µL of reaction) is enough to obtain reliable results in the PCR amplification [23].

There is a variety of commercial reverse transcriptases available, any of them with good performance. Previous test are advised to determine the best conditions for the enzyme, including Mg^{2+} and dNTP concentration. Regarding the primers, the procedure for the RT step is slightly different depending on the chosen version. In this sense, if the 1-step RT-PCR system has been selected, the reaction mixture must be supplemented with both sense and antisense primers, to a final concentration of around 0.5–1 µM, or 10–15 pmol per reaction. For the 2-step version only one primer (sense or antisense) is needed in this first step but, some authors reported better performance by using random primers [202,219–221]. Although many authors have reported an initial denaturation step, previous to addition of the enzyme, at 90°C–98°C [23,208,215] or 65°C–70°C

[203,214,222] for 5 min, others skip this step obtaining good results [161,207,218]. Finally, the enzyme is inactivated by treatment at 95°C–99°C for 2–10 min. If using the 2-step method, the PCR mixture must be added to the RT product (or vice versa). If the 1-step version has been selected, the DNA polymerase must be thermostable to avoid inactivation during the last step. However, this is not really a concern of the user since all one-step systems provide the ready-to-use mixture of both enzymes. PCR protocols are quite similar to that cited earlier, but the diagnostician must use specific references for the virus under assay.

An important subject is the type or types of primers used for the specific amplification. As already indicated, normally sets of primers are selected by specific software on known viral sequence, but must be tested in previous assays to ensure their reliability. Special care must be taken when variants of a virus with large differences can be present. For those cases, some authors prefer the design of primers specific for each type [219], but others prefer the use of degenerate primers to ensure that mismatches will not be responsible of false negatives [115,120,222]. An alternative to ensure the detection (and/or differential diagnosis) of different viruses is the multiplex PCR /RT-PCR, using mixtures of different (normally 2) pairs of primers [207,213,217,221]. However, special care must be taken to optimize the conditions of multiplex amplification, as specificity and sensitivity can be compromised [213].

Finally, for analysis and confirmation of the diagnostic results, different approaches, other than simple gel electrophoresis, can be performed, as already indicated.

Real-time PCR: The protocol to apply this technique for diagnosis is quite similar to that of PCR except for the addition of an extra component constituted by the intercalating dye (SYBR-Green) or the internal labeled probe (TagMan or Molecular Beacon). Therefore, for precise description of the procedure we redirect the readers to those protocols reported elsewhere in the literature for fish and shellfish viruses of both, DNA [170,182,183,224] or RNA [143,171–173,175–178,183,184].

18.10 Nonlethal Methods of Diagnosis

The official directives published by the OIE and other organizations for the surveillance of diseased and healthy populations [3,7,29] are mainly based on the sacrifice of the animals for sampling of internal organisms or tissues. This can be admissible for regular surveillance and monitoring, but not when the populations under study are in-danger species or broodstocks [224]. For some scientists, the available methods of diagnosis are supposed to yield more reliable results if applied onto internal organs. However, this assumption is not necessary true.

For nonlethal sampling and diagnosis there have been different approaches. The detection of antibodies in fish serum has been used (and still is) by many authors, which for some authors might be helpful for the selection of breeders, in some cases even with better detection capacities than PCR [129]. However, the OIE does not accept this as an official method of diagnosis of fish viruses due to the partial lack of knowledge on the immune response of fish to viral infection under certain conditions [7]. Serum has also been employed for viral detection by PCR [80] or by isolation in cell culture. However, the level of virus in serum can be, in some cases, much lower than in other tissues and organs.

For breeders selection, sex products such as sperm or ovarian fluids have been used for diagnosis by means of PCR [129,160], CC isolation, or even flow cytometry [72]. However, the best results have been obtained by the detection of the virus in leukocytes by both, isolation of the virus by cocultivation of the leukocytes in cell monolayers, or by the inoculation of lysed leukocytes in CC [21,23,24,60,137], or by PCR, or RT-PCR detection [23,26,155,158,160,204,225–227]. However, although the best results of viral detection in carrier fish seem to be obtained by PCR-based technologies applied on blood samples [23,204,225], recent studies reported the lower performance of the nonlethal methods, compared with the traditional lethal diagnosis [226]. Therefore, much effort is needed to standardize and validate the methods of diagnosis to be applied on nonlethal sampling.

References

1. Anon 2000. Aquaplan zoning policy guidelines. Agriculture, Fisheries, Forestry. Canberra, Australia, p. 41.
2. EC Council Directive 91/67/EEC [and further amends] concerning the animal health conditions governing the placing on the market of aquatic animals and products.
3. EC Commission Decision 92/532/EEC [and further amends] laying down the sampling plans and diagnostic methods for the detection and confirmation of certain fish diseases.
4. FAO 2004. Surveillance and zoning for aquatic animal diseases. FAO Fisheries Technical paper 451.
5. OIE 2006. Aquatic animal health code. OIE, Paris.
6. Cameron, A. 2004. Principles for the design and conduct of surveys to show presence or absence of infectious diseases in aquatic animals. *Nat. Aquat. Anim. Health-Tech. Work G.- Policy Doc. Aus. Vet.* Animal Health Services. p. 37.
7. OIE 2000b. Manual of diagnostic tests for aquatic animals. OIE, Paris.
8. Dopazo, C.P. and Barja, J.L. 2001. A comparison between polymerase chain reaction and serological techniques for detection of fish viruses. In: *Risk Analysis in Aquatic Animal Health* Rodgers, C.J. (Ed.), World Org. An. Health (OIE), Paris. pp. 271–275.
9. Dohoo, I., Martin, W., and Stryhn, H. 2003. Veterinary epidemiology research. Prince Edward Island. AVC Ic.
10. Cameron, A.R. 2002. Survey toolbox for aquatic diseases—A practical manual and software package. Australian Center for International Agriculture Research (ACIAR). Monograph no. 94, Canberra, Australia.
11. Mulcahy, D. and Batts, W.N. 1987. Infectious hematopoietic necrosis virus detection by separation and incubation of cells from salmonid cavity fluid. *Can. J. Fish. Aquat. Sci.* 44: 1071–1075.
12. McAllister, P.E., Schill, W.B., Owens, W.J., and Hodge, D.L. 1993. Determining the prevalence of infectious pancreatic necrosis virus in asymptomatic brook trout *Salvenilus fontinalis*: A study of clinical samples and processing methods. *Dis. Aquat. Org.* 15: 157–162.
13. Hostnik, P., Barlic-Maganja, D., Strancar, M., Jencic, V., Toplak, I., and Grom, J. 2002. Influence of storage temperature on infectious hematopoietic necrosis virus detection by cell culture isolation and RT-PCR methods. *Dis. Aquat. Org.* 52: 179–184.
14. Agius, C., Richardson, A., and Walker, W. 1983. Further observations on the co-cultivation method for isolating infectious pancreatic necrosis virus from asymptomatic carrier rainbow trout, *Salmo gairdneri*, Richardson. *J. Fish Dis.* 6: 477–480.
15. Hedrick, R.P., McDowell, T., Rosemark, R., Aronstein, D., and Chan, L. 1986. A comparison of four apparatuses for recovering infectious pancreatic necrosis virus from rainbow trout. *Prog. Fish-Cultur.* 48: 47–51.
16. Smail, D.A., Burnside, K., Watt, A., and Munro, E.S. 2003. Enhanced cell culture isolation of infectious pancreatic necrosis virus from kidney tissue of carrier Atlantic salmon (*Salmo salar* L.) using sonication of the cell harvest. *Bull. Eur. Ass. Fish Pathol.* 23: 250–254.
17. Grant, R. and Smail, D.A. 2003. Comparative isolation of infectious salmon anaemia virus (ISAV) from Scotland on TO, SHK-1 and CHSE-214 cells. *Bull. Eur. Ass. Fish Pathol.* 23: 80–85.
18. Agius, C., Mangunwiryo, H., Johnson, R.H., and Smail, D.A. 1982. A more sensitive technique for isolating infectious pancreatic necrosis virus from asymptomatic carrier rainbow trout, *Salmo gairdneri* Richardson. *J. Fish Dis.* 5: 285–292.

19. Griffiths, S. and Melville, K. 2000. Non-lethal detection of ISAV in Atlantic salmon by RT-PCR using serum an mucus samples. *Bull. Eur. Ass. Fish Pathol.* 20: 157–162.

20. LaPatra, S.E. 1996. The use of serological techniques for virus surveillance and certification of finfish. *Ann. Rev. Fish Dis.* 6: 15–28.

21. Munro, E.S., Gahlawat, S.K., and Ellis, A.E. 2004. A sensitive non-destructive method for detecting IPNV carrier Atlantic salmon, *Salmo salar* L., by culture of virus from plastic adherent blood leucocytes. *J. Fish Dis.* 27: 129–134.

22. St-Hilaire, S., Ribble, C., Traxler, G., Davies, T., and Kent, M.L. 2001. Evidence for a carrier state of infectious hematopoietic necrosis virus in chinook salmon *Oncorhynchus tshawytscha*. *Dis. Aquat. Org.* 46: 173–179.

23. Cutrin, J.M., Lopez-Vazquez, C., Olveira, J.G., Castro, S., Dopazo, C.P., and Bandin, I. 2005. Isolation in cell culture and detection by PCR-based technology of IPNV-like virus from leucocytes of carrier turbot, *Scophthalmus maximus* (L.). *J. Fish Dis.* 28: 713–722.

24. Gahlawat, S.K., Munro, E.S., and Ellis, A.E. 2004. A nondestructive test for detection of IPNV-carriers in Atlantic halibut, *Hippoglossus hippoglossus* (L.). *J. Fish Dis.* 27: 233–239.

25. Gray, W.L., Williams, R.J., Jordan, R.L., and Griffin, B.R. 1999. Detection of channel catfish virus DNA in latently infected catfish. *J. Gen. Virol.* 80: 1817–1822.

26. Arakawa, C.K., Deering, R.E., Higman, K.H., Oshima, K.H., O'Hara, P.J., and Winton, J.R. 1990. Polymerase chain reaction (PCR) amplification of a nucleoprotein gene sequence of infectious hematopoietic necrosis virus. *Dis. Aquat. Org.* 8: 165–170.

27. Dopazo, C.P., Hetrick, F.M., and Samal, S.K. 1994. Use of cloned cDNA for diagnosis of infectious pancreatic necrosis virus infections. *J. Fish Dis.* 17: 1–16.

28. Wang, W.-S., Wi, Y.-L., and Lee, J.-S. 1997. Single-tube, non-interrupted reverse transcription PCR for detection of infectious pancreatic necrosis virus. *Dis. Aquat. Org.* 28: 229–233.

29. Thoesen, J. (Ed.). 1994. *Suggested Procedures for the Detection and Identification of Certain Finfish and Shellfish Pathogens*. 4th edn. Fish Health Section, American Fisheries Society. Bethesda, MD.

30. Humphrey J.D. 1995. Australian quarantine policies and practices for aquatic animals and their products: A review for the scientific working party on aquatic animal quarantine. Bureau of Resource Sciences, Canberra, Australia.

31. Aguirre Guzman, G. and Ascendio Valle, F. 2000. Infectious disease in shrimp species with aquaculture potential. *Recent Res. Dev. Microbiol.* 4: 333–348.

32. Brock, J.A. and Main, K. 1994. A guide to the common problems and diseases of cultured *Penaeus vannamei*. *Oceanic Inst.*, Makapuu Point, Honolulu, Hamaii, p. 241.

33. Nylund, A., Krossoy, B., Devold, M., Asphaug, V., Steine, N.O., and Hovland, T. 1998. Outbreak of ISA during first feeding of salmon fry (*Salmo salar*). *Bull. Eur. Assoc. Fish Pathol.* 19: 71–74.

34. Wolf, K. 1988. *Fish Viruses and Fish Viral diseases*. Cornell University Press, Ithaca, NY.

35. Woo, P.T.K. and Bruno, D.W. (Eds). 1999. *Fish Diseases and Disorders, Vol 3: Viral, Bacterial and Fungal Infections*. CABI Publishing, U.K.

36. Groocock, G.H., Getchell, R.G., Wooster, G.A., Britt, K.L., Batts, W.N., Winton, J.R., Casey, R.N., Casey, J.W., and Bowser, P.R. 2007. Detection of viral hemorrhagic septicemia in round gobies in New York. *Dis. Aquat. Org.* 76: 187–192.

37. Kwak, K.T., Gardner, I.A., Farver, T.B., and Hedrick, R.P. 2006. Rapid detection of white sturgeon iridovirus (WSIV) using a polymerase chain reaction (PCR) assay. *Aquaculture* 254: 92–101.

38. McClure, C.A., Hammell, K.L., Stryhn, H., Dohoo, I.R., and Hawkins, L.J. 2005. Application of surveillance data in evaluation of diagnostic tests for infectious salmon anemia. *Dis. Aquat. Org.* 63: 119–127.

39. Opitz, H.M., Bouchard, D., Anderson, E., Blake, S., Nicholson, B., and Keleher, W. 2000. A comparison of methods for the detection of experimentally induced subclinical infectious salmon anaemia in Atlantic salmon. *Bull. Eur. Assoc. Fish Pathol.* 20: 12–22.

40. Wilson, L., McBeath, S.J., Adamson, K.L., Cook, P.F., Ellis, L.M., and Bricknell, I.R. 2002. An alkaline phosphatase-based method for the detection of infectious salmon anaemia virus (ISAV) in tissue culture and tissue imprints. *J. Fish Dis.* 25: 615–619.

41. Evensen, O. and Lorenzen, E. 1997. Simultaneous demonstration of infectious pancreatic necrosis virus (IPNV) and *Flavobacterium psychrophilum* in paraffin-embedded specimens of rainbow trout *Oncorhynchus mykiss* fry by use of paired immunohistochemistry. *Dis. Aquat. Org.* 29: 227–232.

42. Lai, Y.-S., Chiu, H.-C., Murali, S., Guo, I.-C., Chen, S.-C., Fang, K., and Chang, C.-Y. 2001. In vitro neutralization by monoclonal antibodies against yellow grouper nervous necrosis virus (YGNNV) and immunolocalization of virus infection in yellow grouper, *Epinephelus awoara* (Temminck & Schlegel). *J. Fish Dis.* 24: 237–244.

43. Faisal, M. and Ahne, W. 1984. Spring viremia of carp virus (SVCV): Comparison of immunoperoxidase, fluorescent antibody and cell culture isolation techniques for detection of antigen. *J. Fish Dis.* 7: 57–64.

44. Dannevig, B.H., Nilsen, R., Modahl, I., Jankowska, M., Taksdal, T., and Press, C. McL. 2000. Isolation in cell culture of nodavirus from farmed Atlantic halibut *Hippoglossus hippoglossus* in Norway. *Dis. Aquat. Org.* 43: 183–189.

45. Drolet, B.S., Rohovec, J.S., and Leong, J.C. 1993. Serological identification of infectious hematopoietic necrosis virus in fixed tissue culture cells by alkaline phosphatase immunocytochemistry. *J. Aquat. Anim. Health.* 5: 265–269.

46. Hyatt, A.D., Eaton, B.T., Hengstberger, S., and Russel, G. 1991. Epizootic haematopoietic necrosis virus detection by ELISA, immunohistochemistry and immunoelectron-microscopy. *J. Fish Dis.* 14: 605–617.

47. Dannevig, B.H., Olesen, N.J., Jentoft, S., Kvellestad, A., Taksdal, T., and Haastein, T. 2001. The first isolation of a rhabdovirus from perch (*Perca fluviatilis*) in Norway. *Bull. Eur. Ass. Fish Pathol.* 21: 145–153.

48. Dixon, P.F., Hattenberger-Baudouy, A-M., and Way, K. 1994. Detection of carp antibodies to spring viraemia of carp virus by a competitive immunoassay. *Dis. Aquat. Org.* 19: 181–186.

49. Wang, C.S., Tang, K.F.J., Kou, G.H., and Chen, S.N. 1997. Light and electron microscopic evidence of white spot disease in the giant tiger shrimp, *Penaeus monodon* (Fabricius), and the kuruma shrimp, *Penaeus japonicus* (Bate), cultured in Taiwan. *J. Fish Dis.* 20: 323–331.

50. Choi, D.L., Sohn, S.G., Bang, J.D., Do, J.W., and Park, M.S. 2004. Ultrastructural identification of a herpes-like virus infection in common carp *Cyprinus carpio* in Korea. *Dis. Aquat. Org.* 61: 165–168.

51. Hyatt, A.D., Hine, P.M., Jones, J.B., Whittington, R.J., Kearns, C., Wise, T.G., Crane, M.S., and Williams, L.M. 1997. Epizootic mortality in the pilchard *Sardinops sagax* neopilchardus in Australia and New Zealand in 1995. 2. Identification of a herpesvirus within the gill epithelium. *Dis. Aquat. Org.* 28: 17–29.

52. Johnson, S.C., Sperker, S.A., Leggiadro, C.T., Groman, D.B., Griffiths, S.G., Ritchie, R.J., Cook, M.D., and Cusack, R.R. 2002. Identification and characterization of a piscine neuropathy and nodavirus from juvenile Atlantic cod from the Atlantic coast of North America. *J. Aquat. Anim. Health* 14: 124–133.

53. Doyle, A and Griffiths, J.B. (Eds.). 1998. *Cell and Tissue Culture: Laboratory Proceedings*. Wiley, New York.

54. Jakoby, W.B. and Pastan, I.H. (Eds). 1979. *Methods in Enzymology. Vol LVIII: Cell Culture*. Acad. Press Inc., United States. p. 642.

55. Deering, R.E., Arakawa, C.K., Oshima, K.H., O'Hara, P.J., Landolt, M.L., and Winton, J.R. 1991. Development of a biotinylated DNA probe for detection and identification of infectious hematopoietic necrosis virus. *Dis. Aquat. Org.* 11: 57–65.

56. Fendrick, J.L., Groberg, W.J. Jr., and Leong, H.J.C. 1982. Comparative sensitivity of five fish cell lines to wild type infectious haematopoietic necrosis virus from two Oregon sources. *J. Fish Dis.* 5: 87–95.

57. Ogut, H. and Reno, P.W. 2005. In vitro host range of aquatic birnaviruses. *Bull. Eur. Ass. Fish Pathol.* 25: 53–63.

58. Olesen, N.J. and Vestergaard Joergensen, P.E. 1992. Comparative susceptibility of three fish cell lines to Egtved virus, the virus of viral haemorrhagic septicaemia (VHS). *Dis. Aquat. Org.* 12: 235–237.

59. Nelson, R.T., McLoughlin, M.F., Rowley, H.M., Platten, M.A., and McCormick, J.I. 1995. Isolation of a toga-like virus from farmed Atlantic salmon *Salmo salar* with pancreas disease. *Dis. Aquat. Org.* 22: 25–32.

60. Yu, K.K.-Y., Macdonald, R.D., and Moore, A.R. 1982. Replication of infectious pancreatic necrosis virus in trout leucocytes and detection of the carrier state. *J. Fish Dis.* 5: 401–404.

61. Alonso, M.C., Cano, I., Castro, D., Perez-Prieto, S.I., and Borrego, J.J. 2004. Development of an in situ hybridisation procedure for the detection of sole aquabirnavirus in infected fish cell cultures. *J. Virol. Methods* 116: 133–138.

62. Shchelkunov, I.S., Popova, A.G., Shchelkunova, T.I., Oreshkova, S.F., Pichugina, T.D., Zavyalova, E.A., and Borisova, M.N. 2005. First report of spring viraemia of carp virus in Moscow Province, Russia. *Bull. Eur. Assoc. Fish Pathol.* 25: 203–211.

63. Takano, R., Mori, K.-I., Nishizawa, T., Arimoto, M., and Muroga, K. 2001. Isolation of viruses from wild Japanese flounder *Paralichthy olivaceus*. *Fish Pathol.* 36: 153–160.

64. Takano, R., Nishizawa, T., Arimoto, M., and Muroga, K. 2000. Isolation of viral haemorrhagic septi-caemia virus (VHSV) from wild Japanese flounder, *Paralichthys olivaceus*. *Bull. Eur. Ass. Fish Pathol.* 20: 186–192.

65. Varvarigos, P. and Way, K. 2002. First isolation and identification of the infectious pancreatic necro-sis (IPN) virus from rainbow trout *Oncorhynchus mykiss* fingerlings farmed in Greece. *Bull. Eur. Ass. Fish Pathol.* 22: 195–200.

66. Davis, P.J., Laidler, L.A., Perry, P.W., Rossington, D., and Alcock, R. 1994. The detection of infec-tious pancreatic necrosis virus in asymptomatic carrier fish by an integrated cell-culture and ELISA technique. *J. Fish Dis.* 17: 99–110.

67. Dixon, P.F. and Hill, B.J. 1983. Rapid detection of infectious pancreatic necrosis virus (IPNV) by the enzyme-linked immunosorbent assay (ELISA). *J. Gen. Virol.* 64: 321–330.

68. Way, K. 1996. Development of a rapid dipstick-format field test for VHS virus. *Bull. Eur. Ass. Fish Pathol.* 16: 58–62.

69. Whittington, R.J. and Steiner, K.A. 1993. Epizootic haematopoietic necrosis virus (EHNV): Improved ELISA for detection in fish tissues and cell cultures and an efficient method for release of antigen from tissues. *J. Virol. Methods* 43: 205–230.

70. Maheshkumar, S., Goyal, S.M., and Economon, P.P. 1991. Evaluation of a water concentration method for the detection of infectious pancreatic necrosis virus in a fish hatchery. *J. Appl. Ichthyol.* 7: 115–119.

71. Gonzalez, M.P., Sanchez, X., Ganga, M.A., Lopez-Lastra, M., Jashes, M., and Sandino, A.M. 1997. Detection of the infectious hematopoietic necrosis virus directly from infected fish tissues by dot blot hybridization with a non-radioactive probe. *J. Virol. Methods* 65: 273–279.

72. Lupiani, B., Subramanian, K., Hetrick, F.M., and Samal, S.K. 1993. A genetic probe for identification of the turbot aquareovirus in infected cell cultures. *Dis. Aquat. Org.* 15: 187–192.

73. McClenahan, S.D., Beck, B.H., and Grizzle, J.M. 2005. Evaluation of cell culture methods for detec-tion of largemouth bass virus. *J. Aquat. Anim. Health* 17: 365–372.

74. Wang, W.-S., Lee, J.-S., Shieh, M.-T., Wi, Y.-L., Huang, C.-J., and Chien, M.-S. 1996. Detection of infectious hematopoietic necrosis virus in rainbow trout *Oncorhynchus mykiss* from an outbreak in Taiwan by serological and polymerase chain reaction assays. *Dis. Aquat. Org.* 26: 237–239.

75. EC 2000. European Commission Decision SANCO/1965/2000.

76. Sanz, F. and Coll, J. 1992. Techniques for diagnosing viral diseases of salmonid fish. *Dis. Aquat. Org.* 13: 211–223.

77. Ganga, M.A., Gonzalez, M.P., Lopez-Lastra, M., and Sandino, A.M. 1994. Polyacrylamide gel electrophoresis of viral genomic RNA as a diagnosis method for infectious pancreatic necrosis virus detection. *J. Virol. Methods* 50: 227–236.

78. Mikalsen, A.B., Teig, A., Helleman, A.-L., Mjaaland, S., and Rimstad, E. 2001. Detection of infectious salmon anaemia virus (ISAV) by RT-PCR after cohabitant exposure in Atlantic salmon *Salmo salar. Dis. Aquat. Org.* 47: 175–181.

79. Batts, W.N. and Winton, J.R. 1989. Enhanced detection of infectious hematopoetic necrosis virus and other fish viruses by pretreatment of cell monolayers with polyethylene glycol. *J. Aquat. Anim. Health* 1: 284–290.

80. Perez, L., Mas, V., Coll, J., and Estepa, A. 2002. Enhanced detection of viral hemorrhagic septicemia virus (a salmonid rhabdovirus) by pretreatment of the virus with a combinatorial library-selected peptide. *J. Virol. Methods* 106: 17–23.

81. Bruneau, N.N., Thorburn, M.A., and Stevenson, R.M.W. 1999. Use of the Delphi panel method to assess expert perception of the accuracy of screening test systems for infectious pancreatic necrosis virus and infectious hematopoietic necrosis virus. *J. Aquat. Anim. Health* 11: 139–147.

82. Dopazo, C.P. and Barja, J.L. 2002. Diagnosis and identification of IPNV in salmonids by molecular methods. In: *Molecular Diagnosis of Salminid Diseases* Cunningan, C. (Ed.), Kluwer Acad. Publ. Holanda. pp. 23–48.

83. McAllister, P.E. 1997. Susceptibility of 12 lineages of chinook salmon embryo cells (CHSE-214) to four viruses from salmonid fish: Implications for clinical assay sensitivity. *J. Aquat. Anim. Health* 9: 291–294.

84. Dikkeboom, A.L., Radi, C., Toohey-Kurth, K., Marcquenski, S., Engel, M., Goodwin, A.E., Way, K., and Stone, D.M. 2004. First report of spring viremia of carp virus (SVCV) in wild common carp in North America. *Aquat. Anim. Health* Vol: 169–178.

85. Kennedy, J.C. and MacDonald, R.D. 1982. Persistent infection with infectious pancreatic necrosis virus mediated by defective-interfering (DI) virus particles in a cell line showing strong interference but little DI replication. *J. Gen. Virol.* 58: 361–371.

86. Fregeneda-Grandes, J.M. and Olesen, N.J. 2006. Detection of rainbow trout antibodies against viral haemorrhagic septicaemia virus (VHSV) by neutralisation test is highly dependent on the virus isolate used. *Dis. Aquat. Org.* 74: 151–158.

87. Lorenzen, N. and LaPatra, S.E. 1999. Immunity to rhabdoviruses in rainbow trout: The antibody response. *Fish Shellfish Immunol.* 9: 345–360.

88. Winton, J.R. and Einer-Jensen, K. 2002. Molecular diagnosis of infectious hematopoietic necrosis and viral hemorrhagic septicemia. In: *Molecular Diagnosis of Salminid Diseases* Cunningan, C. (Ed.), Kluwer Acad. Publ. Holanda. pp. 49–79.

89. Winton, J.R. 1991. Recent advances in detection and control of infectious hematopoietic necrosis virus in aquaculture. *Annu. Rev. Fish Dis.* 1: 83–93.

90. Enzmann, P.J., Benger, G., Bruchhof, B., and Hoffmeister, K. 1991. Cross-reactions between VHS-V and IHN-V detected by polyclonal immune sera and monoclonal antibodies. In: *Book of Abstracts. Diseases of Fish and Shellfish. Fifth Int. Conf. EAFP*, Budapest, Hungary, p. 36.

91. Ristow, S.S. and Arnzen, J.M. 1989. Development of monoclonal antibodies that recognize a type-2 specific and common epitope on the nucleoprotein of infectious hematopoietic necrosis virus. *J. Aquat. Anim. Health* 1: 119–125.

92. Ariel, E. and Olesen, N.J. 2001. Assessment of a commercial kit collection for diagnosis of the fish viruses: IHNV, IPNV, SVCV and VHSV. *Bull. Eur. Assoc. Fish Pathol.* 21: 6–11.

93. Dixon, P.F. and Longshaw, C.B. 2005. Assessment of commercial test kits for identification of spring viraemia of carp virus. *Dis. Aquat. Org.* 67: 25–29.

94. Arnzen, J.M., Ristow, S.S., Hesson, C.P., and Lientz, J. 1991. Rapid fluorescent antibody tests for infectious hematopoietic necrosis virus (IHNV) utilizing monoclonal antibodies to the nucleoprotein and glycoprotein. *J. Aquat. Anim. Health* 3: 109–113.

95. Sanz, F.A. and Coll, J.M. 1992. Detection of hemorrhagic septicemia virus of salmonid fishes by use of an enzyme-linked immunosorbent assay containing high sodium chloride concentration and two non-competitive monoclonal antibodies against early viral nucleoproteins. *Am. J. Vet. Res.* 53: 897–903.

96. Dominguez, J., Hedrick, R.P., and Sanchez-Vizcaino, J.M.1990. Use of monoclonal antibodies for detection of infectious pancreatic necrosis virus by the enzyme-linked immunosorbent assay (ELISA). *Dis. Aquat. Org.* 8: 157–163.

97. McAllister, P.E. and Schill, W.B. 1986. Immunoblot assay: A rapid and sensitive method for identification of salmonid fish viruses. *J. Wildl. Dis.* 22: 468–474.

98. Mourton, C., Bearzotti, M., Piechaczyk, M., Paolucci, F., Bastide, J.-M., and Kinkelin, P. De. 1990. Antigen-capture ELISA for viral haemorrhagic septicaemia virus serotype I. *J. Virol. Methods* 29: 325–334.

99. Rodak, L., Pospisil, Z., Tomanek, J., Vesely, T., Obr, T., and Valicek, L. 1988. Enzyme-linked immunosorbent assay (ELISA) detection of infectious pancreatic necrosis virus (IPNV) in culture fluids and tissue homogenates of the rainbow trout, *Salmo gairdneri* [*Oncorhynchus mykiss*] Richardson. *J. Fish Dis.* 11: 225–235.

100. Way, K. and Dixon, P.F. 1988. Rapid detection of VHS and IHN viruses by the enzyme-linked immunosorbent assay (ELISA). *J. Appl. Ichthyol.* 4: 182–189.

101. Ristow, S.S., Lorenzen, N., and Joergensen, P.E. 1991. Monoclonal-antibody-based immunodot assay distinguishes between viral hemorrhagic septicemia virus (VHSV) and infectious hematopoietic necrosis virus (IHNV). *J. Aquat. Anim. Health* 3: 176–180.

102. Fenner, B.J., Du, Q., Goh, W., Thiagarajan, R., Chua, H.K., and Kwang, J. 2006. Detection of betanodavirus in juvenile barramundi, *Lates calcarifer* (Bloch), by antigen capture ELISA. *J. Fish Dis.* 29: 423–432.

103. Lu, Y., Tapay, L.M., and Loh, P.C. 1996. Development of a nitrocellulose-enzyme immunoassay for the detection of yellow-head virus from penaeid shrimp. *J. Fish Dis.* 19: 9–13.

104. Dopazo, C.P., Bandín, I., Rivas, C., Cepeda, C., and Barja, J.L. 1996. Antigenic differences among aquareoviruses correlate with previously established gene groups. *Dis. Aquat. Org.* 26: 159–162.

105. Okamoto, N., Sano, T., Hedrick, R.P., and Fryer, J.L. 1983. Antigenic relationships of selected strains of infectious pancreatic necrosis virus and European eel virus. *J. Fish Dis.* 6: 19–25.

106. Hsu, Y.-L., Chiang, S.-Y., Lin, S.-T., and Wu, J.-L. 1989. The specific detection of infectious pancreatic necrosis virus in infected cells and fish by the immuno dot blot method. *J. Fish Dis.* 12: 561–571.

107. Schultz, C.L., McAllister, P.E., Schill, W.B., Lidgerding, B.C., and Hetrick, F.M. 1989. Detection of infectious hematopoietic necrosis virus in cell culture fluid using immunoblot assay and biotinylated monoclonal antibody. *Dis. Aquat. Org.* 7: 31–37.

108. Carr, W.H., Sweeney, J.N., Nunan, L., Lightner, D.V., Hirsch, H.H., and Reddington, J.J. 1996. The use of an infectious hypodermal and hematopoietic necrosis virus gene probe serodiagnostic field kit for the screening of candidate specific pathogen-free *Penaeus vannamei* broodstock. *Aquaculture* 147: 1–8.

109. McBeath, S.J., Ellis, L.M., Cook, P.F., Wilson, L., Urquhart, K.L., and Bricknell, I.R. 2006. Rapid development of polyclonal antisera against infectious salmon anaemia virus and its optimization and application as a diagnostic tool. *J. Fish Dis.* 29: 293–300.

110. Shieh, J.R. and Chi, S.C. 2005. Production of monoclonal antibodies against grouper nervous necrosis virus (GNNV) and development of an antigen capture ELISA. *Dis. Aquat. Org.* 63: 53–60.

111. Vesely, T., Nevorankova, Z., Hulova, J., Reschova, S., and Pokorova, D. 2004. Monoclonal antibodies to nucleoprotein and glycoprotein of the virus of infectious haematopoietic necrosis of salmonids (INHV) and their use in immunoperoxidase test. *Bull. Eur. Ass. Fish Pathol.* 24: 218–230.

112. Shih, H.H. 2002. Detection and titration of white spot syndrome virus using a Blue-Cell ELISA. *J. Fish Dis.* 25: 185–189.

113. Shih, H.H., Wang, C.S., Tan, L.F., and Chen, S.N. 2001. Characterization and application of monoclonal antibodies against white spot syndrome virus. *J. Fish Dis.* 24: 143–150.

114. Lorenzen, N., Olesen, N.J., and Jorgensen, P.E. 1988. Production and characterization of monoclonal antibodies to four egtved virus structural proteins. *Dis. Aquat. Org.* 4: 35–42.

115. McClenahan, S.D., Grizzle, J.M., and Schneider, J.E. 2005. Evaluation of unpurified cell culture supernatant as template for the polymerase chain reaction (PCR) with largemouth bass virus. *J. Aquat. An. Health* 17: 191–196.

116. Nakajima, K., Maeno, Y., Fukudome, M., Fukuda, Y., Tanaka, S., Matsuoka, S., and Sorimachi, M. 1995. Immunofluorescence test for the rapid diagnosis of red sea bream iridovirus infection using monoclonal antibody. *Fish Pathol.* 30: 115–119.

117. Espinoza, J.C. and Kuznar, J. 2002. Rapid simultaneous detection and quantitation of infectious pancreatic necrosis virus (IPNV)- *J. Virol. Methods* 105: 81–85.

118. Arimoto, M., Mushiake, K., Mizuta, Y., Nakai, T., Muroga, K., and Furusawa, I. 1992. Detection of striped jack nervous necrosis virus (SJNNV) by enzyme-linked immunosorbent assay (ELISA). *Fish Pathol.* 27: 191–195.

119. Chen, Z.J., Wang, C.S., and Shih, H.H. 2002. An assay for quantification of white spot syndrome virus using a capture ELISA. *J. Fish Dis.* 25: 249–251.

120. Dixon, P.F., Feist, S., Kehoe, E., Parry, L., Stone, D.M., and Way, K. 1997. Isolation of viral haemorrhagic septicaemia virus from Atlantic herring *Clupea harengus* from the English Channel. *Dis. Aquat. Org.* 30: 81–89.

121. Lewis, D.H. 1986. An enzyme-linked immunosorbent assay (ELISA) for detecting penaeid baculovirus. *J. Fish Dis.* 9: 519–522.

122. Medina, D.J., Chang, P.W., Bradley, T.M., Yeh, M.-T., and Sadasiv, E.C. 1992. Diagnosis of infectious hematopoietic necrosis virus in Atlantic salmon *Salmo salar* by enzyme-linked immunosorbent assay. *Dis. Aquat. Org.* 13: 147–150.

123. Munro, J. and Owens, L. 2006. Sensitivity and specificity of current diagnostic tests for gill-associated virus in *Penaeus monodon. J. Fish Dis.* 29: 649–655.

124. Romestand, B. and Bonami, J.-R. 2003. A sandwich enzyme linked immunosorbent assay (S-ELISA) for detection of MrNV in the giant freshwater prawn, *Macrobrachium rosenbergii* (de Man). *J. Fish Dis.* 26: 71–75.

125. Way, K., Bark, S.J., Longshaw, C.B., Denham, K.L., Dixon, P.F., Feist, S.W., Gardiner, R. et al. 2003. Isolation of a rhabdovirus during outbreaks of disease in cyprinid fish species at fishery sites in England. *Dis. Aquat. Org.* 57: 43–50.

126. Rodak, L., Pospisil, Z., Tomanek, J., Vesely, T., Obr, T., and Valicek, L. 1993. Enzyme-linked immunosorbent assay (ELISA) for the detection of spring viraemia of carp virus (SVCV) in tissue homogenates of the carp, *Cyprinus carpio* L. *J. Fish Dis.* 16: 101–111.

127. Adkison, M.A., Gilad, O., and Hedrick, R.P. 2005. An enzyme linked immunosorbent assay (ELISA) for detection of antibodies to the koi herpesvirus (KHV) in the serum of koi *Cyprinus carpio. Fish Pathol.* 40: 53–62.

128. Crawford, S.A., Gardner, I.A., and Hedrick, R.P. 1999. An enzyme-linked immunosorbent assay (ELISA) for detection of antibodies to channel catfish virus (CCV) in channel catfish. *J. Aquat. Anim. Health* 11: 148–153.

129. Mushiake, K., Nishizawa, T., Nakai, T., Furusawa, I., and Muroga, K. 1994. Control of VNN in striped jack: Selection of spawners based on the detection of SJNNV gene by polymerase chain reaction (PCR). *Fish Pathol.* 29: 177–182.

130. Van Nieuwstadt, A.P., Dijkstra, S.G., and Haenen, O.L.M. 2000. Persistence of herpesvirus of eel *Herpesvirus anguillae*, in farmed European eel *Anguilla anguilla. Dis. Aquat. Org.* 45: 103–107.

131. Whittington, R.J., Philbey, A., Reddacliff, G.L., and Macgown, A.R. 1994. Epidemiology of epizootic haematopoietic necrosis virus (EHNV) infection in farmed rainbow trout, *Oncorhynchus mykiss* (Walbaum): Findings based on virus isolation, antigen capture ELISA and serology. *J. Fish Dis.* 17: 205–218.

132. Ristow, S.S., de Avila, J., LaPatra, S.E., and Lauda, K. 1993. Detection and characterization of rainbow trout antibody against infectious hematopoietic necrosis virus. *Dis. Aquat. Org.* 15: 109–114.

133. LaPatra, S.E., Turner, T., Lauda, K.A., Jones, G.R., and Walker, R. 1993. Characterization of the humoral response of rainbow trout to infectious hematopoietic necrosis virus. *J. Aquat. Anim. Health* 5: 165–171.

134. Huang, B., Tan, C., Chang, S.F., Munday, B., Mathew, J.A., Ngoh, G.H., and Kwang, J. 2001. Detection of nodavirus in barramundi, *Lates calcarifer* (Bloch), using recombinant coat protein-based ELISA and RT-PCR. *J. Fish Dis.* 24: 135–141.

135. Watanabe, K.-I., Nishizawa, T., and Yoshimizu, M. 2000. Selection of brood stock candidates of barfin flounder using an ELISA system with recombinant protein of barfin flounder nervous necrosis virus. *Dis. Aquat. Org.* 41: 219–223.

136. Cunningan, C. (Ed.). 2002. *Molecular Diagnosis of Salminid Diseases.* Kluwer Acad. Publ. Holanda. ELIMINABLE.

137. Durand, S., Lightner, D.V., Nunan, L.M., Redman, R.M., Mari, J., and Bonami, J.-R. 1996. Application of gene probes as diagnostic tools for white spot baculovirus (WSBV) of penaeid shrimp. *Dis. Aquat. Org.* 27: 59–66.

138. Hasson, K.W., Lightner, D.V., Mari, J., Bonami, J.R., Poulos, B.T., Mohney, L.L., Redman, R.M., and Brock, J.A. 1999. The geographic distribution of Taura syndrome virus (TSV) in the Americas: Determination by histopathology and in situ hybridization using TSV-specific cDNA probes. *Aquaculture* 171: 13–26.

139. Hsieh, C.-Y., Wu, Z.-B., Tung, M.-C., Tu, C., Lo, S.-P., Chang, T.-C., Chang, C.-D., Chen, S.-C., Hsieh, Y.-C., and Tsai, S.-S. 2006. In situ hybridization and RT-PCR detection of *Macrobrachium rosenbergii* nodavirus in giant freshwater prawn, *Macrobrachium rosenbergii* (de Man), in Taiwan. *J. Fish Dis.* 29: 665–671.

140. Li, C., Shields, J.D., Small, H.J., Reece, K.S., Hartwig, C.L., Cooper, R.A., and Ratzlaff, R.E. 2006. Detection of *Panulirus argus* Virus 1 (PaV1) in the Caribbean spiny lobster using fluorescence in situ hybridization (FISH). *Dis. Aquat. Org.* 72: 185–192.

141. Wang, C.S., Tsai, Y.J., Chen, S.N. 1996. Detection of white spot disease virus (WSDV) infection in shrimp using in situ hybridization. *J. Invertebr. Pathol.* 72: 170–173.

142. Biering, E. and Bergh, O. 1996. Experimental infection of Atlantic halibut, *Hippoglossus hippoglossus* L., yolk-sac larvae with infectious pancreatic necrosis virus: Detection of virus by immunohistochemistry and in situ hybridization. *J. Fish Dis.* 19: 405–413.

143. Munir, K. and Kibenge, F.S. 2004. Detection of infectious salmon anaemia virus by real-time RT-PCR. *J. Virol. Methods* 117: 37–47.

144. Gregory, A. 2002. Detection of infectious salmon anaemia virus (ISAV) by in situ hybridisation. *Dis. Aquat. Org.* 50: 105–110.

145. Huang, C., Zhang, X., Gin, K.Y., and Qin, Q.W. 2004. In situ hybridization of a marine fish virus, Singapore grouper iridovirus with a nucleic acid probe of major capsid proteíni. *J. Virol. Methods* 117: 123–128.

146. Sano, N., Sano, M., Sano, T., and Hondo, R. 1992. *Herpesvirus cyprini*: Detection of the viral genome by in situ hybridization. *J. Fish Dis.* 15: 153–162.

147. Batts, W.N., Arakawa, C.K., Bernard, J., and Winton, J.R. 1993. Isolates of viral hemorrhagic septicemia virus from North America and Europe can be detected and distinguished by DNA probes. *Dis. Aquat. Org.* 17: 67–71.

148. Rimstad, E., Krona, R., Hornes, E., Olsvik, Oe., and Hyllseth, B. 1990. Detection of infectious pancreatic necrosis virus (IPNV) RNA by hybridization with an oligonucleotide DNA probe. *Vet. Microbiology* 23: 211–219.

149. De la Rosa-Velez, J., Cedano-Thomas, Y., Cid-Becerra, J., Mendez-Payan, J.C., Vega-Perez, C., Zambrano-Garcia, J., and Bonami, J.-R. 2006. Presumptive detection of yellow head virus by reverse transcriptase-polymerase chain reaction and dot-blot hybridization in *Litopenaeus vannamei* and *L. stylirostris* cultured on the Northwest coast of Mexico. *J. Fish Dis.* 29: 717–726.

150. Lu, C.C., Tang, K.F.J., Kou, G.H., and Chen, S.N. 1993. Development of a Penaeus monodon-type baculovirus (MBV) DNA probe by polymerase chain reaction and sequence analysis. *J. Fish Dis.* 16: 551–559.

151. Oreshkova, S.F., Shchelkunov, I.S., Tikunova, N.V., Shchelkunova, T.I., Puzyrev, A.T., and Ilyichev, A.A. 1999. Detection of spring viremia of carp virus isolates by hybridization with non-radioactive probes and amplification by polymerase chain reaction. *Virus Res.* 63: 3–10.

152. Wise, J.A. and Boyle, J.A. 1985. Detection of channel catfish virus in channel catfish, *Ictalurus punctatus* (Rafinesque): Use of a nucleic acid probe. *J. Fish Dis.* 8: 417–424.

153. Saiki, R.K., Scharf, S., Fallona, F.A., Mullis, K.B., Horn, G.T., Erlich, H.A., and Arnheim, N. 1985. Enzymatic amplification of b-globin genomic sequences and restriction site analysis for diagnosis of sickle cell anemia. *Science* 230: 1350–1354.

154. Mjaaland, S., Rimstad, E., and Cunnigham, C. 2002. Molecular diagnosis of infectious salmon anaemia. In: *Molecular Diagnosis of Salmind Diseases* Cunningan, C. (Ed.), Kluwer Acad. Publ. Holanda. pp. 1–22.

155. Baek, Y.-S. and Boyle, J.A. 1996. Detection of channel catfish virus in adult channel catfish by use of a nested polymerase chain reaction. *J. Aquat. Anim. Health* 8: 97–103.

156. Chao, C.-B., Yang, S.-C., Tsai, H.-Y., Chen, C.-Y., Lin, C.S., and Huang, H.-T. 2002. A nested PCR for the detection of grouper iridovirus in Taiwan (TGIV) in cultured hybrid grouper, giant seaperch, and largemouth bass. *J. Aquat. Anim. Health* 14: 104–113.

157. Gray, W.L., Williams, R.J., and Griffin, B.R. 1999. Detection of channel catfish virus DNA in acutely infected channel catfish, *Ictalurus punctatus* (Rafinesque), using the polymerase chain reaction. *J. Fish Dis.* 22: 111–116.

158. Kurita, J., Nakajima, K., Hirono, I., and Aoki, T. 1998. Polymerase chain reaction (PCR) amplification of DNA of red sea bream iridovirus (RSIV). *Fish Pathol.* 33: 17–23.

159. Thiery, R., Raymond, J.C. and Castric, J. 1999. Natural outbreak of viral encephalopathy and retinopathy in juvenile sea bass, *Dicentrarchus labrax*: Study by nested reverse transcriptase-polymerase chain reaction. *Virus Res.* 63: 11–17.

160. Valle, L.D., Zanella, L., Patarnello, P., Paolucci, L., Belvedere, P., and Colombo, L. 2000. Development of a sensitive diagnostic assay for fish nervous necrosis virus based on RT-PCR plus nested PCR. *J. Fish Dis.* 23: 321–327.

161. Cano, I., Ferro, P., Alonso, M.C., Bergmann, S.M., Roemer-Oberdoerfer, A., Garcia-Rosado, E., Castro, D., and Borrego, J.J. 2007. Development of molecular techniques for detection of lymphocystis disease virus in different marine fish species. *J. Appl. Microbiol.* 102: 32–40.

162. Gould, A.R., Hyatt, A.D., Hengstberger, S.H., Whittington, R.J., and Coupar, B.E.H. 1999. A polymerase chain reaction (PCR) to detect epizootic haematopoietic necrosis virus and Bohle iridovirus. *Dis. Aquat. Org.* 22: 211–215.

163. McBeath, A.J.A., Burr, K.L.-A., and Cunningham, C. 2000. Development and use of a DNA probe for confirmation of cDNA from infectious salmon anaemia virus (ISAV) in PCR products. *Bull. Eur. Ass. Fish Pathol.* 20: 130–134.

164. Mjaaland, S., Rimstad, E., Falk, K., and Dannevig, B.H. 1997. Genome characterization of the virus causing infectious salmon anemia in Atlantic salmon (*Salmo salar* L.): An orthomyxo-like virus in a teleost. *J. Virol.* 71: 7681–7686.

165. Quere, R., Commes, T., Marti, J., Bonami, J.R., and Piquemal, D. 2002. White spot syndrome virus and infectious hypodermal and hematopoietic necrosis virus simultaneous diagnosis by miniarray system with colorimetry detection. *J. Virol. Methods* 105: 189–196.

166. Iida, Y. and Nagai, T. 2004. Detection of flounder herpesvirus (FHV) by polymerase chain reaction. *Fish Pathol.* 39: 209–212.

167. Koutna, M., Vesely, T., Psikal, I., and Hulova, J. 2003. Identification of spring viraemia of carp virus (SVCV) by combined RT-PCR and nested PCR. *Dis. Aquat. Org.* 55: 229–235.

168. Dhar, A.K., Roux, M.M., and Klimpel, K.K. 2002. Detection and quantification of infectious hypodermal and hematopoietic necrosis virus and white spot virus in shrimp using real-time quantitative PCR and SYBR green chemistry. *J. Clin. Microbiol.* 39: 2835–2845.

169. Durand, S.V. and Lightner, D.V. 2002. Quantitative real time PCR for the measurement of white spot syndrome virus in shrimp. *J. Fish Dis.* 25: 381–400.

170. Tang, F.F.J. and Lightner, D.V. 2001. Detection and quantification of infectious hypodermal and hematopoietic necrosis virus in penaeid shrimp by real-time PCR. *Dis. Aquat. Org.* 44: 79–85.

171. Chico, V., Gomez, N., Estepa, A., and Perez, L. 2006. Rapid detection and quantitation of viral hemorrhagic septicemia virus in experimentally challenged rainbow trout by real-time RT-PCR. *J. Virol. Methods* 132: 154–159.

172. Overturf, K., LaPatra, S., and Powell, M. 2001. Real-time PCR for the detection and quantitative analysis of IHNV in salmonids. *J. Fish Dis.* 24: 325–333.

173. Purcell, M.K., Hart, S.A., Kurath, G., and Winton, J.R. 2006. Strand-specific, real time RT-PCR assays for quantification of genomic and positive-sense RNAs of the fish rhabdovirus, infectious hematopoietic necrosis virus. *J. Virol. Methods* 132, 18–24.

174. Graham, D.A., Taylor, C., Rodgers, D., Weston, J., Khalili, M., Ball, N., Christie, K.E., and Todd, D. 2006. Development and evaluation of a one-step real-time reverse transcription polymerase chain reaction assay for the detection of salmonid alphaviruses in serum and tissúes. *Dis. Aquat. Org.* 70: 47–54.

175. Hodneland, K. and Endresen, C. 2006. Sensitive and specific detection of Salmonid alphavirus using real-time PCR (TaqMan). *J. Virol. Methods* 131: 184–192.

176. Zhang, H., Wang, J., Yuan, J., Li, L., Zhang, J., Bonami, J.-R., and Shi, Z. 2006. Quantitative relationship of two viruses (MrNV and XSV) in white-tail disease of *Macrobrachium rosenbergii*. *Dis. Aquat. Org.* 71: 11–17.

177. Grove, S., Faller, R., Soleim, K. B., and Dannevig, B. H. 2006. Absolute quantitation of RNA by a competitive real-time RT-PCR method using piscine nodavirus as a model. *J. Virol. Methods* 132, 104–112.

178. Nerland, A.H., Skaar, C., Eriksen, T.B., and Bleie, H. 2006. Detection of nodavirus in seawater from rearing facilities for Atlantic halibut *Hippoglossus hippoglossus* larvae. *Dis. Aquat. Org.* 73: 201–205.

179. Chi, S.C., Shieh, J.R., and Lin, S.-J. 2003. Genetic and antigenic analysis of betanodaviruses isolated from aquatic organisms in Taiwan. *Dis. Aquat. Org.* 55: 221–228.

180. Caipang, C.M.A., Haraguchi, I., Ohira, T., Hirono, I., and Aoki, T. 2004. Rapid detection of a fish iridovirus using loop-mediated isothermal amplification (LAMP). *J. Virol. Methods* 121: 155–161.

181. Wang, X.W., Ao, J.Q., Li, Q.G., and Chen, X.H. 2006. Quantitative detection of a marine fish iridovirus isolated from large yellow croaker, *Pseudosciaena crocea*, using a molecular beacon. *J. Virol. Methods* 133: 76–81.

182. Caipang, C.M., Hirono, I., and Aoki, T. 2003. Development of a real-time PCR assay for the detection and quantification of red reabream iridovirus (RSIV). *Fish Pathol.* 38: 1–8.

183. Dalla-Valle, L., Toffolo, V., Lamprecht, M., Maltese, C., Bovo, G., Belvedere, P., and Colombo, L. 2005. Development of a sensitive and quantitative diagnostic assay for fish nervous necrosis virus based on two-target real-time PCR. *Vet. Microbiol.* 110: 167–179.

184. Graham, D.A., Jewhurst, H., McLoughlin, M.F., Sourd, P., Rowley, H.M., Taylor, C., and Todd, D. 2006. Sub-clinical infection of farmed Atlantic salmon *Salmo salar* with salmonid alphavirus—a prospective longitudinal study. *Dis. Aquat. Org.* 72: 193–199.

185. Compton, J. 1991. Nucleic acid sequence based amplification. *Nature* 350: 91–92.

186. Starkey, W.G., Millar, R.M., Jenkins, M.E., Ireland, J.H., Muir, K.F., and Richards, R.H. 2004. Detection of piscine nodaviruses by real-time nucleic acid sequence based amplification (NASBA). *Dis. Aquat. Org.* 59: 93–100.

187. Starkey, W.G., Smail, D.A., Bleie, H., Muir, K., Ireland, J.H., and Richards, R.H. 2006. Detection of infectious salmon anaemia virus by real-time nucleic acid sequence based amplification. *Dis. Aquat. Org.* 72: 107–113.

188. Notomi, T., Okayama, H., Masubuchi, H., Yonekawa, T., Watanabe, K., Amino, N., and Hase, T. 2000. Loop-mediated isothermal amplification of DNA. *Nucleic Acid Res.* 28: E63.

189. Savan, R., Kono, T., Itami, T., and Sakai, M. 2005. Loop-mediated isothermal amplification: An emerging technology for detection of fish and shellfish pathogens. *J. Fish Dis.* 28: 573–581.

190. Gunimaladevi, I., Kono, T., Venugopal, M.N., and Sakai, M. 2004. Detection of koi herpesvirus in common carp, *Cyprinus carpio* L., by loop-mediated isothermal amplification. *J. Fish Dis.* 27: 583–589.

191. Gunimaladevi, I., Kono, T., LaPatra, S.E., and Sakai, M. 2005. A loop mediated isothermal amplification (LAMP) method for detection of infectious haematopoietic necrosis virus (IHNV) in rainbow trout (*Onchorhynchus mykiss*). *Ach. Virol.* 150: 899–909.

192. Shivappa, R.B., Savan, R., Kono, T., Sakai, M., Emmenegger, E., Kurath, G., and Levine, J.F. 2008. Detection of spring virevia of carp virus (SVCV) by loop-mediated isothermal amplification (LAMP) in koi carp, *Cyprinus carpio* L. *J. Fish Dis.* 31: 249–258.

193. Soliman, H. and El-Matbouli, M. 2006. Reverse-transcription loop-mediated isothermal amplification (RT-LAMP) for rapid detection of viral hemorrhagic septicaemia virus (VHS). *Vet. Microbiol.* 114: 205–213.

194. Kono, T., Savan, R., Sakai, M., and Irame, T. 2004. Detection of white spot syndrome virus in shrimp by loop-mediated isothermal amplification. *J. Virol. Methods* 115: 59–65.

195. Pillai, D., Bonami, J.-R., and Sri Widada, J. 2006. Rapid detection of *Macrobrachium rosenbergii* nodavirus (MrNV) and extra small virus (XSV), the pathogenic agents of white tail disease of *Macrobrachium rosenbergii* (De Man), by loop-mediated isothermal amplification. *J. Fish Dis.* 29: 275–283.

196. Nerette, P., Dohoo, I., and Hammell, L. 2005. Estimation of specificity and sensitivity of three diagnostic tests for infectious salmon anaemia virus in the absence of a gold standard. *J. Fish Dis.* Vol: 89–99.

197. Chomczynski, P. and Sachi, N. 1987. Single step method of RNA isolation by acid guanidinium thiocyanate-phenol chloroform extraction. *Anal. Biochem.* 162: 156–159.

198. Lopez-Lastra, M., Gonzalez, M., Jashes, M., and Sandino, A.M. 1994. A detection method for infectious pancreatic necrosis virus (IPNV) based on reverse transcription (RT)-polymerase chain reaction (PCR). *J. Fish Dis.* 17: 269–282.

199. Rimstad, E., Hornes, E., Olsvik, O., and Hyllseth, B. 1990. Identification of a double-stranded RNA virus by using polymerase chain reaction and magnetic separation of the synthesized DNA segments. *J. Clin. Microbiol.* 28: 2275–2278.

200. Dopazo, C.P., De Blas, I., Miossec, L., Cameron, A.R., Vallejo, A., and Dalsgaard, I. 2006. Common errors in surveillance and monitoring programs on fish populations. In *XI International Symposium on Veterinary Epidemiology and Economics.* 6–11 August 2006, Cairns, Queensland, Australia.

201. Cunningham, C.O. and Hoffs, M.S. 2002. Development of a positive control for detection of infectious salmon anaemia virus (ISAV) by PCR. *Bull. Eur. Assoc. Fish Pathol.* 22: 212–217.

202. Devold, M., Krossoy, B., Aspehaug, V., and Nylund, A. 2000. Use of RT-PCR for diagnosis of infectious salmon anaemia virus (ISAV) in carrier sea trout *Salmo trutta* after experimental infection. *Dis. Aquat. Org.* 40: 9–18.

203. Desselberger, U. 1995. *Medical Virology: A Practical Approach.* IRL Press, Oxford, U.K., p. 214.

204. López-Vázquez, C., Dopazo, C.P., Olveira, J.G., Barja, J.L., and Bandín, I. 2006. Development of a rapid, sensitive and non-lethal diagnostic assay for the detection of viral haemorrhagic septicaemia virus. *J. Virol. Methods* 133: 167–17.

205. Williams, K., Blake, S., Sweeney, A., Singer, J.T., and Nicholson, B.L. 1999. Multiplex reverse transcriptase PCR assay for simultaneous detection of three fish viruses. *J. Clin. Microbiol.* 37: 4139–4141.

206. Wang, W.S., Lee, J.S., Blake, S.L., and Nicholson, B.L. 1995. Developing the polymerase chain reaction technique to detect aquatic birnaviruses. *Taiwan J. Vet. Med. Anim. Husb.* 65: 167–180.
207. Yoganandhan, K., Sri Widada, J., Bonami, J.R., and Sahul Hameed, A.S. 2005. Simultaneous detection of *Macrobrachium rosenbergii* nodavirus and extra small virus by a single tube, one-step multiplex RT-PCR assay. *J. Fish Dis.* 28: 65–69.
208. Nishizawa, T., Mori, K., Nakai, T., Furusawa, I., and Muroga, K. 1994. Polymerase chain reaction (PCR) amplification of RNA of striped jack nervous necrosis virus (SJNNV). *Dis. Aquat. Org.* 18: 103–107.
209. Gilad, O., Yun, S., Andree, K.B., Adkison, M.A., Zlotkin, A., Bercovier, H., Eldar, A., and Hedrick, R.P. 2002. Initial characteristics of koi herpesvirus and development of a polymerase chain reaction assay to detect the virus in koi, *Cyprinus carpio koi. Dis. Aquat. Org.* 48: 101–108.
210. Boyle, J. and Blackwell, J. 1991. Use of polymerase chain reaction to detect latent channel catfish virus. *Am. J. Vet. Res.* 52: 1965–1968.
211. Kim, C.K., Kim, P.K., Sohn, S.G., Sim, D.S., Park, M.A., Heo, M.S., Lee, T.H., Lee, J.D., Jun, H.K., and Jang, K.L. 1998. Development of a polymerase chain reaction (PCR) procedure for the detection of baculovirus associated with white spot syndrome (WSBV) in penaeid shrimp. *J. Fish Dis.* 21: 11–17.
212. Thakur, P.C., Corsin, F., Turnbull, J.F., Shankar, K.M., Hao, N.V., Padiyar, P.A., Madhusudhan, M., Morgan, K.L., and Mohan, C.V. 2002. Estimation of prevalence of white spot syndrome virus (WSSV) by polymerase chain reaction in *Penaeus monodon* postlarvae at time of stocking in shrimp farms of Karnataka, India: A population-based study. *Dis. Aquat. Org.* 49: 235–243.
213. Yang, B., Song, X.-L., Huang, J., Shi, C.-Y., Liu, Q.-H., and Liu, L. 2006. A single-step multiplex PCR for simultaneous detection of white spot syndrome virus and infectious hypodermal and haematopoietic necrosis virus in penaeid shrimp. *J. Fish Dis.* 29: 301–305.
214. Sambrook, J., Fritsch, E.F., and Maniatis, T. 2001. *Molecular Cloning: A Laboratory Manual*, 3rd edn. Cold Spring Harbor Laboratory Press, Cold Spring, NY.
215. Seng, E.K., Fang, Q., Lam, T.J., and Sin, Y.M. 2004. Development of a rapid, sensitive and specific diagnostic assay for fish Aquareovirus based on RT-PCR. *J. Virol. Methods* 118: 111–122.
216. Yuasa, K., Sano, M., Kurita, J., Ito, T., and Iida, T. 2005. Improvement of a PCR Method with the Sph I-5 primer set for the detection of koi herpesvirus (KHV). *Fish Pathol.* 40: 37–40.
217. Barlic-Maganja, D., Strancar, M., Hostnik, P., Jencic, V., and Grom, J. 2002. Comparison of the efficiency and sensitivity of virus isolation and molecular methods for routine diagnosis of infectious haematopoietic necrosis virus and infectious pancreatic necrosis virus. *J. Fish Dis.* 25: 73–80.
218. Taksdal, T., Dannevig, B.H., and Rimstad, E. 2001. Detection of infectious pancreatic necrosis (IPN)-virus in experimentally infected Atlantic salmon parr by RT-PCR and cell culture isolation. *Bull. Eur. Assoc. Fish Pathol.* 21: 214–219.
219. Gagne, N., Johnson, S.C., Cook-Versloot, M., MacKinnon, A.M., and Olivier, G. 2004. Molecular detection and characterization of nodavirus in several marine fish species from the northeastern Atlantic. *Dis. Aquat. Org.* 62: 181–189.
220. Nylund, A., Devold, M., Mullins, J., and Plarre, H. 2002. Herring (*Clupea harengus*): A host for infectious salmon anemia virus. *Bull. Eur. Assoc. Fish Pathol.* 22: 311–318.
221. Miller, T.A., Rapp, J., Wastlhuber, U., Hoffmann, R.W., and Enzmann, P.-J. 1998. Rapid and sensitive reverse transcriptase-polymerase chain reaction based detection and differential diagnosis of fish pathogenic rhabdoviruses in organ samples and cultured cells. *Dis. Aquat. Org.* 34: 13–20.
222. Goodwin, A.E., Khoo, L., LaPatra, S.E., Bonar, C., Key, D.W., Garner, M., Lee, M.V., and Hanson, L. 2006. Goldfish hematopoietic necrosis herpesvirus (Cyprinid Herpesvirus 2) in the USA: Molecular confirmation of isolates from diseased fish. *J. Aquat. Anim. Health* 18: 11–18.
223. Gilad, O., Yun, S., Zagmutt-Vergara, F.J., Leutenegger, C.M., Bercovier, H., and Hedrick, R.P. 2004. Concentrations of a Koi herpesvirus (KHV) in tissues of experimentally infected *Cyprinus carpio koi* as assessed by real-time TaqMan PCR. *Dis. Aquat. Org.* 60: 179–187.

224. Bandín, I. and Dopazo, C.P. 2006. Restocking of salmon in Galician rivers: A health management program to reduce risk of introduction of certain fish viruses. DIPNET Newslett 35.

225. Olveira, J.G., Soares, F., Engrola, L., Dopazo, C.P., and Bandín, I. 2008. Antemorten versus post-mortem methods for detection of betanodavirus in Senegalese sole (*Solea senegalensis*). *J. Vet. Diag. Invest.* 20: 215–219.

226. Munro, E.S. and Ellis, A.E. 2008. A comparison between non-destructive and destructive testing of Atlantic salmon, *Salmo salar* L., broodfish for IPNV – destructive testing is still the best at time of maturation. *J. Fish Dis.* 31: 187–195.

227. Oshima, S., Hata, J.-I., Hirasawa, N., Ohtaka, T., Hirono, I., Aoki, T., and Yamashita, S. 1998. Rapid diagnosis of red sea bream iridovirus infection using the polymerase chain reaction. *Dis. Aquat. Org.* 32: 87–90.

228. Strommer, H.K. and Stone, D.M. 1998. Detection of viral hemorrhagic septicaemia (VHS) virus in fish tissues by semi-nested polymerase chain reaction (PCR). In: Barnes, A.C., Davidson, G.A., Hiney, M.P., and McIntosh, I. (Eds.), *Methodology in Fish*. Disease Research. Fisheries Research Services, Aberdeen.

229. Loevdal, T. and Enger, O. 2002. Detection of infectious salmon anemia virus in sea water by nested RT-PCR. *Dis. Aquat. Org.* 49: 123–128.

230. Grotmol, S., Nerland, A.H., Biering, E., Totland, G.K., and Nishizawa, T. 2000. Characterisation of the capsid protein gene from a nodavirus strain affecting the Atlantic halibut *Hippoglossus hippoglossus* and design of an optimal reverse-transcriptase polymerase chain reaction (RT-PCR) detection assay. *Dis. Aquat. Org.* 39: 79–88.

231. Juan, L., Tiehui, W., Yonglan, Y., Hanqin, L., Renhou, L., and Hongxi, C. 1997. A detection method for grass carp hemorrhagic virus (GCHV) based on a reverse transcription-polymerase chain reaction. *Dis. Aquat. Org.* 29: 7–12.

232. Gou, D.F., Kubota, H., Onuma, M., and Kodama, H. 1991. Detection of salmonid herpesvirus (*Oncorhynchus masou* virus) in fish by Southern-blot technique. *J. Vet. Med. Sci.* 53: 43–48.

233. Grizzle, J.M., Altinok, I., and Noyes, A.D. 2003. PCR method for detection of largemouth bass virus. *Dis. Aquat. Org.* 54: 29–33.

Chapter 19

Marine Toxins

Cara Empey Campora and Yoshitsugi Hokama

Contents

19.1 Introduction 578
19.2 Common Ichthyosarcotoxins 578
    19.2.1 Ciguatoxin 578
        19.2.1.1 Overview 578
        19.2.1.2 Clinical Symptoms 579
        19.2.1.3 Detection Methods 579
    19.2.2 Tetrodotoxin 581
        19.2.2.1 Overview 581
        19.2.2.2 Clinical Symptoms 581
        19.2.2.3 Detection Methods 582
19.3 Common Shellfish Toxins 583
    19.3.1 Saxitoxin—Paralytic Shellfish Poisoning 583
        19.3.1.1 Overview 583
        19.3.1.2 Clinical Symptoms 584
        19.3.1.3 Detection Methods 584
    19.3.2 Okadaic Acid—Diarrhetic Shellfish Poisoning 586
        19.3.2.1 Overview 586
        19.3.2.2 Clinical Symptoms 586
        19.3.2.3 Detection Methods 586
    19.3.3 Domoic Acid—Amnesic Shellfish Poisoning 588
        19.3.3.1 Overview 588
        19.3.3.2 Clinical Symptoms 588
        19.3.3.3 Detection Methods 589

19.3.4 Brevetoxin—Neurotoxic Shellfish Poisoning..590
    19.3.4.1 Overview 590
    19.3.4.2 Clinical Symptoms...590
    19.3.4.3 Detection Methods 591
19.4 Other Toxins..592
    19.4.1 Hepatotoxins—Microcystins..592
        19.4.1.1 Overview 592
        19.4.1.2 Clinical Symptoms...592
        19.4.1.3 Detection Methods 593
    19.4.2 New and Emerging Toxins 594
        19.4.2.1 Pinnatoxins 594
        19.4.2.2 Azaspiracids 594
        19.4.2.3 Gymnodimine 594
        19.4.2.4 Spirolides 594
19.5 Conclusion 594
References 595

19.1 Introduction

Fish poisoning dates back to antiquity. It was cited in Homer's Odyssey in 800 BC and was observed during the time of Alexander the Great (356–323 BC) when armies were forbidden to eat fish in order to avoid the accompanying sickness and malaise that could threaten his conquests [1]. Marine toxins, often the cause of various seafood poisonings, arise naturally from marine algal sources and accumulate through the food chain, ultimately depositing in predator fish- or filter-feeding bivalves destined for mammalian consumption. Such seafood-borne diseases account for a large and growing proportion of all food poisoning incidents, and are associated with several acute and chronic diseases in humans worldwide, which are characterized by gastrointestinal, neurological, and/or cardiovascular disturbances that can persist or recur for many months.

In this chapter, selected marine toxins originating from phytoplankton or their associated bacteria will be discussed in detail with respect to their general mechanisms of action, clinical symptoms, and available bioassay, immunoassay, and analytical detection methods.

19.2 Common Ichthyosarcotoxins

19.2.1 Ciguatoxin

19.2.1.1 Overview

Ciguatera fish poisoning was described as early as 1606 in the South Pacific island chain of New Hebrides [2]. A similar outbreak there and in nearby New Caledonia was reported by the famous English navigator Captain James Cook in 1774 [3], who described the clinical symptoms of his sick crew—symptoms that coincide with the clinical manifestations described today for ciguatera fish poisoning [4,5]. Representing a crude bioassay, viscera from the same fishes given to Cook's crew were also given to pigs, causing their deaths [3].

The term "ciguatera" originated in the Caribbean area to designate intoxication induced by the ingestion of the marine snail, *Turbo livona pica* (called *cigua*), as described by a Cuban ichthyologist. Today, it is widely used to denote the most commonly reported marine toxin disease in the world resulting from the ingestion of certain fishes, primarily reef fish, encountered in the islands of the Caribbean Sea, the Pacific and Indian Oceans, and other tropical and subtropical regions circumglobally between the Tropic of Cancer and the Tropic of Capricorn.

Ciguatera fish poisoning affects between 50,000 and 500,000 people annually, and stems from the consumption of fish containing high levels of ciguatoxins (CTXs), a family of complex, lipid-soluble, highly oxygenated cyclic polyether compounds produced by the benthic marine dinoflagellate, *Gambierdiscus toxicus*. CTXs are small molecular weight toxins (~1111 Da) with some 21 Pacific congeners varying in toxicity elucidated thus far [6–9]. They are biomagnified through the food chain, ultimately causing human and mammalian illness, as they are heat stable, colorless, odorless, and cannot be inactivated through cooking or freezing [7].

The CTXs are the most potent sodium channel toxins known, with the Pacific CTX-1 congener in mice having an intraperitoneal (IP) LD_{50} of $0.25\,\mu g/kg$ [10]. CTXs and the closely related brevetoxins (a family of lipid-soluble polyether toxins produced by the marine dinoflagellate *Karenia brevis*, detailed in Section 19.3.4) are characterized by their ability to cause the persistent activation of voltage-sensitive sodium channels, leading to increased cell Na^+ permeability. As a consequence, Na^+-dependent mechanisms in numerous cell types are modified, leading to increased neuronal excitability and neurotransmitter release, impairment of synaptic vesicle recycling, and induced cell swelling [10].

19.2.1.2 Clinical Symptoms

CTXs cause gastrointestinal and neurological symptoms that typically persist for days to weeks, with common symptoms such as vomiting, diarrhea, nausea, abdominal pain, dysesthesia, pruritus, and myalgia. Severe cases of ciguatera may involve hypotension and bradycardia, although fatalities are rare. Neurological signs may persist for several months or even years [11]. Remarkably, the diagnosis of ciguatera is still largely dependent on the astuteness of the clinician. A history of recent consumption of potentially toxic fish, and at least one neurological sign and one other typical symptom are required to establish the clinical diagnosis. In the absence of a confirmatory laboratory test, a sizable proportion of cases still go undiagnosed and unreported. Treatment is largely empiric and symptomatic. In severe cases, supportive care, particularly monitoring fluid and electrolyte balance, is paramount, and local anesthetics and antidepressants may also be useful in some instances. Following its somewhat serendipitous use for a coma victim in the Marshall Islands who was later diagnosed with severe ciguatera, intravenous mannitol is now the mainstay of therapy [12]. Mannitol, however, is not universally beneficial, and is best when used during the acute phase of severe intoxications.

19.2.1.3 Detection Methods

19.2.1.3.1 Bioassays

A commonly used method to detect CTXs involves the IP injection of mice with the crude extracts of fish [13,14]. Using estimates from known cases of ciguatera fishes obtained by the Hawaii Department of Health and other laboratories, it has been found that 1 mouse unit (MU) = 7–8 ng of CTX [15], which is equivalent to the concentration of toxic extract injected IP that kills a 20 g mouse within 24 h. The general protocol for testing crude fish extract is as follows: Swiss-Webster mice weighing 20–25 g are injected IP with 100 mg of crude fish extract resuspended in 1 mL of 1% Tween 60 in saline. Symptoms displayed by the mouse are observed from 0.5 to 48 h after injection and rated on a scale of 0–5 according to toxicity. Characteristic ionotropic responses to various toxin extracts including CTX have also been established using the guinea pig atrial assay, which involves specialized dissection techniques, requires a small amount of test material, and gives some measure of specificity, as the actions are at the sites of the sodium channel [16]. Other organisms, such as brine shrimp [17], mosquitoes [18,19], chickens [20], and dipteral larvae [21] have been used to screen for CTX, however, most have been found to be nonspecific, nonquantitative, and generally unreliable for routine screening.

Directed cytotoxicity to the sodium channels of neuroblastoma cells has been established for purified CTXs, brevetoxins, saxitoxin (STX), and crude seafood extracts [22]. Using a microplate high-throughput format, this assay takes several days to complete and serves as a valuable tool for marine toxin studies, detecting CTX at subpicogram levels. A fluorescent-based assay detecting sodium channel activators has also been useful in the nonspecific analysis of crude extract [23], and recently, a rapid hemolysis assay based on the neuroblastoma cell bioassay using red cells from the red tilapia (*Sarotherodon mossambicus*) has been developed for the detection of sodium channel-specific marine toxins, including CTX [24].

19.2.1.3.2 Immunoassays

The radioimmunoassay (RIA) [25] and membrane immunobead assay (MIA) [26] are advances in simple, rapid, sensitive, and specific qualitative detection methods for CTX. The MIA is a field usable assay that employs a monoclonal antibody to purified moray eel CTX-1 coated with polystyrene microbeads and a hydrophobic membrane laminated onto a solid plastic support. The membrane binds polyether lipids such as CTX and specifically detects the toxin using the monoclonal antibody to CTX coated with microbeads. The intensity of the color on the membrane correlates to the concentration of toxin on the solid support. This assay has a limit of detection at ~0.032 ng CTX/g fish tissue and has a sensitivity of 91% and specificity of 87%. However, immunochemical methods are subject to cross-reactivity issues with other polyether compounds, and often there is a limited supply of antibody for use.

19.2.1.3.3 Chemical Methods

Because CTX and brevetoxin share a common receptor at the sodium channel receptor site 5, the use of labeled brevetoxin (^{3}H-PbTx-B) allows CTX to be quantified by competitive-binding assay with sodium channel containing proteins using isolated rat brain synaptosomes [27]. This method requires a small amount of fish extract, is rapid and simple, and has a high sensitivity. Best suited for research purposes, this method is likely impractical for large-scale fish screening because of specialized equipment and the use of radiolabeled compounds.

Gradient reverse phase high-performance liquid chromatography/mass spectroscopy (HPLC/MS), fast-atom bombardment tandem mass spectroscopy (MS/MS), and other chemical methods have recently been used to elucidate CTXs and their structures. While CTXs do not possess a useful chromophore for selective spectroscopic detection, they do contain a reactive primary hydroxyl group that can be labeled after a clean-up step. HPLC coupled to fluorescence detection has proven effective when screening for CTX in crude fish extracts [28,29]. HPLC with ionspray MS has shown promise as a confirmatory analytical assay for CTXs in fish flesh [30]. Nuclear magnetic resonance (NMR) has been used to characterize CTXs in fish flesh [31] and wild and cultured *G. toxicus* extracts [7,32], and Lewis and Jones [33] used gradient reverse phase LC/MS methods to identify 11 new P-CTX congeners in a partially purified sample of toxic moray eel viscera. Similarly, LC–ESI-MS/MS (ESI = electrospray ionization) was reported to detect the levels of CTX equivalent to 40 ng/g P-CTX-1 and 100 ng/kg C-CTX-1 in fish flesh [34].

19.2.2 Tetrodotoxin

19.2.2.1 Overview

Tetrodotoxin (TTX), often referred to as puffer fish poisoning, is one of the most potent and common lethal marine poisonings. It occurs primarily in Southeast Asia where fugu (puffer fish fillet) in Japan is considered a delicacy. It is one of the oldest known natural toxins, recorded as early as 2700 BC in Chinese literature describing the toxicity of the puffer [35]. Because there is no cure or antidote, the mortality rate is relatively high, although incidence is steadily declining due in part to increased government regulations and legislation regarding preparation and marketing of aquacultured nontoxic fish. According to the Japanese Ministry of Health and Welfare, there were ~88 deaths due to TTX poisoning in 1965 compared to five deaths in 2001 [35].

TTX concentrates in the liver of bony fish in the order Tetraodontiformes, mainly from the family Tetraodontidae, which includes the puffer fish and toadfish. However, TTX has also been found in xanthid crabs, horse-shoe crabs and their eggs, the blue-ringed octopus, newts, and several other fish species such as marine gobies [36]. TTX has also been found in several bacterial species, including *Shewanella* sp. and *Vibrio* sp., and is believed to be bacterial in origin [35].

TTX is a water-soluble heterocyclic guanidine that blocks Na^+ conductance over the single nanomolar range by binding extracellularly to receptor site 1 of voltage-gated sodium channels. This mechanism of action prevents the access of monovalent cations to the outer pore of the channel and primarily affects the control of peripheral nerve excitability by influencing the generation of action potentials and impulse conduction [37].

19.2.2.2 Clinical Symptoms

The type, the severity, and the range of symptoms of TTX poisoning are dependent on the amount of toxin ingested, and the age and the preexisting health of the victim. The minimum dose for developing TTX poisoning symptoms in humans is ~2 mg of TTX. Early symptoms are sensory, including perioral and distal limb numbness and paresthesia, taste disturbances, dizziness, headache, diaphoresis, and other symptoms such as salivation, nausea, vomiting, diarrhea, and abdominal pain [35]. Mild poisoning cases might include several sensory features and minor gastrointestinal effects. Patients with moderate poisoning may develop distal muscle weakness, weakness of the bulbar and facial muscles, and ataxia and incoordination with normal reflexes [36]. Severe poisoning causes generalized flaccid paralysis, respiratory distress with possible eventual respiratory failure, extreme hypotension, seizures, and loss of deep tendon and spinal reflexes. Although some patients may exhibit impaired mental capabilities, most remain fully conscious for 6–24 h, after which the prognosis for recovery is good. Otherwise, death is caused by cardiovascular effects and ascending paralysis involving the respiratory muscles.

The diagnosis of TTX poisoning is based on the clinical examination and the history of the consumption of toxic organisms. Because TTX may remain detectable and quantifiable in urine using HPLC up to 5 days following exposure [38], testing for exposure immediately after a suspected poisoning is likely to be the most sensitive method of determination. There are no known antidotes or antitoxins to TTX and therefore treatment involves careful observation and supportive care, including serial neurological assessments, and admission to intensive care units so that respiratory failure or cardiac effects are appropriately anticipated and treated. The case reports have suggested the use of neostigmine in an effort to reduce paresthesia and numbness [39,40], although other reports indicate that it has no effect on symptom improvement [41]. The prevention of TTX poisoning, with an emphasis on public education, is the primary method of avoiding illness.

19.2.2.3 Detection Methods

19.2.2.3.1 Bioassays

The mouse bioassay is commonly used to determine the toxicity of TTX in a given sample, as well as the identification of unknown toxin extract when compared to a TTX-specific dose death time relationship curve. Although the mouse bioassay is the animal of choice for such determinations, drawbacks to the method include low accuracy due to inherent individual variation in a biological system, lack of specificity, and the inconvenience and controversy that often accompany the use of live animals for experimentation.

Cell-based bioassays have been employed for the quantitative measurement of the sodium channel blocker TTX, even at low levels (~3 nmol/L). This assay is based on the ability of sodium channel-blocking toxins to antagonize the combined effects of the chemicals veratridine and ouabain on neuroblastoma cell lines. While veratridine at 0.075 mmol/L and ouabain at 1.0 mmol/L cause the cells to round up and die, the presence of TTX counters this effect and the cells exhibit growth. The amount of toxin can then be estimated from the linear relationship of the relative abundance of living cells and the concentration of toxin in the samples. A modified assay that employs a water-soluble tetrazolium salt to quantitate the assay using a microplate reader streamlines the process [42,43]. The sensitivity of this method is much higher than that of the mouse bioassay, however, it is time consuming, requires laboratory expertise, and is not suitable for routine screening.

19.2.2.3.2 Immunoassays

Several attempts to develop immunoassay techniques to detect TTX have been made in recent years with limited success [44–47]. However, recently a monoclonal antibody against TTX has been developed from Balb/c mice immunized with TTX-bovine serum albumin conjugate by which a rapid and highly sensitive enzyme immunoassay capable of monitoring seafood has been established for the quantitative analysis of TTX. It detects concentrations as low as 2–100 ng/mL in 30 min [48]. Using this highly specific monoclonal antibody, immunoaffinity column chromatography methods have also been developed for identification of TTX from the urine of poisoned patients, detecting as low as 2 ng/mL [49].

19.2.2.3.3 Chemical Methods

HPLC methods have been examined for both the qualitative and the quantitative analyses of TTX and its derivatives, including a fluorometric HPLC continuous analyzer first constructed in 1982 [50] and reconfigured in 1989 to improve the detection and the separation of TTX and TTX analogues including 6-*epi*TTX [51]. Reversed-phase HPLC is a fast and efficient method used by many researchers for analyzing TTX and its analogues by using heptanesulfonic acid as a counterion [52,53]. Methods such as thin layer chromatography (TLC) and electrophoresis are useful techniques for detecting the levels down to 2 μg of TTX in laboratories where HPLC and other costly analytical systems are not available [35]. Capillary isotachophoresis is also a rapid, accurate, and potential detection method for TTX, with a quantitative detection limit of ~0.25 μg of TTX [54]. LC–MS is considered an accurate method of detecting TTX [55], combining an HPLC–MS equipped with a 1.5×150 mm column coupled to a mass spectrometer, using acetonitrile (50%, flow rate 70 μL/min) as the mobile solvent. This method has also shown promise in screening biological samples such as blood and urine at a detection limit of 12.5 nM, equivalent to about 3.9 ng/mL [56].

Several other methods including UV spectroscopy [57], gas chromatography–MS [58], infrared spectrometry [59], fast-atom bombardment MS [60], and ESI-time of flight/MS [61] have all been used in the determination of TTX and its derivatives, as well as ^{1}H NMR spectrometry for determining absolute configurations [62].

19.3 Common Shellfish Toxins

Microscopic planktonic algae are critical food sources for filter-feeding bivalve shellfish such as oysters, mussels, scallops, and clams. It is not clear why some microalgal species produce toxins; however, during the past few decades the frequency, intensity, and geographic distribution of toxic compounds produced by marine algae have increased, contributing to the awareness of poisoning events from the ingestion of contaminated shellfish products. The four groups of shellfish toxins and their associated poisonings will be reviewed, namely: STX (paralytic shellfish poisoning [PSP]), okadaic acid (OA) (diarrhetic shellfish poisoning [DSP]), domoic acid (DA) (amnesic shellfish poisoning [ASP]), and brevetoxin (neurotoxic shellfish poisoning [NSP]).

19.3.1 Saxitoxin—Paralytic Shellfish Poisoning

19.3.1.1 Overview

The water-soluble STX and its derivatives, including the gonyautoxins (GNTXs), are responsible for PSP, and are accumulated from dinoflagellates from the genus *Alexandrium* as well as *Pyrodinium bahamense* and *Gymnodinium catenatum* by shellfish filter feeders, primarily mussels, oysters, and clams, and passed through the food chain to humans in tropical and moderate climate zones. The link between shellfish toxicity and dinoflagellates was first identified in the San Francisco Bay in 1927 [63,64] following an outbreak of PSP in the region. The PSP toxins behave pharmacologically similar to TTX in that they bind with nanomolar affinity to receptor site 1 on the sodium channel and are the reversible blockers of voltage-gated sodium channels. Structural differences between the various congeners of STX alters the rates at which they bind and release from the binding site on the sodium channel, and the lifetime of the open channel is reversibly correlated with toxin concentrations and association constants [65].

19.3.1.2 Clinical Symptoms

The outbreaks of PSP occur periodically, attributed in part to poorly understand environmental changes that may be related to "red tides." Individual sensitivity to the toxins determines the level at which PSP toxins cause illness; for example, oral intake causing mild symptoms ranges from 144 to 1660 μg of STX equivalents/person, and fatal intoxications were calculated ranging from 456 to 12,400 μg STX equivalents/person [66]. The fluctuation in the methods of determination and the reconstruction of STX values based on the remaining toxic food sources may contribute to the variations in toxicity reported.

In cases of mild poisoning, clinical symptoms may include a tingling sensation or numbness around the lips within 30 min of ingestion, due to localized absorption of the toxins through the buccal mucous membranes. Gradually these symptoms spread to the face and neck, and prickly sensations in the fingertips and toes as well as headaches, dizziness, nausea, vomiting, and diarrhea are commonly observed. In cases of moderately severe poisoning, paresthesia progresses to the arms and legs, and incoherent speech, motor incoordination, and ataxia are frequent. In severe cases, respiratory difficulties including muscular paralysis are pronounced and death through respiratory paralysis may occur within 2–24 h of ingestion [65]. The overall mortality is reportedly between 1% and 10%, and appears to depend to some degree on medical care, age, and previous health status of the patient [67].

Supportive treatment generally resolves the symptoms, although several weeks or months may pass before the fatigue, tingling, or memory loss is completely resolved. Initial treatments may include gastric lavage to remove unabsorbed toxin, maintenance of adequate ventilation, and fluid therapy to correct acidosis and facilitate renal excretion of the water-soluble toxins. Animal studies have shown that 4-aminopyridine may be useful as a therapeutic antidote for STX intoxication by markedly improving the cardiorespiratory performances in rats and guinea pigs exposed to STX [68,69].

19.3.1.3 Detection Methods

19.3.1.3.1 Bioassays

The detection of STXs is challenging because there are a large number of different but related causative compounds that can be encountered at low levels. The original bioassay for the STXs was a mouse bioassay based on the IP injection of mice [70] and is still in use as the current benchmark technique in food safety, although it cannot distinguish STX from TTX. The refined procedure, standardized by the Association of Official Analytical Chemists (AOAC), produces a rapid and reasonably accurate measurement of total PSP toxins [71]. In most countries, the action level for the closure of a fishery is 400 MU/100 g of shellfish, where 1 MU is defined as the amount of toxin that kills a 20 g mouse in 15 min by IP injection, equivalent to 0.18 μg of STX [72]. The limit of the detection of the assay is ~40 μg STX/100 g shellfish tissue with a precision of ±15%–20% [66]. While alternative organisms have been sought, including houseflies and other insects, they have yet to show the same precision and efficiency as mammalian-based assays, and may be less accurate as the predictors of human oral potency. The drawbacks to the mouse bioassay include maintenance of mice colonies at specific weights, strains, and sizes, lack of linearity between the time of death and the toxin levels, time and labor-intensive procedures, and the use and the sacrifice of animals during the process. To reduce the number of mouse tests in several European countries, a qualitative technique that involves the direct monitoring of toxic algal cells in seawater is often used [73].

A cell bioassay modified by Jellet et al. [74] incorporates the use of an automated microplate reader using mouse neuroblastoma cells, which swell and lyse in the presence of ouabain and veratridine by enhancing sodium ion influx. The addition of STX will block the sodium channel and the cells will remain morphologically normal, and changes can be detected using the absorption of stained cells. This method has a detection limit of about 10 ng STX equivalents/mL of extract, or 2.0 μg STX equivalents/100 g shellfish tissue. This method is a promising screening tool; however, it is recommended that any results measuring close to regulatory limits be reevaluated using another method to confirm. By sequencing, the addition of veratridine, ouabain, and extracted samples to neuroblastoma cells, a hemolysis assay [24] recently developed reportedly detects STXs in concentrations at 0.3 μg/mL, although its value as a practical shellfish-screening tool has not yet been evaluated.

19.3.1.3.2 Immunoassays

Indirect enzyme-linked immunoassays (ELISAs) have been developed and are commercially available for the detection of STX [75] and more recently adapted to detect STX derivatives including neoSTX, GNTX1, and GNTX3 [76]. Such methods appear to be more sensitive than LC and more specific than the mouse bioassay. In addition to indirect ELISAs, direct competitive ELISAs have also been available for the detection of STX and derivatives [77,78] and show excellent correlation between the ELISA data and the mouse assay results, often detectable at concentrations lower than the regulatory limits. However, ELISAs are prone to cross-reactivity, and the difficulty of adequately detecting STX derivatives at low levels limits the use of ELISA as a means of regulating shellfish for PSP toxins.

19.3.1.3.3 Chemical Methods

The alkaline oxidation of PSP toxins yields fluorescent products, allowing determination using fluorometric techniques [79,80] and has given way to the development of a fluorescent sensor that is reportedly selective for STX and not TTX using acridinyl crowns [81]. LC techniques are the most widely used nonbioassay methods for PSP compound determination and are generally based on the separation of toxins by ion-interaction chromatography and use of a postcolumn reactor that oxidizes the column effluent to produce readily detectable derivatives. The methodology developed by the United States Food and Drug Administration was reported to resolve 12 carbamate and sulfocarbamoyl PSP toxins at detection limits with an order of magnitude lower than that of the mouse bioassay, and validation against the mouse bioassay showed good correlation between the two methods at $r > 0.9$ [82]. However, in practice, this method has shown some difficulties in separating STX from the derivative dcSTX and has gone out of use in many European laboratories screening for PSP toxins [83]. Though LC methods are promising, operating such a system requires a considerable amount of skill and time, and may not be robust enough to handle the large numbers of samples that are necessary for screening during a bloom event [84].

MS, specifically LC–MS, has been used for qualitative determination of STX with detection limits five times lower than that of the mouse bioassay [85–87] and variations in the methods have shown promise in confirming accumulation of PSP toxins in mussel and shellfish samples.

19.3.2 Okadaic Acid—Diarrhetic Shellfish Poisoning

19.3.2.1 Overview

DSP is a toxic syndrome that is caused by the consumption of shellfish that has been contaminated with algal toxins produced by marine dinoflagellates belonging to the generas *Dinophysis* spp. and *Prorocentrum* spp. The DSP toxins, which are heat stable polyether lipophilic compounds, can be grouped into three categories based on their unique chemical structures. The first group are acidic toxins including OA and related dinophysistoxin (DTX) derivatives, and are potent phosphate inhibitors, which can cause inflammation of the intestinal tract and diarrhea in humans [88]. The second group is neutral polyether–lactones of the pectenotoxins (PTXs), 10 of which have been isolated. The third group are sulfated polyether compounds called yessotoxins (YTXs), and their derivative 45-hydroxyyessotoxin (45-OH-YTX) [89,90]. Interestingly, the YTXs do not cause diarrhea, but rather attack the cardiac muscle in mice after IP injection, while the desulfated YTX damages the liver [90]. The reevaluation of their toxicities may lead to the removal of these toxins from classification as a DSP toxin, although they currently remain as such [91].

OA is a potent inhibitor of the serine/threonine phosphatases PP1 and PP2A. Because these phosphatases are enzymes responsible for phosphorylation and dephosphorylation of proteins associated with critical metabolic processes within a cell, their dysregulation leads to the specific symptoms associated with DSP. It is suggested that diarrhea in humans is caused by the hyperphosphorylation of proteins that control sodium secretion by intestinal cells or by the increased phosphorylation of junctional moieties that regulate solute permeability, resulting in the passive loss of fluids [90,92].

19.3.2.2 Clinical Symptoms

The clinical symptoms vary depending on the DSP toxin and the intensity depends on the amount of toxin ingested. While rarely fatal, the predominant symptoms from OA and DTX include diarrhea, nausea, vomiting, and abdominal pain within 30 min to several hours after ingestion, and complete recovery is expected within 3 days generally without hospitalization. Intravenous injection of an electrolyte can assist in ameliorating the symptoms. The data indicates that the minimum dose to induce toxic effects in humans is 48 μg of OA and 38.4 μg of DTX1 [70]. The primary clinical result from the ingestion of PTXs is liver necrosis [93], while YTXs can cause cardiac muscle damage when administered intraperitoneally in mice [94].

The prevention of exposure is enforced in many countries including the frequent inspection of seawater around aquaculture facilities and monitoring programs that keep records on the occurrence of toxic phytoplankton and the closures of harvesting areas when toxic algae levels are high. In Europe, the maximum level of OA, DTXs, and PTXs together in edible tissues of molluscs, echinoderms, tunicates, and marine gastropods are 160 μg OA equivalents/kg of shellfish meat, while YTX levels are 1 mg YTX equivalents/kg of meat [95]. Shellfish containing more than 2 μg OA/g hepatopancreas and/or more than 1.8 μg DTX/g hepatopancreas are considered unsafe for human consumption [92].

19.3.2.3 Detection Methods

19.3.2.3.1 Bioassays

The mouse bioassay is a preferred method of analysis for DSP toxins in Europe and Japan, although complementary chemical or immunological analyses may accompany the evaluation [66], and is the officially recognized regulatory method for detection in the European Union (EU) [96]. The mouse bioassay, first developed by Yasumoto et al. [97], involves the extraction of shellfish tissues using acetone, followed by IP injection into a 20 g mouse and survival monitoring for 24–48 h. The toxicity of the sample, expressed in MU/g of whole tissue, is determined as the minimum quantity of toxin capable of killing a 20 g mouse within 24 h after IP injection. In many countries, the regulatory level is set at 0.05 MU/g whole tissue. The disadvantages to this assay include: lack of specificity in that there is no differentiation between the various components of DSP toxins or unknown toxic groups exhibiting ichthyotoxic and hemolytic properties, subjectivity to the time of death in the animals, and the need for routine maintenance of laboratory animals. In addition, the selectivity, specificity, and toxin recovery depend greatly on the selection, the purity, and the ratios of the organic solvents used in the extraction and the clean up.

A semiquantitative method for OA and DTX toxin evaluation is a rat bioassay in which animals are starved, then fed suspect shellfish tissue and observed for signs of diarrhea, fecal consistency, and food refusal. However, this method, officially allowed in the EU, does not detect PTXs and YTXs. An inexpensive, sensitive method for screening OA and some coextracting toxins is a bioassay using small planktonic crustaceans, *Daphnia magna*, and has been reported to measure OA levels 10 times below the threshold of the mouse bioassay method [98].

Cytotoxicity assays using rat hepatocytes and KB cells (a human cell line derived from epidermoid carcinoma) have shown promise for detecting some DSP toxins. The hepatocyte assay is based on morphological changes in the cell, and can differentiate between the diarrhetic DSP toxins and the nondiarrhetic toxins [99]. OA and PTX appear to have a high toxicity on KB cell lines, and thus several different assays using the cell line have been developed using various methods [100].

19.3.2.3.2 Immunoassays

A variety of ELISA kits are commercially available to detect DSP toxins, including the DSP-Check® ELISA test kit (UBE Industries, Japan), used to screen OA and DTX1 at a claimed detection limit of 20 ng/g. While reports about its performance vary, it appears to be more sensitive and specific than LC. The Rougier Bio-Tech® ELISA kit (Montreal, Canada) has undergone extensive comparisons using analytic methods for DSP toxin detection and has been found to be reliable for OA quantification in both mussel and phytoplankton extracts [92]. A direct ELISA developed by Biosense® (Bergen, Norway) for YTX is still being evaluated for efficacy in detecting YTX and its analogues.

Immuno biosensors, which are defined as "a self-consistent bioanalytical device incorporating a biologically active material, either connected to, or integrated within, an appropriate physicochemical transducer, for the purpose of detecting-reversibly and selectively—the concentration or activity of chemical species in any type of sample [101]," have been applied in the development of sensors for DSP toxins. A semiautomated chemiluminescent immunosensor for OA in mussels has already been described [102], and it is expected that such technology will further advance in the coming years. The phosphate inhibition bioassays using colorimetric or fluorometric detection are capable of the quantitative measurement of OA and have been shown to be rapid, accurate, specific, and simple procedures for detecting OA in buffered or complex solutions [103–105].

19.3.2.3.3 Chemical Methods

Chromatography methods are often used to assess DSP toxins. TLC offers a fairly simple method of assessing the acidic DSP toxins at levels of ~1–3 μg of toxin [92], however, these high detection limits can be a limiting factor in the use of TLC. LC methods are commonly used for the determination of OA and DTX1. The original method involves sequential extraction of shellfish tissue with methanol, ether, and chloroform, derivatization with 9-anthryldiazomethane, silica Sep-pak clean up, and determination by HPLC with fluorescence detection [106]. Permutations to the original method have been made to streamline the analysis, including use of various solvents [107], changes to the derivatization reagents including coumarin, luminarine-3, and 9-chloromethylanthracene [108], and adapting the analysis to include the determination of YTXs and PTXs using fluorescent labeling [109,110].

LC combined with ESI-MS can achieve a detection limit of 1 ng/g shellfish tissue, resulting in a fast, sensitive technique for determination of DSP toxins even when analytical standards are not readily available. The interlaboratory studies of a new LC–MS method for the determination of ASP and DSP toxins have obtained consistent results and represents an encouraging alternative to the mouse bioassay [111].

19.3.3 Domoic Acid—Amnesic Shellfish Poisoning

19.3.3.1 Overview

First discovered in Prince Edward Island, Canada in 1987, amnesic or encephalopathic shellfish poisoning (ASP) primarily affects the central nervous system, leading to severe memory loss and confusion. The causative toxin, DA, is a heat stable, water soluble, neuroexcitatory amino acid that acts like the neurotransmitter glutamic acid and is produced by diatoms from the genus *Pseudonitzschia*. Specifically, *Pseudonitzschia pungens f. multiseries*, *P. australis*, and *P. pseudodelicataissima* have been implicated in human and bird intoxications [112,113]. Until the toxic event in Canada, it was thought that phycotoxins were only produced by dinoflagellates, and diatoms were not considered a potential source of toxins.

Cultured blue mussels, soft-shelled clams, razor clams, and some species of scallops have all been shown to potentially contain DA in Canada and from the California coast up to Washington in the United States. In addition, DA was found in the viscera of Dungeness crabs from Oregon and Washington, and some species of anchovies and mackerel have been found to be contaminated with DA after sea lions and water birds died in Central and Northern California and Baja Mexico after ingestion of these fish.

DA is an agonist of the glutamate receptor [114], and binds with high affinity to the glutamate receptors of the quisqualate type, which are targets for neurotransmitters. The receptor serves to conduct Na^+ ion channels in the postsynaptic membrane; DA acts to open these Na^+ channels, leading to Na^+ influx which induces depolarization, with the resulting increased influx of Ca^{2+} ions leading to cell death. DA is about 100 times more potent than glutamate [66].

19.3.3.2 Clinical Symptoms

The diagnosis of ASP is difficult because there have only been a few outbreaks reported, however, a combination of gastrointestinal and neurological features, particularly memory loss and confusion after ingestion of shellfish appear to be common. In the Canadian outbreak, 107 patients were reported to have an acute illness after the ingestion of mussels contaminated with DA [115,116]. Patients presented with gastrointestinal symptoms ~5.5 h after ingestion, including vomiting, abdominal cramps, and diarrhea. Unusual neurological features developed after 48 h including headache, confusion, disorientation, and short-term memory loss correlated with age, mutism, seizures, disordered eye movements, myoclonus, and coma. Hemodynamic instability, cardiac arrhythmias, and respiratory secretions were also noted [116]. Four patients died and 14 were severely affected with ongoing neurological abnormalities, while the remaining patients recovered fully. Treatment is supportive, and symptomatic and/or neurological dysfunction should be carefully monitored and treated accordingly.

19.3.3.3 Detection Methods

19.3.3.3.1 Bioassays

The AOAC approved mouse bioassay for PSP toxins [117] can also be used to detect DA at the concentrations of ~40 μg/g of tissue because the symptoms of ASP in mice are distinguishable from the classic PSP symptoms. The typical sign of DA presence in an extract is a unique scratching of the mouse shoulder by the hind leg, followed by convulsions over an observation period of 4 h. The common regulatory limit for DA is 20 μg DA/g of mussel tissue, and as such the mouse bioassay is not sensitive enough to quantify the toxin for routine screenings.

In vitro assays for detecting the toxin include a competitive receptor-binding assay in which frog (*Rana pipiens*) brain synaptosomes are used and assayed based on binding competition with radiolabeled kainic acid for the kainite/quisqualate glutamate receptor. This assay was further optimized [118] to use a cloned rat GLUR6 glutamate receptor and is suitable for the analysis of DA in seawater extracts from algae and shellfish tissue.

19.3.3.3.2 Immunoassays

An ELISA for DA determination in mussel extracts measuring total DA content including a diastereoisomer and at least two *cis–trans* isomers was developed in 1995 [119] using a polyclonal antiserum raised in mice against an ovalbumin–DA conjugate. A limit of detection was found to be 0.25 mg/mL of extract, representing 0.5 mg DA/g of extracted mussel tissue [119]. The routine monitoring of DA levels in cultured bivalve molluscs can be accomplished through a commercial indirect ELISA originally developed in 1998 [120] where the limit of quantitation is 10 mg/ kg shellfish. According to the manufacturer (Biosense®, Bergen, Norway), method validation between reference laboratories in Scotland, Chile, and New Zealand yielded excellent results.

New antibody-based approaches involve the use of biosensors [121] wherein DA is bound to the surface of a sensor and detected with polyclonal antibodies raised to DA–human serum albumin conjugates with the promising limits of detection. It is expected that biosensor technology will become more refined and effective for use in regulatory situations in the near future.

19.3.3.3.3 Chemical Methods

DA can be determined by TLC as a weak UV-quenching spot that stains yellow following treatment with 1% ninhydrin [122], although normal amino acids present in crude extracts have the potential to interfere, thus a separation step is required. A clean up procedure using strong anion change solid phase extraction, or SAX-SPE, yields fractions that can be used directly in onedimensional TLC. The detection limit of DA using TLC is ~10 mg/g in shellfish tissues and is a useful tool, particularly as a secondary screen following immunoassay detection, or for laboratories that do not have LC available for use.

The liquid and/or ion exchange chromatography can analyze and preparatively isolate DA. Reverse phased LC–UV gives the fastest and the most efficient separations, and has become a preferred analytical technique for the determination of DA in shellfish following an AOAC collaborative study [123]. The detection limit using this method is about 10–80 ng/mL, depending on the sensitivity of the UV detector used. When crude extracts are analyzed without clean-up, the practical limit of quantitation is ~1 mg/g [124], which is suitable for regulatory laboratories concerned with detecting contamination at levels greater than 20 mg/g. The use of fluorescent derivatives can detect DA as low as 15 pg/mL in marine matrices such as seawater and phytoplankton as well as shellfish extracts [125,126]. In addition to other methods, capillary electrophoresis is a relatively simple method that allows for rapid, high-resolution separations and gives comparable precision and accuracy rates when compared with LC.

Electrospray is a technique used to interface LC with MS [127–129]. The interlaboratory studies of the LC–MS method for the determination of ASP toxins in shellfish have been performed and yielded consistent sets of data, and were shown to be a viable alternative to mouse bioassay [111]. The certified materials including a DA calibration solution and a mussel tissue reference material have been developed for ASP to aid in analytical quality assurance through the Certified Reference Materials Programme of the National Research Council, Canada [130].

19.3.4 Brevetoxin—Neurotoxic Shellfish Poisoning

19.3.4.1 Overview

The first documented event of a "red tide" dinoflagellate bloom of *K. brevis* (also known as *Gymnodinium breve* and *Ptychodiscus breve*) was over 100 years ago. Since that time, scientific interest in the mammalian intoxications, massive fish, and bird kills that result from such blooms along the Gulf coast of the United States and in other parts of the world has increased and resulted in advanced research. An unusual feature of *K. brevis* is the formation of toxic aerosols through wave action that can lead to asthma-like symptoms in humans. NSP is caused by brevetoxins (PbTxs), which are tasteless, odorless, heat and acid stable, lipid-soluble, and cyclic polyethers. The molecular structure of the brevetoxins consists of 10–11 transfused rings; their molecular weights are around 900 Da, and 10 brevetoxins have been isolated and identified from field blooms and *K. brevis* cultures [131]. The two major brevetoxins, PbTx-2 and PbTx-3, have been shown to act on receptor site 5 of the voltage-sensitive sodium channel where they bind and cause persistent activation, increased sodium flux, and subsequent depolarization of excitable cells at resting potential.

19.3.4.2 Clinical Symptoms

The toxic effects of brevetoxin can be passed through inhalation and the dermal exposure of aerosolized dinoflagellate particles, and the oral ingestion of raw or cooked shellfish contaminated with brevetoxins. Dermal exposure occurs when the fragile *K. brevis* is broken open during rough surf, releasing the toxins that can cause irritation of the eye and the nasal membranes of the swimmers or those in direct contact with toxic blooms [132,133]. In addition to skin irritations, inhalation of aerosolized red tide brevetoxins may cause respiratory distress, conjunctival irritations, rhinorrhea, nonproductive cough, and bronchoconstriction. Other symptoms such as dizziness, tunnel vision, and skin rashes are also common. The condition is readily reversible in most individuals once they leave the affected area, however, those with asthma or chronic lung conditions have reported more difficulties including prolonged lung disease as a result of exposure

[133,134]. Brevetoxin is thought to cause chronic immunosuppression, possibly mediated through interactions with cysteine cathepsins that are naturally present in immune cells and involved in antigen presentation [135].

The oral ingestion of contaminated shellfish induces a toxic syndrome similar to PSP and ciguatera fish poisoning, although with a lesser degree of severity. The symptoms of brevetoxin through ingestion generally appear within 30 min to 3 h of exposure and may include nausea, vomiting, diarrhea, chills, sweats, reversal of temperature sensation, hypotension, numbness, tingling, paresthesias of lips, face, and extremities, bronchoconstriction, paralysis, and even coma. Fatalities are extremely rare, chronic symptoms as a result of ingestion have not been reported, and treatment is primarily supportive.

19.3.4.3 Detection Methods

19.3.4.3.1 Bioassays

The mouse bioassay involves the IP injection of a crude lipid obtained from a diethylether extraction of shellfish into mice weighing 20 g where 1 MU is defined as the amount of crude toxic residue that on an average will kill 50% of the test animals in 930 min. In practice, a residue toxicity of 20 MU per 100 g shellfish tissue was adopted, and remains as the guidance level for the prohibition of shellfish harvesting [136]. The drawbacks to the mouse assay are that it requires large numbers of animals, uses relatively large amounts of tissue extracts, the results are interpreted subjectively, and it lacks specificity [137].

Mosquito fish (*Gambusia affinis*) bioassays can be conducted in 20 mL seawater (3.5% salinity) using one fish per vessel with toxin added in 0.01 mL ethanol and median lethal doses determined using the tables in Weil from 1952 [138]. The fish bioassay is generally used to determine the potency of either the contaminated seawater or crude and purified toxin extracts [139].

A neuroblastoma cell assay takes advantage of the toxic effects of NSPs and their affinity for voltage-sensitive Na^+ channels. Using this method, the detection limit for PbTxs is 0.25 ng/10 mL tissue extract and can be detected within 4–6 h, though the detection limit can be decreased with an incubation time of 22 h [140]. The detection is based on functional activity rather than on the recognition of a structural component, as is the case of an antibody-based assay and the affinity of a toxin for its receptor is directly proportional to its toxic potency, which can affect the specificity and the sensitivity of this assay.

Fairey et al. [141] reported a further modification of the receptor-binding assay in neuroblastoma cells to a reporter gene assay that utilizes luciferase-catalyzed light generation as an endpoint and a microplate luminometer for quantification. The results indicated that the assay was capable of meeting or exceeding the sensitivity of bioassays for sodium channel active algal toxins.

Van Dolah et al. [142] developed a high-throughput synaptosome-binding assay for brevetoxins using microplate scintillation technology. The microplate assay can be completed within 3 h, has a detection limit of less than 1 ng and can analyze dozens of samples simultaneously. The assay has been demonstrated to be useful for assessing algal toxicity, for purification of brevetoxins, and for the detection of brevetoxins in seafood.

19.3.4.3.2 Immunoassays

A competitive RIA was developed for the detection of PbTx-2 and PbTx-3 at 1 nM concentrations [143], and ELISA methods for brevetoxin detection have since ensued. The modifications to early

ELISA methods have resulted in improved detection and specificity to where the method can be used to screen for brevetoxins in dinoflagellate cells, in shellfish and fish seafood samples, in seawater and culture media, and in human serum samples [144–147] with detection limits ranging from 0.33 pmol for PbTx-3 to 2.5 μg/100 g shellfish meat in spiked oysters.

19.3.4.3.3 Chemical Methods

Using micellar electrokinetic capillary chromatography, brevetoxins were isolated from cell cultures and fish tissue and the method detection limit in fish tissue was ~4 pg/g [148]. The reversed-phase LC–ESI-MS was successfully applied to the separation and the identification of brevetoxins associated with red tide algae [149], and an ionspray LC–MS method was shown to have mass detection limits as low as 10 pg (10 fmol) when using the selected ion monitoring of the $(M + H)^+$ ions. The analyses by LC–MS can be very rapid (as low as 2 min in some cases) and can be completely automated [150]. A fish tissue procedure based on gradient reversed-phase LC/MS/MS was used for the detection of PbTx-2 in fish tissue, and the detection limit in fish flesh using this method was at least 0.2 ng/g [34].

19.4   Other Toxins

19.4.1   *Hepatotoxins—Microcystins*

19.4.1.1   *Overview*

Cyanobacteria, also known as blue-green algae, are Gram-negative photosynthetic prokaryotes that can be found in both terrestrial and aquatic habitats, generally preferring temperatures between 20°C and 25°C [151]. Toxins produced by cyanobacteria differ according to their toxicological properties and chemical structures, which include hepatotoxic cyclic peptides such as microcystins and nodularins, neurotoxic alkaloids, and lipopolysaccharides. Cyanobacterial genera that produce microcystins include *Microcystis, Planktothrix (Oscillatoria), Anabaena, Nostoc, Anabaenopsis,* and *Hapalosiphon* [152], while nodularins are produced by *Nodularia spumigena*, a brackish water cyanobacterium [153].

Currently there are more than 60 variants of microcystin, which differ in toxicity [154,155], however, microcystin-LR is considered the most common in cyanobacteria. Nodularins are structurally similar to microcystins and exert similar toxicities. Microcystins contain five invariant amino acids, namely, D-alanine, D-methylaspartic acid, adda, D-glutamic acid, and *N*-methyldehydroalanine, and two variant L-amino acids. The "adda" amino acid (3-amino-9-methoxy-2,6,8-trimethyl-10-phenyl-4,6-dienoic acid) contributes to the toxicity of the microcystins and the nodularins by inhibiting several eukaryotic processes such as growth, protein synthesis, glycogen metabolism, and muscle contraction, and provides the microcystins with a characteristic absorption wavelength at 238 nm due to the presence of a conjugated diene group in the long carbon chain. This absorption provides a means of analysis after separation using reverse phase chromatography [155].

19.4.1.2   *Clinical Symptoms*

Human exposure to cyanobacterial toxins is mainly through ingestion and direct contact with contaminated waters. In the case of ingestion, drinking water contaminated with toxic blue-green algae has been reported, as has the consumption of fish and blue-green algal products used as food supplements. Swimming in waters where toxic blooms are occurring can lead to dermatitis and gastrointestinal symptoms such as vomiting and diarrhea [154]. Hepatotoxins induce massive hemorrhages, hepatocyte necrosis, disruption of mammalian liver systems, tumor promotion, and adverse kidney effects. Apoptotic and morphological changes have been observed at the cellular level including cell shrinkages, chromosomal breakage, and organelle redistribution.

19.4.1.3   Detection Methods

19.4.1.3.1   Bioassays

Bioassays involving mice, *Artemia salina*, *Sinapis alba* seedlings, and animal cell lines offer simple and rapid screening for microcystins [156], however these methods often lack the specificity necessary for adequate detection and validation. Protein phosphatase inhibition assays (PPIAs) can be radioisotopic and colorimetric, and have been developed based on the ability of microcystins to inhibit serine–threonine protein phosphatase enzymes [157–159]. The detection has also been reported using bioluminescence and fluorogenic substrates [160,161].

19.4.1.3.2   Immunoassays

A specific, sensitive ELISA has been developed using either polyclonal [162] or monoclonal antibodies [163,164] and while this and the PPIA detect microcystins that are below the guideline levels of the World Health Organization (WHO), which recommends that drinking water should have less than 1 μg/L, there are compatibility issues including cross-reactivity of the antibodies with variants, and underlying phosphatase activity in the sample preparation that masks the effects of the toxin.

19.4.1.3.3   Chemical Methods

HPLC retrofitted with UV detection or MS is a powerful tool for the identification of microcystins, capable of providing both quantitative and qualitative data [165]. In general, microcystins are separated on C18 silica column using a gradient of water and acetonitrile, acidified with trifluoroacetic acid or formic acid. Microcystins have characteristic spectra with absorption maxima at either 238 nm due to the "adda" residue, or at 222 nm for microcystins containing tryptophan [166]. One issue in the quantitative analysis of microcystins is the lack of suitable standards, as there are some 60 microcystin variants. In the absence of such standards, variants are often expressed as equivalents of microcystin-LR [167–169].

The HPLC–MS is widely accepted for the qualitative analysis of microcystins of interest. In this method, molecules are converted to desolvated ions, which are resolved based on mass and charge [170]. Other related methods including ESI and MS/MS have been utilized with success. More time-consuming, less specific methods include TLC, gas chromatography–MS, and capillary zone electrophoresis.

Novel approaches for the environmental monitoring of cyanobacterial blooms are developing with the advent of DNA sequencing and polymerase chain reaction. Such sequencing has led to the coding of microcystin genes in several major producers and has enabled the design of primers and probes to specifically detect and identify toxin-producing species in natural samples with low quantities [171,172]. Obstacles to water resource management include the inability to differentiate between toxic and nontoxic cyanobacterial blooms without isolation and testing, as neither strain shows a measurable difference in appearance.

19.4.2   New and Emerging Toxins

19.4.2.1   Pinnatoxins

Pinnatoxins are potent marine toxins common to the bivalve from genus Pinna, common in China and Japan where human intoxication is a regular occurrence [173]. Symptoms include diarrhea and neurological disturbances, and the toxin is thought to be a Ca^{2+} channel activator.

19.4.2.2   Azaspiracids

Azaspiracids, first found in mussels after a toxic incident in the Netherlands, exhibit the symptoms typical of DSP, including nausea, vomiting, diarrhea, and abdominal cramps. However, structural and toxicological studies show that the target organs and the mode of action are distinctly different from those of DSP, PSP, and ASP toxins [174,175].

19.4.2.3   Gymnodimine

Oysters from South Island, New Zealand in 1994 were found to contain a potent compound whose causative organism is *Gymnodinium* sp. This toxin exhibits potent mouse and ichthyotoxicity, with mice dying within 5–15 min following a minimum lethal dose of 450 mg/kg and fish at levels of 250–500 ppb. The structure of the toxin has been resolved through NMR [176].

19.4.2.4   Spirolides

Spirolides were isolated from the digestive glands of shellfish collected near Nova Scotia, Canada and possess an unusual seven-membered cyclic imine moiety that is spirolinked to a cyclohexane ring. The macrocyclic toxins may activate Ca^{2+} channels [176].

19.5   Conclusion

Many varied dynamics characterize the field of algal toxins, posing a challenge for biologists, toxicologists, biochemists, and pharmacologists interested in elucidating the molecular mechanisms and developing more sophisticated detection methods for such toxins. While there have been major advancements in this field in recent years, the increasing incidence of marine toxin poisonings worldwide as well as the continual discovery of new toxins demonstrate the need for the development of additional tools for biotoxin monitoring in seafood intended for mammalian consumption.

Analytical methods that could allow for the accurate estimates of the toxicity of the multiple classes of toxins using a single procedure would be ideal in managing the risks posed by phycotoxins. While such a global approach does not appear likely in the near future, continued efforts toward more rapid, sensitive, specific, and accurate testing methodologies will be encouraged in an effort to monitor marine toxins in the environment.

References

1. Halstead, B.W., *Poisonous and Venomous Marine Animals of the World*, 2nd edn. Darwin Press, Princeton, NJ, 1988, p. 1006.
2. Helfrich, P., Fish poisoning in Hawaii, *Hawaii Med. J.*, 22, 361, 1964.
3. Cook, J., Gurneaux, T., and Hodoes, W., Eds., *A Voyage towards the South Pole and Around the World*, 3rd edn., W. Strahan and T. Cadell, London, 1977.
4. Bagnis, R.A., Clinical aspects of ciguatera (fish poisoning) in French Polynesia, *Hawaii Med. J.*, 20, 25, 1964.
5. Engleberg, N.C. et al., Ciguatera fish poisoning: A major common-source outbreak in the US Virgin Islands, *Ann. Int. Med.*, 98, 336, 1964.
6. Murata, M. et al., Structures of ciguatoxin and its congener, *J. Am. Chem. Soc.*, 111, 8929, 1989.
7. Murata, M. et al., Structures and configurations of ciguatoxin from the moray eel Gymnothorax-javanicus and its likely precursor from the dinoflagellate *Gambierdiscus toxicus*, *J. Am. Chem. Soc.*, 112, 4380, 1990.
8. Lewis, R.J. et al., Ciguatoxin-2 is a diastereomer of ciguatoxin-3, *Toxicon*, 31, 637, 1993.
9. Yasumoto, T. et al., Structural elucidation of ciguatoxin congeners by fast-atom bombardment tandem mass spectroscopy, *J. Am. Chem. Soc.*, 122, 4988, 2000.
10. Lewis, R.J., Ion channel toxins and therapeutics: From cone snail venoms to ciguatoxin, *Ther. Drug. Monit.*, 22, 61, 2000.
11. Chan, T.Y.K., Lengthy persistence of ciguatoxin in the body, *Trans. R. Soc. Trop. Med. Hyg.*, 92, 662, 2000.
12. Palafox, N.A. et al., Successful treatment of ciguatera fish poisoning with intravenous mannitol, *JAMA*, 259, 2740, 1988.
13. Hokama, Y. et al., Assessment of ciguateric fish in Hawaii by immunological mouse toxicity and guinea pig atrial assay, *Memoirs Qld. Museum*, 34, 489, 1994.
14. Kimura, L.H. et al., Comparison of the different assays for the assessment of ciguatoxin in fish tissue: Radioimmunoassay, mouse bioassay and *in vitro* guinea pig atrium assay, *Toxicon*, 20, 907, 1982.
15. Hokama, Y., Ciguatera fish poisoning: Features, tissue, and body effects, in *Reviews in Food and Nutrition Toxicity*, vol. 2, Preedy, V. and Watson, R., Eds., CRC Press, Boca Raton, FL, 2004, p. 43.
16. Miyahara, J.T., Oyama, M.M., and Hokama, Y., The mechanism of cardiotonic action of ciguatoxin, in *Proceedings of the Fifth International Coral Reef Congress*, Gabrie, C. and Salvat, B., Eds., Tahiti, 1985, p. 449.
17. Hungerford, J.M., Seafood toxins and seafood products, *J. AOAC Int.*, 76, 120, 1993.
18. Bagnis, R. et al., Epidemiology of ciguatera in French Polynesia from 1960 to 1984, in *Proceedings of the Fifth International Coral Reef Congress*, Gabrie, C. and Salvat, B., Eds., Tahiti, 1985, p. 475.
19. Bagnis, R. et al., The use of mosquito bioassay for determining toxicity to man of ciguateric fish, *Biol. Bull.*, 172, 137, 1987.
20. Vernoux, J.P. et al., Chick feeding test: A simple system to detect ciguatoxins, *Acta Trop.*, 42, 235, 1985.
21. Labrousse, H. and Matile, L., Toxicological biotest on diptera larvae to detect ciguatoxins and various other toxic substances, *Toxicon*, 34, 881, 1996.
22. Manger, R. et al., Detection of sodium channel toxins: Directed cytotoxicity assays of purified ciguatoxins, brevetoxins, saxitoxins, and seafood extracts, *J. AOAC Int.*, 78, 521, 1995.
23. Louzao, M.C. et al., Detection of sodium channel activators by a rapid fluorimetric microplate assay, *Chem. Res. Toxicol.*, 17, 572, 2004.
24. Shimojo, R.Y. and Iwaoka, W.T., A rapid hemolysis assay for the detection of sodium channel-specific marine toxins, *Toxicology*, 154, 1, 2000.
25. Hokama, Y., Banner, A.H., and Boylan, D.B., A radioimmunoassay for the detection of ciguatoxin, *Toxicon*, 15, 317, 1977.

26. Hokama, Y. et al., A simple membrane immunobead assay for detecting ciguatoxin and related polyethers from human ciguatera intoxication and natural reef fishes, *J. AOAC Int.*, 84, 727, 1998.

27. Trainer, V.L., Baden, D.G., and Catterall, W.A., Detection of marine toxins using reconstituted sodium channels, *J. AOAC Int.*, 78, 570, 1995.

28. Dickey, R.W. et al., Liquid chromatographic-mass spectrometric methods for the determination of marine polyether toxins, *Bull. Soc. Pathol. Exp.*, 85, 514, 1992.

29. Yasumoto, T. et al., A turning point in ciguatera study, in *Toxic Phytoplankton Blooms in the Sea*, Smayda, T.J. and Shimizu, Y., Eds., Elsevier, New York, 1993, p. 455.

30. Lewis, R.J. et al., Ionspray mass spectrometry of ciguatoxin-1, maitotoxin-2 and -3 and related marine polyether toxins, *Nat. Toxins.*, 2, 56, 1994.

31. Lewis, R.J. and Sellin, M., Multiple ciguatoxins in the flesh of fishes, *Toxicon*, 30, 915, 1992.

32. Satake, M. et al., Isolation of a ciguatoxin analog from cultures of *Gambierdiscus toxicus*, in *Toxic Phytoplankton Blooms in the Sea*, Smayda, T.J. and Shimizu, Y., Eds., Elsevier, New York, 1993, p. 575.

33. Lewis, R.J. and Jones, A., Characterization of ciguatoxins and ciguatoxin congeners present in ciguateric fish by gradient reverse-phase high-performance liquid chromatography/mass spectrometry, *Toxicon*, 35, 159, 1997.

34. Lewis, R.J., Jones, A., and Vernoux, J.P., HPLC/tandem electrospray mass spectrometry for the determination of sub-ppb levels of Pacific and Caribbean ciguatoxins in crude extracts of fish, *Anal. Chem.*, 71, 247, 1999.

35. Hwang, D.-F. and Noguchi, T., Tetrodotoxin poisoning, in *Advances in Food and Nutrition Research*, vol. 2, Taylor, S., Ed., Elsevier, San Diego, CA, 2007, p. 141.

36. Isbister, G.K. and Kiernan, M.C., Neurotoxic marine poisoning, *Lancet Neurol.*, 4, 219, 2005.

37. Cestele, S. and Caterall, W.A., Molecular mechanisms of neurotoxin action on voltage-gated sodium channels, *Biochimie*, 82, 997, 2000.

38. O'Leary, M.A., Schneider, J.J., and Isbister, G.K., Use of high performance liquid chromatography to measure tetrodotoxin in serum and urine of poisoned patients, *Toxicon*, 44, 549, 2004.

39. Torda, T.A., Sinclair, E., and Ulyatt, D.B., Puffer fish (tetrodotoxin) poisoning: Clinical record and suggested management, *Med. J. Aust.*, 1, 599, 1973.

40. Sorokin, M., Puffer fish poisoning, *Med. J. Aust.*, 1, 957, 1973.

41. Tibballs, J., Severe tetrodotoxic fish poisoning, *Anaesth. Intensive Care*, 16, 215, 1988.

42. Kogure, K. et al., A tissue culture assay for tetrodotoxin, saxitoxin, and related toxins, *Toxicon*, 26, 191, 1988.

43. Hamasaki, K., Kogure, K., and Ohwada, K., A biological method for the quantitative measurement of tetrodotoxin TTX: Tissue culture bioassay in combination with a water soluble tetrazolium salt, *Toxicon*, 34, 490, 1996.

44. Huot, R.I., Armstrong, D.L., and Chanh, T.C., Protection against nerve toxicity by monoclonal antibodies to the sodium channel blocker tetrodotoxin, *J. Clin. Invest.*, 83, 1821, 1989.

45. Watabe, S. et al., Monoclonal antibody raised against tetrodonic acid, a derivative of tetrodotoxin, *Toxicon*, 27, 265, 1989.

46. Matsumura, K. and Fukiya, S., Indirect competitive enzyme immunoassay for tetrodotoxin using a biotin-avidin system, *J. AOAC Int.*, 75, 883, 1992.

47. Raybould, T.J.G. et al., A monoclonal antibody-based immunoassay for detection tetrodotoxin in biological samples, *J. Clin. Lab. Anal.*, 6, 65, 1992.

48. Kawatsu, K. et al., Rapid and highly sensitive enzyme immunoassay for quantitative determination of tetrodotoxin, *Jpn. Med. Sci. Biol.*, 50, 133, 1997.

49. Kawatsu, K., Shibata, T., and Hamano, Y., Application of immunoaffinity chromatography for detection of tetrodotoxin from urine samples of poisoned patients, *Toxicon*, 37, 325, 1999.

50. Yasumoto, T. et al., Construction of a continuous tetrodotoxin analyzer, *Bull. Jpn. Soc. Sci. Fish.*, 48, 1481, 1982.

51. Yotsu, M., Endo, A., and Yasumoto, T., An improved tetrodotoxin analyzer, *Agric. Biol. Chem.*, 53, 893, 1989.

52. Nagashima, Y. et al., Analysis of paralytic shellfish poison and tetrodotoxin by ion-pairing high performance liquid chromatography, *Nippon Suisan Gakkaishi*, 53, 819, 1987.

53. Arakawa, O. et al., Occurrence of 11-oxotetrodotoxin and 11-nortetrodotoxin-6R-ol in a xanthid crab *Atergatis floridus* collected at Kojima, Ishigaki Island, *Fish. Sci.*, 60, 769, 1994.

54. Shimada, K. et al., Determination of tetrodotoxin by capillary isotachophoresis, *J. Food Sci.*, 48, 665, 1983.

55. Shida, Y. et al., LC/MS of marine toxin-1, in *Proceedings of the 46th Annual Conference on Mass Spectrometry*, Mass Spectrometry Society, Japan, 1998, p. 137.

56. Hwang, P.A. et al., Identification of tetrodotoxin in a marine gastropod *Nassarius glans* responsible for human morbidity and mortality in Taiwan, *J. Food Protect.*, 68, 1696, 2005.

57. Tanu, M.B. and Noguchi, T., Tetrodotoxin as a toxic principle in the horseshoe crab *Carcinoscorpius rotundicauda* collected from Bangladesh, *J. Food. Hyg. Soc. Jpn.*, 40, 426, 1999.

58. Narita, H. et al., Occurrence of tetrodotoxin in a trumpet shellfish "boshubora" *Charonia sauliae*, *Nippon Suisan Gakkaishi*, 47, 935, 1981.

59. Onoue, Y., Noguchi, T., and Hashimoto, K., Tetrodotoxin determination methods, in *Seafood Toxins*, Ragelis, E.P., Ed., American Chemical Society, Washington, DC, 1984, p. 345.

60. Noguchi, T. et al., Tetrodonic acid-like substance: A possible precursor of tetrodotoxin, *Toxicon*, 29, 845, 1991.

61. Tanu, M.B. et al., Occurrence of tetrodotoxin in the skin of a rhacophoridid frog *Polypedates* sp. from Bangladesh, *Toxicon*, 39, 937, 2001.

62. Endo, A. et al., Isolation of 11-nortetrodotoxin-6R-ol and other tetrodotoxin derivatives from the puffer *Fugu nipholes*, *Tetrahedron Lett.*, 29, 4127, 1988.

63. Meyer, K.F., Sommer, H., and Schoenholz, P., Mussel poisoning, *J. Prevent. Med.*, 2, 365, 1928.

64. Sommer, H. and Meyer, K.F., Paralytic shellfish poison, *Arch. Pathol.*, 24, 560, 1937.

65. Mons, M.N., Van Egmond, H.P., and Speijers, G.J.A., Paralytic shellfish poisoning: A review. RIVM Report 288802 005, June 1998.

66. Food and Agriculture Organization of the United Nations (FAO), Food and Nutrition Paper 80, Rome, 2004, p. 278.

67. De Carvalho, M., Jacinto, J., and Ramos, N., Paralytic shellfish poisoning: Clinical and electrophysiological observations, *J. Neurol.*, 245, 551, 1998.

68. Chen, H.M., Lin, C.H., and Wang, T.M., Effects of 4-aminopyridine on saxitoxin intoxication, *Toxicol. Appl. Pharmacol.*, 141, 44, 1996.

69. Chang, F.C. et al., 4-Aminopyridine reverses saxitoxin (STX) and tetrodotoxin (TTX) induced cardiorespiratory depression in chronically instrumented guinea pigs, *Fundam. Appl. Toxicol.*, 38, 75, 1997.

70. Fernandez, M. and Cembella, A.D., Mammalian bioassays, in *Manual on Harmful Marine Microalgae, IOC Manuals and Guides*, vol. 33, Hallegraeff, G.M., Anderson, D.M., Cembella, A.D., Eds., UNESCO, Paris, 1995, p. 213.

71. Hollingworth, T. and Wekell, M.M., Fish and other marine products 959.08 Paralytic shellfish poisoning. Biological Method, Final Action, in *Official Methods of Analysis of the Association of Analytical Chemists*, 15th edn., Helrich, K., Ed., AOAC, Richmond, VA, 1990, p. 881.

72. Schantz, E.J., Historical perspective on paralytic shellfish poisoning, in *Seafood Toxins*, Ragelis, E.P., Ed., American Chemical Society, Washington, DC, 1984, p. 99.

73. Hald, B., Bjergskov, T., and Emsholm, H., Monitoring and analytical programmes on phycotoxins in Denmark, in *Proceedings of the Symposium on Marine Biotoxins*, Fremy, J.M., Ed., Centre National d-Etudes Veterinaires et Alimentaires, Paris, 1991, p. 181.

74. Jellet, J.F. et al., Paralytic shellfish poison (saxitoxin family) bioassays: Automated endpoint determination and standardization of the in vitro tissue culture bioassay, and comparison with the standard mouse bioassay, *Toxicon*, 30, 1143, 1992.

75. Chu, F.S. and Fan, T.S.L., Indirect enzyme-linked immunosorbent assay for saxitoxin in shellfish, *J. AOAC Int.*, 68, 13, 1985.

76. Cembella, A. D. and Lamoureux G., A competitive inhibition enzyme linked immunoassay for the detection of paralytic shellfish toxins in marine phytoplankton, in *Toxic Phytoplankton Blooms in the Sea*, Smayda, T.J. and Shimizu, Y., Eds., Elsevier, Amsterdam, the Netherlands, 1993, p. 857.

77. Chu, F.S. et al., Screening of paralytic shellfish poisoning toxins in naturally occurring samples with three different direct competitive enzyme-linked immunosorbent assays, *J. Agric. Food Chem.*, 44, 4043, 1996.

78. Kawatsu, K. et al., Development and application of an enzyme immunoassay based on a monoclonal antibody against gonyautoxin components of paralytic shellfish poisoning toxins, *J. Food Protect.*, 65, 1304, 2002.

79. Bates, H.A. and Rapoport, H., A chemical assay for saxitoxin, the paralytic shellfish poison, *J. Agric. Food Chem.*, 23, 237, 1975.

80. Bates, H.A. and Rapoport, H., A chemical assay for saxitoxin. Improvements and modifications, *J. Agric. Food Chem.*, 26, 252, 1978.

81. Gawley, R.E. et al., Selective detection of saxitoxin over tetrodotoxin using acridinylmethyl crown ether chemosensor, *Toxicon*, 45, 783, 2005.

82. Sullivan, J.J., Methods of analysis for DSP and PSP toxins in shellfish: A review, *J. Shellfish Res.*, 7, 587, 1988.

83. Van Egmond, H.P. et al., Paralytic shellfish poison reference materials: An intercomparison of methods for the determination of saxitoxin, *Food Add. Contam.*, 11, 39, 1994.

84. Waldock, M.J. et al., An assessment of the suitability of HPLC techniques for monitoring of PSP and DSP on the east coast of England, in *Proceedings of the International Symposium on Marine Biotoxins*, Fremy, J.M. Ed., CNEVA Publication Series, Paris, 1991, p. 137.

85. Quilliam, M.A. et al., Ion-spray mass spectrometry of marine neurotoxins, *Rapid Commun. Mass Spectrom.*, 3, 145, 1989.

86. Pleasance, S. et al., Ionspray mass spectrometry of marine toxins, III. Analysis of paralytic shellfish poisoning toxins by flow-injection analysis, liquid chromatography/mass spectrometry and capillary electrophoresis/mass spectrometry, *Rapid Commun. Mass Spectrom.*, 6, 14, 1992.

87. Jaime, E. et al., Determination of paralytic shellfish poisoning toxins by high-performance ion-exchange chromatography, *J. Chromatogr. A*, 929, 43, 2001.

88. Van Apeldoorn, M.E., Diarrhoeic shellfish poisoning: A review, RIVM/CSR Report 05722A00, August 1998.

89. Draisci, R. et al., First report of pectenotoxin-2 (PTX-2) in algae (*Dinophysis fortii*) related to seafood poisoning in Europe, *Toxicon*, 34, 923, 1996.

90. Van Egmond, H.P. et al., Paralytic and diarrhoeic shellfish poisons: Occurrence in Europe, toxicity, analysis and regulation, *J. Nat. Toxins*, 2, 41, 1993.

91. Quilliam, M.A., General referee reports, committee on natural toxins, phycotoxins, *J. AOAC Int.*, 81, 142, 1998.

92. Hallegraeff, G.M., Anderson, D.M., and Cembella, A.D., Eds., *Manual on Harmful Marine Microalgae, IOC Manuals and Guides*, No. 33, UNESCO, Paris, 1995.

93. Terao, K. et al., Histopathological studies on experimental marine toxin poisoning: The effects in mice of yessotoxin isolated from *Patinopecten yessoensis* and of a desulfated derivative, *Toxicon*, 28, 1095, 1990.

94. Terao, K. et al., A comparative study of the effects of DSP toxins on mice and rats, in *Toxic Phytoplankton Blooms in the Sea*, Smayda, T.J., and Shimitzu, Y., Eds., Elsevier, Amsterdam, the Netherlands, 1993, p. 581.

95. EC, Commission Decision of 15 March 2002, Laying down rules for the implementation of Council Directive 91/492/EEC as regards the maximum levels and the methods of analysis of certain marine biotoxins in bivalve molluscs, echinoderms, tunicates and marine gastropods, *Off. J. Eur. Commun.*, 2002, 62.

96. Mouratidou, T. et al., Detection of marine toxin okadaic acid in mussels during a diarrhetic shellfish poisoning (DSP) episode in Thermaikos Gulf, Greece, using biological, chemical and immunological methods, *Sci. Tot. Environ.*, 366, 894, 2006.

97. Yasumoto, T., Oshima, Y., and Yamaguchi, M., Occurrence of a new type of shellfish poisoning in the Tokohu District, *Bull. Jpn. Soc. Sci. Fish.*, 44, 1249, 1978.

98. Vernoux, J.P. et al., The use of *Daphnia magna* for detection of okadaic acid in mussel extracts, *Food Add. Contam.*, 10, 603, 1993.

99. Aune, T., Yasumoto, T., and Engeland, E., Light and scanning electron microscopic studies on effects of marine algal toxins toward freshly prepared hepatocytes, *J. Toxicol. Environ. Health*, 34, 1, 1991.

100. Amzil, Z. et al., Short-time cytotoxicity of mussel extracts: A new bioassay for okadaic acid detection, *Toxicon*, 30, 1419, 1992.

101. Botrè, F. and Mazzei, F., Inhibition enzymic biosensors: An alternative to global toxicity bioassays for the rapid determination of phycotoxins, *Int. J. Environ. Pollut.*, 13, 173, 2000.

102. Marquette, C.A., Coulet, P.R., and Blum, L.J., Semi-automated membrane based chemiluminiscent immunosensor for flow injection analysis of okadaic acid in mussels, *Anal. Chim. Acta*, 398, 173, 1999.

103. Simon, J.F. and Vernoux, J.P., Highly sensitive assay of okadaic acid using protein phosphatase and paranitrophenyl phosphate, *Nat. Toxins*, 2, 293, 1994.

104. Tubaro, A. et al., A protein phosphatase 2A inhibition assay for a fast and sensitive assessment of okadaic acid contamination in mussels, *Toxicon*, 34, 743, 1996.

105. Vieytes, M.R. et al., A fluorescent microplate assay for diarrheic shellfish toxins, *Anal. Biochem.*, 248, 258, 1997.

106. Lee, J.S. et al., Fluorimetric determination of diarrhetic shellfish toxins by high-performance liquid chromatography, *Agric. Biol. Chem.*, 51, 877, 1987.

107. Aase, B. and Rogstad, A., Optimization of sample clean-up procedure for determination of diarrhetic shellfish poisoning toxins by use of experimental design, *J. Chromatogr. A.*, 764, 223, 1997.

108. GFL, Determination of Okadasäure in mussels with HPLC (L 12.03/04-2), in *Official Collection of Methods under Article 35 of the German Federal Act; Methods of Sampling and Analysis of Foods, Tobacco Products, Cosmetics and Commodity Goods/BgVV*, vol. 1., Köln, Beuth Verlag GmbH, Berlin, 2001.

109. Yasumoto, T. and Takizawa, A., Fluorimetric measurement of Yessotoxins in shellfish by high-pressure liquid chromatography, *Biosci. Biotech. Biochem.*, 61, 1775, 1997.

110. Sasaki, K. et al., Fluorometric analysis of Pectenotoxin-2 in microalgal samples by high performance liquid chromatography, *Nat. Toxins*, 7, 241, 1999.

111. Holland, P. and McNabb, P., Inter-Laboratory Study of an LC-MS Method for ASP and DSP Toxins in Shellfish, Cawthron Report No. 790, Cawthron Institute, Nelson, New Zealand, April 2003.

112. Subba Rao, D.V., Quilliam, M.A., and Pocklington, R., Domoic acid—A neurotoxic amino acid produced by the marine diatom *Nitzscia pungens* in culture, *Can. J. Fish. Aquat. Sci.*, 45, 2076, 1988.

113. Fritz, L. et al., An outbreak of domoic acid and poisoning attributed to the pinnate diatom *Pseudonitzschia australis*, *J. Phycol.*, 28, 439, 1992.

114. Takemoto, T., Isolation and structural identification of naturally occurring excitatory amino acids, in *Kainic Acid as a Tool in Neurobiology*, McGeer, E.G., Olney, J.W., and McGeer, P.L., Eds., Raven Press, New York, 1978, p. 1.

115. Perl, T.M. et al., An outbreak of toxic encephalopathy caused by eating mussels contaminated with domoic acid, *N. Engl. J. Med.*, 322, 1775, 1990.

116. Teitelbaum, J.S. et al., Neurologic sequelae of domoic acid intoxication due to the ingestion of contaminated mussels, *N. Engl. J. Med.*, 322, 1781, 1990.

117. AOAC, Paralytic Shellfish Poison. Biological method. Final action, in *Official Method of Analysis*, 15th edn., Sec 959.08., Hellrich, K., Ed., Association of Official Analytical Chemists (AOAC), Richmond, VA, 1990, p. 881.

118. Van Dolah, F.M. et al., A microplate receptor assay for the amnesic shellfish poisoning toxin, domoic acid, utilizing a cloned glutamate receptor, *Anal. Biochem.*, 245, 102, 1997.

119. Smith, D.S. and Kitts, D.D., Enzyme immunoassay for the determination of domoic acid in mussel extracts, *J. Agric. Food. Chem.*, 43, 367, 1995.

120. Garthwaite, I. et al., Polyclonal antibodies to domoic acid and their use in immunoassays for domoic acid in seawater and shellfish, *Nat. Toxins*, 6, 93, 1998.

121. Traynor, I.M. et al., Detection of the marine toxin domoic acid in bivalve molluscs by immunobiosensor, Poster presented at the *4th International Symposium on Hormone and Veterinary Drug Residue Analysis*, Antwerpen, Belgium, 2002.

122. Quilliam, M.A., Thomas, K., and Wright, J.L.C., Analysis of domoic acid in shellfish by thin-layer chromatography, *Nat. Toxins*, 6, 147, 1998.

123. Lawrence, J.F., Charbonneau, C.F., and Ménard, C., Liquid chromatographic determination of domoic acid in mussels, using AOAC paralytic shellfish poison extraction procedure: Collaborative study, *J. AOAC Int.*, 74, 68, 1991.

124. Lawrence, J.F. et al., Liquid chromatographic determination of domoic acid in shellfish products using the paralytic shellfish poison extraction procedure of the association of official analytical chemists, *J. Chromatogr.* 462, 349, 1998.

125. Pocklington, R. et al., Trace determination of domoic acid in seawater and phytoplankton by high-performance liquid chromatography of the fluorenylmethoxycarbonyl (FMOC) derivative, *Intern. J. Environ. Anal. Chem.*, 38, 351, 1990.

126. Wright, J.L.C. and Quilliam, M.A., Methods for domoic acid, the amnesic shellfish poisons, in *Manual on Harmful Marine Microalgae, IOC Manuals and Guides*, No. 33, Hallegraeff, G.M. et al., Eds., UNESCO, Paris, 1995, p. 113.

127. Hess, P. et al., Determination and confirmation of the amnesic shellfish poisoning toxin, domoic acid in shellfish from Scotland by liquid chromatography and mass spectrometry, *J. AOAC Int.*, 84, 1657, 2001.

128. Powell, C.L. et al., Development of a protocol for determination of domoic acid in the sand crab (*Emerita analoga*): A possible new indicator species, *Toxicon*, 40, 485, 2002.

129 Furey, A. et al., Determination of azaspiracids in shellfish using liquid chromatography/tandem electrospray mass spectrometry, *Rapid Commun. Mass Spectrom.*, 16, 238, 2002.

130. National Research Council Canada (NRC) Institute for Marine Biosciences, 2003. Available at http://www.nrc.ca/imb.

131. Benson, J.M., Thischler, D.L., and Baden, D.G., Uptake, distribution, and excretion of brevetoxin 3 administered to rats by intratracheal instillation, *J. Toxicol. Environ. Health Part A*, 56, 345, 1999.

132. Cembella, A.D. et al., In vitro biochemical and cellular assays, in *Manual on Harmful Marine Microalgae, IOC Manuals and Guides*, No. 33, Hallegraeff, G.M., Anderson, D.M., and Cembella, A.D., Eds., UNESCO, Paris, 1995.

133. Fleming, L.E. and Baden, D.G., *Florida Red Tide and Human Health: Background*, 1999. Available at http://www.redtide.whoi.edu/hab/illness/floridaredtide.html.

134. Watters, M.R., Organic neurotoxins in seafoods, *Clin. Neurol. Neurosurg.*, 97, 119, 1995.

135. Van Dolah, F.M., Roelke, D., and Greene, R.M., Health and ecological impacts of harmful algal blooms: Risk assessment needs. *Hum. Ecol. Risk Assess.*, 7, 1329, 2001.

136. Dickey, R. et al., Monitoring brevetoxins during a *Gymnodinium Breve* red tide: Comparison of sodium channel specific cytotoxicity assay and mouse bioassay for determination of neurotoxic shellfish toxins in shellfish extracts, *Nat. Toxins*, 7, 157, 1999.

137. Hokama, Y., Recent methods for detection of seafood toxins: Recent immunological methods for ciguatoxin and related polyethers, *Food Addit Contam.*, 10, 71, 1993.

138. Weil, C.S., Tables for convenient calculation of median effective dose (LD50 or ED50) and instruction in their use, *Biometrics*, 8, 249, 1952.

139. Viviani, R., Eutrophication, marine biotoxins, human health, *Sci. Total Environ. Suppl.*, 631, 631–662, 1992.

140. Manger, R.L. et al., Tetrazolium-based cell bioassay for neurotoxins active on voltage-sensitive sodium channels: Semiautomated assay for saxitoxin, brevetoxin and ciguatoxins, *Anal. Biochem.*, 214, 190, 1993.

141. Fairey, E.R., Edmunds, J.S.G., and Ramsdell J.S., A cell-based assay for brevetoxins, saxitoxins, and ciguatoxins using a stably expressed c-fos-luciferase reporter gene, *Anal. Biochem.*, 251, 129, 1997.

142. Van Dolah, F.M. et al., Development of rapid and sensitive high throughput pharmacologic assays for marine phycotoxins, *Nat. Toxins*, 2,189, 1994.

143. Trainer, V.L. and Baden, D.G., An enzyme immunoassay for the detection of Florida red tide brevetoxins, *Toxicon*, 29, 1387, 1991.

144. Baden, D.G. et al., Modified immunoassays for polyether toxins: Implications of biological matrixes, metabolic states, and epitope recognition, *J. AOAC Int.*, 78, 499, 1995.

145. Naar, J. et al., Polyclonal and monoclonal antibodies to PbTx-2-type brevetoxins using minute amount of hapten-protein conjugates obtained in a reversed micellar medium, *Toxicon*, 39, 869, 2001.

146. Naar, J. et al., A competitive ELISA to detect brevetoxins from *Karenia brevis* (formerly *Gymnodinium breve*) in seawater, shellfish and mammalian body fluid, *Environ. Health Perspect.*, 10, 179, 2002.

147. Garthwaite, I. et al., Integrated enzyme-linked immunosorbent assay screening system for amnesic, neurotoxic, diarrhetic, and paralytic shellfish poisoning toxins found in New Zealand, *J. AOAC Int.*, 84, 1643, 2002.

148. Shea, D., Analysis of brevetoxins by micellar electrokinetic capillary chromatography and laser-induced fluorescence detection, *Electrophoresis*, 18, 277, 2002.

149. Hua, Y. et al., On-line high-performance liquid chromatography-electrospray ionization mass spectrometry for the determination of brevetoxins in "red tide" algae, *Anal. Chem.*, 67, 1815, 1995.

150. Quilliam, M.A., Liquid chromatography-mass spectrometry: A universal method for analysis of toxins? In *Harmful Algae, Proceedings of the VIII International Conference on Harmful Algae*, Reguera, B., Blanco, J., Fernandez, M., and Wyatt, T., Eds., UNESCO, Paris, 1998, p. 509.

151. Msagati, T.A.M., Siame, B.A., and Shushu, D.D., Evaluation of methods for the isolation, detection and quantification of cyanobacterial hepatoxins, *Aquat. Toxicol.*, 78, 382, 2006.

152. Carmichael, W.W., The cyanotoxins, in *Advances in Botanical Research*, vol. 27, Callow, J.A., Ed., Academic Press Inc., San Diego, CA, 1997, p. 211.

153. Mankiewicz, J. et al., Natural toxins from cyanobacteria, *Acta Biol., Cracov. Bot.*, 45, 9, 2003.

154. Falconer, I.R., Cyanobacterial toxins of drinking water supplies: Cylindrospermopsins and microcystins, in *Cyanobacterial Poisoning of Livestock and People*, CRC Press, Boca Raton, FL, 2005, ch. 5.

155. Falconer, I.R., Is there a human health hazard from microcystins in the drinking water supply?, *Acta Hydrochim. Hydrobiol.*, 33, 64, 2005.

156. Rapala, J. and K. Lahti, Methods for detection of cyanobacterial toxins, in *Detection Methods for Algae, Protozoa and Helminthes in Fresh and Drinking Water, Water Quality Measurement Series*, Palumbo, F., Ziglip, G., Van der Beken, A., Eds., Wiley, New York, 2002, ch. 7.

157. Lambert, T.W. et al., Quantitation of the microcystin hepatotoxins in water at environmentally relevant concentrations with the protein phosphatase bioassay, *Environ. Sci. Technol.*, 28, 753, 1994.

158. Wong, B.S.F. et al., A colorimetric assay for screening microcystin class compounds in aquatic systems, *Chemosphere*, 38, 1113, 1999.

159. Almeida, V.P.S. et al., Colorimetric test for the monitoring of microcystins in cyanobacterial culture and environmental samples from southeast Brazil, *Braz. J. Microbiol.*, 37, 192, 2006.

160. Sugiyama, Y. et al., Sensitive analysis of protein phosphatase inhibitors by the firefly bioluminescence system; application to PP1, *Biosci. Biotechnol. Biochem.*, 60, 1260, 1996.

161. Bouaicha, N. et al., A colorimetric and fluorometric microplate assay for the detection of microcystin-LR in drinking water without preconcentration, *Food Chem. Toxicol.*, 40, 1677, 2002.

162. Metcalf, J.S., Bell, S.G., and Codd, G.A., Production of novel polyclonal antibodies against the cyanobacterial toxin microcystin-LR and their application for the detection and quantification of microcystins and nodularin, *Water Res.*, 34, 2761, 2000.

163. Mikhailov, A. et al., Production and specificity of mono and polyclonal antibodies against microcystins conjugated through N-methyldehydroalanine, *Toxicon*, 39, 477, 2001.
164. Zeck, A. et al., Highly sensitive immunoassay based on a monoclonal antibody specific for [4-arginine] microcystins, *Anal. Chim. Acta*, 441, 1, 2001.
165. Merliuoto, J.A.O., Chromatography of microcystins, *Anal. Chim. Acta*, 352, 277, 1997.
166. Moollan, R.W., Rae, B., and Verbeek, A., Some comments on the determination of microcystin toxins in waters by high performance liquid chromatography, *Analyst*, 121, 233, 1996.
167. Barco, M., Rivera, J., and Caixach, J., Analysis of cyanobacterial hepatotoxins in water samples by microbore reversed-phase liquid chromatography-electrospray ionisation mass spectrometry, *J. Chromatogr. A.*, 959, 103, 2002.
168. Barco, M. et al., Determination of microcystin variants and related peptides present in a water bloom of *Planktothrix (Oscillatoria) rubescens* in a Spanish drinking water reservoir by LC/ESI-MS, *Toxicon*, 44, 881, 2004.
169. McElhiney, J. and Lawton, L.A., Detection of the cyanobacterial hepatotoxins microcystins, *Toxicol. Appl. Pharmacol.*, 203, 219, 2005.
170. Graves, P.R. and Haystead, T.A.J., Molecular biologist's guide to proteomics, *Microb. Mol. Biol. Rev.*, 66, 39, 2002.
171. Rouhiainen, L. et al., Genes coding for hepatotoxic heptapeptides (microcystins) in the cyanobacterium *Anabaena* strain 90, *Appl. Environ. Microb.*, 70, 686, 2004.
172. Ouellette, A.J., Handy, S.M., and Wilhelm, S.W., Toxic *Microcystis* is widespread in Lake Erie: PCR detection of toxin genes and molecular characterization of associated cyanobacterial communities, *Microb. Ecol.*, 51, 154, 2006.
173. Twohig, M., New analytical methods for the determination of acidic polyether toxins in shellfish and marine phytoplankton, MSc thesis, Cork Institute of Technology, Cork, Ireland, 2001.
174. Satake, M. et al., Azaspiracid, a new marine toxin having unique spiro ring assemblies, isolated from Irish mussels, *Mytilus edulis*, *J. Am. Chem. Soc.*, 120, 9967, 1998.
175. Ito, E. et al., Multiple organ damage caused by a new toxin azaspiracid, isolated from mussels produced in Ireland, *Toxicon*, 38, 917, 2000.
176. Gago Martinez, A. and Lawrence, J.F., Shellfish toxins, in *Food Safety*, D'Mello, J.P.F., Ed., CAB International, Wellingford, U.K., 2003, p. 47.

Chapter 20

Detection of Adulterations: Addition of Foreign Proteins

Véronique Verrez-Bagnis

Contents

20.1 Introduction 604
20.2 Electrophoresis 604
20.3 Immunological Techniques 605
20.4 Visible and Near-Infrared Spectrometry 606
20.5 Microscopic Methods 607
20.6 Chromatographic Techniques 607
20.7 DNA Methods 607
    20.7.1 PCR-Sequencing 608
    20.7.2 Species-Specific PCR or Multiplex PCR 608
    20.7.3 Amplified Fragment Length Polymorphism 609
    20.7.4 PCR-Restriction Fragment Length Polymorphism 609
    20.7.5 Real-Time PCR 610
    20.7.6 PCR Lab-on-a-Chip 610
    20.7.7 Commercial PCR Kits for Fish Species Differentiation 611
20.8 Conclusion 611
References 611

20.1   Introduction

Recent food scares such as bovine spongiform encephalopathy (BSE), malpractices of some food producers, religious reasons, and food allergies have tremendously reinforced public awareness in the composition of food products [1]. Therefore, the description and/or labeling of food must be honest and accurate, particularly if the food has been processed removing the ability to distinguish one ingredient from another [2]. There are several ways in which food can be misdescribed: (1) the nondeclaration of processes (e.g., previous freezing or irradiation), (2) substitution of high-quality materials with ones of lower value, (3) overdeclaring a quantitative ingredient declaration, and (4) extending or adulteration of food with a base ingredient, such as water [2]. However, because labels do not provide sufficient guarantee about the true contents of a product, it is mandatory to identify and/or authenticate the components of processed food, thus protecting both consumers and producers from illegal substitutions [3]. Woolfe and Primrose [2] wrote on the needs of methods for detecting misdescription and fraud, that detecting the total substitution of one ingredient is easier than investigating partial substitution or adulteration. In many cases, it is necessary to know the possible adulterant before it can be detected. To decide whether it is adventitious mixing or deliberate substitution, the amount of adulterant present is usually to be confirmed.

Many different chemical and biochemical techniques have been developed for determining the authenticity of food. However, some techniques work well with raw products but lose their discrimination when applied to cooked or highly processed foods [2]. Molecular authentification or molecular traceability, based on the polymerase chain reaction (PCR) amplification of DNA, which has been developed in recent years, offers promising solutions for these issues [1].

In this chapter on seafood product adulteration by addition of foreign proteins, the different techniques used to identify such adulterations will be summarized. In fact, there are, strangely, few studies on seafood adulteration even if a considerable number of fish products may contain muscle or other tissue from one or more fish species. Examples are cooked and sterilized fish commodities such as cakes, pies, pastries, soups, patés, and industrial products such as fish meals. As Ascensio Gil [4] reported there are many forms of adulteration for economic gain such as the addition of undeclared cheaper fish in fish products that are labeled using the names of higher price and quality fish species. This chapter gives the reader, through results of research studies, an idea of the main methods used to detect seafood adulteration by substitution or addition of unlabeled component (foreign proteins).

20.2   Electrophoresis

The identification of fish can be problematic when morphological characteristics, such as the head, the skin, and the fins are removed. A well-used method for identifying raw fish is the characterization of muscle proteins using electrophoresis [5–11]. The methods used depend upon the separation of the muscle proteins (sarcoplasmic proteins and/or myofibrillar proteins) into species-specific profiles which, when compared with those of authentic species obtained under the same electrophoretic conditions, enable the species to be established unequivocally [12]. The identity of raw fish or shellfish is generally determined from the muscular water-soluble or sarcoplasmic proteins obtained by isoelectric focusing (IEF: separation of proteins according to their pI) [5], while cooked fish is analyzed by sodium dodecyl sulfate (SDS) electrophoresis of SDS protein extracts (myofibrillar, connective, and sarcoplasmic proteins) [6,8,10,11]. When such procedures are applied to detect adulteration rather than substitution of one species by another (i.e., mixed species products), their success depend upon the characteristic zones of the component species being identifiable in the profile of mixture. For most species of fish, IEF of the sarcoplasmic proteins is the preferred analytical system as the profiles generally have more species-specific components, with differences between species being much greater than SDS electrophoresis. A report of the European Commission (EC) project "Identification and Quantification of species in marine products" [13] noted that it is possible to identify and to evaluate the concentration of each species in a binary mixture (in the study: saithe and ling, respectively) by IEF if each species contained distinguishing protein bands with measurable intensity. In the same way, Podeszewski and Zarzycki [14] have previously applied the starch-gel electrophoresis on sarcoplasmic protein fractions of fish and demonstrated that this made it possible to demonstrate the presence of another fish species in minced fish meat stated to contain only one species. Under favorable circumstances, it is also possible to detect and identify foreign additions in a mixture of more than two fish species [14]. The IEF method, however, is not suitable for identifying mixtures of crustacean species, as the profiles have few zones and most of them are focused within the same narrow range of pH [12]. However, Craig et al. [12] in a study to detect adulteration of raw reformed breaded scampi (*Nephrops norvegicus*) demonstrated the successful application of SDS acrylamide gel electrophoresis to the identification of scampi and other crustacean species such as tropical shrimp (*Penaeus indicus)* and Pacific scampi (*Metanephrops andamanicus*) when present in reformed scampi products.

Martinez and co-workers [10,15,16] thought that two-dimensional gel electrophoresis (2-DE) could have a major application within food authentification to characterize the species and tissue. They noted that obviously, the first studies carried out on given species will have to deal with 2-DE, sequencing marker proteins, and identification with reference organisms. However, as the databases for food and feed material increase, it is possible that, in the future, the procedure will be made easier, faster, and perhaps cheaper by using, for example, tailor-made peptide chips for each product type.

20.3 Immunological Techniques

Blot hybridization, enzyme-linked immunosorbent assay (ELISA), immunodot, and immunodiffusion tests are immunological techniques which could be used for the detection of adulteration as these techniques are specific and sensitive analytical methods. Immunoassays using antigen-specific antibodies offer a powerful tool for the detection of the added proteins in a complex food protein mixture.

Some research studies have been realized on the implementation of immunological techniques for the detection of seafood product adulteration by the addition of foreign proteins.

Verrez et al. [17,18] have focused on immunological methods such as immunoblots and ELISA to detect the addition of crab meat in surimi (washed fish mince)-based products. Indeed, the label of surimi-based crabsticks sometimes indicates the addition of crab flesh in these products; analytical methods should be implemented to check this assertion. The authors have shown that using antiarginine kinase antibodies (arginine kinase is a cytoplasmic protein present in many invertebrates and absent from vertebrates), crustacean flesh could easily be detected in surimi crab supplemented preparations with a correlation between the level of crab added to surimi and the level of immunological response.

Taylor and Leighton Jones [19] have developed an immunoassay based on a noncompetitive indirect ELISA using antibodies directed to albacore, bonito, skipjack, and yellowfin to detect the adulteration of high-value crustacean tail meat products with lower value white fish.

Another study was done on the development of a dipstick immunoassay for the detection of trace amounts of egg proteins in food [20]. Actually, allergy against egg, for example, can be caused by relatively small amounts of egg proteins and can exhibit typical symptoms, however life-threatening anaphylactic reactions to egg are very rare. In this study, the authors have developed tests with antibodies against egg white and ovalbumin and they have tested different food samples. In nearly all analyzed foods where egg proteins have been declared, they were detected by the dipstick method.

On the contrary, seafood product could also be added to meat products and adulterate them. Indeed, surimi, which is a source of high content in myofibrillar proteins, was expected to become an additive to meat products. In order to distinguish the addition of Alaska pollock surimi to meat products, Dreyfuss et al. [21] have developed a test for rapid identification of pollock surimi in raw meat products. Their test based on the detection by antibodies directed against proteins of Alaska pollock surimi gave positive results with all the finfish species tested (14 species). The test was specific for Alaska pollock surimi at 2% concentration and showed detectable sensitivity to surimi from other finfish at concentration between 2% and 4%, and was 100% accurate in the laboratory trials.

20.4 Visible and Near-Infrared Spectrometry

The origin of the near-infrared (NIR) spectra of agro-food products is the absorption of the NIR light by chemical bounds of organic molecules. Spectra of food products include mainly absorption bands characteristic of the main constituents of the materials (i.e., water, proteins, fat, and carbohydrates) [22]. The main limitation of the NIR technique is its indirect nature as it measures no single target such as a specific molecule, DNA fragment, or proteins. However, NIR-based methods were proposed for the detection of meat and bone meal (MBM) in compound feeds [23,24]. To prevent the transmission to BSE between animals and humans, European authorities [25] prohibited the use of animal meat for feeding to ruminants. This measure includes fish meals, although this disease does not affect fish. The objective of this measure is to prevent adulteration and cross-contamination between fish and land animal meals. That is why analytical methods have been proposed for the determination of animal origin of feeding stuffs.

Murray et al. [24] have developed a method based on a partial least squared (PLS) discriminant analysis, using visible and NIR reflectance spectra. From their results, it seems that visible–NIR reflectance spectroscopy could routinely provide the first line of defense of the food chain against accidental contamination or fraudulent adulteration of fish meal with MBM. In their article, van Raamsdonk et al. [22] noted that a NIR spectrometer coupled to a microscope (NIRM) could also be proposed to tackle the problem of detection of MBM in compound feed; the method can also be used to detect fish meal ingredients. When using a NIR microscope the subjective judgment of the microscopist is replaced by the spectra that can be subjected to statistical analysis.

In another possible application field, Gayo et al. [26,27] have successfully used visible and NIR spectroscopy (Vis/NIR) to detect and quantify species authenticity and adulteration in crabmeat samples. In their studies, visible and NIR spectroscopy have been successfully used to detect the adulteration in crab meat samples adulterated with surimi-based imitation crabmeat and to detect the adulteration of Atlantic blue crab (*Callinectes sapidus*) meat with blue swimmer crab (*Portunus pelagicus*) meat in 10% increments.

20.5   Microscopic Methods

Following the measure to prohibit the use of animal proteins for feeding ruminant including also fish meals, EC directive 2003/126/EC [28] indicates the analytical method to be applied for the detection and characterization of processed animal proteins (PAPs) in feeds. This analytical method is based on a microscopic technique. However, this method is not applicable to fish meals, because some typical structures detected are common to both fish and land animals, and it only gives useful results when bones are present in the sample [29].

Whereas microscopic techniques could be used to check if fish meals are not adulterated by fraudulent addition of terrestrial animal meal. van Raamsdonk et al. [22] have reviewed different proficiency studies and ring trials organized since 2003 for the detection of mammalian PAP in fish meal. The first proficiency study, allowing the participants to apply their own protocol, revealed that microscopic detection of 1 g/kg of mammalian PAP in the presence of 50 g fish meal/kg was realized in 44% of the cases. However, a microscopic detection of 98% can be reached by providing the application of an optimal protocol and a sufficient level of expertise. Recent studies showed that training, application of a decision support system, and use of an improved microscopy protocol resulted in a higher sensitivity. As van Raamsdonk et al. [22] noted, an attractive approach to detect fish meal adulteration by meat meal is the combination of the very low detection level of microscopy with identification by other methods (PCR and immunoassays).

Microscopical techniques could also be used to detect other types of seafood product adulteration than adulteration of fish meals. Thus, Ebert and Islam [30] used histological examinations to identify, as caviar imitations; products labeled "special caviar product manufactured with sturgeon- and salmon roe in the Russian way." The caviar-like, spherical products appeared to be formed out of a homogeneous, unstructured mass and showed none of the fish roe specific biological structures.

20.6   Chromatographic Techniques

There is very little literature on the use of chromatographic techniques to detect adulteration. However, the study of Chou et al. [31] is based on an high-performance liquid chromatography (HPLC) method with electrochemical detection (HPLC-EC). They reported that a major advantage of EC detection is its ability to directly detect peptides and amino acids that exhibit little or no chromogenic or fluorescent properties. In addition, under appropriate chromatographic conditions and simultaneous use of a copper nanoparticle-plated electrode, reliable detection is feasible without sample pretreatment. In their study, Chou et al. [31] have tested the applicability of the method to detect the species in mixtures only on three land animals (beef, pork, and horse meats). However, as they could identify cod, crab, salmon, scallop, and shrimp with this method, it seems possible to apply such technique—that is fast, economic, and reliable for identification of meats from multiple species—to detect adulteration of seafood flesh by meat components.

20.7   DNA Methods

Advances in DNA technologies have led to rapid development of genetic methods mainly based on PCR for fish species identification and for the detection of fish product adulteration as protein analysis are in general not suited to fish species identification in heat-processed matrices [32]. DNA offers advantages over proteins, including stability at high temperature, presence in all tissue types, and greater variation in genetic sequence [33]. Fish and fishery products authentification can be achieved by PCR-based methods (see reviews by Leighton Jones [34], Sotelo et al. [35], Mackie [36], Lockley and Bardsley [32], Asensio Gil [4]). Asensio Gil's [4] work provided an extensive overview on various techniques such as PCR-sequencing, species-specific PCR primers or multiplex PCR, PCR-restriction fragment length polymorphism (PCR-RFLP), PCR-single-stranded conformation polymorphism (PCR-SSCP), random amplification of polymorphic DNA (RAPD), real-time PCR, and PCR lab-on-a-chip. All these PCR-based techniques have high potential because of their rapidity, increased sensitivity, and specificity [32]. Nevertheless, some of these techniques such as RAPD analysis may not be suitable to identify adulterations. RAPD technique is not adapted to detect the species of origin in products containing mixtures of species containing 50%–50% mixtures of species that can interbreed [37]. On the other hand, quantitative PCR tests such as real-time PCR have been widely used for food authentification and quantification. In this section on DNA techniques, only those that could be used to detect adulteration are reviewed.

20.7.1 PCR-Sequencing

PCR-sequencing which is the most direct means of obtaining information from PCR products has been extensively used to identify various fish mainly based on mitochondrial genes such as cytochrome *b* or cytochrome oxidase I (COI) (e.g., see study of Jérôme et al. [38]). In addition fragments of nuclear genes such as α-actinine, 5S ribosomal DNA, rhodopsin, among others, have been sequenced for the discrimination of fish (e.g., see study of Sevilla et al. [39]). Dedicated Internet databases offer the possibility to rely on unknown sequences to reference fish species sequences (e.g., FishTrace database [40] and Fish-BOL [41]).

Even if sequencing is time consuming, it produces large amounts of information that could be used in other PCR-based methods such as PCR-RFLP to fish species identification [42–44]. PCR-sequencing technique seems, therefore, difficult to adapt to check seafood adulteration by substituting partially foreign fish in labeled one fish species product.

20.7.2 Species-Specific PCR or Multiplex PCR

Because this method has the potential to detect qualitative admixture, it is a method adapted to adulteration analysis. Prior sequence knowledge is required in order to design primers and appropriate controls should be included to preclude the possibility of false positive or negative results being obtained [45].

In this idea, Colombo et al. [46] have developed a species-specific PCR to identify frozen and seasoned food labeled as pectinid scallop and suspected to be or to contain vertebrate (in particular teleostean fish).

Multiplex PCR can also be used with the intention to examine fish meal for contamination with mammalian and poultry products. Bellagamba et al. [47] have seeked a method based on three species-specific primer pairs designed for the identification of ruminant, pig, and poultry DNA. The PCR specifically detected mammalian and poultry adulteration in fish meals containing 0.125% beef, 0.125% sheep, 0.125% pig, 0.125% chicken, and 0.5% goat. The multiplex PCR assay for ruminant and pig adulteration in fish meals had a detection limit of 0.25% after optimization.

As different tuna species have different qualities and prices, a fraudulent replacement of valuable species by less valuable ones (e.g., *Katsuwonus pelamis*) may occur. Bottero et al. [48] have developed a multiplex primer-extension assay (PER) to discriminate four closely related species of *Thunnus* (*T. alalunga*, *T. albacores*, *T. obesus*, and *T. thynnus*) and one species of *Euthynnus* genus (*K. pelamis*) in raw and canned tuna. The technique enables the simultaneous and unambiguous identification of the five tuna species.

20.7.3 Amplified Fragment Length Polymorphism

As Atlantic salmon (*Salmo salar*) is highly appreciated in the Chinese market, illegal practices can occur by adulterating or substituting rainbow trout products (*Oncorhynchus mykiss*) of much lower value in China for those of Atlantic salmon. Zhang and Cai [49] have developed a species-specific amplified fragment length polymorphism (AFLP) marker based on AFLP analysis and converted into reliable sequence-characterized amplified regions (SCARs) for constructing a direct and fast method to detect frauds in fresh and processed products of Atlantic salmon being adulterated and substituted by rainbow trout. The SCAR marker could be amplified and visualized in 1% agarose gel in all tested rainbow trout samples and was absent in all salmon samples. Using DNA admixtures, the detection of 1% (0.5 ng) and 10% (5 ng) rainbow trout DNA in Atlantic salmon DNA for fresh and processed samples, respectively, was readily achieved. In another study, Zhang et al. [50] have demonstrated that the detection sensitivity of AFLP-derived SCAR was higher than that of DNA amplicons separation by denaturing gradient gel electrophoresis (DDGE) when analyzing experimental mixtures of Atlantic salmon and rainbow trout. The AFLP-derived SCAR approach was sensitive and demonstrated to be a rapid and reliable method for identifying frauds in salmon products, and it could be extended for the applications of species identification in food industry.

20.7.4 PCR-Restriction Fragment Length Polymorphism

Only three examples of the use of PCR-RFLP in the detection of seafood product adulteration are detailed in this chapter. The first one is the study of Hold et al. [51] who have used this technique to develop a method to differentiate between several different fish species. The method was tested in a collaborative study in which 12 European laboratories participated to ascertain whether the method was reproducible. From a total of 120 tests performed, unknown samples identified by comparison with RFLP profiles of reference species were correctly identified in 96% of cases. They have also tested the ability of this method to analyze mixed and processed fish samples. In all cases, the species contained within mixed samples were correctly identified, indicating the efficacy of the method for detecting fraudulent substitution of fish species in food products.

Horstkotte and Rehbein [52] have tested the usability of fish species identification with RFLP using HPLC for sturgeon, salmon, and tuna samples. Unequivocal species identification was achieved with HPLC, despite nucleotide sequence-depending separation. Separation of DNA fragments by HPLC could be demonstrated to be a fast and reliable alternative to electrophoresis.

In the aim of guaranteeing the composition and security of fish meals, a method based on PCR and length polymorphism, followed by a RFLP was developed by Santaclara et al. [29]. Specific primers for every species were designed and calibrated to generate a PCR product with a specific size when DNA of each land species that can be used for elaboration of meat meals (cow, chicken, pig, horse, sheep, and goat) was present in the sample. This methodology allows verification of the adulteration and cross-contamination of fish meals with the six land species studied.

20.7.5 Real-Time PCR

The specificity and sensitivity of this technique, combined with its high speed, robustness, reliability, and the possibility of automation contribute to the adequacy of the method for quantifying fish species in fishery products [53]. For instance, Asensio Gil [4] reported study of Sotelo et al. [54] who used TaqMan assay for the identification and the quantification of cod. In the same idea, Trotta et al. [55] used real-time PCR for the identification of fish fillets from grouper and common substitute species. They also used conventional multiplex PCR in which electrophoretic migration of different sizes of bands allowed identification of the fish species. These two approaches, real-time PCR and multiplex PCR made possible to discriminate grouper from substitute fish species. Hird et al. [56] have designed real-time PCR primer and probe set for the detection and quantification of haddock. The presence of this fish in concentrations of up to 7% in raw or slightly heat-treated products could be detected. While Lopez and Pardo [57] applied the real-time PCR technology for the identification and the quantification of albacore and yellowfin tuna, the real-time methodology described in their study was suitable to detect the fraudulent presence of yellowfin or even to identify the absence of albacore in cans labeled as white tuna.

As Asensio Gil [4] reported, the accuracy of this technique could be affected by several factors, such as the DNA yield of the samples, which can be variable depending on the strength of the technique used to process the fish product, and by the fact that the sample material could be thermally processed in different ways. Owing to its cost, real-time PCR has only a remarkable interest in analyzing products with an important economic value. However, the enormous utility and possible applications of the real-time PCR will make it affordable for most laboratories in the near future.

20.7.6 PCR Lab-on-a-Chip

This technology that uses microfluidic devices has been recently used for fish species authentification. Dooley et al. [58] used a chip-based capillary electrophoresis system to discriminate mixtures of salmon and trout. Experimental repeatability was less than 3%, allowing species identification without the need to run reference materials with every sample. Using DNA admixtures, the discrimination of 5% salmon DNA in trout DNA was readily achieved. This technology permitted an improvement of a published PCR-RFLP approach for fish species identification by the replacement of the gel-electrophoretic steps by capillary electrophoresis. Dooley et al. [59] used the same methodology for identification of 10 white fish species associated with the U.K. food products. The method was subjected to an interlaboratory study carried out by five U.K. food control laboratories. One hundred percent correct identification of single species samples and six of nine identifications of admixture samples were achieved by all laboratories. The results indicated that fish species identification could be carried out using a database of PCR-RFLP profiles without the need for reference materials. Although this technology is relatively expensive, the cost of the instrumentation and disposable chips are relatively low when compared to that for real-time PCR analysis (cited by Ascensio Gil [4]).

20.7.7 Commercial PCR Kits for Fish Species Differentiation

In the recent years, advances in PCR-based methods have led to rapid development of different commercial kits for fish species identification. There is no literature on the use of these rapid diagnostic kits for the detection of substitution or adulteration in seafood products, but without doubt, these kits could be very useful for screening purposes in inspection programs. The present list below of those commercial kits or commercial proposals is not exhaustive and is only valuable for at present time.

- DNA kit for eight fish species identification provided by Tepnel Biosystems company
- Biofish kit for cod and gadiform species and Biofish kit for Atlantic salmon, sea trout, and rainbow trout commercialized by Biotools company
- FishID kit for the identification of more than 200 fish species developed by Bionostra company
- GeneChip based on DNA microarray technology developed by Biomérieux
- Proposal of Eurofins/GeneScan company to develop methods and analysis kits based on PCR according to a specific request

20.8 Conclusion

Of the wide range of analytical methods available, it is likely that DNA-based techniques will be the favorite approach for determining adulteration, because they are easy to use. However, now, they are generally more expensive than electrophoresis or chromatographic techniques and in some cases, the latter are sufficient to clearly demonstrate an adulteration in seafood products by addition of foreign proteins. Nevertheless, researches into relatively novel techniques such as PCR lab-on-chip and real-time PCR offer the greatest potential for the development on new fish discrimination applications and protocols (Asensio Gil [4]). Quantifying methods are, without doubt, the more adapted to analyze adulteration, and these methods are to be developed.

References

1. Teletchea, F., Maudet, C., and Hanni, C., Food and forensic molecular identification: Update and challenges, *Trends Biotechnol.*, 23, 359, 2005.
2. Woolfe, M. and Primrose, S., Food forensics: Using DNA technology to combat misdescription and fraud, *Trends Biotechnol.*, 22, 222, 2004.
3. Pascal, G. and Mahé, S., Identity, traceability, acceptability and substantial equivalence of food, *Cell. Mol. Biol.*, 47, 1329, 2001.
4. Asensio Gil, L., PCR-based methods for fish and fishery products authentication, *Trends Food Sci. Technol.*, 18, 558, 2007.
5. Mackie, I. M., Identifying species of fish, *Anal. Proc.*, 27, 89, 1990.
6. Scobbie, A. E. and Mackie, I. M., The use of sodium dodecyl sulphate polyacrylamide gel electrophoresis in fish species identification—A procedure suitable for cooked and raw fish, *J Sci. Food Agric.*, 44, 343, 1988.
7. Rehbein, H., Etienne, M., Jerome, M., Hattula, T., Knudsen, L. B., Jessen, F., Luten, J. B. et al., Influence of variation in methodology on the reliability of the isoelectric focusing method of fish species identification, *Food Chem.*, 52, 193, 1995.

8. Rehbein, H., Kundiger, R., Yman, I. M., Ferm, M., Etienne, M., Jerome, M., Craig, A. et al., Species identification of cooked fish by urea isoelectric focusing and sodium dodecylsulfate polyacrylamide gel electrophoresis: A collaborative study, *Food Chem.*, 67, 333, 1999.

9. Etienne, M., Jerome, M., Fleurence, J., Rehbein, H., Kundiger, R., Mendes, R., Costa, H., and Martinez, I., Species identification of formed fishery products and high pressure-treated fish by electrophoresis: A collaborative study, *Food Chem.*, 72, 105, 2001.

10. Piñeiro, C., Barros-Velázquez, J., Pérez-Martín, R. I., Martínez, I., Jacobsen, T., Rehbein, H., Kündiger, R. et al., Development of a sodium dodecyl sulfate-polyacrylamide gel electrophoresis reference method for the analysis and identification of fish species in raw and heat-processed samples: A collaborative study, *Electrophoresis*, 20, 1425, 1999.

11. Etienne, M., Jérôme, M., Fleurence, J., Rehbein, H., Kündiger, R., Mendes, R., Costa, H., Pérez-Martín, R., and Piñeiro-González, C., Identification of fish species after cooking by SDS-PAGE and urea IEF: A collaborative study, *J. Agric. Food Chem.*, 48, 2653, 2000.

12. Craig, A., Ritchie, A. H., and Mackie, I. M., Determining the authenticity of raw reformed breaded scampi (*Nephrops norvegicus*) by electrophoretic techniques, *Food Chem.*, 52, 451, 1995.

13. European Commission, EU Project FAR UP-3-783. 1992–1995, Identification and quantitation of species in marine products, 1992–1995.

14. Podeszewski, Z. and Zarzycki, B., Identification of fish species by testing the minced-meat tissue, *Food/Nahrung*, 22, 377, 1978.

15. Martinez, I. and Friis, T. J., Application of proteome analysis to seafood authentication, *Proteomics*, 4, 347, 2004.

16. Martinez, I., James, D., and Loréal, H., Application of modern analytical techniques to ensure seafood safety and authenticity, FAO Fisheries Technical Paper No. 455, Ed., Food and Agriculture Organization of the United Nations, Rome, Italy, 2005, p. 73.

17. Verrez, V., Benyamin, Y., and Roustan, C., Detection of marine invertebrates in surimi-based products, in *Quality Assurance in the Fish Industry*, Huss, H. H., Jakobsen, M., and Liston, J. (Eds.), Elsevier Science Publisher B.V., Amsterdam, the Netherlands, 1992, p. 441.

18. Verrez-Bagnis, V. and Escriche Roberto, I., The performance of ELISA and dot-blot methods for the detection of crab flesh in heated and sterilized surimi-based products, *J. Sci. Food Agricul.*, 63, 445, 1993.

19. Taylor, W. J. and Leighton Jones, J., An immunoassay for distinguishing between crustacean tailmeat and white fish, *Food Agric. Immunol.*, 4, 177, 1992.

20. Baumgartner, S., Steiner, I., Kloiber, S., Hirmann, D., Krska, R., and Yeung, J., Towards the development of a dipstick immunoassay for the detection of trace amounts of egg proteins in food, *Eur. Food Res. Technol.*, 214, 168, 2002.

21. Dreyfuss, M. S., Cutrufelli, M. E., Mageau, R. P., and McNamara, A. M., Agar-gel immunodiffusion test for rapid identification of pollock surimi in raw meat products, *J. Food Sci.*, 62, 972, 1997.

22. van Raamsdonk, L. W. D., von Holst, C., Baeten, V., Berben, G., Boix, A., and de Jong, J., New developments in the detection and identification of processed animal proteins in feeds, *Anim. Feed Sci. Technol.*, 133, 63, 2007.

23. Garrido-Varo, A., Pérez-Marín, M. D., Guerrero, J. E., Gómez-Cabrera, A., Haba, M. J. D. L., Bautista, J., Soldado, A. et al., Near infrared spectroscopy for enforcement of European legislation concerning the use of animal by-products in animal feeds, *Biotechnologie, Agronomie, Société et Environnement*, 9, 3, 2005.

24. Murray, I., Aucott, L. S., and Pike, I. H., Use of discriminant analysis on visible and near infrared reflectance spectra to detect adulteration of fishmeal with meat and bone meal, *J. Near Infrared Spectrosc.*, 9, 297, 2001.

25. European Commission, Commission decision of 27 March 2002 (2002/248/EC) amending council decision 2000/766/EC and commission decision 2001/9/EC with regard to transmissible spongiform encephalopathies and the feeding of animal proteins, *Off. J. Eur. Communities* (L84 28/3/2002), 71, 2002.

26. Gayo, J. and Hale, S. A., Detection and quantification of species authenticity and adulteration in crabmeat using visible and near-infrared spectroscopy, *J. Agric. Food Chem.*, 55, 585, 2007.

27. Gayo, J., Hale, S. A., and Blanchard, S. M., Quantitative analysis and detection of adulteration in crab meat using visible and near-infrared spectroscopy, *J. Agric. Food Chem.*, 54, 1130, 2006.

28. European Commission, Commission Directive 2003/126/EC of 23 December 2003 on the analytical method for the determination of constituents of animal origin for the official control of feeding stuffs, *Off. J. Eur. Communities*, L 339 24/12/2003, 0078, 2003.

29. Santaclara, F. J., Espiñeira, M., Cabado, A. G., and Vieites, J. M., Detection of land animal remains in fish meals by the polymerase chain reaction-restriction fragment length polymorphism technique, *J. Agric. Food Chem.*, 55, 305, 2007.

30. Ebert, M. and Islam, R., Kaviar und Kaviarimitate: Verfälschungen des teuren Störrogens mit histologischer Methode nachweisbar (Caviar and caviar imitations adulterations of sturgeon roe provable by histological method), *Fleischwirtschaft*, 87, 124, 2007.

31. Chou, C.-C., Lin, S.-P., Lee, K.-M., Hsu, C.-T., Vickroy, T. W., and Zen, J.-M., Fast differentiation of meats from fifteen animal species by liquid chromatography with electrochemical detection using copper nanoparticle plated electrodes, *J. Chromatogr. B*, 846, 230, 2007.

32. Lockley, A. K. and Bardsley, R. G., DNA-based methods for food authentication, *Trends Food Sci. Technol.*, 11, 67, 2000.

33. Mackie, I. M., Authenticity of fish, in *Food Authentification*, Ashurt, P. R. and Dennis, M. J. (Eds.), Blackie Academic and Professional, London, U.K., 1996, p. 140.

34. Leighton Jones, J., DNA probes: Applications in the food industry, *Trends Food Sci. Technol.*, 2, 28, 1991.

35. Sotelo, C. G., Pinciro, C., Gallardo, J. M., and Perez-Martin, R. I., Fish species identification in seafood products, *Trends Food Sci. Technol.*, 4, 395, 1993.

36. Mackie, I. M., Fish speciation, *Food Technol. Int. Eur.*, 1, 177, 1994.

37. Martinez, I. and Malmheden Yman, I., Species identification in meat products by RAPD analysis, *Food Res. Int.*, 31, 459, 1998.

38. Jerome, M., Lemaire, C., Verrez-Bagnis, V., and Etienne, M., Direct sequencing method for species identification of canned sardine and sardine-type products, *J. Agric. Food Chem.*, 51, 7326, 2003.

39. Sevilla, R. G., Diez, A., Noren, M., Mouchel, O., Jerome, M., Verrez-Bagnis, V., Van Pelt, H. et al., Primers and polymerase chain reaction conditions for DNA barcoding teleost fish based on the mitochondrial cytochrome b and nuclear rhodopsin genes, *Mol. Ecol. Notes*, 7, 730, 2007.

40. FishTrace: Genetic catalogue, www.fishtrace.org.

41. Fish Barcode of Life Initiative (FISH-BOL), www.fishbol.org.

42. Quinteiro, J., Sotelo, C. G., Rehbein, H., Pryde, S. E., Medina, I., Pérez-Martin, R. I., Rey-Méndez, M., and Mackie, I. M., Use of mtDNA direct polymerase chain reaction (PCR) sequencing and PCR-restriction fragment length polymorphism methodologies in species identification of canned tuna, *J. Agric. Food Chem.*, 46, 1662, 1998.

43. Sebastio, P., Zanelli, P., and Neri, T. M., Identification of anchovy (*Engraulis encrasicholus* L.) and gilt sardine (*Sardinella aurita*) by polymerase chain reaction, sequence of their mitochondrial cytochrome *b* gene, and restriction analysis of polymerase chain reaction products in semipreserves, *J. Agric. Food Chem.*, 49, 1194, 2001.

44. Ram, J. L., Ram, M. L., and Baidoun, F. F., Authentication of canned tuna and bonito by sequence and restriction site analysis of polymerase chain reaction products of mitochondrial DNA, *J. Agric. Food Chem.*, 44, 2460, 1996.

45. Edwards, M. C. and Gibbs, R. A., Multiplex PCR: Advantages, development, and applications, *PCR Methods Appl.*, 3, S65, 1994.

46. Colombo, F., Trezzi, I., Bernardi, C., Cantoni, C., and Renon, P., A case of identification of pectinid scallop (*Pecten jacobaeus*, *Pecten maximus*) in a frozen and seasoned food product with PCR technique, *Food Control*, 15, 527, 2004.

47. Bellagamba, F., Valfre, F., Panseri, S., and Moretti, V. M., Polymerase chain reaction-based analysis to detect terrestrial animal protein in fish meal, *J. Food Prot.*, 66, 682, 2003.
48. Bottero, M. T., Dalmasso, A., Cappelletti, M., Secchi, C., and Civera, T., Differentiation of five tuna species by a multiplex primer-extension assay, *J. Biotechnol.*, 129, 575, 2007.
49. Zhang, J. and Cai, Z., Differentiation of the rainbow trout (*Oncorhynchus mykiss*) from Atlantic salmon (*Salmon salar*) by the AFLP-derived SCAR, *Eur. Food Res. Technol.*, 223, 413, 2006.
50. Zhang, J., Wang, H., and Cai, Z., The application of DGGE and AFLP-derived SCAR for discrimination between Atlantic salmon (*Salmo salar*) and rainbow trout (*Oncorhynchus mykiss*), *Food Control*, 18, 672, 2007.
51. Hold, G. L., Russell, V. J., Pryde, S. E., Rehbein, H., Quinteiro, J., Vidal, R., Rey-Mendez, M. et al., Development of a DNA-based method aimed at identifying the fish species present in food products, *J. Agric. Food Chem.*, 49, 1175, 2001.
52. Horstkotte, B. and Rehbein, H., Fish species identification by means of restriction fragment length polymorphism and high-performance liquid chromatography, *J. Food Sci.*, 68, 2658, 2003.
53. Heid, C. A., Stevens, J., Livak, K. J., and Williams, P. M., Real time quantitative PCR, *Genome Res.*, 6, 986, 1996.
54. Sotelo, C. G., Chapela, M. J., Rey, M., and Pérez-Martín, R. I., Development of an identification and quantitation system for cod (*Gadus morhua*) using Taqman assay, in *First Joint Trans-Atlantic Fisheries Technology Conference*, Reykjavik, Iceland, 2003, p. 195.
55. Trotta, M., Schonhuth, S., Pepe, T., Cortesi, M. L., Puyet, A., and Bautista, J. M., Multiplex PCR method for use in real-time PCR for identification of fish fillets from grouper (*Epinephelus* and *Mycteroperca* species) and common substitute species, *J. Agric. Food Chem.*, 53, 2039, 2005.
56. Hird, H. J., Hold, G. L., Chisholm, J., Reece, P., Russell, V. J., Brown, J., Goodier, R., and MacArthur, R., Development of a method for the quantification of haddock (*Melanogrammus aeglefinus*) in commercial products using real-time PCR, *Eur. Food Res. Technol.*, 220, 633, 2005.
57. Lopez, I. and Pardo, M. A., Application of relative quantification TaqMan real-time polymerase chain reaction technology for the identification and quantification of *Thunnus alalunga* and *Thunnus albacares*, *J. Agric. Food Chem.*, 53, 4554, 2005.
58. Dooley, J. J., Sage, H. D., Brown, H. M., and Garrett, S. D., Improved fish species identification by use of lab-on-a-chip technology, *Food Control*, 16, 601, 2005.
59. Dooley, J. J., Sage, H. D., Clarke, M. A. L., Brown, H. M., and Garrett, S. D., Fish species identification using PCR-RFLP analysis and lab-on-a-chip capillary electrophoresis: Application to detect white fish species in food products and an interlaboratory study, *J. Agric. Food Chem.*, 53, 3348, 2005.

Chapter 21

Detection of Adulterations: Identification of Seafood Species

Antonio Puyet and José M. Bautista

Contents

21.1 Introduction 616
21.2 Replacement Species and Adulterations 616
    21.2.1 Processed Products Adulteration 619
21.3 Methods Based on Proteins 619
21.4 Methods Based on DNA 620
    21.4.1 Sample Handling and DNA Extraction 620
    21.4.2 DNA Sequencing Methods 621
        21.4.2.1 Standardized Fish Molecular Databases and Barcoding 621
        21.4.2.2 Identification from General Databases 625
    21.4.3 Non-DNA Sequencing Methods 625
        21.4.3.1 RFLP 626
        21.4.3.2 AFLP, RAPD, and Satellite DNA Analysis 626
        21.4.3.3 SSCP and DGGE 628
        21.4.3.4 Selective Amplification 628
        21.4.3.5 Quantitative Methods 629
        21.4.3.6 High-Throughput, Microarray Technologies and Bioinformatics 630
21.5 Future Prospects 631
References 632

21.1 Introduction

In the seafood market, substitution of valuable species for species of lower value is a common practice because it is uncomplicated and has immediate economic reward. In addition, it is favored by the depletion in some areas of highly appreciated species, the high variety of fish species, the global market, the difficult differential diagnosis, and the overall lack of taxonomical expertise.

Studies around the world have shown that up to 75% of a given species from fish samples in the market can be mislabeled [1]. The need for fish species identification in seafood products rely on the consumer's right to make informed choices and to guarantee consumer confidence. In a global trade, misleading or deceptive conduct in the commercialization of fisheries products should be always avoided and tracked if they happen, since the effectiveness of the seafood marketing and promotion could be, otherwise, depreciated.

Also, at the level of public perception, whenever popular fish species are readily available in the marketplace, it supports the idea that there is abundant supply of them from fisheries, misleading the real condition of the stock. Thus, species mislabeling in the catch vessels could also negatively affect stock size assessment since incorrect data reported influences fisheries management. To this respect it should be mentioned that to identify fish correctly is not easy, and substitution or incorrect labeling can be unintended rather than deliberate, but always harms consumer perception and confidence in seafood products.

Finally, it should be emphasized that fish species substitution is not only an economic or ecological fraud but also is a health threat since some seafood species from some areas may elicit maladies to susceptible populations, ranging from allergies [2] to serious illness [3]. For instance, the level of mercury or dioxins in fish species from some fishing grounds [4,5] might promote the medical or governmental advice to consumers, particularly vulnerable population, to limit their consumption, which, in turn, advocate for a correct fish labeling.

With around 30,000 living fish species in the world, these can be only correctly identified by visual inspection when the specimens are undamaged. Nevertheless, even in these cases, professional education in fish identification (fishermen, fishmongers, and restaurateurs) may not be sufficient, particularly if there is a certain extent of overlapping features between taxa, as it frequently occurs in many fish species. Moreover, all processed fish products lose their morphological characteristics early in the processing food chain. Thus, potential misidentification or mislabeling of fish species can be considerable as it has been already reported in some market surveys [1,6], which would be specially significant in fish fillets, cooked food, or fish-transformed products [7].

Although not all fish species are subject to food trade, only in Europe, more than 500 species are currently in the market and 60% of them are caught outside controlled European waters. Similar trade values are found in the American, Australian, and Japanese markets, making the identification process as an essential tool for traceability of the transforming seafood chain.

21.2 Replacement Species and Adulterations

Increasing vulnerability of fish species due to exploitation, climate change, overfishing, and byfishing [8,9] has augmented the commercial replacement of species everywhere. Although there are no wide screening studies on the identification of replacement species in global markets, the rising in the scientific publication of differential diagnostic systems for groups of fish species allows identifying major concerns at present in control laboratories and researchers as depicted in Table 21.1.

Table 21.1   Potential Substitutions of Fish Species in Markets Worldwide According to the Main Differential Diagnostics Methodology Published

Species	Substituted By	Market	Refs.
Cod (*Gadus morhua*)	Pacific cod (*Gadus macrocephalus*), Alaska Pollack (*Theragra chalcogramma*), Saffron cod (*Eleginus gracilis*), Arctic cod (*Arctogadus glacialis*), Southern blue whiting (*Micromesistius australis*), Chilean hake (*Merluccius gayi*), Southern hake (*M. australis*), Longfin codling (*Laemonema longipes*) or Blue grenadier (*Macruronus novaezelandiae*)	Worldwide (Japan, Europe, United States)	[10–12]
Grouper (*Ephinephelus* spp. and *Mycteroperca* spp.)	Nile perch (*Lates niloticus*) and wreck fish (*Polyprion americanus*)	Europe	[13]
Frigate tunas (*Auxis thazard* and *Auxis rochei*)	Skipjack (*Katsuwonus pelamis*), little tuny (*Euthynnus alletteratus*), or yellowfin tuna (*Thunnus albacares*)	Europe	[14]
Red snapper (*Lutjanus campechanus*)	Vermilion snapper (*Rhomboplites aurorubens*), crimson snapper (*L. erythropterus*), or Lane snapper (*L. synagris*)	United States	[15,16]
European anchovy (*Engraulis encrasicolus*)	*Engraulis* spp. (*E. anchoita, E. ringens, E. japonicus, E. mordax*), *Coilia* spp., *Sardina* spp., *Sprattus* spp., and *Sardinella* spp.	Europe	[17,18]
Atlantic salmon (*S. salar*)	Rainbow trout (*Oncorhynchus mykiss*)novaezelandiae	Worldwide	[19]
Pacific salmon (*Oncorhyncus* spp.)	Atlantic salmon (*S. salar*), Brown trout (*Salmo trutta*)	North America	[20]
European perch (*Perca fluviatilis*)	Nile perch (*Lates niloticus*), European pikeperch (*Stizostedion lucioperca*), Sunshine bass (*Morone chrysops x saxatalis*)	Europe	[21]
Albacore (*Thunnus alalunga*)	Skipjack tuna (*Katsuwonus pelamis*), yellowfin (*T. albacares*)	Europe	[22,23]
Mediterranean horse mackerel (*Trachurus mediterraneus*)	Blue jack mackerel (*T. picturatus*)	Europe	[24]
Japanese mackerel (*Scomber japonicus*)	Atlantic mackerel (*Scomber scombrus*)	Japan	[25]

(*continued*)

Table 21.1 (continued)   Potential Substitutions of Fish Species in Markets Worldwide According to the Main Differential Diagnostics Methodology Published

Species	Substituted By	Market	Refs.
European pilchard (*Sardina pilchardus*)	Other pilchards and sardinellas (*Sardinops, Sardinella* spp.)	Europe	[18,26]
European hake (*Merluccius merluccius*)	Deep water hake (*Merluccius paradoxus*), Senegalese hake (*Merluccius senegalensis*), Silver hake (*Merluccius bilinearis*), Chilean hake (*M. gayi*), Argentine hake (*Merluccius hubbsi*), Patagonian grenadier (*Macruronus magellanicus*)	Europe	[27,28]
Sturgeon caviar (*Acipenser sturio*)	Siberian sturgeon eggs (*Acipenser baerii*)	Worldwide	[29]
Surimi (*Theragra chalcograma*)	*Merluccius* spp.	Europe	[6]
Blue mussel (*Mytilus galloprovincialis*)	Green mussel (*Perna* spp.) and others (*Aulacomya, Semimytilus, Brachidontes* and *Choromytilus* spp.)	Europe	[30]
Prawn (*Fenneropenaeus, Penaeus, Parapenaeus, Marsupenaeus, Melicertus, Solenocera, Pleoticus,* and *Aristeomorpha* spp.) and shrimp (*Farfantepenaeus* and, *Litopenaeus* spp.)	Mislabeling	Europe	[32]
Shortfin squids (Family Ommastrephidae: Genera Loligo, Loliolus, Uroteuthis, and Alloteuthis) and longfin squids (family Loliginidae: Genera Todarodes, Illex, Todaropsis, Nototodarus, Dosidicus, and Ommastrephes)	Mislabeling	Europe	[32]
Billfish species: *Makaira nigricans* (blue marlin), *Makaira indica* (black marlin), *Istiophorus platypterus* (sailfish), and *Tetrapturus audax* (striped marlin)	Swordfish (*Xiphias gladius*)	Wordwide	[33]

21.2.1 Processed Products Adulteration

Trade for processed seafood is particularly complex given the large number of species traded, countries involved, and production processes [36]. Protein addition from nondeclared species can be a source of the adulterations in seafood as it also happens in the meat industry [35]. Identification of this manipulation from nonseafood species usually follow specific methodology for detecting mixed DNA from different origin [35–37] that can be applied to groups of species [13]. Nevertheless it should be stressed that addition of artificially synthesized DNA in caviar [29] can be theoretically used to manipulate the detection of the origin, and thus the use of several markers would be recommended to identify foreign DNA in complex food mixtures. Other processed products adulterations are related to the trade of unrecognized parts of fish. This is the case of shark fin where types of fin are mainly described with English common names for sharks [38] in contrast with Chinese market categories that do not correspond to the taxonomic names of shark species. Since shark policy resources lacks species-specific catch and trade data for most fisheries [39], DNA-based species identification techniques have been used to determine the relationship between market category and species, showing that only 14 species made up approximately 40% of the auctioned fin weight in Hong Kong [38]. To follow food labeling regulations, in some instances, quantification of a given fish species in complex food mixtures is required. This has been usually based on the relative content of nitrogen determination, but more recently a model real-time polymerase chain reaction (PCR) has been developed for haddock [40] and tuna species [22] to determine proportion of muscle tissue relative to the amount detected of a single copy gene. This type of methodology is also allowing enforcement of the legislation regarding complex food mixtures containing fish species.

21.3 Methods Based on Proteins

Although DNA technologies are the first choice in the identification of fish species since it is not dependent on the specific condition of the fishery product (e.g., processed or not), several techniques based on protein analysis have also been described, particularly in the most recent past. Protein identification, as a label for species diagnosis, is based on physical and chemical properties of the polypeptide chain: size and net charge of the amino acidic sequence, three-dimensional structure exposure, and immunoreactivity of specific epitopes. Therefore, protein denaturation by heat, chemical additives, or proteolysis can severely modify those native properties.

Separation and characterization of soluble sarcoplasmic proteins by isoelectric focusing (IEF) has been the most frequently described protein method for fish species identification [41–43]. Nevertheless, application of IEF is considered mostly limited to raw fish fillet [46] due to the processing of the fish carcass (involving heating, salting, drying, or smoking) which leads to the loss or modification of species-specific protein fragments and yields indefinite IEF patterns, not allowing clear-cut identifications [45,46] and requiring a faithful set of reference samples [46]. Moreover the identification procedure is generally laborious and requires skilled human resources to strictly follow highly optimized standard operation procedures [46], since reproducibility between laboratories is not always achieved [45]. From an IEF survey of 14 species obtained in a fish market, intraspecific polymorphisms and discrepancies were detected, some of them due to unpredictable band distortions that required special computer-assisted comparison of the IEF gels [42]. Although there are not large standardized sets of IEF standard patterns for seafood species identification, the Regulatory Fish Encyclopedia (RFE: http://www.fda.gov/Food/FoodSafety/Product-SpecificInformation/Seafood/RegulatoryFishEncyclopediaRFE/default.htm) has compiled up to 94 fish species (July 2009) with their respective IEF pattern, the largest deposit at present (see Section 21.4.2.1).

Other protein analytical techniques have also been developed for fish species identification including some attempts based on immunological procedures [47–49]. These methods have a narrow covering to identify different fish species and thus they have a very limited impact on routine analysis. The large number of different fish species and cellular types that could be involved in generating specific antibodies, either polyclonal [49] or monoclonal [47], would also require to analyze in parallel reference samples to verify that the antibodies used in the identification do not cross-react with other or similar species which is not clearly demonstrated with the available antibodies.

Thus, although most of the protein-based methods can be of certain value in some instances, they are not suitable for forensic or certified analysis given that conservation and treatments alter three-dimensional structure and physical characteristics of the proteins, and therefore losing and/or modifying their identification features.

21.4 Methods Based on DNA

21.4.1 Sample Handling and DNA Extraction

All methods described below make use of some kind of PCR amplification and, in consequence, require the extraction of certain amounts of DNA from the sample, either fresh or processed seafood, as template. Nevertheless, in spite of the powerful ability of PCR to detect minute amounts of DNA, several problems may appear associated to its application for species identification in foods. DNA quality in the sample rely upon physical (temperature, moisture, time, mechanic pressure), chemical (pH, oxidizing agents), and biological (endogen nucleases and proteases, manipulation) factors. Depurination, strand break, pyrimidine dimerization, and deoxyribose fragmentation are usual DNA alterations as a consequence of oxidative and hydrolytic damage in the DNA. Thus, in a DNA purification protocol for identification analysis, particular care should be taken to

1. Sample contamination
2. Low DNA yields associated to some species
3. Inhibition of the DNA polymerase activity
4. DNA degradation in processed products

Standard methods of DNA extraction from tissues are usually adequate for application to seafood and seafood products. The sample to be analyzed should be carefully removed from the food product to avoid contamination with other ingredients or particles from other specimens, preferably after rinsing of the surface and separation of an inner portion. Typically, a 5–50 mg muscle tissue sample from fish or shellfish, or equivalent amount from prepared/processed seafood, is finely minced and immersed in an extraction buffer containing detergents and proteinases. Proteins are removed by phenol and/or chloroform extraction and the nucleic acids are subsequently precipitated with ethanol or isopropanol [50]. These methods, fully manual and time consuming have been progressively replaced in the last years by commercial kits that allow simplifying the purification of multiple samples in a short period and have been successfully applied to processed seafood including canned tuna [51]. These kits usually consist of a single DNA-binding chromatography step in spin columns replacing both the protein extraction and the DNA precipitation steps, and can also be applied to the whole process (from tissue to DNA) when including a tissue-extraction initial step.

It should be noticed that the DNA extraction method may be critical for a successful identification of some species. For instance, several species of chondrichthyes, eels, shellfish, and decapods can yield either low amounts or hardly amplifiable DNA. DNA purification for PCR amplification from these samples may require the use of modified or specialized methods. Chelex-100 has been widely used in problematic DNA extractions since it is a styrene–divinylbenzene chelating resin with functional iminodiacetic acid groups that protect against DNA degradation and is recommended for samples where other methods fail [52]. Also the cetyl–trimethyl ammonium bromide (CTAB) method has shown to isolate good quality DNA from diverse and difficult samples and organisms, particularly if those contain carbohydrates. The method was originally devised for plant tissues [53] based on the CTAB properties as cationic detergent that bind to negatively charged molecules like DNA [54]. The procedure use buffered 1% CTAB/0.7 M NaCl/10 mM EDTA to homogenize the tissue (even dehydrated) to be followed by chloroform extraction. Subsequently an isopropanol precipitation step in 1% CTAB/Tris/EDTA without salt is performed [53] where the CTAB/nucleic acids complex precipitate [54] and leave solubilized proteins and polysaccharides in the solution.

Above all, DNA quality may be affected in processed food, particularly during storage and heat treatments [55]. Thus, while high-molecular weight DNA of a length of 20–50 kbp can be isolated from fresh meats, the DNA length is shortened to 15–20 kbp upon storage for only several days. Heating of the sample before extraction leads to a further degradation down to 300 bp for a 10 min cooking at 121°C [56,57].

Low DNA quality can also drive to misidentification when it occurs together with DNA contamination from other species within the sample. In these cases, if only PCR amplification is achieved from the contaminant DNA then, the proband individual is erroneously identified as belonging to the contaminant DNA. In order to detect early the problems of DNA quality in sample preparation, it is highly advisable to use a negative control of reagents used in DNA isolation together with positive and negative controls of PCR amplification in every diagnostic setup. At present, there is not a given DNA isolation method that could be successfully used for any type of sample. The convenience of a given procedure depends upon the nature of the sample and the use that will be given to the isolated DNA, and it should be empirically tested.

21.4.2 DNA Sequencing Methods

Due to the intraspecies genetic variability among fish populations and the lack of complete genetic information for many commercial species, DNA sequencing can be considered as one of the most reliable tools for fish identification. Moreover, comparative genetics of fish species has notably improved due, to a great extent, to the easy PCR amplification of specific DNA sequences in the last decade [58,59] and the subsequent automated and cheap DNA sequencing. Thus, direct DNA sequencing and database search has proved to be a highly accurate method for the unequivocal identification of fish species, subspecies, and even populations.

21.4.2.1 Standardized Fish Molecular Databases and Barcoding

Although current taxonomy and systematics tools permit the classification of practically all fish species, its usefulness is hindered by the lack of efficient and fast reference tools [60,61]. Nevertheless, there are fish identification databases that mainly collect taxonomical and general biological information from worldwide distributed species (e.g., FishBase: www.fishbase.org; The Census of Marine Life: www.coml.org; The FAO Species Identification and Data Programme, SIDP, at www.fao.org, and independent Natural History museums databases) that can be used for taxonomical identification and geographical identification of fish. On the other hand, the utility of DNA sequences for taxonomical purposes is well established at present [62], and even a single specific short sequence has been proposed to be sufficient to differentiate the vast majority of animal species [63], since congeneric species of animals regularly possess enough divergence between nucleotide sequences to ensure easy specific diagnosis. Thus, molecular features to taxon discrimination take advantage of the DNA sequence specificity to identify organisms and therefore, DNA sequences can be taken as identification markers or barcodes in any organism, and particularly in fish. This concept forms the basis for the implementation of databases for biological identifications through the DNA analysis, and is aimed to develop molecular systems based on DNA species-specific profiles or DNA-barcodes, which can be used as unique genetic fingerprint for living beings, allowing further investigations of DNA variation among them. Current studies in this field support the barcoding concept [63–65]. This potential is of particular interest to fisheries products for human consumption as well as to issues related to fisheries management.

For traceability of fish, species identification should be potentially feasible on processed food, including fillets, ready-to-eat dishes, canned fish, etc., and to this respect, DNA-based diagnostic analysis is the most appropriate. Molecular genetics methodologies have widely progressed over the last decade in the area of DNA identification, and specific systems have been developed to obtain DNA fragments of diagnostic significance from most organisms, including fish. To this respect it should be pointed out that advanced and affordable DNA-sequencing equipment as well as private or academic enterprises can generate DNA sequences within 24 h with a relatively low cost that has a downward tendency, making feasible the identification of fish at species level at low cost and in a short time.

Thus, several initiatives have been specifically developed for the identification of fish species based on DNA sequences. The species-diagnostic DNA sequences are online, publicly available and therefore accessible to control laboratories in any part of the world where also sequencing of short DNA fragments can be rapidly obtained at low cost from fish tissue samples.

The FishTrace database (www.fishtrace.org) covers most teleost fish species of commercial, ecological, and zoological interest for the European countries, paying particular emphasis to local data collected in Europe. In addition, FishTrace database provides molecular data, detailed protocols, and tools for the correct identification of fish species [66], standardized photographs taken from fish specimens, otoliths and fish products, and also, a large list of relevant technical publications on taxonomy, distribution, ecology, and biological parameters that have been ad hoc collected for the database.

Moreover, the information collected in FishTrace is connected to a biological reference collection from cataloged fish specimens validated by taxonomists. This additional endeavor allows cross-referring analyses with available vouchers deposited in Natural History Museums around Europe. Thus, the FishTrace database has developed a specific infrastructure in Europe for referencing and comparison of teleost fish sequences, information, and materials.

The Fish DNA Barcode of Life (FishBOL: www.fishbol.org) is a global database effort to collect a reference DNA sequence for all fish species. Also, identifying DNA sequences are derived from voucher specimens identified by taxonomists. At present DNA sequences from more than 6800 species have been deposited (July 2009). In the United States, the National Atmospheric and

Oceanic Administration (NOAA) already makes use of the FishBOL database for species identification purposes into their support for fisheries inspection and control.

FishTrace and FishBOL share common concepts regarding standardization of common sequence information but differ in the standardized sequences chosen to identify fish. It is well accepted that metazoan mitochondrial genomes (mtDNA) are more suitable for the implementation of a microgenomic identification system than nuclear genomes. The usual limits of intraspecific divergence in mitochondrial genes derived from phylogenetic analyses were established between 1% and 2% in general animal species [67]. Fish genomes undergo genetic changes rapidly, often due to polyploidiation, gain of spliceosomal introns, speciation, and gene duplication phenomenon [68–70]. FishBOL focuses on a DNA-based identification system using a relatively small sequence fragment (~600 bp) from the mitochondrial cytochrome *c* oxidase subunit I (COI). This DNA sequence provides sufficient identification labels in terms of nucleotide positions [65] to discriminate even between congeneric fish species, where a 2% sequence divergence is found in 98% of them [65]. Another mitochondrial gene used as DNA label is cytochrome *b* (cyt*b*) which also contains enough resolution to discriminate from the intraspecific to the intergeneric level [71], possesses a phylogenetic performance equivalent to that of COI [72], and has been widely used to identify and develop diagnostic systems for seafood species [10,18,32,33,73,74]. FishTrace database uses a mitochondrial (cyt*b*) and a nuclear (rhodopsin: rhod) sequence to construct a hybrid DNA-barcode that is used to identify European fish species.

The short fragment of the COI sequence proposed as a universal DNA-barcode [63] presents low interspecific divergences, or what is the same, low phylogenetic resolution in some fish families like tunas [65]. In a recent study, ~1% average interspecific Kimura 2-parameter (K2P) distance was obtained from the phylogenetic analysis of 46 tuna COI barcodes [65], while the average interspecific K2P distance obtained from the analysis of FishTrace DNA-barcodes in 29 tuna increased to ~1.7% [66]. These results strengthen the practical efficacy of a DNA-barcode with cyt*b* and rhod to identify fish species. Thus, although it is clear that safer identification labels depend upon the length of the DNA-barcodes, the DNA-barcoding efficiency can also be further improved by the simultaneous use of two genes with different evolutionary rates and genomic locations. This latter approach is used in the barcoding proposed in FishTrace by the use of the complete mitochondrial cyt*b* (1141 bp) and a nuclear fragment (460 bp) of the rhod gene, with independent genetic variation rate for each of them [75]. In fact, both cyt*b* and rhod genes have been widely used as effective molecular markers for fish species identification and for the establishment of unresolved or unknown fish phylogenies [76–80]. The absence of introns in the fish rhodopsin [81] makes easy the PCR amplification of a representative coding sequence for a nuclear gene, and being also conserved in chondrichthyes and tetrapods [81] allows the discrimination of teleosteii from other organism taxa in processed seafood. Moreover, the use of this nuclear gene in parallel with cyt*b* has the advantage of including an internal phylogenetic control for each other, with an increased resolution and guarantees for the identification of fishes to the species level. From the phylogenetic analyses performed within FishTrace, both mitochondrial and nuclear DNA sequence data produce similar phylogenetic tree topologies and congruency with other taxonomical-based phylogenies [82–84]. In addition, the use of two independent genes allows avoiding erroneous ascribing of DNA-barcodes and potential crossover contamination or other errors occurred during the PCR amplification. These errors can be detected since each gene sequence can be independently validated and phylogenetically analyzed to finally perform a morphological cross-checking for testing the reliability of the formed clades. Furthermore, phylogenetic analysis of both the assembled sequences (cyt*b* + rhod) reveal that most recent evolutionary changes are better resolved by the cyt*b*, whereas basal phylogenetic relationships are better defined by the rhod gene, since it shows higher conservation than cyt*b* (less overall changes between taxa).

Some other DNA sequences have also been proposed as identification labels for fish species. Among them ribosomal subunits are the most popular (16S-rRNA [13,85,86]; 12S-rRNA [14]; 18S-rRNA [28,87,90]; 5S-rRNA [27]), but nuclear genes have also been employed for fish species identification [12].

When standardized sequences (COI, cyt*b*, rhod, etc.) from fish species are not available at the time of developing an identification method, the possibility of finding at the general databases of NCBI, EMBL, and CIB-DDBJ* DNA sequences that could be used as identification labels, still exists. With the Taxonomy Browser at the NCBI site it is possible to identify rare fish species from which DNA sequences are available. Moreover, the Tree Tool within the basic local alignment search tool (BLAST; http://blast.ncbi.nlm.nih.gov/Blast.cgi) has also the ability to display a given sequence within a dendogram of closely related sequences to that query sequence uploaded by the user. Although caution should be taken for the correct phylogenetic clustering using this Tree Tool,† an adequate identification of a fish species can be performed provided that a matched sequence is available in the database (see below Section 21.4.2.2 where the DNA identification procedures from general databases are widely covered).

In addition, based on the genetic data entered into the FishTrace or FishBOL databases, tailored molecular identification systems for fish teleost species can be also specifically developed [13]. According to this approach, forensically informative nucleotide sequencing (FINS) [88] has also been adapted to design molecular diagnostic identification of fish [17] and squid [33] species, and thus simplifying the barcoding method and minimizing the potential misassignations by nonsequencing methods like PCR-restriction fragment length polymorphism (RFLP) [10] (see below: Section 21.4.3.1).

The Regulatory Fish Encyclopedia (RFE: http://www.fda.gov/Food/FoodSafety/Product-SpecificInformation/Seafood/RegulatoryFishEncyclopediaRFE/default.htm) is a project launched by the Food and Drug Administration, (FDA) United States, Center for Food Safety and Applied Nutrition (CFSAN) which compile fish species data in several formats to assist accurate identification. This is basically an initiative from administrative bodies focused to identify species substitution and economic deception in the marketplace. At present (July 2009), the RFE contains data on 94 commercially relevant fish species sold in the U.S. market including "chemical taxonomic" information consisting of species-characteristic biochemical patterns for comparison to patterns obtained by an appropriate laboratory analysis of the fish query. These are mainly protein-IEF and DNA-RFLP banding gel patterns. In addition the RFE comprises anatomical information in the form of pictures of the whole fish and their marketed product forms (such as fillets) and unique taxonomic features in a "checklist" format, to aid in identification. Nevertheless, IEF and RFLP patterns from similar or taxonomically closed fish species are not available at present, making practical comparisons between potential substitution species and differential diagnosis not a straightforward analysis. It is expected that in the near future the RFE would be able to include the DNA-barcodes of those species already listed and other relevant for practical analytical purposes [89]. An accurate barcoding procedure would improve species identification, which is essential in determining associated hazards, addressing economic fraud issues, and aiding in foodborne illness outbreak investigations.

---

* NCBI: http://www.ncbi.nlm.nih.gov/; EMBL: www.ebi.ac.uk/; CIB-DDBJ: www.ddbj.nig.ac.jp/.

† The phylogenetic analyses routinely performed in the NCBI Tree Tool are fast but technically limited and do not employ the most advanced (and computer-time consuming) methodology for accurate phylogenetic performance.

21.4.2.2 Identification from General Databases

Most sequencing identification methods are based on the nonstringent PCR amplification of a gene DNA section (the mitochondrial cytochrome *b*, 12S-rRNA, and cytochrome oxidase are the most commonly used target genes) followed by sequencing of the PCR product. The sequence is then compared to the current nucleotide database, and the identity of the specimen is established when the nucleotide identity fulfills a minimum score with one or more known specimens in the database, usually calculated by using the algorithm utilized in the Basic Local Alignment Search Tool (BLAST) software [90]. This approach should be carefully supervised, as nonvalidated databases, like NCBI or EMBL, may lead to DNA identities with entries erroneously labeled for a given species. In addition, BLAST results are typically formatted as a list of database entries arranged by the *Expectation value* (*E*) calculated as the number of different alignments with scores equivalent to or better than the score (sum of substitution and gap scores assigned by the similarity matrix used) that are expected to occur in a database search by chance; the lower the *E* value, the more significant the alignment between two given sequences. However, the *E* value is affected by the length of the alignments. Short alignments with higher identity may yield higher *E* values than longer alignments displaying slightly lower identity. As some database entries contain incomplete gene sequences, identical sequence matches may be overlooked behind lower identity full-length alignments. A more structured approach, which overcomes this inconvenience, should involve a taxonomical study of the sequence, involving the generation of a tree which will fit the input sequence in the appropriate taxonomy clade [92]. The building of phylogenetic trees requires the use of identical length sequences. The method, named FINS, has been applied in a number of applications to fish and other seafood species: the identification of scombrids [91]; sardines, including canned products [18,26]; cephalopods [92]; grey mullet, including processed ovary products [93]; or fish species in surimi [6,73] are just a few examples for the applicability of direct sequencing methods.

Advantages of full sequencing of relatively short DNA sequences (200–400 bp) include the taxonomical identification of the individual down to subspecies or population level, allowing also the ascription of specimens to taxonomy groups even when there is no previous sequence information on a given species. There are, however, several drawbacks for using DNA sequencing as routine food analysis: It is not a straightforward method requiring the use of sophisticated equipment (sequencers) or the use of external facilities providing sequencing services. It also requires some skills to analyze data (database comparison, alignments), and it cannot be directly used on samples containing mixtures of fish species, as mixed PCR products will deliver unreadable sequencing patterns. To facilitate the sequencing and further data analysis, attempts have been made to automate all sample processing, DNA sequencing, and data analysis, with special reference to the identification of fish larvae present in ichthyoplankton [94].

21.4.3 Non-DNA Sequencing Methods

Nonsequencing methods may not always provide forensically reliable information due to the absence of all appropriate control samples from potential substitution species or even because of intraspecific variation in the species. Thus a nonsequencing method can only be forensically applied if all controls have been considered and experimentally evaluated for a given differential identification of a group of fish species and their substitutions [13]. Partial comparison of fish species groups, which in practice can comprise some more species [95,96], are not recommended for fish species identification in control laboratories.

21.4.3.1 RFLP

RFLP is a widespread procedure for species identification in food analysis [104]. The set up of the method requires the previous collection of a significant number of specimens of the species to be identified, the sequencing of the selected target DNA region, and its comparison with the same region sequence obtained from other species, in particular close phylogenetic relatives. Thus, single nucleotide positions showing either low or no intraspecific variations, but which are polymorphic with respect to all other species, can be used as single nucleotide polymorphism (SNP) markers for species identification. To carry out an RFLP analysis, these polymorphisms should be located at a restriction site, allowing the identification of the SNP by variations in the restriction pattern of the PCR product. After a protocol is established and tested, the identification process can be performed routinely by any laboratory skilled in DNA extraction, PCR, and agarose-gel electrophoresis. These methods can be easily adapted to new technical developments, like capillary electrophoresis or Lab-on-a-Chip systems [98], which reduce laboratory manipulations and help to standardize results. The reliability of RFLP methods depends largely on its careful design and testing with target and nontarget species.

RFLP methods have been used to ascertain the origin of mackerels, either by using nontranscribed regions of nuclear DNA [25,99] or a double test using both mitochondrial *cytb* and nuclear 5S rRNA genes [100]. Several *cytb* gene regions have been successfully used as target sequences in PCR-RFLP fish identification methods, including gadoids [101], salmonids [102], or flatfishes [103], whether the p53 gene has been used for salmon and trout [104] and the 16S-rRNA gene for hairtail species [105].

RFLP in short DNA fragments has also been proved useful for the analysis of highly processed food. A diagnostic system set up to identify five different species in canned tuna show that the amplifiable fragment length could not exceed 278 bp due to DNA fragmentation during the sterilization process [23]. Recently, a wider range of tuna species, *Thunnus thynnus, T. alalunga, T. obesus, T. albacares, Euthynnus pelands (Katsuivonus pelamis), E. affinis, Auxis thazard*, and *Sarda orientalis* species in canned tuna could be identified by a combination of five restriction enzymes on two cytochrome *b* fragments [106]. The mitochondrial control region [107] and the 5S rRNA gene [27] have been used to identify several species of hake (*Merluccius* sp.), both in fresh and heat-treated samples. The identification of anchovy species is a representative example for the application of FINS and PCR-RFLP methods to discriminate closely related species. *Engraulis japonicus* and *E. encrasicolus* (Japanese and European anchovies, respectively) display highly similar *cytb* sequences. In addition, both species showed a relatively high genetic diversity, and the existence of subspecies or cryptic species of *E. encrasicolus* has been proposed [17,108,109]. As PCR-RFLP methods may not be fully selective for discrimination between these two species, the additional use of FINS or alternative methods may be needed for identification.

PCR-RFLP methods have been also useful for the authentication of nonfish seafood, like crustaceans [86,110] and mollusks [111,112].

21.4.3.2 AFLP, RAPD, and Satellite DNA Analysis

Amplified fragment length polymorphism (AFLP) is based on the PCR amplification of endonuclease-restricted fragments ligated to synthetic adapters and then amplified using primers which carry selective nucleotides at their 3′ ends [113,114]. After the selection of adequate restriction enzymes, usually applied by pairs, which will yield a wide assortment of genomic DNA fragments upon digestion, and the attachment of linkers to the protruding ends, several sets of selective primers are tested to obtain a pattern of DNA fragments after PCR amplification. Differences among species or populations are visualized by the modification of such amplification patterns, due to the presence of polymorphic sequences located next to the restriction site. The selectivity of the method can be adjusted to identify species, strains, populations, or even lineages in cultured specimens, both in fish and shellfish species, as reported for sturgeon commercial products and interspecific hybrids [115], *Morone* and *Thunnus* species [116], and a wide variety of fish, mollusks, and crustacean species with the purpose of developing an AFLP database [117].

The randomly amplified polymorphic DNA approach (RAPD) has also been widely tested for identification of fish species and populations. The RAPD method relies on the generation of a collection of DNA fragments or fingerprint by using a single or limited number of arbitrary oligonucleotides as primers. The pattern of amplification is expected to be consistent for the set of primers used, the DNA, and the conditions used. The RAPD approach has been reported to be suitable for the identification, among others, of grouper, Nile perch and wreck [118], salmonids [87], and whiting [119]. Some other examples of primers used in RAPD methods for fish authentication have been previously compiled [120]. Both AFLP and RAPD techniques share some advantages such as the relatively low cost, the requirement of only small amounts of DNA, and that they do not require previous knowledge of the species DNA sequences. However, both methods are strongly dependent on the integrity of the DNA. Thus, samples containing highly fragmented DNA, as in thermally treated food, may lead to altered banding in AFLP and RAPD patterns. In addition, samples containing two or more DNA species cannot be easily analyzed by these methods, as the banding would become too complex to discriminate each species.

These two above-mentioned methods are, however, a powerful tool for the identification of sequence characterized amplified regions (SCAR). Species-specific bands identified by AFLP or RAPD are isolated from separating gels, reamplified, and sequenced. These regions can be used subsequently as targets for species identification by using a sequence-specific identification method, as DGGE, selective amplification, RFLP, or array technologies. This approach has been successfully used for the discrimination of trout (*Oncorhynchus mykiss*) and Atlantic salmon (*Salmo salar*) [19].

Tandemly arrayed, highly repetitive DNA sequences on eukaryotic genomes, known as satellite DNA, are highly informative to investigate inheritance patterns and for the identification of intraspecific populations. These sequences range from very short (2–4 bp) repeated sequences (microsatellites), medium sized (10–64 bp) minisatellites, or long (>64 bp) satellite sequences, extensively dispersed throughout eukaryotic chromosomes. Among these, microsatellite sequences have become the most used nuclear markers for genetic analysis in fishes, which are estimated to occur once every 10 kbp [121]. These repeats are inherited, with new variants arising in the population during recombination and segregation by changes in the copy number of the repeat unit. Genetic isolation leads to the fixation of these variants (alleles) in the populations, in which the length of the repeat can be ascertained by PCR amplification with oligonucleotide primers encompassing the neighbor DNA region. Due to its high power of discrimination, microsatellite analysis has become the most powerful tool to study the geographical distribution of populations, as has been reported for sockeye salmon [122], horse mackerel [123], and others [124–126]. It is also one of the preferred methods in aquaculture allowing the genotyping, parentage, stock structure studies, and the traceability of the products (see Chistiakov et al. [127]). One useful application of this approach is the differentiation of wild and hatchery-produced variants of the same species, as are the cases for chinook salmon [142]. The experimental comparison of the performance of microsatellites and SNPs to differentiate wild and farmed Atlantic salmon shows that both approaches produce similar results, being SNPs more suitable for high-throughput applications [129].

Although mainly used for stock and population studies, the technology based on DNA satellites can also be used for species identification: for example, six microsatellite loci allow the identification of up to eight different grouper species [130]. In another report, the feasibility to differentiate 3 species of North Atlantic wolffishes using 16 tetranucleotide and dinucleotide microsatellite markers has also been demonstrated [131]. The use of satellite analysis for routine species identification is however hampered due to the extensive genetic analysis required during the set up of the method. Each microsatellite locus has to be identified and its flanking region sequenced for primer design, and the validation process may become laborious and time consuming.

21.4.3.3   SSCP and DGGE

Although initially developed for the identification of SNPs, single-strand conformation polymorphisms (SSCP) [132] is a useful tool for fish product identification [133]. The method relies on the differences observed in the electrophoretic mobility of short single-stranded DNA which differ in one nucleotide position. After PCR amplification of the region of interest, the resulting double-stranded product is denatured followed by rapid chilling to prevent reannealing of the strands, and separated by electrophoresis under nondenaturing conditions.

An alternate technology to visualize mobility changes based on DNA sequences is denaturing gradient gel electrophoresis (DGGE) [134], which separates single-strand DNA fragments using a gel containing chemicals which break apart the strands of the DNA molecule. Because the amplicon segments are the same length, separation must be made based on the genetic sequences rather than on size. Denaturing gels have an increasing concentration gradient (from top to bottom) of denaturing chemicals. Furthermore, as the identification is based on one or a few electrophoretic bands, these methods can be used on complex samples containing DNA from different species. Recent applications of DGGE methods to fish identification have been reported [19]. Moreover, the comparison of the efficiency obtained by RFLP, SSCP, and DGGE methods to differentiate eight cod-fish species demonstrated that RFLP and SSCP were not able to identify all species tested, but DGGE achieved the best performance [11]. In spite of some advantages, SSCP and DGGE methods are not extensively used due to their lack of robustness and requirements for a skilled interpretation of the band patterns. Thus, accurate experimental conditions have to be maintained to obtain repetitive results, as small changes in temperature, pH, and gel composition may affect the band migration pattern. In addition, these methods are rather time consuming and laborious, and therefore are not well suited for routine monitoring laboratories.

21.4.3.4   Selective Amplification

Under selective amplification we include all PCR methods which deliver a species-specific amplification product under stringent conditions. This may be achieved by selecting short oligonucleotides that match DNA sequences found exclusively in the target species, either as primers for the Taq DNA polymerase, as internal probes in the amplicons, or both. Compared to other PCR-based identification methods, selective amplification is faster and simpler, as it requires little processing after the PCR has been completed. Lockley and Bardsley (2000) [135] used a single-step PCR for discrimination of tuna (*T. thynnus*) and bonito (*Sarda sarda*). The identification may be based on the direct detection of the expected size amplicon in agarose gels after electrophoresis, monitorization of double-chain DNA products by using nonselective fluorochromes like SYBRgreen, or the detection of fluorescence along the PCR using sequence-specific fluorescently labeled DNA probes (see below). Regardless of the method for visualization, PCR-specific methods rely strongly on the adequate selection of target sequences for oligonucleotide hybridization. The exponential increase in sequence information in the last years has assisted greatly to the development of this approach for species identification. Usual targets for PCR amplification from closely related species, like the mitochondrial cyt*b*, COX1, D-loop, 12S-rRNA genes, or nuclear 18S-rRNA gene, may differ in only a few nucleotide positions. In the less-favorable condition, the selective PCR may depend solely on melting temperature (Tm) differences of one oligonucleotide to hybridize with a single-base mismatch in its target sequence. Alternatively, the selective base position can be located at the 3′ end of the primer, preventing the initiation of DNA polymerization after hybridization with the nontarget DNA [136,137].

The selective amplification approach can be expanded in several ways. By combination of several primer pairs, or species-specific primers with a nonselective counterpart, it is possible to identify several species in the same reaction [95,138,139]. Also, nested PCR-based methods have been developed, in which the first reaction is carried out using nonselective or family-specific primers, and the second reaction makes use of species selective primers, yielding PCR products that can be analyzed by RFLP [23,28]. This latter approach increases the discrimination power of the method.

The technology developed for real-time PCR can also be used to increase specificity, reduce the labor required, or automate the analysis. The real-time approach has found two main applications in genetic studies: the detection of SNPs and the quantification of the number of copies of target DNA. The technique relies on the monitorization of the accumulation of products at every PCR cycle, which can be achieved by the addition of an oligonucleotide probe labeled with a reporter and a quencher dye that binds to a target DNA between the flanking primers. The 5′ to 3′ exonuclease activity of the *Taq* DNA polymerase cleaves the probe during PCR, allowing the emission of fluorescence by releasing the reporter from the quencher. In either case, the fluorescent emission would increase exponentially at every PCR cycle [140]. Methods using fluorescent species-specific oligonucleotides probes in real-time PCR have been tested for the identification of eel species based on SNPs [141]. Alternatively, a fluorescent dye (SYBRgreen) which preferentially binds to double-strand DNA can be added to the reaction replacing the oligonucleotide probe [142]. This approach lowers the assay cost and avoids the need to use a specific probe for each species analyzed. In turn, the assay specificity relies exclusively on the primer sequence, which may require confirmation of the PCR product identity after the amplification. A method for identification of grouper and other substitution species based on a multiplex PCR assay using SYBRgreen and a postreaction DNA dissociation analysis has been developed [13]. The amplicons of different size display different melting temperatures, which can be monitorized as a decrease in fluorescence due to the separation of the dye when double-strand DNA dissociates. The identification could be performed based on the appearance of PCR products displaying a melting temperature corresponding to the species analyzed, and can be improved to high throughput screening [13].

21.4.3.5 Quantitative Methods

The development of real-time PCR technology has facilitated significantly the development of techniques aiming to the quantification of animal and vegetal ingredients in processed food, where mixtures of materials from different species are easily found, and there is an increasing need to differentiate purposely added ingredients from trace contaminant materials which may derive from the manufacturing process. The limit of detection of DNA-based qualitative methods may be well below the legal limits for a component to be considered as ingredient, and can therefore lead to conflictive results in the detection of frauds.

As the amount of product accumulated at each PCR cycle is proportional to the initial copy number of the target DNA, it can be quantified using fluorescent monitored real-time PCR by comparison of the cycle at which the unknown and a standard DNA of known concentration reach the same fluorescence [143]. Although most protocols of quantitative real-time PCR for food authentication have been devoted to meat products [37,144,145] or transgenic material (see [146] and references therein), a few examples on applications to fish products have been reported. Thus, the feasibility to quantify albacore and yellowfin in binary mixtures [22] and the detection and precise quantification of haddock in mixed samples [40] are first-rate models for potential development of similar diagnostic systems in other groups of fish species. So far, considering the need for validation tests required for fish species identification, and the fact that quantification of DNA is a valuable data (although may not fully correlate with the amount of raw material or protein in the sample), real-time PCR appears as the most versatile and effective method for quantification of components in food products and mixtures.

21.4.3.6 High-Throughput, Microarray Technologies and Bioinformatics

While most DNA-based methods can be successfully used for the identification of farmed species used in food manufacturing, setting up methods for identification of species from extractive fishing may become extremely laborious due to the very high diversity of species and the presence of intraspecific genetic variations. Compared to specific-amplification, RFLP, and other species-specific closed methods, DNA sequencing is an open method which may provide information on almost any species or variant. However, faster methods suitable for high-throughput analysis would be desirable for routine labs. The application of array technologies, widely used in SNP identification and gene expression applications, can be particularly suited for fish species identification as they have the potential to handle and identify hundreds of species in parallel. In their standard format, DNA microarrays are microscope slides on which oligonucleotides are spotted, whose DNA sequence are complementary to the DNA target sequences. The sample DNA is amplified by PCR using fluorescent-labeled primers flanking the hybridization region. The labeled PCR product hybridizes with the immobilized oligonucleotide on the microarray, and can be detected after washing steps. Despite some inherent drawbacks, as the high cost, methodological difficulties for standardization, and relatively high interlaboratory variability, there are examples of arrays developed for the identification of marine organisms in plankton [147,148] and fish species [149]. In this last report, a prototype assay with 11 commercial fish species were targeted for identification by using a single oligonucleotide probe (23–27 nucleotides long) from the 16S-rRNA gene per species, and a 600 bp fluorescent-labeled PCR product for hybridization. True-positive signals could be differentiated from false-positive due to their higher fluorescent signal, although those signal intensities were heterogeneous. This heterogeneity is common in hybridization array methods, being likely caused by the dependence of hybridization efficiency on several complex parameters like the nucleotide sequence, steric hindrance, secondary structures, and the relative position of the label at the target. This variability often leads to problems of reproducibility both intra- and interlaboratorily, being the main drawback for utilization as a general identification tool. In turn, this method can be the only practical approach for the routine analysis of very complex samples (e.g., plankton), in which a wide variety of known and unknown species can be found. Future improvements on the microarray design, like the use of multiple probe sets for each species to be identified, redundant hybridization using both DNA strands, addition of multiple-labeled target PCR products, and others, may allow a broader utilization of this methodology.

An interesting alternative to the hybridization array approach for high-throughput analysis is the use of primer extension technology. These systems are based on the identification of polymorphisms by using a primer that hybridizes immediately upstream of the SNP, and is extended by just one base in the presence of a fluorescent-labeled dideoxynucleotide. The polymorphisms are detected by the emission of fluorescent signal at located spots (array) [150,151], or at specific peaks after separation by capillary electrophoresis or combined primer extension-capillary electrophoresis (SnaPshot, Applied Biosystems, Foster City, California). Following this approach, the efficiency of a multiplex primer-extension assay for the identification of five tuna species has been recently demonstrated [74]. After amplification of a 132 bp region from the *cytb* gene, the PCR product is used as template for simultaneous single-nucleotide extension using four primers which hybridize next to diagnostic base positions and have different lengths by the inclusion of poly(T) tails at the 5′ end. All four A, G, T, C dideoxynucleotides, each labeled with a different fluorochrome, are used in the reaction, and the resulting products are separated by capillary electrophoresis using a standard sequencing equipment. The resulting pattern of labeled bands allowed the unambiguous identification of five *Thunnus* species. As primer extension protocols require only very short DNA fragments for PCR amplification [152], they appear as a promising alternative for the analysis of highly processed food products.

21.5   Future Prospects

Legislation and law enforcement of fish labeling rely on fast, efficient, and accurate diagnostic analysis of large sets of samples. At present powerful molecular technologies are available, but mostly partial diagnostic systems have only been provided by the scientific community to the control laboratories. Thus in Table 21.2, a suggested practical use of the different methodologies presently available for the identification of seafood species is given. Meanwhile many described

Table 21.2   Recommended Fish Species Identification Methodologies for Different Types of Seafood Samples and Processing

Method	Fresh Single Species	Low Processed (Freezing, Mincing)	Highly Processed (High Temperature, Additives)	Mixed Species
Sequencing	A	A	AR	NA
RFLP	A	A	AR	AR
AFLP, RAPD	A	A	NA	NA
Microsatellite	AR	AR	NA	NA
SSCP	AR	AR	AR	NA
Selective PCR	A	A	AR	A
Array	A	A	AR	A
Quantitative PCR	A	A	AR	A

A, Adequate; AR, adequate with restrictions; NA, not adequate.

molecular methods that show laboratory viability have limited use for routine analysis. In the near future we envisage that high-throughput technologies for DNA analysis associated to the automation of sample handling and processing that is controlled with specialized software and database identification could pave the path for an effective species diagnosis in seafood to overcome adulteration, mislabeling, and realistic fisheries control.

References

1. Marko, P. B., Lee, S. C., Rice, A. M., Gramling, J. M., Fitzhenry, T. M., McAlister, J. S., Harper, G. R., and Moran, A. L., Fisheries: Mislabeling of a depleted reef fish, *Nature* 430 (6997), 309–310, 2004.
2. Chegini, S. and Metcalfe, D. D., Contemporary issues in food allergy: Seafood toxin-induced disease in the differential diagnosis of allergic reactions, *Allergy and Asthma Proceedings* 26 (3), 183–190, 2005.
3. Shinzato, T., Furusu, A., Nishino, T., Abe, K., Kanda, T., Maeda, T., and Kohno, S., Cowfish (Umisuzume, *Lactoria diaphana*) poisoning with rhabdomyolysis, *Internal Medicine* 47 (9), 853–856, 2008.
4. Karl, H. and Ruoff, U., Dioxins, dioxin-like PCBs and chloroorganic contaminants in herring, *Clupea harengus*, from different fishing grounds of the Baltic Sea, *Chemosphere* 67 (9), S90–S95, 2007.
5. Cheung, K. C., Leung, H. M., Kong, K. Y., and Wong, M. H., Residual levels of DDTs and PAHs in freshwater and marine fish from Hong Kong markets and their health risk assessment, *Chemosphere* 66 (3), 460–468, 2007.
6. Pepe, T., Trotta, M., Di Marco, I., Anastasio, A., Bautista, J. M., and Cortesi, M. L., Fish species identification in surimi-based products, *Journal of Agricultural and Food Chemistry* 55 (9), 3681–3685, 2007.
7. Renon, P., Bernardi, C., Malandra, R., and Biondi, P. A., Isoelectric focusing of sarcoplasmic proteins to distinguish swordfish, blue marlin and Mediterranean spearfish, *Food Control* 16 (5), 473–477, 2005.
8. Brander, K. M., Global fish production and climate change, *Proceedings of the National Academy of Sciences of the United States of America* 104 (50), 19709–19714, 2007.
9. Reynolds, J. D., Dulvy, N. K., Goodwin, N. B., and Hutchings, J. A., Biology of extinction risk in marine fishes, *Proceedings of the Royal Society B-Biological Sciences* 272 (1579), 2337–2344, 2005.
10. Akasaki, T., Yanagimoto, T., Yamakami, K., Tomonaga, H., and Sato, S., Species identification and PCR-RFLP analysis of cytochrome b gene in cod fish (order Gadiformes) products, *Journal of Food Science* 71 (3), C190–C195, 2006.
11. Comi, G., Iacumin, L., Rantsiou, K., Cantoni, C., and Cocolin, L., Molecular methods for the differentiation of species used in production of cod-fish can detect commercial frauds, *Food Control* 16 (1), 37–42, 2005.
12. Hubalkova, Z., Kralik, P., Kasalova, J., and Rencova, E., Identification of gadoid species in fish meat by polymerase chain reaction (PCR) on genomic DNA, *Journal of Agricultural and Food Chemistry* 56 (10), 3454–3459, 2008.
13. Trotta, M., Schönhuth, S., Pepe, T., Cortesi, M. L., Puyet, A., and Bautista, J. M., Multiplex PCR method for use in real-time PCR for identification of fish fillets from grouper (*Epinephelus* and *Mycteroperca* species) and common substitute species, *Journal of Agricultural and Food Chemistry* 53, 2039–2045, 2005.
14. Infante, C., Catanese, G., Ponce, M., and Manchado, M., Novel method for the authentication of frigate tunas (*Auxis thazard* and *Auxis rochei*) in commercial canned products, *Journal of Agricultural and Food Chemistry* 52 (25), 7435–7443, 2004.
15. Marko, P. B., Lee, S. C., Rice, A. M., Gramling, J. M., Fitzhenry, T. M., McAlister, J. S., Harper, G. R., and Moran, A. L., Mislabeling of a depleted reef fish, *Nature* 430 (6997), 309–310, 2004.

16. Zhang, J. B., Huang, H., Cai, Z. P., and Huang, L. M., Species identification in salted products of red snappers by semi-nested PCR-RFLP based on the mitochondrial 12S rRNA gene sequence, *Food Control* 17 (7), 557–563, 2006.

17. Santaclara, F. J., Cabado, A. G., and Vieites, J. M., Development of a method for genetic identification of four species of anchovies: *E. encrasicolus*, *E. anchoita*, *E. ringens*, and *E. japonicus, European Food Research and Technology* 223 (5), 609–614, 2006.

18. Jerome, M., Lemaire, C., Bautista, J. M., Fleurence, J., and Etienne, M., Molecular phylogeny and species identification of sardines, *Journal of Agricultural and Food Chemistry* 51 (1), 43–50, 2003.

19. Zhang, J. B., Wang, H. J., and Cai, Z. P., The application of DGGE and AFLP-derived SCAR for discrimination between Atlantic salmon (*Salmo salar*) and rainbow trout (*Oncorhynchus mykiss*), *Food Control* 18 (6), 672–676, 2007.

20. Withler, R. E., Candy, J. R., Beacham, T. D., and Miller, K. M., Forensic DNA analysis of Pacific salmonid samples for species and stock identification, *Environmental Biology of Fishes* 69 (1–4), 275–285, 2004.

21. Berrini, A., Tepedino, V., Borromeo, V., and Secchi, C., Identification of freshwater fish commercially labeled "perch" by isoelectric focusing and two-dimensional electrophoresis, *Food Chemistry* 96 (1), 163–168, 2006.

22. López, I. and Pardo, M. A., Application of relative quantification TaqMan real-time polymerase chain reaction technology for the identification and quantification of *Thunnus alalunga* and *Thunnus albacares, Journal of Agricultural and Food Chemistry* 53 (11), 4554–4560, 2005.

23. Pardo, M. A. and Perez-Villareal, B., Identification of commercial canned tuna species by restriction site analysis of mitochondrial DNA products obtained by nested primer PCR, *Food Chemistry* 86 (1), 143–150, 2004.

24. Karaiskou, N., Apostolidis, A. P., Triantafyllidis, A., Kouvatsi, A., and Triantaphyllidis, C., Genetic identification and phylogeny of three species of the genus *Trachurus* based on mitochondrial DNA analysis, *Marine Biotechnology* 5 (5), 493–504, 2003.

25. Aranishi, F., Rapid PCR-RFLP method for discrimination of imported and domestic mackerel, *Marine Biotechnology* 7 (6), 571–575, 2005.

26. Jerome, M., Lemaire, C., Verrez-Bagnis, V., and Etienne, M., Direct sequencing method for species identification of canned sardine and sardine-type products, *Journal of Agricultural and Food Chemistry* 51 (25), 7326–7332, 2003.

27. Pérez, J. and García-Vázquez, E., Genetic identification of nine hake species for detection of commercial fraud, *Journal of Food Protection* 67 (12), 2792–2796, 2004.

28. Pérez, M., Vieites, J. M., and Presa, P., ITS1-rDNA-based methodology to identify world-wide hake species of the genus *Meriuccius, Journal of Agricultural and Food Chemistry* 53 (13), 5239–5247, 2005.

29. Wuertz, S., Belay, M., and Kirschbaum, F., On the risk of criminal manipulation in caviar trade by intended contamination of caviar with PCR products, *Aquaculture* 269 (1–4), 130–134, 2007.

30. Santaclara, F. J., Espineira, M., Cabado, G., Aldasoro, A., Gonzalez-Lavin, N., and Vieites, J. M., Development of a method for the genetic identification of mussel species belonging to *Mytilus, Perna, Aulacomya*, and other genera, *Journal of Agricultural and Food Chemistry* 54 (22), 8461–8470, 2006.

31. Pascoal, A., Barros-Velazquez, J., Cepeda, A., Gallardo, J. M., and Calo-Mata, P., Survey of the authenticity of prawn and shrimp species in commercial food products by PCR-RFLP analysis of a 16S rRNA/tRNA(Val) mitochondrial region, *Food Chemistry* 109 (3), 638–646, 2008.

32. Santaclara, F. J., Espineira, M., and Vieites, J. M., Genetic identification of squids (families Ommastrephidae and Loliginidae) by PCR-RFLP and FINS methodologies, *Journal of Agricultural and Food Chemistry* 55 (24), 9913–9920, 2007.

33. Hsieh, H. S., Chai, T. J., and Hwang, D. F., Using the PCR-RFLP method to identify the species of different processed products of billfish meats, *Food Control* 18 (4), 369–374, 2007.

34. Allshouse, J., Buzby, J., Harvey, D., and Zorn, D., International trade and seafood safety, In: *International Trade and Food Safety.* Agricultural Economic Report 828, USDA Economic Research Service Buzby J. (Ed.), 2003, pp. 109–124, www.ers.usda.gov/publications/aer828/.

35. Lopez-Andreo, M., Garrido-Pertierra, A., and Puyet, A., Evaluation of post-polymerase chain reaction melting temperature analysis for meat species identification in mixed DNA samples, *Journal of Agricultural and Food Chemistry* 54 (21), 7973–7978, 2006.

36. López-Andreo, M., Lugo, L., Garrido-Pertierra, A., Prieto, M. I., and Puyet, A., Identification and quantitation of species in complex DNA mixtures by real-time polymerase chain reaction, *Analytical Biochemistry* 339, 73–82, 2005.

37. Martin, I., Garcia, T., Fajardo, V., Rojas, M., Hernandez, P. E., Gonzalez, I., and Martin, R., Real-time PCR for quantitative detection of bovine tissues in food and feed, *Journal of Food Protection* 71 (3), 564–572, 2008.

38. Clarke, S. C., Magnussen, J. E., Abercrombie, D. L., McAllister, M. K., and Shivji, M. S., Identification of shark species composition and proportion in the Hong Kong shark fin market based on molecular genetics and trade records, *Conservation Biology* 20 (1), 201–211, 2006.

39. Clarke, S., Understanding pressures on fishery resources through trade statistics: A pilot study of four products in the Chinese dried seafood market, *Fish and Fisheries* 5 (1), 53–74, 2004.

40. Hird, H. J., Hold, G. L., Chisholm, J., Reece, P., Russell, V. J., Brown, J., Goodier, R., and MacArthur, R., Development of a method for the quantification of haddock (*Melanogrammus aeglefinus*) in commercial products using real-time PCR, *European Food Research and Technology* 220 (5–6), 633–637, 2005.

41. Ataman, C., Celik, U., and Rehbein, H., Identification of some Aegean fish species by native isoelectric focussing, *European Food Research and Technology* 222 (1–2), 99–104, 2006.

42. Colombo, M. M., Colombo, F., Biondi, P. A., Malandra, R., and Renon, P., Substitution of fish species detected by thin-layer isoelectric focusing and a computer-assisted method for the evaluation of gels, *Journal of Chromatography A* 880 (1–2), 303–309, 2000.

43. Pineiro, C., Barros-Velazquez, J., Perez-Martin, R. I., and Gallardo, J. M., Specific enzyme detection following isoelectric focusing as a complimentary tool for the differentiation of related Gadoid fish species, *Food Chemistry* 70 (2), 241–245, 2000.

44. Asensio Gil, L., PCR-based methods for fish and fishery products authentication, *Trends in Food Science & Technology* 18 (11), 558–566, 2007.

45. Etienne, M., Jerome, M., Fleurence, J., Rehbein, H., Kundiger, R., Mendes, R., Costa, H. et al., Identification of fish species after cooking by SDS-PAGE and urea IEF: A collaborative study, *Journal of Agricultural and Food Chemistry* 48 (7), 2653–2658, 2000.

46. Rehbein, H., Kundiger, R., Yman, I. M., Ferm, M., Etienne, M., Jerome, M., Craig, A. et al., Species identification of cooked fish by urea isoelectric focusing and sodium dodecylsulfate polyacrylamide gel electrophoresis: A collaborative study, *Food Chemistry* 67 (4), 333–339, 1999.

47. Asensio, L., Gonzalez, I., Rodriguez, M. A., Hernandez, P. E., Garcia, T., and Martin, R., Development of a monoclonal antibody for grouper (*Epinephelus marginatus*) and wreck fish (*Polyprion americanus*) authentication using an indirect ELISA, *Journal of Food Science* 68 (6), 1900–1903, 2003.

48. Fernandez, A., Garcia, T., Asensio, L., Rodriguez, M. A., Gonzalez, I., Lobo, E., Hernandez, P. E., and Martin, R., Identification of the clam species *Ruditapes decussatus* (grooved carpet shell), *Venerupis rhomboides* (yellow carpet shell) and *Venerupis pullastra* (pullet carpet shell) by ELISA, *Food and Agricultural Immunology* 14 (1), 65–71, 2002.

49. Ochiai, Y., Ochiai, L., Hashimoto, K., and Watabe, S., Quantitative estimation of dark muscle content in the mackerel meat paste and its products using antisera against myosin light chains, *Journal of Food Science* 66 (9), 1301–1305, 2001.

50. Sambrook, J. and Russell, D. W., *Molecular Cloning: A Laboratory Manual*, Third edn., CSHL Press, New York, 2001.

51. Chapela, M. J., Sotelo, C. G., Perez-Martin, R. I., Pardo, M. A., Perez-Villareal, B., Gilardi, P., and Riese, J., Comparison of DNA extraction methods from muscle of canned tuna for species identification, *Food Control* 18 (10), 1211–1215, 2007.

52. Walsh, P. S., Metzger, D. A., and Higuchi, R., Chelex 100 as a medium for simple extraction of DNA for PCR-based typing from forensic material, *Biotechniques* 10 (4), 506–513, 1991.

53. Doyle, J. J. and Doyle, J. L., A rapid DNA isolation procedure for small quantities of fresh leaf tissue, *Phytochemistry Bulletin* 19, 11–15, 1987.

54. Spink, C. H. and Chaires, J. B., Thermodynamics of the binding of a cationic lipid to DNA, *Journal of the American Chemical Society* 119 (45), 10920–10928, 1997.

55. Ram, J. L., Ram, M. L., and Baidoun, F. F., Authentication of canned tuna and bonito by sequence and restriction site analysis of polymerase chain reaction products of mitochondrial DNA, *Journal of Agricultural and Food Chemistry* 44, 2460–2467, 1996.

56. Ebbehoj, K. F. and Thomsen, P. D., Differentiation of closely related species by DNA hybridisation, *Meat Science* 30, 359–366, 1991.

57. Chikuni, K., Ozutsumi, K., Koishikawa, T., and Kato, S., Species identification of cooked meats by DNA hybridization assay, *Meat Science* 27, 119–128, 1990.

58. Partis, L. and Wells, R. J., Identification of fish species using random amplified polymorphic DNA (RAPD), *Molecular and Cellular Probes* 10 (6), 435–441, 1996.

59. Rehbein, H., Mackie, I. M., Pryde, S., Gonzales-Sotelo, C., Medina, I., Perez-Martin, R., Quinteiro, J., and Rey-Mendez, M., Fish species identification in canned tuna by PCR-SSCP: Validation by a collaborative study and investigation of intra-species variability of the DNA-patterns, *Food Chemistry* 64 (2), 263–268, 1999.

60. Blaxter, M., Molecular systematics: Counting angels with DNA, *Nature* 421 (6919), 122–124, 2003.

61. Godfray, H. C., Towards taxonomy's "glorious revolution," *Nature* 420 (6915), 461, 2002.

62. Tautz, D., Arctander, P., Minelli, A., Thomas, R. H., and Vogler, A. P., A plea for DNA taxonomy, *Trends in Ecology & Evolution* 18 (2), 70–74, 2003.

63. Hebert, P. D., Cywinska, A., Ball, S. L., and deWaard, J. R., Biological identifications through DNA barcodes, *Proceedings of Biological Sciences* 270 (1512), 313–321, 2003.

64. Blaxter, M. L., The promise of a DNA taxonomy, *Philosophical Transactions of the Royal Society of London Series B: Biological Sciences* 359 (1444), 669–679, 2004.

65. Ward, R. D., Zemlak, T. S., Innes, B. H., Last, P. R., and Hebert, P. D., DNA barcoding Australia's fish species, *Philosophical Transactions of the Royal Society of London Series B: Biological Sciences* 360 (1462), 1847–1857, 2005.

66. Sevilla, R. G., Diez, A., Noren, M., Mouchel, O., Jerome, M., Verrez-Bagnis, V., van Pelt, H., Favre-Krey, L., Krey, G., and Bautista, J. M., Primers and polymerase chain reaction conditions for DNA barcoding teleost fish based on the mitochondrial cytochrome *b* and nuclear rhodopsin genes, *Molecular Ecology Notes* 7 (5), 730–734, 2007.

67. Avise, J. C., Arnold, J., Ball, R. M., Bermingham, E., Lamb, T., Neigel, J. E., Reeb, C. A., and Saunders, N. C., Intraspecific phylogeography—The mitochondrial-DNA bridge between population-genetics and systematics, *Annual Review of Ecology and Systematics* 18, 489–522, 1987.

68. Robinson-Rechavi, M. and Laudet, V., Evolutionary rates of duplicate genes in fish and mammals, *Molecular Biology and Evolution* 18 (4), 681–683, 2001.

69. Robinson-Rechavi, M., Marchand, O., Escriva, H., Bardet, P. L., Zelus, D., Hughes, S., and Laudet, V., Euteleost fish genomes are characterized by expansion of gene families, *Genome Research* 11 (5), 781–788, 2001.

70. Venkatesh, B., Evolution and diversity of fish genomes, *Current Opinion in Genetics and Development* 13 (6), 588–592, 2003.

71. Kocher, T. D., Thomas, W. K., Meyer, A., Edwards, S. V., Paabo, S., Villablanca, F. X., and Wilson, A. C., Dynamics of mitochondrial DNA evolution in animals: Amplification and sequencing with conserved primers, *Proceedings of the National Academy of Sciences USA* 86 (16), 6196–6200, 1989.

72. Zardoya, R. and Meyer, A., Phylogenetic performance of mitochondrial protein-coding genes in resolving relationships among vertebrates, *Molecular Biology and Evolution* 13 (7), 933–942, 1996.

73. Pepe, T., Trotta, M., Di Marco, I., Cennamo, P., Anastasio, A., and Cortesi, M. L., Mitochondrial cytochrome *b* DNA sequence variations: An approach to fish species identification in processed fish products, *Journal of Food Protection* 68 (2), 421–425, 2005.

74. Bottero, M. T., Dalmasso, A., Cappelletti, M., Secchi, C., and Civera, T., Differentiation of five tuna species by a multiplex primer-extension assay, *Journal of Biotechnology* 129 (3), 575–580, 2007.

75. Vawter, L. and Brown, W. M., Nuclear and mitochondrial DNA comparisons reveal extreme rate variation in the molecular clock, *Science* 234 (4773), 194–196, 1986.

76. Zardoya, R. and Doadrio, I., Molecular evidence on the evolutionary and biogeographical patterns of European cyprinids, *Journal of Molecular Evolution* 49 (2), 227–237, 1999.

77. Farias, I. P., Orti, G., Sampaio, I., Schneider, H., and Meyer, A., The cytochrome b gene as a phylogenetic marker: The limits of resolution for analyzing relationships among cichlid fishes, *Journal of Molecular Evolution* 53 (2), 89–103, 2001.

78. Chen, T. Y., Hsieh, Y. W., Tsai, Y. H., Shiau, C. Y., and Hwang, D. F., Identification of species and measurement of tetrodotoxin in dried dressed fillets of the puffer fish, *Lagocephalus lunaris*, *Journal of Food Protection* 65 (10), 1670–1673, 2002.

79. Dettai, A. and Lecointre, G., Further support for the clades obtained by multiple molecular phylogenies in the acanthomorph bush, *Comptes Rendus Biologies* 328 (7), 674–689, 2005.

80. Jimenez, J., Schonhuth, S., Lozano, I. J., González, J. A., Sevilla, R. G., Diez, A., and Bautista, J. M., Morphological, ecological, and molecular analyses separate *Muraena augusti* from *Muraena helena* as a valid species, *Copeia,* 1007 (1), 101–113, 2007.

81. Venkatesh, B., Ning, Y., and Brenner, S., Late changes in spliceosomal introns define clades in vertebrate evolution, *Proceedings of National Academy of Sciences U S A* 96 (18), 10267–10271, 1999.

82. Nelson, J. S., *Fishes of the World*, 4th edn., John Wiley and Sons, Inc., New York, 2006.

83. Stiassny, M. L. J. and Moore, J. A., A review of the pelvic girdle of acanthomorph fishes, with comments on hypotheses of acanthomorph intrarelationships, *Zoological Journal of the Linnean Society* 104 (3), 209–242, 1992.

84. Inoue, J. G., Miya, M., Tsukamoto, K., and Nishida, M., Basal actinopterygian relationships: A mitogenomic perspective on the phylogeny of the "ancient fish," *Molecular Phylogenetics and Evolution* 26 (1), 110–120, 2003.

85. Akasaki, T., Saruwatari, T., Tomonaga, H., Sato, S., and Watanabe, Y., Identification of imported Chirimen at the genus level by a direct sequencing method using mitochondrial partial 16S rDNA region, *Fisheries Science* 72 (3), 686–692, 2006.

86. Brzezinski, J. L., Detection of crustacean DNA and species identification using a PCR-restriction fragment length polymorphism method, *Journal of Food Protection* 68 (9), 1866–1873, 2005.

87. Jin, L., Cho, j. G., Seong, K. B., Park, J. Y., Kong, I. S., and Hong, Y. K., 18 rRNA gene sequences and random amplified polymorphic DNA used in discriminating Manchurian trout from other freshwater salmonids, *Fisheries Science* 72 (4), 903–905, 2006.

88. Bartlett, S. E. and Davidson, W. S., FINS (Forensically informative nucleotide sequencing): A procedure for identifying the animal origin of biological specimens, *BioTechniques* 12 (3), 408–411, 1992.

89. Yancy, H. F., Zemlak, T. S., Mason, J. A., Washington, J. D., Tenge, B. J., Nguyen, N. L., Barnett, J. D. et al., Potential use of DNA barcodes in regulatory science: Applications of the Regulatory Fish Encyclopedia, *Journal of Food Protection* 71 (1), 210–217, 2008.

90. Altschul, S. F., Gish, W., Miller, W., Myers, E. W., and Lipman, D. J., Basic local alignment search tool, *Journal of Molecular Biology* 215 (3), 403–410, 1990.

91. Paine, M. A., McDowell, J. R., and Graves, J. E., Specific identification of western Atlantic Ocean scombrids using mitochondrial DNA cytochrome C oxidase subunit I (COI) gene region sequences, *Bulletin of Marine Science* 80 (2), 353–367, 2007.

92. Chapela, M. J., Sotelo, C. G., Calo-Mata, P., Perez-Martin, R. I., Rehbein, H., Hold, G. L., Quinteiro, J., Rey-Mendez, M., Rosa, C., and Santos, A. T., Identification of cephalopod species (Ommastrephidae and Loliginidae) in seafood products by forensically informative nucleotide sequencing (FINS), *Journal of Food Science* 67 (5), 1672–1676, 2002.

93. Murgia, R., Tola, G., Archer, S. N., Vallerga, S., and Hirano, J., Genetic identification of grey mullet species (Mugilidae) by analysis of mitochondrial DNA sequence: Application to identify the origin of processed ovary products (bottarga), *Marine Biotechnology* 4 (2), 119–126, 2002.

94. Richardson, D. E., Vanwye, J. D., Exum, A. M., Cowen, R. K., and Crawford, D. L., High-throughput species identification: From DNA isolation to bioinformatics, *Molecular Ecology Notes* 7 (2), 199–207, 2007.

95. Asensio, L., Gonzalez, I., Fernandez, A., Cespedes, A., Rodriguez, M. A., Hernandez, P. E., Garcia, T., and Martin, R., Identification of nile perch (*Lates niloticus*), grouper (*Epinephelus guaza*), and wreck fish (*Polyprion americanus*) fillets by PCR amplification of the 5S rDNA gene, *Journal of the AOAC International* 84 (3), 777–781, 2001.

96. Aranishi, F., Okimoto, T., and Izumi, S., Identification of gadoid species (Pisces, Gadidae) by PCR-RFLP analysis, *Journal of Applied Genetics* 46 (1), 69–73, 2005.

97. Meyer, R., Höfelein, C., Lüthy, J., and Candrian, U., Polymerase chain reaction-restriction fragment length polymorphism analysis: A simple method for species identification, *Journal of the AOAC International* 78, 1542–1551, 1995.

98. Dooley, J. J., Sage, H. D., Clarke, M. A. L., Brown, H. M., and Garrett, S. D., Fish species identification using PCR-RFLP analysis and lab-on-a-chip capillary electrophoresis: Application to detect white fish species in food products and an interlaboratory study, *Journal of Agricultural and Food Chemistry* 53 (9), 3348–3357, 2005.

99. Aranishi, F., PCR-RFLP analysis of nuclear nontranscribed spacer for mackerel species identification, *Journal of Agricultural and Food Chemistry* 53 (3), 508–511, 2005.

100. Karaiskou, N., Triantafyllidis, A., and Triantaphyllidis, C., Discrimination of three Trachurus species using both mitochondrial- and nuclear-based DNA approaches, *Journal of Agricultural and Food Chemistry* 51 (17), 4935–4940, 2003.

101. Calo-Mata, P., Sotelo, C. G., Perez-Martin, P. I., Rehbein, H., Hold, G. L., Russell, V. J., Pryde, S. et al., Identification of gadoid fish species using DNA-based techniques, *European Food Research and Technology* 217 (3), 259–264, 2003.

102. Russell, V. J., Hold, G. L., Pryde, S. E., Rehbein, H., Quinteiro, J., Rey-Mendez, M., Sotelo, C. G., Perez-Martin, R. I., Santos, A. T., and Rosa, C., Use of restriction fragment length polymorphism to distinguish between salmon species, *Journal of Agricultural and Food Chemistry* 48 (6), 2184–2188, 2000.

103. Sotelo, C. G., Calo-Mata, P., Chapela, M. J., Perez-Martin, R. I., Rehbein, H., Hold, G. L., Russell, V. J. et al., Identification of flatfish (*Pleuronectiforme*) species using DNA-based techniques, *Journal of Agricultural and Food Chemistry* 49 (10), 4562–4569, 2001.

104. Carrera, E., García, T., Céspedes, A., González, I., Fernández, A., Asensio, L. M., Hernández, P. E., and Martín, R., Identification of smoked Atlantic salmon (*Salmo salar*) and rainbow trout (*Oncorhynchus mykiss*) using PCR-restriction fragment length polymorphism of the p53 gene, *Journal of the AOAC International* 83, 341346, 2000.

105. Chakraborty, A., Aranishi, F., and Iwatsuki, Y., Polymerase chain reaction-restriction fragment length polymorphism analysis for species identification of hairtail fish fillets from supermarkets in Japan, *Fisheries Science* 73 (1), 197–201, 2007.

106. Lin, W. F. and Hwang, D. F., Application of PCR-RFLP analysis on species identification of canned tuna, *Food Control* 18 (9), 1050–1057, 2007.

107. Quinteiro, J., Vidal, R., Izquierdo, V., Sotelo, C. G., Chapela, M. J., Perez-Martin, R. I., Rehbein, H. et al., Identification of hake species (*Merluccius genus*) using sequencing and PCR-RFLP analysis of mitochondrial DNA control region sequences, *Journal of Agricultural and Food Chemistry* 49 (11), 5108–5114, 2001.

108. Bembo, D. G., Carvalho, G. R., Snow, M., Cingolani, N., and Pitcher, T., Stock discrimination among European anchovies *Engraulis encrasicolus,* by means of PCR-amplified mitochondrial DNA analysis., *Fishery Bulletin* 94, 31–40, 1995.

109. Magoulas, A., Castilho, R., Caetano, S., Marcato, S., and Patarnello, T., Mitochondrial DNA reveals a mosaic pattern of phylogeographical structure in Atlantic and Mediterranean populations of anchovy (*Engraulis encrasicolus*), *Molecular Phylogenetics and Evolution* 39 (3), 734–746, 2006.

110. Bossier, P., Wang, X. M., Catania, F., Dooms, S., Van Stappen, G., Naessens, E., and Sorgeloos, P., An RFLP database for authentication of commercial cyst samples of the brine shrimp *Artemia* spp. (International Study on Artemia LXX), *Aquaculture* 231 (1–4), 93–112, 2004.

111. Chapela, M. J., Sotelo, C. G., and Perez-Martin, R. I., Molecular identification of cephalopod species by FINS and PCR-RFLP of a cytochrome *b* gene fragment, *European Food Research and Technology* 217 (6), 524–529, 2003.

112. Fernandez, A., Garcia, T., Gonzalez, I., Asensio, L., Rodriguez, M. A., Hernandez, P. E., and Martin, R., Polymerase chain reaction-restriction fragment length polymorphism analysis of a 16S rRNA gene fragment for authentication of four clam species, *Journal of Food Protection* 65 (4), 692–695, 2002.

113. Vos, P., Hogers, R., Bleeker, M., Reijans, M., Van de lee, T., Hornes, M., Frijters, A., Pot, J., Peleman, J., and Kuiper, M., AFLP: A new technique for DNA fingerprinting, *Nucleic Acids Research* 23, 4407–4414, 1995.

114. Papa, R., Troggio, M., Ajmone-Marsan, P., and Nonnis-Marzano, F., An improved protocol for the production of AFLP markers in complex genomes by means of capillary electrophoresis, *Journal of Animal Breeding and Genetics* 122, 62–68, 2005.

115. Congiu, L., Fontana, F., Patarnello, T., Rossi, R., and Zane, L., The use of AFLP in sturgeon identification, *Journal of Applied Ichtiiology* 18, 286–289, 2002.

116. Han, K. P. and Ely, B., Use of AFLP analyses to assess genetic variation in *Morone* and *Thunnus* species, *Marine Biotechnology* 4 (2), 141–145, 2002.

117. Maldini, M., Marzano, F. N., Fortes, G. G., Papa, R., and Gandolfi, G., Fish and seafood traceability based on AFLP markers: Elaboration of a species database, *Aquaculture* 261 (2), 487–494, 2006.

118. Asensio, L., Gonzalez, I., Fernandez, A., Rodriguez, M. A., Lobo, E., Hernandez, P. E., Garcia, T., and Martin, R., Application of random amplified polymorphic DNA (RAPD) analysis for identification of grouper (*Epinephelus guaza*), wreck fish (*Polyprion americanus*), and Nile perch (*Lates niloticus*) fillets, *Journal of Food Protection* 65 (2), 432–435, 2002.

119. Bektas, Y. and Belduz, A. O., Molecular characterization of the whiting (*Merlangius merlangus euxinus* nordmann, 1840) in Turkish Black Sea coast by RAPD analysis, *Journal of Animal and Veterinary Advances* 6, 739–744, 2007.

120. Bossier, P., Authentication of seafood products by DNA patterns, *Journal of Food Science* 64 (2), 189–193, 1999.

121. Wright, J. M., DNA fingerprintig in fishes, in *Biochemistry and Molecular Biology of Fishes*, Hochachka, P. W. and Mommsen, T. Elsevier, Amsterdam, the Netherlands, 1993, pp. 58–91.

122. Beacham, T. D., Lapointe, M., Candy, J. R., McIntosh, B., MacConnachie, C., Tabata, A., Kaukinen, K., Deng, L. T., Miller, K. M., and Withler, R. E., Stock identification of Fraser River sockeye salmon using microsatellites and major histocompatibility complex variation, *Transactions of the American Fisheries Society* 133 (5), 1117–1137, 2004.

123. Kasapidis, P. and Magoulas, A., Development and application of microsatellite markers to address the population structure of the horse mackerel *Trachurus trachurus*, *Fisheries Research* 89 (2), 132–135, 2008.

124. Zatcoff, M. S., Ball, A. O., and Sedberry, G. R., Population genetic analysis of red grouper, *Epinephelus morio*, and scamp, *Mycteroperca phenax*, from the southeastern U.S. Atlantic and Gulf of Mexico, *Marine Biology* 144 (4), 769–777, 2004.

125. Nielsen, E. E., Hansen, M. M., and Meldrup, D., Evidence of microsatellite hitch-hiking selection in Atlantic cod (*Gadus morhua* L.): Implications for inferring population structure in nonmodel organisms, *Molecular Ecology* 15 (11), 3219–3229, 2006.

126. Yu, H. T., Lee, Y. J., Huang, S. W., and Chiu, T. S., Genetic analysis of the populations of Japanese anchovy (Engraulidae: *Engraulis japonicus*) using microsatellite DNA, *Marine Biotechnology* 4, 471–479, 2002.

127. Chistiakov, D. A., Hellemans, B., and Volckaert, F. A. M., Microsatellites and their genomic distribution, evolution, function and applications: A review with special reference to fish genetics, *Aquaculture* 255 (1–4), 1–29, 2006.

128. Withler, R. E., Rundle, T., and Beacham, T. D., Genetic identification of wild and domesticated strains of chinook salmon (*Oncorhynchus tshawytscha*) in southern British Columbia, Canada, *Aquaculture* 272, S161–S171, 2007.

129. Rengmark, A. H., Slettan, A., Skaala, O., Lie, O., and Lingaas, F., Genetic variability in wild and farmed Atlantic salmon (*Salmo salar*) strains estimated by SNP and microsatellites, *Aquaculture* 253 (1–4), 229–237, 2006.

130. Koedprang, W., Na-Nakorn, U., Nakajima, M., and Taniguchi, N., Evaluation of genetic diversity of eight grouper species *Epinephelus* spp. based on microsatellite variations, *Fisheries Science* 73 (2), 227–236, 2007.

131. McCusker, M. R., Paterson, I. G., and Bentzen, P., Microsatellite markers discriminate three species of North Atlantic wolffishes (*Anarhichas* spp.), *Journal of Fish Biology* 72 (2), 375–385, 2008.

132. Orita, M., Suzuki, Y., Sekiya, T., and Hayashi, K., Rapid and sensitive detection of point mutations and DNA polymorphisms using the polymerase chain-reaction, *Genomics* 5 (4), 874–879, 1989.

133. Rehbein, H., Kress, G., and Schmidt, T., Application of PCR-SSCP to species identification of fishery products, *Journal of the Science of Food and Agriculture* 74 (1), 35–41, 1997.

134. Muyzer, G., Dewaal, E. C., and Uitterlinden, A. G., Profiling of complex microbial-populations by denaturing gradient gel-electrophoresis analysis of polymerase chain reaction-amplified genes-coding for 16s ribosomal-RNA, *Applied and Environmental Microbiology* 59 (3), 695–700, 1993.

135. Lockley, A. K. and Bardsley, R. G., Novel method for the discrimination of tuna (*Thunnus thynnus*) and bonito (*Sarda sarda*) DNA, *Journal of Agricultural and Food Chemistry* 48 (10), 4463–4468, 2000.

136. Kwok, S., Kellog, D. E., Spasic, D., Goda, L., Levenson, C., and Sninsky, J. J., Effects of primer-template mismatches on the polymerase chain reaction: Human immunodeficiency virus type 1 model studies, *Nucleic Acids Research* 18, 999–1005, 1990.

137. Hird, H., Goodier, R., Schneede, K., Boltz, C., Chisholm, J., Lloyd, J., and Popping, B., Truncation of oligonucleotide primers confers specificity on real-time polymerase chain reaction assays for food authentication, *Food Additives and Contaminants* 21 (11), 1035–1040, 2004.

138. Infante, C., Crespo, A., Zuasti, E., Ponce, M., Perez, L., Funes, V., Catanese, G., and Manchado, M., PCR-based methodology for the authentication of the Atlantic mackerel *Scomber scombrus* in commercial canned products, *Food Research International* 39 (9), 1023–1028, 2006.

139. Infante, C. and Manchado, M., Multiplex-polymerase chain reaction assay for the authentication of the mackerel *Scomber colias* in commercial canned products, *Journal of the AOAC International* 89 (3), 708–711, 2006.

140. Holland, P. M., Abramson, R. D., Watson, R., and Gelfand, D. H., Detection of specific polymerase chain reaction product by utilizing the 5′-3′ exonuclease activity of *Thermus aquaticus* DNA polymerase, *Proceedings of the Natural Academy of Sciences* 88, 7276–7280, 1991.

141. Itoi, S., Nakaya, M., Kaneko, G., Kondo, H., Sezaki, K., and Watabe, S., Rapid identification of eels *Anguilla joponica* and *Anguilla anguilla* by polymerase chain reaction with single nucleotide polymorphism-based specific probes, *Fisheries Science* 71 (6), 1356–1364, 2005.

142. Wittwer, C. T., Herrmann, M. G., Moss, A. A., and Rasmussen, R. P., Continuous fluorescence monitoring of rapid cycle DNA amplification, *Biotechniques* 22, 134–138, 1997.

143. Higuchi, R., Fockler, C., Dollinger, G., and Watson, R., Kinetic PCR analysis: Real-time monitoring of DNA amplification reactions, *Bio/Technology* 11, 1026–1030, 1993.

144. Brodmann, P. D. and Moor, D., Sensitive and semi-quantitative TaqMan™ real-time polymerase chain reaction systems for the detection of beef (*Bos taurus*) and the detection of the family *Mammalia* in food and feed, *Meat Science* 65, 599–607, 2003.

145. Sawyer, J., Wood, C., Shanahan, D., Gout, S., and McDowell, D., Real-time PCR for quantitative meat species testing, *Food Control* 14, 579–583, 2003.

146. Rodriguez-Lazaro, D., Lombard, B., Smith, H., Rzezutka, A., D'Agostino, M., Helmuth, R., Schroeter, A. et al., Trends in analytical methodology in food safety and quality: Monitoring microorganisms and genetically modified organisms, *Trends in Food Science & Technology* 18 (6), 306–319, 2007.

147. Rosel, P. E. and Kocher, T. D., DNA-based identification of larval cod in stomach contents of predatory fishes, *Journal of Experimental Marine Biology and Ecology* 267 (1), 75–88, 2002.

148. Kiesling, T. L., Wilkinson, E., Rabalais, J., Ortner, P. B., McCabe, M. M., and Fell, J. W., Rapid identification of adult and naupliar stages of copepods using DNA hybridization methodology, *Marine Biotechnology* 4 (1), 30–39, 2002.

149. Kochzius, M., Nolte, M., Weber, H., Silkenbeumer, N., Hjorleifsdottir, S., Hreggvidsson, G. O., Marteinsson, V. et al., DNA microarrays for identifying fishes, *Marine Biotechnology* 10 (2), 207–217, 2008.

150. Shumaker, J. M., Metspalu, A., and Caskey, C. T., Mutation detection by solid phase primer extension, *Human Mutation* 7 (4), 346–354, 1996.

151. Kurg, A., Tonisson, N., Georgiou, I., Shumaker, J., Tollett, J., and Metspalu, A., Arrayed primer extension: Solid-phase four-color DNA resequencing and mutation detection technology, *Genetic Testing* 4 (1), 1–7, 2000.

152. Sanchez, J. J. and Endicott, P., Developing multiplexed SNP assays with special reference to degraded DNA templates, *Nature Protocols* 1 (3), 1370–1378, 2006.

Chapter 22

Spectrochemical Methods for the Determination of Metals in Seafood

Joseph Sneddon and Chad A. Thibodeaux

Contents

22.1 Introduction 642
22.2 Spectrochemical Methods 642
     22.2.1 Atomic Absorption Spectrometry 642
          22.2.1.1 Theory 642
          22.2.1.2 Instrumentation 643
          22.2.1.3 Cold Vapor Atomic Absorption Spectrometry 646
     22.2.2 Atomic Emission Spectrometry 646
          22.2.2.1 Theory 646
          22.2.2.2 Instrumentation 647
     22.2.3 Inductively Coupled Plasma-Mass Spectrometry 649
     22.2.4 Practice of Analytical Atomic Spectroscopy 650
     22.2.5 Sample Preparation for Metal Determination in Seafood by
          Spectrochemical Methods 651
22.3 Selected Application of Spectrochemical Methods in Seafood 652
Acknowledgments 659
References 659

22.1   Introduction

The surface of the earth is covered by approximately 70% of water (seas, lakes, rivers, etc.) and seafood is a major source of food for the majority of the inhabitants of the earth. It also has a large economic factor for many communities around the world. As well as anthropogenic sources, the seas have been found to be a dumping ground or last refuge for many potential pollutants including metals. The metals in the seas will be taken up by the seafood and can enter the human food cycle potentially causing serious health hazards. Many countries have enacted laws and warnings regarding the minimum concentration of many metals in seafood. This has attracted considerable interest and desire in determining metals in various seafood.

While there are numerous analytical techniques for metal determination such as various electrochemical methods of voltammetry, coulometry, neutron activation analysis, and the like, this chapter will be confined to spectrochemical or atomic spectroscopic techniques. These are the most widely used and accepted techniques for the determination of metals in seafood.

22.2   Spectrochemical Methods

Interaction of energy (light) with matter (gaseous atoms) produce three closely related, yet separate atomic phenomena, namely atomic absorption (AA), atomic emission (AE), and atomic fluorescence (AF). In these techniques, the atoms are detected by optical means. A closely related technique is that of plasma source-mass spectrometry, in particular inductively coupled plasma-mass spectrometry (ICP-MS). In this case, the atoms are detected by mass spectrometry. These techniques are collectively known as atomic spectroscopy or spectrochemical techniques and have detection limits for metals and metalloids ranging from μg/mL to ng/mL and even as low as pg/mL. They have been used to detect metals and metalloids in solids, liquids, and gasses in just about every conceivable matrix including biological, clinical, environmental, food and drugs, petroleum products as well as seafood.

The object of this chapter is to give the reader an overview of spectrochemical techniques, including instrumentation and general analytical performance. It is not intended to be comprehensive or discuss the areas on the fringe of atomic spectroscopy. It is beyond the scope of this chapter to describe in detail these techniques and the reader is referred to a number of texts that provide detailed discussion of these four analytical phenomena [1–4].

This chapter provides an overview of atomic absorption spectrometry (AAS), atomic emission spectrometry (AES) with inductively coupled plasma (ICP) as the excitation source, and ICP-MS. Despite some early promise, atomic fluorescence spectrometry (AFS) has failed to live up to its potential and will not be discussed in this chapter.

An additional technique will be described namely cold vapor (mostly coupled with AAS) as this has extensive use in Hg determination in seafood.

22.2.1   Atomic Absorption Spectrometry

22.2.1.1   Theory

Atomic absorption involves the impingement of light of a specific wavelength onto gaseous atoms. This causes a valence electron in the atom to be raised from a lower energy level to a higher energy level (called an electronic transition). When the energy of the photon is identical to the energy difference between the lower and higher energy level of the atom, then absorption will occur. The intensity of this transition is related to the original concentration of the ground state atoms. This can be represented as follows:

$$T = I/I_o \tag{22.1}$$

where
    T is the transmittance
    I is the intensity of the light source passing through the sample zone
    I_o is the intensity of the light source before it passes through the sample zone

The sample zone or path length, b, is relatively long to maximize the amount of light absorbed by the atoms. The amount of light absorbed will depend on the AA coefficient, k. This value is related to the number of atoms per cm^3 in the atom cell, n the Einstein probability for the absorption process; and the energy difference between the two levels of the transition. In practice these are all constants, which are combined to give one constant, called the absorptivity, a. k is related exponentially to the transmittance as follows:

$$T = I/I_o = e^{-kb} \tag{22.2}$$

In practice, the absorbance, A, is used in AAS and is related log arithmetically to the transmittance as follows:

$$A = -\log T = -\log I/I_o = \log I_o/I = \log 1/T = kb \log e = 0.43\, kb \tag{22.3}$$

The Beer–Lambert law relates A to the concentration of the metal in the atom cell, c, as follows:

$$A = abc \quad \text{or} \quad A = e_o bc \tag{22.4}$$

where
    a is the absorptivity in L/g-cm
    e_o is the molar absorptivity in L/mol-cm
    b is the atom cell width in cm

AAS involves the measurement of the drop in light intensity of I_o to I (depending on the concentration of the metal). Current and modern instrumentation automatically converts the logarithmic value into A. Absorbance is a unit less number, typically, 0.01 to 2.0. In practice, it is better to work in the middle of this range (recommended 0.1–0.3 A) as the precision is poorer at the extremes due to instrumental noise. The most intense transition from the ground state to the first excited state (resonance transition) is the most widely used transition because it is the most sensitive.

The origins of atomic spectra and detailed discussion are available elsewhere [5] AAS was discovered independently by Walsh, Alkemade, and Melatz in the early to mid 1950s.

22.2.1.2 Instrumentation

A typical AA system consists of six basic parts: a light source, atomizer, sample introduction system, wavelength selection device, a detection system, and a readout system. All the components are conveniently packaged in a complete benchtop unit and are connected to a computer for control, sample preparation, data reduction, and printout. There are numerous commercial instrumentation available with cost ranging from a small compact flame AAS of around $10,000.00 to a top-of-the-line multimetal flame/furnace AAS system with automatic sample introduction and data station of around $100,000.00. A detailed description of light sources (hollow cathode lamp and electrodeless discharge lamp), wavelength selection devices (monochromator), sample introduction systems (pneumatic nebulizers), detection systems (photomultiplier tubes, PMT) and readout (connected to external computers) parts are described elsewhere [1–4]. Sample introduction is very important in AAS (and AES and ICP-MS) and is discussed in detail elsewhere [6]. A short discussion on the atomizer is included in this chapter.

22.2.1.2.1 Atomizer

The only widely used and accepted atomizers in AAS are the flame and graphite furnace.

In flame AAS, the sample is (usually) introduced into the flame as a fine mist or aerosol. Flames consist of an oxidant and a fuel. The most widely used flames in AAS are air-acetylene (air is the oxidant and acetylene the fuel) and nitrous oxide-acetylene (nitrous oxide is the oxidant and acetylene the fuel). These flames are called combustion flames. Other flames, called diffusion flames, have been proposed but are not widely used. The primary object of the flame is to dissociate molecules into atoms. Air-acetylene (2500 K) does this readily and efficiently for about 40–50 metals in the periodic table. The other 10–20 metals in the periodic table require the hotter nitrous oxide-acetylene flame (3200 K). A long thin flame is desirable in AAS for maximum sensitivity.

The graphite furnace atom cell or electrothermal atomizer (ETA) for AAS was commercially developed in the late 1960s. Their principal advantage over flame atomizers are the improvement in sensitivity, typically 10–100×, the ability to use microvolumes (2–200 μL) and micromass solid (few mg) sampling, and in situ pretreatment of the sample. However, ETAs are prone to interferences, particularly from alkali and alkaline earth halides and requires a more complex (and subsequently more expensive) system.

The use of an electrically heated tubular furnace was first reported by King in 1905 but for analytical chemistry, the work and system developed by L'Vov around 1960 is regarded as the forerunner of present day ETAs. It consisted of a carbon electrode in which the sample was applied and a carbon tube that could be heated by electrical resistance. The initial design used a supplementary electrode for preheating the furnace, lined the carbon tube with tungsten or tantalum foil to minimize vapor diffusion, and purged the system with argon to prevent oxidation of the carbon. Later work involved direct heating of the sampling electrode by resistance heating and the tube was made of pyrolytic carbon. After heating the tube to an elevated temperature, the sample electrode was inserted into the underside of the tube, and vaporization of the sample was confined to the tube where AAS measurements were made. The system was difficult to operate and the reproducibility could be poor.

In 1967, Massmann described a heated graphite atomizer (HGA) which was commercially developed by the Perkin–Elmer Corporation and proved the forerunner for all current commercial ETAs. An isothermal type furnace system proposed by Woodriff at around the same time was considered more difficult to commercialize although recent work has shown the advantage of atomization under isothermal conditions. The Massmann system was typically 50 mm long and 10 mm diameter graphite tube, which was heated by electrical resistance, typically 7–10 V at 400 amps. An inert gas, usually argon or nitrogen, flowed at a constant rate of around 1.5 L/min and the entire system was enclosed in a water jacket. A microliter sample was deposited through an entry or injection port in the center of the tube and could be heated in three stages by applying variable current to the system; drying to remove the solvent, ashing or pyrolysis to remove the matrix, and finally atomization of the element. Careful control of the temperature was required in order to obtain good reproducibility.

In 1969, West and coworkers developed a rod or filament atomizer. It consisted of a graphite filament of 40 mm in length and 2 mm diameter, supported by water-cooled electrodes, and heated very quickly by the use of current of 70 amps at 10–12 V. Shielding from the air was achieved by a flow of inert gas around the filament. While primarily developed for AAS, West and coworkers showed the potential of the system for AFS.

The West filament was the forerunner for the mini-Massmann atomizer developed commercially by Varian Associates. A commercial system was called the carbon rod atomizer (CRA 63). Its main advantage was in the somewhat simpler and less complex design compared to the HGA, low power requirements (2–3 kW), and fast (~2 s) heating rate. There were differences between this system and the West filament, principally by drilling a hole in a solid cylindrical graphite tube and later using a small cup or crucible between two spring loaded graphite rods. The system was proposed for low microliter volumes, typically 1–20 μL. In general, detection limits and increased interferences were found using the CRA type system compared to the HGA and this type of system has been discontinued from around the mid-1980s and not currently commercially available.

A typical schematic furnace AAS system is shown in Figure 22.1. Most current commercial furnaces are similar to that shown in Figure 22.1.

Graphite furnace AAS has essentially the same instrumentation as flame AAS except for the (a) atom cell, and (b) sample introduction system. An additional need in furnace AAS is faster electronics to process the transient and faster generated signal compared to flame AAS. In practice,

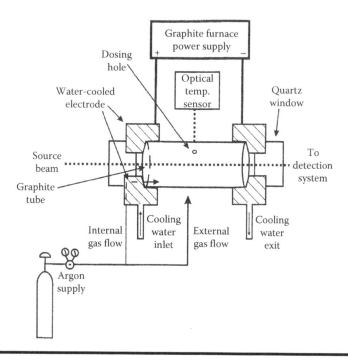

Figure 22.1   Typical GFAAS system.

AAS usually has the fast electronics capability and most commercial systems have the flame and furnace as interchangeable.

Typical volumes used with a graphite furnace are from 1 to 200 μL. This volume can be conveniently introduced to the furnace by manual introduction using a micropipette. There are various dedicated micropipettes available for this range as well as adjustable micropipettes. In the late 1970s, it was suggested that the precision of furnace AAS could be improved by automatic introduction systems. This led to systems which could be added to furnace AAS for automatic sample introduction. It was shown that the precision was not significantly improved by automatic sample introduction but is now an accepted parts of furnace AAS particularly for unattended operation and when numerous samples are required to be analyzed. These systems can be incorporated into sample preparation stations.

A monograph by Butcher and Sneddon [7] provides an in-depth coverage of areas in graphite furnace atomic absorption spectrometry (GFAAS) such as matrix or chemical modification to allow a higher ashing temperature without loss of the more volatile metals such as Cd; background correction techniques, in particularly various geometrical configurations of Zeeman effect (inverse ac longitudinal and inverse transverse ac and dc) for a more accurate measurement; graphite tube design and material that allow a more rapid heating of the furnace which improves the signal from more volatile metals; and platform atomization.

Atomic absorption spectrometry has reached a maturity in the mid-1990s since its initial development in the mid-1950s. Current developments are in refinements and modest improvements, e.g., software.

22.2.1.3 Cold Vapor Atomic Absorption Spectrometry

The unique properties, high toxicity, and large use in many industrial processes of Hg has led to the development of the cold vapor accessory, most widely used in conjunction with AAS. Mercury has an appreciable vapor pressure at room temperature (0.16 Pa at 25°C). Mercury (in the ionic form) and in an acidic medium can be reduced by stannous chloride to produce ground state atomic mercury. After equilibration, the mercury vapor is swept from the reaction vessel with a carrier gas (argon, air, or nitrogen) into the optical path of an AA instrument for determination as a transient signal. Alternatively, a closed system will produce a steady state signal.

The primary advantage of CVAAS for Hg determination is a low detection limit of sub-μg/mL. This can be lowered further using a dedicated commercial cold vapor atomic fluorescence spectrometry system (CV-AFS) where a detection limit in tens of ng/mL is possible. It should be noted that type and concentration of acid, chemical form (inorganic versus organic mercury), matrix components and other components such as the reducing agents can degrade these low detection limits. The use of stannous chloride as a reducing agent will not reduce organomercury compounds. Various pretreatment processes have been developed to overcome this potential problem.

22.2.2 Atomic Emission Spectrometry

22.2.2.1 Theory

Atomic emission spectrometry involves the impingement of an external source of energy on ground state atoms. The radiation from these atoms is observed in AES.

The probability of transitions from the given energy level of a fixed atomic population was expressed by Einstein, in the form of three coefficients, termed transition probabilities, A_{ji} (spontaneous emission), B_{ij} (spontaneous absorption), and B_{ji} (stimulated emission). These can be considered as representing the ratio of the number of atoms undergoing a transition to an upper level to the number of atoms in the initial or lower level and can be represented as follows:

$$N_j = N_o \frac{g_j}{g_o} \exp[-\Delta E/KT] \tag{22.5}$$

where

N_o is the number of atoms in the lower state (or ground state usually for analytical work)
N_j is the number of atoms in the excited or upper level
g_j and g_o are the statistical weights of the jth (upper state) and o (ground state)
ΔE is the difference in energy in Joules between these two states
K is the Boltzmann constant (1.38066×10^{-23} J/K)
T is the absolute temperature.

If self-absorption is neglected, then the intensity of emission, I_{em} is

$$I_{em} = A_{ji} h v_{ji} N_j \tag{22.6}$$

where

h is Planck's constant (6.624×10^{-24} Js)
v_{ji} is the frequency of the transition ($\Delta E = hv$)

Therefore, N is directly related to the concentration of the solutions as follows:

$$N_j = N \frac{g_j}{g_o} \exp[-\Delta E/KT] \tag{22.7}$$

The intensity emission of a spontaneous emission line, I_{em} is related to this equation (sometimes called the Maxwell–Boltzmann equation) as follows:

$$I_{em} = A_{ij} h v_{ji} N \frac{g_j}{g_o} \exp[-\Delta E/KT] \tag{22.8}$$

It can be seen that the atomic emission intensity is dependent on temperature and wavelength. Thus, a higher temperature at longer wavelength would give the most intense atomic emission signal.

A plot of emission intensity against sample concentration will be linear. AES (and AFS) has linearity extending up to five to seven orders of magnitude compared to two to three orders of magnitude of AAS.

22.2.2.2 Instrumentation

The primary components of an atomic emission spectrometer (AES) are similar to that of AAS, although for optimum performance the components are different. The excitation source and atomization source are the same, most frequently a plasma.

A typical setup for ICP-OES is shown in Figure 22.2.

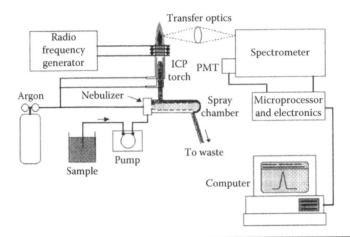

Figure 22.2 **Schematic of a typical inductively coupled plasma-optical emission spectrometry system.**

The higher temperature of the plasma will lead to a richer spectrum with many more lines. In order to separate these lines and prevent or minimize spectral interferences, a high-resolution monochromator is required. The 0.20 nm grating typically used in AAS does not provide the required resolution. The most widely adopted system is the Echelle Monochromator, which uses high-diffraction orders and large angles of diffraction. Resolution is around 0.015 nm compared to around 0.2 nm for a typical AAS monochromator. Simultaneous multimetal determination requires a polychromator.

During the development of the plasma as an excitation source in analytical AES, PMTs were first used. The PMT continues to be used in ICP-OES, particularly for cost reduction and sequential determinations, as shown in Figure 22.2. However, they have very limited use in multichannel systems and AES quickly adopted systems capable of simultaneous multimetal analysis. In the early 1980s, the photodiode arrays (PDAs) and image sensing vacuum tubes, the so-called vidicons were used. A PDA consists of arrays (256, 512, 1024, 2048 elements arranged in linear manner) of photodiodes operated on a charge transfer storage mode. Each diode is sequentially integrated (several μs) after all the diodes have been integrated with the incident radiation. The current generated by each photodiode is proportional to the intensity of the radiation it receives. The sequential measurement of the current can occur many times a second under the control of a microprocessor. This digitized information can be stored in a computer for electronic processing and visual display. Diode array systems are excellent for studying transient signals such as those on the laser-induced plasma with gate delay generator systems. However, it does have somewhat of a limited resolution, usually 1–2 nm. Diode arrays are used in vidicons in the form of a spectrum. These are similar to a small television tube.

The late 1980s through to the present has shown an interest in the use of the charge transfer device (CTD), specifically the charge coupled device (CCD) and to a lesser extent the charge injection device (CID). These are solid-state sensors that have integrated circuits. The charge generated by a photon is collected and stored in a capacitor. A typical pixel arrangement can be 512 × 320 CCD (much larger arrangements such as 2000 × 2000 pixels have been constructed). The capacitor can be reversed biased by a positive voltage applied to the electrode, creating a potential well. The photons striking the array give electron-hole pairs and the electrons can be stored for a short time in the well. The amount of charge accumulated is a direct function of the incident radiation (and time) and is linear. The charge is shifted horizontally and down to a readout preamplifier which results in a scan of each row in series. CCDs are very useful where low levels of radiation are to be detected. At high levels, "blooming" occur that results in curvature of the response. A CID is a two-dimensional array of pixels. The photons generate positive charges below the negative well capacitors. Again, the amount of charge is proportional to the incident radiation. There is a rapid development of the CTD in spectrochemical analysis.

The development of the ICP as an excitation source for analytical AES has been a major advancement in atomic spectroscopy. Its higher temperature has made this source the choice for many atomic spectroscopists work.

The most common plasma source is the ICP, which was first developed in the mid-1960s by Fassel and coworkers at Iowa State University and Greenfield and coworkers at Albright and Wilson Ltd. in England. It became commercially available around the mid-1970s. A typical ICP consists of three concentric quartz tubes. These are frequently referred to as the "outer," "intermediate," and "inner or carrier gas" tubes. The outer tube can be of various sizes in the range 9–27 mm. A two or three turn induction coil surrounds the top of the quartz tube or torch and is connected to a radiofrequency generator. The coil is water-cooled. The argon, typically at a flow rate of about 1–2 L/min is introduced into the torch and the radiofrequency field operated at 4–50 MHz, most typically 27.12 MHz, and a forward power of 1–5 kW, typically 1.3 kW, is applied. An intense magnetic field around the coil is developed and a spark from a Tesla coil is used to produce "seed" electrons and ions in this region. This induced current flowing in a closed circular path, results in great heating of the argon gas and an avalanche of ions is produced. Temperatures in an ICP have been estimated to be around 8000–10,000 K. The high temperature necessitates cooling which is applied using argon to the outer tubes at flow rates as high as 17 L/min. The sample is introduced, usually as an aerosol, through the inner tube and is viewed at a distance of 5–20 mm above the coil. The advantages of the ICP include high temperature, long residence times, presence of no or few molecular species, optically thin, and few ionization interferences.

The last decade has seen a tremendous amount of effort evaluating and understanding the ICP with numerous studies on mechanisms and characterizing variations of the system. The reader is referred to a recent book edited by Montaser and Golightly [8] that describes the current status of the ICP.

Considerable improvement and refinement in plasma source-AES has occurred over the last decade. Improved detection limits have been achieved by rotating the plasma through 90° and the development of the miniature ICP. Considerable effort has been expended in the area of sample introduction (see earlier). Improved software has pushed ICP-AES into a well-established and frequently used technique; particularly for multimetal AES.

22.2.3 Inductively Coupled Plasma-Mass Spectrometry

Since the early to mid-1980s, the ICP has been used as the ion source for mass spectrometry to determine metals. Its advantages include from two to three orders of magnitude improvement in sensitivity compared to traditional ICP-OES, the mass spectra of the metal are very simple and unique giving high specificity, inherent multimetal coverage, and the technique will measure

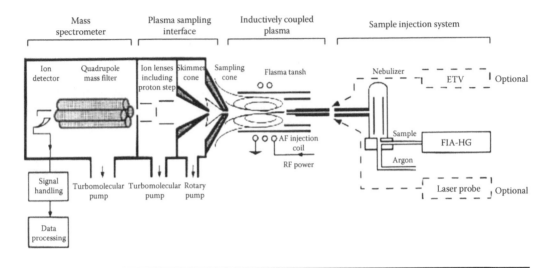

Figure 22.3 Typical Quadrupole ICP-MS system with various sample introduction options. ETV, electrothermal vaporization; LA, laser probe or ablation; FIA-HIG, flow injection analysis-hydride generation.

metal isotopic ratios. Disadvantages include potential spectral interferences from molecular species and the increased cost and complexity of instrumentation.

An inductively coupled plasma-mass spectroscopy (ICP-MS) system consists of the ion source that is the ICP, an interface system, which consists of a sampling cone, a differentially pumped zone, and a skimming zone, ion lenses, a quadrupole mass spectrometer, and a detector. A schematic diagram of a typical ICP-MS is shown in Figure 22.3. Since its commercial availability sample introduction has been a fertile area of research in ICP-MS (as has ICP-OES) and the system in Figure 22.3 has several variations of commercially available sample introduction systems of electrothermal vaporization (ETV), laser ablation (LA), and flow injection analysis-hydride generation (FIA-HG). A less commonly used system involves the use of a high-resolution mass spectrometer as opposed to the quadrupole system. This system shows less resistance to molecular interferences.

A detailed description of the instrumentation and the performance of ICP-MS is described elsewhere [4,8]. Essentially the ICP is in a horizontal position and works under atmospheric or normal pressure. Ions produced by this ICP are introduced to the MS through a small orifice, typically 1 mm diameter. The MS is a low pressure, typically at 10^{-5} to 10^{-6} Torr.

ICP-MS has increasingly become the choice of many spectrochemical analysts.

22.2.4 Practice of Analytical Atomic Spectroscopy

The choice of which analytical atomic spectroscopic technique to use will depend on the needs and expectations of the analyst, and the sample. There are many and varied commercial systems available or the analyst may decide that their needs are best suited to a laboratory-constructed system. The factors to be taken into consideration are the size of sample, whether it is a solid, liquid, or gas; the level to be detected; the accuracy and precision, which is acceptable; availability of a particular system; cost per sample, or the speed of analyses.

Spectrochemical methods are techniques, which depend on the comparison of signals obtained from samples with those obtained from sample standards of known composition. In most cases, these standards are aqueous solutions of the metals of interest. However, the analysis of real samples is complicated by the fact that the metal of interest is present as part of a sample matrix. The matrix can cause interference in the analysis. Therefore, in analytical atomic spectroscopy much attention is paid to the possibility of interferences. This can lead to reduced or poor accuracy. Accuracy can be defined as how close the atomic spectroscopic analysis is to the "correct" answer. In a typical method development, accuracy will be established via several or many ways including standard additions, comparison of the results of the atomic spectroscopic analyses with the results from a different method, recoveries or spikes, or applying the atomic spectroscopic method to standard samples such as those supplied by the National Institutes of Science & Technology (NIST) (Gaithersburg, Maryland). A concern of analytical atomic spectroscopy is precision which can be defined as the repetitive analyses of a particular sample expressed as a percent. Precision will vary with many factors including sample, level to be determined, and choice of instrumentation. Finally, the detection limit is an important factor in analytical atomic spectroscopy. Current atomic spectroscopic techniques have detection limits in the μg/L to μg/mL range. However, lower detection limits are possible using newer and improved techniques in analytical atomic spectroscopy.

The reader is referred to a recent book by Butcher and Sneddon [7] that describes the practice of graphite furnace AAS. Much of the advice and suggestions in the book could be equally applied to many areas of analytical atomic spectroscopy.

22.2.5 Sample Preparation for Metal Determination in Seafood by Spectrochemical Methods

Most analyses are preferentially performed on solution samples. This can be attributed to the desire for a more homogeneous analysis sample, concern over whether a few micro- or milli-grams will be representative of the bulk properties of a large solid sample, improved precision and accuracy is frequently obtained with solution samples (as opposed to solid samples), and the fact that most commercial instrumentation performs at an optimum with solution samples. Therefore, sample preparation remains an integral part of spectrochemical analysis. It is widely used in the preparation of seafood for metal determination by spectrochemical techniques. However, while the metal determination is most frequently performed on solutions, seafood results are most commonly reported as μg/g or in some cases as ng/g.

Sample preparation can be conveniently divided into areas such as classical and microwave. Classical methods involve wet or acid decomposition and involves the use of various mineral acids (HNO_3, H_2SO_4, $HClO_4$, and/or HF), and oxidizing agents (typically H_2O_2) to effect dissolution of the sample. It can be performed on an open or closed system. Microwave digestion has rapidly become the choice for many digestion/dissolutions, particularly in seafood preparation [9]. It involves the use of 2450 MHz electromagnetic radiation to digest samples in a teflon or quartz container. Commercial systems are readily available and are conveniently automated and can digest up to 48 samples simultaneously under controlled temperatures conditions. A recent review describes sample preparation on spectrochemical samples for solid materials [10].

22.3 Selected Application of Spectrochemical Methods in Seafood

In the following section selected applications of spectrochemical methods applied to metal determination in various seafood is presented. This area has been discussed previously [11,12]. It is not meant to be comprehensive but show numerous results of several studies involving various seafood, spectrochemical techniques, and variety of metals determined. The results are summarized in Table 22.1.

Table 22.1   Selected Results of Metals in Seafood

Metals	Samples Analyzed	Method	Comments	Reference
Cu, Fe, and Zn	Crawfish	ICP-AES	Microwave digestion procedure	[14]
As, Cd, and Pb	Tuna, salmon, shrimp, walleye, clams, oysters, and lobster	ICP-AES	Used single microwave digestion procedure	[16]
Cd, Zn, Cu, Al, and Fe	Scallop tissue	ICP-MS and ICP-AES	Used microwave assisted acid digestion	[17]
Na, Al, K, V, Co, Zn, Se, Sr, Ag, Cd, Ni, Pb, Mg, Mn, Fe, Cu, As, Ba, Hg, Ca, and Cr	Flounder, scup, and blue crab	Hg cold vapor and ICP-MS	Used microwave digestion with HNO_3, H_2O_2, and HF	[18]
Cd, Hg, Pb, As, Se, Mn, Cu, and Zn	Brown rice and fish	ICP-MS and AAS	Used open digestion and did a comparison of the techniques	[19]
Al, Bi, Cd, Co, Cu, Ga, Mn, Ni, Pb, V, and Zn	Fish otoliths from the American eel	ICP-MS		[20]
Ag, Co, Cr, and Ni	Algae, crustaceans, and fish	ICP-MS		[21]
Ni, Cu, Zn, Cr, Cd, Pb, and V	Trivela mactroidea clams	ICP-OES and GFAAS	Pb and V done by GFAAS and others done by ICP-OES	[22]
As, Cd, Co, Cr, Cu, Fe, Hg, Mn, Ni, Pd, Se, and Zn	Mollusks	ICP-MS	Used nine kinds of Mollusks through a 1 year period	[23]

Table 22.1 (continued)   Selected Results of Metals in Seafood

Metals	Samples Analyzed	Method	Comments	Reference
Cd, Cr, Cu, Mn, Zn, Hg, Pb, and Ni	Mediterranean mussel	AFS, ICP-MS, and AAS	AFS and ICP-MS used for Pb and Hg, while AAS used for the other metals	[24]
Pb, Cd, Fe, Cu, Mn, and Zn	Fish	GFAAS		[25]
Zn, Cu, Cd, and Pb	Benthic fish	Flame AAS	Looked at 20 different species of fish of the coast of Taiwan	[26]
Fe, Cu, Mn, Zn, and Pb	Fish	Flame and Graphite furnace AAS	Caught fish from the Black Sea and river Yesilimirmak	[27]
Al, Mn, Co, Cu, Mo, Cd, Fe, Zn, Pb, and Hg	Bathymodiolus mussels and grazer shrimp	ICP-MS	Looked at organisms from hydrothermal vents along the mid-Altlantic Ridge	[28]
Hg, Cd, Mn, Pb, and Sn	Trout and Tuna	Zeeman GFAAS		[29]
Co and Cr	Tuna, scallop, mussels, fish, clam, sardine	Flame AAS	Used ultrasound-assisted acid extraction	[30]
Cu, Cd, Pb, Zn, Mn, Fe, Cr, and Ni	Nine fish species	AAS	Microwave digestion used	[31]
Cd, Cu, Fe, Zn, and Pb	Two fish species	AAS		[32]
Cu and Cd	Five fish species	AAS		[33]
Na, Mg, Ca, Fe, Cu, Mn, Ni, Cd, Cr, Pb	Oysters	Flame AAS	Used microwave digestion	[34]
As, Cd, Hg, and Pb	Carrot puree, fish muscle, mushroom, graham flour, scampi, and mussel powder	ICP-MS	Used pressure digestion. Looked at foodstuffs	[35]

Abbreviations: ICP-AES, inductively coupled plasma-atomic emission spectroscopy; ICP-OES, inductively coupled plasma-optical emission spectroscopy; ICP-MS, inductively coupled plasma-mass spectrometry; GFAAS, graphite furnace atomic absorption spectrometry; AAS, atomic absorption spectrometry; AFS, atomic fluorescence spectrometry.

Crawfish or crayfish are widely consumed throughout the world, in particular in Louisiana, People's Republic of China, and Far East, and their metal (and organic) concentrations has generated a great deal of interest [13]. Hagen and Sneddon [14] and Richert and Sneddon [15] have investigated the concentrations of several heavy metals in crawfish following microwave digestion. Hagen and Sneddon determined Cu, Fe, and Zn and found no differences between male and female species. Richert and Sneddon's study was to determine the variation over a season (February through May) and involved monitoring on four separate occasions for two distinct sampling waters: one from natural run off from the nearby fields and other water from an underground source. Concentrations between the two sampling areas were not considered statistically significant.

A single microwave digestion procedure was developed for use with a variety of seafood products by Sheppard et al. [16]. Inductively coupled plasma atomic emission and mass spectrometry were used to determine the levels of As, Cd, and Pb in samples of tuna, salmon, shrimp, walleye, clams, oysters, and lobster. The precision for 10 replicate analyses of clams was 2.1% for As at the 10.0 μg/g level, 5.6% for Pb at the 0.067 μg/g level, and 2.5% for Cd at the 0.079 μg/g level. Acceptable spike recoveries in each of the sample types were achieved using both detection methods. Results from two standard reference materials were in good agreement with certified values.

A multimetal determination method for trace metals in scallop tissue samples was developed by ICP-MS and ICP-AES after microwave assisted acid digestion by Tamaru et al. [17]. A standard reference material of oyster tissue (SRM 1566b Oyster tissue) was analyzed to verify the method. Thirty metals (K, Na, In, Mg, Ca, Fe, Al, Cu, Mn, Ba, Sr, Cd, Ni, Co, Pb, Y, and rare earth elements) in the reference material could be determined. The concentrations of metals obtained were in good agreement with their certified and reference values and had good precision within relative standard deviations (RSD) of 4% except for Ni and Pb. This method was applied to determine concentrations of metals in a scallop tissue sample. Cadmium, In, Cu, Al, and Fe in the scallop tissue sample were obtained at 69.0,148, 19.0, 415, and 221 μg/g level, respectively. In particular, the bio-accumulation factors of Cd, which were estimated between the concentrations in scallop tissue sample and those in shale and seawater, were highest among the 30 metals determined.

A microwave digestion method suitable for the determination of multiple metals in marine species was developed, using cold vapor atomic spectrometry for the determination of Hg, and ICP-MS for all of the other metals by Yang and Swami [18]. An optimized reagent mixture composed of 2 mL of HNO_3, 2 mL of H_2O_2, and 0.3 mL of HF was used in the microwave digestion of about 0.15 g (dry weight) of sample and was found to give the best overall recoveries of metals in two standard reference materials. In the oyster tissue standard reference material (SRM 1566b), recoveries of Na, Al, K, V, Co, In, Se, Sr, Ag, Cd, Ni, and Pb were between 90% and 110%; Mg, Mn, Fe, Cu, As, and Ba recoveries were between 85% and 90%; Hg recovery was 81%; and Ca recovery was 64%. In a dogfish certified reference material (DORM-2), the recoveries of Al, Cr, Mn, Se, and Hg were between 90% and 110%; Ni, Cu, In, and As recoveries were about 85%; and Fe recovery was 112%. Method detection limits of the metals were established. Metal concentrations in flounder, scup, and blue crab samples from coastal locations around Long Island and the Hudson River estuary were determined.

A study was conducted by Oshima et al. [19] to evaluate the applicability of ICP-MS for the determination of metals in brown rice and fish. Cadmium, Pb, Hg, As, Se, Mn, Cu, and Zn were determined by this method. An open digestion with HNO_3 (Method A) and a rapid open digestion with HNO_3 and HF (Method B) were used to solubilize analytes in samples, and these procedures were followed by determination by ICP-MS. Recovery of certified metals from standard reference materials by Method A and Method B ranged from 92% to 110% except for Hg (70%–100%). Analytical results of brown rice and fish samples obtained agreed with those obtained by AAS.

The results of this study demonstrated that quadrupole ICP-MS provides precise and accurate measurements of the metals tested in brown rice and fish samples.

Transition and heavy metals within the calcined otoliths of estuarine fish may represent valuable tracers of environmental exposures, allowing inferences on fatality, habitat use, and exposure to pollution. Accurate measurement of very low concentrations of these metals in otoliths by ICP-MS is often precluded by the interferences of predominant calcium matrix. Arslan and Secor [20] coupled a solid-phase extraction procedure to an ICP-MS instrument to overcome the matrix problems and improve the limits of detection. To test this novel application and the utility of otolith transition and heavy metals as tracers of habitat use, otoliths of American eel (*Anguilla rostrata*) captured from six locations (George Washington Bridge, Haverstraw, Newburgh, Kingston, Athens, and Albany, all in New York State) throughout the Hudson River estuary were analyzed for site specific differences expected due to varying environmental exposure. Several trace metals, including Al, Bi, Cd, Co, Cu, Ga, Mn, Ni, Pb, V, and In, were selectively extracted from otolith solutions and preconcentrated on a microcolumn of chelating resin. The concentrations of all metals in *A. rostrata* otoliths were above the limits of detection that ranged from 0.2 ng/g for Co to 7 ng/g for In. Differences in the metal composition of the otoliths among the groups were significant indicating different levels of exposure to environmental conditions. Discriminant analysis yielded an overall location classification rate of 78%. Aluminum, Bi, Cd, Mn, Ni, and V contributed most to the discriminant function. Samples collected at George Washington Bridge showed 100% discrimination from other locations, and higher levels of many transition and heavy metals, consistent with higher exposure to those metals in the most polluted region of the Hudson River estuary.

Original results concerning Ag, Co, Cr, and Ni determination in marine biotope (sediment and water) and biocenosis (algae, crustaceans, and fish) collected in 2003, 2004, 2005, and 2006 from the Romanian Black seacoast ecosystem are presented by Chirila et al. [21]. The solid samples were carefully prepared (washed and dried) and subjected to dissolution with HNO_3 and H_2O_2 in a Digesdahl device. Metal concentrations were determined by ICP-AES and applied to solid samples (sediment and biota). The levels of Ag varied from ND (not detectable) to 0.20, Co from 0.03–0.65, Cr from 0.49–22.44, and Ni from 0.32–28.13 µg/g. In water, the mean metal concentrations were Ag of 1.07, Co of 0.75 and Ni of 8.68 µg/L.

LaBrecque et al. [22] performed a study using *Trivela mactroidea* clams which were hand-picked directly from the marine sediment at 14 sampling sites along the Venezuelan coast of the state Miranda. Clam soft tissues were washed, dried, and ground into fine powder. For heavy metal analysis, the powder samples were digested with HNO_3 and further with H_2O_2. Determination for Ni, Cu, In, Cr, and Cd were performed by ICP-OES while determination for Pb and V were made by GFAAS. The suitability of the ICP-OES method was assessed by analyzing mussel tissue standard reference material NIST -2976. Trace metal concentrations of 11–49 µg/g for Cu, 55–166 µg/g for In, <1–6.2 µg/g for Cr, 6–15 µg/g for Ni, 2–13.2 µg/g for V, <1–1.9 µg/g for Cd, and <1.5–4.9 µg/g for Pb were determined. These values were significantly lower than those obtained in a study 12 years ago on soft clam tissue from the same area.

Mollusks living in seas can accumulate heavy metals, and may serve as excellent passive biomonitors. During a period of one year, bioaccumulation of As, Cd, Co, Cr, Cu, Fe, Hg, Mn, Ni, Pb, Se, and Zn was examined in nine kinds of mollusks (*Rapana venosa*, *Neverita didyma*, *Scapharca subcrenata*, *Mytilus edulis*, *Amusium*, *Crassostrea talienwhanensis*, *Meretix meretix*, *Ruditapes philippinarum*, and *Mactra veneriformis*). These were collected at eight coastal sites along the Chinese Bohai Se by Wang et al. [23]. Metal concentrations were directly determined by ICP-MS. Two certified reference materials, dogfish muscle (DORM-2) and mussel (GBW 08571), were used to validate the methods, and the recoveries were within 83.72%–112.30% of the certified values.

Bioaccumulation of metals varied strongly among sampling sites and species. Statistical analysis (one-way ANOVA) indicated that different species examined showed different bioaccumulation of metals, and perhaps they could be used as potential biomonitors to investigate the contamination levels of heavy metals. Principal component analysis (PCA) and correlation analysis were used to study the relationships between these heavy metals. The results showed that, in nine mollusks' tissues, there are significant correlations between these metals in the adjacent group or subgroup in the periodic table of elements (metals).

Heavy metal concentrations of Hg, Cd, Pb, Zn, Cu, Ni, Mn, and Cr in *Mytilus galloprovincialis* were investigated by Maanan [24] to provide information on pollution of the Safi coastal area in Morocco, since these metals have the highest toxic potential. The concentration of Hg and Pb was determined by AFS and ICP-MS methods, respectively, while the remaining metals (Cd, Cr, Cu, Mn, In, and Ni) were quantified by AAS. High Pb, Cd, Cr, and Hg levels were registered in tissue samples collected from two stations near the Jon Lihoudi and Safi city, while elevated concentrations of Mn and Zn (14.70–25.30 mg/kg and 570–650 mg/kg dry weight, respectively) were found in mussel specimens from Cap Cantin. The high levels of Ni found near the industrial area are of concern in terms of environmental health need frequent monitoring. The metal concentrations recorded at the clean stations may be considered as useful background levels to which to refer for comparison within the Atlantic coast. *M. galloprovincialis* are suitable biomonitors to investigate the contamination levels of heavy metals pollution face a different human activity in this coastal area of the Atlantic coast.

The concentrations of heavy metals (Pb, Cd, Fe, Cu, Mn, and Zn) in fish samples were determined using GFAAS after dry and wet ashing methods by Tuzen [25]. Different matrix modifiers were used for the stabilization of the analyte. Good accuracy was assured by the analysis of biological reference materials. Recoveries were quantified for all metals studied (~95%). The RSD were less than 7% for all metals.

Taiwanese consume a large amount of marine fish, most of which are collected from the coastal waters around Taiwan. Heavy metals are recognized as one of the most important pollutants, and their accumulations in the organisms were studied and monitored for the safety of seafood consumption in the coastal waters of Taiwan; however, its regulation was overlooked in the eastern region. Huang [26] evaluated the seafood consumption safety of the coastal fisheries in eastern Taiwan and established a baseline reference of the heavy metal levels in the fish of this region for the future monitoring of heavy metal pollution. Indium, Cu, Cd, and Pb concentrations were determined in muscles, gills, intestines, and livers of 20 benthic species of the most common fish caught from the coastal waters of eastern Taiwan using flame atomic absorption spectrometry (FAAS). Indium concentrations were the highest in the tissues, followed by Cu and Cd, and Pb being the lowest except in the gills. Among the tissues, liver showed the highest metal concentrations followed by intestine and gill, and was the lowest in the muscle. The concentrations of In, Cu, Cd, and lead in muscle ranged 2.0–6.2, 0.15–0.81, 0.02–0.12, and <0.02–0.15 μg/g wet weight, respectively The concentrations of the four metals in liver were at 16.9–59.1, 1.4–12.4, 0.11–1.16, and <0.02–1.09 μg/g wet weight, respectively. The concentrations of the heavy metals in the tissues varied significantly among species. Spottyback searobin *Pterygotrigla hemistica* and soldierfish *Myripristis berndti* contained in general higher concentrations of the metals in muscle and liver than other species of fish, respectively The metal concentrations of fish found in this study are similar to the metal levels of the fish caught from slightly polluted waters in other parts of Taiwan, while the metal concentrations in the authors' fish muscle are far below the consumption safety tolerance set by most countries in the world. Therefore, no public health problem would be raised from the consumption of fish from the coastal waters of eastern Taiwan.

The concentrations of trace metals in the fish species from the Black Sea and Yesilirmak River in Turkey were determined by Tuzen et al. [27] using FAAS and GFAAS after microwave digestion. The proposed method showed satisfactory recovery rates, detection limits, and standard deviations. The average metal concentrations (μg/g) of the five species varied in the following ranges: Fe 90.16–102.51; Cu 1.34–1.72; Mn 1.29–9.21; Zn 25.76–112.71; Pb 0.53–1.73; Cd 0.98–2.2; Cr 1.43–1.92, and Ni 2.90–9.36 μg/g, respectively.

Kadar et al. [28] describes several features of the aquatic environment with the emphasis on the total versus filter-passing fraction (FP) of heavy metals in microhabitats of two typical deep-sea vent organisms: the filter-feeder, symbiont-bearing *Bathymodiolus*, and the grazer shrimps *Rimica-ris/Mirocaris* from the Mid-Atlantic Ridge (MAR). The concentration of 10 trace metals: Al, Mn, Co, Cu, Mo, Cd, Fe, Zn, Pb, and Hg was explored highlighting common and distinctive features among the five hydrothermal vent sites of the MAR: Menez Gwen, Lucky Strike, Rainbow, Saldanha, and Menez Hom that are all geochemical different when looking at the undiluted hydrothermal fluid composition. The drop off in the percentage of FP from total metal concentration in mussel and/or shrimp inhabited water samples (in mussel beds at Rainbow, for instance, FP fraction of Fe was <23%, Zn 24%, Al 65%, Cu 70%, and Mn 89%) as compared to noninhabited areas (94% of the Fe, 90% of the Zn, 100% of the other metals was in the FP fraction) may indicate an influence of vent organisms on their habitat's chemistry, which in turn may determine adaptational strategies to elevated levels of toxic heavy metals. Predominance of particulate fraction over the soluble metals, jointly with the morphological structure and elemental composition of typical particles in these vent habitats suggest a more limited metal bioavailability to vent organisms as previously thought. It is evoked that vent invertebrates may have developed highly efficient metal-handling strategies targeting particulate phase of various metals present in the mixing zones that enables their survival under these extreme conditions.

Direct solid sampling Zeeman GFAAS methods were developed by Detcheva and Grobecker [29] for the determination of Hg, Cd, Mn, Pb, and Sn in seafood. All metals except Hg were measured by a third generation Zeeman AAS combined with an automatic solid sampler. In 3-field- and dynamic mode the calibrations concentration range was substantially extended and high amounts of analyte were detectable without laborious dilution of solid samples. The measurements were based on calibrations using certified reference materials of organic matrixes. In this case, solid certified reference materials were not available and calibration by aqueous standard solutions was proved an alternative. No matrix effects were observed under the optimized conditions. Results obtained were in good agreement with the certified values. Solid sampling Zeeman AAS was shown to be a reliable, rapid, and low-cost method for the control of trace metals in seafood.

A rapid and sensitive method was proposed by Yebra-Biurrun and Cancela-Perez [30] for the determination of Cr and Co in seafood samples by FAAS combined with a dynamic ultrasound-assisted acid extraction and an online mini-column preconcentration. The use of dilute HNO_3 as an extractant in a continuous mode at a flow rate of 3.5 mL/min and room temperature was sufficient for quantitative extraction of these trace metals from seafood. A mini-column containing a chelating resin was an excellent device for the quantitative preconcentration of Cr and Co prior to their detection. A flow-injection manifold was used as interface for coupling all analytical steps, which allowed the automation of the whole analytical process. A Plackett–Burman experimental design was used as a multivariate strategy for the optimization of both sample preparation and preconcentration steps. The method was successfully applied to the determination of Cr and Co in seafood samples.

Trace metal content of nine fish species harvested from the Black and Aegean Seas in Turkey were determined by microwave digestion and atomic absorption spectroscopy (MD-AAS) by Uluozlu et al. [31]. Verification of the MD-AAS method was demonstrated by analysis of standard reference material (NRCC-DORM-2 dogfish muscle). Trace metal content in fish samples were 0.73–1.83 µg/g for Cu, 0.45–0.90 µg/g for Cd, 0.33–0.93 µg/g for Pb, 35.4–1 06 µg/g for In, 1.28–7.40 µg/g for Mn 68.6–163 µg/g for Fe, 0.95–1.98 µg/g for Cr, and 1.92–5.68 µg/g for Ni. The levels of Pb and Cd in fish samples were higher than the recommended legal limits for human consumption.

Samples of *Mugil cephalus* and *Mullus barbatus* were collected in the Northeast Mediterranean coast of Turkey to determine the concentrations of Cd, Cu, Fe, Zn, and Pb in the liver, the gill, and the muscle tissues were determined by FAAS [32]. Except for Pb, highest levels of each metal were found in the liver and this was followed by the gill and the muscle in both species. Among the metals analyzed, Cu, Zn, and Fe were the most abundant in the different tissues while Cd and Pb were the least abundant both in *M. cephalus* and *M. barbatus*. Seasonal changes in metal (Cd, Cu, Pb, Fe, and Zn) concentration were observed in the tissues of both species, but these seasonal variations may not influence consumption advisories. In general, the highest concentrations were detected for all metals in summer.

The Cd and Cu levels were determined by Erdogrul et al. [33] in a total of one hundred and twenty six fish samples which belongs to five fish species collected from Sir and Menzelet Dam Lakes in Kahramanmaras Province, Turkey by AAS. The concentrations of heavy metals were expressed as parts per million (ppm) wet weight of tissue. The mean levels of cadmium and copper in the muscle, the liver, and the gill tissues of *Cyprinus carpio* from the Menzelet Dam were found to be 0.27, 0.91, 1.49 and 0.94, 1.2, 1.05, respectively. The mean levels of Cd in the muscle tissues of *Leuciscus cephalus* from the Menzelet Dam were found 0.32 ppm, Cd was not found in tissues of the liver and the gill. The mean levels of Cu in the muscle, the liver and the gill tissues were found as 3.17 ppm, 1.19 ppm, 0.96 ppm, respectively. The mean levels of the Cd and Cu in muscle and gill tissues of *Acanthobrama marmid* from the Sir Dam were found to be 1.28, 2.64 and 0.72, 0.08, respectively. The levels of the Cd and Cu in muscle tissues of *Cyprinus carpio* from the Sir Dam were found 0.87 and 0.02 ppm, respectively. The mean levels of the Cd and Cu in the muscle and gill tissues of *Chondrostoma regium* from the Sir Dam were found to be 0.80, 2.62 and 0.67, 1.34 ppm, respectively. The mean concentrations of Cd in the muscle tissues of *Silurus glanis* were found to be 0.60 ppm. In the muscle of the *Silurus glanis* from the Sir Dam, Cu was not found. The Sir Dam is more polluted than the Menzelet Dam from the point of Cd but less polluted than the Menzelet Dam. From the point of Cu a relationship was determined between species and their habitat region in terms of the levels reflected metal residues. In this study, it was emphasized that the amounts of Cd and Cu in the samples were low, however, seas, lakes, rivers, soil, air, and consumed foods has to be routinely controlled.

A comparison was made between microwave digestion and wet digestion methods for the determination of Na, Mg, Ca, Fe, Cu, Mn, Ni, Cd, Cr, and Pb in oyster with FAAS by Ren et al. [34]. Using microwave digestion method with a closed-vessel, the digestion could be done more rapidly and more effectively. The method could save reagents and display a lower background. It was available for biomonitoring of seawater and analysis of seafood.

Thirteen laboratories participated in an inter-laboratory method performance (collaborative) study on a method for the determination of As, Cd, Hg, and Pb by ICP-MS after pressure digestion including a microwave heating technique [35]. Prior to the study, the laboratories were able to practice on samples with defined metal levels (pretrial test). The method was tested on a total of seven foodstuffs: carrot puree, fish muscle, mushroom, graham flour, simulated diet, scampi, and mussel powder. The metal concentrations in mg/kg dry matter (dm) ranged from 0.06–21.4 for As, 0.03–28.3 for Cd, 0.04–0.6 for Hg, and 0.01–2.4 for Pb. The materials used in the study were presented to the participants as blind duplicates, and the participants were asked to perform single determinations on each sample. The repeatability RSD for As ranged from 3.8% to 24%, for Cd from 2.6% to 6.9%, for Hg from 4.8% to 8.3%, and for Pb from 2.9% to 27%. The reproducibility relative standard deviation for As ranged from 9.0% to 28%, for Cd from 2.8% to 18%, for Hg from 9.9% to 24%, and for Pb from 8.0% to 50%. The HorRat values were less than 1.5 r for all test samples, except for Pb in wheat flour at a level close to the limit of quantitation (0.01 mg/kg). The study showed that the ICP-MS method was satisfactory as a standard method for metal determination in seafood.

Acknowledgments

This work was supported, in part, by Merck undergraduate research program awarded to McNeese State University for 2005–2007. Partial support from Environmental Protection Agency, EPA-R-82958401–1 is gratefully acknowledged.

References

1. J.D. Ingle and S.R. Crouch, *Spectrochemical Analysis*, Prentice Hall, Englewood, NJ, 1988.
2. L.H.J. Lajunen, *Spectrochemical Analysis by Atomic Absorption and Emission*, Royal Society of Chemistry, Cambridge, England, 1992.
3. S.J. Haswell, ed., *Atomic Absorption Spectrometry: Theory, Design and Applications*, Elsevier Science Publishers, Amsterdam, the Netherlands, 1991.
4. J. Nolte, *ICP-Emission Spectrometry–A Practical Guide*, Wiley-VCH, Hoboken, NJ, 2003.
5. I.I. Sobelman, *Atomic Spectra and Radiative Transitions*, second edition, Springer-Verlag, Berlin, Germany, 1992.
6. J. Sneddon, ed., *Sample Introduction in Atomic Spectrometry*, Elsevier Science Publishers, Amsterdam, the Netherlands, 1990.
7. D.J. Butcher and J. Sneddon, *A Practical Guide to Graphite Furnace Atomic Absorption Spectrometry*, John Wiley & Sons, New York, 1997.
8. A. Montaser and D.W. Golightly, eds., *Inductively Coupled Plasmas in Analytical Atomic Spectrometry*, second edition, VCH Publishers, Inc., New York, 1992.
9. H.M. Kingston and S.J. Haswell, eds., *Microwave-Enhanced Chemistry, Fundamentals, Sample Preparation and Applications*, American Chemical Society, Washingon DC, 1997.
10. J. Sneddon, C. Hardaway, K.K. Bobbadi, and A.K. Reddy, Sample preparation of solid samples for metal determination by atomic spectroscopy-an overview and selected recent applications, *Applied Spectroscopy Reviews*, (2006), 41(1), 23–42.
11. J. Sneddon, P.W. Rode, M.A. Hamilton, S. Pingeli, and J.P. Hagen, Determination of metals in seafood, *Applied Spectroscopy Reviews*, (2007), 42(1), 1–16.
12. J. Sneddon, Use of spectrochemical methods for the determination of metals in fish and other seafood in Louisiana. In *The Determination of Chemical Elements in Food: Applications for Atomic and Mass Spectrometry*, S. Caroli, ed., Chapter 14, John Wiley & Sons, Hoboken, NJ, 2007, pp. 437–454.
13. J.C. Richert and J. Sneddon, Determination of inorganics and organics in crawfish, *Applied Spectroscopy Reviews*, (2008), 43, 1–17.
14. J.P. Hagen and J. Sneddon, Determination of copper, iron and zinc in crawfish (*Procambrus clarkii*) by inductively coupled plasma-optical emission spectrometry, *Spectroscopy Letters*, (2009), 42(1), 58–61.
15. J.C. Richert and J. Sneddon, Determination of heavy metals in crawfish (*Procambrus clarkii*) by inductively coupled plasma optical emission spectrometry; a study over the season in Southwest Louisiana, *Analytical Letters*, (2008), 44(17), 3198–3209.

16. B.S. Sheppard, D.T. Heitkemper, and C.M. Gaston, Microwave digestion for the determination of arsenic, cadmium and lead in seafood products by inductively coupled plasma atomic emission and mass spectrometry, *Analyst*, (1994), 119(8), 1683–1686.

17. M. Tamaru, T. Yabutani, and J. Motonaka, Multielement determination of trace metals in scallop tissue samples, *Bunseki Kagaku* (2004), 53(12), 1435–1440.

18. K.X. Yang, and K. Swami, Determination of metals in marine species by microwave digestion and inductively coupled plasma mass spectrometry analysis, *Spectrochimica Acta, Part B: Atomic Spectroscopy* (2007), 62(10), 1177–1181.

19. H. Oshima, E. Ueno, I. Saito, and H. Matsumoto, A comparative study of cadmium, lead, mercury, arsenic, selenium, manganese, copper and zinc in brown rice and fish by inductively coupled plasma-mass spectrometry (ICP-MS) and atomic absorption spectrometry, *Shokuhin Eiseigaku Zasshi* (2004), 45(5), 270–276.

20. L. Arslan, and D.H. Secor, Analysis of trace transition elements and heavy metals in fish otoliths as tracers of habitat use by American eels in the Hudson River estuary, *Estuaries* (2005), 28(3), 382–393.

21. E. Chirila, T. Petisleam, I.C. Popovici, and Z. Caradima, ICP-MS utilization for some trace elements determination in marine samples, Chem. Oep, Constanta, Rom. *Revista de Chimie* (2006), 57(8), 803–807.

22. J.J. LaBrecque, L Benzo, J.A. Alfonso, P.R. Cordoves, M. Quintal, N. Manuelita, C.V. Gomez, and E. Marcano, The concentrations of selected trace elements in clams, Trivela mactroidea along the Venezuelan coast in the state of Miranda, *Marine Pollution Bulletin* (2004), 49(7–8), 664–667.

23. Y. Wang, L. Liang, J. Shi, and G. Jiang, Study on the contamination of heavy metals and their correlations in mollusks collected from coastal sites along the Chinese Bohai Sea, *Environment International* (2005), 31(8), 1103–1113.

24. M. Maanan, Biomonitoring of heavy metals using Mytilus galloprovincialis in Safi coastal waters, Morocco, *Environmental Toxicology* (2007), 22(5), 525–531.

25. M. Tuzen, Determination of heavy metals in fish samples of the middle Black Sea (Turkey) by graphite furnace atomic absorption spectrometry, *Food Chemistry* (2003), 80(1), 119–123.

26. W.-B. Huang, Heavy metal concentrations in the common benthic fishes caught from the coastal waters of eastern Taiwan, *Fenxi* (2003), 11(4), 324–330.

27. M. Tuzen, D. Mend, H.Sari, M. Suicmez, and E. Hasdemir, Investigation of trace metal levels in fish species from the Black Sea and the river Yesilirmak, Turkey by atomic absorption spectrometry, *Fresenius Environmental Bulletin* (2004), 13(5), 472–474.

28. E. Kadar, V. Costa, I. Martins, R.S. Santos, Ricardo and J.J. Powell, Enrichment in Trace Metals (Al, Mn, Co, Cu, Mo, Cd, Fe, Zn, Pb and Hg) of Macro-Invertebrate Habitats at Hydrothermal Vents Along the Mid-Atlantic Ridge, *Hydrobiologia* (2005), 548 191–205.

29. A. Detcheva, and K.H. Grobecker, Determination of Hg, Cd, Mn, Pb, and Sn in seafood by solid sampling Zeeman atomic absorption spectrometry, *Spectrochimica Acta, Part B: Atomic Spectroscopy* (2006), 61B(4), 454–459.

30. M.C. Yebra-Biurrun and S. Cancela-Perez, Continuous approach for ultrasound-assisted acid extraction-minicolumn preconcentration of chromium and cobalt from seafood samples prior to flame atomic absorption spectrometry, *Analytical Sciences*, (2007) 23(8), 993–996.

31. O.D. Uluozlu, M. Tuzen, D. Mendil, and M. Soylak, Trace metal content in nine species of fish from the Black and Aegean Seas, Turkey, *Food Chemistry* (2007), 104(2), 835–840.

32. H.Y. Cogun, A. Yuezereroglu, Oe. Firat, G. Goek, and F. Kargin, Metal concentrations in fish species from the Northeast Mediterranean Sea, *Environmental Monitoring and Assessment* (2006), 121(1–3), 431–438.

33. O. Erdogrul, D. Ates, and D. Ayfer, Determination of cadmium and copper in fish samples from Sir and Menzlet Dam Lake Kahramanmaras, Turkey, *Environmental Monitoring and Assessment* (2006), 117(1–3), 281–290.

34. N. Ren, H.Li, Hong, Q. Zeng, and X. Xijiang, Determination of metal ions in oyster by microwave digestion of sample and flame atomic absorption spectrometry (FAAS), *Huaxue Shijie* (2005), 46(2), 83–85, 108, 117.

35. K. Julshamn, A. Maage, N. Amund, S. Hilde, K.H. Grobecker, L. Jorhem, and P. Fecher, Determination of arsenic, cadmium, mercury, and lead by inductively coupled plasma/mass spectrometry in foods after pressure digestion: NMKL inter-laboratory study, *Journal of AOAC International* (2007), 90(3), 844–856.

Chapter 23

Food Irradiation and Its Detection

Yiu Chung Wong, Della Wai Mei Sin, and Wai Yin Yao

Contents

23.1 Introduction 664
    23.1.1 Foodborne Diseases 664
    23.1.2 High Energy Irradiation for Food Preservation 664
    23.1.3 Development of Food Irradiation 665
    23.1.4 Global Acceptance and Attitudes 669
    23.1.5 Detection Methods for Irradiated Foods 670
23.2 Detection Methods 671
    23.2.1 Electron Spin Resonance Spectroscopy 671
    23.2.2 Analysis of Radiolytic Chemicals 672
        23.2.2.1 2-Alkylcyclobutanones 672
        23.2.2.2 Volatile Hydrocarbons 673
        23.2.2.3 *o-*, *m*-Tyrosine 675
        23.2.2.4 Hydrogen and Carbon Monoxide Gases 675
    23.2.3 DNA Methods 675
    23.2.4 Luminescence 677
    23.2.5 Microbiological Methods 678
23.3 Conclusions 679
Acknowledgment 679
References 679

23.1 Introduction

23.1.1 Foodborne Diseases

Diseases transmitted through contaminated food (or foodborne diseases) are always major social problems recognized by many national and international health authorities. Foodborne microbes such as *Salmonella* spp. and *Escherichia coli* O157:H7 are the primary cause of food poisoning in the United States and other industrialized nations, whereas *Vibrio* spp. especially *V. cholerae*, *V. parahaemolyticus*, and *V. vulnificus* caused a significant number of outbreaks and deaths in Asia and Latin America since the end of the last century [1,2]. In addition, foodborne diseases from parasites such as tapeworms, taxoplasmosis, and trichinosis are also of concern in developing countries. Foodborne diseases were responsible for a total of 76 million cases, 325,000 hospitalizations, and 5,000 deaths annually in the United States [3] and 2.4 million cases, 21,138 hospitalizations, and 718 deaths annually in England and Wales [4], respectively. However, there is a lack of adequate reporting mechanisms in both the developed and developing countries; only a very small proportion (less than 1%–10%) was said to be taken into account in the surveys [5]. The true incidence of such diseases is very difficult to determine and unavoidably results in an underestimated figure. Although, the real picture of the problem has remained a mystery, the situation has been well demonstrated by the World Health Organization (WHO) stating that approximately one in three people worldwide suffer annually from a foodborne disease and 1.8 million die from severe food and waterborne diarrhea [6].

23.1.2 High Energy Irradiation for Food Preservation

Two common ways have long been used to prevent the occurrence of foodborne diseases. Either physical methods such as heating or chemical methods such as adding salt as preservative are known to be extremely simple and effective to get rid of most pathogenic microbes and parasites. These well-accepted methods, however, are not deemed suitable to treat some solid foods, in particular raw meats, seafoods, and fresh fruits, because the texture, flavors, taste, and incurred ingredients would be irreversibly changed during the treatments. Nowadays, with the advancement of food-processing technology, there are more options and combinations of techniques available in food preservation. As described by a number of comprehensive reviews [7–12], the application of ionization radiation is regarded as one of the important new techniques in preserving hygienic quality of food in the food industry. Food irradiation is a process in which food matrices are exposed to high ionization energy gamma-ray produced from radionuclides (usually ^{60}Co or ^{137}Cs sources), fast moving electrons (maximum energy of 10 MeV), or x-rays (maximum of 5 MeV) from machines. These sources are permitted to be used in food treatment processes and have been adopted as a Codex Alimentarius General Standard [13]. However, the relatively low penetrating power of electron beams and the low efficiency of x-ray conversion limited their uses when compared to that of gamma irradiation. At present, cobalt-60, having higher penetrating power than that of cesium-137 is the preferred choice of irradiation source. It is estimated that 120 electron accelerator processing units [14] and more than 200 industrial ^{60}Co irradiation facilities [15] are being operated worldwide for sterilizing medical devices and for food irradiation.

The penetrating high energy could damage the DNA of living cells through energy transfer and significantly reduce the number of bacteria, yeasts, and moulds in food. The mechanisms of these energy transfer actions are well understood and described in detail in the literature [16,17].

The ionization radiation loses its energy to molecules of any matter (such as water, carbohydrates, fats, and proteins in food molecules) and leads to direct breakdown of molecules (primary effect) with the ejection of an electron and formation of a free radical:

$$M \rightarrow •M^+ + e^-$$

Both the electron and the free radical are highly reactive and cause a cascade of further ionization reactions (secondary effect) along the track. A combination of the primary and secondary effect leads to the chemical decomposition of molecules that are exposed to radiation in the medium.

The magnitude of the effects achieved is solely governed by the radiation dose being applied. The International System of Unit for radiation is the Gray (Gy) where 1 Gy is equal to 1 J of energy absorbed per kilogram of food mass. According to the Joint Expert Committee on Food Irradiation (JECFI) [18], there are three defined categories when food is irradiated at low, medium, and high doses:

1. Radurization at below 1 kGy: Prevents sprouting in vegetables, delays ripening in fruits, and inactivates parasites in insects and fish, kills or sterilizes insects in grains, or dried fish and fruits.
2. Radicidation at 1–10 kGy: Also termed as radiation-pasteurization, kills parasites and insects, reduces significantly the number of bacteria, yeasts, and moulds.
3. Radappertization at above 10 kGy: Eliminates all bacteria, achieves a complete sterilization of food.

The potential applications of food irradiation at different doses over a variety of foods including seafood have been extensively studied. Many of the pathogenic microorganisms are evident to be sensitive to ionization radiation [19] and can be conveniently reflected by the D_{10} values. D_{10} values represent the dose necessary to reduce population of 90% of a particular species. D_{10} values are usually under 1 kGy for vegetative cells, yeasts, and moulds, and under 4 kGy for spore-forming species (Table 23.1). As shown in Table 23.2, while the shelf life of unirradiated fishery products is less than 10 days of acceptability at 0°C–2°C, irradiation at optimum dose could extend the shelf life to several weeks [20].

Apart from effective disinfestation, food irradiation also offers a distinct advantage of virtually not raising the temperature of the food being processed. It was estimated that the heat energy absorbed at 10 kGy is equivalent to 10 J/g, an amount of energy needed to increase 1 g of water by 2.4°C [21]. Hence, nutrient losses are often small and are substantially less than other methods of preservation such as canning, drying, and heat pasteurization and sterilization. The technique has been described as possibly the most significant contribution to public health to be made by food science and technology after the pasteurization of milk [24].

23.1.3 Development of Food Irradiation

Although the principle and concept of food irradiation have been known for a long time, the technology was not utilized in the food industry until the 1950s and is still regarded as a "new" technology. The first commercial use of food irradiation was reported in 1957 [10] where a spice manufacturer in Stuttgart, Germany employed electron beam to improve the hygienic quality

Table 23.1 D_{10} **Values of Some Common Food Pathogens**

Organism	Matrix	D_{10} Value (kGy)
Vegetative Cells		
Campylobacter jejuni	Ground turkey	0.19
Escherichia coli (including O157:H7)	Ground beef	0.24–0.31
Listeria monocytogenes	Fish, shrimps	0.15–0.25
Salmonella paratyphi A	Oysters	0.85
Salmonella senftenberg	Liquid whole egg	0.47
Salmonella typhimurium	Roast beef, gravy	0.57
Streptococcus faecium	Shrimp	0.65–1.0
Staphylococcus aureus	Prawn, crabmeat	0.16–0.29
Vibrio cholerae	Clams, fish	0.14
Vibrio parahaemolyticus	Shrimp	0.11
Spores		
Bacillus cereus	Mozzarella cheese	3.6
Clostridium botulinum type E	Beef stew	1.4
Clostridium perfringens	Water	2.1
Yeasts and Moulds		
Aspergillus flavus	Growth culture	1.0
Trichosporon cutaneum	Fresh sausage	1.0

Source: Adapted from Miller, R.B., *Electronic Irradiation of Foods*, Springer Science, New York, 2005; Stewart, E.M., *Biologist*, 51, 91, 2004; Irradiation to control Vibrio infection from consumption of raw seafood and fresh produce, TECDOC Series No. 1213, IAEA, Vienna, Austria, 2001; Foley, D.M., *Food Irradiation Research and Technology*, Blackwell Publishing, Ames, IA, 2006.

of its products. Ironically, the irradiation machine had to be dismantled 2 years later because a new law prohibited such treatment for food. Like many other innovations, the adoption of irradiation for preventing foodborne diseases is slow and lengthy. In particular, the safety issues concerning human consumption of irradiated products have often been questioned. To promote successful implementation of the technique in the control of food pathogens, several national and international food control authorities have extensively studied this irradiation process under a variety of testing conditions over the past few decades. The research on the wholesomeness

Table 23.2 Shelf Life of Some Seafood Items by Irradiation

Items	Radiation Dose (kGy)	Storage Temperature (°C)	Shelf Life (Days)
Catfish	1–2	0	20
Salmon	1.5	2.2	20
Lake trout	3	0.6	26
Whitefish	1.5–3	0	15–29
Yellowperch	3	0.6	40–45
Clams	2	0.6	39
Crabs	2–2.5	0.6	28–42
Lobster	0.75	0	35
Mussels	1.5–2.5	3	42
Oysters	2	0	23
Scallops	0.75	0	28
Freshwater prawns	1.45	0	28
Tropical shrimps	1.5–2	3	42

Source: Adapted from Venugopal, V. et al., *Crit. Rev. Food Sci. Nutr.*, 39, 391, 1999.

and safety of irradiated food is said to be the most extensive undertaking of food scientists in history. One of the most important international authorities is the coalition of international organizations like the Food and Agriculture Organization (FAO) of the United Nations, the International Atomic Energy Agency (IAEA), and the WHO [25]. They had sponsored the establishment of the International Consultative Group on Food Irradiation (ICGFI) which aimed to develop a practical framework for proper application, to evaluate regulations and to give expert advice related to food irradiation. With untiring efforts and unprecedented depth of investigations, the FAO/IAEA/WHO JECFI [18] in 1980 declared that "irradiation of any food commodity up to an overall average dose of 10 kGy causes no toxicological hazard; hence, toxicological testing of food so treated is no longer required." Following the important declaration by JECFI was the issuance of "Codex General Standard for Irradiated Foods" and "Recommended Code for Practice for the Operation of Radiation Facilities Used for the Treatment of Foods" by Codex Alimentarius Commission [13] in 1983. Later in 1997, the joint FAO/IAEA/WHO group conducted further studies on the dose higher than 10 kGy and confirmed it was safe. With the accumulation of reliable scientific information, many nations have shown intense interest in using the technology to treat foodstuffs as one of the effective tools to safeguard public health and safety and also an alternative to reduce the use of banned fumigants such as ethylene oxide. At present, there are more than 50 nations granted clearance for a range of food items (Table 23.3) and an estimated amount of 200,000–500,000 tons of foods is being treated with irradiation every year.

Table 23.3 Countries with Commercial Radiation Processing Facilities

Country	Irradiated Food Type
Algeria	Potato
Argentina	Cocoa, spices
Bangladesh	Dried fish, onion, potato
Belgium	Deep frozen food, dehydrated vegetable, spices
Brazil	Dehydrated vegetables, spices
Canada	Spices
Chile	Dehydrated vegetable, onion, poultry meat, spices
China	Apple, Chinese sausage, dehydrated vegetable, garlic, onion, potato, rice, tomato
Cote d'Ivoire	Cocoa bean, yarns
Croatia	Food ingredients, spices
Czech Republic	Dry food ingredients, spices
Cuba	Beans, onion, potato
Denmark	Spices
Finland	Spices
France	Dried fruits, frozen frog leg, poultry, shrimp, spices, vegetable seasonings
Hungary	Enzymes, onion, spices, wine cork
India	Spices
Indonesia	Rice, spices
Iran	Spices
Israel	Condiments, spices
Japan	Potato
Korea	Garlic powder, spices
Mexico	Dry food ingredients, spices
Netherlands	Egg, frozen and dehydrated vegetable, rice, spices
Norway	Spices
Poland	Garlic, onion
South Africa	Chicken, fish, fruits, meat, onion, potato, processed products, spices

Table 23.3 (continued)   Countries with Commercial Radiation Processing Facilities

Country	Irradiated Food Type
Thailand	Enzymes, fermented pork sausages, onion, spices
Ukraine	Grain
United Kingdom	Spices
United States	Fruits, meat, poultry, spices, vegetable
Former Yugoslavia	Spices

23.1.4 Global Acceptance and Attitudes

The number of nations on the approval list has kept increasing over the past decades, but the degree of acceptance of using radiation processing, to a certain extent, is still varying throughout the world. Application of irradiation as a method of controlling foodborne diseases mainly depends upon consumers' attitude, regulatory actions, and the economic situation. The resistance and reluctance to accept the technique are thought to be comprised of a variety of factors. It has been reviewed [26] that food irradiation is perceived to be associated with nuclear radioactivity and therefore has often been opposed and challenged by numerous antinuclear groups, environmental protection activists, and some other food groups. Among the countless examples, the consumers in Europe Group stated that food irradiation should only be applied if other methods are not available or possible, and it should not be used as a substitute for poor hygiene and is not a low cost method [27]. Second, the general public has limited knowledge of the cause and prevention of foodborne diseases. These factors inevitably have negative impacts on consumer attitudes and eventually the legislative decision of policy makers toward food irradiation. The key to consumer acceptance for irradiated foods is education [28] and it has been shown in a study that there was a drastic change of consumer attitudes in irradiated foods in the United States between 1993 and 2003 [29]. Approximately 76% prefer to buy irradiated pork and 68% prefer to buy irradiated poultry; and more consumers were willing to buy irradiated products in 2003 than in 1993. A variety of recent market surveys also confirmed that a positive shift in attitudes toward irradiated foods could be achieved through the delivery of proper and accurate information to consumers [30,31]. Support from legislator and acceptance from the public and industry has helped the rapid growth of food irradiation in the United States [32] and in the Asia Pacific region [33]. Albeit the contributions of some members of the European Union (EU) like Belgium, France, and the Netherlands, the progress in Europe, meanwhile, is lagging behind [12]. For instance, food irradiation for specific categories of food (fruits, vegetables, cereals, tubers, spices fish, shellfish, and poultry) have been authorized since the early 1990s in the United Kingdom, however, the volume of irradiated foods in the retail markets is almost nonexistent [34]. Furthermore, in 1999, the European Parliament and the Council of EU issued Directives on irradiated food and the permitted commodities have been restricted to dried aromatic herbs, spices, and vegetable seasonings [35,36]. While the use of food irradiation as a distinct application for preventing outbreaks of foodborne diseases in red meats and seafood is gaining popularity across the world, it is rather remote for most of the countries in the continent and results from the European regulations.

23.1.5 Detection Methods for Irradiated Foods

Although policies about food irradiation vary from one country to another, the availability of methods for identifying irradiated food is one of the crucial requisites for legitimate implementation of the irradiation regulations. Reliable methods provide scientific tools for upholding regulatory controls, checking compliance against labeling requirements, facilitating international trade, and reinforcing consumer confidence. However, detection is very difficult as the actual changes that present in irradiated foodstuffs are extremely small within the working irradiation doses of less than 10 kGy; and in many cases, the changes involved are far less than those of classic food treatment processes. It is not surprising that the overall progress on detecting food irradiation was not satisfactory in the early years. Since an agreement on promoting food irradiation adopted by delegates from 57 countries at the International Conference on the Acceptance, Control of and Trade in Irradiation Foods in 1988 [37], intensive research studies and cooperation on detection of irradiation have been supported nationally and internationally. With concerted actions from the Community Bureau of Reference, and the Joint Division of the FAO and the IAEA, a number of detection methods, on the basis of chemical, physical, biological, and microbiological changes, were successfully developed under the cooperation framework. Five methods were adopted as the European standards (EN1784–1788) in 1997; four of them were revised after some years of publication. Another five EN standards have also been published in 2002–2004 (Table 23.4). The methods using electron spin resonance spectroscopy (ESR) and thermoluminescence (TL) for detecting primary radiolytic radicals and the analysis of radiolytic 2-alkylcyclobutanones (2-ACBs) and hydrocarbons are more specific and conclusive, while others are convenient to be used as fast screening methods. This chapter discusses the technical

Table 23.4   EN Protocols for the Detection of Irradiated Foods

Protocols	Title
EN1784:2003	Detection of irradiated food containing fat. GC analysis of hydrocarbons.
EN1785:2003	Detection of irradiated food containing fat. GC/MS analysis of 2-ACBs.
EN1786:1997	Detection of irradiated food containing bone. Method by ESR spectroscopy.
EN1787:2000	Detection of irradiated food containing cellulose by ESR spectroscopy.
EN1788:2001	TL detection of irradiated food from which silicate minerals can be isolated.
EN13708:2002	Detection of irradiated food containing crystalline sugar by ESR spectroscopy.
EN13751:2002	Detection of irradiated food using photostimulated luminescence.
EN13783:2002	Detection of irradiated food using direct epifluorescent filter technique/aerobic plate count.
EN13784:2002	DNA comet assay for the detection of irradiated foodstuffs. Screening method.
EN14569:2004	Microbiological screening for irradiated food using LAL/GNB procedures.

information on some common and validated detection methods that are mostly reported in the literature. There are also a number of other methods available, which have been thoroughly reviewed elsewhere [38–40].

23.2 Detection Methods

23.2.1 Electron Spin Resonance Spectroscopy

Electrons or radicals possess a magnetic moment that arises from their intrinsic spin and motion in the orbit. When placed in a magnetic field, the magnetic moment is proportional to the angular momentum of the electron. The torque exerted then produces a change in angular momentum which is perpendicular to that angular momentum, causing the magnetic moment to process around the direction of the magnetic field. The frequency at the precession is called Larmor frequency. When an external electromagnetic wave of the same frequency as the Larmor frequency is applied, a portion of the energy is absorbed and gives characteristic resonance signals. ESR, or electron paramagnetic resonance (EPR) spectroscopy is a nondestructive physical technique pertaining to the detection of such resonance signals.

ESR had originally been used as a tool for postirradiation dosimetry, dating, and imaging [41–43], and was later proposed to be applicable to detect induced radicals in irradiated foods [44,45]. To allow accurate identification, the ESR signals must be stable or fairly stable during the usual storage time of foodstuff; and must be distinguishable from those of unirradiated substances. Such detection is not suited to foods containing high water content in which induced radicals will be rapidly and significantly removed. Foods having dry and hard compositions, on the other hand, are known to be a favorable environment for stabilizing free radicals, and subsequently prolong their life span. Therefore, confirmation of irradiation status in these materials has been widely determined by ESR including the recent studies on bones [46], crustaceans' shell [47], beans [48], seeds [49], dried fruits [50], and spices [51]. At the early stage of the method development, the majority of the ESR work has been focused on the study of meat bones [52–54]. The ESR spectra were attributed to the CO_3^{3-}, CO_3^-, CO_2^-, CO^- radicals being trapped in the lattices of hydroxyapatite ($Ca_{10}(PO_4)_6(OH)_2$) [15], the major components of calcified materials. At approximately 0.5–1 kGy, the radiation-induced signals are clearly resolvable from those of nonirradiated bones with a characteristic asymmetric singlet (Figure 23.1). The shape of signals is basically similar for all bones and is observed in the region at $g = 2.0010$–2.0050 [53,56,57]. With the ESR method it was also possible to detect mechanically recovered meats, a product where flesh is separated from carcass through mechanical processes, by recovering the bone fragments using alcoholic alkaline hydrolysis [58]. The treatment allowed the ESR detection at an inclusion level of 10% (w/w) but recently the sensitivity has been improved to the level of 0.5% (w/w) [59]. For exoskeletons of crustaceans, such

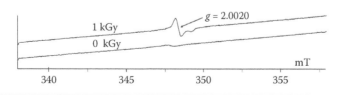

Figure 23.1 ESR spectra of nonirradiated and irradiated fish bone at 1 kGy.

as mollusks [60] and cuttlefish [61], the ESR signals due to the Mn^{2+} and CO_2^- are very intense and allowing a detection limit of less than 0.5 kGy. On the other hand, complex ESR signals are derived from dried cellulose materials [48–51]. The multicomponent signals consist of the predominant peak of crystalline sugars centered at $g = 2.003$ as stipulated in the EN13708 method, and the weak cellulose peak, typically at $g = 2.0045$ [62].

The induced radicals in irradiated foods are stable in most cases and the signal intensities are found to be dose dependent. As a consequence, some workers proposed that ESR could be used as a quantitative procedure where dose–response curves were commonly applied to estimate the original dose in irradiated foods [63–65]. However, the chemical composition [66,67], storage conditions [50,67], and processing treatment [70] could influence the dose response, and one must take into account correction factors in order to obtain a reliable dose estimation.

23.2.2 Analysis of Radiolytic Chemicals

23.2.2.1 2-Alkylcyclobutanones

In the early 1970s, LeTellier and Nawar [71] had isolated and identified a group of cyclic ketones as radiolytic products from pure triglycerides irradiated at 60 kGy in vacuum. The compounds were known as 2-ACBs, having the same carbon number as their respective parent fatty acid molecules, that were formed via ionization and cyclic rearrangement processes (Figure 23.2). These compounds are found to be degraded by oxidation during storage, but their stability is long enough to be detected. The production of radiolytic 2-ACBs is one of the most debated issues for the safety of irradiated food as some experimental studies indicated that these compounds are cancer promoters [71–74], while others claimed no mutagenic and genotoxic effects [75–77]. Despite the arguments, 2-ACBs were not detected in the processes of microwave treatment, oven heating, ultraviolet irradiation, and high pressure treatments and conformed to be used as unique markers for irradiated fat-containing foods [78,79]. Since then, a large amount of research work was initiated to use 2-ACBs to detect irradiated foods. 2-Dodecylcyclobutanone (DCB) and 2-tetradecylcyclobutanone (TCB), which are derived from the abundantly occurring palmitic and stearic acids in food items, are the most studied 2-ACB members. Positive identification of DCB and TCB were reported in irradiated chicken meat [80–82], lamb meat [84], ground beef [82,84], pork [82,85], quenelles [81], cheese [86], egg [80], fruits [81,86], melon seeds [87], fish [82,86,88], and dried shrimp [89]. The radiolytic 2-ACBs were extracted from foods by Soxhlet extraction, then purified, and isolated by column chromatography using Florisil [80,83,86,87,89],

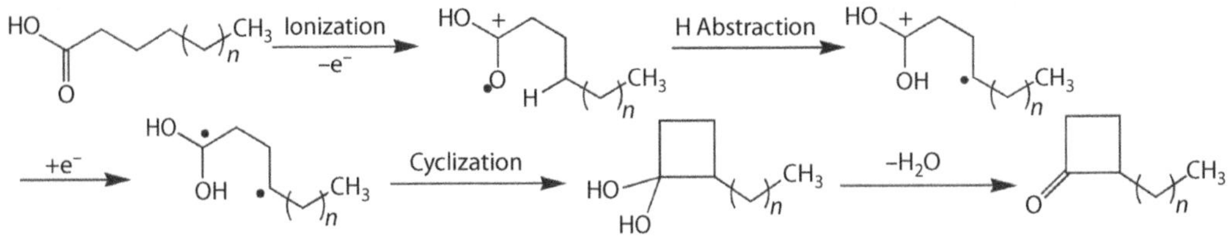

Figure 23.2 Schematic transformation of free fatty acids to 2-ACBs during high energy irradiation. Palmitic acid (*n* = 11) forms 2-dodecylcyclobutanone (DCB) and stearic acid (*n* = 13) forms 2-tetradecylcyclobutanone (TCB).

and detected using gas chromatography (GC)–mass spectrometry (MS), which was adopted as a standard protocol in EN1785 after being validated by a series of interlaboratory comparison studies [90]. Non-EN detection methods such as enzyme-linked immunosorbent assay (ELISA) [91] and TLC [93] have been used for fast screening, but these methods were subject to selectivity and sensitivity problems. In view of the tedious and time-consuming extraction procedures involved in EN1785, some workers proposed to replace the Soxhlet–Florisil chromatography by a supercritical fluid extraction (SFE). A SFE operated at its optimized conditions could extract low levels of DCB and TCB from samples within 30–60 min with good efficiency [81,84,88]. Others recommended the use of a fully automated accelerated solvent extraction method [82] for extracting 2-ACBs.

A typical mass chromatogram of DCB and TCB, usually monitored at m/z 98 and 112 is shown in Figure 23.3. The amount of DCB and TCB produced is proportional to the irradiation dose and is also dependent on the food types and the presence of fatty acid precursors. The limits of detection for DCB and TCB were only at a dose of 0.5–1 kGy for red meats and seafoods [82,88,89] compared to that of 0.1 kGy for mango and papaya [86]. The low sensitivity in the former food types was attributed to the presence of interfering substances that had not been adequately removed. An inclusion of solid phase extraction using cation exchanger impregnated with silver ions after Florisil extraction was found to improve the sensitivity to about 0.1 kGy. Another study using pentafluorophenyl hydrazine as the coupling agent for DCB and TCB [93] was reported to enhance the sensitivity of detection by two to five times for chicken meat and pork samples.

23.2.2.2 Volatile Hydrocarbons

Radiolytic cleavage of fatty acids ($C_{m:n}$), where m is the number of carbon atoms and n is the double bond, at α- and β-position of the carbonyl group leads to the formation of two characteristic

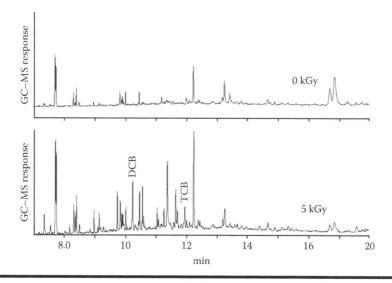

Figure 23.3 **Total ion chromatogram showing detectable quantities of the DCB and TCB in black melon seed at 5 kGy. These radiation-induced compounds were not present in the nonirradiated sample.**

volatile hydrocarbons; one has a carbon atom less than the parent fatty acid ($C_{m-1:n}$) and the other has two carbon atoms less and one extra double bond in position 1 ($C_{m-2::n+1}$). Although volatile hydrocarbons are also found in other nonirradiated foods, the detection of the hydrocarbon couple unambiguously indicated the presence of irradiation treatment [94,95]. Similar to that of 2-ACB, extracted hydrocarbons could be isolated by Florisil chromatography and detected by GC-FID or GC–MS (Figure 23.4). The method was validated through interlaboratory comparison [90,96] and adopted as a standard protocol in EN1784. Pentadecane ($C_{15:0}$) and 1-tetradecene ($C_{14:1}$), heptadecane ($C_{17:0}$) and 1-hexadecene ($C_{16:1}$), and 8-heptadecene ($C_{17:1}$) and 1,7-hexadecadiene ($C_{16:2}$) that respectively were generated from abundant palmitic, stearic, and oleic acids in foods were widely reported in beans [97], cereals [98], nuts [99], eggs [100], meats [101,102], shrimps, and seafood [89,103,104]. The radiolytic hydrocarbons found in those food matrices were relatively stable and varied from a week to several months postirradiation, and the respective concentrations were proportional to the dose applied. A comparative study [105] on investigating the profile of hydrocarbon markers in different foodstuffs showed that while $C_{16:2}$ was commonly detected at the lowest practical dose in dairy products, fruits, beef, pork, chicken, and tuna, only 6,9-heptadecadiene ($C_{17:2}$), owing to the different fatty acids composition, was the suitable marker for dried shrimp at the practical dose of 0.75–2 kGy. Another recent study [107] also showed that only $C_{17:1}$, $C_{16:2}$, $C_{17:2}$, and 1,7,10-hexadecatriene ($C_{16:3}$) were detected at 0.5 kGy in irradiated soybeans as the food consisted mostly of oleic (26%) and linoleic acids (49.6%). Therefore, in order to achieve the best performance of the detection method, it was recommended to check the fatty acid profile of the food type under study as the marker hydrocarbons would vary from one food to another [105].

The disadvantages of using Florisil chromatography were that it is tedious and the procedure was not very efficient in removing all lipid interference prior to detection, hence decreasing the sensitivity. Some workers proposed using argentation chromatography [107] for enrichment of radiolytic unsaturated hydrocarbons from fruits and meats using a silver column; and the use of online coupled LC–GC [109] for fish and prawn samples to improve sample preparation and separation efficiency. Others developed a solid phase microextraction (SPME) and purge and trap

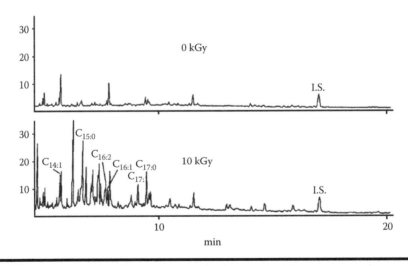

Figure 23.4 **Chromatograms for the hydrocarbons from nonirradiated and 10 kGy-irradiated dried shrimps. (From Kim, K.S. et al., *J. Food Prot.*, 67, 142, 2004. With permission.)**

method [109] to overcome the tedious extraction process. All those modifications might have good potential to enhance the detection capability of EN1784 where further investigation work and research are required.

23.2.2.3   o-, m-Tyrosine

Amino acids are vulnerable to the active radicals induced in irradiation and the end products might be used as markers for identifying irradiated protein-containing foods. An experiment showed that *o*-, *m*-, and *p*-tyrosine were produced upon irradiation of a phenylalanine solution and the yields were proportional to the applied dose [110]. As *o*- and *m*-tyrosine were nonnatural amino acids, some early studies have demonstrated the feasibility using these compounds as potential markers to detect food irradiation [111–114]. At almost about the same time, minute quantities of *o*-tyrosine were found to be present in nonirradiated food and thus diminished their potential usefulness for detecting irradiated foods. However, another research study showed that the background level of *o*-tyrosine determined by high performance liquid chromatography (HPLC) with florescence detection [115] in unirradiated shrimps was 19.3 μg/kg, which was 10 times less than those irradiated at 1 kGy and concluded that *o*-tyrosine was a reasonable marker to detect irradiated shrimps down to 1 kGy dose. The results were in good agreement with another similar study using HPLC with coulometric electrode array detection [116] to detect *o*- and *m*-tyrosine in irradiated shrimps. Unfortunately, the use of tyrosine isomers has not been thoroughly validated and the relevant literature information on other food matrices is very limited.

23.2.2.4   Hydrogen and Carbon Monoxide Gases

Simple molecular inorganic gases produced upon the radiolysis of water and organic components (carbohydrates, lipids, proteins, etc.) in foods were proposed as versatile probes for irradiation detection. Radiolytic hydrogen was detected in pepper [117] by GC and in frozen chicken by a headspace analyzer based on a hydrogen-specific electronic sensor [118]. The latter work also claimed that hydrogen generated during the irradiation of frozen chicken at 5 kGy was measurable after storage for up to at least 6 months. Similar studies on radiolytic carbon dioxide in deboned frozen meats [119], spices, and dry grains [120] showed that this gas could also be well retained after irradiation and could serve as another reliable marker for detection. Another study [121] successfully used microwave heating and headspace GC to measure the level of hydrogen and carbon monoxide in frozen shrimps, cod slices, and deshelled oyster irradiated at 1–8.8 kGy. The radiolytic gases were detected up to 3 months after irradiation. The above studies demonstrated hydrogen and carbon monoxide are convenient for distinguishing some irradiated foods and could provide a rapid screening test. However, no definite conclusions can be drawn from negative results, which should be endorsed by other confirmatory tests.

23.2.3   DNA Methods

Since DNA is vulnerable to high energy irradiation, radiation-induced changes in DNA molecules could be used as a tool to detect the radiation treatment of foods. Upon irradiation, the double helical strands of DNA molecules usually break and form various fragments having lower molecular weight than their parent DNA. Östling and Johanson [122] first reported the experiment by mixing extracted DNA in buffer solution with low-melt agarose and subjecting to electrophoresis analysis. DNA fragments from irradiated cells showed longer migration distance from the nuclei and the unradiated cells presented no or little movement. The migration of the DNA fragments from cells giving the appearance of a comet, and the technique is commonly termed DNA comet assay. Using the approach, other workers [123,124] found that the extension of the comet tail correlated with the degree of cellular DNA damage, i.e., the radiation dose. As a consequence, DNA comet assay has proved its wide application in genotoxicity [125,126], environmental contamination monitoring [127–129], and fundamental research in DNA damage and repair [130,131].

The method was modified to detect irradiated foods [132] with the aid of sets of reference samples at 0–5 kGy for checking migration patterns. Because of the fast analysis time (<1 h), such assay was also applied as a screening to control imported foods in Sweden [133]. Another study [134] on DNA degradation of chilled fresh chicken explicated that the length and shape of the comets obtained could be used to estimate the dose. The observed comet tail was short for unirradiated cells, but it became longer and even separated from the comet head when the dose increased. At a very high dose, almost no DNA was left in the head, and the tail appeared as a cloud (Figure 23.5). Validation of comet assay through interlaboratory comparisons was satisfactory, with an average of over 95% test results correctly identified for seeds, dried fruits, spices [135], chicken, and pork [136] and other foodstuffs [137,138] and the procedure served as a screening in the EN method. However, DNA is known to be naturally degraded at room temperature through the activities of nucleases within the cells. EN 13784 stated that the application of DNA comet assay is limited to determining fresh or frozen foods, and not applicable to foods that have been subjected to various forms of physical and chemical treatments that resulted in DNA fragmentation. Any new type of foodstuffs shall also be tested by the method before unknown samples are analyzed. Since then, a large number of food detections such as pork [139], poultry [140], beef [141], hamburger [142], papaya, and melons [143], using comet assay were reported. An unsuccessful study [144] on halibut, herring, saithe, plaice, and squid, however, might explain limited work on seafood using comet assay.

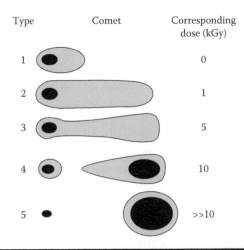

Figure 23.5 Drawing of the comet types observed with increasing DNA degradation (types 1–5) in fresh chicken legs at 2°C–4°C. The darker regions represent higher amount of DNA. For comparison, the doses of ionizing radiation which produce similar comets are given. (From Cerda, H. and Koppen, G., *Z. Lebensm. Unters. Forcsh. A,* **207, 22, 1998. With permission.)**

The use of supercoiled mitochondrial DNA (mtDNA) was proposed [145] as a good alternative for comet assay because mtDNA is well protected by the mitochondrial wall from enzymatic reactions. mtDNA was shown to be very stable during prolonged storage at 4°C, but it was significantly reduced upon irradiation at 2–4 kGy [146]. Therefore, detection of mtDNA could be a good marker for irradiation. However, the method had not been extensively studied for irradiated food by other workers owing to the complication and time-consuming extraction process. Another approach relying on the immunological detection of DNA base changes was also proposed. Dihydrothymidine (DiHT) was known to be produced from thymidine base via radical reaction upon irradiation [147,148]. With the development of monoclonal antibody, novel ELISA assays for the detection of DiHT in prawns at 2 kGy and other irradiated food were reported [149,150]. The potential advantage of ELISA was its usefulness in crude food homogenates, which could reduce the extraction process and offer a rapid screening test. The method, however, would require more validated work in order to be applicable to other foodstuffs.

23.2.4   Luminescence

Electrons can be excited to higher energy states by absorption of ionization radiation, and trapped if the substance has a crystalline structure. When the trapped electrons are released and returned to the ground state, some of the energy appears in the form of light and causes the substance to luminesce. The release process could be stimulated by heat (TL) or light (photostimulated luminescence or pulsed infrared stimulation [PSL]). The trapped electrons can remain in the crystalline lattice for many years, the measurement of the energy emission reveals the ionizing radiation to which the substance had been exposed. The technique of TL has already been used for the radiation dosimetry in archaeological and geological dating [151]. In 1989, it was first reported to identify irradiated spices [152] by measuring the intensity of emitted photons over a range of temperature (glow curve). As an illustrative example shown in Figure 23.6 [153], typical glow curves of irradiated dried fish from 1 to 7 kGy, which were distinguishable from the nonirradiated sample, were observed to peak at about 150°C. The TL signals detected from earlier studies were thought to come from organic materials, but later studies [154,155] indicated the signals originated from the contaminated silicate minerals present in the samples. With the isolation and normalization of minerals, TL was extensively studied in detecting other food items that contain concomitant minerals such as fruits, vegetables, shellfish, shrimps, and prawns [156–160]. The TL method was the first confirmative method adopted in the United Kingdom for detecting irradiated foods and later as a European standard (EN 1788) in 1997, which was revised in 2001.

TL is a very specific method for the confirmation of food irradiation, but it requires the tedious and skillful separation of silicates from the food matrices. These mandated procedures limited the usefulness of TL in routine surveillance examination. Furthermore, the mixture of endogenous inorganic and organic materials in food could inhibit high-temperature TL analysis. Therefore, a novel development was proposed to release the trapped charge carriers from the excited energy level using pulsed infrared stimulation (PSL), which allowed the detection in the presence of interfering organic materials, and eliminated the necessity of isolating inorganic materials [156,157]. With the employment of a simple instrument, the PSL measurement was claimed to produce a fast qualitative analysis for irradiated food within 15–60 s [161]. PSL was shown to be a reliable screening method for herbs, species, and shellfish after the successful outcome of collaborative trials [162–165]; PSL was adopted as a European Standard (EN13751) in 2002.

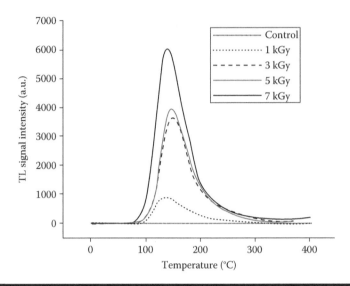

Figure 23.6   Dose-dependent TL glow curves of minerals separated from irradiated sliced-dried Pollack at different doses. (From Kwon, J.H. et al., *Radiat. Phys. Chem.*, 71, 81, 2004. With permission.)

23.2.5   Microbiological Methods

Food irradiation is a process to eliminate/reduce the number of microbial flora that is present, and therefore the change of these microorganisms could be used to detect irradiated food. Through elaborate validations and verifications, two microbiological tests have been adopted by EU as the standard screening methods in 2002 and 2004. EN13783 relies on the difference between aerobic bacteria plate count (APC) and direct epifluorescent filter count (DEFT). APC is a convention method to show the number of viable bacteria capable of forming colonies on agar plate and the unit is expressed as cfu. DEFT is another standard count protocol used for enumerating the total number of live and dead microbes [166]. For unirradiated samples, the counts of APC are often comparable to those in DEFT. Conversely, APC is found to be significantly lower, usually in the range of 3–4 log units in irradiated food such as frozen meats [167], herbs and spices [168,169].

EN14569 is based on two different microbiological techniques, viz., the enumeration of viable Gram-negative bacteria (GNB) and immunological analysis of endotoxins contained in the GNB by *Limulus* Ameobocyte Lysate test (LAL) in the test samples. In general, GNB count is proportional to the LAL results. For a low GNB count with a high LAL result, it might indicate the presence of a large population of dead microbes caused by irradiation. Application of the technique in poultry meats [170,171] was found to be successful.

Since the change of microbe loading in food could also be influenced by other physical and chemical treatments, both microbiological methods are not specific methods and any positive results obtained shall be confirmed by a specific test.

Apart from the two EN standards, turbidimetric measurement for the number of bacteria present in the extracted medium was also proposed [172]. The method was claimed to provide a very good index for food irradiation and was simpler and quicker than that of the DEFT/APC method.

23.3   Conclusions

Owing to the minute change of physiochemical properties in food matrices before and after irradiation, detection of food radiation was considered as an unreachable target two decades ago. With the concerted actions from various national and international organizations such as the Community Bureau of Reference and the Joint Division of the FAO, IAEA, and WHO, we observed an overflow of useful and validated methodologies that are applicable in food irradiation. One of the most noticeable events has been the approval of 10 validated standard methods by European Committee for Standardization from 1996 to 2004. ESR spectroscopy, TL glow curve studies, and the chromatographic analyses of hydrocarbons and 2-ACB methodologies provide reliable and confirmative information of the irradiation status in a wide variety of food items. Whereas, the DNA comet assay, PSL, and microbiological tests could offer a fast and inexpensive alternative for screening a large amount of samples.

The research and development for food irradiation was at its climax in the 1990s but it showed signs of losing momentum in the 2000s. For example, the mandate of ICGFI ended in 2004 and no official arrangement was recommended to continue the mission of this active group. The downfall of ICGFI was an illustration of fading supports from federal governments of most leading countries. However, as stated by Ehlermann, the former Federal Research Centre for Nutrition, Germany, in his review of "Four decades in food irradiation" [173], "After the age of enlightenment it is time for a modern society to counteract myths, ideologies, superstition and to rely on facts and knowledge based on sound science. This must include the appreciation of processing food by ionization irradiation as a valuable and justified technology."

Acknowledgment

The authors would like to express their sincere thanks to Mr. Peter Brown of the Association of Public Analysts, the United Kingdom, for his invaluable technical comments to this manuscript.

References

1. Murano, E.A. and Cross, H.R. Enhancing food safety through irradiation, the International Consultative Group on Food Irradiation (ICGFI), Vienna, Austria, 1999.
2. Kilgen, M. Economic benefits of irradiation of molluscan shellfish in Louisiana, in *Cost-Benefit Aspects of Food Irradiation Processing*, STI/PUB/905, IAEA, Vienna, Austria, 1993.
3. Mead, P.S. et al. Food-related illness and death in the United States, *Emerg. Infect. Dis.*, 5, 607, 1999.
4. Adak, G.K., Long, S.M., and O'Brien, S.J. Trends in indigenous foodborne disease and deaths, England and Wales: 1992 to 2000, *Gut*, 51, 832, 2002.
5. The role of food safety in health and development, the World Health Organization (WHO) Technical Report Series No. 705, Geneva, Switzerland, 1984.
6. Food safety and foodborne illness, WHO Fact Sheet No. 237, Geneva, Switzerland, 2007.
7. Loaharanu, P. Food irradiation: Current and future prospects, in *New Methods of Food Preservation*, Gould, G.W. (Ed.), Dekker, New York, 1995, chap. 5.
8. Rahman, M.S. Irradiation preservation of foods, in *Handbook of Food Preservation*, Rahman, M.S. (Ed.), Dekker, New York, 1999, chap. 13.
9. Dickson, J.S. Radiation inactivation of microorganisms, in *Food Irradiation: Principles and Applications*, Molins, R.A. (Ed.), Wiley Inter science, New York, 2001, chap. 2.

10. Diehl, J.F. Food irradiation—past, present and future, *Radiat. Phys. Chem.*, 63, 211, 2002.
11. Miller, R.B. Introduction to food irradiation, in *Electronic Irradiation of Foods*, Springer Science, New York, 2005, chap. 1.
12. Farkas, J. Irradiation for better foods, *Trends Food Sci. Technol.*, 17, 148, 2006.
13. Codex General Standards for irradiation foods and recommended international code of practice for the operation of radiation facilities used for the treatment of foods, Codex Alimentarius Commission, XV, E-1, Food and Agriculture Organizations (FAO) of the United Nations, Rome, Italy, 1984.
14. Energy applications of radiation processing, TECDOC Series No. 1386, International Atomic Energy Agency (IAEA), Vienna, Austria, 2004.
15. Cleland, M.R. Advances in gamma ray, electron beam, and x-ray technologies for food irradiation, in *Food Irradiation Research and Technology*, Sommers, C.H. and Fan, X. (Eds.), Blackwell Publishing, Ames, IA, 2006, chap. 2.
16. Murano, E.A. Irradiation of fresh meats, *Food Technol.*, 49, 52, 1995.
17. Stewart, E.M. Food radiation chemistry, in *Food Irradiation: Principles and Applications*, Molins, R.A. (Ed.), Wiley Inter-Science, New York, 2001, chap. 3.
18. Wholesomeness of irradiated food, Joint FAO/IAEA/WHO Expert Committee on Food Irradiation, WHO Technical Report Series No. 659, Geneva, Switzerland, 1981.
19. Stewart, E.M. Food irradiation: More pros than cons? *Biologist*, 51, 91, 2004.
20. Venugopal, V., Doke, S.N., and Thomas, P. Radiation processing to improve the quality of fishery products, *Crit. Rev. Food Sci. Nutr.*, 39, 391, 1999.
21. Loaharanu, P. and Murrell, D. A role for irradiation in the control of foodborne parasites, *Trends Food Sci. Nutr.*, 5, 190, 1994.
22. Irradiation to control Vibrio infection from consumption of raw seafood and fresh produce, TECDOC Series No. 1213, IAEA, Vienna, Austria, 2001.
23. Foley, D.M. Irradiation of seafood with a particular emphasis on *Listeria moncytogenes* in ready-to-eat products, in *Food Irradiation Research and Technology*, Sommers, C.H. and Fan, X. (Eds.), Blackwell Publishing, Ames, IA, 2006, chap. 11.
24. Kaperstein, F.K. and Moy, G.G. Public health aspects of food irradiation, *J. Pub. Health Policy*, 14, 149, 1993.
25. Swallow, A.J. Wholesomeness and safety of irradiated foods. in *Nutritional and Toxicological Consequences of Food Processing*. Friedman M. (Ed.), New York: Plenum Press; 1991, pp. 11–31.
26. Osterholm, M.T. and Norgan, A.P. The role of irradiation in food safety, *N. Engl. J. Med.*, 350, 1898, 2004.
27. http://www.foodcomm.org.uk/irradiation_legislation.htm#Euro_cons_orgs.
28. Foley, D.M. Consumer acceptance and marketing of irradiated foods, in *Food Irradiation Research and Technology*, Sommers, C.H. and Fan, X. (Eds.), Blackwell Publishing, Ames, IA, 2006, chap. 5.
29. Johnson, A.M. et al. Consumer attitudes towards irradiated food: 2003 vs. 1993, *Food Prot. Trends*, 24, 408, 2004.
30. Nayga Jr., R.M., Aiew, W., and Nichols, J.P. Information effects on consumers' willingness to purchase irradiated food products, *Rev. Agric. Econ.*, 27, 37, 2005.
31. Rodriguez, L. The impact of risk communication on the acceptance of irradiated food, *Sci. Comm.*, 28, 476, 2007.
32. Chmielewski, A.G. and Haji-Saeid, M. Radiation technologies: Past, present and future, *Radiat. Phys. Chem.*, 71, 16, 2004
33. Luckman, G.J. Food irradiation: Regulatory aspects in the Asia and Pacific region, *Radiat. Phys. Chem.*, 63, 285, 2002.
34. Woolston, J. Food irradiation in the UK and the European Directive, *Radiat. Phys. Chem.*, 57, 245, 2000.
35. Directive 99/2/EC of the European Parliament and of the Council on the approximation of the laws of the member states concerning foods and food ingredients treated with ionizing radiation, *Off. J. Eur. Comm.*, 66, 1999, p. 16.

36. Directive 99/3/EC of the European Parliament and of the Council on the establishment of a community list of foods and food ingredients treated with ionizing radiation, *Off. J. Eur. Comm.*, 66, 1999, p. 24.

37. Stochowicz, W. et al. Accredited laboratory for detection of irradiated foods in Poland, *Radiat. Phys. Chem.*, 63, 427, 2002.

38. McMurray, C.H., Stewart E.M., Gray, R. and Peaece, J. (Eds.), *Detection Methods for Irradiated Foods – Current Status*, Royal Society of Chemistry, London, U.K., 1996.

39. Stewart, E.M. Detection methods for irradiated food, in *Food Irradiation: Principles and Applications*, Molins, R.A. (Ed.), Wiley Inter science, New York, 2001, chap. 14.

40. Marchioni, E. Detection of irradiated foods, in *Food Irradiation Research and Technology*, Sommers, C.H. and Fan, X. (Eds.), Blackwell Publishing, Ames, IA, 2006, chap. 6.

41. Ikeya, M. and Miki, T. ESR Dating of animal and human bones, *Science*, 207, 977, 1980.

42. Desrosiers, M.F. and Schauer, D.A. Electron paramagnetic resonance (EPR) biodosimetry, *Nucl. Instr. Meth. Phys. Res. B*, 184, 219, 2001.

43. Regulla, D.F. ESR spectrometry: A future-oriented tool for dosimetry and dating, *Appl. Radiat. Isot.*, 62, 117, 2005.

44. Dodd, N.J.F. and Swallow, A.J. Use of ESR to identify irradiated food, *Radiat. Phys. Chem.*, 26, 451, 1985.

45. Dodd, N.J.F., Lea, J.S., and Swallow, A.J. ESR detection of irradiated food, *Nature*, 334, 387, 1988.

46. Chawla, S.P. Thomas, P., and Bongirwar, D.R. Factors influencing the yield of radiation induced electron spin resonance (ESR) signal in lamb bones, *Food Res. Int.*, 35, 467, 2002.

47. Köksal, F. and Köseoğlu, R. Detection of free radicals in γ-irradiated seasnail hard tissues by electron paramagnetic resonance, *Appl. Radiat. Isot.*, 59, 73, 2003.

48. Bhushan, B., Bhat, R., and Sharma, A. Status of free radicals in Indian monsooned coffee beans γ-irradiated for disinfestations, *J. Agric. Food Chem.*, 51, 4960, 2003.

49. Eschrig, U. et al. Electron seed dressing of barley—aspects of its verification, *Eur. Food Res. Technol.*, 224, 489, 2007.

50. Yordanov, N.D. and Pachova, Z. Gamma-irradiated dry fruits. An example of a wide variety of long-time dependent EPR spectra, *Spectro. Acta A*, 63, 891, 2006.

51. Ukai, M., Nakamura, H., and Shimoyama, Y. An ESR protocol based on relaxation phenomena of irradiated Japanese pepper, *Spectro. Acta A*, 63, 879, 2006.

52. Desrosiers, M.F. and Simic, M.G. Post-irradiation dosimetry of meat by electron spin resonance spectroscopy of bones, *J. Agric. Food Chem.*, 36, 601, 1988.

53. Raffi, J. et al. ESR analysis of irradiated frog's legs and fishes, *Appl. Radiat. Isot.*, 40, 1215, 1989.

54. Dodd, N.J.F. et al. Factors influencing the yield of free radicals in irradiated chicken, *Int. J. Food Sci. Technol.*, 27, 371, 1992.

55. Bacquet, G. et al. ESR of carbon dioxide radical anion in x-irradiated tooth enamel and A-type carbonated apatite, *Calcif. Tissue Int.*, 33, 105, 1981.

56. Desrosiers, M.F., γ-Irradiated seafoods: Identification and dosimetry by electron paramagnetic resonance spectroscopy, *J. Agric. Food Chem.*, 37, 96, 1989.

57. Stochowicz, W. et al. The EPR detection of foods preserved with the use of ionising radiation, *Radiat. Phys. Chem.*, 46, 771, 1995.

58. Gray, R. and Stevenson, H.M. Detection of irradiated deboned turkey meat using electron spin resonance spectroscopy, *Radiat. Phys. Chem.*, 34, 899, 1989.

59. Marchioni, E. et al. Detection of irradiated ingredients induced in low quantity in non-irradiated matrix. 1. Extraction and ESR analysis of bones from mechanically recovered poultry meat, *J. Agric. Food Chem.*, 53, 3769, 2005.

60. Ziegelmann, B., Bögl, K.W., and Schreiber, G.A. TL and ESR signals of mollusc shells—correlations and suitability for the detection of irradiated foods, *Radiat. Phys. Chem.*, 54, 413, 1999.

61. Duliu, O.G. Electron paramagnetic resonance identification of irradiated cuttlefish (*Sepia officinalis* L.), *Appl. Radiat. Isot.*, 52, 1385, 2000.

62. Yordanov, N.D., Alesksieva, K., and Mansour, I. Improvement of the EPR detection of irradiated dry plants using microwave saturation and thermal treatment, *Radiat. Phys. Chem.*, 73, 55, 2005.
63. Desrosiers, M.F. Estimation of the absorbed dose in radiation-processed food – 2. Test of the ESR response function by an exponential fitting analysis, *Appl. Radiat. Isot.*, 42, 617, 1991.
64. Onori, S. and Pantaloni, M. ESR dosimetry of irradiated chicken legs and chicken eggs, in *Detection Methods for Irradiated Food: Current Status*, McMurray, C.H., Stewart, E.M., Gray, R. and Pearce, J. (Eds.), Royal Society of Chemistry, Cambridge, London, 1996, p. 62.
65. Parlato, A. et al. Application of the ESR spectroscopy to estimate the original dose in irradiated chicken bone, *Radiat. Phys. Chem.*, 74, 1466, 2007.
66. Sin, D.W.M. et al. Identification and stability study of irradiated chicken, pork, beef lamb, fish and mollusk shells by electron paramagnetic resonance (EPR) spectroscopy, *Eur. Food Res. Technol.*, 221, 684, 2005.
67. Stewart, E.M., Stevenson, M.H., and Gray, R. Use of ESR spectroscopy for the detection of irradiated Crustacea, *J. Sci. Food Agric.*, 65, 191, 1994.
68. Chawla, S.P. et al. Detection of irradiated lamb meat with bone: Effect of chilled storage and cooking on ESR signal strength, *Int. J. Food Sci. Technol.*, 34, 41, 1999.
69. Polat, M. and Korkmaz, M. The effect of temperature and storage time on the resonance signals of irradiated pea (*Pisum sativum* L.), *Food Res. Int.*, 36, 857, 2003.
70. Stewart, E.M. et al. The effect of processing treatments on the radiation-induced ESR signal in the cuticle of irradiated Norway lobster (*Nephrops norvegicus*), *Radiat. Phys. Chem.*, 42, 367, 1993.
71. LeTellier, P.R. and Nawar, W.W. 2-Alkycyclobutanones from the radiolysis of triglycerides, *Lipids*, 7, 75, 1972.
72. Delincée, H. and Pool-Zobel, B.L. Genotoxic properties of 2-dodecylcyclobutanone, a compound formed on irradiation of food containing fat., *Radiat. Phys. Chem.*, 52, 39, 1998.
73. Raul, F. et al. Food-borne radiolytic compounds (2-alkylcyclobutanones) may promote experimental colon carcinogenesis, *Nutr. Cancer*, 44, 189, 2002.
74. Marchioni, E. et al. Toxicological study on 2-alkycyclobutanones-results of a collaborative study, *Radiat. Phys. Chem.*, 71, 145, 2004.
75. Sommers, C.H. 2-Dodecylcyclobutanone does not induce mutations in the *Escherichia coli* tryptophan reverse mutation assay, *J. Agric. Food Chem.*, 51, 6367, 2003.
76. Gadgil, P. and Smith, J.S. Mutagenicity and acute toxicity evaluation of 2-dodecylcyclobutanone, *J. Food Sci.*, 69, C713, 2004.
77. Sommers, C.H. and Schiestl, R. 2-Dodecylcyclobutanone does not induce mutations in the Salmonella mutagenicity test or intrachromosomal recombination in *Saccharomyces cerevisiae*, *J. Food Prot.*, 67, 1293, 2004.
78. Stevenson, M.H., Crone, A.V.J., and Hamilton, J.T.G. Irradiation detection, *Nature*, 344, 202, 1990.
79. Boyd, D.R. et al. Synthesis, characterisation and potential use of 2-dodecylcyclobutanone as a marker for irradiated chicken, *J. Agric. Food Chem.*, 39, 789, 1991.
80. Stevenson, M.H. et al. The use of 2-alkylcyclobutanones for the identification of irradiated chicken meat and eggs, *Radiat. Phys. Chem.*, 42, 363, 1993.
81. Horvatovich, P. et al. Supercritical fluid extraction of hydrocarbons and 2-alkylcyclobutanones for the detection of irradiated foodstuffs, *J. Chromatogr. A*, 897, 259, 2000.
82. Obana, H., Furuta, M., and Tanaka, Y. Analysis of 2-alkylcyclobutanones with accelerated solvent extraction to detect irradiated meat and fish, *J. Agric. Food Chem.*, 53, 6603, 2005.
83. Stevenson, M.H. Identification of irradiated food, *Food Technol.*, 48, 141, 1994.
84. Gadgil, P. et al. Evaluation of 2-dodecylcyclobutanone as an irradiation dose indicator in fresh irradiated ground beef, *J. Agric. Food Chem.*, 53, 1890, 2005.
85. Zanaedi, E. et al. Evaluation of 2-alkylcyclobutanones in irradiated cured pork products during vacuum-packed storage, *J. Agric. Food Chem.*, 55, 4246, 2007.

86. Stewart, E.M. et al. 2-Alkylcyclobutanones as a markers for the detection of irradiated mango, papaya, Camembert cheese and salmon meat, *J. Sci. Food Agric.*, 80, 121, 2000.

87. Sin, D.W.M., Wong, Y.C., and Yao, M.W.Y. Analysis of γ-irradiated melon, pumpkin and sunflower seeds by electron paramagnetic resonance spectroscopy and gas chromatography-mass spectrometry, *J. Agric. Food Chem.*, 54, 7159, 2006.

88. Tewik, I.H., Ismail, H.M., and Sumar, S. A rapid supercritical fluid extraction method for the qualitative detection of 2-alkylcyclobutanone in gamma-irradiated fresh and sea water fish, *Int. J. Food Sci. Nutr.*, 50, 51, 1999.

89. Kim, K.S. et al. Analysis of radiation-induced hydrocarbons and 2-alkylcyclobutanones from dried shrimps (*Penaeus aztecus*), *J. Food Prot.*, 67, 142, 2004.

90. Stevenson, M.H., Meier, W., and Kilpatrick, D.J. *A European Collaborative Blind Trial using Volatile Hydrocarbons and 2-Dodecylcyclobutanone to Detect Irradiated Chicken Meat*, report EUR 15969 EN, Commission of the European communities (BCR), Brussels, Belgium, 1994.

91. Elliott, C.T. et al. Detection of irradiated chicken meat by analysis of lipid extracts for 2-substituted cyclobutanones using an enzyme-linked immunosorbent assay, *Analyst*, 120, 2337, 1995.

92. Rahman, R. et al. A rapid method (SFE-TLC) for the identification of irradiated chicken, *Food Res. Int.*, 29, 301, 1996.

93. Sin, D.W.M., Wong, Y.C., and Yao, M.W.Y. Application of pentafluorophenyl hydrazine on detecting 2-dodecylcyclobutanone and 2-tetradecylcyclobutanone in irradiated chicken, pork and mangoes by gas chromatography-mass spectrometry, *Eur. Food Res. Technol.*, 222, 674, 2006.

94. Nawar, W.W. and Balboni, J.J. Detection of irradiation treatment in foods, *J. AOAC Int.*, 53, 726, 1970.

95. Newar, W.W. Volatiles from food irradiation, *Food Rev. Int.*, 21, 45, 1986.

96. Schreiber, G.A. et al. Evaluation of a gas chromatographic method to identify irradiated chicken meat, pork and beef by detection of volatile hydrocarbons, *J. AOAC Int.*, 77, 1201, 1994.

97. Villavicencio, A.C.H. et al. Formation of hydrocarbons in irradiated Brazilian beans: Gas chromatographic analysis to detect radiation processing, *J. Agric. Food Chem.*, 45, 4215, 1997.

98. Hwang, K.T. et al. Detection of alkanes and alkenes for identifying irradiated cereals, *J. Am. Oil Chem. Soc.*, 78, 1145, 2001.

99. Bhattacharjee, P. et al. Hydrocarbons as marker compounds for irradiated cashew nuts, *Food Chem.*, 80, 151, 2003.

100. Hwang, K.T. et al. Hydrocarbons detected in irradiated and heat-treated eggs, *Food Res. Int.*, 34, 321, 2001.

101. Hwang, K.T. Hydrocarbons detected in irradiated pork, bacon and ham, *Food Res. Int.*, 32, 389, 1999.

102. Merino, L. and Cerda, H. Control of imported irradiated frozen meat and poultry using the hydrocarbon method and the DNA comet assay, *Eur. Food. Res. Technol.*, 211, 298, 2000.

103. Morehouse, K.M. and Ku, Y. Gas chromatographic and electron spin resonance investigations of gamma-irradiated shrimp, *J. Agric. Food Chem.*, 40, 1963, 1992.

104. Kwon, J.H. et al. The identification of irradiated seasoned filefish (*Thamnaconus modestus*) by different analytical methods, *Radiat. Phys. Chem.*, 76, 1833, 2007.

105. Miyahara, M. et al. Detection of hydrocarbons in irradiated foods, *J. Health Sci.*, 49, 179, 2003.

106. Hwang, K.T. et al. Effects of roasting, powdering and storing irradiated soybeans on hydrocarbon detection for identifying post-irradiation of soybeans, *Food Chem.*, 102, 263, 2007.

107. Hartmann, M., Ammon, J., and Berg, H. Determination of radiation-induced hydrocarbons in processed food and complex lipid matrices, *Z. Lebensm. Unters. Forsch.* A, 204, 231, 1997.

108. Schulzki, G. et al. Detection of radiation-induced hydrocarbons in irradiated fish and prawns by means of on-line coupled liquid chromatography-gas chromatography, *J. Agric. Food Chem.*, 45, 3921, 1997.

109. Kim, H. et al. Identification of radiolytic marker compounds in irradiated beef extract powder by volatile analysis, *Microchem. J.*, 80, 127, 2005.

110. Solar, S. Reaction of OH with phenylalanine in neutral aqueous solution, *Radiat. Phys. Chem.*, 26, 103, 1985.

111. Karam, L.R. and Simic, M.G. detecting irradiated foods: Use of hydroxyl radical biomarkers, *Anal. Chem.*, 60, 1117A, 1988.

112. Chauqui-Offermanns, N. and McDougall, T. An HPLC method to determine *o*-tyrosine in chicken meat, *J. Agric. Food Chem.*, 39, 300, 1991.

113. Miyahara, M. et al. Identification of boned chicken by determination of *o*-tyrosine and electron spin resonance spectrometry, *J. Health Sci.*, 48, 79, 2002.

114. Hart, R.J., White, J.A., and Reid, W.J. Occurrence of *o*-tyrosine in nonirradiated foods, *Int. J. Food Sci. Technol.*, 23, 643, 1988.

115. Hein, W.G., Simat, T.J., and Steinhart, H. Determination of non-protein bound *o*-tyrosine as a marker for the detection of irradiated shrimps, *Eur. Food Res. Technol.*, 210, 299, 2000.

116. Krach, C., Sontag, G., and Solar, S. Determination of *o*- and *m*-tyrosine for identification of irradiated shrimps, *Z. Lebensm. Unters. Forsch.* A, 204, 417, 1997.

117. Dohmaru, T. et al. Identification of irradiated pepper with the level of hydrogen gas as a probe, *Radiat. Res.*, 120, 552, 1989.

118. Hitchcock, C.H. Determination of hydrogen in irradiated frozen chicken, *J. Sci. Food Agric.*, 68, 319, 1995.

119. Furuta, M. et al. Detention of irradiated frozen meat and poultry using carbon monoxide gas as a probe, *J. Agric. Food Chem.*, 40, 1099, 1992.

120. Furuta, M. et al. Retention of radiolytic CO gas in irradiated pepper grains and irradiation detection of spices and dry grains with the level of stocked CO gas, *J. Agric. Food Chem.*, 43, 2130, 1995.

121. Furuta, M. et al. Detection of irradiated frozen deboned seafood with the level of radiolytic H_2 and CO gases as a probe, *J. Agric. Food Chem.*, 45, 3928, 1997.

122. Östling, O. and Johanson, K.J. Microelectrophoretic study of radiation-induced DNA damages in individual mammalian cells, *Biochem. Biophys. Res. Comm.*, 123, 291, 1984.

123. Fairbairn, D.W., Olive, P.L., and O'Neil, K.L. The comet assay: A comprehensive review. *Mutation Res.*, 339, 37, 1995.

124. McKelvey-Martin, V.J. et al. The single cell gel electrophoresis assay (comet assay): A European review, *Mutation Res.*, 288, 47, 1993.

125. Schindewolf, C. et al. Comet assay as a tool to screen for mouse models with inherited radiation sensitivity, *Mammal. Genome*, 11, 552, 2000.

126. Gedik. C.M. et al. Oxidative stress in humans: Validation of biomarkers of DNA damage, *Carcinogenesis*, 23, 1441, 2002.

127. Hamoutene, D. et al. Use of the comet assay to assess DNA damage in hemocytes and digestive gland cells of mussels and clams exposed to water contaminated with petroleum hydrocarbons, *Mar. Environ. Res.*, 54, 471, 2002.

128. Lee, R. and Kim, G.B. Comet assays to assess DNA damage and repair in grass shrimp embryos exposed to phototoxicans, *Mar. Environ. Res.*, 54, 465, 2002.

129. Taban, I.C. et al. Detection of DNA damage in mussels and sea urchins exposed to crude oil using comet assay, *Mar. Environ. Res.*, 58, 701, 2004.

130. Wada, S. et al. Detection of DNA damage induced by heavy ion irradiation in the individual cells with comet assay, *Nucl. Instr. Meth. Phys. Res. B*, 206, 553, 2003.

131. Collins, A.R. The comet assay for DNA damage and repair, *Mol. Biotech.*, 26, 249, 2006.

132. Cerda, H. Analysis of DNA in fresh meat, poultry and fish. Possibility of identifying irradiated food. Proceedings of a workshop, Strasbourg, Commission of the European Communities, EUR 15012, pp. 5–6, 1993.

133. Nilson, H. and Cerda, H. Analys av livsmedelverksamhetsrapport under peridon 1989/90–1992/93. SLV rapport 17/1993, National Food Administration, Uppsala, Sweden, 1993.

134. Cerda, H. and Koppen, G. DNA degradation in chilled fresh chicken studied with the neutral comet assay, *Z. Lebensm. Unters. Forcsh.* A, 207, 22, 1998.

135. Cerda, H. et al. The DNA "comet assay" as a rapid screening technique to control irradiated food, *Mutation Res.*, 375, 167, 1997.
136. Cerda, H. Detection of irradiated frozen food with the DNA comet assay: Interlaboratory test, *J. Sci. Food Agric.*, 76, 435, 1998.
137. Delincée, H. Detection of irradiated food using simple screening methods, *Food Sci. Technol. Today*, 8, 109, 1994.
138. Delincée, H. Rapid and simple screening tests to detect the radiation of food, *Radiat. Phys. Chem.*, 46, 667, 1995.
139. Araújo, M.M. et al. Identification of irradiated refrigerated pork with the DNA comet assay, *Radiat. Phys. Chem.*, 71, 183, 2004.
140. Villavicencio, A.L.C.H. et al. Identification of irradiated refrigerated poultry with the DNA comet assay, *Radiat. Phys. Chem.*, 71, 187, 2004.
141. Marin-Huachaca, N. et al. Use of the DNA comet assay to detect beef meat treated by ionizing irradiation, *Meat Sci.*, 71, 446, 2005.
142. Delincée, H. Rapid detection of irradiated frozen hamburgers, *Radiat. Phys. Chem.*, 63, 443, 2002.
143. Marin-Huachaca, N. et al. Identification of gamma-irradiated papaya, melon and watermelon, *Radiat. Phys. Chem.*, 71, 191, 2004.
144. Khan, A.A., Khan, H.M., and Delincée, H. "DNA comet assay" – a validity assessment for the identification of radiation treatment of meats and seafood, *Eur. Food Res. Technol.*, 216, 88, 2003.
145. Marchioni, E. et al. Alterations in mitochondrial DNA: A technique for the detection of beef, in *Potential New Methods of detection of Irradiated Food, BCR Information, (Chemical Analysis)*, Report EUR 13331 EN, Commission of the European Communities, Brussels, Belgium, 1991, pp. 11–25.
146. Marchioni, E. et al. Eds., McMurrays, C.H., Stewart E.M., Gray, R., and Pearce, J. Detection of irradiated fresh, chilled and frozen foods by the mitochondrial DNA method, in *Detection Methods for Irradiated Foods – Current Status*, Royal Society of Chemistry, Cambridge, 1996, pp. 355–366.
147. Deeble, D.J. et al. Changes in DNA as a possible means of detecting irradiated food, in *Food Irradiation and the Chemist*, Royal Society of Chemistry, Cambridge, 1990, pp. 57–79.
148. Deeble, D.J. et al. Detection of irradiated food based on DNA bases changes, *Food Sci. Technol. Today*, 8, 96, 1994.
149. Tyreman, A.L. et al. Detection of food irradiation by ELISA, *Food Sci. Technol. Today*, 12, 108, 1998.
150. Tyreman, A.L. et al. Detection of irradiated food by immunoassay-development and optimization of an ELISA for dihydrothymidine in irradiated prawns, *J. Food Sci. Technol.*, 39, 533, 2004.
151. Aitken, M.J. Recent Advances in thermoluminescence dating, *Radiat. Prot. Dosimetry*, 6, 181, 1983.
152. Sanderson, D.C.W., Slater, C., and Cairns, K.J. Detection of irradiated food, *Nature*, 340, 23, 1989.
153. Kwon, J.H. et al. Inter-country transportation of irradiated dried Korean fish to prove its quality and identity, *Radiat. Phys. Chem.*, 71, 81, 2004.
154. Autio, T. and Pinnioja, S. Identification of irradiated foods by the thermoluminescene of mineral contamination, *Z. Lebensm. Unters. Forsh.*, 191, 177, 1990.
155. Autio, T. and Pinnioja, S. Eds., Leonardi, M., Raffi, J.J., and Belliardo, J. Identification of irradiated foods by the thermoluminescene of contaminating minerals, in *Recent Advances on Detection of Irradiated Food*, EUR-14315: Commission of the European Communities, Luxembourg, 1992, pp. 177–183.
156. Sanderson, D.C.W., Carmicheal, L.A., and Naylor, J.D. Recent advances in thermoluminescence and photostimulated luminescence detection methods for irradiated foods, in *Detection Methods for Irradiated Foods – Current Status*, McMurray, C.H., Stewart E.M., Gray, R., and Pearce, J. (Eds.), Royal Society of Chemistry, Cambridge, London, 1996, pp. 124–138.
157. Sanderson, D.C.W. et al. Luminescence detection of shellfish, in *Detection Methods for Irradiated Foods – Current Status*, McMurray, C.H., Stewart E.M., Gray, R., and Pearce, J. (Eds.), Royal Society of Chemistry, Cambridge, London, 1996, pp. 139–148.
158. Schreiber, G.A. et al. Methods for routine control of irradiated food: Determination of the irradiation status of shellfish by thermoluminescence analysis, *Radiat. Phys. Chem.*, 43, 533, 1994.

159. Pinnioja, S. and Pajo, L. Thermoluminescence of minerals useful for identification of irradiated seafood, *Radiat. Phys. Chem.*, 46, 753, 1995.

160. Carmichael, L.A. and Sanderson, D.C.W. The use of acid hydrolysis for extracting minerals from shellfish for thermoluminescence detection of irradiation, *Food Chem.*, 68, 233, 2000.

161. http://www.gla.ac.uk/surrc/luminescene/psl.html.

162. Sanderson, D.C.W., Carmichael, L.A., and Naylor J.D. Photostimulated luminescence and thermoluminescene techniques for the detection of irradiated food, *Food Sci. Technol. Today*, 9, 150, 1995.

163. Sanderson, D.C.W., Carmichael, L.A., and Fisk, S. Establishing luminescence methods to detect irradiated food, *Food Sci. Technol. Today*, 12, 97, 1998.

164. Sanderson, D.C.W., Carmichael, L.A., and Fisk, S. Photostimulated luminescence detection of irradiated shellfish: International interlaboratory trial, *J. AOAC Int.*, 86, 983, 2003.

165. Sanderson, D.C.W., Carmichael, L.A., and Fisk, S. Photostimulated luminescence detection of irradiated herbs, spices, and seasonings: International interlaboratory trial, *J. AOAC Int.*, 86, 990, 2003.

166. Pettipher, G.I. et al. Rapid membrane filtration-epiflourescence microscopy technique for the direct enumeration of bacteria in raw milk, *Appl. Environ. Microbiol.*, 39, 423, 1980.

167. Jones, K. et al. The DEFT/APC screening method for the detection of irradiated, frozen stored foods: A collaborative trial, *Food Sci. Technol. Today*, 10, 175, 1996.

168. Wirtanen, G. et al. Microbiological screening method for indication of irradiation of spices and herbs: A BCR collaborative study, *J. AOAC Int.*, 76, 674, 1993.

169. Oh, K.N. et al. Screening of gamma irradiated spices in Korea by using a microbiological method (DEFT/APC), *Food Chem.*, 14, 489, 2003.

170. Scotter, S.L., Beardwood, K., and Wood, R. *Limulus* ameobocyte lysate test/gram negative bacteria count method for the detection of irradiated poultry: Results of two interlaboratory studies, *Food Sci. Technol. Today*, 8, 106, 1994.

171. Scotter, S.L., Beardwood, K., and Wood, R. Detection of irradiation treatment of poultry meat using the *Limulus* amoebocyte lysate test in conjunction with a Gram negative bacteria count, *J. Assoc. Publ. Anal.*, 31, 163, 1995.

172. Guatam, S., Sharma, A., and Thomas, P. Improved bacterial turbidimetric method for detection of irradiated spices, *J. Agric. Food Chem.*, 46, 5110, 1998.

173. Ehlermann, D.A.E. Four decades in food irradiation, *Radiat. Phys. Chem.*, 73, 346, 2005.

Chapter 24

Veterinary Drugs

Anton Kaufmann

Contents

24.1 Introduction.. 687
24.2 Veterinary Drugs Used in Aquaculture 688
24.3 Origin of Veterinary Drug Residues.. 690
24.4 Analytical Techniques Used for the Analysis of Veterinary Drugs...........691
    24.4.1 Biological Test Systems..691
    24.4.2 Instrumental Analytical Approaches 691
24.5 Veterinary Drugs Used in Aquaculture 695
    24.5.1 Bactericidal Drugs (Antibiotics) 695
    24.5.2 Antimycotica 701
    24.5.3 Tranquilizers 701
    24.5.4 Antiparasitica 702
    24.5.5 Anthelmintic Drugs 702
    24.5.6 Hormones...702
24.6 Future Developments 704
References 705

24.1 Introduction

The term antibiotics was originally reserved for substances, produced by fungi or bacteria, which are capable to kill microorganisms. Semisynthetic or synthetic substances lethal to microorganisms are often included under the term antibiotics. The following agents cannot be considered antibiotics: Anthelmintica (active against parasites like nematodes), antimycotica (against moulds and fungi), and hormones (used to control the sex ratio of males to females of fishes). However, all these substances can be conveniently grouped under the term "veterinary drugs."

24.2 Veterinary Drugs Used in Aquaculture

Aquaculture is a multibillion industry with a production value of more than 70 billion US$ in 2004 [1]. Sixty-nine percent of this production volume is located in China and 22% in the rest of Asia-Pacific [1]. These figures might be misleading, since a country like Ecuador exports almost its entire shrimp production, while only a fraction of the aquaculture output of China is being exported. Still, Asia is strongly dominating the aquaculture industry. This relates not only to the production volume but also to the technology, which is increasingly developed in Asia and then spreads to other production regions. Therefore, it is hardly possible to overemphasize the importance of Asia. From the Western perspective, aquaculture in Asia is still associated with rice fields stocked with fish, which fertilize the rice plants and provide valuable food proteins to the farmers at the same time. Although, such a symbiotic form of aquaculture still exists, it is definitely not responsible for the enormous growth of cultured fish in Asia. Aquaculture is a veritable industry, requiring large investments and enormous know-how to be successful. This is the major reason why aquaculture is hardly practiced in Africa [1]. Tilapia, which came originally from the great African lakes are produced in much larger quantities in the Philippines and Indonesia than in sub-Saharan Africa. This has mostly to do with the lack of infrastructure in sub-Saharan Africa (transportation, reliable electricity supply, skilled manpower, etc.). Many fish farms have more similarities with an industrial site than with a traditional fishpond. There are a number of reasons for this change. First, traditional fishponds permit only a low production volume. Second, lack of oxygen, remaining feeds, and produced excrements, are factors, which limits the number of fishes per volume of water. Third, traditional fishponds are difficult to sterilize as required after an outbreak of a disease. They might be infested by snails, which can be the host of many parasites and diseases that infect fishes and even humans. Using concrete basins is more expensive, but superior in many ways. It is easier to transfer fishes from one basin to another, and they can be more easily cleaned and sterilized if necessary. Maintenance of essential parameters like oxygen concentration, pH, ammonia levels, temperature, and the like can be controlled with less effort, permitting higher fish densities and faster growth rates. Still aquaculture can be a high-risk business. The outbreak of disease has led to huge production losses and even to the collapse of the industry in certain regions [1,2]. Fishes and shrimps are susceptible to a number of diseases caused by bacteria, virus, or fungus. Furthermore, they are prey to many parasites like protozoan, trematodes, and helminths. The infection pressure caused by such bacteria, virus, or parasites might be present for a long time without causing any harm to the fish or shrimps. However, stress factors, such as low oxygen concentration, changes of pH or temperature, might weaken cultured fishes and shrimps and lead to widespread infection with possible lethal consequences. The likelihood of infections increases when production shifts from extensive to semi-intensive and intensive farming. High fish density deprives the fish of space that can lead to lesion. Even minor injuries of the skin or scales can weaken its physical barriers, permitting an infection or penetration of parasites. Fish infected by viral diseases like viral hemorrhagic septicemia (VHS) develop hemorrhages and bleeding gills, making them very susceptible to secondary bacterial infections. New disease or new strains of known disease have regularly appeared in the past and are likely to be a threat in the future [3]. Diseases have spread in the past quickly from one country to another, and have even crossed continents. One reason for such a spread is the increasing specialization within the aquaculture.

Broodstock and postlarvae are most often purchased outside the farm. In some poorer countries like Bangladesh [1] it might be obtained wild-caught. This practice carries the inherent risk of introducing infected juveniles into the farm. A number of species cannot yet be reproduced without using wild-caught broodstock. An example is the black Tiger Prawn (*Penaeus monodon*), cultures of which rely on wild-caught breeders [1]. The risk associated with the introduction of such wild-caught breeders has caused a shift in many countries to the Pacific White Shrimp (*Penaeus vennamei*), which is commercially available as "specific pathogen-free" broodstock.

It is generally accepted that aquaculture diseases are preferably prevented by prophylaxis like proper sanitary conditions, sterilization of ponds, healthy juveniles, immunization, and good aquaculture practices. Still the sudden emergence of diseases can have enormous financial consequences for the involved aquaculturist [4]. As a result, bacterial and parasitic infections are often treated with veterinary drugs.

While there is a significant number of veterinary drugs permitted for treating food-producing animals, such as cattle, swine, and fowl, there are only very few drugs allowed for the use in aquaculture. One major problem is the definition of withdrawal times (time between the end of a treatment and the clearance of the drug from the organism) for aquaculture species. The metabolism and consequently the elimination rate of a drug is dependent on the body temperature of an animal. While the body temperature of mammals and birds are a rather constant parameter, fishes and shrimps are cold blooded. Furthermore, aquaculture is not a traditional way of farming in Europe and the United States. This might be the reason why little pressure was exerted on regulatory agencies to register more veterinary drugs, permitted for the use in aquaculture.

In the past, antibiotics were often used in a way that cannot be called responsible. The focus was on quick results and less on sustainable aquaculture practices. This has led to the emergence of resistance among bacteria [2–4] which forced farmers to increase the concentration of the antibiotic or shift to another drug. Although some drugs are explicitly prohibited in certain drug-producing countries, they have been used because of weak legal enforcement.

Banned drugs like chloramphenicol and nitrofurans have found a widespread use in China and Southeast Asia. In the year 2001/2002, food safety laboratories in Europe discovered that a large percentage of fishes and shrimps from Asia contained significant residue concentrations of these drugs. Consequently, shipments from these countries to Europe were only released after the particular lot has been thoroughly analyzed by an accredited laboratory and declared free of nitrofurans and chloramphenicol. This had enormous consequences to many farmers in Asia, some of which utilized these drugs on a regular basis. The local authorities in the drug-producing countries could not adequately respond because of a lack of technology and trained manpower. Some drug-producing countries had virtually no laboratories to control veterinary drug residues. There were cases where analytical methods designed for the assay of medicated food were used for the determination of veterinary drug residues in fish or shrimps. Since the concentration of veterinary drugs in medicated feed is rather in the low percent range, all animals tested with these insensitive methods showed "no residues." This situation has dramatically changed in many countries where highly sophisticated analytical instruments are now operated by skilled chemists. There was indeed progress and shipments contaminated with chloramphenicol or nitrofuran significantly dropped. However, some 2 years later a new problem arose. Enrofloxacin, a veterinary drug belonging to the group of chinolons was found in many shipments. After removing a number of shipments from the market, "residue-free" products were again rather the rule and not the exception. Soon another weave of contaminated Asian seafood was discovered. It was the malachite green, an antimycotica, which has also a long history of usage in European aquaculture. This development is shown in Table 24.1 that lists the percentage of samples with detectable residues of

Table 24.1 Percentage of Samples Tested Positive for Veterinary Drug Residues by the Official Food Authority of the Canton of Zurich between the Year 2003 and 2006

Product	Year				
	2002	*2003*	*2004*	*2005*	*2006*
Nitrofurans					
Fish Vietnam	62% (*n*=8)	13% (*n*=53)	0% (*n*=2)	0% (*n*=69)	0% (*n*=29)
Shrimps Vietnam	0% (*n*=3)	3% (*n*=38)	—	—	0% (*n*=13)
Shrimps Thailand	8% (*n*=12)	0% (*n*=18)	—	—	0% (*n*=2)
Enrofloxacine					
Fish Vietnam	—	—	0% (*n*=2)	38% (*n*=69)	10% (*n*=29)
Shrimps Vietnam	—	—	—	—	0% (*n*=13)
Shrimps Thailand	—	—	—	—	0% (*n*=2)
Malachite green					
Fish Vietnam	—	—	—	48% (*n*=69)	0% (*n*=29)

Note: The figures indicate the percentage of positive samples for a given drug within the specified product group and year. The number in brackets refers to the number of tested samples.

nitrofurans, enrofloxacin, and malachite green over a period of 5 years. All samples were obtained from the local market in Switzerland; measurements were made at the official food authority of Zurich. Only data from two export countries are listed. This reflects their importance as exporter of seafood to Switzerland as well as the history of previous positive findings. Besides these widespread antibiotics, one case with a very high concentration of sulfadimidine (1140 µg/kg) and one sample with penicillin residues (Dicloxacilline) was discovered.

24.3 Origin of Veterinary Drug Residues

It seems to be logical that veterinary drug residues are the unequivocal proof that such substances have been deliberately used to prevent or treat a disease of the particular aquatic animals. However, there were cases where this "axiom" was violated. The increasing sensitivity of analytical methods is capable in detecting minute amounts of drug concentrations, which can be possibly caused by pre- or postharvest contamination.

Some veterinary drugs have extremely long depletion times. Malachite green (respectively its metabolite leuco-malachite green) was reported to deplete in European eel within 100 days from initial 700 to 15 µg/kg [5]. Such long depletion periods bear the risk that juvenile fish treated with malachite green can still contain residues after being harvested. If such fishes are purchased from an external hatchery, the aquaculturist who raises the juvenile fish might not even be aware of the presence of residues.

Low levels of antibiotics were observed in commercially available feed [6]. Feed mills produce a variety of unmedicated and medicated products. Some antibiotics are permitted for certain animals, which are not used for food production. In such mills, carryover can occur if the equipment is not carefully cleaned after the production of a medicated feed batch.

There have been reports about the contamination of crabs during processing. Workers employed in processing facilities in Southeast Asia were reported to use commercially available hand creams containing chloramphenicol [6].

Analytical chemistry has made great progress concerning sensitivity and selectivity. This has strongly reduced the likelihood of false positive findings. Problems can arise when a veterinary drug undergoes fast and extensive metabolism in the animal. Some of such drugs, e.g., nitrofurans produce relatively small breakdown products which are used as indicator that the parent substance has been administered. Semicarbazide is the indicator (metabolite) of nitrofurazone. This small molecule was found to originate also from other treatments, ingredients, packing material, and natural sources. The details will be given in the section "Nitrofurans." The analysis of metabolites instead of the parent drug is not necessarily an unreliable way of detection. Problems only arise if metabolites that contain less diagnostic information than the parent drug are produced, e.g., as caused by breaking the molecular structure and a corresponding loss of structural information.

24.4 Analytical Techniques Used for the Analysis of Veterinary Drugs

24.4.1 Biological Test Systems

The presence of an antibacterial drug inhibits the growth of microorganisms and can be utilized for their detection [7]. Some tests based on this principle are commercially available, while others require the user to prepare growth media. Generally, a drop of meat extract is put on a growth media plate, which undergoes an incubation period later. Antibacterial drugs will cause an inhibition zone around the position of the drop. The diameter of which, correlates to the microbial inhibition activity of the sample. Such tests are not very selective; however, the use of growth media with different response toward various inhibitors can give information about the type of inhibitor (class of antibiotics) present. The mentioned tests can potentially detect any antibacterial compound, however, they can respond with widely varying sensitivity toward different antibiotic groups. Good sensitivity was reported for penicillins, yet sulfonamides or nitrofurans produce weak or even no inhibition. Inhibition tests are commonly used for large-scale screening. Higher sensitivity and selectivity are obtained by immunoassay test systems [8]. Immunoassay methods like enzyme-linked immunosorbent assays (ELISA) permit the fast and highly selective screening and semiquantification of various veterinary drugs. The technique is well suited for processing large number of samples.

The high selectivity caused by the recognition of an epitope of the analyzed drug is both an advantage and a limitation. Although there are reports about the development of drug class specific techniques [9], ELISA is still recognized as a "one drug one test" approach.

24.4.2 Instrumental Analytical Approaches

The early days of veterinary drug residue analysis has been dominated by microbiological tests. Chemical methods were slower, less sensitive, and more cumbersome. A low concentration of residues has to be detected in the presence of an overwhelming number and concentration of endogenous compounds, calling for some form of sample cleanup and chromatographic or electrophoresis separation prior to the detection. Unlike pesticides, most veterinary drugs are rather large and polar molecules, which in many cases prevent their separation by gas chromatography (GC). Liquid chromatography (LC) is definitely more suited for the analysis of veterinary drugs than GC. An exception remains, the area of hormone analysis, which is still dominated by GC, coupled to mass spectrometry (GC–MS).

There has to be an extraction and cleanup step prior to any separation and quantification. Depending on the group of veterinary drug, different extraction procedures are employed. They might require a polar, nonpolar, highly buffered, or strongly acid environment. This variability creates problems for designing multiresidue methods. Matrix bound drugs such as aminoglycosides require low pH for quantitative extraction. Such an environment will induce degradation of many other drugs such as penicillins and tetracyclines. The proposed alternatives to liquid extraction are matrix solid phase dispersion extraction (MSPD) and accelerated solvent extraction (ASE). MSPD is a manual, rather labor-intensive approach, while ASE is limited to thermostabile analytes. The desorption of incurred drugs from the matrix is not always easy, because some drugs are bound to certain cell compartments. Quantitative or near-quantitative extraction can require one or two re-extraction steps. Drugs from fishes and shrimps are more easily extracted than from matrices such as mammalian kidney or liver. However, shrimp extracts were often found to produce excessive foam or viscous emulsion that can create difficulties in the following cleanup steps. The extract (after centrifugation) can seldom be directly injected into an analytical instrument. Depending on the selectivity of the used analytical technique and the sensitivity requirements, cleanup, and concentration steps are required. Modern methods of veterinary drug cleanup rely heavily on solid phase extraction (SPE). Often used are reversed phase techniques such as C-18 cartridges or more modern polymeric materials such as OASIS. Such novel materials do not suffer from irreproducible silanol activities, causing nonspecific losses of basic analytes. This approach of cleanup is not orthogonal if a reversed phase LC separation follows. However, a reversed phase SPE step removes peptides and proteins, which would otherwise precipitate on the analytical column, reducing separation performance, and cause signal suppression if LC–MS is used. Acid and basic drugs are often purified by the use of an anion or cation exchanger cartridges. Such a cleanup can be highly selective, producing clean extracts. However, it is very difficult to optimize an ion-exchanger cleanup to recover many different drugs with satisfactory recovery rates. Some compounds will only be insufficiently retained, while others might not be quantitatively eluded. The underlying problem can be traced to the different pK values of the analytes, each of which requires a different pH value for best retention and elution. Hence, multimethods covering more than one drug group utilize more often the less selective reversed phase SPE cartridges.

Assay of pharmaceutical formulations are often performed with LC and ultraviolet detection (LC–UV). Such a configuration, however, provides only insufficient selectivity and sensitivity for most trace analysis purposes. More widely used are fluorescence detectors (LC–FL). They permit higher sensitivity and selectivity. Unfortunately only a number of veterinary drugs possess an inherent fluorescence, e.g., chinolons. Most drugs require therefore a pre- or postcolumn derivatization before detection becomes feasible. Still LC–FL is not very satisfactory for some drugs such as aminoglycosides, nitrofurans, nitroimidazoles, and the like. The advent of LC coupled to MS (LC–MS) opened new frontiers for the analysis of veterinary drug residues. Among the many developed interfaces, only electrospray (ESI), atmospheric pressure chemical ionization (APCI), and photoionization are in use today. It was soon discovered that LC–MS is not as selective and sensitive as desired. Adducts produced by the mobile phase and endogenous compounds of equal nominal mass were recognized as relevant interferences. Hence, most of the veterinary drug analysis is performed by LC coupled to tandem mass spectrometry (LC–MS–MS). The mass selection of a precursor ion, the induced fragmentation and the selective monitoring of one or more derived product ions delivered a previously unknown level of selectivity and consequently sensitivity. Within the EU, procedures have been defined (identification points) to prevent the reporting of false positive findings. Sufficient identification points for a positive confirmation can be earned if a drug can be detected by monitoring two independent MS–MS transitions. The area ratio of these transitions and the retention time of the peaks have to correspond to the standard. Confirmations are also possible by other techniques than MS or MS–MS, however, it might require the use of several different methods (ELISA, HPLC–FL, etc.) to earn the required confidence level (identification points). To compare selectivity and sensitivity of various LC technologies, a blank fish sample was processed by a published method [10] and spiked to obtain a concentration of oxytetracycline and tetracycline corresponding to 100 µg/kg, which corresponds to the maximum residue level (MRL) of this drug. This sample was analyzed with different detection techniques. Figure 24.1 gives an impression about the selectivity provided by LC–UV and LC–FL after postcolumn derivatization [10]. Figure 24.2 compares a single ion resolution MS trace obtained from a quadrupole instrument versus a MS–MS

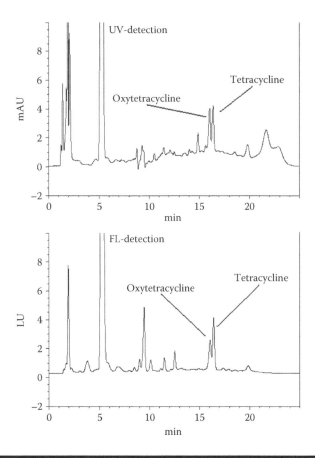

Figure 24.1 Chromatograms of a fish sample containing 100 µg/kg oxy- and tetracycline. Higher selectivity and more stable baseline are obtained by postcolumn derivatization and fluorescence (bottom) detection than by UV (top) detection.

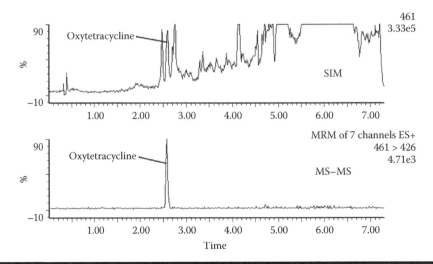

Figure 24.2 Chromatograms of a fish sample containing 100 μg/kg oxy- and tetracycline. Measurements were made by a quadrupole. Single stage quadrupole (top) give lower selectivity and sensitivity than triple quadrupole (bottom). The MS–MS transition depicted represents the MRM trace of oxytetracycline.

trace measured by the same instrument. Instead of using MS–MS as a tool for increasing selectivity, high resolution MS can produce similar results. Figure 24.3 depicts two reconstructed LC-TOF chromatograms.

The chromatogram at the top uses a unit mass resolution of 1 Da. This degree of resolution is typically provided by LC–MS (quadrupole). The bottom chromatogram shows the same time of flight (TOF) chromatogram where a narrower mass window (0.02 Da) was extracted.

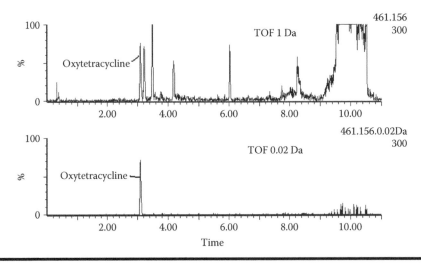

Figure 24.3 Chromatograms of a fish sample containing 100 μg/kg oxy- and tetracycline. Measurements were made by an UPLC-TOF instrument. A 1 Da mass extraction window (corresponding to single stage quadrupole resolution) is shown at the top. The selectivity provided by TOF is shown by the chromatogram depicted at the bottom, where a narrow mass extraction window of 0.02 Da is shown.

Both chromatograms use the same scale. Note signal intensity is not affected, while a significant reduction of endogenous interferences and noise are observed.

LC–MS–MS is a quantitative technique although in many cases it requires sophisticated calibration procedures to avoid or compensate ion source related matrix effects caused by endogenous compounds present in the sample. Ion trap instruments permitting MS–MS experiments have been used less frequently than tandem mass spectrometers, probably due to the larger standard deviation of measured mass peaks and the relatively long dwell times. The older instruments suffered from limited trap capacities (space charging). This might be less an issue with modern instruments. The recent introduction of linear traps has probably opened new opportunities. The more classical sector instruments are considered too bulky, slow, and expensive for most veterinary drug analysis.

Multiresidue methods for veterinary drugs are not yet as widespread as in the field of the pesticide analysis. This is related to the widely varying chemical and physical properties of the different veterinary drugs, as indicated by the chemical structures given in Section 24.5.1. Veterinary drugs cover the whole polarity spectra, making it difficult to ensure a complete extraction from the tissue and separation in one chromatographic run. Some analytes lack functional groups required for a sensitive and selective detection. Others, such as nitrofurans, are covalently bound to tissues and require a liberation step. Still there were intensive efforts to include not only several members of one drug class, but also several different drug classes in one analytical method. Multimethods capable to analyze more than one class of veterinary drugs are almost exclusively based on LC–MS(/MS), specifically by triple quadrupole instruments. A promising approach is the use of emerging high-resolution techniques like TOF or possible Orbitrap. High resolution improves the selectivity as compared to a unit resolution single stage quadrupole. An important advantage of the mentioned high-resolution technique is the provided full scan signal. This permits the postanalysis extraction of any desired mass trace. Triple quadrupole rely on the multireaction mode (MRM) which requires the preanalysis definition of analyte specific transitions (precursor ion, product ion mass, and appropriate collision energy). In the case of multimethods, the setting up of dozens of transitions reduces sensitivity because of the reduced dwell time. This problem can be reduced by defining time programmed MRM acquisition windows. However, drifting analyte peak retention times can complicate such an approach. Readjustments of time windows are required, if one or more peaks move out of a predefined retention time window.

24.5 Veterinary Drugs Used in Aquaculture

24.5.1 Bactericidal Drugs (Antibiotics)

Sulfonamides represent a group of antibacterial drugs that are exclusively produced in a synthetical manner (Figure 24.4).

They are antibacterial and antiprotozoal and show activity against a broad-spectrum of Grampositives and Gram-negatives. These substances have been in use for more than 50 years, which is probably the reason for widespread resistance of animal pathogens to sulfonamides. However, due to their low price, sulfonamides are still widely in use. They are reported to be often administered in combination with "potentizers" such as ormetoprim and trimethoprim that enlarges their therapeutic range.

There are virtually hundreds of different sulfonamide drugs described in the literature, some of which are registered for veterinary use. Sulfonamides are rather stable molecules (pH and temperature) that facilitates their analysis. Significant concentrations of metabolites are not observed,

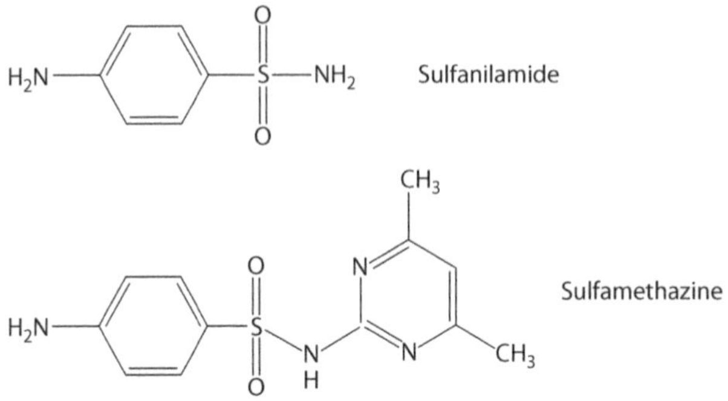

Figure 24.4 Sulfonamides.

unless in urine, where the *N*-acetyl metabolites dominate. There are a number of microbiological tests (ELISA, Charm, etc.) available, which cover one specific sulfonamide or are capable for the detection of generic sulfonamide structures [9]. Extraction with medium polar solvents is exhaustive and detection facilitated by the presence of functional groups, which even permits UV detection. The derivatization of the amino group (pre- or postcolumn) allows sensitive fluorescence detection. LC–MS–MS in the positive ESI mode permits quantification of sulfonamides [11] together with potentiators [12].

Tetracycline is an antibiotic produced by *Streptomyces*. Semisynthetic derivates such as chlortetracycline or doxycycline have enlarged this drug group. Tetracyclines are broad-spectrum drugs with activity against Gram-positives and Gram-negatives. Due to their activity and low price, tetracyclines have been widely used in aquaculture and animal husbandry. Consequently, resistance is a common problem (Figure 24.5).

Tetracyclines are significantly less stable than sulfonamides, they form epimeres as well as complexes with metals, which complicate quantification. There is an extensive set of substance specific microbiological tests available to detect tetracyclines at residue levels. Extraction of tetracyclines is based on acid aqueous buffers containing complex agents such as oxalic acid and EDTA. UV detection is often unsatisfactory, fluorescence after postcolumn derivatization improves specificity and sensitivity [13]. Again, LC–MS–MS (positive ESI mode) is currently the most often employed instrumental analytical technique.

Chinolones are a group of synthetically produced bactericidal drugs. They are active against a broad spectrum of animal pathogens, especially Gram-negatives. Chinolones are a relative

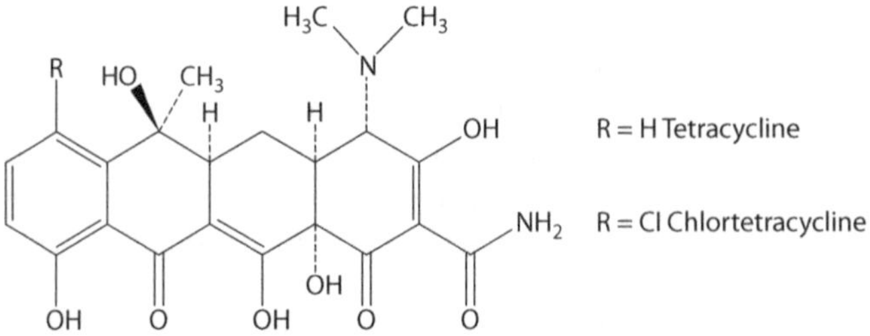

Figure 24.5 Tetracyclines.

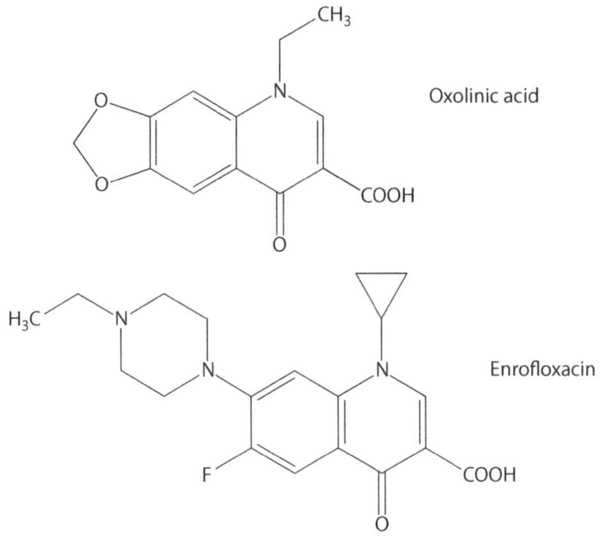

Figure 24.6 Chinolones.

new group of bactericidal drugs, many of which are reserved for the use of human treatment (Figure 24.6).

Residues of enrofloxacin and its metabolite ciprofloxacin were found in many shrimps and fish samples from Southeast Asia in the years 2005 and 2006.

Chinolones are relative stable molecules, most of which show intensive fluorescence. Hence, LC fluorescence detection is often employed while LC–MS–MS (positive ESI mode) is preferred for confirmation purposes [14].

Penicillin is a substance produced by the mould *Penicillium notatum* and *Penicillium chrysogenum*. It was the first discovered antibiotic and its activity extends against Gram-positives and Gram-negatives. Its long use has led to extensive resistance. A number of pathogens produce beta-lactamase that inhibits the action of penicillin. A number of semisynthetic penicillin derivates are available such as the aminopenicillins (amoxicillin, ampicillin) and the more modern cephalosporins. Cephalosporins are not affected by beta-lactamase. Several generations of these drugs are now available, most of which are reserved for human use only (Figure 24.7).

Penicillins are very labile molecules, easily degraded by enzymatic or chemical hydrolysis. Qualitative and semiquantitative detection can be achieved by a variety of microbiological test systems. Extraction is preferably done with polar solvents. Deproteinization achieved by adding acetonitrile or tungstate, e.g., is an important step of every analytical method. The amphoteric character of penicillins can cause severe asymmetrical LC peaks in insufficiently buffered mobile phases. Detection requires derivatization or LC–MS–MS [7].

Nitrofurans are a purely synthetic class of antibiotics. They show a wide spectrum of activities against many microorganism. Their uses have been banned or strongly limited due to their possible carcinogenic and mutagenic potential. While most nitrofurans were originally developed for the human use, one such as Furanace (Nifurpirinol) was designed as a chemotherapeutic for fish. This drug is not anymore permitted for the use of fish intended for human consumption, yet it is widely used in the ornamental fish business.

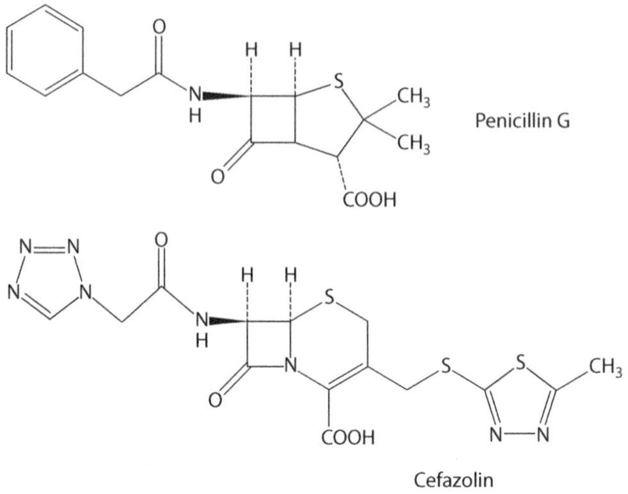

Figure 24.7 Penicillins and cephalosporines.

Nitrofurans fed to animals are quickly metabolized and escape analytical detection [6] if the parent drug is searched. This is probably the reason why these banned drugs have not been detected for a long time, although they were heavily used. Only the discovery that metabolites are covalently bound to tissue proteins permitted the liberation, derivatization, and consequently detection of nitrofuran residues [15]. The detectable metabolites, which are considered as markers, are void of the nitrofuran structure typical to the parent drug as shown in Figure 24.8.

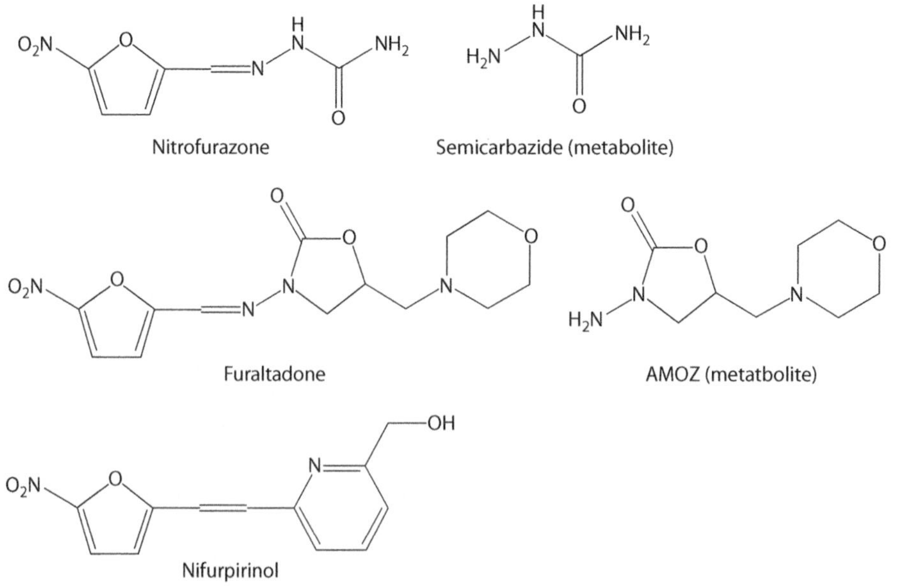

Figure 24.8 Nitrofurans including their metabolites.

These relatively small metabolites were initially not expected to be possible endogenous matrix components. This assumption turned out to be wrong in the case of nitrofurazone that produces semicarbazide as marker molecule. Semicarbazide was shown to be produced by the hypochlorite treatment of food [16]. It is a degradation product of azodicarbonamide used as flour improving agent and as blowing agent used for gaskets of certain food jars [16]. Furthermore, semicarbazide was reported to be an endogenous compound in Finnish crayfish [17].

All the published nitrofuran methods employ a very similar approach for the derivatization and quantification. Samples are hydrolyzed and derivatized by the use of hydrochloric acid and nitrobenzaldehyde. Processing includes liquid or SPE which is followed by LC–MS–MS separation and detection. The required performance limit of 1 μg/kg leaves almost only MS as detection technique. Quantification is preferably done by using isotopic labeled standards which are now readily commercially available for four different nitrofurans.

Chloramphenicol is a very effective antibiotic, which has a long history of use in human and animal treatment. The drug was reported to have very serious, life-threatening side effects. Although, this occurs very seldom, there is apparently no safe concentration of the drug. This led to the ban of this drug for human and veterinary use in most countries of the world. Florfenicol, a derivative of chloramphenicol was developed and registered for the use in aquaculture (Figure 24.9).

Chloramphenicol is a small, rather polar molecule. The major analytical challenge is posed by the low minimum required performance limit (MRPL) of 0.3 μg/kg, which has to be met by the utilized analytical method. Commercial ELISA tests are widely used for screening while confirmation was done after derivatization by GC–MS [18]. LC–MS–MS (negative ESI) seems to replace GC–MS since it does not require derivatization and produces excellent sensitivities [19,20].

Macrolides and lincosamides are a group of semisynthetic antibiotics that are active against Gram-positives. They are often used against microorganisms having developed a resistance against penicillins. Macrolides are often derivatized and detected by LC–FL, however, LC–MS–MS seems to produce best results in terms of selectivity and sensitivity [21] (Figure 24.10).

Aminoglycosides are polar compounds which are used against aerobic Gram-negative bacteria. Aminoglycosides were reported to show a narrow therapeutic range, due to their toxicity against

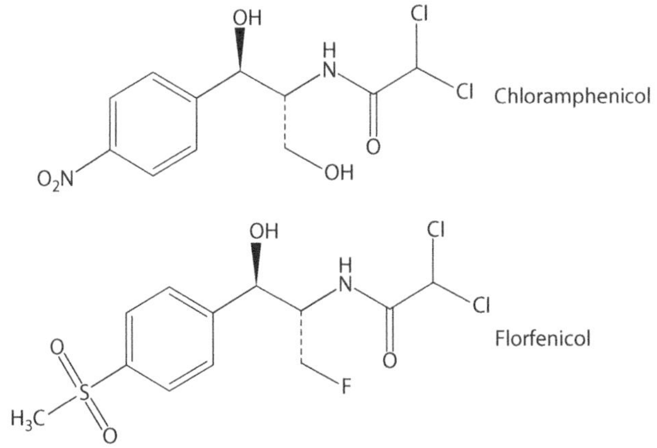

Figure 24.9 Phenicols.

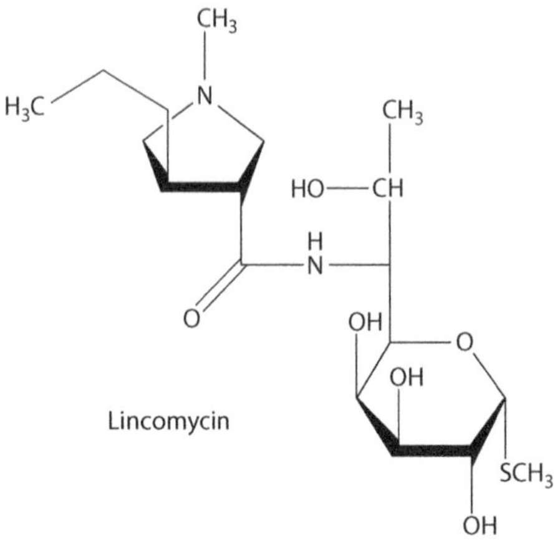

Figure 24.10   Lincomycin.

the animal treated. Amikacin and gentamycin were used for ornamental fish and in aquaculture. Aminoglycosides require harsh (low pH) extraction conditions to be released from the sample. Most published methods use cation exchange SPE for sample processing. Aminoglycosides degrade on GC columns and elute unretained from most LC columns. Moreover, they have no chromophoric groups, which make derivatization compulsory. Only LC–MS–MS permitted the development of multimethods, covering more than one or two aminoglycosides [22]. Chromatographic retention is often achieved by using reversed phase columns conditioned with volatile ion pair agents (perfluorinated carbonic acids) or HILIC columns with relatively high volatile salt (ammonia formiate) concentrations. MS–MS is vastly superior to other detection techniques like derivatization and fluorescence detection (Figure 24.11).

Nitroimidazoles, such as nitrofurans are banned drugs, which are suspected to be human carcinogens and mutagens. Still these drugs (metronidazole and ipronidazole) were used in aquaculture for prophylactic and therapeutic treatments of diseases. Nitroimidazoles undergo extensive metabolism and are preferably detected as hydroxylated metabolites. LC–MS–MS permits the quantification and confirmation of nitrimidazole residues without requiring derivatization [23] (Figure 24.12).

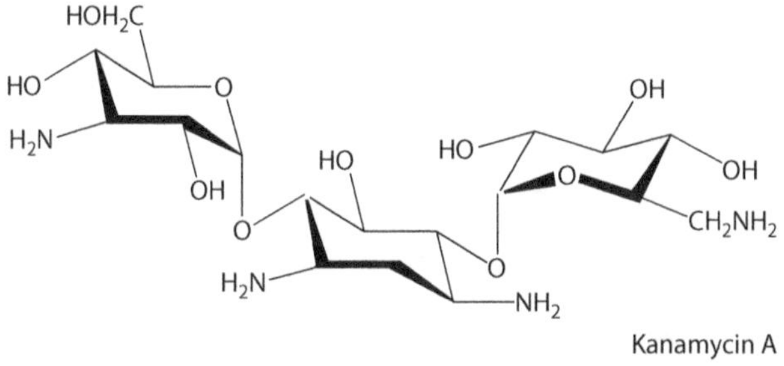

Figure 24.11   Aminoglycosides.

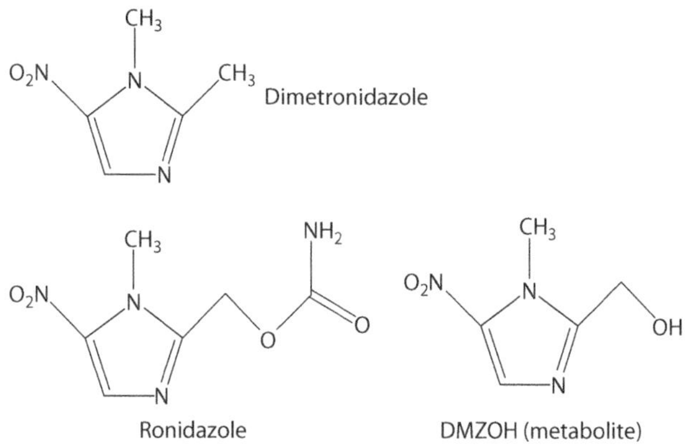

Figure 24.12 Nitroimidazoles including metabolites.

24.5.2 Antimycotica

Fungal and protozoal infections of fish can be treated by the use of triphenylmethane dyes.

Known members of these families are malachite green and crystal violet. Although effective and often used in aquaculture, these compounds are not registered in most countries. Residues of malachite green are very persistent and can be detected months after the application of the drug is stopped [5]. Malachite green is generally detected in the leuco form. Older analytical methods using LC–UV detection, utilize the chromophoric changes when oxidizing leuco-malachite green to malachite green. The required low detection limits are best achieved by LC–MS–MS (Figure 24.13).

24.5.3 Tranquilizers

Tranquilizers are used in aquaculture to sedate fishes and reduce mortality during transport and handling procedures. Reported was the use of benzocaine and tricaine [24] in aquaculture. The residue target organ for tranquilizers applied to mammals is the kidney. Since this organ from fish

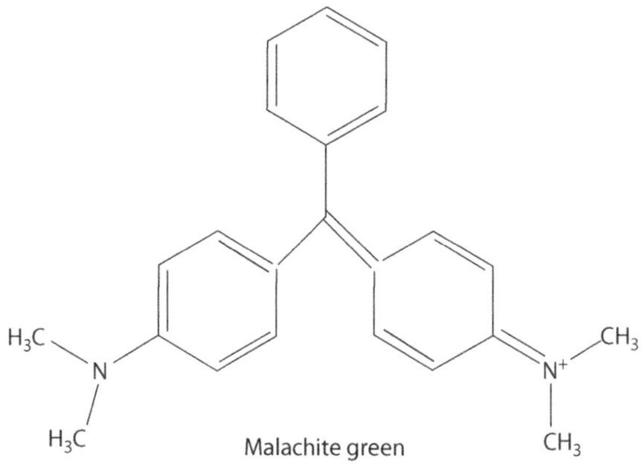

Figure 24.13 Malachite green.

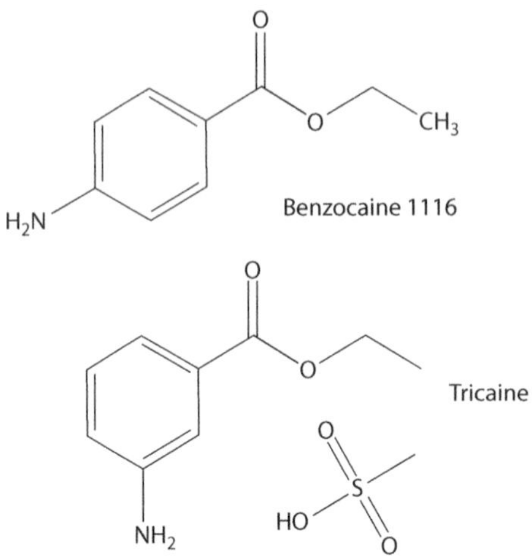

Figure 24.14 Tranquilizers.

or shrimps is not eaten and mostly not available to the residue laboratory, analysis has to focus on muscle tissues where degradation occurs within hours. Detection is only possible if the drug has been applied directly prior catching, slaughtering, and freezing (Figure 24.14).

24.5.4 Antiparasitica

Avermectins act against parasites like sea lice. Members of this group (emamectin and ivermectin) were used to treat salmons and trout against such parasites. The marker residue for emamectin benzoate is emamectin B_1a that has been detected by LC–FL [25] (Figure 24.15).

24.5.5 Anthelmintic Drugs

Benzimidazoles, as for example albendazole, are active against a broad-spectrum of intestinal helminth infection. These drugs were developed to treat mammalians, however activity against fish parasites is observed as well.

Albendazole is extensively metabolized in fish, forming albendazole sulfoxide, albendazole sulfone, and albendazole-2-aminosulfone. Quantification is hampered by a lack of commercial availability of reference substances. LC–FL was reported to be used for the determination of albendazole [26] (Figure 24.16).

24.5.6 Hormones

Hormones are used in aquaculture for sex reversal of newly hatched fry. Tilapia males are known to grow faster and larger than females [27], hence sex reversal is of economical interest. Such treatment is applied on juvenile fishes, hence residues of the applied hormones are not anymore likely to detect when the fishes reach the market. Besides, there remains the possibility that hormones are used as growth promoters. The analysis of hormones was a domain of GC–MS, however, newer papers increasingly often report LC–MS–MS with APCI interfaces [27] (Figure 24.17).

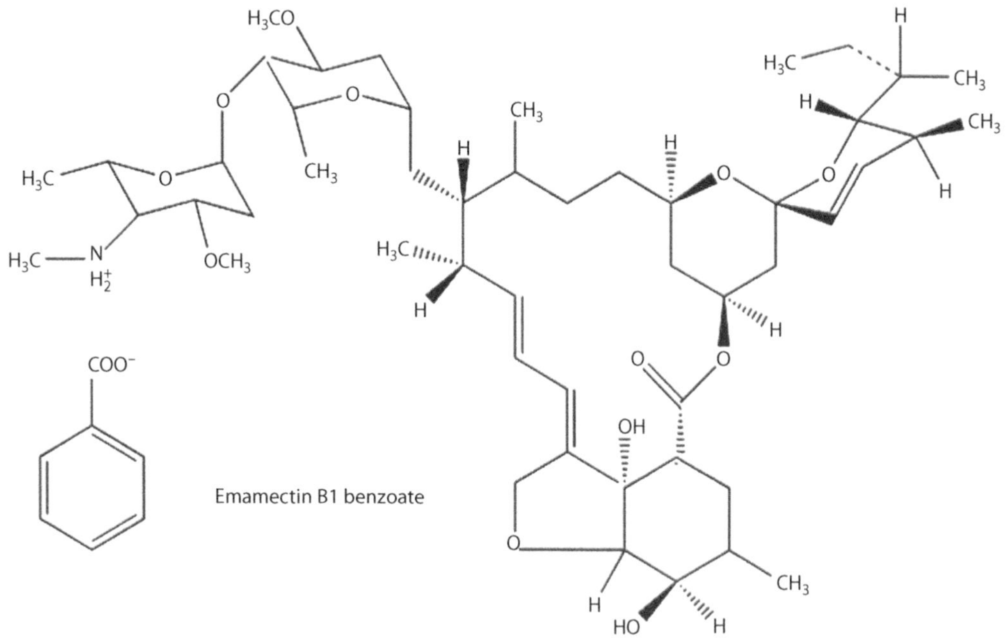

Figure 24.15   Avermectins (emamectin B1).

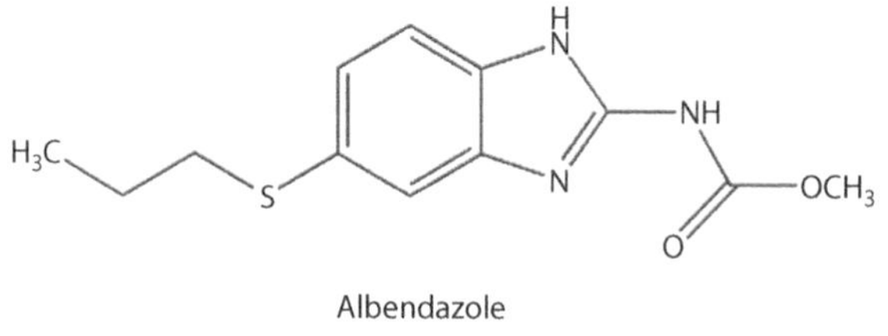

Figure 24.16   Benzimidazoles (albendazole).

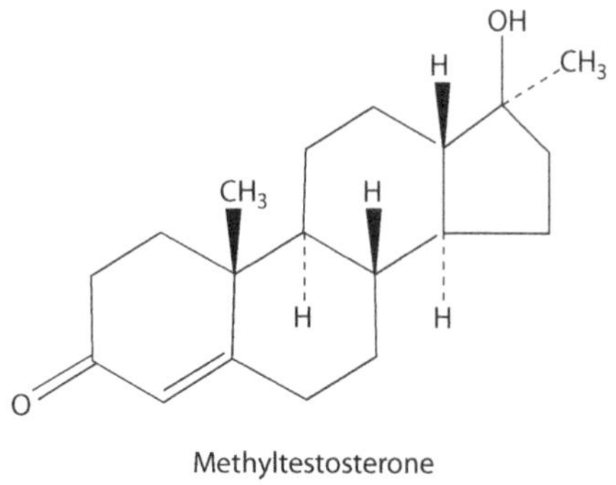

Figure 24.17   Hormones (methyltestosterone).

24.6 Future Developments

Aquaculture is probably still in its infancy. There is an enormous potential for growth in such markets where consumers are not yet accustomed to fish or shrimp or do not have yet the purchase power to add such food onto their plates. On the other hand, the current fish harvest from the world oceans cannot probably be further increased. Overfishing will likely even lead to shrinking harvests in the near future. All these are factors that point to the increasing importance of aquaculture. Aquaculture will likely serve different markets. There is the fish supply for poor countries, where fish is primarily a source of essential proteins. On the other hand, there are developed countries, which will likely develop a fancy for high end products, including exclusive aquatic species or fish produced under a certified organic environment. It is obvious that producers supplying these two market segments will have different opinions concerning the use of veterinary drugs. Therefore, it is to be expected that products produced for these two segments, will not always be consequently separated. A producer who caters for the European and U.S. market might be unable to supply the volume of fish as specified by contracts with his customers. Consequently, he or she might be tempted to fill the gap with fish coming from his neighbor's ponds, which were intended for the local market. Considering the fact that processing plants cater to the local and the export markets, there is the likelihood that some fishes or shrimps intended for the low-end market will end up as a high-end shipments. Therefore, there is a continuing need to control antibiotic residues in the production regions as well as in the importing countries.

With the more widespread use of aquaculture, the treat for lethal infections will stay or even increase. It is very difficult to envision the possibilities of future prevention or therapy methods. Certainly, increasing research will increase our knowledge of how aquatic species should be kept and fed. Improved breeding will lead to species that are less likely to develop disease. This might even include genetically modified fishes and shrimps. Likewise, there will be also further developments in the field of immunization. A marked progress for the immunization of fish has been achieved, however, immunization of shrimps was not very successful. There is the speculation that shrimps have a rather "primitive" immune system, which prevents the successful development of immunization strategies [4].

Progress is also to be expected in the field of residue analysis. Multimethods covering several groups of veterinary drugs are not yet commonly used. The widely different chemical and physical properties of the various drugs make it difficult to analyze many analytes by a single analytical method. Progress has to be made on two different fronts. First a generic extraction procedure has to be established, which quantitatively liberates incurred veterinary drugs from the tissue, followed by a cleanup intended to remove matrix compounds like peptides, proteins, and fats, without otherwise strongly affecting analyte recoveries. Detection will most likely rely on LC–MS–MS or LC-TOF. Modern LC–MS–MS is suitable for multiresidue methods. However, the more analytes are monitored, the shorter the MS dwell times have to be chosen. This does not only affect sensitivity, but also decreases reproducibility due to higher variation of the peak areas. TOF does not show such limitations. However, the current available resolution provides mass selectivity, which is above LC–MS, but still below LC–MS–MS. Technical developments and engineering improvements might ease the limitations for both MS–MS and TOF, or might even permit the use of ultra high-resolution MS in the routine residue analysis environment.

References

1. Nomura I., State of the world aquaculture 2006, FAO Fisheries technical paper 500, FOA, Rome, Italy, 2006.
2. Serrano P., Responsible use of antibiotics in aquaculture, FOA Fisheries technical paper 469, FOA, Rome, Italy, 2005.
3. Pillay TV.R. and Kutty M.N., *Aquaculture Principles and Practices*, 2nd edn., Blackwell Publishing, Oxford, U.K., 2005.
4. Stickney R.R., *Aquaculture: An Introductory Text*, CABI Publishing, Oxfordshire, U.K., 2005.
5. Kuiper R.V., Scherpenisse P., and Bergwerff A.A., Persistence of residues of malachite green in European eel after water-born exposure of juvenile eels, *Poster Euroresidue IV*, Veldhoven, the Netherlands, 2000.
6. Kennedy G., Antibiotic residues in aquaculture systems, *Lecture at Annual Conference of the Shellfish Association of Great Britain*, London, 2004.
7. Oka H., Nakszawa H., Harada K., and MacNeil J., *Chemical Analysis for Antibiotics Used in Agriculture*, AOAC International, Airlington, TX, 1995.
8. Kurtz D., Skerritt J., and Stanker L., *New Frontiers in Agrochemical Immunoassay*, AOAC International, Airlington, TX, 1995.
9. Franek M., Diblikove I., Broad-specificity immunoassays for sulfonamide detection: Immunochemical strategy for generic antibodies and competitors, *Anal. Chem.* 78, 1559, 2006.
10. Kaufmann A. and Pacciarelli B., Bestimmung von Rückständen von Tetracyclinen in Lebensmitteln, *Mitt. Lebenms. Hyg.*, 90, 167, 1999.
11. Bogialli S., Curini R., Di Corcia A., Nazzari M., and Saperi R., A liquid chromatography-mass spectrometry assay for analyzing sulfonamide antibioticals in callte and fish muscle tissues, *Anal. Chem.*, 75, 1798, 2003.
12. Potter R.A., Burns B.G., Van De Riet J.M., North D.H., and Darvesh R., Simultaneous determination of 17 sulfonamides and the potentiators ormetoprim and trimethoprim in salmon muscle with liquid chromatography with tandem mass spectrometry detection, *JAOAC*, 90, 343, 2007.
13. Brillantes S., Tanasomwang V., Thongrod S., and Dachanantawitaya N., Oxytetracycline residues in Gian Freshwater Prawn, *J. Agric. Food Chem.*, 49, 4995, 2001.
14. Schneider M., Vazquez-Moreno L., and Barraza R., Multiresidue determination of fluoroquinolones in shrimp by liquid chromatography-fluorescence-mass spectrometry, *JAOAC*, 88, 1160, 2005.
15. Leitner A., Zöllner P., and Lindner W., Determination of the metabolites of nitrofuran antibiotics in animal tissue by high-performance liquid chromatography-tandem mass spectrometry. *J. Chromatogr. A*, 939, 49, 2001.
16. Hoenicke K., Gatermann R., Hartig L., and Mandix M., Formation of semicarbazide (SEM) in food by hypochlorite treatment: Is SEM a specific marker for nitrofurazone abuse?, *Food Addit. Contam.*, 21 526, 2004.
17. Saari L. and Peltonen K., Novel source of semicarbazide: Levels of semicarbazide in cooked crayfish samples determined by LC/MS/MS, *Food Addit. Contam.*, 21, 825, 2004.
18. Shen H. and Jiang H., Screening determination and confirmation of chloramphenicol in seafood, meat and honey using ELISA, HPLC-UVD, GC-ECD, GC-MS-EI-SIM and GCMS-NCI-SIM methods, *Anal. Chim. Acta*, 535, 33, 2005.
19. Kaufmann A. and Butcher P., Quantitative liquid chromatography/tandem mass spectrometry determination of chloramphenicol in food using sub-2 μm particulate high-performance liquid chromatography columns for sensitivity and speed, *Rapid Commun. Mass Spectrom.*, 19, 3694, 2005.
20. Van de Riet J., Potter R., and Christine-Fougere M., Simultaneous determination of residues of chloramphenicol, thiamphenicol, florfenicol, and florfenicol amine in farmed aquatic species, *JAOAC*, 86, 510, 2003.

21. Horie M., Harumi T., and Kazuo T., Determination of macrolide antibiotics in meat and fish by liquid chromatography-electrospray mass spectrometry. *Anal. Chim. Acta*, 492, 187, 2003.

22. Kaufmann A. and Maden K., Determination of 11 aminoglycosides in meat and liver by liquid chromatography with tandem mass spectrometry, *JAOAC*, 88, 1118, 2005.

23. Mottier P., Huré I., Gremaud E., and Guy P., Analysis of four 5-nitroimidazoles and their corresponding hydroxylated metabolites in egg, processed egg, and chicken meat by isotope dilution liquid chromatography tandem mass spectrometry, *J. Agric. Food Chem.*, 54, 2018, 2006.

24. Scherpenisse P. and Bergwerff A., Determination of residues of tricaine in fish using liquid chromatography tandem mass spectrometry, *Anal. Chim. Acta*, 586, 407, 2007.

25. Kim-Kang H., Bova A., Crouch L., and Wislocki P., Tissue distribution, metabolism, and residue depletion study in atlantic salmon following oral administration of [³H] emamectin benzoate, *J. Agric. Food Chem.*, 52, 2108, 2004.

26. Shaikh B., Rummel N., and Reimschuessel R., Determination of albendazole and its major metabolites in the muscle of atlantic salmon, tilapia, and rainbow trout by high performance liquid chromatography with fluorescence detection, *J. Agric. Food Chem.*, 51, 3254, 2003.

27. Chu P., Lopez M., Serfling S., and Gieseker C., Determination of 17α-methyltestosterone in muscle tissues of tilapia, rainbow trout, and salmon using liquid chromatography-tandem mass spectrometry, *J. Agric. Food Chem.*, 54, 3193, 2006.

Chapter 25

Analysis of Dioxins in Seafood and Seafood Products

Luisa Ramos Bordajandi, Belén Gómara,
and María José González

Contents

25.1 Introduction 708
25.2 Sample Pretreatment and Recovery Studies 709
    25.2.1 Sample Storage 709
    25.2.2 Spiking and Recovery Studies 709
25.3 Extraction Methods 710
    25.3.1 Soxhlet Extraction 710
    25.3.2 Solid Phase Extraction and Matrix Solid Phase Dispersion 710
    25.3.3 Supercritical Fluid Extraction 711
    25.3.4 Accelerated Solvent Extraction 711
    25.3.5 Microwave Oven 712
25.4 Cleanup Methods 712
    25.4.1 Lipid Removal 713
    25.4.2 Isolation of Uncommon Chemical Interferences 713
25.5 Fractionation/Group Separation 713
25.6 Automation of Extraction and Cleanup 714
25.7 Instrumental Determination 715
    25.7.1 GC Congener Separation 715
    25.7.2 GC Detectors 717
25.8 Bioanalytical Screening Methods 718
References 719

25.1 Introduction

Polychlorinated dibenzo-*p*-dioxins (PCDDs) and dibenzofurans (PCDFs), two groups of Persistent Organic Pollutants (POPs), are structurally related chlorinated aromatic hydrocarbons which are generally referred to as "dioxins." They are of great concern due to the extreme toxicity of the 2,3,7,8 chlorine substituted congeners and their presence in all compartments of the environment. PCDD/Fs are formed as by-products of a wide variety of chemical industry and combustion processes that contain chlorine and chlorinated aromatic hydrocarbon sources [1]. Due to their low water solubility, hydrophobicity, and resistance to degradation, these substances are found in a wide range of biological samples, and tend to accumulate in animal and human adipose tissues through the food web [2]. Among the 210 possible congeners, seven 2,3,7,8-substituted PCDDs and 10 PCDFs are generally considered the most persistent and toxic PCDD/F congeners, since they have toxic properties similar to 2,3,7,8-tetrachlorodibenzo-*p*-dioxin (TCDD), which is the most toxic congener of these compounds [3–5].

For the general population, dietary intake is the main route of PCDD/F exposure, contributing to more than 90% of the daily exposure [6,7]. Public concern over the adverse health effects of these toxicants at this time has been intensified by a number of dioxin contamination incidents involving food and feedstuffs [8–10]. Recent reports concerning toxicological aspects have led to a revaluation of the tolerable daily intake (TDI) of dioxins [4] and have prompted wide-ranging efforts and the tightening of regulations to reduce dioxin release into the environment [11]. To prevent the health risk from dioxin exposure, the European Commission has recently established maximum permissible levels of dioxins and dioxin-like polychlorinated biphenyls (PCBs) in foods [12]; the minimal risk level (MRL) for fish and seafood is 8 pg of toxic equivalents (TEQs)/g fresh weight (including dioxin-like PCBs), except for eels (*Anguilla anguilla*) and their products, where it is 12 pg of TEQs/g fresh weight.

As required by the current legislation regulating dioxins in foodstuffs, including seafood and seafood products, large numbers of samples have been analyzed, and a great deal of data concerning PCDD/F levels in foodstuffs is now available. All these studies show that human dietary dioxin intake has been decreasing in recent years, and that seafood and seafood products have received special attention due to their widespread consumption by the population and their high dioxin contents [13–25].

Although seafood and seafood products exhibit higher dioxin levels than any other food category [14,16,19,21,24,25], they are present in sub-ppb levels and their analysis is complex and challenging. At present, there is a need for cost-efficient, reliable, and rapid analytical alternatives to expensive methods involving the use of gas chromatography coupled with high-resolution mass spectrometry (GC-HRMS) so that food items can be routinely monitored to detect contamination at an early stage.

Usually, the analysis of these compounds in fatty seafood tissues requires three main steps: extraction of the target analytes, cleanup of the extract obtained, and GC separation [26]. Several extraction and cleanup procedures are described in the literature, and which one is chosen depends on individual analytical laboratories. Soxhlet (SOX) [15,27,28], solid-phase extraction (SPE) [29], matrix solid phase dispersion (MSPD) [20,30], supercritical fluid extraction (SFE) [31,32], microwave-assisted solvent extraction (MASE) [33], and accelerated solvent extraction (ASE) (also named pressurized liquid (PLE) [34,35] have all been employed as extraction methods. Cleanup procedures, including open column chromatography on activated Florisil®, alumina, silica, carbon, and size exclusion chromatography (SEC) [26,36–39] are also currently used. Automatic online procedures combining extraction and cleanup methods to obtain extracts ready for

GC analysis with the maximum extraction efficiency and overcoming matrix-related interferences have always been a target, but they are still being developed [40].

Over the last few years, the European Union (EU) has initiated a large-scale research project to develop new analytical methodologies for the determination of dioxins (most of them including dioxin-like PCBs) in food matrices to serve as alternatives to GC-HRMS. This last technique is taken as the benchmark for accurate and specific determination of these compounds in food samples as described in Environmental Protection Agency (EPA) and EU official methods [41–43]. GC-HRMS provides enough specificity and selectivity at concentration levels down to femtograms per gram for the analysis of these compounds, but it is a relatively expensive technique and requires qualified personnel. Because of that, alternative techniques such as GC coupled to ion trap mass spectrometry (GC-ITMS), working in tandem mode (MS/MS) [44–46], and also comprehensive two-dimensional GC (GC × GC) coupled to microelectron capture detection (μECD) [47] and time of flight (TOF) MS [48], have recently been validated for acceptance as an alternative to GC-HRMS. In addition, bioanalytical methods have improved considerably in sensitivity and selectivity to the extent that they can be used as screening methods to determine the total quantities of dioxin-like compounds [49].

Although most analytical methods for measuring dioxins in seafood and seafood products include dioxin-like PCBs, this chapter specifically focuses on those targeting the 17 toxic 2,3,7,8-PCDD/Fs in seafood and seafood products. Attention has been paid to both, methods that are in current use and methods that have recently been developed for each step of the analysis from sample preparation to instrumental determination of these congeners.

25.2 Sample Pretreatment and Recovery Studies

25.2.1 Sample Storage

Seafood and seafood products collected in the field are usually preserved by freezing immediately, either in the field, on board ship or at the laboratory. Whenever possible the seafood should be dissected immediately and the individual tissues stored in individual packs of approximately the size required for analysis to minimize thawing of subsampling material. Seafood tissues are first macerated and then freeze-dried or ground with sodium sulfate and silica to reduce the water content and rupture cell walls and these are the most commonly used pretreatment procedures for seafood tissue matrices [20,29,50]. It should be noted that the concentration of the 2,3,7,8-PCDD/F congeners is generally at femtogram per gram. It is therefore necessary to analyze samples containing around 6 g of fat, which require large amounts of fresh sample (from 600 g for mussels [1% fat] to 30–40 g for salmon [20% fat]). In any case, that is much more than what is usually employed for the analysis of other POPs. For this reason, almost all sample pretreatment methods involve freeze-drying, which completely eliminates the water content and drastically reduces the sample size. On the other hand, the freeze-drying step takes 48 h, which considerably increases the total analysis time.

25.2.2 Spiking and Recovery Studies

In PCDD/Fs analysis it is mandatory to use isotope dilution mass spectrometry (IDMS) for the final quantitative determination of the target 2,3,7,8-PCDD/Fs and the recoveries of the total analysis (extraction + cleanup + analytical determination) [41,43]. IDMS is the most elegant way to overcome the whole problem of sample recovery and quantification. The seventeen $^{13}C_{12}$ 2,3,7,8-PCDD/F labeled isotopes are added to the sample at a known concentration prior to extraction, as extraction standards for quantification. Two more $^{13}C_{12}$-PCDD congeners (1,2,3,4-TCDD and 1,2,3,7,8,9-HxCDD) are added to the extract at a known concentration prior to analytical determination, as instrumental standards. The ratio of the labeled and native compounds is measured by MS and automatically accounts for any losses in the procedure. Although it is not necessary to calculate the recoveries for quantification purposes, they are calculated as a quality parameter from the ratio of the labeled congeners in the extraction and recovery standards [13–22,24,25,43].

25.3 Extraction Methods

The purpose of extraction step is to remove the bulk of the sample matrix and to transfer the fraction containing the analytes to a suitable solvent. Extraction techniques for fish and seafood are generally based on the assumption that lipophilic compounds such as PCDD/Fs predominantly occur in the fat fraction of the food matrix, and they are based on general methods for isolation of the lipid fraction from the sample matrix. Conventional extraction methods of extraction are SOX [15,27,28], SPE [29], MSPD [20,30], and SFE [31,32]. The need to change the nature of the solvent, the amount of solvent used, and the time required to undertake an extraction from a food matrix has driven the development of various techniques in recent years to challenge SOX extraction, which has been a long-standing and proven technique. Substantial progress has been made toward developing improved techniques such as MASE [33] and the most popular ASE or PLE [34,35]. Finally, other methods such as dialysis [51] have also been tested.

Saponification under alkaline conditions (in the presence of ethanol and KOH) [52] followed by extraction with organic solvents is often employed for the analysis of large amounts (up to 100 g) of fat. However, this method is known to lead to degradation of dioxins in proportion to their chlorine content, and in the case of PCDFs to production of lower chlorinated PCDFs and ethoxy-PCDFs as artifacts [53]. Finally, it is worth noting that although the approach has not yet been studied extensively, some applications have already demonstrated the potential of sonication (USE) for food dioxin analysis [54].

25.3.1 Soxhlet Extraction

One of the most frequently used liquid–solid extraction methods, developed in the late nineteenth century, is still routinely used for extraction of dioxins from seafood tissues [15,27,28]. However, the technique has a number of drawbacks, the most important of which are the large volume of solvent (200 mL for 100 g of tissues), the long extraction time (more than 18 h), the generation of dirty extracts that require extensive cleanup, and the impossibility of automation. In order to overcome these, alternative extraction strategies have been developed, offering analysts a choice of newer techniques such as SPE [29], MSPD [20,30], and more recently ASE [35].

25.3.2 Solid Phase Extraction and Matrix Solid Phase Dispersion

SPE is today a classic extraction system, thanks mainly to the popularization of SPE cartridges, which have been successfully applied to biological human fluids [29]. However, in the case of solid samples, SPE is less popular and has almost never been used to extract dioxins from seafood tissues because of the large amount of sample needed. On the contrary, MSPD, using open conventional glass chromatography columns, is very often used in routine analysis of seafood samples [20,22,24,30]. In MSPD, the sample is mixed or blended with an appropriate sorbent (e.g., C_{18}, silica) until a homogeneous mixture is obtained; this mixture is packed into a column, from which the analytes of interest are eluted with a suitable organic solvent [20,24,30]. The extraction and first cleanup step are performed simultaneously, and most of the artifacts are eliminated. Because a large amount of sample is needed, the method compares unfavorably with SOX in terms of the amount of solvent required (around 400 mL).

25.3.3 Supercritical Fluid Extraction

SFE is another classic method for seafood dioxin analysis, but not as popular as SOX and MSPD. SFE has attracted intense interest during the past 20 years, mainly for extraction of solid samples, because it offers short extraction times and minimum use of organic solvents [26,55]. Carbon dioxide (CO_2) is mostly used as the extraction solvent because of its moderate critical temperature (31°C) and pressure (73 atm). In the 1990s, SFE instruments became available, enabling larger sample sizes and rendering it more suitable for wider applications. For seafood and seafood matrixes, fat retainers such as Florisil and silica are usually introduced in the extraction thimble to achieve a fat-free extract. Some applications for seafood dioxin determinations have been published [31,32,56]. Although SFE extraction is automated and offers a short extraction time and minimum use of organic solvents with no additional cleanup step before GC–MS, it is not widely used because of the large number of parameters that have to be optimized, especially in the analyte collection chamber, and the high cost of the equipment.

Similarly to SPE, due to the large amounts of samples required (5–10 g of lipid equivalents) to be able to reach the low levels at which dioxins are present in food samples, the use of SFE for this purpose is scarce.

25.3.4 Accelerated Solvent Extraction

One of the most recent extraction methods used instead of SOX and MSPD is ASE or PLE, an alternative to the classic extraction methods [26,34,57,58]. It uses conventional liquid solvents at high pressures (150–200 psi) and temperatures (50°C–200°C) to extract solid samples quickly, and with much less solvent than conventional techniques. Seafood samples are placed in extraction cells, which are filled with an extraction solvent and heated. The sample is statically extracted for 5–10 min, with the expanding solvent vented to a collection vial. Following this period, compressed nitrogen is used to purge the remaining solvent into the same vial. The entire procedure is completed in 10–20 min per sample, and uses only 15–20 mL of solvent. Of special interest are applications dealing with selective extraction procedures, where integrated cleanup strategies are used to combine extraction and cleanup or fractionation to further simplify all the sample-preparation steps [34]. Fish and seafood have been the food matrices most commonly investigated using a fat retainer (alumina, silica, and Florisil) in the extraction thimble to achieve a fat-free extract; with satisfactory results for dioxins and dioxin-like PCBs [59–62]. As part of the DIFFERENCE project [63], Wiberg et al. [35] evaluated traditional extraction techniques vs. alternative techniques such as ASE for PCDD/F and dioxin-like PCB determinations in food and feed, including certified reference materials. They demonstrated that ASE is more of a quantitative extraction process than other conventional techniques. The ASE method in combination with HRMS detection meets the quality criteria for official control of dioxins in foodstuffs [43]. One of the recent developments is the combination of ASE with integrated carbon fractionation [64], in which dioxins can be fractionated and obtained in backward elution, and only a small, miniaturized multilayer silica column cleanup is required after ASE and before detection. Some attempts have also been made to combine ASE with automated cleanup systems, in particular the power-prep FMS system (which is discussed further in Section 25.4), and to construct a fully automated system (extraction plus cleanup); however, the results have not been satisfactory because ASE, as a dynamic system, requires the incorporation of a concentration phase prior to prep-FMS, rendering automation virtually impossible and considerably increasing the analysis times [65].

More extensive information about this technique can be found in the literature, where there are some reviews dealing exclusively with ASE for dioxins in foods [34] and biological matrices [66].

25.3.5 Microwave Oven

MASE in the analysis of dioxins has been only recently introduced, and there are no published studies in which MASE was used for dioxin seafood extraction. In recent years, MASE has attracted growing interest, as it allows rapid extraction of solutes from solid samples by employing microwave energy as a source of heat, with an extraction efficiency comparable to that of classic techniques. The partitioning of the analytes from the sample matrix to the later extractant depends on the temperature and the nature of the extractant. Unlike conventional systems, microwaves heat the entire sample simultaneously without heating the vessel; thus the solution reaches its boiling point very rapidly and the extraction time is very short [33]. In view of the good results of MASE in the extraction of PCBs and DDTs [67] from biological tissues with only 8 mL of ethyl-acetate (1:1, v:v), this technique is very attractive for dioxin analysis in food samples. Its main drawbacks are the loss of more volatile solutes if the temperature of the vessel rises rapidly; and that the vessels need to be cooled to room temperature after extraction before they can be opened, which increases the overall extraction time. In addition, it is not possible to automate the procedure to incorporate cleanup steps.

25.4 Cleanup Methods

Analytical procedures for determination of PCDD/Fs in seafood samples involve sophisticated and tedious cleanup methods. Several steps are usually required to remove the bulk of coextractants (including lipids) in order to end up with an extract containing only PCDD/Fs, in which the analytes can be detected at the ultra trace levels at which they occur in seafood and seafood samples. The choice of a particular sequence of steps will depend very much on the analytical system that is finally used. Sample extraction, cleanup, and GC method together form a delicately balanced combination, each contributing to the ultimate specificity and selectivity. For the determination of dioxins, nearly all established schemes involve combinations of cleanup methods developed for the analysis of PCBs and organochlorinated pesticides (OCPs) (solid–liquid adsorption chromatography using Florisil, silica, and alumina, gel permeation chromatography [GPC], and high-performance liquid chromatography [HPLC]) in combination with an active carbon step to isolate the specific fraction containing the dioxins without chemical interferences. In many of the methods used today, the sample extraction and cleanup steps are combined "online" or "at-line," and some are automated.

25.4.1 Lipid Removal

Lipid removal is the first step in the cleanup process, and some other interferences are also usually eliminated. Several different methods have been used to remove lipids, including destructive methods such as sulfuric acid [28] or sodium hydroxide treatment [52], and nondestructive methods such as GPC [68] and dialysis [51,69]. GPC, which is sometimes referred to as SEC has been successfully applied to seafood tissues for POPs analysis using SX-3 Biobeds (200–400 mesh) in a range of column sizes and solvents. It can be fully automated and, unlike adsorption chromatography, it is also more suitable for the isolation of unknown contaminants on whose polarity or chemical functionality there is little information. The method can also handle a large mass of lipid in each sample (e.g., columns of ca. 500 × 25 mm ID can handle up to 500 mg of lipids) compared to adsorption columns that are limited to 50 mg of lipids per g of adsorbent [68,70]. Dialysis with semipermeable membranes (SPMs) in an organic solvent can separate other similar POPs from lipids. The method can eliminate more than 20 g of lipids in a single membrane with acceptable recoveries of internal standards, practically irrespective of the amount and type of lipid dialyzed. The method has been successfully used for dioxin analysis in a large variety of seafood species [69]. Although it is efficient, simple, and versatile and does not entail excessive solvent use, this procedure is not very often used for dioxin analysis because it is very time consuming (72 h). In addition, online coupling, either with extraction or with the following cleanup steps, is not possible.

25.4.2 Isolation of Uncommon Chemical Interferences

For dioxin analysis in a seafood matrix, an additional purification step is necessary to eliminate other interferences (including any other lipids). A combination of adsorbents (neutral, basic and acid alumina, silica, modified silica with acids and basics, and Florisil in multilayer or one-layer columns) and solvents with different polarities and dielectric constants are used to eliminate interferences [15,22,24,27]. It is well known to experts that the application of the extract to a strongly basic adsorbent (potassium or cesium hydroxides) silica gel with a low-polarity solvent hexane is very effective for removing trace residues of acidic compounds such as phenolic and carboxylic acids, and sulfonamide compounds [71]. On the other hand, the sulfuric acid-impregnated silica gel (20%–40%, w/w) is very effective in removing numerous types of compounds by dehydration, acid-catalyzed condensation, and oxidation reactions [72]. Alumina (basic, acid, and neutral) and Florisil are used, at different activation grades, mainly to eliminate all other lipids and other coextractants [37,39]. The literature gives no indication of preferences for any specific adsorbent or solvent, the choice of which depends more on the laboratory's preferences than on performance. Cleaning up of seafood samples for dioxin analysis is a laborious and tedious task, which has to be validated. The combination of adsorbents and solvents chosen to obtain a clean extract without any dioxin loss before the GC–MS analysis is up to each laboratory.

25.5 Fractionation/Group Separation

Normally, a group separation is necessary before final analysis of dioxins by GC-HRMS. At this stage, the cleanup extract may contain other similar organohalogen compounds such as PCBs. With the exception of non*ortho* PCBs, dioxins are present at substantially lower concentrations than the other POPs, and it is therefore necessary to separate dioxins from the bulk of POPs. The methods available for the isolation of POPs into separate fractions prior to GC analysis are based on the spatial planarity of dioxins to separate them as a distinct fraction.

The available methods for fractionation have been extensively reviewed [36,73]. Open liquid chromatography columns of Florisil [37,74]; alumina [39] active carbon [30,75,76]; *and* graphitic carbon [38] are among the most widely used methods. In recent years, HPLC with either porous graphitic carbon (PGC) [77,78] or active carbon [5] and PYE (2-(1-pyrenyl) ethyldimethylsilylated silica gel) columns [79,80] has become more popular thanks to the inherent advantages of HPLC.

Concejero et al. [81] studied the feasibility of employing four different carbons, Amoco PX-21, Carbosphere and Carbopack B and C, and one HPLC stationary phase, PYE (typically used for PCB and PCDD/F fractionation) for environmental studies. Recoveries for fractionation of the target compounds with all the sorbents studied were generally good and reproducibility satisfactory. All were able to isolate PCDD/Fs from PCBs, which could interfere in the final determination of the former by GC-HRMS. As a result, Carbopack B (as SPE cartridges) and PYE were considered the most valuable alternatives for simultaneous fractionation of PCDD/Fs and of the different classes of PCBs typically investigated in environmental studies. An additional merit of this HPLC stationary phase is the possible of automates.

It is worth noting that Immunoaffinity Chromatography (IAC) using mono- and polyclonal antibodies specifically developed to recognize 2,3,7,8-CDD/Fs was considered a very attractive technique in the 1990s. Thanks to the good results achieved in the cleanup of aqueous samples (water and blood) for dioxin analysis [82], it was initially thought to be very promising. However, because of the need to inject fat-free extracts, the variability of the results and the presence of cross-reactions, only a few years later it had been forgotten as an alternative cleanup process for dioxin analysis.

25.6 Automation of Extraction and Cleanup

In view of the extreme difficulty and tediousness of the extraction + cleanup process in seafood dioxin analysis, there have been many attempts at automation, but so far no one has come up with an automated procedure for simultaneous extraction and cleaning up.

The first attempt at a semiautomated at-line extraction/cleanup procedure was made by Smith et al. [71]. They developed a method for dioxin analysis of biological (including fish) tissues in two steps. In the first step, the extraction and a first cleanup step, using active carbon, were performed simultaneously. In the second step, the extract was applied to a second series of adsorbents contained in two tandem columns. Based on this general scheme, in 1997 was developed a semiautomated method for online extraction plus cleanup and fractionation of PCBs and PCDD/Fs [30]. Up to 6 g of fat can be extracted with this method, which is very useful in the case of fish, and in fact it has been successfully used to determine dioxins in seafood and seafood products [83,84]. However, despite its good performance, the method has not been widely used because there is no commercially available apparatus.

The efficiency of the automated Power-Prep system (FMS, Waltham, MA) in purifying sample extracts for dioxin analyses has already been demonstrated in recent years for different type of matrices, including food [40,85]. The multistep procedure is based on the use of disposable multilayer silica columns, basic alumina and PX-21 carbon columns, which can be combined to suit the target analytes. This means that dioxins and non*ortho* PCBs can be isolated in a fraction with good recoveries, and several samples can be analyzed in parallel, even ones with high fat contents [86]. This method has become increasingly popular over the years; as the only commercially available apparatus, it is gradually making its way into all laboratories that perform large numbers of analyses. Some efforts have been made [65] to couple in an online ASE extraction step, but for the moment no satisfactory results have been achieved.

While any laboratory can choose the method that best suits it for extraction and cleanup of seafood dioxins, there is no doubt that ASE as an extraction method and the Power-prep system for cleanup, both commercial products, are the only ones that, although not fully automated, permit large numbers of samples to be analyzed in the shortest possible time. With this combination, it is possible to handle 10 samples at once in both the extraction and the purification steps, and to deal with any food health emergency due to dioxin contamination of foodstuffs.

25.7 Instrumental Determination

As noted earlier, the choice of analytical procedure for extraction plus cleanup is up to each laboratory, if the analyte recoveries that they achieve are within the range laid down by the EU directive [43].

However, in the case of instrumental determination of the seventeen 2,3,7,8-PCDD/F congeners, the EU directive requires the use of GC-HRMS, which was the only method able to reliably determine dioxins at levels appropriate for food analysis. Some other methods, such as DR CALUX® bioassay and GC–MS/MS (ion trap detector [ITD]), are only officially accepted for screening purposes.

The basic requirements for acceptance of analytical requirements [43] are high sensitivity (10^{-12} g), selectivity, accuracy, and low limits of detection. The most important specific requirements are recovery control by the addition of $^{13}C_{12}$-PCDD/Fs as standards. The recoveries of the individual internal standards should be between 50% and 130%; the GC separation of the isomers should be <25% peak to peak; the identification should be performed according to EPA Method 1613 revision B and EU official method [43] using isotope dilution RGC/HRMS; and the difference between upper bound (not detected at limit of detection) and lower bound (not detected equal to 0) determination levels, should not exceed 20% for foodstuffs with about 1 pg WHO-TEQ/g fat (only PCDD/F), and 25%–40% for foodstuffs with about 0.5 pg WHO-TEQ/g fat.

Following a number of dioxin contamination incidents involving foodstuffs [8–10], there has been a tremendous increase in the demand for PCDD/Fs measurements in foodstuffs, including seafood and seafood products. Because of this, alternative and relatively inexpensive techniques such as GC-ITMS, working in tandem mode (MS/MS) [44,45], or GC×GC μECD [47] and GC×GC-ToFs [48], have recently been developed and validated in order to be accepted as an alternative to GC-HRMS for instrumental determination of PCDD/Fs.

25.7.1 GC Congener Separation

High-resolution gas chromatographic methods for analysis of PCDDs and PCDFs have been developed extensively in the last two decades and continue to progress today. GC isomer-specific separation of all 136 tetra- to octa-PCDD/Fs on a series of nine fused-silica capillary GC columns containing silicone stationary phases of diverse polarity (100% methyl, 5% phenyl methyl, 50% phenyl methyl, 50% methyl trifluoropropyl, 50%, 75%, 90%, and 100% cyanopropyl, and liquid crystalline smectic) was studied by Ryan et al. [87]. They showed that all 136 PCDD/F compounds, including the biologically important 2,3,7,8-substituted congeners, could be separated from each other mostly with two stationary phases. More recent studies have focused on separation of the seventeen 2,3,7,8-PCDD/F congeners from closely co-eluting isomers. Almost all methods found in the literature for seafood analysis use DB-5 stationary phase (5% diphenyl 95% dimethyl polysiloxane) (J&W Scientific, Folsom, CA) or equivalent [15,17,20,27]. Since the DB-5MS (J&W Scientific Folsom, CA) stationary phase (5% Silphenylene Silicone copolymer or Si-Arylene) has become commercially available, most laboratories use these products for dioxin analysis because their thermal stability is much improved as compared to DB-5 [88]. However, either of the two GC stationary phases can completely separate all seventeen 2,3,7,8-PCDD/F congeners, particularly the 2,3,7,8-TCDF. The EPA and European Standard methods [41–43] recommend the use of a second polar GC stationary phase such as DB-255 (50% cyanopropyl-methyl 50% phenylmethylsiloxane), DB-Dioxin (44% methyl, 28% phenyl, 20% cyanopropyl polixiloxane) (J&W Scientific, Folsom, CA), Supelco SP-2330 (100% cyanopropyl polysiloxane), or equivalent as a complementary tool. Most recently, Fishman et al. [89], evaluated 13 different GC columns: HP-5MS (Agilent technology), Rtx-5MS and Rtx-Dioxin2 (Restek, Bellefonte, PA), Supelco Equity 5 and SP-2331 (Supelco, Bellefonte, PA), Factor Four VF-5MS and CP-Sil 8 CB LowBleed/MS (Varian, Walnut Creek, CA), DB-5, DB-5MS, DB-225, DB-XLB (J&W Scientific, Folsom, CA), ZB-5MS, and ZB-5UMS (Phenomenex, Torrance, CA) for separation of the 17 toxic dioxin congeners. Their conclusion was similar to that of Ryan et al. (15 years ago): all dioxins can be separated from closely eluting isomers using either of two sets of nonpolar and polar stationary phase combinations. On the other hand, all efforts made to improve the separation among target isomers in some specific stationary phases (i.e., 1,2,3,7,8- and 1,2,3,6,7-penta-chlorinated dioxins) using the Rtx-Dioxin-2 (Restek Bellefonte, PA) have failed [90].

In this context, multidimensional GC techniques such as heart-cut multidimensional GC (heart-cut MDGC) and lately comprehensive two-dimensional GC (GC×GC) are regarded as powerful alternatives to the one-dimensional GC to solve coelutions between dioxins and other similar compounds present in the extract.

Heart-cut MDGC allows coeluting congeners on a precolumn to be transferred to a second capillary column with a different selectivity, improving the separation of the selected regions. A number of applications for PCDD/Fs analysis can be found in the literature [91]. However, when dealing with such complex mixtures, the number of heart cuts that can be made in one analytical run is limited to avoid coelutions in the second column. The need of reinjecting several times the same extract makes this technique time consuming, and is therefore a major drawback. Recently, GC×GC has been recognized as a powerful chromatographic technique for the resolution of complex mixtures. In this case, a modulation process transfers the entire effluent from the first column into the second one as consecutive narrow bands. Compared to heart-cut MDGC, a much higher peak capacity is obtained since the whole extract is subjected to two independent chromatographic separations by two sequential GC columns, without the need of reinjecting several times the same extract. In addition, the focusing effect that takes place during the modulation yields an increase in the signal-to-noise ratio, improving the limits of detection [92]. Since its introduction at the beginning of the 1990s by Liu and Phillips [93], the number of applications of GC×GC in the environmental and food fields has grown exponentially thanks to the advances in the instrument setup, such as more robust interfaces (modulators) between the first and second capillary columns and the possibility of coupling to a number of detection systems. The combination of GC×GC with microelectron detection (GC×GC-μECD) has been regarded as a promising technique for the determination of PCDD/Fs. Besides providing sufficient selectivity and sensitivity for their reliable determination at low levels in complex matrices, it would be a more cost-efficient option than GC-HRMS, the confirmatory method for the official control of dioxins in food [43]. The selection of the stationary phase of the first and second dimension columns is a critical step. A number of column combinations have been tested for the complete separation of the 17 priority PCDD/Fs isomers and the 12 WHO-PCBs (that have assigned a toxic equivalency factor [TEF] value) from each other and from other compounds potentially present in the extracts. Korytár et al. [94] found that the combination of a nonpolar stationary phase such as DB-XLB, in the first dimension, and the liquid-crystalline LC-50 (J&K Environmental, Milton, ONT, Canada), as second dimension, provided the complete separation of the 29 priority congeners, as well as from matrix constituents. In a further study, Danielsson et al. [47] used the same column combination (DB-XLB × LC-50) for the analysis of food samples, including fish, and compared the results with those obtained by GC-HRMS. The TEQ data correlated well between the two methods, pointing that although a more intensive validation should be performed to propose GC × GC-µECD as complementary/confirmatory method, it has a great potential as screening method, providing not only the TEQs' value but also a profile of the congener distribution in the samples [95].

Certainly, the coupling of GC × GC with MS has additional advantages to µECD, including the possibility of using isotope dilution for quantification, although increasing the costs [96]. Up to now most of the studies have been carried out using ToF-MS. Focant et al. [48] explored the possibilities of GC × GC-ToF-MS with the column combination Rtx-500 × BPX-50. The TEQ results obtained compared favorably to those obtained by GC-HRMS for seafood samples, although lower limits of detection would be desirable. On the other hand, the introduction in the market of rapid-scanning quadrupole MS instruments that have a lower cost than ToF-MS systems is promising, enabling also the possibility of using electron-capture negative ionization (ECNI) instead of electronic impact (EI) that would, in many cases, enhance the analyte detectability [97].

Improvements in the general setup of GC × GC system and advances in the MS detectors will lead this technique to a full establishment in routine analysis laboratories for dioxins analysis, if more user-friendly software for visualization and data treatment are available.

25.7.2 GC Detectors

As noted earlier, HRMS is mandatory for dioxin analysis in foodstuffs (including seafood and seafood products); however, due to the high cost of acquisition and maintenance, there has been research into suitable alternatives to HRMS over the last few years. HRMS was first used for TCDD determination in seafood samples in 1973 with detection limits in excess of 3 ppt [98]. Since then, improvements in mass spectrometers have made it possible to quantify all PCDDs and PCDFs in the sub-parts per trillion range.

Almost all recently investigated MS alternatives are based on tandem mass spectrometry (MS/MS) because this mode of operation theoretically provides higher signal to noise ratios than lower-resolution MS working in selective ion monitoring (SIM) mode. Of the different instruments capable of performing MS/MS experiments, ITD are the most widely studied. However, ITD instruments are considerably less sensitive than HRMS instruments, and therefore this instrumentation is difficult to use for dioxin analysis in seafood samples, which usually present low dioxin concentrations. During the last decade, GC-ITD working in tandem operation mode (MS/MS) has been successfully used for the analysis of PCDD/Fs and related compounds in environmental samples (with relatively high concentrations of dioxins) such as sewage effluents [99,100], atmospheric aerosols [101], and fly ashes [100]. However, until now very few papers dealing with the analysis of PCDD/Fs in food samples have been published, and almost all of them analyzed fatty foods that present concentrations above 2 pg/g [102]. In the case of seafood and seafood products, GC-ITD(MS/MS) has produced comparable results to that of HRMS for PCDD/F determinations in seafood presenting total PCDD/F concentrations higher than 3 pg/g (fresh weight) [103,104]. In other works, Malavia et al. [44,45], compared GC-HRMS with GC-ITD (MS/MS) for the determination of dioxins and furans in vegetable and seafood oils and seafood tissue samples. The study was done within the framework of the European research project DIFFERENCE [63], and the results obtained with both instruments were comparable and within the consensus values. Nevertheless, the limits of detection obtained for GC-ITD(MS/MS) were between 0.07 and 0.20 pg/g of oil, whereas modern HRMS is two or three orders of magnitude lower. The main advantages of using ITD(MS/MS) included rapid determinations and low cost, simplicity of operation (once the method is developed) and maintenance, and high selectivity for dioxin isomers. The main disadvantage of ITD reported by other authors [105], such as the low reproducibility of quantification due to excessive ions coexisting with dioxins in the trap, have been solved in the modern and more recent GC-ITD(MS/MS) instrumentation with the ion source outside the trap, as has been demonstrated by Malavia et al. [46]. In their paper, the authors concluded that it is only possible to achieve reliable results for PCDD/F determinations at concentrations close to the maximum residue levels established by the EU for food by using external ionization. Other systems capable of performing MS/MS experiments, such as triple-quadrupole and hybrid MS instruments have also been proposed as suitable alternatives for dioxin analysis in foods. However, now they are not being used at laboratories for routine analysis of PCDD/Fs due to the high cost of acquisition, not being a real low-cost alternative to HRMS for dioxins analysis.

At the same time, some authors have explored the possibilities of the ToFMS analyzer for dioxin analysis [106]. Although the ToFMS analyzer allows simultaneous sampling and measurement of all ions across the mass range and full-spectrum sensitivity is comparable to a quadrupole instrument in the SIM mode, its limits of detection (in the pg range) make dioxin analysis in seafood samples very difficult. In fact the literature records a large variety of environmental applications such as screening for PCBs, pesticides, and brominated flame retardants in biological sample [107,108], but applications to dioxins are still rare. In addition, as mentioned earlier, at present GC×GC–ToF-MS are neither cheaper nor easier to use than HRMS, and so they are still not a real alternative to HRMS.

25.8 Bioanalytical Screening Methods

Several dioxin food incidents [8–10], and also new EU regulations [12], highlight the need for screening methods for food and feed materials. This need is even more acute during an incident, first to rapidly locate the source and second to reserve often-limited GC-HRMS capacity for confirmation of suspect samples.

Several bioanalytical detection methods (BDMs) for measuring dioxin-like activity have been developed since the early 1990s. These methods are based on the ability of key biological molecules to recognize a unique structural property of dioxins or to respond to dioxins in a specific way. Most bioassays are based on the assumption that dioxin compounds act through the aryl hydrocarbon receptor (AhR) signal transduction pathway. The biological methods include biomarkers (e.g., wildlife/human effects) [109], whole animal exposures (in vivo, laboratory exposure) [3], cell- or organ-based bioassays (e.g., EROD, in vitro luciferase) [110], and protein binding assays (e.g., ligand binding as well as immunoassays) [111,112].

All these methods entail an estimation of the TEQs present in the sample, so that unless an interference-free dioxin fraction is obtained, the TEQs calculated in this way may be dioxin-like PCBs or any other compound that responds to dioxins in a similar way.

Of all the methods mentioned, the one that has achieved most popularity and is accepted by the new EU regulations, as a screening method, is the one called DR CALUX bioassay. This method uses genetically modified rat or mouse hepatoma cells which respond to chemicals that activate the AhR. The recombinant CALUX cells contain a stably transfected AhR responsive firefly luciferase reported gene, which responds to dioxins and also to dioxin-like chemicals. At present, the different cell lines are commercialized and sold as the DR CALUX assay; one is based on modified rat H4IIE hepatoma cells (GudLuc1.1), and the other is based on modified H1L6.1 mouse hepatoma cells. The rat cells appear to be more sensitive, showing a response at TCDD concentrations below 1 pM [113]. Recently, there have been a number of international validation studies, such as ring trials on different foodstuffs (including seafood) under the EU-sponsored DIFFERENCE project [114]. The results for both methodologies were comparable and within the consensus values.

Bovee et al. [115] were the first to show their utility for screening milk fat around the existing limit of 6 pg TEQ/fat. Based on this work, the test was validated for other food matrices, including seafood and seafood products [116]. However, all of them stress that exhaustive cleanup (similar to that necessary for GC-HRMS) is essential to assure accurate results, which means that one of the advantages of using biological analysis—rapidity—is lost. Clearly therefore, there are some issues that still need improvement, chiefly relating to cleanup procedure.

References

1. Travis, C.C. and Hattemer-Frey, H.A., Human exposure to dioxins, *Sci. Total Environ.*, 104, 97, 1991.
2. Ormerod, S.J., Tyler, S.J., and Juttner, I., Effects of point source PCB contamination on breeding performance and post-fledging survival in the dipper *Cinclus cinclus*, *Environ. Pollut.*, 110, 505, 2000.
3. van den Berg, M. et al., Toxic equivalency factors (TEFs) for PCBs, PCDDs and PCDFs for humans and wildlife, *Environ. Health Perspect.*, 106, 775, 1998.
4. van Leeuwen, F.X.R. et al., Dioxins: WHO's tolerable daily intake (TDI) revisited, *Chemosphere*, 40, 1095, 2000.
5. Lundgren, K., van Babel, B., and Tysklind, M., Development of a high-performance liquid chromatography carbon column method for the fractionation of dioxin-like polychlorinated biphenyls, *J. Chromatogr. A*, 962, 79, 2002.
6. Sweetman, A.J. et al., Human exposure to PCDD/Fs in the U.K.: The development of a modelling approach to give historical and future perspectives, *Environ. Int.*, 26, 37, 2000.
7. Hays, S.M. and Aylward, L.L. Dioxin risk in perspective: Past, present, and future, *Regul. Toxicol. Pharm.*, 37, 202, 2003.
8. Malisch, R., Increase of the PCDD/F-contamination of milk, butter and meat samples by use of contaminated citrus pulp, *Chemosphere*, 40, 1041, 2000.
9. Bernard, A. et al., The Belgian PCB/Dioxin crisis: Analysis of the food chain contamination and health risk evaluation, *Environ. Res. Section A*, 88, 1, 2002.
10. Llerena, J.J. et al., A new episode of PCDDs/PCDFs feed contamination in Europe: The choline chloride, *Organohalogen Compd.*, 51, 283, 2001.
11. Commission Regulation (EC) No. 76/2000/EC, Setting maximum levels for dioxins in emissions of municipal waste incinerators, *Official J. Eur. Commun.*, L 321/91–100, 2000.

12. Council Regulation (EC) No. 199/2006/EC amending Regulation (EC) No. 466/2001 setting maximum levels for certain contaminants in foodstuffs as regards dioxins and dioxin-like PCBs, *Official J. Eur. Commun.*, L 32/34, February 2006.

13. Liem, A.-K.D. et al., Dietary intake of dioxins and dioxin-like PCBs by the general population of ten European countries. Results of EU-SCOOP Task 3.2.5. (Dioxins), *Organohalogen Compd.*, 48, 13, 2000.

14. Kiviranta, H. et al., Dietary intakes of polychlorinated dibenzo-*p*-dioxins, dibenzofurans and polychlorinated biphenyls in Finland, *Food Addit. Contam.*, 18, 945, 2001.

15. Kiviranta, H., Ovaskainen, M.-L., and Vartiainen, T., Market basket study on dietary intake of PCDD/Fs, PCBs, and PBDEs in Finland, *Environ. Intern.* 30, 923, 2004.

16. Focant, J.-F. et al., Levels and congener distribution of PCDDs, PCDFs and non-*ortho* PCBs in Belgian foodstuff. Assessment of dietary intake, *Chemosphere*, 48, 167, 2002.

17. Abad, E. et al., Study on PCDDs/PCDFs and co-PCBs content in food samples from Catalonia (Spain), *Chemosphere*, 46, 1435, 2002.

18. Karl, H., Ruoff, U., and Blüthgen, A., Levels of dioxins in fish and fishery products on the German market, *Chemosphere*, 49, 765, 2002.

19. Llobet, J.M. et al., Human exposure to dioxins through the diet in Catalonia, Spain: Carcinogenic and non-carcinogenic risk, *Chemosphere*, 50, 1193, 2003.

20. Bordajandi, L.R. et al., Study on PCBs, PCDD/Fs, organochlorine pesticides, heavy metals and arsenic content in freshwater fish species from the River Turia (Spain), *Chemosphere*, 53, 163, 2003.

21. Bordajandi, L.R. et al., Survey of persistent organic pollutants (PCBs, PCDD/Fs, PAHs), heavy metals (Cu, Cd, Zn, Pb, Hg) and arsenic in food samples from Huelva (Spain): Levels, congener distribution and health implications, *J. Agric. Food Chem.*, 52, 992, 2004.

22. Knutzen, J. et al., Polychlorinated dibenzofurans/dibenzo-*p*-dioxins (PCDF/PCDDs) and other dioxin-like substances in marine organisms from the Grenland fjords, S. Norway, 1975–2001: Present contamination levels, trends and species specific accumulation of PCDF/PCDD congeners, *Chemosphere*, 52, 745, 2003.

23. Fernández, M.A. et al., Temporal trends in PCDD, PCDF and non-*ortho* PCB concentrations in Spanish commercial dairy products from 1993 to 2001, *Organohalogen Compd.*, 56, 485, 2002.

24. Fernández, M.A. et al. Dietary intakes of polychlorinated dibenzo-*p*-dioxins, dibenzofurans and dioxin-like polychlorinated biphenyls in Spain, *Food Addit. Contam.*, 21, 983, 2004.

25. Baars, A.J. et al., Dioxins, dioxin-like PCBs and non-dioxin-like PCBs in foodstuffs: Occurrence and dietary intake in The Netherlands, *Toxicol. Lett.*, 151, 51, 2004.

26. Ahmed, F.E., Analysis of polychlorinated biphenyls in food products, *Trends Anal. Chem.*, 22, 170, 2003.

27. Choi, D. et al., Determining dioxin-like compounds in selected Korean food, *Chemosphere*, 46, 1423, 2002.

28. Abad, E. et al., Evidence for a specific pattern of polycholrinated dibenzo-*p*-dioxins and dibenzofurans in bivalves, *Environ. Sci. Technol.*, 37, 5090, 2003.

29. Chang, R.R., Jarman, W.M., and Hennings, J.A., Sample cleanup by solid phase extraction for the ultratrace determination of polychlorinated dibenzo-p-dioxins and dibenzofurans in biological samples, *Anal. Chem.*, 65, 2420, 1993.

30. Krokos, F. et al., congener-specific method for the determination of *ortho* and non-*ortho* polychlorinated biphenyls, polychlorinated dibenzo-*p*-dioxins and polychlorinated dibenzofurans in foods by carbon-column fractionation and gas chromatography-isotope dilution mass spectrometry, *Fresen. J. Anal. Chem.*, 357, 732, 1997.

31. van Babel, B. et al., Development of a solid phase carbon trap for simultaneous determination of PCDDs, PCDFs, PCBs and pesticides in environmental samples using SFE-LC, *Anal. Chem.*, 68, 1279, 1996.

32. Miyawaki, T., Kawashima, A., and Honda, K. Development of supercritical carbon dioxide extraction with a solid phase trap for dioxins in soils and sediments, *Chemosphere*, 70, 648, 2008.

33. Camel, V., Microwave-assisted solvent extraction of environmental samples, *Trends Anal. Chem.*, 19, 229, 2000.

34. Björklund, E. et al., New strategies for extraction and clean up of persistent organic pollutants from food and feed samples using selective pressurized liquid extraction, *Trends Anal. Chem.*, 25, 318, 2006.

35. Wiberg, K. et al., Pressurized liquid extraction of polychlorinated dibenzo-p-dioxins, dibenzofurans and dioxin-like polychlorinated biphenyls, from food and feed samples, *J. Chromatogr. A*, 1138, 55, 2007.

36. Hess, P. et al., Critical review of the analysis of non- and mono-*ortho*-chlorobiphenyls, *J. Chromatogr. A*, 703, 417, 1995.

37. Ramos, L., Hernández, L.M., and González, M.J., Elution pattern of pattern of planar CBs and 2,3,7,8-PCDD/Fs on chromatographic adsorbents and factors affecting the mechanism of retention, possibilities of selective separation of both families, *J. Chromatogr. A*, 759, 127, 1997.

38. Molina, L. et al., Separation of non-*ortho* polychlorinated biphenyls congeners on pre-packed carbon tubes. Application to analysis in sewage sludge and soils samples, *Chemosphere*, 40, 921, 2000.

39. Liu, H.X. et al., Separation of polybrominated diphenyl ethers, polychlorinated biphenyls, polychlorinated dibenzo-p-dioxins and dibenzo-furans in environmental samples using silica gel and florisil fractionation chromatography, *Anal. Chem. Acta*, 557, 314, 2006.

40. Pirard, C., Focant J.-F., and Pauw, E.D., An improved clean-up strategy for simultaneous analysis of polychlorinated dibenzo-p-dioxins (PCDD), and polychlorinated dibenzofurans (PDF), and polychlorinated biphenyls (PCB) in fatty food samples, *Anal. Bioanal. Chem.*, 372, 373, 2002.

41. US EPA Method 1613 Revision B, Tetra-through Octa-Chlorinated Dioxins and Furans by Isotope Dilution HRGC/HRMS, US EPA, Washington, DC, April 2002.

42. US EPA Method 1668, Toxic polychlorinated biphenyls by isotope dilution high resolution gas chromatography/high resolution mass spectrometry, US EPA, Washington, DC, 1999.

43. Commission Regulation (EC) No. 1883/2006/EC laying down methods of sampling and analysis for the official control of the levels of nitrates in certain foodstuffs, *Official J. Eur. Commun.*, L 364/32, December 2006.

44. Malavia, J. et al., Analysis of polychlorinated dibenzo-p-dioxins, dibenzofurans, and dioxin-like polychlorinated biphenyls in vegetable oil samples by gas chromatography-ion-trap tandem mass spectrometry, *J. Chromatogr. A*, 1149, 321, 2007.

45. Malavia, J. et al., Ion-trap tandem mass spectrometry for the analysis of polychlorinated dibenzo-p-dioxins, dibenzofurans, and dioxin-like polychlorinated biphenyls in food, *J. Agric. Food Chem.*, 55, 10531, 2007.

46. Malavia, J., Santos, F.J., and Galceran, M.T., Comparison of gas chromatography–ion-trap tandem mass spectrometry systems for the determination of polychlorinated dibenzo-p-dioxins, dibenzofurans and dioxin-like polychlorinated biphenyls, *J. Chromatogr. A*, 1186, 302, 2008.

47. Danielsson, C. et al., Trace analysis of polychlorinated dibenzo-p-dioxins, dibenzofurans and WHO polychlorinated biphenyls in food using comprehensive two-dimensional gas chromatography with electron-capture detection, *J. Chromatogr. A*, 1086, 61, 2005.

48. Focant, J.-F. et al., Comprehensive two-dimensional gas chromatography with isotope dilution time-of-flight mass spectrometry for the measurement of dioxins and polychlorinated biphenyls in foodstuffs: Comparison with other methods, *J. Chromatogr. A*, 1086, 45, 2005.

49. Hoogenboom, L. et al., The CALUX bioassay: Current status of its application to screening food and feed, *Trends Anal. Chem.*, 25, 410, 2006.

50. Eskilsson, C.S. and Björklung, E., Analytical-scale microwave-assisted extraction. *J. Chromatogr. A*, 902, 227, 2000.

51. Hess, P. and Wells, D.E., Evaluation of dialysis as a technique for the removal of lipids prior to the GC determination of *ortho* and non-*ortho* chlorobiphenyls, using 14C-labelled congeners, *Analyst*, 126, 829, 2001.

52. Otaka, H. and Hashimoto, S., Fast matrix digestion with ethanolic alkali plus pyrogallol for polychlorinated dibenzo-*p*-dioxins, polychlorinated dibenzofurans and coplanar polychlorinated biphenyls analysis in biological samples, *Anal. Chem Acta*, 509, 21, 2004.

53. Ryan, J.J. et al., The effect of strong alkali on the determination of polychlorinated dibenzofurans (PCDFs) and polychlorinated dibenzo-*p*-dioxins (PCDDs), *Chemosphere*, 18, 135, 1989.

54. Lanbropoulou, D.Q., Konstantinou, I.K., and Albanis, T.A., Sample pretreatment method for the determination of polychlorinated biphenyls in bird livers using ultrasonic extraction followed by headspace solid-phase microextraction and gas chromatography-mass spectrometry, *J. Chromatogr. A*, 1124, 97, 2006.

55. Smith R.M., Supercritical fluids in separation science-the dreams, the reality and the future, *J. Chromatogr. A*, 856, 83, 1999.

56. van der Velde, E.G. et al., SFE as clean-up technique for ppt-levels of PCBs in fatty samples, *Organohalogen Compd.*, 27, 247, 1996.

57. Suchan, P. et al., Pressurized liquid extraction in determination of polychlorinated biphenyls and organochlorinated pesticides in fish samples, *Anal. Chim. Acta*, 520, 193, 2004.

58. Björklund, E., von Holst, C., and Anklam, E., Fast extraction, clean up and detection methods for rapid analysis and screening of seven indicator PCBs in food matrices, *Trends Anal. Chem.*, 21, 39, 2002.

59. Sporring, S. and Björklund, E., Selective accelerated solvent extraction of PCBs from food and feed samples, *Organohalogen Compd.*, 60, 1, 2003.

60. Sporring, S. and Björklund E., Selective pressurized liquid extraction of polychlorinated biphenyls from fat-containing food and feed samples: Influence of cell dimensions, solvent type, temperature and flush volume, *J. Chromatogr. A*, 1040, 155, 2004.

61. Haglund, P. et al., Hyphenated techniques for dioxin analysis: LC-LC-GC-ECD, GCXGC-ECD, and selective PLE with GC-HRMS or bioanalytical detection, *Organohalogen Compd.*, 66, 376, 2004.

62. Bernsmann, T. and Fürst, P., Comparison of accelerated solvent extraction (ASE) with integrated sulphuric acid clean up and soxhlet extraction for determination of PCDD/PCDF, dioxin-like PCB and indicator PCB in feeding stuffs, *Organohalogen Compd.*, 66, 159, 2004.

63. European Commission DIFFERENCE. Project G6RD-CT-2001-00623. www.dioxins.nl

64. Nording, M. et al., Monitoring dioxins in food and feedstuffs using accelerated solvent extraction with a novel integrated carbon fractionation cell in combination with CAFLUX bioassay, *Anal. Bioanal. Chem.*, 381, 1472, 2005.

65. Focant, J.-F. et al., Integrated PLE-multi step automated clean up and fractionation for the measurement of dioxins and PCBs in food and feed, *Organohalogen Compd.*, 67, 261, 2005.

66. Focant, J.-F., Pirard, C., and De Pauw, E., Automated sample preparation-fractionation for the measurement of dioxins and related compounds in biological matrices: A review, *Talanta*, 63, 1101, 2004.

67. de Boer, J., Trends in chorobiphenyl contents in livers of Atlantic cod (*Gadus morhua*) from the North Sea, 1979–1987, *Chemosphere*, 17, 1811, 1988.

68. De Boer, J. and Lau, R.J., Developments in the use of chromatographic techniques in marine laboratories for the determination of halogenated contaminants and polycyclic aromatic hydrocarbons, *J. Chromatogr. A*, 1000, 223, 2003.

69. Strandberg, B., Bergqvist, P.A., and Rappe, C., Dyalisis with semipermeable membrane as an efficient lipid removal method in the analysis of bioaccumulative chemicals, *Anal. Chem.*, 70, 528, 1998.

70. Ahmed, F.E., Analysis of pesticides and their metabolites in foods and drinks, *Trends Anal. Chem.*, 20, 649, 2001.

71. Smith, L.M., Stalling, D.L., and Johnson, J.L., Determination of part-per-trillion levels of polychlorinated dibenzofurans and dioxins in environmental samples, *Anal. Chem.*, 56, 1839, 1984.

72. Lamparski, L.L., Nestrick, T.J., and Stehl, R.H., Determination of part-per-trillion concentrations of 2,3,7,8-tetrachlorodibenzo-*p*-dioxins in fish, *Anal. Chem.*, 51, 1453, 1979.

73. Creaser, C.S., Krokos, F., and Startin, J.R., Analytical methods for the determination of non-*ortho* substituted chlorobiphenyls: A review, *Chemosphere*, 24, 1981, 1992.

74. Harrad, S.J. et al., A method for the determination of PCB congeners 77, 126 and 169 in biotic and abiotic matrices, *Chemosphere*, 24, 1147, 1992.

75. Kannan, N. et al., A comparison between activated charcoals and multidimensional GC in the separation and determination of (non-*ortho* Cl substituted) toxic chlorobiphenyls, *Chemosphere*, 23, 1055, 1991.

76. van der Velde, E.G. et al., Analysis and occurrence of toxic planar PCBs, PCDDs and PCDFs in milk by use of carbosphere activated carbon, *Chemosphere*, 28, 693, 1994.

77. Creaser, C.S. and Al-Haddad, A., Fractionation of polychlorinated-biphenyls, polychlorinated dibenzo-*p*-dioxins and polychlorinated dibenzofurans on porous graphitic carbon, *Anal. Chem.*, 61, 1300, 1989.

78. de Boer, J. et al., Non-*ortho* and mono-*ortho* substituted chlorobiphenyls and chlorinated dibenzo-p-dioxins and dibenzofurans in marine freshwater fish and shellfish from the Netherlands, *Chemosphere*, 26, 1823, 1993.

79. Ramos, L., Hernández, L.M., and González, M.J., Simultaneous separation of coplanar and chiral polychlorinated biphenyls by off-line pyrenil-silica HPLC/HRGC. Enantiomeric ratios of chiral PCBs by HRGC/LRMS (SIM), *Anal. Chem.*, 71, 70, 1999.

80. Diaz-Ferrero et al., Study of dioxins, furans and polychlorinated biphenyl fractionation on HPLC using a pyrenil column for their analysis in meat and fish samples, *Afinidad*, 62, 433, 2005.

81. Concejero, M.A. et al., Suitability of several carbon sorbents for the fractionation of various sub-groups of toxic polychlorinated biphenyls, polychlorinated dibenzo-*p*-dioxins and polychlorinated dibenzofurans, *J. Chromatogr A*, 917, 227, 2001.

82. Concejero, M.A. et al., Different retention of dioxin-like compounds and organochlorinated insecticides on an immunochromatographic column. Interpretation and applicability, *J. Sep. Sci.*, 27, 1101, 2004.

83. Jiménez, B. et al., Levels of PCDDs and PCDFs in oil components of the Spanish diet, *Chemosphere*, 32, 461, 1996.

84. Serrano, R. et al., Congener-specific determination of polychlorinated biphenyls in shark and grouper livers from the northwest African Atlantic Ocean, *Arch. Environ. Contam. Toxicol.*, 38, 217, 2000.

85. Eljarrat, E. et al., Evaluation of an automated clean-up system for the isotope-dilution high resolution mass spectrometry analysis of PCB, PCDD and PCDF in food, *Fresen. J. Anal. Chem.*, 371, 983, 2001.

86. Focant, J.-F. et al., Fast clean-up for polychlorinated dibenzo-*p*-dioxins, dibenzofurans and coplanar polychlorinated biphenyls analysis of high-fat-content biological samples, *J. Chromatogr. A*, 925, 20, 2001.

87. Ryan, J.J. et al., Gas chromatographic separations of all 136 tetra- to octapolychlorinated dibenzo-p-dioxins and polychlorinated dibenzofurans on nine different stationary phases, *J. Chromatogr. A*, 541, 131, 1991.

88. Abad E., Caixach, J., and Rivera, J., Application of DB-5ms gas chromatography column for the complete assignment of 2,3,7,8-substituted polychlorodibenzo-*p*-dioxins and polychlorodibenzo-furans in samples from municipal waste incinerator emissions, *J. Chromatogr. A*, 786, 125, 1997.

89. Fishman, V.N., Martin, G.D., and Lamparski, L.L., Comparison of a variety of gas chromatographic columns with different polarities for the separation of chlorinated dibenzo-*p*-dioxins and dibenzo-furans by high-resolution mass spectrometry, *J Chromatogr A*, 1139, 285, 2007.

90. Cochram, J. et al., Retention time profiling for all 136 tetra-through octa-chlorinated dioxins and furans on a unique, low-bleed, thermally-stable gas chromatography column, *Organohalogen Compd.*, 69, 115, 2007.

91. Schomburg, G., Husmann, H., and Hubinger, E., Multidimensional separation of isomeric species of chlorinated hydrocarbons such as PCB, PCDD, and PCDF, *J. High Res. Chromatogr.*, 5, 395, 1985.

92. Dallüge, J., Beens, J., and Brinkman, U.A.Th., Comprehensive two-dimensional gas chromatography: A powerful and versatile analytical tool, *J. Chromatogr. A*, 1000, 69, 2003.

93. Liu, Z. and Phillips, J.B., Comprehensive two-dimensional gas chromatography using an on-column thermal modulator interface, *J. Chromatogr. Sci.*, 29, 227, 1991.

94. Korytár, P. et al., Separation of seventeen 2,3,7,8-substituted polychlorinated dibenzo-*p*-dioxins and dibenzofurans and 12 dioxin-like polychlorinated biphenyls by comprehensive two-dimensional gas chromatography with electron-capture detection, *J. Chromatogr. A*, 1038, 189, 2004.

95. Haglund, P. et al., GCxGC-ECD: A promising method for the determination of dioxins and dioxin-like PCBs in food and feed, *Anal. Bioanal. Chem.*, 390, 1815, 2008.

96. Mondello, L. et al., Comprehensive two-dimensional gas chromatography-mass spectrometry: A review, *Mass Spectrom. Rev.*, 27, 101, 2008.

97. Korytár P. et al., Quadrupole mass spectrometer operating in the electron-capture negative ion mode as detector for comprehensive two-dimensional gas chromatography, *J. Chromatogr. A*, 1067, 255, 2005.

98. Baughman, R. and Meselson, M., An analytical method for detecting TCDD (dioxin): Levels of TCDD in samples from Vietnam, *Environ. Health Perspect.*, 5, 27, 1973.

99. Küchler, T. and Brzezinski, H., A comparison of GC-MS/MS for the analysis of PCDD/Fs in sewage effluents, *Chemosphere*, 40, 213, 2000.

100. Fabrellas, B. et al., Analysis of dioxins and furans in environmental samples, *Chemosphere*, 55, 1469, 2004.

101. Mandalakis, M., Tsapakis, M., and Stephanou, E.G., Optimization and application of high-resolution gas chromatography with ion trap tandem mass spectrometry to the determination of polychlorinated biphenyls in atmospheric aerosols, *J. Chromatogr. A*, 925, 183, 2001.

102. Eppe, G. et al., PTV-LV-GC/MS/MS as screening and complementary method to HRMS for the monitoring of dioxin levels in food and feed, *Talanta*, 63, 1135, 2004.

103. Grabic, R., Novák, J., and Pacáková V., Optimization of a GC-MS/MS method for the analysis of PCDDs and PCDFs in human and fish tissue, *J. High Resolut. Chromatgr.*, 23, 595, 2000.

104. Haywar, D.G. et al., Quadrupolo ion storage tandem mass spectrometry and high resolution mass spectrometry: Complementary application in the measurements of 2,3,7,8-chlorine substituted polychlorinated dibenzo-*p*-dioxins, dibenzofurans, in US Foods, *Chemosphere*, 43, 407, 2001.

105. Eljarrat, E. and Barceló, D., Congener-specific determination of dioxins and related compounds by gas chromatography coupled to LRMS, HRMS, MS/MS and ToFMS, *J. Mass Spectrom.*, 37(11), 1105, 2002.

106. Focant, J.-F. et al., Recent advances in mass spectrometric measurement of dioxins. *J. Chromatogr. A*, 1067, 265, 2005c.

107. van Babel, B. et al., Fast screening for PCBs, pesticides and brominated flame retardants in biological samples by SFE-LC in combination with GC-ToF, *Organohalogen Compd.*, 40, 293, 1999.

108. Dallüge, J., Roose, P., and Brinkman, U.A.Th., Evaluation of a high-resolution time of flight mass spectrometer for gas chromatography determination of selected environmental contaminants, *J. Chromatogr. A*, 965, 207, 2002.

109. Scheter, A., *Dioxins and Health*, Plenum, New York, 1994.

110. Jones, J.M., Anderson, J.W., and Tukey, R.H., Using the metabolism of PAHs in a human cell line to characterize environmental samples, *Environ. Toxicol. Pharmacol.*, 8, 119, 2000.

111. Diaz-Ferrero, J. et al., Bioanalytical methods applied to endocrine disrupting polychlorinated biphenyls, polychlorinated dibenzo-*p*-dioxins and polychlorinated dibenzofurans. A review, *Trends Anal. Chem.*, 16, 563, 1997.

112. Seidel, S.D. et al., Ah receptor-based chemical screening bioassays: Application and limitations for the detection of Ah receptors agonists, *Toxicol. Sci.*, 55, 107, 2005.

113. Goeyens, L. et al., Comparison of the rat and mouse cell lines commercially available for CALUX bioassays, *Organohalogen Compd.*, 66, 608, 2004.

114. van Loco, J. et al., The international validation of bio- and chemical-analytical screening methods for dioxins and dioxin-like PCBs: The DIFFERENCE project rounds 1 and 2, *Talanta*, 63, 1169, 2004.

115. Bovee, T.F.H. et al., Validation and use of the CALUX-bioassay for the determination of dioxins and PCBs in bovine milk, *Food Addit. Contam.*, 15, 863, 1998.

116. van Leeuwen, S.P.J. et al., Polychlorinated dibenzo-*p*-dioxins, dibenzofurans and biphenyls in fish from the Netherlands: Concentrations, profiles and comparison with DR CALUX® bioassay results, *Anal. Bioanal. Chem.*, 389, 321, 2007.

Chapter 26

Environmental Contaminants: Persistent Organic Pollutants

Monia Perugini

Contents

26.1 Introduction 728
26.2 Characterization of PAHs 728
    26.2.1 PAHs Methods of Extraction 729
        26.2.1.1 Saponification 729
        26.2.1.2 Soxhlet Extraction 730
        26.2.1.3 Sonication Method 730
        26.2.1.4 Pressurized Liquid Extraction 730
        26.2.1.5 Supercritical Fluid Extraction 731
    26.2.2 Clean-Up of Extracts 731
        26.2.2.1 Solid-Phase Extraction 732
        26.2.2.2 Gel Permeation Chromatography 732
    26.2.3 Chromatographic Analysis 733
        26.2.3.1 High-Performance Liquid Chromatography 733
        26.2.3.2 Gas Chromatography 734
26.3 Characterization of PCBs 734
    26.3.1 PCBs Methods of Extraction 735
        26.3.1.1 Soxhlet Extraction 735
        26.3.1.2 Sonication Method and Liquid–Liquid Partitioning 736
        26.3.1.3 Pressurized Liquid Extraction 736
        26.3.1.4 Supercritical Fluid Extraction 737

26.3.2 Clean-Up of Extracts 737
26.3.3 Chromatographic Analysis.. 738
   26.3.3.1 Gas Chromatography with Electron Capture Detector and Gas
        Chromatography–Mass Spectrometry... 738
   26.3.3.2 Gas Chromatography–High-Resolution Mass Spectrometry................739
References 740

26.1 Introduction

Persistent organic pollutants (POPs) are organic compounds of natural or anthropogenic origin characterized by low water and high lipid solubility, resulting in bioaccumulation in fatty tissues of living organisms. They are widespread contaminants of the marine ecosystem mainly because of their depositions from the atmosphere, able to be transported for long distances from their point of origin, or as result of river wastewater transport, surface runoff, industrial development, or agricultural activities. It is important to highlight that all pollutants, whether in air or on land tend to end up in the ocean [1]; furthermore, closed or semienclosed seas are particularly exposed to the pollution risk. In recent years there has been a growing interest in these pollutants, in particular for their impact on human health. The risks posed by POPs for human health have become of increasing concern and are actually object of a worldwide agreement among several governments, including measures to reduce or eliminate their release in the environment. Their environmental presence is of particular gravity because of their toxicity, bioavailability, and persistence. The seas and costal areas are, generally, final recipients for terrestrial wastewaters containing both anthropological and natural origin pollutants; furthermore, at the same time they are very important economic and aquatic resources. POPs' presence poses serious adverse effects on the marine ecosystem because they affect all organisms from primary to secondary producer levels up until the top levels of the seafood chain. Polychlorinated biphenyls (PCBs) and polycyclic aromatic hydrocarbons (PAHs) have been listed as priority pollutants by the United Nations Environment Programme (UNEP) because of their potential carcinogenicity, mutagenicity, and toxicity to aquatic organisms and humans. This chapter will focus primarily on the sample preparation and analytical methods of determination of PAHs and PCBs that represent two classes of pollutants very often detected in fish and shellfish. The intent of this chapter is to provide an exhaustive summary of analytical methods, carefully for the sampling, preparation, and analysis techniques.

26.2 Characterization of PAHs

Polycyclic aromatic hydrocarbons (PAHs) are a class of compounds consisting of at least two or more fused aromatic rings of carbon and hydrogen atoms. The chemical properties of PAHs depend on their number of rings and molecular mass. In general they are solids having high melting (60°C–450°C) and boiling (200°C–600°C) points, showing low degrees of volatility, and are rather inert lipophilic compounds, which easily dissolve in organic solvents. There have been identified over 100 PAHs in the environment that occur as complex mixtures, of which the composition may differ by source. Of these 100 PAHs 16 were classified as "priority pollutants" according to the U.S. Environmental Protection Agency (EPA). PAHs are produced by natural and anthropogenic activities as products of incomplete pyrolysis from organic materials [2]. Forest fires, domestic heating, combustion of fossil fuels as gasoline, coal, and diesel fuel, industrial activities as petroleum, refining processes and catalytic cracking, rural and urban sewage sludge, smoking food processes, and tobacco and cigarette smoke represent only a few PAHs sources. Aquatic organisms that metabolize PAHs to little or no extent, such as algae, mollusks, and the more primitive invertebrates (protozoans, porifers, and cnidaria) accumulate high concentrations of PAHs, whereas fish and higher invertebrates, which metabolize PAHs, accumulate little or no PAHs. Biomagnification of PAHs has not been observed in aquatic systems and would not be expected to occur because most organisms have a high biotransformation potential for PAHs. Organisms at higher trophic levels in food chains show the highest potential for biotransformation.

The general concern for this class of compounds is due to their mutagenic and carcinogenic activity. Dihydrodiols and epoxide derivatives, products of the liver by PAHs metabolism, form covalent adducts with DNA and proteins that begin a mutagenic process in the cells. The analytical choice for determining PAHs depends on the purpose of the measurement: carcinogenic PAHs are of interest in studies of human health, but those widespread in the environment may be of interest in ecotoxicological studies. The quantification of PAHs is particularly advantageous when their profiles can be correlated with sources and effects.

Many extraction, purification techniques, and combinations have been described and validated, but no single scheme is commonly recognized as "the best" for seafood samples, because all of these display advantages and disadvantages. Chromatographic techniques, such as high-performance liquid chromatography (HPLC) and gas chromatography (GC) are common methods of PAHs detection. The intent of this section is both to provide an exhaustive list of analytical methods to detect PAHs in marine organisms and to compare their efficiency.

26.2.1 PAHs Methods of Extraction

Generally PAH levels are lower in fish musculature than in the liver or in the soft tissues of mollusks, because they have the ability to metabolize and excrete PAHs to water-soluble compounds. All solid samples require homogenization before their extraction. Shells of bivalves are washed, using distilled water, to remove external impurities, in order to avoid the contamination of edible parts. Instead, in fish, the skin and bone are removed before homogenization. The efficiency of PAHs extraction depends both on sample preparations and the polarity of the solvents used [3]. Generally when samples are totally soluble in the organic solvents the recovery of PAHs is high. Furthermore, it is rational to prefer the use of certified reference materials (CRM) rather than add PAHs standards to the samples prior to extraction, because they remain unbound and are easier to extract.

26.2.1.1 Saponification

Saponification is the classical isolation method of PAHs from lipophilic matrices and protein-rich foods [4–7]. The possibility of alkaline saponification in handling wet samples directly assists the recovery of the more volatile PAHs. In marine organisms, due to the presence of insoluble fats and proteins, alkaline digestion, using aqueous, methanolic, or ethanolic potassium hydroxide solutions (KOH) for the hydrolysis of lipids, is necessary [8]. However, methanol in the presence of acid or basic catalyst can convert the fats in fatty acid methyl esters which are difficult to remove from a PAH fraction. The normality of the solution can range from 0.5 to 6 N and the length of saponification varies from 2 to 24 h, depending on the characteristics of the sample. Lean tissues take less time than adipose tissues. Reflux on a water bath, can improve the PAHs recovery and speed the length of extraction. Cyclohexane, hexane, pentane, and isooctane are the solvents of election for the following liquid–liquid partition. To protect PAHs from light photodegradation the samples can be covered with aluminum foils. Saponification is an easy method not requiring sophisticated instruments but a long period of analysis, about 3–5 h. Moreover, the presence of alcohol in the hydrolytic solution can interfere with the alkylated PAHs derivatives, and harsh alkaline digestive treatment could have a partial effect of decomposition on the more labile PAHs [5,9].

26.2.1.2   Soxhlet Extraction

Soxhlet extraction represents a common method in routine laboratories [10–14].

Generally extraction is performed using cellulose extraction thimbles filled with fresh homogenized samples and anhydrous sodium sulfate and covered with glass wool. It is possible to use dried samples, but the drying process can determine loss of low-molecular weight PAHs. Physicochemical properties and toxicity are facts considered when choosing extraction solvents. For the extraction, it is possible to use several organic solvents such as acetone, hexane, or methanol, but mixtures of hexane–acetone (1:1, v/v) or chloroform–methanol (2:1, v/v) are the most suitable. The total time of extraction is about 6–8 h. It is possible to cover the Soxhlet apparatus with aluminum foil to avoid access of daylight. The Soxhlet extraction often uses large volumes of organic solvents but its efficiency, considering recoveries of analytes and repeatability, is still the method of choice for many studies [15]. For the high molecular PAHs extraction this method achieves the best recovery.

26.2.1.3   Sonication Method

Ultrasonication with solvents, as an alternative to Soxhlet extraction, has advantages in terms of reduced time of extraction. In fact this process lasts only 20–30 min. The reproducibility and the recovery efficiency, above all for the lower PAHs, is practically equivalent to all the other techniques, particularly for solid samples. Homogenized samples, dried with anhydrous sodium sulfate, can be extracted using several extraction solvents including chloroform, hexane–dichloromethane (1:1, v/v), or hexane–acetone (1:1, v/v). As reported in literature this last mixture is the most efficient isolation solvent [16,17]. The ultrasonic procedure can be repeated to improve the efficiency of the extraction. To support the best performance and avoid a decrease of efficiency of the ultrasonic extraction the probe requires frequent replacement.

26.2.1.4   Pressurized Liquid Extraction

Pressurized liquid extraction (PLE) represents a modern and alternative extraction technique, at elevated pressures and temperatures, enabling the reduction of solvent quantities and the time of analysis, the improvement in precision of the analyte recovery, and avoids the contamination of samples. Accelerated solvent extraction system (ASE) is the Dionex trade name for the instrument that uses this technique. Many different solvents or mixtures can be employed for extraction. Hexane, toluene, hexane–acetone (1:1, v/v), and dichloromethane–acetone (1:1, v/v) are the most widely used organic solvents. Considering the hydrophobicity of the majority of PAHs one would expect higher extraction efficiency for less polar mixtures as hexane–acetone (3:1 or 4:1, v/v) but, by using these solvents the recovery for hydrophobic compounds is lower, because the solvent is immiscible with water and enables penetrating into the wet sample [18]. The use of the hexane–acetone (1:1, v/v) mixture is recommended by EPA method 3545A for extraction of semivolatile organics, OCPs, and PCBs [19]. The extraction can also be performed at several temperatures within a range of 60°C–200°C, but usually an oven temperature of 100°C is considered the optimal condition. Lower temperatures enable the extraction of low molecular PAHs and the use of temperatures higher than 140°C increases the recovery of single analytes but decreases the selectivity of extraction, and the waxes and pigments matrix can interfere with analytical determination. With ASE it is also possible to select the number of static cycles. One extraction cycle of 5–7 min or two of these cycles are the best choice to achieve maximum efficiency of extraction. Using more or longer cycles increase the risk of the waxes and pigments extraction making more difficult the handling of extracts. Extraction is performed using homogenized fresh fish samples mixed with drying agents, such as sodium sulfate anhydrous or diatomaceous earth. A cellulose paper is placed at the bottom of extraction cells before the sample homogenates are loaded. The final step is the setup of instrument conditions such as heating-up time, number of cycles, oven temperature, flush volume, and purge time. The length of this last parameter does not influence the PAHs recovery. Although the repeatability of PLE extraction is comparable with the classic techniques like the Soxhlet and both methods are able to give good recoveries, however, only the PLE displays a low solvent consumption, a short time of analysis, and a higher sample number. The main disadvantages are represented in the high cost of ASE as compared to equipment used for the Soxhlet or the extraction enhanced by sonication.

26.2.1.5 Supercritical Fluid Extraction

Supercritical fluid extraction (SFE) is an expanding analytical technique that has gained attention as a rapid alternative to conventional liquid extraction. The main advantages of SFE include non-toxicity, cost-effectiveness, high separation efficiencies, and short analysis time. Carbon dioxide (CO_2) is the most used supercritical material for its ease of manipulation, good solvent strength, compatibility with solutes, lack of toxicity, and also because it is nonflammable, noncorrosive, odorless, and inexpensive. This technique can also be directly coupled with on-column GC. SFE is often used for PAH extraction from sediments, while its use for processing biotic samples is limited, probably because the sample composition (fat and moisture) influences the robustness of this technique. Compared with the Soxhlet extraction, SFE gives the same results in terms of accuracy and precision but reduces the use of organic solvents and extraction time.

26.2.2 Clean-Up of Extracts

Especially in marine organisms, the PAHs are associated with substances that interfere with their separation and identification, and extracts may necessitate additional clean-up before analysis and quantification, above all, if the alkaline saponification has not been used as an extraction step. Lipids and pigments may be the main interfering substances in the analysis of PAHs in biological samples. The object of the clean-up is to remove these coextracted materials in order to extend the column lifetime and to improve detection and quantification limits. Solid-phase extraction (SPE) and gel permeation chromatography (GPC) are two nondestructive techniques very often applied to purify the samples. Some authors [20] carry out the clean-up using concentrated sulfuric acid, but this method is advised against PAHs because they are compounds of low chemical stability and can be partially destroyed by this treatment.

26.2.2.1 Solid-Phase Extraction

This method is a sample treatment technique which passes a liquid sample through a sorbent. For determining PAHs in seafoods or solid samples solid-phase extraction (SPE) is used after the extraction processes. It is a technique of common use in many laboratories because it does not require large quantities of solvents, has a short period of analysis, can be automated, and several kinds of commercial sorbents are available. Recently SPE is carried out in online mode, coupled with HPLC or GC, but usually is used in offline mode. At present, conventional chromatographic columns are substituted by prepacked commercial cartridges, which have advantages in terms of time, solvents consumed, and reproducibility. Disks are also available, but cartridges are more commonly used because the commercial availability of disks is reduced. The classical sorbents, alumina, Florisil, silica gel, or C_{18}-bonded silica are widely used for PAH purification, but the choice of the sorbent depends on the selectivity of the final detection step. The solvents widely used for the elution of PAHs are acetone, acetonitrile, methanol, toluene, dichloromethane, tetrahydrofurane, or mixtures of these. The mixtures represent a better combination to gain best recoveries as the high molecular PAH recovery is generally higher with nonpolar solvents, while less polar solvents ensure higher recovery for low molecular PAHs. An other critical parameter to consider is the solvent for reconstituting the extracts before the SPE and its concentration. The hydrophobicity of PAHs may lead to adsorption problems, and if the organic solvent is weak or low concentrations are used it is difficult to get a full solubilization of the high molecular PAHs. The best method, to gain good recoveries for all compounds before the SPE should be the optimization of these parameters. Some authors suggest to avoid the complete evaporation of extracts after SPE because this procedure may lead to a loss of more volatile PAHs. At last, conditioning of SPE cartridge, flow-rate elution, drying of SPE cartridge, and PAH concentration in the samples are other factors affecting the recovery of these pollutants.

26.2.2.2 Gel Permeation Chromatography

This technique is a chromatographic process of the separation of molecules in which impurities and target contaminants are separated based on their hydrodynamic volume when a solution flows through a packed bed of porous gels. The underlying principle of GPC is that particles of different sizes will elute through a stationary phase at different rates. The collected fractions are examined by chromatographic instruments to determine the concentration of the particles eluted. The GPC procedure may employ several types of gels for the purification such as polystyrene divinylbenzene copolymer gels Bio-Beads S-X3, S-X12, XAD-2, Envirogel, and Phenogel. The stationary phase may also interact in undesirable ways with a particle and influence retention times, though great care is taken by column manufacturers to use stationary phases that are inert and minimize this issue. Proper column packing is important to maximize resolution: an overpacked column can collapse the pores in the beads, resulting in a loss of resolution. An underpacked column can reduce the relative surface area of the stationary phase accessible to smaller species, resulting in those species spending less time trapped in pores. The optimization of GPC mobile phase is an other important parameter to consider. The classical mobile phases used for the elution of samples are organic solvents such as chloroform, toluene, benzene, or dichloromethane. Considering workplace hazards as well as ecological aspects for the chloroform several authors prefer to employ safer solvent mixtures as dichloromethane–cyclohexane or acetate–cyclohexane. The GPC provides good purification but when the sample is a rich tissue it is important to inject a small amount of sample extract to avoid the column saturation.

26.2.3 Chromatographic Analysis

PAHs are now routinely identified and quantified by HPLC or GC. Each technique is rather expensive, and requires qualified operating personnel but both have a number of relative advantages. They are considered necessary in order to analyze "real" samples for a large number of PAHs with accuracy and precision.

26.2.3.1 High-Performance Liquid Chromatography

Liquid chromatography is a very sensitive technique and an excellent detection method for PAHs. Usually the analysis is carried out at ambient temperature, avoiding a thermal decomposition of heat-sensitive compounds and using a guard precolumn in order to require less clean-up than GC. Furthermore, HPLC is more suited to the analysis of high molecular PAHs than GC. The main advantages of HPLC derive from the capabilities of the detectors. Those most widely used for PAHs are ultraviolet (UV), diode array detectors (DAD), and fluorescence detectors (FLD). Especially the latter provides very high selectivity and sensitivity, showing detection limits at least one order of magnitude lower than those obtained with ultraviolet detectors. Furthermore, UV and DAD can be employed but more clean-up is required. DAD can be also used to confirm peaks and moreover, additional information on isomeric structure can be obtained from the spectra seen during the elution phase [21]. The specificity of FLD allows the determination of individual PAHs in the presence of other nonfluorescing impurities, but the solvents have to be oxygen-free to avoid the quenching of fluorescence of some PAHs, e.g., pyrene [7]. In addition, since different PAHs have different absorptivity or different fluorescence spectral characteristics at given wavelengths, the detector can be optimized for maximal response to specific compounds. In particular, wavelength-programmed fluorescence detection, to measure changes in excitation and emission wavelengths during a chromatographic run is being used for the analysis of marine samples. Due to the ring differences among PAHs the selection of an appropriate detection wavelength is very important. Acenaphthylene is not detected by the fluorescence detector because it does not emit any fluorescence [22]. In recent years, several selective HPLC columns for PAH separation are available on the market. The packing material considered most suitable for separating PAHs consists of reversed-phase columns, of silica particles chemically bonded to linear C_8 or C_{18} hydrocarbon chains. With these columns, the mobile phase should be more polar than the stationary phase. Typically, the elution program is in the gradient elution technique, and the mobile phase consists of mixtures of acetonitrile and water or methanol and water. The use of an isocratic separation technique is also possible, but separation time is generally too long and some peaks, like acenaphthene and fluorene can overlap.

As the efficiency of separation that can be achieved with HPLC columns is much lower than that with capillary GC, HPLC is generally less suitable for separating samples containing complex PAH mixtures, but more suitable for the separation of isomeric compounds. Several isomers including the chrysene–triphenylene and benzo[b]fluoranthene–benzo[k]fluorathene pairs are difficult to separate efficiently using the usual capillary gas chromatographic columns and can be identified by HPLC because this instrument offers a higher column selectivity [23]. Furthermore, the analysis of PAHs in complex matrices can be carried out using HPLC–mass spectrometry (LC–MS), which is also a very helpful detector for the characterization of thermally unstable compounds and begins to be used in routine analysis. According to the U.S. Environmental Protection Agency, HPLC is suitable, in particular, for lower molecular mass compounds like naphthalene, acenaphthene, and acenaphthylene, for which the detection limits can be relatively high.

26.2.3.2 Gas Chromatography

PAHs from aquatic organisms can also be analyzed by GC, that is considered an excellent method for the analysis of complex matrices. The signal is related linearly to the carbon mass of PAHs, and the chromatogram shows the quantitative composition of the sample directly. The greatest separation of these compounds can be obtained using columns with high efficiency in the order of 50,000–70,000 height equivalent to a theoretical plate (HETP). Fused silica capillary columns often with nonpolar phases, nowadays commercially available, are making it possible to analyze very complex mixtures containing more than 100 PAHs. The most widely used stationary phases are the methylpolylsiloxanes: especially SE-54 (5% phenyl-, 1% vinyl-substituted) and SE-52 (5% phenyl-substituted), but SE-30 and OV-101 (unsubstituted), OV-17 (50% phenyl-substituted), Dexsil 300 (carborane-substituted), and their equivalent phases are also used. Chemically bonded phases are used increasingly because they can be rinsed to restore column performance and undergo little "bleeding" at high temperatures of analysis (about 300°C) that are required for determining high-boiling-point compounds. Splitless or cold on-column injection is necessary to gain sensitivity in trace analysis. The latter is preferred as it allows better reproducibility and reduces discrimination against the high molecular PAHs which is difficult to avoid entirely when using splitless injection [7]. Although the flame ionization detector (FID) was the most commonly applied detector in GC in the 1980s and was used because of its excellent linearity, sensitivity, and reliability, a range of more selective and sensitive detectors as the electron capture detector and MS have led to its replacement. Because FIDs are nonselective and are subject to background interferences from other carbonaceous sources the samples must be highly purified. GC–MS, conducted in electron impact (EI) mode, has gained wide acceptance and represents the analytical method of choice for identifying minor as well as major PAHs. By using the selected ion mode method (SIM) the singular compounds can be identified at concentrations of at least 100 times lower than is possible by HPLC and it is possible to simplify the time-consuming clean-up process. Identification of single PAHs in SIM mode is helped by means of available libraries of reference spectra that can be used to match the spectra obtained and control the purity of a compound. As isomeric compounds often have indistinguishable spectra, however, the final assignment must also be based on retention. Capillary GC is an excellent technique to separate and determine PAHs in complex mixtures: it presents a high column efficiency in terms of plate number but is not more suitable for the separation of isomeric compounds because it does not show a great selectivity.

A number of unconventional instruments and techniques based on spectroscopic principles have been developed as possible alternatives to the chromatographic methods for PAHs. Most of them are, however, expensive, require skilled personnel, and are not yet considered useful for the practicing analyst [24].

26.3 Characterization of PCBs

PCBs constitute a family of environmental persistent pollutants of synthetic organic compounds that have mainly been used in electrical equipment as dielectric insulating media. The use of these compounds is now restricted, but because of their wide usage in the past and of their high stability in the environment they are so widely distributed that detectable levels can be found in marine organisms, from mollusks to fish. PCBs have been linked with subtle subchronic effects such as reduced male fertility and long-term behavioral and non-*ortho* and mono-*ortho* PCBs have been assessed as having dioxin-like effects. PCBs are extremely persistent in the environment and possess the ability to accumulate in the food chain. These compounds are highly insoluble in water and tend to accumulate in body fat. Human exposure is probably dominated by the accumulation through the food chain of the PCBs present in environmental reservoirs. Determination of PCBs in marine organisms generally consists of three steps: sample extraction, purification, and chromatographic separation, identification, and quantification. The extraction methods are generally set up to maximize the extraction of all analytes and the clean-up is performed to improve the selectivity of the extraction removing lipids and interfering compounds. Low selectivity of extraction method yields considerable amounts of undesirable coextractives. For a correct evaluation of data it is important to report the PCB concentrations in an adequate manner or on wet weight or on fat weight. In the case of marine organisms, since an equilibrium partitioning of PCBs between fats in the organisms and water is established, lipid normalized data is used because it allows to compare the PCBs concentrations in the several species of fish, independently of their lipid content. In general, in biotic samples, the lipids are usually extracted together with PCBs. PCBs can be determined using GC techniques with electron capture detection, though more sophisticated methods, such as GC coupled with mass spectrometry (GC–MS), can be used to identify the individual congeners and to improve the comparability of the analytical data from different sources. Higher specificity than ECD and higher sensitivity of conventional GC–MS techniques can be performed using the high-resolution mass spectrometry (HRMS). This powerful system is the reference method for the determination of trace level of non-*ortho* and mono-*ortho* PCBs in various environmental matrices [25].

The accuracy in determining PCB levels is highly variable and matrix dependent. Many factors including the water solubility, volatility, and biodegradability of individual PCBs, will alter the composition of a commercial PCB preparation introduced as a pollutant into the environment. Thus, the composition of PCB extracts from environmental matrices will vary widely and often do not resemble any commercial mixture.

26.3.1 PCBs Methods of Extraction

Sample preparation methods for PCB determination involve several steps for exhaustive extraction of the analytes and subsequent clean-up steps for removal of the lipids. PCB levels in fish or shellfish are very high above all in those organisms located at the top of the food chain, because of their ability to accumulate these compounds. Like PAHs the biotic samples require washing with distilled water and homogenization process before the extraction.

26.3.1.1 Soxhlet Extraction

This method is a traditional procedure, largely used in the laboratory and reported also by EPA [26]. The fresh samples, with high moisture content, have to be dried before the extraction to perform a better penetration of solvent into the sample matrix. All samples have to be mixed with anhydrous sodium sulfate to form a flowing powder. The nature of the extraction solvent influences the efficiency of the procedure. Nonpolar or semipolar solvents as pentane, hexane, dichloromethane, acetone, toluene, diethyl ether, or polar–apolar solvent mixtures are usually selected as extraction solvents because they lead to better extraction of most of the PCB congeners present in marine organisms. For fish and shellfish the hexane–acetone (4:1, v/v or 1:1, v/v), methylene chloride–acetone (1:1), and hexane–dichloromethane (1:1, v/v) mixtures are the most employed and give the best recovery for PCBs and lipid content. Use of hexane–acetone mixture generally reduces the amount of interferences that are extracted and improves signal-to-noise ratio.

Soxhlet extraction method achieves very high recoveries, superior to those reported in the samples extracted with organic solvents, but requires large volumes of highly purified organic solvents and consumes long time. To perform an efficient extraction of PCBs using the Soxhlet technique the total extraction time requires approximately 6–8 h. In addition, some volatile compounds may be lost unless efficient condensers are used [27].

26.3.1.2 Sonication Method and Liquid–Liquid Partitioning

Both methods allow a very simple and fast PCBs extraction, but the main disadvantage is that the extracts are very dirty because they contain a lot of coextracted components and require a more accurate clean-up. Hexane or acetone–hexane mixture [28] (1:1, v/v) are commonly used to extract PCBs by sonication, however this can yield the formation of emulsions which may cause a loss of compounds. The time of extraction is about 15–20 min per cycle but the process is normally repeated two more times with fresh solvent.

26.3.1.3 Pressurized Liquid Extraction

In recent years this technique has been applied to marine matrices and has shown high recovery in the extraction of PCBs, when compared with conventional methods. Furthermore, it provides cleaner extracts in shorter time and smaller solvent consumption. The most important variables affecting the efficiency of the PLE process are the nature of the solvent, the temperature of the extraction, and the extraction time.

Most applications dealing with extraction of PCBs make use of hexane because it provides a quantitative extraction of most PCB congeners, also this is a more toxic solvent than other linear alkanes [29–32]. *N*-Heptane, *n*-pentane, toluene, or hexane–acetone (1:1, v/v), and hexane–dichloromethane (1:1, v/v or 4:1, v/v) mixtures are further extraction solvents often used for the isolation of PCBs from abiotic or biotic samples like mussel and oyster [33]. The best efficiency of PCBs extraction in fish samples is achieved by mixtures of low-polar and high-polar solvents as boiling point, polarity, and specific density influence the penetration into the sample matrix allowing a more efficient extraction of analytes than single solvents and the complete extraction of lipids. This fact is very important because the PCB concentrations, in fish and shellfish, should be standardized to the lipid content. An other important aspect to consider is the influence of extraction temperature. Higher temperatures decrease the viscosity of solvents allowing their better penetration into the sample and enhance the extraction efficiency [34]. Using a temperature range of 90°C–100°C it is possible to achieve elevated recoveries but it needs to consider that the extraction efficiency for lower chlorinated PCBs can be increased with higher temperatures, in the range 90°C–120°C. The selection of the number of static cycles is another important parameter for achieving quantitative extraction. Two static cycles (5–7 min) offer the practical solution to achieve maximum efficiency of extraction, as reported in literature [35,36]. Using three or more cycles increases the total extraction time and the solvent consumption but does not improve the efficiency of PCBs extraction.

Although the repeatability of PLE extraction is comparable with the Soxhlet and both methods are able to give good recoveries, however, only the PLE displays a low solvent consumption, a short time of analysis, and a better efficiency of extraction for lower chlorinated PCBs. The cost of this equipment is the main disadvantage if compared to Soxhlet or batch extraction enhanced by sonication. Nevertheless, PLE shows the best performances for extraction of PCBs from fish and shellfish, better than those of classic extraction methods or Soxhlet procedure.

With the PLE system it is also possible to perform an online clean-up, introducing the suitable fat retainer directly into the extraction cell. This technique is used for the extraction of PCBs from fatty samples in order to avoid further sample treatment before the GC analyses and to save the time spent on sample handling. Florisil, basic alumina, neutral alumina, acidic alumina, and sulfuric acid-impregnated silica are the main employed sorbent retaining lipids. Silica gel impregnated with sulfuric acid is considered the best choice for fat removal because of the clearness of the extracts and because it is much less sensitive to high temperatures [29,37].

26.3.1.4 Supercritical Fluid Extraction

SFE has received increasing attention and popularity as a technique used in the extraction and clean-up procedures for the determination of organic pollutants in complex matrices. The application of SFE to the isolation of PCBs from fish tissues is documented, also if the high lipid contents of these organisms, when coextracted may block SFE restrictors. The introduction of heated, adjustable restrictors have ameliorated this problem [38].

The use of a single fluid, generally carbon dioxide, the selectivity in the extraction of different classes of compounds, the possibility of extraction at high temperatures, the consumption of very limited volumes of organic solvents and a very limited contamination from the laboratory environment, for the use of a sealed system are the main advantages of SFE. Nevertheless, this technique requires two different sequential extraction steps when, on the same sample, the compounds investigated present a different polarity. Sample size is an other critical parameter when low concentrations of analytes are present. However, lyophilization of the tissues prior to extraction conserves considerable vessel volume by eliminating the need for inclusion of drying agents, such as sodium sulfate anhydrous or diatomaceous earth. The extraction conditions have to be generally set up to maximize the extraction of the analytes. Bøwadt et al. [39] found that, in SFE analysis of lyophilized fish tissue, high extraction temperatures yield scarcely better extraction and also imply a less pure extract, with more lipids and interfering compounds. Mild temperatures (60°C–70°C) avoid this problem, however when these are used on fish species with fat content higher than 8%–10% yielded an extract that is injectable only on a split–splitless injector because of the presence of lipids.

Inclusion of alumina in the SFE extraction vessel eliminates the need for any additional off-line lipid purification since the combination of selective SFE extraction and alumina retains more than 99% of the lipids. SFE extracts generated could be collected in autosampler vials and injected directly onto a gas chromatograph. Compared with the many hours necessary with the conventional procedures this clean-up procedure takes only 30 min and greatly reduces manipulations. In addition to the parameters discussed above, the selection of one extraction technique in preference to another is usually made on the basis of initial capital cost, operating costs, amount of organic solvent, and sample weight.

26.3.2 Clean-Up of Extracts

The procedures followed to purify sample extracts are common to all regardless of matrix. The first purification treatment consists of the elimination of coextracted materials, as lipids and pigments, which can be performed by either destructive or nondestructive methods. In the first case the lipids are removed by oxidation reactions with concentrated sulfuric acid that is very efficient for the most unreactive chemical groups like PCBs. The sulfuric acid can be directly added to the extracts or immobilized on silica layers allowing an on-column removal. When it is added to the extracts the solution has to be centrifuged. This method cannot be used to clean-up extracts for other target analytes, as it will destroy most organic chemicals including the pesticides aldrin, dieldrin, endrin, endosulfan (I and II), and endosulfan sulfate.

Adsorption chromatography on Florisil, alumina, or silica columns is a nondestructive method that allows the removal of coextracted materials and the elimination of polar interferences. This approach is performed in order to separate PCBs from other organochlorine compounds. It is possible to use columns with 1–2 cm of anhydrous sodium sulfate to the top, but before the elution the sodium sulfate and the cartridge have to be wet and rinsed by adding hexane. The first fraction eluted, generally with hexane, contains PCBs and some DDTs, whereas the second fraction, eluted with ethylether in hexane, contains the remaining DDTs and other organochlorine compounds. The ease of separation appears to depend on the characteristics of the absorbent, of the eluting solvent, and of the sample extract. The option of using standard column chromatography techniques or solid-phase extraction cartridges depends on the amount of interferences in the sample extract and the degree of clean-up required. The cartridges require less elution solvent and less time, however, their clean-up capacity is drastically reduced in comparison with standard chromatographic columns.

As a nondestructive clean-up method the GPC was used for removal of lipid from fish sample extracts. Using a column of Bio-Beads S-X3 with dichloromethane/hexane (1:1) as the mobile phase, the GPC clean-up gives an excellent efficiency for the lipid removal. Furthermore, it is capable of separating high boiling material from the sample analytes avoiding the contamination of injection ports and column heads, prolonging column life, stabilizing the instrument, and reducing column reactivity.

26.3.3 Chromatographic Analysis

Presently the most widely practiced technique for the PCB determination is capillary GC–ECD. Although ECD has many advantages it is unable, using a single column, to differentiate between coeluting PCBs and interferences, and cannot resolve PCB congener pairs like 77/110 [40,41]. Furthermore, its ability to identify individual PCBs relies on retention time alone and may suffer from limited selectivity in cases of very complicated samples.

The use of mass spectrometric detectors not only enhances selectivity but, by the use of selected ion monitoring (SIM) and isotope ratios, provides qualitative information to supplement that supplied by GC retention time. Although the two techniques provide very low detection limits, quantification is complicated with both because detector responses vary significantly with molecular structures. Thus, individual calibration for each compound of interest is required in order to achieve quantitative data of acceptable accuracy. GC–HRMS is an other technique generally employed to solve some specific problems in different GC–MS applications.

26.3.3.1 Gas Chromatography with Electron Capture Detector and Gas Chromatography–Mass Spectrometry

GC–ECD is the most popular technique owing to the relatively low costs, whereas the high selectivity of GC–MS is superior in the presence of abundant electron-capturing coextractives. Selection of appropriate chromatographic columns is of the major importance for correctly identifying and quantifying PCBs. Capillary GC columns, currently in use, are made of fused silica, chemically bonded with various stationary phases, to achieve a range of different selectivities toward complex samples. In general, packed columns have been replaced by capillary columns, because of their far superior efficiency, but the use of a single column does not allow the separation of all congeners in a single chromatographic run. The use of two columns with different polarities provides different elution patterns enabling separation of coeluting congeners. Furthermore, the approach of using a second column is important to confirm the identity of the compounds. Identification of PCB congeners in the sample is performed by comparing the retention times of the peaks with those of the peaks in standard chromatograms. The width of the retention time window used to make identifications should be based upon measurements of actual retention time variations of standards over the course of a day and should be carefully established to minimize the occurrence of both false positive and false negative results.

If the response for a peak exceeds the working range of the system, a dilution of the extract is required. If the measurement of the peak response is prevented by the presence of interferences, further clean-up is required. The quality and utility of the analytical data depend critically on the validity of the sample and the adequacy of the sampling. For PCB determination an extensive quality assurance program is required and, furthermore intercalibration studies are recommended. If the internal standard calibration procedure is being used, the internal standard must be added to the sample extract and mixed thoroughly before injection into the gas chromatograph.

When capillary columns are used with temperature programming, almost all PCB isomers and congeners normally present in samples can be identified. The injector temperature set from 250°C to 280°C and the injections made in the splitless mode are the more suitable conditions for PCB analysis. GC oven temperature can be programmed in order to elute all PCBs during the temperature gradient, but the ramp conditions change depending on the column used. Using a GC–MS the transfer line temperature of the GC–MS interface the ion source temperature can be set at 280°C and 260°C, respectively, and a selected ion monitoring can be performed. GC–MS plays an important role in the identification and quantification of PCBs in complex samples and is one of the most attractive and powerful techniques for routine analysis due to its good sensitivity, selectivity, and versatility. The GC separation usually provides isomer selectivity, while the MS shows compound class and homologue specificity.

The MS fragmentation pattern provides unambiguous identification by comparing an unknown electron ionization MS spectrum with library spectra. For the identification of unknown peaks the MS conditions at which spectra have been obtained must be similar and the GC separation must be sufficiently efficient to obtain a clean mass spectrum [42].

26.3.3.2 Gas Chromatography–High-Resolution Mass Spectrometry

Non-*ortho* and mono-*ortho* PCBs are specific compounds whose determination is mainly performed by GC–HRMS, to provide the required sensitivity and selectivity for analysis. The use of HRMS is based on enhancing the selectivity of the MS as a detector by increasing resolution. HRMS presents a very high capacity to remove the contribution of matrix interfering compounds in the determination of the analytes. Using SIM at a mass resolution of 10,000, the presence of matrix components in the extracts does not interfere and detection at a high level of mass accuracy can be performed. Very high sensitivity and powerful identification capability of HRMS have made this technique the reference method for determination of many POPs at sub pg/g concentrations.

However, HRMS systems are relatively expensive and require specialized laboratory infrastructure to run effectively [25].

References

1. Williams, C., Combatting marine pollution from land-based activities: Australian initiatives, *Ocean Coast. Manage.*, 33, 87, 1996.
2. Bye, T., Romundstad, P.R., Ronneberg, A., and Hilt, B., Health survey of former workers in a Norwegian coke plant: Part 2. Cancer incidence and cause specific mortality, *Occup. Environ. Med.*, 55, 622, 1998.
3. Moret, S., Conte, L., and Dean, D., Assessment of polycyclic aromatic hydrocarbon content of smoked fish by means of a fast HPLC/HPLC method, *J. Agric. Food Chem.*, 47, 1367, 1999.
4. Dafflon, O. et al., Le dosage des hydrocarbures aromatiques polycycliques dans le poisson, les produites carnès et le fromage par chromatographie liquide à haute performance, *Trav. Chim. Aliment. Hyg.*, 86, 534, 1995.
5. Vassilaros, D.L. et al., Capillary gas chromatographic determination of polycyclic aromatic compounds in vertebrate fish tissue, *Anal. Chem.*, 54, 106, 1982.
6. Chen, B.H. and Lin, Y.S., Formation of polycyclic aromatic hydrocarbons during processing of duck meat, *J. Agric. Food Chem.* 45, 1394, 1997.
7. De Boer, J. and Law, R.J., Developments in the use of chromatographic techniques in marine laboratories for the determination of halogenated contaminants and polycyclic aromatic hydrocarbons, *J. Chromatogr., A*, 1000, 223, 2003.
8. Bartle, K.D. Analysis and occurrence of PAHs in food, in *Food Contaminants: Sources and Surveillance*, Creaser, C.S. and Purchase R., Eds., Royal Society of Chemistry, Cambridge, U.K., 1991, pp. 41–60.
9. Lebo, J.A. et al., Determination of monocyclic and polycyclic aromatic hydrocarbons in fish tissue, *J. Assoc. Off. Anal. Chem.*, 74, 538, 1991.
10. Chen, B.H., Wang, C.Y., and Chiu, C.P., Evaluation of analysis of polycyclic aromatic hydrocarbons in meat products by liquid chromatography, *J. Agric. Food Chem.*, 44, 2244, 1996.
11. Jaouen-Madoulet, A.J. et al., Validation of the analytical procedure for polychlorinated biphenyls, coplanar polychlorinated biphenyls and polycyclic aromatic hydrocarbons in environmental samples, *J. Chromatogr., A*, 886, 153, 2000.
12. Anyakora, C. et al., Determination of polynuclear aromatic hydrocarbons in marine samples of Siokolo fishing settlement, *J. Chromatogr., A*, 1073, 323, 2005.
13. Vives, I. and Grimalt, J.O., Method for integrated analysis of polycyclic aromatic hydrocarbons and organochlorine compounds in fish liver, *J. Chromatogr., B*, 768, 247, 2002.
14. Vives, I. et al., Polycyclic aromatic hydrocarbons in fish from remote and high mountain lakes in Europe and Greenland, *Sci. Total Environ.*, 324, 67, 2004.
15. Pensado, L. et al., Application of matrix solid-phase dispersion in the analysis of priority polycyclic aromatic hydrocarbons in fish samples, *J. Chromatogr. A*, 1077, 1103, 2005.
16. Shu, Y.Y. et al., Analysis of polycyclic aromatic hydrocarbons in sediment reference materials by microwave-assisted extraction, *Chemosphere*, 41, 1709, 2000.
17. Lopez-Avila, V., Young, R., and Teplitsky, N.L., Microwave-assisted extraction as an alternative to Soxhlet, sonication, and supercritical fluid extraction, *J. Assoc. Off. Anal. Chem.*, 79, 142, 1995.
18. Janská, M. et al., Appraisal of "classic" and "novel" extraction procedure efficiencies for the isolation of polycyclic aromatic hydrocarbons and their derivatives from biotic matrices, *Anal. Chim. Acta*, 520, 93, 2004.
19. U.S. EPA SW-846, *Update III: Test Methods for Evaluating Solid Waste, Method 3545: Fed. Reg. Vol. 62*, 114:32451 U.S. GPO, Washington, D.C., 1997.
20. Guangdi, W. et al., Accelerated solvent extraction and gas chromatography/mass spectrometry for determination of polycyclic aromatic hydrocarbons in smoked food samples, *J. Agric. Food Chem.*, 47, 1062, 1999.
21. Dong, M.W. and Greenberg, A., Liquid chromatographic analysis of polynuclear aromatic hydrocarbons with diode array detection, *J. Liq. Chromatogr. Relat. Technol.*, 11, 1887, 1988.

22. Stołyhwo, A. and Sikorski, Z.E., Polycyclic aromatic hydrocarbons in smoked fish—a critical review, *Food Chem.,* 91, 303, 2005.

23. Wise, S.A., Bonnett, W.J., and May, W.E., Normal- and reverse-phase liquid chromatographic separations of polycyclic aromatic hydrocarbons, in *Polynuclear Aromatic Hydrocarbons: Chemistry and Biological Effects*, Bjorseth, A. and Dennis, A.J., Eds., Battelle Press, Columbus, OH, 1980, pp. 791–806.

24. Vo-Dinh, T., Significance of chemical analysis of polycyclic aromatic compounds and related biological systems, in *Chemical Analysis of Polycyclic Aromatic Compounds*, Vo-Dinh, T., Ed., Wiley, New York, 1989, pp. 1–30.

25. Verenitch, S.S. et al., Ion-trap tandem mass spectrometry-based analytical methodology for the determination of polychlorinated biphenyls in fish and shellfish. Performance comparison against electron-capture detection and high-resolution mass spectrometry detection, *J. Chromatogr. A*, 1142, 199, 2007.

26. Schantz, M.M. et al., Comparison of supercritical fluid extraction and Soxhlet extraction for the determination of polychlorinated biphenyls in environmental matrix standard reference materials, *J. Chromatogr. A*, 816, 213, 1998.

27. Hess, P. et al., Critical review of the analysis of non- and mono-ortho-chlorobiphenyls, *J. Chromatogr., A*, 703, 417, 1995.

28. Kitamura, K. et al., Effective extraction method for dioxin analysis from lipid-rich biological matrices using a combination of pressurized liquid extraction and dimethyl sulfoxide/acetonitrile/hexane partitioning, *Anal. Chim. Acta,* 512, 27, 2004.

29. Sporring, S. and Bjorklund, E., Selective pressurized liquid extraction of polychlorinated biphenyls from fat-containing food and feed samples. Influence of cell dimensions, solvent type, temperature and flush volume, *J. Chromatogr. A,* 1040, 155, 2004.

30. Focant, J.F. et al., Fast clean-up for polychlorinated dibenzo-*p*-dioxins, dibenzofurans and coplanar polychlorinated biphenyls analysis of high-fat-content biological samples, *J. Chromatogr., A*, 925, 207, 2001.

31. Bjorklund, E. et al., New strategies for extraction and clean-up of persistent organic pollutants from food and feed samples using selective pressurized liquid extraction, *Trends Anal. Chem.* 25, 318, 2006.

32. Ramos, J.J. et al., Miniaturised selective pressurised liquid extraction of polychlorinated biphenyls from foodstuffs, *J. Chromatogr. A,* 1152, 254, 2007.

33. Hubert, A. et al., Accelerated solvent extraction—more efficient extraction of POPs and PAHs from real contaminated plant and soil samples, *Rev. Anal. Chem.,* 20, 101, 2001.

34. Richter, B.E. et al., Accelerated solvent extraction: A technique for sample preparation, *Anal. Chem.,* 68, 1033, 1996.

35. Schantz, M.M., Nichols, J.J., and Wise, S.A., Evaluation of pressurized fluid extraction for the extraction of environmental matrix reference materials, *Anal. Chem.,* 69, 4210, 1997.

36. Bjorklund, J. et al., Pressurized fluid extraction of polychlorinated biphenyls in solid environmental samples, *J. Chromatogr. A,* 836, 285, 1999.

37. Björklund, E., Muller, A., and von Holst, C., Comparison of fat retainers in accelerated solvent extraction for the selective extraction of PCBs from fat-containing samples, *Anal. Chem.,* 73, 4050, 2001.

38. Hale, R.C. and Gaylor, M.O., Determination of PCBs in fish tissues using supercritical fluid extraction, *Environ. Sci. Technol.,* 29, 1043, 1995.

39. Bówadt, S. et al., Supercritical fluid extraction of polychlorinated biphenyls from lyophilized fish tissue, *J. Chromatogr., A*, 675, 189, 1994.

40. Harrad, S.J. et al., A method for the determination of PCB congeners 77, 126 and 169 in biotic and abiotic matrices, *Chemosphere* 24, 1147, 1992.

41. Ayris, S. et al., GC/MS procedures for the determination of PCBs in environmental matrices, *Chemosphere* 35, 905, 1997.

42. Santos, F.J. and Galceran, M.T., Modern developments in gas chromatography–mass spectrometry based-environmental analysis, *J. Chromatogr. A,* 1000, 125, 2003.

Chapter 27

Biogenic Amines in Seafood Products

Claudia Ruiz-Capillas and Francisco Jiménez-Colmenero

Contents

27.1 Introduction..743
27.2 Toxicity of Biogenic Amines in Seafood..744
27.3 Biogenic Amines as Quality Index in Seafood...746
27.4 Legal Limits of Biogenic Amines in Seafood...747
27.5 Factors Influencing the Formation of Biogenic Amines in Seafood 748
    27.5.1 Raw Material 748
    27.5.2 Microorganisms...749
    27.5.3 Processing and the Storage Conditions of Seafood...750
27.6 Determination of Biogenic Amines in Seafood 753
    27.6.1 Extraction Process..753
    27.6.2 Determination Process..753
Acknowledgments 756
References 756

27.1 Introduction

Biogenic amines (BA) are biologically active low-molecular-weight basic nitrogenous compounds (Figure 27.1) which are present in the great majority of foods, including fish and fishery products. According to their chemical structure, they can be classified as aromatic amines (histamine, tyramine, serotonin, β-phenylalanine, and tryptamine), aliphatic diamines (putrescine and cadaverine),

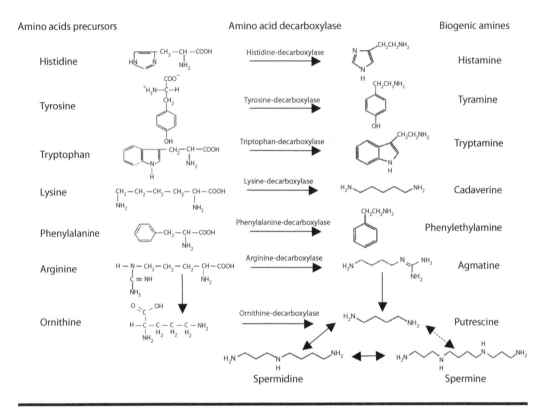

Figure 27.1 Amino acid precursors and amino acid decarboxylase enzymes in the formation of BA in seafood.

or aliphatic polyamines (agmatine, spermidine, and spermine) [1]. Animal, plant, and microorganism metabolism form some of them (putrescine, spermidine, and spermine) naturally. In the case of animals or plants, cadaverine and agmatine are also produced naturally. Some of these amines play important roles in many human and animal physiological functions. They are necessary for normal cell growth and play an important role in nucleic acid regulation and protein synthesis, and possibly also in the stabilization of membranes [1,2]. On the other hand, BA are produced by decarboxylation of free amino acids (FAA) from the action of microbial amino acid decarboxylase enzymes (Figures 27.1 and 27.2) [1,3–6], which is of particular interest in fish and fishery products as these are extremely perishable. Biogenic amines are important for two reasons. Firstly, the intake of foods containing high concentrations of BA can present a health hazard [4,6–8]; and secondly, they may have a role as indicators of quality and/or acceptability in some foods such as seafood [5,9–13].

27.2   Toxicity of Biogenic Amines in Seafood

The consumption of food containing high concentrations of BA has been associated with toxic effects and constitutes a potential health hazard. The compounds mainly implicated in these toxic effects are histamine and tyramine. Histamine is the most significant biogenic amine in fish and fish products. It is the main component in "scombroid poisoning" or "histamine poisoning" caused by consumption of fish containing high levels of histamine and/or other BA. This foodborne intoxication was originally called "scombroid poisoning" because it was primarily associated with

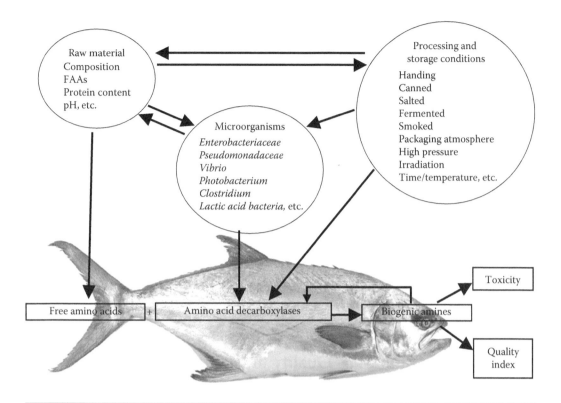

Figure 27.2   Factors affecting the formation of BA in seafood.

the consumption of fish of the Scombridae and Scomberesocidae families such as tuna, mackerel, bonito, bluefish, and the like. These species contain high levels of free histidine in their muscle that is decarboxylased to histamine [7,8]. However, the term is misleading in that fish from several species of nonscombroid fish such as bluefish, herring, sardine, anchovy, pink salmon, redfish yellowtail, marlin, sailfish, amberjack, mahi-mahi, and the like, have often been implicated in cases of scombroid poisoning in different countries such as Japan, Great Britain, Australia, USA, Taiwan, and the like [7,8,14]. Non scombroid species also contain high levels of free histamine in their muscle tissue [15], hence this illness came to be called "histamine poisoning." However, histamine does not appear to be the only causative agent of scombroid poisoning, since it is not itself toxic when taken orally [7,8]. Other amines such as putrescine and cadaverine are also implicated in this illness as they enhance the toxicity of histamine [3,16]. The most common symptoms of histamine poisoning involve the cardiovascular system; histamine has a vasodilating effect, producing low blood pressure, reddening of the skin, headaches, edemas, and rashes and a burning or peppery taste in the mouth typical of allergic reactions, diarrhea, and abdominal cramps [4,7,8]. These symptoms usually disappear within several hours without medical intervention. More severe symptoms (e.g., respiratory distress, swelling of the tongue and throat, and blurred vision) can occur and require medical treatment with antihistamines. In many cases, these symptoms disappear too quickly for the poisoning to be naturally associated with the consumption of food containing BA, and in some cases they are difficult to distinguish from symptoms of other illnesses; for example, the symptoms of food allergy and histamine poisoning are similar. However, the toxicity of BA depends on the ability of individuals to metabolize normal dietary intakes of BAs [4,16]. This detoxification system includes specific enzymes such as monoamine oxidase (MAO; EC 1.4.3.4), diamine oxidase (DAO; EC 1.4.3.6), and polyamine oxidase (PAO; EC 1.5.3.11). However, the system is susceptible to some risk factors involving these enzymes, such as dietary consumption of foods containing high levels of BA (as in the case of spoiled or fermented fish), alcohol, gastrointestinal diseases, or in case of amino oxidase activity due to intake of certain drugs that can act as monoamine oxidase inhibitors (MAOI). It is important to note that large quantities of these drugs are consumed in Europe as antidepressants [17]. Alcohol and acetaldehyde seem to promote the transportation of these amines through the intestinal wall and augment their toxic potential [4,8]. The presence of other amines such as putrescine and cadaverine or spermidine, phenylethylamine, agmatine, and spermine, can also contribute to histamine toxicity. Phenylethylamine is a known inhibitor of the enzymes diamine oxidase and histamine methyl-transferase. Putrescine and cadaverine can inhibit the intestinal enzyme MAO, thus promoting the absorption and/or reducing the catabolism of this amine and enhancing its toxicity even when the histamine concentration is not so high [4]. Therefore, the effect of a given amount of histamine in seafood may be greater or lesser depending on the amount of enhancers that are present and on the efficiency of the individual's detoxification system [7]. Moreover, certain BA, essentially putrescine and cadaverine, can react with nitrites to form toxic or carcinogenic compounds such as nitrosamines [5]. However, this is much less important in seafood products than in meat products.

27.3 Biogenic Amines as Quality Index in Seafood

Although the control of BA in seafood has frequently been undertaken because of their involvement in food poisoning [4,8], the determination of these BA has also been proposed as a quality index to detect fish spoilage and/or defective preparation (Table 27.1) [9–13,18–25]. BAs are used as quality control indices because they undergo change during fish processing and storage. They are found at very low levels in fresh fish; their formation tends to increase during storage and is associated with bacterial spoilage [3,19,26].

The biogenic amine index of Mietz and Karmas [9] is the most widely used criterion for assessing spoilage of fish on the basis of biogenic amine (histamine, cadaverine, putrescine, spermidine, and spermine) contents (Table 27.1). The limit of fish acceptability for this quality index was set at 10. Later on, other authors [22] proposed an index calculated from the sum of the contents of histamine, tyramine, cadaverine, and putrescine, which showed good correlations with both time in storage and sensory evaluation, for assessment of tuna. Other individual BAs have also been used as quality indices in fish (Table 27.1). The Food and Drug Administration (FDA) [25] recommended not only histamine to indicate defect action levels, but also other scientific data to judge fish freshness, such as the presence of other BAs associated with fish decomposition.

Some BAs such as agmatine and cadaverine have been used as fish freshness indicators (Table 27.1), reflecting the changes taking place in fish at the outset of storage prior to the point of onset of appreciable microbial spoilage [10,12,18,19]. Both agmatine and cadaverine have been associated with the autolytic changes responsible for loss of freshness that take place in fish muscle before the onset of microbial spoilage. Some authors suggest that this be defined as the beginning of the formation of some BAs and other spoilage compounds such as total volatile basic nitrogen or trimethylamine nitrogen, which have traditionally been used as indices to assess the spoilage of refrigerated fish [10,27].

In addition, BA can be useful as indicators of poor-quality raw material in preserved fish products, e.g., canned fish, because they are thermally stable compounds [21]. A good correlation has

Table 27.1   Quality Index of Biogenic Amines in Seafood

Seafood	Biogenic Amines	References
Rockfish, salmon, lobster, and shrimp	Quality index = (histamine + putrescine + cadaverine)/(1 + spermidine + spermine)	[9]
Hake	Cadaverine and agmatine	[10]
Hake in protective atmospheres	Cadaverine and agmatine	[11]
Squid	Agmatine	[12,19]
Salmonoid	Cadaverine	[13]
Trout	Putrescine, cadaverine, and histamine	[18]
Skipjack tuna	Putrescine and cadaverine	[20]
Canned tuna	Histamine	[21]
Tuna	Quality index = (histamine + tyramine + putrescine + cadaverine)	[22]
Cold smoked salmon	Cadaverine, histamine, putrescine, and tyramine	[23]
Skipjack and bigeye tuna	Cadaverine	[24]
	Cadaverine + histamine	
Scombridae and *Clupeidae* families	Histamine	[25]

been found between sensory evaluation and levels of putrescine and cadaverine in canned skipjack tuna [20]. Jorgensen et al. [23] found that the BAs correlated well with sensory analysis in cold-smoked salmon, but they thought that the BAs are not necessarily the causal agents of spoilage off-flavors. BA have also been used as a quality index in fish and fish products treated with other technologies such as protective atmospheres (vacuum, modified atmosphere) [11]. Cadaverine and agmatine have been proposed as a control index for whole chilled hake, gutted and stored in controlled atmospheres [11].

27.4   Legal Limits of Biogenic Amines in Seafood

The ingestion of BA can pose a risk to consumers, and therefore there have been various attempts to set limits for safe human consumption, particularly in the case of histamine. The European Union has set [25] a legal limit for histamine in certain fish species taking into account the technology used on them. The maximum permitted average concentration of histamine in fresh fish products of the *Scombridae* and *Clupeidae* families is 100 mg/kg and up to 400 mg/kg in cured products from the same families [25]. The FDA [15,28] has set this histamine level at 50 mg/kg, above which it is considered a potential health hazard. In Australia the legal limit of histamine concentration within which it is regarded as safe for health is 200 mg/kg [29], and in South Africa the limit is 100 mg/kg [29]. However, as noted earlier, the relationship between the level of histamine and the toxicity of a fish sample is not clear and the presence of other enhancers of histamine toxicity needs to be taken into account.

27.5 Factors Influencing the Formation of Biogenic Amines in Seafood

The formation of BA in seafood is caused mainly by microbial enzymatic decarboxylation of certain FAA (Figure 27.1). Such formation is affected by the substrate and the enzyme, as well as by factors influencing these such as the raw material, microorganisms affecting the amount and type of enzymes, and technological processing and storage conditions which directly affect the enzyme, the substrate, and the reaction medium (Figure 27.2) [3,5].

27.5.1 Raw Material

Biogenic amines in the muscle. The raw material is decisive in determining the level and formation of BA in the final product. Initially, the muscle of freshly caught fish contains certain concentrations of BA, mainly polyamines (spermidine and spermine and in some cases putrescine and agmatine), which are characteristic of the fish species and type of muscle concerned [10,11,30]. In the case of tuna, for instance, initial concentrations of BA are generally higher in white than in red muscle. The exceptions are putrescine, levels of which are similar in both muscle types, and spermidine, levels of which are higher in red than in white muscle [30–34]. Wendakoon et al. [31] found higher biogenic amine concentrations in red than in white muscle of herring. Agmatine has only been found in trace amounts in freshly caught squid [19,35]. The BA with the highest concentrations reported in raw hake are putrescine and spermidine, with levels of 3.08 and 2.71 mg/kg, respectively. Initial concentrations of histamine, cadaverine, and agmatine are very low (less than 1 mg/kg) and tyramine concentrations are less than 1 mg/kg, or in some cases not detectable in hake [10,36].

Free amino acids in muscle. Fish raw material is the natural source of FAA from which BA are produced (Figures 27.1 and 27.2). The FAA levels in fish are high when compared to terrestrial animals since the primary function of these compounds in aquatic organisms is to serve as osmoregulators. The level and the type of these FAA in fish depend on the fish (family, species, etc.) and the muscle type (red or white). Fish contain varying proportions of red and white muscle, which perform different physiological functions in the live animal [21]. Members of the Scombridae and Scomberosocidae families contain higher concentrations of free histidine (which is decarboxylated to histamine) (Figure 27.1) than other fish species [30–34,36–40]. The two muscle types in tuna contain different levels of histidine, with higher concentrations in white than in red muscle [30]. In the case of other FAA biogenic amine precursors in tuna muscle, there are no significant differences in concentration between white and red muscle; in the case of lysine, red muscle contains higher concentrations than white muscle [30].

Differences in muscle composition affect the chemical changes taking place during storage, and consequently also in the formation of BA [30,38]. FAA in general are formed in fish as a result of muscle proteolysis, and decreases in FAA concentrations are associated with greater consumption of these amino acids by the microorganisms that use them as a growth substrate, and hence with greater spoilage [21], giving rise to a variety of compounds including BA [30,38]. Initial free amino acid contents have generally been observed to decrease during fish storage. Ruiz-Capillas and Moral [30] observed that histidine concentrations decreased significantly in white muscle but remained stable in red muscle. A correlation has been observed between the amino acid histidine and the biogenic amine histamine during storage of tuna, but only in the white muscle. Other authors have likewise found no correlation between FAA and BA [30,31,40,41]. However, Stede and Stockemer [42] found a relationship between the formation of BA and FAA in cod and haddock.

Muscle as a reaction medium. The muscle is also the medium where decarboxylation takes place and is therefore necessary in determining essential factor in enzymatic activity: pH, ionic strength, substrate concentration, inhibitors, and the like [5]. The relatively low pH of sardine muscle possibly favors histidine decarboxylation [41].

27.5.2 Microorganisms

Biogenic amines are mainly produced by the decarboxylation of certain FAA by microbial enzymes [3,4,21,43]. It is important to identify which bacteria possess amino acid decarboxylase activity in order to estimate the risk of biogenic amine production in seafood and to prevent its build up in seafood products. Several Gram positive and Gram negative bacteria are implicated in the formation of BA. The most widely studied decarboxylase bacteria in the case of fish are the ones involved in histamine formation because of their implication in foodborne intoxication. In the case of fish, histamine decarboxylase enzymes are generally found in bacteria belonging to the *Enterobacteriaceae* family and to genera of *Pseudomonas, Clostridium, Vibrio,* and *Photobacterium* spp. [19,21,26,29,35,43–50]. *Acinetobacter, Aeromonas, Staphylococcus,* and *Bacillus* spp. have also been reported to have the potential to produce histamine by decarboxylase activity [51–54]. While histamine formation in raw fish products is caused mostly by Gram negative enteric bacteria, histamine in fermented products such as fish sauces is produced by Gram positive lactic acid bacteria (LAB). Tyrosine decarboxylase activity is more common among Gram positive bacteria [55]. Thapa et al. [56] showed that the LAB isolated from traditional processed fish products from the Eastern Himalayas did not produce BA. However, other authors have identified LAB as responsible for the production of some BA in squid kept in ice and in sterile cold-smoked salmon [35,57].

The variation observed in the decarboxylase ability of different species is extremely wide. *Proteus vulgaris, P. mirabilis, Enterobacter aerogenes, Enterobacter cloacae, Serratia fonticola, S. liquefaciens, Citrobacter freundii, Vibrio alginolyticus, Acinetobacter lowffi, Plesiomonas shigelloides, Pseudomonas putida, P. fluorescens, Photobacterium phosphoreum, Photobacterium damselae,* and *Raoultella planticola* have all been identified as histamine-formers in fish [35,47,49,50,51,58–65].

Differences in the ability to produce BA have even also been observed between strains of the same species. *Morganella (Proteus) morganii, Klebsiella pneumoniae,* and *Hafnia alvei* have all been isolated from fish blamed for scombroid poisoning [47,50]. Most of them are also involved in the production of other BA besides histamine, such as *P. phosphoreum,* which is also capable of producing agmatine and cadaverine [35,61].

The microorganism with the strongest histidine decarboxylase activity is *M. morganii.* The next most active appears to be *K. pneumoniae, H. alvei,* and some strains of *E. cloacae* and *E. aerogenes,* which produce more than 500 mg/kg of histamine, always depending on the storage conditions, especially temperature [3,41,47].

These biogenic amine-forming microorganisms may constitute part of the endogenous microbiota associated with the microflora of the fish or may be introduced by contamination during processing and storage of these fish. In freshly caught fish, bacterial contamination is located initially on the skin and gills; from there, these microorganisms invade the fish muscle and grow rapidly in response to a number of factors relating to processing and the storage conditions such as temperature, time, and the like.

27.5.3 Processing and the Storage Conditions of Seafood

In general, levels of BA increase during processing and storage of seafood, and this increase depends on many factors associated with processing and storage conditions. These factors affect the main elements implicated in biogenic amine formation, such as FAA or microbial enzyme decarboxylase (Figures 27.1 and 27.2) [4,21]. Therefore, in order to limit the formation of BA in seafood it is not enough to have suitable raw material; it is also necessary to optimize the processing and storage conditions. Biogenic amine formation in seafood products has been studied with reference to the processing and technological practices.

Handling. The handling of fish is decisive for the formation of BA, and the result clearly depends on the kind of fish [40,66,67]. Haaland et al. [66] reported that the formation of cadaverine and putrescine was higher in ungutted mackerel than in fillets during the storage. Fernández-Salguero and Mackei [67] reported that histamine, cadaverine, and putrescine were produced more rapidly in haddock fillets than in the whole gutted fish, and ungutted fish spoiled more rapidly than fillets.

Refrigeration. Storage time and temperature are decisive factors in the production of BA. These factors affect microorganism growth and hence production of the amino acid descarboxylase enzyme necessary for the production of BA. Generally the levels of BA, except the polyamines spermidine and spermine (which usually remain constant), increase progressively throughout fish storage in ice. The increase is smaller if the fish had been efficiently cooled on ice to 0°C–2°C [9–13,31,32]. The behavior in each case is dependent on the fish species and the type of muscle. In tuna, pronounced increases of histamine have been recorded at days 18–25 in white muscle, but only on the last day of storage in the case of red muscle. In the case of cadaverine and agmatine, changes again occur later in red than in white tuna muscle [30]. Other authors [19,12] also observed that agmatine was the biogenic amine in the highest concentration in squid stored in ice. In high quality lean fish such as redfish, haddock, plaice, or hake stored in ice at 0°C–4°C, the levels of BA (histamine, putrescine, cadaverine, and tyramine) remained below 10 mg/kg. However, when storage time is prolonged, BA concentrations increase. Hake kept at 0°C–2°C for 33 days attained peak cadaverine concentrations (72.14 mg/kg) at the end of storage as compared with 13.47 mg/kg of agmatine and 7.12 mg/kg of putrescine, or even with levels of histamine and tyramine, which were less than 3.5 mg/kg by the end of storage [10]. These low levels of BA are the result of very low microbial growth and decarboxylase activity at low storage temperatures [21,42].

Increasing temperature during storage has also been studied as a factor in the formation of BA. Frank et al. [45] reported that biogenic amine formation is more closely related to the activity of mesophilic than of psychrotrophic bacteria, which could explain the fact that formation of those amines was more extensive at 20°C. Klausen and Huss [68] reported that *M. morganii* grows well at 15°C or above and that growth is greatly reduced at below 10°C. These authors confirmed that *M. morganii* could produce large amounts of histamine (600–1400 mg/kg) in mackerel stored at low temperature (0°C–5°C) following storage at higher temperatures (10°C–25°C). However, *Vibrio* and *Photobacterium* spp. are probably primarily responsible for histamine production at lower temperatures [26,54,64] since the optimum temperature for these bacteria is below 10°C. Ababouch et al. [41] showed that the rate of histamine formation in sardines was greater at ambient temperature than in ice storage. After 24 h of storage at ambient temperature, histamine, cadaverine, and putrescine reached levels of 2350, 1050, and 300 mg/kg, respectively. Similar results have been reported by other authors in various fish species [11,13,22,31,51,67,69,70]. Veciana-Nogués et al. [22] found similar profiles for evolution of BA in tuna at three storage temperatures (0°C, 8°C, and 20°C). Levels of the polyamines spermidine and spermine generally remain constant during storage and are independent of temperature. This is because the formation of these amines is scarcely affected by spoilage [22,10].

On the other hand, decreases in histamine in advanced fish spoilage have been reported in association with the growth of microorganisms presenting histaminolytic activity [8,22,41]. Veciana-Nogués et al. [22] reported considerable decreases in histamine and cadaverine after 3 days of storage at 20°C, at which time of spoilage was far advanced, whereas no decreases in histamine, putrescine, cadaverine, and tyramine were observed when samples were stored at 0°C or 8°C. Ababouch et al. [41] observed that for longer periods of storage of sardine at ambient temperature (over 24 h), histamine levels began to decrease, indicating proliferation of bacteria presenting histaminase activity.

Therefore, the most effective way to prevent BA formation in fish is to keep it at a low temperature throughout processing and storage. It is commonly accepted that storage at or near 0°C limits the formation of amines by substantially retarding bacterial growth and the activity of decarboxylating enzymes [21].

In general, frozen storage does not affect microbial growth or enzymatic activity and therefore does not affect the production of biogenic amine. However, some authors [47,40] have isolated histamine-producing bacteria from frozen fish. The presence of amines in frozen fish is indicative of the characteristics of the raw material prior to thawing [21].

Canning. Biogenic amines are stable to thermal processing, and therefore the presence of BA in canned products indicates that the fish has been microbially spoiled before heating [21,6]. Fernández-Salguero and Mackie [71] reported very low levels (less than 2 mg/kg) of BA in canned tuna, mackerel, and sardine.

Fermentation. Fermented fish products are particularly rich in histamine, followed by phenylethylamine [72]. Stages in fermentation are sometimes carried out without adequate temperature control; and higher fermentation temperatures increase biogenic amine concentrations, especially histamine. A considerable increase in putrescine, histamine, and tyramine contents of fermented sardines has been associated with increased concentrations of halotolerant and halophilic histamine-forming bacteria (*Staphylococcus*, *Micrococcus*, *Vibrio*, and *Pseudomonas*).

Salting. Biogenic amine contents of salted fish generally vary considerably [73]. The average concentration of each biogenic amine in salted mackerel sold in retail markets and supermarkets in Taiwan was less than 3 mg/100 g [74]. Higher levels of histamine have also been detected in some samples, and minute amounts of spermidine, phenylethylamine, agmatine, and spermine have been detected in some tested mackerel samples. Rodríguez-Jerez et al. [54] assessed histidine decarboxylase activity and production of putrescine and cadaverine bacterial isolates from ripened semipreserved Spanish anchovies. They found the highest levels of histidine decarboxylase activity in *M. morganii*, which is also a producer of putrescine, and cadaverine. Other species such as *S. epidermidis*, *S. xylosus*, *K. oxytoca*, *E. cloacae*, *Pseudomonas cepaciae*, and *Bacillus* spp. were also implicated in the production of these BA. Other authors have identified halotolerant *Staphylococcus* spp., *Vibrio* spp., and *Pseudomonas III/IV-NH* as histamine-formers in salted fish [52–54,75,76]. Tsai et al. [74] identified *Pantoea* spp., *Pantoea agglomerans*, and *E. cloacae* as histamine-producing bacteria in salted mackerel.

Smoking. Biogenic amines have also been studied in smoked fish. The smoking process usually commences with a drying phase. This phase should be kept short, as prolonged exposure to ambient temperature may lead to unwanted microbiological growth and to formation of histamine in susceptible species. Shalaby [6] observed levels of putrescine, cadaverine, and spermine between 1–16 mg/kg and low levels of tyramine, spermidine, and histamine (1–8 mg/kg) in smoked herring. Tryptamine and phenylethylamine were not detected in any of the samples analyzed.

Protective atmospheres. This technology, used as a coadjuvant to chilled storage, affects biogenic amine formation in various fish species (tuna, hake cod, sardine, salmon, etc.) [11,26,30, 33,36,39,77,78]. In general, the application of protective atmospheres reduces the production of BA except spermidine and spermine. This effect is mainly due to the way that the mixture of gases in the atmosphere acts on microbial growth and hence on the amino acid descarboxylase enzyme (Figure 27.2). The effect of this technology also depends on the kind of atmosphere used (controlled, modified, etc.), the type of biogenic amine and the fish species [11,26,30,33,36,39,77–79]. Ruiz-Capillas and Moral [30] showed that a controlled atmosphere with a gas mix containing a high concentration of CO_2 was most effective in reducing BA in bigeye tuna. However, high levels of CO_2 (60%) in the controlled atmosphere were not sufficient to inhibit the production of BA in gutted hake kept in refrigeration, while the high O_2 concentration (40%) in the controlled atmosphere had an inhibiting effect on the production of BA in hake [11]. The combination of different protective atmospheres has also been found to be effective in reducing BA levels. Hake bulk-stored in a controlled atmosphere (40% CO_2:40% O_2:20% N_2) for the first 12 days and then packed in trays with modified atmospheres and the same mixture of gases also exhibited lower levels of BA, except for agmatine, throughout storage [36]. However, these protective atmospheres are only effective if the product is kept in refrigerated storage. An appropriate combination of low temperature and atmosphere, then, potentates the inhibiting effect of CO_2, retarding the growth of spoilage microorganisms in this fish [11,26,33,77,80]. Emborg et al. [26] showed that the spoilage of modified atmosphere packaged (MAP) cod is caused by growth and metabolism of the CO_2-resistant bacterium *P. phosphoreum*. This specific spoilage organism grows to high levels in different MAP fish but is inactivated by freezing at—20°C, and the shelf life of thawed MAP cod can be substantially prolonged in this way [64]. Jorgensen et al. [23] also found *P. phosphoreum* to be primarily responsible for the production of BA in vacuum-packed cold-smoked salmon, where agmatine, cadaverine, histamine, and tyramine were formed at 5°C.

High-pressure treatment. There has hardly been any research into the effect of high-pressure treatment on the formation and evolution of BA in seafoods. Paarup et al. [35] observed that the onset of formation of agmatine and other BA was delayed by increasing pressure in vacuum-packed squid. The application of moderate pressures (150–200 MPa) reduces the rate of agmatine formation, whereas higher pressures (300 and 400 MPa) delay the onset of production of this amine. Pressurization at 400 MPa inhibits histamine formation and keeps putrescine formation low, while higher concentrations of tyramine have been detected in squid pressurized at 300 and 400 MPa. Fujii et al. [81] also reported absence of histamine in minced mackerel meat pressurized at 200 MPa during chilled storage. It has been suggested that *P. phosphoreum* is responsible for biogenic amine production, mainly agmatine and histamine, in pressurized squid, while *Carnobacterium* spp. has been identified as responsible for the production of tyramine [35].

Irradiation. The effect of irradiation on the formation of BA has been studied in Atlantic horse mackerel during chilled storage [82]. Histamine in the irradiated mackerel (even at 1 kg) was undetectable at the end of 23 days when the fish had spoiled. This effect was associated with a majority of Gram negative anaerobic bacteria, since around 10%–18% was Gram positive bacteria in the irradiated samples.

27.6 Determination of Biogenic Amines in Seafood

Considering the importance of BA in fish and fish products for legal, toxicological, and quality purposes, it is essential to have accurate analytical methods. Biogenic amines in different foods, including fish and fish products, have traditionally been determined by means of standard chromatographic techniques such as thin layer chromatography, gas chromatography, capillary electrophoresis, flow injection analysis, and high-performance liquid chromatography (HPLC) [83,84]. Positive confirmation using mass spectrometry after either HPLC [85] or gas chromatographic separation [86] has also been reported for other food. Enzyme-based amperometric biosensors using histamine oxidase have also been developed for the determination of histamine [87].

The determination of BA from the fish matrix frequently presents problems for a variety of reasons. The major BA in seafood (normally not less than nine), which are present in a wide range of concentrations, are usually determined simultaneously. Moreover, this kind of sample is very complex, containing high protein levels and a wide variety of fat contents (0.5%–30%). For these reasons, most methods for determining BA frequently involve preliminary steps to extract these compounds from the fish matrix and subsequent separation and quantification steps.

27.6.1 Extraction Process

Sample preparation, or extraction of BA from the fish matrix, is a crucial step in the analysis. Many different solvents have been used to extract BAs from fish and fish products, including hydrochloric acid, trichloroacetic acid (TCA), perchloric acid (PCA), and other organic solvents such as methanol, dichloromethane, acetone, and acetonitrile [88,89]. Of these solvents the most commonly used are PCA and TCA because of their effect on protein precipitation, which makes them highly effective biogenic amine extractors for fish and fish products. Extraction commonly involves a first step where 5–15 g of fish muscle is homogenized with 10–50 mL of the acids at different concentrations (5%, 6%, and 7.5%); this homogenate is then centrifuged and the extract filtered [10,22]. The precipitate is washed with more acid, centrifuged, and filtered again. The acid extract made with TCA can also be used for other chemical determinations such as trimethylamine, volatile base nitrogen, ammonia, urea, and some FAA [10,90].

In some cases, this first step in the determination of BA in fish and fish products may include purification or cleanup of the final extract based on ion exchange resins and a solid phase. HLPC normally have a small clean-separation ion exchange (guard) column which is set up online prior to the separation column used to determine BA [10,91].

27.6.2 Determination Process

Of the available methods for determination of BA in seafood, the most widely used and frequently reported for the separation and quantification of BA in seafood are chromatographic procedures, especially HPLC with an ion exchange column or a reverse-phase column using ion pairs to separate BA. This procedure offers high resolution, sensitivity, and versatility. Moreover, the sample treatments are generally simple.

It is well known that BAs respond poorly to detection systems due to low volatility and lack of chromophores. Many BA occurring in food exhibit neither satisfactory absorption nor significant fluorescence properties. A chemical derivatization is therefore usually performed to increase their sensitivity. Covalent labeling with chromophores or fluorophores normally greatly improves detection sensitivity and detection limits. Thanks to UV–VIS and fluorescence detection [84].

There are many known derivatization reagents. Among them dansyl and dabsyl chloride, benzoyl chloride, fluoresceine, 9-fluorenylmethyl chloroformate, *o*-phthalaldehyde (OPA), and naphthalene-2,3-dicarboxaldehyde [84]. Of these, OPA and dansyl chloride are the most widely used. Dansyl chloride forms stable compounds after reaction with both primary and secondary amino groups and the products are more stable than those formed using OPA. This last reagent reacts rapidly with primary amines in the presence of a reducing agent such as *N*-acetylcysteine, 2-mercaptoethanol, or thiofluor, which is a stable solid substitute for 2-mercaptoethanol during the preparation of the OPA reagent. It forms a more stable and longer-lasting fluorophore with OPA than does 2-mercaptoethanol while possessing the same fluorescence properties. Under basic conditions (pH > 9) and at ambient temperature, the reaction is generally complete in 1–30 s. The products of this reaction, 1-alkyl-2-alkylthio-substituted isoindoles, exhibit optimal excitation at 330 nm and maximum emission at 465 nm [10,91,92]. Moreover, OPAs are faster and much simpler for purposes of sample pretreatment, which can be fully automated using an autosampler and is more sensitive because florescence detection is used rather than spectrophotometric detection as in the case of dansyl chloride [91]. On the other hand, the OPA derivative is unstable and the fluorescence intensity diminishes quickly, especially in alkaline media. This problem requires strict control of reaction times, but it can be solved using postcolumn derivatization or automatic precolumn derivatization.

Biogenic amine derivatization may be performed before (precolumn), during (on-column) or after (postcolumn) chromatographic separation [84,85]. Automatic online postcolumn derivatization is the most common procedure as it offers a number of advantages: it entails less handling, thus reducing the likelihood of interferences or artefacts in the sample, the analysis time is shorter and derivatization occurs at the same time, thus enhancing the reproducibility and sensitivity of the analysis [91]. However prederivatization involves more sample preparation steps, which can produce problems in the analysis later on. In addition, postderivatization is usually performed with an internal standard. Postderivatization for the quantification of BA entails comparison with an external calibration standard composed of the different BA. Dansyl chloride is normally used for precolumn derivatization with reverse-phase separation coupled with UV detection [33,80] and OPAs are used in ion-pair reverse-phase HPLC with post and precolumn derivatization coupled with fluorescence detection [10,22,70,91,92].

Flow injection analysis (FIA) is a new, fast, and simple method with low operating costs for the determination of BAs in seafood. FIA coupled with automated OPA derivatization has been used for histamine determination in canned tuna [93,94]. In the FIA method, all reagents are added automatically. Flow rates and OPA reactor volumes afford the required pH and reaction timing. This system employs three channels, using an anion-exchange column to eliminate sample matrix interferences. Selectivity for histamine versus interfering compounds appears to be based on differences in the reaction rates with OPA since histamine reacts quicker than the remainder compounds. Interfering substances such as ascorbate are fully electrooxidized, and the signals are removed upstream of the detector. This system is based on the AOAC method for determining histamine in seafood [95].

Most recently, an FIA method has been reported which uses an enzyme electrode to detect histamine. Takagi and Shikata [96] developed a new FIA method using a histamine dehydrogenase-based electrode, which they used to determine histamine in fish samples. Histamine dehydrogenase is immobilized in an osmium-derivatized redox polymer, poly(1-vinylimidazole) complexed with Os(4,4′-dimethylbipyridine)2Cl2 (PVI-dmeOs) film on a glassy carbon electrode. This electrode exhibits high selectivity to histamine and is not sensitive to other primary amines including common BA, putrescine, cadaverine, and tyramine. This is an effective method for rapid and efficient laboratory histamine testing, particularly in laboratories analyzing large numbers of samples. Not only does this allow more rapid analysis without sample cleanup, but also operator dependence is reduced by automation and the instrument completes each determination step in <1 min [96].

An enzymatic method has also been developed for the determination of histamine in fish. This one is based on the reaction of diaminoxidase (DAO) with histamine to yield hydrogen peroxide which, coupled with horseradish peroxidase, converts a reduced dye to its oxidized form. The color change is used to quantify the histamine in the sample [97]. However, this technique has a draw-back in the low specificity of the enzyme, which can react with other BA such as putrescine and tyramine, and as in the case of FIA, only one biogenic amine can be determined at a time.

More recently, this enzymatic reaction has been used to develop a biosensor with integrated pulsed amperometric detection for determination of BA in salted anchovy samples. The probe is based on a platinum electrode, which senses the hydrogen peroxide produced by the reaction cata-lyzed by the enzyme diamine oxidase (DAO). This is obtained from different sources (microorgan-isms, plants, and animal tissue) with different enzymatic activities such as seeds of cicer, porcine kidney, pea lentil, and the like. The DAO is immobilized on the electrode surface. The conditions selected were immobilization of the enzyme on a nylon-net membrane using glutaraldehyde as cross-linking agent and phosphate buffer at pH 8.0. Carelli et al. [98] also developed an ampero-metric biosensor for determination of total biogenic amine content using commercial diamino oxidase (from porcine kidney) as the biocomponent, entrapped by glutaraldehyde onto an elec-trosynthesized bilayer film. In order to minimize both fouling and interference caused by direct electrochemical oxidation of both the analytes (i.e., BA) and the common interferents usually present in food products, the performances of Pt and Au electrodes and of several electroproduced antiinterferent mono- and bilayer films were tested. Although the commercial DAO presented very low activity, the biosensor displayed high response sensitivity in flow experiments, short response times, a good linear response, and low detection limits. The antiinterference characteristics are so good that the biosensor can be used in screening analysis of seafood products [98].

A capillary electrophoresis method with conductometric detection of BA has been reported [99]. Clear separation of six BAs (cadaverine, putrescine, agmatine, histamine, tryptamine, and tyramine) from other components of acidic sample extract was achieved within 10 min. The advantages of this capillary electrophoresis method include low laboriousness (no derivatization step or sample cleaning, dilution or acidic extraction, and filtration only), adequate sensitivity, low running costs (no separation column, only an empty capillary), and speed of analysis and environmental friendliness (small amounts of water-based diluted electrolyte for analysis). The disadvantages are nonselectivity of conductometric detection and higher detection limits in the case of salty samples [99].

Interest in "portable" procedures for analysis of BA that would be capable of rapidly screening fishery products has led to the development of commercial test kits, which have been proposed for histamine determination and for hazard analysis and critical control point (HACCP) applications. A number of these commercial test kits have been compared with the AOAC method by analyzing samples of tuna and mahi-mahi. These kits are based on an enzymatic immunoassay [100,101]. Take for example the ALERT® kit and the Veratox histamine kit. Both kits are direct competi-tive enzyme-linked immunosorbent assays (ELISA). They are used for quantitative analysis of histamine in scombroid fish species such as tuna, bluefish, and mahi-mahi. Histamine is extracted from a sample using a quick water extraction process. Free histamine in the sample and controls competes with enzyme-labeled histamine (conjugate) for the antibody-binding sites. After a wash step, the substrate reacts with the bound enzyme conjugate to produce a blue color. A microwell reader is used to yield optical densities. Control optical densities are used to form a standard curve, and sample optical densities are plotted against the curve to calculate the exact concentration of histamine. The results are read using a microwell reader at 450–650 nm. The method offers simplicity, rapidity, and relatively low cost in comparison with other methodologies such as HPLC, and no previous derivatization is required [101].

Acknowledgments

This research was also supported under projects AGL2003-00454, AGL2007-61038/ALI of the Plan Nacional de Investigación Científica, Desarrollo e Innovación Tecnológica (I+D+I), the Consolider CSD2007–00016, Ministerio de Ciencia y Tecnología.

References

1. Smith, T.A. Amines in food, *Food Chem.*, 6, 169, 1980.
2. Bardócz, S. et al. The importance of dietary polyamines in cell regeneration and growth, *Brit. J. Nutr.*, 73, 819, 1995.
3. Halász, A. et al. Biogenic amines and their production by microorganisms in food, *Trends Food Sci. Tech.*, 5, 42, 1994.
4. Bardócz, S. Polyamines in food and their consequences for food quality and human health, *Trends Food Sci. Tech.*, 6, 341, 1995.
5. Ruiz-Capillas, C. and Jiménez-Colmenero, F. Biogenic amines in meat and meat products, *Crit. Rev. Food Sci.*, 44, 489, 2004.
6. Shalaby, A.R. Significance of biogenic amines to food safety and human health, *Food Res. Int.*, 29, 675, 1996.
7. Lehane, L. and Olley, J. Histamine fish poisoning revisited, *Int. Food. Microbiol.*, 58, 1, 2000.
8. Taylor, S.L. Histamine food poisoning: Toxicology and clinical aspects, *Crit. Rev. Toxicol.*, 17, 91, 1986.
9. Mietz, J.L. and Karmas, E. Polyamine and histamine content of rockfish, salmon, lobster and shrimp as an indicator of decomposition, *J. Assoc. Offic. Anal. Chem.*, 61, 139, 1977.
10. Ruiz-Capillas, C. and Moral, A. Production of biogenic amines and their potential use as quality control indices for hake (*Merluccius merluccius L.*) stored in ice, *J. Food Sci.*, 66, 1030, 2001.
11. Ruiz-Capillas, C. and Moral, A. Effect of controlled atmospheres enriched with O_2 in formation of biogenic amines in chilled hake (*Merluccius merluccius L.*), *Eur. Food Res. Technol.*, 212, 546, 2001.
12. Yamanaka, H., Shiomi, K., and Kikuchi, T. Agmatine as a potential index for freshness of common squid, *J. Food Sci.*, 52, 936, 1987.
13. Yamanaka, H., Shiomi, K., and Kikuchi, T. Cadaverine as a potential index for decomposition of salmonoid fishes, *J. Food Hyg. Soc. Jpn.*, 30, 170, 1989.
14. Gessner, B., Hokama, Y., and Isto, S. Scombrotoxicosis-like illness following the ingestion of smoked salmon that demonstrated low histamine levels and high toxicity on mouse bioassay, *Clin. Infect. Dis.*, 23, 1316, 1996.
15. Food and Drug Administration. Fish and fisheries products hazards and controls guidance. 3rd ed. Scombrotoxin (histamine) formation: A chemical hazard, Available at http://www.cfsan.fda.gov/~comm/haccp4g.html (accessed 5 November, 2008).
16. Rice, S.L., Eitenmiller, R.R., and Koehler, P.E. Biologically active amines in food: A review, *J. Milk Food Technol.*, 39, 353, 1976.
17. Sattler, J. et al. Food induced histaminosis as an epdemiological problem: Plasma histamine elevation and haemodynamic alterations after oral histamine administration and blockade of diamine oxidase (DAO), *Agents Actions*, 23, 361, 1988.

18. Dawood, A.A. et al. The occurrence of non-volatile amines in chilled-stored rainbow trout (*Salmo irideus*), *Food Chem.*, 27, 33, 1988.

19. Paarup, T. et al. Sensory, chemical and bacteriological changes during storage of iced squid (*Todaropsis eblanae*), *J. Appl. Microbiol.*, 92, 941, 2002.

20. Sims, G.G., Farn, G., and York, R.Y. Quality index for tuna correlation of sensory attributes with chemical indices, *J. Food Sci.*, 57, 1112, 1992.

21. Huss, H.H. Quality and quality changes in fresh fish, FAO. Fisheries Tecnical., Paper. 348, 1995.

22. Veciana-Nogués, M.T., Mariné Font, A., and Vidal Carou, M.C. Biogenic amines as hygienic quality indicators of tuna. Relationships with microbial counts, ATPrelated compounds, volatile and organoleptic changes, *J. Agric. Food Chem.*, 45, 2036, 1997.

23. Jorgensen, L.V., Dalgaard, P., and Huss, H.H. Multiple compound quality index for cold-smoked salmon (*Salmo salar*) developed by multivariate regression of biogenic amines and pH, *J. Agric. Food Chem.*, 48, 2448, 2000.

24. Rossi, S. et al. Biogenic amines formation in bigeye tuna steaks and whole skipjack tuna, *J. Food Sci.*, 67, 2056, 2002.

25. Commission Regulation (EC) No. 2073/2005 of 15 November 2005 on microbiological criteria for foodstuffs, 2005.

26. Emborg, J. et al. Microbial spoilage and formation of biogenic amines in fresh and thawed modified atmosphere packed salmon (*Salmo salar*) at 2°C, *J. Appl. Microbiol.*, 92, 790, 2002.

27. Ruiz-Capillas, C. and Moral, A. Correlation between biochemical and sensory quality indices in hake stored in ice, *Food Res. Int.*, 34, 441, 2001.

28. Food and Drug Administration (FDA). 2001, In *Fish and Fishery Products Hazards and Controls Guidance*, Food and Drug Administration, Center for Food Safety and Applied Nutrition, Office of Seafood, Washington, DC, 83–102.

29. Auerswald, L., Morren, C., and Lopata, AL. Histamine levels in seventeen species of fresh and processed South African seafood, *Food Chem.*, 98, 231, 2006.

30. Ruiz-Capillas, C. and Moral, A. Free amino acids and biogenic amines in red and white muscle of tuna stored in controlled atmospheres, *Amino Acids.*, 26, 125, 2003.

31. Wendakoon, C.N., Murata, M., and Sakaguchi, M. Comparison of non-volatile amine formation between the dark and white muscles of mackerel during storage, *Nippon Suisan Gakk.*, 56, 809, 1990.

32. Watanabe, H., Yamanaka, H., and Yamanakawa, H. Post mortem biochemical changes in the muscle of disk abalone during storage, *Nippon Suisan Gakk.*, 58, 2081, 1992.

33. López-Gálvez, D., De la Hoz, L., and Ordóñez, J.A. Effect of carbon dioxide and oxygen enriched atmospheres on microbiological and chemical changes in refrigerated tuna (*Thunnus alalunga*) steaks, *J. Agric. Food Chem.*, 30, 435, 1995.

34. Du, W.X. et al. Development of biogenic amines in yellowfin tuna (*Thunnus albacore*): Effect of storage and correlation with decarboxylase-positive bacterial flora, *J. Food Sci.*, 67, 292, 2002.

35. Paarup, T. et al. Sensory, chemical and bacteriological changes in vacuum-packed pressurised squid mantle (*Todaropsis eblanae*) stored at 4°C, *Int. J. Food Microbiol.*, 74, 1, 2002.

36. Ruiz-Capillas, C., and Moral, A. Formation of biogenic amines in bulk stored chilled hake (*Merluccius merluccius, L.*) packed under atmospheres, *J. Food Prot.*, 64, 1045, 2001.

37. Wei, C.I. et al. Bacterial growth and histmamine production on vacuum packaged tuna, *J. Food Sci.*, 55, 59, 1990.

38. Ochiai, Y., Aleman-Polo, J.M., and Hashimoto, K. Postmortem diffusion of taurine and free histidine between ordinary and dark muscle of mackerel during ice storage, *Nippon Suisan Gakk.*, 56, 1017, 1990.

39. Ruiz-Capillas, C., and Moral, A. Effect of controlled and modified atmospheres on the production of biogenic amines and free amino acids during storage of hake, *Eur. Food Res. Technol. A.*, 214, 476, 2002.

40. Mendes, R., Goncalves, A., and Nunes, M.L. Changes in free amino acids and biogenic amines during ripening of fresh and frozen sardine, *J. Food Biochem.*, 23, 295, 1999.

41. Ababouch, L.H. et al. Quality changes in sardines (*Sardina pilchardus*) stored in ice and at ambient temperatures, *Food Microbiol.*, 13, 123, 1996.

42. Stede, M. and Stockemer, J. Biogenic amines in marine fish, *Lebensm Wiss. Technol.*, 19, 283, 1986.

43. Silla, M.H. Biogenic amines: Their importance in foods, *Int. J. Food Microbiol.*, 29, 213, 1996.

44. Gram, L. Fish spoilage bacteria–problems and solutions, *Curr. Opin. Biotechnol.*, 13, 262, 2002.

45. Frank, H.A. et al. Identification and decarboxylase activities of bacteria isolated from decomposed mahimahi (*Coryphaena hippurus*) after incubation at 0 and 32°C, *Int. J. Food Microbiol.*, 2, 331, 1985.

46. Taylor, S.L. and Sumner, S.S. (eds.), Determination of histamine, putrescine, and cadaverine, In *Seafood Quality Determination*, Kramer & Liston, Amsterdam, the Netherlands, 1986, 235.

47. Taylor, S.L. and Speckhard, M.W. Inhibition of bacteria histamine production by sorbate and other antimicrobial agents, *J. Food Prot.*, 47, 508, 1984.

48. Arnold, S.H., Price, R.J., and Browen, W.D. Histamine formation by bacteria isolated from skipjack tuna *Katsuwonus plamis*, *Bull. Jpn. Soc. Sci. Fish.*, 46, 991, 1980.

49. López-Sabater, E.I. et al. Evaluation of histidine decarboxylase activity of bacteria isolated from sardine (*Sardina pilchardus*) by an enzymatic method, *Lett. Appl. Microbiol.*, 19, 70, 1994.

50. Eitenmiller, R.R. et al. Production of histidine decarboxylase and histamine by *Proteus morganii*, *J. Food Prot.*, 44, 815, 1981.

51. Middlebrooks, B.L. et al. Effects of storage time and temperature on the microflora and amine development in Spanish mackerel (*Scomberomorus maculatus*), *J. Food Sci.*, 53, 1024, 1988.

52. Yatsunami, K. and Echigo, T. Changes in the number of halotolerant histamine-forming bacteria and contents of non-volatile amines in sardine meat with addition of NaCl, *Bull. Jpn. Soc. Sci. Fish.*, 59, 123, 1993.

53. Hernández-Herrero, M.M. et al. Halotolerant and halophilic histamine-forming bacteria isolated during the ripening of salted anchovies, *J. Food Prot.*, 62, 509, 1999.

54. Rodríguez-Jerez, J.J. et al. Histamine, cadaverine and putrescine forming bacteria from ripened Spanish semipreserved anchovies, *J. Food Sci.*, 59, 998, 1994.

55. Bover-Cid, S. and Holzapfel, W.H. Improved screening procedure for biogenic amine production by lactic acid bacteria, *Int. J. Food Microbiol.*, 53, 33, 1999.

56. Thapa, N., Pal, J., and Tamang, J.P. Phenotypic identification and technological properties of lactic acid bacteria isolated from traditionally processed fish products of the Eastern Himalayas, *Int. J. Food Microbiol.*, 107, 33, 2006.

57. Jorgensen, L.V., Huss, H.H., and Dalgaard, P. The effect of biogenic amine production by single bacterial cultures and metabiosis in cold-smoked salmon, *J. Appl. Microbiol.*, 89, 920, 2000.

58. López-Sabater, E.I. et al. Sensory quality and histamine formation during controlled decomposition of tuna (*Thunnus thynnus*), *J. Food Prot.*, 59, 167, 1996.

59. Tsai et al. Histamine related hygienic qualities and bacteria found in popular commercial scombroid fish fillets in Taiwan, *J. Food Prot.*, 67, 407, 2004.

60. Yoshinaga, D.H. and Frank, H.A. Histamine-producing bacteria in decomposing skipjack tuna (*Katsuwonus pelamia*), *Appl. Environ. Microbiol.*, 44, 447, 1982.

61. Okuzumi, M. et al. *Photobacterium histaminum* sp. nov., a histamine-producing marine bacterium, *Int. J. Syst. Bacteriol.*, 44, 631, 1994.

62. Ryser, E.T., Marth, E.H., and Taylor, S.L. Histamine production by psychrotrophic pseudomonas isolated from tuna fish, *J. Food Prot.*, 47, 378, 1984.

63. Kanki, M. et al. *Klebsiella pneumoniae* produces no histamine: *Raoultella planticola* and *Raoultella ornithinolytica* strains are histamine producers, *Appl. Environ. Microbiol.*, 68, 3462, 2002.

64. Dalgaard, P. et al. Importance of *Photobacterium phosphoreum* in relation to spoilage of modified atmosphere packed fish products, *Lett. Appl. Microbiol.*, 24, 373, 1997.

65. Kanki, M. et al. *Photobacterium phosphoreum* caused a histamine fish poisoning incident, *Int. J. Food Microbiol.*, 92, 79, 2004.

66. Haaland, H., Arnesen, E., and Njaa, L.R. Amino-acid-composition of whole mackerel (*Scomber scombrus*) stored anaerobically at 20°C and at 2°C, *Int. J. Food Sci. Technol.*, 25, 82, 1990.

67. Fernández-Salguero, J. and Mackei, I.M. Comparative rates of spoilage of fillets and whole fish during storage of haddock (*Melanogammus aeglefinus*) and herring (*Clupea arengus*) as determined by the formation of non-volatile and volatile amines, *Int. J. Food Sci. Technol.*, 22, 385, 1987.

68. Klausen, N.K. and Huss, H.H. Rapid method for detection of histamine-producing bacteria, *Int. J. Food Microbiol.*, 5, 137, 1987.

69. Nagayama, T. et al. Non-volatile amines formation and decomposition in abusively stored fishes and shelfishes, *J. Food Hyg. Soc. Jpn.*, 31, 362, 1985.

70. Baixas-Nogueras, S. et al. Volatiles and nonvolatile amines in Mediterranean Hake as a function of their storage temperature, *J. Food Sci.*, 66, 83, 2001.

71. Fernández-Salguero, J. and Mackie, I.M. Technical note: Preliminary survey of the content of histamine and other higher amines in some samples of Spanish canned fish, *Int. J. Food Sci. Tech.*, 22, 409, 1987.

72. Fardiaz, D. and Markakis, P. Amine in fermented fish paste, *J. Food Sci.*, 44, 1562, 1979.

73. Lee, H. et al. Histamine and other biogenic amines and bacterial isolation in retail canned anchovies, *J. Food Sci.*, 70, C145, 2005.

74. Tsai, Y.H. et al. Occurrence of histamine and histamine-forming bacteria in salted mackerel in Taiwan, *Food Microbiol.*, 22, 461, 2005.

75. Yatsunami, K. and Echigo, T. Isolation of salt tolerant histamine-forming bacteria from commercial rice-bran pickle sardine, *Bull. Jpn. Soc. Sci. Fish.*, 57, 1723, 1991.

76. Yatsunami, K. and Echigo, T. Occurrence of halotolerant and halophili histamine-forming bacteria in red meat fish products, *Bull. Jpn. Soc. Sci. Fish.*, 58, 515, 1992.

77. Emborg, J., Laursen, B.G., and Dalgaard, P. Significant histamine formation in tuna (*Thunnus albacares*) at 2 C—effect of vacuum- and modified atmosphere-packaging on psychrotolerant bacteria, *Int. J. Food Microbiol.*, 101, 263, 2005.

78. Suzuki, S., Noda, J., and Takama, K. Growth and polyamine production of *Alteromonas* ssp. in fish meat extracts under modified atmosphere, *Bull. Fac. Fish. Hokkaido Univ.*, 41, 213, 1990.

79. Randell, K. et al. Quality of whole gutted salmon in various bulk packages, *J. Food Quality*, 22, 483, 1999.

80. Lannelongue M. et al. Microbiological and chemical changes during storage of swordfish (*Xiphias gladius*) steaks in retail packages containing CO_2-enriched atmospheres, *J. Food Prot.*, 45, 1197, 1982.

81. Fujii, T. et al. Changes in freshness indexes and bacterial flora during storage of pressurized mackerel, *J. Food Hyg. Soc. Jpn.*, 35, 195, 1994.

82. Mendes, R. et al. Effect of low-dose irradiation and refrigeration on the microflora, sensory characteristics and biogenic amines of Atlantic horse mackerel (*Trachurus trachurus*), *Eur. Food Res. Technol.*, 221, 329, 2005.

83. Teti, D., Visalli, M., and McNair, H. Analysis of polyamines as markers of (patho) physiological conditions, *J. Chromatogr. B.*, 781, 107, 2002.

84. Önal, A. A review: Current analytical methods for the determination of biogenic amines in foods, *Food Chem.*, 103, 1475, 2007.

85. Gosetti, F. et al. High performance liquid chromatography/tandem mass spectrometry determination of biogenic amines in typical Piedmont cheeses, *J. Chromatogr. A.*, 1149, 151, 2007.

86. Awan, M.A., Fleet, I., and Thomas, C.L.P. Determination of biogenic diamines with a vaporization derivatisation approach using solid-phase microextraction gas chromatography-mass spectrometry, *Food Chem.*, 111, 462, 2008.

87. Draisci, R. et al. Determination of biogenic amines with an electrochemical biosensor and its application to salted anchovies, *Food Chem.*, 62, 225, 1998.

88. Lapa-Guimarães, J. and Pickova, J. New solvent systems for thin-layer chromatographic determination of nine biogenic amines in fish and squid, *J. Chrom. A.*, 1045, 223, 2004.

89. Moret, S. and Conte, L.S. High-performance liquid chromatographic evaluation of biogenic amines in foods—An analysis of different methods of sample preparation in relation to food characteristics, *J. Chrom. A.*, 729, 363, 1996.

90. Ruiz-Capillas, C. and Horner, W.F.A. Determination of trimethylamine nitrogen and total volatile basic nitrogen in fresh fish by flow injection analysis, *J. Sci. Food Agric.*, 79, 1982, 1999.

91. Tracy, M.L., Pickering M.V., and Verhulst, T. Cation exchange analysis of foods and beverages for biogenic amines, *Food Test. Anal.*, 1, 48, 1995.

92. Tapia-Salazar, M., Smith, T.K., and Harris, A. High performance liquid chromatographic method for detection of biogenic amines in feedstuffs, complete feeds and animal tissues, *J. Agric. Food Chem.*, 48, 1708, 2000.

93. Gutiérrez, C., Rubio, S., Gómez-Hens, A., and Valcárcel, M. Determination of histamine by derivative synchronous fluorescence spectrometry, *Anal. Chem.*, 59, 769, 1987.

94. Hungerford, M., Hollingworth, T.A., and Wekell, M.M. Automated kinetics-enhanced flow-injection method for histamine in regulatory laboratories: Rapid screening and suitability requirements, *Anal. Chim. Acta.*, 438, 123, 2001.

95. Association of Official Analytical Chemists, Histamine in seafood: Fluorometric method [35.1.32 method 977.13], In *Official Methods of Analysis*, 17th edn., AOAC International, Gaithersburg, MD, 2003.

96. Takagi, K. and Shikata, S. Flow injection determination of histamine with a histamine dehydrogenase-based electrode, *Anal. Chim. Acta.*, 505, 189, 2004.

97. Lerke, P.A., Martina P.N., and Henry, B.C. Screening test for histamine in fish, *J. Food Sci.*, 48, 155, 1983.

98. Carelli, D., Centonze, D., Palermo, C., Quinto, M., and Rotunno, T. An interference free amperometric biosensor for the detection of biogenic amines in food products, *Biosens. Bioelectron.*, 23, 640, 2007.

99. Kvasnicka, F. and Voldrich, M. Determination of biogenic amines by capillary zone electrophoresis with conductometric detection, *J. Chromatogr.*, 1103, 145, 2006.

100. Staruszkiewicz, W.F. and Rogers, P.L. *Performance of Histamine Test Kits for Applications to Seafood.* Presentation on October 26, 2001. 4th World Fish Inspection & Quality Control Congress, Vancouver, BC.

101. Rogers, P.L. and Staruszkiewicz, W.F. http://seafood.ucdavis.edu/pubs/histamine.htm (accessed December, 10th 2008).

102. Neogen, (http://www.neogen.com/FoodSafety/V_Index.html) (accessed December, 10th 2008).

Chapter 28

Detection of GM Ingredients in Fish Feed

Kathy Messens, Nicolas Gryson, Kris Audenaert, and Mia Eeckhout

Contents

28.1 Introduction..762
    28.1.1 Genetically Modified Organisms..762
    28.1.2 GMOs in Fish Feed..762
    28.1.3 International Regulations 763
28.2 GMO Analysis...764
    28.2.1 Introduction 764
    28.2.2 DNA-Based Detection..764
        28.2.2.1 Extraction of DNA...764
        28.2.2.2 Qualitative Conventional PCR....................................764
        28.2.2.3 Multiplex PCR 766
        28.2.2.4 Quantitative PCR..766
    28.2.3 Protein-Based Detection 767
28.3 Comparison of DNA and Protein Methods 769
    28.3.1 Protein-Based Methods...769
    28.3.2 DNA-Based Methods 770
28.4 Recent Developments in GMO Detection 771
    28.4.1 Microarray Technology..771
    28.4.2 Biosensors 772
28.5 Conclusions..772
References 773

28.1 Introduction

28.1.1 Genetically Modified Organisms

Genetically modified organisms (GMOs) can be defined as organisms whose genetic constitution has been altered by gene technology. The genetic material is changed in a way that is not possible by reproduction or natural recombination (transgene, genetically altered, or GMOs). Gene technology makes the insertion of new properties possible in an efficient and direct way. For this purpose, a coding DNA sequence is brought to expression, e.g., in the genome of a plant. This gene transfer is possible through a physical process (injection, gene gun) or a biological process (plasmids, viruses). It is possible to select a property in different kinds of organisms and insert it in the crop of interest enabling the enlargement of genetic variations.[1]

Gene technology has mainly been used to produce agriculturally improved plant varieties. The majority of plants commercialized are either herbicide tolerant (canola, sugar beet, chicory, soybean, flax, alfalfa, tobacco, rice, wheat, and maize), produce their own insecticide (*Bt* cotton, potato, maize, and tomato) or both (cotton, maize). Next to this, genes of interest can be related to delayed ripening, altered amino acid and fatty acid composition, starch hydrolysis, male sterility, virus and lepidopteran resistance.[2] Genes have also been inserted that speed up the growth rate or lead to the synthesis of new proteins, leading to the development of so-called bio-factories, which produce substantial quantities of pharmaceuticals.[3]

In May 1994, the first genetically modified (GM) product for food use, the FlavrSavr™ tomato, was approved for commercial sale in the United States. Since then, the amount of GM crops has increased significantly. Nowadays, more than 80 different GM plants are commercially available for food and feed purposes all over the world. Next to the aforementioned GMOs, the list of approved GM crops worldwide also includes carnation, creeping bent grass, lentil, papaya, plum, squash, and sunflower.[2] At present 60% of all soybeans on the world market are GM and almost one-fourth of all maize is GM, with numbers increasing every year. However, growth of GM soy and GM cotton in the United States is decreasing due to the growing market for ethanol.[4–6]

Although genetic engineering has emerged as one of the most powerful transforming technologies, no higher animals have been exploited as GMOs until now.[7] To date, the only animals that have a chance of being genetically engineered for commercial purposes are fish species. The possible applications are medical research, increased food/feed production, and development of specific ornamental traits. The main constrain on the commercial use of these GMOs is the absence of laws regulating production, importation, and consumption of transgenic fish.[8]

The commercial pressure groups focus on developing rapidly growing fish with a highly efficient food conversion, but many questions remain on benefits and risks. Transgenic fish that escape into natural ecosystems could turn into an invasive species eliminating native fish populations since they often possess increased fitness or breeding capacities. The use of sterile GMOs might provide a solution for this problem. In spite of doubts on the commercial use of GMO fish, a number of transgenic fish are being developed on a lab scale for scientific research. Salmon, trout, tilapia, bass, catfish, and flounder have been genetically engineered for faster growth, higher disease resistance, and a better temperature tolerance.[8]

28.1.2 GMOs in Fish Feed

Fish feed formulation will vary according to nutritional requirements of the species. Fish feed ingredients may be derived from GMOs with agronomical desirable traits.[9] Soybean meal is universally available and has one of the best amino acids profiles of all protein-rich plant feedstuffs to meet most of the essential amino acid requirements of fish. Some fish such as young salmon, find soybean meal unpalatable while others, such as channel catfish, readily consume diets containing up to 50% soybean meal. For Egyptian sole (*Solea aegyptiaca*) juveniles for example, a diet with 56.2% fish meal, 5% maize gluten meal, and 4% of soy oil has been taken as a reference in replacement trials.[9] In this study, growth performance was evaluated when fish meal was replaced by soybean meal (48% protein) up to an amount of 30%, with a subsequent reduction of fish meal down to 37.1%. According to the performance of sole during an experiment of 87 days, the amino acid profile of the soybean meal diets did not seem to influence the protein utilization negatively and this despite to a different content of certain essential amino acid such as methionine. Therefore, the authors suggest that soybean meal and even soy protein concentrates could be good replacers for fish meal.[9] This could suggest an increasing use of soybean products in fish feed in the future. Meals from cottonseed and peanut have also been used in fish feeds in the United States, as well as meal from canola seeds for salmonids and lupin flour in feeds for rainbow trout.

The use of GMO-derived products in fish feed does not influence the animal performance. Studies performed for catfish, rainbow trout, and salmon show that there is no difference in animal performance or the composition of the end product between animals fed with conventional crops and those fed with grain, silage, or byproducts derived from GM crops.[10]

28.1.3 International Regulations

GM crops have become part of the global feed and food market. In the United States, three independent authorities are involved in the regulation of the release of GM plants and their use in foodstuffs: Animal and Plant Health Inspection Service (APHIS), Food and Drug Administration (FDA), and Environmental Protection Agency (EPA). The authorization procedure is simple and based on the principle of substantial equivalence, which means that a GM product is in essence not distinct from a its conventional counterpart. The GM risk assessment focuses on human, animal, and environmental safety and there is no requirement for traceability or labeling of approved GMOs. Product tracing should only be considered in cases of food safety concerns, which is obtained through the governmental establishment of food performance standards for food producers and processors.[11,12]

The current European-based GM legislation is more complex and includes pre-authorization safety assessments[13] by the European Food Safety Authority,[14] availability of validated detection methods, reference materials, and thresholds for labeling,[15,16] post-market monitoring, and post-marketing traceability requirements.[17,18] More and more emphasis is also set on the coexistence of GM crops next to the conventional and organic production systems throughout the entire supply chain.[19–22] Gene-stacked GMOs, which are the result of the crossbreeding between two GMO lines, require separate authorization in the EU.

European Regulations 1829/2003 and 1830/2009/EC mandate the labeling and traceability of GMOs in the EU.[15,17] Only in cases where the content of the authorized GMO ingredient is below 0.9% and in cases of accidental of adventitious presence of this GMO in the product, the labeling requirement may be omitted. Seed legislation requires that GM varieties have to be authorized in accordance with EU GMO legislation, in particular with Directive 2001/18/EEC before they are included in the Common Catalogue and marketed in the EU. If the seed is intended for use in food or feed, it can also be authorized in accordance with the GM food and feed Regulation. Similar labeling legislations exist in countries such as Russia (0.9%), Brazil (1%), and Japan (5%), whereas in Canada and the Philippines, GMO products require no labeling[23] All these legislations, together with the Cartagena protocol,[24] which has been established to preserve the worldwide biosafety, enforce the need for the identification and traceability of GMOs worldwide. For this purpose, methods to sample, detect, identify, quantify, and trace GMOs and derived products are necessary.

GM crops can be detected either by searching for the altered DNA, by detecting the newly expressed proteins or by assessing the presence of the trait (bioassays). This chapter explores the existing methods based on DNA and protein detection. Advantages and limitations of both detection methods are discussed.

28.2 GMO Analysis

28.2.1 Introduction

The manner in which to discriminate between GM versus nonmodified products is in most cases based on the presence of the newly introduced genes. Besides protein- and DNA-based methods, the so-called genetic analyses, biological (phenotypical), and chemical methods also exist. This chapter will focus on the genetic analyses. Methods have been developed either based on the detection of DNA using the polymerase chain reaction (PCR), or based on protein detection using enzyme-linked immunosorbent assays (ELISA). These methods however vary in their reliability, robustness and reproducibility, cost, complexity, and speed. Although PCR-based methods are known to be highly sensitive, the use of protein-based methods is in some cases more obvious.

28.2.2 DNA-Based Detection

28.2.2.1 Extraction of DNA

A first step in the DNA-based detection methods is the extraction of suitable DNA from the sample. In the context of GMO analysis, where GMO quantification is necessary according to legislation, it is of utmost importance to extract DNA from samples, which are homogeneous and representative for an entire lot or batch. DNA extraction methods may be based on the precipitation of DNA in a test tube, on the binding of DNA to a (silica) resin in an extraction column or on a combination of both. The choice for a specific extraction method may depend on the characteristics of the sample. For instance, for the extraction of DNA from soybean oil, a hexane-based extraction method may be used,[25] whereas the extraction from particulate material may be performed using CTAB, as proposed by the European Committee for Standardization.[26] The evaluation of the efficiency of different DNA extraction methods has shown that, due to the variability and complexity of food and feed products, the choice for a particular method should be done on a case-by-case basis.[27–30] The suitability of a method may be tested by a DNA amplification test, which amplifies a target gene specific for the species under investigation (species- or taxon-specific PCR). For example, in the case of soybean the detection of the *lectin* gene is used as an endogenous control,[31] for maize the *zein* gene[32] or the *invertase* gene[31] may be targeted (Figure 28.1).

28.2.2.2 Qualitative Conventional PCR

Depending on the sequences selected for PCR amplification, the detection of GMOs can be categorized into different levels of specificity: screening methods, gene-specific, construct-specific, and line- or event-specific methods (Figure 28.1).

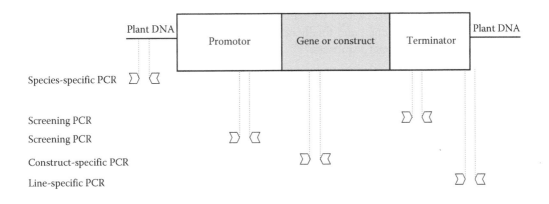

Figure 28.1 Novel gene construct integrated into a plant genome and the PCR amplification styles.

PCR screening methods target the regulatory elements that are used in the transformation process, e.g., the cauliflower mosaic virus (CaMV) 35S promotor, the nopalin synthase (*nos*) terminator, and the kanamycin resistance marker gene (*nptII*), which are present in many of the transgenic plants that have been developed so far. However, certain limitations of this method have been described. Because the CaMV 35S promotor, *3′-nos* terminator, and *nptII* sequence may occur naturally, their detection does not necessarily prove that a GMO is present.[33] Moreover, this screening system does not allow to discriminate between different transgenic lines and to identify them, since only the presence of one specific element is shown. GMO mixtures cannot be detected.[34]

Construct-specific PCR methods have also been derived for a range of different GMOs. The construct-specific PCR is based on the use of primer pairs that span the boundary of two (or more) adjacent introduced genetic elements (e.g., promotor, target gene(s), and terminator) or that are specific for the altered gene sequence.[35] This construct-specific PCR will identify a specific insert and lines transformed with different inserts can be discriminated. However, lines with an identical insert cannot be distinguished and it is impossible to distinguish nonauthorized from authorized GMOs.

Line-specific PCR detection methods target the junction at the integration site between the plant genome and the inserted DNA. This method was developed by the use of anchored PCR.[35–37] Through the amplification of the left and the right border of the insertion site (i.e., the junction between the insert and the plant DNA), a typical fingerprint can be obtained. In the latter PCR one primer, labeled with a radioactive marker, is complementary to the DNA, which is integrated in the target genome, while the other is complementary to the adapter linked to a restriction site generated by a frequent cutting enzyme. As a result, a specific junction that spans the boundary between plant DNA and inserted DNA will be amplified. This junction fragment will be unique for each transgenic line. After sequencing this junction fragment, the obtained data of the flanking plant DNA makes it possible to design line-specific primers. The created line-specific primer and the anchored primer can then be used to perform line-specific PCR.[35,38,39] All the line-specific detection methods for the EU-authorized GMOs have been validated by the Community Reference Laboratory (CRL) of the Joint Research Center (Ispra, Italy) and are available on the net (http://gmo-crl.jrc.it/statusofdoss.htm). In addition, different methods are available to confirm the PCR results: specific cleavage of the amplification product by restriction endonuclease digestion; hybridization with a DNA probe specific for the target sequence; direct sequencing of the PCR product; and finally nested PCR, in which two sets of primers bind specifically to the amplified target sequences.[37,40]

28.2.2.3 Multiplex PCR

The specific requirements for fast, multiple, and high-throughput GMO determination is fed by the increasing amount of transgenic events, the co-occurrence of several GMOs in one given matrix, and the customs formalities of countries with restricted import of GMOs. Each of these are a driving force to accelerate cost-effective methods assessing multiple GMO presence. Multiplex PCR methods based on the simultaneous amplification of different sequences save considerable time and effort by decreasing the number of reactions.[23,41] The different size of the amplicons leading to differential migration through agarose or polyacrylamide gels makes it possible to differentiate between the different GMOs present. Capillary electrophoresis in combination with different fluorochromes for each amplicon is a useful alternative for size-based differentiation.[42] A plethora of successful examples in which multiplex PCR has been applied successfully in detection of GMO are available for maize, canola, and soybean.[43–45] In a study by Hernandez et al.,[44] a multiplex PCR able to detect several GM maize lines proved to be 100% event-specific. Although this technique has the advantage of being time-saving, multiplex PCR often involves the amplification of larger amplicons and is therefore sometimes less suitable to assess GMO presence in processed food or feed since DNA degradation precludes the survival of long stretches of intact DNA. Furthermore, multiplex PCR is less sensitive compared to simplex (conventional) PCR methods due to competition between amplicons, depletion of reagents in the PCR tube, and the production of aspecific PCR products. In this context, Morisset et al.[46] published a review on alternative DNA amplification methods that could be used in GMO detection to overcome these problems.

28.2.2.4 Quantitative PCR

Because in several countries maximum limits are set for the (accidental) presence of GMOs in food and feed products, a critical aspect of the GMO analysis is the quantification. Initially quantitative competitive PCR was developed. The principle of this method is the co-amplification of internal DNA standards together with target DNA and their subsequent quantification. Therefore, DNA standards were constructed in a similar manner to the construction of the target DNA, but these are distinct from the GMO DNA by specific sequence insertions. The system is calibrated by co-amplification of mixtures of GMO DNA and corresponding amounts of conventional DNA. Such standards are commercially available and contain known amounts of standard DNA. Determination of the so-called equivalence point is the basis for quantification.[47] The system has been tested on several samples such as soy and maize.[48–50]

PCR–ELISA uses the strategy of real-time PCR and can be quantitative when the PCR is stopped before a significant decrease in amplification efficiency occurs (i.e., before the plateau phase is reached). Then ELISA can be used to quantify the relatively low amounts of PCR products.[51,52]

Real-time PCR (qPCR), an online method for determining ratios of transgenic to nonmodified DNA components, is the method of choice at the moment to detect and quantify DNA. This real-time PCR allows the quantification of DNA by measuring the kinetics of the PCR amplification instead of an endpoint measurement. This continuous measurement is based on the accumulation of a fluorescence signal. A signal is created by the binding of a fluorogene component to the amplified DNA and is proportional to the amounts of PCR products generated. There exist different strategies: the detection is either based on the measurement of the accumulation of DNA-binding dyes (e.g., SYBR Green) or the binding of specific labeled probes (e.g., TaqMan probe). During PCR, the accumulation of fluorescence can then be measured online. With the first method, the DNA-binding dye does not discriminate between the target and any co-amplified nontarget sequence, e.g., primer dimers. Still it is possible to verify the identity of the amplified dsDNA by melting curve analysis. The reassociation of the dsDNA will result in the creation of a peak signal and the specific temperature profile of the peak is characteristic of a specific sequence. Nevertheless, in the presence of more than one GMO in a sample, it is difficult to quantify each specific target DNA sequence. With probes, the reliability of the real-time PCR increases significantly. These probes can be divided into different categories: hydrolysis and hybridization probes. Hydrolysis probes emit fluorescence upon degradation and require that the two fluorophores are brought into close proximity by hybridization and/or altered configuration of the probes. Hybridization probes emit fluorescence upon specific hybridization to the target amplicon sequence and require that the two fluorochromes are sufficiently separated in space by hydrolysis and/or altered configuration of the probes.[53] The most commonly used real-time probe is the TaqMan probe, a hydrolysis probe where reporter and quencher are located on the same probe within a short distance apart and where the hydrolysis of the probe results in the release of the reporter from the quencher. This results in a fluorescence signal, emitted by the reporter dye.

In the field of GMO analysis, the quantification requires simultaneous assessment of the recombinant and of a species- or taxon-specific reference gene. The quantification standards are usually certified reference material or cloned plasmid standards.[23,54] The method is well suited for automation and high throughput of samples and can be used for raw, processed, and even mixed products.[31,55,56]

28.2.3   Protein-Based Detection

The majority of protein detection methods are based on immunoassays, i.e., they rely on the interaction between a specific antibody and its antigen. Due to this strong interaction, immunoassays are highly specific and generally samples only need a simple preparation before analysis.

The Western blot technique is based on the separation of total protein on a gel followed by staining of the target protein with a labeled antibody. The method is highly specific, which provides qualitative results for determining whether a sample contains the target protein below or above a predetermined threshold level.[57]

The ELISA is the most common type of immunoassay. It covers any enzyme immunoassay involving an enzyme-labeled immunoreactant (antigen or antibody) and an immunosorbent (antigen or antibody bound to a solid support). Therefore, several variants of the ELISA-method exist, with the sandwich assay being the most widely used and most flexible type of ELISA. The detections of GM proteins in feed, based on commercially available ELISA kits, are mostly sandwich assays, although some are competitive ELISA assays.[58]

For the sandwich ELISA, antibodies specific to the target protein are bound to the surface of typically a microtiter well plate. When the solution containing the test material is added, the antibody will work as a capture molecule for the target protein. Following an incubation period, a washing step removes all unbound components. Then a second specific antibody, chemically bound to an enzyme that catalyzes a color reaction, is added. If the target protein is present, the second labeled antibody binds to it and any unbound labeled antibody is washed away. Enzyme substrate is then added to yield a colored product. The intensity of the signal produced is proportional to the amount of target protein present.[57]

In a competitive ELISA assay, the wells of the plate are coated with the target protein. A solution containing a limited number of first antibody together with the test sample is added. A competition for the first antibody will then occur between the target protein in the sample and the target protein coated on the wells. The antibodies that are not bound to the antigens bound to the well will be washed away. A second antibody–enzyme complex is then added which specifically binds the antigen–antibody complex. After a second washing step, an enzyme substrate is added, resulting in color production. For the competitive assay, the intensity of the color is inversely proportional to the concentration of the GM protein in the sample.[59]

Some commercial ELISA plate kits are supplied with calibrators (known concentrations of the target analyte in solution) and a negative control. These standards are run concurrently with each sample set and allow a standard curve to be set up. By using the spectrophotometer for all samples and all standards at the same time, a quantitative interpretation can be performed. The protein concentration in the sample can then be calculated from the standard curve. A semiquantitative interpretation can be made by comparing the color of the sample against the standards without the use of a spectrophotometer and determining the concentration range of the sample.[58]

The lateral flow strip is a variation on ELISA, using strips rather than microtiter well plates. The test strip is placed in an Eppendorf vial containing the test solution. The sample migrates up the strip by capillary action. As it moves up the strip, the sample passes through a zone that contains mobile antibodies, usually labeled with colloidal gold. This labeled antibody binds to the target protein, if present in the sample. The antibody–protein complex then continues to move up the strip through a porous membrane that contains two capture zones. The first capture zone of immobilized antibodies binds the antibody–protein complex and the gold becomes visible as a red line. The second capture zone acts as a control zone and contains antibodies specific for untreated antibodies coupled to the color reagent. If there is no target GM protein present, only one single line will appear at the control zone. A result is called positive when both the control line and the line indicating presence of target protein change color.[58] The test is simple, cheap, and fast, and does not require any laboratory facilities or technical expertise. Van den Bulcke et al.[60] compared the use of this strip test with PCR analysis for the detection of GM soy in feed samples. In most cases, matches were found between the obtained results. However, in cases of low level of GM soy the PCR analysis gave more positive results, due to its higher sensitivity. The strip test however remains a good tool in the traceability of GMOs early in the chain. Some other drawbacks are discussed later in this chapter.

Another format of immunoassays uses magnetic beads as a solid surface. The principle is the same as the ELISA-format, with the magnetic particles being coated with capture antibodies and the reaction being performed in a test tube. The target protein, bound to the magnetic particles, can be separated using a magnet. The advantages include superior kinetics, as the particles are free to move in the reaction solution, and increased precision, due to uniformity of the particles.[52]

A new evolution in protein-based assays is high-throughput technologies that have gained a great deal of attention over the last several years. The high throughput is obtained by miniature and highly sensitive microfluidic devices.[61] Although very small volumes are used, the technique remains reliable compared to the traditional immunoassays.[62] The device consists of a flat surface, usually glass or silicon, onto which narrow bands of antigen are deposited. Perpendicular to these stripes, a second set of lines is engraved. When diluted proteins flow through this second set of small channels, induced by capillary forces, they bind with their specific antigen, and a mosaic pattern of tiny squares occurs. This can then be analyzed using fluorescence microscopy.[62] The advantages of this technique are numerous. One sample can be screened for several proteins in one test and several assays can be performed simultaneously. The amount of sample needed is reduced to nanoliter range, thereby reducing the time to perform the analysis. In addition, this technique is highly suitable for automation. However, the equipment needed is expensive and skilled personnel are required for the analysis of results. Therefore, this technique is currently not used for routine analysis of feed samples.

28.3   Comparison of DNA and Protein Methods

28.3.1   Protein-Based Methods

GMOs can be detected on basis of the altered genome, thus by DNA analysis, or on basis of the novel protein. The protein-based assays comprise a very large and diverse group of assays and are commercially very successful. The strong interaction between the antibody and the antigen is translated into high sensitivity assays and antibody specificity minimizes sample preparation. Their cost, ease-of-use, and flexible test format have resulted in wide-scale use in highly diverse markets. The cost per test of an immunoassay compared to other analytical methods is low and a number of test formats exist, which require little or no training to perform the analysis. For instance, the one-step "strip" tests can be performed by untrained personnel and give yes/no type results in minutes. The time-to-result may be one of the most critical attributes of a test method in applications where a large lot of material needs to be screened before being pooled with other lots or proceeded to the next stage in a process. For on-site testing, the lateral flow test is generally the most cost-effective solution.[63]

Immunoassays can yield quantitative results in about an hour or they can be incorporated into fully automated instruments capable of running hundreds of samples an hour. Quantification of proteins is easier than quantification of DNA if one looks at the expression unit of GMO levels. Quantitative results from protein analyses are expressed on a weight/weight basis (molar concentrations), while those from DNA assays represent genome equivalents. The influence of gene copy numbers makes DNA quantification more complex and the results more uncertain.[37]

Despite the advantages protein-based methods offer, many drawbacks need to be considered when it comes to their application for the detection of GMOs. In those cases, the DNA-based method is preferred. One of the difficulties of an immunoassay system is experienced at the outset. Generating a specific antibody against the antigen of concern can be a slow, difficult, and time-consuming process, and requires many skills and a lot of experience. This is where the use of recombinant antibody technology could offer benefits. Using this technique, it will be easier to select antibodies with rare properties and to manipulate the characteristics of already existing antibodies.[64] Once a specific (monoclonal or polyclonal) antibody with high affinity for its target protein is generated, careful standardization and testing for unexpected cross-reactivity must be performed. Nonetheless, once established, the flexibility and turnaround time for immunoassay systems is excellent.[65]

Since the binding between antibody and antigen is based on the native structure of the target, any conformational change in the epitope of the antigen renders the assay ineffective. Food and feed processing, such as heating or exposure to strong acids or alkalis, can cause this conformational change, by which detection of the protein will no longer be possible. Therefore, protein detection using immunoassays is limited to grains, raw materials, and unprocessed products.[66] Furthermore, the accuracy and precision of the assay can be affected by substances present in feed matrices such as surfactants, phenolic compounds, fatty acids, endogenous phosphatases, or enzymes that may inhibit the specific antigen–antibody interaction.[64–67] Another disadvantage of the immunoassays is due to the fact that they rely on the expression of the newly introduced gene, e.g., genetic modification does after all not always result in the production of a new protein or the expression level of the introduced DNA can be too low for detection. Moreover, the protein levels can vary from tissue to tissue. As a result, the transgenic protein might not be present in the part of the plant that is used in feed production. For example, the endotoxin *cry*1A of GM maize *Bt*-176 is expressed in the green tissues and in the pollen of the plant. Since the novel protein is not expressed in the maize kernels, protein analysis of the kernels will not reveal the presence of any foreign protein.[68] Expression is furthermore influenced by external factors such as weather, soil, and other cultivation conditions.[69] This complicates the quantification of the GMO by protein-based detection methods.

Quantification by the use of immunoassays is already a complex subject, since the assays generate absolute values, such as the total amount of a novel protein present. To comply with GMO labeling legislation however, not the absolute quantity but the relative quantity of the GM trait is important, i.e., the relative ratio of the GM trait to the conventional counterpart of the same ingredient (e.g., percentage of GM soy out of total soy in the feed product). The relative protein-based quantification is only possible if a species- or taxon-specific protein is measured simultaneously[70] or if the sample exists of one single ingredient and an appropriate reference material for the standard dose response curve is available.

Since there are no structures common to all GM proteins or groups of proteins and a single antibody will only bind to one particular protein, the immunoassays are less suited for a general GM screening.[35] In the future, it is likely that tests will be developed that detect multiple novel proteins using a single lateral flow strip. Moreover, protein-based methods cannot distinguish between GM varieties with a different genetic construct but the same expressed protein.[38,71,72]

28.3.2   DNA-Based Methods

The suitability of the PCR technique to detect genetic modifications in feedstuffs is determined by the quality and quantity of the DNA still present in the final product. Various factors can contribute to the degradation of the DNA such as heat treatments, nuclease activity, and low pH. In a typical feed-pelleting process, the feedstuff is first moisture conditioned, then subjected to pressures, and finally, with the aid of the associated heat generated, forced through die openings. The larger particles or pellets formed are easier to handle, more palatable, and usually result in improved feeding results when compared to the unpelleted feed. The extrusion of cereals aims at the complete rupture of the starch granules by a combination of moisture, heat, pressure, and mechanical shear.

Although extensive research has already been performed on the effect of food processing on the stability and detection of DNA,[73–76] research on feed products is poor.[77,78] Gawienowski et al.[74] and Aulrich, Pahlow, and Flachowsky[79] have investigated the effect on ensilage of maize, showing that DNA is degraded due to a pH drop with time. Dry and steam heating under laboratory conditions by Forbes et al.[80] and Chiter, Forbes, and Blair[81] indicated that temperatures above 95°C for at least 5 min completely fragment DNA, whereas physical extrusion highly fragments DNA. Results indicate that DNA of sufficient quality and quantity remained detectable after steam conditioning at 95°C, allowing subsequent GMO analysis (personal communication). However, the extrusion of feed samples, where temperatures up to 135°C were applied, led to the degradation of

DNA. A lower moisture content even emphasizes the destruction of DNA. A significant decrease in the average DNA fragment length may render the detection of the inserted DNA sequence impossible.

Other processing steps may remove DNA from the sample, as during the refining of glucose syrup, soy lecithin, or oil.[82,83] In these cases, DNA is depleted to very small quantities, rendering detection impossible with the current official ISO detection methods.[75,84,85] Furthermore, potential problems will arise with respect to adequate sampling (homogeneity and representativity) and PCR inhibitors in the feed matrix. Inhibition of the DNA polymerase by co-purification of proteins, fats, polysaccharides, polyphenolics, and other components present in highly processed feedstuffs is a major problem in the preparation of DNA for PCR.[86]

These circumstances may not only influence GMO screening, construct-, gene-, and line-specific PCRs, but may render the quantification of GMOs inaccurate. Although validated by the Community Reference Laboratory (CRL, Ispra, Italy), these methods are rarely tested and optimized for DNA samples extracted from processed food or feed samples.[87,88] Problems encountered with the quantification of GMOs are translated in high limits of detection and quantification. Furthermore, as GMO quantification is based on the amplification of two DNA fragments, different PCR amplification efficiencies for those fragments may divert the calculated GMO content from its true value.[88] Another problem encountered with GMO quantification is due to the ploidy of the material under investigation, as GMO percentages should be expressed in terms of haploid genome equivalents. Since the true level of zygosity or ploidy of the material to be analyzed is not always known, a high degree of measurement uncertainty is associated with quantitative analytical estimates.[33] All these inconveniences result in a standard deviation of 20%–30% for quantitative GMO analysis.

28.4 Recent Developments in GMO Detection

Conventional methods for GMO detection have their own limitations. Qualitative PCR reactions are time-consuming and technically demanding, quantitative PCRs require internal standards and protein-derived methods often suffer from protein denaturation.[89] Therefore, new technologies are under development in order to overcome these shortcomings.

28.4.1 *Microarray Technology*

From an economic point of view, microarray technology is very promising by viewing its feature of simultaneous identification of multiple GMOs. In addition compared to multiplex PCR, it has the advantage of being a more flexible tool not being hampered by the range of GMOs, the amplicon size or the number of amplicons in the analysis. This is a direct consequence of the fact that microarrays are two-dimensional compared to classical gel electrophoresis rendering it more discriminative. The principle of microarray analysis is the attachment of nucleic acids to a solid support (often glass), which is subsequently probed by labeled nucleic acid molecules. Although microarray technology has its main applications in transcriptome analysis, some specific arrays were developed for GMO detection and characterization. Several low-density arrays allowing the simultaneous detection of nine GMO events including the GMO screening elements (35S promotor, *nos* terminator, and *nptII*) were developed on a laboratory scale.[90,91]

The enormous data obtained after a microarray analysis however is probably one of the only restraints that has hampered this method being applied routinely. To our knowledge, only a few studies resulted in the commercially available microarray device.[92] A commercial kit, DualChip® GMO has been developed by Eppendorf array technology by coupling multiplex PCR to microarray hybridization and as such combining the best of both worlds. The systems detects and identifies GM events by screening multiple genetic elements, namely, *Bt*-176 maize, Mon 810 maize, *Bt*-11 sweet maize, Mon 531 cotton, GA21 maize, and Roundup Ready™ soja GMO events.[89]

28.4.2 Biosensors

The major restraint on the use of PCR-derived techniques in commercial samples is undoubtedly the highly technical demand. The use of biosensors could be a useful alternative to detect GMOs in the future. The principle of a sensor is that a change in the probe environment results in a specific change in a physical property of the probe, which in turn is converted into an electrical signal. The development of different types of biosensors is in progress. In the work by Mannelli et al.[93,94] two different types of sensors were developed. A bulk acoustic wave affinity biosensor for GMO detection was developed. This device consisted of a DNA probe immobilized on the sensor surface while the target sequence was free in solution and available for hybridization.[93] Probe immobilization was monitored by a surface plasmon resonance (SPR) system. The same group described a piezoelectric sensor based on the immobilization of single-stranded DNA probes on the surface of a quartz crystal microbalance device.[94]

Since its development, SPR technology has been successfully used for the detection of Roundup Ready® soybean.[95] In this application, household and transgene sequences were mounted on the sensor chips and were shown to detect the transgene in real-time formats. In another approach, biotinylated PCR products containing 0.5% and 2% of *Bt*-176 sequences were mounted on a chip. After immobilization, the authors succeeded to discriminate between both *Bt*-176 concentrations making them conclude that biosensor technology has the capacity to be as efficient as real-time PCR.

Even though biosensors may not be able to entirely replace the use of PCR for the accurate determination of GMO in products, they can nonetheless be useful in the preliminary stages of control to identify samples to be subjected to successive analyses.[23]

28.5 Conclusions

A wide variety of protein- and DNA-based formats are available to analyze products for the presence of GMOs. Immunoassays are very suitable for the detection of GMOs at critical control points early in the feed production chain, i.e., on raw materials, unprocessed ingredients, and simple food or feed matrices (single ingredients products or products with only one GMO ingredient). Application on processed and complex matrices is limited. Thanks to the higher stability of DNA compared to proteins, PCR-based methods are currently the methods of choice in the field of GMO detection and quantification. In cases where no protein or DNA can be detected in the feed sample, enforcement of the legislation should be supported by traceability measures.

It is to be expected that in the near future, an increasing number of GM crops will be released on the market, producing an increased variety of genetic constructs present in food and feed. To challenge the entrance of these new GM products, high throughput technologies will be required. The multiplex amplification method, possibly in combination with microarray format detection, will be a very interesting option in order to decrease the number of assays needed for efficient identification and quantification of feedstuffs containing GM material.

With respect to the current quantification of GMOs, which is mostly based on real-time PCR quantification, still work has to be done to improve the accuracy of the available methods. Other issues that will need attention in the following years are the detection of unauthorized GMOs, the detection and quantification of stacked genes (because no distinction can be made between the gene-stacked GMO and a mixture of its two parental GMOs), the optimization of extraction methods and sampling procedures, and the development of reliable field tests for all GMOs. This could be obtained by the use of lateral flow test that are able to detect different GMOs at the same time. Some tests are already commercially available.

References

1. Gryson, N., Messens, K., and Dewettinck, K., Genetically modified organisms in food industry, in *Handbook of Food Science, Technology and Engineering*, Hui, Y.H., Ed., CRC Press, Boca Raton, FL, 2006.
2. Agbios, *Agriculture and Biotechnology Strategies*, Canada Inc. GMO database, Available from: http://www.agbios.com. 2009.
3. Ivarie, R., Competitive bioreactor hens on the horizon, *Trends Biotechnol.*, 24, 99, 2006.
4. James, C., *Executive Summary of Global Status of Commercialised Biotech/GM Crops*, ISAAA Briefs 34, Ithaca, NY, 2005.
5. James, C., *Executive Summary of Global Status of Commercialised Biotech/GM crops*, ISAAA Briefs 37, Ithaca, NY, 2007.
6. Greiner, R. and Konietzny, U., Presence of genetically modified maize and soy in food products sold commercially in Brazil from 2000 to 2005, *Food Control*, 19, 499, 2008.
7. Maga, E.A., Genetically engineered livestock: Closer than we think? *Trends Biotechnol.*, 23, 533, 2005.
8. Varzakas, T.H., Arvanitoyannis, I.S., and Baltas, H., The politics and science behind GMO acceptance, *Crit. Rev. Food Sci.*, 47, 335, 2007.
9. Bonaldo, A. et al., Influence of dietary soybean meal levels on growth, feed utilization and gut histology of Egyptian sole (*Solea aegyptiaca*) juveniles, *Aquaculture*, 12(2), 580, 2006.
10. Anonymous, Safety and nutritional assessment of GM plants and derived food and feed: The role of animal feeding trials, *Food Chem. Toxicol.*, 46, S1, 2008.
11. Gryson, N., Rotthier, A., Messens, K., and Dewettinck, K., Regulation of genetically-modified crops in the European Union, *Lipid Technol.*, 17(1), 11, 2005.
12. USDA-APHIS (US Department of Agriculture, Animal and Plant Health Inspection Service), *Factsheet on Biotechnology Regulatory Services, USDA's Biotechnology Deregulation Process*. Fabruary 2006, Available from http://www.aphis.usda.gov/.
13. European Commission, Directive 2001/18/EC of the European Parliament and of the Council of 17 April 2001 on the deliberate release into the environment of genetically modified organisms and repealing Council Directive 90/220/EEC, *Off. J. Eur. Commun.*, L106, 1, 2001.
14. European Commission, Regulation (EC) 178/2002 of the European Parliament and of the Council of 28 January 2002 laying down the general principles and requirements of food law, establishing the European Food Safety Authority and laying down procedures in matters of food safety, *Off. J. Eur. Commun.*, L31, 1, 2002.
15. European Commission, Regulation (EC) No 1829/2003 of the European Parliament and of the Council of 22 September 2003 on genetically modified food and feed, *Off. J. Eur. Commun.*, L268, 1, 2003.

16. European Commission, Commission Regulation (EC) No 641/2004 on detailed rules for the implementation of Regulation (EC) No 1829/2003 of the European Parliament and of the Council as regards the application for the authorisation of new genetically modified food and feed, the notification of existing products and adventitious or technically unavoidable presence of genetically modified material which has benefited from a favourable risk evaluation, *Off. J. Eur. Commun.*, L102, 14, 2004.

17. European Commission, Regulation (EC) No 1830/2003 of the European Parliament and of the Council of 22 September 2003 concerning the traceability of food and feed products produced from genetically modified organisms and amending Directive 2001/18/EC, *Off. J. Eur. Commun.*, L268, 24, 2003.

18. European Commission, Commission Regulation (EC) No 65/2004 of 14 January 2004 establishing a system for the development and assignment of unique identifiers for genetically modified organisms, *Off. J. Eur. Commun.*, L10, 5, 2004.

19. European Commission, Commission recommendation 2003/556/EC of 23 July 2003 on guidelines for the development of national strategies and best practices to ensure the co-existence of genetically modified crops with conventional and organic farming, *Off. J. Eur. Commun.*, L189, 36, 2003.

20. European Commission, Commission decision 2005/463/EC of 21 June 2005 establishing a network group for the exchange and coordination of information concerning coexistence of genetically modified, conventional and organic crops, *Off. J. Eur. Commun.*, L164, 50, 2005.

21. Gryson, N. et al., Co-existence and traceability of GM and non-GM products in the feed chain, *Eur. Food Res. Technol.*, 226, 81, 2007.

22. Gryson, N. and Eeckhout, M., Co-existence of GM and non-GM crops: Vision of the European Commission anno 2007 (Co-existentie van GGO en non-GGO gewassen: Visie van de Europese Commissie anno 2007), *De Molenaar*, 1, 24, 2008.

23. Marmiroli, N. et al., Methods for detection of GMOs in food and feed, *Anal. Bioanal. Chem.*, 392, 369, 2008.

24. Cartagena protocol on biosafety to the convention on biological diversity. Montreal, Quebec, Canada. Available from: http://www.biodiv.org/doc/legal/cartagena-protocol-en.pdf.

25. Wurz, A. et al., DNA-Extraktionsmethode für den Nachweis gentechnisch veränderter Soja in Sojalecithin, *Deut. Lebensm.-Rundsch.*, 94(5), 159, 1998.

26. International Organization for Standardization, ISO 21571, *Foodstuffs—Methods of Analysis for the Detection of Genetically Modified Organisms and Derived Products—Nucleic Acid Extraction*, ISO, Geneva, Switzerland, 2005.

27. Di Pinto, A. et al., A comparison of DNA extraction methods for food analysis, *Food Control*, 18, 76, 2007.

28. Gryson, N., Messens, K., and Dewettinck, K., Evaluation and optimalisation of five different extraction methods for soya DNA in chocolate and biscuits. Extraction of DNA as a first step in GMO analysis, *J. Sci. Food Agr.*, 84, 1357, 2004.

29. Holden, M.J. et al., Evaluation of extraction methodologies for corn kernel (*Zea mays*) DNA for detection of trace amounts of biotechnology-derived DNA, *J. Agric. Food Chem*, 51, 2468, 2003.

30. Mafra, I., Ferreira, I., and Oliveira, M., Food authentication by PCR-based methods, *Eur. Food Res. Technol.*, 227, 649, 2008.

31. Berdal, K.G. and Holst-Jensen, A., Roundup Ready soybean event-specific real-time quantitative PCR assay and estimation of the practical detection and quantification limits in GMO analyses, *Eur. Food Res. Technol.*, 213, 432, 2001.

32. Peano, C., Samson, M.C., Palmieri, L. Gulli, M., and Marmiroli, N., Qualitative and quantitative evaluation of the genomic DNA extracted from GMO and non-GMO foodstuffs with four different extraction methods, *J. Agric. Food Chem.*, 52, 5829, 2004.

33. Miraglia, M. et al., Detection and traceability of genetically modified organisms in the food production chain, *Food Chem. Toxicol.*, 42, 1157, 2004.

34. Pietsch, K. et al., Screeningsverfahren zur identifizierung gentechnisch veränderter pflantelicher Lebensmittel, *D. Lebensm.-Rundsch.*, 93, 35, 1997.

35. Griffiths, K. et al., *Review of Technologies for Detecting Genetically Modified Materials in Commodities and Food*. Department of Agriculture, Fisheries and Forestry, Australia, 126, 2002.

36. Windels, P. et al., Characterisation of the Roundup Ready soybean insert. *Eur. Food Res. Technol.*, 213(2), 107, 2001.

37. Anklam, E. and Neumann D.A., Method development in relation to regulatory requirements for detection of GMOs in the food chain,. *J. AOAC Int.,* 85, 754, 2002.

38. Gachet, E. et al., Detection of genetically modified organisms (GMOs) by PCR: A brief review of methodologies available, *Trends Food Sci. Technol.*, 9, 380, 1999.

39. Meyer, R., Development and application of DNA analytical methods for the detection of GMOs in food, *Food Control*, 10, 391, 1999.

40. Tengs, T. et al., Microarray-based method for detection of unknown genetic modifications, *BMC Biotecnol.*, 7, 91, 2007.

41. Mafra, I. et al., Comparative study of DNA extraction methods for soybean derived food products, *Food Control*, 19, 1183, 2008.

42. Nadal, A. et al., A new PCR-CGE (size and color) method for simultaneous detection of genetically modified maize events, *Electrophoresis*, 27, 3879, 2006.

43. Germini, A. et al., Determination of transgenic material on the Italian food market using a new multiplex PCR method, *Ital. J. Food Sci.* 17, 371, 2005.

44. Hernandez, M. et al., Interlaboratory transfer of a PCR multiplex method for simultaneous detection of four genetically modified maize lines: *Bt*11, MON810, T25, and GA21, *J. Agric. Food Chem.*, 53, 3333, 2005.

45. James, D. et al., Reliable detection and identification of genetically modified maize, soybean, and canola by multiplex PCR analysis, *J. Agric. Food Chem.*, 51, 5829, 2003.

46. Morisset, D. et al., Alternative DNA amplification methods to PCR and their application in GMO detection: A review, *Eur. Food Res. Technol.*, 227, 1287, 2008.

47. Raeymackers, L., Quantitative PCR: Theoretical considerations with practical implications, *Anal Biochem.*, 214, 582, 1993.

48. Wurz, A. et al., Quantitative analysis of genetically modified organisms in processed food by PCR-based methods, *Food Control*, 10, 385, 1999.

49. Zimmerman, A., Lüthy, J., and Pauli, U. Event specific transgene detection in *Bt*11 corn by quantitative PCR at the integration site, *Lebensm. Wiss. Technol.*, 33, 210, 2000.

50. Studer, E. et al., Quantitative competitive PCR for the detection of genetically modified soybean and maize, *Z. Lebensm. Unters. For. A*, 207, 207, 1998.

51. Landgraf, A., Reckmann, B., and Pingoud, A., Direct analysis of polymerase chain reaction products using Enzyme-Linked Immunosorbent Assay techniques, *Anal. Biochem.*, 198, 86, 1991.

52. Ahmed, F.E., Detection of genetically modified organisms in food, *Trends Biotechnol.*, 20, 215, 2002.

53. Holst-Jensen, A., Ronning, S.B., Lovseth, A., and Berdal, K.G., PCR technology for screening and quantification of genetically modified organisms (GMOs), *Anal. Bioanal. Chem.*, 375, 985, 2003.

54. Taverniers, I., Windels, P., Vaitilingom, M., Milcamps, A., Van Bockstaele, E., Van den Eede, G., and De Loose, M., Event-specific plasmid standards and real-time PCR methods for transgenic *Bt*11, *Bt*176, and GA21 maize and transgenic GT73 canola, *J. Agric. Food Chem.*, 53, 3041, 2005.

55. Greiner, R., Konietzny, U., and Villavicencio, A.L.C.H., Qualitative and quantitative detection of genetically modified maize and soy in processed foods sold commercially in Brazil by PCR-based methods, *Food Control*, 16, 753, 2005.

56. Hernàndez, M. et al., A specific real-time quantitative PCR detection system for event MON810 in maize YieldGard® based on the 3′-transgene integration sequence, *Transgenic Res.*, 12, 179, 2003.

57. Lipton, C.R. et al., Guidelines for the validation and use of immunoassays for determination of introduced proteins in biotechnology enchanced crops and derived food ingredients, *Food Agric. Immunol.*, 12, 153, 2000.

58. Rotthier, A. et al., Protein-based detection of GM ingredients, in *Advances in Food Diagnostics*, Nollet, L. and Toldrá, F., Eds., Blackwell Publishing, Ames, IA, 10, 199, 2007.

59. Yeung, J. Enzyme-linked immunosorbent assays (ELISAs) for detecting allergens in foods, in *Detecting Allergens in Food*, Koppelman, S. and Hefle, S., Eds., Woodhead Publishing in Food Science and Technology, Cambridge, U.K., 109, 2006.

60. Van de Bulcke, M. et al., Detection of genetically modified plant products by protein strip testing: An evaluation of real-life samples, *Eur. Food Res. Technol.*, 225, 49, 2007.

61. Burns, M.A. et al., Microfabricated structures for integrated DNA analysis, *Proc. Natl. Acad. Sci. USA*, 93(11), 5556, 1996.

62. Bernard, A., Michel, B., and Delamarche, E., Micromosaic immunoassays, *Anal. Chem.*, 73, 8, 2001.

63. Stave, J.W., Detection of new or modified proteins in novel foods derived from GMO - future needs, *Food Control*, 10, 367, 1999.

64. Brett, G.M. et al., Design and development of immunoassays for detection of proteins, *Food Control*, 10, 401, 1999.

65. Bindler, G. et al., Report of the CORESTA Task Force Genetically Modified Tobacco: Detection methods, *CORESTA*, 1, 1998.

66. Anklam, E. et al., Analytical methods for detection and determination of genetically modified organisms in agricultural crops and plant-derived food products, *Eur. Food Res. Technol.*, 214, 3, 2002.

67. Spiegelhalter, F., Lauter, F.R., and Russel, J.M., Detection of genetically modified food products in a commercial laboratory, *J. Food Sci.*, 66, 634, 2001.

68. Popping, B. and Broll, H., Detection of genetically modified foods—Past and future, *Actual. Chim.*, 11, 3, 2001.

69. Wilson, R.F., Hou, C.T., and Hildebrand, D.F., Eds., *Dealing with Genetically Modified Crops*, AOCS Press, Champaign, IL, 2001.

70. Van Duijn, G. et al., Detection of genetically modified organisms in foods by protein- and DNA-based techniques: Bridging the methods, *J. AOAC Int.*, 85, 787, 2002.

71. Lüthy, J., Detection strategies for food authenticity and genetically modified foods, *Food Control*, 10, 359, 1999.

72. Kok, E.J. et al., DNA methods: Critical review of innovative approaches, *J. AOAC Int.*, 85, 797, 2002.

73. Chen, Y. et al., Degradation of endogenous and exogenous genes of Roundup-Ready soybean during food processing, *J. Agric. Food Chem.*, 53, 10239, 2005.

74. Gawienowski, M.C. et al., Fate of maize DNA during steeping, wet-milling, and processing, *Cereal Chem.*, 76, 371, 1999.

75. Gryson, N., Stability and detection of genetically modified soy in processed foods, PhD dissertation, 2006.

76. Gryson, N., Messens, K., and Dewettinck, K., PCR detection of soy ingredients in bread, *Eur. Food Res. Technol.*, 227, 345, 2008.

77. Alexander, T.W. et al., A review of the detection and fate of novel plant molecules derived from biotechnology in livestock production, *Anim. Feed Sci. Technol.*, 133, 31, 2007.

78. Alexander, T.W. et al., Impact of feed processing and mixed ruminal culture on the fate of recombinant EPSP synthase and endogenous canola plant DNA, *FEMS Microbiol. Lett.*, 214, 263, 2002.

79. Aulrich, K., Pahlow, G., and Flachowsky, G., Influence of ensiling on the DNA-degradation in isogenic and transgenic corn, *Proc. Soc. Nutr. Physiol.*, 13, 112, 2004.

80. Forbes, J.M. et al., Effect of feed processing conditions on DNA fragmentation, Scientific Report No. 376 to the Ministry of Agriculture, Fisheries and Food, United Kingdom, 1998.

81. Chiter, A., Forbes, J.M., and Blair, G.E., DNA stability in plant tissues: Implications for the possible transfer of genes from genetically modified food, *FEBS Lett.*, 481, 164, 2000.

82. Gryson, N. et al., Detection of DNA during the refining of soy bean oil, *J. Am. Oil Chem. Soc.*, 79, 171, 2002.

83. Gryson, N., Messens, K., and Dewettinck, K. Influence of different oil refining parameters and sampling size on the detection of GM-DNA in soybean oil, *J. Am. Oil Chem. Soc.*, 81, 231, 2004.

84. International Organization for Standardization, ISO 21569, Foodstuffs—Methods of Analysis for the Detection of Genetically Modified Organisms and Derived Products—Qualitative Nucleic Acid Based Methods, ISO, Geneva, Switzerland, 2005.

85. International Organization for Standardization, ISO 21570, Foodstuffs—Methods of Analysis for the Detection of Genetically Modified Organisms and Derived Products—Quantitative Nucleic Acid Based Methods, ISO, Geneva, Switzerland, 2005.

86. Gryson, N., Dewettinck, K., and Messens K., Influence of cocoa components on the PCR detection of soy lecithin, *Eur. Food Res. Technol.*, 226, 247, 2007.

87. Trapman, S. and Emons, H., Reliable GMO analysis, *Anal. Bioanal. Chem.*, 381, 72, 2005.

88. Weighardt, F., GMO quantification in processed food and feed, *Nat. Biotechnol.*, 25, 1213, 2007.

89. Lasken, R.S. and Egholm, M., Whole genome amplification: Abundant supplies of DNA from precious samples or clinical specimens, *Trends Biotechn.*, 21, 531, 2003.

90. Leimanis, S. et al., A microarray-based detection system for genetically modified (GM) food ingredients, *Plant Mol. Biol.* 61, 123, 2006.

91. Xu, X.D. Rapid and reliable detection and identification of GM events using multiplex PCR coupled with oligonucleotide microarray, *J. Agric. Food Chem.*, 53, 3789, 2005.

92. Birch, L. et al., Evaluation of LabChip™ technology for GMO analysis in food, *Food Control*, 12, 535, 2001.

93. Mannelli, I. et al., Bulk acoustic wave affinity biosensor for genetically modified organisms detection. *IEEE Sens. J.*, 3, 369, 2003.

94. Mannelli, I. et al., Quartz crystal microbalance (QCM) affinity biosensor for genetically modified organisms (GMOs) detection, *Biosens. Bioelectron.*, 18, 129, 2003.

95. Feriotto, G. et al., Biosensor technology and surface plasmon resonance for real-time detection of genetically modified Roundup Ready soybean gene sequences, *J. Agric. Food Chem.*, 50, 955, 2002.

MILK AND DAIRY FOODS

MILK AND DAIRY PRODUCTS

Chapter 29

Microbial Flora

Effie Tsakalidou

Contents

29.1 Introduction..781
29.2 Microbial Ecology of Dairy Products..782
29.3 Methods of Microbiological Analysis in Dairy Products 783
     29.3.1 Culture-Dependent Methods...783
          29.3.1.1 Classical and Advanced Phenotypic Methods....................783
          29.3.1.2 Molecular Methods 785
     29.3.2 Culture-Independent Methods 788
References 790

29.1 Introduction

The production of fermented dairy products, such as cheese, yogurt, and fermented milks is one of the oldest methods practiced by man for the preservation of a highly perishable and nutritional foodstuff like milk. The first fermented dairy products were produced by an accidental combination of events. The ability of a group of bacteria, now known as lactic acid bacteria (LAB), to grow in milk and produce enough acid to reduce the pH of milk, caused the coagulation of proteins, thus fermenting the milk. An alternative mechanism was also recognized from an early date, in which the proteolytic enzymes were observed to modify the milk proteins, causing them to coagulate under certain circumstances.

The need for an inoculum was understood and usually a sample from the previous production, also known as back-slopping, was retaining as an inoculum. With the discovery of microorganisms, it became possible to improve the products and the fermentation processes by using well-characterized starter cultures. LAB, yeasts, and molds are the dominant starter cultures used in the production of fermented foods, in general, with a market size of approximately US$ 250 million. Among them, LAB constitute the majority in volume and value of the commercial starter cultures, with the largest part being used in the dairy industry [68].

The primary metabolic actions of microorganisms in dairy fermentations include their ability to ferment carbohydrates and, to a lesser degree, to degrade the proteins and fats present in the raw material. This leads to the production of a broad range of compounds, such as organic acids, peptides, and free fatty acids, along with many volatile and nonvolatile low-molecular mass compounds. Other metabolites, such as antimicrobial compounds (e.g., bacteriocins), exopolysaccharides, bioactive peptides, vitamins, and enzymes are also often produced. In this way, the starter cultures enhance the shelf-life and microbial safety, improve the texture, and shape the nutritious properties and the pleasant sensory profile of the end product. This contribution is further responsible for the differences observed between the products of different brands and thereby adds significantly to the value of the product [95].

In a special category of the so-called "probiotic" cultures, the primary activity is a positive impact on the human health by promoting physiological processes and/or stimulating the host's immune responses [155]. Over the past 15 years, considerable advances were made in the development and conceptualization of novel health-related functional dairy products. Scientific progress in the nutritional and biological sciences has been a major drive in these developments and has contributed significantly to an increase in the consumer awareness on the link between nutrition and health [61].

29.2 Microbial Ecology of Dairy Products

The transformation of milk to a fermented dairy product, especially to cheese, involves a complex and dynamic microbial ecosystem, in which numerous biochemical reactions occur.

LAB are the major microbial group involved in cheese manufacturing. These are divided into the starters and the secondary flora. The primary function of the starter bacteria that mainly belong to the genera *Lactococcus* and *Lactobacillus*, along with *Streptococcus thermophilus*, is to produce acid during the fermentation process. However, they also contribute to cheese ripening, where their enzymes are involved in proteolysis and conversion of amino acids into flavor compounds. The secondary flora consists of adventitious microorganisms from the environment, which contaminate the milk or cheese curd during manufacture and ripening. This group includes numerous species of LAB, such as *Lactobacillus*, *Enterococcus*, *Pediococcus*, and *Leuconostoc*. These microorganisms may become the dominant viable microorganisms in cheese. The numerous hydrolytic enzymes produced by both the starter and the secondary flora affect the proteolysis and lipolysis during cheese ripening and thus contribute to cheese maturation [11].

Propionic acid bacteria grow in many cheese varieties during ripening, and are the characteristic microflora associated with Swiss-type cheeses, such as Emmental, Gruyere, Appenzell, and Comte. The classical propionic acid bacteria are the most important with respect to cheese microbiology, and five species are currently recognized, namely *P. freudenreichii*, *P. jensenii*, *P. thoenii*, *P. acidipropionici*, and *P. cyclohexanicum* [157].

Smear-ripened cheeses are characterized by the development of a smear of bacteria and yeast on the surface of the cheese during ripening. It is generally believed that *Brevibacterium linens* is the major bacterium growing on the surface of the smear-ripened cheeses. Recent studies have indicated that several micrococci (*M. luteus*, *M. lylae*, *Kocuria kristinae*, and *K. roseus*), staphylococci (*St. equorum*, *St. vitulus*, *St. xylosus*, *St. saptrophyticus*, *St. lentus*, and *St. sciuri*), and coryneform bacteria (*Arthrobacter citreus, A. globiformis, A. nicotianae, B. imperiale, B. fuscum, B. oxydans, B. helvolum, Corynebacterium ammoniagenes, C. betae, C. insidiosum, C. variabilis, Curtobacterium poinsettiae, Microbacterium imperiale*, and *Rhodococcus fascians*) are also found on the surface of these cheeses. However, the stage of ripening at which these bacteria are involved is not yet clear [11].

In certain types of cheeses, molds comprise a major part of the cheese microbiota. The mold-ripened cheeses include the mold surface-ripened cheeses, mainly represented by the French varieties like Camembert and Brie, with *Penicillium camemberti* being the dominant microorganism, and the blue-veined cheeses, such as the French Roquefort and the Italian Gorgonzolla, where *Penicillium roqueforti* is grown within the cheese curd. The surface of the French cheeses, St. Nectaire and Tome de Savoie, is covered by a complex fungal flora containing *Penicillium, Mucor, Cladosporium, Geotrichum, Epicoccum*, and *Sporotrichum*, while *Penicillium* and *Mucor* have been reported on the surface of the Italian cheese, Taleggio, and *Geotrichum* on that of Robiola [66].

Yeasts are found in a wide variety of cheeses. However, in most cases, their role in cheese ripening is not fully understood [52]. Fox et al. [53] summarized the yeasts found in several different cheeses. They found that *Debaryomyces hansenii* is, by far, the dominant yeast occurring in nearly all the cheeses, including Weinkase, Romadour, Limburger, Tilsit, Roquefort, Cabrales, Camembert, and St. Nectaire. The next most-important species include *Kluyveromyces lactis, Yarrowia lipolytica*, and *Trichospora beigelii*. However, whether a progression in the species of yeast occurs during the ripening is not clear, as, in many of the relevant studies, the stage of ripening at which the yeasts were isolated was not defined. In addition, many commercial smear-cheese preparations were observed to contain yeast species, such as *Geotrichum candidum, Candida utilis, Debaryomyces hansenii*, and *Kluyveromyces lactis*.

Specific yeast species are essential for the typical characteristics of certain fermented milks, such as kefir, koumis, viili, and longfil. The FAO/WHO food standards defines Kefir starter culture as being composed of Kefir grains, *L. kefiri*, species of the genera *Leuconostoc* (*L. mesenteroides* and *L. cremoris)*, *Lactococcus* (*L. lactis*), *Streptococcus* (*S. thermophilus*), and the acetic acid bacterium, *Acetobacter* (*A. aceti*). It also contains *Kluyveromyces marxianus, Saccharomyces unisporus, S. cerevisiae, S. exiguous, Candida*, and *Torulopsis*. However, this definition does not cover the full composition of the Kefir grains' microbiota, and does not list *L. kefiranofaciens, L. kefirgranum, L. parakefir, L. brevis, L. acidophilus, L. casei, L. helveticus, L. lactis, L. bulgaricus*, and *L. cellobiosus*, which are thought to be present in a Kefir starter (www.codexalimentarius.net).

29.3 Methods of Microbiological Analysis in Dairy Products

29.3.1 Culture-Dependent Methods

29.3.1.1 Classical and Advanced Phenotypic Methods

Routine methods to enumerate the microorganisms in dairy products are based on conventional microbial techniques. These rely on the enumeration of various microbial groups by using selective growth media and growth conditions. Several selective media have been developed and reported in the literature [29,35,147]. Moreover, some media have been established as international standards through a collaborative work between the International Organization of Standardization (ISO) and the International Dairy Federation (IDF) [78–81]. However, this approach has several drawbacks, such as the *de facto* limited selectivity of most growth media as well as the discrepancies between genotypic and phenotypic identifications [86]. A major disadvantage is the failure to recover strains that cannot be cultured using existing methods, or strains that are metabolically active and viable, but have entered a nonculturable state.

The enumeration of microorganisms is generally followed by the random isolation of a representative number of colonies, usually by considering the appearance of the colonies. For bacterial identification, the basic techniques applied include gram staining, catalase and oxidase reactions, as well as cell morphology and physiology. The LAB can be divided into rods (*Lactobacillus* and *Carnobacterium*) and cocci (all other genera), while the genus *Weissella* includes both rods and cocci [6]. It should be stressed, however, that the growth conditions and the growth stage of the cells may seriously affect the cell morphology. The physiological tests include the ability of the isolates to produce CO_2 from glucose and the ability to grow at different temperatures, NaCl concentrations, and pH conditions [56,69,85]. Beyond this and despite a number of serious disadvantages, such as interlaboratory variation, strain-to-strain variation, and number of characteristics tested, carbohydrate fermentation patterns are still very useful as a classical phenotypic approach. Nowadays, well-standardized commercially available kits, such as API 20STREP, API 50CH, API ZYM, and Rapid ID32STREP (bioMerieux, Marcy-l'Etoile, France) or BIOLOG GP (BIOLOG Inc., Hayward, CA), usually accompanied with a database and the respective software, are available. Meanwhile, older phenotypic methods, like serological grouping as well as the determination of cell wall composition, chemotaxonimic markers, and electrophoretic mobility of lactic acid dehydrogenase, have been practically abandoned [123].

On the other hand, advanced phenotypic methods have been developed and successfully used for the identification down to species level. These include the sodium dodecyl sulfate-polyacrylamide gel electrophoresis (SDS-PAGE) of the whole cell proteins, the Fourier transform infrared (FT-IR) spectroscopy of intact cells, and the analysis of the cellular fatty acids. All these methods demand a well-structured database with patterns of well-characterized reference and type strains, and the appropriate mathematical device for the numerical analysis of the patterns obtained, especially when dealing with large numbers of isolates.

The comparison of the whole cell protein patterns obtained by SDS-PAGE offers the advantages of being a fairly fast and easy identification method, which when performed under highly standardized conditions, produces a good level of taxonomic resolution at species or subspecies level [120–122]. In the early days as well as when applied on LAB, it was possible to solve specific identification problems for lactococci [43,82], *Lactobacillus kefir* and *Lactobacillus reuteri* [36], and leuconostocs [8,37] using SDS-PAGE. The method has been widely used for studying the biodiversity of LAB in various dairy products [30,57,58,96,152,158].

Originally introduced by Naumann et al. [111], FT-IR spectroscopy is a vibrational spectroscopic technique with high resolution power, capable of distinguishing the microbial cells at different taxonomic levels [65,72,88,148], and acts a valuable tool for rapid screening of environmental isolates [149]. The identification is achieved by calculating the overall difference between a test spectrum and all the reference spectra. A test strain is assigned to the source of the nearest reference spectrum. However, such a procedure is univariate and does not consider the patterns of individual differences at different wavelengths, leaving lots of information stored in the spectra unused. Therefore, for the differentiation of closely related species within the same genus, advanced multivariate methods for data analysis are required [103,128]. The FT-IR spectroscopy and cluster analysis have been successfully applied to differentiate and identify LAB [2,3,41,64,91], to follow the evolution of *Lactococcus* strains during ripening in Brie cheese [93], and to identify yeast isolates from Irish smear-ripened cheeses [106].

The fatty-acid composition of the microorganisms depends on several factors, such as growth temperature, pH, oxygen tension, growth phase, medium composition, and salt concentration [142].

Under highly standardized growth conditions, gas chromatographic analysis of cellular fatty acids can be used in the chemotaxonomy of LAB [150]. Indeed, several studies have shown the validity of the method for distinguishing between different species of lactococci [83,140], lactobacilli [31,83,129,131], carnobacteria [25], leuconostocs [151], and *Weissella* [137].

29.3.1.2 Molecular Methods

As an alternative to the phenotypic methods, the need of molecular methods for taxonomic purposes was recognized early. The data of considerable taxonomic importance have been derived from the molecular biology studies of DNA, particularly the determination of DNA base composition and the percentage of the DNA similarity, using DNA–DNA hybridization. The latter technique has been extensively applied in the past for the taxonomy of LAB [122]. Although DNA–DNA hybridization values remain the "gold standard" for defining the bacterial species [159], nowadays, it is mainly applied to describe new species. Similarly, other early molecular techniques, such as DNA–rRNA hybridization and 16S rRNA cataloguing have been replaced by the rRNA sequence analysis and rRNA sequencing, respectively, owing to the fact that the earlier techniques are too laborious and time-consuming [123].

In recent years, the increasing availability of the sequences of the 16S rRNA, 23S rRNA genes, and the 16S–23S rRNA intergenic spacer regions or genes encoding enzymes [26,112] allowed the design of numerous primers and probes, and thus, the development of PCR-based, reproducible, easily automated, and rapid molecular methods for the identification of microbial species of interest in the field of dairy products. Although DNA-based methods provide complementary information on the biodiversity of dairy products as well as the temporal and spatial distribution, microbial ecology studies that compare the ratios of culturable cells with both total cells and active cells have indicated the usefulness of culturability in assessing the succession of microbial communities [55]. Furthermore, the importance of the information on the quantitative populations of microorganisms, furnished by culture-dependent methods resides on the fact that microorganisms affect the food ecosystem according to their biochemical reactivity and peak populations. Therefore, a combination of both types of methods, the so-called polyphasic taxonomy, would give more complete information on the microbial complexity of the fermented dairy products.

Table 29.1 gives a representative overview from recent taxonomical studies performed on the microbiota of various dairy products using PCR-based techniques. The overview certainly does not cover the numerous literature data accumulated so far. However, it tries to deal with the main genera/species found in the fermented dairy products. The molecular techniques that either serve as typing methods or culture-independent techniques are described in more detail in the subsequent paragraphs.

Among the PCR-based techniques, randomly amplified polymorphic DNA (RAPD)-PCR analysis is widely recognized as a rapid and reliable method for intra- and interspecific differentiation of most of the food-associated bacterial species. Moreover, its resolving power can be easily enhanced by increasing the number of primers used to randomly amplify the bacterial genome [145]. RAPD-PCR analysis has been used to estimate the diverse *Lactobacillus* strains in the Centre National de Recherches Zootechniques collection [145], to establish the correct nomenclature and classification of strains of *L. casei* subsp. *casei* [38], to type *L. plantarum* strains in Cheddar cheese [92], *Lactococcus lactis* isolated from raw milk used to produce Camembert [100], nonstarter LAB in mature Cheddar cheese [51] and Italian ewe's milk cheeses [30], enterococci in Italian dairy products [105], natural whey starter cultures in Mozzarella cheese [32], bifidobacteria [156], and dairy propionic acid bacteria [135].

Table 29.1   PCR-Based Techniques Used in the Identification of Microorganisms in Dairy Products

Product	Genus/Species	Identification Method	Reference
Cheese	*Propionibacterium* species	Genus-specific PCR	[136]
Milk		Species-specific PCR	
Cheese	*Enterococcus* species	Specific and random amplification (SARA)-PCR	[87]
Cheese	*Enterococcus* species	Species-specific PCR	[101]
	Lactobacillus species		
	Streptococcus thermophilus		
Cheese	*Streptococcus macedonicus*	Species-specific PCR	[118]
Fermented milk	*Lactobacillus* species	Real-time quantitative PCR	[54]
	Streptococcus thermophilus		
Cheese	*Lactobacillus* species	Species-specific PCR	[116]
Cheese	*Enterococcus* species	Species-specific PCR	[134]
Raw milk	*Lactobacillus* species		
	Lactococcus lactis		
	Streptococcus thermophilus		
Cheese	*Enterococcus* species	(GTG)$_5$-rep-PCR	[143]
Cheese	*Enterococcus* species	(GTG)$_5$-rep-PCR	[158]
Fermented milk	*Lactobacillus* species		
Raw milk	*Lactococcus* species		
Sour cream	*Leuconostoc* species		
Fermented milk	*Lactobacillus* species	Amplified ribosomal DNA restriction analysis (ARDRA)	[24]
	Bifidobacterium species		
	Streptococcus thermophilus		
Fermented milk	*Bifidobacterium lactis*	Species-specific PCR	[144]
	Lactobacillus species		
	Streptococcus thermophilus		

Terminal restriction fragment length polymorphism (T-RFLP) is a method that analyzes the variation among 16S rRNA genes from different bacteria and gives information about the microbial community structure. It is based on the restriction endonuclease digestion of fluorescent end-labeled PCR products. The individual terminal restriction fragments (T-RFs) are separated by gel electrophoresis and the fluorescence signal intensities are quantified. Depending on the species composition of the microbial community, distinct profiles (T-RF patterns) are obtained, as each fragment represents each species present. A relative quantitative distribution can be obtained by this method, as the fluorescence intensity of each peak is proportional to the amount of genomic DNA present for each species in the mixture. Nevertheless, PCR bias could negatively affect the quantification of the real composition of the microbial community [62]. Although 16S rRNA offers the benefits of robust database and well-characterized phylogenetic primers, the T-RFLP approach should not be limited to ribosomal gene markers. The accumulating set of new sequences from various genes from less conserved DNA regions and the high quality of information provided, namely the exact base-pair length of the T-RFs generated, could allow the comparison of profiles for any gene system of interest. The method has been proven suitable for the rapid and routine identification of the classical propionibacteria [130], for the characterization of LAB in Kefir [99], and lactobacilli in Provolone del Monaco cheese [13]. Sánchez et al. [138] assayed the ability of the T-RFLP analysis coupled with RT-PCR in monitoring the population dynamics of the metabolically active fraction of well-defined microbial communities, such as dairy defined-strain starters.

Multilocus sequence-typing (MLST) has emerged as a new powerful DNA-typing tool for the evaluation of intraspecies genetic relatedness [98]. It relies on DNA sequence analysis of usually five to eight internal, ~500 bp fragments of housekeeping genes, and has shown a high degree of intraspecies discriminatory power for bacterial and fungal pathogens [21,132]. The overwhelming advantage of MLST over other molecular-typing methods is that sequence data are truly portable between laboratories, permitting a single expanding global database per species on the World Wide Web site, thus, enabling exchange of molecular-typing data for global epidemiology via the Internet [98]. MLST has been successfully applied for defining the genomic subpopulations within the species *Pediococcus acidilactici* [104], for the genotypic characterization of *Lactobacillus casei* strains isolated from different ecological niches, in cheeses from different geographical locations [19], and for the identification of *Enterococcus* [109] and *Lactobacillus* [15,110] strains derived from humans, animals and food products, in milk, yogurt, and cheese.

Standard gel electrophoresis techniques are not capable of effectively separating very large molecules of DNA, which, when migrating through a gel, essentially move together in a size-independent manner. Schwartz and Cantor [141] developed pulsed field gel electrophoresis (PFGE) by introducing an alternating voltage gradient to improve the resolution of larger DNA molecules. The analysis of chromosomal-DNA restriction-endonuclease profiles using PFGE, by either field-inversion gel electrophoresis or counter-clamped homogenous electric field electrophoresis, is currently considered as the most reliable typing method and the golden standard for epidemiological studies [48,146]. However, the need for specialized equipment and the lack of standardized electrophoresis conditions and interpretation criteria of the PFGE profiles still limit the more extensive application, especially in long-term studies. PFGE has been used for typing bacteria in smear cheeses [17,73,75,106], for elucidating the genotypic heterogeneity of enterococci [12,84,125,126,153], lactococci [34,124], lactobacilli [16,28,39,63], *Streptococcus thermophilus* [115], and *Staphylococcus* [74] in dairy products. Gelsomino et al. [59] applied PFGE to determine the impact of the consumption of cheese containing enterococci on the composition of the enterococcal flora of the feces in healthy humans. Furthermore, Leite et al. [94] characterized *L. monocytogenes* from cheese and clinical isolates, which were collected in partially overlapping dates from the same geographical area, and using PFGE analysis, examined whether there was any clonal relationship between the cheese and the clinical isolates.

As far as taxonomic studies of cheese yeasts are concerned, a combination of physiological and morphological characteristics has been used traditionally [7,89]. However, in the last decade, molecular approaches have been developed that have overcome the inherent variability of phenotypic tests. Analysis of the coenzyme Q system and the monosaccharide pattern of cell walls [124], random amplification of polymorphic DNA (RAPD) microsatellite analysis [5,60,102,124], RFLP of transcribed and spacer sequences of ribosomal DNA [1,9,18,47,67], chromosome polymorphism determined by PFGE [119], and sequencing of the 18S rRNA gene [20,154] have been used in the classification and typing of yeast species.

29.3.2 Culture-Independent Methods

In the last decade, it was shown that classical microbiological techniques do not accurately detect the microbial diversity. It is well documented, for example, that stressed or injured cells do not recover in the selective media and that cells present in low numbers are very often inhibited by microbial populations numerically more abundantly [77]. As a consequence, an increasing interest in the development and use of culture-independent techniques has emerged. A variety of new methods have been developed to directly characterize the microorganisms in particular habitats without the need for enrichment or isolation [70]. Typically, these strategies examine the total microbial DNA or RNA derived from mixed microbial populations to identify individual constituents. This approach eliminates the necessity for strain isolation, thereby negating the potential biases inherent to the microbial enrichment. Studies that employed such direct analysis have repeatedly demonstrated a tremendous variance between cultivated and naturally occurring species, thereby dramatically altering our understanding of the true microbial diversity present in various habitats [76].

A culture-independent method for studying the diversity of microbial communities is the analysis of PCR products generated with primers homologous to relatively conserved regions in the genome, by using denaturing-gradient gel electrophoresis (DGGE) or temperature-gradient gel electrophoresis (TGGE). These approaches allow the electrophoretic separation of DNA molecules that are of the same length, but have different nucleotide sequences. Hence, they have the potential to provide information about variations in the target genes in a bacterial population. By adjusting the primers used for amplification, both the major and minor constituents of microbial communities can be characterized, and were first used to detect single-base DNA sequence variations [50]. In DGGE, PCR-amplified double-stranded DNA is subjected to electrophoresis under denaturing conditions achieved by a solvent gradient, and migration depends on the degree of DNA denaturation. In TGGE, a temperature gradient rather than a solvent gradient is used to denature the DNA [14].

Although the techniques are reliable, reproducible, rapid, and inexpensive [108], their main limitation is that the community fingerprints they generate do not directly translate into taxonomic information, for which a comparative analysis of the sequences from excised and reamplified DNA fragments to 16S rDNA sequences reported in nucleotide databases, is necessary. More information about the identity of the community members could be obtained by hybridization analysis of DGGE/TGGE patterns with taxon-specific oligonucleotide probes to the hypervariable regions of the 16S rRNA. Both DGGE and TTGE are now methods of choice for environmental microbiologists, and have been used to determine the genetic diversities of natural microbial communities, such as the communities in biofilms [107], hot springs [139], biodegraded wall painting [133], and fermented foods, such as fermented maize dough [4] and sausages [22].

Recent studies on the microbial diversity in different types of cheeses have made use of these culture-independent methods. The dynamics of bacterial communities have been analyzed using DGGE in evaluating the microbial diversity of natural whey cultures from water-buffalo Mozzarella cheese production [27,44], during the production of an artisanal Sicilian cheese [127], in the elucidation of the bacterial community structure and location in Stilton cheese [45], in studying the microbial succession during the manufacture of traditional water-buffalo mozzarella cheese [46], and in studying the microbial diversity and succession during the manufacture and ripening of the traditional, Spanish, blue-veined Cabrales cheese [10].

Ogier et al. [113] applied TTGE to describe the diversity of LAB in commercial dairy products by setting up a bacterial database that allows rapid identification of the unknown bands. This database essentially included bacteria with a low G + C content genome, i.e., numerous LAB and a few dairy *Staphylococcus* species. In 2004, Ogier et al. [114] modified their approach to expand the bacterial database to other species of dairy interest, including psychrotrophic and spoilage bacteria, pathogens, and bacteria present on the cheese surface. As one of the limitations of TTGE is poor resolution of species having high G + C content genomes, they combined TTGE and DGGE, which is more suitable for these bacterial species. Henri-Dubernet et al. [71] applied TGGE for the assessment of the lactobacilli-community biodiversity and evolution during the production of Camembert. Lafarge et al. [90] studied the evolution of the bacterial community in raw milk upon conservation at 4°C by using both TGGE and DGGE, and both the methods were also used for the elucidation of the bacterial biodiversity occurring in traditional Egyptian soft Domiati cheese [42].

Single-stranded conformational polymorphism analysis (SSCP) is the electrophoretic separation of single-stranded nucleic acids based on subtle differences in sequence, often a single base pair, which results in a different secondary structure and a measurable difference in mobility through an electrophoresis gel. However, similar to the DGGE/TGGE analyses, SSCP provides community fingerprints that cannot be phylogenetically assigned. SSCP analysis on gel has been successfully applied to monitor the dynamics of bacterial population in anaerobic bioreactor [160] or in hot composting [117], or to study the fungal diversity in soils [97]. Duthoit et al. [40] were the first to apply SSCP in a dairy product. Using this method, they effectively described the ecosystem of the registered designation of the origin of Salers cheese, an artisanal cheese produced in France. SSCP was also applied to investigate the microbial community composition and dynamics during the production of a French soft, red-smear cheese [49]. In addition, Delbès and Montel [33] designed and applied a *Staphylococcus*-specific SSCP-PCR analysis to monitor *Staphylococcus* populations' diversity and dynamics during the production of raw milk cheese.

Fluorescence *in situ* hybridization (FISH) represents a new non-PCR-based culture-independent technique in the field of food fermentations. Despite its considerable background knowledge, its application in studying the distribution of microbial populations in food has been limited. In the field of dairy products, this method was successfully used for elucidating the bacterial community structure and location in Stilton cheese [45] as well as other cheese varieties, such as cottage cheese, Kefalotiri, Hallumi, Stracchino, and Mozzarella [23].

References

1. Alvarez-Martin P., A.B. Florez, T.M. López-Díaz, and B. Mayo (2007) Phenotypic and molecular identification of yeast species associated with Spanish blue-veined Cabrales cheese. *Int. Dairy J.* 17, 961–967.
2. Amiel C., L. Mariey, M.C. Curk-Daubie, P. Pichon, and J. Travert (2000) Potential of Fourier infrared spectroscopy (FTIR) for discrimination and identification of dairy lactic acid bacteria. *Le Lait* 80, 445–449.
3. Amiel C., L. Mariey, C. Denis, P. Pichon, and J. Travert (2001) FTIR spectroscopy and taxonomic purpose: Contribution to the classification of lactic acid bacteria. *Le Lait* 81, 249–255.
4. Ampe F., N. ben Omar, C. Moizan, C. Wacher, and J.P. Guyot (1999) Polyphasic study of the spatial distribution of microorganisms in Mexican pozol, a fermented maize dough, demonstrates the need for cultivation independent methods to investigate traditional fermentations. *Appl. Environ. Microbiol.* 65, 5464–5473.
5. Andrighetto C., E. Psomas, N. Tzanetakis, G. Suzzi, and A. Lombardi A (2000) Randomly amplified polymorphic DNA (RAPD) PCR for the identification of yeasts isolated from dairy products. *Lett. Appl. Microbiol.* 30, 5–9.
6. Axelsson L. (2004) Lactic acid bacteria: Classification and physiology. In *Lactic Acid Bacteria-Microbiological and Functional Aspects*, pp. 1–66, S. Salminen, A. von Wright, and A. Ouwehand, Eds., Marcel Dekker, New York.
7. Barnett J.A., R.W. Payne, and D. Yarrow (2000) *Yeasts: Characteristics and Identification.* Cambridge University Press, Cambridge, U.K.
8. Barreau C. and G. Wagener (1990) Characterization of *Leuconostoc lactis* strains from human sources. *J. Clin. Microbiol.* 28, 1728–1733.
9. Belen Florez A. and B. Mayo (2006) Microbial diversity and succession during the manufacture and ripening of traditional, Spanish, blue-veined Cabrales cheese, as determined by PCR-DGGE. *Int. J. Food Microbiol.* 110, 165–171.
10. Belén Flórez A., P. Álvarez-Martín, T.M. López-Díaz, and B. Mayo (2007) Morphotypic and molecular identification of filamentous fungi from Spanish blue-veined Cabrales cheese, and typing of *Penicillium roqueforti* and *Geotrichum candidum* isolates. *Int. J. Food Microbiol.* 17, 350–357.
11. Beresford T.P., N.A. Fitzsimons, N.L. Brennan, and T.M. Cogan (2001) Recent advances in cheese microbiology. *Int. Dairy J.* 11, 259–274.
12. Bertrand X., B. Mulin, J.F. Viel, M. Thouverez, and D. Talon (2000) Common PFGE patterns in antibiotic-resistant *Enterococcus faecalis* from humans and cheeses. *Food Microbiol.* 17, 543–551.
13. Blaiotta G., V. Fusco, D. Ercolini, M. Aponte, O. Pepe, and F. Villani1 (2008) *Lactobacillus* strain diversity based on partial *hsp60* gene sequences and design of PCR-restriction fragment length polymorphism assays for species identification and differentiation. *Appl. Environ. Microbiol.* 74, 208–215.
14. Børresen-Dale A.L., S. Lystad, and A. Langerød (1997) Temporal temperature gradient gel electrophoresis (TTGE) compared with denaturing gradient gel electrophoresis (DGGE) and constant denaturing gel electrophoresis (CDGE) in mutation screening. *Bioradiation* 99, 12–13.
15. Borgo F., G. Ricci, P.L. Manachini, and M. Grazia Fortina (2007) Multilocus restriction typing: A tool for studying molecular diversity within *Lactobacillus helveticus* of dairy origin. *Int. Dairy J.* 17, 336–342.
16. Bouton Y., P. Guyot, E. Beuvier, P. Tailliez, and R. Grappin (2002) Use of PCR-based methods and PFGE for typing and monitoring homofermentative lactobacilli during Comté cheese ripening. *Int. J. Food Microbiol.* 76, 27–38.
17. Brennan N.M., A.C. Ward, T.P. Beresford, P.F. Fox, M. Goodfellow, and T.M. Cogan (2002) Biodiversity of the bacterial flora on the surface of a smear cheese. *Appl. Environ. Microbiol.* 68, 820–830.
18. Caggia D., C. Restuccia, A. Pulvirenti, and P. Giudici (2001) Identification of *Pichia anomala* isolated from yoghurt by RFLP of the ITS region. *Int. J. Food Microbiol.* 71, 71–73.

19. Cai H., B.T. Rodríguez, W. Zhang, J.R. Broadbent, and J.L. Steele (2007) Genotypic and phenotypic characterization of *Lactobacillus casei* strains isolated from different ecological niches suggests frequent recombination and niche specificity. *Microbiology* 153, 2655–2665.

20. Cappa F. and P.S. Cocconchelli (2001) Identification of fungi from dairy products by means of 18S rRNA analysis. *Int. J. Food Microbiol.* 69, 157–160.

21. Chan M.S., M.C.J. Maiden, and B.G. Spratt (2001) Database-driven multi locus sequence typing (MLST) of bacterial pathogen. *Bioinformatics* 17, 1077–1083.

22. Cocolin L., M. Manzano, C. Cantoni, and G. Comi (2001) Denaturing gradient gel electrophoresis analysis of the 16S rRNA gene V1 region to monitor dynamic changes in the bacterial population during fermentation of Italian sausages. *Appl. Environ. Microbiol.* 67, 5113–5121.

23. Cocolin L., A. Diez, R. Urso, K. Rantsiou, G. Comi, I. Bergmaier, and C. Beimfohr (2007) Optimization of conditions for profiling bacterial populations in food by culture-independent methods. *Int. J. Food Microbiol.* 120, 100–109.

24. Collado M.C. and M. Hernández (2007) Identification and differentiation of *Lactobacillus*, *Streptococcus* and *Bifidobacterium* species in fermented milk products with bifidobacteria. *Microbiol. Res.* 162, 86–92.

25. Collins M.D., J.A.E. Farrow, B.A. Phillips, S. Ferusu, and D. Jones (1987) Classification of *Lactobacillus divergens*, *Lactobacillus piscicola* and some catalase-negative, asporogenous, rodshaped bacteria from poultry in a new genus, *Carnobacterium*. *Int. J. System. Bacteriol.* 37, 311–316.

26. Collins M.D., U. Rodriguez, C. Ash, M. Aguirre, J.E. Farrow, A. Martinezmurcia, B.A. Philips, A.M. Williams, and S. Wallbanks (1991) Phylogenetic analysis of the genus *Lactobacillus* and related lactic acid bacteria as determined by reverse-transcriptase sequencing of 16S ribosomal-RNA. *FEMS Microbiol. Lett.* 77, 5–12.

27. Coppola S., G. Blaiotta, D. Ercolini, and G. Moschetti (2001) Molecular evaluation of microbial diversity occurring in different types of Mozzarella cheese. *J. Appl. Microbiol.* 90, 414–420.

28. Christiansen P., M.H. Petersen, S. Kask, P.L. Møller, M. Petersen, E.W. Nielsen, F.K. Vogensen, and Y. Ardö (2005) Anticlostridial activity of *Lactobacillus* isolated from semi-hard cheeses. *Int. Dairy J.* 15, 901–909.

29. Dave R.I. and N.P. Shah (1996) Evaluation of media for selective enumeration of *Streptococcus thermophilus*, *Lactobacillus delbrueckii* ssp. *bulgaricus*, *Lactobacillus acidophilus* and bifidobacteria, *J. Dairy Sci.* 79, 1529–1536.

30. De Angelis M., A. Corsetti, N. Tosti, J. Rossi, M.R. Corbo, and M. Gobbetti (2001) Characterization of non-starter lactic acid bacteria from Italian ewe cheeses based on phenotypic, genotypic and cell wall protein analyses. *Appl. Environ. Microbiol.* 67, 2011–2020.

31. Decallonne J., M. Delmee, P. Wauthoz, M. El Lioui, and R. Lambert (1991). A rapid procedure for the identification of lactic acid bacteria based on the gas chromatographic analysis of the cellular fatty acids. *J. Food Prot.* 54, 217–224.

32. de Candia S., M. De Angelis, E. Dunlea, F. Minervini, P.L.H. McSweeney, M. Faccia, and M. Gobbetti (2007) Molecular identification and typing of natural whey starter cultures and microbiological and compositional properties of related traditional Mozzarella cheeses. *Int. J. Food Microbiol.* 119, 182–191.

33. Delbès C. and M.-C. Montel (2005) Design and application of a *Staphylococcus*-specific single strand conformation polymorphism-PCR analysis to monitor *Staphylococcus* populations diversity and dynamics during production of raw milk cheese. *Lett. Appl. Microbiol.* 41, 169–174.

34. Delgado S. and B. Mayo (2004) Phenotypic and genetic diversity of *Lactococcus lactis* and *Enterococcus* spp. strains isolated from Northern Spain starter-free farmhouse cheeses. *Int. J. Food Microbiol.* 90, 309–319.

35. de Man J.C., M. Rogosa, and M.E. Sharpe (1960) A medium for the cultivation of lactobacilli. *J. Appl. Bacteriol.* 23, 130–135.

36. Dicks L.M.T. and H.J.J. van Vuuren (1987) Relatedness of heterofermentative *Lactobacillus* species revealed by numerical analysis of total soluble cell protein patterns. *Int. J. Syst. Bacteriol.* 37, 437–440.

37. Dicks L.M.T., H.J.J. van Vuuren, and F. Dellaglio (1990) Taxonomy of *Leuconostoc* species, particularly *Leuconostoc oenos*, as revealed by numerical analysis of total soluble protein patterns, DNA base compositions and DNA-DNA hybridizationa. *Int. J. Syst. Bacteriol.* 40, 83–91.

38. Dicks L.M.T., E.M. Du Plessis, F. Dellaglio, and E. Lauer (1996) Reclassification of *Lactobacillus casei* subsp. *casei* ATCC 393 and *Lactobacillus rhamnosus* ATCC 15820 as *Lactobacillus zeae* nom. rev., designation of ATCC 334 as the neotype of *L. casei* subsp. *casei*, and rejection of the name *Lactobacillus paracasei*. *Int. J. Syst. Bacteriol.* 46, 337–340.

39. Dimitrov Z., M. Michaylova, and S. Mincova (2005) Characterization of *Lactobacillus helveticus* strains isolated from Bulgarian yoghurt, cheese, plants and human faecal samples by sodium dodecylsulfate polyacrylamide gel electrophoresis of cell-wall proteins, ribotyping and pulsed field gel fingerprinting. *Int. Dairy J.* 15, 998–1005.

40. Duthoit F., J.J. Godon, and M.C. Montel (2003) Bacterial community dynamics during production of registered designation of origin Salers cheese as evaluated by 16S rRNA gene single-strand conformation polymorphism analysis. *Appl. Environ. Microbiol.* 69, 3840–3848.

41. Dziuba B., A. Babuchowski, D. Nałęcz, and M. Niklewicz (2007) Identification of lactic acid bacteria using FTIR spectroscopy and cluster analysis. *Int. Dairy J.* 17, 183–189.

42. El-Baradei G., A. Delacroix-Buchet, and J.-C. Ogier (2007) Biodiversity of bacterial ecosystems in traditional Egyptian Domiati cheese. *Appl. Environ. Microbiol.* 73, 1248–1255.

43. Elliot J.A., M.D. Collins, Pigott N.E., and R.R. Facklam (1991) Differentiation of *Lactococcus lactis* and *Lactococcus garviae* from humans by comparison of whole cell protein patterns. *J. Clin. Microbiol.* 29, 2731–2734.

44. Ercolini D., G. Moschetti, G. Blaiotta, and S. Coppola (2001) The potential of a polyphasic PCR-DGGE approach in evaluating microbial diversity of natural whey cultures from water-buffalo Mozzarella cheese production: Bias of "culture dependent" and "culture independent" approaches. *Syst. Appl. Microbiol.* 24, 610–617.

45. Ercolini D., P.J. Hill, and E.R. Dodd (2003) Bacterial community structure and location in Stilton cheese. *Appl. Environ. Microbiol.* 69, 3540–3548.

46. Ercolini D., G. Mauriello, G. Blaiotta, G. Moschetti, and S. Coppola (2004) PCR-DGGE fingerprints of microbial succession during a manufacture of traditional water buffalo mozzarella cheese. *J. Appl. Microbiol.* 96, 263–270.

47. Esteve-Zarzoso B., C. Belloch, F. Uruburu, and A. Querol (1999) Identification of yeasts by RFLP analysis of the 5.8S rRNA gene and the two ribosomal internal transcribed spacers. *Int. J. Syst. Bacteriol.* 49, 329–337.

48. Facklam R., M.G. Carvallo, and L. Teixeira (2002) History, taxonomy, biochemical characteristics and antibiotic susceptibility testing of enterococci. In *The Enterococci: Pathogenesis, Molecular Biology and Antibiotic Resistance*, M. Gimore, Ed., ASM Press, Washington, D.C.

49. Feurer C., T. Vallaeys, G. Corrieu, and F. Irlinger (2004) Does smearing inoculum reflect the bacterial composition of the smear at the end of the ripening of a French soft, red-smear cheese? *J. Dairy Sci.* 87, 3189–3197.

50. Fischer S.G. and L.S. Lerman (1983) DNA fragments differing by single base-pair substitutions are separated in denaturing gradient gels: Correspondence with melting theory. *Proc. Natl. Acad. Sci.* 80, 1579–1583.

51. Fitzsimons N.A., T.M. Cogan, S. Condon, and T. Beresford (1999) Phenotypic and genotypic characterization of non-starter lactic acid bacteria in mature cheddar cheese. *Appl. Environ. Microbiol.* 65, 3418–3426.

52. Fleet G.H. and M.A. Mian (1987) The occurrence and growth of yeasts in dairy products. *Int. J. Food Microbiol.* 4, 145–155.

53. Fox P.F., T.P. Guinee, T.M. Cogan, and P.L.H. McSweeney (2000) *Fundamentals of Cheese Science.* Aspen Publishers, Inc., Gaithersburg, MD.

54. Furet J.P., P. Quénée, and P. Tailliez (2004) Molecular quantification of lactic acid bacteria in fermented milk products using real-time quantitative PCR. *Int. J. Food Microbiol.* 97, 197–207.

55. Garland J.L., K.L. Cook, J.L. Adams, and L. Kerkhof (2001) Culturability as an indicator of succession in microbial communities. *Microb. Ecol.* 42, 150–158.

56. Garvie E.I. (1986) Genus *Leuconostoc* van Tieghem 1878, 198[AL], emend. Mut. Chat. Hucker and Peterson 1930, 66[AL]. In *Bergey's Manual of Systematic Bacteriology*, Vol. 2, pp. 1071–1075, P.A. Sneath, N.S. Mair, M.E. Sharpe, and J.G. Holt, Eds., The Williams and Wilkins Co., Baltimore, MD.

57. Gatti M., M.E. Fornasari, and E. Neviani (2001) Differentiation of *Lactobacillus delbrueckii* subsp. *bulgaricus* and *Lactobacillus delbrueckii* subsp. *lactis* by SDS-PAGE of cell-wall proteins. *Lett. Appl. Microbiol.* 32, 352–356.

58. Gatti M., C. Lazzi, L. Rossetti, G. Mucchetti, and E. Neviani (2003) Biodiversity of *Lactobacillus helveticus* strains present in natural whey starter used for Parmigiano Reggiano cheese. *J. Appl. Microbiol.* 95, 463–470.

59. Gelsomino R., M. Vancanneyt, T.M. Cogan, and J. Swings (2003) Effect of raw-milk cheese consumption on the enterococcal flora of human faeces. *Appl. Environ. Microbiol.* 69, 312–319.

60. Gente S., N. Desmasures, J.M. Panoff, and M. Gueguen (2002) Genetic diversity among *Geotrichum candidum* strains from various substrates studied using RAM and RAPD-PCR. *J. Appl. Microbiol.* 92, 491–501.

61. Gibney M.J., M. Walsh, L. Brennan, H.M. Roche, B. German, and B. van Ommen (2005) Metabolomics in human nutrition: Opportunities and challenges. *Am. J. Clin. Nutr.* 82, 497–503.

62. Giraffa G. and E. Neviani (2001) DNA-based, culture-independent strategies for evaluating microbial communities in food-associated ecosystems. *Int. J. Food Microbiol.* 67, 19–34.

63. Giraffa G., C. Andrighetto, C. Antonello, M. Gatti, C. Lazzi, G. Marcazzan, A. Lombardi, and E. Neviani (2004) Genotypic and phenotypic diversity of *Lactobacillus delbrueckii* subsp. *lactis* strains of dairy origin. *Int. J. Food Microbiol.* 91, 129–139.

64. Goodacre R., E.M. Timmins, P.J. Rooney, J.J. Rowland, and D.B. Kell (1996) Rapid identification of *Streptococcus* and *Enterococcus* species using diffuse reflectance-absorbance Fourier transform infrared spectroscopy and artificial neural networks. *FEMS Microbiol. Lett.* 140, 233–239.

65. Goodacre R., E.M. Timmins, R. Burton, N. Kaderbhai, A.M. Woodward, D.B. Kell, and P.J. Rooney, P.J. (1998) Rapid identification of urinary tract infection bacteria using hyperspectral whole-organism fingerprinting and artificial neural networks. *Microbiology* 144, 1157–1170.

66. Gripon J.C. (1999) Mould-ripened cheeses. In *Cheese Chemistry, Physics and Microbiology*, Vol. 2, pp. 111–136, P.F. Fox, Ed., Chapman & Hall, London, U.K.

67. Guillamon J.M., J. Sabate, E. Barrio, J. Cano, and A. Querol (1998) Rapid identification of wine yeasts species based on RFLP analysis of ribosomal internal transcribed spacer (ITS) region. *Arch. Microbiol.* 169, 387–392.

68. Hansen E.B. (2002) Commercial bacterial starter cultures for 1 fermented foods of the future. *Int. J. Food Microbiol.* 78, 119–131.

69. Hardie J.M. (1986) Genus *Streptococcus* Rosenbach 1884, 22[AL]. In *Bergey's Manual of Systematic Bacteriology*, Vol. 2, pp. 1043–1071, P.A. Sneath, N.S. Mair, M.E. Sharpe, and J.G. Holt, Eds., The Williams and Wilkins Co., Baltimore, MD.

70. Head I.M., J.R. Saunders, and R.W. Pickup (1998) Microbial evolution, diversity and ecology: A decade of ribosomal RNA analysis of uncultivated microorganisms. *Microb. Ecol.* 35, 1–21.

71. Henri-Dubernet S., N. Desmasures, and M. Guéguen (2004) Culture-dependent and culture-independent methods for molecular analysis of the diversity of lactobacilli in "Camembert de Normandie" cheese. *Le Lait* 84, 179–189.

72. Holt C., D. Hirst, A. Sutherland, and F. MacDonald (1995) Discrimination of species in the genus *Listeria* by Fourier transform infrared spectroscopy and canonical variate analysis. *Appl. Environ. Microbiol.* 61, 377–378.

73. Hoppe-Seyler T.S., B. Jaeger, W. Bockelmann, W.H. Noordman, A. Geis, and K.J. Heller (2003) Identification and differentiation of species and strains of *Arthrobacter* and *Microbacterium barkeri* isolated from smear cheeses with amplified ribosomal DNA restriction analysis (ARDRA) and pulsed field gel electrophoresis (PFGE). *Syst. Appl. Microbiol.* 26, 438–444.

74. Hoppe-Seyler T.S., B. Jaeger, W. Bockelmann, W.H. Noordman, A. Geis, and K.J. Heller (2004) Molecular identification and differentiation of *Staphylococcus* species and strains of cheese origin. *Syst. Appl. Microbiol.* 27, 211–218.

75. Hoppe-Seyler T.S., B. Jaeger, W. Bockelmann, A. Geis, and K.J. Heller (2007) Molecular identification and differentiation of *Brevibacterium* species and strains. *Syst. Appl. Microbiol.* 30, 50–57.

76. Hugenholtz P. and N.R. Pace (1996) Identifying microbial diversity in the natural environment: A molecular phylogenetic approach. *Trends Biotechnol.* 14, 90–97.

77. Hugenholtz P., B.M. Goebel, and N.R. Pace (1998) Impact of culture-independent studies on the emerging phylogenetic view of bacterial diversity. *J. Bacteriol.* 180, 4765–4774.

78. ISO7889/IDF117 (2003) Yogurt – enumeration of characteristic microorganisms–colony count technique at 37°C.

79. ISO9232/IDF146 (2003) Yogurt – enumeration of characteristic microorganisms (*Lactobacillus delbreuckii* subsp. *bulgaricus* and *Streptococcus thermophilus*).

80. ISO17792/IDF180 (2006) Milk, milk products and mesophilic starter cultures–enumeration of citrate lactic acid bacteria – colony count technique at 25°C.

81. ISO20128/IDF192 (2006) Milk products – enumeration of presumptive *Lactobacillus acidophilus* on a selective medium – colony count technique at 37°C.

82. Jarvis A.W. and J.M. Wolff (1979) Grouping of lactic streptococci by gel electrophoresis of soluble cell extracts. *Appl. Environ. Microbiol.* 37, 391–398.

83. Johnsson T., P. Nikkil, L. Toivonen, H. Rosenqvist, and S. Laakso (1995) Cellular fatty acid profiles of *Lactobacillus* and *Lactococcus* strains in relation to the oleic acid content of the cultivation medium. *Appl. Environ. Microbiol.* 61, 4497–4499.

84. Jurkovič D., L. Križková, M. Sojka, M. Takáčová, R. Dušinský, J. Krajčovič, P. Vandamme, and M. Vancanneyt (2007) Genetic diversity of *Enterococcus faecium* isolated from Bryndza cheese. *Int. J. Food Microbiol.* 116, 82–87.

85. Kandler O. and N. Weiss (1986) Genus *Lactobacillus* Beijerinck 1901, 212[AL]. In *Bergey's manual of systematic bacteriology*, Vol. 2, pp. 1209–1234, P.A. Sneath, N.S. Mair, M.E. Sharpe, and J.G. Holt, Eds., The Williams and Wilkins Co., Baltimore, MD.

86. Kelly W. and L. Ward (2002) Genotypic vs. phenotypic biodiversity in *Lactococcus lactis*. *Microbiol.* 148, 3332–3333.

87. Knijff E., F. Dellaglio, A. Lombardi, S. Biesterveld, and S. Torriani (2001) Development of the specific and random amplification (SARA)-PCR for both species identification of enterococci and detection of the *van*A gene *J. Microbiol. Methods* 43, 233–239.

88. Kuemmerle M., S. Scherer, and H. Seiler (1998) Rapid and reliable identification of fermentative yeasts by Fourier-transform infrared spectroscopy. *Appl. Environ. Microbiol.* 64, 2207–2214.

89. Kurtzman C.P. and J.W. Fell (1998). *The Yeasts: A Taxonomic Study* (4th ed.), Elsevier Science Publishers, Amsterdam, The Netherlands.

90. Lafarge V., J.-C. Ogier, V. Girard, V. Maladen, J.-Y. Leveau, A. Gruss, and A. Delacroix-Buchet (2004) Raw cow milk bacterial population shifts attributable to refrigeration. *Appl Environ. Microbiol.* 70, 5644–5650.

91. Lai S., R. Goodacre, and L.N. Manchester (2004) Whole-organism fingerprinting of the genus *Carnobacterium* using Fourier transform infrared spectroscopy (FT-IR). *Syst. Appl. Microbiol.* 27, 186–191.

92. Lane C.N. and P.F. Fox (1996) Contribution of starter and added lactobacilli to proteolysis in Cheddar cheese during ripening. *Int. Dairy J.* 6, 715–728.

93. Lefier D., H. Lamprell, and G. Mazerolles (2000) Evolution of *Lactococcus* strains during ripening in Brie cheese using Fourier transform infrared spectroscopy. *Le Lait* 80, 247–254.

94. Leite P., R. Rodrigues, M. Ferreira, G. Ribeiro, C. Jacquet, P. Martin, and L. Brito (2006) Comparative characterization of *Listeria monocytogenes* isolated from Portuguese farmhouse ewe's cheese and from humans. *Int. J. Food Microbiol.* 106, 111–121.

95. Leroy F. and L. DeVuyst (2004) Lactic acid bacteria as functional starter cultures for the food fermentation Industry. *Trends Food Sci. Technol.* 15, 67–78.

96. Lombardi L., M. Gatti, L. Rizzotti, S. Torriani, C. Andrighetto, and G. Giraffa (2004) Characterization of *Streptococcus macedonicus* strains isolated from artisanal Italian raw milk cheeses. *Int. Dairy J.* 14, 967–976.

97. Lowell J.L. and D.A. Klein (2001) Comparative single-strand conformation polymorphism (SSCP) and microscopy-based analysis of nitrogen cultivation interactive effects on the fungal community of a semiarid steppe soil. *FEMS Microbiol. Ecol.* 36, 85–92.

98. Maiden M.C.J., J.A. Bygraves, E. Feil, J. Morelli, J.E. Russell, R. Urwin, Q. Zhang, et al. (1998) Multilocus sequence typing: A portable approach to the identification of clones within populations of pathogenic microorganisms. *Proc. Nat. Acad. Sci.* 95, 3140–3145.

99. Mainville I., N. Robert, By. Lee, and E.R. Farnworth (2006) Polyphasic characterization of the lactic acid bacteria in kefir. *Syst. Appl. Microbiol.* 29, 59–68.

100. Mangin I., D. Corroler, A. Reinhardt, and M. Gueguen (1999) Genetic diversity among dairy lactococcal strains investigated by polymerase chain reaction with three arbitrary primers. *J. Appl. Microbiol.* 86, 514–520.

101. Mannu L., G. Riu, R. Comunian, M.C. Fozzi, and M.F. Scintu (2002) A preliminary study of lactic acid bacteria in whey starter culture and industrial Pecorino Sardo ewes' milk cheese: PCR-identification and evolution during ripening. *Int. Dairy J.* 12, 17–26.

102. Marcellino N., E. Beuvier, R. Grappin, M. Gueguen, and D.R. Benson (2001) Diversity of *Geotrichum candidum* strains isolated from traditional cheese making fabrications in France. *Appl. Environ. Microbiol.* 67, 4752–4759.

103. Mariey L., J.P. Signolle, C. Amiel, and J. Travert (2001) Discrimination, classification, identification of microorganisms using FTIR spectroscopy and chemometrics. *Vibrat. Spectr.* 26, 151–159.

104. Mora D., M.G. Fortina, C. Parini, D. Daffonchio, and P.L. Manachini (2000) Genomic subpopulations within the species *Pediococcus acidilactici* detected by multilocus typing analysis: Relationships between pediocin AcH/PA-1 producing and non-producing strains. *Microbiology* 146, 2027–2038.

105. Morandi S., M. Brasca, C. Andrighetto, A. Lombardi, and R. Lodi (2006) Technological and molecular characterisation of enterococci isolated from north–west Italian dairy products. *Int. Dairy J.* 16, 867–875.

106. Mounier J., R.Gelsomino, S. Goerges, M. Vancanneyt, K. Vandemeulebroecke, B. Hoste, S. Scherer, J. Swings, G.F. Fitzgerald, and T.M. Cogan (2005) Surface microflora of four smear-ripened cheeses. *Appl. Environ. Microbiol.* 71, 6489–6500.

107. Muyzer G.E., C. de Waal, and A.G. Uitterlinden (1993) Profiling of complex microbial populations by denaturing gradient gel electrophoresis analysis of polymerase chain reaction-amplified genes coding for 16S rRNA. *Appl. Environ. Microbiol.* 59, 695–700.

108. Muyzer G. (1999) DGGE/TGGE a method for identifying genes from natural ecosystems. *Curr. Opin. Microbiol.* 2, 317–322.

109. Naser S.M., F.L. Thompson, B. Hoste, D. Gevers, P. Dawyndt, M. Vancanneyt, and J. Swings (2005) Application of multilocus sequence analysis (MLSA) for rapid identification of *Enterococcus* species based on *rpoA* and *pheS* genes. *Microbiology* 151, 2141–215.

110. Naser S.M., P. Dawyndt, B. Hoste, D. Gevers, K. Vandemeulebroecke, I. Cleenwerck, M. Vancanneyt, and J. Swings (2007) Identification of lactobacilli by *pheS* and *rpoA* gene sequence analyses. *Int. J. Syst. Evol. Microbiol.* 57, 2777–2778.

111. Naumann D., D. Helm, and C. Schultz (1994) Characterization and identification of micro-organisms by FT-IR spectroscopy and FT-IR microscopy. In *Bacterial Diversity and Systematics*, pp. 67–85, F.G. Priest, A. Ramos Cormenzana, and B.J. Tindall, Eds., Plenum, NY.

112. Nour M. (1998) 16S-23S and 23S-5S intergenic spacer regions of lactobacilli: Nucleotide sequence, secondary structure and comparative analysis. *Res. Microbiol.* 149, 433–448.

113. Ogier J.C., O. Son, A. Gruss, P. Tailliez, and A. Delacroix-Buchet (2002) Identification of the bacterial microflora in dairy products by temporal temperature gradient gel electrophoresis. *Appl. Environ. Microbiol.* 68, 3691–3701.

114. Ogier J.-C., V. Lafarge, V. Girard, A. Rault, V. Maladen, A. Gruss, J.-Y. Leveau, and A. Delacroix-Buchet (2004) Molecular fingerprinting of dairy microbial ecosystems by use of temporal temperature and denaturing gradient gel electrophoresis. *Appl. Environ. Microbiol.* 70, 5628–5643.

115. O'Sullivan T.F. and G.F. Fitzgerald (1998) Comparison of *Streptococcus thermophilus* strains by pulse field gel electrophoresis of genomic DNA. *FEMS Microbiol. Lett.* 168, 213–219.

116. Østlie H.M., L. Eliassen, A. Florvaag, and S. Skeie (2005) Phenotypic and PCR-based characterization of the microflora in Präst cheese during ripening. *Int. Dairy J.* 15, 911–920.

117. Peters S., S. Koschinsky, F. Schwieger, and C.C. Tebbe (2000) Succession of microbial communities during hot composting as detected by PCR-single-strand-conformation-polymorphism-based genetic profiles of small-subunit rRNA genes. *Appl. Environ. Microbiol.* 66, 930–936.

118. Papadelli M., E. Manolopoulou, G. Kalantzopoulos, and E. Tsakalidou (2003) Rapid detection and identification of *Streptococcus macedonicus* by species-specific PCR and DNA hybridisation. *Int. J. Food Microbiol.* 81, 233–241.

119. Petersen K.M. and L. Jesperen (2004) Genetic diversity of the species *Debaryomyces hansenii* and the use of chromosome polymorphism for typing of strains isolated from surface-ripened cheeses. *J. Appl. Microbiol.* 97, 205–213.

120. Piraino P., A. Ricciardi, G. Salzano, T. Zotta, and E. Parente (2006) Use of unsupervised and supervised artificial neural networks for the identification of lactic acid bacteria on the basis of SDS-PAGE patterns of whole cell proteins. *J. Microbiol. Methods* 66, 336–346.

121. Pot B., C. Hertel, W. Ludwig, P. Deschee-Maeker, K. Kersters, and K.H. Schleifer (1993) Identification and classification of *Lactobacillus acidophilus*, *L. gasseri* and *L. johnsonii* strains by SDS-PAGE and rRNA-targeted oligonucleotide probe hybridization. *J. Gen. Microbiol.* 139, 513–517.

122. Pot B., P. Vandamme, and K. Kersters (1994a) Analysis of electrophoretic whole-organism protein fingerprints. In *Chemical Methods in Prokaryotic Systematics*, pp. 493–521, M. Goodfellow and A.G. O'Donnell, Eds., Wiley, Chichester, U.K.

123. Pot B., W. Ludwig, K. Kersters, and K.-H. Schleifer (1994b) Taxonomy of lactic acid bacteria. In *Bacteriocins of Lactic Acid Bacteria*, pp. 13–90, L. De Vuyst and E.J. Vandamme, Eds., Blackie Academic and Professional Glasgow, U.K.

124. Prillinger H., O. Molnar, F. Eliskases-Lechner, and K. Lopandic (1999) Phenotypic and genotypic identification of yeasts from cheese. *Ant. Leeuwen.* 75, 267–283.

125. Psoni L., C. Kotzamanides, C. Andrighetto, A. Lombardi, N. Tzanetakis, and E. Litopoulou-Tzanetaki (2006) Genotypic and phenotypic heterogeneity in *Enterococcus* isolates from Batzos, a raw goat milk cheese. *Int. J. Food Microbiol.* 109, 109–120.

126. Psoni L., C. Kotzamanidis, M. Yiangou, N. Tzanetakis, and E. Litopoulou-Tzanetaki (2007) Genotypic and phenotypic diversity of *Lactococcus lactis* isolates from Batzos, a Greek PDO raw goat milk cheese. *Int. J. Food Microbiol.* 114, 211–220.

127. Randazzo C.L., S. Torriani, A.D.L. Akkermans, W.M. de Vos, and E.E. Vaughan (2002) Diversity, dynamics, and activity of bacterial communities during production of an artisanal Sicilian cheese as evaluated by 16S rRNA analysis. *Appl. Environ. Microbiol.* 68, 1882–1892.

128. Rebuffo C.A., J. Schmitt, M. Wenning, F. von Stetten, and S. Scherer (2006) Reliable and rapid identification of *Listeria monocytogenes* and *Listeria* species by artificial neural network-based Fourier transform infrared spectroscopy. *Appl. Envirom. Microbiol.* 72, 994–1000.

129. Rementzis J. and J. Samelis (1996) Rapid GC analysis of cellular fatty acids for characterizing *Lactobacillus sake* and *Lact. curvatus* strains of meat origin. *Lett. Appl. Microbiol.* 23, 379–384.

130. Riedel K.-H.J., B.D. Wingfield, and T.J. Britz (1998) Identification of classical *Propionibacterium* species using 16S rDNA-restriction fragment length polymorphisms. *Syst. Appl. Microbiol.* 21, 419–428.

131. Rizzo A.F., H. Korkeala, and I. Mononen (1987) Gas chromatography analysis of cellular fatty acids and neutral monosaccharides in the identification of lactobacilli. *Appl. Environ. Microbiol.* 53, 2883–2888.

132. Robles J.C., L. Koreen, S. Park, and D.S. Perlin (2004) Multilocus sequence typing is a reliable alternative method to DNA fingerprinting for discriminating among strains of *Candida albicans. J. Clin. Microbiol.* 42, 2480–2488.

133. Roelleke S., G. Muyzer, C. Wawer, G. Wanner, and W. Lubitz (1996) Identification of bacteria in a biodegraded wall painting by denaturing gradient gel electrophoresis of PCR-amplified gene fragments coding for 16S rRNA. *Appl. Environ. Microbiol.* 62, 2059–2065.

134. Rossetti L. and G. Giraffa (2005) Rapid identification of dairy lactic acid bacteria by M13-generated, RAPD-PCR fingerprint databases. *J. Microbiol. Methods* 63, 135–144.

135. Rossi F., S. Torriani, and F. Dellaglio (1998) Identification and clustering of dairy propionibacteria by RAPD-PCR and CGE-REA methods. *J. Appl. Microbiol.* 85, 956–964.

136. Rossi F., S. Torriani, and F. Dellaglio (1999) Genus- and species specific PCR based detection of dairy propionibacteria in environmental samples by using primers targeted to the genes encoding16S rRNA. *Appl. Environ. Microbiol.* 65, 4241–4244.

137. Samelis J., J. Rementzis, E. Tsakalidou, and J. Metaxopoulos (1998) Usefulness of rapid GC analysis of cellular fatty acids for distinguishing *Weissella viridenscens, Weissella paramesenteroides, Weissella helenica* and some non-identifiable, arginine-negative *Weiissella* strains of meat origin. *Syst. Appl. Microbiol.* 21, 260–265.

138. Sánchez J.I., L. Rossetti, B. Martínez, A. Rodríguez, and G. Giraffa (2006) Application of reverse transcriptase PCR-based T-RFLP to perform semi-quantitative analysis of metabolically active bacteria in dairy fermentations. *J. Microbiol. Methods* 65, 268–277.

139. Santegoeds C.M., S.C. Nold, and D.M. Ward (1996) Denaturing gradient gel electrophoresis used to monitor the enrichment culture of aerobic chemoorganotrophic bacteria from a hot spring cyanobacterial mat. *Appl. Environ. Microbiol.* 62, 3922–3928.

140. Schleifer K.H., J. Kraus, C. Dvorak, R. Kilpper-Balz, M.D. Collins, and W. Fischer (1985) Transfer of *Streptococcus luctis* and related streptococci to the genus *Lactococcus* gen. nov. *Syst. Appl. Microbiol.* 6, 183–195.

141. Schwartz D.C. and C.R. Cantor (1984) Separation of yeast chromosome-sized DNAs by pulsed field gradient gel electrophoresis. *Cell* 37, 67–75.

142. Smith D.D. Jr. and S.J. Norton (1980) *S*-Adenosylmethionine, cyclopropane fatty acid synthase, and the production of lactobacillic acid in *Lactobacillus plantarum. Arch. Biochem. Biophys.* 205, 564–570.

143. Švec P., M. Vancanneyt, M. Seman, C. Snauwaert, K.Lefebvre, I. Sedláček, and J. Swings (2005) Evaluation of (GTG)₅-PCR for identification of *Enterococcus* spp. *FEMS Microbiol. Lett.* 247, 59–63.

144. Tabasco R., T. Paarup, C. Janer, C. Peláez, and T. Requena (2007) Selective enumeration and identification of mixed cultures of *Streptococcus thermophilus, Lactobacillus delbrueckii* subsp. *bulgaricus, L. acidophilus, L. paracasei* subsp. *paracasei* and *Bifidobacterium lactis* in fermented milk. *Int. Dairy J.* 17, 1107–1114.

145. Tailliez P., P. Quenee, and A. Chopin (1996) Estimation de la diversite' parmi les souches de la colection CNRZ: application de la RAPD a' un groupe de lactobacilles. *Lait* 76, 147–158.

146. Tanskanen E.I., D.L. Tulloch, A.J. Hillier, and B.E. Davidson (1990) Pulsed-field gel electrophoresis of *Sma*I digests of lactococcal genomic DNA, a novel method of strain identification. *Appl. Environ. Microbiol.* 56, 3105–3111.

147. Terzaghi B.E. and W.E. Sandine (1975) Improved medium for lactic streptococci and their bacteriophages. *Appl. Environ. Microbiol.* 29, 807–813.

148. Timmins E.M., S.A. Howell, B.K. Alsberg, W.C. Noble, and R. Goodacre (1998) Rapid differentiation of closely related *Candida* species and strains by pyrolysis-mass spectrometry and Fourier transform-infrared spectroscopy. *J. Clin. Microbiol.* 36, 367–374.

149. Tindall B.J., E. Brambilla, M. Steffen, R. Neumann, R. Pukall, R.M. Kroppenstedt, and E. Stacke-brandt (2000) Cultivatable microbial diversity: Gnawing at the Gordian knot. *Environ. Microbiol.* 2, 310–318.

150. Tornabene T.G. (1985) Lipid analysis and the relationship to chemotaxonomy. In *Methods in Microbiology*, Vol. 18, pp. 209–234, G. Gottschalk, Ed., Academic Press, London.

151. Tracey R.P. and T.J. Britz (1989) Cellular fatty acid composition of *Leuconostoc oenos*. *J. Appl. Bacteriol.* 66, 445–456.

152. Tsakalidou E., E. Manolopoulou, E. Kabaraki, E. Zoidou, B. Pot, K. Kersters, and G. Kalantzopoulos (1994) The combined use of whole-cell protein extracts for the identification (SDS-PAGE) and enzyme activity screening of lactic acid bacteria isolated from traditional Greek dairy products. *Syst. Appl. Microbiol.* 17, 444–458.

153. Vancanneyt M., A. Lombardi, C. Andrighetto, E. Knijff, S. Torriani, K.J. Björkroth, C.M.A.P. Franz, et al. (2002) Intraspecies genomic groups in *Enterococcus faecium* and their correlation with origin and pathogenicity. *Appl. Environ. Microbiol.* 68, 1381–1391.

154. Vasdinyei R. and T. Deák (2003) Characterization of yeast isolates originating from Hungarian dairy products using traditional and molecular identification techniques. *Int. J. Food Microbiol.* 86, 123–130.

155. Vasiljevic T. and N.P. Shah (2008) Probiotics—From Metchnikoff to bioactives. *Int. Dairy J.*, 18, 714–728.

156. Vincent D., D. Roy, F. Mondou, and C. Dery (1998) Characterization of bifidobacteria by random DNA amplification. *Int. J. Food Microbiol.* 43, 185–193.

157. Vorobjeva L.I. (1999). Economic and medical applications. In *Propionibacteria*, pp. 209–243, L.I. Vorobjeva, Ed., Kluwer Academic Publishers, The Netherlands.

158. Zamfir M., M. Vancanneyt, L. Makras, F. Vaningelgem, K. Lefebvre, B. Pot, J. Swings, and L. De Vuyst (2006) Biodiversity of lactic acid bacteria in Romanian dairy products. *Syst. Appl. Microbiol.* 29, 487–495.

159. Zeigler D.R. (2003) Gene sequences useful for predicting relatedness of whole genomes in bacteria. *Int. J. Syst. Evolut. Microbiol.* 53, 1893–1900.

160. Zumstein E., R. Moletta, and J.J. Godon (2000) Examination of two years of community dynamics in an anaerobic bioreactor using fluorescence polymerase chain reaction (PCR) single-strand conformation polymorphism analysis. *Environ. Microbiol.* 2, 69–78.

Chapter 30

Spoilage Detection

Maria Cristina Dantas Vanetti

Contents

30.1 Introduction 799
30.2 Detection of Microbial Spoilage of Milk and Dairy Products 800
    30.2.1 Detection Methods of Spoilage Microorganisms 802
    30.2.2 Sensorial Detection of Spoilage of Milk and Dairy Products 804
    30.2.3 Microbial Metabolites as Markers of Milk
           and Dairy Products Spoilage 804
    30.2.4 Volatile Compounds as Markers of Milk
           and Dairy Products Spoilage 805
30.3 Detection of Chemical and Physical Spoilage of Milk 806
30.4 Modeling Spoilage 806
30.5 Future Trends 807
References 807

30.1 Introduction

With the current world production and distribution systems in the food industry, there is a real need for high-quality, extended shelf-life products. The dairy industry must optimize and improve the processes that result in products that meet the consumers' demands for foods having high quality, nutrition, with functionality, wholesomeness, with less fat and salt, and safe. Despite the development of the dairy industry in the last century, premature spoilage of milk continues to be a problem and causes considerable environmental and economic losses.

Spoilage is a subjective term used to describe the deterioration of foods' texture, color, odor, or flavor as well as the development of slime to the point where the foods are unsuitable for human consumption. Off-odors and off-flavors are a common cause of spoilage of dairy products, and the economic consequences can be serious. Some spoilage is inevitable, and a variety of factors cause the deterioration of milk and dairy products, including some factors that are mainly physical or chemical, while others are due to the actions of enzymes or microorganisms. These factors are interrelated and dependent on intrinsic product properties, e.g., pH, water activity, endogenous enzymes, and starter cultures, cross-contamination during milking and processing in combination with the presence of oxygen and temperature abuse.

Modern dairy processing utilizes various preservation treatments that result in an assortment of dairy products having vastly different tastes and textures and a complex spoilage microbiota. Despite the complexity of spoilage, detection needs to be fast and accurate, and it may involve detailed microbiological, sensory, and chemical analysis to determine the specific spoilage organism or the actual cause. Rapid and effective means of identifying the potential of spoilage of milk and dairy products and being able to instigate remedial action with little delay are, therefore, essential and advantageous in reducing product food loss.

Numerous methods to detect spoilage have been proposed that aim to determine concentrations of spoilage microorganisms or compounds produced by them or by reactions of food components. However, many of these methods are considered inadequate because they are time-consuming, labor-intensive, and/or do not reliably give consistent results [14]. Sensory and microbiological analyses are most widely used to serve these purposes in today's industry. While sensory analysis is appropriate and, indeed, essential for product development, its reliance on highly trained panels to minimize subjectivity makes it costly and, therefore, unattractive for the other, more routine requirements [11]. It is essential to adopt objective techniques, such as microbiological or chemical analysis, that are less expensive and more convenient. Consequently, for a limited number of foods, various chemical and biochemical markers for spoilage have been proposed and used to measure the quality or degree of spoilage [50].

Microbiological methods, at least in their traditional form, give retrospective information that is satisfactory for product development, but less so for the other requirements [11]. Although traditional methods of estimating bacterial populations offer many advantages for quality control in the dairy industry, they do not provide results quickly enough to allow for intervention. Despite this, they are, to date, routinely used as the main means to detect spoilage of milk and dairy products. To understand the changes that occur in milk and dairy products due to the microbial growth and metabolism and, therefore, to establish the microbial survey for spoilage detection, it is necessary to know this microbiota in specific conditions of the product.

30.2   Detection of Microbial Spoilage of Milk and Dairy Products

Numerous microorganisms, including bacteria, yeasts, and molds, constitute the complex ecosystem present in milk and dairy products, and, in most situations, they quite frequently are associated with product spoilage. Even before spoilage becomes obvious, microorganisms have begun the process of breaking down milk constituents for their own metabolic needs.

Microbial spoilage of milk often involves the degradation of carbohydrates, proteins, and fats by the microorganisms or their enzymes. The metabolic diversity of microorganisms associated with the complexity of food composition requires a more complete understanding of the chemical and physiological characteristics of these organisms in milk, which may lead to the development of better methods of detection and prevention.

Milk, as it leaves the udder of healthy animals, normally contains low numbers of microorganisms, typically ranging from several hundred to a few thousand colony-forming units per milliliter (CFU/mL). This contaminant microbiota is quite limited and consists predominantly of gram-positive bacteria belonging to micrococci and lactococci groups and *Corynebacterium bovis* [43]. These bacteria are generally mesophilic, and their growth and concomitant spoilage of milk are inhibited if the milk is immediately refrigerated and stored at temperatures below 4°C. Without prompt refrigeration, milk spoilage occurs due to the conversion of lactose to lactic acid by mesophilic contaminants. The development of lactic acid in milk is accompanied by an odor usually described as "sour" due to the production of very small amounts of acetic and propionic acids [44]. Pasteurization will not improve the flavor of raw milk if acid has already developed. The acidity determination is a fundamentally important test for the industry because it indicates the convenience or inconvenience of using the milk. The AOAC official method of number 947.05 for determining the acidity of milk established the titrimetric procedure with 0.1 M NaOH and phenolphthalein as an indicator [1]. The spectrophotometric method (number 437.05) for lactic acid in milk is also described [1]. The acidity of the milk can also be determined routinely and quickly by the Alizarol test.

Nonaseptically drown milk usually contains a diverse group of bacteria capable of growing over a wide range of storage temperatures. These contaminants originate from contact surface, soil, dust, water, bedding, manure, feed, milking equipment, and milk handlers. This contamination of raw milk will affect not only the shelf-life of dairy products but also the technological and economical aspects of milk processing. Refrigerating raw milk is universally acceptable for extending the shelf-life and eliminating spoilage by mesophilic bacteria. However, the growth of psychrotrophic microorganisms is permitted, mainly gram-negative bacteria, which produce heat-resistant extracellular enzymes such as proteases and lipases that further damage milk and milk products. Psychrotrophic microorganisms are defined as those that can grow at 7°C or below, within 7–10 days incubation, regardless of their optimal growth temperature [19]. In most countries, changes in the procedures for collecting milk on farms and in management practices at dairies lead to a fluid milk plant processing raw milk 2–5 days old. At this time, psychrotrophic bacteria will develop and generate a variety of defects in dairy products.

Although psychrotrophic bacteria are in a small part of fresh collected milk, they compose up to 80% of the population of raw refrigerated milk, and *Pseudomonas* spp. are the most important of the psychrotrophs that dominate the microbiota of raw or pasteurized milk at the time of storage. This genus is represented by species with the shortest generation times at 0°C–7°C. *Pseudomonas fragi, Pseudomonas fluorescens*, and *Pseudomonas putida* are the most common species, and they are recognized as producers of proteolytic and lipolytic thermostable enzymes. Other genera of gram-negative psychrotrophic bacteria include *Achromobater, Aeromonas, Alcaligenes, Chromobacterium, Flavobacterium, Serratia*, and *Enterobacter*. Thermoduric bacteria are those that survive pasteurization, and they are represented mainly by gram-positive bacteria in the genera *Bacillus* and *Clostridium* spp. and the nonsporeformers genera *Arthrobacter, Microbacterium, Streptococcus*, and *Corynebacterium* that are involved in spoilage. Some psychrotrophic *Bacillus* spp. secrete heat-resistant extracellular proteases, lipases, and phospholipases (lecithinase) that are of comparable heat resistance as those of pseudomonas. *Bacillus cereus* frequently isolated from milk has been examined carefully because of its "bitty cream" defect and potential enterotoxin production. Some *Enterococcus* isolates can grow at 7°C and have demonstrable proteolytic activity. These bacteria constitute only a minor population of the microbiota in raw milk, but their number may be proportionally higher in pasteurized milk because of their resistance to pasteurization temperatures.

Yeasts and molds are a common cause of spoilage of fermented dairy products because of the low pH usually found in these products. Low water activity in some hard cheeses, sweetened condensed milk, and butter can also favor yeasts and molds spoilage.

30.2.1 Detection Methods of Spoilage Microorganisms

Once they are of crucial importance in milk spoilage, psychrotrophic populations could be determined for milk shelf-life prediction or to determine spoilage. Standard plate count procedures are traditionally used to detect psychrotrophic bacteria in milk and milk products, but these techniques require plates to be incubated at 7°C for 7–10 days [19]. This method is time-consuming, labor-intensive, and does not leave time for intervention, but it is still applicable to raw and pasteurized milk, cream, and cottage cheese [19]. Several variations of time and temperature of incubation of the conventional plate count procedure were proposed as 16 h at 17°C followed by 3 days at 7°C [10] and 25 h at 21°C for milk and cream [29].

These quantitative methods for psychrotrophs require careful interpretation, since there is no agreement about the number of this group of bacteria that cause milk spoilage. The number of psychrotrophs required to produce off-flavors varies among species, and it is determined not only by the growth rate at the storage temperature but also by the proteolytic and lipolytic activity and heat resistance of the enzymes. Some authors defend the theory that there is a significant correlation between the initial count of psychrotrophs and the storage life of raw milk at refrigeration temperatures. Generally, high levels of psychrotrophic bacteria in raw milk are required to contribute sufficient quantities of heat-stable proteases and lipases to cause the breakdown of protein and fat after pasteurization. The number of psychrotrophs generally required to initiate spoilage in milk is about 10^6 CFU/mL [4,39]. For *Pseudomonas* sp., 2.7×10^6 to 9.3×10^7 CFU/mL were required to produce off-flavors, while for *Alcaligenes* sp., 2.2×10^6 to 3.6×10^7 CFU/mL were needed [5]. The development of off-flavors, including bitterness and texture problems in cheese caused by proteases from psychrotrophs, has been reported, but only when psychrotroph counts in milk were 2×10^6 to 5×10^8 CFU/mL. However, milk spoilage by psychrotrophs was reported in the range of populations of 10^2–10^9 per mL [46]. Gelation of UHT milk can result from the activity of proteolytic enzymes of psychrotrophs at counts from 10^4 to 10^8 CFU/mL [5]. Milk spoilage observed in counts as low as 10^2 CFU/mL makes it unclear whether psychrotroph counts can be used as an index in the determination of milk quality or shelf-life from a sensory standpoint. The results of Duyvesteyn et al. [15] showed that the psychrotrophs count at the sensory end of shelf-life is poorly correlated with the sensory shelf-life of milk; therefore, they suggest that the best way to determine the sensory endpoint of milk is by sensory testing and not by plate count method.

Despite this controversy regarding the number of microbial contaminants for milk spoilage, *Pseudomonas* spp. are considered the most important causative agent, and detection and enumeration of these bacteria is useful to establish contamination and potential spoilage microbiota. Current methods of identification and enumeration of *Pseudomonas* spp. in milk involve plating milk or dairy samples onto *Pseudomonas* selective media, e.g., cetrimide, fucidin, cephaloridine (CFC) agar [17], and confirmation of well-isolated colonies by biochemical methods. One major problem associated with commercially available *Pseudomonas* selective media is insufficient selectivity for the genus *Pseudomonas* [17]. Indeed, culture and identification assay require time to produce results, and underestimation of bacterial numbers sometimes occurs because the conventional techniques could not recover sublethally injured cells that may occur in heat-treated products such as pasteurized milk.

Another alternative is to test for groups of microorganisms that are of particular significance in milk spoilage such as proteolytic and lipolytic bacteria or yeasts and molds in fermented milk and hard cheeses. Proteolytic bacteria can be determined by plating samples on skim milk agar or standard caseinate agar and lipolytic bacteria on spirit blue agar [19].

Enumeration of lipolytic microorganisms is not usually performed as a routine analysis but only when a problem arises, and the results can indicate whether the particular lipid-related problem is of microbial or nonmicrobial origin. Considering that microbial lipases are often heat-resistant while the producer microorganisms are not, enzymes of microbial origin can be found in the absence of viable cells.

Count of yeasts and molds by conventional plating method using agar media added to antibiotic or acid for bacteria inhibition is time-consuming and at least 5 days incubation is suggested [19].

Although conventional microbiological methods can identify the spoilage potential of the microbiota found in milk and dairy products, they are time-consuming. To overcome these limitations, molecular biological, biochemical, and immunological techniques have been applied for the rapid and specific detection of microorganisms [21]. Rapid and simple culture-independent methods are required for the detection of proteolytic psychrotrophic bacteria in milk, once it is considered the most important spoilage microbiota. Several culture-independent methods are used for the detection of bacteria in food. Molecular approaches based on direct analyses of DNA or RNA in its environment without microbial enrichment has allowed more precise descriptions of microbial dynamics in complex ecosystems [31]. PCR is one of the most useful techniques because of its high sensitivity, and most research that has applied PCR to milk analysis has focused on pathogen detection [2,13,20,32,40].

Improvements in molecular diagnostic methods are largely dependent on the identification of suitable DNA sequences to use as targets for species' identification and enumeration [35]. The *apr* gene encodes for alkaline metalloprotease in *Pseudomonas* and other related bacteria and was used to detect proteolytic *Pseudomonas* in milk by PCR [37]. A detection limit assay indicated that the *apr* gene could be directly amplified from pasteurized milk contaminated with 10^8 CFU/mL of *P. fluorescens* and with 10^5 CFU/mL in reconstituted skim milk powder if cells were recovered for DNA extraction before amplification [37]. This could reduce the time for detection of proteolytic bacteria in raw milk, allowing the processor to decide about the best use of raw milk during processing. However, an improvement in sensitivity of the assay and a reduction in the cost of the reagents and equipments would seem to be required before this goal could be achieved. Moreover, the sensitivity of PCR assays may be further improved when combined with immunocapture.

Total cell numbers in milk can be obtained with flow cytometry analysis, and this method is currently used by many dairies to determine milk quality [51]. The currently applied flow cytometry techniques do not provide information about the number and identity of potential pathogens or spoilage microorganisms that might be present in milk. This limitation could be eliminated by combining flow cytometry with fluorescent in situ hybridization (FISH) that utilizes fluorescently labeled DNA oligonucleotide probes to detect specific sequences of ribosomal RNA (rRNA) [24]. FISH is another rapid technique considered for the detection and enumeration of *Pseudomonas* spp. in milk [24,30]. The numbers of respiring *Pseudomonas* cells as determined by FISH using fluorescent redox dye 5-cyano-2,3-ditolyl tetrazolium chloride (CTC) staining (CTC-FISH) were almost the same or higher than the numbers of colony counts as determined by the conventional culture method.

New highly sensitive and specific microbial methods based on immunological assay have already been developed for the detection of pathogenic microorganisms, and many are available commercially. Enzyme-linked immunosorbent assay (ELISA) was used to detect *P. fluorescens* in milk and has a sensitivity of 10^5 CFU/mL [22]. Polyclonal antibodies were produced against a pool of *P. fluorescens* strains isolated from milk and, using immunodot blot, the limit of detection was 10^5 CFU/mL [34]. However, more research needs to be done to develop a polyclonal antibody to recognize many genera of psychrotrophs associated with milk spoilage.

More rapid techniques to detect spoilage microorganisms in foods continue to be evaluated (e.g., epifluorescent microscopy and electrical impedance). Like the traditional methods, they also presuppose that the specific spoilage organisms are known and detectable by the chosen technique.

30.2.2 Sensorial Detection of Spoilage of Milk and Dairy Products

Despite the importance of microorganisms in food spoilage, the definition and assessment of spoilage relies on sensory evaluation [23]. A sensory evaluation technique, such as descriptive analysis, is useful in obtaining objective data from human subjects and can be used to characterize aromas and differentiate milks on quality aspects [8]. Although suitable panel methods and statistical examination by humans are fairly reliable, daily and real-time tasting of foods is very laborious.

The most common defect observed in milk and associated with psychrotrophs is an "unclean" flavor, but aroma characteristics of spoiled milk differ by the specific spoilage microorganisms and fat content of the milk. Milks containing *P. fragi* were high in fruity attributes, while those with *P. fluorescens* and *P. putida* exhibited proteolytic aromas. Whole milks were high in rancid/cheesy aromas regardless of the organism. Unpleasant aromas are characteristic of spoiled milk, and a more complete understanding of bacteria-induced spoilage is necessary for the development of shelf-life prediction procedures.

30.2.3 Microbial Metabolites as Markers of Milk and Dairy Products Spoilage

An alternative or ancillary method to microbiological and sensorial analyses involves the measurement of chemical changes associated with microbial spoilage of foods [11]. However, its application has not been as intensively researched as the microbiological and sensory methods in routine use today.

Growth of psychrotrophic bacteria in raw milk during cold storage results in simultaneous production of various heat-stable proteolytic and lipolytic enzymes that are resistant to pasteurization and ultrahigh temperatures used to treat UHT milk. Many enzymes resistant to the heat processes applied in the manufacture of processed milk and dairy products, particularly proteases and lipases, are from *Pseudomonas* and *Bacillus* species. These enzymes may, therefore, cause spoilage of the final products during storage.

Proteolysis in milk occurs also due to the activity of the native milk's proteases such as plasmin, a serine protease that enters milk from the blood in the form of plasminogen. Other proteases may be secreted from mammary tissue cells, blood plasma, or leucocytes.

The peptides produced as a result of proteolysis usually give rise to bitter flavors, and reactions of the released amino acids produce browning on heating. Furthermore, proteolytic enzymes strongly contribute to spoilage off-flavor development, decreased yield during the cheese production, milk heat-stability loss, gelation of UHT-sterilized milk, and reduced shelf-life of dairy products [6,12,16]. Proteolytic activity is the main cause of UHT milk spoilage, causing bitterness and gelation problems [12]. As low levels of this enzyme are sufficient to cause undesirable amounts of protein degradation in UHT milk during storage at room temperature, sensitive methods for their detection have been sought by the dairy industry. However, no method has been universally adopted for this purpose [12].

Methods for measuring the extent of proteolysis in milk by bacterial proteases include analysis of the peptides produced and/or quantifying them by the external standard. Early methods for the detection of protease activity in milk were based on measuring increases in the levels of tyrosine- or tryptophan-containing peptides using the Folin–Ciocalteau reagent [26]. Later, methods using reagents, such as fluorescamine, trinitrobenzene sulfonic acid (TNBS), and o-phthaldialdehyde (OPA), were developed to detect changes in the levels of α-amino groups [7,27,38]. In the last decade, more sensitive assays have been developed, such as enzyme-linked bioluminescent, fluorescent, immunological, and radiometric assays. The possible responsible proteases could be indicated by examining the peptide cleavage by capillary reversed-phase high-performance liquid chromatography (RP-HPLC) and identified by matrix-assisted laser desorption/ionization-time of flight tandem mass spectrometry (MALDI-TOF MS/MS) [52].

Other products resulting from protein catabolism are the biogenic amines such as putrescine, cadaverine, histamine, and tyramine, which are commonly produced during fermentation or spoilage of high protein products by decarboxylation of the amino acids through substrate-specific enzymes produced by microorganisms. Starter cultures or contaminant microorganisms in milk and cheese production processes can present decarboxylase activity. At higher concentrations, biogenic amines may have unwanted health consequences for consumers. Several methods to analyze biogenic amines in food based on thin layer chromatography, liquid chromatography, gas chromatography (GC), biochemical assays, and capillary electrophoresis have so far been described, but the complexity of the real matrices is the most critical in terms of obtaining adequate recoveries for all amines [41]. In cheese, a direct correlation between microorganism counts and the content of biogenic amines is difficult to find because the amine-producing abilities of different bacteria differ widely [25,28,48]. However, a positive correlation between the concentration of the biogenic amine cadaverine and Enterobacteriaceae counts in hard and semihard cheeses was determined [36], but, at this time, this analysis is not adopted as a definitive test for spoilage detection in this product.

As proteolysis can be due to the presence of native protease plasmin and other proteases liberated from somatic cells, the count of somatic cells (SCC) is indicated as a marker for proteolytic potential present in milk. In the past, fluid milk processors have not focused much on milk SCC, but now this view is changing, as it is known that increased SCC is correlated with increased amounts of the heat-stable protease plasmin and lipase in milk. When processing raw milk that has a low bacterial count, and in the absence of microbial growth in pasteurized milk, enzymes associated with high SCC will cause protein and fat degradation during refrigerated storage and produce off-flavors [3]. Using high SCC milk for cheese-making causes compromised on sensory quality. The detected sensory defects were predominantly "rancid" and "bitter," which were consistent with the increased proteolysis and lipolysis observed in the high SCC milks [33]. Somatic cells in milk have been determined by using direct microscopic count or electronically by flow cytometry [19].

Lipolysis occurs due to the action of natural or microbial lipolytic enzymes that are able to hydrolyze triglycerides, a milk fat constituent, in the fatty acids of small chains such as butyric, caproic, caprylic, and capric acid, which are mainly responsible for off-flavors in milk and for rancidity in cheese [6]. Free fatty acids with short chain acids (C4–C8) give rise mainly to rancid flavors, while the middle length chains (C10–C12) give rise to most of soapy, unclean, or bitter flavors. Microorganisms that produce lipolytic enzymes are important in the dairy industry because they can produce rancid flavors and odors in milk and dairy products that make these foods unacceptable to consumers [9]. Lipolytic enzymes produced by psychrotrophs are more important than proteases in relation to the development of defects of flavor in cheese because proteases are soluble in water and lost in the whey, while lipases are adsorbed in the fatty globules and retained in cheese mass [18].

30.2.4 Volatile Compounds as Markers of Milk and Dairy Products Spoilage

All the analyses described so far require extracts of foods. A less invasive and more rapid means for monitoring spoilage is the detection of volatile compounds produced by spoilage bacteria. At least some of the problems inherent in sampling are thereby avoided, and the food itself is not disturbed. Specific volatile compounds have been identified and related to the growth of several microorganisms in biological samples, and these results promise to be useful for early diagnosis of food spoilage. To identify the individual volatile components, the headspace sampling techniques are usually coupled to GC/MS. Thus, a relatively simple, rapid technique with great resolving power is available for routine troubleshooting of spoilage problems.

The range of end-products of microbial growth that have the potential for use in the determination of shelf-life is far wider than that for substrates. Of particular interest in the determination of volatile biomolecules is the commercial availability of the so-called "electronic noses." The electronic nose instrumentation was developed in the early 1980s, and it can perform odor detection continually without being subject to individual sensitivity. Since then, the analyses of volatile compounds have been of increasing interest, and many studies have been dedicated to the improvement of odor measurements. This technology aims to mimic the mammalian sense of smell by producing a composite response unique to each odorant. It consists of an array of gas sensors with different selectivity patterns, a signal-collecting unit, and pattern recognition software applied to a computer. Multivariate statistics were used to create models that detect the spoilage markers.

With this technique, volatiles are detected, but not identified, through their relatively nonspecific adsorption to electronic sensors (e.g., gas-sensitive metal oxide semiconductor field effect transistors and conducting organic polymers). The responses are analyzed within the instrument using pattern recognition techniques such as artificial neural networks and results printed out in real time [11]. The electronic nose system could distinguish among the volatile profiles of different microbial species inoculated in milk-based media after 2 and 5 h of incubation.

A high correlation was established between the complex mixtures of volatile compounds formed and the shelf-life of the refrigerated milk as determined by sensory analysis [49].

In the last years, there has been interest in using similar concepts of the electronic nose in aqueous solutions. This system, denominated of "electronic tongue," is related to the sense of taste in similar ways as the electronic nose is related to olfaction, and it is composed of several kinds of lipid/polymer membranes for transforming information about taste substances into electric signals, which are analyzed by a computer [47]. The taste sensor may be applicable for quality control in the food industry and help assess taste objectively.

30.3   Detection of Chemical and Physical Spoilage of Milk

Milk has a high content of both protein and reducing sugar, and its close-to-neutral pH favors the occurrence of the Maillard reaction that causes the formation of off-flavor and color changes during storage that impair product quality. Additionally, dairy products, in particular, are very sensitive to light oxidation that results in the development of off-flavors, discoloration and, the decrease in nutritional quality. Products of Maillard reaction and oxidation are measured by chemical means (e.g., GC and HPLC, loss of lysine availability, advanced glycosylation end-products, and fluorescence spectroscopy).

30.4   Modeling Spoilage

Several intrinsic and extrinsic factors determine whether spoilage microorganisms will be successful in utilizing the nutrients in a food. These include water activity and types of solutes, pH, storage, and processing temperature, oxygen and carbon dioxide levels, solid or liquid state of food, available nutrients and preservatives, and competing microbiota [14]. The knowledge of microbial responses to these conditions enables objective evaluation and prediction of the spoilage process. Predicting spoilage involves the accumulation of knowledge on microbial behavior in foods and its distillation into mathematical models, based on and validated by actual experimental data. These models can provide useful information for product development and modification, shelf-life estimates, processing requirements, and quality assurance programs. Depending on their objective, models are constructed to focus on the probability of growth/no growth, time required to initiate growth, growth rate, or survival of spoilage organisms under a particular set of parameters. Inactivation and destruction of microbes exposed to different preservatives or preservation techniques can also be modeled. However, models cannot incorporate every factor that may affect the spoilage process, and processors should validate models for their own products to account for different variables [14].

30.5   Future Trends

Although food spoilage is a huge economical problem worldwide, it is obvious that the mechanisms and interaction leading to food spoilage are very poorly understood. Understanding microbial food spoilage is a multidisciplinary task that is required to provide a scientific basis for better preservation methods. The spoilage of some foods is not just a function of cell biomass but a complex process whereby the spoilage may be regulated by bacterial communication signals such as acylated homoserine lactones (AHLs). These molecules allow cells to control many of their functions such as surface colonization and motility, production of exopolymers, production of antibiotics, biofilm development, bioluminescence, cell differentiation, competence for DNA uptake, growth, pigment production, conjugal plasmid transfer, sporulation, toxin production, virulence gene expression, and production of a range of hydrolytic enzymes [45]. AHL-production is common among psychrotrophic bacteria isolated from milk, and indicate that quorum sensing may play an important role in the spoilage of this product [42]. However, our knowledge about the influence of the different spoilage organisms and bacterial pathogens is still very limited from a microbial cell-signaling point of view. Such understanding of spoilage processes and their regulation may allow the development of more targeted, and often milder, food preservation techniques.

Another future application of particular interest is food spoilage detection by sensors integrated into the food packaging. These sensors would eliminate the need for inaccurate expiration dates and provide real-time status of food freshness. Furthermore, it is expected that some advances in nanotechnology will improve the portability, sensitivity, and speed of detection of food spoilage.

References

1. AOAC—Association of Official Analytical Chemists. 2006. *Official Methods of Analysis*, 18th edn., Washington, DC.
2. Aurora, R., Prakash, A., Prakash, C., Rawool, D.B., Barbuddhe, S.B. 2008. Comparison of PI PLC based assays and PCR along with *in vivo* pathogenicity tests for rapid detection of pathogenic *Listeria monocytogenes*. *Food Control*, 19:641–647.
3. Barbano, D.M., Ma, Y., Santos, M.V. 2006. Influence of raw milk quality on fluid milk shelf life. *Journal of Dairy Science*, 89(E. Suppl.):E15–E19.

4. Birkeland, S.E., Stepaniak, L., Sørhaug, T. 1985. Quantitative studies of heat-stable proteinase from *Pseudomonas fluorescens* P1 by the enzyme-linked immunosorbent assay. *Applied and Environmental Microbiology*, 49:382–387.

5. Champagne, C.P., Laing, R.R., Roy, D., Mafu, A.A., Griffiths, M.W. 1994. Psychrotrophs in dairy products: their effects and their control. *Critical Reviews in Food Science and Nutrition*, 34:1–30.

6. Chen, L., Daniel, R.M., Coolbear, T. 2003. Detection and impact of protease and lipase in milk and milk powders. *International Dairy Journal*, 13:255–275.

7. Church, F.C., Swaisgood, H.E., Porter, D.H., Catignani, G.L. 1983. Spectrophotometric assay using *o*-phthaldialdehyde for determination of proteolysis in milk and isolated milk proteins. *Journal of Dairy Science*, 66:1219–1227.

8. Claassen, M., Lawless, H.T. 1992. Comparison of descriptive terminology systems for sensory evaluation of fluid milk. *Journal of Food Science*, 57: 596–600.

9. Cousin, M.A. 1982. Presence and activity of psychrotrophic microorganisms in milk and dairy products: A review. *Journal of Food Protection*, 45:172–207.

10. Cousin, M.A., Jay, J.M., Vasavada, P.C. 2001. Psychrotrophic microorganisms. In Dowes, F.P. and Ito, K. (eds.), *Compendium of Methods for the Microbiological Examination of Foods*, 4th edn., American Public Health Association, Washington, DC, pp. 159–166.

11. Dainty, R.H. 1996. Chemical/biochemical detection of spoilage. *International Journal of Food Microbiology*, 33:19–33.

12. Datta, N., Deeth, H.C. 2001. Age gelation of UHT milk—A review. *Institution Chemical of Engineers*, 79:197–210.

13. Desmarchelier, P.M., Bilge, S.S., Fegan, N., Mills, L., Vary J.C. Jr., Tarr, P.I. 1998. A PCR specific for *Escherichia coli* O157 based on the *rfb* locus encoding O157 lipopolysaccharide. *Journal of Clinical Microbiology*, 36:1801–1804.

14. Doyle, E.M. 2007. *Microbial Food Spoilage—Losses and Control Strategies A Brief Review of the Literature*, University of Wisconsin, Madison, WI.

15. Duyvesteyn, W.S., Shimoni, E., Labuza, T.P. 2001. Determination of the end of shelf-life for milk using Weibull hazard method. *Lebensm.-Wiss. u.-Technology*, 34:143–148.

16. Fairbairn, D.J., Law, B.A. 1986. Proteinases of psychrotrophic bacteria: Their production, properties, effects and control. *Journal of Dairy Research*, 53:139–177.

17. Flint, S., Hartley, N. 1996. A modified selective medium for the detection of *Pseudomonas* species that cause spoilage of milk and dairy plants. *International Dairy Journal*, 6:223–230.

18. Fox, P.F. 1989. Proteolysis during cheese manufacture and ripening. *Journal of Dairy Science*, 72:1379–1400.

19. Frank, J.F., Yousef, A.E. 2004. Tests for groups of microorganisms. In Wehr, H.M and Frank, J.F. (eds.), *Standard Methods of the Examination of Dairy Products*, 17th edn., American Public Health Association, Washington, DC, pp. 227–247.

20. Gao, A., Mutharia, L., Raymond, M., Odumeru, J. 2007. Improved template DNA preparation procedure for detection *Mycobacterium avium* subsp. *paratuberculosis* in milk by PCR. *Journal of Microbiological Methods*, 69:417–420.

21. Giraffa, G., Neviani, E. 2001. DNA-based, culture-independent strategies for evaluating microbial communities in food-associated ecosystems. *International Journal of Food Microbiology*, 67:19–34.

22. González, I., Martin, R., Garcia, T., Morales, P., Sanz, B., Hernández, P.E. 1994. Detection of *Pseudomonas fluorescens* and related psychrotrophic bacteria in refrigerated meat by sandwich ELISA. *Journal of Food Protection*, 57:710–714.

23. Gram, L., Ravn, L., Rasch, M., Bruhn, J.B., Christensen, A.B., Givskov, M. 2002. Food spoilage-interactions between food spoilage bacteria. *International Journal of Food Microbiology*, 78:79–97.

24. Gunasekera, T.S., Dorsch, M.R., Slade, M.B., Veal, D.A. 2003. Specific detection of *Pseudomonas* spp. in milk by fluorescence in situ hybridization using ribosomal RNA directed probes. *Journal of Applied Microbiology*, 94:936–945.

25. Halasz, A., Barath, A., Simon-Sarkadi, L., Holzhapeel, W. 1994. Biogenic amines and their production by microorganisms in food. *Trends in Food Science and Technology*, 5:42–46.

26. Hull, M.E. 1947. Studies on milk proteins. II. Colorimetric determination of the partial hydrolysis of the proteins in milk. *Journal Dairy Science*, 30:881–894.

27. Humbert, G., Guingamp, M.F., Kouomegne, R., Linden, G. 1990. Measurement of proteolysis in milk and cheese using trinitrobenzene sulphonic acid and a new dissolving reagent. *Journal of Dairy Research*, 57:143–148.

28. Innocente, N., D'Agostin, P. 2002. Formation of biogenic amines in a typical semihard Italian cheese. *Journal of Food Protection*, 65:1498–1501.

29. IDF—International Dairy Federation. 1991. Determination of free fatty acids in milk and milk products. Brussels: IDF (*FIL-IDF Standard no. 265*).

30. Kitaguchi, A., Yamaguchi, N., Nasu, M. 2005. Enumeration of respiring *Pseudomonas* spp. in milk within 6 hours by fluorescence in situ hybridization following formazan reduction. *Applied and Environmental Microbiology*, 71:2748–2752.

31. Lafarge, V., Ogier, J.C., Girard, V., Maladen, V., Leveau, J.Y., Gruss, A., Delacroix-Buchet, A. 2004. Raw cow milk bacterial population shifts attributable to refrigeration. *Applied and Environmental Microbiology*, 70:5644–5650.

32. Lindqvist, L., Norling, B., Thisted Lambertz, S. 1997. A rapid sample preparation method for PCR detection of food pathogens based on buoyant density centrifugation. *Letters in Applied Microbiology*, 24:306–310.

33. Ma, Y., Ryan, C., Barbano, D.M., Galton, D.M., Rudan, M., Boor, K. 2000. Effects of somatic cell count on quality and shelf-life of pasteurized fluid milk. *Journal of Dairy Science*, 83:1–11.

34. Machado, A.D.S. 2006. Atividade proteolítica de *Pseudomonas fluorescens* em biofilmes e detecção das células por anti-soro policlonal. DSc Tese. Departamento de Microbiologia, Universidade Federal de Viçosa.

35. Marco, M.L., Wells-Bennik, M.H.J. 2008. Impact of bacterial genomics on determining quality and safety in the dairy production chain. *International Dairy Journal*, 18:486–495.

36. Marino, M., Maifreni, M., Moret, S., Rondinini, G. 2000. The capacity of Enterobacteriacee species biogenic amines in cheese. *Letters in Applied Microbiology*, 31:169–173.

37. Martins, M.L., Araújo, E.F., Mantovani, H.C., Moraes, C.A., Vanetti, M.C.D. 2005. Detection of the *apr* gene in proteolytic psychrotrophic bacteria isolated from refrigerated raw milk. *International Journal of Food Microbiology*, 102: 203–211.

38. McKellar, R.C. 1981. Development of off-flavors in ultra-high temperature and pasteurized milk as a function of proteolysis. *Journal of Dairy Science*, 64:2138–2145.

39. Matta, H., Punj, V., Kanwar, S.S. 1997. An immuno-dot blot assay for detection of thermostable protease from *Pseudomonas* sp. AFT-36 of dairy origin. *Letters in Applied Microbiology*, 25:300–302.

40. O' Gradya, J., Sedano-Balba, S., Maherb, M., Smith, T., Barry, T. 2008. Rapid real-time PCR detection of *Listeria monocytogenes* in enriched food samples based on the *ssrA* gene, a novel diagnostic target. *Food Microbiology*, 25:75–84.

41. Önal, A. 2007. Current analytical methods for the determination of biogenic amines in foods. *Food Chemistry*, 103:1475–1486.

42. Pinto, U.M., Viana, E.S., Martins, M.L., Vanetti, M.C.D. 2007. Detection of acylated homoserine lactones in gram-negative proteolytic psychrotrophic bacteria isolated from cooled raw milk. *Food Control*, 18:1322–1327.

43. Ryser, E. 1999. Microorganisms of importance in raw milk. *Michigan Dairy Review*, 8:7–9.

44. Shipe, W.F., Bassette, R., Deane, D.D., Dunkley, W.L., Hammond, E.G., Harper, W.J., Kleyn, D.H., Morgan, M.E., Nelson, J.H., Scanlan, R.A. 1978. Off flavors of milk: Nomenclature, standards, and bibliography. *Journal of Dairy Science*, 61:857–858.

45. Smith, J.L., Fratamico, P.M., Novak, J.S. 2004. Quorum sensing: A primer for food microbiologists. *Journal of Food Protection*, 67:1053–1070.

46. Tekinson, O.C., Tothwell, J.A. 1974. A study on the effect of storage at 5°C on the microbial flora of heat-treated marked cream. *Journal of Society of Dairy Technology*, 27:57–59.
47. Toko, K. 1996. Taste sensor with global selectivity. *Materials Science and Engineering* C 4:69–82.
48. Valsamaki, K., Michaelidou, A., Polychroniadou, A. 2000. Biogenic amine production in Feta cheese. *Food Chemistry*, 71:259–266.
49. Vallejo-Cordoba, B., Nakai, S. 1994. Keeping quality assessment of pasteurized milk by multivariate analysis of dynamic headspace gas chromatographic data. 1. Shelf life prediction by principal component regression. *Journal of Agricultural and Food Chemistry*, 42: 989–993.
50. Velt, J.H.J.H. 1996. Microbial and biochemical spoilage of foods: An overview. *International Journal of Food Microbiology*, 33:1–18.
51. Walte, H.G., Suhren, G., Reichmuth, J. 2005. Bacteriological raw milk quality: Factors influencing the relationship between colony forming units and Bactoscan-FC counts. *Milchwissenschaft-Milk Science International*, 60:28–31.
52. Wedholm A., Møller, H.S., Lindmark-Mansson, H., Rasmussen, M.D., Andrén, A., Larsen, L.B. 2008. Identification of peptides in milk as a result of proteolysis at different levels of somatic cell counts using LC MALDI MS/MS detection. *Journal of Dairy Research*, 75:76–83.

Chapter 31

PCR-Based Methods for Detection of Foodborne Bacterial Pathogens in Dairy Products

Ilex Whiting, Nigel Cook, Marta Hernández,
David Rodríguez-Lázaro, and Martin D'Agostino

Contents

31.1 Introduction 812
31.2 PCR: Principles and Applications 812
31.3 Critical Features of a PCR-Based Method 813
31.4 PCR Methods for Foodborne Pathogens in Dairy Products 813
      31.4.1 *Salmonella* 814
      31.4.2 *Listeria monocytogenes* 816
      31.4.3 *Enterobacter sakazakii* 816
      31.4.4 *Mycobacterium avium* subsp. *paratuberculosis* 818
31.5 Future Perspective 818
References 819

31.1 Introduction

Microbiological quality control programs are being increasingly applied throughout the milk production chain to minimize the risk of infection in the consumer. The benefits of adopting the latest advancements of molecular microbial diagnostics in routine food analysis are becoming increasingly apparent [32], as they possess inherent advantages over the traditional microbiological culturing techniques, such as shorter time to results, excellent detection limits, specificity, and potential for automation. Several molecular detection techniques have been devised in the last two decades, such as nucleic acid sequence-based amplification (NASBA) [12,31]. The technique that has had the maximum development as a practical food analytical tool is the polymerase chain reaction (PCR) [16,24]. A number of PCR-based methods for detection of pathogens in dairy products have been published; there are also methods marketed commercially. This chapter will focus only on open-formula methods published in the scientific literature. Because of their transparency, such methods have the potential for adoption as international standards [16].

31.2 PCR: Principles and Applications

Kleppe et al. first described in 1971 the principles of PCR, but it was in 1985, with the introduction of thermostable DNA polymerase [35,36], when the first experimental data were published in collaboration with Dr. Kary Mullis who was awarded the Nobel Prize in Chemistry in 1993. This technique has been applied in different areas owing to its versatility, specificity, and sensitivity and has b32 [32]. PCR is a simple, versatile, sensitive, specific, and reproducible technique that amplifies a DNA fragment exponentially, and its principle is based on the mechanism of DNA replication in vivo: double-stranded DNA (dsDNA) is denatured to single-stranded DNA (ssDNA), duplicated, and this process is repeated along the reaction.

A subsequent advancement in PCR has been the development of real-time (RTi) PCR in 1996. It allows monitoring of the synthesis of new amplicon molecules by using fluorescence during the cycling that can be used to quantify the initial amounts of template DNA molecules. Data are therefore collected throughout the PCR process and not just at the end of the reaction (as it occurs in conventional PCR). The major advantages of RTi-PCR are the closed-tube format (that avoids risks of carryover contamination), fast and easy-to-perform analysis, the extremely wide dynamic range of quantification (more than eight orders of magnitude), and the significantly higher reliability of the results when compared with conventional PCR. Fluorescence can be produced during RTi-PCR by an unspecific detection strategy independent of the target sequence using unspecific fluorescent molecules when bound to dsDNA (e.g., ethidium bromide, YO-PRO-1, or SYBR Green I), or by sequence-specific fluorescent oligonucleotides (hydrolysis and hybridization probes). The hydrolysis probes are cleaved by 5′-3′ exonuclease activity during the elongation phase of primers. One of the most used are the TaqMan® probes that are double-labeled oligonucleotides with a reporter fluorophore at the 5′ end and a quencher internally or at the 3′ end, which absorbs the fluorescence of the reporter dye because of its proximity allowing the physical phenomenon defined as "fluorescence resonance energy transfer" (FRET). In contrast to hydrolysis probes, hybridization probes are not hydrolyzed during PCR and the fluorescence is generated by a change in its secondary structure during the hybridization phase, which results in an increase in the distance separating the reporter and the quencher dyes.

31.3   Critical Features of a PCR-Based Method

The main features that an ideal PCR-based analytical method should possess are defined high-performance characteristics, efficient sample preparation, and appropriate controls.

The principal criteria and parameters for PCR performance as a diagnostic tool are defined in the International Standard ISO 22174 "Microbiology of food and animal feeding stuffs—Polymerase chain reaction (PCR) for the detection of food-borne pathogens—General requirements and definitions" [5]. The ideal PCR assay should be fully specific (able to detect only the desired targets) and possess an excellent analytical sensitivity, e.g., be able to detect 10^0–10^1 targets per reaction. In addition, there are some other critical parameters for food analysts: accuracy, precision, and robustness. Accuracy describes the veracity of the test results [38], and can be defined as closeness of agreement between a test result and the accepted reference value [1,28]. Similar terms are trueness and relative accuracy [2]. Precision describes the reproducibility of the test results [38], and can be defined as the closeness of agreement between independent test results obtained under stipulated conditions of repeatability and reproducibility [1,40]. Finally, robustness is the reproducibility by other laboratories using different batches and brands of reagents and validated equipment [17].

The most critical aspect for a PCR-based method is appropriate sample preparation. Bacterial pathogens need in many instances only to be present in low numbers in a foodstuff to pose a hazard to the consumer. The target pathogen or its nucleic acid must be concentrated out of the foodstuff (normally 25 mL or g) into an appropriate volume for a PCR (usually 1–10 μL). This is normally achieved by increasing the number of target cells by incubating the food sample in a nutrient broth (enrichment), and chemical extraction of target nucleic acids. In many foods, clinical and environmental matrices, some components may influence the effectiveness of a PCR [33], and can inhibit the reaction preventing a signal even when targets are present. The use of an enrichment step prior to bacterial nucleic acids extraction allows not only the concentration of target bacteria but also the dilution of inhibitory substances that can affect the subsequent analytical steps. In addition, as only living bacterial cells can grow, an enrichment step can be adapted for viability studies, and therefore can guarantee against false-positive results by residual nucleic acids. However, the accuracy of the use of enrichment for detecting only viable bacteria will depend on the background of DNA of dead cells in the food sample.

In PCR diagnostics, internal amplification controls (IACs) are essential to identify false-negative results [17,18]. The IAC is a nontarget nucleic acid sequence present in every reaction, which is coamplified simultaneously with the target sequence [32]. Few published noncommercial assays have included an IAC. The IAC is an absolutely essential feature [6], and any method that does not contain one has no practical value in actual food analysis, since without an IAC, negative results cannot be accepted as unambiguously signifying that the original sample did not contain the target microorganism. In a reaction with an IAC, a control signal will always be produced when there is no target sequence present. When no IAC signal is observed, this means that the reaction has failed. This review therefore will include only those published methods that contain an IAC.

31.4   PCR Methods for Foodborne Pathogens in Dairy Products

This section provides brief descriptions of a selection of the currently available PCR-based methods for detection of main foodborne pathogens in dairy products: *Salmonella, Enterobacter sakazakii, Mycobacterium avium* subsp. *paratuberculosis,* and *Listeria monocytogenes.* Other

Table 31.1 PCR-Based Methods for the Principal Foodborne Pathogens in Dairy Products

Bacterium	Target Sequence	Matrix	Sample Preparation	LOD	Reference
Salmonella	ttrRSBCA gene	Milk	Enrichment	≈1 CFU/25 mL	[25]
	invA gene	Milk	Enrichment	≈1 CFU/25 mL	[29]
	invA, prt, fliC-d, and viaB genes	Milk	Enrichment	480 CFU/10 mL	[20]
E. sakazakii	tRNA-glu and 23S rRNA region	Infant formula	Enrichment	≈1 CFU/25 mL	[15]
	16S RNA	Pure culture	(Enrichment) chelex	5 genome equivalent/ reaction	[21]
	palE	Infant formula	Enrichment	100 cells/mL	[19]
M. avium paratuberculosis	IS9000	Milk	Direct DNA extraction	100 cells/20 mL	[30]
L. monocytogenes	prfA	Milk	Enrichment	20 cells/20 mL	[13]
	Prú	Cheese and milk	Enrichment	≈1 CFU/25 mL	[34]
	ssrA gene	Soft cheese, milk	Enrichment	1–10 genome equivalents/ reaction	[21]

important microbial pathogens in dairy products such as *Staphylococcus aureus* do not have specific open-formula PCR-based methods including IAC, which is a principal control that should be included in each analytical method, and therefore they will not be discussed in this section. Table 31.1 summarizes the principal analytical features of all the described methods.

31.4.1 Salmonella

Malorny et al. [25] developed a robust RTi-PCR method for detection of *Salmonella enterica* and *S. bongori* in different meat products. The target of the RTi-PCR assay was the *ttrRSBCA* gene, required for the tetrathionate respiration in this bacterium, which is located near the *Salmonella* pathogenicity island 2 at centisome 30.5. The platform used by the authors was the DNA Engine Opticon 2 System (MJ Research, South San Francisco, CA). This method was able to identify 110

Salmonella strains correctly, and not to detect 87 non-*Salmonella* strains. They sourced 46 raw milk samples obtained from one farm in France. The samples were cooled at 4°C for no longer than 24 h before investigation. The traditional enrichment method for the detection of *Salmonella* in artificially and naturally contaminated samples was performed according to International Standard ISO 6579:2003, which is the internationally accepted traditional culture method to detect *Salmonella* in foodstuffs [3]. A 25 mL sample of raw milk was homogenized in 225 mL of buffered peptone water (BPW) by mixing. All samples were preenriched for 20 h at 37°C without shaking. DNA was extracted from 1 mL aliquots of the resulting cultures, by Chelex 100 resin (Biorad, Munich, Germany). The diagnostic sensitivity (the proportion of culture-positive samples that test positive in the PCR assay) was 100%, and the diagnostic specificity (the proportion of culture-negative samples that test negative in the PCR assay) was 100%. In addition, Malorny et al. [26] demonstrated the robustness of the assays based on amplification of *ttrRSBCA* and *invA* sequences, by validating them in a multicenter collaborative trial conducted in Germany. Thirteen laboratories analyzed samples of artificially contaminated milk powder by the PCR-based methods (using various thermocycling instruments) in parallel with the standard culture-based method EN ISO 6579:2003 [3]. The trial demonstrated that the PCR-based methods were repeatable, reproducible, and produced results, which were highly comparable with those obtained by the standard method. The work of Malorny and coworkers provides an excellent example of how to take a PCR-based method from development to implementation.

Malorny and coworkers had also previously devised a conventional PCR assay for *Salmonella* based on targeting sequences of the *invA* gene, and validated its analytical accuracy in two collaborative trials [22,23]. Perelle et al. [29] adapted this assay to RTi-format using the LightCycler platform (Roche Diagnostics, Basel, Switzerland). They evaluated the selectivity of the new RTi-PCR method using 84 *Salmonella* and 44 non-*Salmonella* strains, obtaining 100% selectivity with the RTi-PCR assay. Finally, they artificially contaminated 25 mL of milk with different concentrations of *Salmonella* (0, 1–5, 5–10, 10–20, 20–200 CFU/25 g), and diluted them tenfold in BPW, and subsequently they were incubated 18 h at 37°C. One milliliter of enrichment was used for the bacterial DNA extraction using the InstaGene Matrix (Bio-Rad Laboratories, Germany). Simultaneously, they analyzed the enrichments by the standard culture-based method ISO 6579 [4]. There was 100% agreement between the results obtained by the two methods.

Kumar et al. [20] devised a multiplex PCR method for detection of *S. typhi*, based on amplification of specific regions of the *invA*, *prt*, *fliC-d*, and *viaB* genes. An IAC, which coamplified with *prt* primers, was also included in the assay. They proposed that a multiplex format would mediate more reliable detection than uniplex PCR when analyzing food and environmental samples, where a range of bacterial types would be present. Detection of PCR products was performed conventionally by gel electrophoresis. 13 *Salmonella* and 16 non-*Salmonella* strains were used to evaluate the selectivity of the PCR assay. All *Salmonella* (*invA* PCR positive) and non-*Salmonella* strains (*invA* PCR negative) were identified correctly. In addition, only the *S. typhi* strains were PCR positive for the four genes tested (*invA*, *prt*, *fliC-d*, and *viaB*). The detection probability of the assay was found to be 20% at a concentration of 10^3 CFU/mL (50 CFU/reaction) and 100% at a concentration of 10^4 CFU/mL (500 CFU/reaction) when pure cultures were used. To evaluate the capacity of the system for the detection of *S. typhi* in food samples, 10 mL samples of milk were artificially contaminated with cultures of *S. typhi* containing various cell concentrations (10^3–10^{-1} CFU/mL). After 18 h enrichment in BPW (dilution 1:10) at 37°C, a 1 mL aliquot was taken for nucleic acid extraction by boiling. Detection of artificially contaminating *S. typhi* was achieved down to 480 CFU/10 mL original milk sample.

31.4.2 *Listeria monocytogenes*

D'Agostino et al. [13] developed a conventional PCR assay for *L. monocytogenes*, containing an IAC. The assay is based on amplification of *prfA* gene sequences [37]. It has a 99% detection probability of 7 cells per reaction. When tested against 38 *L. monocytogenes* strains and 52 non-target strains, the PCR assay was 100% inclusive (positive signal from target) and 100% exclusive (no positive signal from nontarget). The assay was incorporated within a method for the detection of *L. monocytogenes* in raw milk. The method comprises 24 h enrichment in half-Fraser broth followed by 16 h enrichment in a medium, which can be added directly into the PCR. The performance characteristics of this PCR-based method were evaluated in a collaborative trial involving 13 European laboratories. A specificity value, or percentage correct identification of uncontaminated milk samples, of 81.8% was obtained. Sensitivity or correct identification of milk samples inoculated with between 20 and 200 *L. monocytogenes* cells per 25 mL was 89.4%. This method has the advantage of being fully compatible with the standard procedure for analysis of foodstuffs for *L. monocytogenes*, ISO 11290-1 [41], and is most suitable as a screening method. PCR-positive results can be confirmed by completing the standard procedure on the same sample, by following the steps after half-Fraser enrichment.

The *prfA* primer set was subsequently used in RTi-PCR format by Rossmanith et al. [34]. They tested the selectivity of the new method using 100 *L. monocytogenes* isolates, 30 non-*monocytogenes Listeria* spp. isolates, and 29 non-*Listeria* isolates, and they obtained that the method was 100% selective. The theoretical detection limit was 1 genome equivalent per PCR reaction and the practical detection limit was about 5 genome equivalents per PCR. The RTi-assay was incorporated in a method involving the ISO 11290-1 [41] primary and secondary enrichments followed by a DNA extraction step, to analyze samples of cheese and milk. It was able to detect down to 7.5 CFU/25 mL of artificially contaminated raw milk, and 1 CFU/15 g of artificially contaminated green-veined cheese.

O'Grady et al. [27] developed a RTi-PCR method for the detection of *L. monocytogenes* in naturally and artificially contaminated cheese and milk samples after 30 enrichment steps. Its target was the *ssrA* gene encoding for tmRNA, which rescues stalled ribosomes and clears the cell of incomplete polypeptides. The detection strategy was based on FRET hybridization probes using the Lightcycler (Roche) as the RTi-PCR platform. The method was fully specific, with a limit of detection (LOD) of 1–10 genome equivalents. For its application in food analysis, in three independent experiments, 25 g or mL of different dairy products (soft cheese and milk) were independently added to 225 mL of half-Fraser broth (Oxoid, Hampshire, U.K.), and homogenized in a stomacher for 2 min. Subsequently, the samples were incubated at 30°C for 22 h with shaking, and then 100 µL were added to 10 mL Fraser broths, respectively, and incubated at 37°C for 4 h with shaking. Finally, 1.5 mL aliquots of the secondary enrichment cultures were used for the DNA isolation using the Bacterial Genomic DNA purification Kit (Edge BioSystems, Gaithersburg, MD). The PCR method detected *L. monocytogenes* in all artificially contaminated samples, and did not detect any in the control samples. These results were confirmed by culturing the samples.

31.4.3 *Enterobacter sakazakii*

Malorny and Wagner [21] developed and validated in-house a TaqMan RTi-PCR for the specific detection of *E. sakazakii*. The target of the RTi-PCR assay was *E. sakazakii*-specific region of the 16S rRNA gene and the platform used was DNA Engine Opticon 2 System (MJ Research). The specificity of the system was evaluated using 27 *E. sakazakii* and 141 non-*E. sakazakii* isolates, which were identified correctly. The RTi-PCR system can detect robustly as little as 10^3 *E. sakazakii* CFU/mL (corresponding to 5 genome equivalent per reaction). The authors did not evaluate the method using actual food samples, but they concluded that the assay could be a practical tool for the detection of *E. sakazakii* in powdered infant formula (PIF) after cultural enrichment.

The International Standard Organisation (ISO) and the International Dairy Federation (IDF) recently jointly adopted a technical specification [9], defining a method for the detection of *E. sakazakii* in PIF. Derzelle and Dilasser [15] evaluated a RTi-PCR-based assay and an automated nucleic acid extraction method that can be used in combination with the ISO–IDF enrichment steps for the routine examination of naturally contaminated PIF. Infant formula powders from three different commercial brands were inoculated with *E. sakazakii* strains ATCC 29544 or ATCC 51329 at four levels of contamination (1–5, 5–10, 10–20, and 20–200 CFU/25 g) plus negative control. Twenty five grams of PIF were dissolved in 225 mL of BPW and then inoculated with diluted *E. sakazakii* culture. The artificial contaminations were carried out in triplicate, except the blank, which was in duplicate. Samples were analyzed in parallel by the conventional ISO–IDF (TS 22964/RM 210) method and by RTi-PCR after a common cultural enrichment. The DNA region located between the tRNA-glu and 23S rRNA genes was selected as a target for detecting the *E. sakazakii* species. Primers ESFor and ESRevB were demonstrated to amplify a 158 bp fragment in all 35 strains of *E. sakazakii* tested with no cross-reaction with other non-*E. sakazakii* bacterial strains. Exclusivity was performed on a total of 139 non-*E. sakazakii* *Enterobacteriaceae*. Forty-five non-*Enterobacteriaceae* strains were chosen. All 184 non-*E. sakazakii* strains tested were negative by PCR and/or RTi-PCR while a positive IAC signal was always detected. The FRET RTi-PCR was combined with a robust nucleic acid extraction procedure, the MagNA Pure LC automated DNA extraction system. A total of 41 samples were suspected to be naturally contaminated, with *E. sakazakii* including infant formulae and samples from the production environment of infant formulae factories. These were investigated using the ISO cultural method and RTi-PCR in parallel. Twenty-two samples were positive for *E. sakazakii* by the ISO–IDF method and 23 were positive by RTi-PCR, providing more than 97.5% concordance between methods. One sample tested positive by PCR and negative by the culture method and it had a very low amplification value (mean C_T cycle to threshold value above 35.00). This value, largely higher than those found for the other positive samples (i.e., 19.15–26.82 cycles), indicated a lower *E. sakazakii* cell density in the enriched sample. The detection limit was approximately 1–5 equivalent genome(s) per reaction for the strain ATCC 29544 (18 cells per PCR tube when combined with DNA extraction step), and was approximately 25 copies (180 cells per PCR tube when combined with DNA extraction step) for the phylogenetically more distinct strain ATCC 51329. The enrichment procedures recommended by the ISO–IDF (TS 22964/RM 210) method allowed detection of an initial contamination level of 1 cell per 100 g of PIF.

Krascsenicsová and coworkwers [19] have recently developed a RTi-5'-nuclease PCR for the specific detection and quantification of *E. sakazakii*. The PCR system targeted a sequence of *E. sakazakii*-specific *palE* gene and the platform used was the PTC-200 thermal cycler coupled to a Chromo 4 continuous fluorescence detector (MJ Research, Waltham, MA). It was 100% selective as determined using 54 *E. sakazakii* and 99 non-*E. sakazakii* strains. The analytical sensitivity was 4×10^1 CFU/mL in 90% of the PCR replicates when pure cultures were used. In addition, the results obtained using the RTi-PCR system were highly linear in the range of 1×10^8–1×10^1 CFU/mL. Finally, they artificially contaminated powdered infant milk formula with tenfold dilutions of *E. sakazakii*. Subsequently, they followed the two-step enrichment ISO standard [9]. The detection limit was 1×10^2 CFU/mL.

31.4.4 Mycobacterium avium subsp. paratuberculosis

Rodríguez-Lázaro et al. [30] developed a RTi-PCR assay for quantitative detection of *M. avium* subsp. *paratuberculosis*. The assay amplifies sequences from the IS900 insertion element, which is specific for this bacterium. The assay was tested against 18 isolates of *M. avium* subsp. *paratuberculosis*, 17 other mycobacterial strains, and 25 nonmycobacterial strains and was fully selective. It was capable of detecting <3 genomic DNA copies with 99% probability or alternatively, using cells directly in the reaction, 12 cells can be detected with 99% probability. To allow the detection of *M. avium* subsp. *paratuberculosis* in milk, 20 mL samples were incubated at 37°C for 30 min with 11% Triton X-100 and 1% trypsin, followed by centrifugation at 2000 g for 30 min and subsequent nucleic acid extraction from the pellet. Harnessed to this sample treatment, the assay was able to consistently detect 10^2 *M. avium* subsp. *paratuberculosis* in 20 mL artificially contaminated semiskimmed milk.

Tasara and Stephan [39] developed a light cycler-based RTi-PCR assay that targets the F57 sequence for the detection of *M. avium* subsp. *paratuberculosis*. The system was 100% selective in correctly identifying 10 *M. avium* subsp. *paratuberculosis* and 33 non-*M. avium* subsp. *paratuberculosis* strains. The analytical sensitivity of the system was 100 *M. avium* subsp. *paratuberculosis* cells per ml when 10 mL of milk samples artificially contaminated was used. Finally, they evaluated the method in naturally contaminated milk. Eighty milk samples were collected from a dairy herd with a history of paratuberculosis. Sixteen pooled samples were prepared from the 80 raw milk samples; each 10 mL sample was made up of 2 mL samples from five different cows. Two of the 16 pooled samples were found to be positive for *M. avium* subsp. *paratuberculosis*. Later, the same research team analyzed a total of 100 individual farm raw milk bulk tank samples on three occasions during August and September 2005 [11]. Among the 100 bulk tank milk samples that were tested, three samples (3%) were positive for MAP F57.

Ayele and collaborators [10] did an extensive study of the presence of *M. avium* subsp. *paratuberculosis* in bottles and cartons (244) of commercially pasteurized cow's milk in retail outlets throughout the Czech Republic. Milk samples were brought to the Veterinary Research Institute in Brno, Czech Republic, processed, inoculated onto Herrold's egg yolk slants, and incubated for 32 weeks. Colonies were characterized by standard techniques and confirmed by PCR based on the IS900. *M. avium* subsp. *paratuberculosis* was cultured and confirmed by PCR from 4 of 244 units (1.6%) of commercially pasteurized retail milk.

31.5 Future Perspective

There needs to be a focused drive toward taking proven methods from the scientist's laboratory and implementing them in actual use in the analyst's laboratory. However, further developments are needed for an effective implementation of amplification techniques in food microbiology. Among the main issues that must be addressed for the effective adoption of molecular techniques by food analysis laboratories are the development of rational and easy-to-use strategies for sample treatment and greater automation of the whole analytical process.

Although most of the published molecular-based methods for foodborne detection in dairy products possess a very high potential for its application in routine food analysis laboratories and even for being adopted as standard methods, none of them have been implemented effectively in food microbiology so far. This is particularly surprising when the capacity of these technologies for screening and identifying new agents and specific forms found in food environments such as viable but not culturable forms or the high performance for bacterial typing (from a taxonomic point of view and from a capacity for drug resistance). However, there are multifaceted reasons for that: the classical reasons are based on the cost of the equipment and reagents required and the difficulty of finding adequately trained personnel. However, a wider offer for new platforms for RTi-PCR is available each day (from only two or three platforms in the late-1990s to more than 20 available in the market currently), and the ample number of different biotechnology companies offering DNA polymerase and fluorescent probes. In addition, more than 10 years have passed since the first publication of RTi-PCR in 1996, and now there are many more and better trained analysts who can develop these methods.

Thus, the factors impairing the adoption of these methods principally include the lack of international validation of these methods in comparison with the microbiological standards and the lack of trust of these methods within the food industry. The absolute prerequisite for successful adoption of molecular-based diagnostic methodology is international validation and subsequent standardization [14,16,24]. Most analysts still regard the conventional "gold standard" culture-based methods as the only accepted method. Therefore, any molecular-based method should be shown to work at least as well as the corresponding conventional method, by direct comparison of the analytical performance of each, on identical food samples. There is an international standard guideline for performing this validation [4]. Standard guidelines regarding the use of PCR for the detection of foodborne pathogens have also been established [6–8].

A clear feedback obtained from the food industry is the lack of trust of molecular-based methods. This fact is exacerbated especially if the results are positive, as they need to wait for a classical confirmation. Therefore, the potential advantage of the molecular-based methods is lost in the waiting time. It is obvious that further steps need to be taken to guarantee and reinforce the value of the analytical results obtained using these methods.

Finally, a determined effort to communicate and promote dialog between the researcher and the analyst is necessary, to encourage and mediate adoption of fit-for-purpose methodology. Ideally, this effort requires the establishment of a solid international infrastructure for taking promising PCR-based analytical methods through development and validation and finally delivering them for use. The foundation of this scenario awaits support from international funding agencies.

References

1. Anonymous. (1993). Statistics. Vocabulary and symbols. Part 1: Probability and general statistical terms (ISO 3534-1:1993). International Organization for Standardization, Geneva, Switzerland.
2. Anonymous. (1994). Accuracy (trueness and precision) of measurement and results (ISO 5725-1:1994). International Organization for Standardization, Geneva, Switzerland.
3. Anonymous. (2003a). Microbiology of food and animal feeding stuffs. Horizontal method for the detection of *Salmonella* (ISO 6579:2003). International Organization for Standardization, Geneva, Switzerland.
4. Anonymous. (2003b). Microbiology of food and animal feeding stuffs–Protocol for the validation of alternative methods (ISO 16140:2003). International Organization for Standardization, Geneva, Switzerland.
5. Anonymous. (2005a). Microbiology of food and animal feeding stuffs—Polymerase chain reaction (PCR) for the detection of food-borne pathogens—General requirements and definitions (ISO 22174:2005). International Organization for Standardization, Geneva, Switzerland.
6. Anonymous. (2005b). Microbiology of food and animal feeding stuffs—Polymerase chain reaction (PCR) for the detection of food-borne pathogens—Performance testing for thermal cyclers (ISO/TS 20836). International Organization for Standardization, Geneva, Switzerland.

7. Anonymous. (2006a). Microbiology of food and animal feeding stuffs—Polymerase chain reaction (PCR) for the detection of food-borne pathogens—Requirements for sample preparation for qualitative detection (ISO 20837). International Organization for Standardization, Geneva, Switzerland.

8. Anonymous. (2006b). Microbiology of food and animal feeding stuffs—Polymerase chain reaction (PCR) for the detection of food-borne pathogens—Requirements for amplification and detection for qualitative methods (ISO 20838). International Organization for Standardization., Geneva, Switzerland.

9. Anonymous. (2006c). Milk and milk products — Detection of *Enterobacter sakazakii* (ISO/TS 22964:2006). International Organization for Standardization, Geneva, Switzerland.

10. Ayele, W.Y., Svastova, P., Roubal, P., Bartos, M., and Pavlik, I. (2005). *Mycobacterium avium* subspecies *paratuberculosis* cultured from locally and commercially pasteurized cow's milk in the Czech Republic. *Applied and Environmental Microbiology* **71**, 1210–1214.

11. Bosshard, C., Stephan, R., and Tasara, T. (2006). Application of an F57 sequence-based real-time PCR assay for *Mycobacterium paratuberculosis* detection in bulk tank raw milk and slaughtered healthy dairy cows. *Journal of Food Protection* **69**, 1662–1667.

12. Cook, N. (2003). The use of NASBA for the detection of microbial pathogens in food and environmental samples. *Journal of Microbiological Methods* **53**, 165–174.

13. D'Agostino, M., Wagner, M., Vazquez-Boland, J.A., Kuchta, T., Karpiskova, R., Hoorfar, J., Novella, S. et al. (2004). A validated PCR-based method to detect *Listeria monocytogenes* using raw milk as a food model–towards an international standard. *Journal of Food Protection* **67**, 1646–1655.

14. D'Agostino, M. and Rodriguez-Lazaro, D. (2009) Harmonization and validation of methods in food safety – "FOOD-PCR" a case study. In: *Global Issues in Food Safety and Technology* (G. Barbosa-Canovas et al., Eds.), Academic Press, New York, Chapter 13.

15. Derzelle, S. and Dilasser, F. (2006). A robotic DNA purification protocol and real-time PCR for the detection of *Enterobacter sakazakii* in powdered infant formulae. *BMC Microbiology* **6**, 100.

16. Hoorfar, J. and Cook, N. (2003). Critical aspects of standardization of PCR. In: *Methods in Molecular Biology: PCR Detection of Microbial Pathogens* (K. Sachse and J. Frey, Eds.), Humana Press, Totowa, pp. 51–64.

17. Hoorfar, J., Cook, N., Malorny, B., Rådström, P., De Medici, D., Abdulmawjood, A., and Fach, P. (2003). Diagnostic PCR: Making internal amplification control mandatory. *Journal of Clinical Microbiology* **41**, 5835.

18. Hoorfar, J., Malorny, B., Abdulmawjood, A., Cook, N., Wagner, M., and Fach, P. (2004). Practical considerations in design of internal amplification control for diagnostic PCR assays. *Journal of Clinical Microbiology* **42**, 1863–1868.

19. Krascsenicsová, P., Trnčíková, T., and Kaclíková, E. (2008). Detection and quantification of *Enterobacter sakazakii* by real-time 5′-nuclease polymerase chain reaction targeting the *palE* gene. *Food Analytical Methods* **1**, 85–94.

20. Kumar, S., Balakrishna K., and Batra, H.V. (2006). Detection of *Salmonella enterica* serovar Typhi (S. Typhi) by selective amplification of *invA*, *viaB*, *fliC-d* and *prt* genes by polymerase chain reaction in mutiplex format. *Letters in Applied Microbiology* **42**, 149–154.

21. Malorny, B. and Wagner, M. (2005). Detection of *Enterobacter sakazakii* strains by real-time PCR. *Journal of Food Protection* **68**, 1623–1627.

22. Malorny, B., Hoorfar, J., Hugas, M., Heuvelink, A., Fach, P., Ellerbroek, L., Bunge, C., Dorn, C., and Helmuth, R. (2003a). Interlaboratory diagnostic accuracy of a *Salmonella* specific PCR-based method. *International Journal of Food Microbiology* **89**, 241–249.

23. Malorny, B., Hoorfar, J., Bunge, C., and Helmuth, R. (2003b). Multicenter validation of the analytical accuracy of *Salmonella* PCR: Towards an international standard. *Applied and Environmental Microbiology* **69**, 290–296.

24. Malorny, B., Tassios, P.T., Rådström, P., Cook, N., Wagner, M., and Hoorfar, J. (2003c). Standardization of diagnostic PCR for the detection of foodborne pathogens. *International Journal of Food Microbiology* **83**, 39–48.

25. Malorny, B., Paccassoni, E., Fach, P., Bunge, C., Martin, A., and Helmuth, R. (2004). Diagnostic real-time PCR for detection of *Salmonella* in food. *Applied and Environmental Microbiology* **70**, 7046–7052.

26. Malorny, B., Mäde, D., Teufel, P., Berghof-Jäger, C., Huber, I., Anderson, A., and Helmuth, R. (2007). Multicenter validation study of two blockcycler- and one capillary-based real-time PCR methods for the detection of *Salmonella* in milk powder. *International Journal of Food Microbiology* **117**, 211–218.

27. O' Grady, J., Sedano-Balbas, S., Maher, M., Smith, Y., and Barry, T. (2008). Rapid real-time detection of *Listeria monocytogenes* in enriched food samples based on the *ssrA* gene, a novel diagnostic target. *Food Microbiology* **25**, 75–84.

28. Paoletti, C. and Wighardt, F. (2002). Definition of pre-validation performance requirements. In: 4th Meeting of the European Network of GMO Laboratories – ENGL, April 29–30, 2002. Joint Research Centre, Ispra, Italy.

29. Perelle, S., Dilasser, F., Malorny, B., Grout, J., Hoorfar, J., and Fach, P. (2004). Comparison of PCR-ELISA and LightCycler real-time PCR assays for detecting *Salmonella* spp. in milk and meat samples. *Molecular and Cellular Probes* **18**, 409–420.

30. Rodríguez-Lázaro, D., D'Agostino, M., Herrewegh, A., Pla, M., Cook, N., and Ikonomopoulos, J. (2005). Real-time PCR-based methods for detection of *Mycobacterium avium* subsp. *paratuberculosis* in water and milk. *International Journal of Food Microbiology* **101**, 93–94.

31. Rodríguez-Lázaro, D., Hernández, M., D'Agostino, M., and Cook, N. (2006). Application of nucleic acid sequence-based amplification (NASBA) for the detection of viable foodborne pathogens: Progress and challenges. *Journal of Rapid Methods and Automation in Microbiology* **14**, 218–236.

32. Rodríguez-Lázaro, D., Lombard, B., Smith, H., Rzezutka, A., D'Agostino, M., Helmuth, R., Schroeter, A. et al. (2007). Trends in analytical methodology in food safety and quality: Monitoring microorganisms and genetically modified organisms. *Trends in Food Science and Technology* **18**, 306–309.

33. Rossen, L., Nøskov, P., Holmstrøm, K., and Rasmussen, O.F. (1992). Inhibition of PCR by components of food samples, microbial diagnostic assays and DNA-extraction solution. *International Journal of Food Microbiology* **17**, 37–45.

34. Rossmanith, P., Krassnig, M., Wagner, M., and Hein, I. (2006). Detection of *Listeria monocytogenes* in food using a combined enrichment/real-time PCR method targeting the *prfA* gene. *Research in Microbiology* **157**, 763–771.

35. Saiki, R.K., Scharf, S., Faloona, F., Mullis, K.B., Horn, G.T., Erlich, H.A., and Arnheim N. (1985). Enzymatic amplification of beta-globin genomic sequences and restriction site analysis for diagnosis of sickle cell anemia. *Science* **230**, 1350–1354.

36. Saiki, R.K., Gelfand, D.H., Stoffel, S., Scharf, S.J., Higuchi, R., Horn, G.T., Mullis, K.B., and Erlich, H.A. (1988). Primer-directed enzymatic amplification of DNA with a thermostable DNA polimerasa. *Science* **239**, 487–491.

37. Simon, M.C., Gray, D.I., and Cook, N. (1996) DNA extraction and PCR methods for the detection of *Listeria monocytogenes* in cold-smoked salmon. *Applied and Environmental Microbiology* **62**, 822–824.

38. Skoog, D.A. and Leary, J.J. (1992). *Principles of Instrumental Analysis*, Saunders College Publishing, London, U.K.

39. Tasara, T. and Stephan, R. (2005). Development of an F57 sequence-based real-time PCR assay for detection of *Mycobacterium avium* subsp. *paratuberculosis* in milk. *Applied and Environmental Microbiology* **71**, 5957–5968.

40. Thompson, M., Ellison, S.L.R., and Wood, R. (2002), Harmonised guidelines for single-laboratory validation of methods of analysis (IUAPC Technical Report). *Pure and Applied Chemistry* **74**, 835–855.

41. Anonymous. (1997). Microbiology of food and animal feeding stuffs. Horizontal method for the detection and enumeration of *Listeria monocytogenes*. Detection method (ISO 11290-1:1997). International Organization for Standardization, Geneva, Switzerland.

Chapter 32

Mycotoxins and Toxins

Carla Soler, José Miguel Soriano, and Jordi Mañes

Contents

32.1 Mycotoxins in Dairy Food 824
    32.1.1 Introduction 824
    32.1.2 Mycotoxin Analysis 824
        32.1.2.1 Analytical Quality Assurance 824
        32.1.2.2 Laboratory Precautions.. 827
        32.1.2.3 Sample Preparation 827
        32.1.2.4 Extraction Procedures.. 827
        32.1.2.5 Cleanup Methods 828
        32.1.2.6 Screening Tests... 828
        32.1.2.7 Quantitative Methods... 829
        32.1.2.8 Detection Systems 829
32.2 Toxins in Dairy Food...839
    32.2.1 Introduction 839
        32.2.1.1 Principal Bacterial Toxins in Dairy Products 839
    32.2.2 Analysis of Bacterial Toxins in Dairy Foods 840
        32.2.2.1 Biological Assays... 840
        32.2.2.2 Immunological Tests 841
        32.2.2.3 Phenotypic Assays...842
        32.2.2.4 Biologic–Immunologic–Phenotypic Combination Studies 843
32.3 Future Trends.. 844
References 844

32.1 Mycotoxins in Dairy Food

32.1.1 Introduction

Mycotoxins are products, together with antibiotics, of secondary metabolism of molds, with a molecular weight ranging from ca. 200 to 500 Da that cause undesirable effects, called mycotoxicoses, when animals or humans are exposed to them [1]. Many species of fungi produce mycotoxins in feedstuffs, either preharvest or postharvest, during storage, transport, processing, or feeding. Contamination of feeds with mycotoxins results in significant economic losses in animal husbandry, as well as in undesirable trade barriers [2]. Three sources of mycotoxins are identified in ruminant diets: (i) the contamination of energy-rich concentrates (cereal grains, corn gluten, etc.) with aflatoxins, ochratoxins, ergot alkaloids, and trichothecenes [3,4], (ii) exposition to different classes of mycotoxins (lolitrem–paxilline group, ergovaline, and other ergot alkaloids) that occur in forages [5], and (iii) the consumption of preserved feeding stuffs, such as silage, hay, and straw [6,7]. Kiessling et al. [8] observed that ruminant animals develop mycotoxicoses less frequently, as the rumen flora acts as a first line of defense against mycotoxins. However, various mycotoxins pass the blood–milk barrier or are converted into metabolites that retain their biological activity, and are till date very important because they might impair the milk quality and the use of milk for dairy products such as yoghurt and cheese [9–12]. Other mycotoxins analyzed in dairy and dairy products are zearalenone (ZEN), α-zearalenol (α-ZEL), β-zearalenol (β-ZEL), α-zearalanol (α-ZAL), β-zearalanol (β-ZAL), fumonisin B1 (FB1) and B2 (FB2), T-2 toxin and HT-2 toxin, T-2 triol, diacetoxyscirpenol (DAS), 15-monoacetoxyscirpenol (MAS), deoxynivalenol (DON), 3-acetyldeoxynivalenol (3-AcDON), 15-acetyldeoxynivalenol (15-AcDON), deepoxy-deoxynivalenol (DOM-1), cyclopiazonic acid, fusarenon-X (FUS-X), nivalenol (NIV), neosolaniol (NEO), ergovaline, and slaframine (alkaloidal compound produced by *Rhizoctonia leguminicola*) [16–21]. Figure 32.1 presents the chemical structures of several mycotoxins analyzed in dairy and dairy products. However, the number of studies about the mycotoxins in these matrices is limited than those with other foods. Figure 32.2 shows the percentages of articles cited from 1997 to 2007 related to mycotoxins in dairy and dairy products, and those obtained from the ISI Web of Science, mainly on aflatoxin M1 (AFM1) and ochratoxin A (OTA). AFM1 is a hydroxylated derivative of aflatoxin B1 (AFB1), which occurs in the milk of lactating animals. Several authors [13–15] reflected that the percentage range (1%–3%) of the AFB1 initially present in the animal feedstuff appearing as AFM1 in milk is not real, owing to the day-to-day variations among the animals and milking processes. Ochratoxins is a group of mycotoxins produced by some species of *Aspergillus* and *Penicillium*, and OTA is the most important toxin of this family. OTA is metabolized by rumen microorganisms into a less toxic metabolite called ochratoxin α (OTα) that is excreted in milk. The principal difference between OTα and other ochratoxins is the lack of phenylalanine group in the chemical structure.

32.1.2 Mycotoxin Analysis

32.1.2.1 Analytical Quality Assurance

Basically, two components are used in the analytical quality assurance [34]. First, the use of certified reference materials (CRM) whenever possible, owing to the fact that they are stable and homogeneous products containing certified amounts of mycotoxin(s) of interest [35]. They should be routinely used as much as possible. These CRMs have developed with the coordination of the Standards, Measurements and Testing Programme (also called European Union's Community Bureau of Reference in the past) [36,37]. The characteristics of CRM for mycotoxins are shown in Table 32.1.

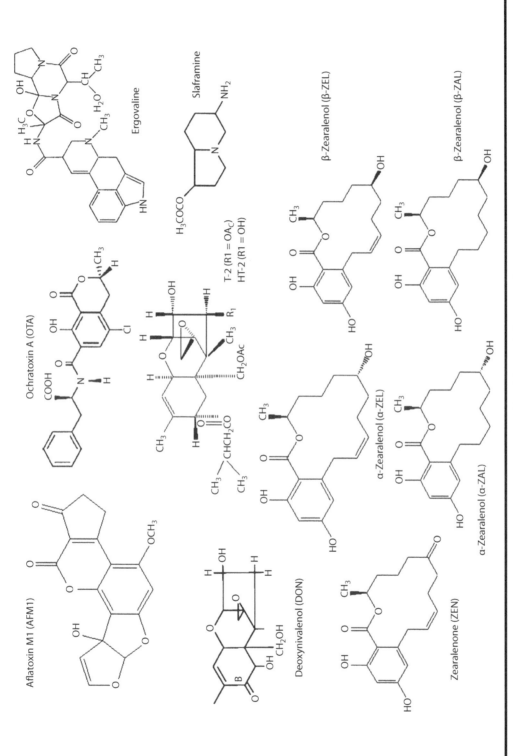

Figure 32.1 **Chemical structures of several mycotoxins analyzed in dairy and dairy products.**

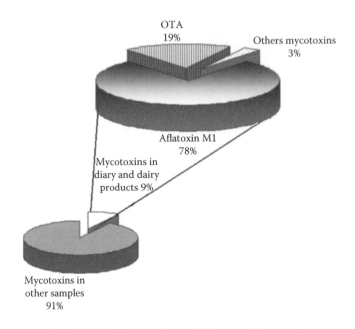

Figure 32.2 **Percentages of articles published in the last 10 years about mycotoxins in dairy and dairy products. (Obtained from the ISI Web of Science.)**

Table 32.1 **Certified Dairy and Dairy Products Reference Materials for AFM1**

Number	Matrix	Certified Value (µg/kg)	Uncertainty (µg/kg)
CRM 282	Full-cream milk powder	<0.05	Not reported
CRM 283	Full-cream milk powder	0.09	+0.04; −0.02
CRM 285	Full-cream milk powder	0.76	±0.05
RM	Chloroform	Information value: 9.93	Not reported

The second component in the analytical quality assurance is proficiency testing, a key issue in achieving external quality control for national reference laboratories conducted by he European Union's Community Reference Laboratory for Milk and Milk Products in France in 1996 [38] and 1998 [39]. Conclusion of the organizers demonstrated that the network of national reference laboratories had shown good analytical competency for the determination of AFM1 in milk, including the very low concentrations of this mycotoxin in the distributed samples. Furthermore, other body involved in proficiency testing is Food Analysis Proficiency Assessment Scheme (FAPAS), in the United Kingdom Ministry of Agriculture Fisheries and Food (UK MAFF).

32.1.2.2 Laboratory Precautions

There are several safety measures in the laboratory which include precautions in handling myco-toxins and in the decontamination and destruction of laboratory wastes. At the outset, mycotoxins are extremely toxic chemicals, while crystalline standards are highly electrostatic and can disperse in the work area. The handling of standards, either in powder or concentrated form and reference standard working solutions, must be carried out with extreme care under a hood, and the face of the operator should be protected with an appropriate mask and the operator's hands should be protected with latex gloves (and not vinyl gloves). Risk assessment should be carried out before any starting any work, as continuous exposure even to low concentrations of mycotoxins presents a potential chronic hazard to the analysts who may be working with these mycotoxins for many years, because they may be carcinogenic, genotoxic, or immunosuppressant. Working solutions should aim for zero-operator exposure. Furthermore, the glassware must be scrupulously cleaned by immersion into a powerful oxidant (bleach or sulfochromic mixture). A flask with 10%–20% aqueous solution of sodium hypochlorite must be kept near the bench in case of emergency, during the analysis. Also, it is recommended that the work areas be decontaminated overnight with sodium hypochlorite at the end of the workday and the surfaces should be thoroughly washed and checked for neutrality before starting a new analysis [34].

32.1.2.3 Sample Preparation

Liquid samples such as milk are usually more homogeneous; however, when these are cloudy, myc-otoxins can be unequally distributed between the liquid and solid phases. In solid food commodities such as cheese, mycotoxins may be distributed in a very heterogeneous manner. According to the Directive and a Decision of the Commission of the European Union [40,41], a minimum of 9.5 kg should be collected from a batch of milk mixed by manual or mechanical means and should be composed of at least five increments, and the batch is accepted only if the concentration of AFM1 does not exceed the permitted limit. If the analysis is not carried out immediately, then the samples must be stored at 4°C–8°C, and never at frozen temperature, owing to the possibility, for example, of aflatoxins to bind to the milk proteins and other components, which could affect the mycotoxin recovery [28].

32.1.2.4 Extraction Procedures

The purpose of extraction is to remove as much of the mycotoxin from the dairy matrix as possible, into a solvent suitable for subsequent cleanup and determination. For AFM1 and OTA, the use of centrifugation and filtration is a key in the extraction to obtain defatted milk, while for other matrices, other options are used to extract these compounds. González-Osnaya et al. [31] proposed a methodology for the extraction of OTA in milk, which is short and easy to perform, based on the mix of 2 mL of milk and 2 mL of methanol. This helps the aggregation of the casein micelles by dehydration, and hence, structures of average sizes as large as 9 μm are precipitated [42,43]. With the separation of these particles, by filtration and/or centrifugation, the upper cream layers are discarded and the cleaner extracts are obtained. Seeling et al. [17] proposed a simple extraction for nine mycotoxins with a mixture of water and ethanol, and Sorensen et al. [19] applied a ternary mixture (sulfuric acid 18%, hexane, and acetonitrile) for the extraction of 18 mycotoxins. As dairy products are a more complex matrix, the addition of diatomaceous earth is required for the extraction of AFM1 [28].

32.1.2.5 Cleanup Methods

The cleanup step in the analysis of mycotoxins in dairy foods consists of one of the two following approaches: (a) the use of solid-phase extraction (SPE) columns or (b) the use of immunoaffinity columns (IACs).

Few articles have been published regarding SPE, reporting that the technique is useful to regenerate the cartridge for further analysis and is cheaper than IACs. However, the disadvantages of SPE are that it obtains low repeatability within a single batch and/or low reproducibility of different batches. For milk, a C18 Sep-Pak cartridge has been used by Imerman and Stahr [21] as well as Fremy and Chu [44] for the determination of slaframine and AFM1, respectively, and a silica gel cartridge Sep-Pak® has been used by Prasongsidh et al. [18], for the determination of cyclopiazonic acid. Durix et al. [20] used 100 mg of Ergosil® on a small column with the chloroform extract, the impurities were removed by washing the column with 3 mL of acetone–chloroform (75:25), and the ergovaline was eluted with 1.5 mL of methanol.

The use of IACs help to considerably increase the reliability of the results owing to its high selectivity, the possibility of analyzing more than one sample simultaneously, and reduce the time of analysis. The IACs must be stored between 4°C and 8°C, but brought to room temperature before analysis. The extract is purified with IAC-containing antibodies specific to mycotoxins and previously conditioned with phosphate-buffered saline (PBS). Subsequently, the IAC is washed with water primarily to remove the impurities and mycotoxin is separated from the antibody by passing methanol for OTA, and methanol, acetonitrile, or a mixture of both for AFM1 in dairy and dairy products. For milk, the IACs are used for AFM1 and OTA. In 1987, Mortimer et al. [45] applied IACs for the first time in the cleanup procedure of AFM1 in milk. Since then, it has been the election procedure for practically all laboratories.

32.1.2.6 Screening Tests

Five methods of screening tests have been used to analyze the presence of AFM1 in milk: thin-layer chromatography (TLC), radioimmunoassay (RIA), enzyme-linked immunoabsorbent assay (ELISA), electronic nose, and Charm Rapid One Step Assay (ROSA).

Screening methods based on TLC are available and applied for AFM1 in milk [45–48], but they are used in only a few laboratories, as they do not provide an adequate limit of quantification (LOQ). In 2004 [26], the Food and Agriculture Organization (FAO), as the collaborative center, the International Dairy Federation (IDF; Committee on Organic Contaminants E501), the International Union of Pure and Applied Chemistry (IUPAC; Commission on the Chemistry of Food), and the International Atomic Energy Agency (IAEA), with 14 laboratories representing 11 countries participating in the trial, validated a method combining immunoaffinity cleanup to TLC for the determination of AFM1 in milk. This study reflected a variation of the recovery rate, from 32% to 120%, and these were used to correct the data. This wide variation of the recovery rate suggested two crucial steps in the protocol, such as matrix sample reconstitution and extract evaporation. Nowadays, this method is the standard ISO/DIS 14674-IDF 190 entitled "Milk and milk powder-determination of aflatoxin M1 content-clean-up by immunoaffinity chromatography and determination by TLC" [49].

Saitanu [50] and Offiah as well as Adesiyun [51] used the RIAS technique for the routine investigations of AFM1 in milk from Thailand as well as Trinidad and Tobago, respectively.

ELISA has been more often used when compared with the other immunochemical procedures [52]. A review of the ELISA as a method for the detection of mycotoxins in dairy and dairy products indicated that, till date, it has been applied mainly for AFM1 (Ridascreen®) [53,54], and Fast AFM1, produced by R-Biopharm (Germany), has been very frequently used in several studies [55–57]. Kim et al. [58] used a direct competitive ELISA for the determination of AFM1 in pasteurized milk, infant formula, powdered milk, and yoghurt. According to the kit supplier, the sample aliquot must be centrifuged and a little quantity of the supernatant is used in the Microplate Reader. The detection limit of this test is 245 ng/L [55].

Benedetti et al. [59] and Barbiroli et al. [53] applied a commercial electronic nose as an innovative screening methodology for simple and rapid detection of AFM1 in a large number of ovine and caprine milk samples, and confirmed that the analysis of the electronic nose data offered substantial assistance in creating clusters that allow recognition of samples at different contamination levels.

The use of ROSA Safe-Level AFM1 quantitative lateral-flow method has been validated in an interlaboratory study of 21 public health, state agriculture, and industry laboratories in the United States testing raw commingled bovine milk. The average intralaboratory repeatability was 11% and the average interlaboratory reproducibility was 13% for the fortified sample pairs. Liquid chromatography (LC) analysis of the study samples by five laboratories showed 38% false negatives with 500 and 550 ppt samples [60].

32.1.2.7 Quantitative Methods

Methods used for mycotoxin in dairy products are mainly based on LC with octadecyl as the stationary phase. Some mycotoxins in milk, such as cyclopiazonic acid [18] and trichothecenes [17] are analyzed by capillary electrophoresis (CE) and gas chromatography (GC), respectively (Table 32.2). The CE with diode array detector (DAD) has been used for the determination of cyclopiazonic acid in milk (absorbance at 220 nm wavelength) [18]. On the other hand, GC is used for trichothecenes, because they have nonfluorescent and weak UV–vis absorption properties that require derivatization to facilitate detection, and makes the determination of trace levels unreliable. For OTA and ergovaline, the method applied is LC followed by fluorescence detection (FLD) and mass spectrometry detection (MSD). For official controls, the analytical methods for determining AFM1 in milk should be able to detect traces of this mycotoxin, i.e., at the ng/kg level. This performance criterion is satisfied by the use of an immunoaffinity cleanup step, followed by LC and fluorimetric detection [61], standardized by the IDF (IDF 171:1995) [62] and ISO and CEN (EN ISO 14501:1998) [63], and validated by Dragacci et al. (2001) with minor modifications as the AOAC official method 2000.08 [38].

32.1.2.8 Detection Systems

To date, the use of a fluorescence detector is mainly applied for AFM1 and OTA owing to the fact that it is a more sensitive and selective technique. The operating conditions range from an excitation wavelength of 360–365 and 274–334 nm, respectively, and an emission wavelength of 430–440 and 440–464 nm, respectively (Table 32.3). The typical chromatograms of AFM1 by LC–FLD are shown in Figure 32.3. Some authors carried out the formation of the mycotoxin derivative to confirm AFM1 and OTA in several matrices including dairy and dairy products. The AFM1 in positive samples has been confirmed by the formation of the AFM1 hemiacetal derivative (AFM2a), as reported by Takeda [64]. This method is carried out as follows: 50 µL of the sample eluate is evaporated to dryness at 40°C under a gentle stream of nitrogen. Subsequently,

Table 32.2   Analysis of Mycotoxins in Milk

Mycotoxin	Extraction Technique	Cleanup Technique	Separation	Detection		Recovery (%)	LD (ng/L)	Reference
AFM1	Centrifugation to obtain defatted milk	IAC	LC	C_{18}-RP Select/B (250 × 4.6 mm, 5 μm), H_2O/ CH_3CN/CH_3OH (65:15:20) at 1 mL/min	FLD	λ_{exc} 360 nm	n.r.	n.r.	[22]
						λ_{em} 430 nm			
AFM1	Centrifugation to obtain defatted milk and filtration (Whatman No. 4)	IAC (Aflaprep) eluted with CH_3CN/ CH_3OH (3/2), followed by H_2O	LC	Phenomenex Prodigy C_{18} (250 × 4.6 mm, 3 μm), H_2O/ CH_3CN/CH_3OH (50:30:20) at 0.8 mL/min	FLD	λ_{exc} 360 nm	90.7–113.5	3	[23]
						λ_{em} 430 nm			
AFM1	Centrifugation with NaCl to obtain defatted milk	IAC (AflaM1) eluted with CH_3OH	LC	Bio-Sil C_{18} (150 × 4.6 mm, 3 μm), CH_3CN/H_2O/CH_3OH (25:50:25) at 1 mL/min	FLD	λ_{exc} 360 nm	68.3–90.1	250	[24]
						λ_{em} 440 nm			
AFM1	Centrifugation to obtain defatted milk and filter (Whatman No. 4)	IAC (AflaM1) eluted with CH_3OH	LC		FLD		—	—	[25]

Analyte	Sample preparation	Cleanup	Separation	Detection	Recovery (%)	LOD	Reference
			Zorbax SB C$_{18}$ (150 × 4.6mm, 3μm), CH$_3$CN/H$_2$O/CH$_3$COOH (25:75:1) at 1mL/min	λ$_{exc}$ 365 nm, λ$_{em}$ 435 nm			
AFM1	Centrifugation to obtain defatted milk and filter (Whatman No. 4)	IAC (AflaPrep) eluted with CH$_3$CN/CH$_3$OH (3/2) and CH$_3$OH	TLC	UV 365 nm	32–120	—	[26]
AFM1	—	—	TLC	The presence was confirmed by derivative formation on TLC plates (hexane–trifluoroacetic acid) (4:1)	84.6–88.0	2000	[27]
AFM1	—	IAC (AflaPrep) eluted with CH$_3$OH	LC; C$_{18}$ (250 × 4.6mm), CH$_3$OH/CH$_3$CN/H$_2$O (20:20:60) at 0.8mL/min	FLD, λ$_{exc}$ 366 nm, λ$_{em}$ 440 nm	98.5	56	[28]
AFM1	—	—	LC; Column TSK-GEL® C$_{18}$ (250 × 4.6mm, 3μm), CH$_3$CN/CH$_3$OH/H$_2$O (20:20:60) at 1mL/min	FLD, λ$_{exc}$ 360 nm, λ$_{em}$ 440 nm	68–81	50 (LOQ)	[29]

(continued)

Table 32.2 (continued)   Analysis of Mycotoxins in Milk

Mycotoxin	Extraction Technique	Cleanup Technique	Separation	Detection	Recovery (%)	LD (ng/L)	Reference
AFM1	Shaken with sulfuric acid 18% (v/v) (pH 2.0), hexane, and CH₃CN, removed the organic layer	For nonbound mycotoxins, Oasis, eluted with CH₃OH	LC	MS	76–108	20–150	[19]
OTA							
ZEN							
α-ZEL, β-ZEL			Hypersil ENV (150 × 4.6 mm, 5 μm) for ESI in PI mode, H₂O, and CH₃OH acidified with CH₃COOH at 100 μL/min	Single quadrupole using ESI in PI mode (T-2 toxin, HT-2 toxin, T-2 triol, DAS, MAS, FB1, FB2, AFM1) and in NI mode (DON, DOM-1, 3-AcDON, 15-AcDON, OTA, ZEN, α-ZEL, β-ZEL, α-ZAL β-ZAL)			
α-ZAL, β-ZAL							
FB1, FB2							
T-2 toxin							
HT-2 toxin							
T-2 triol							
DAS			Luna C₁₈ (150 × 4.6 mm, 5 μm) for ESI in NI mode, H₂O, and CH₃OH at 100 μL/min				
MAS							
DON							
3-AcDON							
15-AcDON							
DOM-1							

Analyte	Extraction	Cleanup	Separation	Detection	Recovery (%)	LOD	Reference
Cyclopiazonic acid	Alkalization with CH$_3$OH–NaHCO$_3$, acidified with HCl, extraction with CHCl$_3$	Silica gel cartridge Sep-Pak® and eluted with CHCl$_3$–CH$_3$OH	CE bare fused silica capillary-extended light path (50 μm i.d. × 64.5 cm and 150 μm i.d. bubble, 60 cm effective length and alignment interface)	DAD 220 nm	—	—	[18]
DON	H$_2$O/CH$_3$CH$_2$OH (90/10)	ChemElut and Mycosep columns	GC	MS	91 (DON)	0.1[a] (DON)	[17]
FUS-X					96 (FUS-X)	0.16[a] (FUS-X)	
NIV					44 (NIV)	0.25[a] (NIV)	
3-AcDON					104 (3-AcDON)	0.22[a] (3-AcDON)	
15-AcDON					97 (15-AcDON)	0.18[a] (15-AcDON)	
DAS					104 (DAS)	0.30[a] (DAS)	
NEO					109 (NEO)	0.33[a] (NEO)	
HT2					106 (HT2)	0.22[a] (HT2)	
T2					111 (T2)	0.16[a] (T2)	

(continued)

Table 32.2 (continued) Analysis of Mycotoxins in Milk

Mycotoxin	Extraction Technique	Cleanup Technique	Separation	Detection	Recovery (%)	LD (ng/L)	Reference
Ergovaline	Deproteinization with acetone. Centrifuged. Acetone evaporated and aqueous residue adjusted to pH 9 and extracted with $CHCl_3$	100 mg Ergosil®, eluted with CH_3OH	LC Zorbax C_{18} (150 × 4.6 mm, 3.5 μm), CH_3CN–$(NH_4)CO_3$ (2 mM) (36.5:63.5) at 1 mL/min	FLD	99.8	200	[20]
OTA	Shaken H_3PO_4– NaCl solution and $CHCl_3$, separated the organic layer, adjusted to pH 7.6 and separated the aqueous layer	IAC® eluted with methanol	LC	FLD	89.4 (OTA)	5 (LOQ)	[16]
OTB			Nucleodur C_{18} (125 × 4.6 mm, 5 μm), 10 mL/L of CH_3COOH: CH_3OH at 1 mL/min	λ_{exc} 274 nm λ_{em} 440 nm	115 (OTB)		
OTα					19.8 (OTα)		
ÓTA	Centrifugation to obtain defatted milk	OchraTest, IAC®, eluted with CH_3OH	Symmetry C_{18} (150 × 4.6 mm, 3.5 μm), CH_3CN:H_2O:CH_3COOH (55:45:0.5) at 0.6 mL/min	FLD λ_{exc} 333 nm λ_{em} 460 nm	34–47	—	[30]

Analyte	Sample preparation	Cleanup	Separation	Column/mobile phase	Detection	Wavelengths	Recovery	LOD	Ref.
OTA	Centrifugation to obtain defatted milk, diluted with water solution containing NaCl and NaHCO$_3$	OchraTest IAC®, eluted with CH$_3$OH	LC	Symmetry C$_{18}$ (150 × 4.6mm, 3.5μm), CH$_3$CN:H$_2$O:CH$_3$COOH (55:45:0.5) at 0.6mL/min	FLD	λ_{exc} 333 nm λ_{em} 460 nm	75.6	—	[30]
OTA	Centrifugation to obtain defatted milk, diluted with PEG 8000 (1%) and NaHCO$_3$ (2%) in water	OchraTest IAC®, eluted with CH$_3$OH	LC	Symmetry C$_{18}$ (150 × 4.6mm, 3.5μm), CH$_3$CN:H$_2$O:CH$_3$COOH (55:45:0.5) at 0.6mL/min	FLD	λ_{exc} 333 nm λ_{em} 460 nm	62.3	—	[30]
OTA	Centrifugation to obtain defatted milk	OchraPrep IAC®, eluted with CH$_3$OH	LC	Symmetry C$_{18}$ (150 × 4.6mm, 3.5μm), CH$_3$CN:H$_2$O:CH$_3$COOH (55:45:0.5) at 0.6mL/min	FLD	λ_{exc} 333 nm λ_{em} 460 nm	89.8% (skimmed milk) 71.1% (whole milk)	0.5	[30]
OTA	Shaken with CH$_3$OH and filter a nylon acrodisk (0.45μm)	—	LC	Phenomenex C$_{18}$ (150 × 4.6mm, 5μm), CH$_3$CN:H$_2$O:CH$_3$COOH (50:49:1) at 0.4mL/min	FLD	λ_{exc} 334 nm λ_{em} 464 nm	93	10	[31]

(continued)

Table 32.2 (continued)   Analysis of Mycotoxins in Milk

Mycotoxin	Extraction Technique	Cleanup Technique	Separation	Detection		Recovery (%)	LD (ng/L)	Reference
Slaframine	C_{18} cartridge. After passing the sample, cleaned with water and eluted with CH_3OH:H_2O (75:25). Re-extracted with H_2O NaCl + 10% Na_2CO_3 (pH 10) + CH_2Cl_2	—	LC Hamilton PRP-1 (250 × 4.1 mm, 10 μm), CH_3CN/20 mM NaH_2BO_3 containing 10 mM triethylamine (35:65) at 1 mL/min	FLD	λ_{exc} 365 nm λ_{em} 400 nm	91	n.r.	[21]

See abbreviations of mycotoxins in the text.

n.r., not reported; FLD, fluorescence detection; CE, capillary electrophoresis; DAD, diode array detector; LD, limit of detection; LOQ, limit of quantification; LC, liquid chromatography; GC, gas chromatography.

[a] μg/kg.

Table 32.3 Analysis of Mycotoxins in Dairy Products

Mycotoxin	Sample	Extraction Technique	Cleanup Technique	Separation	Detection	Recovery (%)	LD (ng/L)	Reference
AFM1	Cheese	Centrifugation with CH_2Cl_2 and celite, filter (Whatman No. 1), redissolved in CH_3OH, H_2O, and n-hexane (30:50:20), and the aqueous phase separated for cleanup technique	IAC (Aflaprep) eluted with CH_3CN	LC — Column TSK-GEL C_{18} (250 × 4.6 mm, 5 μm), CH_3CN: CH_3OH:H_2O (20:20:60) at 1 mL/min	FLD — λ_{exc} 360 nm λ_{em} 440 nm	75.2–88.4	n.r.	[32]
AFM1	Cheese	Blended with CH_2Cl_2 and diatomaceous earth, evaporated and redissolved with CH_3OH, H_2O, and n-hexane and, the aqueous phase separated for cleanup technique	IAC (AflaPrep) eluted with CH_3OH	LC — C_{18} (250 × 4.6 mm), CH_3OH/CH_3CN/H_2O (20:20:60) at 0.8 mL/min	FLD — λ_{exc} 366 nm λ_{em} 440 nm	98.5	56	[28]
AFM1	Yoghurt	Extraction with CH_2Cl_2, evaporated and redissolved in CH_3OH and water. n-Hexane is used to eliminate fat	IAC (Aflaprep) eluted with CH_3CN	LC — LiChrospher 100 RP-18 (250 × 4.6 mm, 5 μm), CH_3CN/H_2O (25:75) at 0.8 mL/min	FLD — λ_{exc} 360 nm λ_{em} 435 nm	88–99	10 ng/kg	[33]
AFM1	Curd	Shaken CH_3Cl, celite, and saturated NaCl and removed the organic layer	IAC (AflaM1) eluted with CH_3OH	LC — Zorbax SB C_{18} (150 × 4.6 mm, 3 μm), CH_3CN/H_2O/CH_3COOH (25:75:1) at 1 mL/min	FLD — λ_{exc} 365 nm λ_{em} 435 nm	n.r.	n.r.	[25]

See abbreviations of mycotoxins in the text.

n.r., not reported; FLD, fluorescence detection; LC, liquid chromatography; IAC, immunoaffinity column; LD, limit of detection.

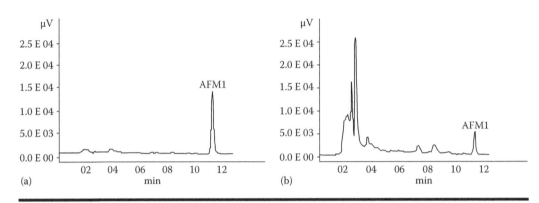

Figure 32.3 LC-FLD chromatograms of (a) AFM1 standard solution at 0.004 µg/mL; (b) naturally contaminated pasteurized milk with an AFM1 concentration of 0.004 µg/L. (Reprinted from Zinedine, A. et al., *Int. J. Food Microbiol.*, 114, 25, 2007. With permission.)

n-hexane (20 µL) and TFA (5 µL) are added to the residue and the mixture is vortexed and allowed to stand at 40°C for 20 min. The mixture is again evaporated to dryness under a gentle stream of nitrogen, reconstituted in 50 µL of mobile phase, and reinjected into the LC system. For OTA, the method used is based on the methyl-ester formation according to Zimmerli and Dick [65]. The procedure consists of adding 2.5 mL of methanol and 0.1 mL of concentrated hydrochloric acid to 200 µL of OTA residue. The vial is closed and kept overnight at room temperature, and the reaction mixture is evaporated to dryness and the residue is redissolved in mobile phase. The LC with MSD has been introduced for the analysis of mycotoxins in dairy and dairy products, but the number of articles published is limited when compared with the other detectors. Figure 32.4 demonstrates a chromatogram for AFM1 analyzed by LC–MS.

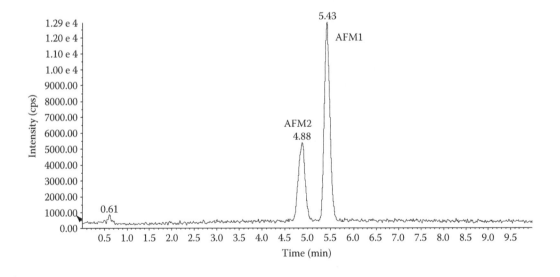

Figure 32.4 LC/MS/MS chromatogram of an ewe-milk AFM1 contaminated sample. (Reprinted from Bognanno, M. et al., *Mol. Nutr. Food Res.*, 50, 300, 2006. With permission.)

32.2 Toxins in Dairy Food

32.2.1 Introduction

Food poisoning owing to bacterial toxins can be caused by the ingestion of exotoxins that are formed in the food, or by the ingestion of food containing large numbers of bacterial cells that subsequently release endotoxins in the gastrointestinal tract. Early methods for the assay of bacteria toxins were based on in vivo or in vitro tests. Later, immunological test were developed based on techniques such as gel diffusion, but these tests were laborious and difficult to apply to foods. Now, a number of rapid test kits are available which give results within hours and are much simpler to perform and interpret, than bioassays.

32.2.1.1 Principal Bacterial Toxins in Dairy Products

The high level of nutrients in milk makes it an especially suitable growth medium for various bacteria, including those belonging to the families of *Enterobacteriaceae*, *Streptococcaceae*, and *Bacillaceae*. In fact, these microorganisms can reach high population densities following contamination during milk processing in dairy farms and dairy industry [66]. In particular, bacterial toxins are an important cause of a variety of human and animal diseases.

The most important bacterial toxins involved in outbreaks or food poisonings from dairy products are as follows.

32.2.1.1.1 Staphylococcus aureus

Staphylococcus aureus is considered as the third most important cause of diseases in the world among the reported foodborne illnesses [67–69]. This food poisoning is caused by consuming foods containing the enterotoxins produced by the strains of *S. aureus*.

Staphylococci can multiply rapidly in many foods, but milk is a good substrate for *S. aureus* growth, and milk and milk products have been the source of many staphylococcal food poisonings [70,71]. *S. aureus* can gain access to milk either by direct excretion from udders with clinical or subclinical staphylococcal mastitis, or by the contamination from the environment during handling and processing of raw milk [72,73]. Although pasteurization kills *S. aureus* cells, the thermostable staphylococcal enterotoxins (SEs) remain and generally retain their biological activity [74].

Traditionally, classic SE-type antigens have been recognized: SEA, SEB, SEC1, SEC2, SEC3, SED, and SEE [75]. During the 1990s, new SEs (SEG, SHE, SEI, and SEJ) were reported and their genes described [76–78]. More recent data resulting from partial or complete genome sequence analyses have led to the description of further "new" genes: *sek*, *sel*, *sem*, *sen*, *seo*, *sep*, *seq*, *ser*, and *seu* [79–82].

Detection of SEs in implicated foods is essential to confirm staphylococcal food poisoning, but the detection of *S. aureus* and SEs in food is often difficult [83]. In an outbreak of gastroenteritis owing to chocolate milk, Evenson et al. [84] determined that ingestion of 100–200 ng of enterotoxins can induce symptoms of food poisoning, and foods implicated with staphylococcal food poisoning typically contain about 0.5–10 μg of toxin per 100 g of food. Therefore, the sensitivity of any detection method needs to be below this level.

32.2.1.1.2 Shiga Toxin-Producing Escherichia coli

Escherichia coli is a genetically heterogeneous group of bacteria whose members are typically nonpathogens that are a part of the normal microflora of the intestinal tract of humans and animals

[85]. However, certain subsets of this bacteria cause enteric diseases. One of these subsets called Shiga toxin-producing *E. coli* (STEC) includes strains of *E. coli* that produce at least two potent phage-encoded cytotoxins called Shiga toxins (Stxs) [86]. The STEC are also called verotoxin (VT)-producing *E. coli*. The names Stx, derived from the similarity to a cytotoxin produced by *Shigella dysenteriae* serotype 1 [87], and VT, based on the cytotoxicity for Vero cells [88] are used interchangeably. In fact, STEC has emerged as an important global health threat and is recognized as an important pathogen of human diarrhea capable of causing life-threatening conditions, like hemolytic-uremic syndrome (HUS) [89].

Fecal contamination during the milking process, along with poor hygienic practices is known to account for the presence of STEC in raw milk [90]. The possibility of transmission through the consumption of raw milk [91] as well as raw-milk dairy products, such as cheese [92–94] and yoghurt [95], has been repeatedly documented as responsible for the outbreaks and sporadic cases of illnesses.

32.2.1.1.3 *Bacillus cereus*

Bacillus cereus is a ubiquitous spore-forming bacterium that is a common cause of food poisoning. Moreover, *B. cereus* is associated with spoilage problems in the dairy industry [96,97], including defects such as off-flavors, sweet curdling, and bitty cream [98]. In addition, *B. cereus* has been also associated with the outbreaks of food poisoning. This microorganism is responsible for 1% and 25% of food-poisoning outbreaks worldwide of known etiology. However, surprisingly, only few reports of food poisoning caused by *B. cereus* from milk and cream have been reported.

Their main contamination route to milk is via teats contaminated by soil and feces or bedding material, and to some extent via feed [99]. Milking equipment can also be a contamination source; silos tanks, pasteurizers, and packing machines may lead to further contamination of the milk and milk-containing products [100–105].

Bacillus cereus can cause two types of food-poisoning diseases: the diarrheal and emetic syndromes. The emetic syndrome is caused by only one heat-stable toxin (cereulide) that is formed in food [104]. The diarrheal syndrome is caused by several different heat-labile toxins formed by the vegetative bacteria [105]. Toxin production is strongly dependent on the culture medium and bacterial growth conditions, and the ability of toxin produced by *B. cereus* in milk under different dairy processing and storage conditions is not known.

The identification and widespread incidence of toxin-producing strains from a variety of food, including pasteurized dairy products led to renewed interest in methods for the detection of *B. cereus* toxins. The presence of diarrheal strains in milk is well known and there are several polymerase chain reaction (PCR) methods and immunological kits available for the detection of these strains [106,107]. However, the study of emetic strains in food chain has been hampered by the lack of suitable detection methods. In fact, at present, there are no commercially available rapid test kits for the detection of the emetic toxin, owing to difficulties in purification and characterization of the toxin [108].

32.2.2 Analysis of Bacterial Toxins in Dairy Foods

32.2.2.1 Biological Assays

Although modern assay methods are a rapid and convenient means of testing for bacterial toxins, they do not provide information on the biological activity of the toxin. Rasooly et al. [109] studied the in-vitro T-cell proliferation of human and rat lymphocytes in response to the concentrations of

SEA of *S. aureus*. They demonstrated that the T-cell response to SEA correlated well with increasing amounts of this toxin in the studied food matrix, with the exception of milk. In milk, proliferation at 10 ng/well was lower than that for the sample with 1 ng/well, but the difference between the two values was not statistically significant. This correlation may not be feasible because, the presence of milk in the SEA sample increased the efficiency of SEA heat inactivation. This suggests that the presence of SEA, even at low concentration levels, produces the T-cell proliferation, and this fact could be employed for the detection of this toxin in milk.

In the case of STEC, a range of in vivo and in vitro tests have been described including ileal loop [110], rabbit skin, Chinese hamster ovary (CHO) [111], and suckling mouse assay [112], but they were not suitable for routine use. Simpler alternatives to the biological assays based on immunological techniques and nucleic-acid hybridization were needed [108].

Owing to the difficulties in the determination of *B. cereus* emetic toxin, early biological assay methods for screening for *B. cereus* emetic toxin involved feeding to rhesus monkeys. However, by European legislation, effective from June 1, 2007, whole animals are not allowed for food testing. Consequently, in-vitro assays were used for toxin detection in food. Cereulide is observed to cause vacuolation of the mitochondria in HEp-2 [113]. Another biological technique to determine this emetic toxin is the sperm-based bioassay, which is based on the loss of mobility of boar sperm cells upon exposure to the emetic toxin [114]. Biological assays for the detection of *B. cereus* diarrheal toxin include the rabbit ileal loop test and the vascular permeability reaction test [108].

These methods are laborious and very expensive, and rather difficult to perform in food products on a routine basis, and thus, may not be easily accessible to food industry laboratories [115].

32.2.2.2 Immunological Tests

Immunological assays are much simpler and cheaper than biological assays, and have therefore been widely adopted [108]. The currently available methods for bacterial toxins detection are based on microbiological cultures of milk and milk products, and have been developed according to three methods: ELISA, enzyme-linked immunofiltration assay (ELIFA), and reversed passive latex agglutination (RPLA) [116].

ELISA is probably the most widely used immunoassay. In ELISA, the target antigen is captured by incubating the test sample in specific antibody-coated wells. The bound antigen is detected by reacting with another enzyme-specific substrate to form a colored or fluorescent product. The amount of label present at completion of the assay (and color) is directly proportional to the target analyte concentration. A rapid alternative to ELISA is the ELIFA, in which the filtration of the test sample through a high-affinity membrane accelerates the reaction between the analyte and the ligand immobilized on the membrane, reducing the total assay time to 1 h [108].

The RPLA can be used to detect soluble antigens in food extracts or culture filtrates by a simple latex agglutination assay. The antibody is attached to the latex particles and allowed to react with the soluble antigen. If the antigen is present in the sample, agglutination occurs owing to the formation of molecular lattice and a diffuse layer is formed at the base of the well. The assay is simple and rapid to perform, but it is relatively expensive and gives only semiquantitative results. For example, the SET-RPLA kit (Oxoid) is one of the most widely used commercial kits for SEs. This kit is a latex-based immunological test, in which visible cross-linking of antibody-coated latex particles occurs in the presence of SEs, allowing simultaneous detection of SEs A, B, C, and D in food extracts and culture filtrates. The initial studies showed that nonspecific reactions were obtained when analyzing cheese, making the kit unsuitable for the analysis of dairy products [117,118]. However, Rose et al. [119] applied this test to a variety of dairy products, demonstrating that the nonspecific reactions could be reduced by the addition of 10 nmol/L of hexametaphosphate to the diluent, without affecting the ability to detect SEs in these products. The sensitivity of the SET-RPLA was demonstrated to be 0.25 ng/mL.

Several commercial immunoassays exist for the detection of STEC in pure cultures of *E. coli*, although there is limited literature based on this. As it will be described later, almost all the studies have been carried out by PCR.

However, there are several disadvantages associated with microbiological cultures, such as time consumption, cost, and detection limits higher than the level required for bacterial intoxications [116].

Two commercial kits also exist for the rapid detection of *B. cereus* diarrheal toxin in food and cultures. The BCET-RPLA kit was evaluated by Granum [106] in dairy products and compared with the results of Western immunoblot and vascular permeability reaction. They concluded that the BCET-RPLA is a very simple and reliable method for the detection of *B. cereus* diarrheal toxin. However, other authors obtained results referring that the immunological activity measured in BCET-RPLA does not correlate with the biological activity [119]. For example, Day et al. [120] detected the enterotoxin in the culture supernatants of 13 strains of *B. cereus* using the other commercial kit, TECRA-kit, but only 6 strains were detected with the BCET-RPLA. One of the seven strains that were negative in the BCET-RPLA had previously been shown to produce diarrheal toxin in monkey feeding test, and four of the other six had been implicated with food-poisoning outbreaks [121,122].

32.2.2.3 Phenotypic Assays

PCR is a highly specific and sensitive method for amplifying nucleic-acid sequences exponentially. PCR assays for the detection of the toxin-encoding genes in bacteria have been developed, but none have yet been commercialized [123].

This technique has been often experimented in milk and cheese for the direct detection of *S. aureus* [124,125] and has been introduced as a simple technique for the detection of enterotoxigenic strains [67,126]. Although the PCR-based approach is specific, highly sensitive, and rapid, it can only demonstrate the presence of enterotoxin genes in *S. aureus* isolates rather than the production of the SEs protein [125].

A number of nucleic acid-based assays have been described for the detection of STEC. In fact, PCR is the method of choice to determine this type of bacterial toxins. Vivegnis et al. [127] achieved the growth of the bacteria on McConkey agar from raw milk cheese and used PCR to detect Stx genes. For each PCR-positive sample, isolated colonies were subsequently identified through a biochemical test (API 20) and a complementary indole production test. These authors concluded that the production of Stx was not sufficient to cause the disease, as other factors are thought to contribute to the virulence of this bacteria; Stx genes were detected in 17 cheese samples, but the toxin-producing strains could be isolated only from 5 of them. This low isolation level can probably be related to the loss of Stx genes in-vitro by some STEC strains or to the unfavorable proportion of STEC versus other *E. coli* strains. To overcome this difficulty, colony blot or DNA/DNA hybridization assay can be used to detect and isolate STEC [128]. Similarly, Rey et al. [129] and Caro et al. [130] determined the occurrence of STEC in different Spanish dairy products. In the first study [129], a total of 502 dairy products were examined for STEC using genotypic (PCR) methods. As in the previous work, the prevalence of STEC in milk was low, and the authors hypothesized that this circumstance would be related to the fact that milk carries a number of immune factors (principally IgA) and nonimmune factors (e.g., lactoferrin) that specifically hinder the adherence and subsequent proliferation of STEC on certain cell substrates. By contrast, serotype O157:H7 showed a high prevalence owing to its resistance and survival in refrigerated milk tanks, and resistance to acid pH and high NaCl concentrations.

In the second study [130], a total of 83 raw-milk cheese samples were examined for virulence genes using PCR. The obtained results and conclusions were similar to those obtained by the above-mentioned authors.

32.2.2.4 Biologic–Immunologic–Phenotypic Combination Studies

Sometimes, the use of a combination of techniques is needed. The PCR demonstrates the presence of genes capable to produce the toxin, and the immunological methods indicate the serotype of the toxin. Hence, coupling of these two techniques can give the global information of a possible toxin outbreak.

To demonstrate the capability of the strain to produce an amount of SE protein sufficient to induce a disease, the bioassay or immunological methods for the detection of SEs protein must be used [116]. Some studies on this topic are reported in the literature. Morandi et al. [131] compared the results obtained by PCR and SET-RPLA, and concluded that the PCR technique revealed a higher number of potential enterotoxin-producing strains. Indeed, in a high percentage of isolates where classical SEs production was identified by SET-RPLA (A, B, C, and D), the presence of other strains (*g*, *j*, *i*, *h*, and *l*) was confirmed by PCR technique. Jørgensen et al. [72] isolated samples of bovine and caprine bulk milk and raw milk products, and tested these isolated samples for SEs production by SET-RPLA, and for SE genes by PCR. They concluded that the most commonly toxin detected in these products was SEC and *sec*.

In the same way, Normanno et al. [68] evaluated the occurrence of *S. aureus*, characterized the isolated strains based on their production of SEs and antimicrobial-resistance pattern, and biotyped the isolated strains from milk, dairy, and meat products. For these purposes, they used the SET-RPLA to detect the enterotoxin production (SEA to SED) and a PCR to screen from *sea* to *sed* genes. Loncarevic et al. [132] used SET-RPLA to test the SEA to SED of *S. aureus* from raw milk and raw milk products, and PCR for the identification and characterization of the same isolates as tested with SET-RPLA.

Das et al. [133] determined the distribution, virulence-gene profile, and phenotypes of STEC strains within a dairy farm in India. The milk samples were inoculated into the EC medium, and after incubation, each enriched culture was directly tested by multiplex PCR. The colony that yielded a positive result was further confirmed for the presence of Stx by Stx-PCR. For a highly sensitive Stx detection, Bead-ELISA and Vero cells assay for determining cytotoxic effects were performed. As reported in other studies, the isolation rate of STEC from the PCR-positive samples was low (only two from the fresh milk). With regard to the virulence-gene profiles, most of the strains harbored only Stx1. Moreover, of the 30 strains examined, 27 were found to be cytotoxic to Vero cells. Out of these 27 strains, only 5 showed positivity for Stx in Bead-ELISA.

Borge et al. [134] investigated the growth, sporulation, and germination of a selection of toxin-producing *B. cereus* strains, isolated from dairy and meat products using PCR analysis of coding regions of enterotoxin genes, and evaluated the cytotoxicity to Vero cells. In the same way, Beattie and Williams [135] studied the factors that affect toxin formation by *B. cereus* in the fermentation process of dairy products. For this purpose, diarrheal enterotoxin was detected using CHO cells. The presence of the toxin in the culture's supernatant fluids could be detected by measurement of the total metabolic activity of the CHO cells. Enterotoxin was also determined using commercially available BCET-RPLA and TECRA immunoassays kits. These authors concluded that the immunological activity measured in BCET-RPLA did not correlate with the biological activity.

Furthermore, Arnesen et al. [136] employed the PCR technique to determine the genes encoding the enterotoxins of *B. cereus* in different dairies, and the cell cultures assays to measure the cytotoxicity toward Vero cells, to discriminate pathogenic *B. cereus* group strains from the nonpathogenic ones. Similarly, Te Giffel et al. [137] carried out an investigation to determine the level of *B. cereus* in pasteurized milk by sampling and testing the milk stored in household refrigerators. Using immunoblotting, Vero cells assays, and PCR, the samples were examined and were found to produce the toxin.

On the other hand, some authors combined PCR, to prove the presence of toxin genes, and immunoassays, to show the production of the toxin. For example, Svensson et al. [138], to characterize the hazard posed by *B. cereus* in the milk-production chain, tested the *B. cereus* group from farms, silo tanks, and production lines for pasteurized milk for toxin-production potential, using PCR to detect the presence of toxin genes. The toxin production was measured with the two commercial kits, TECRA and BCET-RPLA.

The study of Svensson et al. [139] went one step further in the analysis of *B. cereus* emetic toxin in dairy products, in which the phenotypic methods, RAPD-PCR, as well as the sperm test were applied to determine the cytotoxicity. The quantitative analysis for cereulide was carried out by liquid chromatography–ion trap-mass spectrometry (LC–IT-MS) at dairy farms and dairy plants.

32.3 Future Trends

Although immunoassay-based methods are sensitive and widely used for measuring protein toxins in food matrices, there is a need for the methods that can directly confirm the molecular identity of the toxin in situations where immunoassay tests yield a positive result.

The applications of the techniques, such as HPLC or GC coupled with MS have been scarcely applied to identify and characterize bacterial toxins and mycotoxins. The complicated food matrix, high cost of the equipments, low sample throughput, and amount of work involved, along with the length of time taken to achieve a result and the level of experience needed for the analysis, preclude the use of these techniques as routine procedures [121].

A method using HPLC–MS/MS has been developed to identify SEB in apple juice. The approach employs ultrafiltration to remove low-molecular weight components from the sample, after which, the remaining high-molecular weight fraction, containing the protein, is digested with trypsin. The authors indicated that this analysis cannot be applied to a large number of foods. The results showed that it was probably generally applicable to food matrices with low concentrations of soluble proteins, but there were difficulties with high-protein matrices such as milk. Measurement of SEB in milk using this approach is limited at present to ppm levels, principally owing to the suppression by the large number of peptides produced upon digestion of the milk proteins. The authors proposed two possibilities: (i) more selective sample extraction approaches (immunomagnetic method using antibodies for SEB, followed by extraction/digestion, or (ii) direct digestion of the antibody beads and reduction of the suppression of the target analyte signals by milk peptides through the use of multidimensional separations (ion exchange combined with reversed-phase LC), to fractionate the sample further prior to MS analysis [140].

References

1. Fink-Gremmels, F. Mycotoxins in cattle feeds and carry-over to dairy milk: A review. *Food Addit. Contam.* 25, 172, 2008.
2. Wu, F. Mycotoxin reduction in Bt corn: Potential economic, health, and regulatory impacts. *Transgenic Res.* 15, 277, 2006.

3. Nawaz, S., Scudamore, K.A., Rainbird, S.C. Mycotoxins in ingredients of animal feeding stuffs: I. Determination of *Alternaria* mycotoxins in oilseed rape meal and sunflower seed meal. *Food Addit. Contam.* 14, 249, 1997.

4. Scudamore, K.A., Nawaz, S., Hetmanski, M.T. Mycotoxins in ingredients of animal feeding stuffs: II. Determination of mycotoxins in maize and maize products. *Food Addit. Contam.* 15, 30, 1998.

5. Cheeke, P.R. Endogenous toxins and mycotoxins in forage grasses and their effects on livestock. *J. Animal Sci.* 73, 909, 1995.

6. O'Brien, M., O'Kiely, P., Forristal, P.D., Fuller, H.T. Fungi isolated from contaminated baled grass silage on farms in the Irish Midlands. *FEMS Microbiol. Lett.* 247, 131, 2005.

7. Mansfield, M.A., Kuldau, G.A. Microbiological and molecular determination of mycobiota in fresh and ensiled maize silage. *Mycologia.* 99, 269, 2007.

8. Kiessling, K., Pettersson, H., Sandholm, K., Olsen, M. Metabolism of aflatoxin, ochratoxin, zearalenone and three trichothecenes by intact rumen fluid, rumen protozoa and rumen bacteria. *Appl. Environ. Microbiol.* 47, 1070, 1984.

9. Prandini, A. et al. On the occurrence of aflatoxin M1 in milk and dairy products. *Food Chem. Toxicol.* 47, 984, 2009.

10. Fernández, M., Ferrer, E. Aflatoxin M1. In: *Food Mycotoxins*, Soriano, J.M., Ed. Díaz de Santos, Madrid, 2007, Chapter 9.

11. Henry, S.H. et al. Aflatoxin M1. In: *Safety Evaluation of Certain Mycotoxin in Food*, Joint FAO/WHO Expert Committee on Food Additives (JECFA), Ed. WHO, Geneva, 2001, Chapter 1.

12. Zinedine, A., González-Osnaya, L., Soriano, J.M., Moltó, J.C., Idrissi, L., Mañes, J. Presence of AFM1 in pasteurized milk from Morocco. *Int. J. Food Microbiol.* 114, 25, 2007.

13. Van Egmond, H.P. *Mycotoxins in Dairy Products.* Elsevier Science Publishing Co., Ltd., New York, 1989.

14. Veldman, A., Meijs, J.A.C., Borggreve, G.J., Heeresvan der Tol, J.J. Carry-over of aflatoxin from cows' food to milk. *Anim. Prod.* 55, 163, 1992.

15. Barbieri, G., Bergamini, C., Ori, E., Pesca, P. Aflatoxin M₁ in Parmesan cheese: HPLC determination. *J. Food Sci.* 54, 1313, 1994.

16. Boudra, H., Morgavi, D.P. Development and validation of a HPLC method for the quantification of ochratoxins in plasma and raw milk. *J. Chromatogr. B* 843, 295, 2006.

17. Seeling, K. et al. Effects of Fusarium toxin-contaminated wheat and feed intake level on the biotransformation and carry-over of deoxynivalenol in dairy cows. *Food Addit. Contam.* 23, 1008, 2006.

18. Prasongsidh, B.C., Kailasapathy, K., Skurray, G.R., Bryden, W.L. Stability of cyclopiazonic acid during storage and processing of milk. *Food Res. Int.* 30, 793, 1997.

19. Sorensen, L.K., Elbaek, T.H. Determination of mycotoxins in bovine milk by liquid chromatography tandem mass spectrometry. *J. Chromatogr. B*, 820, 183, 2005.

20. Durix, A. et al. Analysis of ergovaline in milk using high-performance liquid chromatography with fluorescence detection. *J. Chromatogr. B*, 729, 255, 1999.

21. Imerman, P.M., Stahr, H.M. New, sensitive high-performance liquid chromatography method for the determination of slaframine in plasma and milk. *J. Chromatogr. A*, 815, 141, 1998.

22. Decastelli, L. et al. Aflatoxins occurrence in milk and feed in Northern Italy during 2004–2005. *Food Control*, 18, 1263, 2007.

23. Diaz, G.J., Espitia, E. Occurrence of aflatoxin M1 in retail milk samples from Bogotá, Colombia. *Food Addit. Contam.* 23, 811, 2006.

24. Bognanno, M. et al. Survey of the occurrence of aflatoxin M1 in ovine milk by HPLC and its confirmation by MS. *Mol. Nutr. Food Res.* 50, 300, 2006.

25. Battacone, G. et al. Transfer of aflatoxin B1 from feed to milk and from milk to curd and whey in dairy sheep feed artificially contaminated concentrates. *J. Dairy Sci.* 88, 3063, 2005.

26. Grosso, F., Fremy, J.M., Bevis, S., Dragacci, S. Joint IDF-IUPAC-IAEA (FAO) interlaboratory validation for determining aflatoxin M1 in milk by using immunoaffinity clean-up before thin-layer chromatography. *Food Addit. Contam.* 21, 348, 2004.

27. Atanda, O. et al. Aflatoxin M1 contamination of milk and ice cream in Abeokuta and Odeda local governments of Ogun State, Nigeria. *Chemosphere*, 68, 1455, 2007.

28. Elgerbi, A.M., Aidoo, K.E., Candlish, A.A.G., Tester, R.F. Occurrence of aflatoxin M1 in randomly selected North African milk and cheese samples. *Food Addit. Contam.* 21, 592, 2004.

29. Tajkarimi, M. et al. Seasonal study of aflatoxin M1 contamination in milk in five regions in Iran. *Int. J. Food Microbiol.* 116, 346, 2007.

30. Bascarán, V., Hernández de Rojas, A., Chouciño, P., Delgado, T. Analysis of ochratoxin A in milk after direct immunoaffinity column clean-up by high-performance liquid chromatography with fluorescence detection. *J. Chromatogr. A*, 1167, 95, 2007.

31. González-Osnaya, L., Soriano, J.M., Moltó, J.C, Mañes, J. Simple liquid chromatography assay for analyzing ochratoxin A in bovine milk. *Int. J. Food Microbiol.* 108, 272, 2008.

32. Kamkar, A., Karim, G., Shojaee Aliabadi, F., Khaksar, R. Fate of aflatoxin M1 in Iranian white cheese processing. *Food Chem. Toxicol.* 46, 2236, 2008.

33. Martins, L.M., Marina M.H. Aflatoxin M1 in yoghurts in Portugal. *Int. J. Food Microbiol.* 91, 315, 2004.

34. Gilbert, J. Validation of analytical methods for determining mycotoxins in foodstuffs. *Trends Anal. Chem.* 21, 468, 2002.

35. Van Egmond, H.P., Wagstaffe, P.J. Development of milk powder reference materials certified for aflatoxin M_1 content (Part 1). *J. AOAC Int.* 70, 605, 1987.

36. Boenke, A. BCR- and M&T-activities in the area of mycotoxin analysis in food and feedstuffs. *Nat. Toxins* 3, 243, 2006.

37. Standard Measurement and Testing (SMT). Standard Measurement Testing Project SMT-CT96-2045. European Communities, Brussels, 1996.

38. Dragacci, S., Grosso, F., Gilbert, J. Immunoaffinity column cleanup with liquid chromatography for determination of aflatoxin M1 in liquid milk: Collaborative study. *J. AOAC Int.* 84, 437, 2001.

39. Grosso, F., Dragacci, S., Lombard, B. Second community proficiency testing for the determination of aflatoxin M1 in milk (November 1998). Report aflatoxin SSA/TOMI/FG 9901, Paris, May 1999.

40. European Communities (EC). Commission Regulation No. 1525/98. *Off. J. Eur. Com.* L201, 93, 1998.

41. European Communities (EC). Commission Decision 91/180/EEC. *Off. J. Eur. Com.* L93, 1, 1991.

42. Agboola, S.O., Dalgleish, D.G. Enzymatic hydrolysis of milk proteins used for emulsion formation. 1. Kinetics of protein breakdown and storage stability of the emulsions. *J. Agric. Food Chem.* 44, 3631, 1996.

43. Huppertz, T., Fox, P.F., Kelly, A.L. Degradation of ochratoxin A by a ruminant. *Appl. Environ. Microbiol.* 32, 443, 2004.

44. Fremy, J.M., Chu, F.S. A direct ELISA for determining aflatoxin M1 at ppt levels in various dairy products. *J. Assoc. Off. Anal. Chem.* 67, 1098, 1984.

45. Mortimer, D.N., Gilbert, J., Shepherd, M. J. Rapid and highly sensitive analysis of aflatoxin M1 in liquid and powdered milks using an affinity column cleanup. *J. Chromatogr.* 407, 393, 1987.

46. de Sylos, C.M., Rodriguez-Amaya, D.B., Carvalho, P.R. Occurrence of aflatoxin M_1 in milk and dairy products commercialized in Campinas, Brazil. *Food Addit. Contam.* 13, 169–172, 1996.

47. Paul, P., Thurm, V. Determination and occurrence of aflatoxin M1 in milk and dried milk products. *Nahrung.* 27, 877, 1983.

48. van Egmond, H.P., Stubblefield, R.D. Improved method for confirmation of identity of aflatoxins B1 and M1 in dairy products and animal tissue extracts. *J. Assoc. Off. Anal. Chem.* 64, 152, 1981.

49. International Organization for Standardization (ISO). Milk and milk powder-determination of aflatoxin M1 content-clean-up by immunoaffinity chromatography and determination by TLC. ISO/DIS 14674-IDF 190, 2005.

50. Saitanu, K. Incidence of aflatoxin M_1 in Thai milk products. *J. Food Prot.* 60, 1010, 1997.

51. Offiah, N., Adesiyun, A. Occurrence of aflatoxins in peanuts, milk, and animal feed in Trinidad. *J. Food Prot.* 70, 771, 2007.

52. Frémy, J.M., Chu, F.S. Immunochemical methods of analysis for aflatoxin M$_1$. In: *Mycotoxins in Dairy Products*, van Egmond, H.P, Ed., Elsevier Applied Science, London, 1989, pp. 97–125.

53. Barbiroli, A. et al. Binding of aflatoxin M1 to different protein fractions in ovine and caprine milk. *J. Dairy Sci.* 90, 352, 2007.

54. Rodríguez V.M.L., Calonge D.M.M., Ordóñez E.D. ELISA and HPLC determination of the occurrence of aflatoxin M1 in raw cow's milk. *Food Addit. Contam.* 20, 276, 2003.

55. Sassahara, M., Pontes Netto, D., Yanaka, E.K. Aflatoxin occurrence in foodstuff supplied to dairy cattle and aflatoxin M1 in raw milk in the North of Paraná state. *Food Chem. Toxicol.* 43, 981, 2005.

56. López, C. et al. Distribution of aflatoxin M1 in cheese obtained from milk artificially contaminated. *Int. J. Food Microbiol.* 64, 211, 2001.

57. Unusan, N. Occurrence of aflatoxin M1 in UHT milk in Turkey. *Food Chem. Toxicol.* 44, 1897, 2006.

58. Kim, E.K., Shon, D.H., Ryu, D., Park, J.W., Hwang, H.J., Kim, Y.B. Occurrence of aflatoxin M1 in Korean dairy products determined by ELISA and HPLC. *Food Addit. Contam.* 17, 59, 2000.

59. Benedetti, S., Iametti, S., Bonomi, F., Mannino, S. Head space sensor array for the detection of aflatoxin M1 in raw ewe's milk. *J. Food Prot.* 68, 1089, 2005.

60. Salter, R., Douglas, D., Tess, M., Markovsky, B., Saul, S.J. Interlaboratory study of the Charm ROSA Safe Level Aflatoxin M1 Quantitative lateral flow test for raw bovine milk. *J. AOAC Int.* 89, 1327, 2006.

61. Tuinstra, L.G., Roos, A.H., van Trijp, J.M. Liquid chromatographic determination of aflatoxin M1 in milk powder using immunoaffinity columns for cleanup: Interlaboratory study. *J. AOAC Int.* 76, 1248, 1993.

62. International Dairy Federation (IDF). Protocol IDF study aflatoxin M1. IDF 171, 1995.

63. International Organization for Standardization (ISO). Milk and milk powder. Determination of aflatoxin M1 content. Clean-up by immunoaffinity chromatography and determination by high-performance liquid chromatography. ISO 14501, 1998.

64. Takeda, N. Determination of aflatoxin M1 in milk by reversed phase high performance liquid chromatography. *J. Chromatogr.* 288, 484, 1984.

65. Zimmerli, B., Dick, R. Determination of ochratoxin A at the ppt level in human blood, serum, milk and some foodstuffs by high-performance liquid chromatography with enhanced fluorescence detection and immunoaffinity column cleanup: Methodology and Swiss data. *J. Chromatogr. B.* 666, 85, 1995.

66. Bartoszewicz, M., Hansen, B.M., Swiecicka, I. The members of the *Bacillus cereus* group are commonly present contaminants of fresh and heat-treated milk. *Food Microbiol.* 25, 588, 2008.

67. Asperger, H., Zangerl, P. *Staphylococcus aureus*. In: *Encyclopaedia of Dairy Sciences*, Vol. 4, Roginsky, H., Fuquay, J.W., Fox, P.F., Eds, Academic Press, London, 2003, p. 2563.

68. Normanno, G. et al. Occurrence, characterization and antimicrobial resistance of enterotoxigenic *Staphylococcus aureus* isolated from meat and dairy products. *Int. J. Food Microbiol.* 115, 290, 2007.

69. Boerema, J.A., Clemens, R., Brightwell, G. Evaluation of molecular methods to determine enterotoxigenic status and molecular genotype of bovine, ovine and human and food isolates of *Staphylococcus aureus*. *Int. J. Food Microbiol.* 107, 192, 2006.

70. Davis, J.G. Microbiology of cream and dairy desserts. In: *Dairy Microbiology*, Vol. 2, R.K. Robinson, Ed., The Microbiology of Milk Products, Applied Science Publisher, Barking, 1981, p. 157.

71. De Buyser, M.L. et al. Implication of milk and milk products in food-borne diseases in France and in different industrialised countries. *Int. J. Food Microbiol.* 67, 1, 2001.

72. Jørgensen, H.J., Mørk, T., Rørvik, L.M. The occurrence of *Staphylococcus aureus* on a farm with small-scale production of raw milk cheese. *J. Dairy Sci.* 88, 3810, 2005.

73. Scherrer, D. et al. Phenotypic and genotypic characteristics of *Staphylococcus aureus* isolates from raw bulk-tank milk samples of goats and sheep. *Vet. Microbiol.* 101, 101, 2004.

74. Jablonsky, L.M., Bohach, G.A. *Staphylococcus aureus*. In: *Food Microbiology Fundamentals and Frontiers*, Doyle, M.P., Beuchat, L.R., Montville, T.J., Eds. American Society for Microbiology, Washington, DC, 1997, p. 353.

75. Bergedoll, M.S. et al. The staphylococcal enterotoxins: Similarities. *Contrib. Microbiol. Immunol.* 1, 390, 1973.

76. Munson, S.H. et al. Identification and characterization of staphylococcal enterotoxins type I and G from *Staphylococcus aureus. Infect. Immun.* 66, 3337, 1998.

77. Su, Y.C., Wong, A.C. Identification and purification of a new staphylococcal enterotoxins. *Appl. Environ. Microbiol.* 61, 1438, 1995.

78. Zhang, S., Iandolo, J.J., Steward, G.C. The enterotoxin D plasmid of *Staphylococcus aureus* encodes a second enterotoxin determinant (*sej*). *FEMS Microbiol. Lett.* 168, 227, 1998.

79. Fitzgerald, J.R. et al. Characterization of a putative pathogenicity island from *Staphylococcus aureus encoding* multiple superantigens. *J. Bacteriol.* 183, 63, 2001.

80. Jarraud, S. et al. *egc,* a highly prevalent operon of enterotoxin gene, forms a putative nursery of superantigens in *Staphylococcus aureus. J. Immunol.* 166, 669, 2001.

81. Letetre, C. et al. Identification of a new putative enterotoxin SEU encoded by the *egc* cluster of *Staphylococcus aureus. J. Appl. Microbiol.* 95, 38, 2003.

82. Omoe, K. et al. Detection *seg, seh* and *sei* genes in *Staphylococcus aureus* isolates and determination of enterotoxins productivities of *S. aureus* isolates harbouring *seg, seh* and *sei* genes. *J. Clin. Microbiol.* 40, 857, 2002.

83. Meyrand, A. et al. Evaluation of an alternative extraction procedure for enterotoxin determination in dairy products. *Lett. Appl. Microbiol.* 28, 411, 1999.

84. Evenson, M.L. et al. Estimation of human dose of staphylococcal enterotoxin A from a large outbreak of staphylococcal food poisoning involving chocolate milk. *Int. J. Food Microbiol.* 7, 311, 1988.

85. Gyles, C.L. Shiga toxin-producing *Escherichia coli*: An overview. *J. Anim. Sci.* 85, 45–62, 2007.

86. Doyle, M.P., Padhye, N.V. *Escherichia coli.* In: *Foodborne Bacterial Pathogens,* Doyle, M.P., Ed., Marcel Dekker, New York, 1989, p. 236.

87. O'Bryen, A.D. et al. Production of *Shigella dysenteriae* type 1-like cytotoxin by *Escherichia coli. J. Infect. Dis.* 146, 763, 1982.

88. Konowalchuk, J., Speirs, J.I., Stavric, S., Vero response to a cytotoxin of *Escherichia coli. Infect. Immun.* 18, 775, 1977.

89. Paton, J.C., Paton, A.W. Pathogenesis and diagnosis of Shiga toxin producing *Escherichia coli* infections. *Clin. Microbiol. Rev.* 11, 450, 1998.

90. Griffin, P.M., Tauxe, R.V. The epidemiology of infections caused by *Escherichia coli* O157:H7, other entero-hemorrhagic *E. coli* and associated haemolytic uremic syndrome. *Epidemiol. Rev.* 13, 60, 1991.

91. Lahti, E. et al. Use of phenotyping and genotyping to verify transmission of *Escherichia coli* O157:H7 from dairy farms. *Eur. J. Clin. Microbiol.* 21, 189, 2002.

92. Karmali, M.A. Infection by verocytotoxin-producing *Escherichia coli. Clin. Microbiol. Rev.* 2, 15, 1989.

93. Deschenes et al. Cluster of cases of haemolytic uremic syndrome due to unpasteurised cheese. *Pediatr. Nephrol.* 10, 203, 1996.

94. Altekruse, S.F. et al. Cheese-associated outbreaks of human illness in the United States, 1973 to 1992; sanitary manufacturing practices protect consumers. *J. Food Prot.* 61, 1405, 1998.

95. Morgan D. et al. Verotoxin producing *Escherichia coli* O157 infections associated with the consumption of yoghurt. *Epidemiol. Infect.* 111, 181, 1993.

96. Anderson, A., Rönner, U., Granum, P.E. What problems does the food industry have with the spore-forming pathogens *Bacillus cereus* and *Clostridium perfringens? Int. J. Food Micobiol.* 28, 45, 1995.

97. Larsen, H.D., Jørgensen, K. Growth of *Bacillus cereus* in pasteurized milk. *Int. J. Food Microbiol.* 46, 173, 1999.

98. Meer, R.R. et al. Psychrotrophic *Bacillus* spp. in fluid milk products: A review. *J. Food Prot.* 54, 969, 1991.

99. Christiansson, A., Bertilsson, J., Svensson, S., *Bacillus cereus* spores in raw milk: Factors affecting the contamination of milk during the grazing period. *J. Dairy Sci.* 82, 305, 1999.

100. Svensson, B. et al. Characterisation of *Bacillus cereus* isolated from milk silo tanks at eight different dairy plants. *Int. Dairy J.* 14, 17, 2004.

101. Svensson, B. et al. Involvement of pasteurizer in the contamination of milk by *Bacillus cereus* in a commercial dairy plant. *J. Dairy Res.* 67, 455, 2000.

102. Eneroth, Å. et al. Contamination of pasteurized milk by *Bacillus cereus* in the filling machine. *J. Dairy Res.* 68, 189, 2001.

103. Lin, A. et al. Identification of contamination sources of *Bacillus cereus* in pasteurized milk. *J. Food Microbiol.* 43, 159, 1998.

104. Melling, J. Capel, B.J. Characteristics of *Bacillus cereus* emetic toxin. *FEMS Microbiol. Lett.* 4, 133, 1978.

105. Granum, P.E., Brynestad, S., Kramer, J.M. Analysis of enterotoxin production by *Bacillus cereus* from dairy products, food poisoning incidents and non-gastrointestinal infections. *Int. Dairy J.* 17, 269, 1993.

106. Granum, P.E. *Bacillus cereus.* In: *Food Microbiology: Fundamentals and Frontiers,* M.P. Doyle, Ed. AMS Press, Washington, DC, 2001, p. 373.

107. In't Veld, P.H. et al. Detection genes encoding for enterotoxins and determination of the production of enterotoxins by HBL blood plates and immunoassays of psychrotropic strains of *Bacillus cereus* isolated from pasteurized milk. *Int. J. Food Microbiol.* 64, 63, 2001.

108. Pimbley, D.W., Patel, P.D. A review of analytical methods for the detection of bacterial toxins. *J. Appl. Microbiol. (Symp. Suppl.)* 84, 98, 1998.

109. Rasooly, L. et al. In vitro assay of *Staphylococcus aureus* Enterotoxin A activity in food. *Appl. Environ. Microbiol.* 63, 2361, 1997.

110. Burges, M.N. et al. Biological evaluation of a methanol soluble, heat stable *Escherichia coli* enterotoxin in the rabbit ileal loop. *Infect. Immun.* 21, 526, 1978.

111. Guerrant, R.L. et al. Cyclic adenosine monophosphate and alteration of Chinese hamster ovary cell morphology: A rapid, sensitive *in-vitro* assay for the enterotoxins of *Vibrio cholerae* and *Escherichia coli. Infect. Immun.* 10, 32, 1974.

112. Lovett, J., Peeler, J.T. Detection of *Escherichia coli* enterotoxins by mouse adrenal cells and suckling mouse assays: Collaborative study. *J. Assoc. Off. Anal. Chem.* 67, 946, 1984.

113. Ehling-Schulz, M., Fricker, M., Scherer S. Identification of emetic toxin producing *Bacillus cereus* strains by a novel molecular assay. *FEMS Microbiol. Lett.* 232, 189, 2004.

114. Anderson, M.A. et al. Sperm bioassay for rapid detection of cereulide-producing *Bacilllus cereus* in food and related environments. *Int. J. Food Microbiol.* 94, 175, 2004.

115. Ehling-Schulz, M., Fricker, M., Scherer S. *Bacillus cereus,* the causative agent of an emetic type of food-borne illness. *Mol. Nutr. Food Res.* 48, 479, 2004.

116. Cremonesi, P. et al. Detection of enterotoxigenic *Staphylococcus aureus* isolates in raw milk cheese. *Lett. Appl. Microbiol.* 45, 586, 2007.

117. Park, C.E., Szabo, R. Evaluation of the reserved passive latex agglutination (RPLA) test kit for detection of staphylococcal enterotoxins A, B, C and D in food. *Can. J. Microbiol.* 32, 723, 1986.

118. Wieneke, A.A., Gilbert, R.J. Comparison of four methods for detection of staphylococcal enterotoxins in foods from outbreaks of food poisoning. *Int. J. Food Microbiol.* 4, 135, 1987.

119. Rose, S.A., Bankes, P., Stringer, M.F. Detection of staphylococcal enterotoxins in dairy products by the reversed passive latex agglutination kit. *Int. J. Food Microbiol.* 8, 65, 1989.

120. Day, T.L. et al. A comparison of ELISA and RPLA for detection of *Bacillus cereus* diarrhoeal enterotoxins. *J. Appl. Bacteriol.* 77, 9, 1994.

121. Brett, M.M. Kits for detection of some bacterial food poisoning toxins: Problems, pitfalls and benefits. *J. Appl. Microbiol. (Symp. Suppl.)* 84, 110, 1998.

122. McKillip, J.L. Prevalence and expression of enterotoxins in *Bacillus cereus* and other *Bacillus* spp., a literature review. *A. Van Leeuw.* 77, 393, 2000.

123. Hill, W.E., Olsvik, O. Detection and identification of foodborne pathogens by the polymerase chain reaction: Food safety applications. In: *Rapid Analysis Techniques in Food Microbiology,* Patel, P.D., Ed. Academic Press, London, 1994, p. 268.

124. Kim, C.H. et al. Optimisation of the PCR for detection of *Staphylococcus aureus nuc* gene in bovine milk. *J. Dairy Sci.* 84, 74, 2001.

125. Tamarapu, S., McKillip, J.L., Drake, M. Development of a multiplex polymerase chain reaction assay for detection and differentiation of *Staphylococcus aureus* in dairy products. *J. Food Protect.* 64, 664, 2001.

126. Know, N.H. et al. Application of extended single-reaction multiplex polymerase chain reaction for toxin typing of *Staphylococcus aureus* isolates in South Korea. *Int. J. Food Microbiol.* 97, 137, 2004.

127. Vivegnis, J. et al. Detection of Shiga-like toxin producing *Escherichia coli* from raw milk cheese produced in Wallonia. *Biotechnol. Agron. Soc. Environ.* 3, 159, 1999.

128. Padhyle, N.V., Doyle, P.M. *Escherichia coli* O157:H7: Epidemiology, pathogenesis, and methods for detection in foods. *J. Food Protect.* 55, 555, 1992.

129. Rey, J. et al. Prevalence, serotypes and virulence genes of Shiga toxin-producing *Escherichia coli* isolated from ovine and caprine milk and other dairy products in Spain. *Int. J. Food Microbiol.* 107, 212, 2006.

130. Caro, I., Garcia-Armesto, M.R. Occurrence of Shiga toxin-producing *Escherichia coli* in a Spanich raw ewe's milk cheese. *Int. J. Food Microbiol.* 116, 410, 2007.

131. Morandi, S. et al. Detection of classical enterotoxins and identification of enterotoxins genes in *Staphylococcus aureus* from milk and dairy products. *Vet. Microbiol.* 124, 66, 2007.

132. Loncarevic, S. et al. Diversity of *Staphylococcus aureus* enterotoxin types within single samples of raw milk and raw milk products. *J. Appl. Microbiol.* 98, 344, 2005.

133. Das, S.C. et al. Dairy farm investigation on Shiga toxin-producing *Escherichia coli* (STEC) in Kolkata, India with emphasis on molecular characterization. *Epidemiol. Infect.* 133, 617, 2005.

134. Borge, G.I. et al. Growth and toxin profiles of *Bacillus cereus* isolated from different food sources. *Int. J. Food Microbiol.* 69, 237, 2001.

135. Beattie, S.H., Williams, A.G. Growth and diarrhoeagenic enterotoxin formation by strains of *Bacillus cereus* in vitro in controlled fermentation and in situ in food products and a model food system. *Food Microbiol.* 19, 329, 2002.

136. Arnesen, L.P.S., O'Sullivan, K., Granum, P.E. Food poisoning potential of *Bacillus cereus* strains from Norwegian dairies. *Int. J. Food Microbiol.* 116, 292, 2007.

137. Te Giffel, M.C. et al. Isolation and characterization of *Bacillus cereus* from pasteurized milk in household refrigerators in the Netherland. *Int. J. Food Microbiol.* 34, 307, 1997.

138. Svensson, B. et al. Toxin production potential and detection of toxin genes among strains of the *Bacillus cereus* group isolated along the dairy production chain. *Int. Dairy J.* 17, 1201, 2007.

139. Svensson, B. et al. Occurrence of emetic toxin producing *Bacillus cereus* in dairy production chain. *Int. Dairy J.* 16, 740, 2006.

140. Callahan, J.H. et al. Detection, confirmation, and quantification of staphylococcal enterotoxin B in food matrixes using liquid chromatography-mass spectrometry. *Anal. Chem.* 78, 1789, 2006.

Chapter 33

Detection of Adulterations: Addition of Foreign Lipids and Proteins

Saskia M. van Ruth, Maria G. E. G. Bremer, and Rob Frankhuizen

Contents

33.1 General Introduction 852
    33.1.1 Milk and Adulteration 852
    33.1.2 Milk Products and Adulteration 852
    33.1.3 Lipids, Proteins, and Authentication Testing 852
33.2 Lipids 853
    33.2.1 Introduction 853
    33.2.2 Authentication Testing 853
        33.2.2.1 Fatty Acid Analysis...853
        33.2.2.2 Triacylglycerol Analysis 855
33.3 Proteins 856
    33.3.1 Introduction 856
    33.3.2 Authentication Testing 856
        33.3.2.1 Immunoassays 856
        33.3.2.2 Mass Spectrometry 857
        33.3.2.3 Near-Infrared Spectroscopy 858
33.4 Conclusions..860
References 861

33.1 General Introduction

Product authenticity and authentication are emerging topics in the food sector. It is a major concern not only for consumers, but also for producers and distributors. Regulatory authorities, food processors, retailers, and consumer groups are all interested in ensuring that foods are correctly labeled. Food adulteration has been practiced forever, but has become more sophisticated in the recent past. Foods or ingredients, most likely to be targets for adulteration, include those which are of high value and which undergo a number of processing steps before they appear on the market. With the European harmonization of the agricultural policy and the emergence of the international markets, authentication of such food products requires more attention. This trend is the result of efforts made by regional authorities, as well as producers to protect and support local productions [1].

33.1.1 Milk and Adulteration

Milk is a biologically complex fluid, constituted mainly of water, proteins, lactose, fat, and inorganic compounds. The majority of these substances have important nutritional and technological properties. According to its solubility at pH 4.6 and 20°C, the protein fraction can be divided into caseins that are insoluble at this pH, and whey proteins that are soluble. Caseins are, quantitatively, the most important protein components. This protein complex, known as a micelle, comprises four different caseins (α_{s1}-, α_{s2}-, β-, and κ-caseins) that are held together by noncovalent interactions, and appear as a highly stabilized dispersion in milk [2].

Adulteration of sheep's milk with cow's milk is relatively common owing to seasonal fluctuations of the availability of sheep's milk, the higher price of sheep's milk than cow's milk, and the opportunity to use the overproduction of cow's milk without loss of profit [3]. Consumers allergic to cow's milk may suffer severely if they ingest, e.g., ovine or caprine milk fraudulently extended with bovine milk or whey.

33.1.2 Milk Products and Adulteration

The quality of milk plays a very important role in the production of all types of cheeses, affecting both cheese yield and characteristics of the cheese. In regions with high production costs, agriculture must produce food of superior quality. The products can be labeled according to the specific conditions that characterize their origin and/or the processing technology. Animal feeding is one of the elements that are often considered as important by cheese-makers. The relationships between the origin of cheeses and the type of pasture have been extensively highlighted [4].

An example of a susceptible cheese product from an adulteration perspective is Italian Mozzarella. The seasonal increase in the market demand for Italian Mozzarella cheese occurring every summer and, on the other hand, the limited productions of buffalo milk may induce fraudulent addition of bovine milk during the manufacture of Mozzarella [5].

33.1.3 Lipids, Proteins, and Authentication Testing

The fat of milk is often regarded as superior to other fats, because of its sensory properties. Therefore, its adulteration has always been a serious problem because of the economic advantages taken by partly replacing the high-priced milk fat with low-priced fats without labeling the product accordingly.

Milk proteins are probably the best characterized of all the food proteins. However, the existence of genetic and nongenetic polymorphism as well as the application of technological treatments complicate their quantitative determination. Modifications such as heat denaturation or proteolysis, common in the manufacture of many dairy products, give rise to complex, insoluble, new compounds, and smaller peptides and amino acids, and their analysis is not easy to perform. In addition, information on the occurrence and amount of a particular protein or derived compound is extremely useful in the assessment of processing and adulterations [6].

Virtually, all components present in the complex physicochemical system of milk contribute information that is valuable for authentication testing. Traditional analytical strategies to uncover adulteration and guarantee quality have relied on wet chemistry to determine the amount of a marker compound or compounds in a suspected material and the subsequent univariate comparison of the value(s) obtained with those established for equivalent material of known provenance [7]. This approach suffers from a number of disadvantages, namely, the ever-increasing range of analytes that must be included in any test procedure and the limited knowledge of the range of each constituent in normal lots of the substance. Accordingly, these ranges may be expected to vary with the breed, feed, season, geographic source, dairy products processing procedure, etc. It is often not possible to make a definitive statement on the authenticity or otherwise of a material, even after its examination for a large suite of single marker compounds. Hence, there is a continuing demand for new, rapid, nondestructive, cost-efficient methods for direct quality measurements in food and food ingredients. Spectroscopic techniques, including the near-infrared (NIR), mid-infrared (MIR), front face fluorescence spectroscopy (FFFS), stable isotope, and nuclear magnetic resonance (NMR) have been examined to assess their suitability for the determination of the quality and/or geographical origins of dairy products [8]. This chapter presents a brief overview of the techniques for the detection of foreign lipids and proteins in milk and milk products. It includes some classical techniques, as well as some of the reported approaches adopted for the determination of the identity and quality of fats and proteins with application of a chemometric strategy.

33.2 Lipids

33.2.1 Introduction

Adulteration of butters has a history reaching back to ancient times. As early as in 1877, the bureau of the Leipzig Pharmaceutical Union, offered a prize of 800 marks for the discovery of a sure and practical method for the detection of adulteration of butter by other fatty substances [9]. In the following section, two widely applied approaches for the detection of milk fat adulteration are discussed. Alternatives to these techniques like differential scanning calorimetry, infrared (IR) spectroscopic techniques, proton-transfer reaction mass spectrometry (MS) were proposed, but have not yet found wide applications. These rapid techniques would be widely accepted, as they do not need lengthy sample preparations, have a high throughput, and are nondestructive testing methods.

33.2.2 Authentication Testing

33.2.2.1 Fatty Acid Analysis

The fat of milk from all sources contains short-chain fatty acids, presumably because these are more easily absorbed by the young animal. The fatty acid profile is characteristic for each oil and fat. It is influenced by several factors, such as breed, feeding, season, climate, geographical origin, and technological variables. The composition can be changed by refining and fat modifications like fractionation, hydrogenation, and interesterification. Furthermore, the fatty acid content has been modified over the years by using conventional methods of breeding [10]. Butyric acid (C4) is fairly the characteristic fatty acid of milk fat. Hence, methods for estimation of milk fat content have usually relied in some way on the amount of butyric acid (C4). This was initially by means of the Reichert value, where the water-soluble/steam-distillable acids are determined. Generally, the percentage of butyric acid present in milk fat is usually taken as about 3.6%. However, as the value can vary over a range (2.40%–4.22%, w/w [11]), this determination is of no use in finding the complete authenticity of the milk fat. Other traditional physicochemical methods to verify the authenticity of milk fat include the iodine value (a measure of the total unsaturation of a fat) or Polenske value (titrimetric determination of steam-volatile, but water-insoluble fatty acids). Unfortunately, these univariate, relatively simple methods are successful only in detecting massive adulteration of milk fat or even its substitution by another fat. James and Martin first determined the extended fatty acid composition of milk by packed column gas chromatography (GC) in 1956 [12]. The packed column GC was later replaced by capillary GC. The main problem concerning butter analysis is the reliable determination of the short-chain fatty acids (C4–C8), and this is problem is more pronounced when split injection is used.

Multiple fatty acid analysis of a fat is nowadays a relatively routine analytical procedure. After methylation of the fat using reaction with boron trifluoride/methanol, boron trichloride/methanol, methanolic hydrogen chloride solution, diazomethane, or, if free fatty acids are not present, alkaline catalysts such as sodium methoxide/methanol, the prepared methyl esters are analyzed by GC on a polar column. The high polarity of the column is required to completely separate the saturated and unsaturated fatty acids. Milk fat does have a very characteristic fatty acid composition, and contains about 15 major fatty acids and several hundred minor fatty acids [13]. One might think that this would mean that authentication would be relatively easy from just the fatty acid composition. However, the fatty acid composition is not just complicated, but is also very variable.

Fatty acid compositions can be compared with univariate purity criteria specified by the FAO/WHO Codex Committee on Fats and Oils [14]. The admixture of a certain amount of foreign fat with a high concentration of a particular fatty acid in its spectrum would shift the concerned fatty acid out of the range, which is normally encountered in the genuine fat or oil. To increase the sensitivity of the fatty acid approach for the purity testing of milk fat various ratios of different fatty acids have been proposed as authenticity criteria. Antonelli et al. [15] suggested the use of the fatty acids, butyric acid and enanthic acid, for butter authentication in concentrated butters (butter oils). By using a combination of four fatty acid ratios (C18:0/C8:0 < 7.63; C14:0/C18:0 > 1.02, (C6:0 + C8:0 + C10:0 + C12:0)/C18:0 > 0.95; C18:1/C18:0 < 2.34 for genuine milk fat), the detection of an addition of 10% beef suet to milk fat was possible. Furthermore, differentiation of milk fat from different species based on fatty acid profiling is also possible. The ratio of C14:1/C15:0 is 1.00 in cow's milk fat while it is 0.20 in sheep [16]. Instead of using a univariate approach, the information content of the total fatty acid can be more efficiently explored by multivariate data analysis [17]. The art of extracting chemically relevant information from the data produced in chemical experiments by means of statistical and mathematical tools is called chemometrics. It is an indirect approach to the study of the effects of multivariate factors and hidden patterns in complex data sets. Chemometrics is routinely used for: (a) exploring patterns of association in data, and (b) preparing and using multivariate classification models. A partial least square-discriminant analysis (PLS-DA) plot (Figure 33.1) of the first two dimensions of a four-component model,

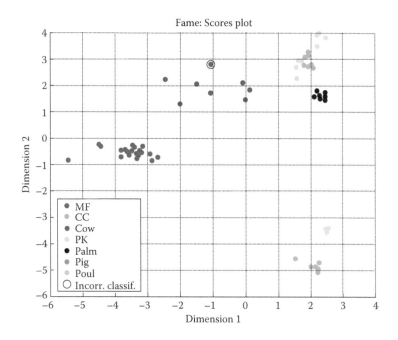

Figure 33.1 Scores plot of the first two dimensions of PLS-DA on the fatty acid composition data of milk fat (MF) and six other fats and oils.

predicting the identity of a variety of animal fats and vegetables oils (including milk fat) by their fatty acid compositions illustrates this approach. This multivariate approach adopted by the authors shows that the milk fats are fairly separated from the other fats and oils, with beef tallow most closely positioned in the proximity of the milk fats.

33.2.2.2 Triacylglycerol Analysis

When it was discovered that extreme variations of composition precluded the use of univariate fatty acid analysis for milk fat authentication, the possibility of analysis of whole triglycerides (TGs) was investigated. TGs are more difficult to separate and analyze satisfactorily by GC, owing to their high molecular weight and corresponding volatility. TG composition, as determined by measuring the carbon numbers of the TG fraction, is affected by many of the same factors as is fatty acid composition. Early attempts to detect certain foreign fats from TG analyses using regression procedures could theoretically detect the levels of 4%–7% of some oils, but could not handle mixtures [18]. Precht [19] subsequently developed a formula for detection of some adulterations in milk fat based on TG composition analysis. Pure milk was characterized by the presence of C40, C42, and C44 TGs. Again, the formula was limited to some potential adulterants and mixtures which increased the detection levels considerably. Furthermore, alternative computational models applied were equally effective [20].

Milk fat TGs can also be analyzed by other techniques. The application of high-performance liquid chromatography (HPLC) in normal and reversed-phase mode, thin-layer chromatography, and supercritical fluid chromatography have been reported. Detection systems include ultraviolet (UV), refractive index (RI), and evaporative light-scattering detector (ELSD). The major advantage of HPLC is that it is possible to separate TGs at ambient or slightly elevated temperatures, thereby obviating thermal stress on thermolabile long-chain polyunsaturated TGs. However, high-molecular mass TGs are not easy to elute from HPLC columns owing to their insolubility in a number of popular mobile phases. A further disadvantage is that commonly available HPLC detectors are only compatible with isocratic elution, or the detector response is influenced by the unsaturation of the separated substances which renders quantification unreliable. Today, GC using capillary columns coated with high-temperature polarizable phenylmethylsilicone stationary phases has been shown to be as effective as the other techniques. In comparison with HPLC, capillary GC yields higher resolution. However, to date, complete resolution of any molecular species is not attainable, neither by GC nor by HPLC. A combination of complementary chromatographic techniques would therefore be required to elucidate the entire TG pattern of milk fat [21].

33.3 Proteins

33.3.1 Introduction

For unethical farmers and dairy manufacturers, it may be attractive to adulterate high-priced milk products with less expensive protein sources, such as the low-priced soy, pea and soluble wheat proteins (SWPs), and bovine rennet whey (BRW). The latter is a low-priced by-product obtained during cheese production [22]. As mentioned earlier, apart from the economical/quality loss, adulteration of dairy products with other proteins can cause severe problems for allergic individuals as they are inadvertently exposed to allergenic proteins. Furthermore, when applied to feed, change in protein composition may affect digestibility. Therefore, detection methods for milk product adulteration which can be routinely employed by food/feed control authorities and food/feed processors are required. Several protein-targeted methods have been developed based on sodium dodecyl sulfate polyacrylamide gel electrophoresis (SDS-PAGE), capillary zone electrophoresis, colorimetry, chromatography, immunoassays and immunoblotting, biosensors, near-infrared spectroscopy (NIRS), and more recently MS. Immunoassays, MS, and NIRS will be described in greater detail in the following sections.

33.3.2 Authentication Testing

33.3.2.1 Immunoassays

In an immunoassay, the detection of a target molecule (antigen) is based on the specific antigen–antibody binding. The applied antibodies, Y-shaped 150 kDa proteins containing two antigen-binding sites, must possess high affinity and selectivity for the antigen to allow the detection of trace amounts of the antigen and to avoid false-positive test results, especially in complex matrices like dairy products. Nowadays, different immunoassay formats are available; classical formats like enzyme-linked immunosorbent assays (ELISA) and lateral flow devices (LFDs), and novel formats like biosensors and microsphere-based flow cytometric systems. The classical methods are relatively inexpensive, fast for small (LFDs) or large numbers of samples (ELISA), and easy-to-use without the need of expensive equipment. However, with these methods, only one target molecule can be detected simultaneously. Although biosensors and microsphere-based flow cytometric systems require relatively costly equipment (most of) these methods have the major advantage of the fact that they can detect several target molecules simultaneously. Furthermore, these new methods are time-efficient with sample analysis duration of a few minutes only.

For the detection of species adulteration, ELISA and biosensor applications based on the detection of caseins, whey proteins, or immunoglobulin G have been reported [23–25]. In addition, commercial immunoassays are also available (e.g., R-Biopharm AG, Darmstadt, Germany). For the detection of plant proteins in milk (powders), only an immunoblotting procedure [26], an ELISA [27], a biosensor application [28], and recently, a microsphere-based flow cytometric system [29], have been described. Both biosensor and microsphere-based methods can detect proteins from three different plants simultaneously. Until recently, only two biosensor immunoassays had been described for the immunochemical detection of the adulteration of milk powders with rennet whey [22,25]. Recent developments include a strip test (Operon S.A., Zaragoza, Spain) for the detection of BRW in milk and milk powders. Other new developments include an inhibition ELISA [30], reported by the authors recently. The inhibition ELISA is suitable for the detection of BRW in milk and milk powders with a detection limit of 0.1% (w/w), using a monoclonal antibody that recognizes caseinomacropeptide (CMP) as a marker. The signal (absorbance) is inversely proportional to the CMP concentration in the sample. The CMP concentrations are calibrated against standards of known BRW concentrations. A typical calibration curve is shown in Figure 33.2.

33.3.2.2 Mass Spectrometry

MS is an analytical technique used for the identification of analytes based on the accurate measurement of their molecular masses. Ionized analytes, produced in the ionization source of the mass spectrometer, are separated by their mass-to-charge ratio (m/z). In a mass spectrum, the m/z values are plotted against their intensities to reveal the different (ionizable) components in the sample and their molecular masses. In general, proteins and peptides are ionized by electrospray ionization (ESI) and matrix-assisted laser desorption ionization (MALDI). These methods are "soft" ionization methods indicating that fragmentation of the protein or peptide ions scarcely occur, enabling the mass measurement of intact proteins and peptides. MALDI time-of-flight (TOF) MS is fast and relatively easy-to-use. However, online coupling to sample pretreatment and separation techniques is still a challenge. ESI-MS, on the other hand, can be conveniently coupled with liquid chromatography (LC), which greatly improves the quality of the spectra.

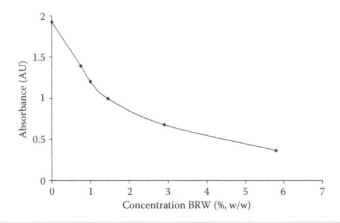

Figure 33.2 **Calibration curve of BRW powder in milk powder obtained with inhibition ELISA.**

To obtain amino acid sequence information of peptides and proteins, after enzymatic digestion of proteins, tandem mass spectrometry (MS/MS) is used. Inside the mass spectrometer, the peptide ions are fragmented and the *m/z* values of the fragments are plotted in the MS/MS spectra. These spectra are compared with a spectra database and the amino acid sequence is confirmed when a good match between the measured and theoretical spectra is obtained.

In the field of dairy adulterations, MALDI-TOF MS has been used to identify the presence of cow's milk in ewe and water-buffalo milk, as well as in cheeses. Identification is based on the protein profiles of the samples [31] or the use of α-lactalbumin and β-lactoglobulin as molecular markers [32,33]. However, Chen et al. [34] presented a more specific high-performance HPLC–ESI-MS method. The bovine milk protein identification procedure is based on the use of both retention time and molecular mass derived from multiple charged molecular ions. Furthermore, an HPLC–ESI/MS method was also developed for the detection of rennet whey in "traditional butter" based on the monitoring of two multicharged ions [35]. For a similar application, the detection of rennet whey in dairy powders, a more specific LC–ESI MS/MS method has been reported, which is based on the measurement of the fixed transition from a precursor ion (a specific CMP fragment) to a product ion [36].

For the identification of plant proteins in milk powder, a technique based on the determination of the amino acid sequence of the plant proteins by nano-LC ESI MS/MS has been developed [37]. Plant proteins are concentrated using a borate buffer and are subsequently digested with trypsin. The peptide mixture is analyzed using LC coupled with a quadrupole (Q) TOF MS instrument. Subsequently, the obtained tandem mass spectra are matched with those included in the National Center for Biotechnology Information (NCBI) database for identification purposes. An example of a tandem mass spectrum is shown in Figure 33.3.

33.3.2.3 Near-Infrared Spectroscopy

The IR is based on the concept of specific frequency vibration of atom-to-atom bonds within the molecules. Therefore, mid-IR absorption peaks are unique for specific bond pairs in a particular molecular environment. NIRS is based on the molecular overtone and combination vibrations of the fundamental vibrations occurring in the mid-IR region. Several molecular bonds (O–H in

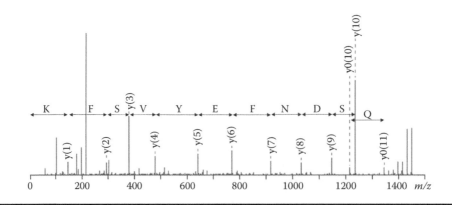

Figure 33.3 Tandem mass spectra of soy peptide SQSDNFEYVSFK. The corresponding amino acids are indicated.

water, C–H in carbohydrates and oils, and N–H in proteins) absorb NIR light (1100–2500 nm) at well-defined wavelengths. The absorbance level at these specific wavelengths is generally proportional to the quantity of that constituent in the material [38].

An advantage of NIR light is that it can be transmitted through a reasonably thick sample as the molar absorptivity in the NIR region is quite small. Such diffuse transmittance measurements have particularly proven to be useful in the analysis of liquids, slurries, suspensions, and pastes [39]. On the contrary, NIR absorption bands also often overlap and are strongly influenced by light-scattering effects. The latter is mainly caused by solid particles in a sample. Food materials are organically complex and are present in a multiplicity of physical forms. The complexity of the factors determining NIR spectra demand the use of mathematical models for NIRS data interpretation.

In combination with multiple linear regression (MLR) analysis, the NIRS technique was applied for the quantification of major food components in dairy products, as early as in the 1980s [40]. The NIRS spectra were calibrated against the data obtained by classical wet chemistry procedures. Precision of NIRS in this type of application is limited to a great extend to the precision of the reference methods used for calibration. The representativeness of the calibration sample sets is fairly challenging, which is partly owing to the sample preparation issues. Particle size, homogeneity, temperature, and presentation of the sample require standardization. From the calibration sets, regression equations can be generated to determine the major constituents of milk, milk powder, casein, butter, and cheese with an accuracy similar to that obtained with the wet chemistry methods [41].

With the development of chemometrics, NIRS has received more scientific attention and has generally become more popular [42]. Over the last few years, many NIRS applications in the dairy field, including online applications, have been reported [43–50]. In 2006, the International Organization for Standardization (ISO) and the International Dairy Federation (IDF) jointly published an International Guidance for the application of NIRS to milk product analysis [51].

For authentication purposes, the full NIR spectral data set needs to be considered, as plant and milk protein spectra differ considerably (Figure 33.4). Multivariate data analysis techniques, such as PLS analysis, are employed to enable the complete use of the spectral data. DA and spectral matching methods are applied for discrimination the products that differ considerably, whereas principle component analysis (PCA) is used for spectral identification and differentiation of fairly similar products [52,53]. For detailed comparison of the spectra for the detection of adulterations, multivariate classification models have been established from the full spectra of training sets, considering the natural and processing-induced variance. The identity and authenticity of an unknown sample can subsequently be established in a single analysis by comparing their NIR fingerprint spectra with the collection of NIR spectra of training samples. An example of a plot of the first two dimensions of a PLS-DA model based on NIR spectra of pure skimmed milk powder and butter milk powder, predicting the identity of the samples is presented in Figure 33.5. When samples appear out of the range compared with their unadulterated counterparts, additional analyses can be carried out to determine the identity of the suspicious sample.

If the identity is determined, then a unique "databank" can be constructed which can be used with increasing certainty to authenticate the dairy products. Comprehensive descriptions and references on spectra–structure correlations and additional technical details of applications for NIR spectra have been reported extensively [54–58]. NIR continues to provide a valuable measurement technique, applicable to both quantification and identification of dairy products, for use as a control technique, as well as a real-time process-monitoring technique.

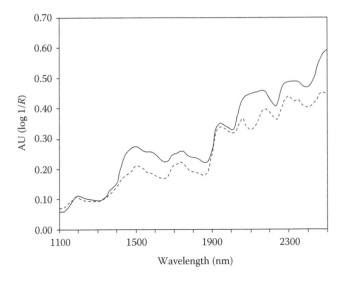

Figure 33.4   Typical NIR absorbance spectra of skimmed milk powder (—) and soy powder (- - -) displayed in log(1/R) form.

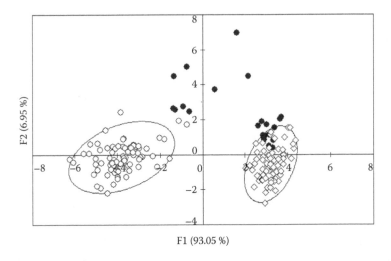

Figure 33.5   Discriminant plot determined by discriminant factors 1 (F1) and 2 (F2) for calibration sets of skimmed milk powder (◊) and butter milk powder (○), and the scores of 16 suspicious samples (●). Five samples lie outside both the databases with 99% confidence limit and are highly suspected.

33.4   Conclusions

The brief overview on techniques being applied for the detection of foreign lipids and proteins in dairy products illustrates the complex task encountered in authentication. Approximately a few hundred of the thousands of chemical compounds identified in dairy products are used in the authentication process. Although single analyte methods are suitable for particular purposes, the ultimate solution may be rather sought in a multivariate approach, combining information of several analytes in a multidimensional space.

References

1. Karoui, R., Mazerolles, G., and Dufour, E., Spectroscopic techniques coupled with chemometric tools for structure and texture determinations in dairy products: A review, *Int. Dairy J.*, 13, 607, 2003.
2. Ausar, S.F. et al., Characterization of casein micelle precipitation by chitosans, *J. Dairy Sci.*, 84, 361, 2001.
3. Colak, H. et al., Detection of presence of cow's milk in sheep's cheeses by immunochromatography, *Food Control*, 17, 905, 2006.
4. Bosset, J.O. et al., Comparison of some highland and lowland Gruyère type cheese of Switzerland: A study of their potential PDO/AOC/AOP characteristics. Authenticity and adulteration of food-the analytical approach, *Proc. Eur. Food Chem.*, 220, 395, 1999.
5. Enne, G. et al., High-performance liquid chromatography of governing liquid to detect illegal bovine milk's addition in water buffalo Mozzarella: Comparison with results from raw milk and cheese matrix, *J. Chromatogr. A*, 1094, 169, 2005.
6. Recio, I., Amigo, L., and López-Fandiño, R., Assessment of the quality of dairy products by capillary electrophoresis of milk proteins, *J. Chromatogr. B*, 697, 231, 1997.
7. Downey, G., Authentication of food and food ingredients by near infrared spectroscopy, *J. Near Infrared Spectrosc.*, 4, 47, 1996.
8. Karoui, R. and De Baerdemaeker, J., A review of analytical methods coupled with chemometric tools for the determination of the quality and identity of dairy products, *Food Chem.*, 102, 621, 2007.
9. Prize for a method of detecting adulterations of butter, *Dingler's Polytech. J.*, 2, 223, 1877.
10. Kamm, W. et al., Authenticity assessment of fats and oils, *Food Rev. Int.*, 17, 249, 2001.
11. Jee, M., Milk fat and other animal fats, in *Oils and Fats Authentication*, Jee, M., ed., CRC Press, Boca Raton, FL, 2002, 115.
12. James, A.T. and Martin, A.J.P., Gas-liquid chromatography: The separation and identification of the methyl esters of saturated and unsaturated acids from formic acid to n-octadecanoic acid, *Biochem. J.*, 63, 144, 1956.
13. Hettinga, D., Butter, in *Bailey's Industrial Oil and Fat Products*, 5th edn., Hui, Y.H., ed., Wiley and Sons, New York, 1996, p. 1.
14. Codex Alimentarius Commission, Report of the Fourteenth Session of the Codex Committee on Fats and Oils, *Alinorm* 95/17, FAO/WHO, Rome, 1995.
15. Antonelli, A., Conte, L.S., and Lercker, G., Applications of capillary gas chromatography to the quality control of butter and related products, *J. Chromatogr.*, 552, 273, 1991.
16. Matter, L. et al., Characterization of animal fats via the GC patterns of FAME mixtures obtained by transesterification of the triglycerides, *Chromatograpia*, 27, 31, 1989.
17. van Ruth, S.M. et al., Prediction of the identity of fats and oils by their fatty acid, triacylglycerol and volatile compositions using PLS-DA, *Food Chem.*, submitted.
18. Lercker, G. et al., Milk fat. Gas chromatographic analysis of triglycerides in assessment of authenticity of butter, *Scienza Tecnica Lattiero Caseria*, 43, 95, 1992.
19. Precht, D., Detection of foreign fat in milk fat, I. Qualitative detection by triacylglycerol formulae, *Z. Lebensm. Unters. Forsch.*, 194, 1, 1992.
20. Lipp, M., Determination of the adulteration of butter fat by its triglyceride composition obtained by GC. A comparison of the suitability of PLS and neural networks, *Food Chem.*, 55, 389, 1996.
21. Buchgraber, M. et al., Triacylglycerol profiling by using chromatographic techniques, *Eur. J. Lipid Sci. Technol.*, 106, 621, 2004.
22. Haasnoot, W., Detection of adulterants in milk products, *Biacore J.*, 5, 12, 2005.

23. Haasnoot, W. et al., Fast biosensor immunoassay for the detection of cows' milk in the milk of ewes and goats, *J. Dairy Res.*, 71, 322, 2004.

24. Hurley, I.P. et al., Application of immunological methods for the detection of species adulteration in dairy products, *Int. J. Food Sc. Technol.*, 39, 873, 2004.

25. Haasnoot, W., Marchesini, G.R., and Koopal, K., Spreeta-based biosensor immunoassay to detect fraudulent adulteration in milk and milk powder, *J. AOAC Int.*, 89, 849, 2006.

26. Ventling, B. and Hurley, W.L., Soy protein in milk replacers identified by immunoblotting, *J. Food Sci.*, 54, 766, 1988.

27. Hewedy, M.M. and Smith, C.J., Modified immunoassay for the detection of soy milk in pasteurized skimmed bovine milk, *Food Hydrocolloids*, 3, 485, 1990.

28. Haasnoot, W. et al., Direct biosensor immunoassay for the detection of nonmilk proteins in milk powder, *J. Agric. Food Chem.*, 49, 5201, 2001.

29. Haasnoot W. and Du Pré, J.G., Luminex-based triplex immunoassay for the simultaneous detection of soy, pea, and soluble wheat proteins in milk powder, *J. Agric. Food Chem.*, 55, 3771, 2007.

30. Bremer, M.G.E.G. et al., Enzyme-linked-immunosorbent-assay for the detection of bovine rennet whey powder in milk powder and buttermilk powder, *Int. Dairy J.*, 18, 294, 2008.

31. Fanton, C. et al., Matrix-assisted laser desorption/ionization mass spectrometry in the dairy industry. 2. The protein fingerprint of ewe cheese and its application to detection of adulteration by bovine milk, *Rapid Commun. Mass Spectrom.*, 12, 1569, 1998.

32. Cozzolino, R. et al., Identification of adulteration in milk by matrix-assisted laser desorption/ionization time-of-flight mass spectrometry, *J. Mass Spectrosc.*, 36, 1031, 2001.

33. Cozzolino, R. et al., Identification of adulteration in water buffalo mozzarella and in ewe cheese by using whey proteins as biomarkers and by matrix-assisted laser desorption/ionization mass spectrometry, *J. Mass Spectrosc.*, 37, 985, 2002.

34. Chen, R.-K. et al., Quantification of cow milk adulteration in goat milk using high-performance liquid chromatography with electrospray ionization mass spectrometry, *Rapid Commun. Mass Spectrosc.*, 18, 1167, 2004.

35. De Noni, I. and Resmini, P., Identification of rennet-whey solids in "traditional butter" by means of HPLC/ESI-MS of non-glycosylated caseinomacropeptide A, *Food Chem.*, 93, 65, 2005.

36. Mollé, D. and Léonil, J., Quantitative determination of bovine k-casein macropeptide in dairy products by liquid chromatography/electrospray coupled to mass spectrometry (LC-ESI/MS) and liquid chromatography/electrospray coupled to tandem mass spectrometry (LC-ESI/MS/MS), *Int. Dairy J.*, 15, 419, 2005.

37. Luykx, D.M.A.M. et al., Identification of plant proteins in adulterated skimmed milk powder by high-performance liquid chromatography-mass spectrometry, *J. Chromatogr. A*, 1164, 189, 2007.

38. Stark, E., Luchter, K., and Margoshes, M., Near-infrared analysis (NIRA): A technology for quantitative and qualitative analysis, *Appl. Spectrosc.*, 22, 335, 1986.

39. Davies, A.M.C. and Grant, A., Near-infrared analysis of food, *Int. J. Food Sci. Technol.*, 22, 191, 1987.

40. Frankhuizen, R. and Veen van der, N.G., Determination of major and minor constituents in milk powders and cheese by near-infrared reflectance spectroscopy, *Neth. Milk Dairy J.*, 39, 191, 1985.

41. Frankhuizen, R., NIR analysis of dairy products, in *Handbook of Near-Infrared Analysis*, 2nd edn., Burns, D.A. and Ciurczak, E.W., eds., Marcel Dekker Inc., New York, 2001, Chapter 18.

42. Cen, H. and He, Y., Theory and application of near infrared reflectance spectroscopy in determination of food quality, *Trends Food Sci. Technol.*, 18, 72, 2007.

43. Rodriquez-Otero, J.L. et al., Determination of fat, protein and total solids in cheese by near-infrared reflectance spectroscopy, *J. AOAC Int.*, 78, 802, 1995.

44. Sørensen, L.K. and Jepsen, R., Detection of cheese batches exposed to clostrium tyrobutyricum spoilage by near infrared spectroscopy, *J. Near Infrared Spectrosc.*, 5, 91, 1997.

45. Laporte, M.F. and Paquin, P., Near-infrared analysis of fat, protein and casein in cow's milk, *J. Agric. Food Chem.*, 47, 2600, 1999.

46. Reeves, J.B. and Kessel van, J.S., Determination of ammonium-N, moisture, total C and total N in dairy manures using a near infrared fibre-optic spectrometer, *J. Near Infrared Spectrosc.*, 8, 151, 2000.

47. Blazquez, C. et al., Prediction of moisture, fat and inorganic salts in processed cheese by near infrared reflectance spectroscopy and multivariate data analysis, *J. Near Infrared Spectrosc.*, 12, 149, 2004.

48. Sinelli, N. et al., A preliminary study using Fourier transform near infrared spectroscopy to monitor the shelf-life of packed industrial ricotta cheese, *J. Near Infrared Spectrosc.*, 13, 293, 2005.

49. Feten, G. et al., Using NIR spectroscopy for the prediction of free amino acids during cheese ripening, *NIR News*, 18, 4, 2007.

50. Frankhuizen, R., Cruijsen, J.M.M., and Kok, H.L., On-line determination of moisture content during cheese processing, in *Book of Abstracts, Fourth International Workshop on Water in Food*, Bernreuther, A., ed., EC-DG JRC-IRMM, Brussels, 2006, 59.

51. Milk products—Determination of total solids, fat, protein and moisture contents—Guidance for the application of near infrared spectrometry, *ISO21543/IDF210*, 2006.

52. Martens, H. and Naes, T., *Multivariate Calibration*, John Wiley and Sons, Chichester, 1992, 1.

53. Mark, H. and Workman, J., *Statistics in Spectroscopy*, 2nd edn., Elsevier, Amsterdam, 2003, 1.

54. Weyer, L.G. and Lo, S.-C., Spectra-structure correlations in the near-infrared, in *Handbook of Vibrational Spectroscopy*, vol. 3, Chalmens, J.H. and Griffiths, P.R., eds., Wiley and Sons, New York, 2002, 1817.

55. Maraboli, A., Cattanea, T.M.O., and Giangiacomo, R., Detection of vegetable proteins from soy, pea and wheat isolates in milk powder by near infrared spectroscopy, *J. Near Infrared Spectrosc.*, 10, 63, 2002.

56. Pillonel, L. et al., Cheese Authenticity. Determining the geographic origin of emmental cheese, *NIR News*, 15, 14, 2004.

57. Tsenkova, R. et al., Near infrared spectra of cow's milk for milk quality evaluation: Disease diagnosis and pathogen identification, *J. Near Infrared Spectrosc.*, 14, 363, 2006.

58. Mouazen, A.M. et al., Feasibility study on using visible-near infrared spectroscopy coupled with a factorial discriminant analysis technique to identify sheep milk from different genotypes and feeding systems, *J. Near Infrared Spectrosc.*, 15, 359, 2007.

Chapter 34

Detection of Adulterations: Identification of Milk Origin

Golfo Moatsou

Contents

34.1 Introduction 865
34.2 Methods of Detection Based on Proteins 866
    34.2.1 Electrophoretic Methods.. 866
        34.2.1.1 Caseins 866
        34.2.1.2 Whey Proteins 868
    34.2.2 Chromatographic Methods.. 869
        34.2.2.1 Caseins 869
        34.2.2.2 Whey Proteins 869
    34.2.3 Antibody-Based Analytical Methods 870
34.3 Methods of Detection Based on Milk Fat 874
34.4 DNA-Based Methods of Detection—Species-Specific PCR 874
References 880

34.1   Introduction

The use of a nondeclared milk type in the manufacture of dairy products is characterized as a fraud. The most common practice is the partial or total substitution of a milk kind by another of lower commercial value. The adulteration of cheesemilk, related to both cheese quality and legal requirements, is a rather frequent problem. The increased demand for genuine and accurately labeled traditional products necessitates the protection against adulteration. Moreover, adulteration may affect the health of consumers with nondeclared allergic problems. Traditional cheeses produced in the Mediterranean countries accepted by consumers worldwide, e.g., Feta, Manchego, and Pecorino, are made from ovine milk or from its mixtures with caprine milk. The composition of cheesemilk affects the characteristics and the organoleptic properties of the final product. The seasonal production and the higher prices of caprine milk, especially of the ovine milk, than the bovine milk are the main reasons for the admixture of cheesemilk with bovine milk. In addition, the existence of mixed flocks of goats and ewes can result in the accidental or fraudulent substitution of ovine milk by caprine. Fraudulent addition of bovine milk also occurs during the manufacture of "Mozzarella di Bufala" cheese that is normally made from raw water-buffalo milk. Cheesemilk adulteration is carried out using raw milk, heated, condensed milk, milk powder, or caseinates from milk kinds that are not declared on the product label.

The confirmation of the milk kinds used for the manufacture of dairy products and the determination of their relative percentages is based on milk components. The objective of the analytical methods used for this purpose is the detection of substances or the determination of abnormalities in the composition which cannot be assigned to any of the materials used during the manufacture of dairy products. Detection is based on milk constituents (protein, fat) or on DNA from the somatic-cell content of the milk. The analytical methods for the identification of milk origin in dairy products have been reviewed by Ramos and Juárez [1], and more recently by De la Fuente and Juárez [2]. There are also reviews regarding the immunological techniques [3,4] and polymerase chain reaction (PCR) techniques [5].

34.2   Methods of Detection Based on Proteins

Milk proteins are the basis of many analytical methods that have been applied successfully for the detection of adulteration. The objective of analyses is the separation and the subsequent detection and quantification of a homologous protein fraction from different milk kinds. Casein and whey proteins are analyzed for this purpose. It has to be taken into consideration that during cheese-making, a great part of whey proteins is removed from the cheese curd with the whey. Moreover, the thermal treatments applied to cheesemilk or cheese curd may denature part of them (e.g., immunoglobulins). Therefore, casein seems to be more appropriate than whey proteins in terms of the detection of cheesemilk adulteration. However, whey proteins are tolerant to hydrolysis during ripening. In addition, heat treatments applied in cheesemilk, usually lower than or equal to high-temperature short-time (HTST) pasteurization, do not affect them.

34.2.1   Electrophoretic Methods

34.2.1.1   Caseins

Urea-polyacrylamide gel electrophoresis (PAGE) of milk caseins under alkaline conditions [6] is one of the oldest methods used for the detection of bovine milk in milk mixtures. Detection is based on the higher mobility of bovine α_{s1}-casein than that of its ovine and caprine counterparts. It is a rather simple method, but its sensitivity is limited by the sensitivity of the staining techniques used for the visualization of protein bands. Urea-PAGE has been found appropriate for analyzing milk mixtures [7,8]. In the majority of cheese varieties, α_{s1}-casein is hydrolyzed to a variety of smaller peptides during ripening, which are not retained in the polyacrylamide gel matrix. Therefore, this method is not always suitable for cheese fraud detection. It has been successfully applied in Halloumi cheese, a cheese variety with limited proteolysis [9]. More recently, detection of the rather high percentages of 10% and 20% of bovine milk in ovine-milk cheeses by urea-PAGE has been reported [10]. Among the casein fractions of cheese, para-κ-casein seems to be the most advantageous for the detection of cheesemilk origin, because it is not affected substantially by proteolysis during cheese ripening. However, the differences among the amino acid sequences of ovine, caprine, and bovine para-κ-caseins are rather limited. Cationic PAGE of para-κ-casein has been used for the detection and determination of bovine milk in ovine yoghurt after the treatment of yoghurt caseins with rennet. A detection limit of 1% of bovine milk has been reported [11].

Isoelectric focusing (IEF) methods have been proposed since 1986 for the detection of milk kind, especially for the detection of cheesemilk adulteration [12–14]. IEF of γ-caseins resulting from the hydrolysis of paracasein fraction of milk or cheese with plasmin is the basis of the official reference method of European communities, for the detection of bovine milk or caseinates in ovine and caprine milk products [15]. Detection is based on the different focusing points of bovine γ_2- and γ_3-caseins when compared with their ovine and caprine counterparts. Standard samples with 0% and 1% bovine milk are analyzed simultaneously with the unknown samples. The quantification of bovine milk is based on the intensity of γ_2- and γ_3-casein bands when compared with the respective bands of the standard samples. The method is sensitive and appropriate for cheese, as it is not affected by the thermal treatment of milk or cheese curd and by the extent of proteolysis. However, it cannot detect mixtures of ovine and caprine milk. Moreover, it is not accurate in quantitative terms, when the adulteration level is high, because there are declinations as the band intensity increases. The detection of γ-caseins and peptides along the IEF profiles using immunoblotting with polyclonal antibodies against β-casein has also been reported [16,17]. IEF of para-κ-casein for the detection of ovine, caprine, and bovine protein in the cheesemilk was first used by Addeo et al., and it has also been applied in model cheeses [8,12,13,18]. In general, this method has been found suitable for the detection of bovine para-κ-casein in hard-pressed and young mold-ripened cheeses. However, a peptide with p*I* similar to that of bovine para-κ-casein in matured cheese profiles can cause false-positive results in Roquefort cheese. Moreover, a band with p*I* similar to caprine para-κ-casein has been detected within the profile of ovine cheese. Therefore, IEF of para-κ-casein has not been found suitable for the detection of small quantities of either ovine or caprine para-κ-casein, but it could be used for the estimation of high percentages of adulteration (i.e., >10% of goat milk).

Capillary zone electrophoresis (CZE) has been extensively used since 1990. It is a rather fast, sensitive, and easily operated method used for the separation of casein fraction of both milk and cheese [19,20]. Following the studies about casein fraction and milk-protein polymorphism, and taking into consideration the different migration times of α_{s1}-casein of different milk kinds [21,22], the presence of up to 8% of bovine casein in milk mixtures has been detected [23]. The detection limit has been effectively improved up to 1% of bovine milk in caprine milk by means of an uncoated capillary tube [24]. Identification and quantitative determination of milk kind in binary and ternary mixtures based on particular predictor variables, i.e., the peak areas of bovine and ovine α_{s1}-casein, bovine, ovine and caprine κ-casein, bovine β-casein A1 and A2, and ovine and caprine β_1- and β_2-casein, has been reported [25]. This approach involves the principle component regression and partial least-squares regression with the mean square errors in prediction being <2.4% in all the cases. Characteristic capillary electrophoresis (CE) patterns of cheese from ovine, caprine, and bovine milk have been presented, considering the peaks of caprine para-κ-casein and bovine β-casein as indicatives of the milk kind [26].

The quantification of β-lactoglobulin and para-κ-casein in the capillary electropherograms has been related to the determination of milk origin in fresh cheeses [27]. Caprine para-κ-casein and bovine α_{s1}-casein peaks have been used to detect caprine and/or bovine milk in ovine Halloumi cheese. The detection limit has been found to be 2% and 5% for caprine and bovine milk, respectively, by means of stepwise multiple linear regression analysis [28]. Similarly, multivariate statistical techniques have been utilized for the prediction of the percentages of ternary mixtures of bovine, ovine, and caprine milk using the areas of 13 selected peaks of the capillary electropherograms of an unripened cheese; the root square error has been estimated at 2.2% [29]. De la Fuente and Juárez [2] suggested that instrumental developments such as mass spectrometry (MS) detection and new injection devices will improve the performance of the CE technique.

34.2.1.2   Whey Proteins

Native PAGE of whey protein fraction has been found suitable for the detection of 10%–40% bovine milk in ovine or caprine milk submitted to thermal treatment commonly used in cheesemaking, i.e., at 74°C for 30 s [30]. The detection has been based on the greater electrophoretic mobility of bovine β-lactoglobulin when compared with that of ovine β-lactoglobulin. Native PAGE of whey proteins has also been used for the differentiation of bovine, ovine, caprine, and mare milk [8]. Technological parameters of cheesemaking, such as type of rennet, pressing, and ripening time have not interfered with the detection of bovine milk in ovine Machengo, Roquefort, and Serra da Estrella cheeses [31,32]. Stepwise multiple linear regression and principle components regression applied to the results of PAGE analysis of cheese whey fraction have been used for the prediction of the percentages of ovine, caprine, and bovine cheesemilk [33]. The addition of 1% ultrahigh temperature (UHT) bovine milk or heat-denatured bovine whey proteins in cheese made from ovine, caprine, and buffalo milk has been detected by means of immunoblotting [34]. Adequate sample preparation is necessary to obtain denatured whey proteins from the casein fraction, and polyclonal antibovine β-lactoglobulin antibodies against native and denatured β-lactoglobulin have been utilized.

Detection of adulteration by means of IEF of acid or cheese whey fraction has been reported, based on the different isoelectric point of bovine β-lactoglobulin when compared with the caprine and ovine counterparts [8,32–37]. The results were similar to that of the more simple native PAGE method.

There are several studies on the analysis of whey protein fraction of milk and cheese using CE with alkaline buffer, to detect nondeclared bovine milk in milk mixtures or cheeses. However, genetic variability of milk proteins and the possible heat treatment of one of the milk kinds used in the mixture can limit the efficacy of the method. The detection of 5% bovine milk in buffalo milk and buffalo mozzarella cheese has been based on bovine β-lactoglobulin A or α-lactalbumin [38]. Bovine milk in ovine milk and cheese has been detected using the bovine β-lactoglobulin B as a marker, with a detection limit of 0.5% in milk mixtures and 2% in cheeses [39]. The same method has been used for the determination of bovine milk in caprine milk and cheese by means of the ratio of bovine β-lactoglobulin A to caprine α-lactalbumin areas. Detection limits of 2% and 4% for milk and cheese, respectively, and the quantification results have been presented [40]. A quick CE procedure in isoelectric acidic buffers, not requiring coated capillaries, has been developed for the analysis of milk and cheese whey protein fraction. One percent of bovine milk has been detected using the ratios of areas of bovine α-lactalbumin or β-lactoglobulin to those of their ovine or caprine counterparts [41]. Partial least-squares multivariate regression applied to the results obtained by CE analysis in isoelectric acidic buffers of an ethanol–water extract of cheese, has predicted the content of bovine milk in caprine and ovine cheeses with relative standard deviation of about 6%–7% [42].

In conclusion, electrophoretic methods are not always effective for the accurate quantification of adulteration in cheese. The main difficulties arise from the differences in the casein contents of the milk used in mixtures [43], as well as from the various technological conditions applied in cheese manufacture (i.e., cheesemaking conditions, heat treatment of cheesemilk, cheese ripening).

34.2.2 Chromatographic Methods

34.2.2.1 Caseins

The chromatographic methods for the identification of milk origin have the objective to separate individual caseins and whey proteins of different animal species which exist in a dairy sample. Despite the fact that many samples can be analyzed at the same time by means of PAGE or IEF, the chromatographic methods are more advantageous as they can be fully automated.

Anion-exchange fast protein liquid chromatography (FPLC®, Amersham Biosciences, part of GE Healthcare, Piscataway, NJ) or high-performance liquid chromatography (HPLC) methods have been applied for bovine milk detection in milk mixtures using the difference between elution times of bovine and ovine/caprine α_{s1}-casein. A detection limit of about 2%–4% has been reported, similar to the electrophoretic detection [8,44]. However, the rapid hydrolysis of α_{s1}-casein during cheese ripening limits the efficacy of these methods regarding cheesemilk adulteration. However, α_{s1}-I peptide (α_{s1}-CN f24-199) has been used as a marker for fraud detection in Gouda cheese; the quantification results depend upon the maturity of the cheese [44]. Addition of bovine milk as low as 1% in Halloumi cheese has been determined after casein hydrolysis by plasmin, by means of a strong anion-exchange HPLC column [45].

Reversed-phase HPLC (RP-HPLC) profiles have been used for the detection of bovine milk in ovine cheeses at different stages of ripening using the area of α-casein peaks as a marker; the method has been found to be less sensitive than PAGE, as adulteration equal to or higher than 20% could be detected [10]. Hydrophobic interaction chromatography of the casein fraction has been applied for the analysis of binary mixtures of bovine/caprine and bovine/ovine milks and the cheeses subsequently made. The ratios of individual casein peaks have been proposed as possible markers for the detection of adulteration [46].

The cation-exchange HPLC method of para-κ-casein of Mayer et al. [18] has given very promising semiquantitative results in model Camembert cheeses of different ages. The new element of this method is the differentiation between ovine and caprine cheesemilk using the para-κ-casein peaks. The same method has been used for Tilsit and Halloumi cheese [8,47].

34.2.2.2 Whey Proteins

Chromatographic analysis of whey proteins has been extensively used for the detection of adulteration. Anion-exchange FPLC distinguishes α-lactalbumin and β-lactoglobulin from bovine and ovine milk [48]. More promising results has been given by an RP-HPLC method, detecting 10% of bovine in ovine milk using bovine β-lactoglobulin A as indicator [49,50]. The RP-HPLC has been also applied successfully for whey protein separation, especially for β-lactoglobulin of Mozzarella cheese and pickle, and detection limits of 1% and 2% of bovine milk have been reported [51–53]. Nevertheless, ovine and caprine milks cannot be differentiated using these methods. Moreover, the chromatograms of whey fraction of cheese varieties with extended proteolysis are expected to be very complex, because they also include a variety of medium- and small-sized peptides resulting from casein hydrolysis, apart from whey proteins. RP-HPLC of caseinomacropeptide (CMP) has been proposed for the detection of bovine milk in caprine milk after treatment with rennet. The detection limit of bovine milk in caprine milk has been found to be 2.5%, and the intensity of heat treatment does not affect the results [54].

Furthermore, new methods based on the determination of α-lactalbumin and β-lactoglobulin masses by means of MS have been developed. Matrix-assisted laser desorption/ionization time-of-flight mass spectrometry (MALDI-TOFMS) analysis of whey fraction of Mozzarella cheese made from binary mixtures of buffalo, bovine, and ovine milk has resulted in detection limits of 2% and 5% for ovine and bovine milk, respectively [55]. The presence of 5% bovine milk in goat milk has been detected by means of RP-HPLC/electrospray ionization mass spectrometry (ESI-MS) using β-lactoglobulin as a molecular marker [56]. The high resolving power and the sensitivity of MS methods make them a very promising tool for the rapid detection of the adulteration of milk and dairy products.

34.2.3 Antibody-Based Analytical Methods

Methods based on antigen–antibody precipitation reactions for differentiating the proteins in milk from different species have been presented since 1901 [1,3]. Initially, rabbit antiserums to milk whey proteins or blood serum proteins were used for the detection of bovine milk. Subsequently, more specific antiserums were prepared by elimination of cross-reacting antibodies. Levieux [57,58] prepared an antiserum against bovine immunoglobulin by immunizing goats or sheep to overcome the adsorption step. The methods of radial immunodiffusion or inhibition of hemagglutination were used with detection limits of about 1%. Both the methods have been commercialized as a patent (CV test for detection of bovine IgG_1 and BC test for caprine IgG_1), and have been applied to many cheese varieties. The heat resistance of caseins makes them suitable for the elaboration of immunological methods, although preservation of their epitopes depends on protein hydrolysis during cheese ripening. A method that uses rabbit antiserum against whole bovine casein has been presented with a detection limit of 0.1% of bovine casein in ovine casein [59].

Since 1990, enzyme-linked immunosorbent assay (ELISA) has been the most frequently used immunoassay for the detection of milk from different species. It is a simple, sensitive, rapid, reliable, and versatile assay system for the quantifications of antigen and antibodies. Furthermore, ELISAs are usually performed in commercially available 96-well microtitre plates that allow the use of small sample volumes and the rapid simultaneous analysis of high sample numbers. A wide range of ELISA configurations have been developed for the detection of milk adulteration. Apart from the configuration of the assay, the methods are differentiated by the type of antibody used, i.e., monoclonal or polyclonal, and by the protein fractions used for the antibodies production, i.e., caseins or whey proteins or peptides corresponding to defined regions of milk proteins (Table 34.1).

Various ELISAs use polyclonal antibodies against whey proteins. The detection of bovine milk in ovine or caprine milk with detection limits of 0.1% [60], and the detection of caprine milk in ovine milk with detection limit of 0.5% have been reported [61,62]. The response of the assay has been found to be lower in sterilized milks. Polyclonal antibodies against heat-denatured β-lactoglobulin have been used for the detection of very low quantities, i.e., 0.1%–0.2%, of bovine milk in ovine and caprine cheese [63].

The high degree of identity among the homologous caseins from different species complicates their differentiation by immunological methods. Furthermore, proteolysis during cheese ripening

Table 34.1 ELISA Methods Used for the Detection of Adulteration in Milk and Dairy Products

Type of Antibodies	ELISA Format	Detection Limit	Quantitative Determination	Ref.
Polyclonal antibodies against				
Bovine whey proteins	Indirect	1% bovine milk in ovine milk	1%–50%	[60]
Caprine whey proteins	Sandwich	0.5% caprine milk in ovine milk	0.5%–100%	[61]
Caprine whey proteins	Indirect	1% caprine milk in ovine milk	1%–100%	[62]
Heat-denatured β-lactoglobulin	Indirect competitive	0.1%–0.2% bovine milk in cheese	N/A	[63]
Bovine caseins	Indirect	1% bovine milk in ovine milk and cheese	1%–50%	[64]
Caprine caseins	Sandwich	1% caprine milk in ovine milk and cheese	1%–100%	[67]
Bovine γ_3-casein	Indirect competitive	0.1% bovine milk in ovine and caprine cheese	N/A	[68]
Chemically synthesized bovine κ-CN f139–152	Competitive	0.25% raw or heated bovine milk in ovine or caprine milk and cheese	0.25%–64%	[69]
Chemically synthesized bovine α_{s1}-CN f140–149	Competitive	0.125% raw or heated bovine milk in ovine milk	0.125%–64% for milk 0.5%–25% for cheese	[70]
Monoclonal antibodies against				
Bovine β-lactoglobulin (MAbs 17 and 102)	Sandwich "two site"	10 ppm bovine milk in ovine or caprine milk	N/A	[73]
Bovine β-lactoglobulin (MAbs 88N and 117N)	Sandwich "two site"	0.03% bovine milk in caprine milk	N/A	[74]
Bovine IgG (MAb BG-18)	Indirect competitive	0.1% bovine milk in caprine, ovine, or buffalo milk	0.1%–10%	[75]

(*continued*)

Table 34.1 (continued)   ELISA Methods Used for the Detection of Adulteration in Milk and Dairy Products

Type of Antibodies	ELISA Format	Detection Limit	Quantitative Determination	Ref.
Bovine IgG (MAb BG-18)	Indirect competitive	0.001% bovine milk in ovine and buffalo milk	0.01%–100%	[76]
	Indirect sandwich	0.01% bovine milk in caprine milk		
Bovine β-casein (MAb AH4)	Indirect	0.5% of raw bovine milk in ovine or caprine milk	0.5%–100%	[77]
		1% of thermal-treated bovine milk in ovine or caprine milk		
Bovine β-casein (MAb AH4)	Immunostick	>1% of bovine milk in ovine milk	N/A	[78]
		>0.5% bovine cheese in ovine cheese		
Bovine β-casein (MAb AH4)	Sandwich	0.5% bovine cheese in ovine cheese	0.5%–10%	[79]
Caprine α_{s2}-casein (MAb B2B)	Indirect	0.5% caprine milk in ovine milk	0.5%–15%	[81]
Caprine α_{s2}-casein (MAb B2B)	Competitive indirect	0.25% caprine milk in ovine milk	0.25%–15%	[82]
Caprine α_{s2}-casein (MAb B2B)	Indirect	1% caprine cheese in ovine cheese	1%–15%	[82]
Caprine α_{s2}-casein (MAb B2B)	Competitive indirect	0.5% caprine cheese in ovine cheese	0.5%–25%	[82]

N/A, not available.

can alter their antigenicity characteristics. However, the immunological reactivity of caseins is not affected by heat treatment. Therefore, they can be used in immunological methods for detecting milk mixtures in heat-treated products. Polyclonal antibodies against caseins have been used for the detection of 1% bovine milk or 1% caprine milk in ovine milk and in cheeses made from milk mixtures [64–67]. In addition, polyclonal antibodies against γ_3-casein have been used for the detection of 0.1% bovine milk in ovine or caprine cheese, and the method has not been affected by the intensity of heat treatment of bovine milk [68].

Polyclonal antibodies against synthetic peptides that correspond to defined regions of milk proteins and are conjugated to carriers have also been used as antigens. In addition, polyclonal antibodies against the chemically synthesized 139–152 peptide of bovine κ-casein (κ-CN f139–152) have been used for the detection of bovine CMP in milk and cheese made from mixtures of ovine or caprine milk with raw or heated (115°C for 15 min) bovine milk; the detection limit has been estimated at 0.25% bovine milk [69]. Assays using polyclonal antibodies against the chemically synthesized bovine fragment α_{s1}-CN f140–149, which is two amino acids longer than the ovine α_{s1}-casein deletion, have detected the presence of bovine milk in ovine and caprine milk and cheese with detection limits of 0.125% and 0.5%, respectively [70,71]. Preliminary experiments have indicated that polyclonal antibodies raised against β-CN f1–28 4P phosphopeptide released from bovine β-casein by plasmin might be suitable for the detection of bovine casein in fresh dairy products of ovine and caprine milk [72].

Hybridoma technology for continuous production of monoclonal, monospecific antibodies (MAbs) of consistent specificity against milk proteins or against their fractions could eliminate cross-reactivity between the proteins from different species. Monoclonal antibodies against bovine β-lactoglobulin have been used in very sensitive ELISAs for the detection of bovine milk in ovine or caprine milk; very low detection limits ranging from 0.0001% to 0.03% have been reported [73,74]. Furthermore, very low levels of bovine milk, i.e., 0.1% in caprine, ovine, and buffalo milk can be detected by means of an ELISA that uses a commercial MAb against bovine IgG. However, the assay does not detect bovine IgG in UHT or in reconstituted nonfat dried milk, owing to the denaturation of the target epitope by heat treatment required to produce such products [75]. Using the same antibody, 0.001% bovine milk adulteration of ovine or buffalo milk, 0.01% bovine milk adulteration of goat milk, 0.001% bovine milk in caprine cheese, and 0.01% bovine in ovine and buffalo soft cheese have been detected [76]. A monoclonal antibody against bovine β-casein has been used for the detection of bovine milk in ovine and caprine milk, and for the detection of bovine cheese in ovine cheese. A detection limit of 0.5% bovine milk or cheese has been reported, which is not affected by the intensity of heat treatment of bovine milk [77–80]. The detection of 0.25% of caprine milk in ovine milk and 0.5% of bovine cheese in ovine cheese has been carried out by means of ELISAs that use an MAb against caprine α_{s2}-casein [81–83].

Immunoassays using monoclonal antibodies against bovine κ-casein have been applied in an automated optical biosensor, for the detection of bovine milk in caprine and ovine milk with a detection limit of 0.1% and a measurement range of 0.1%–10% bovine milk. They are proposed as fast control system of raw milk prior to manufacture of milk products [84,85].

Furthermore, ELISA techniques for the detection of milk or cheese adulteration have been commercialized. Bovine IgG can be detected in milk and cheese (e.g., RIDASCREEN® CIS, R-Biopharm AG, Darmstadt, Germany; RC-BOVINO®, Zeu-Inmunotec SL, Saragosa, Spain) with detection limits of 0.1% and 0.5%, respectively. In addition, using a similar procedure, caprine IgG can be detected in ovine milk with a detection limit of about 1% (e.g., RIDASCREEN GIS, R-Biopharm AG; RC-CAPRINO®, Zeu-Inmunotec SL). However, these methods are not accurate, if bovine milk treated with UHT has been used. In this case, assays based on antibodies that recognize caseins are adequate. A test involving MAbs recognizing bovine γ_1-, γ_2-, γ_3-, and β-caseins (e.g., RIDASCREEN Casein, R-Biopharm AG) with a detection limit of 0.5% of bovine casein in cheese has also been commercialized. Another approach is the development of fast immunochromatographic tests that use antibodies against bovine IgGs and have a detection limit of 0.5% or 1% of bovine milk in milk or cheese, respectively (e.g., RIDA® QUICK CIS, R-Biopharm AG; IC-BOVINO, Zeu-Inmunotec SL). A similar commercial test detects caprine milk in ovine milk or cheese (IC-CAPRINO, Zeu-Inmunotec SL).

In conclusion, ELISAs can be applied for quantitative determination of adulteration. Their simplicity and sensitivity make them very practical for routine controls of dairy products.

34.3 Methods of Detection Based on Milk Fat

The genetically controlled biosynthesis of milk fat results in differences in the composition of milk fat triglycerides and fatty acid profiles of each milk kind. Gas chromatography methods have been applied for the determination of triglyceride profiles or the ratios of individual fatty acids. Relationships among particular fatty acids have been proposed since 1963, as indexes for the detection of bovine milk in ovine and caprine milk with a high detection limit of 15%–20% [1]. The ratio of lauric:capric fatty acids (12:10), which is much higher in bovine than in caprine or ovine milk has been used to determine the presence of bovine milk in ovine or caprine cheeses [86]. More recently, 10 NMR parameters of ^{13}C NMR spectra of triglycerides have been used for distinguishing milks from different animal species [87].

However, detection based on milk fat composition is not reliable, as triglyceride profiles in milk fat are also affected by the environmental factors, such as animal nutrition and season of the year. Furthermore, the addition of skimmed milk from different animal species cannot be detected. Finally, fatty acids profiles can change during cheese ripening owing to their modifications resulting in various aromatic substances.

34.4 DNA-Based Methods of Detection—Species-Specific PCR

PCR is an amplification procedure for generating large quantities, over a million fold, of a specific DNA sequence *in vitro*. As described by Glick and Pasternak [88], a typical PCR process entails 30 or more successive cycles, each one consisting of three successive steps, i.e., denaturation at 95°C, renaturation at ~55°C, and *in vitro* DNA synthesis at ~75°C. The essential components are the target sequence in a DNA sample from 100 to ~35,000 bp in length, two synthetic oligonucleotide primers in a vast molar excess, which are complementary to regions on the opposite strand that flank the target DNA sequence, a thermostable DNA polymerase (e.g., Taq DNA polymerase), and four deoxyribonucleotides (dNTPs). Specific PCR procedures for the detection of a species-specific nucleotide sequence in food of animal origin have been developed. PCR techniques are more advantageous than ELISA techniques in terms of sensitivity and suitability for analyzing processed products. Accordingly, they have been considered as a promising tool of dairy research for the detection of milk adulteration.

Lipkin et al. [89] proved that milk samples can serve as a convenient source of purified DNA owing to their somatic-cell content, consisting mainly of leucocytes, as well as epithelial cells. This DNA can serve as a substrate for the amplification of specific DNA sequences using PCR. Milk and dairy products are subjected to various treatments and processes to have an extended shelf-life or to develop special characteristics. Heat treatments of various intensities, condensation, drying, rennet or acid coagulation, fermentation, and cheese ripening can substantially change the environment of the dairy food, whereas somatic cells and DNA molecules are relatively stable under these conditions. The PCR methods reported for the detection of adulteration in milk and dairy products are presented in Table 34.2.

The first step for the application of molecular genetic techniques is the isolation from dairy samples of genomic DNA, free of inhibitors. The adequate protocols have to be optimized to extract a high quantity of DNA efficiently without affecting its integrity. A cell pellet from milk samples is obtained by centrifugation [89,90]. The cell lysis is carried out with the appropriate extraction buffer, followed by treatment with chloroform/methanol, and finally, DNA can be concentrated by ethanol precipitation [80,91–96] or by adsorption on silica particles [97]. The DNA can be extracted from the cell pellet using silica spin columns [98]. A procedure based on

Table 34.2 PCR Methods Used for the Detection of Adulteration in Milk and Dairy Products

PCR Format/Dairy Samples	Species Detection	Detection Limit	Quantitative Detection	Ref.
PCR-RFLP				
Ovine and caprine cheese samples	Bovine DNA in cheese	0.5%	N/A	[97]
Commercial Mozzarella and Feta cheese samples	Bovine DNA in cheese	N/A	N/A	[103]
Commercial cheese samples	Bovine DNA in cheese	1%	N/A	[104]
Experimental Feta cheese and yoghurt prepared from binary mixtures of ovine and bovine milk	Bovine DNA in cheese and yoghurt	1% for cheese / 2.5% for yoghurt	N/A	[110]
PCR-LCR-EIA				
Experimental binary mixtures of bovine milk with ovine, caprine, and buffalo milk	Bovine DNA in milk and cheese	5%	N/A	[116]
Commercial cheese samples				
Species-Specific PCR (Simplex PCR)				
Experimental mixtures of caprine and bovine milk	Bovine DNA in milk	0.1%	N/A	[91]
Commercial cheeses made with nonpasteurized, pasteurized, or UHT-treated milk mixtures (ovine/bovine/caprine), in different ripening stages	Bovine DNA in cheese	0.1%	N/A	[101]
Commercial cheese samples made from ovine, caprine, bovine milk, or milk mixtures (bovine/ovine, bovine/caprine)	Bovine DNA in cheese	N/A	N/A	[115]
Experimental Mozzarella cheese made from mixtures of bovine and buffalo milk	Bovine DNA in cheese	1.5%	N/A	[105]
Commercial Mozzarella cheese samples				
Experimental binary raw, pasteurized, and sterilized milk mixtures (bovine/caprine and ovine/caprine)	Bovine DNA in ovine or caprine milk	0.1%	N/A	[92]

(continued)

Table 34.2 (continued)   PCR Methods Used for the Detection of Adulteration in Milk and Dairy Products

PCR Format/Dairy Samples	Species Detection	Detection Limit	Quantitative Detection	Ref.
Experimental binary mixtures of raw, pasteurized, and sterilized milk	Bovine DNA in buffalo milk and cheese	0.1%	N/A	[93]
Experimental binary mixtures of buffalo Mozzarella and bovine Mozzarella cheese				
Experimental binary mixtures of raw, pasteurized, and sterilized milk (caprine/ovine)	Caprine DNA in ovine milk	0.1%	N/A	[94]
Mozzarella cheese	Bovine DNA in buffalo cheese	0.5%	N/A	[108]
Experimental Camembert cheeses made from binary mixtures of bovine/caprine or bovine/ovine milk	Bovine DNA in ovine and caprine cheese	0.5%	N/A	[8]
Experimental binary mixtures of caprine and ovine cheese	Caprine DNA in ovine cheese	1%	N/A	[111]
Commercial cheese samples				
Experimental binary mixtures of bovine and caprine or ovine cheese	Bovine DNA in ovine and caprine cheese	1%	N/A	[80]
Reference cheese made from binary mixtures of cheesemilk				
Commercial cheese samples				
Experimental mixtures of bovine and buffalo milk	Bovine DNA in buffalo milk and cheese	0.1%	N/A	[112]
Commercial Mozzarella cheese samples				
Duplex PCR				
Experimental mixtures of bovine and buffalo milk and commercial buffalo Mozzarella samples	Simultaneous detection of bovine and buffalo DNA in cheese and milk	1% bovine milk or 1% buffalo milk	N/A	[102]

Table 34.2 (continued)   PCR Methods Used for the Detection of Adulteration in Milk and Dairy Products

PCR Format/Dairy Samples	*Species Detection*	*Detection Limit*	*Quantitative Detection*	*Ref.*
Commercial Mozzarella cheese samples	Simultaneous detection of bovine and buffalo DNA in cheese	N/A	N/A	[107]
Experimental cheeses made from binary mixtures of bovine and ovine milk	Simultaneous detection of bovine and ovine DNA in cheese	0.1% bovine milk	1%–50%	[98]
Commercial cheese samples				
Experimental cheeses made from binary mixtures of bovine and caprine milk	Simultaneous detection of bovine and caprine DNA in cheese	0.1% bovine milk	1%–60%	[99]
Commercial cheese samples				
Multiplex PCR				
Experimental cheeses made from mixtures of bovine, ovine, and caprine milk	Simultaneous detection of bovine, ovine, and caprine DNA in cheese	0.5% bovine milk	N/A	[104]
Commercial cheese samples				
RT-PCR				
Experimental and commercial Mozzarella cheese	Bovine DNA in cheese	0.1%	0.6%–20%	[109]
Experimental binary mixtures of raw and pasteurized bovine and ovine milk	Bovine DNA in milk	0.5%	0.5%–10%	[95]
Experimental binary mixtures of raw and pasteurized caprine and ovine milk	Bovine DNA in milk	0.6%	0.5%–10%	[96]
Commercial bovine and caprine milk	Bovine DNA	35 pg bovine DNA	N/A	[100]
Commercial bovine, caprine, and water-buffalo cheese samples				

N/A, not available.

resin with selective affinity has been proposed, to avoid the step of purification of somatic cells from other milk components [90]. Several adequate protocols for DNA purification are provided with commercial kits. The DNA extraction from cheese starts with the preparation of a cheese homogenate in Tris–HCl buffer at pH 7.5–8.0, in the presence of guanidium isocyanate, EDTA, 2-mercaptoeathanol, or sodium dodecyl sulfate. After addition of chilled ethanol, further purification is carried out by spin column or by adsorption to silica particles followed by repetitive washing steps [97–100]. Often, after cheese-sample digestion, lysis by proteinase K is carried out and the lysate can be purified by chloroform addition or ethanol precipitation, or/and by means of a spin column, according to the instructions of the kit manufacturer [93,101–105].

The second step is the efficient PCR amplification of an appropriate target DNA sequence. Initially, DNA-based methods used nuclear DNA. The target for PCR amplification was a *Bos taurus* β-casein region [97]. However, mitochondrial (mt) DNA has been found to be more suitable than nuclear DNA for PCR amplification. The reasons are that the copies of mt DNA in a cell are about 1000 times more than those of nuclear DNA, they have an appropriate length, and contain a great number of point mutations defining differences among the species [101,106]. Amplification of mitochondrial cytochrome *b* DNA sequences by PCR has been used for the identification of milk kind in dairy samples [8,91,100,102,103,105–107,109,110]. In addition, D-loop region [8,101], sequence of mitochondrial cytochrome oxidase I subunit [108], and cytochrome oxidase II [8] have been amplified. There are several reports about PCR targeting the mitochondrial-encoded gene for 12S rRNA [8,80,92–96,111]. In addition, PCR targeting both the mitochondrial 12S rRNA and 16S rRNA has been also reported [98,104]. Another procedure based on two targets, i.e., mitochondrial cytochrome *b* and nuclear growth hormone (GH) genes has been presented [109]. Single-copy nuclear genes can be used to avoid problems resulting from the large variability in mt DNA copy number among the species and individuals of the same species. Very recently, nuclear κ-casein gene has been used for the simultaneous detection of DNA from bovine, ovine, caprine, buffalo milk, and dairy products [112].

Various PCR formats have been put into practice for the detection of milk kind in dairy products (Table 34.2). PCR-restriction fragment length polymorphism analysis (PCR-RFLP), i.e., digestion of PCR products with restriction enzymes has been reported using β-casein [97] and cytochrome *b* primers [103,110], and it is proposed as a qualitative rather than quantitative method. Apart from simplex PCR, duplex PCRs have been configured with the objective to identify two milk kinds in a single PCR assay. For this purpose, specific primers for the two different targets are included in the reaction mixture [98–100,102,107]. Moreover, multiplex PCR using primers for the mitochondrial 12S and 16S rRNA genes of ewe's, goat's, and cow's milk in a single PCR assay has been presented [104].

The third step of molecular techniques is the detection of amplicons. Amplification products are resolved by agarose gel electrophoresis calibrated by simultaneous analysis of a molecular weight marker containing fragments of known sizes. Gels are stained with ethidium bromide, and UV light is used for their visualization. The quantification of milk kind in the dairy samples is based on the intensity of the relevant amplicon band. In the simplex PCR format, the quantification is related to one band that results from the amplification of the target sequence assigned to the nondeclared milk kind [92–94,101,111]. In the duplex PCR, the intensities of the bands of both the targets are used to normalize the calculation or to detect two milk kinds simultaneously in a dairy product. It is proposed as a simple and accurate quantitative approach to overcome variations that might occur during sample preparation, because the quantity of the target fragment is related to the sum of the quantity of the two targets [98,99].

Real-time PCR (RT-PCR) procedures have been successfully applied for the quantification of a target sequence in a dairy sample in a simplex [95,96,109] or a duplex format [100]. Detection is by fluorescence continuously monitored during PCR amplification. Therefore, gel electrophoresis is not required. For RT-PCR, a TaqMan® (Roche Molecular Systems, Inc, Pleasanton, CA) fluorogenic probe, labeled with a fluorophore (reporter) at the 5′ end and a nonfluorescent chromophore (quencher) at the 3′ end, is used. The middle nucleotides of the probe are complementary to the target DNA. As they hybridize to the target DNA sequence between the flanking primers, the exonuclease activity of the Taq DNA polymerase releases the reporter fluorophore molecule from the probe. As a result, fluorophore is not quenched, i.e., it fluoresces, and the fluorescence resulting from the accumulation of PCR product is continuously monitored. The increase in the fluorescence is proportional to the amount of amplicons produced during PCR [88,95,96,113]. The detection of adulterant species is based on the calculation of the threshold cycle (Ct), i.e., the cycle at which statistically significant fluorescence is detected above the background. López-Calleja et al. [95,96] have used a TaqMan probe, designed to hybridize in a mammalian PCR system as well, which serves as an endogenous control and amplifies any mammalian DNA from the sample. The Ct value in the mammalian PCR system is used to normalize the results. Loparelli et al. [109] have used the single-copy nuclear GH gene PCR system as a reference marker to check the reliability of RT-PCR quantification of species-specific DNA. The superiority of RT-PCR in terms of detection limit is shown in Table 34.2. Furthermore, RT-PCR does not require post-PCR processing steps and many samples can be analyzed in a single run. However, till date, there exist cost limitations with regard to instrumentation and TaqMan chemistry.

Another approach is the oligonucleotide microarray hybridization analysis of PCR products from the mitochondrial cytochrome *b* gene DNA that has been applied to cheese samples [114]. The fluorophor-labeled PCR products are detected by hybridization to an oligonucleotide microarray carrying a set of characteristic sequences covalently immobilized on the activated probe glass slides. When PCR products are hybridized to the immobilized probe set, distinct signals are detected assigned to the corresponding species-specific probes. The post-PCR procedure is short and the method has detected up to three different species in cheeses.

As somatic cell counts (SCC) of milk samples are the source of DNA for the various PCR approaches, the effect of processes such as thermal treatments of milk or cheese and cheese ripening has been examined. It is expected that DNA yield depends on the total number of somatic cells in the samples [102]. The SCC for raw milk has been found to be double the value than that of heat-treated milk, without affecting the sensitivity of PCR used [92–94]. The sensitivity of the methods is strongly influenced by the number of PCR cycles [90,91,99,106]. However, addition of preservatives in milk and refrigeration or freezing up to 200 days does not interfere with PCR amplification [89]. Furthermore, DNA suitable for PCR can be extracted from milk powder or even bovine caseinate [8].

As shown in Table 34.2, the results regarding cheese samples are very promising. Nevertheless, very low level of amplification of cheese DNA has been reported and has been attributed to the low integrity of the cheese DNA [99]. The presence of inhibitory substances can affect DNA amplification; the existence of such a problem can be checked using an internal control in each PCR reaction. Furthermore, calf rennet used in cheesemaking has not influenced the results regarding genuine ovine and caprine cheeses [80,110].

In conclusion, DNA-based techniques could be appropriate control methods for the detection of adulteration of milk and dairy products manufactured from adulterated milk. Their main advantage is that they can be applied in heat-treated milk and dairy products, in which particular protein fractions may be denatured. The same is true for cheese. During cheese ripening, extensive proteolysis may occur but the mammary somatic cells are not affected. Molecular techniques have been found to be sensitive, whereas duplex and RT-PCR assays can provide reliable estimation of nondeclared milk kinds in dairy products. However, it has to be taken into consideration that SCC of milk, which is the source of DNA, is not controlled, as it is affected by the animal species, and by the genetic and physiological factors for the same species.

References

1. Ramos, M. and Juárez, M., Chromatographic, electrophoretic and immunological methods for detecting mixtures of milk from different species, IDF Bulletin No. 202, Brussels: International Dairy Federation, 1986, pp. 175–187.
2. De la Fuente, M.A. and Juárez, M., Authenticity assessment of dairy products, *Crit. Rev. Food Sci. Nutr.*, 45, 563, 2005.
3. Moatsou, G. and Anifantakis, E., Recent developments in antibody-based analytical methods for the differentiation of milk from different species, *Int. J. Dairy Technol.*, 56, 133, 2003.
4. Hurley, I.P., Ireland, H.E., Coleman, R.C., and Willimas, J.H.H., Application of immunological methods for the detection of species adulteration in dairy products, *Int. J. Food Sci. Technol.*, 39, 873, 2004.
5. Mafra, I., Ferreira, I.M.P.L.V.O., and Oliveira, M.B.P.P., Food authentication by PCR-based methods, *Eur. Food Res. Technol.*, 227, 649, 2008.
6. Andrews, A.T., Proteinases in normal bovine milk and their action on caseins, *J. Dairy Res.*, 50, 45, 1983.
7. Furtado, M.M., Detection of cow milk in goat milk by polyacrylamide gel electrophoresis, *J. Dairy Sci.*, 66, 1822, 1983.
8. Mayer, H.K., Milk species identification in cheese varieties using electrophoretic, chromatographic and PCR techniques, *Int. Dairy J.*, 15, 595, 2005.
9. Kaminarides, S.E., Kandarakis, I.G., and Moschopoulou, E., Detection of bovine milk in ovine Halloumi cheese by electrophoresis of αs1-casein, *Aust. J. Dairy Technol.*, 50, 58, 1995.
10. Veloso, A.C.A., Teixera, N., Peres, A.M., Mendonça, A., and Ferreira, I.M.P.L.V.O., Evaluation of cheese authenticity and proteolysis by HPLC and urea-polyacrylamide gel electrophoresis, *Food Chem.*, 87, 289, 2004.
11. Kaminarides, S.E. and Koukiasa, P., Detection of bovine milk in ovine yoghurt by electrophoresis of para-κ-casein, *Food Chem.*, 78, 53, 2002.
12. Addeo, F., Anelli, G., and Chianese, L., Gel isoelectric focusing of cheese proteins to detect milk from different species in mixture, In: *IDF Bulletin 202*, International Dairy Federation, Eds, Brussels, Belgium, 1986, pp. 191–192.
13. Addeo, F., Moio, L., Chianese, L., and Stingo, C., Improved procedure for detecting bovine and ovine milk mixtures in cheese by isoelectric focusing of para-κ-casein, *Milchwissenschaft*, 45, 221, 1990.
14. Addeo, F., Moio, L., Chianese, L., Stingo, C., Resmini, P., Berner, I., Karause, I., Di Luccia, A., and Bocca, A., Use of plasmin to increase the sensitivity of the detection of bovine milk in ovine cheese by gel isoelectric focusing of γ_2-caseins, *Milchwissenschaft*, 45, 708, 1990.
15. Commission Regulation, Reference method for the detection of cows' milk and caseinate in cheeses from ewes', goats' and buffaloes' milk, EC regulation No 213/2001 of 9 January 2001, *Off. J. Eur. Commun.*, L37, 51–60, 2001.
16. Addeo, F., Nicolai, M.A., Chianese, L., Moio, L., Spagna, M.S., Bocca, A., and Del Giovine, L., A control method to detect bovine milk in ewe and water buffalo cheese using immunoblotting, *Milchwissenschaft*, 50, 83, 1995.
17. Moio, L., Chianese, L., Rivemale, M., and Addeo, F., Fast detection of bovine milk in Roquefort cheese with Phastsystem® by gel electrophoretic focusing and immunoblotting, *Le Lait*, 72, 87, 1992.

18. Mayer, H.K., Heidler, D., and Rockenbauer, C., Determination of the percentages of cows', ewes' and goats' milk in cheese by isoelectric focusing and cation-exchange HPLC of γ- and para-κ-caseins, *Int. Dairy J.*, 10, 619, 1997.

19. Recio, L., Amigo, L., and López-Fandiño, R., Assessment of the quality of dairy products by capillary electrophoresis of milk proteins, *J. Chromatogr. A*, 697, 231, 1997.

20. Corradini, C. and Cavazza, A., Application of capillary zone electrophoresis (CZE) and micellar electrokinetic chromatography (MEKC) in food analysis, *Ital. J. Food Sci.*, 4, 299, 1998.

21. De Jong, N., Visser, S., and Olieman, C., Determination of milk proteins by capillary electrophoresis, *J. Chromatogr. A*, 652, 207, 1993.

22. Cattaneo, T.M.P., Nigro, F., Toppino, P.M., and Denti, V., Characterization of ewe's milk by capillary zone electrophoresis, *J. Chromatogr. A*, 721, 345, 1996.

23. Cattaneo, T.M.P., Nigro, F., and Greppi, G.F., Analysis of cow, goat and ewe milk mixtures by capillary zone electrophoresis (CZE): Preliminary approach, *Milchwissenschaft*, 51, 616, 1996.

24. Lee, S.-J., Chen, M-C., and Lin C.-W., Detection of cows' milk in goats' milk by capillary zone electrophoresis, *Aust. J. Dairy Technol.*, 56, 24, 2001.

25. Molina, E., Martín-Álvarez, J., and Ramos, M., Analysis of cows', ewes' and goats' milk mixtures by capillary electrophoresis: Quantification by multivariate regression analysis, *Int. Dairy J.*, 9, 99, 1999.

26. Molina, E., De Frutos, M., and Ramos, M., Capillary electrophoresis characterization of the casein fraction of cheeses made from cows', ewes' and goats' milks, *J. Dairy Res.*, 67, 209, 2000.

27. Miralles, B., Ramos, M., and Amigo, L., Characterization of fresh cheeses by capillary electrophoresis, *Milchwissenschaft*, 60, 278, 2005.

28. Recio, I., García-Risco, M.R., Amigo, L., Molina, E., Ramos, M., and Martin-Álvarez, P.J., Detection of milk mixtures in Halloumi cheese, *J. Dairy Sci.*, 87, 1595, 2004.

29. Rodriguez-Nogales, J.M. and Vázquez, F., Application of electrophoretic and chemometric analysis to predict the bovine, ovine and caprine milk percentages in Panela cheese, an unripened cheese, *Food Control*, 18, 580, 2007.

30. Calvo, M.M., Amigo, L., Olano, A., Martin, P.J., and Ramos, M., Effect of thermal treatments on the determination of bovine milk added to ovine or caprine milk, *Food Chem.*, 32, 99, 1989.

31. Amigo, L., Ramos, M., and Martin-Álvarez, P.J., Effect of technological parameters on electrophoretic detection of cow's milk in ewe's milk cheeses, *J. Dairy Sci.*, 74, 1482, 1991.

32. Amigo, L., Ramos, M., Calahau, L., and Barbosa, M., Comparison of electrophoresis, isoelectric focusing, and immunodiffusion in determination of cow's and goat's milk in Serra da Estrela cheeses, *Le Lait*, 72, 95, 1992.

33. Molina, E., Ramos, M., and Martin-Álvarez, P.J., Prediction of the percentages of cows', goats' and ewes' milk in "Iberico" cheese by electrophoretic analysis of whey proteins, *Z. Lebensm. Unters. Forsch.*, 201, 331, 1995.

34. Molina, E., Fernández-Fournier, A., De Frutos, M., and Ramos, M., Western blotting of native and denatured bovine β-lactoglobulin to detect addition of bovine milk in cheese, *J. Dairy Sci.*, 79, 191, 1996.

35. Addeo, F., Moio, L., Chianese, L., and Di Luccia, A., Detection of bovine milk in ovine milk or cheese by gel electrophoretic focusing of β-lactoglobulin: Applications and limitations, *Ital. J. Food Sci.*, 1, 45, 1989.

36. Rispoli, S. and Saugues, R., Isoélectroocalisation des lactosérums de fromages de mélange brebis-vache sur gel polyacrylamide—Application à la recherche et au dosage du lait de vache dans les fromages de brebis, *Le Lait*, 69, 211, 1989.

37. Rispoli, S., Rivemale, M., and Saugues, R, Mise en évidence et évaluation de la quantité de lait de vache dans les fromages de brebis par isoélectrofocalisation des lactosérums. Application au cas de fromages très protéolysés: fromages type Roquefort, *Le Lait*, 71, 501, 1991.

38. Cartoni, G.P., Coccioli, E., Jasionowska, R., and Masci, M., Determination of cow milk in buffalo milk and mozzarella cheese by capillary electrophoresis of the whey protein fractions, *Ital. J. Food Sci.*, 10, 127, 1998.

39. Cartoni, G.P., Coccioli, E., Jasionowska, R., and Masci, M., Determination of cow milk in ewe milk and cheese by capillary electrophoresis of the whey protein fractions, *Ital. J. Food Sci.*, 10, 317, 1998.

40. Cartoni, G.P., Coccioli, E., Jasionowska, R., and Masci, M, Determination of cows' milk in goats' milk and cheese by capillary electrophoresis of the whey protein fractions, *J. Chromatogr. A*, 846, 135, 1999.

41. Herrero-Martínez, J.M., Simó-Alfonso, E.F., Ramis-Ramos, G., Gelfi, C., and Righetti, P.G., Determination of cow's milk in non-bovine and mixed cheeses by capillary electrophoresis of whey proteins in acidic isoelectric buffers, *J. Chromatogr. A*, 878, 261, 2000.

42. Herrero-Martínez, J.M., Simó-Alfonso, E.F., Ramis-Ramos, G., Gelfi, C., and Righetti, P.G., Determination of cow's milk and ripening time in nonbovine cheese by capillary electrophoresis of the ethanol-water protein fraction, *Electrophoresis*, 21, 633, 2000.

43. Park, Y.W., Juárez, M., Ramos, M., and Haenlein, G.F.W., Physico-chemical characteristics of goat and sheep milk, *Small Ruminant Res.*, 68, 88, 2007.

44. Haasnoot, W., Venema, D.P., and Elenbaas, H.L., Determination of cow milk in the milk and cheese of ewes and goats by fast protein liquid chromatography, *Milchwissenschaft*, 41, 642–645, 1986.

45. Volitaki, A.J. and Kaminarides, S.E., Detection of bovine milk in ovine Halloumi cheese by HPLC analysis of cheese caseins hydrolysed by plasmin, *Milchwissenschaft*, 56, 207–210, 2001.

46. Bramanti, E., Sortino, C., Onor, M., Beni, F. and Raspi, G., Separation and determination of denatured α_{s1}-, α_{s2}-, β- and κ-caseins by hydrophobic interaction chromatography in cows', ewes' and goats' milk, milk mixtures and cheeses, *J. Chromatogr. A*, 994, 59, 2003.

47. Moatsou, G., Hatzinaki, A., Psathas, G., and Anifantakis, E., Detection of caprine casein in ovine Halloumi cheese, *Int. Dairy J.*, 14, 219, 2004.

48. Laezza, P., Nota, G., and Addeo, F., Determination of bovine and ovine milk in mixtures by fast ion-exchange chromatography of whey proteins, *Milchwissenschaft*, 46, 559–561, 1991.

49. De Frutos, M., Cifuentes, A., and Diez-Masa, J.C., Amigo, L and Ramos, M., Application of HPLC for the detection of proteins in whey mixtures from different animal species, *J. High Resolut. Chromatogr.*, 14, 289, 1991.

50. De Frutos, M., Cifuentes, A., Amigo, L., Ramos, M., and Diez-Masa, J.C., Rapid analysis of whey proteins from different animal species by reversed-phase high-performance liquid chromatography, *Z. Lebensm. Unters. Forsch.*, 195, 326, 1992.

51. Pellegrino, L., Tirelli, A., and Masotti, F, Detection of cow milk in non-bovine cheese by HPLC of whey proteins. Note 2: Application to ewe's milk cheeses, *Scienza e Tecnica Lattiero Casearia*, 43, 297, 1992.

52. Ferreira, I.M.P.L.V.O. and Cacote, H., Detection and quantification of bovine, ovine and caprine milk percentages in protected denomination of origin cheeses by reversed-phase high-performance liquid chromatography of beta-lactoglobulins, *J. Chromatogr. A.*, 1015, 111, 2003.

53. Enne, G., Elez, D., Fondrini, F., Bonizzi, I., Feligini, M., and Aleandri, R., High-performance liquid chromatography of governing liquid to detect illegal bovine milk's addition in water buffalo Mozzarella: comparison with results from raw milk and cheese matrix, *J. Chromatogr. A.*, 1094, 169, 2005.

54. Moatsou, G., Kandarakis, I., and Fournarakou, M., Detection of bovine milk in caprine milk by reversed-phase HPLC of caseinomacropeptides, *Milchwissenschaft*, 58, 274, 2003.

55. Cozzolino, R., Passalacqua, S., Salerni, S., and Garozzo, D., Identification of adulteration in water buffalo mozzarella and in ewe cheese by using whey proteins as biomarkers and matrix-assisted laser desorption/ionization mass spectrometry, *J. Mass Spectrom.*, 37, 985, 2002.

56. Chen, R.K., Chang, L.-W., Chung, Y.-Y., Lee, M.-H., and Ling, Y.-C., Quantification of cow milk adulteration in goat milk using high-performance liquid chromatography with electrospray ionization mass spectrometry, *Rapid Commun. Mass Spectrom.*, 18, 1167, 2004.

57. Levieux, D., Detection immunologique des malanges de laits de diverses especes, *Proc. Int. Congr. Paris*, 15ST, 1978.

58. Levieux, D., The development of a rapid and sensitive method based on hemagglutination inhibition for the measurement of cows' milk in goats' milk, *Ann. Res. Vétérinaires*, 11, 151, 1980.

59. Aranda, P., Oría, R., and Calvo, M. Detection of cow's milk in ewe's cheese by an immunoblotting method, *J. Dairy Res.*, 55, 121, 1988.

60. Garcia, T., Martin, R., Rodriguez, E., Morales, P., Hernández, P., and Sanz, B., Detection of bovine milk in ovine milk by an indirect enzyme-linked immunosorbent assay, *J. Dairy Sci.*, 73, 1489, 1990.

61. García, T., Martín, R., Morales, P., González, I., Sanz, B., and Hernández, P., Sandwich ELISA for detection of caprine milk in ovine milk, *Milchwissenschaft*, 48, 563, 1993.

62. García, T., Martín, R., Morales, P., González, I., Sanz, B., and Hernández, P., Detection of goats' milk in ewes' milk by an indirect ELISA. *Food Agric. Immunol.*, 6, 113, 1994.

63. Beer, M., Krause, I., Stapf, M., Schwarzer, C., and Klostermeyer, H., Indirect competitive enzyme-linked immunosorbent assay for the detection of native and heat-denatured bovine β-lactoglobulin in ewes' and goats' milk cheese, *Zeitschrift Lebensmittel Unters. Forsch.*, 203, 21, 1996.

64. Rodríguez, E., Martín, R., García, T., Hernández, P.E., and Sanz, B., Detection of cows' milk in ewes' milk and cheese by an indirect enzyme-linked immunosorbent assay (ELISA), *J. Dairy Res.*, 57, 197, 1990.

65. Rodríguez, E., Martín, R., García, T., Azcona, J.L., Sanz, B., and Hernández, P.E., Indirect ELISA for detection of goats' milk in ewes' milk and cheese. *Int. J. Food Sci. Technol.*, 26, 457, 1991.

66. Rodriguez, E., Martin, R., García, T., Gonzàlez, I., Morales, P., Sanz, B., and Hernández, P.E., Detection of cows' milk in ewes' milk cheese by a sandwich enzyme-linked immunosorbent assay (ELISA), *J. Sci. Food Agric.*, 61, 175, 1993.

67. Rodríguez, E., Martín, R., García, T., Morales, P., Gonzàlez, I., Sanz, B., and Hernández, P.E., Sandwich ELISA for detection of goats' milk in ewes' milk and cheese, *Food Agric. Immunol.*, 6, 105, 1994.

68. Richter, W., Krause, I., Graf, C., Sperrer, I., Schwarzer, C., and Klostermeyer, H., An indirect competitive ELISA for the detection of cows' milk and caseinate in goats' and ewes' milk and cheese using polyclonal antibodies against bovine γ-caseins. *Zeitschrift Lebensmittel Unters. Forsch.*, 204, 21, 1997.

69. Bitri, L., Rolland, M.P., and Besançon, P., Immunological detection of bovine caseinomacropeptide in ovine and caprine dairy products, *Milchwissenschaft*, 48, 367, 1993.

70. Rolland, M.P., Bitri, L., and Besançon, P., Polyclonal antibodies with predetermined specificity against bovine αs1-casein: Application to the detection of bovine milk in ovine milk and cheese, *J. Dairy Res.*, 60, 413, 1993.

71. Rolland, M.P., Bitri, L., and Besançon, P., Monospecificity of the antibodies to bovine αs1-casein fragment 140–149: Application to the detection of bovine milk in caprine dairy products, *J. Dairy Res.*, 62, 83, 1995.

72. Pizzano, R., Nicolai, M.A., Padovano, P., Ferranti, P., Barone, F., and Addeo, F., Immunochemical evaluation of bovine β-casein and its 1–28 phosphopeptide in cheese during ripening, *J. Agric. Food Chem.*, 48, 4555, 2000.

73. Levieux, D. and Venien, A., Rapid, sensitive, two-site ELISA for the detection of cows' milk in goats' or ewes' milk using monoclonal antibodies., *J. Dairy Res.*, 61, 91, 1994.

74. Negroni, L., Bernard, H., Clement, G., Chatel, J.M., Brune, P., Frobert, Y., Wal, J.M., and Grassi, J., Two-site enzyme immunometric assays for determination of native and denatured β-lactoglobulin, *J. Immunol. Methods*, 220, 25, 1998.

75. Hurley, I.P., Coleman, R.C., Ireland, H.E., and Williams J.H.H., Measurement of bovine IgG by indirect competitive ELISA as a means of detecting milk adulteration, *J. Dairy Sci.*, 87, 543, 2004.

76. Hurley, I.P., Coleman, R.C., Ireland, H.E., and Williams J.H.H., Use of sandwich ELISA for the detection and quantification of adulteration of milk and soft cheese, *Int. Dairy J.*, 16, 805, 2006.

77. Anguita, G., Martín, R., García, T., Morales, P., Haza, A.I., González, I., Sanz, B., and Hernández, P.E., Indirect ELISA for detection of cows' milk in ewes' and goats' milks using a monoclonal antibody against bovine β-casein, *J. Dairy Res.*, 62, 655, 1995.

78. Anguita, G., Martín, R., García, T., Morales, P., Haza, A.I., González, I., Sanz, B., and Hernández, P.E., Immunostick ELISA for detection of cows' milk in ewes' milk and cheese using a monoclonal antibody against bovine β-casein, *J. Food Protection*, 59, 436, 1996.

79. Anguita, G., Martín, R., García, T., Morales, P., Haza, A.I., González, I., Sanz, B., and Hernández, P.E., Detection of bovine casein in ovine cheese using deoxigenated monoclonal antibodies and a sandwich ELISA, *Milchwissenschaft*, 52, 511, 1997.

80. López-Calleja, I.M., González, I., Fajardo, V., Martin, I., Hernández, P.E., García, T., and Martín, R., Application of an indirect ELISA and a PCR technique for detection of cows' milk in sheep's and goats' milk cheeses, *Int. Dairy J.*, 17, 87, 2007.

81. Haza, A.I., Morales, P., Martín, R., García, T., Anguita, G., González, I., Sanz, B., and Hernández, P.E., Development of monoclonal antibodies against caprine α_{s2}-casein and their potential for detecting the substitution of ovine milk by caprine milk by an indirect ELISA, *J. Agric. Food Chem.*, 44, 1756, 1996.

82. Haza, A.I., Morales, P., Martín, R., García, T., Anguita, G., González, I., Sanz, B., and Hernández, P.E., Use of monoclonal antibody and two enzyme-linked immunosorbent assay formats for detection and quantification of the substitution of caprine milk for ovine milk, *J. Food Prot.*, 60, 973, 1997.

83. Haza, A.I., Morales, P., Martín, R., García, T., Anguita, G., González, I., Sanz, B., and Hernández, P.E., Detection and quantification of goat's cheese in ewe's cheese using a monoclonal antibody and two ELISA formats, *J. Sci. Food Agric.*, 79, 1043, 1999.

84. Haasnoot, W., Smits, N.G.E., Kemmers-Voncken, A.E.M., and Bremer, M.G.E.G., Fast biosensor immunoassays for the detection of cows' milk in the milk of ewes and goats, *J. Dairy Res.*, 71, 322, 2004.

85. Haasnoot, W., Marchesini, G.R., and Koopai, K., Spreeta-based biosensor immunoassays to detect fraudulent adulteration in milk and milk powder, *J. AOAC Int.*, 89, 849, 2006.

86. Iverson, J.L. and Sheppard, A.J., Detection of adulteration in cow, goat and sheep cheeses utilizing gas-liquid chromatographic fatty acid data, *J. Dairy Sci.*, 72, 1707, 1989.

87. Andreotti, G., Trivellone, E., Lamanna, R., Di Luccia, A., and Motta, A., Milk identification of different species: 13C NMR Spectra of triglycerols from cows' and buffaloes' milks, *J. Dairy Sci.*, 83, 2432, 2000.

88. Glick, B.R., and Pasternak, J.J., PCR, In: *Molecular Biotechnology*, 3rd edn., Glick, B.R. and Pasternak, J.J., Eds, ASM Press, Washington, DC, 2003, Chapter 5.

89. Lipkin, E., Shalom, A., Khatib, H., Soller, M., and Friedmann, A., Milk as a source of deoxyribonucleic acid and as a substrate for the polymerase chain reaction, *J. Dairy Sci.*, 76, 2025, 1993.

90. Amills, M., Francino, O., Jansa, M., and Sanchez, A., Isolation of genomic DNA from milk samples by using Chelex resin, *J. Dairy Res.*, 64, 231, 1997.

91. Bania, J., Ugorski, M., Polanowski, A., and Adamczyk, E., Application of polymerase chain reaction for detection of goats' milk adulteration by milk of cow, *J. Dairy Res.*, 68, 333, 2001.

92. López-Calleja, I., González Alonso, I., Fajardo, V., Rodríguez, M.A., Hernández, P.E., García, T., and Martín, R., Rapid detection of cows' milk in sheeps' and goats' milk by a species specific polymerase chain reaction technique, *J. Dairy Sci.*, 87, 2839, 2004.

93. López-Calleja, I., González Alonso, I., Fajardo, V., Rodríguez, M.A., Hernández, P.E., García, T., and Martín, R., PCR detection of cows' milk in water buffalo milk and mozzarella cheese, *Int. Dairy J.*, 15, 1122–1129, 2005.

94. López-Calleja, I., González Alonso, I., Fajardo, V., Martin, I., Hernández, P.E., García, T., and Martín, R., Application of polymerase chain reaction to detect adulteration of sheep's milk with goats' milk, *J. Dairy Sci.*, 88, 3115, 2005.

95. López-Calleja, I.M., González Alonso, I., Fajardo, V., Martin, I., Hernández, P.E., García, T., and Martín, R., Real-time TaqMan PCR for quantitative detection of cows' milk in ewes' milk mixtures, *Int. Dairy J.*, 17, 729, 2007.

96. López-Calleja, I., González Alonso, I., Fajardo, V., Martin, I., Hernández, P.E., García, T., and Martín, R., Quantitative detection of goats' milk in sheep's milk by real-time PCR, *Food Control*, 18, 1466, 2007.

97. Plath, A., Krause, I., and Einspanier, R., Species identification in dairy products by three different DNA-based techniques, *Z. Lebensm. Unters. Forsch.* A, 205, 437, 1997.

98. Mafra, I., Ferreira, I.M.P.L.V.O., Faria, M.A., and Oliveira, B.P.P., A novel approach to the quantification of bovine milk in ovine cheeses using duplex polymerase chain reaction method, *J. Agric. Food Chem.*, 52, 4943, 2004.

99. Mafra, I., Roxo, Á., Ferreira, I.M.P.L.V.O., Beatriz, M., and Oliveira, P.P., A duplex polymerase chain reaction for the quantitative detection of cows' milk in goats' milk cheese, *Int. Dairy J.*, 17, 1132, 2007.

100. Zhang, C.L., Fowler, M.R., Scott, N.W., Lawson, G., and Slater, A., A TaqMan real-time PCR system for the identification and quantification of bovine DNA in meats, milks and cheeses, *Food Control*, 18, 1149, 2007.

101. Maudet, C. and Taberlet, P., Detection of cows' milk in goats; cheeses inferred from mitochondrial DNA polymorphism, *J. Dairy Res.*, 68, 229, 2001.

102. Rea, S., Chikuni, K., Branciari, R., Sangamayya, R.S., Ranucci, D., and Avellini, P., Use of duplex polymerase chain reaction (duplex PCR) technique to identify bovine and water buffalo milk used in making mozzarella cheese, *J. Dairy Res.*, 68, 689, 2001.

103. Branciari, R., Nijman, I.J., Plas, M.E., Di Antonio, E., and Lenstra, J.A., Species origin of milk in Italian Mozzarella and Greek Feta cheese, *J. Food Prot.*, 63, 408, 2000.

104. Bottero, M.T., Civera, T., Nucera, D., Rosati, S., Sacchi, P., and Turi, R.M., A multiplex polymerase chain reaction for the identification of cows', goats' and sheeps' milk in dairy products, *Int. Dairy J.*, 13, 277, 2003.

105. Di Pinto, A., Conversano, C., Forte, V.T., Novello, L., and Tantillo, G.M., Detection of cow milk in buffalo "Mozzarella" by polymerase chain reaction (PCR) assay, *J. Food Qual.*, 27, 428, 2004.

106. Herman, L., Determination of the animal origin of raw food by species-specific PCR, *J. Dairy Res.*, 68, 429, 2001.

107. Bottero, M.T., Civera, T., Anastasio, A., Turi, R.M., and Rosati, S., Identification of cow's milk in "buffalo" cheese by duplex polymerase chain reaction, *J. Food Prot.*, 65, 362, 2002.

108. Feligini, M., Bonizzi, I., Curik, V.C., Parma, P., Greppi, G.F., and Enne, G. Detection of adulteration in Italian mozzarella cheese using mitochondrial DNA templates as biomarkers, *Food Technol. Biotechnol.*, 43, 91, 2005.

109. Lopparelli, R.M., Cardazzo, B., Balzan, S, Giaccone, V., and Novelli, E., Real-time TaqMan polymerase chain reaction detection and quantification of cow DNA in pure water buffalo mozzarella cheese: method validation and its application on commercial samples. *J. Agric. Food Chem.*, 55, 3429, 2007.

110. Stefos, G., Argyrokastritis, A., Bizelis, I., Moatsou, G., Anifantakis, E., and Rogdakis, E., Detection of bovine mitochondrial DNA specific sequences in Feta cheese and ovine yogurt by PCR-RFLP, *Milchwissenschaft*, 59, 509, 2004.

111. Díaz, I.,L.-C., González Alonso, I., Fajardo, V., Martin, I., Hernández, P., García Lacarra, T., and Martín de Santos, R., Application of a polymerase chain reaction to detect adulteration of ovine cheeses with caprine milk, *Eur. Food Res. Technol.*, 225, 345, 2007.

112. Reale, S., Campanella, A., Merigioli, A., and Pilla, F., A novel method for species identification in milk and milk-based products, *J. Dairy Res.*, 75, 107, 2008.

113. Woolfe, M. and Primrose, S., Food forensics: Using DNA technology to combat misdescription and fraud, *Trends Biotechnol*, 22, 222, 2004.

114. Peter, C., Brünen-Nieweler, C., Camman, K., and Börchers, T., Differentiation of animal species in food by oligonucleotide microarray hybridization, *Eur. Food Res. Technol.*, 219, 286, 2004.

115. Calvo, J.H., Osta, R., and Zaragoza, P., Species-specific amplification for detection of bovine, ovine and caprine cheese, *Milchwissenschaft*, 57, 444, 2002.

116. Klotz, A. and Einspanier, R., Development of a DNA-based screening method to detect cow milk in ewe, goat and buffalo milk and dairy products using PCR-LCR-EIA technique, *Milchwissenschaft*, 56, 67, 2001.

Chapter 35

Analysis of Antibiotics in Milk and Its Products

Jian Wang

Contents

35.1 Introduction..887
35.2 Analytical Methods..888
    35.2.1 Screening Methods...889
    35.2.2 Confirmatory Methods...890
        35.2.2.1 Sample Preparation...890
        35.2.2.2 Liquid Chromatography and Ultraperformance Liquid
                Chromatography Separation...891
        35.2.2.3 Mass Spectrometry...896
        35.2.2.4 LC–MS Confirmatory Criteria...897
35.3 Method Validation and Measurement Uncertainty...900
35.4 Conclusions...901
References..901

35.1 Introduction

Antibiotics, also known as antimicrobial, antibacterial, or anti-infective agents, include synthetic compounds such as sulfonamides and natural compounds such as penicillins, tetracyclines, and some macrolides. The term antibiotics originally meant only natural substances produced by

---

Disclaimer: The list of test kits, biosensors, SPE cartridges, and analytical columns described or mentioned in this chapter is by no means exhaustive and that any commercially available items cited do not in any way constitute an endorsement by the author.

bacteria or fungi, but now it is often used to refer to both synthetic and natural compounds. Antibiotics are used in both human medicine and veterinary practice. In the livestock industry and fish farming, antibiotics are employed for therapeutic (disease control), prophylactic (disease prevention), and subtherapeutic (growth promotion) purposes. Consequently, if the withdrawal time after treatment is not respected, or if antibiotics are not used correctly, it could lead to the presence of antibiotic residues in foods of animal origin, which in turn may provoke allergic reactions in some hypersensitive individuals, or cause the problem of drug-resistant pathogenic bacterial strains [1–3]. Some antibiotics such as chloramphenicol, and nitrofurans and their metabolites are associated with serious toxic effects in humans causing bone marrow depression and aplastic anemia [4], and/or mutagenic and carcinogenic effects [5]. Therefore, they are not allowed to be present in food. To ensure the safety of food for consumers and to facilitate the interest of international trade, the U.S. Food and Drug Administration (FDA) [6], European Union (EU) [7,8], Canada [9], FAO/WHO [10], and other international regulatory bodies have established the relevant regulations and maximum residue limits (MRLs) to monitor the level of approved antibiotics present in food. For example, the EU Council Regulation (EEC) 2377/90 [7], which describes the procedure for the establishment of MRLs for veterinary medicinal products in food of animal origin, controls the use of veterinary drugs. The EU Council Directive 96/23/EC [11] regulates and implements the residue control limits, which are set under the EEC 2377/90, of pharmacologically active compounds, i.e., substances having anabolic effects, banned or unauthorized substances, veterinary drugs or antibiotics, environmental contaminants, etc. The Directive 96/23/EC divides all chemical residues into Group A compounds, which comprise banned substances such as chloramphenicol and nitrofurans, and Group B compounds, which comprise all registered veterinary drugs and other compounds with MRLs. Analytical methods often focus on the detection of antibiotic residues in raw materials, which serves as an effectively preventative measure to ensure that no residues beyond authorized or permitted levels are transferred into their end-products through food processing. However, this does not mean that there is no need to monitor antibiotic residues in final food products. Chloramphenicol and nitrofurans (Group A), and aminoglycosides, β-lactams, macrolides, sulfonamides, tetracyclines, quinolones, etc. (Group B) are common antibiotics that have been monitored and investigated actively in bovine, ovine, and/or caprine milk (raw material) [3,6,8,12–14], and occasionally in its products such as yogurt [15], cheese [16], and milk powder [17,18].

35.2 Analytical Methods

Analytical methods for the detection and/or determination of antibiotic residues in milk and its products fall in two categories: (1) screening methods such as microbial inhibition tests, rapid test kits, etc., and (2) confirmatory methods including gas chromatography with electron capture, flame ionization, or mass spectrometry (MS) detection; and liquid chromatography (LC) with ultraviolet (UV), fluorometric or electrochemical detection, or MS. Antibiotics are predominantly LC-amenable compounds, and therefore they are likely to be determined by LC techniques. According to the European Commission Decision 2002/657/EC, confirmatory methods for organic residues or contaminants shall provide information on the chemical structure of the analyte. Consequently, methods based only on chromatographic analysis without the use of spectrometric detection are not suitable on their own for use as confirmatory methods. However, if a single technique lacks sufficient specificity, the desired specificity shall be achieved by analytical procedures consisting of suitable combinations of clean-up, chromatographic separation(s), and spectrometric detection [19].

35.2.1 Screening Methods

Common screening methods or techniques include microbial inhibition assay, rapid test kits, surface plasmon resonance (SPR) technology biosensor, enzyme-linked immunosorbent assay (ELISA), etc. Screening tests have advantages of easy-to-use, low cost, and high sample throughput, but they lack specificity and sometimes display a relatively high false-positive rate. There are numerous commercially available microbial inhibition and rapid test kits that are able to detect antibiotics in milk at or below the FDA safe tolerance levels or the EU MRLs. Those kits include various Charm test kits (both inhibition assay and rapid test) [20], Copan milk test (inhibition assay) [21], Delvotest SP and SP-NT (inhibition assay) [22], BetaStar and Penzym test kits (rapid test) [23], SNAP (rapid test) [24], etc. Microbial inhibition assays are nonspecific, and test for a broad spectrum of antibiotics. Microbial inhibition tests are time-consuming and it may take a few hours to complete a test. In contrast, rapid test kits, which are based on microbial receptor, enzymatic, or immunological assay, are fast and a test could be done in a few minutes [25]. The rapid test kits are somewhat selective, and can detect a specific family of antibiotics per kit or assay.

An SPR technology biosensor has proven to be a rapid and sensitive technique to detect chemical contaminants in food or milk. SPR biosensors are designed to be operated in real time and be able to detect single or multiple antibiotic residues in a sample with minimum sample preparation. A classical SPR device employs the immobilization of antibody, antigen, or other receptors to a sensor chip, and then measures the minute changes in the refractive index as a shift in the angle of total absorption of light incident on a metal layer carrying the receptors. An SPR biosensor has been investigated for its applications on the determination of sulfonamides [26], chloramphenicol [27], penicillins or beta-lactams [25,28,29], streptomycin [30], and tetracycline [31] in milk. Biacore Q with Qflex kits, which is dedicated for food applications, is capable of determining antibiotics including sulfadiazine (SDZ), sulfamethazine (SMZ), streptomycin, chloramphenicol, sulfonamides, and tylosin in various foods [32]. For example, the Qflex streptomycin kit can detect streptomycin and dihydrostreptomycin in bovine milk with limits of detection (LOD) at 28 μg/L. The Qflex chloramphenicol kit is validated for bovine milk to detect chloramphenicol with LOD at 0.03 μg/L (ppb). The test is straightforward with a relatively high sample throughput such that it may only take 8–10 h to analyze 40 milk samples. The Qflex kits are developed and validated for certain matrices, but there is the potential to validate the kit reagents for other matrices following certain method development and validation procedures.

ELISA that uses microtitration plates is a valuable technique to detect antibiotics in milk because of its high sensitivity, simplicity, and ability to screen a large number of small-volume samples. The test, however, could be time-consuming. Some studies have reported on the development of various ELISA kits for the detection of antibiotics in milk including tetracyclines [33], β-lactams or penicillins [34], aminoglycosides [35], chloramphenicol [36], fluoroquinolones [37], and sulfonamides [38]. A recent study demonstrated the applicability of a commercial ELISA kit to analyze 11 beta-lactams (nafcillin, ampicillin, amoxicillin, piperacillin, azlocillin, cloxacillin, penicillin G, dicloxacillin, oxacillin, metampicillin, and penicillin V) in milk with LOD below the EU MRLs [34].

35.2.2 Confirmatory Methods

Although LC with UV, fluorometric or electrochemical detection can be used to determine antibiotics, LC–MS has largely superseded other detection approaches and has become an important technique for quantifying and confirming antibiotic residues in milk and its products with respect to its sensitivity and specificity [3,12–14,39,40].

35.2.2.1 Sample Preparation

Sample preparation serves as a critical step to extract and concentrate antibiotic residues into an aqueous buffer or solution that is suitable for LC–MS injection. Sample preparation for milk or its products with a sample size ranging from 1 to 5 g usually involves deproteinization and removal of fat and other interferences. Proteins can be precipitated using acetonitrile in a sample-to-solvent ratio of 1:2 to 1:5 [41–44], methanol [45], trichloroacetic acid (20% in water or methanol) [46–49], 5-sulfosalicylic acid [50], acetic acid [51], or sodium tungstate [52]. Deproteinization can also be achieved by means of ultrafiltration using a cut-off mass filter device [53–55]. Fat or lipids are removed using hexane and/or through centrifugation [3,14,52,54,56]. After deproteinization and/or removal of fat, further sample cleanup and/or concentration are necessary for reproducible chromatograms and improved mass spectrometric sensitivity using solid-phase extraction (SPE), liquid–liquid extraction (LLE) or liquid–liquid partitioning (LLP), matrix solid-phase dispersion (MSPD), etc. The SPE has been adopted as routine because of its advantage that sample extracts are further cleaned up, and therefore interferences are removed and matrix effects are reduced. Moreover, analytes are concentrated to achieve sensitivity with LODs at sub μg/kg. Commonly used SPE cartridges include hydrophilic–lipophilic balanced (HLB) [43,44,47,49,52,56], Strata-X [55], C18 [41,46,53], cation exchange (i.e., for aminoglycosides) [48,57], and anion exchange (i.e., for quinolone) [56]. The HLB cartridges, which are made from a copolymer of hydrophilic *N*-vinylpyrrolidone and lipophilic divinylbenzene reversed-phase sorbents, have been used widely as a result of its good retention and highly reproducible recoveries of acidic, basic, and neutral compounds, whether polar or nonpolar. Strata-X cartridges, which have functionality similar to HLB, provide comparable results in retaining these analytes. Generally, SPE is performed off-line by passing sample aqueous extracts through SPE cartridges (30–500 mg) that are placed onto a regulated vacuum manifold. Antibiotics are retained or trapped on the cartridges. After washing with water or buffer, they are eluted with a few milliliters of an appropriate organic solvent such as methanol or acetonitrile with or without pH adjustment, depending on the SPE binding mechanism. MSPD uses a solid supporting material such as sand mixed with samples that subsequently are packed into extraction cells, and antibiotics are extracted or eluted with heated water. MSPD has been recently reported to extract aminoglycosides, tetracyclines, quinolones, macrolides, and lincomycin from milk, yogurt, and cheese [15,16,58,59].

The adjustment of pH and/or addition of chelating agents prior to the SPE or during the extraction step are necessary to prevent the degradation of some antibiotics, and to enhance extraction efficiency based on antibiotic chemical properties and SPE bonding mechanism. For example, tetracyclines and sulfonamides require pH 2–4 [47,60,61], while macrolides and β-lactams prefer a slight basic buffer, i.e., pH 8 or 8.5 [40,43,44,52,53] to maintain the stability and/or to increase the hydrophobicity on reversed-phase SPE cartridges in relating to their pK_as. Tetracyclines tend to form a strong complex with cations or metals (Ca^{2+} and Mg^{2+} ions), and bind to protein and silanol groups. Therefore, chelating agents such as McIlvain buffer or EDTA, Na_2EDTA, citric acid, oxalic acid, etc., are used to prevent the chelation of tetracyclines with metals or others to improve the extraction efficiency [16,47,60,62].

Under most circumstances, the respective parent antibiotics are targeted as marker residues; however, there are a few exceptions. First, nitrofurans including furazolidone, furaltadone, nitrofurazone, and nitrofurantoin are metabolized rapidly to 3-amino-2-oxazolidinone (AOZ), 3-amino-5-morpholinomethyl-2-oxazolidinone (AMOZ), semicarbazide (SC), and 1-aminohydantoin (AH) in animals, which are bound to proteins. Parent nitrofurans may not be detected in most food products. Therefore, an analytical method for the determination of nitrofurans often focuses on the detection of protein-bound residues or active side chains (Table 35.1). The extraction procedure requires an overnight acid hydrolysis and simultaneous derivatization of the released side chains with 2-nitrobenzaldehyde (2-NBA) to form their nitrophenyl derivatives that are able to be analyzed by LC-MS [5]. Second, when ceftiofur is administered parenterally to lactating dairy cattle, it is metabolized quickly to desfuroylceftiofur, which then forms a variety of metabolites and conjugates or is bound to proteins. Therefore, the EEC 2377/90 sets a MRL of 100 μg/kg for the sum of all residues retaining the β-lactam structure expressed as desfuroylceftiofur. The extraction involves the release of desfuroylceftiofur from the various conjugated forms with a reducing agent such as dithioerythritol followed by derivatization with iodoacetamide to form an acetamide derivative, i.e., desfuroylceftiofur acetamide, which is stable and suitable for LC-MS analysis [63,64]. Third, tetracyclines are susceptible to conformational degradation to their 4-epimers in aqueous solution and even during the sample preparation as a function of pH and temperature [16,47,49,60,62,65]. Therefore, quantification of both tetracyclines and their 4-epimers residues remains a challenge. Fourth, a few antibiotics are required to be monitored as the sum of the parent compound and its metabolite, examples of which include cephapirin and deacetylcephapirin, spiramycin and neospiramycin, enrofloxacin, and ciprofloxacin in milk [7,8].

35.2.2.2 Liquid Chromatography and Ultraperformance Liquid Chromatography Separation

Generally, the chromatographic separation of antibiotics relies on the use of reversed-phase columns prior to a mass spectrometer. A conventional LC with a C_{18}-modified silica stationary phase is a practical choice [3,12–14], but more recently an ultraperformance liquid chromatography (UPLC) with sub-2 μm particle C18 columns has been reported with respect to its application for antibiotic analysis as well [66]. UPLC or other fast chromatography with sub-2 μm particle columns is a novel separation technology that has gained popularity in analytical chemistry. Particularly, when coupled with mass spectrometers capable of performing high-speed data acquisition, UPLC offers significant advantages in resolution, speed, and sensitivity [67,68]. The mobile phase composition, concentration, and pH are critical to the optimal ionization and chromatographic separation of antibiotics. Acetonitrile and methanol are two common organic solvents used as LC or UPLC mobile phases. Formic acid (0.1%), ammonium acetate, or ammonium formate (10–20 mM) can be employed as a mobile phase modifier. Heptafluorobutyric acid (HFBA, <20 mM), pentafluoropropionic acid (PFPA, <20 mM), or trifluoroacetic acid (TFA, <0.1%), which are ion-pair reagents, are used for the benefits of improved chromatographic peak shape and extended retention of polar analytes such as lincomycin and aminoglycosides on reversed-phased LC stationary phases [48,50,57,59]. However, it is known that ion-pair reagents such as HFBA, PFPA, and TFA could cause electrospray ionization (ESI) ion suppression, resulting in a significant loss in signal owing to ion-pairing effects in the ESI process, especially for compounds containing nitrogen atoms [69]. Therefore, ion-pair reagents should be avoided, if possible, to achieve better sensitivity for trace antibiotic detection. Alternatively, hydrophilic interaction chromatographic (HILIC) columns, which have the separation mechanism different from that of the reversed-phase

Table 35.1   Examples of LC–MS Analysis of Antibiotic Residues in Milk and its Products

| Class | Compound | Matrix | Sample Preparation | | LC | |
			Extraction	Cleanup	Column	Mobile Phase
	Chloramphenicol	Milk	Deproteinization by acetonitrile	LLE with chloroform	Purospher Star RP-18 column, 55 × 4 mm, 3 μm	0.15% formic acid and methanol
Nitrofurans	Furazolidone (side chain: AOZ)	Milk	Overnight incubation and derivatization with 0.125 M HCl and 2-NBA	SPE with HLB and LLE with ethyl acetate	Inertsil ODS, 150 × 2.1 mm, 3.5 μm	20 mM ammonium acetate and methanol
	Furaltadone (side chain: AMOZ)					
	Nitrofurazone (side chain: SC)					
	Nitrofurantoin (side chain: AH)					
Aminoglycosides	Dihydrostreptomycin	Milk and milk powder	Deproteinization by 5% 5-sulfosalicylic acid		Alltima C18, 150 × 2.1 mm, 5 μm	6.4 mM ammonium formate, 1.9 mM pentafluoropropionic acid (PFPA) and acetonitrile
	Streptomycin					
β-Lactams	Amoxicillin	Milk	Deproteinization by acetonitrile and extraction with 0.1 M phosphate buffer (pH 8.5)	SPE with HLB	Luna C18(2), 250 × 4.6, 5 μm	1% acetic acid and methanol
	Ampicillin					
	Cephapirin					
	Cloxacillin					
	Penicillin G					
Macrolides	Spiramycin	Milk	Deproteinization by acetonitrile and extraction with 0.1 M phosphate buffer (pH 8)	SPE with HLB	Acquity UPLC BEH C18, 100 × 2.1 mm, 1.7 μm and YMC ODS-AQ S-3, 50 × 2 mm	UPLC: 10 mM ammonium acetate and acetonitrile. HPLC: 0.1% formic acid and acetonitrile
	Erythromycin					
	Neopsiramycin					

MS				Sensitivity		
Type	Ionization	Mass or Transitions	Calibration	LOD or CCα	LOQ or CCβ	Reference
QqLIT	ESI⁻	321 → 152, 194, 257	Matrix-matched. Deuterated (d5) chloramphenicol used as an internal standard	CCα = 0.02 µg/kg	CCβ = 0.04 µg/kg	Ronning et al. [4]
QqQ	ESI⁺	Monitored as nitrophenyl derivatives. 3-[(2-Nitro-benzylidene)-amino]-oxazolidin-2-one (NPAOZ). 236 → 134, 104, 149	Matrix-matched. SC hydrochloride-¹³C,¹⁵N2 (SC + 3), 3-amino-2-oxazolidinone-d4 (AOZ-d4), and AMOZ-d5 used as internal standards	0.1 ng/g		Chu and Lopez [5]
		Monitored as nitrophenyl derivatives. 5-Morpholin-4-ylmethyl-3-[(2-nitrobenzylidene)-amino]-oxazolidin-2-one (NPAMOZ). 335 → 128, 262, 291		0.1 ng/g		
		Monitored as nitrophenyl derivatives. 2-Nitro-benzaldehyde-semicarbazone (NPSC). 209 → 166, 192, 134		0.2 ng/g		
		Monitored as nitrophenyl derivatives. 1-[(2-Nitro-benzylidene)-amino]-imidazolidine-2,4-dione (NPAH). 249 → 104, 134, 178		0.2 ng/g		
QqQ	ESI⁺	584 → 263, 246	Matrix-matched.	CCα = 0.22 µg/kg	CCβ = 0.26 µg/kg	van Bruijnsvoort et al. [50]
		582 → 263, 246		CCα = 0.23 µg/kg	CCβ = 0.28 µg/kg	
QIT	ESI⁺	366 → 349	Matrix-matched.	1 ng/mL		Holstege et al. [43]
		350 → 160, 191, 333		0.2 ng/mL		
		424 → 292, 320, 333		0.8 ng/mL		
		458 → 182, 299, 330		2 ng/mL		
		357 → 181, 198, 229		1 ng/mL		
QqTOF and QqQ	ESI⁺	Q-TOF: 843.5218 MS/MS: 843 → 174, 142	Matrix-matched. Roxithromycin used as an internal standard	UPLC/Q-TOF: 0.8–1.0 µg/kg. LC/MS/MS: 0.1–0.2 µg/kg		Wang and Leung [66]
		Q-TOF: 734.4690 MS/MS: 734 → 158, 576		UPLC/Q-TOF: 0.2–0.5 µg/kg. LC/MS/MS: 0.01 µg/kg		
		Q-TOF: 699.4432 MS/MS: 699 → 174, 142		UPLC/Q-TOF: 1.0 µg/kg. LC/MS/MS: 0.1–0.2 µg/kg		

(continued)

Table 35.1 (continued) Examples of LC–MS Analysis of Antibiotic Residues in Milk and its Products

Class	Compound	Matrix	Sample Preparation		LC	
			Extraction	Cleanup	Column	Mobile Phase
	Oleandomycin					
	Tilmicosin					
	Tylosin A					
Sulfonamides	Sulfadiazine (SDZ)	Milk	Deproteinization by 20% trichloroacetic acid	SPE with HLB	SymmetryShield RP18, 150 × 2.1 mm, 3.5 μm	1 mM oxalic acid and acetonitrile
	Sulfathiazole (STZ)					
	Sulfamethazine (SMZ)					
	Sulfamethoxy-pyridazine (SMP)					
	Sulfamethoxazole (SMX)					
	Sulfadimethoxine (SDM)					
Tetracyclines	Tetracycline (TC) and 4-epi-tetracycline (4-epi-TC)	Milk	Deproteinization by 20% trichloroacetic acid	SPE with HLB	Alltima C18, 150 × 2.1 mm, 3 μm	1% formic acid and a mixture of acetonitrile and methanol
	Oxytetracycline (OTC) and 4-epi-oxytetracycline (4-epi-OTC)					
	Chlortetracycline (CTC) and 4-epi-chlortetracycline (4-epi–CTC)					
Quinolones	Ciprofloxacin	Milk	MSPD		Alltima C18, 250 × 4.6 mm, 5 μm	Water and methanol acidified with formic acid
	Danofloxacin					
	Enrofloxacin					
	Flumequine					
	Marbofloxacin					

MS			Calibration	Sensitivity		Reference
Type	Ionization	Mass or Transitions		LOD or CCα	LOQ or CCβ	
		Q-TOF: 688.4272 MS/MS: 688 → 158, 544		UPLC/Q-TOF: 0.2–0.5 μg/kg. LC/MS/MS: 0.01 μg/kg		
		Q-TOF: 869.5738 MS/MS: 869 → 174, 132		UPLC/Q-TOF: 1.0 μg/kg. LC/MS/MS: 0.2–0.5 μg/kg		
		Q-TOF: 916.5270 MS/MS: 916 → 174, 145		UPLC/Q-TOF: 0.2 μg/kg. LC/MS/MS: 0.01–0.02 μg/kg		
Q	ESI+	251, 156	Matrix-matched.	0.75 ng/mL	1.12 ng/mL	Koesukwiwat et al. [47]
		256, 108		1.27 ng/mL	4.16 ng/mL	
		279, 124		1.47 ng/mL	5.10 ng/mL	
		281, 126		0.87 ng/mL	3.00 ng/mL	
		254, 156		0.84 ng/mL	2.68 ng/mL	
		311, 156		0.48 ng/mL	0.61 ng/mL	
QqQ	ESI+	445 → 410, 427	Matrix-matched. Demethylchlortetracycline used as an internal standard	TC: 7.8 μg/L. 4-epi-TC: 10 μg/L	TC: 8.8 μg/L. 4-epi-TC: 12.2 μg/L	De Ruyck and De Ridder [49]
		461 → 426, 444		OTC: 25 μg/L. 4-epi-OTC: 17.5 μg/L	OTC: 29.4 μg/L. 4-epi-OTC: 20.6 μg/L	
		479 → 444, 462		CTC: 7.5 μg/L. 4-epi-CTC: 5 μg/L	CTC: 9.1 μg/L. 4-epi-CTC: 7.1 μg/L	
QqQ	ESI+	332 → 288, 314	Matrix-matched. Lomefloxacin used as an internal standard		Between 0.3 and 1.5 ng/mL	Bogialli et al. [58]
		358 → 314, 340			Between 0.3 and 1.5 ng/mL	
		360 → 316, 342			Between 0.3 and 1.5 ng/mL	
		262 → 202, 244			1.5 ng/mL	
		363 → 320, 72			0.3 ng/mL	

columns, can be used for better column retention of lincomycin and aminoglycosides without using any ion-pairing reagents [70]. Oxalic acid (1 mM), a chelating agent, is a mobile phase modifier for LC–MS analysis of tetracyclines to reduce peak tailing and to maintain their stability [47]. However, oxalic acid is not a good choice for ESI because of its low volatility, and therefore extra maintenance is needed to avoid clogging the capillary needle and to reduce ion source contamination through ion source cleaning, splitting the LC flow, and/or the use of a divert value. It has been reported that some reversed-phase columns such as Alltima C_{18} or Atlantis dC_{18}, with the use of 0.1% formic acid and a mixture of methanol/acetonitrile as mobile phases, are suitable for LC–MS analysis of tetracyclines and their respective epimers where Gaussian distribution peak shape along with baseline resolution is able to be obtained [49,62].

35.2.2.3 Mass Spectrometry

Electrospray ionization (ESI) and atmospheric pressure chemical ionization (APCI) are two common LC–MS interfaces. ESI is applicable to polar and medium nonpolar analytes covering a very broad mass range, and therefore, it has become a popular LC–MS interface. Matrix effects can be a major challenge in LC–MS quantitative work, especially when ESI is used as the interface. Matrix can either enhance or suppress ionization of antibiotics, and its effects vary from sample to sample. Matrix effects are able to be estimated or determined by comparing the responses of analytes in the solvent or buffer to those in the presence of matrices [71]. The uses of isotopically labeled standards, matrix-matched standard calibration curves, and standard addition are general approaches utilized to overcome matrix effects and to improve accuracy of the method. Owing to a lack of deuterium-labeled standards for each individual antibiotic, matrix-matched standard calibration, with or without the use of a chemical analog as an internal standard, is the most common approach for reducing the matrix effects. The method of standard addition is effective in compensating for matrix effects, but the procedure can be very tedious and often requires additional sample preparation.

Mass spectrometers with various designs, performances, and functions that are currently available for the analysis of antibiotic residues include single quadrupole (Q), triple-quadrupole (QqQ), quadrupole ion trap (QIT), quadrupole linear ion trap (LIT), time-of-flight (TOF), and quadrupole time-of-flight (QqTOF) [3,12,14,39,40,72]. Examples of LC–MS analyses of antibiotic residues in milk and its products are presented in Table 35.1. A Q mass spectrometer is applicable for the analysis of antibiotics [47,73], but it has been replaced by a tandem mass spectrometer for improved sensitivity and specificity. A QqQ mass spectrometer operated in the multiple-reaction monitoring (MRM) mode is the most sensitive and common tool for quantifying and confirming antibiotics [5,15–17,42,44,49–54,58,59,66]. In general, a product ion spectrum (Figure 35.1a) is first acquired using a reference standard where MRM transitions are defined and selected to perform LC/MS/MS analysis (Figure 35.1b). A QIT mass spectrometer with its capability to perform MS/MS and MS^n experiment and its relatively high sensitivity in scan mode makes it a valuable instrument to quantify and characterize antibiotics [41,43,48,57,74]. A triple-quadrupole linear ion-trap (QqLIT) mass analyzer has some novel functions that combine the advantages of a QqQ mass spectrometer and an ion-trap mass spectrometer within the same platform without compromising on the performance of either mass spectrometer [4]. A QqLIT mass spectrometer is very valuable for the determination of antibiotic residues because of its capability to perform MRM and acquire product ion scan spectra at low concentrations in one single run. TOF and QqTOF mass spectrometers, as a result of their high sensitivity in full-scan mode, medium-range high resolution, and accurate mass measurement capability, are emerging tools for screening, quantification, confirmation, and identification of antibiotics and their degradation products or

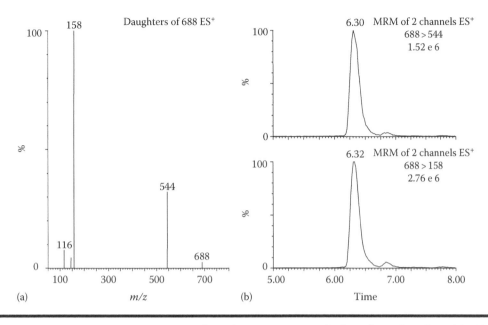

Figure 35.1 (a) An ESI–MS/MS product ion spectrum of oleandomycin. (b) LC/MS/MS chromatograms of a blank raw milk sample fortified with oleandomycin (5 mg/kg). The mass spectrum and chromatograms are unpublished data and are obtained from the author's previous research project, Calgary Laboratory, Canadian Food Inspection Agency. Instrumental parameters are described in the paper by Wang and Leung [66].

metabolites [66]. Figure 35.2 shows an example that uses the UPLC/QqTOF MS (full-scan) to screen six macrolides spiked in a blank milk sample, and extracted ion chromatograms are based on the accurate mass with the mass error window set at 50 mDa. Moreover, a QqTOF mass spectrometer can be operated in QqTOF MS/MS mode, which provides accurate mass product ion spectra for unequivocal confirmation of antibiotics in complex matrices, which eliminates false-positives and avoids ambiguous data interpretation. In general, QqQ, QIT, QqLIT, and TOF or QqTOF mass spectrometers are complementary to each other for the determination of antibiotic residues in milk and its products.

35.2.2.4 *LC–MS Confirmatory Criteria*

LC–MS confirmatory criteria are well defined in the Decision 2002/657/EC [19] and the "Guidance for industry—Mass spectrometry for confirmation of the identity of animal drug residues" from the Center for Veterinary Medicine of the U.S. FDA [75]. The confirmatory characteristics generally include retention time and ion ratio with certain tolerances (Table 35.2). The retention time of an analyte in a chromatographic run should match that of the calibration standard within a specified relative retention time window, i.e., typically ±2.5%. The relative abundances of two or more transitions should fall in the maximum permitted tolerances of the comparison standard. The FDA guidance recommends that the relative abundance ratio be within ±10% absolute when two transitions are monitored. If three or more transitions are monitored, the relative abundance ratios should match the comparison standard within ±20% absolute. The Decision 2002/657/EC has set relative abundance criteria that are dependent on the relative intensities of the two

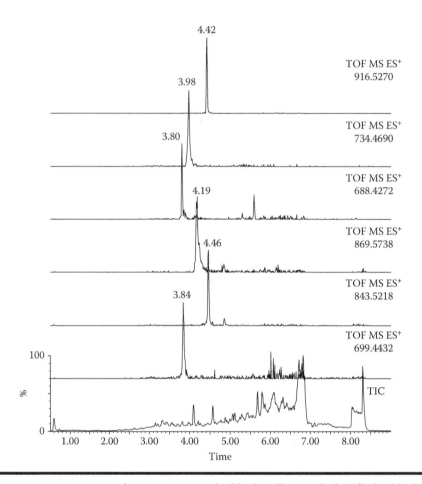

Figure 35.2 **UPLC QTOF MS chromatograms of a blank milk sample fortified with six macro-lides (5 mg/kg per analyte). TIC, total ion current. From bottom to top: 1, neospiramycin I; 2, spiramycin I; 3, tilmicosin; 4, oleandomycin; 5, erythromycin; 6, tylosin A. The chromatograms are unpublished data and are obtained from the author's previous research project, Calgary Laboratory, Canadian Food Inspection Agency. Instrumental parameters are described in the paper by Wang and Leung [66].**

Table 35.2 **Maximum Permitted Tolerances for MS Relative Ion Intensity**

Relative Intensity (% of Base Peak)	EU[a] (Relative) (%)	FDA[b] (Absolute)
>50	±20	
>20–50	±25	Two transitions: ±10%
>10–20	±30	More than two transitions: ±20%
≤10	±50	

[a] Criterion set by the Commission Decision 2002/657/EC [19].
[b] Criterion set by U.S. FDA [75].

transitions, and has also established an identification points (IPs) system to confirm organic residues and contaminants in live animals and animal products (Table 35.3) [19]. Regarding the assignment of IPs, for a low-resolution mass spectrometer (LRMS) with unit mass resolution such as QqQ, QIT, or QqLIT, one precursor ion and one transition product are assigned 1 and 1.5 IPs, respectively. According to the Decision 2002/657/EC, for the confirmation of substances such as chloramphenicol and nitrofurans listed in the Group A (banned substances) in the Directive 96/23/EC [11], a minimum of 4 IPs are required. For the confirmation of substances listed in the Group B (substances with established MRLs), a minimum of 3 IPs are required. Typically, two or three transitions (equivalent to 4 or 5.5 IPs), which are adequate for confirmation, are feasibly

Table 35.3   Mass Resolution, Mass Accuracy, and IPs

MS Technique	IPs Obtained for Each Ion[a]
LRMS	1
LR-MS^n precursor ion	1
LR-MS^n product ion or transition products	1.5
HRMS	2
HR-MS^n precursor ion	2
HR-MS^n product ion or transition products	2.5
Mass Accuracy	*IPs Obtained for Each Ion[b]*
Error higher than 10 mDa[b] or ppm[c]	
Single ion	1
Precursor ion	1
Product ion or transition products	1.5
Error between 2 and 10 mDa or ppm	
Single ion	1.5
Precursor ion	1.5
Product ion or transition products	2
Error below 2 mDa or ppm	
Single ion	2
Precursor ion	2
Product ion or transition products	2.5

[a] Criterion proposed by the Commission Decision 2002/657/EC [19].
[b] Criterion proposed by Hernandez et al. [76] and mass error in mDa.
[c] Criterion proposed by Wang and Leung [66] and mass error in ppm.

obtained using a low-resolution tandem mass spectrometer (Table 35.1). TOF and QqTOF are able to achieve as high as 15,000 FWHM resolution with mass accuracy less than 5 ppm, which provides specificity for target compound confirmation and permits assigning possible elemental compositions for unknown identification. In the Decision 2002/657/EC, high-resolution mass spectrometry (HRMS) is defined as the resolution that shall typically be greater than 10,000 for the entire mass range at 10% valley. This definition does not take into account mass accuracy and disadvantages TOF instruments for the confirmation of a chemical contaminant. Therefore, the criterion for IPs assignment (Table 35.3) based on mass measurement accuracy rather than on resolving power has been proposed, which uses either absolute [76] or relative mass errors [66]. The latter has an advantage that the IP rating criterion is consistent across a mass range or is independent of mass. Thus, for substances with established MRLs, at least two ions need to be monitored to achieve a minimum of 3 IPs for satisfactory confirmation of the compound's identity with mass errors that are between 2 and 10 mDa or ppm. Using in-source fragmentation or collision-induced dissociation with a low and high fragmentation or collision energy, a TOF or QqTOF instrument could acquire fragment-rich spectra, and therefore, additional IPs are able to be assigned for confirmation.

35.3 Method Validation and Measurement Uncertainty

Method validation is critical to ensure that a newly developed analytical method is reliable in routine practice. The method performance parameters that generally need to be evaluated include accuracy or recovery, trueness that uses certified reference materials, precision (repeatability, intermediate precision, and reproducibility), calibration curves or functions, analytical range, decision limit (CCα), detection capability (CCβ), LOD, limit of quantification (LOQ), specificity, ruggedness, analyte stability, estimation of measurement uncertainty, etc. The Decision 2002/657/EC [19] is a well-established EU legislative document that describes criteria and procedures for the validation of both screening and confirmatory methods to ensure the quality and comparability of analytical results generated by official laboratories. It is a guide that has been widely adopted and frequently cited in the food science community. In addition, there are many other scientific references [77–79] that can be followed for method validation.

Measurement uncertainty is an important aspect of an analytical method associated with its performance. It can be estimated using either in-house validation data or quality control data. Measurement uncertainty is defined as a parameter, associated with the result of a measurement, which characterizes the dispersion of values that could reasonably be attributed to the measurand [80]. It is commonly interpreted as an interval within which the true value lies with a probability. The measurement uncertainty associated with a result is an essential part of quantitative results. It provides comparability and reliability among accredited laboratories nationally and internationally. It is a root-cause analysis that helps one to identify the key factors contributing to large variations of a method. Many accreditation bodies have been requiring uncertainty values or estimations when a laboratory implements the ISO standard 17025 [81]. Uncertainty can be estimated either by calculating all the sources of uncertainty whenever possible using the "bottom-up" approach proposed by ISO [80] and EURACHEM/CITAC Guide [82], or through the commonly known "top-down" approach including the nested data analysis using information from interlaboratory study [83] and method validation results [44,84,85]. The main contributions of uncertainty for LC–MS analysis of antibiotics mainly result from the repeatability of the measurement and matrix effects [44].

35.4   Conclusions

Antibiotics have been widely used in veterinary practice, but are stringently controlled under regulations to ensure the safety of food supply. Antibiotic residues in milk and its products can be determined using either screening methods or confirmatory methods. Screening methods have advantages of easy-to-use, low cost, and high sample throughput, but they lack specificity. Confirmatory methods are required to confirm the positive findings, and LC–MS is the important technique for both quantification and confirmation. Organic solvents such acetonitrile and methanol, and acids are often used to precipitate milk proteins. Hexane and/or centrifugation are employed to remove lipids and fat. Further sample cleanup and concentration through SPE, LLE, etc., are necessary to minimize interferences, to reduce matrix effects, and to improve sensitivity. Analytical methods are required to be validated, and measurement uncertainty should be estimated to ensure the quality and comparability of scientific data generated by analytical laboratories.

References

1. Florea, N.F. and Nightingale, C.H., Review of the pharmacodynamics of antibiotic use in animal food production, *Diagn. Microbiol. Infect. Dis.*, 49, 105, 2004.
2. Mathur, S. and Singh, R., Antibiotic resistance in food lactic acid bacteria—A review, *Int. J. Food Microbiol.*, 105, 281, 2005.
3. Stolker, A.A.M. and Brinkman, U.A.T., Analytical strategies for residue analysis of veterinary drugs and growth-promoting agents in food-producing animals—A review, *J. Chromatogr. A*, 1067, 15, 2005.
4. Ronning, H.T., Einarsen, K., and Asp, T.N., Determination of chloramphenicol residues in meat, seafood, egg, honey, milk, plasma and urine with liquid chromatography-tandem mass spectrometry, and the validation of the method based on 2002/657/EC, *J. Chromatogr. A*, 1118, 226, 2006.
5. Chu, P.S. and Lopez, M.I., Determination of nitrofuran residues in milk of dairy cows using liquid chromatography-tandem mass spectrometry, *J. Agric. Food Chem.*, 55, 2129, 2007.
6. FDA Center for Veterinary Medicine: Database of Approved Animal Drug Products, http://dil.vetmed.vt.edu/.
7. Council Regulation (EEC) No 2377/90 of 26 June 1990 laying down a community procedure for the establishment of maximum residue limits of veterinary medicinal products in foodstuffs of animal origin, *Off. J. Eur. Communities*, L224, 18, 1990.
8. European Medicines Agency: Veterinary Medicines—Maximum Residue Limits (MRL), http://www.emea.europa.eu/htms/vet/mrls/mrlop.htm.
9. Health Canada, Drugs & Health Products, Veterinary Drugs, Maximum Residue Limits, http://www.hc-sc.gc.ca/dhp-mps/vet/mrl-lmr/mrl-lmr_versus_new-nouveau_e.html.
10. FAO/WHO, CODEX alimentarius: Veterinary Drug Residues in Food Maximum Residue Limits, http://www.codexalimentarius.net/mrls/vetdrugs/jsp/vetd_q-e.jsp.
11. Council Directive 96/23/EC of 29 April 1996 on measures to monitor certain substances and residues thereof in live animals and animal products and repealing Directives 85/358/EEC and 86/469/EEC and Decisions 89/187/EEC and 91/664/EEC, *Off. J. Eur. Communities*, L125, 10, 1996.
12. Kennedy, D.G., McCracken, R.J., Cannavan, A., and Hewitt, S.A., Use of liquid chromatography-mass spectrometry in the analysis of residues of antibiotics in meat and milk, *J. Chromatogr. A*, 812, 77, 1998.
13. Schenck, F.J. and Callery, P.S., Chromatographic methods of analysis of antibiotics in milk, *J. Chromatogr. A*, 812, 99, 1998.
14. Di Corcia, A. and Nazzari, M., Liquid chromatographic-mass spectrometric methods for analyzing antibiotic and antibacterial agents in animal food products, *J. Chromatogr. A*, 974, 53, 2002.

15. Bogialli, S., Di Corcia, A., Lagana, A., Mastrantoni, V., and Sergi, M., A simple and rapid confirmatory assay for analyzing antibiotic residues of the macrolide class and lincomycin in bovine milk and yoghurt: Hot water extraction followed by liquid chromatography/tandem mass spectrometry, *Rapid Commun. Mass Spectrom.*, 21, 237, 2007.

16. Bogialli, S., Coradazzi, C., Di Corcia, A., Lagana, A., and Sergi, M., A rapid method based on hot water extraction and liquid chromatography-tandem mass spectrometry for analyzing tetracycline antibiotic residues in cheese, *J. AOAC Int.*, 90, 864, 2007.

17. Guy, P.A., Royer, D., Mottier, P., Gremaud, E., Perisset, A., and Stadler, R.H., Quantitative determination of chloramphenicol in milk powders by isotope dilution liquid chromatography coupled to tandem mass spectrometry, *J. Chromatogr. A*, 1054, 365, 2004.

18. Anderson, K.L., Lyman, R.L., Moats, W.A., Hansen, A.P., and Rushing, J.E., Comparison of microbial receptor assay and liquid chromatography for determination of penicillin G and amoxicillin in milk powder, *J. AOAC Int.*, 85, 546, 2002.

19. Commission Decision of 12 August 2002 implementing Council Directive 96/23/EC concerning the performance of analytical methods and the interpretation of results. 2002/657/EC., *Off. J. Eur. Communities*, L221, 8, 2002.

20. Charm Sciences Inc., http://www.charm.com/.

21. Copan Italia S.p.A., http://www.copanswabs.com/.

22. DSM Food Specialties, http://www.dsm.com/en_US/html/dfs/dairy-products-tests-delvotest.htm.

23. Chr. Hansen A/S, http://www.chr-hansen.com/.

24. IDEXX Laboratories, Inc., http://www.idexx.com/dairy/snap/.

25. Gustavsson, E. and Sternesjo, A., Biosensor analysis of beta-lactams in milk: Comparison with microbiological, immunological, and receptor-based screening methods, *J. AOAC Int.*, 87, 614, 2004.

26. Gaudin, V., Hedou, C., and Sanders, P., Validation of a Biacore method for screening eight sulfonamides in milk and porcine muscle tissues according to European decision 2002/657/EC, *J. AOAC Int.*, 90, 1706, 2007.

27. Gaudin, V. and Maris, P., Development of a biosensor-based immunoassay for screening of chloramphenicol residues in milk, *Food Agric. Immunol.*, 13, 77, 2001.

28. Gustavsson, E., Degelaen, J., Bjurling, P., and Sternesjo, A., Determination of beta-lactams in milk using a surface plasmon resonance-based biosensor, *J. Agric. Food Chem.*, 52, 2791, 2004.

29. Sternesjo, A. and Gustavsson, E., Biosensor analysis of beta-lactams in milk using the carboxypeptidase activity of a bacterial penicillin binding protein, *J. AOAC Int.*, 89, 832, 2006.

30. Ferguson, J.P., Baxter, G.A., McEvoy, J.D., Stead, S., Rawlings, E., and Sharman, M., Detection of streptomycin and dihydrostreptomycin residues in milk, honey and meat samples using an optical biosensor, *Analyst*, 127, 951, 2002.

31. Moeller, N., Mueller Seitz, E., Scholz, O., Hillen, W., Bergwerff, A.A., and Petz, M., A new strategy for the analysis of tetracycline residues in foodstuffs by a surface plasmon resonance biosensor, *Eur. Food Res. Technol.*, 224, 285, 2007.

32. General Electric Company, http://www.biacore.com/.

33. Zhang, Y., Lu, S., Liu, W., Zhao, C., and Xi, R., Preparation of anti-tetracycline antibodies and development of an indirect heterologous competitive enzyme-linked immunosorbent assay to detect residues of tetracycline in milk, *J. Agric. Food Chem.*, 55, 211, 2007.

34. Fitzgerald, S.P., O'Loan, N., McConnell, R.I., Benchikh el, O., and Kane, N.E., Stable competitive enzyme-linked immunosorbent assay kit for rapid measurement of 11 active beta-lactams in milk, tissue, urine, and serum, *J. AOAC Int.*, 90, 334, 2007.

35. Abuknesha, R.A. and Luk, C., Enzyme immunoassays for the analysis of streptomycin in milk, serum and water: Development and assessment of a polyclonal antiserum and assay procedures using novel streptomycin derivatives, *Analyst*, 130, 964, 2005.

36. Fodey, T., Murilla, G., Cannavan, A., and Elliott, C., Characterisation of antibodies to chloramphenicol, produced in different species by enzyme-linked immunosorbent assay and biosensor technologies, *Anal. Chim. Acta*, 592, 51, 2007.

37. Van Coillie, E., De Block, J., and Reybroeck, W., Development of an indirect competitive ELISA for flumequine residues in raw milk using chicken egg yolk antibodies, *J. Agric. Food Chem.*, 52, 4975, 2004.

38. Spinks, C.A., Schut, C.G., Wyatt, G.M., and Morgan, M.R., Development of an ELISA for sulfachlorpyridazine and investigation of matrix effects from different sample extraction procedures, *Food Addit. Contam.*, 18, 11, 2001.

39. Kotretsou, S.I., Determination of aminoglycosides and quinolones in food using tandem mass spectrometry: A review, *Crit. Rev. Food Sci. Nutr.*, 44, 173, 2004.

40. Gonzalez de la Huebra, M.J., and Vincent, U., Analysis of macrolide antibiotics by liquid chromatography, *J. Pharm. Biomed. Anal.*, 39, 376, 2005.

41. Heller, D.N. and Ngoh, M.A., Electrospray ionization and tandem ion trap mass spectrometry for the confirmation of seven beta-lactam antibiotics in bovine milk, *Rapid Commun. Mass Spectrom.*, 12, 2031, 1998.

42. Daeseleire, E., De Ruyck, H., and Van Renterghem, R., Confirmatory assay for the simultaneous detection of penicillins and cephalosporins in milk using liquid chromatography/tandem mass spectrometry, *Rapid Commun. Mass Spectrom.*, 14, 1404, 2000.

43. Holstege, D.M., Puschner, B., Whitehead, G., and Galey, F.D., Screening and mass spectral confirmation of beta-lactam antibiotic residues in milk using LC-MS/MS, *J. Agric. Food Chem.*, 50, 406, 2002.

44. Wang, J., Leung, D., and Lenz, S.P., Determination of five macrolide antibiotic residues in raw milk using liquid chromatography-electrospray ionization tandem mass spectrometry, *J. Agric. Food Chem.*, 54, 2873, 2006.

45. Moats, W.A., Liquid chromatographic approaches to antibiotic residue analysis, *J. Assoc. Off. Anal. Chem.*, 73, 343, 1990.

46. Cinquina, A.L., Robertia, P., Giannettia, L., Longoa, F., Draiscib, R., Fagioloa, A., and Brizioli, N.R., Determination of enrofloxacin and its metabolite ciprofloxacin in goat milk by high-performance liquid chromatography with diode-array detection Optimization and validation, *J. Chromatogr. A.*, 987, 221, 2003.

47. Koesukwiwat, U., Jayanta, S., and Leepipatpiboon, N., Validation of a liquid chromatography-mass spectrometry multi-residue method for the simultaneous determination of sulfonamides, tetracyclines, and pyrimethamine in milk, *J. Chromatogr. A*, 1140, 147, 2007.

48. Cherlet, M., De Baere, S., and De Backer, P., Quantitative determination of dihydrostreptomycin in bovine tissues and milk by liquid chromatography-electrospray ionization-tandem mass spectrometry, *J. Mass Spectrom.*, 42, 647, 2007.

49. De Ruyck, H. and De Ridder, H., Determination of tetracycline antibiotics in cow's milk by liquid chromatography/tandem mass spectrometry, *Rapid Commun. Mass Spectrom.*, 21, 1511, 2007.

50. van Bruijnsvoort, M., Ottink, S.J., Jonker, K.M., and de Boer, E., Determination of streptomycin and dihydrostreptomycin in milk and honey by liquid chromatography with tandem mass spectrometry, *J. Chromatogr. A.*, 1058, 137, 2004.

51. Ghidini, S., Zanardi, E., Varisco, G., and Chizzolini, R., Residues of beta-lactam antibiotics in bovine milk: Confirmatory analysis by liquid chromatography tandem mass spectrometry after microbial assay screening, *Food Addit. Contam.*, 20, 528, 2003.

52. Dubois, M., Fluchard, D., Sior, E., and Delahaut, P., Identification and quantification of five macrolide antibiotics in several tissues, eggs and milk by liquid chromatography-electrospray tandem mass spectrometry, *J. Chromatogr. B*, 753, 189, 2001.

53. Riediker, S. and Stadler, R.H., Simultaneous determination of five beta-lactam antibiotics in bovine milk using liquid chromatography coupled with electrospray ionization tandem mass spectrometry, *Anal. Chem.*, 73, 1614, 2001.

54. Riediker, S., Rytz, A., and Stadler, R.H., Cold-temperature stability of five beta-lactam antibiotics in bovine milk and milk extracts prepared for liquid chromatography-electrospray ionization tandem mass spectrometry analysis, *J. Chromatogr, A*, 1054, 359, 2004.

55. Marazuela, M.D. and Moreno-Bondi, M.C., Multiresidue determination of fluoroquinolones in milk by column liquid chromatography with fluorescence and ultraviolet absorbance detection, *J. Chromatogr. A*, 1034, 25, 2004.

56. Lara, F.J., Garcia-Campana, A.M., Ales-Barrero, F., Bosque-Sendra, J.M., and Garcia-Ayuso, L.E., Multiresidue method for the determination of quinolone antibiotics in bovine raw milk by capillary electrophoresis-tandem mass spectrometry, *Anal. Chem.*, 78, 7665, 2006.

57. Heller, D.N., Clark, S.B., and Righter, H.F., Confirmation of gentamicin and neomycin in milk by weak cation-exchange extraction and electrospray ionization/ion trap tandem mass spectrometry, *J. Mass Spectrom.*, 35, 39, 2000.

58. Bogialli, S., D'Ascenzo, G., Di Corcia, A., Lagana, A., and Nicolardi, S., A simple and rapid assay based on hot water extraction and liquid chromatography-tandem mass spectrometry for monitoring quinolone residues in bovine milk, *Food Chem.*, 108, 354, 2008.

59. Bogialli, S., Curini, R., Di Corcia, A., Lagana, A., Mele, M., and Nazzari, M., Simple confirmatory assay for analyzing residues of aminoglycoside antibiotics in bovine milk: Hot water extraction followed by liquid chromatography-tandem mass spectrometry, *J. Chromatogr. A*, 1067, 93, 2005.

60. Anderson, C.R., Rupp, H.S., and Wu, W.H., Complexities in tetracycline analysis-chemistry, matrix extraction, cleanup, and liquid chromatography, *J. Chromatogr. A*, 1075, 23, 2005.

61. Wen, Y., Zhang, M., Zhao, Q., and Feng, Y.Q., Monitoring of five sulfonamide antibacterial residues in milk by in-tube solid-phase microextraction coupled to high-performance liquid chromatography, *J. Agric. Food Chem.*, 53, 8468, 2005.

62. Khong, S.P., Hammel, Y.A., and Guy, P.A., Analysis of tetracyclines in honey by high-performance liquid chromatography/tandem mass spectrometry, *Rapid Commun. Mass Spectrom.*, 19, 493, 2005.

63. Becker, M., Zittlau, E., and Petz, M., Quantitative determination of ceftiofur-related residues in bovine raw milk by LC-MS/MS with electrospray ionization, *Eur. Food Res. Technol.*, 217, 449, 2003.

64. Makeswaran, S., Patterson, I., and Points, J., An analytical method to determine conjugated residues of ceftiofur in milk using liquid chromatography with tandem mass spectrometry, *Anal. Chim. Acta*, 529, 151, 2005.

65. Blanchflower, W.J., McCracken, R.J., Haggan, A.S., and Kennedy, D.G., Confirmatory assay for the determination of tetracycline, oxytetracycline, chlortetracycline and its isomers in muscle and kidney using liquid chromatography-mass spectrometry, *J. Chromatogr. B*, 692, 351, 1997.

66. Wang, J. and Leung, D., Analyses of macrolide antibiotic residues in eggs, raw milk, and honey using both ultra-performance liquid chromatography/quadrupole time-of-flight mass spectrometry and high-performance liquid chromatography/tandem mass spectrometry, *Rapid Commun. Mass Spectrom.*, 21, 3213, 2007.

67. Churchwell, M.I., Twaddle, N.C., Meeker, L.R., and Doerge, D.R., Improving LC-MS sensitivity through increases in chromatographic performance: Comparisons of UPLC-ES/MS/MS to HPLC-ES/MS/MS, *J. Chromatogr. B*, 825, 134, 2005.

68. Guillarme, D., Nguyen, D.T., Rudaz, S., and Veuthey, J.L., Recent developments in liquid chromatography—Impact on qualitative and quantitative performance, *J. Chromatogr. A*, 1149, 20, 2007.

69. Gustavsson, S.A., Samskog, J., Markides, K.E., and Langstrom, B., Studies of signal suppression in liquid chromatography-electrospray ionization mass spectrometry using volatile ion-pairing reagents, *J. Chromatogr. A*, 937, 41, 2001.

70. Peru, K.M., Kuchta, S.L., Headley, J.V., and Cessna, A.J., Development of a hydrophilic interaction chromatography-mass spectrometry assay for spectinomycin and lincomycin in liquid hog manure supernatant and run-off from cropland, *J. Chromatogr. A*, 1107, 152, 2006.

71. Matuszewski, B.K., Constanzer, M.L., and Chavez-Eng, C.M., Strategies for the assessment of matrix effect in quantitative bioanalytical methods based on HPLC-MS/MS, *Anal. Chem.*, 75, 3019, 2003.

72. Gentili, A., Perret, D., and Marchese, S., Liquid chromatography-tandem mass spectrometry for performing confirmatory analysis of veterinary drugs in animal-food products, *Trends Anal. Chem.*, 24, 704, 2005.

73. Huang, J.F., Zhang, H.J., and Feng, Y.Q., Chloramphenicol extraction from honey, milk, and eggs using polymer monolith microextraction followed by liquid chromatography-mass spectrometry determination, *J. Agric. Food Chem.*, 54, 9279, 2006.

74. Klagkou, K., Pullen, F., Harrison, M., Organ, A., Firth, A., and Langley, G.J., Fragmentation pathways of sulphonamides under electrospray tandem mass spectrometric conditions, *Rapid Commun. Mass Spectrom.*, 17, 2373, 2003.

75. Guidance for industry. *Mass Spectrometry for Confirmation of the Identity of Animal Drug Residues. Final Guidance.* U.S. Department of Health and Human Services Food and Drug Administration Center for Veterinary Medicine. May 1 2003, http://www.fda.gov/cvm/guidance/guide118.pdf., 2003.

76. Hernandez, F., Ibanez, M., Sancho, J.V., and Pozo, O.J., Comparison of different mass spectrometric techniques combined with liquid chromatography for confirmation of pesticides in environmental water based on the use of identification points, *Anal. Chem.*, 76, 4349, 2004.

77. The Fitness for purpose of analytical methods. A laboratory guide to method validation and related topics, http://www.eurachem.org/guides/valid.pdf, 1998.

78. Guidance for industry. *Validation of Analytical Procedures: Methodology. Final Guidance.* U.S. Department of Health and Human Services, Food and Drug Administration, Center for Veterinary Medicine, http://www.fda.gov/cvm/Guidance/guida64.pdf, 1999.

79. Thompson, M., Ellison, S.L.R., and Wood, R., Harmonized guidelines for single-laboratory validation of methods of analysis, *Pure Appl. Chem.*, 74, 835, 2002.

80. *Guide to the Expression of Uncertainty in Measurement*, International Standards Organization (ISO), Geneva, 1993.

81. ISO/IEC 17025:2005, *General Requirements for the Competence of Testing and Calibration Laboratories*, 2005.

82. EURACHEM/CITAC Guide. *Quantifying Uncertainty in Analytical Measurement*, 2nd edn., http://www.eurachem.org/guides/QUAM2000-1.pdf, 2000.

83. Dehouck, P., Vander Heyden, Y., Smeyers Verbeke, J., Massart, D.L., Crommen, J., Hubert, P., Marini, R.D., Smeets, O.S.N.M., Decristoforo, G., Van de Wauw, W., De Beer, J., Quaglia, M.G., Stella, C., Veuthey, J.L., Estevenon, O., Van Schepdael, A., Roets, E., and Hoogmartens, J., Determination of uncertainty in analytical measurements from collaborative study results on the analysis of a phenoxymethylpenicillin sample, *Anal. Chim. Acta*, 481, 261, 2003.

84. Maroto, A., Boque, R., Riu, J., and Rius, F.X., Measurement uncertainty in analytical methods in which trueness is assessed from recovery assays, *Anal. Chim. Acta*, 440, 171, 2001.

85. Barwick, V. and Ellison, S.L.R., Measurement uncertainty: Approaches to the evaluation of uncertainties associated with recovery, *Analyst*, 124, 981–990, 1999.

Chapter 36

Chemical Contaminants: Phthalates

Jiping Zhu, Susan P. Phillips, and Xu-Liang Cao

Contents

36.1 Introduction 907
36.2 Levels of Phthalates in Dairy Products and Other Foods 910
    36.2.1 Phthalates in Infant Milk 910
    36.2.2 Milk for General Consumption 916
    36.2.3 Phthalates in Dairy Products 917
    36.2.4 Phthalates in Nondairy Food Products 917
    36.2.5 Phthalates in Total Diet 918
36.3 Migration of Phthalates into Milk and Other Dairy Products 920
36.4 Analytical Methods for the Detection of Phthalates in Dairy Products 921
    36.4.1 Avoidance of Sample Contamination 921
    36.4.2 Sample Pretreatment, Extraction, and Cleanup 921
    36.4.3 Instrumental Conditions 924
36.5 Concluding Remarks 925
References 926

36.1 Introduction

Without the naturally occurring chemicals that underlay all growth and development, there would be no life on our planet. Chemicals manufactured from combinations of the basic elements are inseparable from modern life. Society, as we know, would be unrecognizable without them. Many consumer products used daily contain or are made of synthetic materials. Food, itself, contains many naturally occurring chemicals. Other chemicals, either natural or synthetic, are often added to food to modify or improve its quality or to preserve it. However, some chemicals that are not intended for consumption are being introduced into the food chain, and their presence is not desirable and can be harmful to consumers. They are called chemical contaminants. The source of these chemical contaminants in food are numerous and sometimes not apparent. Their presence could be a result of using contaminant-containing materials in food production (for example, the use of pesticides/fungicides), food handling (such as containers for food storage and food wrappings/packaging materials), or food preparation (such as heating). Among them, there is a class of chemical contaminants called phthalic acid diesters, commonly known as phthalates.

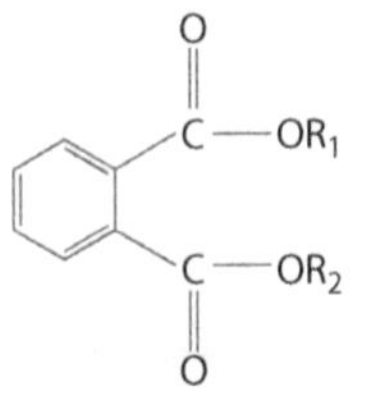

Scheme 36.1   General structure of phthalates.

Structurally, phthalates are composed of a common basic moiety of 1,2-benzenedicarboxylic acid group, in which the two acidic groups are linked to two alcohols to form diesters (Scheme 36.1). The most well-known and widely used phthalate is di(2-ethylhexyl)phthalate (DEHP, $R_1 = R_2 = -CH_2CH(CH_2CH_3)CH_2CH_2CH_2CH_3$). Other common phthalates being environmentally monitored include dimethyl phthalate (DMP, $R_1 = R_2 = -CH_3$), diethyl phthalate (DEP, $R_1 = R_2 = -CH_2CH_3$), di-*n*-butyl phthalate (DnBP, $R_1 = R_2 = -CH_2CH_2CH_2CH_3$), benzylbutyl phthalate (BBzP, $R_1 = -CH_2CH_2CH_2CH_3$, $R_2 = -CH_2C_6H_5$), and di-*n*-octyl phthalate (DnOP, $R_1 = R_2 = -CH_2CH_2CH_2CH_2CH_2CH_2CH_2CH_3$). Their physical properties are summarized in Table 36.1.

Phthalates are semivolatile organic compounds with vapor pressures of the six commonly monitored phthalate ranging from 1.00×10^{-7} Pa to 2.63×10^{-1} Pa and boiling points from 282°C to 428°C. Their other physical–chemical properties have been described by Cousins and Mackay [1]. Using a quantitative structure–property relationship (QSPR) method, air–water (K_{AW}), octanol–water (K_{OW}), and octanol–air (K_{OA}) partition coefficients at 25°C were estimated. These coefficients represent the distribution of phthalates in air and water, in octanol and water, and in octanol and air under equilibrium conditions, respectively. The estimated coefficients of six commonly monitored phthalates are included in Table 36.1.

Among these coefficients, K_{OW} values are important for evaluating the distribution of a chemical in various tissues and organs in humans and animals. Octanol is a straight chain fatty alcohol with eight carbon atoms and a molecular structure of $CH_3(CH_2)_7OH$. It is lipophilic ("fat loving") and is considered to be the representative of body's fatty tissues and fluids such as milk, while water is hydrophilic ("water loving") and represents the body's less fatty fluids such as blood and urine. Log K_{OW} values of the six commonly monitored phthalates range from 1.61 to 9.46, of which DEHP has a value of 7.73 (Table 36.1). This means that for every molecule of DEHP stored in water, there will be $10^{7.73}$ molecules in octanol. The fat-loving nature of phthalates such as DEHP is an important property explaining its presence in dairy products, particularly those with a relatively high fat content.

Phthalates are widely used in today's society. DEHP is predominantly found in polyvinyl chloride (PVC), a thermoplastic polymer, to make it softer and more flexible. Besides being used in some food-packaging materials, DEHP is also a component of medical product containers, intravenous tubing, medical equipment, plastic toys, vinyl upholstery, shower curtains, adhesives, and coatings. Phthalates with smaller R_1 and R_2 groups such as DEP and DBP are mainly used as solvents in products such as perfumes and pesticides.

Extensive use of phthalate esters in both industrial processes and consumer products has resulted in the ubiquitous presence of these chemicals in the environment. They have been

Table 36.1 Physical Properties of Phthalates

Name	Dimethyl Phthalate	Diethyl Phthalate	Di-n-Butyl Phthalate	Butyl Benzyl Phthalate	Di(2-Ethylhexyl) Phthalate	Di(n-Octyl) Phthalate
Abbreviation	DMP	DEP	DnBP	BBzP	DEHP	DnOP
CAS No.	131-11-3	84-66-2	84-74-2	85-68-7	117-81-7	117-84-0
Structural formula	$C_{10}H_{10}O_4$	$C_{12}H_{14}O_4$	$C_{16}H_{22}O_4$	$C_{19}H_{20}O_4$	$C_{24}H_{38}O_4$	$C_{24}H_{38}O_4$
Molecular weight	194.2	222.2	278.4	312.4	390.6	390.6
Melting point (°C)	5.5	−40	−35	−35	−47	−25
Boiling point (°C)	282	298	340	370	385	428
VP(Pa) (25°C)	2.63×10^{-1}	2.83×10^{-3}	2.17×10^{-4}	1.17×10^{-4}	1.19×10^{-6}	1.00×10^{-7}
Specific gravity (20°C)	1.192	1.118	1.042	1.111	0.986	0.978
Log K_{OW} (25°C)	1.61	2.54	4.27	4.7	7.73	9.46
Log K_{OA} (25°C)	7.01	7.55	8.54	8.78	10.53	11.52
Log K_{AW} (25°C)	−5.4	−5.01	−4.27	−4.08	−2.8	−2.06

Source: Cousins, I. and Mackay, D., *Chemosphere*, 41, 1389, 2001.

Partition coefficients of air–water (K_{AW}), octanol–water (K_{OW}), and octanol–air (K_{OA}) are from Cousins and Mackay [1].

detected in water and soil [2–6], consumer products [7–10], medical devices [11], marine ecosystems [12], indoor air [13,14], and indoor dust [15,16]. Some phthalate esters and their monoester metabolites (called monophthalates) have also been detected in human urine [17–21] and amniotic fluid [22] samples.

Toxicology and epidemiology studies indicated that in animals and humans, phthalates can mimic hormones and have endocrine disrupting properties [23]. Studies on rodents have shown that phthalate esters are estrogenic and are associated with adverse reproductive effects [24–27]. Phthalate esters have been linked to premature breast development observed in very young Puerto Rican girls [28]. An inverse linear association between phthalate metabolite levels in urine and observed mobility, concentration, and normal morphology of sperm in American men suggested that these chemicals have estrogen-like activity in humans [29]. In a case-controlled study, BBP in house dust was found to be associated with asthma and allergic symptoms in children [30].

36.2 Levels of Phthalates in Dairy Products and Other Foods

Dairy products are generally defined as foodstuffs produced from milk. These products include milk, milk powder, butter, cheese, and yoghurt among others. They are usually high-energy yielding food products. Raw milk used to make processed dairy products such as cheese generally comes from cows, but occasionally from other mammals such as goats or sheep. Owing to their large K_{OW} values, that is, their fat solubility, phthalates are found primarily in fat-containing dairy products. Table 36.2 summarizes the levels of phthalates reported in dairy products and other foods consumed by infants or the general population. The products are grouped into infant milk, general retail milk, dairy products, baby food, general food, and dietary food. DnBP and DEHP are the two most commonly detected phthalates in dairy products although other phthalates are occasionally present as well.

36.2.1 Phthalates in Infant Milk

Yano et al. [31] has measured 27 baby milk powders purchased from supermarkets or local open markets in several cities in 11 European, North American, and Asian countries for the presence of DnBP (15–77 µg/kg) and DEHP (34–281 µg/kg). Each sample was analyzed in triplicate and the average value of the three was used for reporting the concentrations. Infant exposure to these two phthalates was then estimated based on a daily intake of 700 mL of milk and was found to be below the European Commission Scientific Committee's tolerable daily intake. Although the authors did not report how the milk powder was mixed prior to analysis, based on the estimation of infant daily intake, we believe this study used reconstituted milk solutions.

Another study examined levels of five phthalates, namely DBP, BBP, DEHP, di-*n*-nonyl phthalate (DINP), and di-*n*-decyl phthalate (DIDP) in reconstituted infant formula ($n=6$) and liquid infant formula ($n=2$) [32]. Only DEHP was detected in the samples. The levels of DEHP in liquid infant formula were in the range of 10–23 µg/kg; however, DEHP levels in reconstituted infant formula (37–138 µg/kg) were much higher. The results from the latter study were in agreement with another earlier German study, which reported DnBP (<20–85 µg/kg) and DEHP (<50–196 µg/kg) in infant milk [33].

Casajuana and Lacorte reported mean values of five phthalates (DMP, DEP, DnBP, BBP, and DEHP) in one powdered infant formula packed in a metal can [34]. DEP, DnBP, and DEHP were the three major phthalates and the concentrations of DMP and BBP were much lower at 1–2 µg/kg. While the levels of DnBP (18 µg/kg) and DEHP (20 µg/kg) were similar to or lower than those in other studies, the high DEP (76 µg/kg) levels reported in this study were rather surprising. In fact, this is the only study that reported the presence of DEP in commercial milk.

Owing to the importance of human milk in the early stage of infant growth and the potential health impact of chemical contaminants in human milk on infant development, several studies have been conducted to measure various chemical contaminants in human milk [33,35–37] and to estimate the potential intake of these chemical contaminants by breast-fed infants [36,37]. A German study identified DnBP (10–50 µg/kg) and DEHP (10–20 µg/kg, except for one at 110 µg/kg) in five human milk samples [35]. Similar results were also obtained in an earlier German study in which ranges of DnBP (<20–51 µg/kg) and DEHP (<50–160 µg/kg) were reported from five human milk samples [33]. More extensive studies on phthalates in human milk have been recently reported from Canada [36] and Sweden [37]. The Canadian study

Table 36.2 Levels of Phthalates (µg/kg) in Dairy Products and Other Foods

| Type of Dairy Products | Level (µg/kg) | | | | | | | Size (n=) | Country | Ref. | Year of Publication |
	DMP	DEP	DnBP	BBP	DEHP	DnOP				
Infant milk											
Liquid infant formula			<9	<4	10–23	<5	6	Different countries	[32]	2006
Reconstituted baby formula			15–77		34–281		27	11 countries	[31]	2005
Reconstituted baby formula			<20–85		<50–196		8	Germany	[33]	1998
Reconstituted baby formula			<9	<4	37–138	<5	6	Different countries	[32]	2006
Reconstituted baby formula				Up to 10	Up to 60		11	Denmark	[43]	2000
Reconstituted baby formula			n.d. to 30		10–20		5	Germany	[35]	2000
Reconstituted baby formula[a]	1.4	76	18	1.2	20		2	Spain	[34]	2004
Human milk			<20–51		<50–160		5	Germany	[33]	1998
Human milk			10–50		10–20, 110		5	Germany	[35]	2000
Human milk[a]	n.d.	0.31	0.87	n.d.	222	n.d.	86	Canada	[36]	2006
Human milk[a]		0.3	2.8	0.75	17	1.1	50	Sweden	[37]	2008
General retail milk											
Raw milk			<9	<4	7–30	<5	18	Denmark	[32]	2006

(continued)

Table 36.2 (continued) Levels of Phthalates (µg/kg) in Dairy Products and Other Foods

Type of Dairy Products	Level (µg/kg)							Size (n=)	Country	Ref.	Year of Publication
	DMP	DEP	DnBP	BBP	DEHP	DnOP					
Raw milk			n.d. to 30		100–150			3	Germany	[35]	2000
Raw milk (hand milking)[a]	n.d.	0.6	6.39	n.d.	16	n.d.		6	Canada	[49]	2005
Raw milk (machine milking)[a]	n.d.	0.63	5.79	n.d.	215	n.d.		6	Canada	[49]	2005
Retail milk			<9	<4	13–27	<5		4	Denmark	[32]	2006
Retail milk (<0.1%–3% fat)					<10–50			11	Spain	[41]	1994
Retail milk (<1% fat)					20–40			5	Norway	[41]	1994
Retail milk (1% fat)					50			3	Norway	[41]	1994
Retail milk (3% fat)					60–380			9	Norway	[41]	1994
Retail milk in glass bottle (2.3%–4.2% fat)					10–90			16	United Kingdom	[41]	1994
Retail milk			n.d. to 50		n.d. to 40			5	Germany	[35]	2000
Retail milk (container type 1)[a]	1.3–1.7	36–72	7.3–9.5	1.1–2.9	15–25			4	Spain	[34]	2004
Retail milk (container type 2)[a]	0.97–1.2	71–85	40–50	1.2–2.9	23–27			4	Spain	[34]	2004
Retail milk, skim (no fat)					20–25			2	Norway	[48]	1990
Retail milk, whole					50			1	United Kingdom	[48]	1990

					n	Country	Reference	Year
Retail milk[a]		70	210		50	Italy	[44]	1986
Retail milk, skim (no fat)	n.d.	n.d.	10		1	Canada	[42]	1995
Retail milk, whole (3.3% fat)	n.d.	n.d.	100		1	Canada	[42]	1995
Retail milk (2% fat)	n.d.	n.d.	40		1	Canada	[42]	1995
Dairy Products								
Yoghurt with fruit	<9	<4	15–37	<5	3	Denmark	[32]	2006
Retail butter			2500–7400		10	United Kingdom	[41]	1994
Retail butter spread/margarine			1200–2400		10	United Kingdom	[41]	1994
Retail cheese			200–16,800		25	United Kingdom	[41]	1994
Retail cream			200–2700		10	United Kingdom	[41]	1994
Retail cream (35% fat)			1060–1670		5	Norway	[41]	1994
Cream (31%–33% fat)			480, 550		2	Spain	[41]	1994
Retail cream (no %fat indicated)	n.d. to 70		180–320		6	Germany	[35]	2000
Retail milk, evaporated (7.6% fat)	n.d.		130		1	Canada	[42]	1995
Ice cream (16.0% fat)	n.d.		820		1	Canada	[42]	1995
Butter (80% fat)	1500	6400	3400		1	Canada	[42]	1995

(continued)

Table 36.2 (continued) Levels of Phthalates (µg/kg) in Dairy Products and Other Foods

Type of Dairy Products	Level (µg/kg)						Size (n=)	Country	Ref.	Year of Publication
	DMP	DEP	DnBP	BBP	DEHP	DnOP				
Cheese, Cheddar (32.6% fat)			n.d.	1600	2200		1	Canada	[42]	1995
Cheese, cottage (3.0% fat)			n.d.	n.d.	70		1	Canada	[42]	1995
Cheese, processed (17.7% fat)			n.d.	n.d.	1100		1	Canada	[42]	1995
Cream (17.1% fat)			n.d.	n.d.	1200		1	Canada	[42]	1995
Yoghurt (8.6% fat)			n.d.	600	70		1	Canada	[42]	1995
Baby Food										
Processed baby food			10–55		52–210		7	Germany	[33]	1998
Processed baby food			up to 40	up to 5	up to 630		11	Denmark	[43]	2000
Processed baby food			n.d. to 30		10–20		5	Germany	[35]	2000
Processed baby food[a]			n.d.		750		50	Italy	[44]	1986
General Food[b]										
Packaged food (nuts)			120–570		80–220		3	Germany	[35]	2000
Packaged food (cheese)[a]			840		1080		20	Italy	[44]	1986
Packaged food (fruit jam)[a]			n.d.		170		20	Italy	[44]	1986

Packaged food (potato chips)[a]	2800		350	20	Italy	[44]	1986
Packaged food (salted meat)[a]	1090		2380	20	Italy	[44]	1986
Packaged food (vegetable soups)[a]	2060		2090	20	Italy	[44]	1986
Cucumbers			242–347	30	China	[45]	2007
Tomatoes			311–517	30	China	[45]	2007
Total Diet							
Energy equivalent diet	90–190	17–19	110–180	29	Denmark	[43]	2000
Hospital duplicate diet (7-day, 1999)	n.d. to 7	0.1–5.0	46–478	21	Japan	[46]	2003
Hospital duplicate diet (7-day, 2001)	2–7.1	0.6–2.8	77–103	21	Japan	[46]	2003

a Mean values.
b The data on general food in Canadian food basket survey [42] are not included. See Section 36.2.5 and Table 36.3.

analyzed 21 mothers during the first 6 months of their breast-feeding and collected a total of 86 milk samples. DnBP and DEHP were detected in almost all samples with a mean value of 0.87 and 222 μg/kg for DnBP and DEHP, respectively. DEP was only detected in 15 of the 86 samples, with a mean value of 0.31 μg/kg. The Canadian study also indicated that the levels of phthalates in human milk did not decrease during lactation as is the case for other persistent chemical contaminants [38–40]. The Swedish study reported measurements of phthalates in human milk samples collected from 42 breast-feeding women [37]. One milk sample was collected from each woman when her baby was 14–20 days old (median, 17 days). DEHP and BBP were detected in all samples, whereas a much lower detection frequency was obtained for DEP (8/42), DnBP (12/42), and DOP (10/42). Compared with the Canadian data, the Swedish data showed a lower mean value of DEHP (17 μg/kg). The higher mean value of DnBP (2.8 μg/kg) reported in the Swedish study might have contributed to the higher number of nondetectable (n.d.) values, where 1.5 μg/kg (half of the detection limit) was used to calculate the mean. Mean values of other phthalates were 0.30 μg/kg for DEP, 0.75 μg/kg for BBP, and 1.1 μg/kg for DOP, respectively.

36.2.2 Milk for General Consumption

Besides the milk for consumption by infants, concentrations of phthalates in other milk and milk products have been reported. Measurements of 18 raw milk samples from Denmark showed the presence of DEHP in the range of 7–30 μg/kg, while DnBP, BBP, and DnOP were below the detection limit [32]. One hundred to 150 μg/kg of DEHP and up to 30 μg/kg of DnBP were detected in three raw milk samples from Germany [35]. The milking methods used in these two studies, however, were not specified.

The fat content in retail milk samples ranged from less than 0.1% in skim milk to approximately 3% in whole milk. Levels of phthalates in retail milk samples have been reported by various research groups in Spain, Norway, United Kingdom, Germany, Italy, and Canada. Again, DEHP was the most frequently detected major phthalate in these samples. The levels of DEHP in retail milk samples among the different studies ranged from a few μg/kg to several hundred μg/kg. The levels of DEHP in four Danish retail milk samples ranged from 13 to 27 μg/kg, similar to the levels in Danish raw milk samples measured in the same study [32]. However, the German study reported lower DEHP levels in retail milk (n.d. to 40 μg/kg) compared with raw milk (100–150 μg/kg) [35], most likely due to partial removal of fat in the production of retail milk. Sharman et al. have measured DEHP and total phthalates in a number of samples of retail milk from Norway (DEHP: 20–130 μg/kg), Spain (DEHP: up to 50 μg/kg), and United Kingdom (DEHP: up to 90 μg/kg) [42]. In addition to DnBP (7–50 μg/kg) and DEHP (15–27 μg/kg), the levels of DEP (36–72 μg/kg) and other phthalates were reported in another study on Spanish retail milk [34].

Phthalate levels in retail milk seem to vary considerably. In addition to the source and handling of the raw milk from which the retail milk is produced, packaging and manufacturing processes may be the contributing factors as well. Casajuana and Lacorte reported mean values of five phthalates (DMP, DEP, DnBP, BBP, and DEHP) in retail milk packed in two types of containers (type 1: Tetra Brik and type 2: high-density polyethylene (HDPE)) [34]. DEP, DnBP, and DEHP were the three major phthalates identified. The concentrations of DMP and BBP were much lower at 1–3 μg/kg. There was a significant difference in DnBP levels (7–9 vs. 40–50 μg/kg) in the milk packed in Tetra Brik and HDPE, respectively.

36.2.3 Phthalates in Dairy Products

Many dairy products have been analyzed for phthalates. These have included yoghurt, butter, cheese, margarine, cream, and evaporated milk powder. The fat content in different milk products varies greatly, ranging from a few percent in yoghurt to 80% in butter. The levels of phthalates found in milk products are also listed in Table 36.2. The levels of phthalates in these dairy products were much higher on a per weight basis than those of milk owing to their higher fat content. For example, DEHP was found in 25 U.K. retail cheese samples at levels ranging from 200 to 16,800 μg/kg [41], among which, the highest DEHP levels were found in mild cheddar (16,800 μg/kg), Pompadom with herbs (14,900 μg/kg), and Old Amsterdam (7,500 μg/kg). DEHP levels in 10 U.K. cream samples were 200–2700 μg/kg, whereas in 10 butter samples their levels were 2500–7400 μg/kg [41]. The levels of DEHP in yoghurt were much lower [32,35]. DEHP is the predominant phthalate in milk products reported in these studies and DnBP was occasionally detected at low levels. The Canadian study, however, showed high levels of BBP in butter and Cheddar cheese at 6400 and 1600 μg/kg, respectively. The former product also contained a high level of DnBP (1500 μg/kg) [42].

The levels of DEHP in milk products such as cheese and butter were directly linked to their fat content. It was demonstrated that the levels of phthalates in retail milk were proportional to the percentage of fat in the milk [41]. For example, DEHP levels in the Norwegian retail milk (<1%, 1%, and 3%) and cream (35%) had a linear relationship with fat content (Figure 36.1a). Such fat dependency was also evident in the Spanish samples, where both DEHP and fat content were measured in dairy products (Figure 36.1b). The Canadian study on the measurement of phthalates in milk products provided further evidence of the relationship between DEHP levels in milk products and their fat content (Figure 36.1c) [42].

36.2.4 Phthalates in Nondairy Food Products

The presence of phthalates in foods extends well beyond the dairy products and includes baby food and packaged foods such as meats, vegetables, and fruits. Analysis of seven baby food samples sold in Germany revealed a concentration range of 10–55 μg/kg of DnBP and 52–210 μg/kg of DEHP, respectively [33]. A second German study also found phthalates in five baby food samples (DEHP in the range of 10–20 μg/kg and DnBP in the range of n.d. to 30 μg/kg) [35]. Eleven baby food samples sold in Denmark showed a maximum concentration of 40 μg/kg DnBP, 5 μg/kg BBP, and 360–630 μg/kg DEHP [43].

DEHP was detected in 93% of 50 Italian baby food samples with a mean value of 750 μg/kg and a maximum value of 3400 μg/kg; DnBP, on the other hand, was not detected in these samples. In general, as with dairy products, DEHP and DnBP are the two major phthalate congeners detected in baby foods. The only other phthalate detected in baby food is BBP, which is present at much lower concentrations (up to 5 μg/kg) [44]. The same Italian study also measured phthalates in several types of packaged foods including cheese, salted meat, vegetable soup, fruit jam, and potato chips. The mean values of DEHP in these samples ranged from 170 μg/kg in fruit jams to 2380 μg/kg in salted meat, while the range of DnBP was in the range of n.d. to 1580 μg/kg.

Measurement of phthalates in three different nut samples showed a range of DEHP levels at 80–220 μg/kg and DnBP in the range of 120–570 μg/kg [35]. A Chinese study on 30 cucumbers and 30 tomatoes purchased from the market detected 242–347 μg/kg of DnBP and 311–517 μg/kg of DEHP with a 100% detection frequency [45].

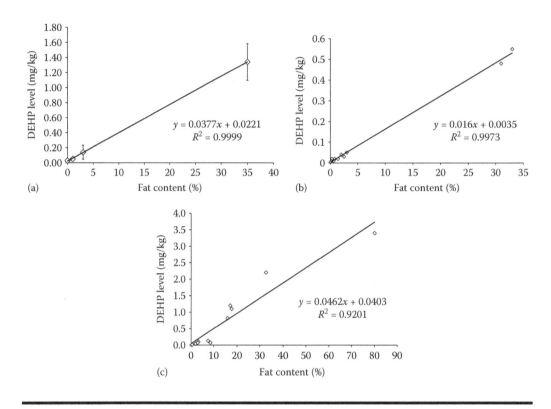

Figure 36.1 Fat content and DEHP levels in dairy products. (a) Norwegian milk products collected at various points in the distribution chain [41]; error bar indicates one standard deviation obtained among samples at each fat content level, (b) retail milk and cream samples obtained from Spain [41], and (c) diary products of total diet samples in Canada [42].

36.2.5 Phthalates in Total Diet

A total diet study consists of purchasing foods commonly consumed at the retail level, processing them for consumption, often combining the foods into food composites, homogenizing them, and analyzing them for the chemicals of interest. It is the most reliable way to estimate the dietary intake of chemicals by large population groups and is supported and recommended by the World Health Organization (WHO). Although total diet studies have been conducted in various countries for many different chemical contaminants, phthalates were determined in total diet samples only occasionally in a few countries. A total diet study conducted in Denmark based on 29 samples generated low to high ranges of mean values for DnBP (90–190 µg/kg), BBP (17–19 µg/kg), and DEHP (110–118 µg/kg), with a detection frequency of 8/29 for DnBP, 9/29 for BBP, and 6/29 for DEHP [43]. The diet sample in the study was not the second prepared portion of the food consumed, rather it was constituted from 24 h daily diet, and the phthalates levels were corrected to an equivalent of 10 MJ of energy, which is considered to be the mean energy intake for adults in Denmark. The Japanese duplicate diet studies conducted in three hospitals in 1999 and in 2001 yielded similar mean values of DEHP, i.e., 46–478 µg/kg and 77–103 µg/kg for the 1999 and 2001 samples, respectively, but much lower mean DnBP levels (n.d. to 7 µg/kg in 1999 samples and 2–7.1 µg/kg in 2001 samples) in the composite food [46].

In Canada, 99 of the 112 food-based total diet composites were collected in 1986 for the analysis of phthalates and other contaminants in 11 dairy products (Table 36.2) as well as 12 samples of meat, poultry, and fish, 19 cereal products, 18 vegetables, 16 fruits, and 23 miscellaneous [42]. Among meat, poultry, and fish products, DEHP was detected in ground beef (100 μg/kg), cured pork (500 μg/kg), poultry (2600 μg/kg), cold cut luncheon meat (200 μg/kg), canned luncheon meat (200 μg/kg), freshwater fish (100 μg/kg), and canned fish (100 μg/kg), whereas DnBP was detected in freshwater fish (500 μg/kg). Among vegetables and fruits, DEHP was detected in cabbage coleslaw (140 μg/kg), fresh tomato (90 μg/kg), cucumber and pickle (0.17 μg/kg), canned citrus fruit (50 μg/kg), and plum/prune (70 μg/kg), whereas DnBP was found in baked potato (630 μg/kg), banana (120 μg/kg), blueberry (90 μg/kg), and pineapple (50 μg/kg), and DEP in canned citrus fruit (40 μg/kg) and blueberry (730 μg/kg). Phthalates were also detected in 17 of the 19 cereal products (Table 36.3). In addition to DnBP and DEHP, a number of cereal products were found to contain DEP, while one product (crackers) contained BBzP (480 μg/kg). Other

Table 36.3   Phthalates Levels (μg/kg) Found in 19 Cereal Products in a Canadian Total Diet Survey

	DEP	*DBP*	*BBzP*	*DEHP*
White bread		90		680
Whole wheat bread	50	100		1500
Rolls and biscuits				1100
Wheat flour		1900		
Cakes				
Cookies		620		1500
Danish pastry and doughnut				3400
Crackers	1200	600	480	
Pancake				120
Cooked wheat cereal	170	300		
Cooked oatmeal cereal		100		
Corn cereal	40	100		820
Wheat and bran cereal	190	500		20
Rice				60
Apple pie	2200	40		80
Blueberry pie	1300			1000
Pizza				1200
Pasta		30		140

Source: Page, D.B. and Lacroix G.M., *Food Addit. Contam.*, 2, 129, 1995.

sampled foods that contained phthalates were margarine (DnBP, 640 μg/kg; DEHP, 1240 μg/kg), white sugar (DnBP, 200 μ/kg), instant pudding (410 μg/kg), chocolate bars (DnBP, 5300 μg/kg; DEHP, 510 μg/kg), soft drinks (90 μg/kg), muffins (1000 μg/kg), gelatin dessert (90 μg/kg), and canned meat soup (100 μg/kg). Levels of DEHP in the samples from the 1996 total diet study in Canada were in general lower or the same as those found in the 1986 study. For example, DEHP was not detected in 13 food composites compared with DEHP levels from 0.02 to 0.14 μg/g in the same food composites in 1986, while DEHP levels in butter (3.2 μg/g in 1996 vs. 3.4 μg/g in 1986) and ice cream (0.83 μg/g in 1996 vs. 0.82 μg/g in 1986) were about the same [47].

36.3 Migration of Phthalates into Milk and Other Dairy Products

Phthalates are not intentionally added to food but are there as contaminants from some other processes or sources. Phthalates in food may arise from contamination during the production and preparation of food. One of the main sources of such contamination results from the migration of phthalates from products that are in contact with food during food processing. A number of studies have been conducted on the migration of DEHP from PVC tubing in the machine milking used on dairy farms [48–51]. PVC tubing contains up to 40% DEHP by weight. A Norwegian study showed that there was a clear difference in DEHP levels between raw milk collected by hand milking (about 5 μg/kg) and by machine milking involving PVC tubing (30 μg/kg in milking chamber and 50 μg/kg in collection tank) [48]. This finding is corroborated by Canadian data showing that the average DEHP value in the raw milk collected by hand milking from six cows in a Canadian dairy farm was 16 μg/kg and this value increased to 215 μg/kg when the same cows were machine milked using PVC tubing. The DEP (0.6 μg/kg) and DnBP (6 μg/kg) values were not affected by the method of milking used [49].

These studies have demonstrated that although it is not the sole source of contamination, PVC tubing used on dairy farms contributes to the levels of DEHP in raw milk and ultimately in the retail milk and milk products. Both a clear time-dependent leaching of DEHP from PVC tubing into milk and a correlation between the leaching rate and the percentage of DEHP in the PVC tubes have been reported [50]. The same study also pointed out that retail milk from the United Kingdom still contained 35 μg/kg of DEHP after DEHP plasticizers were no longer used in milking machines. A feeding study on the postsecretory migration of DEHP from cows using deuterated DEHP indicated that the secretory contamination of milk is negligibly low (3.2 ng/kg) compared with the ubiquitous background contamination [51].

Apart from the PVC tubing used in machine milking, other contamination sources must also be considered. Food-packaging materials have been evaluated for their contribution to the phthalate levels of foods they contain. Up to 11 mg/kg of DnBP and up to 61 mg/kg of DEHP in cardboard and paper used as food containers in four European countries were reported [7]. The migration of phthalates from packaging materials into food was also reported [52]. The paper bags intended for packaging and marketing sugar contained up to 450 mg/kg of DiBP and 200 mg/kg of DnBP, respectively; these phthalates were thought to originate from adhesives used in the joints of the packaging. The migration from packaging to sugar after 4 months storage was estimated at 74% for DiBP and 57% for DBP. Measurements of phthalates in retail packaging materials (plastic foil, paper, cardboard, and aluminum foil with color printing) used for various foods (sweets, wafers, meat, and milk products, frozen foods, vegetables, dry ready-to-cook products, and potato chips) were carried out in the Czech Republic. DnBP and DEHP were found in all 42 tested materials at concentrations up to 1 g/kg [53]. A study of Brazilian food-packaging materials acquired on the retail market also showed the presence of DEHP and other plasticizers in PVC films manufactured for domestic use, squeeze bags for honey, and wrapping for soft milk candy at around 20%–35% (w/w). DEHP in packaging closure seals for fatty foods such as palm oil, coconut milk, and soft high-fat cheese was 18%–33% [53]. Page and Lacroix [42] have measured phthalates and di-2-ethylhexyl adipate (DEHA) plasticizers in Canadian food-packaging materials and food samples during the 1985–1989 survey. They showed wide use of DEHP-containing or DEHA-containing food-packaging materials and demonstrated the potential link between the presence of chemicals in packaging materials and in food. They also observed that some phthalates could migrate from aluminum foil paper laminates into foods wrapped with this packaging material. Retail samples of butter and margarine wrapped in aluminum foil paper laminate were found to contain DBP, BBP, and DEHP at levels up to 10.6, 47.8, and 11.9 µg/g, respectively [54].

36.4 Analytical Methods for the Detection of Phthalates in Dairy Products

36.4.1 Avoidance of Sample Contamination

The widespread presence of phthalates poses tremendous challenge to analytical chemists in their efforts to create contamination-free environment in the laboratory. Like the ubiquitous presence of phthalates in the environment, phthalates are common contaminants in the laboratory environment. They are present in air, solvents, and many tools and containers used in the laboratory. Among them, DEHP is usually the predominant phthalate in the blank. Various measures have been taken to reduce the phthalate contamination and achieve a low and stable blank level. The common precautious measures include rinsing all glassware coming in contact with samples with organic solvents such as methanol, acetone, and hexane [32,41–43], heating the glassware to a high temperature [31,35], or having the combination of heat treatment followed by solvent rinse [33,53]. In addition, septa, caps and sample vials [41], glass wool [52], and utensils for sample preparation [42] are generally solvent rinsed before use.

Other measures to reduce phthalate contamination include careful selection and prescreening of solvents [32], distillation of purchased solvents in an all-glass device [42,53], and avoiding using Latex or vinyl gloves when handling samples [33]. When adsorbent is to be used in sample clean up, it should be decontaminated by heating prior to use [32,54]. In some cases, sodium chloride for sample handling is heat-treated prior to use as well [55].

Another approach to reduce background levels of phthalates during sample treatment is solvent-less sample preparation such as use of solid-phase microextraction (SPME) techniques followed by gas chromatography/mass spectrometry (GC/MS) analysis [49]. In SPME, phthalates in the sample are pushed out into the headspace by heat in an enclosed vial. They are then adsorbed by the extraction fiber without involvement of any organic solvent in the sample treatment.

36.4.2 Sample Pretreatment, Extraction, and Cleanup

Sample preparation methods employed in the analysis of milk and other products are summarized in Table 36.4. Liquid milk samples were processed as they were received. Powdered milk was first

Table 36.4   Examples of Measurement Methods Employed in the Analysis of Phthalates in Dairy Products

Matrix	Preparation/Extraction	Clean Up	Analysis	Ref.
Milk	Two grams of sample extracted with acetonitrile/hexane (not mixable)	Centrifuge, filtration	GC/MS	[31]
Milk	1.5 mL of milk mixed with 1.5 mL of methanol. Extracted with hexane/*t*-butyl methyl ether	Shake with acetonitrile. Acetonitrile solution was cleaned up further with silica gel column eluted with ethyl acetate/hexane	LC/MS/MS	[32]
Milk	Ten milliliters of milk mixed with 10 mL of methanol and sonicated for 10 min afterwards, 80 mL of water were added	C-18 cartridge and further Florisil cartridge	GC/MS	[34]
Milk	Five grams of samples extracted using headspace SPME	Not needed	GC/MS	[49]
Milk and cream	Sample mixed with water and acetone, extracted with DCM, and filtered	DCM extract further cleaned up by GPC with cyclohexane	GC/MS	[35]
Milk, cream, butter, and cheese	Ten grams of sample mixed with 5 mL of methanol, 3 mL of hexane and ca 300 mg of KOH, and shaken well. Repeated two more times with hexane	Further cleaned up by GPC (Biobeads SX3) with DCM/cyclohexane 1:1	GC/MS	[41]
Milk, baby food	Ten grams of sample mixed with 20 mL of water. Extracted with cyclohexane	Further cleaned up by GPC (Biobeads SX3) with DCM/cyclohexane 1:1	GC/MS	[33]

reconstituted by mixing with water according to the manufacturer's instructions. For example, Casajuana and Lacorte reported mixing milk powder with water in a 1:10 ratio (3 g of powdered milk in 30 mL of water) [34]. Solid or semisolid samples such as baby food were also mixed with water before the extraction process [33].

Different combinations of organic solvents were used for the extraction of phthalates from milk. Owing to the presence of water content in the sample, generally a mixture of water-soluble solvents (methanol, acetonitrile) and water-insoluble solvents (hexane, dichloromethane [DCM], cyclohexane) was used for the initial extraction. The extracts were then either centrifuged and filtered for subsequent instrumental analysis [31], or subjected to further clean up procedures, usually through an adsorbent column (C-18, silica gel, Florisil, etc.) for low fat milk samples [32,34] or a gel permeation column (GPC) in the case of samples with higher fat content such as cream and butter [33,35,41].

Feng et al. reported measurements of phthalates in milk using a headspace (HS) SPME technique, in which the adsorption fiber was suspended in the headspace of the vial containing milk samples allowing evaporated phthalates to be concentrated in the fiber [49]. Salt was added to the sample to aid the extraction efficiency.

The following paragraphs describe briefly some examples of extraction and cleanup methods for milk samples as examples.

Method 1 [31]: To 2 g of baby milk powder in a test tube, 1 mL of a mixture of isotope-labeled internal standards (DnBP-d_4 and DEHP-d_4) in acetonitrile (as internal standard) were added. The tube was capped and kept in a refrigerator overnight. The following day, 4.5 mL of acetonitrile, saturated with hexane, was added to the tube. After vigorous mixing (shaking and sonication), the mixture was centrifuged at 4000 rpm for 20 min and the upper solution was separated from the solid then filtered. The filtered solution was washed once with a small volume of hexane saturated with acetonitrile before undergoing GC/MS analysis.

Method 2 [32]: To 1.5 mL of milk in a 10 mL centrifuge tube, 50 μL of internal standard mixture and 1.5 mL of methanol were added. After mixing, 2.0 mL of hexane and 2.0 mL of *tert*-butyl methyl ether were added and the mixture was shaken vigorously for 1 min followed by centrifugation at 1500 rpm for 2 min. The hexane/ether phase was transferred to another tube and the extraction was repeated once. The combined extract was evaporated to dryness at 70°C under nitrogen flow and redissolved in 3.0 mL of hexane. An aliquot (2.0 mL) of the hexane solution was shaken with 2.0 mL of acetonitrile for 1 min. The hexane phase was removed and the remaining acetonitrile phase was shaken with 1 mL of hexane. After removing the hexane phase completely, the acetonitrile solution was evaporated at 70°C under nitrogen flow and redissolved in 0.5 mL of acetonitrile for LC/MS/MS analysis.

Method 3 [33]: 10 g of sample spiked with internal standard DnBP-d_4 and DnOP-d_4 were mixed with 20 mL of water and 60 mL of cyclohexane. The mixture was shaken for 3 min and left to stand for at least 2 h. Afterwards, 60 mL of acetone were added to the mixture and again shaken for 3 min. The whole mixture was transferred to a centrifuge tube. The solution was carefully collected using a pipette and the residual was then mixed with 20 mL of cyclohexane once more. Both solutions were combined and evaporated to approximately 0.5–1 mL. The separation of phthalates from fat was achieved by GPC. The preparative GPC comprised of a Biobeds SX-3 column with an internal diameter of 16 mm and a length of 52 cm. The eluting solvent was a mixture of 50:50 cyclohexane/DCM at a flow rate of 3.0 mL/min. The phthalate fraction (eluting at 16–32 min) was collected and concentrated to 1 mL for GC/MS analysis.

Method 4 [34]: 10 mL of reconstituted milk spiked with 4-*n*-nonylphenol as surrogate was mixed with 10 mL of methanol and sonicated for 10 min. Afterwards, the mixture was diluted with 80 mL of high-performance liquid chromatography (HPLC) grade water and passed through a 0.5-g C-18 SPE cartridge that had been conditioned with 12 mL of 4:1 DCM/hexane, 12 mL of methanol, and 12 mL of water. After preconcentration, the adsorbent was rinsed with 15 mL of water and dried under vacuum. The trapped compounds were desorbed with 12 mL of 4:1 methylene chloride/hexane and the volume reduced for loading onto a 5 g Florisil cartridge that had been preconditioned with 60 mL of methanol and 60 mL of 4:1 DCM/hexane. The Florisil cartridge was eluted with 40 mL of 4:1 DCM/hexane. Notice that the cartridge was further eluted with 40 mL of ethyl acetate to recover the more polar compounds (bisphenol A, nonylphenol, and bisphenol A diglycidyl ether) as the cotarget analytes under the same study. The eluent was evaporated to near dryness and reconstituted with ethyl acetate to a final volume of 0.3 mL with DEHP-d_4 being added as an internal standard.

Method 5 [49]: 5 g of milk were weighed into a 15 mL SPME vial using a pipette. A magnetic stirring bar and 2.5 g of sodium chloride were then added into the vial and the vial was closed with the vial cap. The vial was tightly closed to avoid possible leakage of gas when the vial was heated. The vial was then placed into a preheated oil bath (90°C) on a hot plate. The stirring speed was adjusted to ensure that the solution was well stirred. After 2 min, the SPME needle was punched through the cap into the headspace of the vial and the fiber (PDMS-100 μm) was pushed out from the protection needle to start headspace sampling. The SPME holder was placed at a height that would result in the tip of the inserted fiber being suspended about 1.5 cm above the milk sample. After the sampling was finished, the fiber was retracted into the protection needle. The needle was then removed from the sampling vial and inserted into a clean vial to protect the fiber from exposure to laboratory air. The needle was then immediately inserted into the GC injection port for GC/MS analysis.

36.4.3 Instrumental Conditions

Owing to the semivolatile nature of phthalates, instruments used in their analysis are either GC or HPLC. MS is almost a nominal detector nowadays for the measurements of phthalates. Although LC/MS is popular for the analysis of phthalate metabolites, such as the monoester of the phthalate, GC/MS is more common in the analysis of phthalates themselves.

For GC/MS analysis (Table 36.5), prepared samples were injected into the GC/MS at an injection temperature around 250°C (240°C–260°C reported from several studies). The commonly used GC capillary column for the separation of phthalates is a fused-silica capillary column containing 5% phenyl and 95% methyl polysiloxane with a column length of either 30 or 60 m. There are several different brands of such columns available including HP-5, DB-5, ZB-5, XTI-5, etc. The oven temperature programs varied greatly depending on the complexity of the samples and number of target analytes monitored. For example, the rise of oven temperature ranged from 6°C/min in one study to 15°C/min in another. Usually, the phthalates can be well separated under these oven temperature programs. Figure 36.2 shows a typical GC/MS chromatogram of six commonly monitored phthalates.

The separated phthalates were detected by a mass spectrometer usually operated under the selected ion-monitoring (SIM) mode. One target ion (T-ion) and two qualifier ions (Q-ion) were selected for each of the target phthalates. The base peak of *m/z* 149 was monitored as the T-ions for all phthalates except for DMP that had a base peak of *m/z* 163. The other characteristic fragments of phthalates are listed in Table 36.5. In practice, one has to consider the relative abundance of these fragments in selecting qualifier (Q-) ions. There were some variations in selecting Q-ions for phthalates among various studies. For example, Feng et al. selected Q-ions of *m/z* 77 and 194 for DMP; *m/z* 177 and 104 for DEP; *m/z* 223 and 104 for DBP; *m/z* 91 and 206 for BBP; *m/z* 167 and 279 for DEHP; *m/z* 279 and 104 for DOP [49], while Casajuana and Lacorte selected Q-ions

Table 36.5 GC/MS Operation Conditions for the Analysis of Phthalates in Dairy Products

Injection vol/temp	Column	Oven Temperature	MS	Quantification	Ref.
1 μL/260°C	ZB-05 (0.25 mm × 30 m × 0.25 μm)	50°C (1 min), 15°C/min to 270°C (5 min)	NA	Comparing peak area with corresponding isotopic standard	[31]
1 μL/NA	DB-5 (0.25 mm × 60 m × 0.25 μm)	140°C (2 min), 10°C/min to 340°C	SIM	Relative response factor to d_4-labeled standards	[33]
1 μL/240°C	XTI-5 (0.25 mm × 30 m × 0.25 μm)	90°C (1 min), 8°C/min to 250°C, 4°C/min to 280°C (5 min)	SIM	Relative response factor to d_4-labeled standards	[43]
2 μL/250°C	HP-5MS (0.25 mm × 30 m × 0.25 μm)	60°C (1 min), 6°C/min to 175°C, 3°C/min to 280, 7°C/min to 300°C	SIM	Relative response factor to d_4-labeled standards	[34]
SPME/280°C	DB-5 (0.25 mm × 30 m × 0.25 μm)	55°C (1 min), 15°C/min to 280°C (15 min)	SIM	Relative response factor to d_4-labeled standards	[49]

of 77 and 135 for DMP, *m/z* 177 and 105 for DEP; *m/z* 223 and 76 for DBP; *m/z* 91 and 206 for BBP; and *m/z* 167 and 279 for DEHP [34].

36.5 Concluding Remarks

Phthalates are present in foods and particularly in dairy products, owing to the high fat content of the latter. Oral intake is a major contributor to the total exposure of humans to this group of chemical contaminants. A recent study on sources of human exposure to phthalates among Europeans indicated a high proportion contributed by food, across all age groups, for DnBP and DEHP [56]. The contribution of foods to total DEHP exposure range from about 50% in infants and toddlers to almost 100% in teens and adults. The Canadian study on phthalates in human milk following a 6 month postpartum period also indicated a continuous human exposure to DnBP and DEHP, as there was no decrease in concentration levels in breast milk observed over the lactation period [36]. The migration of phthalates from phthalate-containing products and packages into food is one of the major sources of phthalates found in food. A reduction in phthalate concentrations in foods will significantly reduce the total human exposure to these chemicals.

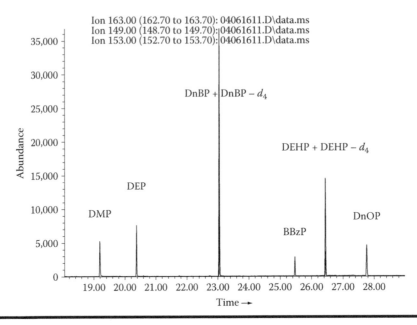

Figure 36.2 GC/MS chromatogram of six common phthalates and two isotope-labeled phthalates.

References

1. Cousins, I. and Mackay, D. Correlating the physical–chemical properties of phthalate esters using the 'three solubility' approach. *Chemosphere*, 41, 1389, 2001.
2. do Nascimento Filho, I. et al. Identification of some plasticizers compounds in landfill leachate. *Chemosphere*, 50, 657, 2003.
3. Peñalver, A. et al. Determination of phthalate esters in water samples by solid-phase microextraction and gas chromatography with mass spectrometric detection. *J. Chromatogr. A*, 872, 191, 2000.
4. Fauser, P. and Thomsen, M. Sensitivity analysis of calculated exposure concentrations and dissipation of DEHP in a topsoil compartment: The influence of the third phase effect and dissolved organic matter (DOM). *Sci. Total Environ.*, 296, 89, 2002.
5. Suzuki, T. et al. Monitoring of phthalic acid monoesters in river water by solid-phase extraction and GC–MS determination. *Environ. Sci. Technol.*, 35, 3757, 2001.
6. Luke-Betlej, K. et al. Solid-phase microextraction of phthalates from water. *J. Chromatogr. A*, 938, 93, 2001.
7. Aurela, B., Kulmala, H., and Söderhjelm, L. Phthalates in paper and board packaging and their migration into Tenax and sugar. *Food Addit. Contam.*, 16, 571, 1999.
8. Uhde, E. et al. Phthalic esters in the indoor environment—Test chamber studies on PVC-coated wallcoverings. *Indoor Air*, 11, 150, 2001.
9. Wilkinson, C.F. and Lamb IV, J.C. The potential health effects of phthalate esters in children's toys: A review and risk assessment. *Regul. Toxicol. Pharmacol.*, 30, 140, 1999.
10. Bouma, K. and Schakel, D.J. Migration of phthalates from PVC toys into saliva simulant by dynamic extraction. *Food Addit. Contam.*, 19, 602, 2002.
11. Inoue, K. et al. The validation of column-switching LC/MS as a high-throughput approach for direct analysis of di(2-ethylhexyl) phthalate released from PVC medical devices in intravenous solution. *J. Pharmaceut. Biomed. Anal.*, 31, 1145, 2003.

12. Lin, Z.-P. et al. Determination of phthalate ester congeners and mixtures by LC/ESI-MS in sediments and biota of an urbanized marine inlet. *Environ. Sci. Technol.*, 37, 2100, 2003.
13. Otake, T., Yoshinaga, J., and Yanagisawa, Y. Analysis of organic esters of plasticizer in indoor air by GC–MS and GC-FPD. *Environ. Sci. Technol.*, 35, 3099, 2001.
14. Zhu, J. et al. Phthalates in indoor air of Canadian residences. In *Proceedings of the ISIAQ 7th International Conference*; 2003; Vol. 1, pp. 542–547.
15. Rudel, R.A. et al. Phthalates, alkylphenols, pesticides, polybrominated diphenyl ethers, and other endocrine-disrupting compounds in indoor air and dust. *Environ. Sci. Technol.*, 37, 4543, 2003.
16. Butte, W. and Heinzow, B. Pollutants in house dust as indicators of indoor contamination. *Rev. Environ. Contam. Toxicol.*, 175, 1, 2002.
17. Koch, H.M. et al. Internal exposure of the general population to DEHP and other phthalates—Determination of secondary and primary phthalate monoester metabolites in urine. *Environ. Res.*, 93, 177, 2003.
18. Brock, J.W. et al. Phthalate monoesters levels in urine of young children. *Bull. Environ. Contam. Toxicol.*, 68, 309, 2002.
19. Anderson, W.A.C. et al. Determination of isotopically labeled monoesterphthalates in urine by high performance liquid chromatography–mass spectrometry. *Analytes*, 127, 1193, 2002.
20. Blount, B.C. et al. Quantitative detection of eight phthalate metabolites in human urine using HPLC-APCI-MS/MS. *Anal. Chem.*, 72, 4127, 2000.
21. Blount, B.C. et al. Levels of seven urinary phthalate metabolites in a human reference population. *Environ. Health Persp.*, 108, 979, 2000.
22. Silva, M.J. et al. Determination of phthalate metabolites in human amniotic fluid. *Bull. Environ. Contam. Toxicol.*, 72, 1226, 2004.
23. Jobling, S. et al. A variety of environmentally persistent chemicals, including some phthalate plasticizers, are weakly estrogenic. *Environ. Health Persp.*, 103, 582, 1995.
24. Higuchi, T.T. et al. Effects of dibutyl phthalate in male rabbits following in utero, adolescent, or postpubertal exposure. *Toxicol. Sci.*, 72, 301, 2003.
25. Fukuwatari, T. et al. Elucidation of the toxic mechanism of the plasticizers, phthalic acid esters, putative endocrine disrupters: Effects of dietary di(2-ethylhexyl)phthalate on the metabolism of tryptophan to niacin in rats. *Biosci. Biotechnol. Biochem.*, 66, 705, 2002.
26. Lamb IV, J.C. et al. Reproductive effects of four phthalic acid esters in the mouse. *Toxicol. Appl. Pharmacol.*, 88, 255, 1987.
27. Poon, R. et al. Subchronic oral toxicity of di-*n*-octyl phthalate and di(2-ethylhexyl) phthalate in the rat. *Food Chem. Toxicol.*, 35, 225, 1997.
28. Colón, I. et al. Identification of phthalate esters in the serum of young Puerto Rican girls with premature breast development. *Environ. Health Persp.*, 108, 895, 2000.
29. Duty, S.M. et al. Phthalate exposure and human semen parameters. *Epidemiology*, 14, 269, 2003.
30. Bornehag, C.-G. et al. The association between asthma and allergic symptoms in children and phthalates in house dust: A nested case–control study. *Environ. Health Persp.*, 112, 1393, 2004.
31. Yano, K. et al. Phthalates levels in baby milk powder sold in several countries. *Bull. Environ. Contam. Toxicol.*, 74, 373, 2005.
32. Soerensen, L. Determination of phthalate in milk and milk products by liquid chromatography/tandem mass spectrometry. *Rapid Commun. Mass Spectrom.*, 20, 1135, 2006.
33. Gruber, L., Wolz, G., and Piringer, O. Analysis of phthalates in baby food. *Deutsche Lebensmittel-Rundashau*, 94, 177, 1998.
34. Casajuana, N. and Lacorte, S. New methodology for the determination of phthalate esters, bisphenol A, bisphenol A diglycidyl ether, and nonylphenol in commercial whole milk samples. *J. Agric. Food Chem.*, 52, 3702, 2004.
35. Bruns-Weller, E. and Pfordt, J. Determination of phthalic acid esters in foods, mother's milk, dust and textiles. *Umweltwissenschaften und Schadstoff-Forschung*, 12, 125, 2000.

36. Zhu, J. et al. Phthalate esters in human milk: Concentration variations over a six-month postpartum time. *Environ. Sci. Technol.*, 40, 5276, 2006.

37. Högberg, J. et al. Phthalate diesters and their metabolites in human breast milk, blood and urine as biomarkers of exposure in vulnerable populations. *Environ. Health Persp.*, in press.

38. Schecter, A., Ryan J.J., and Papke, O. Decrease in levels and body burden of dioxins, dibenzofurans, PCBS, DDE, and HCB in blood and milk in a mother nursing twins over a thirty-eight month period. *Chemosphere*, 27, 1807, 1998.

39. LaKind, J.S. et al. Methodology for characterizing distributions of incremental body burdens of 2,3,7,8-TCDD and DDE from breast milk in North American nursing infants. *J. Toxicol. Environ. Health* Part A 59, 605–639, 2000.

40. Yakushiji T. Contamination, clearance, and transfer of PCB from human milk. *Environ. Contam. Toxicol.*, 101, 139, 1988.

41. Sharman, M. et al. Levels of di-(2-ethylhexyl) phthalate and total phthalate ester in milk, cream, butter and cheese. *Food Addit. Contam.*, 11, 357, 1994.

42. Page, D.B. and Lacroix G.M. The occurrence of phthalate ester and di-2-ethylhexyl adipate plasticizers in Canadian packaging and food sampled in 1985–1989: A survey. *Food Addit. Contam.*, 2, 129, 1995.

43. Petersen, J.H. and Breindahl, T. Plasticizers in total diet samples, baby food and infant formulae. *Food Addit. Contam.*, 17, 133, 2000.

44. Cocchieri, R.A. Occurrence of phthalate esters in Italian packaged foods. *J. Food Protect.*, 49, 265, 1986.

45. Wang, M. et al. Matrix solid-phase dispersion and gas chromatography/mass spectrometry for the determination of phthalic acid esters in vegetables. *Chin. J. Chromatogr.*, 25, 577, 2007.

46. Tsumura, Y. et al. Estimated daily intake of plasticizers in 1-week duplicate diet samples following regulation of DEHP-containing PVC gloves in Japan. *Food Addit. Contam.*, 20, 317, 2003.

47. Castle, L., Gibert, J., and Eklund, T. Migration of plasticizer from poly(vinyl chloride) milk tubing. *Food Addit. Contam.*, 7, 591, 1990.

48. Feng, Y.L., Zhu, J., and Sensenstein, R. Development of a headspace solid phase microextraction method combined with gas chromatography mass spectrometry for the determination of phthalate esters in cow milk. *Anal. Chim. Acta*, 538, 41, 2005.

49. Ruuska, R.M. et al. Migration of contaminants from milk tubes and teat liners. *J. Food Protect.*, 50, 316, 1987.

50. Blüthgen, A. Organic migration agents into milk at farm level (illustrated with diethylhexylphthalate). *Bull. IDF*, 356, 39, 2003 [German].

51. Lopez-Espinosa, M.-J. et al. Oestrogenicity of paper rand cardboard extracts used as food containers. *Food Addit. Contam.*, 24, 95, 2007.

52. Gajduskova, V., Jarosova, A., and Ulrich, R. Occurrence of phthalic acid esters in food packaging materials. *Potravinarske Vedy*, 14, 99, 1996.

53. de A. Freire, M.T., Santana, I.A., and Reyes, F.G.R. Plasticizers in Brazilian food-packaging materials acquired on the retail market. *Food Addit. Contam.*, 23, 93, 2006.

54. Page, B.D. and Lacroix, G.M. Studies into the transfer and migration of phthalate esters from aluminium foil paper laminates to butter and margarine. *Food Addit. Contam.*, 9, 197, 1992.

55. Tsumura, Y. et al. Eleven phthalate esters and di(2-ethylhexyl) adipate in one-week duplicate diet samples obtained from hospitals and their estimated daily intake. *Food Addit. Contam.*, 18, 449, 2001.

56. Wormuth, M. et al. What are the sources of exposure to eight frequently used phthalic acid esters in Europeans? *Risk Anal.*, 26, 803, 2006.

Chapter 37

Environmental Contaminants

Sara Bogialli and Antonio Di Corcia

Contents

37.1 Introduction 929
    37.1.1 Persistent Organic Pollutants 930
    37.1.2 Pesticides 931
37.2 Regulations 932
37.3 Analysis of Pesticides and POPs in Dairy Foods 933
    37.3.1 Extraction 933
        37.3.1.1 Analyte Extraction from Liquid Foodstuff (Milk) 933
        37.3.1.2 Analyte Extraction from Semiliquid and Solid Matrices.......................935
    37.3.2 Cleanup 941
    37.3.3 Identification and Quantitation 943
        37.3.3.1 Capillary GC with Selective Detectors 943
        37.3.3.2 Selection of the HRGC Column... 943
        37.3.3.3 Injection Devices 944
    37.3.4 Selective Detectors for HRGC.. 944
    37.3.5 The Mass Spectrometric Detector 945
References 946

37.1 Introduction

Milk may include more than the nutrients required for humans that are well studied and well documented. Less well-understood and less well-studied is the composition of milk as it reflects other ingestants of the cattle, such as environmental contaminants, that are ingested, inhaled, or absorbed through the skin or mucous membranes. This pool of substances is absorbed into the bloodstream of lactating animals or stored in their bones or fat and reach the target organ, the breast, during active lactation.

Organic contaminants in dairy food can be divided into four categories:

1. Veterinary drugs
2. Toxins produced by fungi and bacteria
3. Pesticides
4. Persistent organic pollutants (POPs)

In this chapter, we will illustrate analytical methodologies elaborated for detecting pesticide residues and POPs in milk and milk derivatives. Figure 37.1 shows the chemical structures of some representative environmental contaminants.

It is known that milk, like other fatty matrices, is one of the most important routes of excretion for lipophilic pesticides, i.e., organochlorine and organophosphorous insecticides, and POPs, like polychlorinated biphenyls (PCBs), polychlorinated dibenzodioxins (PCDDs), polychlorinatedibenzofurans (PCDFs), and polycyclic aromatic hydrocarbons (PAHs). In addition to the physiology of milk production, there are other factors that affect the body's burden of xenobiotics and the amount excreted in the milk; the influence of cattle residence, industrial or not industrial, and proximity to unusual exposures, spills, or accidents. When cattle live in a relatively clean environment, the diet may be the only source of contaminants.

37.1.1 Persistent Organic Pollutants

POPs include industrial chemicals and by-products of certain manufacturing processes and waste incineration such as PCBs, PAHs, and dioxins. The term "dioxins" is often used in a confusing way. In toxicological consideration, and also in the present chapter, the term is used to designate the PCDDs, the PCDFs, and the coplanar (dioxin-like) PCBs, since these classes of compounds show the same type of toxicity. In addition to environmental pollution, PAHs can contaminate

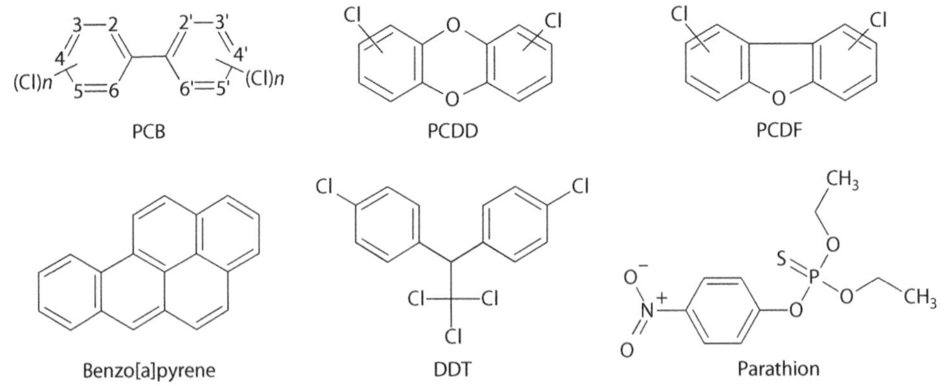

Figure 37.1 General structures of PCBs, PCDDs, PCDFs, and chemical structures of some representative compounds belonging to the class of PAHs, benzo[a]pyrene, organochlorine insecticides (DDT) and organophosphate pesticides (parathion).

foods during smoking processes, and heating and drying processes that allow combustion products to come into direct contact with food.

The characteristics that make POPs chemicals unique also make them a serious global environmental pollutant because they (a) persist in the environment for decades; (b) concentrate in fatty tissues and bioaccumulate as they move up the food chain; (c) travel long distances in global area and water currents, generally moving from tropical and temperate regions to concentrate in the northern latitudes; (d) have been linked with serious health effects in humans and other living organisms, even at very low exposures.

In just a few decades, POPs have spread throughout the global environment to threaten human health and damage land and water ecosystems. All living organisms on the Earth now carry measurable levels of POPs in their tissues. POPs have been found in sea mammals at levels high enough to qualify their bodies as hazardous waste under U.S. law [1] and evidence of POPs contamination in human blood and breast milk has been documented worldwide.

Despite their hazards, these chemicals continue to be produced, used, and stored in many countries. Even where national bans or other controls exist, these restrictions are often poorly enforced, and, in any case, they cannot protect citizens from exposure to POPs who have migrated from other regions where these chemicals are still in use.

37.1.2 Pesticides

The term "pesticide" is used to indicate any substance, preparation, or organism used for destroying pests. This broad definition covers substances used for many purposes, including insecticides, herbicides, fungicides, nematocides, acaricides, and lumbricides. According to their chemical nature, the most important classes of pesticides are the following: organochlorines (OCs), organophosphates (OPs), carbamates, triazines, phenoxyacids, phenylureas and sulfonylureas, acetoanilides, benzimidazoles, and pyrethroids.

Today, there are more than 1800 basic chemicals that are used as active ingredients of pesticides dispersed in approximately 33,600 formulations. Over the last 20 years, in the United States alone, about 15 Mton of pesticides were used for pest control. This situation has urged local governments to enact more and more restrictive regulations for banning some dangerous pesticides and lowering the maximum admissible concentrations of pesticides in drinking water and foodstuffs.

Possible sources of contamination of milk are (a) foodstuffs containing high levels of pesticide residues from postharvest treatment or contamination, for instance, by drift during commercial aerial application; (b) foodstuffs manufactured from plant material that has been treated during the growing season with insecticides; (c) use of insecticides directly on the animal against disease vectors; (d) use of insecticides in stables (treatment against flies); (e) hygienic treatments against insects in milk-processing factories.

Contamination of milk from source (a) and (b) with a pesticide depends on the stability of the compound, its mode of application, the duration of intake or exposure, and its metabolic fate in the animal. Contamination from source (c) is more important, especially in tropical countries where the use of insecticides (cattle dipping) is necessary to protect the health and productivity of animals. Spraying of stables frequently leads to contamination of the milking equipment, and treatment of factory premises against cockroaches and other insects may introduce significant quantities of pesticides into the milk products. Although ingested pesticides are not excreted as such in milk, some of the previously described routes can lead to contamination of milk and dairy products by (bio)degradation products of pesticides. Indeed, maximum residue limits (MRLs) for transformation products of certain classes of insecticides have been set by several organizations such as FAO-Codex Alimentarius [1] and European Union (EU) [2].

Among pesticides, the class of OC insecticides, e.g., DDT, aldrin, dieldrin, toxaphene, chlordane, heptachlor, and others, has received special attention by regulatory laboratories entrusted to monitor contamination levels of dairy foods, especially milk. This is so because characteristics and fate of OCs in the environment are very similar to those of POPs.

37.2 Regulations

The Codex Alimentarius is the global reference point for consumers, food producers and processors, national food control agencies, and the international food trade. The Codex Alimentarius standards, guidelines, and recommendations are internationally acknowledged as the best-established measures to protect human health from risks arising from contaminants in foods. The respective committees in the Codex Alimentarius for chemical contaminants are the Codex Committee on Pesticide Residues (CCPR) and the Joint Meeting on Pesticide Residues (JMPR) for the pesticide residue monitoring, while the Codex Committee on Food Additives and Contaminants (CCFAC) and the Joint FAO/World Health Organization (WHO) Committee on Food Additives (JECFA) are charged with the management of other chemical contaminants.

To include a specific contaminant in a list of forbidden or regulated compounds in foodstuff, generally these Committees take into account the following criteria:

■ The compound has been often detected in at least one commodity at a significant concentration by reliable analysis.
■ The compound has proved or is suspected to be toxicologically adverse to human or animal health at the concentration observed in the foodstuff.
■ The foodstuff is widespread and it shares in the total intake of the contaminant of interest.

Therefore, the proposed safety level is an arrangement among toxicological data, analytical performances, and trade demands. These threshold values in foodstuff are established by the Food and Drug Administration (FDA) in the United States and by the European Community (EC) in EU and are known as MRLs. In the case of contaminants that are considered to be genotoxic carcinogens or in cases where current exposure of the population is close to or exceeds the tolerable intakes, maximum levels should be set as the levels that are as low as reasonably achievable (ALARA).

With regard to dioxins and PCBs, in 2001 maximum levels were set on EC level only for dioxins and not for dioxin-like PCBs, given the very limited data available at that time on the prevalence of dioxin-like PCBs. Since 2001, however, more data on the presence of dioxin-like PCBs have become available; therefore, maximum levels for the sum of dioxins and dioxin-like PCBs have been set in 2006 [3], as this is the most appropriate approach from a toxicological point of view. Each congener of dioxins or dioxin-like PCBs exhibits a different level of toxicity. To enable the sum up of the toxicity of these different congeners, the concept of toxic equivalency factors has been introduced to facilitate risk assessment and regulatory control. This means that the analytical results relating to all the individual dioxin and dioxin-like PCB congeners of toxicological concern are expressed in terms of a quantifiable unit, namely the TCDD toxic equivalent (TEQ). On this basis, the Scientific Committee on Food fixed a tolerable weekly intake of 14 pg WHO-TEQ/kg bw for dioxins and dioxin-like PCBs.

Milk is the main feed for infants, a vulnerable group. As a consequence, EC has set very strict MRLs in this matrix [3,4]. No specific MRLs related to pesticide residues has been set by the FDA and the EC in dairy products, maybe because quality control tests are supposed to be carried out on raw milk. Conversely, MRLs of POPs have been set in some milk derivates.

Some helpful databases are available on the websites of government agencies, i.e., FAO/WHO, FDA, or EC [5].

37.3 Analysis of Pesticides and POPs in Dairy Foods

37.3.1 *Extraction*

37.3.1.1 *Analyte Extraction from Liquid Foodstuff (Milk)*

Methods for the extraction of toxicants in liquid foodstuffs exploit the partitioning of analytes between the aqueous phase and a water-immiscible solvent (liquid–liquid extraction [LLE]) or a sorbent material (solid-phase extraction [SPE]). Conventional LLE is still the most diffused in many laboratories. However, SPE and, to a lesser extent, solid-phase microextraction (SPME) are constantly gaining popularity in regulatory laboratories.

37.3.1.1.1 Liquid–Liquid Extraction

Depending on the nature of the analytes, the methodology for their extraction from milk involves single organic solvents or mixtures of them.

After extraction and phase separation, the organic phase is often dried with Na_2SO_4. The extractant is then removed by using a rotary evaporator in a water bath. Often, solvent substitution is needed to make the final extract more compatible to gas chromatography (GC) or high-performance liquid chromatography (LC) analysis.

Table 37.1 shows some selected applications involving LLE for analyzing OCs and OPs insecticides and POPs in milk.

37.3.1.1.2 Solid-Phase Extraction

The SPE technique was first introduced in the mid-1970s as an alternative to LLE. It became commercially available in 1978, and now SPE cartridges and disks are available from many suppliers. Conventional SPE is generally performed by passing liquid samples through a cartridge filled with a solid sorbent. Analytes are eluted from the cartridge with an appropriate organic solvent or mixture of solvents. Typical sorbents for SPE are silica chemically modified with a C_{18} alkyl chain, commonly referred to as C-18; highly cross-linked polystyrene-divinylbenzene copolymers (PS-DVB), commonly referred to as PRP-1, Envichrom P or Lichrolut; hydrophobic/hydrophilic copolymers, commercially referred to as Oasis; graphitized carbon blacks (GCBs), commonly referred to as Carbopack or Carbograph. All these materials are commercially available in medical-grade polypropylene housing and polyethylene frits. This technique is widely applied to low-viscosity liquid matrices. Depending on the type of the sorbent and the final destination of the extract, various solvents or solvent mixtures are used to re-extract pesticides from sorbent cartridges. For both C-18 and PS-DVB, methanol or acetonitrile is the eluent of choice, when analyzing using LC. With C-18 cartridges and GC instrumentation, ethyl acetate is usually preferred. With GCB cartridges, a CH_2Cl_2/CH_3OH (80:20, v/v) mixture offers quantitative desorption of base/neutral pesticides having a broad range of polarity.

Table 37.1 Selected Applications Using LLE for the Analysis of Pesticides and POPs in Milk

Analytes	Solvent	Quantitation Technique	Ref.
PCBs	Boiling hexane	GC–MS	[6]
PCBs	Acetone/hexane (2:1)	GC–MS	[7]
OP pesticides	ethylacetate	GC–PPD	[8]
OP pesticides	CH_3CN	GC–PPD	[9]
OC pesticides	Acetone/CH_3CN/hexane (2:2:15)	GC–ECD	[10]
OC pesticides	Conc H_2SO_4 + hexane	GC–ECD	[11]
Dioxins	Methylene chloride/hexane (1:1)	GC–MS	[12]

PCBs, polychlorinated biphenyls; GC, gas chromatography; MS, mass spectrometry; OP, organophosphate; PPD, phosphorous photometric detector; OC, organochlorine; ECD, electron capture detector.

Table 37.2 lists some selected SPE-based methods for extracting pesticides and POPs from milk.

37.3.1.1.3 Solid-Phase Microextraction

In the early 1990s, a new technique for extracting analytes from liquid samples was introduced, the so-called SPME. Figure 37.2 shows a typical SPME device. A 0.05–1 mm i.d. uncoated fiber or coated with suitable immobilized liquid phase (in the second case, this technique should be more correctly called liquid-phase microextraction) is immersed into a continuously stirred liquid sample. After equilibrium is reached (a good exposure time takes 15–25 min), the fiber is introduced into the injection port of a gas chromatograph, where analytes are thermally desorbed and analyzed. Positive features of this technique are that the technique is rapid, very simple, and it does

Table 37.2 Selected Applications Using Solid-Phase Extraction for the Analysis of Pesticides and POPs in Milk

Analytes	Sorbent	Eluant	Ref.
Dioxins	25 g C-18 cartridge	Hexane	[13]
OCs	0.5 g C-18 cartridge	CH_3CN/light petroleum	[14]
Herbicides	0.5 g carbograph 4	CH_2Cl_2/MeOH (8:2)	[15]
Triazines	0.5 g carbograph 1	CH_2Cl_2/CH_3CN (6:4)	[16]
OCs	1 g C-18	Hexane	[17]

OCs, organochlorines.

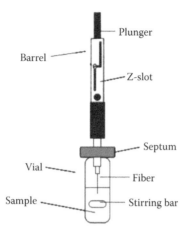

Figure 37.2 SPME device.

not use any solvent. In addition, this technique requires small sample volumes (2–5 mL) and all the sample extract is injected into the analytical column.

Table 37.3 lists some selected SPME-based methods for extracting pesticides and POPs from milk.

37.3.1.2 Analyte Extraction from Semiliquid and Solid Matrices

The extraction and recovery of trace organic material from semiliquid and solid matrices is often the slowest and the most error-prone step in an analytical method. The conventional liquid extraction techniques for solids and semisolid materials (Soxhlet) have two main disadvantages. The first, large volumes of organic solvent are required, which can lead to sample contamination and

Table 37.3 Selected Applications of the Solid-Phase Microextraction Technique for Analyzing Pesticides and POPs in Milk

Analytes	Extractant	Method of Analysis	Ref.
Pesticides	PDMS/DVB-coated fiber	GC-ECD	[18]
PCBs	Uncoated fiber	GC-ECD	[19]
OP insecticides	Uncoated fiber	GC-NPD	[20]
Pesticides	Uncoated fiber	GC-MS/MS	[21]
Triazine herbicides	PDMS/DVB-coated fiber	GC-MS	[22]

PDMS/DVB, polydimethylsiloxane/divinylbenzene; GC, gas chromatography; ECD, electron capture detector; PCBs, polychlorinated biphenyls; OP, organophosphate; NPD, nitrogen photometric detector; MS/MS, tandem mass spectrometer.

"losses" due to volatilization during concentration steps. The second, to achieve an exhaustive extraction may require several hours (6–14 h). With the development of sophisticated instrumentation with detection limits in the picogram and femtogram levels, pressure is finally felt within the analytical community to develop and validate sample preparation procedures that can be used to rapidly isolate trace level organics from complex matrices.

37.3.1.2.1 Soxhlet Extraction

In spite of being time-consuming, Soxhlet extraction still continues to be largely used, as it is included in several official methods. The most salient advantages of Soxhlet are that the sample is repeatedly brought into contact with fresh portion of the solvent, thereby aiding displacement of the distribution equilibrium, and that no filtration is needed. The drawbacks involved in the use of this traditional extraction technique are the inability to provide agitation, which would help process acceleration, and the constant heat applied to the leaching cavity. This heat is dependent on the solvent boiling point and could be insufficient to break some matrix–analyte bonds. A microwave-assisted Soxhlet extractor (Soxtec) has been proposed and commercialized without noticeable success, even though the literature quotes applications in which a saving time (1 instead of 4 h) is achieved using Soxtec instead of conventional Soxhlet.

Recently, Soxhlet extraction has also been applied to liquid dairy products (milk and yoghurt) after dispersing food sample on a suitable solid material following a procedure similar to that used with the matrix solid-phase dispersion (see later).

Table 37.4 lists some selected Soxhlet extraction-based methods for extracting pesticides and POPs from dairy products.

37.3.1.2.2 Liquid-Phase Extraction

Although manually shaking a finely dispersed solid sample with a suitable solvent can be effective in many cases, blending the sample in the presence of the solvent in high-speed homogenizer machines or ultrasonication baths ensures extensive sample disruption and a better analyte extraction. This technique is called liquid-phase extraction (LPE) or liquid–solid extraction (LSE). So

Table 37.4 Selected Applications of Soxhlet Extraction for the Analysis of Pesticides and POPs in Solid Dairy Products

Analytes	Matrix	Extractant	Ref.
Dioxins	Yoghurt, cheese, milk	Toluene, 24 h	[23]
Dioxins	Powder milk	Pentane/DCM (1:1), 12 h	[24]
OCs, PCBs	Yoghurt	Cyclohexane/acetone (1:1), 14 h	[25]
PAHs	Yoghurt, cheese, butter	Cyclohexane/DCM (1:1), 4 h	[26]
Dioxins	Powdered milk	Acetone/hexane (1:1), 16 h	[27]
PCBs, dioxins	Cheese, butter	Hexane/DCM (1:1), 16 h	[28]

DCM, methylene chloride; OCs, organochlorines; PCBs, polychlorinated biphenyls; PAHs, polycyclic aromatic hydrocarbons.

far, this technique is the most popular for extracting contaminants in foodstuffs. Water-miscible solvents, such as acetone, acetonitrile, and methanol, are now widely used as they are effective in extracting both polar and nonpolar toxicants. Ethyl acetate with added anhydrous sodium sulfate is an alternative extractant. Its use offers advantages in that no subsequent partition step is required and the extract can be used directly in gel permeation chromatographic (GPC) cleanup. Each extraction system offers distinct advantages, some of which depend on the way in which the extraction/partition steps are integrated into the cleanup/determination steps of the analytical method. Other factors that may influence the choice of one solvent over another is solvent consumption, which is related to cost, health, and disposal problems.

Table 37.5 shows selected extraction procedures for analyzing contaminants in solid foods by LPE.

37.3.1.2.3 Pressurized Solvent Extraction

This method, also called accelerated solvent extraction (ASE), has been used since 1995. Pressurized solvent extraction (PLE) is an extraction under elevated pressure and temperature (Figure 37.3). It represents an effective extraction technique with the advantages of shorter extraction times and lower consumption of solvents when compared with LPE and Soxhlet. It allows the universal use of solvents or solvent mixtures with different polarities and individually variable pressures of 5–200 atm to maintain the extraction solvent in a liquid state, and temperatures ranging from room temperature up to 200°C to accelerate extraction.

In general, the extraction efficiency of PLE is influenced by both extraction pressure and temperature, which are the operation parameters of PLE. The solvent volume can be reduced because the solubility increases with temperature. In addition, sample matrix effects also affect the extraction efficiency. Therefore, the extraction behavior of PLE is not plain and optimization of operating conditions is laborious. Another weakness of PLE is that, when using hydrophobic organic solvents, the presence of relatively high water percentages in the sample strongly decreases analyte extraction efficiency, as water hinders contact between the solvent and the analyte. When

Table 37.5 Selected Applications of LSE for the Analysis of Pesticides and POPs in Solid Dairy Products

Analytes	Matrix	Extractant	Ref.
PCBs, OCs	Butter	Boiling hexane	[29]
PCBs	Cheese	Petroleum ether	[30]
OCs	Cheese, butter	Chloroform	[31]
Dioxins	Cheese	Hexane/DCM	[32]
PAHs	Cheese	Cyclohexane	[33]
PAHs	Butter, cheese	Cyclohexane/DCM	[26]
PAHs	Smoked cheese	Cyclohexane	[34]

PCBs, polychlorinated biphenyls; OCs, organochlorines; DCM, methylene chloride; PAHs, polycyclic aromatic hydrocarbons.

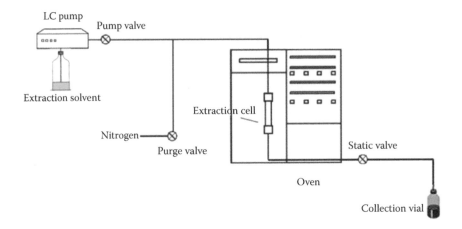

Figure 37.3   Schematic of a pressurized fluid extraction apparatus.

extracting analyte from food samples, a remedy adopted by many searchers is that of adding anhydrous sodium sulfate in the extraction cell or to adopt a preliminary lyophilization step.

Table 37.6 shows some PLE-based extraction procedures involved in the analysis of pesticides and POPs in milk, cheese, and butter.

37.3.1.2.4   Matrix Solid-Phase Dispersion

Matrix solid-phase dispersion (MSPD) is a process for the extraction of target compounds from solid matrices and was introduced in 1989. Later, this technique has been applied also to liquid and semisolid matrices. MSPD combines the aspects of several analytical techniques, performing sample disruption while dispersing the components of the sample on a solid support, thereby generating a chromatographic material that possesses a particular character for the extraction of compounds from the dispersed sample. The MSPD technique involves the use of abrasives blended with the sample by means of a mortar and pestle or by a related mechanical device (Figure 37.4). The shearing forces generated by the blending process disrupt the sample architecture and provide a more finely divided material for extraction. Some procedures use abrasives that also possess the properties of a drying agent, such as anhydrous Na_2SO_4 or silica, producing a material that is finely divided but also quite dry for subsequent extraction as described.

Table 37.6   Selected Applications of the Pressurized Liquid Extraction Technique to the Analysis of Pesticides and POPs in Solid Dairy Products

Analytes	Matrix	Extractant	Ref.
Dioxins	Cheese, butter	Hexane (P: 10 MPa)	[13]
Pesticides	Powdered milk	CH_3CN (T: 100°C, P: 10 MPa)	[35]
PCBs	Powdered milk	Hexane (T: 100°C, P: 10 MPa)	[36]

PCBs, polychlorinated biphenyls.

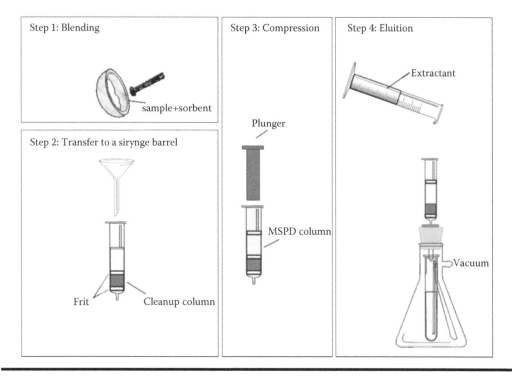

Figure 37.4 **Schematic representation of a typical MSPD extraction procedure.**

Once the MSPD process is complete, the material is transferred to a column generally consisting of a syringe barrel, with two frits inserted on the top and bottom of the MSPD column. The principles of performing good chromatography always apply: one should avoid channels in the column and not overcompress or compact the material. When 0.5 g of the sample is mixed with 2 g of the solid support, evidences from several studies indicate that most target analytes are eluted in the first 4 mL of extractant. Many MSPD procedures have also employed the use of co-columns to obtain further fractionation and to assist in extract cleanup. A co-column material (e.g., Florisil, silica, alumina) can be packed in the bottom of a cartridge containing usually C-18 as support material. Such columns may be literally stacked so as to collect and fractionate the sample as it elutes from the MSPD column.

Over classical sample treatment procedures, MSPD offers distinct advantages in that (a) the analytical protocol is drastically simplified and shortened; (b) the possibility of emulsion formation is eliminated; (c) last but not the least, the extraction efficiency of the analytes is enhanced as the entire sample is exposed to the extractant.

Recently, to achieve faster and more efficient extraction of target compounds from various biological matrices, MSPD with heated or pressurized extractants has been proposed by using PLE or laboratory-made instrumentations.

Recently, hot water has been successfully used for extracting target compounds from biological matrices by using an instrumentation similar to that for PLE. However, MSPD with heated water differs from PLE in that extraction can also be performed in the dynamic mode instead of only in the static one. Water is an environmentally acceptable solvent and it is cost-effective. The polarity of water decreases as the temperature is increased. This means that selective extraction of polar and medium polar compounds can be performed by suitably adjusting the water temperature. In addition to the advantages mentioned earlier, heated water provides sufficiently clean extracts needing little manipulation (pH adjustment and filtration) before injection into a reversed-phase LC column.

Table 37.7 shows some selected applications of MSPD to the extraction of contaminants from dairy products.

37.3.1.2.5 Supercritical Fluid Extraction

Recently, supercritical fluid extraction (SFE) with CO_2 as extractant has been deployed in the analytical field for extracting a variety of pesticides from solid matrices. Over sonication, blending, and Soxhlet extraction, definite advantages of SFE are listed here:

1. A nontoxic, nonflammable, inexpensive fluid, such as CO_2, is used.
2. Selective extraction can be performed by suitably modifying the density of the supercritical fluid. Increasing the density of the fluid increases the extraction yield of high-molecular weight compounds. The density of the fluid can be varied by varying its temperature and pressure.
3. Faster extraction: Extraction by SFE is a matter of minutes, instead of hours. Compared with the conventional solvents, the low viscosity of the supercritical fluid helps rapid penetration into the core of the solid matrix and extraction of analytes. In addition, the high-solute diffusivities into the supercritical fluid results in rapid removal of the analytes from the matrix by a decreased mass transfer resistance.

SFE is conceptually simple to perform. A pump is used to supply a known pressure of the supercritical fluid to an extraction vessel, which is thermostated at a temperature above the critical temperature of the supercritical fluid. During the extraction, the analytes are removed from the bulk sample matrix into the fluid and swept into a decompressing region. Here, the supercritical fluid becomes a gas and is vented, while analytes abandoning the gas are collected in a vial containing a

Table 37.7 Selected Applications of MSPD for the Analysis of Pesticides and POPs in Liquid and Solid Dairy Products

Analytes	Matrix	Support	Extractant	Ref.
OP insecticides	Milk	Hydromatrix	LP/MeCN/EtOH (100:25:5)	[37]
PCBs	Butter	Florisil	Hexane/DCM (9:1)	[38]
OP insecticides	Milk	C-18	CH_3CN	[39]
Dioxins	Milk	Silica/Na_2SO_4	Hexane/acetone (1:1)	[26]
PCBs and PBBs	Cheese	Silica	Hexane/acetone (1:1)	[40]
PCBs	Powdered milk	Silica/Na_2SO_4	Hexane/acetone (1:1)	[41]
Carbamates	Milk	Crystobalite	Water heated at 90°C	[42]

OP, organophosphate; hydromatrix, diatomaceous earth. LP/MeCN/EtOH, light petroleum/ CH_3CN/ethanol. Florisil, magnesium silicate; DCM, methylene chloride; C-18, octade-cyl-bonded silica; PCBs, polychlorinated biphenyls; PBBs, polybrominated biphenyls.

small volume of a suitable solvent. A variation of this scheme is that of substituting the collecting liquid at the outlet of the extractor with a sorbent cartridge.

The extraction of hydrophobic compounds from complex matrices containing sugar, proteins, and fat can be achieved almost quantitatively, but polar molecules give poor recovery rates. The recovery of these compounds can be improved significantly by the addition to CO_2 of a modifying solvent, such as methanol or acetonitrile.

In spite of its unique elevated sensitivity, the interest in SFE has substantially decreased in the last years. This is due to the high dependence of the extraction conditions on the sample, leading to fastidious optimization procedures and difficulty in using this technique routinely.

Table 37.8 shows some application of SFE to the analysis of contaminants in dairy products.

37.3.2 Cleanup

Once an extract has been obtained, a cleanup process for isolating the analytes from coextracted compounds is necessary prior to the final determination step. With liquid samples, simultaneous sample extraction and cleanup can be sometimes accomplished by a single SPE cartridge with a suitable washing step prior to analyte elution.

When extracting target compounds from food samples, sugars, pigments, lipids, denaturized proteins, and other naturally occurring compounds are typical interferences. These endogenous compounds are to be removed from the extracts, as they interfere with the analysis in the following ways:

1. Coextractives can produce large and tailed peaks overlapping those for the analytes.
2. Even when using specific detectors, the presence of great amounts of coextractives can saturate the detector or somehow modify the detector response.
3. With GC analysis, sugars, lipids, and proteins are thermally decomposed and the relative degradation products accumulate on the first part of the column provoking a rapid deterioration of the chromatographic performance.

These problems can be, in large part, resolved by an extract cleanup step. The techniques most frequently used are adsorption and GPC by using disposable high-efficiency cartridges filled with small particle size of the fractionating materials.

Adsorption chromatography: Adsorption chromatography on silica, alumina, Florisil (a synthetic magnesium silicate), carbon, C-18, and copolymer sorbents is widely used for cleanup of

Table 37.8 Selected Applications of SFE for the Analysis of Pesticides and POPs in Liquid and Solid Dairy Products

Analytes	Matrix	Dispersant	Extractant	Ref.
PCBs	Powdered milk	—	CO_2 (50°C, 13–23 MPa)	[43]
OCs and OPs	Cheese	Hydromatrix	CO_2 (80°C, 69 MPa)	[44]
OCs and OPs	Butter	Hydromatrix	CO_2 + 3% CH_3CN (60°C, 28 MPa)	[45]

PCBs, polychlorinated biphenyls; OCs, organochlorines; OP, organophosphates; Hydromatrix, diatomaceous earth.

many pesticides in both official and proposed methods. The first three materials progressively retard elution according to the increasing polarity. The reverse occurs with the last three sorbents. For medium- and low-polar pesticides, Florisil and alumina have been largely used in older standard methods. However, these cleanup procedures fail to fractionate polar pesticides and pesticide metabolites from coextractives. Carbon columns are used in several European standard multiresidue methods. Carbon strongly adsorbs lipids abundantly present in dairy foods, which are low polar in nature. Thus, this sorbent is particularly well suited for the purification of medium and highly polar pesticides.

Many analytical procedures used for analyzing low-polarity target compounds, i.e., PCBs, dioxins, PAHs, and OCs, include a cleanup step using adsorption columns packed with polar sorbents, such as silica, alumina, and Florisil. These columns provide good cleanup only when they are eluted with solvent mixtures of low polarity, eluting nonpolar and low-polar analytes and leaving more polar coextractives in the column. The more the eluting solvent polarity is increased, the greater will be the portion of interfering compounds eluted and the less effective will be the cleanup.

GPC: In GPC, compounds are eluted according to their molecular sizes, the smallest ones being more retarded than the largest ones. GPC is especially used for separating pesticides and POPs from complex heavy molecules, such as lipids and proteins. The most used chromatographic material is Bio-Beads SX-3 (a styrene divinylbenzene resin). When GC analysis is the final determination step, cleanup by GPC is particularly attractive, as ethyl acetate can be used as the eluent. This solvent is well compatible with GC detectors.

Table 37.9 shows selected cleanup procedures used for analyzing contaminants in dairy foods.

Table 37.9   Selected Cleanup Procedures Using Adsorption (AC) and GPC for the Analysis of Pesticides and POPs in Liquid and Solid Dairy Products

Analytes	Matrix	Sorbent(s)	Eluent	Ref.
Dioxins	Yoghurt, cheese, milk	AC on silica and alumina columns	Hexane	[23]
Dioxins	Milk, cheese, butter	GPC on SX-3 Bio-Beads and AC on silica, alumina, and carbon columns	EtAc/cycloC$_6$ (GPC) toluene (AC)	[13]
OCs, PCBs	Yoghurt	AC on magnesium silicate	Hexane	[25]
OCs	Milk	AC on magnesium silicate	1% CH$_3$OH in hexane	[10]
Dioxins	Cheese	AC on silica, alumina and carbon columns	Toluene	[32]
PCBs	Cheese	AC on magnesium silicate	Hexane	[30]
Dioxins	Milk	AC on carbon (Carbosphere)	EtAc/hexane (1:1)	[12]
PCBs	Dairy foods	GPC on SX-3 Bio-Beads	EtAc/CycloC$_6$ (1:1)	[46]

SX-3, polystyrene gel; EtAc/cycloC$_6$, ethylacetate/cyclohexane; OCs, organochlorines; PCBs, polychlorinated biphenyls.

37.3.3 Identification and Quantitation

Once the final extract has been obtained, several chromatographic instrumentations are available for identifying and measuring analyte concentrations in the final extract:

1. GC with selective detectors
2. GC or LC coupled to mass spectrometry

GC is still the most commonly used technique for analyzing POPs and pesticides in dairy food. However, several classes of pesticides are thermally labile and thus not amenable to GC methods, unless affording not easily viable derivatization procedures. LC does not suffer from these limitations and can be used to analyze virtually any nongaseous analyte.

37.3.3.1 Capillary GC with Selective Detectors

Before affording the final step of the analysis by GC instrumentation, reliable analysis can be achieved by adding an internal standard to the final extract. An internal standard is defined by U.S. Environmental Protection Agency (USEPA) as "(a) pure analyte(s) added to a solution in known amount(s) and used to measure the relative response of other method analytes that are components of the same solution. The internal standard must be an analyte that is not a sample component." Almost all modern GC methods of analysis use internal standard calibration.

Although some methods still involves the use of packed columns, GC with capillary columns, commonly referred to as high resolution (HR) GC, with selective detectors or coupled to a specific detector, such as mass spectrometry, has become the staple for analyzing contaminants in food samples. Well-established advantages of this technique are

1. The high-resolution power of a 25 m length capillary column enables rapid screening of more than 100 analytes in less than 40 min.
2. The introduction of fused-silica capillary columns with bonded stationary phases has improved column inertness and ruggedness of HRGC.
3. Availability of cheaper, more reliable, stable highly selective detectors, such as electron capture detector (ECD), nitrogen–phosphorous thermoionic detector (NPD), and flame photometric detector (FPD).
4. The introduction of new injection devices has improved reproducibility and reliability of analyses by HRGC. Moreover, some injection systems allow introduction of large volumes of the final extract, thus improving the sensitivity of the analysis.
5. The low cost of GC instrumentation. The affordability of a benchtop HRGC/mass spectrometric (MS) equipment for unequivocal detection of target compounds has made this hyphenated technique very appealing to regulatory laboratories.

37.3.3.2 Selection of the HRGC Column

Fused-silica capillary column with bonded liquid phases are by far the most preferred because of their higher inertness, stability, and flexibility. To a first view, the choice of the column appears difficult, since there are many manufacturers, each one producing many different columns differing in length, internal diameter, nature of the liquid phase, and film thickness coating the capillary wall. Indeed, only few liquid phases are of effective use for the analysis of contaminants in dairy foods. Columns coated with nonpolar liquid phases, such as 5% phenyl/95% methylsilicone are usually the first choice. These columns offer low bleed and sufficient chemical inertness. For confirmational analysis, more polar columns (phenyl/cyanopropyl) should be used. Regarding the other parameters, a 25 m × 0.25 mm i.d. having a 0.15–0.33 µm film thickness is a good selection for screening purposes.

37.3.3.3 Injection Devices

Sample introduction in HRGC is a complex and critical process. An ideal injection device should have the following properties:

1. To focus analytes in the very first part of the column.
2. Do not discriminate compounds on the basis of their chemical and physical characteristics.
3. Do not decompose or adsorb any one of the mixture components.

Among the commercially available injection devices for introducing relatively large volumes (1–3 µL) of the final extract into a capillary column, the splitless/splitter and the programmed temperature vaporization devices [47] are the most used ones for detecting pesticides and POPs in food.

37.3.4 Selective Detectors for HRGC

Among the various commercially available selective GC detectors, the electron capture and the FPDs are largely the most used ones for detecting OC and OP insecticides, respectively, in dairy products.

Electron capture detector. The ECD is the oldest selective detector for GC analysis. Its fame relies on the fact that it was successfully used to demonstrate the ubiquitous distribution of chlorinated pesticides. These works had huge impact on the scientific world and on public opinion, giving rise to the great interest in the fate of the environment due to human activities. Over the past 40 years, the main modifications of the ECD have been that of (a) replacing tritium adsorbed on a palladium β-ray electron source with a ^{63}Ni source, which can be heated at elevated temperatures; (b) adopting pulsed voltage instead of constant voltage. The latter modification has greatly expanded the linear dynamic range of the detector (10^4 against 10^2). The response mechanism and a sketch of this detector have been reported elsewhere [47].

The most positive feature of the ECD is its extreme sensitivity for compounds bearing more than one halogen atoms, such as organochlorine pesticides. For this class of pesticides, the ECD is the detector of choice. A defect of the ECD is its relatively poor selectivity. To a greater or lesser extent, the ECD responds to a wide range of compounds. Moreover, the ECD sensitivity for monohalogenated compounds is not higher than that of the popular flame ionization detector.

Flame photometric detector. The FPD is a selective detector for sulfur- and phosphorous-containing compounds. The response mechanism and a sketch of this detector have been reported elsewhere [47]. The FPD is a robust and reliable detector but suffers from some limitations and defects:

1. It is not very sensitive for sulfur compounds.
2. By operating the detector in the P-mode, large amounts of sulfur compounds that were coeluted can interfere with the analysis of phosphorous compounds.
3. The abundance of some coeluted or even nearby eluted coextractives able to absorb the radiations emitted by S or P can result in false-negative or, in the best case, in analyte underestimation.

While the FPD operating in the S-mode is of limited utility for pesticides, this detector is largely used for analyzing OP pesticides.

37.3.5 The Mass Spectrometric Detector

A serious weakness of chromatographic methods based on conventional detectors is that they lack sufficient specificity for identifying target compounds in complex biological matrices. Because of legal implications, health agencies in many countries rely on detection by MS for unambiguous confirmation of the presence of contaminants in food.

It is beyond the scope of this chapter to illustrate the principles and theory of mass spectrometry. Here, the authors will describe the information that can be obtained when using GC or LC coupled to a MS detector. Compounds eluted from the GC or LC column enter the MS ion-generating source, where molecules can be ionized by different mechanisms, according to the particular ion source adopted. Under certain conditions, a series of structure-significant fragment ions having characteristic mass-to-charge (m/z) ratios can be formed, in addition to the "molecular ion." By scanning the MS over a defined m/z range, these ions are recorded by a photomultiplier or an electron multiplier and a resulting mass spectrum is obtained, which displays m/z vs. relative abundance.

A GC–MS instrumentation is relatively inexpensive, as it requires only a source to ionize analytes. With LC–MS, both an interface and an ion source are needed to evaporate the liquid mobile phase and produce gas-phase ions. Although the youngest device introduced for LC–MS, the electrospray ion source (ESI) is today the only commercially available interface. Using GC as separation technique, MS acquisition data are usually obtained by electron impact (EI) ionization, a "hard" ionization technique able to produce several daughter ions, in addition to the molecular ion. With ESI, unlike EI ionization, gas-phase ions are softly generated, leading to the formation of $[M + H]^+$ (or cationized ions, usually $[M + Na]^+$) or $[M – H]^-$, even for the most labile and nonvolatile compounds, and confirmatory daughter ions can be obtained by a subsequent collision-induced decomposition (CID) process either using a single quadrupole or a triple quadrupole.

When analyzing target compounds, MS data acquisition with a single-quadrupole mass spectrometer is usually performed in the selective ion monitoring by monitoring the molecular ion plus two characteristic fragment ions for each analyte. Under this condition, the MS instrument affords the maximum sensitivity as no detector time is wasted to collect any other ion formed.

Tandem mass spectrometry (MS/MS) is a method involving two stages of mass analysis in conjunction with a chemical reaction that causes a change in the mass of the molecular ion. This can be done by coupling two physically distinct parts of the instrument (triple quadrupole, $Q_1q_2Q_3$). Briefly, the molecular ion of a given compound is selected by the first quadrupole (Q_1), the second quadrupole (q_2) drives the molecular ion into a cell where the collision of the molecular ion with an inert gas generates characteristic fragment ions that are monitored by the Q_3. This very selective acquisition mode is called selected reaction monitoring (SRM) and affords extremely high selectivity and sensitivity, especially when the MS/MS instrument is coupled to a fractionation device, such as a chromatographic column (LC or GC–tandem MS). The main advantage of using MS/MS in the SRM mode is the discrimination against the chemical noise, which can arise from different sources (matrix compounds, column bleed, and contamination from an ion source).

When analyzing extremely low amounts of dioxins in complex matrices, the low-resolution (unit mass) quadrupole may fail to detect one or more of the analyte ion signals if they are overlapped by those relative to abundant matrix components at concentrations several orders of magnitude higher than those of the analytes. The analysis of dioxins is further complicated by the

Table 37.10   Selected Applications of GC and LC with Selective detectors to the Analysis of Pesticides and POPs in Dairy Products

Analytes	Column/Stationary Phase	Detector	Ref.
Dioxins	HRGC (60 m × 0.25 mm i.d.)/DB 5	HRMS	[28]
Dioxins	HRGC (30 m × 0.25 mm i.d.)/RTX-5SIL-MS	HRMS	[13]
OC pesticides	HRGC (30 m × 0.25 mm i.d.)/CP-Sil 5 CB	ECD	[25]
OC pesticides	PC (2 m)/1.5% OV-17 + 1.95% OV-210 on Chromosorb	ECD	[10]
PCBs, OC pesticides	HRGC (50 m × 025 mm i.d.)/CPSil8	MS	[29]
OP pesticides	HRGC (25 m × 0.2 mm i.d.)/HP-1	FPD	[8]
OP pesticides	HRGC (15 m × 0.53 mm i.d.)/SPB-608	FPD	[37]
Multiclass pesticides	HRGC (30 m × 0.25 mm i.d.)/HP-5MS	MS/MS	[35]
PAHs	HRGC (30 m × 0.25 mm i.d.)/HP-5MS	MS	[26]
Carbamate insecticides	LC (25 cm × 4.6 mm i.d.)/C-18	MS/MS	[42]

HRGC, gas chromatography with capillary column; DB 5, (5%-phenyl)-methylpolysiloxane; HRMS, high-resolution mass spectrometry with a magnetic sector; RTX-5-SIL-MS, 5% diphenyl 95% dimethylsiloxane; OC, organochlorine; Sil 5 CB, dimethylopolysiloxane; ECD, electron capture detector; PC, packed column; OV-17, phenyl methyl, 50% phenyl silicone; OV-210, 50% trifluoropropyl methylsilicone; PCBs, polychlorinated biphenyls; CPSil8, methylsiloxane; OP, organophosphates; HP-1, dimethylpolysiloxane; FPD, flame photometric detector; SPB-608, 65% dimethyl-35% diphenyl polysiloxan;. HP-5MS, 5% diphenyl 95% dimethylsiloxane; PAHs, polycyclic aromatic hydrocarbons.

existence of many isomers (i.e., 75 PCDDs and 135 PCDFs). Since dioxins differ in toxicity by several orders of magnitude, the separation and positive identification/quantification of each dioxin in a biological matrix is a crucial task. Nevertheless, analysis of dioxin traces in biological matrices can be afforded only by coupling HRGC to a mass spectrometer equipped with a magnetic sector mass analyzer (mass resolution 10,000, HRMS). Compared with low-resolution MS (quadrupole), HRMS offers much higher selectivity and sensitivity.

Table 37.10 shows selected analytical methods based on GC or LC with selective detectors for detecting contaminants in dairy food.

References

1. Joint FAO/WHO Food Standards Program—Codex Alimentarius Commission—Codex Alimentarius, Vol. II, Pesticide Residues in Food. Food and Agriculture Organization and World Health Organization, Rome, 1993.
2. European Union Council Directive 93/57/CEE of June 29, 1993, *Off. J. Eur. Commun.* N. L211/1 of 23/8/1993.

3. Commission Regulation (EC) No 1881/2006 of 19 December 2006 setting maximum levels for certain contaminants in foodstuffs (text with EEA relevance).

4. Commission Regulation (EC) No 178/2006 of 1 February 2006 amending Regulation (EC) No 396/2005 of the European Parliament and of the Council to establish Annex I listing the food and feed products to which maximum levels for pesticide residues apply.

5. FAO database: http://www.codexalimentarius.net/mrls/pestdes/jsp/pest_q-e.jsp; EC portal on foodsafety, http://ec.europa.eu/food/food/chemicalsafety/contaminants/index_en.htm

6. Thomas, G.O. et al. Development and validation of methods for the trace determination of PCBs in biological matrices. *Chemosphere*, 36, 2447, 1998.

7. Newsome, W.H. and Davies, D. Determination of PCB metabolites in Canadian human milk. *Chemosphere*, 33, 559, 1996.

8. Salas J.H. et al. Organophosphorus pesticide residues in Mexican commercial pasteurized milk. *J. Agric. Food Chem.*, 51, 4468, 2003.

9. Fenske, R.A. et al. Assessment of organophosphorous pesticide exposures in the diets of preschool children in Washington State exposure. *J. Expo. Anal. Environ. Epidemiol.*, 12, 21, 2002.

10. John, P.J., Bakore, N., and Bhatnagar, P. Assessment of organochlorine pesticide residue levels in dairy milk and buffalo milk from Jaipur City, Rajasthan, India. *Environ. Intern.*, 26, 231, 2001.

11. Barkatina, E.N. et al. Organochlorine pesticide residues in breast milk in the Republic of Belarus. *Bull. Environ. Contam. Toxicol.*, 60, 231 1998.

12. Baars, A.J. et al. Dioxins, dioxin-like PCBs and non-dioxin-like PCBs in foodstuffs: Occurrence and dietary intake in the Netherlands. *Toxicol. Lett.*, 151, 51, 2004.

13. Focant, J.-F. et al. Levels and congener distributions of PCDDs, PCDFs and non-ortho PCBs in Belgian foodstuffs. Assessment of dietary intake. *Chemosphere*, 48, 167, 2002.

14. Di Muccio, A. et al. Selective, on-column extraction of organochlorine pesticide residues from milk. *J. Chromatogr. A*, 456, 143, 1986.

15. Bogialli, S. et al. Development of a multiresidue method for analyzing herbicide and fungicide residues in bovine milk based on solid-phase extraction and liquid chromatography–tandem mass spectrometry. *J. Chromatogr. A*, 1102, 1, 2006.

16. Laganà, A., Marino, A., and Fago, G. Evaluation of double solid-phase extraction system for determining triazine herbicides in milk. *Chromatographia*, 38, 88, 1994.

17. Pico, Y. et al. Determination of organochlorine pesticide content in human milk and infant formulas using solid phase extraction and capillary gas chromatography. *J. Agric. Food Chem.*, 43, 1610, 1995.

18. Fernandez-Alvares, M. et al. Development of a solid-phase microextraction gas chromatography with microelectron-capture detection method for a multiresidue analysis of pesticides in bovine milk. *Anal. Chim. Acta*, 617, 37, 2008.

19. Kowalski, C.H. et al. Neuro-genetic multioptimization of the determination of polychlorinated biphenyl congeners in human milk by headspace solid phase microextraction coupled to gas chromatography with electron capture detection. *Anal. Chim. Acta*, 585, 66, 2007.

20. De Lourdes Cardeal, Z. et al. Analysis of organophosphorus pesticides in whole milk by solid phase microextraction gas chromatography method. *J. Environ. Sci. Health*, 585, 66, 2006.

21. González-Rodríguez, M.J. et al., Determination of pesticides and some metabolites in different kinds of milk by solid-phase microextraction and low-pressure gas chromatography-tandem mass spectrometry. *Anal. Bioanal. Chem.*, 382, 164, 2005.

22. Basheer, C. and Lee, H.K. Hollow fiber membrane-protected solid-phase microextraction of triazine herbicides in bovine milk and sewage sludge samples. *J. Chromatogr. A*, 1047, 189, 2004, 189.

23. Bocio, A. and Domingo, J.L. Daily intake of polychlorinated dibenzo-*p*-dioxins/polychlorinated dibenzofurans (PCDD/PCDFs) in foodstuffs consumed in Tarragona, Spain: A review of recent studies (2001–2003) on human PCDD/PCDF exposure through the diet. *Environ. Res.*, 97, 1, 2005.

24. Focant, J.F. et al. Fast clean-up for polychlorinated dibenzo-*p*-dioxins, dibenzofurans and coplanar polychlorinated biphenyls analysis of high-fat-content biological samples. *J. Chromatogr.*, 925, 207, 2001.

25. Zhang, H. et al. A survey of extractable persistent organochlorine pollutants in chinese commercial yogurt. *J. Dairy Sci.*, 89, 1413, 2006.

26. Siegmund, B., Weiss, R., and Pfanhauser, W. Sensitive method for the determination of nitrated polycyclic aromatic hydrocarbons in the human diet. *Anal. Bioanal. Chem.*, 375, 175, 2003.

27. Ramos, L. et al. Comparative study of methodologies for the analysis of PCDDs and PDCFs in powdered full-fat milk. PCB, PCDD and PCDF levels in commercial samples from Spain. *Chemosphere*, 38, 2577, 1999.

28. Schecter, A. et al. Levels of dioxins, dibenzofurans, PCB and DDE congeners in pooled food samples collected in 1995 at supermarkets across the United States. *Chemosphere*, 34, 1437, 1997.

29. Kalantzi, O.J. et al. The global distribution of PCBs and organochlorine pesticides in butter. *Environ. Sci. Technol.*, 35, 1013, 2001.

30. Santos, J.S. et al. Assessment of polychlorinated biphenyls (PCBs) in cheese from Rio Grande do Sul, Brazil. *Chemosphere*, 65, 1544, 2006.

31. Barkatina, E.N. et al. Organochlorine pesticide residues in basic food products and diets in the Republic of Belarus. *Bull. Environ. Contam. Toxicol.*, 63, 235, 1999.

32. Hayward, D.G. Determination of polychlorinated dibenzo-p-dioxin and dibenzofuran background in milk and cheese by quadrupole ion storage collision induced dissociation MS/MS. *Chemosphere*, 34, 929, 1991.

33. Anastasio, A. et al. Levels of Benzo[*a*]pyrene (BaP) in "Mozzarella di Bufala Campana" cheese smoked according to different procedures. *J. Agric. Food Chem.*, 52, 4452, 2004.

34. Guillen, M.D. and Sopelana, P. Occurrence of polycyclic aromatic hydrocarbons in smoked cheese. *J. Dairy Sci.*, 87, 556, 2004.

35. Mezcua, M. et al. Determination of pesticides in milk-based infant formulas by pressurized liquid extraction followed by gas chromatography tandem mass spectrometry. *Anal. Bioanal. Chem.*, 389, 1833, 2007.

36. Müller, A., Björklund, E., and von Holst, C. On-line clean-up of pressurized liquid extracts for the determination of polychlorinated biphenyls in feedingstuffs and food matrices using gas chromatography–mass spectrometry. *J. Chromatogr. A*, 925, 197, 2001.

37. Di Muccio, A. et al. Selective solid-matrix dispersion extraction of organophosphate pesticide residues from milk. *J. Chromatogr. A*, 754, 497, 1996.

38. Ramil Criado, M. et al. Application of matrix solid-phase dispersion to the determination of polychlorinated biphenyls in fat by gas chromatography with electron-capture and mass spectrometric detection. *J. Chromatogr. A*, 1056, 187, 2004.

39. Schenck, F.J. and Wagner, R. Screening procedure for organochlorine and organophosphorus pesticide residues in milk using matrix solid phase dispersion (MSPD) extraction and gas chromatographic determination. *Food Addit. Contam.*, 12, 535, 1995.

40. Gómara, B. et al. Fractionation of chlorinated and brominated persistent organic pollutants in several food samples by pyrenyl-silica liquid chromatography prior to GC–MS determination. *Anal. Chim. Acta*, 565, 208, 2006.

41. Ramos, L. et al. Levels of PCDDs and PCDFs in farm cow's milk located near potential contaminant sources in Asturias (Spain). Comparison with levels found in control, rural farms and commercial pasteurized cow's milks. *Chemosphere*, 35, 2167, 1997.

42. Bogialli, S. et al. Simple and rapid assay for analyzing residues of carbamate insecticides in bovine milk: Hot water extraction followed by liquid chromatography–mass spectrometry. *J. Chromatogr. A*, 1054, 351, 2004.

43. Ramos, L. et al. Study of the distribution of the polychlorinated biphenyls in the milk fat globule by supercritical fluid extraction. *Chemosphere*, 41, 881, 2000.

44. Hopper, ML. Multivessel supercritical fluid extraction of food items in total diet study. *J. AOAC Int.* 78, 1072, 1995.
45. Hopper, ML. Automated one-step supercritical fluid extraction and clean-up system for the analysis of pesticide residues in fatty matrices. *J. Chromatogr. A*, 840, 93, 1999.
46. Zuccato, E. et al. Level, sources and toxicity of polychlorinated biphenyls in the Italian diet. *Chemosphere*, 38, 2753, 1999.
47. Di Corcia, A. Pesticide trace analysis, in *Encyclopedia of Environmental Analysis and Remediation*, Meyers, R.A., Ed., John Wiley & Sons, Inc., New York, 1998, pp. 3555–3598.

Chapter 38

Allergens

Virginie Tregoat and Arjon J. van Hengel

Contents

38.1 Introduction..952
38.2 Characteristics of Milk Allergy 952
38.3 Identification and Characterization of Milk Allergens....................................953
    38.3.1 Double Blind Placebo-Controlled Food Challenge...............................953
    38.3.2 Skin Prick Test, RAST/EAST Inhibition and Allergen Microarrays....................953
    38.3.3 Patch Tests..954
    38.3.4 Allergen Recognition 954
38.4 Effects of Food Processing on Milk Allergenicity 954
    38.4.1 Heat Treatment...955
    38.4.2 Fermentation 956
    38.4.3 Ripening..956
    38.4.4 Enzymatic Hydrolysis 957
    38.4.5 Homogenization 957
38.5 Analytical Tools for the Detection of Milk Allergens in Food Products..........957
    38.5.1 Immunodetection 958
        38.5.1.1 Immunoprecipitation and Immunodiffusion 958
        38.5.1.2 RAST/EAST Inhibition 958
        38.5.1.3 Enzyme-Linked Immunosorbent Assay 958
        38.5.1.4 Lateral-Flow Immunoassays (LFIAs) or Dipsticks 959
        38.5.1.5 Biosensor and Surface Plasmon Resonance 960
        38.5.1.6 Western Blotting..960
    38.5.2 Proteomic Techniques...961

38.6 Detection of Milk Allergens in Dairy Foods and Other Food Products:
Hidden Allergens 961
38.7 Conclusion.. 962
References 963

38.1   Introduction

Milk is one of the most widely consumed foods, especially during the earliest stages of our life. During later stages, both milk and milk-derived dairy products remain important for human nutrition. This nutritional value as well as the abundant functional properties (e.g., foaming, emulsifying) of milk constituents make them highly attractive ingredients for the food industry [1,2].

However, milk is also well known for being allergenic. Milk allergy affects around 2% of children and 0.1%–0.5% of the adult population [3,4]. Like the other food allergies, it can induce mild to severe reactions that can even be fatal [5]. Since no treatment exists to cure food allergy, only a strict avoidance of the offending food (in this case, milk and its derived forms) can prevent an allergic reaction from occurring [6]. This stresses the necessity for allergic individuals to be aware of the presence of allergenic ingredients in food products. Accurate labeling in combination with good manufacturing practices should help the allergic consumer to avoid unintended exposure to milk. To assist the allergic consumer, the European Commission issued directive 2007/68/ EC [7], which stipulates that milk, as one of the major allergenic foods, has to be declared on the label of food products when used as an ingredient. This also holds true for milk-derived products. To support this legislation, accurate and sensitive analytical methods are required to detect milk allergens and to monitor their presence in food products even at trace levels [8].

This chapter focuses on the characteristics of milk allergy and the variety of allergens present in milk. In addition to this, the effects of food processing, as applied during the manufacture of dairy products, are discussed in relation to allergenicity and the detection of milk allergens. Finally, the techniques with which the detection of milk allergens can be achieved are described along with their application for the detection of hidden milk allergens in food products.

38.2   Characteristics of Milk Allergy

Milk allergy results from a hypersensitivity of the immune system to milk proteins that should normally be tolerated. This is potentially due to the immaturity of the immune system or its failure [9]. Sensitization occurs when the immune system reacts aberrantly during a primary contact with milk proteins by the production of specific antibodies (Immunoglobulin E or IgE) that bind to immune cells (mast cells, basophils). A second exposure to the allergic food results in an activation of the immune cells, which release inflammatory mediators leading to the allergic reaction. Milk allergy is characterized by two types of allergic reactions: (1) an immediate IgE-dependent reaction that occurs within minutes after contact with the allergen, and (2) a delayed reaction appearing after several hours and mainly mediated by immune cells (degranulation) [10]. Milk allergy induces a spectrum of clinical symptoms involving the skin (hives, eczema, and swelling), gastrointestinal tract (nausea, vomiting, diarrhea, and stomach cramps), respiratory tract (runny nose, nasal congestion, wheezing, and coughing), and in more severe cases, anaphylaxis [11]. Milk allergy should not be confused with milk intolerance that does not involve the immune system despite similar symptoms [12]. Milk intolerance refers mainly to lactose intolerance attributed to the lack of lactase, the enzyme needed for the digestion of the milk sugar lactose [13].

Milk allergy predominantly affects children, the majority of whom outgrow this allergy by the age of 5 years. However, around 20% of the affected children remain allergic (persistent allergic patients) and in some cases, milk allergy can develop after childhood [14]. More than 90% of children who are allergic to cow's milk also react to goat's milk and sheep's milk [15–17]. However, there are examples of isolated allergies to goat and sheep milk without cross-reaction to bovine milk [18,19] or vice versa [20].

38.3 Identification and Characterization of Milk Allergens

Typically, the allergenicity of milk is triggered by its proteins [21]. The identification of milk allergenic proteins is established by the determination of their reactivity toward milk allergic patients in *in vitro* as well as *in vivo* tests.

38.3.1 *Double Blind Placebo-Controlled Food Challenge*

The most reliable *in vivo* test to assess the capacity of milk proteins to trigger allergic reactions is the double blind placebo-controlled food challenge (DBPCFC) test [22]. The suspected milk allergic patients are orally challenged with milk protein extracts and the provocation symptoms emerging after the ingestion are studied under strict clinical conditions [23]. The power of this test resides in its capability to trigger allergic reactions in people; however, since this can threaten the health of the allergic individuals, it is tended to be supplanted by other tests. Defining the threshold at which milk proteins induce an allergic reaction is difficult, since this varies considerably from patient to patient and from protein to protein. For sensitive allergic individuals, tiny amounts, in the order of $5\,\mu g$ or 0.1 mL of milk, have been reported to trigger allergic reactions in DBPCFC tests [24,25]. A sorbet containing trace levels of whey proteins as low as $8.8\,\mu g/mL$ has been reported to elicit systemic reactions in a milk allergic individual after ingestion of only $120–180\,\mu g$ of the offending food [26].

38.3.2 *Skin Prick Test, RAST/EAST Inhibition and Allergen Microarrays*

Skin prick test and radio-allergosorbent/enzyme-allergosorbent (RAST/EAST) tests are among the most popular *in vivo* and *in vitro* assays for diagnosing a food allergy [27]. Those qualitative tests, which are based on the immunoreaction of food-specific IgE from the blood of allergic patients with the allergenic food, provide an identification of the allergenic compounds to which the individual reacts.

In skin prick tests, a very small amount of extracted cow's milk proteins (the allergens, e.g., caseins [CNs], β-lactoglobulin [β-LG], and α-lactalbumin [α-LA]) are introduced under the outer layer of the skin. The weal size of the localized reddening and swelling that develops when a milk allergic reaction occurs provides an indication of the severity of the allergic reaction [28].

RAST and EAST are *in vitro* tests that analyze the blood of individuals suspected to have a food allergy to assess the level of IgE antibodies that recognize milk proteins or their derived peptides [29]. Microarrays are emerging techniques based on the same principle as RAST, but offer the possibility to simultaneously measure the reaction of IgE antibodies from allergic patients with a battery of immobilized food allergens (proteins or derived peptides) on a chip [30]. Recently, such a sensitive microarray assay was used to study the immune response to milk and purified milk proteins [31]. Also, a peptide microarray immunoassay has been developed for milk allergens (CNs and β-LG), which was used to map allergenic milk-derived peptide epitopes responsible for triggering the allergic responses [32]. This tool might be useful for developing hypoallergenic formulae, containing milk-derived ingredients devoid of epitopes that are known to trigger allergic reactions.

38.3.3 Patch Tests

A number of infants and the majority of adults with milk allergy do not have freely circulating IgE specific to milk proteins that can be highlighted by skin prick tests and *in vitro* blood tests. Patch tests are employed in such cases, which are based on the application of a patch containing milk allergens on the skin of the patient's back [33]. Such a commercially available noninvasive epicutaneous delivery system has been designed to diagnose allergy to cow's milk protein in infants, children, or adults with delayed allergic reactions [34,35].

38.3.4 Allergen Recognition

Employing the *in vivo* and *in vitro* tests described earlier, it has become apparent that nearly all milk proteins (more than 30 so far, including all CNs and the whey proteins β-LG, α-LA, bovine serum albumin [BSA], and lactoferrin [LF]) can trigger allergic responses [36]. However, the majority of allergic reactions are attributed to the most abundant milk proteins (αs1-CN and β-LG) [37]. The allergenicity of milk proteins resides in specific amino acid sequences within the protein, called epitopes, which are recognized by IgE antibodies. Epitopes can be conformational (domains of proteins made up of nonadjacent amino acids that depend on the three-dimensional structure) or linear (continuous amino acid sequences, that only depend on the primary structure). Epitope mapping that has been performed for the main allergenic milk proteins revealed multiple allergenic epitopes within each protein as well as a high heterogeneity among allergic individuals concerning the epitopes to which they react [38–40]. While β-LG is likely to be the main elicitor of milk allergy (80%) in children and infants [40,41], CNs are apparently the major cause for allergic reactions in adults and persistent allergic patients [42] exposed to milk or to its derived products such as cheese [20,43]. This is potentially linked to the structural characteristics of the allergenic proteins. The poor three-dimensional structures of CNs favor the existence of linear epitopes that may participate in the persisting allergy, while whey proteins with their globular structure are characterized by conformational epitopes [43–45]. After denaturation or digestion of the protein into small fragments, the conformational epitopes can no longer bind the antibody in contrast to linear epitopes as depicted in Figure 38.1.

38.4 Effects of Food Processing on Milk Allergenicity

Milk is usually submitted to different technological processes to improve its safety and shelf-life before consumption. Alternatively, it is transformed into a variety of dairy products (e.g., yoghurts, cheese, ice cream, butter, and cream) [46]. Those manufacturing procedures can modify the structure of milk proteins, which might alter their immunodominant epitopes and thereby modulate allergenicity [47]. The effects are likely to be process dependent. Theoretically, allergenicity can

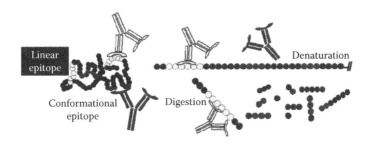

Figure 38.1 Antibody detection of conformational and linear epitopes within a native, denatured, or digested allergenic protein.

decrease because of the destruction of epitopes or it can increase because of the formation of new epitopes or an improved accessibility of cryptic or hidden epitopes after allergen denaturation. From the large number of industrial processes used to manufacture dairy products or specialized foods, only few were investigated for their impact on allergenicity [9,48,49]. Heat treatment is a basic process that milk undergoes before its consumption or its transformation into derived products, while degradation of milk allergens occurs during fermentation, ripening, and enzymatic hydrolysis, which are all processes that are frequently employed by the food industry. All those processes are very likely to impact allergenicity and are discussed here.

38.4.1 Heat Treatment

Thermal treatment is known to induce physicochemical changes in milk constituents [50]. The stability of milk allergenic proteins submitted to heat treatment differs according to the structure of the protein, the intensity of the thermal treatment [51], and the animal species the milk originated from. For instance, the heat stability of caprine and ovine milks is lower than that of bovine milk [52]. During heating, proteins with a globular tertiary structure (especially BSA, Ig, and β-LG) lose their conformational structure during unfolding. This is illustrated by the decrease in recognition by specific antibodies [53,54] or by the modification of the charge state distribution of β-LG as analyzed by electrospray ionization mass spectrometry (ESI-MS) [55]. Higher degrees of protonation were observed in whey protein solutions after increased heat exposure, reflecting the opening up of the molecule, but the presence of other components present in milk was shown to partly protect β-LG from denaturation [55]. The thermal treatment of BSA, Ig, and β-LG is associated with an alteration of their conformational epitopes that can no longer be recognized by IgE resulting in a reduction of allergenicity [56]. The antigenicity of milk allergens depends on the conditions of thermal treatment. Below 90°C, the allergenicity of milk proteins such as β-LG increases when submitted to pasteurization most likely caused by the unmasking of cryptic epitopes [57]. This is confirmed by the stronger allergic response after oral challenge (DBPCFC) of cow's milk allergic children and adults with pasteurized milks (15 s, 75°C) when compared with raw milk [58]. Inversely, heating at temperatures above 90°C drastically decreases the allergenicity of milk as shown by an impaired IgE binding [53], which is most likely to result from a combination of loss of conformational epitopes and a masking of sequential epitopes [59]. In fact, the denaturation of milk proteins that is relatively negligible with pasteurization (20% denaturation of whey proteins) is still incomplete after ultrahigh temperature (UHT) treatment (60% denaturation of whey proteins) and only boiling (100°C 10 min) represented a treatment strong enough to annihilate prick test reactivity of BSA and β-LG. But, even this only partially reduces the allergenicity of α-LA and CNs (50%–66% of IgE binding) [47]. The fact that CNs do not possess a highly structured configuration and that they have predominantly linear epitopes explains their thermostability and their persistence after thermal processing [60]. The maintenance of immunoreactivity of heat-treated milk can also be based on coaggregation and complexation of whey proteins with the CN micelles [61,62], or by the emergence of Maillard products (e.g., lactosylated milk proteins). Both processes might lead to the formation of neoepitopes [63,64].

38.4.2 Fermentation

Other technological processes used in the manufacture of dairy products such as fermentation also affect the allergenicity of milk allergens [53]. Fermented milk products like yoghurt or kefir produced from cow's and ewe's milk are recognized to have beneficial effects on the immune system. Such effects are linked to the presence of viable bacteria that improve gastrointestinal immunity as well as to milk-derived bioactive peptides emerging during proteolysis [65]. Clinical reports have suggested that consumption of fermented foods, such as yoghurt, might reduce the development of allergies, possibly via a mechanism of immune regulation [66] and the presence of tolerogenic peptides emerging from the degradation of cow's milk proteins by lactic acid bacteria [67]. The changes of milk protein profiles by the action of yoghurt bacteria (*Lactobacillus delbrueckii* ssp. *bulgaricus* and *Streptococcus salivarius* ssp. *thermophilus*) were analyzed to identify the emerging peptides [68]. Proteolytic activity during fermentation in kefir manufacturing [69,70] involves the degradation of β-CN followed by αs-CN, which is mediated by proteinases originating from lactic acid bacteria [71]. The extent of proteolysis of milk proteins varies according to the lactic acid bacteria employed for fermentation [72]. *L. delbrueckii* ssp. *bulgaricus* is able to eliminate more than 99% of the antigenicity of α-LA and β-LG. However, despite this drastic diminution of IgE binding, the allergenicity of the product is maintained as observed by provocation tests [73]. *S. salivarius* ssp. *thermophilus* is also able to efficiently diminish the immunoreactivity of α-LA (99.95%) and β-LG (91.26%). The same effect was observed for a whole panel of lactic acid bacteria where a remaining antigenicity of around 10% was detected [74]. Fermentation of milk from bovine species for the production of yoghurt yields a variety of peptides [75] some of which contain epitopes that are recognized by milk allergic patients [38–40]. Fermented milk products are therefore likely to remain allergenic for consumers with a milk allergy.

38.4.3 Ripening

During cheese-making and ripening, proteolysis takes place to form free amino acids from large water-insoluble peptides, as well as medium-sized and small soluble peptides [76]. Currently, complex food matrices like cheese are subjected to proteomic analyses, which provide insight into the multitude of milk-derived proteins (e.g., CNs and whey proteins), their degradation products (peptides), and to the microbial-produced proteins (enzymes) in this type of food [77]. The proteolysis of milk proteins (mainly CNs) has also been monitored with capillary electrophoresis or HPLC techniques coupled to mass spectrometry aiming at the detection of bioactive peptides [78,79]. But, such studies do not report on the residual allergenicity of these products. Despite a continuous hydrolysis of milk proteins during cheese ripening, alteration of the allergenicity of cheese during ripening seems to be limited [80].

38.4.4 Enzymatic Hydrolysis

The use of enzymatic hydrolysis of milk products is widespread within the food industry and often aims to reduce allergenicity by enzymatic degradation of milk proteins to obtain nutritional substitutes for milk allergic children [81]. In addition to this, it can be employed to generate milk protein-derived peptides with bioactive properties [82]. A variety of hydrolyzed milk formulae based on CNs or whey with different degrees of hydrolysis (partial or extensive) are commercially available. The antigenicity of those formulae is profoundly reduced [83]. But, even if the majority of the extensively hydrolyzed formulae developed for milk allergic children are well tolerated, their consumption is known to have triggered allergic reactions in several cases [84]. A study on different hypoallergenic formulae supposed to be deprived of "antigenic binding sites" has shown that β-LG traces could be detected in CN-based hydrolysates, indicating that during precipitation of CNs, contamination with whey proteins occur [85]. Caprine milk hydrolysates have also been developed and marketed since its proteins show a better gastrointestinal digestion than cow's milk proteins. This faster and stronger degradation of caprine milk proteins, especially β-LG, is likely to be caused by differences in the tertiary structure and physicochemical properties [86,87]. New investigations to decrease the allergenicity of milk allergenic proteins have focused on enzymatic hydrolysis under high pressure, which is suggested to be more effective; but it seems that depending on the conditions, the antigenicity can be intensified [88].

38.4.5 Homogenization

Homogenization is often employed for the manufacture of dairy products such as ice cream or fluid milk [89]. By destroying milk fat globules into smaller droplets under pressure, homogenization induces profound modifications in the structure of milk, which potentially affects allergenicity [90]. Homogenization of milk seems to increase its allergenicity, which is potentially due to the exposure of milk allergenic proteins at the surface of the fat globules [89].

38.5 Analytical Tools for the Detection of Milk Allergens in Food Products

The analytical tools that have been developed to detect milk allergens in food products either target (allergenic) proteins or DNA. DNA-based methods for the detection of milk traces in food products are hardly used, since milk contains relatively little DNA (compared with a rather high protein content) and such methods are not specific for milk, but would detect meat as well. A panel of screening methods available for the detection of milk allergen proteins in food products is based on immunoassays. Such assays usually employ animal-produced antibodies raised against the allergenic proteins [91]. Furthermore, proteomic techniques are used to confirm the presence of milk allergens in food products, and to identify milk protein/peptide sequences even after food processing [92]. Those analytical tools have been described extensively in several reviews that focus on the detection of food allergens [93,94]. The availability of methods capable of detecting milk allergen traces in food products at levels that are relevant to improve the protection of the health of allergic consumers is very important, and therefore an overview of commonly used methods is presented here.

38.5.1 Immunodetection

Immunochemical methods developed to detect traces of milk allergens and dairy products in food products are based on the recognition of milk allergenic proteins by specific antibodies raised against those milk proteins. Binding of allergens and antibodies leads to the formation of an allergen–antibody complex that is subsequently detected.

38.5.1.1 Immunoprecipitation and Immunodiffusion

The detection and quantitation of milk allergenic proteins were initially assessed by radial immunodiffusion techniques [95]. The sensitivity of this methodology was subsequently improved and currently radial immunodiffusion kits are commercially available for the specific quantitative measurement of native β-LG, α-LA, BSA, and LF in milk and dairy products from species like cow, goat, sheep, and camel.

Briefly, as illustrated in Figure 38.2, fixed concentrations of anti β-LG, α-LA, BSA, or LF antibodies are incorporated into an agar gel. Standards of diluted milk protein (C1, C2...) as well as test samples (C?) are deposited in holes in the gel and their proteins diffuse in the gel. The antibodies in the gel bind their target proteins and at the equilibrium, a precipitation ring is formed with a diameter that is proportional to the concentration of milk allergenic proteins present in the sample. Monitoring the progress of denaturation and hydrolysis of milk proteins during industrial processing (i.e., heat treatment and proteolysis) is feasible with a limit of detection (LOD) around 1 μg/mL in a measurement range between 1.5 and 12 μg/mL for α-LA [96].

38.5.1.2 RAST/EAST Inhibition

RAST and EAST have been utilized to estimate the presence and level of milk allergens in food products [18]. As illustrated in Figure 38.3, solid-phase-attached milk allergens and free milk allergens from the test sample compete for binding to human IgE. Subsequently, IgE bound to immobilized milk allergens is detected by labeled-antibodies (radiolabeled [RAST] or enzymatically labeled [EAST]). IgE binding of allergens from the test sample leads to the reduction of signal intensity, which is proportional to the level of milk allergenic protein present in the food sample. An LOD of around 1 mg/kg can be achieved with this methodology [8].

38.5.1.3 Enzyme-Linked Immunosorbent Assay

Enzyme-linked immunosorbent assay (ELISA) is the type of method that is most commonly employed to detect trace amounts of food allergens in industrial food products. It is usually based

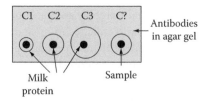

Figure 38.2 Principle of radial immunodiffusion.

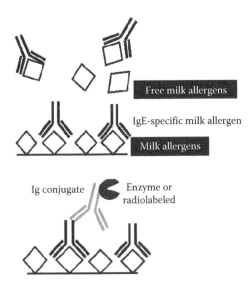

Figure 38.3   Principle of EAST and RAST inhibition.

on immobilized antibodies of animal origin that were raised against milk proteins. The milk proteins in test samples are bound and immobilized, which allows their detection by means of a second (labeled) antibody (Figure 38.4). Several commercial kits as well as in-house developed ELISAs are available to detect and measure the amount of milk allergens present in a food matrix [97]. ELISAs for the detection of milk traces are usually directed against CNs, β-LG, or total milk and use either a sandwich configuration (as described earlier) or a competitive detection. LODs for such kits generally range from below 1 to 7.5 ppm [97].

38.5.1.4   Lateral-Flow Immunoassays (LFIAs) or Dipsticks

Lateral-flow immunochromatographic test systems, also called dipsticks, have been developed to provide food manufacturers with easy-to-use (on site) fast qualitative tests for the detection of milk proteins in food products. Specific antibodies (raised against milk proteins) are attached to stained

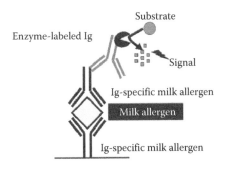

Figure 38.4   Principle of sandwich ELISA.

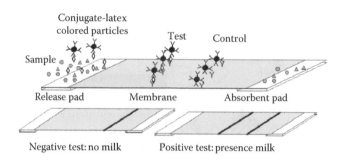

Figure 38.5 Principle of dipstick device and interpretation.

latex beads and deposited at the extremity of a nitrocellulose membrane (Figure 38.5). They bind to milk proteins in a sample extract and migrate as a complex along the membrane driven by capillary forces (Figure 38.5). The complexes are captured by a secondary milk allergen-specific antibody immobilized on the test line of the membrane. This leads to the appearance of a colored line reflecting the presence of milk protein.

Dipsticks for the detection of milk traces are commercially available and claim to have a sensitivity of around 5 ppm.

38.5.1.5 Biosensor and Surface Plasmon Resonance

Surface plasmon resonance (SPR) immunoassays represent an emerging and attractive technology for the food industry, since it monitors in real time the presence of milk allergen traces in food products. Also with this methodology, detection is based on recognition of milk proteins by antibodies. Binding of milk proteins to antibodies immobilized on a sensor chip leads to variation in the measurement of a refractive index that allows quantification of the milk content in the samples. Simultaneous quantification of CNs (αs_1, β, and κ) in dairy products [98] in their intact form can be assessed with an optical immunosensor technique in a fast and sensitive manner (LOD 0.87 μg/mL), and is adapted to the analysis of raw [99] and drinking milk [100]. The use of optical biosensors allows the detection of milk proteins at levels around 1–12.5 mg/kg in food samples [101]. The residual immunogenicity of food products submitted to different processes [102] such as heat treatment can be effectively estimated for the main whey proteins with LODs of 13, 27, and 20 ng/mL for α-LA, β-LG A, and B, respectively [103]. In processed complex food matrices (baby food products like crème dessert and fruit yoghurt), β-LG could be specifically and rapidly identified with a biosensor at concentrations ranging from 500 μg/mL to 2 mg/mL, similar to the immunochemically detectable β-LG content of the products [104].

38.5.1.6 Western Blotting

With this technique, milk allergens present in a sample are separated by gel electrophoresis in one or two dimensions and electrotransferred onto a membrane. Subsequently, specific antibodies are employed to reveal the presence of milk proteins. The antibodies employed can originate from serum of allergic patients [36], or can be raised in animals. The residual antigenicity of food and especially of hypoallergenic formulae can be assessed by this technique [105].

38.5.2 Proteomic Techniques

Proteomic techniques are more and more used to detect and confirm the presence of milk allergens in food products. Indeed, this approach allows an unambiguous identification of milk proteins in food matrices, which cannot be achieved with immunological methods like ELISA owing to potential cross-reactivities of antibodies. Another advantage of proteomic techniques resides in their ability to detect potentially allergenic milk-derived peptides that emerge during food processing. Proteomic techniques are usually based on a combination of separation and identification techniques. Separation of milk protein or peptide mixtures (e.g., hypoallergenic formulae) is generally achieved either by electrophoresis or chromatography. This is then followed by their unambiguous amino acid sequence identification with mass spectrometry [106,107]. For this, the separated sample (milk proteins/peptides) entering in the mass spectrometer is ionized with matrix-assisted laser desorption/ionization (MALDI) or electrospray ionization (ESI) and the resulting ions are propelled into the mass analyzer by an electric field resolving the ions by their mass-to-charge ratio [108].

High-resolution two-dimensional gel electrophoresis (2-DE) resolves milk proteins according to their isoelectric point in a first dimension (isoelectric focusing [IEF]) and their relative molecular weight in a second dimension (SDS-PAGE) [109] before being digested *in situ* into peptides and identified by mass spectrometry [77]. This technique has been used for the detection and characterization of milk allergens in commercial milk powder [106] and for monitoring the proteolysis during cheese ripening [110,111].

Liquid separation techniques like liquid chromatography and capillary electrophoresis are applied as separation methods that offer a variety of separation principles (size exclusion, reverse phase, ion exchange, IEF, etc.). Reversed-phase chromatography constitutes the method of choice for the separation of allergens preceding mass spectrometry. Mass characterization of milk proteins and peptides and their sequence identification have been determined with LC–MS methodology, which allowed their detection in complex food matrices, after hydrolysis, fermentation [112], or during the cheese-making processes [79]. Capillary electrophoresis represents an alternative high-resolution separation technique for the analysis of milk proteins in food products and their quantification [113] and has been proven to be useful to rapidly resolve milk allergens from different matrices including milk, milk powders, hypoallergenic formulae, dairy products, and cheeses [114–118].

A limitation of proteomic techniques resides in the fact that the analysis of complex mixtures such as milk hydrolysates or cheese can be difficult to interpret without prefractionation steps. This is due to the relatively low number of allergen-derived ions compared with all detectable ions, but also to the fact that short peptides (below five amino acids) cannot be clearly attributed to their mother protein(s).

38.6 Detection of Milk Allergens in Dairy Foods and Other Food Products: Hidden Allergens

Milk and its derivatives (e.g., whey proteins and CNs) are more and more incorporated as ingredients into a wide range of nondairy food products because of their broad functional properties [119]. Whey proteins (β-LG and α-LA), for instance, find their application in meat, reformed fish products as gelling additives, or can replace skim milk in ice cream, or even fat or whole egg in dairy and nondairy dessert products (e.g., meringue) owing to foaming and whipping properties

[2]. Milk powders having a high-nutritional value can supplement food, beverages, cereals, and specific nutritional products (e.g., sports drinks and infant formulae) [120,121]. A large variety of essentially nondairy products like bakery products, pastry, chocolate, sausages, hot dogs, tuna, ham, meringue, and many more products have been reported to trigger severe allergic reactions and were demonstrated to contain milk proteins by ELISA analyses [122,123]. Functional foods that are entering the market, products that contain milk protein-derived ingredients valued for their new functionalities (e.g., as biopreservative for fresh cut vegetables), or nutraceuticals could be threatening for the milk allergic population [124,125]. Probiotics that are added to food products for their potential ability to decrease allergy are also not always safe for milk allergic patients who can react to remains of the media on which the probiotics were grown (whey protein and CN) [126]. Furthermore, the ubiquity of milk proteins in food products will be reinforced by the appearance on the market of health benefit products supplemented with milk-derived peptides [127]. This strengthens the necessity to be able to detect the presence of milk proteins or milk-derived peptides in food products. Some of the methods mentioned above have been tested and optimized for this purpose. Immunological assays were applied for testing for traces of β-LG in infant formulae [128]. Furthermore, a series of nonmilk-containing products (fruit juices, fruit juice bars, sorbets, and dark chocolate) as well as food products that were suspected to have triggered allergic reactions were evaluated for the presence of CN employing a sandwich ELISA test, detecting CN levels that varied from 0.5 ppm (LOD) up to 40,000 ppm [129]. A competitive ELISA that is more suited to detect smaller proteolytic fragments was also successfully used to detect the presence of CN in foodstuffs (flour mix, instant potato, soup, and spice mix) with a limit of quantification (LOQ) around 1 mg/kg [130].

So far, only a single validation study of ELISA methods for the detection of milk proteins in food products has been reported. In Japan, an interlaboratory study investigating three types of ELISA kits reported the detection of milk proteins spiked into food products (sausages, sauces, cookies, and cereals) [131]. Besides immunochemical detection, proteomic techniques have been developed to assess and confirm the presence of milk allergens in food products. An LC–MS method has been set up for the detection and quantification of whey proteins (β-LG and α-LA) in mixed fruit juices at concentrations ranging from 5 to 40 μg/mL. This method was shown to have an LOD of 1 μg/mL and an LOQ of 4 μg /mL [132]. Another LC–MS method was developed to detect CNs in spiked cookies and was able to detect 1.25 ppm CN. This method is based on the detection of two peptides derived from αs1-CN (FFVAPFPEVFGK; YLGYLEQLLR) that were identified as markers for the presence of milk in food matrices [133]. Techniques like capillary electrophoresis have been employed to detect whey proteins in soybean dairy-like products with an LOD of 0.6 and 1.0 μg/g for α-LA and β-LG, respectively [134]. The further development of methods based on capillary electrophoresis might advance the detection of milk and dairy traces in food products.

38.7   Conclusion

Milk proteins constitute a very rich source of nutrients with a wide variety of functional properties and are utilized to manufacture a multitude of food products. However, a proper assessment of the allergenicity and a correct declaration of milk-derived ingredients on the label of food products are of paramount importance to prevent a nightmare for milk allergic consumers [135]. The panel of technological treatments referred to in this chapter can unfortunately not guarantee the elimination of allergenic components, while contamination with milk allergens is also a cause for concern

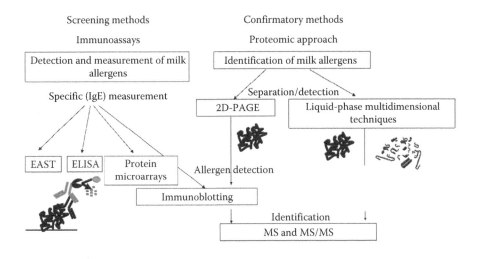

Figure 38.6 Analytical tools for the detection and identification of milk allergens.

[136]. It is therefore crucial that the labeling of food products is clear without ambiguity to help consumers to protect their health. To support this, a number of highly sensitive methods that are described earlier and depicted in Figure 38.6 are available to determine the presence of milk components in food matrices. The availability of such methods is crucial to detect and estimate the level of contamination of food products with allergenic ingredients, to identify mislabeling or adulteration practices, and finally to protect the allergic consumer.

References

1. Pihlanto, A. et al., Bioactive peptides and proteins, *Adv. Food Nutr. Res.*, 47, 175, 2003.
2. Chatterton, D.E.W. et al., Bioactivity of [beta]-lactoglobulin and [alpha]-lactalbumin—Technological implications for processing, *Int. Dairy J.*, 16(11), 1229, 2006.
3. Woods, R.K. et al., Prevalence of food allergies in young adults and their relationship to asthma, nasal allergies, and eczema, *Ann. Allergy Asthma Immunol.*, 88(2), 183, 2002.
4. Pelto, L. et al., Milk hypersensitivity in young adults, *Eur. J. Clin. Nutr.*, 53, 620, 1999.
5. Bock, S.A. et al., Fatalities due to anaphylactic reactions to foods, *J. Allergy Clin. Immunol.*, 107(1), 191, 2001.
6. Isolauri, E., The treatment of cow's milk allergy, *Eur. J. Clin. Nutr.*, 49(Suppl. 1), 49, 1995.
7. Kyprianou, M., Commission directive 2007/68/EC, *Off. J. Eur. Union*, 11, 11, 2007.
8. Poms, R.E. and Anklam, E., Tracking and tracing for allergen-free food production chain, in *Allergy Matters—New Approaches to Allergy Prevention and Management*, Gilissen, L.J.W.J. et al., Eds., Wageningen UR Fontis Series. Springer, Wageningen, 2006, p. 79.
9. El-Agamy, E.I., The challenge of cow milk protein allergy, *Small Ruminant Res.*, 68(1–2), 64, 2007.
10. Crittenden, R.G. and Bennett, L.E., Cow's milk allergy: A complex disorder, *J. Am. Coll. Nutr.*, 24(Suppl. 6), 582, 2005.
11. Sampson, H.A., Food allergy, *J. Allergy Clin. Immunol.*, 111(2, Suppl. 2), 540, 2003.
12. Wilson, J., Milk intolerance: Lactose intolerance and cow's milk protein allergy, *Newborn Infant Nurs. Rev.*, 5(4), 203, 2005.
13. Høst, A., Cow's milk protein allergy and intolerance in infancy. Some clinical, epidemiological and immunological aspects, *Pediatr. Allergy Immunol.*, 5(5 Suppl.), 1, 1994.

14. Olalde, S. et al., Allergy to cow's milk with onset in adult life, *Ann. Allergy.*, 62(3), 185a, 1989.

15. Spuergin, P. et al., Allergenicity of alpha-caseins from cow, sheep, and goat, *Allergy*, 52(3), 293, 1997.

16. Restani, P. et al., Cross-reactivity between milk proteins from different animal species, *Clin. Exp. Allergy.*, 29(7), 997, 1999.

17. Bellioni-Businco, B. et al., Allergenicity of goat's milk in children with cow's milk allergy, *J. Allergy Clin. Immunol.*, 103(6), 1191, 1999.

18. Lamblin, C. et al., Allergie aux laits de chevre et de brebis sans allergie associee au lait de vache, *Rev. Fr. Allergol.*, 41(2), 165, 2001.

19. Paty, E. et al., Allergie au lait de chevre et de brebis sans allergie associee au lait de vache, *Rev. Fr. Allergol.*, 43(7), 455, 2003.

20. Barnig, C. et al., Allergy to cow milk proteins without accompanying allergy to sheep milk proteins in an adult, *Rev. Fr. Allergol.*, 45(8), 608, 2005.

21. Wal, J.M., Bovine milk allergenicity, *Ann. Allergy Asthma Immunol.*, 93(Suppl. 3), 2, 2004.

22. Baehler, P. et al., Distinct patterns of cow's milk allergy in infancy defined by prolonged, two-stage double-blind, placebo-controlled food challenges, *Clin. Exp. Allergy*, 26(3), 254, 1996.

23. Ragno, V. et al., Allergenicity of milk protein hydrolysate formulae in children with cow's milk allergy, *Eur. J. Pediatr.*, 152(9), 760, 1993.

24. Bindslev-Jensen, C. et al., Can we determine a threshold level for allergenic foods by statistical analysis of published data in the literature? Hypothesis paper, *Allergy*, 57(8), 741, 2002.

25. Morisset, M. et al., Thresholds of clinical reactivity to milk, egg, peanut and sesame in immunoglobulin E-dependent allergies: Evaluation by double-blind or single-blind placebo-controlled oral challenges, *Clin. Exp. Allergy*, 33(8), 1046, 2003.

26. Laoprasert, N. et al., Anaphylaxis in a milk-allergic child following ingestion of lemon sorbet containing trace quantities of milk, *J. Food Prot.*, 61(1), 1522, 1998.

27. Majamaa, H. et al., Cow's milk allergy: Diagnostic accuracy of skin prick and patch tests and specific IgE, *Allergy*, 54(4), 346, 1999.

28. Mauro, C. et al., Correlation between skin prick test using commercial extract of cow's milk protein and fresh milk and food challenges, *Pediatr. Allergy Immunol.*, 18(7), 583, 2007.

29. Niggemann, B. et al., In vivo and in vitro studies on the residual allergenicity of partially hydrolysed infant formulae, *Acta Paediatr.*, 88(4), 394, 1999.

30. Zhu, H. and Snyder, M., Protein chip technology, *Curr. Opin. Chem. Biol.*, 7(1), 55, 2003.

31. Gaudin, J.C. et al., Assessment of the immunoglobulin E-mediated immune response to milk-specific proteins in allergic patients using microarrays, *Clin. Exp. Allergy*, 38(4), 686, 2008.

32. Wang, J. et al., Determination of epitope diversity in cow's milk hypersensitivity using microarray immunoassay, *J. Allergy Clin. Immunol.*, 117(2, Suppl. 1), S39, 2006.

33. de Boissieu, D., Comment utiliser le Diallertest(R)?, *Journal de Pediatrie et de Puericulture*, 19(4–5), 149, 2006.

34. de Boissieu, D. et al., The atopy patch tests for detection of cow's milk allergy with digestive symptoms, *J. Pediatr.*, 142(2), 203, 2003.

35. Kalach, N. et al., A pilot study of the usefulness and safety of a ready-to-use atopy patch test (Diallertest) versus a comparator (Finn Chamber) during cow's milk allergy in children, *J. Allergy Clin. Immunol.*, 116(6), 1321, 2005.

36. Natale, M. et al., Cow's milk allergens identification by two-dimensional immunoblotting and mass spectrometry, *Mol. Nutr. Food Res.*, 48(5), 363, 2004.

37. Bernard, H. et al., Specificity of the human IgE response to the different purified caseins in allergy to cow's milk proteins, *Int. Arch. Allergy Immunol.*, 115(3), 235, 1998.

38. Chatchatee, P. et al., Identification of IgE- and IgG-binding epitopes on alpha(s1)-casein: Differences in patients with persistent and transient cow's milk allergy, *J. Allergy Clin. Immunol.*, 107(2), 379, 2001.

39. Chatchatee, P. et al., Identification of IgE and IgG binding epitopes on beta- and kappa-casein in cow's milk allergic patients, *Clin. Exp. Allergy*, 31(8), 1256, 2001.
40. Jarvinen, K.M. et al., IgE and IgG binding epitopes on alpha-lactalbumin and beta-lactoglobulin in cow's milk allergy, *Int. Arch. Allergy Immunol.*, 126(2), 111, 2001.
41. Besler, M., Determination of allergens in foods, *Trends Anal. Chem.*, 20(11), 662, 2001.
42. Docena, G.H. et al., Identification of casein as the major allergenic and antigenic protein of cow's milk, *Allergy*, 51(6), 412, 1996.
43. Jarvinen, K.M. et al., B-cell epitopes as a screening instrument for persistent cow's milk allergy, *J. Allergy Clin. Immunol.*, 110(2), 293, 2002.
44. Vila, L. et al., Role of conformational and linear epitopes in the achievement of tolerance in cow's milk allergy, *Clin. Exp. Allergy*, 31(10), 1599, 2001.
45. Clement, G. et al., Epitopic characterization of native bovine β-lactoglobulin, *J. Immunol. Methods*, 266(1–2), 67, 2002.
46. Paschke, A. and Besler, M., Stability of bovine allergens during food processing, *Ann. Allergy Asthma Immunol.*, 89(6), 16, 2002.
47. Besler, M. et al., Stability of food allergens and allergenicity of processed foods, *J. Chromatogr. B. Biomed. Sci. Appl.*, 756(1–2), 207, 2001.
48. Sathe, S.K. et al., Effects of food processing on the stability of food allergens, *Biotechnol. Adv.*, 23, 423, 2005.
49. Heyman, M., Evaluation of the impact of food technology on the allergenicity of cow's milk proteins, *Proc. Nutr. Soc.*, 58, 587, 1999.
50. Iametti, S. et al., Modifications of high-order structures upon heating of b-lactoglobulin: Dependence on the protein concentration, *J. Agric. Food Chem.*, 43, 53, 1995.
51. Busti, P. et al., Thermal unfolding of bovine [beta]-lactoglobulin studied by UV spectroscopy and fluorescence quenching, *Food Res. Int.*, 38(5), 543, 2005.
52. Raynal-Ljutovac, K. et al., Heat stability and enzymatic modifications of goat and sheep milk, *Small Ruminant Res.*, 68(1–2), 207, 2007.
53. Ehn, B.M. et al., Modification of IgE binding during heat processing of the cow's milk allergen beta-lactoglobulin, *J. Agric. Food Chem.*, 52(5), 1398, 2004.
54. Chen, W.L. et al., A novel conformation-dependent monoclonal antibody specific to the native structure of {beta}-lactoglobulin and its application, *J. Dairy Sci.*, 89(3), 912, 2006.
55. Monaci, L. and van Hengel, A.J., Effect of heat treatment on the detection of intact bovine beta-lactoglobulins by LC mass spectrometry, *J. Agric. Food Chem.*, 55(8), 2985, 2007.
56. Kinsella, J. and Whitehead, D., Proteins in whey: Chemical, physical, and functional properties, *Adv. Food Nutr Res.*, 33, 343, 1989.
57. Bleumink, E. and Young, E., Identification of the atopic allergen in cow's milk, *Int. Arch. Allergy Appl. Immunol.*, 24, 521, 1968.
58. Høst, A. and Samuelsson, E.-G., Allergic reactions to raw, pasteurized, and homogenized/pasteurized cow milk: A comparison. A double-blind placebo-controlled study in milk allergic children, *Allergy*, 43(2), 113, 1988.
59. Kleber, N. et al., The antigenic response of ß-lactoglobulin is modulated by thermally induced aggregation, *Eur. Food Res. Technol.*, 219(2), 105, 2004.
60. Lee, Y.H. Food-processing approaches to altering allergenic potential of milk-based formula, *J. Pediatr.*, 121(5 Pt 2), 47, 1992.
61. Corredig, M. and Dalgleish, D.G., The mechanisms of the heat-induced interaction of whey proteins with casein micelles in milk, *Int. Dairy J.*, 9(3–6), 233, 1999.
62. Oldfield, D.J. et al., Effect of preheating and other process parameters on whey protein reactions during skim milk powder manufacture, *Int. Dairy J.*, 15(5), 501, 2005.
63. Bleumink, E. and Berrens, L., Synthetic approaches to the biological activity of [beta]-lactoglobulin in human allergy to cows' milk, *Nature*, 212(5061), 541, 1966.

64. Chen, W.L. et al., {beta}-Lactoglobulin is a thermal marker in processed milk as studied by electrophoresis and circular dichroic spectra, *J. Dairy Sci.*, 88(5), 1618, 2005.

65. Smacchi, E. and Gobbetti, M., Bioactive peptides in dairy products: Synthesis and interaction with proteolytic enzymes, *Food Microbiol.*, 17(2), 129, 2000.

66. Cross, M.L. et al., Anti-allergy properties of fermented foods: An important immunoregulatory mechanism of lactic acid bacteria? *Int. Immunopharmacol.*, 1(5), 891, 2001.

67. Sutas, Y. et al., Suppression of lymphocyte proliferation in vitro by bovine caseins hydrolyzed with *Lactobacillus casei* GG-derived enzymes, *J. Allergy Clin. Immunol.*, 98, 216, 1996.

68. Fedele, L. et al., Matrix-assisted laser desorption/ionization mass spectrometry for monitoring bacterial protein digestion in yogurt production, *J. Mass Spectrom.*, 34, 1338, 1999.

69. Tamime, A.Y. and Deeth, H.C., Yogurt: Technology and biochemistry, *J. Food Prot.*, 43, 939, 1980.

70. Thomas, T.D. and Pritchard, G.G., Proteolytic enzymes of dairy starter cultures, *FEMS Microbiol. Lett.*, 46(3), 245, 1987.

71. Kabadjova-Hristova, P. et al., Evidence for proteolytic activity of Bacilli isolated from kefir grains, *Biotechnol. Biotechnol. Eq.*, 20(2), 89, 2006.

72. Wroblewska, B. et al., The effect of selected microorganisms on the presence of immunoreactive fractions in cow and goat milks, *Pol. J. Food Nutr. Sci.*, 4(3), 21, 1995.

73. Jedrychowski, L. and Wroblewska, B., Reduction of the antigenicity of whey proteins by lactic acid fermentation, *Food Agric. Immunol.*, 11(1), 91, 1999.

74. Kleber, N. et al., Screening for lactic acid bacteria with potential to reduce antigenic response of [beta]-lactoglobulin in bovine skim milk and sweet whey, *Innovat. Food Sci. Emerg. Tech.*, 7(3), 233, 2006.

75. Schieber, A. and Bruckner, H., Characterization of oligo- and polypeptides isolated from yoghurt, *Eur. Food Res. Technol.*, 210, 310, 2000.

76. Hau, J. and Bovetto, L., Characterisation of modified whey protein in milk ingredients by liquid chromatography coupled to electrospray ionisation mass spectrometry, *J. Chromatogr. A.*, 926(1), 105, 2001.

77. Manso, M.A. et al., Application of proteomics to the characterisation of milk and dairy products, *Int. Dairy J.*, 15(6–9), 845, 2005.

78. Alli, I. et al., Identification of peptides in cheddar cheese by electrospray ionization mass spectrometry, *Int. Dairy J.*, 8(7), 643, 1998.

79. Piraino, P. et al., Use of mass spectrometry to characterize proteolysis in cheese, *Food Chem.*, 101(3), 964, 2007.

80. Faccia, M. et al., Influence of type of milk and ripening time on proteolysis and lipolysis in a cheese made from overheated milk, *Int. J. Food Sci. Technol.*, 42(4), 427, 2007.

81. Høst, A. and Halken, S., Hypoallergenic formulas—When, to whom and how long: After more than 15 years we know the right indication!, *Allergy*, 59(Suppl. 78), 45, 2004.

82. Korhonen, H. and Pihlanto, A., Bioactive peptides: production and functionality, *Int. Dairy J.*, 16(9), 945, 2006.

83. Chirico, G. et al., Immunogenicity and antigenicity of a partially hydrolyzed cow's milk infant formula, *Allergy*, 52(1), 82, 1997.

84. Wróblewska, B. et al., Immunoreactive properties of peptide fractions of cow whey milk proteins after enzymatic hydrolysis, *Int. J. Food Sci. Technol.*, 39(8), 839, 2004.

85. Puerta, A. et al., Immunochromatographic determination of [beta]-lactoglobulin and its antigenic peptides in hypoallergenic formulas, *Int. Dairy J.*, 16(5), 406, 2006.

86. Almaas, H. et al., In vitro digestion of bovine and caprine milk by human gastric and duodenal enzymes, *Int. Dairy J.*, 16(9), 961, 2006.

87. Park, Y.W. et al., Physico-chemical characteristics of goat and sheep milk, *Small Ruminant Res.*, 68(1–2), 88, 2007.

88. Bonomi, F. et al., Reduction of immunoreactivity of bovine beta-lactoglobulin upon combined physical and proteolytic treatment, *J. Dairy Res.*, 70(1), 51, 2003.

89. Michalski, M.-C. and Januel, C., Does homogenization affect the human health properties of cow's milk?, *Trends Food Sci. Technol.*, 17(8), 423, 2006.

90. Garcia-Risco, M.R. et al., Modifications in milk proteins induced by heat treatment and homogenization and their influence on susceptibility to proteolysis, *Int. Dairy J.*, 12(8), 679, 2002.

91. Steinhart, H. et al., Introducing allergists to food chemistry, *Allergy*, 56(Suppl. 67), 9, 2001.

92. Alomirah, H.F. et al., Applications of mass spectrometry to food proteins and peptides, *J. Chromatogr. A*, 893(1), 1, 2000.

93. Poms, R.E. et al., Methods for allergen analysis in food—a review, *Food Addit. Contam.*, 21, 1, 2004.

94. Van Hengel, A.J. et al., Analysis of food allergens and practical applications, in *Food Toxicants Analysis: Techniques, Strategies and Developments*, Pico, Y., Ed. Elsevier, Amsterdam, 2007, p. 189.

95. Guidry, A.J. and Pearson, R.E., Improved methodology for quantitative determination of serum and milk proteins by single radial immunodiffusion, *J. Dairy Sci.*, 62(8), 1252, 1979.

96. Levieux, D. et al., Caprine immunoglobulin G, beta-lactoglobulin, alpha-lactalbumin and serum albumin in colostrum and milk during the early post partum period, *J. Dairy Res.*, 69(3), 391, 2002.

97. Monaci, L. et al., Milk allergens, their characteristics and their detection in food: a review, *Eur. Food Res. Technol.*, 223(2), 149, 2006.

98. Muller-Renaud, S. et al., Quantification of beta-casein in milk and cheese using an optical immunosensor, *J. Agric. Food Chem.*, 52(4), 659, 2004.

99. Dupont, D. et al., Determination of the heat-treatment undergone by milk by following the denaturation of alpha-lactalbumin with a biosensor, *J. Agric. Food Chem.*, 52, 677, 2004.

100. Dupont, D. and Muller-Renaud, S., Quantification of proteins in dairy products using an optical biosensor detection of milk allergens-specific IgE in patient sera using biosensor techniques, *J. AOAC Int.*, 89(3), 843, 2006.

101. Malmheden Yman, I. et al., Food allergen detection with biosensor immunoassays, *J. AOAC Int.*, 89(3), 856, 2006.

102. Haasnoot, W. et al., Fast biosensor immunoassays for the detection of cows' milk in the milk of ewes and goats, *J. Dairy Res.*, 71(3), 322, 2004.

103. Karamanova, L. et al., Immunoprobes for thermally-induced alterations in whey protein structure and their applications to the analysis of thermally-treated milks, *Food Agric. Immunol.*, 15, 77, 2003.

104. Hohensinner, V. et al., A 'gold cluster-linked immunosorbent assay': Optical near-field biosensor chip for the detection of allergenic [beta]-lactoglobulin in processed milk matrices, *J. Biotechnol.*, 130(4), 385, 2007.

105. Restani, P. et al., Use of immunoblotting and monoclonal antibodies to evaluate the residual antigenic activity of milk protein hydrolysed formulae, *Clin. Exp. Allergy*, 26(10), 1182, 1996.

106. Galvani, M. et al., Two-dimensional gel electrophoresis/matrix-assisted laser desorption/ionisation mass spectrometry of a milk powder, *Rapid Commun. Mass Spec.*, 14(20), 1889, 2000.

107. O'Donnell, R. et al., Milk proteomics, *Int. Dairy J.*, 14(12), 1013, 2004.

108. Leonil, J. et al., Application of chromatography and mass spectrometry to the characterization of food proteins and derived peptides, *J. Chromatogr. A*, 881(1–2), 1, 2000.

109. Roncada, P. et al., Identification of caseins in goat milk, *Proteomics*, 2(6), 723, 2002.

110. Chin, H.W. and Rosenberg, M., Monitoring proteolysis during cheddar cheese ripening using two-dimensional gel electrophoresis, *J. Food Sci.*, 63(3), 423, 1998.

111. Molina, E. et al., Characterisation of the casein fraction of Ibérico cheese by electrophoretic techniques, *J. Sci. Food Agric.*, 82(10), 1240, 2002.

112. Hernandez-Ledesma, B. et al., Identification of bioactive peptides after digestion of human milk and infant formula with pepsin and pancreatin, *Int. Dairy J.*, 17(1), 42, 2007.

113. Veledo, M.T. et al., Development of a method for quantitative analysis of the major whey proteins by capillary electrophoresis with on-capillary derivatization and laser-induced fluorescence detection, *J. Sep. Sci.*, 28(9–10), 935, 2005.

114. Vallejo-Cordoba, B., Rapid separation and quantification of major caseins and whey proteins of bovine milk by capillary electrophoresis, *J. Capill. Electrophor.*, 4(5), 219, 1997.

115. Gutierrez, J.E. and Jakobovits, L., Capillary electrophoresis of alpha-lactalbumin in milk powders, *J. Agric. Food Chem.*, 51(11), 3280, 2003.

116. Veledo, M.T. et al., Analysis of trace amounts of bovine beta-lactoglobulin in infant formulas by capillary electrophoresis with on-capillary derivatization and laser-induced fluorescence detection, *J. Sep. Sci.*, 28(9–10), 941, 2005.

117. Miralles, B. et al., Improved method for the simultaneous determination of whey proteins, caseins and para-kappa-casein in milk and dairy products by capillary electrophoresis, *J. Chromatogr. A*, 915(1–2), 225, 2001.

118. Pappa, E.C. et al., Application of proteomic techniques to protein and peptide profiling of Teleme cheese made from different types of milk, *Int. Dairy J.*, 18(6), 605, 2008.

119. de Wit, J.N., Nutritional and functional characteristics of whey proteins in food products, *J. Dairy Sci.*, 81(3), 597, 1998.

120. Cantani, A., Hidden presence of cow's milk proteins in foods, *J. Investig. Allergol. Clin. Immunol.*, 9(3), 141, 1999.

121. Koppelman, S.J. et al., Anaphylaxis caused by the unexpected presence of casein in salmon, *Lancet*, 354(9196), 2136, 1999.

122. Gern, J.E. et al., Allergic reactions to milk-contaminated "nondairy" products, *N. Engl. J. Med.*, 324(14), 976, 1991.

123. Malmheden-Yman, I. and Eriksson, I., Analysis of food proteins for verification of contamination or mislabelling, *Food Agric. Immunol.*, 6, 167, 1994.

124. Martin-Diana, A.B. et al., Whey permeate as a bio-preservative for shelf life maintenance of fresh-cut vegetables, *Innovat. Food Sci. Emerg. Tech.*, 7(1–2), 112, 2006.

125. Semo, E. et al., Casein micelle as a natural nano-capsular vehicle for nutraceuticals, *Food Hydrocolloid.*, 21(5–6), 936, 2007.

126. Tiger Lee, T.-T. et al., Contamination of probiotic preparations with milk allergens can cause anaphylaxis in children with cow's milk allergy, *J. Allergy Clin. Immunol.*, 119(3), 746, 2007.

127. Lucas, A. et al., Probiotic cell counts and acidification in fermented milks supplemented with milk protein hydrolysates, *Int. dairy J.*, 14(1), 47, 2004.

128. Mariager, B. et al., Bovine β-lactoglobulin in hypoallergenic and ordinary infant formulas measured by indirect competitive ELISA using monoclonal and polyclonal antibodies, *Food Agric. Immunol.*, 6, 73, 1994.

129. Hefle, S. and Lambrecht, D., Validated sandwich enzyme-linked immunosorbent assay for casein and its application to retail and milk-allergic complaint foods, *J. Food Prot.*, 67(9), 1933, 2004.

130. Sletten, G.B. et al., A comparison of time-resolved fluoroimmunoassay and ELISA in the detection of casein in foodstuffs, *Food Agric. Immunol.*, 16(3), 235, 2005.

131. Akiyama, H. et al., Inter-laboratory evaluation studies for development of notified ELISA methods for allergenic substances (milk), *J. Food Hyg. Soc. Jpn.*, 45(3), 120, 2004.

132. Monaci, L. and van Hengel, A.J., Development of a method for the quantification of whey allergen traces in mixed-fruit juices based on liquid chromatography and mass spectrometry detection, *J. Chromatrogr. A*, 1192(1), 113, 2008.

133. Weber, D. et al., Development of a liquid chromatography-tandem mass spectrometry method using capillary liquid chromatography and nanoelectrospray ionization-quadrupole time-of-flight hybrid mass spectrometer for the detection of milk allergens, *J. Agric. Food Chem.*, 54(5), 1604, 2006.

134. Garcia-Ruiz, C. et al., Analysis of bovine whey proteins in soybean dairy-like products by capillary electrophoresis, *J. Chromatogr. A*, 859(1), 77, 1999.

135. Henning, D.R. et al., Major advances in concentrated and dry milk products, cheese, and milk fat-based spreads, *J. Dairy Sci.*, 89(4), 1179, 2006.

136. Levin, M.E. et al., Anaphylaxis in a milk-allergic child after ingestion of soy formula cross-contaminated with cow's milk protein, *Pediatrics*, 116(5), 1223, 2005.

Index

A

Acanthocephalans, 523
Accelerated solvent extraction (ASE), 364, 711–712, 730, 937
Adenosine 5′-triphosphate (ATP), 8; *see also* ATP bioluminescence methods
Aerobic standard plate count (SPC), 5, 7
Aeromonas genus, 473
Aflatoxins
    analysis methods, 90–91
    origin and nature, 89
    in poultry, 100–103
    structure and chemical properties, 89–90
Aflatoxins B1
    mycotoxin analysis, processed meat, 105
    structure, 80
AFLP, *see* Amplified fragment length polymorphism
Agar plate count methods, *see* Aerobic standard plate count
Aliphatic diamines, 400
Aliphatic polyamines, 400
Allergens, milk, *see* Milk allergens
Alteromonas nigrifaciens, 469
Amino acids analysis, 172–174
Aminocyclitol ring, 261–262
Aminogenic microorganisms, 402
Aminoglycosides, 260–262, 699–700
Amnesic shellfish poisoning (ASP), 588–590
Amphenicols, 262–266
Amplified fragment length polymorphism (AFLP), 609, 625–626
Analytical atomic spectroscopy, 650–651
Animal species identification
    capillary electrophoresis, 188
    DNA methods
        extraction, 188
        fingerprinting, 194
        multiplex PCR, 194–196
        PCR product determination, 190–192
        PCR reaction design, 189
        real-time PCR, 194, 196–198
        species-specific PCR amplification, 192–193
        universal mtDNA primers, 189–190
    ELISA, 188
    immunochemical methods, 188
Anion exchange liquid chromatography, 146
Anisakid nematodes, 515, 518–523
Anisakis simplex, 515, 518–523
Anthelmintic drugs, 702–703
Antibacterial residues, *see* Antimicrobial residues, muscle tissues
Antibiotics, fish and seafoods, *see* Veterinary drugs
Antibiotics, milk and its products
    confirmatory methods
        LC-MS confirmatory criteria, 897–900
        liquid chromatography (LC), 891–896
        mass spectrometry, 896–897
        sample preparation, 890–891
        ultraperformance liquid chromatography separation (ULCS), 891–896
    method validation and measurement uncertainty, 900
    screening methods, 889
Antibiotics, muscle tissues; *see also* Antimicrobial residues, muscle tissues
    aminoglycosides, 260–262
    amphenicols, 262–266
    vs. antibacterials, 250
    beta-lactams: penicillins and cephalosporins, 266, 270–278
    definition and role, 250–251
    macrolides and lincosamides, 272, 279–283
    novobiocin and tiamulin, 316, 320–321
    polyether antibiotics, 316, 318–319
    polypeptide antibiotics, 309, 315–317
    tetracyclines, 308–314
Antibody detection diagnosis, 549–550
Antimicrobial residues, muscle tissues; *see also* Antibiotics, muscle tissues
    analytical control methods, 253
    applications, chromatographic methods
        aminoglycosides, 260–262
        amphenicols, 262–266

beta-lactams: penicillins and cephalosporins, 266, 270–278
carbadox and olaquindox, 322–324
macrolides and lincosamides, 272, 279–283
nitrofurans, 280, 284–289
nitroimidazoles, 285, 290–293
novobiocin and tiamulin, 316, 320–321
polyether antibiotics, 316, 318–319
polypeptide antibiotics, 309, 315–317
quinolones, 290, 294–300
sulfonamide antiinfectives, 301–307
tetracyclines, 308–314
confirmatory analysis, chromatographic methods
chemical analysis, 257
chromatographic methodologies, 256–257
detection modes, 259–260
sample preparation, 257–258
separation modes, 258–259
screening analysis, biological methods
microbiological methods, 253–254
radioimmunological Charm II test®, 254
surface plasmon resonance-based biosensor immunoassay (SPR-BIA), 255
strategies for screening and confirmation, 252–253
Antiparasitica, 702–703
Antisera, 545
Aquaculture, 688–690; *see also* Veterinary drugs
anthelmintic drugs, 702–703
antiparasitica, 702–703
bactericidal drugs (antibiotics), 695–701
hormones, 702–703
tranquilizers, 701–702
AromaScan, 14
Aromatic monoamines, 400
Artificial digestion, *Trichinella* inspection, 62–63
Ascaris suum, 69–70
Atomic absorption spectrometry (AAS)
cold vapor atomic absorption spectrometry, 646
instrumentation, 643–646
theory, 642–643
Atomic emission spectrometry (AES)
instrumentation, 647–649
theory, 646–647
Atomizers, 644–646
ATP, *see* Adenosine 5′-triphosphate
ATP bioluminescence methods
microbial foodborne pathogens, 25–26
muscle food spoilage prediction, 8–9
Azaspiracids, 594

B

Bacillus cereus, 42–44, 840
Bactericidal drugs (antibiotics), 695–701
Bactometer®, 9
Beta-lactams: penicillins and cephalosporins, 266, 270–278

Bioassays
brevetoxin detection, 591
ciguatoxin detection, 579–580
DR CALUX, 719
hepatotoxins detection, 593
okadaic acid detection, 586–587
POP determination, meat
cell-based, 387
PCR, 387
reliability and applicability, 387–388
screening and confirmatory (HRGC-HRMS) analysis, 388
saxitoxin detection, 584–585
tetrodotoxin detection, 582
Biogenic amines (BA), meat and meat products
aminogenic microorganisms, 402
detection and determination
chromatographic quantification procedures, 407–413
extraction and cleanup, 406–407
rapid screening procedures, 409, 414
food safety and food quality issues, 400–401
hygienic quality evaluation
meat freshness, 404–405
of raw materials, 405–406
occurrence
in cooked meat products, 403
in cured meat products, 403
in fermented meat products, 403–404
in fresh meat and fresh meat products, 402
origin and classification, 400
Biogenic amines (BA), dairy foods, 840–841
Biogenic amines (BA), seafood products
determination
capillary electrophoresis method, 755
derivatization, 754
diaminoxidase (DAO) reaction, 755
extraction process, 753
flow injection analysis (FIA), 754–755
portable procedures, 755–756
formation
amino acid precursors and amino acid decarboxylase enzymes, 743–745
microorganisms, 749–750
processing and storage conditions, 750–752
raw material, 748–749
legal limits, 747–748
as quality index, 746–747
toxicity, 744–746
Bioinformatics, 630–631
Biologically active amines, *see* Biogenic amines, dairy foods; Biogenic amines, meat and meat products; Biogenic amines, seafood products
Biologic-immunologic-phenotypic combination studies, 843–844
Bioluminescence methods, *see* ATP bioluminescence methods

Biosensors, 147–148, 235–236, 587, 772, 889, 960
Blue-green algae, *see* Cyanobacteria
Bovine cysticercosis, 67
Bovine rennet whey (BRW), 856–857
Brevetoxin-neurotoxic shellfish poisoning
  clinical symptoms, 590–591
  detection methods
    bioassays, 591
    chemical methods, 592
    immunoassays, 591–592
  overview, 590
Brochothrix genus, 472
Butter adulteration, 853–854; *see also* Dairy adulteration
    detection

C

Campylobacter jejuni, 47–49
Capillary electrophoresis (CE) method, 755
Carbadox, 322–324
Carbon rod atomizer (CRA 63), 645
Caseins, 852, 866–869
Cation-exchange HPLC method, 869
Cell-based bioassays, 387, 582
Cell culture isolation, fish and shellfish
    cell lines and cell culture, 541–542
    cell line selection, 542
    performance, 543–544
    viral isolation, 542–543
Cephalosporines, 698
Cephalosporins, 266, 270–278
Chemical contaminants, phthalates, *see* Phthalates, dairy
    products
Chinolones, 696–697
Chloramphenicol, 255, 262, 689, 699, 888
Chromatographic methods/analysis
    anion exchange liquid chromatography, GMO, 146
    antimicrobial residues, 256–257
    biogenic amines (BA), 407–413
    gas chromatography (*see* Gas chromatography)
    gel permeation chromatography, 732
    GMO, meat, 146
    growth promoters, 236
    HPLC method (*see* High-pressure liquid
        chromatography method)
    immunoaffinity, 234
    liquid chromatography (LC), 257–260, 891–896
    marine toxins, 587–588
    meat adulteration detection
        amino acids analysis, 172–174
        peptides analysis, 173–174
        whole proteins analysis, 173–176
    polychlorinated biphenyls (PCBs)
        gas chromatography-high-resolution mass
            spectrometry (GC-HRMS), 739
        gas chromatography-mass spectrometry
            (GC-MS), 738–739

    gas chromatography with electron capture
        detector (GC-ECD), 738–739
    polycyclic aromatic hydrocarbons (PAHs)
        gas chromatography (GC), 734
        high-performance liquid chromatography
            (HPLC), 733
    protein detection, milk, 869–870
    seafood adulteration detection, 607
    selected ion monitoring (SIM) GC-MS
        chromatogram, 237–238
    thin-layer chromatography (TLC), 236, 447, 828
    ultraperformance liquid chromatography separation
        (ULCS), 891–896
Ciguatera fish poisoning, 578–579
Ciguatoxin (CTX)
    clinical symptoms, 579
    detection methods
        bioassays, 579–580
        chemical methods, 580–581
        immunoassays, 580
    overview, 578–579
Citrinin
    analytical methods, 93
    origin and nature, 92–93
    physicochemical properties, 93
    in poultry, 104
    in processes meat, 105, 107
    structure, 80
Clonorchis sinensis, 510, 513–514
Clostridium perfringens, 44–47
Cold vapor atomic absorption spectrometry, 646
Colony forming units (CFU), 480
Culture methods, muscle foods
    food spoilage prediction methods, 5
    microbial foodborne pathogens, 22–23
        detection methods, 24
        most probable number, 24
        plate count, 23–24
Cyanobacteria, 592
Cylopiazonic acid (CPA)
    analysis methods, 94
    origin and nature, 93
    physicochemical properties, 93
    in poultry, 104
    in processes meat, 107
Cytophaga, 471–472

D

Dairy adulteration detection
    lipids
        fatty acid analysis, 853–855
        triacylglycerol analysis, 855–856
    proteins
        immunoassays, 856–857
        mass spectrometry (MS), 857–858
        near-infrared spectroscopy (NIRS), 858–860

Dairy products
  allergens (*see* Milk allergens)
  adulteration detection (*see* Dairy adulteration detection)
  antibiotics (*see* Antibiotics, milk and its products)
  chemical contaminants (*see* Phthalates, dairy products)
  culture-dependent microbiological analysis methods
    classical and advanced phenotypic methods, 783–785
    molecular methods, 785–788
  culture-independent microbiological analysis methods, 788–790
  environmental contaminants (*see* Persistent organic pollutants, dairy products; Pesticides, dairy products)
  microbial ecology, 782–783
  pathogen detection (*see* Microbial foodborne pathogens, dairy products)
  spoilage prediction (*see* Milk and dairy products spoilage prediction)
  toxins (*see* Mycotoxin analysis, dairy foods)
Decision limit (CCα), 232
DEHP, 908, 910, 916–921, 923–925
Denaturing gradient gel electrophoresis (DGGE), 628
Deoxynivalenol (DON), 79, 83–84, 98–99
Detection capability (CCβ), 232
Detection limit (DL), 533, 557–558
Dexamethasone detection, 243
DGGE, *see* Denaturing gradient gel electrophoresis
Diarrhetic shellfish poisoning (DSP), 586–588
Dioxin-like polychlorobiphenyls (DL-PCB)
  analytical methods, 369, 378
    cleanup and fractionation, 379
    extraction techniques, 378–379
    HRGC-HRMS instrumental analysis, 379–380
    pretreatment, 378
  examples, 370–377
Dioxins, *see* Polychlorodibenzofurans; Polychlorodibenzo-p-dioxins
Dioxins analysis, seafood and seafood products
  automation of extraction and cleanup, 714–715
  bioanalytical screening methods, 718–719
  cleanup methods
    isolation of uncommon chemical interferences, 713
    lipid removal, 713
  extraction methods
    accelerated solvent extraction (ASE), 711–712
    microwave oven, 712
    solid phase extraction (SPE) and matrix solid phase dispersion (MSPD), 710–711
    Soxhlet extraction, 710
    supercritical fluid extraction (SFE), 711
  fractionation/group separation, 713–714
  instrumental determination
    GC congener separation, 715–717
    GC detectors, 717–718

sample pretreatment and recovery studies
    sample storage, 709
    spiking and recovery studies, 709–710
Diphyllobothriosis, 515
Diphyllobothrium pacificum, 515–517
Dipsticks, 959–960
Direct epifluorescent filtration technique (DEFT)
  aerobic plate count (APC), 219
  microbial foodborne pathogens detection, muscle foods, 25
  muscle food spoilage prediction, 6–8
Disease-resistant plants, 126
DL-PCB, *see* Dioxin-like polychlorobiphenyls
DNA-based methods, GMO detection
  DNA extraction methods, 131–134
  PCR-based assay formats, 130–131
    applications, 136, 143
    enzyme-linked immunosorbent assay (ELISA), 136
    qualitative methods, 132, 134
    quantitative competitive polymerase chain reaction (QC-PCR), 134–135
    real-time, 135–140
DNA comet assay, 218–219
DNA microarray technology, 146–147
DNase activity, 479–480
Domoic acid (DA)-amnesic shellfish poisoning (ASP)
  clinical symptoms, 588–589
  detection methods
    bioassays, 589
    chemical methods, 589–590
    immunoassays, 589
  overview, 588

E

Echinococcus spp., 71
Electronic nose, 14
Electronic transition, 642
Electron microscopy, 541
Electron spin resonance (ESR) spectroscopy, food irradiation identification
  fish and seafoods, 671–672
  muscle foods, 220
    foods containing bone, 214–215
    foods containing cellulose, 215
    foods containing crystalline sugar, 216
Electrophoresis, 437604–605
Electrophoretical separation, 169–170
Electrophoretic techniques, meat adulteration detection, 158–164
Electrothermal atomizer (ETA), *see* Graphite furnace atom cell
Endogenous/natural amines, 400
Enrofloxacine, 689–690, 697
Enterobacter sakazakii, 814, 816–817

Environmental contaminants, *see* Persistent organic pollutants, dairy products; Persistent organic pollutants, fish and seafood; Pesticides, dairy products

Enzyme-linked immunosorbent assay (ELISA)
    animal species identification, 188
    dairy adulteration detection, 856–857
    for fish antibody detection, 549
    foodborne pathogens detection, 25
    GMO detection
        PCR, 136
        protein-based methods, 143–144
        traceability, 128
    indirect ELISA (iELISA), 548
    marine toxin
        domoic acid (DA) detection, 589
        ELISA kits, 587
        indirect ELISA, 585
    milk allergens, 958–959
    milk origin identification, 870–873
    mycotoxin analysis
        aflatoxin, 90–91
        fumonisin quantification, 89
        type B trichothecenes, 85
    sandwich ELISA (swELISA), 548
    Trichinella spp. detection, 67

Escherichia coli O157:H7
    confirmation, 34–35
    cultural enumeration method, 34
    detection, 32–34

F

Fasciola hepatica and other liver flukes, 70–71
Fat detection, milk, 874
Filament atomizer, 645
Fingerprinting, 194
Fish and seafoods
    adulteration (*see* Seafood adulteration detection)
    antibiotics (*see* Veterinary drugs)
    biogenic amines (BA)
        determination process, 754–756
        formation, 743–745, 748–752
        legal limits, 747–748
        as quality index, 746–747
        toxicity, 744–746
    dioxins analysis (*see* Dioxins analysis, seafood and seafood products)
    environmental contaminants (*see* Persistent organic pollutants, fish and seafood)
    food irradiation (*see* Food irradiation identification, fish and seafoods)
    food spoilage prediction (*see* Seafood spoilage prediction)
    metals determination (*see* Spectrochemical methods, seafood metals determination)
    parasites (*see* Parasites, fish and seafoods)

pathogens detection (*see* Seafood pathogens detection)
safety
    allergic and immunosuppressed persons, recommendations, 525
    consumers and restaurateurs, recommendations, 525
    primary production and handling, 523–524
    thermal processing, 524
    toxins (*see* Marine toxins)

Fish poisoning, *see* Marine toxins
Flame atomic absorption spectrometry, 644
Flavobacterium, 471–472
Flow cell cytometry, 11
Flow injection analysis (FIA), 754–755
Fluorescent in situ hybridization (FISH), 13
Food irradiation identification, fish and seafoods
    countries with commercial radiation processing facilities, 668
    development of technology, 665–669
    DNA methods, 675–677
    electron spin resonance spectroscopy, 671–672
    EN standards, 670–671
    foodborne diseases, 664
    food preservation, high energy irradiation, 664–667
    global acceptance and attitudes, 669
    luminescence, 677–678
    microbiological methods, 678
    radiolytic chemicals analysis
        2-alkylcyclobutanones, 672–673
        hydrogen and carbon monoxide gases, 675
        o-, m-tyrosine, 675
        volatile hydrocarbons, 673–675

Food irradiation identification, muscle foods
    agarose electrophoresis, 221
    applications, 221–223
    DNA Comet Assay, 218–219
    electron spin resonance (ESR) spectroscopy, 220
        foods containing bone, 214–215
        foods containing cellulose, 215
        foods containing crystalline sugar, 216
    EU legislation, 210
    European standards, 211
    gas chromatographic analysis
        hydrocarbons (EN1784), 210, 212
        mass spectrometric analysis, 2-alkylcyclobutanones (EN1785), 212–213
    half-embryo test, 221
    microbiological changes measurement
        direct epifluorescent filter technique/aerobic plate count (DEFT/APC), 219
        Limulus amebocyte lysate/gram-negative bacteria test, 220
    photostimulated luminescence (PSL), 217–218
    reasons, 209–210
    thermoluminescence detection, 217

Food preservation, high energy irradiation, 664–667
Food safety, 251–252

Foreign proteins addition
    dairy products (*see* Dairy adulteration detection)
    processed meat (*see* Meat adulteration detection)
    seafoods (*see* Seafood adulteration detection)
Fourier transform infrared (FT-IR) spectroscopy, 10
Fumonisin B1, 79
Fumonisins
    analysis methods, 88–89
    origin and nature, 87–88
    physicochemical properties, 88
    in poultry, 100

G

Gamma irradiated seafood, 474
Gas chromatography (GC), 435–436
    food irradiation identification, muscle foods
        hydrocarbons (EN1784), 210, 212
        mass spectrometric analysis,
            2-alkylcyclobutanones (EN1785), 212–213
    vs. HPLC, 456–457
    nitrosamines, sample preparation, 435–436
    PAH, 448–455, 734
    PCB, 738–739
    Type A trichothecenes detection, 85
Gas chromatography-high-resolution mass spectrometry
    (GC-HRMS), 739
Gas chromatography-mass spectrometry (GC-MS),
    738–739
Gas chromatography with electron capture detector
    (GC-ECD), 738–739
Gel electrophoresis, 131
Gel permeation chromatography, 732
Genetically modified organism (GMO), meat
    crops, 126–127
    detection
        anion exchange liquid chromatography, 146
        biosensors, 147–148
        chromatographic techniques, 146
        DNA-based methods, 130–143
        DNA microarray technology, 146–147
        NIR spectroscopy, 145
        protein-based methods, 143–145
        visible/NIR (vis/NIR) spectroscopy, 145
    production for food and feed, 126–127
    traceability
        analytical methods, 128–129
        legislative framework, 127–128
    transgenic material, 129–130
Genetically modified organisms (GMO), fish feed
    biosensors, 772
    definition, 762
    DNA-based detection
        extraction, 764
        multiplex PCR, 766
        qualitative conventional PCR, 764–766
        quantitative PCR, 766–767

DNA *vs.* protein methods
    DNA-based methods, 770–771
    protein-based methods, 769–770
    formulation, 762–763
    international regulations, 763–764
    microarray technology, 771–772
    protein-based detection, 767–769
Giardia duodenalis, 509–510
Gnathostoma spp., 523
Graphite furnace atom cell, 644–646
Green meat, 72
Gross signs, 537
Growth promoters
    cleanup methods
        extraction procedures, 234
        immunoaffinity chromatography, 234
        molecular recognition, 235
    confirmatory analytical methods
        LC-DAD, 243
        LC-MS/MS, 237, 239–242
        selected ion monitoring (SIM) GC-MS
            chromatogram, 237–238
    control, 231–233
    dexamethasone detection, 243
    sampling and sample preparation
        meat samples, 234
        samples from animal farms, 233
    screening methods
        biosensors, 235–236
        chromatographic techniques, 236
        immunological techniques, 235
Gymnodimine, 594

H

Half-embryo test, 221
Heated graphite atomizer (HGA), 644–645
Herbicide-tolerant plants, 126
Heterocyclic amines, 400
High-performance thin-layer chromatography
    (HPTLC), 236
High-pressure liquid chromatography (HPLC)
    method
    aflatoxins analysis, 90
    antimicrobial analysis, 256
    cation-exchange method, 869
    citrinin analysis, 93, 104
    coupled with TEA detector (HPLC-TEA), 436
    with electrochemical detection (HPLC-EC), 607
    marine toxins, 583
    ochratoxins A analysis, 92
    PAH, 448, 456
    POP characterization, 733
    RP-HPLC, 869–870
    triacylglycerol analysis, 855–856
    whole proteins analysis, 174–176

Histopathology, 537–540
HRGC column selection, 943–944
H₂S production detection, 479
Hydrophobic grid membrane filter, 25

I

Ichthyosarcotoxin
    ciguatoxin (CTX)
        clinical symptoms, 579
        detection methods, 579–581
        overview, 578–579
    tetrodotoxin (TTX)
        clinical symptoms, 581–582
        detection methods, 582–583
        overview, 581
Immunoassays, 170–172
Immunobiosensors, 587
Immunodiffusion, 164–168, 958
Immunodot, 547
Immunofluorescense antibody test, 548
Immunohistochemistry (IHC), 537, 540–541, 547–548
Immunological methods
    meat adulteration detection
        electrophoretical separation, 169–170
        immunoassays, 170–172
        immunodiffusion, 164–168
        indirect hemagglutination, 168
        serology, 164
    seafood adulteration detection, 605–606
Immunomagnetic separation (IMS), 25
Immunoprecipitation, 958
Impedance/conductance technique, muscle foods
    food spoilage prediction, 9
    microbial foodborne pathogens, 26
Indirect hemagglutination, 168
Inductively coupled plasma-mass spectrometry
        (ICP-MS), 649–650
Inductively coupled plasma-optical emission
        spectrometry (ICP-OES) system, 648
Insect-protected plants, 126
Interlaboratory study, 232
Irradiated ingredients detection, muscle foods, *see* Food
        irradiation identification, muscle foods

L

Lactic acid bacteria (LAB), 781–782
Lactobacillus genus, 472–473
LAL, *see Limulus* amoebocyte lysate assay
Lateral-flow immunoassays (LFIAs), 959–960
Limulus amoebocyte lysate/gram-negative bacteria
        test, 220
Limulus amoebocyte lysate assay (LAL), 9–10
Lincomycin, 700
Lincosamides, 272, 279–283, 699

Liquid chromatography (LC), 257–260, 891–896; *see
        also* High-pressure liquid chromatography
        method
Liquid-liquid extraction (LLE), 432–433, 933
Liquid-liquid partitioning, 736
Liquid media-based most probable number (MPN)
        technique, 5
Liquid-phase extraction (LPE), 936–937
Liquid smoke flavors (LSF), 446–447
Liquid-solid extraction (LSE), 936
Listeria monocytogenes, 487, 496–498, 814, 816
    confirmation scheme, 30–32
    cultural enumeration method, 29–30
    detection, 27–29
Litmus milk, 478
Liver trematodes, 70
Loop-mediated isothermal amplification (LAMP), 554
Luciferin, 8
Luminescence, food irradiation identification, 677–678

M

Macrolides, 272, 279–283, 699
Magnetic stirrer digestion, 63–64
Malachite green, 690, 701
Malthus®, 9
Marine toxins
    azaspiracids, 594
    gymnodimine, 594
    hepatotoxins-microcystins
        clinical symptoms, 592–593
        detection methods, 593–594
        overview, 592
    ichthyosarcotoxin (*see also* Ichthyosarcotoxin)
        ciguatoxin, 578–581
        tetrodotoxin (TTX), 581–583
    pinnatoxins, 594
    shellfish toxins (*see also* Shellfish toxins)
        brevetoxin-neurotoxic shellfish poisoning,
            590–592
        domoic acid-amnesic shellfish poisoning,
            588–590
        okadaic acid-diarrhetic shellfish poisoning,
            586–588
        saxitoxin-paralytic shellfish poisoning, 583–585
    spirolides, 594
Mass spectrometric detector, 945–946
Matrix solid-phase dispersion (MSPD), 432–433,
        710–711, 938–940
Meat; *see also* Processed meat
    adulteration detection (*see* Meat adulteration
        detection)
    antibiotics (*see* Antibiotics, muscle tissues)
    antimicrobial residues
        analytical control methods, 253
        strategies for screening and confirmation, 252–253

biogenic amines
    aminogenic microorganisms, 402
    in cooked meat products, 403
    in cured meat products, 403
    detection and determination, 406–414
    in fermented meat products, 403–404
    in fresh meat and fresh meat products, 402
    hygienic quality evaluation, 404–406
food spoilage prediction (*see* Muscle food spoilage
    prediction methods)
GMO (*see* Genetically modified organism, meat)
irradiated ingredients detection (*see* Food irradiation
    identification, muscle foods)
liquid smoke flavors (LSF), 446–447
nitrosamines
    chemistry, 422–424
    formation and occurrence, 424
    toxicological aspects, 425
PAH (*see* Polycyclic aromatic hydrocarbons)
pathogen detection (*see* Microbial foodborne
    pathogens, muscle foods)
POP determination (*see* Persistent organic pollutants
    determination, meat)
smoked meat, 445–446
smoking principles, PAH
    alternatives to traditional procedures, 443
    traditional procedures, 442–443
toxins (*see* Mycotoxin analysis, poultry and processed
    meat)
Meat adulteration detection
    detection methods
        chemical methods, 158
        chromatographic methods, 172–176 (*see
            also* Chromatographic methods, meat
            adulteration)
        DNA analysis, 176–177
        electrophoretic techniques, 158–164
        immunological methods, 164–172 (*see also*
            Immunological methods, meat adulteration
            detection)
    reasons
        exploitation, low-quality meats, 157
        fat content reduction, 156–157
        health benefits, 157
        stabilization and sensory improvement, 156
    types
        milk products, 158
        soybean proteins, 157–158
        wheat gluten, 158
Membrane immunobead assay (MIA), marine
    toxin, 580
Metal hydroxide-based bacterial concentration
    technique, 25
Microarray technology, 630–631, 771–772
Microbial flora, dairy products
    culture-dependent microbiological analysis
        methods

        classical and advanced phenotypic methods,
            783–785
        molecular methods, 785–788
    culture-independent microbiological analysis
        methods, 788–790
    microbial ecology, 782–783
Microbial foodborne pathogens, dairy products
    critical features, 813
    Enterobacter sakazakii, 814, 816–817
    future perspective, 818–819
    Listeria monocytogenes, 814, 816
    Mycobacterium avium subsp. paratuberculosis,
        814, 818
    principles and applications, 812
    Salmonella, 814–815
Microbial foodborne pathogens, muscle foods
    Bacillus cereus, 42–44
    Campylobacter jejuni, 47–49
    Clostridium perfringens, 44–47
    culture methods
        detection methods, 24
        most probable number, 24
        plate count, 23–24
    Escherichia coli O157:H7
        confirmation, 34–35
        cultural enumeration method, 34
        detection, 32–34
    Listeria monocytogenes
        confirmation scheme, 30–32
        cultural enumeration method, 29–30
        detection, 27–29
    rapid microbiological methods
        detection and enumeration methods, 25–27
        target microorganism/toxin concentrating
            step, 25
    Salmonella spp., 35–38
    Staphylococcus aureus, 38–40
    Yersinia enterocolitica, 40–42
Microwave-assisted solvent extraction (MASE), 712
Milk; *see also* Dairy products
    for general consumption, 916
    infant milk, 910–916
    phthalates, 910, 916, 920
Milk allergens
    characteristics, 952–953
    detection tools, 963
        biosensor and surface plasmon resonance, 960
        ELISA, 958–959
        hidden allergens, 961–962
        immunoprecipitation and immunodiffusion, 958
        lateral-flow immunoassays (LFIAs) or dipsticks,
            959–960
        proteomic techniques, 961
        RAST/EAST inhibition, 958
        Western blotting, 960
    food processing effects
        enzymatic hydrolysis, 957

fermentation, 956
heat treatment, 955–956
homogenization, 957
ripening, 956
identification, 963
double blind placebo-controlled food
challenge, 953
patch tests, 954
recognition, 954
skin prick test, RAST/EAST inhibition and
allergen microarrays, 953–954
Milk and adulteration
composition, 852
lipids, proteins, and authentication test, 852–853
(*see also* Dairy adulteration detection)
products, 852
Milk and dairy products spoilage prediction
chemical and physical spoilage, milk, 806
future trends, 807
microbial metabolites as markers, 804–805
microorganisms, 802–803
modeling, 806–807
psychrotrophic bacteria, 801
sensorial detection, 804
volatile compounds as markers, 805–806
yeasts and molds, 801
Milk origin identification
DNA-based methods, 874–880
fat detection, 874
protein detection
antibody-based analytical methods, 870–873
chromatographic methods, 869–870
electrophoretic methods, 866–869
Mini-Massmann atomizer, 645
Minimum required performance limit (MRPL), 232
Modified atmosphere (MA) seafood storage, 473–474
Molecular methods, 11–13
Moraxella and *Acinetobacter,* 470–471
Most probable number, 24
Muscle food spoilage prediction methods
ATP bioluminescence methods, 8–9
culture-based methods, 5
developmental methods
flow cell cytometry, 11
molecular methods, 11–13
direct epifluorescent filtration technique (DEFT),
6–8
electrical methods, 9
electronic nose, 14
Limulus amoebocyte lysate assay, 9–10
time-temperature integrators, 14–15
Mycobacterium avium subsp. paratuberculosis, 814, 818
Mycotoxin analysis, dairy foods
analytical quality assurance, 824–826
cleanup methods, 828
detection systems, 829–838
extraction procedures, 827

laboratory precautions, 827
quantitative methods, 829
sample preparation, 827
screening tests, 828–829
Mycotoxin analysis, poultry and processed meat
aflatoxins
analysis methods, 90–91
origin and nature, 89
in poultry, 100–103
in processed meat, 105
structure and chemical properties, 89–90
citrinin
analytical methods, 93
origin and nature, 92–93
physicochemical properties, 93
in poultry, 104
in processes meat, 105, 107
cylopiazonic acid (CPA)
analysis methods, 94
origin and nature, 93
physicochemical properties, 93
in poultry, 104
in processes meat, 107
EU regulation, 81–83
fumonisins
analysis methods, 88–89
origin and nature, 87–88
physicochemical properties, 88
in poultry, 100
ochratoxins A (OTA)
analysis methods, 91–92
origin and nature, 91
physicochemical properties, 91
in poultry, 101
in processes meat, 105–106
in poultry muscle and tissue, 79–80, 94–101, 104
in processed meat, 70–80, 104–107
toxicity, 78–80
trichothecenes
analytical methods, 84–85
origin and nature, 81, 84
in poultry, 94, 97
structure and physicochemical properties, 84
zearalenone (ZEA)
analytical methods, 86–87
origin and nature, 86
in poultry, 97
structure and physicochemical properties, 86

N

NDL-PCB, *see* Non-dioxin-like polychlorobiphenyls
Nematodes, 515, 518–523
Neurotoxic shellfish poisoning (NSP), 590–592
Neutralization, 547
NIR spectroscopy, 145
Nitrofurans, 280, 284–289, 690, 697–699

Nitroimidazoles, 285, 290–293, 700–701
Nitrosamines
    chemistry, 422–424
    determination steps, 428–430
    formation and occurrence, meat and meat
        products, 424
    regulatory aspects, 425–426
    sample preparation
        distillation and clean-up procedures, 427–428
        electrophoresis, 437
        gas chromatography (GC), 435–436
        HPLC-TEA, 436
        matrix solid-phase dispersion and liquid-liquid
          extraction, 432–433
        solid-phase microextraction (SPME), 433–434
        solvent extraction, 432
        supercritical fluid extraction, 434–435
    toxicological aspects, 425
Non-dioxin-like polychlorobiphenyls (NDL-PCB)
    examples, 381–382
    extraction, 380, 383
    instrumental analysis, 383–384
    pretreatment, 380
Nonlethal diagnosis methods, 562–563
Novobiocin, 316, 320–321
Nucleic acid hybridization (NAH), 550–551, 559–560
Nucleic acid sequence based amplification (NASBA),
    553–554

O

Ochratoxins A (OTA)
    analysis methods, 91–92
    origin and nature, 91
    physicochemical properties, 91
    in poultry, 101
    in processes meat, 105–106
    structure, 80
Okadaic acid (OA)-diarrhetic shellfish poisoning
    (DSP)
    clinical symptoms, 586
    detection methods
        bioassays, 586–587
        chemical methods, 587–588
        immunoassays, 586
    overview, 586
Olaquindox, 322–324
Organochlorine pesticides
    cleanup, 368
    examples, 365–366
    extraction techniques, 364, 367
    instrumental analysis
        HRGC (ECD), 368–369
        HRGC-MS, 369
    multiresidue methods, 364
    pretreatment, 360–361

P

PAGE, *see* Polyacrylamide gel electrophoresis
PAH, *see* Polycyclic aromatic hydrocarbons
Parafilaria bovicola, 72
Paralytic shellfish poisoning (PSP), 583–585
Parasites
    fish and seafoods
        acanthocephalans, 523
        anisakid nematodes, 515, 518–523
        Capillaria philippinensis, 523
        cestodes, 515–517
        further developments, 525–526
        Gnathostoma spp., 523
        protozoa, 509–510
        safety, 523–525 (*see also* Fish and seafoods,
          safety)
        trematodes, 510–515
    meat
        Ascaris suum, 69–70
        Echinococcus spp., 71
        Fasciola hepatica and other liver flukes, 70–71
        future research, 72–73
        Parafilaria bovicola, 72
        Sarcocystis spp., 69
        Taenia spp., 67–68
        Toxoplasma gondii, 68–69
        Trichinella spp., 60–67
Pathogens
    dairy products (*see* Microbial foodborne pathogens,
        dairy products)
    fish and seafoods (*see* Seafood pathogens detection)
    muscle foods (*see* Microbial foodborne pathogens,
        muscle foods)
PBDEs, *see* Polybrominated diphenyl ethers
PCDD, *see* Polychlorodibenzo-p-dioxins
PCDF, *see* Polychlorodibenzofurans
Penicillins, 266, 270–278, 697–698
Peptides analysis, 173–174
Persistent organic pollutants (POPs), dairy products
    chemical structure, 930–931
    cleanup process, 941–942
    detectors for HRGC, 944–945
    extraction from liquid foodstuff (milk)
        liquid-liquid extraction (LLE), 933
        solid-phase extraction (SPE), 933
        solid-phase microextraction (SPME), 934–935
    extraction from semiliquid and solid matrices
        liquid-phase extraction (LPE), 936–937
        matrix solid-phase dispersion (MSPD), 938–940
        pressurized solvent extraction (PLE), 937–938
        Soxhlet extraction, 936
        supercritical fluid extraction (SFE), 940–941
    identification and quantitation
        capillary GC with selective detectors, 943
        HRGC column selection, 943–944
        injection device, 944

mass spectrometric detector, 945–946
regulations, 932–933
Persistent organic pollutants (POPs) determination, meat
 bioassays
  cell-based bioassays, 387
  PCR, 387
  reliability and applicability, 387–388
 chemical methods
  NDL-PCBs and PBDEs, 380–384
  organochlorine pesticides, 363–369
  PCDDs and PCDFs, and DL-PCB, 369–380
  PFAS, 384–386
 definition, 350
 laboratory safety, 350–351
 organochlorine pesticides, 360
 PCDD and PCDF, 360–361
 polybrominated diphenyl ethers (PBDE), 361
 polychlorobiphenyls (PCB), 360–361
 polyfluorinated alkylated substances (PFAS),
  362–363
 Stockholm convention, 351–360
Persistent organic pollutants (POPs), fish and seafood,
 728; *see also* Polychlorinated biphenyls;
 Polycyclic aromatic hydrocarbons
Pesticides, dairy products
 cleanup process, 941–942
 definition, 931–932
 detectors for HRGC, 944–945
 extraction from liquid foodstuff (milk)
  liquid-liquid extraction (LLE), 933
  solid-phase extraction (SPE), 933
  solid-phase microextraction (SPME), 934–935
 extraction from semiliquid and solid matrices
  liquid-phase extraction (LPE), 936–937
  matrix solid-phase dispersion (MSPD), 938–940
  pressurized solvent extraction (PLE), 937–938
  Soxhlet extraction, 936
  supercritical fluid extraction (SFE), 940–941
 identification and quantitation
  capillary GC with selective detectors, 943
  HRGC column selection, 943–944
  injection device, 944
 mass spectrometric detector, 945–946
 regulations, 932–933
Petrifilm®, 5
PFAS, *see* Polyfluorinated alkylated substances
Phenicols, 699
Photobacterium genus, 472
Photostimulated luminescence (PSL), 217–218
Phthalates, dairy products
 in dairy products, 917–918
 detection methods
  instrumental conditions, 924–926
  sample contamination avoidance, 921
  sample pretreatment, extraction, and cleanup,
   921–924
  GC/MS chromatogram, 924, 926

 for general consumption, 916
 in infant milk, 910–916
 levels, 910–920
 migration into milk and other dairy products, 920–921
 nondairy food products, 917
 physical properties, 908–909
 structure, 908
 in total diet, 918–920
Pinnatoxins, 594
Plaque neutralization test, 549
Plate count, 23–24
Poisoning, *see* Marine toxins
Polyacrylamide gel electrophoresis (PAGE), 866
Polybrominated diphenyl ethers (PBDE)
 commercial products, 380, 383
 examples, 381–382
 extraction, 380, 383
 instrumental analysis, 383–384
 pretreatment, 380
Polychlorinated biphenyls (PCBs), 360–361, 734–735
 chromatographic analysis
  gas chromatography-high-resolution mass
   spectrometry (GC-HRMS), 739
  gas chromatography-mass spectrometry
   (GC-MS), 738–739
  gas chromatography with electron capture
   detector (GC-ECD), 738–739
 clean-up of extracts, 737–738
 extraction methods
  pressurized liquid extraction (PLE), 736–737
  sonication method and liquid-liquid
   partitioning, 736
  Soxhlet extraction, 735–736
  supercritical fluid extraction (SFE), 737
Polychlorinated dibenzo-*p*-dioxins and dibenzofurans
 (PCDD/F) analysis, *see* Dioxins analysis,
 seafood and seafood products
Polychlorodibenzofurans (PCDF), 360–361
 analytical methods, 369, 378
  cleanup and fractionation, 379
  extraction techniques, 378–379
  HRGC-HRMS instrumental analysis, 379–380
  pretreatment, 378
 bioassays, 387–388
 examples, 370–377
Polychlorodibenzo-p-dioxins
 analytical methods, 369, 378
  cleanup and fractionation, 379
  extraction techniques, 378–379
  HRGC-HRMS instrumental analysis, 379–380
  pretreatment, 378
 bioassays, 387–388
 examples, 370–377
Polychlorodibenzo-p-dioxins (PCDD), 360–361
Polyclonal antisera, 545
Polycyclic aromatic hydrocarbons (PAH)
 behavior in organism, 443–444

chromatographic analysis, seafood
    gas chromatography (GC), 734
    high-performance liquid chromatography
        (HPLC), 733
clean-up of extracts, seafood, 731
    gel permeation chromatography, 732
    solid-phase extraction, 732
extraction methods, seafood
    pressurized liquid extraction (PLE), 730–731
    saponification, 729–730
    sonication method, 730
    Soxhlet extraction, 730
    supercritical fluid extraction (SFE), 731
GC *vs.* HPLC, 456–457
legislative aspects and international
        normalization, 444
occurrence, 457
preseparation procedures, meat
    gas chromatography (GC), 448–455
    high-pressure liquid chromatography (HPLC)
        method, 448, 456
    thin-layer chromatography (TLC), 447
sample preparation, meat
    liquid smoke flavors (LSF), 446–447
    smoked meat, 445–446
smoking principles, meat
    alternatives to traditional procedures, 443
    traditional procedures, 442–443
Polyether antibiotics, 316, 318–319
Polyfluorinated alkylated substances (PFAS)
    examples, 362–363
    extraction and cleanup, 384–385
    instrumental identification and determination,
        385–386
    observations, 386
    principal MS/MS transition, 386
    sampling and sample storage, 384
Polymerase chain reaction (PCR) method
    amplification from a unique target molecule,
        551, 553
    animal species identification
        fingerprinting, 194
        multiplex PCR, 194–196
        PCR product determination, 190–192
        PCR reaction design, 189
        PCR-RFLP, 190–192
        real-time PCR, 194, 196–198
        species-specific PCR amplification, 192–193
        universal mtDNA primers, 189–190
    bioassays, persistent organic pollutants (POPs)
        determination, 387
    commercial kits, 556
    fish and seafood, 560–562
    GMO detection, DNA-based methods, 130–131
        applications, 136, 143
        enzyme-linked immunosorbent assay
            (ELISA), 136

qualitative methods, 132, 134
quantitative competitive polymerase chain
        reaction (QC-PCR), 134–135
real-time, 135–140
GMO detection, fish feed
    multiplex PCR, 766
    qualitative conventional PCR, 764–766
    quantitative PCR, 766–767
microbial foodborne pathogens, dairy products
    critical features, 813
    Enterobacter sakazakii, 814, 816–817
    future perspective, 818–819
    Listeria monocytogenes, 814, 816
    Mycobacterium avium subsp. paratuberculosis,
        814, 818
    principles and applications, 812
    Salmonella, 814–815
microbial foodborne pathogens, muscle foods
    followed with denaturing gradient gel
        electrophoresis (DGGE), 26–27
    Escherichia coli O157:H7 detection, 34
    Listeria monocytogenes, 28–30
    microbial foodborne pathogens, muscle
        foods, 25
    Salmonella spp., 38
microorganisms identification in dairy products,
        785–787
milk origin identification, 874–880
muscle food spoilage prediction, 11–13
nested PCR (Nt-PCR), 551
quantitative PCR (Qt-PCR), 553
real-time PCR (Rt-PCR), 553, 562
seafood adulteration detection
    commercial kits for fish species
        differentiation, 611
    lab-on-a-chip, 610
    multiplex, 608–609
    PCR-restriction fragment length polymorphism,
        609–610
    real-time, 610
    sequencing, 608
    species-specific, 608–609
seafood pathogens detection, 488–490
sequence amplification, 551–552
SYBR-Green Rt-PCR, 553
Trichinella spp. detection methods, 65–66
Polypeptide antibiotics, 309, 315–317
Poultry muscle and tissue; *see also* Meat
    Campylobacter jejuni, 47
    GMO detection (*see* Genetically modified organism,
        meat)
    mycotoxin analysis, 79–80, 94–101, 104
Precision, 232
Pressurized liquid (solvent) extraction (PLE), 730–731,
        736–737, 937–938
Processed meat; *see also* Meat
    foreign proteins detection methods

chemical methods, 158
chromatographic methods, 172–176 (*see also*
      Chromatographic methods, meat adulteration
      detection)
DNA analysis, 176–177
electrophoretic techniques, 158–164
immunological methods, 164–172 (*see also*
      Immunological methods, meat adulteration
      detection)
mycotoxin analysis, 79–80, 104–107
Propionic acid bacteria, 782
Protein-based methods, GMO detection
   antibody-based assay formats
      enzyme-linked immunosorbent assay (ELISA),
         143–144
      lateral flow assay technology, 144
   applications, 144–145
Protein detection, milk
   antibody-based analytical methods, 870–873
   chromatographic methods
      caseins, 869
      whey protein, 869–870
   electrophoretic methods
      caseins, 866–868
      whey protein, 868–869
Proteolysis, 478–479
Pseudomonas fragi, 469
Pseudomonas genus, 466–469
Pseudomonas perolens, 469
Psychrobacter immobilis isolates, genetic transformation
      assay, 476–478

Q

Quinolones, 290, 294–300

R

Radio-allergosorbent/enzyme-allergosorbent
      (RAST/EAST) tests, 953–954
Radioimmunoassay (RIA), marine toxin, 580
Randomly amplified polymorphic DNA approach
      (RAPD), 627
RAPD, *see* Randomly amplified polymorphic DNA
      approach
Rapid microbiological methods
   detection and enumeration methods, 25–27
   target microorganism/toxin concentrating step, 25
Recovery, 232
Repeatability and reproducibility (R&R), 533–534
Reproducibility, 232
Reversed passive latex agglutination, 26
Reversed-phase HPLC (RP-HPLC), 869–870
RFLP, 626
Rod atomizer, *see* Filament atomizer
Ruggedness, 232

S

Salmonella, 814–815
Salmonella spp., 35–38, 487, 498–500
Saponification, 729–730
Sarcocystis spp., 69
Satellite DNA analysis, 627–628
Saxitoxin (STX)-paralytic shellfish poisoning (PSP)
   clinical symptoms, 584
   detection methods
      bioassays, 584–585
      chemical methods, 585
      immunoassays, 585
   overview, 583
Seafood adulteration detection
   chromatographic techniques, 607
   DNA methods
      amplified fragment length polymorphism, 609
      commercial PCR kits for fish species
         differentiation, 611
      PCR lab-on-a-chip, 610
      PCR-restriction fragment length polymorphism,
         609–610
      PCR-sequencing, 608
      real-time PCR, 610
      species-specific PCR or multiplex PCR, 608–609
   electrophoresis, 604–605
   immunological techniques, 605–606
   microscopic methods, 607
   visible and near-infrared spectrometry, 606
Seafood pathogens detection
   L. monocytogenes, 487, 496–498
   PCR-based method, 488–490
   Salmonella spp., 487, 498–500
   Vibrio cholerae, 486, 495–496
   Vibrio parahaemolyticus, 486, 488–493
   Vibrio vulnificus, 486, 493–495
Seafoods, *see* Fish and seafoods
Seafood species identification
   differential diagnostic systems, 616–618
   for different types of seafood samples, 631–632
   DNA sequencing methods
      identification from general databases, 625
      standardized fish molecular databases and
         barcoding, 621–624
   non-DNA sequencing methods
      AFLP, RAPD, and satellite DNA analysis,
         626–628
      microarray technologies and bioinformatics,
         630–631
      quantitative methods, 629–630
      RFLP, 626
      selective amplification, 628–629
      SSCP and DGGE, 628
   processed products adulteration, 619
   protein analysis, 619–620
   sample handling and DNA extraction, 620–621

Seafood spoilage prediction
  chemical causes, 465
  identity confirmation tests
    DNase activity, 479–480
    H₂S production detection, 479
    litmus milk, 478
    proteolysis, 478–479
    psychrobacter immobilis isolates, genetic
      transformation assay, 476–478
  microbiology
    gamma irradiation, 474
    modified atmosphere (MA) storage, 473–474
  quality assessment assays
    refractive index of eye fluid, 466
    trimethylamine (TMA), 465–466
    vacuum distillation procedure, 466
  taxonomy
    genera *Flavobacterium* and *Cytophaga,* 471–472
    genera *Moraxella* and *Acinetobacter,* 470–471
    genus *Aeromonas,* 473
    genus *Alteromonas,* 469
    genus *Brochothrix,* 472
    genus *Lactobacillus,* 472–473
    genus *Photobacterium,* 472
    genus *Pseudomonas,* 466–469
    genus *Shewanella,* 469–470
    genus *Vibrio,* 473
  varius bacterial counts determination, 474–475
    fluorescent pseudomonads enumeration,
      475–476
    genus *Pseudomonas* selective enumeration, 475
Selected ion monitoring (SIM) GC-MS chromatogram,
      237–238
Sensitivity, 533
Serology, 164
Shelf-life testing, 9
Shellfish toxins; *see also* Fish and seafoods; Virus and
      viral diseases, fish and shellfish
  brevetoxin-neurotoxic shellfish poisoning (NSP)
    clinical symptoms, 590–591
    detection methods, 591–592
    overview, 590
  domoic acid (DA)-amnesic shellfish poisoning (ASP)
    clinical symptoms, 588–589
    detection methods, 589–590
    overview, 588
  okadaic acid (OA)-diarrhetic shellfish poisoning
      (DSP)
    clinical symptoms, 586
    detection methods, 586–588
    overview, 586
  saxitoxin (STX)-paralytic shellfish poisoning (PSP)
    clinical symptoms, 584
    detection methods, 584–585
    overview, 583
Shewanella genus, 469–470
Shewanella putrefaciens, 470

Shiga toxin-producing *Escherichia coli* (STEC),
      839–840
Short-wavelength-near-infrared (SW-NIR) diffuse
      reflectance spectroscopy, 10
SimPlate®, 5
Single-strand conformation polymorphisms
      (SSCP), 628
Skimmed milk powder, 859–860
Smoked meat, 445–446
Sodium dodecyl sulfate-polyacrylamide gel
      electrophoresis (SDS-PAGE), 159–163,
      169–170
Solid-phase extraction (SPE), 710–711, 732, 933
Solid-phase microextraction (SPME), 433–434,
      934–935
Sonication method, 730, 736
Soxhlet extraction, 730, 735–736, 936
Soy powder, 860
SPC, *see* Aerobic standard plate count
Specificity, 232, 533
Spectrochemical methods, seafood metals
      determination
  analytical atomic spectroscopy, 650–651
  applications
    Cd and Cu levels, 658
    crawfish or crayfish, 654
    direct solid sampling Zeeman GFAAS
      methods, 657
    heavy metal concentrations, 656
    microwave digestion method, 654
    mollusks, 655–656
    multimetal determination method, 654
    selected results, 652–653
    Taiwanese, 656
    trace metals concentrations, 657
    transition and heavy metals, 655
  atomic absorption spectrometry
    cold vapor atomic absorption spectrometry, 646
    instrumentation, 643–646
    theory, 642–643
  atomic emission spectrometry
    instrumentation, 647–649
    theory, 646–647
  inductively coupled plasma-mass spectrometry,
      649–650
  sample preparation, 651
Spirolides, 594
Spoilage prediction methods, *see* Muscle food spoilage
      prediction methods
SSCP, *see* Single-strand conformation polymorphisms
Staphylococcus aureus, 38–40, 839
Stockholm convention, POP, 351–360
Sulfonamide antiinfectives, 301–307
Sulfonamides, 695–696
Supercritical fluid extraction (SFE), 434–435, 711, 731,
      737, 940–941
Surface plasmon resonance, 960

T

Taenia spp., 67–68
TEMPO system, 5
Tetracyclines, 308–314, 696
Tetrodotoxin (TTX)
　　clinical symptoms, 581–582
　　detection methods
　　　　bioassays, 582
　　　　chemical methods, 583
　　　　immunoassays, 582
　　overview, 581
Thermoluminescence detection, 217
Thin-layer chromatography (TLC), 236, 447, 828
Tiamulin, 316, 320–321
Time-temperature integrators (TTI), 14–15
Toxins
　　dairy foods (*see also* Mycotoxin analysis, dairy foods)
　　　　Bacillus cereus, 840
　　　　biological assays, 840–841
　　　　biologic-immunologic-phenotypic combination
　　　　　　studies, 843–844
　　　　future trends, 844
　　　　immunological tests, 841–842
　　　　phenotypic assays, 842–843
　　　　Shiga toxin-producing *Escherichia coli* (STEC),
　　　　　　839–840
　　　　Staphylococcus aureus, 839
　　poultry and processed meat (*see* Mycotoxin analysis,
　　　　poultry and processed meat)
Toxoplasma gondii, 68–69
Tranquilizers, 701–702
Trichinella spp.
　　direct detection methods
　　　　digestion methods, 62–65
　　　　histology, 65–66
　　　　trichinoscopic examination, 62
　　indirect detection methods., 66–67
　　polymerase chain reaction (PCR) method, 65–66
Trichinoscopy, *Trichinella* inspection, 62
Trichothecenes
　　analytical methods, 84–85
　　origin and nature, 81, 84
　　in poultry, 94, 97
　　structure and physicochemical properties, 84
TTI, *see* Time-temperature integrators
Turkey X disease, 93
Type A trichothecenes, 85
Type B trichothecenes, 85

U

Ultraperformance liquid chromatography separation
　　(ULCS), 891–896
Universal mtDNA primers, 189–190
Urea-polyacrylamide gel electrophoresis (PAGE),
　　866–867

V

Varius bacterial counts determination, 474–475
　　fluorescent pseudomonads enumeration, 475–476
　　genus *Pseudomonas* selective enumeration, 475
Veterinary drug residue regulatory control, 251–252
Veterinary drugs; *see also* Growth promoters
　　aminoglycosides, 699–700
　　analysis
　　　　biological test systems, 691
　　　　instrumental analytical approaches, 691–695
　　aquaculture, 688–690
　　　　anthelmintic drugs, 702–703
　　　　antiparasitica, 702–703
　　　　bactericidal drugs (antibiotics), 695–701
　　　　hormones, 702–703
　　　　tranquilizers, 701–702
　　cephalosporines, 698
　　chinolones, 696–697
　　chloramphenicol, 699
　　developments, 704
　　enrofloxacine, 689–690, 697
　　lincomycin, 700
　　lincosamides, 699
　　macrolides, 699
　　nitrofurans, 690, 697–699
　　nitroimidazoles, 700–701
　　penicillin, 697–698
　　phenicols, 699
　　sulfonamides, 695–696
　　tetracyclines, 696
Vibrio cholerae, 486, 495–496
Vibrio genus, 473
Vibrio parahaemolyticus, 486, 488–493
Vibrio vulnificus, 486, 493–495
Viral neutralization, 544–545
Virus and viral diseases, fish and shellfish
　　aquatic animal diseases, 536
　　cell culture isolation
　　　　cell lines and cell culture, 541–542
　　　　cell line selection, 542
　　　　performance, 543–544
　　　　viral isolation, 542–543
　　clinical, histological, and microscopical
　　　　techniques
　　　　electron microscopy, 541
　　　　gross signs, 537
　　　　histopathology, 537–540
　　　　immunohistochemistry (IHC), 537, 540–541
　　definition, 532–533
　　molecular diagnosis
　　　　critical steps and critical factors, 555–559
　　　　protocols, 559–562
　　　　scientific basis, 550–554
　　nonlethal methods, 562–563
　　sample processing, 534–536
　　serological and immunological techniques

advantages and disadvantages, 544–547
antibody detection, 549–550
procedures, 547–549
scientific basis, 544
validation, 533–534
Visible and near-infrared spectrometry
seafood adulteration detection, 606
Visible/NIR (vis/NIR) spectroscopy, 145

W

Whey protein, 868–870
Whole proteins analysis, 173–176
Within-laboratory reproducibility, 232

Y

Yersinia enterocolitica, 40–42

Z

Zearalenone (ZEA)
analytical methods, 86–87
EU regulation, 79, 83, 86
origin and nature, 86
in poultry, 97
structure and physicochemical properties,
79, 86